QUALITATIVE RESEARCH & EVALUATION METHODS

FOURTH EDITION

To
Calla Quinn,
my first grandchild,
born during the gestation of this book,
the fifth generation in our family line to carry
the middle name Quinn,
two before me, two after me,
and the first female to do so.
Whatever the future holds, uncertain as it inherently is,
new opportunities worthy of, and in need of,
in-depth qualitative inquiry and illumination
are sure to emerge.

QUALITATIVE RESEARCH & EVALUATION METHODS

FOURTH EDITION

Integrating Theory and Practice

Michael Quinn Patton

Los Angeles | London | New Delhi
Singapore | Washington DC

Los Angeles | London | New Delhi
Singapore | Washington DC

FOR INFORMATION:

SAGE Publications, Inc.
2455 Teller Road
Thousand Oaks, California 91320
E-mail: order@sagepub.com

SAGE Publications Ltd.
1 Oliver's Yard
55 City Road
London EC1Y 1SP
United Kingdom

SAGE Publications India Pvt. Ltd.
B 1/I 1 Mohan Cooperative Industrial Area
Mathura Road, New Delhi 110 044
India

SAGE Publications Asia-Pacific Pte. Ltd.
3 Church Street
#10-04 Samsung Hub
Singapore 049483

Acquisitions Editor: Vicki Knight
Assistant Editor: Katie Bierach
Editorial Assistant: Yvonne McDuffee
Production Editor: David C. Felts
Copy Editor: QuADS Prepress (P) Ltd.
Typesetter: C&M Digitals (P) Ltd.
Proofreaders: Sally Jaskold, Scott Oney
Indexer: Molly Hall
Cover Designer: Candice Harman
Marketing Manager: Nicole Elliott

Copyright © 2015 by SAGE Publications, Inc.

Printed in the United States of America.

Library of Congress Cataloging-in-Publication Data

Patton, Michael Quinn.

[Qualitative evaluation methods]

Qualitative research & evaluation methods : integrating theory and practice / Michael Quinn Patton.—Fourth edition.

pages cm

Revised edition of the author's Qualitative evaluation methods. Includes bibliographical references and index.

ISBN 978-1-4129-7212-3 (alk. paper)

1. Social sciences—Methodology. 2. Evaluation research (Social action programs) I. Title. II. Title: Qualitative research and evaluation methods.

H62.P3218 2014

001.4'2—dc23   2014029195

This book is printed on acid-free paper.

14 15 16 17 18 10 9 8 7 6 5 4 3 2 1

Brief Contents

Contents

Preface

> For a very long time everyone has decried the futility of prefaces—yet everyone keeps writing them. We all know that readers (an already optimistic plural) skip them, which should itself be valid reason not to write any more.
>
> —Preface to *Fortunio* (a romantic tale, 1836) by Theophile Gautier (1811–1872), French poet, dramatist, novelist, journalist, and art and literary critic

This quotation opens a preface by Professor Clement Moisan, University of Laval Quebec, Canada, to a qualitative analysis of literary prefaces by Steven Tötösy de Zepetnek (1993), University of Alberta, Edmonton, Canada, for his doctoral dissertation. Thus, *the preface is a phenomenon* that constitutes a literary genre that manifests substantial variation. That variation can be studied, coded, classified, and labeled, thereby generating a typology of prefaces, which is precisely what Tötösy de Zepetnek did through "analysis of the systemic dimensions of the preface typologies and of the systemic data of the prefaces." His analysis included examining how prefaces are produced, their content, how they are received, and how they have come to be viewed as a genre. His comparative analysis examined different kinds of prefaces for different kinds of works. I found myself especially drawn to a type of preface he labels "preemptive," in which the author attempts to predispose and engage the readers by anticipating their reactions. I can imagine, for example, a reader of this preface reacting, "Enough with analyzing prefaces already. Get on with the preface."

But not yet, Dear Reader (a style of engagement aimed at endearing the reader to the author, and vice versa), for there is more. The preface is but one manifestation of a larger phenomenon, any writing that precedes and/or explains a work of writing, music, or art. Thus, in addition to prefaces, there are forewords, epigraphs, introductions, preambles, prologues, preludes, overtures, invocations, and, my personal favorite, the *prolegomenon*. To which, Dear Patient Reader, I now turn.

Prolegomenon Purposes: The Journey to This Fourth Edition

> Lately it occurs to me: What a long, strange trip it's been.
>
> —Robert C. Hunter American lyricist who collaborated with the Grateful Dead and Bob Dylan, among many others

Ironically, writing a preface comes at the end of the writing journey. Your beginning, my ending.

I aim to do four things in this prefatory, prological prolegomenon. First, to explain and justify why the book is so long. And no, despite initial evidence here to the contrary, it's not because of a proliferation of long, redundant phrases like *prefatory, prological prolegomenon*. Nor, rumors to the contrary, is it because authors are paid by the word (Writers beta, 2012). It's because in the decade since the third edition of this book came out (Patton, 2002), qualitative methods have flourished. Why, how so, and with what implications (they are huge), I'll address momentarily.

A second purpose involves telling you how this fourth edition is different from (and so much better than) the third edition. I should perhaps note in this regard that this book is about both research and evaluation methods. The judgment that this new edition is so much better than the last is an exemplar of a form called "self-evaluation," which should alert those of you who may be new to the field to the fact that self-evaluations are held in very high esteem and considered extraordinarily credible.

The third function of a preface is to acknowledge and thank those who have helped me along the way. I have not been alone on this long, strange journey.

Finally, and very seriously, there is the book's dedication to elaborate and luxuriate in.

From the First, to Second, to Third, to Fourth Edition

> **What's past is prologue.**
>
> —William Shakespeare
> Antonio to Sebastian, *The Tempest*

The first edition of this book appeared in 1980. At the time, there was very little literature dedicated to the nuts and bolts of how to do qualitative inquiry, especially as an approach to program evaluation. The second edition, in 1990, began the process, continued in this latest revision, of integrating theory and practice. The third edition, published in 2002, noted the growth of interest in qualitative methods and observed that "the upshot of all the developmental work in qualitative methods is that there is now as much variation among qualitative researchers as there is between qualitatively and quantitatively oriented scholars and evaluators" (Patton, 2002, p. xxii). Thus, the primary purpose of the third edition was to sort out the major alternative perspectives in the diversity of qualitative approaches that had emerged and examine the influences of that diversity on applications, especially but not exclusively in program evaluation, policy analysis, and action research, and across applied social sciences generally—which brings us to this fourth edition.

In doing this revision, I reviewed well over 1,000 new qualitative resources published in the past decade: the latest books on qualitative methods, recent qualitative studies, scholarly and applied journal articles, program evaluations, case studies, monographs, and dissertations. The volume of qualitative research and evaluation has exploded exponentially. More important, in my judgment, the quality has been enhanced by deeper reflections on the nature and variety of qualitative inquiry, more methodological and analytical options and sophistication, and greatly expanded outlets for publication of qualitative works. To do even minor justice to the diverse flowering of qualitative inquiry has meant a substantial increase in the length of this edition.

What's New in This Edition?

- *Substantive highlights*
  o Chapter 1 opens with extensive new examples of how qualitative inquiry contributes to our understanding of the world. These examples are aimed at illuminating the nature, niche, value, and fruit of qualitative inquiry.
  o Chapter 2 provides up-to-date references and examples for the 12 strategic themes of qualitative inquiry.
  o Chapter 3, on the variety of qualitative inquiry frameworks (paradigmatic, philosophical, and theoretical orientations), includes updated references and examples, with new sections on pragmatism and generic qualitative inquiry as distinct approaches, as well as new directions in and more in-depth treatment of realism, systems theory, and complexity theory as relevant to qualitative inquiry.
  o Chapter 4 has added examples of practical purposes, concrete questions, and actionable answers, with special attention to international research and evaluation studies and findings. (The year 2015 has been designated by the United Nations as the International Year of Evaluation.)
  o Chapter 5 has expanded the crucial qualitative design discussion of purposeful sampling from 16 options in the third edition to 40 distinct case selection approaches in this new edition. Mixed-methods designs have become increasingly important and get much added attention in this edition.
  o Chapter 6, on qualitative observation and fieldwork, includes attention to how technological developments have increased data collection options.
  o Chapter 7, on qualitative interviewing, examines new opportunities afforded by social media, the Internet, and social networking, as well as the challenges of remote and virtual interviewing. Distinct forms and approaches to interviewing that derive from different theoretical orientations are discussed and illustrated.
  o Chapter 8, on qualitative data analysis and reporting, includes more in-depth treatment of case study creation and cross-case analysis; examines the trend toward greater use of visual representations (visualization as a powerful communication tool) and the emergence of principles-focused evaluation as a new, complexity-based approach; and provides extensive discussion of causal inference in qualitative analysis.
  o Chapter 9 has updated and deepened ways of enhancing the credibility and utility of qualitative findings with systematic ways of thinking about and addressing rigor in qualitative studies, and innovative approaches to extrapolating qualitative findings and dealing with issues of generalizing results. The third edition included

review of five distinct sets of criteria for judging the quality of qualitative studies; this new edition adds two new sets of criteria and revises the former sets.

- **Modules:** To make the book more manageable, the nine lengthy chapters are now organized into more digestible modules, 82 in all.
- **Exhibits:** Creating exhibits involves summarizing discussion of complex issues into major points; bringing coherence, succinctness, and closure to a topic; and making that summary readily available for ease of review. The third edition had 59 exhibits. This new edition has 148. Here's a foreshadowing of 10 new exhibits, 1 per chapter plus a bonus, which will also provide a preview of some of the new material in this edition.
  - **Exhibit 1.9** Qualitative Inquiry Pioneers (Prepare to be surprised by who's been considered for this unique, world premier honor role. Let me know who you think has been inappropriately left out.)
  - **Exhibit 2.6** Twelve Core Strategies of Qualitative Inquiry: Interdependent and Interactive
  - **Exhibit 3.19** Distinguishing and Understanding Alternative Inquiry Frameworks: Cross-Cutting Themes
  - **Exhibit 4.11** Reflective Practice Guide
  - **Exhibit 5.15** Outline for a Qualitative Inquiry Design Proposal
  - **Exhibit 6.3** Ethical Issues in Qualitative Research on Internet Communities
  - **Exhibit 7.15** Ten Examples of Variations in Cross-Cultural Norms That Can Affect Interviewing and Qualitative Fieldwork
  - **Exhibit 8.18** Ten Approaches to Qualitative Causal Analysis
  - **Exhibit 8.27** Mixed Methods Challenges and Solutions
  - **Exhibit 9.14** Twelve Perspectives on and Approaches to Generalization of Qualitative Findings
- **More case study exemplars:** One of the most common requests I get is for more examples of case studies and qualitative reporting. Both are important. Both require space. I've added both, which has contributed to the increased size of this fourth edition.
- **Ruminations:** In celebration of this new edition, I have indulged in one personal rumination per chapter. These are issues that have persistently engaged, sometimes annoyed, occasionally haunted, and often amused me over more than 40 years of qualitative research and evaluation practice. In these ruminations, I state my case on the issue and make

my peace with it. Examples are dismissing qualitative case studies as merely "anecdotal"; poor-quality site visits in program evaluations; untrained interviewers unaware of the skills and rigor involved in high-quality qualitative interviewing; and avoiding qualitative research rigor mortis.

I originally called these "rants," but a couple of external reviewers said that label evoked painful memories of certain excruciating faculty meetings, which inclined them to dismiss the rant without even reading it. I considered other terms: *fulminations*, *tirades*, *declamations*, and one that almost made the final cut, *tub-thumbing*: engaging in impassioned utterance. I quite like that. But I eventually settled on *rumination*: "the act of pondering."

- **Sidebars:** Sidebars are boxed items of interest that supplement the text with examples and extended quotations from knowledgeable qualitative theorists and practitioners. They are a way of highlighting experts' insights, case study exemplars, supplementary readings, and additional resources. More than 100 new sidebars have been added throughout the book.
- **Updated references:** As I reviewed the vast new literature on qualitative methods and the thousands of qualitative and mixed-methods studies that have been published in the past decade, and having tracked the literature over the past 40 years, my sense is that more qualitative works have appeared in the past decade than in all the preceding years combined. Simply skimming the references will give you a sense of the amount and diversity of qualitative research and evaluation that has been and is being published. I am indebted to Jean Gornick for helping me track, update, and organize the references, and even more for her personal support throughout the writing.
- **Unique chapter symbols:** Each chapter has a unique symbol that sets the stage for the chapter's content and is then used to introduce modules in that chapter. For example, the symbol for Chapter 1 is the wisdom knot.

Nyansapo, "wisdom knot," Adinkra (West Africa) symbol of wisdom, ingenuity, intelligence, and patience. This symbol conveys the idea that a wise person has the capacity to select the best means to achieve a goal. Being wise means knowing how to apply broad knowledge, learning, and experience for practical purposes (Willis, 1998).

- **Halcolm:** There's new wisdom from Halcolm (pronounced "How come," as in "Why"), my internal philosophical alter ego and muse, who pipes in every so often to remind us that qualitative inquiry is grounded in fundamental philosophical underpinnings about how and why the world works as it does. Halcolm takes the form of an elderly sage in new graphic comics at the end of each chapter. Art teacher and cartoonist Andrew Wales created these comic renditions of Halcolm parables, for which I am deeply grateful. They serve as a meditative transition between chapters. (See more of Andrew's creative work at http://andrewwales.blogspot.com/.)

- *New cartoons:* Cartoons invite us to see things in a different way, both less and more seriously at the same time. This edition includes new cartoons by Mark Rogers and Chris Lysy (www.freshspectrum .com), both practicing program evaluators who bring insightful humor to what many experience as a largely humorless enterprise. Claudius Ceccon, a distinguished Brazilian political cartoonist and father-in-law of Brazilian evaluator Thomaz Chianca, has also contributed five provocative and evocative cartoons.
- *Application exercises:* For the first time, this new edition includes practice exercises at the end of each chapter. These include opportunities to apply the content of a chapter to your own arenas of interest and expertise.

Style Note: On Using Quotations

> All books fail to be everything their authors hoped.
>
> —Scott Sandage (2014)
> Cultural historian

I often use quotations to introduce new sections, like this one. Or Shakespeare's "What's past is prologue," used earlier. I think of such quotations as garnishes, seasoning, and a bit of *amuse-bouche* (a French gourmet tradition of serving an appetizer that is not on the menu but, when served, is done so without charge and entirely at the chef's discretion and preference). For the most part, these are not scholarly quotations, nor are they usually referenced. In the spirit of the gastronomic metaphors offered here, they are *palate cleansers* as you move from one topic to another.

Some people, I am told, find such quotations annoying (including one external reviewer of this book for whom I have direct evidence of disdain). Well, you know, you don't have to eat the garnish. If you don't like it, skip it. Like spam or unwelcome e-mails that you instantly delete, move past them quickly. I offer the same counsel with regard to Halcolm stories, cartoons, fables, sidebars, extended examples, personal reflections, and my ruminations. Some view them as distractions. Exercise *control*. Chose the *alternate* path. *Delete* from your reading. They're there for the many folks who write and tell me that those are their favorite parts of the book and what keeps them going through the more traditional academic and methodological stuff.

Qualitative inquiry is fundamentally about capturing, appreciating, and making sense of diverse perspectives. Different people resonate to different aspects of a book, both content and style. I've included a variety of ways of engaging readers in hopes of hitting on an approach that works for you, Dear Reader. But in the end, I confess, as the author I include cartoons, quotations, Halcolm stories, and fables because they amuse and enlighten me.

Acknowledgments and Collaborations

> Your corn is ripe today; mine will be so tomorrow.
>
> 'Tis profitable for us both, that I should labour with you today, and that you should aid me tomorrow.
>
> —**David Hume** (1711–1776)
> Scottish philosopher

My initial foray into qualitative writing was entirely due to the persuasive powers of Sara Miller McCune, cofounder with her husband, George, of SAGE Publications. She had shepherded my first book, *Utilization-Focused Evaluation* (1978), into print. Based on that book's advocacy of applying the criterion of utility to methods decisions, she urged me to write a qualitative companion. Her vision and follow-through have made SAGE Publications the leading publisher of both program evaluation and qualitative inquiry books. In 2015, SAGE celebrates its 50th anniversary as this book celebrates its 35th. Vicki Knight, my current SAGE editor, provided ongoing support throughout the arduous and overly long process of completing this new edition. She arranged excellent external reviewers who provided helpful and timely feedback to improve the final manuscript, for which I am grateful and you can be as well, for they were especially useful in clarifying obfuscations, eliminating redundancies, and increasing overall coherence and integration.

Only one person has been a constant companion and fellow traveler on this long, strange trip through all four editions, my friend and colleague Malcolm Gray. In the 1970s, Malcolm cocreated and codirected a wilderness leadership development program in the southwestern United States. He came across an essay on evaluation that I had written and contacted me about evaluating the program, which I did through participant observation. It is one of the examples featured in this book. We subsequently became hiking companions, logging many miles and weeks in the Grand Canyon, about which I wrote a book (Patton, 1999). Malcolm has engaged in a great deal of qualitative inquiry over the years and teaches qualitative methods seminars to doctoral students at Capella University. Based on his research and teaching, our discussions about long-enduring issues and new directions have had a major influence on every edition of this book, including most certainly this one.

On the other end of the time continuum is the most recent contributor to the book, Matthew Cameron-Rogers, a graduate student at the University of Melbourne. He took on the gargantuan task of reading the entire near-final manuscript for coherence, redundancy, nonsensical passages, missing words and phrases, typographical errors, and other authorial sins of both commission and omission. He did so quickly and thoroughly, providing valuable feedback. I predict a brilliant career ahead for him in qualitative research. You heard it here first.

I opened this preface with reference to a study of prefaces. I also examined other prefaces in search of inspiration. That's how I came across Bob Stake's prefactory exemplar in his important and influential book *Multiple Case Study Analysis* (2006), cited often herein. Bob lists every single graduate student he's ever advised: seven-plus pages of names in small font. I felt like I was viewing a celebratory version of the Vietnam War Memorial in Washington, D.C., this one honoring survivors and thrivers rather than those who perished. Bob has created a precedent impossible to emulate, at least for me. I began to list the many colleagues, students, workshop participants, and evaluation clients to whom I am indebted and who deserve acknowledgment for their contributions to my understanding and writing over the years. The task soon felt overwhelming. Every yin deserves its yang, so I'm going to the opposite extreme and not acknowledging anyone else by name. Now that I have reached this fourth edition and traveled many qualitative miles over many, many years, the list of those to whom I am indebted is too long and the danger of leaving out important influences too great for me to include such traditional acknowledgments here. I can only refer the reader to the references and stories in the book as a starting point.

Special thanks to current and former University of Minnesota graduate students who helped with final proofreading: Gifty Amarteiifio, Hanife Cakici, Michaelle L. Gensinger, Melissa Haynes, Mary Karlsson, David Milavetz, Anna Kiel Martin, Nora Murphy, and Gayra Ostgaard.

I would also like to acknowledge the following reviewers of this edition: Susan S. Manning, University of Denver; Eva Mika, Northcentral University; Allison Zippay, Rutgers University; Karina Dancza, Canterbury Christ Church University; C. Victor Fung, University of South Florida; Virginia E. Hines, Ferris State University; Kathleen A. Bolland, the University of Alabama; Dan Kaczynski, Central Michigan University; Suzanne M. Leland, University of North Carolina at Charlotte; and Michael P. O'Malley, Texas State University.

Dedication

I close this preface with some personal history. Qualitative inquiry is personal. The researcher is the instrument of inquiry. This theme will be reiterated throughout the book, including at the beginning of the first chapter and in the closing of the final chapter. The redundant emphasis is intentional, for the personal and interpersonal nature of qualitative inquiry is its great strength, a source of direct experiential insight. It is also what sparks controversy among those whose very definition of research involves excluding the personal and interpersonal as potential sources of bias. In this prolegomenon, I am foreshadowing that discussion and debate, which we'll examine in depth along the way. All I'll say at this point is that what is going on in your life during qualitative fieldwork may well become part of your methodological documentation, for it can affect both data collection and analysis, and therefore deserves attention and reflection.

This book is dedicated to Calla Quinn Campbell-Burke, who was born during the book's development. Calla is my first grandchild, my daughter Charmagne Elise Campbell-Patton's first child. Charmagne has become a qualitative evaluator in her own right, which bodes well (from my perspective) for passing on qualitative genes.

The writing of each edition of this book has been marked in my memory by significant family events. I began the first edition just after my middle son, Julius Quinn Campbell, now a biomechanical engineer, was born. I often dictated portions of the book walking with him in a backpack. The second edition was written during the year when my oldest son, Brandon Quinn Tchombiano Patton, was in his senior year in high school, a momentous year in our family's life as he prepared to go off to college. The third edition was written as Charmagne was finishing college, completing the educational journey from preschool through college for all three of my children. This fourth edition has been dominated by the birth of my first grandchild. Indeed, I was in the midst of conducting a webinar that included qualitative evaluation when she was born, and those participants from around the world became part of the experience as I took a break to rush to the hospital and meet Calla for the first time.

I find that writing about qualitative methods deepens and enhances my observational acuity. Qualitative methods are not just for conducting research and evaluation. Observation and interviewing are life skills. Qualitative methods offer windows through which to take in the dynamic unfolding of the world around us. Nothing makes me more mindful and appreciative of the benefits and joys of observation than observing the development of my granddaughter and her awakening to the world.

End of the Beginning

It is said among authors that books don't get finished or completed. At some point, an author just has to stop writing. That time has arrived.

I would add only that I suspect this fourth edition will be the last print copy of this book. Should I have the good fortune to be around to write a fifth edition, given the trends in publishing, I would expect it to be entirely digital. The first edition was written on a manual typewriter. This edition was written on the cloud. Qualitative methods and reporting have been developing in new directions reflecting larger changes in society and technology. Thus has it ever been.

Thus will it be going forward. What a long, strange trip it's been. But the journey is not over. This book is a marker along the way. Those of you reading and using what is here will be part of creating the future of qualitative methods. I'll be watching what you do as I gather material to document the next leg of the trip.

Cover Art Credit

The cover art, titled *Fable II*, is a wood sculpture by the artist Shaftaï (http://www.shaftai.com/).

Born and raised in the United States and having lived in France for 40 years, Shaftaï works essentially in wood and metal. Rather abstract, his works bring to mind natural forms, dynamic movement, or unusual constructions. Using direct-carving technique, his relationship to a sculpture evolves as the piece takes shape, much like emergent qualitative fieldwork and analysis. He says,

> The success of a sculpture can be judged by its capacity to surprise the artist during its creation and the spectator when he sees it. If the spectator wonders what he's looking at, while finding pleasure in sensual or aesthetic contemplation, his imagination sets the base for his connection to the work, giving that relationship a chance to last. If the piece can be immediately "figured out" or reduced to some form of "meaning" (a title, a reference to reality, a function), the mind will not stay focused on it for long. There being no clear answer to "What is it?," the sculpture says alive.

Shaftaï and I were Peace Corps volunteers together in Burkina Faso in the late 1960s.

—Michael Quinn Patton
May 15, 2014
Pine City, Minnesota

About the Author

Michael Quinn Patton is an independent consultant with more than 40 years' experience conducting applied research and program evaluations. He lives in Minnesota, where, according to the state's poet laureate, Garrison Keillor, "all the women are strong, all the men are good-looking, and all the children are above average." It was this interesting lack of statistical variation in Minnesota that led him to qualitative inquiry despite the strong quantitative orientation of his doctoral studies in sociology at the University of Wisconsin. He was on the faculty of the University of Minnesota for 18 years, including 5 years as director of the Minnesota Center for Social Research, where he was awarded the Morse-Amoco Award for innovative teaching. Readers of this book will not be surprised to learn that he has also won the University of Minnesota storytelling competition.

He has authored six other SAGE books: *Utilization-Focused Evaluation, Creative Evaluation, Practical Evaluation, How to Use Qualitative Methods for Evaluation, Essentials of Utilization-Focused Evaluation,* and *Family Sexual Abuse: Frontline Research and Evaluation.* He has edited or contributed articles to numerous books and journals, including several volumes of *New Directions in Program Evaluation,* on subjects as diverse as culture and evaluation, how and why language matters, HIV/AIDS research and evaluation systems, extension methods, feminist evaluation, teaching using the case method, evaluating strategy, utilization of evaluation, and valuing. He is the author of *Developmental Evaluation: Applying Complexity Concepts to Enhance Innovation and Use* and coauthor of *Getting to Maybe: How the World Is Changed,* a book that applies complexity science to social innovation. His creative nonfiction book, *Grand Canyon Celebration: A Father–Son Journey of Discovery,* was a finalist for Minnesota Book of the Year.

He is a former president of the American Evaluation Association and recipient of both the Alva and Gunnar Myrdal Award for Outstanding Contributions to Useful and Practical Evaluation and the Paul F. Lazarsfeld Award for Lifelong Contributions to Evaluation Theory from the American Evaluation Association. The Society for Applied Sociology presented him the Lester F. Ward Award for Outstanding Contributions to Applied Sociology.

He is on the faculty of The Evaluators' Institute and teaches workshops for the American Evaluation Association's professional development courses and Claremont University's Summer Institute. He is a founding trainer for the International Program for Development Evaluation Training, sponsored by The World Bank and other international development agencies each summer in Ottawa, Ohio.

He has conducted applied research and evaluation on a broad range of issues, including antipoverty initiatives, leadership development, education at all levels, human services, the environment, public health, medical education, employment training, agricultural extension, arts, criminal justice, mental health, transportation, diversity initiatives, international development, community development, systems change, policy effectiveness, managing for results, performance indicators, and effective governance. He has worked with organizations and programs at the international, national, state, provincial, and local levels and with philanthropic, not-for-profit, private sector, international agency, and government programs. He has worked with people from many different cultures and perspectives.

He has three children—a musician, an engineer, and a nonprofit organization development and evaluation specialist—and one granddaughter. When not evaluating, he enjoys exploring the woods and rivers of Minnesota with his partner, Jean—kayaking, cross-country skiing, and snowshoeing—and occasionally hiking in the Grand Canyon. He enjoys watching the seasons change from his office overlooking the Mississippi River in Saint Paul and his home in the north woods of Minnesota.

Finally, a note on *Halcolm,* Patton's philosophical creation, who gives you another perspective on his approach as author and storyteller. Halcolm made his debut in the first edition of this book (1980) as a qualitative inquiry muse and Sufi/Zen teaching master who offered stories that probed the deeper philosophical underpinnings of how we come to know what we know—or think we know. Halcolm's musings, like his name (pronounced slowly), lead us to ponder "how come?" Halcolm was inspired by a combination of the character Mulla Nasrudin from Sufi stories (Shah, 1972, 1973) and science fiction writer Robert Heinlein's (1973) immortal character Lazarus Long, the oldest living member of the human race, who travels through time and space offering wisdom to mere mortals. Part muse and part alter ego, part literary character and part scholarly inquirer, Halcolm's occasional appearances in this research and evaluation text remind us to ponder what we think is real, question what we think we know, and inquire into *how come* we think we know it.

List of Exhibits

PART 1

Framing Qualitative Inquiry
Theory Informs Practice, Practice Informs Theory

- Psychometricians try to measure *it*.
  Experimentalists try to control *it*.
  Interviewers ask questions about *it*.
  Observers watch *it*.
  Participant-observers do *it*.
  Statisticians count *it*.
  Evaluators value *it*.
  Qualitative inquirers find meaning in *it*.
- When in doubt, observe and ask questions.
  When certain, observe at length and ask many
  more questions.
- In a world where much is transient, inquiry
  endures.
- Gigo's Law of Deduction: Garbage in, garbage out.
  Halcolm's Law of Induction: No new experience,
  no new insight.
- Qualitative inquiry cultivates the most useful of all
  human capacities:
  The capacity to *learn*.

- Innovators are told, "Think outside the box."
  Qualitative scholars tell their students, "Study
  the box. Observe it. Inside. Outside. From inside
  to outside, and from outside to inside. Where
  is it? How did it get there? What's around it?
  Who says it's a 'box'? What do they mean?
  Why does it matter? Or does it? What is *not* a
  'box'? Ask the box questions. Question others
  about the box. What's the perspective from
  inside? From outside? Study diagrams of the
  box. Find documents related to the box. What
  does *thinking* have to do with the box anyway?
  Understand *this* box. Study another box. And
  another. Understand *box*. Understand. Then, you
  can think inside *and* outside the box. Perhaps.
  For awhile. Until it changes. Until you change.
  Until outside becomes inside—again. Then, start
  over. Study the box."

There is no burden of proof. There is only the world to
experience and understand. Shed the burden of proof to
lighten the load for the journey of experience.

—From Halcolm's *Laws of Inquiry*

The Nature, Niche, Value, and Fruit of Qualitative Inquiry

Nyansapo, "wisdom knot," Adinkra (West Africa) symbol of wisdom, ingenuity, intelligence, and patience. This symbol conveys the idea that a wise person has the capacity to select the best means to achieve a goal. Being wise means knowing how to apply broad knowledge, learning, and experience for practical purposes (Willis, 1998).

Book Overview and Chapter Preview

Part 1 of this book—this journey deep into qualitative inquiry—provides an overview of qualitative methodology in four chapters—on (1) the nature, niche, value, and fruit of qualitative inquiry; (2) strategic themes in qualitative inquiry; (3) a variety of qualitative inquiry frameworks (paradigmatic, philosophical, and theoretical orientations); and (4) practical and actionable qualitative applications. Part 2 covers qualitative designs and data collection, with chapters on (5) design options, (6) fieldwork and observation, and (7) in-depth interviewing. Part 3 completes the book with chapters on (8) qualitative analysis and (9) enhancing the quality and credibility of qualitative studies.

In this first chapter, Module 1 presents examples of how qualitative inquiry contributes to our understanding of the world. Module 2 examines what makes qualitative data *qualitative*. Module 3 provides an overview of the issues involved in making methods decisions. Module 4 concludes the chapter with a summary of the fruit of qualitative methods, that is, a look at what comes out of qualitative studies.

A thick tree grows from a tiny seed.

A tall building arises from a mound of earth.

A journey of a thousand miles starts with one step.

—Lao-tzu
Philosopher and poet of ancient China

QUALITATIVE WISDOM

A Portuguese professional from Barcelona was driving in a remote area of his country when he came upon a sizable herd of sheep being driven along the country road by a shepherd. Seeing that he would be delayed until the sheep could be turned off the road, he got out of the car and struck up a conversation with the shepherd.

"How many sheep do you have?" he asked.

"I don't know," responded the young man. The professional was embarrassed for having exposed what he assumed was the young shepherd's lack of formal schooling, and therefore his inability to count such a large number. But he was also puzzled.

"How do you keep track of the flock if you don't know how many sheep there are? How would you know if one was missing?"

The shepherd, in turn, seemed puzzled by the question. Then he explained, "I don't need to count them. I know each one, and I know the whole flock. I would know if the flock was not whole."

How Qualitative Inquiry Contributes to Our Understanding of the World

This opening chapter will offer an overview of the nature, niche, value, and fruit of qualitative inquiry. In the spirit of the Adinkra *Nyansapo*, symbol of wisdom, our journey together through various purposes for and contributions of qualitative inquiry aims to enhance your capacity to select the best methods and design to achieve a particular research or evaluation purpose. This chapter will offer a sampling of findings from qualitative studies. In this regard, it will be like a wine tasting, meant to introduce possibilities and support developing a more sophisticated palate, or like appetizers, as an opening to the fuller feast yet to come in later chapters.

In this chapter, we are especially attentive to the fruit of qualitative inquiry. It is important to know what qualitative data yield, what findings look like, and how they are produced, so that you will know what you are seeking to find out and produce when you undertake your own qualitative inquiry. Let's begin, then, with seven ways in which qualitative inquiry contributes to our understanding of the world. The first contribution is illuminating meaning.

Illuminating Meanings: From Birth to Death and In-Between

> Being a person is the activity of meaning-making.
>
> —Robert Kegan (1982, p. 11)
> Developmental psychologist
> Harvard University

What makes us different from other animals is our capacity to assign meaning to things. The essence of being human is integrating and making sense of experience (Loevinger, 1976). Language has developed, and continues to develop, as a uniquely human way to express meaning (Halliday, 1978)—and to disguise meaning. As Shakespeare observed in *Measure for Measure*, "It oft falls out, to have what we would have, we speak not what we mean."

Qualitative research inquires into, documents, and interprets the meaning-making process. Let me illustrate how this occurs and explain why it is so important—indeed, why it is the core of qualitative inquiry and analysis. I'll begin with a personal example. During the writing of this book, my first grandchild was born, and this book is dedicated to her. The hospital records document her weight, height, health, and *Apgar* score—activity (muscle tone), pulse, grimace (reflex response), appearance, and respiration. The mother's condition, length of labor, time of birth, and hospital stay are all documented. These are physiological and institutional metrics. When aggregated across many babies and mothers, they provide trend data about the beginning of life—birthing. But nowhere in the hospital records will you find anything about what the birth of Calla Quinn *means*. Her name is recorded but not why it was chosen by her parents and what it means to them. Her existence is documented but not what she means to our family, what decision-making process led up to her birth, the experience and meaning of the pregnancy, the family experience of the birth process, and the familial, social, cultural, political, and economic context that is essential to understanding what her birth means to family and friends in this time and place. A qualitative case study of Calla's birth would capture and interpret the story and meaning of her entry into the world from the perspectives of those involved in and touched by her coming into our lives. This might, or might not, include the fact that at the moment she was born I was in the midst of conducting a webinar on qualitative evaluation and those participants from around the world became part of the experience as I took a break to rush to the hospital and meet Calla for the first time. Several participants subsequently sent me e-mails that Calla's birth made the webinar more meaningful for them. This example of the meaning of her birth as a potential qualitative case study was born during that webinar.

I open with this personal story for another reason. Qualitative inquiry is personal. The researcher is the instrument of inquiry. What brings you to an inquiry matters. Your background, experience, training, skills, interpersonal competence, capacity for empathy, cross-cultural sensitivity, and how you, as a person, engage in fieldwork and analysis—these things undergird the credibility of your finings. Reflection on how your data collection and interpretation are affected by who you are, what's going on in your life, what you care about, how you view the world, and how you've chosen to study what interests you is a part of qualitative methodology. The obligation and commitment to acknowledge and take into account the personal and interpersonal nature of qualitative inquiry will be a recurring theme of this book. I've been at this for more than 40 years. My granddaughter's birth infused my writing with new energy and urgency as I imagined addressing a new generation of qualitative researchers and evaluators.

So let us turn now to the other end of the human existential continuum. Systematically gathering data on deaths began in the Black Plague, when, from 1347 to 1351, a third or more of all Europeans died. From that time, England began tracking deaths, eventually developing the death certificate, which specifies cause of death.

> Of the roughly fifty million people who will die this year, approximately half will get a death certificate. That figure includes every fatality in every developed nation on earth: man, woman, child, infant. The other half, death's dark matter, expire in the world's poorest places, which lack the medical and bureaucratic infrastructures for end-of-life documentation. (Schulz, 2014, p. 32)

The death certificate has become a crucial source of epidemiological data documenting trends in causes of death, which has influenced policymaking, research priorities, and allocation of public health resources. Epidemiological studies go beyond death certificates to estimate deaths caused by poverty, low levels of education, smoking, obesity, and inactivity—causes in the same range as deaths from heart attacks and cancer (Galea, Tracy, Hoggatt, DiMaggio, & Karpati, 2011). But as with birth certificates, death certificates and epidemiological studies do not capture what the death of someone means to those touched by that death. Only an in-depth case study can even begin to do that. To understand how humans face death and make sense of dying under the most extreme conditions, Viktor E. Frankl (2006), a neurologist, psychiatrist, and Holocaust survivor, studied *the search for meaning* in World War II concentration and death camps. The capacity to find meaning in suffering and death, he concluded, was the key to survival.

So aggregate statistics on mortality reveal causes of death but don't tell us how people find meaning in dying and how cultures make sense of death. That kind of inquiry is the focus of the *anthropology of death*, a specialized area of cross-cultural and cross-institutional inquiry.

> The anthropology of death takes as its task to understand the phrase: "All humans die," yet in every culture, each dies in their own way. . . .

Death is an intensely emotional and often taboo subject, so that studying death raises special dilemmas and emotional challenges for the fieldworker. . . . [Anthropologist] Hortense Powdermaker, working in a matrilineal society in New Ireland, described her own extreme distress when she began taking field notes at her first funeral. She imagined how intrusive such an ethnographic presence would be in her own house of mourning. She, however, discovered to her great surprise that these non-literate people felt no such intrusions, but rather that her writing added prestige to the ritual. They demanded her presence at every subsequent funeral, even long after she had constructed a complete account of the funeral process.

> Many ethnographers have discussed the emotional strain of participating closely in the grief of others. . . . Perhaps the most moving account is by Rosaldo who connects how his overwhelming grief at the accidental death of his wife, (also an eminent anthropologist) helped him understand more deeply [the] headhunter's rage of the *Ilongot* in the Philippines. (Abramovitch, 2014, p. 1)

An exemplar of an in-depth qualitative inquiry into death is a study by Karen Martin (2007) of sudden infant death syndrome (SIDS), the leading cause of death among apparently healthy infants between the ages of one week and one year. Her case studies document the painful experience of bereaved parents, who frequently blame themselves for their baby's death. She looks at how parents grieve, the meanings and casual explanations they attribute to a SIDS death, the effects of their grief on family relationships, and the strategies they use to cope and carry on.

The anthropology of death includes how different cultures explain, talk about, and deal with death. Americans appear to have particular difficulty dealing with death. Debate about the wisdom and costs of extending life a few months with hugely expensive medical technology, surgery, and drugs has become not only a difficult matter of medical ethics but also a volatile political issue (Brown, 2014). These are matters for qualitative inquiry.

We construct and attach meaning to births. We try to make sense of death—and culture tells us how to do so. Now let's look at five diverse examples between birth and death of how qualitative inquiry contributes to understanding human *meaning making*.

- *Bodily meaning making:* Qualitative inquirers have studied the meanings attached to male and female bodies, disabled and injured bodies, bodies of different colors and sizes, how and why people adorn their bodies (e.g., with jewelry, tattoos, piercings, intentional scars), and how and why they mutilate them (e.g., by circumcision, female genital mutilation, cutting off limbs in war, scalping, cannibalism, and sexual abuse).

Because our bodies serve as a site of meaning making within our culture, they also serve as a site of scholarly investigation. . . .

Wanda Pillow discovered the centrality of the body in her research . . . on pregnant teenagers and their experiences. . . . The body, the changing body, the experience of the pregnant body, structural responses to girls' changing bodies, and the perceptions of others toward girls' pregnant bodies became central to her research. In fact, without focusing on the body, it would not have been possible to understand much of the experience of this population. Accordingly, Pillow modified her research to focus on the bodies and bodily experiences of the girls she was studying. In other words, she developed a *body-centered methodology.* . . . A shift to the body allowed her to ask and answer research questions that would otherwise be impossible to address. Likewise, she was able to access knowledge that would otherwise remain invisible. (Hesse-Biber & Leavy, 2006a, pp. xx–xxi)

- **Evaluative meaning making:** Evaluation involves making judgments about what is meaningful. One important form of evaluation is assessing students' academic achievement. Magolda and King (2012) interviewed nearly 2,000 college students to find out how they make learning meaningful. They found that many students fail to achieve complex learning goals because they rely too heavily on others' opinions about what to believe, who to be, and how to relate to others. In other words, peer pressure trumps individual meaning making, especially early in the college experience. Over time, successful students learn to decide for themselves what is meaningful, what Magolda and King call "self-authorship." They conclude that understanding and assessing students' meaning making is essential for interpreting students' academic performance and other behaviors and should inform the design of new programs and services.
- **What objects mean:** Humans attach meaning to things, what anthropologists call *material culture.* Art, food, toys, jewelry, land, cars, perfume, clothes . . . anything can become meaningful to those people within a setting who attach value to it. The dictum that "an Englishman's home is his castle" attaches special meaning, legal protections, and

social status to one's place of abode. National flags are symbols full of meaning. Music has meaning. The Olympic medal presentations combining flags and music evoke strong emotions. Qualitative inquiry includes studying the meaning making associated with things as diverse as Smartphones, Facebook, and hair dye (Berger, 2014).

- **Meaning in meaninglessness:** Social groups are typically defined by their shared meaning making. In an ironic twist, some groups find meaning around a commitment to meaninglessness. Nihilism is a philosophical assertion that life has no meaning. Nihilists find common meaningfulness in asserting meaninglessness. How and why this occurs, and its effects on those involved, is a matter well suited for qualitative inquiry. Distinguished British philosopher and author Aldous Huxley (1894–1963) studied and reflected on the political and moral attraction of a philosophy of meaninglessness in England during the 1920s and 1930s. He interpreted it as essentially a means of liberation from conservative morality and politics, resisting being told what to believe by the powerful (religious, corporate, and government leaders). Adherents of a philosophy of meaninglessness justified their political and erotic revolt by denying that the world had any meaning at all (Huxley, 1937).
- **Qualitative interpretation as meaning making:** Qualitative inquiries study how people and groups construct meaning. In so doing, qualitative methodology devotes considerable attention to how qualitative analysts determine what is meaningful. Qualitative analysis involves interpreting interviews, observations, and documents—the data of qualitative inquiry—to find substantively meaningful patterns and themes. Doing so is an act of interpretation. Distinguished qualitative methodologist Robert Stake (2010) explains what this means:

Interpretation is an act of composition. The interpreter takes descriptions and makes them more complex, drawing upon a few conceptual relationships. He or she might take the term *work* and give it muscle, durability, remuneration, and self-respect. These can be some of the larger meanings of *work.* He or she might take an episode observed at the workplace and give it personality, history, tension, and implication. The best interpretations will be logical extensions of the simple description but also will include contemplative, speculative, even aesthetic extension. The reader would be deceived if allowed to think that these interpretations had been agreed upon, certified in some way. They are contributions of the researcher, written so as to make it clear they are personal interpretations. All people make interpretations. All research requires interpretations. Qualitative research relies heavily on interpretive perceptions throughout the planning, data gathering, analysis, and write-up of the study. (p. 55)

The first contribution of qualitative inquiry, then, is illuminating meanings and how humans engage in meaning making—in essence, making sense of the world. Science fiction author Piers Anthony could have been talking about the challenge of qualitative inquiry when he observed, "All things make sense; you just have to fathom how they make sense."

Studying How Things Work

Michael Scriven is a founder of the transdisciplinary profession of evaluation. Research can involve studying how anything works. Program evaluation involves studying how a program works and what results it gets to render a judgment about its effectiveness. Scriven (1998) tells about being invited to evaluate a computer-based approach used by the counseling center at the University of California at Irvine. He accepted, and then things got interesting:

> I ran three of my graduate students through the program, and its disastrous failings emerged readily. From the administrator's desk, dazzled by the computers, these failings—of content as well as of the machinery—were invisible. In any case, they refused payment in order to not have my critical report in their files. I said I would be happy not to charge them and instead use it as the theme for my next published article. So they called and said they had appointed a negotiator. I called the negotiator and asked if he was empowered to negotiate to the full amount of the contract and he said, "Absolutely." So I said fine, that I would not charge them since they did not think it worth paying for, but I would use the example in every future speech that I made on a related topic. (p. 13)

In that story are two examples of how things work. First is a glimpse into how the counseling center's program worked—or rather didn't work. The second story is how negotiating settlement of the evaluation contract worked.

Students of anatomy examine how the body works. Social scientists study how human groups and institutions work. The contribution of qualitative research and evaluation to understanding *how things work* is highlighted by the opposite phenomenon expressed in the title of Nigerian Chinua Achebe's (1994) classic story of the clash between Western and traditional African values during and after the colonial era: *Things Fall Apart*.

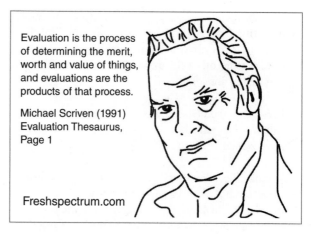

Evaluation is the process of determining the merit, worth and value of things, and evaluations are the products of that process.

Michael Scriven (1991) Evaluation Thesaurus, Page 1

Freshspectrum.com

SOURCE: From Scriven, M. (1991). *Evaluation thesaurus* (4th ed., p. 1). Thousand Oaks, CA: Sage. Used by permission of cartoonist Chris Lysy.

Robert Stake, quoted earlier about the centrality of interpretation in making sense of the world, subtitled his book on qualitative research *Studying How Things Work*. Here's what Stake (2010) says it means:

> Understanding the social and professional worlds around us comes from paying attention to what people are doing and what they are saying. Some of what they do and say is unproductive and silly, but we need to know that, too. A lot of what people do is motivated by their love for their families and a desire to help people, and we need to know that, too. We won't just ask them. We will look closely to see how their productivity and love are manifested. I put "Studying How Things Work" in the title . . . to help you improve your ability to examine how things are working. Most of the things I have in mind are small things—small but not simple, such as classrooms and offices and committees. But also gerundial things, nursing and mainstreaming and fund-raising, in particular situations. And some special things, such as ordering chairs for a classroom, and "labor and delivery," and personal privacy. (p. 2)

The possibilities for studying how things work is vast. How does culture work? How do families work? Small groups? Universities? Movements? Systems? What is meant by qualitatively studying how things work is getting inside the phenomenon of interest to get detailed, descriptive data and perceptions about the variations in what goes on and the implications of those variations for the people and processes involved. A major way to do that is to capture people's stories about how things work.

Capturing Stories to Understand People's Perspectives and Experiences

> The universe is made of stories, not atoms.
>
> —Muriel Rukeyser (1913–1980)
> American poet and political activist
>
> Stories make us human.
>
> —Jonathan Gottschall (2012)
> *The Storytelling Animal*

If you want to know *how much* children can read, give them a reading test. If you want to know what reading means to them, you have to talk with them, listen to them, and hear their stories about the stories they love. Exhibit 1.1 gives examples of the kinds of questions you might ask.

These are qualitative inquiry questions aimed at getting an in-depth, individualized, and contextually sensitive understanding of reading for each child interviewed. Of course, the actual questions asked would have to be adapted to the child's age and language skills, the school and family situation, and the purpose of the inquiry. But regardless of the precise wording and sequence of the questions, the purpose is to hear children talk about reading in their own words; find out about their reading behaviors, attitudes, and experiences; and get them to tell stories that illuminate what reading means to them. You might talk to groups of kids about reading as a basis for developing more in-depth, personalized questions for individual interviews. While doing fieldwork (actually visiting schools and classrooms), you would observe children reading and the interactions between teachers and children around reading. You would also observe what books and reading materials are there in a classroom and how they are arranged, handled, and used. In a comprehensive inquiry, you would also interview teachers and parents to get their perspective on the meaning and practice of reading, both for children and for themselves, as models their children are likely to emulate.

In analyzing your classroom observations and interviews with children, parents, and teachers, you would provide illustrative case examples of variations in reading practices and what it means to those interviewed. You would report and explain any patterns or themes that emerged in the responses to your interview questions. For example, eight-year-old boys I interviewed told me that good readers are bad at sports.

EXHIBIT 1.1 Open-Ended Interview Questions About Reading

If you want to know how much *children can read, give them a reading test. If you want to know what reading means to them, you have to talk with them. Here are examples of open-ended interview questions about reading:*

Tell me about something you're reading now.

What do you like to read in school? How does reading relate to other subjects in school? What do you read on your own, outside school? When do you read?

What do you like about reading? What don't you like?

Tell me about reading in your family. What do people in your family say about reading? What do your friends say about it?

Some kids seem to be good readers, and some have trouble reading. Why is this, do you think? From what you've seen, what's the difference between good readers and not so good readers?

What about you, how would you describe your level of reading? Why?

How is your reading today different from a year ago, if at all?

Why do you think so much emphasis is placed on reading in school? When you think about the future, how important do you think reading will be to what you do? Why?

Tell me one of your favorite stories. What makes it a favorite?

They believed that a little reading was okay, but if you read too much, it interferes with your muscles getting stronger. They were careful to read just enough to do okay in school but not so much as to hurt their aspirations to become good athletes. Teachers had heard this, they told me, but didn't take it seriously. The boys I talked with took it very seriously. The eight-year-old girls thought the boys were just dumb and had silly and stupid ideas.

The results from your in-depth qualitative inquiry into reading could be used to adapt and improve approaches to reading, both in school and at home. Understanding how students view reading is critical to dealing with the development of reading skills as well as supporting positive attitudes about reading and motivation to get better at it.

Discovery of a 4,000-Year-Old Prehistoric Cave Painting of a Wooly Mammoth Encounter: An Early Example of a Qualitative Report

Making sense of and communicating research findings involves "the art of storytelling" (Hastings & Domegan, 2014, p. 117). Capturing and understanding diverse perspectives, observing and analyzing behaviors in context, looking for patterns in what human beings do and think—and examining the implications of those patterns—these are some of the basic contributions of qualitative inquiry.

Elucidating How Systems Function and Their Consequences for People's Lives

Why do people do what they do even when it doesn't seem to make sense to an outsider? Sometimes the answer lies within the individual enmeshed in a person's background, personality, upbringing, worldview, and conditioned behaviors. Qualitative research often inquires into the stories of individuals to capture and understand their perspectives, as just discussed. But often the answer to why people do what they do is found not just within the individual but, rather, within the systems of which they are a part: social, family, organizational, community, religious, political, and economic systems.

Atul Gawande (2007), a Harvard Medical School surgeon, tells of visiting the Walter Reed military hospital early in the Iraq war. He participated in a session interpreting eye injury statistics. The doctors were having considerable success saving some soldiers from blindness, a positive outcome. But digging deeper, the doctors asked why so many severe eye injuries were occurring. Interviewing their patients, they learned that the young soldiers weren't wearing their protective goggles because they were considered too ugly and uncool. They recommended that the military switch to "cooler-looking Wiley X ballistic eyewear. The soldiers wore their eye gear more consistently and the eye-injury rate dropped immediately" (p. A23). By asking these kinds of deeper questions about what's really going on and inquiring into assumptions about why things are happening, qualitative researchers and evaluators contribute to knowledge about what works, what doesn't, and why.

Let's place this example in a larger context. Those involved in solving problems repeatedly share frustrations about the stream of ever more sophisticated ambulances sent to accidents at the bottom of a cliff, when simply building a fence at the top would prevent those accidents. Why doesn't the fence get built? A qualitative researcher would likely find a number of different perspectives and explanations and learn that building a fence involves political jurisdictions, financial priorities, environmental issues, contextual complexities, and conflicts about who is involved in such a decision and who actually decides what kind of fence is to be built by whom, as well as some interesting and important local history and stories about the cliff, the road that runs along it, and the people who use that road, maybe even stories about the fence or fences that used to be there. What might begin as a seemingly simple question about why a fence hasn't been built becomes, with in-depth inquiry, a case study of complex system dynamics. Moreover, social, cultural, and political perspectives about how to solve problems all come into play. In reviewing numerous such stories, distinguished Australian action research scholar and practitioner Yolande Wadsworth (2010) has commented that they are reminders about our repeated tendency to go for the short-term quick fix rather than examine, come to understand, and take action to change *how a system is functioning* that creates the very problems being addressed. In-depth qualitative inquiry can illuminate system and systemic issues and potential solutions.

Understanding Context: How and Why It Matters

When qualitative inquirers study how systems function and the consequences of system dynamics, they include attention to context. Context refers to what's going on around the people, groups, organizations, communities, or systems of interest. If we're studying family farming systems in northwest Minnesota or eastern Burkina Faso, we'll need to attend to larger-context issues such as climate change, political and economic trends, and public health threats, such as the obesity epidemic in the United States or the spread of HIV/AIDS in Africa.

The theme of the 2009 annual conference of the American Evaluation Association (AEA) was "Context and Evaluation." AEA president Debra Rog (2009) articulated the challenge of taking context seriously:

> As evaluators, we recognize that context matters.... Context has multiple layers and is dynamic, changing over time. Increasingly, we are aware of the need to shape our methods and overall approach to the context. Each of the dimensions within the context of an evaluation influences the approaches and methods that are possible, appropriate, and likely to produce actionable evidence. In tandem, some evaluations embrace context and include it within the study, rather than simply attempting to control its effects. Attention to context helps to produce findings that are generalizable and useful to a broader set of stakeholders outside the local decision-making context.

When new cases of polio emerged in a remote area of India, epidemiologists accompanied the revaccination team and interviewed people in the area to find out how and why some children had missed being vaccinated in the recent campaign. They learned that some Muslim mothers had resisted the vaccination and had hidden their children because they'd heard rumors of a Hindu plot to sterilize their boys (Gawande, 2004). Understanding the resistance to vaccination or other health practices, and developing approaches to overcome that resistance, requires an in-depth understanding of the cultural, social, and political systems within a particular context.

Context includes attention to and understanding of important nuances of culture, politics, economy, history, geography, resources, and institutions. Qualitative inquiry makes attention to context a priority both for data collection and for reporting findings. This means documenting diversity and the contextual factors that explain particular variations even while identifying cross-cutting patterns and themes.

Bottom line: *Sensitivity to context is central in qualitative inquiry and analysis.* Sounds reasonable, I trust. Perhaps even matter-of-fact. But maybe a bit abstract. So let's make it concrete. What is the context for your reading this book? For me, or anyone, to understand how you engage with this book—indeed, how you engage with these very words as you read them—I would need to know the context within which you are reading. Are you taking a required course and are required to read this? Are you doing, or considering doing, a qualitative thesis or dissertation? Are you undertaking an evaluation project that includes qualitative methods? Are you new to qualitative inquiry, or are you an experienced researcher reading this as a refresher or to see how these methods have developed in the past few years? Do you come to this book with a strong quantitative methods background? Are you working or studying in a context that values qualitative data or one of skepticism about the value and credibility of qualitative findings? Are you just studying qualitative methods at this point for possible future use, or are you engaged in or about to engage in a qualitative study? If the latter, are you doing so alone or as part of a team? These are questions that illuminate the *context* within which you are reading this book. If I am to understand your reading of this book, and your reactions to it, I need to know the context within which you are reading it.

Identifying Unanticipated Consequences

> That there will be unanticipated consequences is the one sure thing we can anticipate.
>
> —Halcolm

The great delusion of our times is that we can control what happens. Politicians routinely promise to bring about changes in things such as the economy, over which they have little or no control. People running programs of all kinds establish objectives and implementation strategies and then follow the admonition to *plan your work and work your plan.* Good advice. But things seldom work out quite as planned. I've seen literacy programs aimed at helping high school dropouts learn to read that made them hate reading. I've seen juvenile justice programs aimed at rescuing delinquents from a potential life of crime that propelled them inadvertently on a pathway to becoming hardcore criminals. I've seen interventions that intended to support welfare recipients become economically self-sufficient that contributed to multigenerational poverty. I've seen international agricultural development programs aimed at increasing food production and income from cash crops turn fragile soils into desert through the introduction of inappropriate technological and management schemes. None of

Context Matters: The Example of the Paris Declaration on Aid Effectiveness

The 2005 Paris Declaration on Aid Effectiveness was a landmark agreement between aid donors and developing countries aimed at reforming aid processes and increasing aid results. The Paris Declaration Principles were endorsed by more than 150 countries and organizations, including the more developed aid donor countries, such as the United States, developing countries from around the world, and international development institutions, such as the World Bank, the United Nations Development Group, and the Organization for Economic Co-operation and Development. The five principles are as follows:

1. *Ownership:* Developing countries set their own strategies for poverty reduction, improve their institutions, and tackle corruption.
2. *Alignment:* Donor countries align behind these objectives and use local systems.
3. *Harmonization:* Donor countries coordinate, simplify procedures, and share information to avoid duplication.
4. *Results:* Developing countries and donors shift focus to development results, and results get measured.
5. *Mutual accountability:* Donors and partners are accountable for development results. (OECD, 2005)

An independent evaluation examined what difference, if any, the Paris Declaration made to development processes and results (Wood et al., 2011). The scope of the evaluation was immense: case studies in 22 developing countries and in-depth reviews in 18 donor agencies as well as studies on special themes such as health sector aid.

The evaluation was conducted in two phases over four years, between 2007 and 2011. The case studies used mixed methods, including interviews with "key knowledgeables" (people in a position to know firsthand about the subject of an inquiry), reviews of documents and official reports, and surveys of people involved with aid initiatives. The evaluation received the 2012 Outstanding Evaluation Award from the American Evaluation Association.

I became involved as the evaluator of the evaluation, to independently assess the rigor of the evaluation processes and render judgment about the credibility and utility of the findings (Patton & Gornick, 2011a). I observed two international meetings of those involved from developing countries and donor agencies engaged in international development assistance. As the findings were reported and discussed, one theme dominated: *Context matters.* The five reform principles provided a shared global commitment and a standard inquiry framework, but Indonesia is different from Vietnam, Colombia is different from the Cook Islands, and Afghanistan is different from South Africa. Development aid from Sweden is handled differently than aid from the United States, just as Australia, Switzerland, and Japan have different priorities, processes, and relationships with recipient countries.

Thus, the evaluation of the Paris Declaration not only included 22 country case studies but also offered cross-cutting judgments, for example, that country ownership (Principle 1 above) showed significant progress over the past decade, while mutual accountability (Principle 5) has lagged. Qualitative inquiry balances the particular (specific case studies in context) with the general (findings that cut across cases and contexts).

---

these programs were run by uncaring or incompetent people. Quite the contrary. In each case, I found the program leaders and staff to be committed, motivated, hardworking, and deeply engaged in bringing about change but, in the end, ineffective. Indeed, they were not just ineffective but counterproductive, having the opposite of their intended effects.

Qualitative inquiry is especially valuable for identifying unintended consequences and side effects. If all

a program evaluator looks at is whether the intended outcomes are attained, especially using standard performance indicators such as reading tests, employment statistics, and health outcome data, then other, unintended effects will be missed. To find unanticipated effects, you have to go into the field where things are happening, observe what is really going on, interview program participants about what they're experiencing, and find out through *open inquiry* what is happening, both intended *and* unintended. Consider these examples:

- A program for chronically unemployed men wanted to improve its diagnosis of participants' needs, so an extensive battery of assessment tests was added to the intake process. But the men coming into the program were so turned off by all the testing that the drop-out rate soared. The tests were supposed to communicate a deep concern with understanding the particular needs of each man. What the men experienced was quite different. The tests made them feel stupid, reminded them of years of failure in school, and came across as impersonal, mechanistic, and alienating.

- A parent education program for young, poor, first-time mothers aimed to increase their knowledge and skills as parents. Those outcomes were attained, and the mothers were grateful. But when asked what was the most important thing they got out of the program, they said that it wasn't the knowledge and skills attained, as important as those were, but the new relationships with other mothers. They came into the program feeling isolated, lonely, abandoned, and afraid. They came out with friends, playmates for their children, and a network of support. Those were unintended side effects from the perspective of the program funders and staff. They were what made the program worthwhile from the perspective of the mothers.

- The economy of Kiribati, an island nation in the central Pacific, depends on coconut oil and fishing. But overfishing threatened the island's future. The government began subsidizing the coconut oil industry to increase the islanders' income from coconut production and reduce overfishing. The idea was that if people spent more time growing coconuts, they would spend less time fishing. What happened? Fishing increased dramatically, and the reef fish population dropped precipitously, putting the whole ecosystem at risk. It turned out that paying people more to do coconut agriculture actually increased fishing because as people earned more money making coconut oil, they could work less to support themselves and spend more leisure time fishing. They didn't just fish for income. They fished because they liked to fish, and so having more income from coconut production gave them more time to fish (Walsh, 2009).

- Economists and financial analysts spent decades building sophisticated statistical models to predict market behavior and manage both the domestic and the global economy, yet those models did not predict the financial crisis of 2008 that wiped out $50 trillion in global wealth and increased poverty and human suffering around the world. The sophisticated prediction models used by financial planners and government policymakers unexpectedly contributed to and deepened the crisis by promulgating a false sense of security that what actually unfolded could never happen (Brooks, 2010). In-depth interviews with the financial sector, government, and corporate leaders, who, in deep crisis mode, were trying to figure out what to do to avert a worldwide economic collapse, consistently reported a sense of disbelief and shock: "I don't know how this happened." "This was never supposed to happen." "This can't be happening." But it did happen. And to this day, the full story of what happened and its long-term implications for the global economy are still unfolding.

- A social service program was created in Detroit to support young inner-city African Americans living with AIDS. Interviews with participants discovered that some of these destitute young people, when they heard about the services provided, had intentionally contracted HIV/AIDS to become eligible for social services. Despite knowing that it was a potentially fatal disease, they felt so hopeless that the risk felt worth taking to have someone care for them for once (Tourigny, 1998).

- The mission of the United States Forest Service highlighted prevention of forest fires. The Smokey Bear campaign was the centerpiece of a highly effective public campaign to prevent forest fires—by setting aside national forests and parks with the aim of protecting them from fire. The unanticipated consequence has been that the successful prevention of forest fires created a widespread habitat for fires that are much more destructive and devastating than the smaller, natural fires of the past. "So," according to modern forest ecologists, "instead of a few dozen trees per acre, the Southwestern mountains of New Mexico, Arizona, Colorado and Utah are now choked with trees of all sizes, and grass and shrubs. Essentially, it's fuel. And now fires are burning bigger and hotter. They're not just damaging forests—they're wiping them out. [In 2011] more than 74,000 wildfires burned over 8.7 million acres in the U.S." (Joyce, 2012, p. 1).

The particular niche and contribution of qualitative methods in uncovering unanticipated consequences come from the *openness of inquiry*: asking open-ended interview questions, doing fieldwork in a way that is open to whatever turns up, studying documents to discover patterns that are hidden in the details, and observing with open eyes and an open mind. There is a lot of rhetoric in research and evaluation about the importance of looking for unintended consequences, but the rhetoric rings hollow unless the design includes sufficient time, money, and investigator skills to do fieldwork and undertake genuinely open-ended inquiry to find out what is actually happening. You can't find this out from surveys and performance indicators because you don't know what you don't know, so you can't ask questions about it or measure it. The dirty little secret in much of research and evaluation is that the designs do not give serious attention to the emergent and unexpected because those who design studies are primarily interested in testing their predetermined hypotheses and analyzing established indicators rather than openly inquiring into the complex and dynamic ways in which the real world unfolds.

The mind-set that is critical in open inquiry is to expect the unexpected, look for it, and see where it leads you. That is the nature, niche, and value of qualitative inquiry. And, as noted in the previous section, context comes into play. As the famed and infamous actress Angelina Jolie explained in summarizing her youth, "You're young, you're drunk, you're in bed, you have knives; shit happens."

Stuff happens everywhere. Qualitative inquiry documents the stuff that happens among real people in the real world in their own words, from their own perspectives, and within their own contexts; it then makes sense of the stuff that happens by finding patterns and themes among the seeming chaos and idiosyncrasies of lots of stuff. Woody Allen, award-winning filmmaker, once witticized, "If you want to make God laugh, tell him about your plans." Qualitative researchers and evaluators document the plans and study the consequences of attempting to carry out those plans, both intended and unintended.

Making Case Comparisons to Discover Important Patterns and Themes

> The Internet is a big deal, but electricity was bigger.
>
> Building a great company requires adherence to principles predating both.
>
> —Jim Collins (2000)
> "The Timeless Physics of Great Companies"

Jim Collins identifies principles of organizational effectiveness by comparing successful and unsuccessful companies. He is one of the most influential management scholars and consultants of the twenty-first century. His books have sold more than 10 million copies worldwide.

- *Built to Last: Successful Habits of Visionary Companies* compares businesses that have endured over time with those that have failed (Collins & Porras, 2004).
- *Good to Great: Why Some Companies Make the Leap . . . and Others Don't* (Collins, 2001a), translated into 35 languages, compares matched pairs of companies that started out in the same industries with similar profiles (market share, size, and profitability). One in each pair remained "good," while the other went on to become "great" (market share dominance, sustained profit growth, and high prestige).
- *How the Mighty Fall* presents case studies of how some great companies self-destruct (Collins, 2009).
- *Great by Choice: Uncertainty, Chaos, and Luck—Why Some Thrive Despite Them All* compares companies that successfully adapted to complex dynamic environments with those that failed to adapt—and thus failed (Collins & Hansen, 2011).

The cornerstone of Collins's research method is conducting in-depth case studies that allow drawing systematic contrasts between successful and unsuccessful companies.

> The critical question is not "What do successes share in common?" or "What do failures share in common?" The critical question is "What do we learn by studying the contrast between success and failure?" Think of it this way: Suppose you wanted to study what makes gold medal winners in the Olympic Games. If you only studied the gold medal winners by themselves, you'd find that they all had coaches. But if you looked at the athletes that made the Olympic team, but never won a medal, you'd find that they also had coaches! The key question is, "What systemically distinguishes gold medal winners from those who never won a medal?"
>
> Our comparison method has proven to be the key for calling into question powerfully entrenched myths and discerning fundamental principles that apply over long stretches of time and across a wide range of circumstances. (Collins, 2012, p. 1)

Qualitative Inquiry Contributions

In this module, we have looked at seven kinds of knowledge-generating contributions that can flow from qualitative inquiry:

1. Illuminating meaning
2. Studying how things work
3. Capturing stories to understand people's perspectives and experiences

4. Elucidating how systems function and their consequences for people's lives

5. Understanding context: how and why it matters

6. Identifying unanticipated consequences

7. Making case comparisons to discover important patterns and themes across cases

Exhibit 1.2 summarizes these contributions.

As the book unfolds, we will examine many more contributions of qualitative inquiry to knowledge, both theory and practice, and the interconnections between the two. But we're just getting going. Let's step back and more explicitly define the nature and niche of qualitative methods.

EXHIBIT 1.2   The Contributions of Qualitative Inquiry: Seven Examples

QUALITATIVE CONTRIBUTION	INQUIRY FOCUS
1. *Illuminating meanings*	Qualitative inquiry studies, documents, analyzes, and interprets how human beings construct and attach meanings to their experiences. Birth, death, learning—indeed, any and all human experiences—are given meaning by those involved. Interviews and observations reveal those meanings and their implications.
2. *Studying how things work*	Program evaluations study what participants in programs experience, the outcomes of those experiences, and how program experiences lead to program outcomes. More generally, qualitative inquiry can illuminate how any human phenomenon unfolds as it does: how churches, social groups, political campaigns, community events, and social media work—and the effects on those who participate.
3. *Capturing stories to understand people's perspectives and experiences*	An in-depth case study tells *the story* of a person, group, organization, or community. There's a starting point (baseline); events unfold; some point of closure is reached. The story, well-documented and well told, opens a window into the world of the case(s) studied.
4. *Elucidating how systems function and their consequences for people's lives*	Systems involve complex interdependent dimensions that interact in ways that affect the people in those systems. Family systems, cultural systems, organizational systems, political systems, economic systems, community systems: qualitative inquiry systematically gathers perspectives on what happens within systems, and how what happens has implications for those involved. The results are systems stories and insights.
5. *Understanding context: how and why it matters*	Context refers to what's going on around the people, groups, organizations, communities, or systems of interest. If someone wants to understand what brings you to this book, the context within which you are reading (school, job, project, professional development, team, workshop) will be critical to illuminate and understand. People's lives and events unfold within larger, enveloping contexts. For qualitative inquiry and analysis, contextual sensitivity is central.
6. *Identifying unanticipated consequences*	Leaders, planners, social innovators, managers, politicians, change agents, community organizers, evaluators—the list goes on and on—strive to attain their intended goals. The modern world is highly goal oriented. But things seldom go as planned. Much of what was intended never occurs, and things that are never intended, and never even imagined, do occur. The open-ended fieldwork of qualitative inquiry documents both intended and unintended consequences of change processes.
7. *Making case comparisons to discover important patterns and themes across cases*	Comparisons involve analyzing both similarities and differences. We learn and deepen our understanding of phenomena of all kinds by drawing contrasts and making comparisons. Case studies provide rich data for teasing out what cases have in common and what sets them apart: successes versus failures, those who are resilient and those who are not, those who have long marriages and those who have multiple divorces, those who engage with qualitative methods and those who insist that only numbers count. Comparisons illuminate the enormous diversity of humanity even as we seek and find patterns across that diversity.

> Not everything that counts can be counted, and not everything that can be counted counts.
>
> —William Bruce Cameron (1963)
> Sociologist

Qualitative inquiry includes collecting quotes from people, verifying them, and contemplating what they mean. For an in-depth qualitative inquiry into the Cameron quote above, including its origin, varied versions over time, and diverse attributions, including Albert Einstein and many others, see Garson O'Toole (2010). My extensive use of quotations throughout this book treats them as both examples of qualitative data and sources of insight.

Qualitative reports describe and interpret something—whatever was studied. The data are words, stories, observations, and documents. Qualitative findings are based on three kinds of data: (1) in-depth, open-ended interviews; (2) direct observations; and (3) written communications. Interviews yield direct quotations from people about their experiences, opinions, feelings, and knowledge. Data from observations consist of detailed descriptions of people's activities, behaviors, actions, and the full range of interpersonal interactions and organizational processes that are part of observable human experience. Written communications are a rich source of data. Finding, studying, and analyzing documents of all kinds are a part of qualitative inquiry. For example, qualitative data can include excerpts, quotations, or entire passages from organizational, clinical, or program records; memoranda and correspondence; social media postings; official publications and reports; personal diaries; and open-ended written responses to questionnaires and surveys (see Exhibit 1.3).

The data for qualitative analysis typically come from fieldwork. During fieldwork the researcher spends time in the setting under study—a program, an organization, a community, or wherever situations of importance to a study can be observed, people interviewed, and documents analyzed. The researcher makes firsthand observations of activities and interactions, sometimes engaging personally in those activities as a "participant observer."

EXHIBIT 1.3 Three Kinds of Qualitative Data

1. *Interviews*

Open-ended questions and probes yield in-depth responses about people's experiences, perceptions, opinions, feelings, and knowledge. Data consist of verbatim quotations with sufficient context to be interpretable.

2. *Observations and fieldwork*

Fieldwork descriptions of activities, behaviors, actions, conversations, interpersonal interactions, organizational or community processes, or any other aspect of observable human experience are documented. Data consist of field notes: rich, detailed descriptions, including the context within which the observations were made.

3. *Documents*

Written materials and documents from organizational, clinical, or program records; social media postings of all kinds; memoranda and correspondence; official publications and reports; personal diaries, letters, artistic works, photographs, and memorabilia; and written responses to open-ended surveys are collected. Data consist of excerpts from documents captured in a way that records and preserves the context.

For example, an evaluator might participate in all or part of the program under study, participating as a regular program member, client, or student. The qualitative researcher talks with people about their experiences and perceptions. More formal individual or group interviews may be conducted. Relevant records and documents are examined. Extensive field notes are collected through these observations, interviews, and document reviews. The voluminous raw data in these field notes are organized into readable narrative descriptions, with major themes, categories, and illustrative case examples extracted through content analysis. The themes, patterns, understandings, and insights that emerge from fieldwork and subsequent analysis are the fruit of qualitative inquiry.

Mixed Methods

Qualitative findings may be presented alone or in combination with quantitative data. Research and evaluation studies employing multiple methods, including combinations of qualitative and quantitative data, are common. At the simplest level, a questionnaire or interview that asks both fixed-choice (closed) questions and open-ended questions is an example of how quantitative measurement and qualitative inquiry are often combined. Here's an example.

Quantitative survey question (fixed, scaled response categories):

How satisfied are you with the quality of public transportation in your area?

1. Very satisfied
2. More satisfied than dissatisfied
3. More dissatisfied than satisfied
4. Very dissatisfied

Qualitative follow-up question (open-ended):

1. What are you especially satisfied with?
2. What are you especially dissatisfied with?

Mixed methods yield both statistics and stories. Such studies report how many people fall into categories of interest and provide quotations and stories to elucidate what the numbers mean. See Exhibit 1.4 for an example.

The Quality of Qualitative Data

The quality of qualitative data depends to a great extent on the methodological training, skill, sensitivity, and integrity of the researcher. Systematic and rigorous observation involves far more than just being present and looking around. Skillful interviewing requires much more than just asking questions. Credible content analysis demands considerably more than just reading to see what's there. Generating meaningful and useful qualitative findings through observation, interviewing, and content analysis requires discipline, knowledge, training, practice, creativity, and hard work.

This chapter provides an overview of qualitative inquiry. Later chapters examine how to choose among

---

EXHIBIT 1.4 **Mixed-Methods Example**

Statistical data. Roughly 30% of entering freshmen in the United States are first-generation college students, and 24% (4.5 million) are both first-gens and low income. Nationally, 89% of low-income first-gens leave college within six years, without a degree. More than a quarter leave after their first year—four times the dropout rate of higher-income second-generation students (Ramsey & Peale, 2010, p. 1).

Qualitative case quotation. "I did okay in high school, and everybody always said I'd go to college, but I didn't even know you had to apply to go to college. Nobody in my family had ever been to college. I thought it was like going from primary school to high school, that the school authorities would just tell me where I'd be going. Looking back, it's hard to believe that we were so ignorant, but sometimes you don't know what you don't know. We knew nothing about college at all. Me or my family. Nothing. Every bit of it is new to me and my family. It's pretty scary, but I'm figuring it out" (Quote from Mike, a freshman at a large public university).

---

the many options available within the broad range of qualitative methods, theoretical perspectives, and applications; how to design a qualitative study; how to use observational methods and conduct in-depth, open-ended interviews; and how to analyze qualitative data to generate findings. To set the stage for those more detailed methodological discussions, it will be helpful to have a better sense of the kinds of findings that emerge from qualitative studies. If you want to get somewhere, it's helpful to have a sense of the destination. So let's look at the findings from some classic qualitative studies.

Qualitative Findings: Themes, Patterns, Concepts, Insights, Understandings

> Newton and the apple. Freud and anxiety. Jung and dreams. Piaget and his children. Darwin and Galapagos tortoises. Marx and England's factories. Whyte and street corners. What are you obsessed with understanding?
>
> —Halcolm

Mary Field Belenky and her colleagues set out to study women's ways of knowing. They conducted extensive interviews with 135 women from diverse backgrounds probing how they thought about knowledge, authority, truth, themselves, life changes, and life in general. They worked as a team to group similar responses and stories together, informed partly by previous research but ultimately basing the analysis on their own collective sense of what categories best captured what they found in the narrative data. They argued with each other about which responses belonged in which categories. They created and abandoned categories. They looked for commonalities and differences. They worked hard to honor the diverse points of view they found, while also seeking patterns across stories, experiences, and perspectives. One theme emerged as particularly powerful: "Again and again women spoke of 'gaining voice'" (Belenky, Clinchy, Goldberger, & Tarule, 1986, p. 16).

Voice versus silence emerged as a central metaphor for informing variations in ways of knowing. After painstaking analysis, they ended up with the five categories of knowing summarized in Exhibit 1.5, a framework that became very influential in women's studies and represents one kind of fruit from qualitative inquiry.

One of the best-known and most influential books on organizational development and management is *In Search of Excellence: Lessons From America's Best-Run Companies* by Peters and Waterman (1982). The authors based the book on case studies of 62 highly regarded companies. They visited companies, conducted extensive interviews, and studied corporate documents. From that massive amount of data, they extracted eight attributes of excellence: (1) a bias for action; (2) being close to the customer; (3) autonomy and entrepreneurship; (4) productivity through people; (5) hands-on, value-driven work; (6) sticking to the knitting; (7) simple form and lean staff; and (8) simultaneous loose and tight properties. Their book devotes a chapter to each theme, with case examples and implications. Their research helped launch the quality movement, which has now moved from the business world to not-for-profit organizations and government. This study also illustrates a common qualitative sampling strategy: studying a relatively small number of special cases that are successful at something and therefore a good source of lessons learned.

Stephen Covey (1989) used this same sampling approach in doing case studies of "highly effective people." He identified seven habits these people practice: (1) being proactive; (2) beginning with the end in mind; (3) putting first things first; (4) thinking win–win; (5) seeking first to understand, then seeking to

EXHIBIT 1.5   Women's Ways of Knowing: An Example of Qualitative Findings

- ❏ *Silence.* A position in which women experience themselves as mindless and voiceless and subject to the whims of external authority.

- ❏ *Received knowledge.* Women conceive of themselves as capable of receiving, even reproducing knowledge from external authorities but not capable of creating knowledge on their own.

- ❏ *Subjective knowledge.* A perspective from which truth and knowledge are conceived as personal, private, and subjectively known or intuited.

- ❏ *Procedural knowledge.* Women are invested in learning and apply objective procedures for obtaining and communicating knowledge.

- ❏ *Constructed knowledge.* Women view all knowledge as contextual, experience themselves as creators of knowledge, and value both subjective and objective strategies for knowing.

SOURCE: Belenky et al. (1986, p. 15).

be understood; (6) synergizing, or engaging in creative cooperation; and (7) self-renewal. Kurtzman and Goldsmith (2010) studied highly effective leaders and identified patterns in how they get organizations to "achieve the extraordinary."

What these influential books have in common is distilling a small number of important "lessons" from a huge amount of data based on outstanding exemplars. It is common in qualitative analysis for mounds of field notes and months of work to reduce to a small number of core themes. The quality of the insights generated is what matters, not the number of such insights. For example, in an evaluation of 34 programs aimed at people in poverty, we found a core theme that separated more effective from less effective programs: How people are treated affects how they treat others. If staff members are treated autocratically and insensitively by management, with suspicion and disrespect, staff will treat clients the same way. Contrariwise, responsiveness reinforces responsiveness, and empowerment breeds empowerment. These insights became the centerpiece of subsequent cross-project, collaborative organizational and staff development processes.

Jim Collins and his colleagues have continued and updated this methodological approach, conducting in-depth case studies of high-performing organizations. Their findings have been reported in a series of best-selling books mentioned earlier in this chapter in the section on making case comparisons to discover important patterns and themes. In their study of companies that thrive under conditions of uncertainty and chaos (Collins & Hansen, 2011), they began with an initial list of 20,400 companies and screened for small companies that achieved "spectacular results": at least 10-fold growth over 15+ years through good times and bad. *Only 7 companies made it into the final study sample.* Their in-depth case studies revealed and explained three critical success factors: (1) fanatic discipline, (2) productive paranoia, and (3) empirical creativity. They also debunked "entrenched myths" such as the following:

- Successful leaders are not bold, risk-taking visionaries.
- High performance is not distinguished by innovation.
- Acting quickly and making fast, real-time decisions is not an effective way of dealing with rapid change.
- Radical internal change is not an effective response to turbulent external environments (Collins & Hansen, 2011, pp. 9–10).

A different kind of qualitative finding is illustrated by Angela Browne's book *When Battered Women Kill* (1987). Browne conducted in-depth interviews with 42 women from 15 states who were charged with a crime in the death or serious injury of their mates. She was often the first to hear these women's stories. She used one couple's history and vignettes from nine others, representative of the entire sample, to illuminate the progression of an abusive relationship from romantic courtship to the onset of abuse through its escalation until it was ongoing and eventually provoked a homicide. Her work helped lead to legal recognition of *battered women's syndrome* as a legitimate defense, especially in offering insight into the common outsider's question: Why doesn't the woman just leave? Getting an insider perspective on the debilitating, destructive, and all-encompassing brutality of battering reveals that question for what it is: the facile judgment of one who hasn't been there. The effectiveness of Browne's careful, detailed, and straightforward descriptions and quotations lies in their capacity to take us inside the abusive relationship. Offering that inside perspective powers qualitative reporting.

This quick sampling of findings from classic qualitative studies is like a wine tasting, meant to introduce possibilities and support developing a more sophisticated palate or, like appetizers, as an opening to the fuller feast yet to come. Many important scholars have contributed to knowledge through qualitative inquiry. Many breakthroughs in our knowledge of how the

---

QUALIA

Qualia refers to what we, as humans, subjectively add to our physical, sensory experience of the world through consciousness. Consciousness may fundamentally involve processing *qualia* as a neurological capacity (Ramachandran & Blakeslee, 1998). *Qualia* are studied, and debated, at the intersection between philosophy of mind and brain science (Tye, 2013).

Clarence Irving Lewis (1929) coined the term *qualia* in 1929 in his book *Mind and the World Order*:

> There are recognizable qualitative characters of the given, which may be repeated in different experiences, and are thus a sort of universals; I call these "qualia." But although such qualia are universals, in the sense of being recognized from one to another experience, they must be distinguished from the properties of objects . . . because it is purely subjective. (p. 3)

Qualia cannot be measured and standardized; they can only be experienced and reported. *Qualia* are fundamentally, inherently, neurologically, essentially, and epistemologically qualitative. If this intrigues you and you're looking for a philosophical debate to get engaged with and mired in, see *Qualia* (2013).

---

world works and why it works as it does have emerged from qualitative studies. Exhibit 1.9, at the end of this chapter (p. 41), provides examples of distinguished and prestigious qualitative research pioneers throughout history and across a broad range of disciplines. The next section discusses some of the different purposes of and audiences for qualitative inquiry.

Different Purposes of and Audiences for Qualitative Studies: Research, Evaluation, Dissertations, and Personal Inquiry

As the title of this book indicates, qualitative methods are used in both research and evaluation. But because the purposes of research and evaluation are different, the criteria for judging qualitative studies can vary depending on a study's purpose. This point is important. It means that one can't judge the appropriateness of the methods in any study or the quality of the resulting findings without knowing the study's purpose, agreed-on uses, and intended audiences. Evaluation and research typically have different purposes, expected uses, and intended users. Dissertations add yet another layer of complexity to this mix. Let's begin with research.

Qualitative Research

Research aims to generate or test theory and contribute to knowledge. Research findings describe how the world works and why it works as it does. Such knowledge, and the theories that undergird knowledge, may subsequently inform action and evaluation, but action is not the primary purpose of fundamental research. Qualitative inquiry is especially powerful as a source of *grounded theory*, theory that is inductively generated from fieldwork, that is, theory that emerges from the researcher's observations and interviews out in the real world rather than in the laboratory or the academy. The primary audiences for research are other researchers and scholars, as well as policymakers and others interested in understanding some phenomenon or problem of interest. The research training, methodological preferences, and scientific values of those who use research will affect how valuable and credible they find the empirical and theoretical findings of qualitative studies.

Qualitative Dissertations

Dissertations and graduate theses offer special insight into the importance of attention to audience. Savvy graduate students learn that to complete a degree program, the student's committee must approve the work. The particular understandings, values, preferences, and biases of committee members come into play in that approval process. They will, in essence, evaluate the student's contribution, including the quality of the methodological procedures followed and the analysis done. Qualitative dissertations, once quite rare, have become increasingly common as the criteria for judging qualitative contributions to knowledge have become better understood and accepted. But those criteria are not absolute or universally agreed on. As we shall see, there are many varieties of qualitative inquiry and multiple criteria for judging quality, many of which remain disputed.

Qualitative Evaluations

Program evaluation is the systematic collection of information about the activities, characteristics, and outcomes of programs to make judgments about the program, improve program effectiveness, and/or inform decisions about future programming. Policies, organizations, and personnel can also be evaluated. Evaluative research, quite broadly, can include any effort to judge or enhance human effectiveness through systematic data-based inquiry. Human beings are engaged in all kinds of efforts to make the world a better place. These efforts include assessing needs, formulating policies, passing laws, delivering programs, managing people and resources, providing therapy, developing communities, changing organizational culture, educating students, intervening in conflicts, and solving problems. In these and other efforts to make the world a better place, the question of whether the people involved are accomplishing what they want to accomplish arises. When one examines and judges accomplishments and effectiveness, one is engaged in evaluation.

Qualitative methods are often used in evaluations because they tell the *program's story* by capturing and communicating the *participants' stories*. Evaluation case studies have all the elements of a good story. They tell what happened when, to whom, and with what consequences. Many examples in this book are drawn from program evaluation, policy analysis, and organizational development. The purpose of such studies is to gather information and generate findings that are *useful*. Understanding the program's and participants' stories is useful to the extent that they illuminate the processes and outcomes of the program for those who must make decisions about the program. In *Essentials of Utilization-Focused Evaluation* (Patton, 2012a), I have presented a comprehensive approach to doing evaluations that are useful, practical, ethical, accurate, and accountable. The primary criterion for judging such evaluations is the extent to which the intended users actually use the findings for decision making and program improvement. The methodological implication of this criterion is that the intended users must value the findings and find them credible. They must be interested in the stories, experiences, and perceptions of the program participants, beyond simply knowing how many came into the program, how many completed it, and how many did what afterward. Qualitative findings in evaluation illuminate the people behind the numbers and put faces on the statistics to deepen understanding and inform decision making.

Your Personal Interest and Passion as a Basis for Qualitative Inquiry

While the preceding discussion of evaluation, research, and dissertations has emphasized taking into account external audiences and consumers of qualitative studies, it is also important to acknowledge that *you* may study something because *you* want to understand it. As my children grew to adulthood, I found myself asking questions about coming of age in modern society, so I undertook a personal inquiry that became a book (Patton, 1999). But I didn't start out to write a book. I started out trying to understand my own experience and the experiences of my children. That is a form of qualitative inquiry.

TOP TEN PIECES OF ADVICE TO A GRADUATE STUDENT CONSIDERING A QUALITATIVE DISSERTATION

The following query was posted on an Internet listserv devoted to discussing qualitative inquiry:

I am a new graduate student thinking about doing a qualitative dissertation. If you could give just one bit of advice to a student considering qualitative research for a dissertation, what would it be?

The responses below came from different people. I've combined some responses, edited them (while trying to maintain the flavor of the postings), and arranged them for coherence.

1. *Be sure that a qualitative approach fits your research questions and interests.* (Chapter 2 will help with this by presenting the primary themes of qualitative inquiry.)

2. *Study qualitative inquiry.* There are lots of different approaches and a lot to know. So it's not a matter of just using qualitative methods but using a particular framework for undertaking a qualitative study. (Chapter 3 covers different qualitative theoretical orientations and approaches.)

3. *Find a dissertation adviser who will support you in doing qualitative research.* Otherwise, it can be a long, tough haul. A dissertation is a big commitment. There are other practical approaches to using qualitative methods that don't involve all the constraints of doing a dissertation, things like program evaluation, action research, and organizational development. You can still do lots of great qualitative work without doing a dissertation. But if you can find a supportive adviser and committee, then, by all means, go for it. (Chapter 4 covers particularly appropriate practical applications of qualitative methods.)

4. *Really work on design.* Qualitative designs follow a different logic from quantitative research, especially with regard to purposeful sampling and conducting in-depth case studies. This is not the same as questionnaires and tests and experiments. You can combine designs, like *quant* and *qual* approaches, but integrating both kinds of data can be challenging. Either way, you have to understand what's unique about qualitative designs. (Chapter 5 covers qualitative designs.)

5. *Practice interviewing and observation skills.* Practice! Practice! Practice! Do lots of interviews. Spend a lot of time doing practice fieldwork observations. Get feedback from someone who's really good at interviewing and observations. There's an amazing amount to learn. And it's not just head stuff. Qualitative research takes skill. Don't make the mistake of thinking it's easy. The better I get at it, the more I realize how bad I was when I started. (Chapters 6 and 7 cover the skills of qualitative inquiry.)

6. *Figure out how to do qualitative analysis before you gather data.* I've talked with lots of advanced grad students who rushed to collect data before they knew anything about analyzing it—and lived to regret it big time. This is true for statistical data, but somehow people seem to think that qualitative data are easy to analyze. No way. That's a big-time NO WAY. And don't think that the new software will solve the problem. Another big-time NO WAY. You, that is, YOU, still have to analyze the data. (Chapter 8 covers analysis.)

7. *Be sure that you're prepared to deal with the controversies of doing qualitative research.* People on this listserv are constantly sharing stories about people who don't "get" qualitative research and put it down. Don't go into it naively. Understand the paradigms and politics. (Chapter 9 deals with the paradigms, politics, and ways of enhancing the credibility of qualitative inquiry.)

8. *Do it because you want to and are convinced that it's right for you.* Don't do it because someone told you it would be easier. It's not. Try as hard as possible to pick/negotiate dissertation research questions that have to do with some passion/interest in your professional life. Qualitative research is time-consuming, intimate, and intense; you will need to find your questions interesting if you want to be at all sane during the process—and still sane at the end.

9. *Find a good mentor or support group.* Or both. In fact, find several of each. If you can, start a small group of peers in the same boat, so to speak, to talk about your research together on a regular basis—you can share knowledge, brainstorm, and problem solve, as well as share in each other's successes, all in a more relaxed environment that helps take some of the edge off the stress (e.g., you might have potluck meals at different homes). This can be tremendously liberating (even on a less than regular basis). Take care of yourself.

10. *Prepare to be changed.* Looking deeply at other people's lives will force you to look deeply at yourself.

Additional resources for graduate students:

- *Completing Your Qualitative Dissertation: A Road Map From Beginning to End* (Bloomberg & Volpe, 2012)
- *The Qualitative Dissertation: A Guide for Students and Faculty* (Piantanida & Garman, 2009)
- *A Practical Guide to the Qualitative Dissertation* (Biklen & Casella, 2007)
- *Stretching Exercise for Qualitative Researchers* (Janesick, 2011)

While doing interviews with recipients of MacArthur Foundation Fellowships (popularly called "Genius Awards"), I was told by a social scientist that her fieldwork was driven by her own search for understanding and that she disciplined herself to not even think about publication while engaged in interviewing and observing because she didn't want to have her inquiry affected by attention to external audiences. She wanted to know because she wanted to *know*, and she had made a series of career and professional decisions that allowed her to focus on her personal inquiry without being driven by the traditional academic admonition to "publish or perish." She didn't want to subject herself to or have her work influenced by external criteria and judgment.

Thus far in this first chapter, we've looked at how qualitative inquiry contributes to our understanding of the world (Module 1) and what makes qualitative data *qualitative* (Module 2). We turn now to making methods decisions.

WHAT BRINGS YOU TO YOUR CHOSEN INQUIRY? THE INTERSECTION OF PERSONAL INTEREST AND SOCIETAL DYNAMICS

In his classic book *The Sociological Imagination*, C. Wright Mills (1959/2000) challenged scholars in the human disciplines to develop a point of view and a methodological attitude that would allow them to examine how the private troubles of individuals, which occur within the immediate world of experience, are connected to public issues and to public responses to these troubles. Mills's sociological imagination was biographical, international, and historical.

Mills wanted [social science] to make a difference in the lives people lead. . . .

I want a critical methodology that enacts its own version of the sociological imagination. Like Mills, my version of the imagination is moral and methodological. And like Mills, I want a discourse that troubles the world, understanding that all inquiry is moral and political. . . . in the connection between critical inquiry and social justice.

—Norman K. Denzin (2010, p. 9)
The Qualitative Manifesto: A Call to Arms

This is what undergirds and energizes the work of qualitative inquiry pioneer Norman K. Denzin, among the most prolific and eclectic of all qualitative methodologists. He took inspiration from the vision and imagination of C. Wright Mills. What is the source of your inspiration? What motivates you? What is your vision of the contribution you want to make? These questions are fundamental to inquiry because the answers get you through the long, hard slog of fieldwork, arduous data analysis, and disputes about conclusions.

Enter not through this portal of inquiry

if thee be languorous, lackadaisical,

timorous, or without vision.

—Halcolm

> You cannot successfully determine beforehand which side of the bread to butter.
>
> —The Law of the Perversity of Nature

The implication of thinking about purpose and audience in designing studies is that methods, no less than knowledge, are dependent on context. No rigid rules can prescribe what data to gather to investigate a particular interest or problem. There is no recipe or formula in making methods decisions. The widely respected social science methodologist and psychometrician Lee J. Cronbach observed that designing a study is as much art as science. It is "an exercise of the dramatic imagination" (Cronbach, 1982, p. 239). In research as in art there can be no single, ideal standard. Beauty no less than "truth" is in the eye of the beholder, and the beholders of research and evaluation can include a plethora of stakeholders: scholars, policymakers, funders, program managers, staff, program participants, journalists, critics, and the general public. Any given design inevitably reflects some imperfect interplay of resources, capabilities, purposes, possibilities, creativity, and personal judgments by the people involved. Research, like diplomacy, is the art of the possible. Exhibit 1.6 provides a set of questions to consider in the design process, regardless of type of inquiry. With that background, we can turn to consideration of the relative strengths and weaknesses of qualitative and quantitative methods.

EXHIBIT 1.6 Some Guiding Questions and Options for Making Methods Decisions

1. What are the purposes of the inquiry?

   - *Research.* Contribution to knowledge
   - *Evaluation.* Program improvement and decision making
   - *Dissertation.* Demonstrating doctoral-level scholarship
   - *Personal inquiry.* Finding out for oneself

2. Who are the primary audiences for the findings?

   - Scholars, researchers, academicians
   - Program funders, administrators, staff, participants
   - Doctoral committee
   - Oneself, friends, family, lovers

3. What questions will guide the inquiry?

   - Theory-derived, theory-testing, and/or theory-oriented questions
   - Practical, applied, action-oriented questions and issues
   - Academic degree or discipline/specialization priorities
   - Matters of personal interest and concern, even passion

4. What data will answer or illuminate the inquiry questions?

   - *Qualitative.* Interviews, field observations, documents
   - *Quantitative.* Surveys, tests, experiments, secondary data
   - *Mixed methods.* What kind of mix? Which methods are primary?

5. What resources are available to support the inquiry?

   - Financial resources
   - Time
   - People resources
   - Access, connections

6. What criteria will be used to judge the quality of the findings?

   - *Traditional research criteria:* Rigor, validity, reliability, generalizability
   - *Evaluation standards.* Utility, feasibility, propriety, accuracy
   - *Nontraditional criteria.* Trustworthiness, diversity of perspectives, clarity of voice, credibility of the inquirer to primary users of the findings

Methods Choices: Contrasting Qualitative and Quantitative Emphases

> The key to making a good forecast is weighing quantitative and qualitative information appropriately.
>
> —Nate Silver (2012, p. 100)

Thinking about design alternatives and methods choices leads directly to consideration of the relative strengths and weaknesses of qualitative and quantitative data. The approach here is pragmatic. Some questions lend themselves to numerical answers, and some don't. If you want to know how much people weigh, use a scale. If you want to know if they're obese, measure body fat in relation to height and weight and compare the results with population norms. If you want to know what their weight *means* to them, how it affects them, how they think about it, and what they do about it, you need to ask them questions, find out about their experiences, and hear their stories. A comprehensive and multifaceted understanding of weight in people's lives requires both their numbers and their stories. Doctors who look only at test results and don't also listen to their patients are making judgments with inadequate knowledge.

Qualitative methods facilitate study of issues in depth and detail. Approaching fieldwork without being constrained by predetermined categories of analysis contributes to the depth, openness, and detail of qualitative inquiry. Quantitative methods, on the other hand, require the use of standardized measures so that the varying perspectives and experiences of people can be fit into a limited number of predetermined response categories to which numbers are assigned.

The advantage of a quantitative approach is that it's possible to measure the reactions of a great many people to a limited set of questions, thus facilitating comparison and statistical aggregation of the data. This gives a broad, generalizable set of findings presented succinctly and parsimoniously. By contrast, qualitative methods typically produce a wealth of detailed information about a much smaller number of people and cases. This increases the depth of understanding of the cases and situations studied but reduces generalizability.

Validity in quantitative research depends on careful instrument construction to ensure that the instrument measures what it is supposed to measure. The instrument must then be administered in an appropriate, standardized manner according to prescribed procedures. The focus is on the measuring instrument—the

test items, survey questions, or other measurement tools. In qualitative inquiry, the researcher is the instrument. The credibility of qualitative methods, therefore, hinges to a great extent on the skill, competence, and rigor of the person doing the fieldwork—as well as the things going on in a person's life that might prove to be a distraction. Qualitative methodology pioneers Egon Guba and Yvonna Lincoln (1981) have commented on this aspect of qualitative research:

> Fatigue, shifts in knowledge, and cooptation, as well as variations resulting from differences in training, skill, and experience among different "instruments," easily occur. But this loss in rigor is more than offset by the flexibility, insight, and ability to build on tacit knowledge that is the peculiar province of the human instrument. (p. 113)

Because qualitative and quantitative methods involve differing strengths and weaknesses, they constitute alternative, but not mutually exclusive, strategies for research. Both qualitative and quantitative data can be collected in the same study. To further illustrate these contrasting approaches and provide concrete examples of the fruit of qualitative inquiry, the rest of this chapter presents select excerpts from actual studies.

Comparing Two Kinds of Data: An Example

The Technology for Literacy Center was a computer-based adult literacy program in Saint Paul, Minnesota. It operated out of a storefront facility in a lower-socioeconomic area of the city. In 1988, after three years of pilot operation, a major funding decision had to be made about whether to continue the program. Anticipating the funding decision, a year earlier, local foundations and public schools supported a "summative evaluation" to determine the overall outcomes and cost-effectiveness of the center. The evaluation design included both quantitative and qualitative data.

The quantitative testing data showed great variation. The statistics on average achievement gains masked the great differences among the participants. The report concluded that, while testing showed substantial achievement test gains for the treatment group versus the control group, the more important finding concerned the highly individualized nature of student progress. The report concluded that "the data on variation in achievement and instructional hours lead to a very dramatic, important and significant finding: *there is no average student at TLC* (Patton & Stockdill, 1987, p. 33).

This finding highlights the kind of program or treatment situation where qualitative data are particularly helpful and appropriate. The Technology for Literacy Center has a highly individualized program in which learners proceed at their own pace based on specific needs and interest. The students come in at very different levels, with a range of goals, participate in widely varying ways, and make very different gains. Average gain scores and average hours of instruction provide a parsimonious overview of aggregate progress. Adding case studies helps funders understand individual variation and what that variation means. To get at the meaning of the program for individual participants, the evaluation included case studies and qualitative data from interviews.

Individual Case Examples

One case was the story of Barbara, a 65-year-old black grandmother who came to Minnesota after a childhood in the Deep South. She worked as a custodian and house cleaner and was proud of never having been on welfare. She was the primary breadwinner for a home with five children spanning three generations, including her oldest daughter's teenage children, for whom she had cared since her daughter's unexpected death from hepatitis. During the week, she seldom got more than three hours of sleep each night. At the time of the case study, she had spent 15 months in the literacy program and progressed from not reading at all (second-grade level) to being a regular library user (and testing a grade level higher than where she began). She developed an interest in black history and reported being particularly pleased at being able to read the Bible on her own. She described what it was like not being able to read:

> Where do you go for a job? You can't make out an application. You go to a doctor, and you can't fill out the forms, and it's very embarrassing. You have to depend on other people to do things like this for you. Sometimes you don't even want to ask your own kids because it's just like you're depending too much on people, and sometimes they do it willingly, and sometimes you have to beg people to help....

> All the progress has made me feel lots better about myself because I can do some of the things I've been wanting to do and I couldn't do. It's made me feel more independent to do things myself instead of depending on other people to do them for me.

A second contrasting case tells the story of Sara, a 42-year-old Caucasian woman who dropped out of school in the 10th grade. She worked as an office manager and tested at 12th-grade level on entry to the program. After 56 hours of study over 17 days, she passed the exam to receive a Graduate Equivalency Degree (GED), making her a high school graduate. She immediately entered college. She said that the decision to return for her GED was

> an affirmation, as not having a diploma had really hurt me for a long time.... It was always scary wondering if somebody actually found out that I was not a graduate that they would fire me or they wouldn't accept me because I hadn't graduated. The hardest thing for me to do was tell my employer. He is very much into education, and our company is education oriented. So the hardest thing I ever had to do was tell him I was a high school dropout. I needed to tell because I needed time to go and take the test. He was just so understanding. I couldn't believe it. It was just wonderful. I thought he was going to be disappointed in me, and he thought it was wonderful that I was going back. He came to graduation.

These short excerpts from two contrasting cases illustrate the value of detailed, descriptive data in deepening our understanding of individual variation. Knowing that each woman progressed about one grade level on a standardized reading test is only a small part of a larger, much more complex picture. Yet, with more than 500 people in the program, it would be overwhelming for funders and decision makers to attempt to make sense of 500 detailed case studies (about 5,000 double-spaced pages). Statistical data provide a succinct and parsimonious summary of major patterns, while select case studies provide depth, detail, and individual meaning.

Open-Ended Interview Responses

Another instructive contrast is to compare closed-ended questionnaire results with responses to open-ended group interviews. Questionnaire responses to quantitative, standardized items indicated that 77% of the adult literacy students were "very happy" with the Technology for Literacy Center program; 74% reported learning "a great deal." These and similar results revealed a general pattern of satisfaction and progress. But what did the program mean to students in their own words?

To get the perspective of students, I conducted focus group interviews. I asked students to describe the program's outcomes in personal terms. I asked, "What difference has what you are learning made in your lives?" Here are some of the responses:

- I love the newspaper now and actually read it. Yeah, I love to pick up the newspaper now. I used to hate it. Now I *love* the newspaper.

- I can follow sewing directions. I make a grocery list now, so I'm a better shopper. I don't forget things.
- Yeah, you don't know how embarrassing it is to go shopping and not be able to read the wife's grocery list. It's helped me out so much in the grocery store.
- Helps me with my medicine. Now I can read the bottles and the directions! I was afraid to give the kids medicine before because I wasn't sure.
- I don't get lost anymore. I can find my way around. I can make out directions, read the map. I work in construction, and we change locations a lot. Now I can find my way around. I don't get lost anymore!
- Just getting a driver's license will be wonderful. I'm 50. If I don't get the GED but if I can get a license . . . ! I can drive well, but I'm scared to death of the written test. Just getting a driver's license . . . , a driver's license.
- Now I read outdoor magazines. I used to just read the titles of books—now I read the books!
- I was always afraid to read at school and at church. I'm not afraid to read the Bible now at Bible class. It's really important to me to be able to read the Bible.
- I can fill out applications now. You have to know how to fill out an application in this world. I can look in the Yellow Pages. It used to be so embarrassing not to be able to fill out applications, not to be able to find things in the Yellow Pages. I feel so much better now. At least my application is filled out right, even if I don't get the job, at least my application is filled out right.
- I'm learning just enough to keep ahead of my kids. My family is my motivation. Me and my family. Once you can read to your kids, it makes all the difference in the world. It helps you to want to read and to read more. When I can read myself, I can help them read so they can have a better life. The kids love it when I read to them.

These focus group interview excerpts provide some qualitative insights into the individual, personal experiences of adults learning to read. The questionnaire results (77% satisfied) provided data on statistically generalizable patterns, but the standardized questions only tap the surface of what it means for the program to have had "great perceived impact." The much smaller sample of open-ended interviews adds depth, detail, and meaning at a very personal level of experience. The next example will show that qualitative data can yield not only deeper understanding but also political action as the depth of participants' feelings is revealed.

DISCOVERING QUALITATIVE INQUIRY

Distinguished adult education scholar Malcolm Knowles created the discipline of *andragogy* (adult learning, in contrast to pedagogy, child learning). In his autobiography, *The Making of an Adult Educator* (1989), he listed his discovery of qualitative methods as an alternative way of studying and evaluating adult learning as one of the eight most important episodes of his life, right there alongside his marriage. He reported that it completely changed how he viewed the world and opened up critically important new ways of understanding how adults learn.

The Power of Qualitative Data

In the early 1970s, the school system of Kalamazoo, Michigan, implemented a new accountability system. It was a complex system that included using standardized achievement tests administered in both fall and spring, criterion-referenced tests developed by teachers, performance objectives, teacher peer ratings, student ratings of teachers, parent ratings of teachers, principal ratings of teachers, and teacher self-ratings.

The Kalamazoo accountability system began to attract national attention. For example, the *American School Board Journal* reported in April 1974 that "Kalamazoo schools probably will have one of the most comprehensive computerized systems of personnel evaluation and accountability yet devised" (p. 40). In the first of a three-part series on Kalamazoo the *American School Board Journal* asserted, "Take it from Kalamazoo: *a comprehensive, performance-based system of evaluation and accountability can work*" (p. 32).

Not everyone agreed with that positive assessment, however. The Kalamazoo Education Association charged that teachers were being demoralized by the accountability system. Some school officials, on the other hand, argued that teachers did not want to be accountable. In the spring of 1976, the Kalamazoo Education Association, with assistance from the Michigan Education Association and the National Education Association, sponsored a survey of teachers to find out the teachers' perspective on the accountability program (Perrone & Patton, 1976).

The education association officials were interested primarily in a questionnaire consisting of standardized items. One part of the closed-ended questionnaire provided the teachers with a set of statements with which they could agree or disagree. The questionnaire results showed that the teachers felt that the accountability system was largely ineffective and inadequate. For example, 90% of the teachers disagreed with the school administration's published statement: "The Kalamazoo accountability system is designed to personalize and individualize education"; 88% reported that the system does *not* assist teachers to become more effective; 90% responded that the accountability system has *not* improved educational planning in Kalamazoo; 93% believed that "accountability as practiced in Kalamazoo creates an undesirable atmosphere of anxiety among teachers"; and 90% asserted, "The accountability system is mostly a public relations effort." Nor did teachers feel that the accountability system fairly reflected what they did as teachers since 97% of them agreed that "accountability as practiced in Kalamazoo places too much emphasis on things that can be quantified, so that it misses the results of teaching that are not easily measured."

It is relatively clear from these statements that most of the teachers who responded to the questionnaire were negative about the accountability system. When school officials and school board members reviewed the questionnaire results, however, many of them immediately dismissed those results by arguing that they had never expected teachers to like the system, teachers didn't really want to be accountable, and the teachers' unions had told their teachers to respond negatively anyway. In short, many school officials and school board members dismissed the questionnaire results as biased, inaccurate, and the result of teacher union leaders telling teachers how to respond in order to discredit the school authorities.

The same questionnaire included two open-ended questions. The first was placed midway through the questionnaire, and the second came at the end of the questionnaire.

1. Please use this space to make any further comments or recommendations concerning any component of the accountability system.

2. Finally, we'd like you to use this space to add any additional comments you'd like to make about any part of the Kalamazoo accountability system.

A total of 373 teachers (70% of those who responded to the questionnaire) took the time to respond to one of these open-ended questions. All of the comments made by the teachers were typed verbatim and included in the report. These open-ended data filled 101 pages. When the school officials and school board members rejected the questionnaire data, rather than argue with them about the meaningfulness of the teacher responses to the standardized items, we asked them to turn to the pages of open-ended teacher comments and simply read at random what the teachers said. Examples of the comments they read, and could read on virtually any page in the report, are reproduced below in six representative responses from the middle pages of the report.

Teacher Response No. 284: I don't feel that fear is necessary in an accountability situation. The person at the head of a school system has to be human, not a machine. You just don't treat people like they are machines!

The superintendent used fear in this system to get what he wanted. That's very hard to explain in a short space. It's something you have to live through to appreciate. He lied on many occasions and was very deceitful. Teachers need a situation where they feel comfortable. I'm not saying that accountability is not good. I am saying the one we have is lousy. It's hurting the students—the very ones we're supposed to be working for.

Teacher Response No. 257: This system is creating an atmosphere of fear and intimidation. I can only speak for the school I am in, but people are tense, hostile, and losing their humanity. Gone is the goodwill and team spirit of administration and staff, and I believe this all begins at the top. One can work in these conditions but why, if it is to "shape up" a few poor teachers. Instead, it's having disastrous results on the whole faculty community.

Teacher Response No. 244: In order to fully understand the oppressive, stifling atmosphere in Kalamazoo, you have to "be in the trenches"—the classrooms. In 10 years of teaching, I have never ended a school year as depressed about "education" as I have this year. If things do not improve in the next 2 years, I will leave education. The Kalamazoo accountability system must be viewed in its totality and not just the individual component parts of it. In toto, it is oppressive and stifling.

In teaching government and history, students often asked what it was like to live in a dictatorship. I now know firsthand.

The superintendent, with his accountability model and his abrasive condescending manner, has managed in three short years to destroy teacher morale and effective creative classroom teaching.

Last evening, my wife and I went to an end of the school year party. The atmosphere there was strange—little exuberance, laughter, or release. People who in previous years laughed, sang, and danced were unnaturally quiet and somber. Most people went home early. The key topic was the superintendent, the school board election, and a millage campaign. People are still tense and uncertain.

While the school board does not "pay us to be happy," it certainly must recognize that emotional stability is necessary for effective teaching to take place. The involuntary transfers, intimidation, coercion, and top to bottom "channelized" communication in Kalamazoo must qualify this school system for the list of "least desirable" school systems in the nation.

Teacher Response No. 233: I have taught in Kalamazoo for 15 years and under five superintendents. Until the present superintendent, I found working conditions to be enjoyable, and teachers and administration and the Board of Education all had a good working relationship. In the past 4 years—under the present superintendent—I find the atmosphere deteriorating to the point where teachers distrust each other and teachers do not trust administrators at all! We understand the position the administrators have been forced into and feel compassion for them—however, we still have no trust! Going to school each morning is no longer an enjoyable experience.

Teacher Response No. 261: A teacher needs some checks and balances to function effectively; it would be ridiculous to think otherwise—if you are a concerned teacher. But in teaching, you are not turning out neatly packaged little mechanical products, all alike and endowed with the same qualities. This nonsensical accountability program we have here makes the superintendent look good to the community. But someone who is in the classroom dealing with all types of kids, some who cannot read, some who hardly ever come to school, some who are in and out of jail, this teacher can see that and the rigid accountability model that neglects the above-mentioned problems is pure "BULLSHIT!"

Teacher Response No. 251: "Fear" is the word for "accountability" as applied in our system. My teaching before "Accountability" is the same as now. "Accountability" is a political ploy to maintain power. Whatever good there may have been in it in the beginning has been destroyed by the awareness that each new educational "system" has at its base a political motive. Students get screwed.... The bitterness and hatred in our system are incredible. What began as "noble" has been destroyed. You wouldn't believe the new layers of administration that have been created just to keep this monster going.

Our finest compliment around our state is that the other school systems know what is going on and are having none of it. Lucky people. Come down and *visit in hell* sometime.

Face Validity and Credibility

What was the impact of the qualitative data collected from the teachers in Kalamazoo? You will recall that many of the school board members initially dismissed the standardized questionnaire responses as biased, rigged, and the predictable result of the union's campaign to discredit school officials. However, after reading through a few pages of the teachers' own personal comments, after hearing about the teachers' experiences with the accountability system in their own words, the tenor of the discussion about the evaluation report changed. School board members could easily reject what they perceived as a "loaded" questionnaire. They could not so easily dismiss the anguish, fear, and depth of concern revealed in the teachers' own reflections. The teachers' words had face validity and credibility. Discussion of the evaluation results shifted from an attack on the measures used to the question "What do you think we should do?"

Not long after the evaluation report, the superintendent resigned. The new superintendent and school board used the evaluation report as a basis for starting afresh with the teachers. A year later, teacher association officials reported a new environment of teacher–administration cooperation in developing a mutually acceptable accountability system. The evaluation report did not directly cause these changes. Many other factors were involved in Kalamazoo at that time. However, the qualitative information in the evaluation report revealed the full scope and nature of teachers' feelings about what it was like to work in the atmosphere created by the accountability system. The depth of those feelings as expressed in the teachers' own words became part of the impetus for change in Kalamazoo.

The Purpose of Open-Ended Responses

Direct quotations are a basic source of raw data in qualitative inquiry, revealing respondents' depth of emotion, the ways they have organized their world, their thoughts about what is happening, their experiences, and their basic perceptions. The task for the qualitative researcher is to provide a framework within which people can respond in a way that represents accurately and thoroughly their points of view about the world, or that part of the world about which they are talking, for example, their experience with a particular program being evaluated.

I have included the Kalamazoo evaluation findings as an illustration of qualitative inquiry because open-ended responses on questionnaires represent the most elementary form of qualitative data. There are severe limitations to open-ended data collected in writing on questionnaires, limitations related to the writing skills of respondents, the impossibility of probing or extending responses, and the effort required of the person completing the questionnaire. Yet, even at this elementary level of inquiry, the depth and detail of feelings revealed in the open-ended comments of the Kalamazoo teachers illustrate the fruit of qualitative methods.

The major way in which qualitative researchers seek to understand the perceptions, feelings, experiences, and knowledge of people is through in-depth, intensive interviewing, not just open-ended items on questionnaires. Chapter 7 on interviewing will present ways of gathering high-quality information from people. Effective interviewing techniques, skillful questioning, and the capacity to establish rapport are keys to obtaining credible and useful data through interviews. A particularly strong type of qualitative inquiry combines fieldwork observations with in-depth interviewing.

Combining Observations and Interviews

Qualitative data can include both direct observation and interview data. Sometimes a longitudinal study begins with observations and then continues with follow-up interviews. That was the design of a study of female college students. Sociologists Elizabeth A. Armstrong and Laura T. Hamilton, with a team of researchers, embedded themselves in a freshman dormitory at a major midwestern state university, observed the young women throughout their years in college, and then continued to gather data about their lives after college through in-depth interviews. They originally expected to learn a lot about romance and sex in college, but as often happens with open-ended qualitative fieldwork, what emerged as most important turned out to be different from what was expected. They ended up documenting the powerful effects of social class and socioeconomic status on college experiences and outcomes. They studied the culture of status seeking centered on sororities and found that the most well-resourced and seductive route to "success" was a "party pathway" anchored in the Greek sorority system, a system supported and encouraged by the university administration. This *party pathway*, they found, exerted influence over the academic and social experiences of all students but in different ways: It benefited the affluent, elite, and well-connected women but seriously disadvantaged the majority of women (Armstrong & Hamilton, 2013).

Inquiry by Observation

What people say is a major source of qualitative data, whether what they say is obtained verbally through an interview or in written form through document analysis or survey responses. There are limitations, however, to how much can be learned from what people say. To understand fully the complexities of a situation, direct participation in and observation of the phenomenon of interest is a particularly fruitful method. The ideal observation captures context, the unfolding of events over time, and critical interactions, and it includes talking with those involved in the activities observed.

Observational data, especially participant observation, permit a program evaluator to understand a program or treatment to an extent not entirely possible using only the insights obtained through interviews. Of course, not everything can be directly observed or experienced—and participant observation is a highly labor-intensive and, therefore, a relatively expensive research strategy. Chapter 6 will present strategies and practices for high-quality fieldwork observations, including both participant and nonparticipant approaches.

Certain Really Discriminating People Like Nothing Better Than to Relax on the Beach With a Good, In-depth, and Detailed Qualitative Study in Hand

The purpose of qualitative observation is to take the reader into the setting observed. This means that observational data must have depth and detail. The data must be descriptive—sufficiently descriptive that the reader can understand what occurred and how it occurred. The observer's notes become the eyes, ears, and perceptual senses for the reader. The descriptions must be thorough without being cluttered by irrelevant minutiae and trivia. The basic criterion to apply to a well-documented observation is the extent to which the observation permits the reader to enter and understand the situation under study.

THROUGH THE EYES OF A CHILD

Dr. Nancy Boxill, a child psychologist, studied homeless families in Atlanta.

I know of a small boy eight years old who sat alone on a park bench five or six hours every day for almost a week. He alternately played with the pigeons, watched the passing people, made patterns in the air with his feet and legs, or looked blankly into space. On the fourth day of his visit to this bench, a friend of mine asked this boy why he sat there every day. He replied that his mother brought him there in the mornings telling him to wait there while she looked for a job and a place for them to stay. There is no place else for him to go. When asked what he did all day he simply said that he watched and he waited. He watched the pigeons and the people. He made a game of guessing where each had to go. He said that mostly he just waited for his mother to come at the end of the day so they could wait together until the night shelter opened. (Boxill, 1990, p. 1)

I want to provide an observation example to illustrate what such a descriptive account is like. The observation I've selected describes a two-hour session of mothers participating in an early-childhood parent education program. The purpose of the program was to increase the skills, knowledge, and confidence of participants as well as provide a support group for first-time mothers. In funding the program, the legislators emphasized that they did not want parents to be told how to rear their children. Rather, the purpose of the parent education sessions was to increase the options available to parents so that they could make conscious choices about their own parenting styles and increase their confidence about the choices they make. Parents were also to be treated with respect and to be recognized as the primary educators of their children—in other words, the early-childhood educators were not

to impose their expertise on parents but, instead, make clear that parents are the real experts regarding their own children.

We made site visits to see the programs in action and observe the parenting discussions. Descriptions of these sessions became the primary data of the evaluation. Our descriptions provided feedback to the legislature about whether their policy guidance was in fact being followed. In essence, in our role as evaluation observers, we became the eyes and ears of the legislature and the state program staff, permitting them to understand what was happening in various parent sessions throughout the state. Descriptive data about the sessions also provided a mirror for the staff who conducted those sessions, a way of looking at what they were doing to see if that was what they wanted to be doing.

Exhibit 1.7 provides a description from one such session. The criterion I invite you to apply in reading this observation is the extent to which sufficient data are provided to take you, the reader, into the setting and permit you to make your own judgment about the nature and quality of parent education being provided.

The Raw Data of Qualitative Inquiry

In Exhibit 1.7, the description of the parenting session is aimed at permitting the reader to understand what occurred in the session. The qualitative data are descriptive. Pure description and quotations are the raw data of qualitative inquiry. Description is meant to take the reader into the setting. The data do not include judgments about whether what occurred was good or bad, appropriate or inappropriate, or any other interpretive judgments. The data simply describe what occurred. State legislators, program staff, parents, and others used this description, and descriptions like this from other program sites, to discuss what they wanted the programs to be and do. The descriptions helped them make explicit *their own* judgmental criteria. So do anecdotes when systematically collected and analyzed. (See MQP Rumination # 1, page 31.)

Integrating Qualitative Inquiry Methods

Thus far, the examples of observation and interviewing in this chapter have been presented as separate and distinct from each other. In practice, they are often fully integrated approaches. Becoming a skilled observer is essential even if you concentrate primarily on interviewing, because every face-to-face interview also involves and requires observation. The skilled interviewer is thus also a skilled observer, able to read nonverbal messages, sensitive to how the interview setting can affect what is said, and carefully attuned to

(Continued on p. 33)

EXHIBIT 1.7   Observation Description Illustrated: A Discussion for Mothers of Two-Year-Olds

Context

Mothers in an early-childhood parent education program in rural Minnesota are discussing the issues they face as parents. The program operates out of a small classroom in the basement of a church. The toddler center is directly overhead on the first floor, so that noises made by the children these mothers have left upstairs can be heard during the discussion. The room is just large enough for the 12 mothers, one staff person, and me to sit along three sides of the room. The fourth side is used for a movie screen. Some mothers are smoking. (The staff person told me afterward that smoking had been negotiated and agreed on among the mothers.) The seats are padded folding chairs plus two couches. A few colorful posters with pictures of children playing decorate the walls. Small tables are available for holding coffee cups and ashtrays during the discussion. The back wall is lined with brochures on child care and child development, and a metal cabinet in the room holds additional program materials.

The Session Begins

The mothers watch a 20-minute film about nursery school children. The film forms the basis for getting a discussion started about "what two-year-olds do." Louise, a part-time staff person in her early 30s, who has two young children of her own, one of them a two-year-old, leads the discussion. Louise asks the mothers to begin by picking out from the film things that their own children do and talk about the way some of the problems with children were handled in the film. For the most part, the mothers share happy, play activities their children like:

"My Johnny loves the playground just like the kids in the film."

"Yeah, mine could live on the playground."

The focus of the discussion turns quickly to what happens as children grow older, how they change and develop. Louise comments: "Don't worry about what kids do at a particular age. Like don't worry that your kid has to do a certain thing at age two or else he's behind in development or ahead of development. There's just a lot of variation in the ages at which kids do things."

The discussion is free-flowing and, once begun, is not directed much by Louise. The mothers talk back and forth to each other, sharing experiences about their children. A mother will bring up a particular point, and other mothers will talk about their own experiences as they want to. For example, one of the topics is the problem a mother is having with her child urinating in the bathtub. Other mothers share their experiences with this problem, ways of handling it, and whether or not to be concerned about it. The crux of that discussion seems to be that it's not a big deal and not something that the mother ought to be terribly concerned about. It is important not to make it a big deal for the child; the child will outgrow it.

The discussion turns to things that two-year-olds can do around the house to help their mothers. This is followed by some discussion of the things that two-year-olds can't do, and some of their frustrations in trying to do things. There is a good deal of laughing, sharing of funny stories about children, and sharing of frustrations about children. The atmosphere is informal, and there is a good deal of intensity in listening. The mothers seem especially to pick up on things that they share in common about the problems they have with their children.

Another issue from another mother is the problem of her child pouring out her milk. She asks, "What does it mean?" This question elicits some suggestions about using water aprons and cups that don't spill and other mothers' similar problems, but the discussion is not focused and does not really come to much closure. The water apron suggestion brings up a question about whether or not a plastic bag is okay. The discussion turns to the safety problems with different kinds of plastic bags. About 20 minutes of discussion have now taken place. (At this point, one mother leaves because she hears her child crying upstairs.)

The discussion returns to giving children baths. Louise interjects, "Two-year-olds should not be left alone in the bathtub." With reference to the earlier discussion about urinating in the bathtub, a mother interjects that urine in the bathwater is probably better than the lake water her kids swim in. The mother with the problem of urination in the bathtub says again, "It really bugs me when he urinates in the bathtub." Louise responds, "It really is your problem, not his. If you can calm yourself down, he'll be okay."

During a lull in the discussion, Louise asks, "Did you agree with everything in the movie?" The mothers talk a bit about this and focus on an incident in the movie where one child bites another. They share stories about problems they've had with their children biting. Louise interjects, "Biting can be dangerous. It is important to do something about biting." The discussion turns to what to do. One

(Continued)

(Continued)

mother suggests biting the child back. Another mother suggests that kids will work it out themselves by biting each other back. The mothers get very agitated; more than one mother talks at a time. Louise asks them to "cool it," so that only one person talks at a time. (The mother who had left returns.)

The discussion about biting leads to a discussion about child conflict and fighting in general, for example, the problem of children hitting each other or hitting their mothers. Again, the question arises about what to do. One mother suggests that when her child hits her, she hits him back, or when her child bites her, she bites him back. Louise interjects, "Don't model behavior you don't like." She goes on to explain that her philosophy is that you should not do things as a model for children that you don't want them to do. She says that works best for her; however, other mothers may find other things that work better for them. Louise comments that hitting back or biting back is a technique suggested by Dreikurs. She says she disagrees with that technique, "but you all have to decide what works for you." (About 40 minutes have now passed since the film, and 7 of the 11 mothers have participated, most of them actively.)

A New Issue Emerges

Another mother brings up a new problem. Her child is destroying her plants, dumping plants out, and tearing them up. "I really get mad." She says that the technique she has used for punishment is to isolate the child. Then she asks, "How long do you have to punish a two-year-old before it starts working?" This question is followed by intense discussion, with several mothers making comments. (This discussion is reproduced in full to illustrate the type of discussion that occurred.)

Mother No. 2:	"Maybe he needs his own plant."
Mother No. 3:	"Maybe he likes to play in the dirt. Does he have his own sand or dirt to play in around the house?"
Mother No. 4:	"Oatmeal is another good thing to play in."
Louise:	"Rice is another thing that children like to play in, and it's clean, good to use indoors."
Mother No. 5:	"Some things to play in would be bad or dangerous. For example, powdered soap isn't a good thing to let kids play in."
Mother No. 2:	"Can you put the plants where he can't get at them?"

Mother with the problem:	"I have too many plants; I can't put them all out of the way."
Louise:	"Can you put the plants somewhere else or provide a place to play with dirt or rice?" (The mother with the problem kind of shakes her head: "No." Louise goes on.) "Another thing is to tell the kid the plants are alive, to help him learn respect for living things. Give him his own plant that he can get an investment in."
Mother with the problem:	"I'll try it."
Mother No. 3:	"Let him help with the plants. Do you ever let him help you take care of the plants?"
Mother No. 6:	"Some plants are dangerous to help with."

Louise reaches up and pulls down a brochure on plants that are dangerous and says she has brochures for everyone. This is followed by a discussion of childproofing a house as a method of child rearing versus training the child not to touch things.

Session Ends

The time had come for the discussion to end. The mothers stayed around for about 15 minutes, talking to each other informally, going up and getting their children, and getting them dressed. Some brought them back down. They seemed to have enjoyed themselves and continued talking informally. One mother with whom Louise had disagreed about the issue of whether it was all right to bite or hit children back stopped to continue the discussion. Louise said, "I hope you know that I respect your right to have your own views on things. I wasn't trying to tell you what to do. I just disagreed, but I definitely feel that everybody has a right to their own opinion. Part of the purpose of the group is for everyone to be able to come together and appreciate other points of view and understand what works for different people."

The mother said that she certainly didn't feel bad about the disagreement and that she knew that some things that worked for other people didn't work for her and that she had her own ways but that she really enjoyed the group.

Louise cleaned up the room, and the session ended.

Anecdote as Epithet

In celebration of the fourth edition of this book (the first was in 1980), I am indulging in one personal rumination per chapter. These are issues that have persistently engaged, sometimes annoyed, occasionally haunted, and often amused me over more than 40 years of research and evaluation practice. Here's where I state my case on the issue and make my peace.

 An anecdote is nothing more than a short story about something. But *anecdote* has become an *epithet*. "That's just anecdotal" is a common way of dismissing qualitative data. For example, an article on science attacking qualitative case studies insisted, "The plural of *anecdote* is not *evidence*" (Benson, 2013, p. 11).

Sometimes. Cherry-picked anecdotes to supposedly "prove" a predetermined position come across as what they are: argumentative advocacy, not evidence. But the systematic, intentional, and careful recording of purposefully sampled anecdotes (stories) can become evidence when rigorously captured and thoughtfully analyzed. Suppose you're doing fieldwork and ask eight knowledgeable people about an event you've heard about. They give you 15 anecdotes. Look for patterns across the anecdotes. When in doubt about the veracity of a particular anecdote, check it out with others. That's qualitative inquiry using anecdotes as data. But even single anecdotes can be informative. Consider this example.

Fred Shapiro, editor of the *Yale Dictionary of Quotations*, has the job of tracking down and verifying the original source of widely used quotations. He traced the quote "The plural of *anecdote* is not *evidence*" to Raymond Wolfinger, political scientist, University of California, Berkeley, and e-mailed him for confirmation. Wolfinger responded with an anecdote:

I said "The plural of anecdote **IS** data" sometime in the 1969–70 academic year while teaching a graduate seminar at Stanford. The occasion was a student's dismissal of a simple factual statement—by another student or me—as a mere anecdote. The quotation was my rejoinder. Since then I have missed few opportunities to quote myself. The only appearance in print that I can remember is Nelson Polsby's (1984) accurate quotation and attribution in an article in *Political Science and Politics*.

Shapiro goes on to note, "What is interesting about this saying is that it seems to have morphed into its opposite '*Data* is not the plural of *anecdote*.'"

Etymology

The word *anecdote* originally meant "secret or private stories," from the French *anecdote*, derived from the Greek *anekdota*, "things unpublished" or "not given out." The word entered our language to denote stories that weren't given out to the public, meaning the hidden accounts that didn't make the authorized biography, the formal minutes of the meeting, or the official version of events. Diaries are a classic source of anecdotal material. Fadiman (2000) traces the etymology and history of, and the changing connotations associated with, the word *anecdote*, including the contemporary scholarly enterprise of collecting and publishing entertaining, insightful, inspirational, and titillating anecdotes. Having studied anecdotes, he takes on the question of their value.

Men of high philosophic mind have valued the anecdote less for its capacity to divert, more for its power to reveal character. This value was first classically formulated by Plutarch, quoted by Boswell: "Nor is it always in the most distinguished achievements that men's virtues or vices may be best discerned; but very often an action of small note, a short saying, or a jest, shall distinguish a person's real character more than the greatest sieges, or the most important battles."

From anecdotes, thought Prosper Merirnee, one "can distinguish a true picture of the customs and characters of any given period." Nietzsche was confident that "three anecdotes may suffice to paint a picture of a man." Isaac D'Israeli, whose *Dissertation on Anecdotes* affords a perfect reflection of his time's anecdotal preferences, thought anecdotes accurate indices to character: "Opinions are fallible, but not examples." Says Ralph Waldo Emerson: "Ballads, bons mots, and anecdotes give us better insights into the depths of past centuries than grave and voluminous chronicles." His contemporary William Ellery Channing agreed: "One anecdote of a man is worth a volume of biography." (p. xxiii)

Dubious? Consider what this single anecdote reveals about a person, a people, and a colonial empire. It is told

(Continued)

(Continued)

by English author J. R. Ackerley from his travels in India sometime around 1930.

> Talking of snakes, Mrs. Montgomery told me that once she nearly trod upon a *krait*—one of the most venomous snakes in India. "I was going back in the evening to my bungalow, preceded by a servant who was carrying a lamp. Suddenly he stopped and said, '*Krait, Mem-sahib!*'—but I was far too *ill* to notice what he was saying, and went straight on. Then the servant did a thing absolutely without precedent in India—he touched me! He put his hand on my shoulder and pulled me back. Of course *if* he hadn't done that I should undoubtedly have been killed; but I didn't like it all the same, and got rid of him soon after."

Fadiman, as an anecdote collector and scholar, is careful not to overgeneralize from a single anecdote. He has commented that, like a statistic taken out of context, a single anecdote, unless measured against the whole record of a life, may be a damned lie. But he ruminates that a reasonable and diverse number of anecdotes drawn from different circumstances and phases of a life may give us "an imperfect yet authentic sense of character." In the end, though, he comes to see anecdotes as data, though he doesn't use that word. Anecdotes are naturally occurring, readily available, and insight-generating data. He concludes thus about the overall value of anecdotes:

> If one were asked to name the *kind* of book that within one set of covers most adequately reflects the sheer multifariousness of human personality, it might well be a book of anecdotes. . . .

> A reasonably ample gathering of anecdotes, drawn from many times and climes, may reconcile us to our human nature by showing us that, for all its faults and stupidities, it can boast a diversity to which no other animal species can lay claim. (Fadiman, 2000, p. xxiv)

Anecdotal Evidence

In the course of discussing this rumination while doing fieldwork together, colleague Jamie Radner shared two anecdotes with me that illustrate and illuminate anecdotes as data and evidence. With his kind permission (and a caution that these are vivid memories but, being decades old, may be off in one or more detail), I share them here:

> Decades ago I worked for Amnesty International (AI). At the heart of AI's effectiveness was generating impartial, reliable, credible evidence about what was going on in all kinds of brutal places. Our biggest staff unit and biggest investment was therefore our research department, packed with smart, well-trained PhDs. A core research method was collecting stories from refugees. Definitely, anecdotes in both the old and current senses. These were then rigorously analyzed to assess where there was independent corroboration coming through, to push for inconsistencies, to discover patterns (ah, that word—brings back memories). Was the result good evidence? You bet. Good enough, I liked to think, to at least occasionally save lives. And this brings to mind . . .

> Again, decades ago, a colleague working for the Peace Corps in Africa described a scene to me in a remote village in a brutally run country. There was a long line of villagers waiting at some makeshift station. He asked what was up and learned that they were waiting to talk, one by one, to a representative from AI. Everyone knew that such talk, in such a place, was very dangerous. So at a certain point he asked a villager he knew why so many people were running this risk. I'll never forget the answer: "We trust Amnesty International."

Anecdotes as Hypotheses

Scientific observations often begin as anecdotes, for example, Newton getting inspired by observing a falling apple, stimulating him to ponder what we now know as gravity, or so goes the legendary anecdote. Anecdotes yield hypotheses and questions. Was that anecdote an interesting but idiosyncratic story, or is it part of a pattern? Qualitative inquirers examine multiple anecdotes for patterns, insights, and meaning.

Nicholas G. Carr (2010), an American writer who has published books and articles on technology, business, and culture, has labeled *anti-anecdotalism* as antiscience.

> We live anecdotally, proceeding from birth to death through a series of incidents, but scientists can be quick to dismiss the value of anecdotes. "Anecdotal" has become something of a curse word, at least when applied to research and other explorations of the real. . . . The empirical, if it's to provide anything like a full picture, needs to make room for both the statistical and the anecdotal.

> The danger in scorning the anecdotal is that science gets too far removed from the actual experience of life, that it loses sight of the fact that mathematical averages and other such measures are always abstractions. (Carr, 2014)

Anti-Anti-Anecdotalism

So here's how I now respond to the dismissive "It's just an anecdote."

> It's just an anecdote when told in isolation and heard by amateurs. But I'm a professional anecdote collector. If you know how to listen, systematically collect, and rigorously analyze anecdotes, the patterns revealed are windows into what's going on in the world. It's true that to the untrained ear, an anecdote is just a casual story, perhaps amusing, perhaps not. But to the professionally trained and attuned ear, an anecdote is scientific data—a note in a symphony of human experience. Of course, you have to know how to listen. (Smile knowingly.)

On the other hand, this entire rumination may be viewed as evidence that I have moved into an advanced stage of *anecdotage.* "In youth we sow our wild oats, in old age our tame anecdotes" (Fadiman, 2000, p. xiv).

And a final cautionary note: Don't confuse anecdotes with *anecdata*: "information which is presented as if it is based on serious research but is in fact based on what someone thinks is true" (Davidson, 2014; Macmillan, 2014).

Rumination Exercise: Practice Analyzing Anecdotes

For Valentine's Day, the editor of *The New York Times Book Review* collected anecdotes from writers in a variety of genres about books that have taught them about love. Read these anecdotes as data. What themes do you find across the stories? ("A Sentimental Education," 2014).

---

the nuances of the interviewer–interviewee interaction and relationship.

Likewise, interviewing skills are essential for the observer because, during fieldwork, you will need and want to talk with people, whether formally or informally. Participant observers gather a great deal of information through informal, naturally occurring conversations. Understanding that interviewing and observation are mutually reinforcing qualitative techniques is a bridge to understanding the fundamentally people-oriented nature of qualitative inquiry.

Sociologist John Lofland posited four people-oriented mandates in collecting qualitative data. First, the qualitative methodologist must get close enough to the people and situation being studied to personally understand in depth the details of what goes on. Second, the qualitative methodologist must aim at capturing what actually takes place and what people actually say: the perceived facts. Third, qualitative data must include a great deal of pure description of people, activities, interactions, and settings. Fourth, qualitative data must include direct quotations from people, both what they speak and what they write down.

> The commitment to get close, to be factual, descriptive and quotive, constitutes a significant commitment to represent the participants in their own terms. This does not mean that one becomes an apologist for them, but rather that one faithfully depicts what goes on in their lives and what life is like for them, in such a way that one's audience is at least partially able

to project themselves into the point of view of the people depicted. They can "take the role of the other" because the reporter has given them a living sense of day-to-day talk, day-to-day activities, day-to-day concerns and problems. . . .

> A major methodological consequence of these commitments is that the qualitative study of people *in situ* is a *process of discovery.* It is of necessity a process of learning what is happening. Since a major part of what is happening is provided by people in their own terms, one must find out about those terms rather than impose upon them a preconceived or outsider's scheme of what they are about. It is the observer's task to find out what is fundamental or central to the people or world under observation. (Lofland, 1971, p. 4)

Personal Engagement in Qualitative Inquiry

In qualitative inquiry, the person conducting interviews and engaging in field observations is the instrument of the inquiry. The inquirer's skills, experience, perspective, and background matter. The personal nature of qualitative inquiry will be a recurring theme of this book. Qualitative inquiry provides *a point of intersection between the personal and the professional.* Exhibit 1.8 offers a concrete example of inquiry into that personal/professional intersection while contrasting cognitive styles and inviting you to assess your own cognitive style and personal/professional qualitative inquiry intersection.

MAPPING EXPERIENCES:
Our Own as Well as Those of Others

Qualitative inquiry offers opportunities not only to learn about the experiences of others but also to examine the experiences that you, the inquirer, bring to the inquiry, experiences that will, to some extent, affect what is studied and help shape, for better or worse, what is discovered. Qualitative inquiry includes examining and understanding how who we are can shape what we see, hear, know, and learn during fieldwork and subsequent analysis. In that sense, qualitative inquiry can be thought of as mapping experiences, our own as well as those of others.

Imagine a map… drawn from your memory instead of from the atlas. It is made of strong places stitched together by the vivid threads of transforming journeys. It contains all the things you learned from the land and shows where you learned them. …

Think of this map as a living thing, not a chart but a tissue of stories that grows half-consciously with each experience. It tells where and who you are with respect to the earth, and in times of stress or disorientation it gives you the bearings you need in order to move on. We all carry such maps within us as sentient and reflective beings, and we depend upon them unthinkingly, as we do upon language or thought. … And it is part of wisdom, to consider this ecological aspect of our identity. (Tallmadge, 1997, p. 14)

EXHIBIT 1.8   Cognitive Inquiry Styles: PowerPoint Versus Story

PowerPoint constitutes a powerful and widely used presentation tool, but embedded within it may be a way of thinking that reduces knowledge to bullet points. Edward R. Tufte (2006), professor emeritus of political science, computer science, statistics, and graphic design at Yale, argues that the "cognitive style" of PowerPoint weakens verbal and spatial reasoning. His view is emphatic: *PowerPoint Is Evil. Power Corrupts. PowerPoint Corrupts Absolutely* (Tufte, 2003). Victims of poor PowerPoint presentations refer to being caught in "PowerPoint hell." Angela R. Garber (2001), a communications expert, is credited with the phrase "death by PowerPoint." *Dilbert* cartoonist Scott Adams has warned of *PowerPoint poisoning*.

PowerPoint is an effective presentation tool when expertly used. That is not in question. What undergirds these vociferous critiques of PowerPoint is the exalted position it has been thrust into as a symbol of bullet-point simplicity. The tool has been cloaked in a mantle that caricatures simple-mindedness. What does this have to do with qualitative inquiry? Just this: Qualitative inquiry and reporting resist bullet-point simplicity in favor of contextualized

complexity and offer an alternative to the Information Age trend of reducing knowledge to numbers, bullet points, tweets, and text messages.

To illustrate the contrast, which itself is oversimplified and intentionally provocative, let's compare two inquiry frameworks using the life of Dr. Will Wilson (a pseudonym), who teaches qualitative methods. For this thought experiment, the left-hand column below is his life story in 20 bullets. In the right-hand column are the qualitative inquiry questions that would form the basis for a deep, rich, thick, complex, and contextualized story. The contrasting frameworks constitute not only different cognitive styles but also different ways of engaging the world, different forms of understanding, different commitments of time and attention, and different tolerances for ambiguity and complexity. These differences are partly a matter of taste, but taste flows from habit, experience, and behavioral reinforcement as much as, and even more so than, any inherent predilection. This exercise invites you to assess your current taste preferences and habits of mind while aspiring to whet your appetite for the deep engagement that is the essence of qualitative inquiry.

The Story of Dr. Will Wilson

Research Question: **What Experiences Have Informed Your Engagement With and Approach to Qualitative Inquiry?**

LIFE EXPERIENCES IN 20 BULLET POINTS	TEN SETS OF QUESTIONS TO FRAME AN IN-DEPTH QUALITATIVE INQUIRY
<ul><li>65-year-old white male</li><li>Grew up in rural Missouri</li><li>Navy service during the Vietnam War</li><li>Degrees in history, social psychology, bio-cultural anthropology, and demography</li><li>Dissertation used mathematical modeling to study human population dynamics, fertility, and human behavior</li><li>Became a full-time independent evaluation consultant after 20 years at a major state university</li><li>Father (of two children) and grandfather; divorced</li><li>Avid hiker, having logged more than three years in wilderness areas and national parks over his lifetime—and still counting</li><li>Helped build a cabin he has in western Wisconsin</li><li>Broke his leg hiking in 1989</li><li>Serious car accident in 1998, which was life threatening (he was a passenger)—lost his left arm</li><li>Heart attack and quadruple bypass surgery in 2004</li><li>Spent a year working on a development project in West Africa in 2006</li><li>Lost right eye to a blood clot in 2008</li><li>Hip replacement in 2010</li><li>Needs and uses hearing aids</li><li>Has published two qualitative methods guidebooks, one on interviewing and one on analysis</li><li>Hiking philosophy: "Take what nature gives you."</li><li>Life philosophy: "Take what life gives you."</li><li>Epitaph offered by his last serious relationship: "Too unsettled and too feral, but admittedly ever adapting to the end."</li></ul>	<ul><li>☐ What was it like growing up in rural Missouri? What of Missouri do you find still in you? How have those early experiences shaped your research focus and approach?</li><li>☐ What led you to Navy service, and how did that service affect your subsequent life journey and scholarly career?</li><li>☐ How have your diverse areas of academic study affected your understanding of the world? To what extent, if at all, do you still draw on those foundational disciplines in your current qualitative evaluation work and workshops?</li><li>☐ How did you transition from a mathematical modeler to a qualitative inquirer and teacher? What are your seminal experiences with and core approach to qualitative inquiry?</li><li>☐ How did you become an evaluator? What is the nature and focus of your evaluation practice? What have you learned about program effectiveness in conducting many evaluations?</li><li>☐ How did you come to be such an avid hiker? How has the time spent in the wilderness shaped your perspective on inquiry and life?</li><li>☐ You have spent time in the desert, in the mountains, on the plains, in the woods, and on the ocean. How do you experience those different environments? More generally, how does context affect your perspective?</li><li>☐ You've lost an arm and an eye and had other medical challenges. What have those experiences been like for you? How have they affected you? How have you adapted to those losses and health challenges?</li><li>☐ As a father, grandfather, former husband, and veteran of several subsequent relationships, what have you learned about family, parenting, and relationships?</li><li>☐ How, if at all, have your hiking and life philosophies influenced your approach to qualitative inquiry?</li></ul>

Interrelationship Between the Two Columns

This exercise was positioned as a contrast between two competing cognitive frameworks: bullet-point simplicity versus in-depth qualitative complexity. But looked at from a different angle, the 20 bullet-point facts can be treated as an outline for creating the interview questions that guide the in-depth qualitative inquiry. The bullet points are the bare-bones skeleton of Dr. Wilson's life. The qualitative inquiry puts flesh, visage, expression, and personhood on that skeleton. Each informs the other. They are complementary, interconnected cognitive frameworks, a metaphor for integrated mixed methods.

> Wine exists in as many varieties as there are people who produce it. Variations in technique, climate, grape, soil, and culture ensure that wine is, to the ordinary drinker, the most unpredictable of drinks, and to the connoisseur the most intricately informative, responding to its origins like a game of chess to its opening move.
>
> —Roger Scruton (2009)
> *I Drink Therefore I Am:*
> *A Philosopher's Guide to Wine*

You may have noticed that the phrase "the fruit of qualitative methods" appears often throughout this chapter. Subsequent chapters in this book discuss how to collect, analyze, and use qualitative data, but this opening chapter has aimed at giving you a taste of the fruit of qualitative methods. As noted at the beginning, this chapter has been a bit like a qualitative wine tasting, if you will: a chance to cultivate taste as much as judge what is to your liking. It is important to know what qualitative data yield and what findings look like so that you will know what you are seeking to produce when you undertake your own qualitative inquiry. It will also be important to consider criteria for judging the quality of qualitative data. Wines come to market distinguished by type (e.g., red, white, rose, and sparkling), serving a variety of purposes (e.g., fine dining, celebrations, as accompaniment to particular foods, for cooking, for medicinal purposes), and varying in quality, though judges of quality differ in their judgments. Likewise, qualitative studies vary by type, purpose, data-gathering processes, analytical techniques, reporting formats, and quality. These variations are the territory we'll cover in subsequent chapters. Before doing so, here's a brief review of this initial foray into the qualitative vineyard.

Chapter Summary and Conclusion

Module 1 How Qualitative Inquiry Contributes to Our Understanding of the World

Module 2 What Makes Qualitative Data *Qualitative*

Module 3 Making Methods Decisions

Module 4 The Fruit of Qualitative Methods: Chapter Summary and Conclusion

Chapter Review

The fruit of qualitative inquiry emerges from the three kinds of qualitative data:

1. *Interpersonal interviews:* They ask open-ended questions and probe for in-depth responses about people's experiences, perceptions, opinions, feelings, and knowledge; interview data consist of verbatim quotations with sufficient context to be interpretable. Exhibit 1.1 offers examples of open-ended interview questions.

2. *Fieldwork observations:* They describe activities, behaviors, actions, conversations, interpersonal interactions, organizational or community processes, or any other aspect of observable human experience; data consist of field notes—rich, detailed descriptions, including the context within which the observations were made. Exhibit 1.7 presents an example of an observation of a session in an early-childhood parent education program.

3. *Documentation:* This includes any kind of written material from organizational, clinical, or program records; social media postings of all kinds; memoranda and correspondence; official publications and reports; personal diaries, letters, artistic works, photographs, and memorabilia; and written responses to open-ended surveys. The qualitative data consist of excerpts from documents captured in a way that records and preserves context.

This chapter compared and contrasted these qualitative kinds of data with quantitative data and gave examples of mixed methods that integrate qualitative and quantitative methods and findings. (See Exhibit 1.4, p. 15 for details.)

In Module 1, we examined five examples of the qualitative inquiry contributions summarized in Exhibit 1.2 (p. 13):

1. Capturing stories to understand people's perspectives and experiences

2. Elucidating how systems function and their consequences for people's lives

3. Understanding context: how and why it matters

4. Identifying unanticipated consequences

5. Making case comparisons to discover important patterns and themes across cases.

Throughout the chapter, I've provided examples of qualitative findings, such as teachers' reactions to the Kalamazoo accountability system and research on women's ways of knowing (Exhibit 1.5, p. 16).

Exhibit 1.6 (p. 21) provides a list of guiding questions and options for making methods decisions. To place our contemporary inquiries and methods into the broadest possible historical context, Exhibit 1.9, at the end of this chapter (pp. 41–43), lists the contributions of qualitative inquiry pioneers across the full panorama of disciplines and inquiry traditions going back to ancient Greece.

Form Follows Function, Design Follows Purpose

Beginning this book with examples of the fruit of qualitative inquiry follows the basic logic of design: Start with what you want to produce and achieve, the outcomes and results you seek, and then work backward to figure out what processes you must follow and what steps you must take to get where you want to be at the end. It turns out, however, that the planned path and the path actually taken can be quite different as you navigate uncharted parts of the terrain and overcome unexpected obstacles along the way. Start to finish is rarely, if ever, a simple, linear path. Be prepared for some major forks in the road, detours, emergent opportunities, disappointments, and thrills. For qualitative inquiry takes you into the world to experience and document the world, and the world, being multidimensional, multilayered, complex, dynamic, and enveloping, will take you to places both planned and unplanned. It's an amazing journey because the world is an amazing place, offering much to discover, much to ponder, and much to understand. To help determine if you are ready, reflect on what you understand to be the fruit of qualitative inquiry. The ancient Sufi story, in graphic comic format, that ends this chapter (pp. 38–40) is aimed at stimulating that reflective process.

EXHIBIT 1.9 Qualitative Inquiry Pioneers

Below are examples of inquisitive minds who contributed to our knowledge of the world through direct observation, interviewing, document analysis, fieldwork, open-ended inquiry, systematic analysis, and careful reflection into and on the nature of things. No such list can be definitive, and any such list will be controversial. The purpose here is to place what we now understand to be—and label as—qualitative inquiry within the broadest possible historical context. As has been reiterated periodically by scholars throughout history, we surely stand on the shoulders of giants (Bernard of Chartres, 12th century; Sir Isaac Newton, 17th century). This is a purposeful sampling of those giants, the purpose being to demonstrate and remind that qualitative inquiry is deeply grounded in scientific inquiry across time, space, discipline, culture, and knowledge epoch.

QUALITATIVE INQUIRER	DISCIPLINE	FOCUS AND CONTRIBUTION
1. Herodotus (484–425 BCE)	History	He systematically studied the Greco-Persian Wars through interviews and documents capturing in-depth geographical, social, political, and cultural information.
2. Aristotle (384–322 BCE)	Philosophy	His natural philosophy examined phenomena of the natural world using qualitative observations to reason inductively about the essence of things.
3. Herophilos (335–280 BCE)	Anatomy and medicine	He observed and documented the actual structure of the human body. In studying the brain, he was the first to differentiate the cerebrum and the cerebellum.
4. Plutarch (46–120)	History	Case studies of famous Greeks and Romans, arranged in pairs to analyze contrasting patterns of virtues and vices.
5. Alhazen (965–1040)	Optics	He observed and described the structure of the eye and experimented with image formation, the processes of seeing, and the nature of vision. He also documented the annual Nile floods in hopes of discovering and engineering controls.
6. Shen Kuo (1031–1095)	Geology	He observed and analyzed Chinese land formations, soil erosion, inland marine fossils, and the deposition of silt, which led to a theory of gradual climate change, stimulated in part by the observation of ancient petrified bamboos preserved underground in a dry northern habitat that would not support bamboo growth at the time of his observations.
7. Roger Bacon (1214–1294)	Linguistics	Fluent in several languages, he documented the corruption of religious texts and Greek philosophy by mistranslations and misinterpretations. He championed direct study of nature over reliance on religious authority.
8. Marco Polo (1254–1324)	Geography and anthropology	He made systematic observations of nature, geography, and culture in his extensive travels through Asia.
9. Nicolaus Copernicus (1473–1543)	Astronomy	He observed the motions of celestial objects, concluding that the Sun, not Earth, was at the center of the galaxy, which led to the Copernican Revolution in scientific inquiry.
10. Galileo Galilei (1564–1642)	Astronomy	He is known for his discovery of the four largest satellites of Jupiter, observation and analysis of sunspots, and confirmation of the phases of Venus.

(Continued)

(Continued)

QUALITATIVE INQUIRER	DISCIPLINE	FOCUS AND CONTRIBUTION
11. Isaac Newton (1642–1727)	Physics	He developed a theory of color from the observation that a prism divides white light into the range of colors in the visible spectrum. He converted qualitative observations about gravity and motion into mathematical laws.
12. Charles Darwin (1809–1882)	Biology	He observed and compared species and fossils, which led to the theory of evolution.
13. Karl Marx (1818–1883)	Political economy	He provided detailed descriptions of the lives of the poor in the early years of urban, industrialized England. His qualitative observations portrayed the harsh living conditions and impoverished lives of industrial workers.
14. Henry Gray (1827–1861)	Anatomy	His painstaking and methodical dissections led to the breakthrough publication of the carefully illustrated and documented *Gray's Anatomy*.
15. Max Weber (1864–1920)	Sociology	He described cultural influences embedded in religion and observed and analyzed the nature and functions of bureaucracy.
16. Émile Durkheim (1858–1917)	Sociology	He described the division of labor in society and studied and compared the social and cultural lives of aboriginal and modern societies.
17. Sigmund Freud (1856–1939)	Psychiatry	He conducted clinical case studies of people with psychological problems and investigations of the unconscious.
18. Carl Jung (1875–1961)	Psychology	He conducted clinical case studies of personality, dreaming, and the collective unconscious.
19. Franz Boas (1858–1942)	Cultural anthropology	He studied the relation between the life of a people and their physical environment.
20. William Isaac Thomas (1863–1947)	Sociology	He studied the lives and culture of American immigrants by collecting oral and written reports from Chicago's Polish community as well as from Poles in their native land. His qualitative data included newspaper items, records found in immigrant organizations, personal letters, and diaries. He documented how what is defined by people as real has real consequences, a premise known as the Thomas theorem.
21. Bronisław Malinowski (1884–1942)	Cultural anthropology	His ethnography of the Trobriand Islands led to influential theories of reciprocity and exchange. He advanced systematic anthropological fieldwork methods.
22. George Herbert Mead (1863–1931)	Social psychology	He is known for his in-depth observation of human interactions and how humans create social meanings.
23. Ruth Benedict (1887–1948)	Cultural anthropology	She studied patterns of culture: the relationships between personality, art, language, and culture.
24. Jean Piaget (1896–1980)	Child development and education	He conducted interviews with and observations of children as they developed and matured. He focused on the processes of the qualitative development of knowledge and understanding.

QUALITATIVE INQUIRER	DISCIPLINE	FOCUS AND CONTRIBUTION
25. Louis Leakey (1903–1972) and Mary Leakey (1913–1996)	Paleoanthropology	Starting work in Olduvai Gorge, Tanzania, in 1951, they found an ancient bog where animals had been trapped and butchered. This led to major fossil discoveries that advanced understandings of human evolution, including *Zinjanthropus boisei* and *Homo habilis.*
26. Alexander Haley (1921–1992)	Oral history	He pioneered publishing in-depth interviews with prominent and often controversial public figures, including Malcolm X, Muhammad Ali, Miles Davis, Martin Luther King Jr., Melvin Belli, Sammy Davis Jr., Jim Brown, Johnny Carson, and Quincy Jones.
27. William Foote Whyte (1914–2000)	Urban sociology	A pioneer in participant observation, his study of a Boston slum inhabited by immigrants from Italy was published as *Street Corner Society*, a qualitative classic. He also studied industrial and agricultural workers in Venezuela, Peru, Guatemala, and Spain.
28. Jane Goodall (1934–)	Primatology	She made in-depth, longitudinal observations (45years) of the social and family interactions of wild chimpanzees in Gombe Stream National Park, their native habitat.
29. Studs Terkle (1912–2008)	Oral history	He captured and popularized oral histories of ordinary working-class Americans.
30. Oliver Saks (1933–)	Neurologist	He conducted case studies of people with neurological disorders that provided breakthrough insights into the nature of those disorders, how to treat them, and their effects on people's lives.
31. Gregory Bateson (1904–1980)	Ecological anthropology, cybernetics, and systems theory	He illuminated relationships as central to human experience at multiple levels of analysis and crossed disciplines and fields to integrate knowledge about functional and dysfunctional interrelationships. He was a pioneer in using abductive inference for holistic qualitative analysis.
32. Howard Becker (1928–)	Sociologist and qualitative methodologist	He is a pioneer in teaching and writing about qualitative inquiry as a credible method for systematic and rigorous study of social phenomena.
33. Norman K. Denzin and Yvonna Lincoln	Qualitative epistemologists and methodologists	They are pioneers in identifying, documenting, nurturing, and expanding the range and breadth of qualitative inquiry. They are editors of the first edition of *Handbook of Qualitative Research* (1994) and three subsequent editions (2000, 2005, and 2011); editors of three editions of *Collecting and Interpreting Qualitative Materials* (2008); editors of *The Qualitative Inquiry Reader* (2002), *Turning Points in Qualitative Research* (2003), and *Handbook of Critical and Indigenous Methodologies* (with Linda Tuhiiwai Smith, 2008); and founders and editors of the first qualitative methods journal, *Qualitative Inquiry.*

APPLICATION EXERCISES

1. This chapter opened with the story of a Portuguese sheep herder. Identify and explain at least three things that story illustrates about the nature of qualitative inquiry.

2. The first contribution of qualitative inquiry discussed is illuminating meanings (pp. 3–6). Write a case study of an event, experience, or encounter that had meaning for you. First, simply describe what happened, what you experienced, and how the story unfolded in enough detail that a reader knows what occurred. Then, analyze the experience or event for *meaning*. Finally, reflect on your experience of meaning making. Comment on your experience of and thoughts about interpreting a case study (personal story) to extract meanings.

3. Early in the chapter, there is a section on how qualitative inquiry elucidates "how systems function and their consequences for people's lives" (see p. 8). Identify a *system* that you have some knowledge about. Discuss and explain how a qualitative inquiry could help elucidate how that system functions and its system dynamics.

4. The chapter highlights a number of important qualitative studies that have contributed to our knowledge. (See especially the section on qualitative findings (pp. 15–17) as well as Exhibit 1.9, on qualitative inquiry pioneers (pp. 41–43). Find a major qualitative study in your own field of interest. Why did the study use qualitative methods? What was the importance of the contribution made to knowledge of the study you located?

5. My rumination on anecdotes included an application exercise, which, in case you neglected it earlier, I repeat here. For Valentine's Day, *The New York Times* collected anecdotes from writers in a variety of genres about books that have taught them above love. Read these anecdotes as data. What themes do you find across the stories? Practice analysis ("A Sentimental Education," 2014).

6. This chapter ends with an ancient Sufi story, "The Fruit of Qualitative Methods" (pp. 38–40), in which a scholar is seeking to experience and understand fruit. Earlier in the chapter, in the section "The Power of Qualitative Data," the case example of what happened in Kalamazoo schools is presented. Qualitative analysis often involves looking across different stories for common themes and patterns. Identify and discuss at least two patterns or themes that are common in those two stories. This is a creative analysis exercise. There is no particular right answer. What connections can you make between those two quite different stories, the scholar seeking fruit and the Kalamazoo teachers reacting to the school district's accountability system?

Strategic Themes in Qualitative Inquiry

明打狼　闇打狐狸

Hunt foxes stealthily and wolves openly.
Adapt strategies appropriately to meet different challenges.

Strategic Wisdom

Strategos is a Greek word meaning "the thinking and action of a general." What it means to be strategic is epitomized by the greatest of Greek generals, Alexander. He conducted his first independent military operation in northern Macedonia at age 16. He became the ruler of Macedonia after his father, Philip, was assassinated in 336 BCE. Two years later, he embarked on an invasion of Persia and conquest of the known world. In the Battle of Arbela, he decisively defeated Darius III, king of kings of the Persian Empire, despite being outnumbered five to one (250,000 Persians against Alexander and fewer than 50,000 Greeks).

Alexander's military conquests are legendary. What is less known and little appreciated is that his battlefield victories depended on in-depth knowledge of the psychology and culture of the ordinary people and military leaders in the opposing armies. He included in his military intelligence information about the beliefs, worldviews, motivations, and patterns of behavior of those he faced. Moreover, his conquests and subsequent rule were more economic and political in nature than military. He used what we would now understand to be psychological, sociological, and anthropological insights. He understood that lasting victory depended on the goodwill of and alliances with non-Greek peoples. He carefully studied the customs and conditions of the people he conquered and adapted his policies—politically, economically, and culturally—to promote good conditions in each locale, so that the people were reasonably well disposed toward his rule (Garcia, 1989).

In this approach, Alexander had to overcome the arrogance and ethnocentrism of his own training, his culture, and Greek philosophy. Historian C. A. Robinson Jr. (1949) explained that Alexander was brought up on Plato's theory that all non-Greeks were barbarians, enemies of the Greeks by nature; and Aristotle taught that all barbarians (non-Greeks) were slaves by nature. But "Alexander had been able to test the smugness of the Greeks by actual contact with the barbarians, . . . and experience had apparently convinced him of the essential sameness of all people" (p. 136).

In addition to being a great general and an enlightened ruler, Alexander appears to have been an extraordinary ethnographer, a qualitative inquirer par excellence, using observations and firsthand experience to systematically study and understand the peoples he encountered and to challenge his own culture's prejudices. Thinking strategically, enhancing your powers of observation, becoming an ever more astute interviewer—these are not just research methods but life skills and competencies for more deeply experiencing and understanding the world and for engaging effectively in it.

Chapter Preview: The Purpose and Nature of Strategic Principles

> You've got to think about big things while you're doing small things, so that all the small things go in the right direction.
>
> —Alvin Toffler (1928–)
> Futurist

A well-conceived strategy, by providing overall direction, constitutes a framework for decision making and action. It permits seemingly isolated tasks and activities to fit together, integrating separate efforts toward a common purpose. Specific study design and methods decisions are best made within an overall strategic framework. The effectiveness and impacts of the strategic approach taken can then be evaluated for quality, credibility, and utility (Mintzberg, 2007; Patrizi & Patton, 2010).

Reviewing the great variety of approaches to strategic planning requires a *Strategy Safari Through the Wilds of Strategic Management* (Mintzberg, Lampel, & Ahlstrand, 2008). This chapter offers a strategy safari for qualitative inquiry. We'll review 12 major strategic principles that, taken together, constitute a comprehensive and coherent strategic framework for qualitative inquiry, including fundamental assumptions and epistemological ideals. Exhibit 2.1 summarizes these strategic principles in three modules:

Module 5 Strategic design principles for qualitative inquiry

Module 6 Strategic principles guiding data collection and fieldwork

Module 7 Strategic principles for qualitative analysis and reporting findings

EXHIBIT 2.1	Twelve Core Strategies of Qualitative Inquiry
DESIGN STRATEGIES	
1. *Naturalistic inquiry*	Studying real-world situations as they unfold naturally; nonmanipulative and noncontrolling; *openness* to whatever emerges (lack of predetermined constraints on findings)
2. *Emergent design flexibility*	Open to adapting inquiry as understanding deepens and/or situations change; avoids getting locked into rigid designs that eliminate responsiveness; pursues new paths of discovery as they emerge
3. *Purposeful sampling*	Cases for study (e.g., people, organizations, communities, cultures, events, and critical incidences) are selected because they are "information rich" and illuminative, that is, they offer useful manifestations of the phenomenon of interest; sampling, then, is aimed at insight about the phenomenon, not empirical generalization from a sample to a population
DATA COLLECTION AND FIELDWORK STRATEGIES	
4. *Qualitative data*	Observations that yield detailed, thick description; inquiry in depth; interviews that capture direct quotations about people's personal perspectives and experiences; case studies; careful document review
5. *Personal experience and engagement*	The inquirer has direct contact with and gets close to the people, situation, and phenomenon under study; your personal experiences and insights as the inquirer (instrument of qualitative inquiry) are an important part of the inquiry and critical to understanding the phenomenon under study
6. *Empathic neutrality and mindfulness*	An empathic stance in interviewing seeks vicarious understanding without judgment (neutrality) by establishing rapport and showing openness, sensitivity, respect, awareness, and responsiveness; in data collection (observation and interviewing), it means being fully present: *mindful*
7. *Dynamic systems perspective*	Attention to process; assumes change as ongoing whether the focus is on an individual, an organization, a community, or an entire culture; therefore, the inquiry is mindful of and attentive to system and situation dynamics

ANALYSIS AND REPORTING STRATEGIES

8. *Unique case orientation*	Assumes that each case is special and unique; the first level of analysis is being true to, respecting, and capturing the details of the individual cases being studied; cross-case analysis follows from and depends on the particularity and quality of in-depth individual case studies
9. *Inductive analysis and creative synthesis*	Analysis begins with immersion in the details and specifics of the inquiry to discover important patterns, themes, and interrelationships; exploration and attention to what emerges is followed by confirmatory inquiry; analysis from the particular to the general is guided by analytical principles rather than by rules, and it ends with a creative synthesis
10. *Holistic perspective*	The *whole* phenomenon under study is understood as a complex system that is more than the sum of its parts; inquiry focuses on and captures complex interdependencies and system dynamics that cannot meaningfully be reduced to a few discrete variables and linear, cause–effect relationships
11. *Context sensitivity*	The inquiry places findings in a social, historical, and temporal context; in analysis and interpretation, the qualitative inquirer is careful about, even dubious of, the possibility or meaningfulness of generalizations across time and space; the inquiry emphasizes thoughtful comparative case analyses and extrapolating patterns for possible transferability and adaptation in new settings
12. Reflexivity: perspective and voice	The qualitative analyst owns and is reflective about her or his own voice and perspective; a credible voice conveys authenticity and trustworthiness; the inquirer's focus becomes balanced—understanding and depicting the world *authentically* in all its complexity while being self-analytical, politically aware, and reflexive in consciousness; this reiterates that the qualitative inquirer is the instrument of inquiry

This module covers three design strategies: (1) naturalistic inquiry, (2) emergent design flexibility, and (3) purposeful sampling.

Naturalistic Inquiry

> The best [doctors] seem to have a sixth sense about disease. They feel its presence, know it to be there, perceive its gravity before any intellectual process can define, catalog, and put it into words. Patients sense this about such a physician as well: that he is attentive, alert, ready; that he cares. No student of medicine should miss observing such an encounter. Of all the moments in medicine, this one is most filled with drama, with feeling, with history.
>
> —Michael LaCombe
> *Annals of Internal Medicine* (1993)
> (Quoted in Mukherjee, 2010, p. 128)

A health researcher observes a doctor visiting her patients in a hospital. An anthropologist studies initiation rites among the Gourma people of Burkina Faso in West Africa. A sociologist observes interactions among bowlers in their weekly league games. An evaluator participates fully in a leadership training program she is documenting. A naturalist studies bighorn sheep beneath Powell Plateau in the Grand Canyon. A policy analyst interviews people living in public housing in their homes. An agronomist observes farmers' spring planting practices in rural Minnesota. What do these researchers have in common? They are *in the field* studying the real world as it unfolds.

Qualitative designs are naturalistic to the extent that the research takes place in real-world settings and the researcher does not attempt to affect, control, or manipulate what is unfolding naturally.

NATURALISTIC INQUIRY

Naturalistic inquiry is based on the notion that context is essential for understanding human behavior, and acquiring knowledge of human experience outside of its natural context is not possible. Conducting research in participants' natural environments is essential. Researchers must meet participants where they are, in the field, so that data collection occurs while people are engaging in their everyday practices. Research conducted in the field allows investigators to observe participants in action in an effort to obtain a more complete understanding of the phenomenon under investigation. During the process of engaging in naturalistic inquiry, the researcher becomes the instrument for collecting data. Human beings as data collecting instruments are necessary because only humans can gather and evaluate the meaning of complex interactions. Attending to these processes in the field is necessary because the complexity of human interaction is available only in the settings of everyday life, not in a controlled laboratory setting or through created instruments.

—Jillian A. Tullis Owen (2008, p. 547)
"Naturalistic Inquiry"

Observations take place in real-world settings, and people are interviewed with open-ended questions in places and under conditions that are comfortable for and familiar to them.

Egon Guba (1978), in his classic treatise on naturalistic inquiry, identified two dimensions along which types of scientific inquiry can be described: (1) the extent to which the scientist manipulates some phenomenon in advance to study it and (2) the extent to which constraints are placed on outputs, that is, the extent to which *predetermined* categories or variables are used to describe the phenomenon under study. He then defined "naturalistic inquiry" as a "discovery-oriented" approach that minimizes investigator manipulation of the study setting and places no prior constraints on what the outcomes of the research will be. Naturalistic inquiry contrasts with controlled experimental designs and laboratory studies where the investigator controls study conditions by manipulating, changing, or holding constant external influences and where a very limited set of outcome variables are measured. Open-ended, conversation-like interviews as a form of naturalistic inquiry contrast with questionnaires that have predetermined response categories. It's the difference between asking, "Tell me about your reactions to the program?" versus "How satisfied were you with the program?"

1. Very satisfied

2. Somewhat satisfied

3. Not at all satisfied

In the simplest form of controlled experimental inquiry, the researcher enters the program at two points in time, pretest and posttest, and compares the treatment group with some control group on a limited set of standardized measures. Such designs assume a single, identifiable, isolated, and measurable treatment. Moreover, such designs assume that, once introduced, the treatment remains relatively constant.

While there are some narrow, carefully controlled, and standardized treatments that fit this description, in practice, human interventions (programs) are often quite comprehensive, variable, and dynamic—changing as practitioners learn what does and does not work, developing new approaches, and realigning priorities. This, of course, creates difficulty for controlled experimental designs that need specifiable, standardized, unchanging treatments aimed at producing specifiable, predetermined outcomes. Controlled experimental evaluation designs require controlling program adaptation and improvement so as not to interfere with the rigor of the research design. One contribution of qualitative inquiry in experimental designs is to document the extent to which treatment implementation unfolds as planned and, if there are variations in treatment implementation, to help interpret the implications of intervention variations for the observed and measured outcomes.

Under real-world conditions, where programs are subject to change and redirection, naturalistic inquiry replaces the fixed-treatment/outcome emphasis of the controlled experiment with a dynamic, process orientation that documents actual operations and impacts over a period of time. The qualitative evaluator sets out to understand and document the day-to-day reality of participants in the program, making no attempt to manipulate, control, or eliminate situational variables or program developments but accepting the complexity of a changing program reality. The data of the evaluation include whatever emerges as important to understanding participants' experiences.

However, the distinction is not as simple as being in the field versus being in the laboratory; rather, the degree to which a design is naturalistic falls along a continuum with completely open fieldwork on one end and completely controlled laboratory control on the other end, but with varying degrees of researcher control and manipulation between these endpoints. For example, the very presence of the researcher, asking questions or, as in the case of formative program evaluation, providing feedback, can be an intervention that reduces the "natural" unfolding of events. Unobtrusive observations can minimize data collection as an intervention, but opportunities for unobtrusive observation are limited, and covert observations raise ethical issues that we'll address in Chapter 6.

Let me offer two examples to illustrate variations in the naturalistic inquiry design strategy. In evaluating a wilderness-based leadership training program, I participated fully in the 10-day wilderness experience, guided in my observations by nothing more than the sensitizing concept of "leadership." The only "unnatural" elements of my participation were that (a) everyone knew I was taking notes to document what happened and (b) at the end of each day, I conducted open-ended, conversational interviews with the staff and participants. While this constitutes a relatively pure naturalistic inquiry strategy, my presence, note taking, and interviews must be presumed to have altered somewhat the way the program unfolded. I know, for example, that my debriefing interviews with staff in the evenings got them thinking about the things they were doing that led to some changes in how they conducted the training.

The second example comes from the fieldwork of Beverly Strassmann among the Dogon people in the village of Sangui in the Sahel, about 120 miles south of Timbuktu in Mali, West Africa (Gladwell, 2000). Her study focused on the Dogon tradition of having menstruating women stay in small, segregated adobe huts at the edge of the village. She observed the comings and goings of these women and obtained urine samples from them to be sure they were menstruating. The women slept in the isolation huts, but during the day, they went about their normal activities. For 736 consecutive nights, Strassmann kept track of all the women who used the huts. This allowed her to collect statistics on the frequency and length of menstruation among the Dogon women, but with a completely naturalistic inquiry strategy, illustrating how both quantitative and qualitative data can be collected within a naturalistic design strategy.

Emergent Design Flexibility

> There's the inquiry idea you begin with, the newborn idea. Then the inquiry design emerges and evolves as you think about it and discuss it with others. At some point, for some sooner, for others later, the inquiry design reaches the formal proposal stage. Then you start data collection and the fieldwork unfolds and new opportunities emerge and you pursue those opportunities, so the design further evolves. That becomes the actual design you have implemented by the time you cease data collection. Then as you immerse yourself in analysis, review the data you've collected and how you collected it, the design you actually implemented will become clearer. Only retrospectively will you finally know what your design was. Prospective designs articulate possibilities. Retrospective descriptions of what actually occurred constitute the actual design. All of which means carefully documenting the design process throughout the inquiry journey.
>
> —From Halcolm's *Methodology as Journey*

In the 10-day wilderness leadership training program I evaluated, the 20 participants unexpectedly split into two subgroups on the first day. I had to make an in-the-field, on-the-spot decision about which group to join and how to get interviews with the others at a later time. The original design had planned for follow-up questionnaires to be sent to the 20 participants six months after the training experience, but as the program drew to a close, the participants rebelled against the idea of surveys as being too shallow to be meaningful and insisted on open-ended interviews, in which they could tell their stories and reflect on their experiences.

Naturalistic inquiry designs cannot usually be completely specified in advance of fieldwork. While the design will specify an initial focus, plans for observations, and initial guiding interview questions, the naturalistic and inductive nature of the inquiry makes it both impossible and inappropriate to specify operational variables, state testable hypotheses, or finalize either instrumentation or sampling schemes. A naturalistic design unfolds or emerges as the fieldwork unfolds.

> The call for an emergent design by naturalists is not simply an effort on their part to get around the "hard thinking" that is supposed to precede an inquiry; the desire to permit events to unfold is not merely a way of rationalizing what is at bottom "sloppy inquiry." The design specifications of the conventional paradigm [in which details of the design are rigidly determined in advance of data collection and carefully adhered to] form a procrustean bed of such a nature as to make it impossible for the naturalist to lie in it—not only uncomfortably, *but at all.* (Lincoln & Guba, 1985, p. 225)

Design flexibility stems from the open-ended nature of naturalistic inquiry as well as pragmatic considerations. Being open and pragmatic requires a high tolerance for ambiguity and uncertainty as well as trust in the ultimate value of what inductive analysis will yield. Such tolerance, openness, and trust create special problems for dissertation committees and funders of evaluation or research. How will they know what will result from the inquiry if the design is only partially specified? The answer is that they won't know with any certainty. All they can do is look at the results of similar qualitative inquiries, inspect the reasonableness of the overall strategies in the proposed design, and consider the capacity of the researcher to fruitfully undertake the proposed study.

As with other strategic themes of qualitative inquiry, the extent to which the design is specified in advance is a matter of degree. Doctoral students doing qualitative dissertations will usually be expected to present fairly detailed fieldwork proposals and interview schedules so that the approving doctoral committee and institutional review board can guide the student and be sure that the proposed work will lead to satisfying degree requirements. Many funders will fund only detailed proposals. As an ideal, however, the qualitative researcher needs considerable flexibility and openness. The fieldwork approach of anthropologist Brackette F. Williams represents the ideal of emergence in naturalistic inquiry. She has focused on issues of cultural identity and social relationships. Her work has included in-depth study of ritual and symbolism in the construction of national identity in Guyana (1991) and the ways in which race and class function in the national consciousness of the United States. In 1997, she received a five-year MacArthur fellowship (popularly called a "Genius Award"), which allowed her to pursue a truly emergent, naturalistic design in her fieldwork on the phenomenon of killing in America. I had the opportunity to interview her about her work and, with her permission, am including several excerpts from that interview throughout this chapter to illustrate the actual scholarly implementation of some of the strategic ideals of qualitative inquiry. Here, she describes the necessity of an open-ended approach to her fieldwork because her topic is broad and she needs to follow wherever the phenomenon takes her:

> I'm tracking something—killing—that's moving very rapidly in the culture. Every time I talk to someone, there's another set of data, another thing to look at. Anything that happens in America can be relevant, and that's the exhausting part of it. It never shuts off. You listen to the radio. You watch television. You pass a billboard with an advertisement on it. There's no such thing as something irrelevant when you're studying something like this or maybe just studying the society that you're in. You don't always know exactly how it's going to be relevant, but somehow it just strikes you and you say to yourself: I should document the date when I saw this and where it was and what was said because it's data.

> I don't follow every possible lead people give me. But generally, it is a matter in some sense of opportunity sampling, of serendipity, whatever you want to call it. I key into things that turn out to be very important six months later.

> I did a lot of impromptu interviewing in the first year in places like airports to formulate a protocol of questions and issues to pursue. It was general sampling to get a sense of what I wanted to know. At other times, it's just to get a general opinion from John Q. Public about a question that I've gotten all kinds of official responses to, but I want to know what people in general think. In an airport, I may get an opportunity to talk to 5 or 10 people. If I have several stops, I may get 15 or 20 by the time I come home.

> I fashion the research as I want to fashion it based on what I think this week as opposed to what I thought last week. I don't follow some proposal. I don't have in mind that this has to be a book that's going to have to come out a certain way. I'm following where the data take me, where my questions take me.

Few qualitative studies are as fully emergent and open-ended as the fieldwork of Williams. Her work exemplifies the ideal of *emergent design flexibility*.

SIDEBAR

EMERGENT METHODS

Emergent design flexibility not only includes openness, responsiveness, and adaptability within a particular study, but also emergent methods more generally are at the cutting edge of how we inquire into and make some sense of complexity.

> Emergent methods arise as a means of accessing answers to complex research questions and revealing subjugated knowledge. These research techniques are particularly useful for discovering knowledge that lies hidden, that is, difficult to tap into because it has not been part of the dominant culture or discourse.

—Sharlene Nagy Hesse-Biber and Patrcia Leavy (2008, p. v)
Handbook of Emergent Methods

> Emergent research methods are the logical conclusion to paradigm shifts, major developments in theory, and new conceptions of knowledge and the knowledge-building process. As researchers continue to explore new ways of thinking about and framing knowledge construction, so, too, do they develop new ways of building knowledge, accessing data, and generating theory. In this sense, new methods and methodologies are theory driven and question driven. Emergent methods often arise in order to answer research questions that traditional methods may not adequately address. Evolving theoretical paradigms in the disciplines have opened up the possibilities of the development of innovative methods to get at new theoretical perspectives.

—Sharlene Nagy Hesse-Biber and Patrcia Leavy (2006a, p. xi)
Emergent Methods in Social Research

Purposeful Sampling

> In an information-rich world, the wealth of information means a dearth of something else: a scarcity of whatever it is that information consumes. What information consumes is rather obvious: it consumes the attention of its recipients. Hence a wealth of information creates a poverty of attention and a need to allocate that attention efficiently among the overabundance of information sources that might consume it.
>
> —Herbert Simon (1971, pp. 40–41)
> Nobel laureate in economics
>
> Since you can't study everything and everyone, focus on something important and someone from whom you can learn a great deal about that matter of importance. Choose wisely, with purpose. Time is fleeting. Pay attention.
>
> —Halcolm

In 1940, eminent sociologist Kingsley Davis published what was to become a classic case study, the story of Anna, a baby kept in nearly total isolation from the time of her birth until she was discovered at age six. She had been deprived of human contact, had acquired no language skills, and had received only enough care to keep her barely alive. This single case, horrifying as was the abuse and neglect, offered a natural experiment to study socialization effects and the relative contributions of nature and nurture to human development. In 1947, Davis published an update on Anna and a comparison case of socialization isolation, the story of Isabelle. These two cases offered considerable insight into the question of how long a human being could remain isolated before "the capacity for full cultural acquisition" was permanently damaged (Davis, 1940, 1947). The cases of Anna and Isabelle are examples of purposeful case sampling (also called *purposive* case selection).

Perhaps in nothing is the difference between quantitative and qualitative designs better illuminated than in the different strategies, logics, and purposes that distinguish statistical probability sampling from qualitative purposeful sampling. Qualitative inquiry typically focuses on relatively small samples, even single cases ($n = 1$) like Anna or Isabelle, selected *purposefully* to permit inquiry into and understanding of a phenomenon *in depth*. Quantitative methods typically aim for larger samples selected randomly, to generalize with confidence from the sample to the population that it represents. Not only are the techniques for sample selection different, but also the very logic of each approach is distinct because the purpose of each strategy is different. Mixed methods can incorporate both quantitative and qualitative sampling strategies when sufficient time and resources are available to do both.

WISE ADVICE ON CASE SELECTION

SIDEBAR

If you are going to do a case study, you are likely to devote a significant portion of your time to it. What is to be avoided is committing much of your time and resources and then finding that the case study will not work out. Therefore, in using the case study method, your goal should be to select your case study carefully. Try to spot unrealistic or uninformative case studies as early as possible.

More ambitiously, try to select a significant or "special" case or cases for your case study. The more significant your case, the more likely your case study will contribute to the research literature or to improvements in practice (or to the completion of a doctoral dissertation). Conversely, devoting your efforts to a fairly "mundane" case study may not even produce an acceptable study (or dissertation). If you do not have access to a special case, the recommended approach is to consider any candidate for your case study with great care and forethought, even if the process takes more time than you would have anticipated.

Set your goals high. You may only have a once-in-a-lifetime opportunity to contribute to case study research. Also, consult actively with your peers and colleagues about your selection. Choose the most significant case possible. Your success might result in an exemplary study (or dissertation). Your study might present new theoretical or practical themes. It also might capture . . . a case of lasting relevance—if not value—decades later.

—Robert K. Yin (2004, pp. 3–4)
The Case Study Anthology

Unusual clinical cases in medicine and psychology, instructive precisely because they are unusual, offer many examples of purposeful sampling. Neurologist Oliver Sacks presents a number of such cases in his widely read and influential books, *The Man Who Mistook His Wife for a Hat* (1985) and *Hallucination* (2012), the very titles of which suggest the uniqueness of the cases and issues examined. While one cannot generalize from single cases or very small samples, one can learn from them—and learn a great deal—often opening up new territory for further research, as was the case with Piaget's detailed and insightful observations of his own two children. Andrew Solomon (2012) has studied a great diversity of families with children he describes as living "far from the tree," that is, outliers in the sense that they are substantially different from their parents: children manifesting autism, schizophrenia, Down syndrome, dwarfism, deafness, or transgender identity, and prodigies, among others. Malcolm Gladwell, in his book *Outliers* (2008), reports on people with unusual, outside-the-mainstream advantages and success stories, like Bill Gates, the founder of Microsoft and one of the world's wealthiest men.

The logic and power of statistical probability sampling derives from its purpose: generalization. The logic and power of qualitative purposeful sampling derives from the emphasis on in-depth understanding of specific cases: *information-rich cases*. Information-rich cases are those from which one can learn a great deal about issues of central importance to the purpose of the research; thus the term *purposeful sampling*. For example, if the purpose of an evaluation is to increase the effectiveness of a program in reaching lower–socioeconomic status groups, one may learn a great deal by focusing in depth on understanding the needs, interests, and incentives of a small number of carefully selected poor families. Chapter 5 will present a variety of strategies for purposefully selecting information-rich cases.

It is important to add that not all qualitative studies involve small samples. Lenore Manderson (2011) based her inquiry into the effects of amputations and other surgical intrusions that permanently change the body on interviews with 100 people. Exhibit 2.2 presents how she framed her inquiry. Solomon's (2012) study of children living "far from the tree" (exceptional and special-needs children) is based on interviews with more than 300 families over 10 years.

EXHIBIT 2.2   From Able-Bodied to Disabled

Some reading this book have no doubt experienced the loss of a limb, the removal of a cancerous organ, or other surgical procedure or life event that left them disabled. If that has been your experience, you would have been a candidate for inclusion in Lenore Manderson's (2011) study of "surgery, bodily boundaries, and the social self." Most readers are likely to have escaped such an experience, at least so far, and so to understand what it is like would benefit from hearing how those who have gone from being able-bodied to "disabled" have coped. Here is how Manderson framed her inquiry:

The primary empirical data of *Surface Tensions* are narratives of illness, injury, surgery and recovery, elicited in interviews conducted often on more than one occasion with some one hundred remarkable men and women. . . . I explore how people make sense of who they are when the surface of the body is profoundly changed. I ask what occurs when a person experiences a dramatically physical, often very obvious loss, as in the case of the loss of a limb or its functionality. When a woman's sense of being female is so invested in her sexed body, how does she reconstruct "being feminine" after mastectomy, when surfaces dense with gendered meaning are excised? And how, at the same time, does she make sense of the disease when the challenge to her health is internal and cannot be monitored precisely, even though contemporary technologies extend the clinical gaze to the body's interior? What difference does it make to the self when surfaces are reconstructed and the inner workings of the body are brought to the surface—when elimination must be managed with colostomy bags, for example? And what of the self, when the surface is stable but the inner mechanics of the body change—when, as with a kidney transplant, the body becomes host to an organ vital to survival and once part of someone else? Can the self be whole or undamaged when the body . . . has undergone so much change? The interviews on which I draw in this book illustrate how this disorganization is understood and managed by those directly affected.

—Lenore Manderson (2011, p. 44)

This module covers 4 more of the 12 core strategies of qualitative inquiry, with a focus on data collection and fieldwork strategies: (4) qualitative data, (5) personal experience and engagement, (6) empathic neutrality and mindfulness, and (7) the dynamic systems perspective.

Qualitative Data

> Human interconnections consist of molecules we call stories.
>
> —Halcolm

Qualitative data consist of quotations, observations, excerpts from documents, and entries from social media. The first chapter provided several examples of qualitative data. Deciding whether to use naturalistic inquiry or an experimental approach is a design issue. This is different from deciding what kind of data to collect (qualitative, quantitative, or some combination), although design and data alternatives are clearly related. Qualitative data can be collected in experimental designs where participants have been randomly divided into treatment and control groups. Likewise, some quantitative data may be collected in naturalistic inquiry approaches. Nevertheless, controlled experimental designs predominantly aim for statistical analyses of quantitative data, while qualitative data are the primary focus in naturalistic inquiry. This relationship between design and measurement will be explored at greater length in Chapter 5.

Qualitative data describe. They take us, as readers, into the time and place of the observation so that we know what it was like to have been there. Thick description with contextual details captures and communicates someone else's experience of the world in his or her own words. Qualitative data tell a story. In the excerpt below, from my interview with Brackette Williams, she tells the story of checking out a childhood memory. This story gives us insight into the nature of her naturalistic inquiry and open-ended interviewing, shows how a critical incident can be a purposeful sample, and, in the story itself, offers something of the flavor of qualitative data.

I was down in Texas interviewing last March, thinking about my research and interviewing people, and there was a childhood memory that I had of an electrocution of a man that was the son of a woman who lived across the field from us. Now a rumor about this had always been in the back of my mind. Whenever I'd hear about a death penalty case over the years, I would think about this man having been electrocuted. I thought he was electrocuted because he raped this white woman. So I'm sitting in my cousin's kitchen after I had done some of these interviews and another woman, an older woman who was a relative of hers, came in and the conversation goes around. I happen to mention this memory of mine. I asked, "Is that just something that I concocted out of having read a book or something, but it never happened?" She answered, "Oh, no, it happened. You only have one part of the story wrong. He didn't rape her. He looked at her."

You know, you read about these things in history books, and then all of a sudden, it's like a part of a world that you existed in. These things happened around you and yet somehow there was so much of a distance, you couldn't touch it. I knew about this man all my life, but in all the reading and all the history books, I couldn't touch that. *Doing this project the way I'm doing it allows me to touch things that otherwise I would never touch.*

Direct Personal Experience and Engagement: Going Into the Field

> Objects in a mirror may appear closer than they are.
>
> —Warning on passenger-side car mirrors
>
> People in the field may appear more distant than they are. Or closer. Or both at the same time. In every case these appearances are not what's important. Get closer. And closer. And closer. To get beyond appearances.
>
> —From Halcolm's *Fieldwork Advice*

SIDEBAR

STORIES AS QUALITATIVE DATA

Richard Krueger, University of Minnesota, pioneered using focus groups for program evaluation, created the *Focus Group Kit* (1997), and coauthored the most widely used practical guide to focus groups (Krueger & Casey, 2008). In recent years, he has been writing about and conducting workshops on stories as central to qualitative inquiry: collecting, analyzing, and presenting stories. Here's why he thinks stories are the core of qualitative inquiry:

> I believe in the power of stories. I have spent much of my career listening to people tell their stories in focus groups and individual interviews. People's stories have made me laugh, made me cry, made me angry, and kept me awake at night. Quantitative data have never once led me to shed a tear or spend a sleepless night. Numbers may appeal to my head, but they don't grab my heart. I believe that if you want people to do something with your evaluation findings, you have to grab their attention, and one way to do that is through stories.

> Stories can help evaluators get their audience's attention, communicate emotions, illustrate key points or themes, and make findings memorable.... Stories give researchers new ways to understand people and situations and tools for communicating these understandings to others.

> They help us understand. Stories provide insights that can't be found through quantitative data. A story helps us understand motivation, values, emotions, interests, and factors that influence behavior. Stories can give us clues about why an event might occur or how something happens.

> Stories help us interpret quantitative data. Stories can also be used to amplify and communicate quantitative data. For example, a monitoring system might detect a change in outcomes, but what prompted the change usually can't be found in these data systems. Stories from clients and staff can help us understand factors that change lives and influence people.

> They help share what we learned. Stories help communicate evaluation findings. Evidence suggests that people have an easier time remembering a story than recalling numerical data. The story is sticky, but numbers quickly fade away. Evaluation data are hard to remember. The story provides a framework that helps the reader or listener remember salient facts.

> Stories help communicate emotions. Statistical and survey data tend to dwell on the cold hard facts: Stories are different. They show us the challenges people face, how people feel, and how they respond to situations. Stories tug at our hearts and affect our outlook. Evaluators tend to be apprehensive about emotional messages for valid reasons. Emotional messages can ignore important facts, be fabrications, or overlook empirical data. But instead of avoiding these emotional messages, [qualitative data combines] the emotional aspect of stories with empirical data to address the concerns of both the heart and the head.

—Richard A. Krueger (2010, pp. 404–405)
Using Stories in Evaluation

StoryCorps is a resource for stories: http://storycorps.org/listen/

The quotation from Williams that closed the last section exemplifies the *personal* nature of qualitative fieldwork. Getting close to her subject matter, including using her own experiences, from both childhood and her day-to-day adult life, illustrates the all-encompassing and ultimately personal nature of in-depth qualitative inquiry. Traditionally, social scientists have been warned to stay distant from those they studied in order to maintain "objectivity." But that kind of detachment can limit your openness to and understanding of the very nature of what you are studying, especially where meaning making and emotion are part of the phenomenon. Look closely at what Williams says about the effects of immersing herself personally in her fieldwork, even while visiting relatives: *"Doing this project the way I'm doing it allows me to touch things that otherwise I would never touch."*

Fieldwork is the central activity of qualitative inquiry. *Going into the field* means having direct personal contact with the people under study in their own environments—getting close to the people and situations being studied to personally understand the realities and minutiae of daily life, for example, life as experienced by participants in a welfare-to-work program. The inquirer gets close to the people under study through physical proximity for a period of time as well as through development of closeness in the social sense of shared experience, empathy, and confidentiality. That many quantitative methodologists fail to ground their findings in personal qualitative understanding poses what sociologist John Lofland (1971) has called a major contradiction between their public insistence on the adequacy of statistical portrayals of other humans and their personal, everyday dealings with and judgments about other human beings. His classic observation about how we all make sense of the world through direct personal experience, though a half-century old, still gets at the heart of the value of fieldwork.

> In everyday life, statistical sociologists, like everyone else, assume that they do not know or understand very well people they do not see or associate with very much. They assume that knowing and understanding other people require that one see them reasonably often and in a variety of situations relative to a variety of issues. Moreover, statistical sociologists, like other people, assume that in order to know or understand others one is well advised to give some conscious attention to that effort in face-to-face contacts. They assume, too, that the internal world of sociology—or any other social world—is not understandable unless one has been part of it in a face-to-face fashion for quite a period of time. How utterly paradoxical, then, for these same persons to turn around and make, by implication, precisely the opposite claim about people they have never encountered face-to-face—those people appearing as numbers in their tables and as correlations in their matrices! (p. 3)

Qualitative inquiry means going into the field—into the real world of programs, organizations, neighborhoods, street corners—and getting close enough to the people and circumstances there to capture what is happening. This makes possible description and understanding of *both* externally observable behaviors and internal states (worldview, opinions, values, attitudes, and symbolic constructs). The qualitative emphasis on striving for depth of understanding, in context, includes capturing inner perspectives. "The inner perspective assumes that understanding can only be achieved by actively participating in the life of the observed and gaining insight by means of introspection" (Bruyn, 1963, p. 226).

Actively participating in the life of the observed means going where the action is, getting one's hands dirty, participating where possible in actual program activities, and getting to know program staff and participants on a personal level—in other words, getting personally engaged so as to use all of one's senses and capacities, including the capacity to experience emotion no less than cognition. Such engagement stands in sharp contrast to the professional comportment of some in the field, for example, supposedly objective evaluators, who purposely project an image of being cool, calm, external, and detached. Such detachment is presumed to reduce bias. However, qualitative methodologists question the necessity and utility of distance and detachment, asserting that without

VERSTEHEN: MEANINGFUL UNDERSTANDING

An American Indian prayer warns, "Great Spirit, grant that I may not criticize my neighbor until I have walked a mile in his moccasins." The issue in fieldwork is avoiding judgment so as to be open to deep and meaningful understanding of another.

Verstehen involves the capacity to see things from another's perspective. The nineteenth-century German sociologist Max Weber pioneered this approach. *Verstehen* refers to understanding the meaning of action from the actor's point of view, metaphorically entering into the shoes of the other. Adopting this research stance requires respecting a person interviewed or observed as a fellow human being rather than as an abstract object of study. It also implies that unlike objects in the natural world, human actors are not simply the product of the pulls and pushes of external forces. Individuals are seen to create the world by organizing their own understanding of it and by giving it meaning. To do research on people without taking into account the meanings they attribute to their actions or environment is to treat them like objects.

The *verstehen* perspective emphasizes that human beings can and must be understood in a manner different from other objects of study, because we create purposes and experience emotions. We make plans, construct cultures, and hold values that affect our behavior. Our feelings and behaviors are influenced by consciousness, deliberation, and the capacity to think about the future. As human beings, we each live in a world that has special meaning to us, and because we give our behavior meaning, our human actions can and must be studied interpersonally. In this regard, behavioral and social sciences need methods different from those used in agricultural experimentation and physical sciences because human beings are different from plants and nuclear particles. You can't interview a stalk of corn or an electron. The *verstehen* tradition stresses understanding that focuses on the meaning-making capacity of humans, the contextual importance of social interactions, an empathetic understanding based on interpersonal experience, and attention to the connections between mental states and behavior. Interpretation of what we observe flows from empathetic introspection and reflection based on direct observation of and interaction with people.

> *Verstehen* thus entails a kind of empathic identification with the actor. It is an act of psychological reenactment—getting inside the head of an actor to understand what he or she is up to in terms of motives, beliefs, desires, thoughts, and so on. (Schwandt, 2000, p. 192)

empathy and sympathetic introspection derived from personal encounters the observer cannot fully understand human behavior. Understanding comes from trying to put oneself in the other person's shoes, from trying to discern how others think, act, and feel.

In an enduringly relevant classic study, educational evaluator Edna Shapiro (1973) studied young children in classrooms in the National Follow Through Program using both quantitative and qualitative methods. It was her closeness to the children in those classrooms that allowed her to see that something was happening that was not captured by standardized tests. She could see differences in children, observe their responses to diverse situations, and capture the varying meanings they attached to common events. She could feel their tension in the testing situation and their spontaneity in the more natural classroom setting. Had she worked solely with data collected by others or only at a distance, she would never have discovered the crucial differences in the classroom settings she studied—differences that actually allowed her to evaluate the innovative program in a meaningful and relevant way. Where standardized tests showed no differences between classrooms using different approaches, her direct observations documented important and significant program impacts.

It is important to note that the admonition to engage directly and personally in fieldwork is in no way meant to deny the usefulness of quantitative methods. Rather, it means that statistical portrayals must always be interpreted and given human meaning. I once interviewed an evaluator of federal health programs, who expressed frustration at trying to make sense out of statistical data from more than 80 projects after site visit funds had been cut out of the evaluation: "There's no way to evaluate something that's just data. You know, you have to go look."

Going into the field and having personal contact with program participants is not the only legitimate way to understand human behavior. For certain questions and for situations involving large groups, distance is inevitable, perhaps even helpful, but to get at deeper meanings and preserve context, face-to-face interaction is both necessary and desirable. This returns us to a recurrent theme of this book: matching research methods to the purpose of a study, the questions being asked, and the resources available.

In thinking about the issue of closeness to the people and situations being studied, it is useful to remember that many major contributions to our understanding of the world have come from scientists' personal experiences. One finds many instances where closeness to sources of data made key insights possible—Piaget's closeness to his children, Freud's proximity to

and empathy with his patients, Darwin's closeness to nature, and even Newton's intimate encounter with an apple. In short, closeness does not make bias and loss of perspective inevitable; and distance is no guarantee of objectivity.

Empathic Neutrality and Mindfulness

> The idea of acquiring an "inside" understanding—the actors' definitions of the situation—is a powerful central concept for understanding the purpose of qualitative inquiry.
>
> —Thomas A. Schwandt (2000, p. 102)

Since naturalistic inquiry involves fieldwork that puts one in close contact with people and their problems, what is to be the researcher's cognitive and emotional stance toward those people and problems? No universal prescription can capture the range of possibilities, for the answer will depend on the situation, the nature of the inquiry, and the perspective of the researcher. But, thinking strategically, I offer the phrase "empathic neutrality" as a point of departure. It offers a middle ground between becoming too involved, which can cloud judgment, and remaining too distant, which can reduce understanding. What is empathic neutrality? In essence, it is *understanding a person's situation and perspective without judging the person—and communicating that understanding with authenticity to build rapport, trust, and openness.*

Let's examine each idea separately and then the combination, *emphatic neutrality.* We'll start with neutrality.

Neutrality

Methodologists and philosophers of science debate what the researcher's stance should be vis-à-vis the people being studied. Critics of qualitative inquiry have charged that the approach is too *subjective*, in large part because the researcher is the instrument of both data collection and data interpretation, and because a qualitative strategy includes having personal contact with and getting close to the people and situation under study. From the perspective of advocates of a supposedly value-free social science, subjectivity is the very antithesis of scientific inquiry.

Objectivity has been considered the strength of the scientific method. The primary methods for achieving objectivity in science have been conducting blind experiments and quantification. "Objective tests" gather data through instruments that, in principle, are not dependent on human skill, perception, or even presence. Yet it is clear that tests and questionnaires are designed by human beings and, therefore, are subject to the intrusion of the researcher's biases by the very questions asked. Unconscious bias in the skillful manipulation of statistics to prove a hypothesis in which the researcher believes is hardly absent from hypothetical-deductive inquiry.

Part of the difficulty in thinking about the fieldwork stance of the qualitative inquirer is that the terms *objectivity* and *subjectivity* have become so loaded with negative connotations and subject to acrimonious debate that neither of the terms any longer provides useful guidance. These terms have been politicized beyond utility. To claim the mantle of *objectivity* in the postmodern age is to expose oneself as embarrassingly naive. The ideals of absolute objectivity and value-free science are impossible to attain in practice and of questionable desirability in the first place since they ignore the intrinsically social nature and human purposes of research. On the other hand, *subjectivity* has such negative connotations in the public mind that to admit being subjective may undermine one's credibility with audiences unfamiliar with philosophy of science debates. In short, the terms *objectivity* and *subjectivity* have become ideological ammunition in the methodological paradigms debate. My pragmatic solution is to avoid using either word and to stay out of futile debates about subjectivity *versus* objectivity. Qualitative research in recent years has moved toward preferring terms such as *trustworthiness* and *authenticity*. Evaluators aim for balance, fairness, and neutrality (Patton, 2012a). Chapter 9 will discuss these terms and the stances they imply at greater length. At this point, I simply want to note the strategic nature of the issue of inquirer stance and add empathic neutrality to the emerging lexicon that attempts to supersede the hot-button term *objective* and the epithet *subjective*.

Any research strategy ultimately needs credibility to be useful. No credible research advocates distortion of data to serve the researcher's vested interests and prejudices. Both qualitative/naturalistic inquiry and quantitative/experimental inquiry seek honest, meaningful, credible, and empirically supported findings. Any credible research strategy requires that the investigator adopt a stance of openness, being careful to fully document methods of inquiry and their implications for resultant findings. This simply means that the investigator does not set out to prove a particular perspective or manipulate the data to arrive at predisposed propositions. The neutral investigator enters the research arena with no axe to grind, no theory to prove (to test but not to prove), and no predetermined results to support. Rather, the investigator's commitment is to understand the world as it unfolds, be true to complexities and multiple perspectives as they emerge, and be balanced in reporting both confirming and disconfirming evidence with regard to any conclusions offered.

Neutrality is not an easily attainable stance, so all credible research strategies include techniques for helping the investigator become aware of and deal with selective perception, personal biases, and theoretical predispositions. Qualitative inquiry, because the human being is the instrument of data collection, requires that the investigator carefully reflect on, deal with, and report potential sources of bias and error. Systematic data collection procedures, rigorous training, multiple data sources, triangulation, external reviews, and other techniques to be discussed in this book are aimed at producing high-quality qualitative data that are credible, trustworthy, authentic, balanced about the phenomenon under study, and fair to the people studied.

The livelihood of evaluators and researchers depends on their integrity and credibility. Independence and neutrality, then, are serious issues.

Empathy

> Regarding the pain of others requires more than just a pair of eyes. It necessitates an act of the imagination: a willingness to think or feel oneself into the interior of another's experience.
>
> —Leslie Jamison (2014)
> *The Empathy Exams*

Neutrality does not mean detachment. It is on this point that qualitative inquiry makes a special contribution. Qualitative inquiry depends on, uses, and enhances the researcher's direct experiences in the world and insights about those experiences. This includes learning through *empathy*. So, having discussed the neutrality part in the phrase "emphatic neutrality," let's now turn to a closer look at empathy.

Humanistic psychologist Clark Moustakas has described the nonjudgmental empathic stance as *"being-in"* another's world—immersing oneself in another's world by listening deeply and attentively so as to enter into the other person's experience and perception.

> I do not select, interpret, advise, or direct. . . . Being-In the world of the other is a way of going wide open, entering in as if for the first time, hearing just what is, leaving out my own thoughts, feelings, theories, biases. . . . I enter with the intention of understanding and accepting perceptions and not presenting my own view or reactions. . . . I only want to encourage and support the other person's expression, what and how it is, how it came to be, and where it is going. (Moustakas, 1995, pp. 82–83)

Empathy develops from interpersonal interaction with the people interviewed and observed during fieldwork. Empathy involves being able to take and understand the stance, position, feelings, experiences, and worldview of others. Put metaphorically, empathy is "like being able to imagine a life for a spider, a maker's life, or just some aliveness in its wide abdomen and delicate spinnerets so you take it outside in two paper cups instead of stepping on it" (Dunn, 2000, p. 62). Empathy combines cognitive understanding with affective connection, and in that sense it differs from sympathy, which is primarily emotional.

Nor can the capacity for empathy be assumed. University of Michigan studies of incoming students show that "today's students score about 40 percent lower in measures of empathy than students did 30 years ago" (Brooks, 2014, p. A25). Empathy may need to be cultivated, and nurturing empathy begins with valuing it.

Empathy as an inquiry stance is rooted in the phenomenological doctrine of *verstehen*, discussed earlier in a sidebar (p. 56), which undergirds much of qualitative inquiry. *Verstehen* means *understanding at a deep level*, grounded in the unique human capacity to make sense of the world, which has profound implications for how we can study our fellow human beings. The *verstehen* doctrine presumes that since human beings have a unique type of consciousness, as distinct from other forms of life, the study of human beings will be different from the study of other forms of life and nonhuman phenomena. In this regard, the capacity for empathy is one of the major assets available for human inquiry into human affairs. *Verstehen* is primarily cognitive understanding of another; empathy is emotional understanding, *feeling* what it's like for another.

A qualitative strategy of inquiry proposes an active, involved role for the social scientist. This increases the opportunity to generate insight, which deepens social knowledge. Insight emerges from being close to, even sometimes on the inside of, the phenomena being studied. This is quite a different scientific process from that envisioned by the classical, experimental approach to science, but it is still an empirical, that is, data-based, scientific perspective. The qualitative perspective

> in no way suggests that the researcher lacks the ability to be scientific while collecting the data. On the contrary, it merely specifies that it is crucial for validity—and, consequently, for reliability—to try to picture the empirical social world as it actually exists to those under investigation, rather than as the researcher imagines it to be. (Filstead, 1970, p. 4)

This is the reason for the importance of qualitative approaches such as participant observation, in-depth interviewing, detailed description, and case studies.

Empathetic Neutrality

Having discussed empathy and neutrality separately, let's turn our attention back to the phrase that combines these ideas. On first encountering the phrase *empathetic neutrality*, it may appear to be an oxymoron, combining contradictory ideas. While empathy describes a stance toward the people we encounter in fieldwork, calling on us to communicate interest, caring, and understanding, neutrality suggests a stance toward their thoughts, emotions, and behaviors, a stance of being nonjudgmental. Neutrality can actually facilitate rapport and help build a relationship that supports empathy by disciplining the researcher to be open to the other person and nonjudgmental in that openness. Rapport and empathy, however, must not be taken for granted, as Radhika Parameswaran (2001) found in doing fieldwork among young middle-class women in urban India who read Western romance fiction.

> Despite their eventual willingness to share their fears and complaints about gendered social pressures, I still wonder whether these young women would have been more open about their sexuality with a Westerner who might be seen as less likely to judge them based on cultural expectations of women's behavior in Indian society. The well-known word rapport, which is often used to signify acceptance and warm relationships between

informants and researchers, was thus something I could not take for granted despite being an insider; all I could claim was an imperfect rapport. (p. 69)

Evaluation presents special challenges for rapport and neutrality as well. After fieldwork, an evaluator may be called on to render judgments about a program as part of data interpretation and formulating recommendations, but during fieldwork, the focus should be on rigorously observing and interviewing to understand the people and situation being studied. This nuanced relationship between neutrality and empathy will be discussed further in both data collection and analysis chapters.

Empathic Neutrality Grounded in Mindfulness

> Mindfulness is wise attention. Mindlessness is lazy inattention. We have the capacity for and experience of both. Each is part of the human condition and human potential. Which one prevails in a given moment or over time is, paradoxically, a matter of mindfulness. Practice choosing wisely.
>
> —From Halcolm's *Mindfulness Meditations*

Mindfulness involves being focused in the moment, being attentive to what's going on, without distraction, and maintaining attentiveness on a moment-to-moment basis. Mindfulness is best known as a core discipline of Zen Buddhism (Hanh, 1999), but the practice has application well beyond the meditative life of monks. In qualitative inquiry, when interviewing, mindfulness means that you are completely focused on the interaction with the person or people being interviewed. Likewise, in observation, your mind becomes immersed in the setting, the situation, and what is happening so that you can be present to and *see* what is unfolding. Mindfulness is *presence*, which creates the opening to empathy, and is intrinsically nonjudgmental. To achieve empathic neutrality in qualitative inquiry, then, requires mindfulness. You can't hear what you're not taking in. You can't observe what you're not seeing.

A Dynamic, Developmental Perspective

> There is nothing permanent except change.
>
> —Heraclitus
> Philosopher, Ancient Greece
>
> There is a time for everything, and a season for every activity under the heavens.
>
> —Ecclesiastes 3:1

A questionnaire is like a photograph. A qualitative study is like a documentary film. Both offer images. The photograph captures and freezes a moment in time, like recording a respondent's answer to a survey question at a moment in time. The film offers a fluid sense of development, movement, and change.

Qualitative evaluators, for example, conceive of programs as dynamic and developing, with "treatments" changing in subtle but important ways as staff learn what does and doesn't work, as clients move in and out, and as conditions of delivery are altered in response to changing conditions in the program's environment. This kind of adaptive approach to tracking program dynamics and participant outcomes is called *developmental evaluation* (Patton, 2011). The purpose is documenting and understanding dynamic program processes and their effects on participants so as to provide information for ongoing program development. In contrast, an experimental design for an evaluation typically conceives of the program as a fixed thing, like a measured amount of fertilizer applied to a crop—*a* treatment, *an* intervention—which has predetermined, measurable outcomes. Inconsistency in the treatment, instability in the intervention, changes in the program, variability in program processes, and diversity in participants' experiences undermine the logic of an experimental design because these developments—all natural, even inevitable, in real-world programs—call into question what the "treatment" or experiment actually is.

Naturalistic inquiry assumes the ever-changing world posited by the observation in the ancient Chinese proverb that one never steps into the same river twice. Change is a natural, expected, and inevitable part of human experience, and documenting change is a natural, expected, and intrinsic part of fieldwork. Rather than trying to control, limit, or direct change, naturalistic inquirers expect change, anticipate the

MQP Rumination # 2

Confusing Empathy With Bias

I am offering one personal rumination per chapter. These are issues that have persistently engaged, sometimes annoyed, occasionally haunted, and often amused me over more than 40 years of research and evaluation practice. Here's where I state my case on the issue and make my peace.

Researchers and evaluators are admonished to stay rational and independent. Don't get emotional. Feelings are the enemy of rationality and objectivity. Emotions and feelings lead to caring—and caring is a primary source of bias. Stay distant and unfeeling. Caring emerges from connecting to people, an empathic sense of interdependence rather than independence. So avoid connection and caring, eschew empathy, maintain rationality and independence, and you can avoid bias, the greatest of scientific failings.

I hear this view expounded regularly when qualitative findings are attacked for being biased because the researcher or evaluator got close to the people studied and took on the responsibility of communicating their point of view. Early in my career, I was admonished in a public forum by a distinguished university professor who disagreed with the qualitative findings on an innovative education program:

> Your results can't be trusted because you went native. You obviously spent lots of time with them. You totally bought into what those people told you. You've lost all objectivity. You call it empathy. True scientists call it bias.

At that time, I had no ready response. Today, I do, and I will share it at the end of this rumination.

As I've experienced versions of this confusion between empathy and bias over the years, I get the sense that the vociferousness of the attack, and it is often quite vehement, stems from a deep-seated fear of emotions and human connection by those dismissing qualitative data.

So What About the Role of Emotions in Scientific Inquiry?

Brian Knutson is a professor of psychology and neuroscience at Stanford University. Knutson (2014) makes the case that scientific inquiry should incorporate emotions as a source of data and insight into the nature of the human experience.

> The absence of emotion pervades modern scientific models of the mind. In the most popular mental metaphors of social science, mind as reflex (from

behaviorism) explicitly omits emotion and mind as computer (from cognitivism) all but ignores it. Even when emotion appears in later theories, it is usually as an afterthought—an epiphenomenal reaction to some event that has already passed. But over the past decade, the rising field of affective science has revealed that emotions can precede and motivate thought and behavior.

> Emerging physiological, behavioral, and neuroimaging evidence suggests that emotions are proactive as well as reactive. Emotional signals from the brain now yield predictions about choice and mental health symptoms, and may soon guide scientists to specific circuits that confer more precise control over thought and behavior. Thus, the price of continuing to ignore emotion's centrality to mental function could be substantial. By assuming the mind is like a bundle of reflexes, a computer program, or even a self-interested rational actor, we may miss out on significant opportunities to predict and control behavior—both in individuals and groups.

> Literally and figuratively, we should stop relegating emotion to the periphery, and move emotion to the center—where it belongs.

An Anecdote About Empathy and Bias

MQP Rumination # 1 (Chapter 1, pp. 31–33) concerned undervaluing anecdotes as a form of potentially useful data. So let me share an anecdote that illustrates the importance of human empathy as a source of understanding and making sense of the world. It is an anecdote about the nature of bias told by Tom Griffiths, Professor of Psychology and Cognitive Science, University of California, Berkeley, and Director of the Institute of Cognitive and Brain Sciences.

> It's easy to discover the biases that have been built into speech recognition software. I once left my office for a meeting, locking the door behind me, and came back to find a stranger had broken in and typed a series of poetic sentences into my computer. Who was this

(Continued)

(Continued)

person, and what did the message mean? After a few spooky, puzzling minutes, I realized that I had left my speech recognition software running, and the sentences were the guesses it had produced about what the rustling of the trees outside my window meant. But the fact that they were fairly intelligible English sentences reflected the biases of the software, which didn't even consider the possibility that it was listening to the wind rather than a person. (Griffiths, 2014)

Cultivating Empathic Skills and Appreciating the Appropriate Use of Bias

Computers, at least so far, lack the capacity for empathy—or even bias. Biased human beings import bias into software. The distinguished philosopher of science and evaluation research pioneer Michael Scriven concluded his volume on *Hard-Won Lessons in Program Evaluation* (1993) with astute observations about both empathy and bias. First, empathy:

The most difficult problems with program evaluation are not methodological or political but psychological. . . . What is lacking is the ability to see the point of view of those on the receiving end of the evaluation [intended beneficiaries of the program]—*the lack of empathic skills* [italics added]—and that is just as important a failing. (p. 87)

Scriven (1993) also commented on the common fallacy of defining bias as a lack of belief in or concern about something:

Preference and commitment do not entail bias.

It is crucial to begin with a clear idea of the difference between bias in the sense of prejudice, which means a tendency to error, and bias in the . . . sense of preference, support, endorsement, acceptance, or favoring of one side of an issue. Only the first of these senses is derogatory, and in the legal context the term bias is restricted to the first sense. From none of the synonyms for the second sense can one infer prejudice, because the preference, support, and so on may be justified. It is insulting, and never tolerated in a court

of law where these matters are of the essence, to treat someone who has preferences as if they are thereby biased (and hence not a fair witness). It is especially absurd in the science, mathematics, engineering, and technology (SMET) area to act as if belief in [something] shows bias. Bias must be shown, either by demonstrating a pattern of error or by demonstrating the presence of an attitude that definitely and regularly produces error. . . .

People with knowledge about an area are typically people with views about it; the way to avoid panels of ignoramuses or compulsive fence sitters is to go for a balance of views, not an absence of views. (pp. 79–80)

Emotion and Reason

When I was in graduate school, we were constantly warned that emotion was the enemy of reason. Now, based on the latest research on how we as humans make decisions, brain research, and cognitive science, we know that emotion is not opposed to reason; our emotions assign value to things and are the basis of reason (Brooks, 2011; Patton, 2013). "Emotive traits" like "empathetic sensitivity" are not barriers to scientific inquiry about the human experience; rather, the capacity for empathy enhances, enriches, and deepens human understanding (Brooks, 2011, 2014).

Beyond Defensiveness

Nowadays, in the face of attacks on my qualitative findings as biased because I got close enough to people to feel empathetic, I assume the stance of an old man feigning calm and a statesman-like attitude, rather than displaying the passion and defensiveness of youth, and I say,

I'm sorry you feel that way. Oops! I didn't mean to use the verb *feel*. But it must be a terrible thing to be so afraid of feelings and human connections. How much of human experience you miss by staying so doggedly and dogmatically in your head. But I certainly understand why you can't relate to and don't understand my findings. You detect bias. I detect empathic atrophy. Such a tragic loss. My condolences.

likelihood of the unanticipated, and are prepared to go with the flow of change. One gets this sense of pursuing change in the comment by Williams cited earlier: "I'm tracking something—killing—that's moving very rapidly in the culture." Part of her inquiry task is to track cultural changes the way an epidemiologist tracks a disease. As a result, reading a good qualitative case study gives the sense of reading a good story. It has a beginning, middle, and ending—though not necessarily an end.

Strategic Principles for Qualitative Analysis and Reporting Findings

This module covers the final five core strategies for qualitative inquiry focused on qualitative analysis strategies: (8) unique case orientation, (9) inductive analysis and creative synthesis, (10) holistic perspective, (11) context sensitivity, and (12) voice, perspective, and reflexivity.

Unique Case Orientation

> If we wish to know about a man, we ask "what is his story—his real, inmost story?"—for each of us is a biography, a story. Each of us is a singular narrative, which is constructed, continually, unconsciously, by, through, and in us—through our perceptions, our feelings, our thoughts, our actions; and, not least, our discourse, our spoken narrations. Biologically, physiologically, we are not so different from each other; historically, as narratives—we are each of us unique.
>
> —Oliver Sacks (1985, p. vii)
> *The Man Who Mistook His Wife for a Hat and Other Clinical Tales*

"Six windows on respect" is how Harvard sociologist Sara Lawrence-Lightfoot (2000, p. 13) described the six detailed case studies in her book *Respect*. The cases, each a full chapter, offer different perspectives on the meaning and experience of respect in modern society. We enter the worlds of a nurse-midwife, a pediatrician, a teacher, an artist, a law school professor, and a pastoral therapist/AIDS activist. Before drawing themes and contrasts from this small, purposeful sample and before naming the six perspectives they represent, Lawrence-Lightfoot had the task of constructing the unique cases to tell these distinct stories. Her first task, then, was to undertake the "art and science of portraiture" (Lawrence-Lightfoot & Davis, 1997). From these separate portraits, she fashions a metaphoric stained-glass mosaic that depicts and illumines *respect*.

STUDYING EXITS

The diversity, meaning, and impacts of people separating from each other

For two years I sought out and listened to people tell their big and memorable stories of leaving. As we talked together, some were in the midst of composing their exits, anticipating and planning their departures, anxious and excited about moving on. Others had exited long ago and used our dialogues as an opportunity for reflection—revisiting the ancient narratives that had changed the course of their lives, discovering new ways of interpreting and making sense of their journeys. Some interviewees told tales of forced exits; others spoke about designing and executing their planned departures. Still others found it hard to determine whether the impetus for their exits came from within—a decision motivated by them, within their jurisdiction and control—or whether their leave-takings were a response to subtle pressuring from friends and families, covert warnings from bosses, or influenced by the social prescriptions, norms, and rhythms deemed appropriate by our institutional cultures.

In all our conversations I followed the lead of my interviewees as they decided where to begin their stories, chose the central arc of their exit narratives, and rehearsed the major transformational moments of their departures. I listened carefully to the talk and the silences, the text and the subtexts of their narrations. I was attentive to those revelations that surprised them, to those discoveries that disoriented them, to the places where they feared to tread. I pushed for the details of long-buried memories. I stopped my probing when I felt myself crossing the boundaries of resistance and vulnerability. We took breaks, went for walks, and drank lots of water to hydrate us through what one storyteller called "the desert of my despair." For many—in fact, most—these were emotional encounters, filled with weeping and laughter, breakthroughs and breakdowns, curiosity and discovery.

—Sara Lawrence-Lightfoot (2012)
Harvard University
EXIT: The Endings That Set Us Free

I undertook a study of a national fellowship award program that had had more than 600 recipients over a 20-year period. A survey had been done to get the fellows' opinions about select issues, but the staff wanted more depth, richness, and detail to really understand the patterns of fellowship use and impact. With a team of researchers, we conducted 40 in-depth, face-to-face interviews and wrote case studies. Through inductive analysis, we subsequently identified distinct enabling processes and impacts and created a framework that depicted the relationships between status at the time of the award, enabling processes, and impacts. But the heart of the study was always the 40 case studies. To read only the framework analysis without reading the case studies would be to lose much of the richness, depth, meaning, and contribution of qualitative data. That is what is meant by the *unique case orientation* of qualitative inquiry.

Case studies are particularly valuable in program evaluation when the program is individualized, so the evaluation needs to be attentive to and capture individual differences among participants, diverse experiences of the program, or unique variations from one program setting to another. A case can be a person, an event, a program, an organization, a time period, a critical incident, or a community. Regardless of the unit of analysis, a qualitative case study seeks to describe that unit in depth and detail, holistically, and in context.

Inductive Analysis and Creative Synthesis

In solving a problem of this sort, the grand thing is to be able to reason backwards. This is a very useful accomplishment, and a very easy one, but people do not practice it much.

—Sir Arthur Conan Doyle (1887)
(Quoted by Sherlock Holmes in *A Study in Scarlet*)

Benjamin Whorf's development of the famous "Whorf hypothesis"—that language shapes our experience of the environment and that words shape perceptions and actions, a kind of linguistic relativity theory (Schultz, 1991)—provides an instructive example of inductive analysis. Whorf was an insurance investigator assigned to look into explosions in warehouses. He discovered that truck drivers were entering "empty" warehouses smoking cigarettes and cigars. The warehouses, it turned out, often contained invisible but highly flammable gases. He interviewed the truck drivers and found that they associated the word *empty* with "harmless" and acted accordingly. From these specific observations and findings, he *inductively* formulated his influential theory about language and perception, which has informed a half-century of communications scholarship (Lee, 1996).

Qualitative inquiry is particularly oriented toward exploration, discovery, and inductive logic. Inductive analysis begins with specific observations and builds toward general patterns. Categories or dimensions of analysis emerge from open-ended observations as the inquirer comes to understand patterns that exist in the phenomenon being investigated.

Inductive analysis contrasts with the hypothetical-deductive approach of experimental designs, which requires the specification of main variables and the statement of specific research hypotheses *before* data collection begins. A specification of research hypotheses based on an explicit theoretical framework means that general constructs provide the framework for understanding specific observations or cases. The investigator must then decide in advance what variables are important and what relationships among those variables can be expected.

The strategy of induction allows meaningful dimensions to emerge from the patterns found in the cases under study, without presupposing in advance what those important dimensions will be. The qualitative analyst seeks to understand the multiple interrelationships among the dimensions that emerge from the data, without making prior assumptions or specifying hypotheses about the linear or correlative relationships among the narrowly defined, operationalized variables. For example, an inductive approach to program evaluation means that understanding the nature of the intervention emerges from direct observations of program activities and interviews with participants. In general, theories about what is happening in a setting are grounded in and emerge from direct field experience rather than being imposed a priori, as is the case in formal hypothesis and theory testing.

In practice, these approaches are often combined in mixed-methods designs. Some inquiry questions may be determined deductively, while others are left sufficiently open to permit inductive analysis based on participants' responses. While the quantitative/experimental approach is largely hypothetical-deductive and the qualitative/naturalistic approach is largely inductive, a mixed-methods study can include elements of both strategies. Indeed, over a period of inquiry, an investigation may flow from inductive approaches, to find out what the important questions and variables are (exploratory work), to deductive

EXHIBIT 2.3 **A Classic Mixed-Methods Inquiry Sequence**

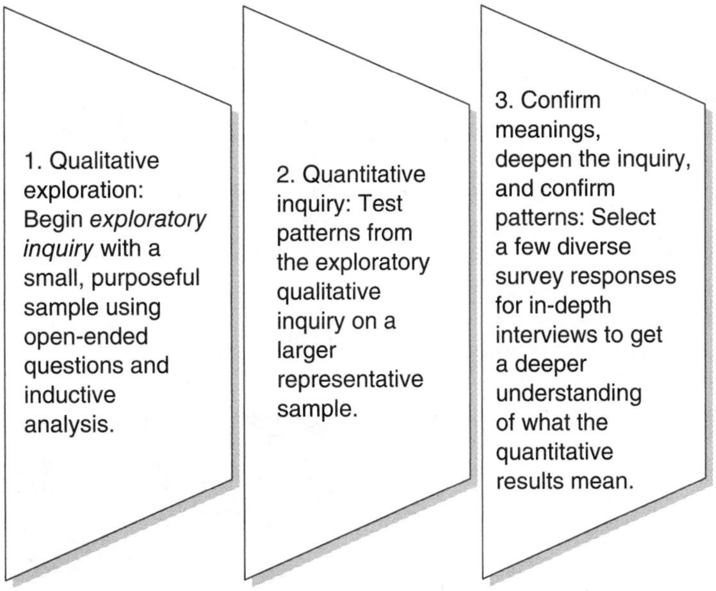

1. Qualitative exploration: Begin *exploratory inquiry* with a small, purposeful sample using open-ended questions and inductive analysis.

2. Quantitative inquiry: Test patterns from the exploratory qualitative inquiry on a larger representative sample.

3. Confirm meanings, deepen the inquiry, and confirm patterns: Select a few diverse survey responses for in-depth interviews to get a deeper understanding of what the quantitative results mean.

hypothesis-testing or outcome measurement, aimed at confirming and/or generalizing exploratory findings, and then back again to inductive analysis to look for rival hypotheses and unanticipated or unmeasured factors. This mixed-methods sequence is depicted graphically in Exhibit 2.3.

Anthropologist Russell Bernard has described how the interaction of inductive and deductive strategies unfolds in fieldwork:

> When I started working with the Ñähñu Indians of central Mexico, for example, I wondered why so many parents wanted their children not to learn how to read and write Ñähñu in school. As I became aware of the issue, I started asking everyone I talked to about it. With each new interview, pieces of the puzzle fell into place. This was a really, really inductive approach. After a while, I came to understand the problem: It's a long, sad story, repeated across the world by indigenous people who have learned to devalue their own cultures and reject their own languages in the hope that this will help their children do better economically. After that, I started right off by asking people about my hunches—for example, about the economic penalty of speaking Spanish in Mexico with an identifiable Indian accent. In other words, I switched to a really, really deductive approach.

It's messy, but this paradigm for building knowledge—the continual combination of inductive and deductive research—is used by scholars across the humanities and the sciences alike and has proved itself, over thousands of years. If we know anything about how and why stars explode or about how HIV is transmitted or about why women lower their fertility when they enter the labor market, it's because of this combination of effort. Human experience—the way real people experience real events—is endlessly interesting because it is endlessly unique, and so, in a way, the study of human experience is always exploratory and is best done inductively. (Bernard, 2013, p. 12)

Just as writers report different creative processes, so too qualitative analysts have different ways of working. While software programs now exist to facilitate working with large amounts of narrative data, and while substantial guidance can be offered about the steps and processes of content analysis, making sense of multiple interview transcripts and pages of field notes cannot be reduced to a formula or even a standard series of steps. There is no equivalent of a statistical significance test or factor score to tell the analyst when results are important or what quotations fit together under the same theme. Finding a way to *creatively synthesize* and present findings is one of the challenges of qualitative analysis, a challenge that will be explored at length in Part III of this book. For the moment, I can offer a flavor of that challenge with another excerpt from my interview with Williams. Here, she describes part of her own unique analytic process.

> My current project follows up work that I have always done, which is to study categories and classifications and their implications. Right now, as I said, the focus of my work is on killing and the desire to kill and the categories people create in relation to killing. Part of it right now focuses on the death penalty, but mainly on killing. My fascination is with the links between category distinctions, commitments, and the desire to kill for those commitments. That's what I study.
>
> I track categories, like "serial killers" or "death row inmates." The business of constantly transforming people into acts and acts into people is part of the way loyalties, commitments, and hatreds are generated. So I'm a classifier. I study classification—theories of classification. A lot of categories have to do with very abstract things; others have to do with very concrete things like skin color. But ultimately, the classification of a kill is what I'm focusing on now. I've been asking myself lately, for the chapter I've been working on, *Is there a fundamental difference, for example, in the way we classify to kill?* Consider the percentage of people classified as *death worthy*—the way we classify to justify the death penalty.
>
> As I write, moving back and forth between my tapes and my interviews, I don't feel that I have to follow some fixed outline or that I have to code things to come out a certain way. Sometimes I listen to a tape and I start to think that I should rewrite this part of this chapter. I had completely forgotten about this tape. It was done in early '98 or late '97 and maybe I hadn't listened to it or looked at the transcript for a while, and I've just finished a chapter or section of a chapter. I pull that tape off the shelf. I listen to it. I go back to the transcript and I start writing again. I start revising in ways that it seems to me that tape *demands*.

As Williams describes her analysis and writing process, she offers insight into what it means when qualitative researchers say they are "working to be true to the data" or that their analytical process is "data driven." She says, "I start revising in ways that it seems to me that tape *demands*." It is common to hear qualitative analysts say that, as they write their conclusions, they keep going back to the cases, rereading the field notes and listening again to the interviews.

Inductive analysis is built on a solid foundation of specific, concrete, and detailed observations, quotations, documents, and cases. As thematic structures and overarching constructs emerge during analysis, the qualitative analyst keeps returning to fieldwork observations, interview transcripts, social media entries, and relevant documents, working from the bottom up, staying grounded in the foundation of case write-ups, and thereby examining emergent themes and constructs in light of what they illuminate about the case descriptions on which they are based. That is inductive analysis.

Holistic Perspective

> The child's life is an integral, a total one. He passes quickly and readily from one topic to another, as from one spot to another, but is not conscious of transition or break. There is no conscious isolation, hardly conscious distinction. The things that occupy him are held together by the unity of the personal and social interests which his life carries along. . . . [His] universe is fluid and fluent; its contents dissolve and reform with amazing rapidity. But after all, it is the child's own world. It has the unity and completeness of his own life.
>
> —John Dewey (1859–1952)
> American philosopher, psychologist, and educational reformer
> *The Child and the Curriculum* (1956, pp. 5–6)

Holography is a method of photography in which the wave field of light scattered by an object is captured as an interference pattern. When the photographic record—the hologram—is illuminated by a laser, a three-dimensional image appears. Any piece of a hologram will reconstruct the entire image. This has become a metaphor for thinking in new ways about the relationships between parts and wholes. The interdependence of flora, fauna, and the physical environment in ecological systems offers another metaphor for what it means to think and analyze holistically. Or consider this holistic wisdom from the great Greek physician Hippocrates (460–377 BCE): "It is more important to know what sort of person has a disease than to know what sort of disease a person has."

Researchers and evaluators analyzing qualitative data strive to understand a person, organization, community, phenomenon, or program as a *whole*. This means that a description and interpretation of a person's social environment or an organization's external context is essential for an overall understanding of what has been observed during fieldwork or said in an interview. This holistic approach assumes that the whole is understood as a complex system that is greater than the sum of its parts. The analyst searches for the totality or unifying nature of particular settings—the *gestalt*. Psychotherapist Fritz Perls (1973) used the term *gestalt* to evoke a holistic perspective in psychology. He used the example of three sticks that are just three sticks until one places them together to form a triangle; then, they are much more than the three separate sticks combined; they form a new whole.

> A gestalt may be a tangible thing, such as a triangle, or it may be a situation. A happening such as a meeting of two people, their conversation, and their leave-taking would constitute a completed situation. If there were an interruption in the middle of the conversation, it would be an incomplete gestalt. (Brown, 1996, p. 36)

The strategy of seeking gestalt units and holistic understandings in qualitative analysis contrasts with the logic and procedures of studies conducted in the analytical tradition of "Let's take it apart and see how it works." The quantitative/experimental approach, for example, requires operationalization of independent and dependent variables with a focus on their statistical covariance. In program evaluation, this means that outcomes must be identified and measured as specific variables. Treatments and programs must also be conceptualized as discrete, independent variables. The characteristics of program participants must be measured by standardized, quantified dimensions. Sometimes, the variables of interest are derived from program goals, for example, student achievement test scores, recidivism statistics for a group of juvenile delinquents, and sobriety rates for participants in chemical dependency treatment programs. At other times, the variables measured are indicators of a larger construct. For example, community well-being may be measured by such rates for delinquency, infant mortality, divorce, unemployment, suicide, and poverty. These variables are statistically manipulated or added together in some

linear fashion to test hypotheses and draw inferences about the relationships among separate indicators or the statistical significance of the differences between measured levels of the variables for different groups. The essential logic of this approach is as follows: (a) key program outcomes and processes can be represented by separate independent variables, (b) these variables can be quantified, and (c) the relationships among these variables are best portrayed statistically.

QUALITATIVE INQUIRY AS A HOLISTIC PROCESS

A holistic approach views research as a process rather than an event. In this regard, adopting a holistic approach means the researcher views all research approaches, from topic selection to final representation, as interrelated. This differs from an event-oriented approach, which views choices as a set of sequential steps. . . . In addition, it is not just the resulting information or research findings that we learn; the process itself becomes a part of the learning experience. In this regard and others, qualitative approaches to social inquiry foster personal satisfaction and growth.

—Sharlene Nagy Hesse-Biber and Patrcia Leavy (2011, pp. 7–8)
The Practice of Qualitative Research, 2nd ed.

The primary critique of this logic by qualitative/ naturalistic evaluators is that such an approach (a) oversimplifies the complexities of real-world programs and participants' experiences, (b) misses major factors of importance that are not easily quantified, and (c) fails to portray a sense of the program and its impacts as a "whole." To support holistic analysis, the qualitative inquirer gathers data on multiple aspects of the setting under study to assemble a comprehensive and complete picture of the social dynamic of the particular situation or program. This means that, at the time of data collection, each case, event, or setting under study, though treated as a unique entity, with its own particular meaning and its own constellation of relationships emerging from and related to the context within which it occurs, is also thought of as a window into the whole. Thus, capturing and documenting history, interconnections, and system relationships is part of fieldwork.

The advantages of using quantitative variables and indicators are parsimony, precision, and ease of analysis. Where key elements can be quantified with validity, reliability, and credibility and where necessary statistical assumptions can be met (e.g., linearity, normality, and independence of measurement), then statistical portrayals can be quite powerful and succinct. The advantages of qualitative portrayals of holistic settings and impacts are that greater attention can be given to nuance, setting, interdependencies, complexities, idiosyncrasies, and context.

> Holism is partly about *how you see*. Breaking a thing down into its constituent parts is one way to understand something—but not the only way. In fact, that can be [a] terribly limiting approach. Instead of narrowing your focus, why not expand it instead? By zooming out and bringing more into the frame, an observer can perceive phenomena that only appear in the whole, not in the parts. (This approach has been called *expansionism* in contrast to *reductionism*.) . . .
>
> In other words, a whole (or system) is both *construed* and *discovered*. So when people say that they want to be holistic, they usually mean that they aspire to expand their range of vision. They want to include more parts and, in that way, perceive a whole that is larger than what they could see before. . . . [Yet] no whole that an observer has construed and discovered can ever be considered complete. And for that reason, holism is necessarily an ongoing aspiration. It implies a disciplined commitment to continue questioning. It is endless pursuit of a broader and richer understanding. It is a radically *open* approach. (Coursen, 2014, p. 1)

Qualitative sociologist Irwin Deutscher (1970) commented that, despite the totality of our personal experiences as living, working human beings, social scientists have tended to focus their research on parts *to the virtual exclusion of wholes*:

> We knew that human behavior was rarely if ever directly influenced or explained by an isolated variable; we knew that it was impossible to assume that any set of such variables was additive (with or without weighting); we knew that the complex mathematics of the interaction among any set of variables . . . was incomprehensible to us. In effect, although we knew they did not exist, we defined them into being. (p. 33)

While many would view this intense critique of variable analysis as too extreme, the reaction of many program staff to scientific research is like the reaction of Copernicus to the astronomers of his day: "With them," he observed,

A Holistic Perspective

> it is as though an artist were to gather the hands, feet, head, and other members for his images from diverse models, each part excellently drawn, but not related to a single body, and since they in no way match each other, the result would be monster rather than man. (Kuhn, 1970, p. 83)

How many program staffs have complained of the evaluation research monster?

It is no simple task to undertake holistic analysis. The challenge is "to seek the essence of the life of the observed, to sum up, to find a central unifying principle" (Bruyn, 1966, p. 316). Again, Shapiro's work (1973) in evaluating innovative follow-through classrooms is instructive. She found that standardized test results could not be interpreted without understanding the larger cultural and institutional context in which the individual child is situated. Taking context seriously, the topic of the next section, is an important element of holistic analysis.

I opened Chapter 1 with an illuminative example of holistic understanding that is worth repeating here. A Portuguese colleague told of driving in a remote area of his country when he came upon a sizable herd of sheep being driven along the road by a shepherd. Seeing that he would be delayed until the sheep could be turned off the road, he got out of the car and struck up a conversation with the shepherd.

"How many sheep do you have?" he asked.

"I don't know," responded the young man.

Surprised at this answer, the traveler asked, "How do you keep track of the flock if you don't know how many sheep there are? How would you know if one was missing?"

The shepherd seemed puzzled by the question. Then, he explained, "I don't need to count them. I know each one and I know the whole flock. I would know if the flock was not whole."

This epitomizes holism. "I would know the flock as a whole." Qualitative inquiry focused on a classroom of students strives to know the class *as a whole*. Qualitative inquiry into a family's life seeks to understand the family class *as a whole*. Qualitative inquiry about a community aims to represent the community *as a whole*.

Context Sensitivity

> The Latin word *contextus* means "to join together" or "to weave together."
> —Dahler-Larsen and Schwandt (2012, p. 75)

Context envelops and completes the whole. Without attention to and inclusion of context, qualitative findings are like a fine painting without a frame. To understand context, let's move now from sheep to elephants.

One of the classic tales used to illustrate the relationship between parts and wholes is the story of the nine blind people and the elephant. Each person touches only one part of the elephant and therefore knows only that part. The person touching the ears thinks an elephant is like a large, thin fan. The person touching the tail thinks the elephant is like a rope. The person touching the trunk thinks of a snake. The legs feel like tree trunks, the elephant's side like a tall wall. And so it goes. The holistic point is that one must put all of these perspectives together to get a full picture of what an elephant actually looks like.

But such a picture will still be limited, even distorted, if the only place you ever see an elephant is in the zoo or at the circus. To understand the elephant—how it developed, how it uses its trunk, why it is so large—you must see it in the African savannah or an Asian jungle. In short, you must see it *in context* as part of an ecological *system* in relation to other flora and fauna, in its natural environment.

When we say to someone, "You've taken my comment out of context," we are saying, "You have distorted what I said," changed its meaning by omitting critical context.

In Victor Hugo's great classic, *Les Misérables*, we first encounter Jean Valjean as a hardened criminal and common thief; then, we learn that he was originally sentenced to five years in prison for stealing a loaf of bread for his sister's starving family. That adds context for his "crime" and changes our understanding. The battle over standardized sentencing guidelines in the criminal justice system is partly a debate about how much to allow judges sway in taking into account context and individual circumstances in pronouncing sentences.

Naturalistic inquiry not only describes context in reporting findings but also highlights and deciphers context when interpreting findings. Social psychology experiments under laboratory conditions strip the observed actions from context. But that is the point of such laboratory experiments—to generate findings that are context-free. The scientific ideal of generalizing across time and space is the ideal of identifying principles that do not depend on context. In contrast, qualitative inquiry elevates context as critical to understanding. Portraitist Sara Lawrence-Lightfoot (1997) explains why she finds context "crucial to the documentation of human experience and organizational culture":

> By context, I mean the setting—physical, geographic, temporal, historical cultural, aesthetic—within which action takes place. Context becomes the framework, the reference point, the map, the ecological sphere; it is used to place people and action in time and space and as a resource for understanding what they say and do. The context is rich in clues for interpreting the experience of the actors in the setting. We have no idea how to decipher or decode an action, a gesture, a conversation, or an exclamation unless we see it embedded in context. (p. 41)

Context also affects how an inquiry is conducted. As president of the American Evaluation Association in 2009, Debra Rog made "Context and Evaluation" the theme of the annual conference. In doing so, she sought to move attention to context "from background to foreground" by focusing on "how best to match designs and methods to particular program and policy contexts to produce the most useful and actionable evidence" (Rog, 2012, p. 26).

> Context-sensitive evaluation eschews a methods-first orientation and suggests that a context-first approach for evaluation is more appropriate. The perspective is that the evaluator needs an understanding of which approaches to evaluation are most appropriate for particular contexts. Much like the question we strive to answer in our evaluations, "What works best for whom under what conditions?" context-sensitive evaluation practice asks "What evaluation approach provides the highest quality and most actionable evidence in which contexts?" The answer to this question requires balancing, at a minimum, attention to context, stakeholder needs, and rigor. Accomplishing this balance likely entails understanding the many context issues that affect an evaluation and its evaluand, actively involving the range of stakeholders in

the process, and drawing on a portfolio of method-ological and analytic strategies to accommodate the context issues and needs in the most rigorous way possible. (pp. 26–27)

While Rog was focusing on selecting evaluation methods to fit program and policy contexts, moving from a methods-first orientation to a context-first approach applies to any kind of inquiry. What kinds of methods are most appropriate for what kinds of questions? That issue is the core of this chapter as we examine the 12 strategic dimensions that char-acterize and distinguish qualitative inquiry. What is the purpose of the inquiry? Who will be assess-ing the rigor of the inquiry, using what standards and criteria, to judge the credibility of the findings? How will qualitative inquiry be received in the context in which the study will be conducted? Exhibit 2.4, adapted and expanded from Rog (2012, p. 28), depicts the interrelated arenas of con-text that come into play in determining the appro-priateness of a particular inquiry approach.

Reflexivity: Perspective and Voice

γνῶθι σεαυτόν ("Know thyself" in Greek)

—Inscription in ancient Temple of Apollo at Delphi

Let me acknowledge immediately that the term *reflexivity* reeks of academic jargon. In everyday con-versation, we don't say, "I'm in a reflexive mood today. I've set aside time to engage in some serious reflexiv-ity." Such an assertion would likely evoke a profoundly unimpressed and skeptical "Whatever." So why not just use the word *reflection*? Reflexivity encompasses reflection—indeed, mandates reflection—but it means to take the reflective process deeper and make it more systematic than is usually implied by the term *reflec-tion*. It may sound pretentious and can elicit negative feedback for sounding academic and highfalutin, but the purpose is not pomposity. The term *reflexivity* is meant to direct us to a particular kind of reflection grounded in the in-depth, experiential, and interper-sonal nature of qualitative inquiry.

In science generally, a reflexive relationship is bidi-rectionally interactive and interdependent. Cause and effect are circular, interconnected, and mutually influ-encing. I affect you, and you affect me. The interviewer affects the interviewee, and the interviewee affects the interviewer. Fieldworkers enter a place in which they

REFLEXIVITY

Reflexivity is self-critical sympathetic introspection and the self-conscious analytical scrutiny of the self as researcher. Indeed reflexivity is critical to the conduct of fieldwork; it induces self-discovery and can lead to insights and new hypotheses about the research questions. A more reflexive and flexible approach to fieldwork allows the researcher to be more open to any challenges to their theoretical position that fieldwork almost inevitably raises.

—Kim V. L. England (1994)
Professional geographer

observe what is going on, they describe what they see and hear, they interact with people in the situation being studied, and these interactions have effects, both on those studied and on the observers. But how do we know what those effects are? How do we figure out how who we are affects what we see, how we see what we see, and how others respond to our being there, observing, asking questions, and taking notes?

The term *reflexivity* has entered the qualitative lexicon as a way of emphasizing the importance of deep introspection, political consciousness, cul-tural awareness, and ownership of one's perspective. Reflexivity calls on us to think about how we think and inquire into our thinking patterns even as we apply thinking to making sense of the patterns we observe around us. Reflexivity involves "*interpreta-tion of interpretation* and the launching of a critical self-exploration of one's own interpretations . . . , a consideration of the perceptual, cognitive, theo-retical, linguistic, (inter)textual, political and cul-tural circumstances that form the backdrop to—as well as impregnate—the interpretations" (Alvesson & Sköldberg, 2009, p. 9). Being reflexive involves self-questioning and self-understanding, for "all understanding is self-understanding" (Schwandt, 1997a, p. xvi). To be reflexive, then, is to under-take an ongoing examination of what I know and how I know it, "to have an ongoing conversation about experience while simultaneously living in the moment" (Hertz, 1997, p. viii). Reflexivity reminds the qualitative inquirer to be attentive to and con-scious of the cultural, political, social, linguistic, and economic origins of one's own perspective and voice as well as the perspective and voices of those one interviews and those to whom one reports.

Reflexivity turns mindfulness inward. Earlier, I discussed mindfulness as a pathway to empathic neu-trality. Here, reflexive mindfulness is the pathway to

EXHIBIT 2.4 Contextual Sensitivity and Assessment

Premises

1. Context sensitivity and assessment affect choices about what methods are appropriate (Conner, Fitzpatrick, & Rog, 2012).

2. Attending to and understanding context sets the stage for studying context and takes it into account in interpreting findings.

3. How the inquirer is involved in a particular context will affect how the inquirer understands context. Researchers and evaluators "do not simply identify and respond to contextual factors, but by virtue of their actions are always constructing, relating to, engaging in, and taking part in some reconstruction of the context in which they operate" (Dahler-Larsen & Schwandt, 2012, p. 84).

4. Contexts are often complex dynamic systems. Static depictions of context as fixed will reduce the inquiry's emergent design flexibility and misrepresent the meaning and effects of context when interpreting and presenting findings (Patton, 2012a).

5. Contextualism makes the first priority of an inquiry understanding perspectives, behaviors, relationships, processes, outcomes, and knowledge within the context or contexts studied.

Contextually Situating and Framing an Inquiry

- ***Purpose context:*** Why is the study being done? Who will judge its rigor and credibility? By what standards?

- ***Inquiry focus context:*** From what inquiry traditions, discipline, knowledge arena, interests, and issues are the inquiry questions derived? Within what larger context is the inquiry framed? To what extent are particular theoretical, philosophical, epistemological, or methodological contexts critical to understanding the inquiry?

- ***Location context:*** Where does the inquiry take place (e.g., physical location, organizational entity, virtual community; one site or multiple sites and levels), and how does location affect both inquiry methods and interpretation of findings?

- ***Broader context:*** This refers to sensitivity to organizational, social, cultural, historical, political, and demographic dynamics and trends (Rog, 2012, pp. 28–30).

- ***Relationship context:*** What is the relationship of the inquirer (researcher or evaluator) to the people studied? To what extent, if at all, are those studied involved as participants in the inquiry? What role does the inquirer play, for example, when undertaking participant observation?

- ***Other relevant contextual arenas:*** The contextual arenas in the graphic below are meant to be suggestive, not definitive. For example, a study involving a team of several researchers or evaluators might add a *team context*, or the funding of the inquiry may be such that a should be included. Part of the point of a contextual assessment is to identify and attend to those arenas of context that are important for a particular inquiry—and to be aware that the relative importance of contextual arenas may change over time as the inquirers' engagement with and understanding of context unfolds and evolves and as the various contextual arenas are affected by broader contextual trends and dynamics.

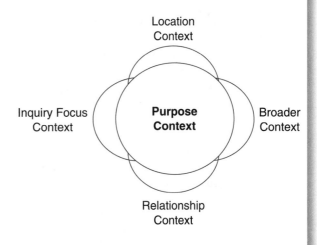

Location Context

Inquiry Focus Context

Purpose Context

Broader Context

Relationship Context

---

self-awareness. To excel in qualitative inquiry requires keen and astute self-awareness. It turns out that people who excel in all kinds of activities share the quality of being self-aware and using that awareness to adapt to whatever presents itself in the course of taking action (Sweeney & Gosfield, 2013). Exhibit 2.5, on the next page, depicts the mindfulness of reflexive triangulation.

Reflexivity Meets Voice

Reflexivity leads both to understanding one's perspective and to owning that perspective. That ownership of perspective is where voice intersects reflexivity. The reflexive voice is the first-person active voice, "I." Contrast that voice to the traditional third-person

EXHIBIT 2.5   Reflexive Questions: Triangulated Inquiry

Those studied (participants):

How do they know what they know? What shapes and has shaped their worldview? How do they perceive me? Why? How do I know? How do I perceive them?

Reflexive Screens: culture, age, gender, class, social status, education, family, political praxis, language, values

Those who receive the study (audience):

How do they make sense of what I give them? What do they bring to the findings I offer? How do they perceive me? How do I perceive them?

Myself (as qualitative inquirer):

*What do I know?
How do I know what I know?
What shapes and has shaped my perspective?
With what voice do I share my perspective?
What do I do with what I have found?*

passive voice of academia epitomized by this educational evaluation abstract:

This study will delineate the major factors that affect school achievement. Instruments were selected to measure achievement based on validity and reliability criteria. Decisions were made about administering the tests in conjunction with administrators taking into account time and resource constraints. A regression model was constructed to test relationships between various background variables and demonstrated achievement. School records were reviewed and coded to ascertain student's background characteristics. Data were obtained on 120 students from four classrooms. The extraction of significant predictor variables is the purpose of the final analysis. Interviews were conducted with teachers and principals to determine how test scores were used. The analysis concludes with the researcher's interpretations. The researcher wishes to thank those who cooperated in this study.

This journal article abstract represents academic writing as I was taught to do it in graduate school. This writing style still predominates in scholarly journals and books. No human being is visible in this writing. The passive voice reigns. Instruments were selected; decisions were made; a model was constructed; records were reviewed and coded; data were obtained;

predictor variables were extracted; and interviews were conducted. The warmth of thanks is extended by a role, the researcher: "The researcher wishes to thank those who cooperated." The third person, passive voice communicates a message: This work is about procedures, not people. This academic style has historically been employed to project a sense of objectivity, control, and authority. The overall impression is mechanical, robot-like, distant, detached, systematic, and procedural. The research is the object of attention. Any real, live human being, subject to all the usual foibles of being human, is barely implied, generally disguised, hidden away, and kept in the background.

Contrast that academic voice with my explanation of how I analyzed a 10-day coming-of-age experience with my son in the Grand Canyon. Here's an excerpt in which I describe the analytical process.

I'm not sure when the notion first took hold of me that articulating alternative coming of age paradigms might help elucidate our Canyon experience. Before formally conceptualizing contrasting paradigm dimensions, I experienced them as conflicting feelings that emanated from my struggle to sort out what I wanted my son's initiation to be, while also grappling with defining my role in the process. I suppose the idea of alternative paradigms first emerged the second night as I paced the narrow beach where White Creek intersects

Shinumo and pondered the Great Unconformity [a geologic reference] as metaphor for the gap between tribal approaches to initiation and coming of age for contemporary youth. In the weeks and months after our Canyon experience, far from languishing in the throes of retox as I expected, the idea of contrasting paradigms stayed with me, as did the Canyon experience. I started listing themes and matching them with incidents and turning points along the way. The sequence of incidents became this book and the contrasting themes became the basis for this closing chapter, a way for me to figure out how what started out as an initiation become a humanist coming of age celebration. (Patton, 1999, p. 332)

The traditional academic voice (third-person passive) may still be used because students haven't been offered and/or don't know that there's an alternative. One reviewer of an earlier draft of this section responded that the academic style

"Not again with the reflexivity stuff. It was a great interview. They loved us."

is employed by people who learned it or think they learned it and have not read the American Psychological Association style manual, which clearly says not to be mysterious. They may be thoughtlessly following bad models, not deliberately trying to portray objectivity, control, and authority.

The contrast between the traditional academic voice and the personal voice of qualitative analysis recalls philosopher and theologian Martin Buber's influential distinction between "I–It" and "I–Thou" relationships. An "I–It" relationship regards other human beings from a distance, from a superior vantage point of authority, as objects or subjects, as things in the environment to be examined and placed in abstract cause–effect chains. An "I–Thou" perspective, in contrast, acknowledges the humanity of both self and others and implies relationship, mutuality, and genuine dialogue.

The perspective that the researcher brings to a qualitative inquiry is part of the context for the findings. You as a human being are the instrument of qualitative methods. You as a real, live person make observations, take field notes, ask interview questions, and interpret responses. Self-awareness, then, can be an asset in both fieldwork and analysis. Developing appropriate self-awareness can be a form of "sharpening the instrument" (Brown, 1996, p. 42). The methods section of a qualitative study reports on the researcher's training, preparation, fieldwork procedures, and analytical processes. This is both the strength and the weakness of qualitative methods, the strength in that a well-trained, experienced, and astute observer adds

value and credibility to the inquiry while an ill-prepared, inexperienced, and imperceptive observer casts doubt on what is reported. Judgments about the significance of findings are thus inevitably connected to the researcher's credibility, competence, thoroughness, and integrity. Those judgments, precisely because they are acknowledged as inevitably personal and perspective dependent to some extent, invite response and dialogue rather than just acceptance or rejection.

Writing in the first person, active voice communicates the inquirer's self-aware role in the inquiry: "I started listing themes and matching them with incidents and turning points along the way." Not the passive: Themes were listed and matched to incidents and turning points along the way. Judith Brown (1996) captured the importance of the first-person voice in the title of her book *The I in Science: Training to Utilize Subjectivity in Research*. By subjectivity, she means "the domain of experiential self-knowledge" (p. 1). Voice reveals and communicates that domain.

Voice Is More Than Grammar

The issue is not just first-person active versus third-person passive voice. A credible, authoritative, authentic, and trustworthy voice engages the reader through rich description, thoughtful sequencing, appropriate use of quotes, and contextual clarity, so that the reader joins the inquirer in the search for meaning. And there are choices of voice: the didactic voice of the teacher; the searching, logical voice of the sleuth; the narrator voice of the storyteller; the personal voice of the autoethnographer; the doubting voice of the skeptic; the intimacy of the insider's voice; the detachment of the outsider's voice; the searching voice of uncertainty; and the excited voice of discovery, to offer but a few examples. Just as point of view and voice have become focal points of writing engaging fiction, so too must qualitative writers learn about, take into account, and communicate perspective and voice. Balancing critical and creative analyses, description and interpretation, or direct quotation and synopsis also involves issues of perspective, audience, purpose, and voice. No rules or formula can tell a qualitative analyst precisely what balance is right or which voice to use, only that finding both balance and voice is part of the work and challenge of qualitative inquiry. This is what Lewis (2001) has acknowledged as "the difficulty of trying to situate the I in narrative research" (p. 109).

In addition to finding voice, the critical and creative writing involved in qualitative analysis and synthesis challenges the inquirer to *own your voice and perspective.* Here, we owe much to classic feminist theory for highlighting and deepening our understanding of the intricate and implicate relationships between language, voice, and consciousness (e.g., England, 1994; Gilligan, 1982; Minnich, 2004). We are challenged by postmodern critiques of knowledge to be clear about and own our authorship of whatever we propound, be self-reflective, acknowledge biases and limitations,

and honor multiple perspectives (Mabry, 1997), while "accepting incredulity and doubt as modal postmodern responses to all attempts to explain ourselves to ourselves" (Schwandt, 1997b, p. 102). From struggles to locate and acknowledge the inevitably political and moral nature of evaluative judgments, we are challenged to connect voice and perspective to *praxis*—acting in the world with an appreciation for and recognition of how those actions inherently express social, political, and moral values (Schwandt, 1989, 2000)—and to personalize evaluation (Kushner, 2000), by not only owning our own perspective but also taking seriously the responsibility to communicate authentically the perspectives of those we encounter during our inquiry. These represent some of the more prominent contextual forces that have elevated the importance of owning voice and perspective in qualitative analysis.

The practical side of owning your voice and perspective comes in reporting findings. Qualitative analysis doesn't have the equivalent of a statistical significance test. Determining substantive significance requires judgment, which makes it personal. It can take considerable self-awareness and confidence to report thus: I coded these 40 interviews; these are the themes I found; here is what I think they mean; and here is the process I undertook to arrive at those meanings. The latter statement calls for, even demands, a sense of one's own perspective, analytical process, and voice.

Integrating the 12 Strategic Qualitative Principles in Practice: Chapter Summary and Conclusion

This chapter has presented and discussed 12 qualitative principles as strategic ideals:

1. Make real-world observations through *naturalistic inquiry*.

2. Stay open, responsive, and flexible through *emergent designs*.

3. Focus on information-rich cases through *purposeful sampling*.

4. Describe what you observe in detail, and illuminate people's verbatim perspectives and meanings through rich, in-depth qualitative *data*.

5. Deepen your understanding through *direct personal experience* with and engagement in what you are studying.

6. Balance your critical and creative sides, your cognitive and affective processes, through a stance of *empathetic neutrality*.

7. Cultivate sensitivity to *dynamic processes and systems*.

8. Appreciate and do justice to each person, organization, community, or situation you study through attention to the particular: a *unique case orientation*.

9. Generate insight and understanding through bottoms-up *inductive analysis*.

10. Broaden your analysis and interpretation through *contextual sensitivity*.

11. Integrate data through a *holistic perspective*.

12. Communicate authenticity and trustworthiness through *reflexivity*: Own your voice and perspective.

Exhibit 2.1, at the beginning of this chapter (pp. 46–47), introduced and summarized these 12 strategic principles. Exhibit 2.6 depicts them graphically.

EXHIBIT 2.6 **Twelve Core Strategies of Qualitative Inquiry: Interdependent and Interactive**

This graphic aims to show the 12 core strategies as interrelated, interdependent, and interacting rather than as isolated, disconnected dimensions.

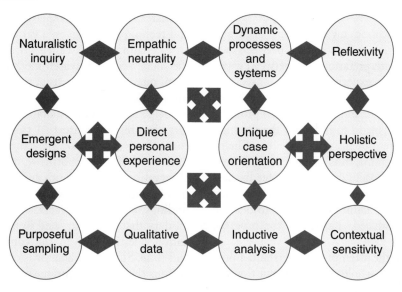

From Strategic Ideals to Practical Choices

> However beautiful the strategy, you should occasionally look at the results.
>
> —Sir Winston Churchill (1874–1965)

The 12 qualitative principles are strategic ideals, not absolute and universal characteristics of qualitative inquiry. They provide a direction and framework for developing specific designs and concrete data collection tactics. Ideally, a pure qualitative inquiry strategy includes all of these themes and dimensions. For example, in an ideal naturalistic inquiry, the approach is highly emergent and inductive. In practice, however, the fieldwork moves from "discovery mode" to "verification mode" (Guba, 1978). As fieldwork begins, the inquirer is open to whatever emerges from the data, a discovery or inductive approach. Then, as the inquiry reveals patterns and major dimensions of interest, the investigator will begin to focus on verifying and elucidating what appears to be emerging—a more deductive approach to data collection and analysis. In essence, what is discovered may be verified by going back to the world under study and examining the extent to which the emergent analysis fits the phenomenon. Glaser and Strauss (1967), in their classic framing of grounded theory, described what it means for results to fit and work.

> By "fit," we mean that the categories must be readily (not forcibly) applicable to and indicated by the data under study; by "work," we mean that they must be meaningfully relevant to and be able to explain the behavior under study. (p. 3)

Discovery and verification mean moving back and forth between induction and deduction, between data gathering and data interpretation, and between experience and reflection on experience.

In program evaluation in particular, the evaluator may, through feedback of initial findings to program participants and staff, begin to influence the program quite directly and intentionally (given the job of helping improve the program), thus moving away from a purely naturalistic approach. As evaluative feedback is used to improve the program, the evaluator may then move back into a more naturalistic stance to observe how the feedback-induced changes in the program unfold.

In the same vein, the attempt to understand a program or treatment as a whole does not mean that the evaluator never becomes involved in component analysis or in looking at particular variables, dimensions, and parts of the phenomenon under study. Rather, it means that the qualitative inquirer consciously works back and forth between parts and wholes, aiming to understand specific variables and dimensions while also capturing the complex, interwoven constellations of variables in a sorting-out and then putting-back-together process. While staying true to a strategy that emphasizes the importance of a holistic picture of the program, the qualitative evaluator recognizes that certain periods of fieldwork may focus on component, variable, and less-than-the-whole kinds of analysis.

The practice and practicalities of fieldwork also mean that the strategic mandate to "get close" to the people and setting under study is neither absolute nor fixed. The degree of closeness to and involvement with the people falls along a continuum. The personal styles and capabilities of evaluators will permit and necessitate variance along these dimensions. Variations in types of programs and evaluation purposes will affect the extent to which an evaluator can or ought to get close to the program staff and participants. Moreover, closeness is likely to vary over the course of an evaluation. At times, the evaluator may become totally immersed in the program experience. These periods of immersion may be followed by times of withdrawal and distance (for personal as well as methodological reasons), to be followed still later by new experiences of immersion in and direct experience with the program.

Nor is it necessary to be a qualitative methods purist. Qualitative data can be collected and used in conjunction with quantitative data. Today's evaluators must be sophisticated about matching research methods to the nuances of particular questions and the idiosyncrasies of specific stakeholder interests. Such an evaluator needs a large repertoire of research methods and techniques to use on a variety of problems. Thus, an evaluator may be called on to use any and all of the social science research methods, including analyses of quantitative data from a program's management information system, questionnaires administered to participants, secondary data available in the community that reveals contextual trends, cost–benefit and cost-effectiveness analyses, standardized tests, experimental designs, unobtrusive measures, participant observation, and in-depth interviewing. The evaluation researcher works with the intended users of the findings to design an evaluation that includes any and all data that will help shed light on important evaluation questions, given the constraints of resources and time. Such an evaluator is committed to research designs that are relevant, meaningful, understandable, and able to produce useful results that are valid, reliable, and believable. On many occasions, a variety of data collection techniques and design approaches may be used together. Multiple methods and diverse kinds of data can contribute to methodological rigor. The ideal in evaluation designs is methodological appropriateness, design flexibility, and situational responsiveness in the service of utility (Patton, 2012a), not absolute allegiance to some ideal standard of methodological orthodoxy and purity.

A Sufi story about the wise fool Mulla Nasrudin illustrates the importance of understanding the connections between strategic ideals and practical tactics in real-world situations. Real-world situations seldom resemble the theoretical ideals taught in the classroom.

Ideal Conditions for Qualitative Inquiry: A Cautionary Tale

In his youth, Nasrudin received training in a small monastery noted for its excellence in the teaching of martial arts. Nasrudin became highly skilled in self-defense, and after two years of training, both his peers and his teachers recognized his superior abilities.

Each day, it was the responsibility of one of the students to go to the village market to beg for alms and food. It happened that a small band of three thieves moved into the area. They observed how the monastery obtained food daily and began hiding along the path the students had to take back to the monastery. As a student returned, laden with food and alms, the thieves would attack. After three days of such losses, the monastery's few supplies were exhausted. It was Nasrudin's turn to go to the village market. His elders and peers were confident that Nasrudin's martial arts skills were more than sufficient to overcome the small band of thieves.

At the end of the day, Nasrudin returned ragged, beaten, and empty-handed. Everyone was amazed. Nasrudin was immediately brought before the Master. "Nasrudin," he asked, "how is it that with all your skill in our ancient arts of defense, you were overcome?"

"But I did not use the ancient arts," replied Nasrudin.

All present were dumfounded. An explanation was demanded.

"All of our competitions are preceded by great and courteous ceremony," Nasrudin explained.

We have learned that the opening prayers, the ceremonial cleansing, the bow to the East—these are essential to the ancient ways. The ruffians seemed not to understand the necessity for these things. I didn't find the situation ideal enough to use the methods you have taught us, Master.

On more than one occasion, researchers or evaluators have told me of their belief in the potential utility of qualitative methods, but they tell me, "I just haven't found the ideal situation in which to use them."

Ideal situations are rare, but we will consider throughout this book the questions and conditions in which qualitative strategies and methods offer advantages. In Chapter 4, I will present a range and variety of situations and inquiry problems that particularly lend themselves to qualitative inquiry. Chapter 5 will then discuss in more detail some of the methodological trade-offs involved in adapting the strategic ideals of qualitative methods to the practical realities of conducting research and evaluation in the field. Chapters devoted to observation, interviewing, analysis, and enhancing the quality and credibility of qualitative studies follow the design chapter. To lay the groundwork for in-depth review of applications and methods, the next chapter examines alternative theoretical frameworks that are closely associated with and used to guide qualitative inquiry.

A Case Study Exemplar That Demonstrates the Application and Integration of the 12 Strategic Qualitative Principles of Inquiry

I close this chapter with a case example that shows how all 12 strategies were applied in one major qualitative inquiry, a case study of Henrietta Lacks. It is the story of how her cancer cells, taken without her knowledge or permission, transformed medical research (Skloot, 2010). It's an extraordinary exemplar of an in-depth qualitative inquiry that took 10 years of fieldwork, involved more than 1,000 hours of interviews, and included reviews of hundreds of documents. And, by the way, all that work resulted in a best-selling book, *The Immortal Life of Henrietta Lacks.*

Now, let me admit that not many qualitative studies become best-selling books, but some do (e.g., Collins, 2001a; Gladwell, 2008; Sacks, 1985, 2012), and this one merits special attention because of its extraordinarily important subject matter and its rigorous application of the full range of qualitative strategies. So I urge you to study Exhibit 2.7 carefully and then, if you haven't already done so, read the book. It's an inspiration about what careful, diligent, and dedicated fieldwork can yield.

EXHIBIT 2.7   Case Example Applying the 12 Strategic Themes of Qualitative Inquiry: Research Into the Immortal Cells of Henrietta Lacks

Henrietta Lacks was a poor African American tobacco farmer who grew up in the South. In 1951, dying of cervical cancer, cells from her tumor were taken for research, without her knowledge or permission, by an oncologist at Johns Hopkins Hospital in Baltimore, where she was being treated. Because of the unique characteristics of those cells, they could be cultured, sustained, reproduced, and distributed to other researchers—the first tissue cells discovered with these extraordinary characteristics. The cells were dubbed *HeLa cells*, the nomenclature created by the first two letters of her first and last names. They were also dubbed "immortal" in that they continue to this day to be cultured in labs around the world, reproduced in the trillions, distributed commercially, and used globally in medical research. Research using the HeLa cells has contributed to developing the polio vaccine, breakthroughs in cancer treatments, and the discovery of drugs to treat HIV/AIDS, to mention but three of a large number of significant breakthroughs. Yet, although HeLa cells became well-known and greatly prized by medical researchers around the world, no one knew about the woman whose cells had become so critical to medical research.

In 1988, Rebecca Skloot learned about HeLa cells as a student and became curious about their origins. She subsequently undertook an in-depth qualitative inquiry into the story of Henrietta Lacks. Her inquiry illustrates and illuminates the 12 strategic themes presented in this chapter. Examining the themes through Skloot's research will hopefully bring them to life, demonstrate their importance and relevance, and serve as a dramatic summary of this chapter. The results of Skloot's extensive qualitative inquiry were published as *The Immortal Life of Henrietta Lacks*, which became a best-selling book (Skloot, 2010).

QUALITATIVE INQUIRY STRATEGY	MANIFESTATION OF STRATEGIES IN SKLOOT'S CASE STUDY RESEARCH ON HENRIETTA LACKS
1. *Naturalistic inquiry*	Skloot began by locating and studying the few available documents and publications she could find. She used medical researchers as key informants for background and then began an arduous process of connecting with Henrietta Lacks's family members, including her husband, siblings, and children. She went into the field where Henrietta Lacks had lived, interviewed her neighbors, and visited her house, church, and cemetery. She tracked down old medical records and got access to archives. She followed the story as it had naturally unfolded, going backward to establish the complete history. At the same time, she was gathering data in real time, through participant observation, about the current experiences, perceptions, opinions, behaviors, and feelings of family, community members, and medical researchers using HeLa cells.
2. *Emergent design flexibility*	Her original design called for interviews with surviving family members, but the family was suspicious and resistant. Phone calls were not returned. Scheduled interviews became no-shows. Promises to meet were not kept. She went to Henrietta Lacks's small town and drove around trying to find people who would talk to her. Once she established contact and credibility with the first family member, other opportunities for interviews and field visits emerged. Toward the end of her fieldwork, the daughter of Henrietta Lacks became a close confidante and coresearcher, joining the inquiry on an extensive journey by car to several places and institutions that emerged as important to the story. Interviews and site visits to institutions provided leads on important, previously unknown documents, including medical records. Those documents led to new interviews and new institutions to visit in ways that could not have been anticipated at the beginning.
3. *Purposeful sampling*	This information-rich, purposefully chosen sample is one person, Henrietta Lacks, an *n* of 1 case study. To study the life and times of Henrietta Lacks, snowball sampling of key informants is the central strategy in this emergent design. Each person identified and interviewed yields leads to additional key informants. Documenting in depth and in detail, the story of Henrietta Lacks illuminates patterns, issues, and developments in medical research, social class, race, education, institutional politics, university policies, science, family dynamics, the legal system, the commercialization of medical discoveries, bioethics, cellular biology, cancer, and African American history, culture, and politics.

QUALITATIVE INQUIRY STRATEGY	MANIFESTATION OF STRATEGIES IN SKLOOT'S CASE STUDY RESEARCH ON HENRIETTA LACKS
4. *Qualitative data*	The book is the result of more than 1,000 hours of open-ended interviews, most of them tape-recorded and transcribed; 10 years of fieldwork; detailed descriptions of field visits to important places, institutions, and communities—and interactions with and observations of the people encountered in the field; and hundreds of archival documents, medical records, photos, personal diaries, videotapes, scientific reports, and both published and unpublished articles. The book is filled with direct quotations, detailed descriptions of experiences and observations, and direct citations from relevant documents.
5. *Personal experience and engagement*	Rebecca Skloot eventually became deeply involved with the Lacks family. She developed a close friendship with Deborah Lacks, Henrietta's daughter. At times, during fieldwork, she personally encountered hostility, anger, suspicion, confusion, fear, danger, attacks on her motives, requests for help, requests for money, appreciation, gratitude, and thanks. She directly engaged and sometimes confronted health authorities, medical researchers, university officials, legal experts, commercial interests, scientists, and journalists. At no point was she accused of being aloof and disengaged.
6. *Empathic neutrality and mindfulness*	Building trust was critical to gaining access to family and community members. She had to communicate her understanding of the family's anger and suspicion about having been kept in the dark about their mother for years, having been denied information, lied to, misinformed, and condescended to. As she established credibility and trust through her passion for getting at and telling Henrietta Lacks's story, she insisted on maintaining the integrity and independence of her research role. Note taking was omnipresent. Everything was documented. She was ever mindful of respecting family and community members, dealing openly with the strong feelings that regularly emerged during interviews and interactions in the field, and honoring the expertise of medical researchers, while also insisting on the independence of her inquiry and her obligation to present all sides of the story. In the end, this stance contributed to her credibility and deepened her access to family members and the medical research community. Her stance of empathic neutrality comes to the fore for the book's reader when she reviews the pros and cons of how human tissue samples are currently taken and stored and opines, "How you should feel about all this isn't obvious?" (p. 316).
7. *Dynamic systems*	During the 10 years of the research, the situation of the Lacks family members was in constant flux. Marriages; divorces; deaths in the family; illnesses; encounters with the criminal justice system, including one family member in and out of jail; financial woes; housing moves; and neighborhood changes. There were also dramatic changes in medical research, legal rulings, and bioethics; medical breakthroughs; waxing and waning interest in Henrietta Lacks among tissue researchers; changes in race relations and government initiatives around poverty that affected the family; and a major international cancer research conference to be held to honor Henrietta Lacks and her children scheduled for September 13, 2001, but overwhelmed by and canceled because of the 9/11 attacks on the World Trade Center and Pentagon—and never rescheduled.
8. *Unique case orientation*	Rebecca Skloot's commitment from the beginning and throughout was to capture and do justice to the life and story of Henrietta Lacks. Henrietta's cells, the HeLa cells, were and remain unique. Henrietta's story is and remains unique. The story illuminates many issues, historical developments, societal conditions, racial tensions, and medical research trends, but all these are the backstory. Always front and center is Skloot's commitment to document, understand, and tell Henrietta's unique story. She is not trying to generalize or extrapolate. She's focused on a single, high-quality case study.
9. *Inductive analysis and creative synthesis*	When Rebecca Skloot begins her research, she doesn't know what she'll find. She doesn't formulate hypotheses. She's not testing theory. She's investigating a case, seeing where the inquiry takes her, and making sense of it as it unfolds. And it unfolds in layers, like unpeeling an onion, going deeper and deeper to reach the core of the story. She works from the details

(Continued)

(Continued)

QUALITATIVE INQUIRY STRATEGY	MANIFESTATION OF STRATEGIES IN SKLOOT'S CASE STUDY RESEARCH ON HENRIETTA LACKS
	of events, quotations, relationships, and observations to construct the dramatic rendering that is the book. The creative synthesis is mostly chronological, but sometimes jumps forward and backward in time, and ultimately centers on cryptic quotations and phrases that become chapter titles, each providing a compelling window into the story. Chapter 5 "Blackness Be Spreadin All Inside" Chapter 11 "The Devil of Pain Itself" Chapter 16 "Spending Eternity in the Same Place" Chapter 19 "The Most Critical Time on This Earth Is Now" Chapter 20 "The HeLa Bomb" Chapter 31 "HeLa, Goddess of Death"
10. *Holistic perspective*	The story of Henrietta Lacks is one of connections, relationships, interconnections, and interdependencies: connections to family, community, and God; relationships with authorities, cousins, friends, her husband, her children, and her doctors; interconnections between her past, her place of origin, her slave ancestors, and her lived experience as a mother, wife, cook, homemaker, matriarch, farmer, Christian, medical patient, and cancer victim; and interdependencies between her multiple roles, her intertwined relationships, her secrets, her painted toenails, and the various institutional and community settings she inhabited along the way. These many crosscutting dimensions become integrated as the story unfolds and we come to know Henrietta Lacks and her immortal HeLa cells.
11. *Context sensitivity*	The whole only becomes complete, or at least as complete as it can be, when context is added. The historical context includes slavery, the Great Depression, World War II, the South and segregation, African American experiences of poverty, inadequate education, hard-to-access health care, and racism. The context includes the Johns Hopkins Hospital and University; the state of cancer research; the ethical norms for conducting research in the 1950s; evolving scientific knowledge about the human body, especially cells; the legal system; the emergence of a commercial market for buying and selling tissue samples and genetic material; and the emergence of HIV/AIDs. The story of Henrietta Lacks only makes sense in the context of the times in which she lived, the history that preceded her, and the issues and societal trends that emerged as the lives of her children and grandchildren unfolded.
12. *Reflexivity: perspective and voice*	Rebecca Skloot was a young, unmarried, middle-class, agnostic white woman studying a poor, married, African American, religious woman and her family. She openly discusses the challenges of being outside the culture she was studying, the mistakes she made, the confusions she encountered and engendered, and the advantages, at times, of being forced to ask for explanations of what was outside her experience. She reflects on the multiple roles that she plays over the years of the inquiry: researcher, documenter, interviewer, participant-observer, investigative journalist, book author, biographer, family friend, and advocate, among others. She went into debt to pursue the research. She shares her motivations, doubts, frustrations, little triumphs, and commitment to persevere without making her perspective intrusive. She speaks in a clear voice, using the first-person active to reflect, "The Lackses challenged everything I thought I knew about faith, science, journalism, and race" (p. 7). She calls the book "a work of nonfiction" (p. ix). I call it an exemplary qualitative inquiry.

Conclusion: Evaluating the Box

This chapter, like all chapters in this book, closes with a Halcolm story in the form of a graphic comic. The cartoon story illustrates the importance of how one looks at things. This chapter has presented 12 qualitative lenses through which one can view, study, and inquire into the world—including something as simple as a box.

APPLICATION EXERCISES

1. ***Naturalistic inquiry study:*** You are a purposeful sample of one (*n* = 1). Undertake a naturalistic inquiry study of a week in your life. At the beginning of the week, plan what you will do during the week. Then, (a) document what you did that unfolded as planned and (b) document what happened that wasn't planned or expected. What emerged in the natural course of living and working and interacting with people that captures and provides insights into you and your life? As an illustration of *emergent design flexibility*, what interactions and/or events occurred during the week that would provide opportunities for interviews with people you encountered whom you would not have anticipated being key informants at the start of the week. For example, an old friend called, and you got together. That friend would now be a potential key informant to be interviewed to get insight into your week. Likewise, an event or problem that you experienced that was unanticipated could become a mini case study of how your life unfolds. Describing that unanticipated event or problem in some depth would be an example of emergent design flexibility.

2. ***Qualitative data:*** Pick a topic that interests you, and search the web for qualitative studies on that topic. For example, let's say that you're interested in how college students perceive "hooking up." A web search of "hooking up qualitative study" at the time of this writing yielded several interesting examples, including the following:

   - "Hooking Up and Sexual Risk Taking Among College Students: A Health Belief Model Perspective"

     Hooking up with friends, strangers, and acquaintances is a popular way for college students to experience sexual intimacy without investing in relationships. Because hooking up often occurs in situations in which prophylactics against sexually transmitted infections (STIs) are not available or in which students' judgment is impaired, it can involve risky behaviors that compromise student health. . . . Based on semistructured interviews with 71 college students about their hooking-up experiences . . . , results demonstrate why students' assessments of their own and their peers' susceptibility to STIs are often misinformed. (Downing-Matibag & Geisinger, 2009, p. 1)

   - "A Plan to Reboot Dating": This article cites a qualitative study by "the feminist sociologist" Lisa Wade of Occidental College, who interviewed 44 of her freshman students (33 of them women) and reported that most of them were "overwhelmingly disappointed with the sex they were having in hook ups. This was true of both men and women, but was felt more intensely by women" (Smith, 2012).

   - "Hooking Up": An ecological psychology perspective on the sexual behaviors of women and the potential effects on sexually transmitted infections. Four

(Continued)

(Continued)

females were interviewed, having been recruited through networking and word-of-mouth strategies at the Boston University undergraduate main campus (Power, 2012).

Through your Internet search, identify three qualitative studies on your topic of interest. What kinds of qualitative data were collected? What kind of purposeful sample was studied? What was the core finding of each study? To what extent and in what ways do the three studies provide consistent findings that illuminate your topic of interest? To what extent are the three studies quite different windows into your topic of interest? What do these studies tell you about the possibilities for qualitative inquiry on your topic of interest?

3. *Contextual sensitivity:* Describe an event or experience in your life *without* context. Now, add at least two layers of context: (1) the local context within which the event or experience occurred and (2) the larger societal/cultural context. What difference does adding context make to understanding the event or experience?

4. On page 81 is a cartoon titled "The Box." Using the format in Exhibit 2.7 (pp. 78–80), describe how you would address the 12 strategic principles if you were to study the box qualitatively. (1) How would you engage in naturalistic inquiry of the box? (2) What would constitute emergent design flexibility in studying the box? (3) How would you select a purposeful sample? (4) What qualitative data would you collect? (5) What personal experience and engagement would be part of your inquiry? (6) How would empathic neutrality come into play? (7) How would you adopt a dynamic systems perspective? (8) What would constitute unique case orientation? (9) How would you engage in inductive analysis and creative synthesis? How would you be (10) holistic and (11) contextually sensitive? (12) How would you convey your own voice, perspective, and reflexivity?

5. Exhibit 2.6 (p. 75) depicts the 12 core strategies of qualitative inquiry as interdependent and interactive. What's the message in that graphic? Express the graphic—and its implications for qualitative inquiry—in words. Tell what the graphic means and why it's important.

CHAPTER 3

Variety of Qualitative Inquiry Frameworks

Paradigmatic, Philosophical, and Theoretical Orientations

Personal coat of arms of Sir Isaac Newton:

Two shinbones in *saltaire argent* crossed in silver

—Michon (2013)

**I can calculate the motion
of heavenly bodies,
but not the madness of people.**

—Sir Isaac Newton (1642–1726)

Chapter Preview

There is no definitive way to categorize the various philosophical and theoretical perspectives that have influenced and that distinguish the types of qualitative inquiry. This chapter distinguishes theoretical perspectives by their foundational questions. We're going to examine, compare, and contrast 16 different theoretical approaches to quality inquiry:

1. Ethnography
2. Autoethnography
3. Reality-testing, foundationalist epistemologies: positivism, postpositivism, empiricism, and objectivism
4. Grounded theory
5. Realism
6. Phenomenology
7. Heuristic inquiry
8. Social constructionism and constructivism
9. Narrative inquiry
10. Ethnomethodology
11. Symbolic interaction
12. Semiotics
13. Hermeneutics
14. Systems theory
15. Complexity theory
16. Pragmatism and generic qualitative inquiry

We'll then look at 10 qualitative inquiry themes that cut across these 16 distinct and diverse theoretical orientations. That will set the stage for the next chapter, where we examine concrete and practical applications of qualitative methods. In effect, these two chapters, Chapters 3 and 4, look at theory and practice, first theory (this chapter), then practice (the next chapter), to integrate them: *theory informing practice* and *practice informing theory*.

To begin, I'm going to review the historic debate between qualitative and quantitative methods. The debate has waxed and waned over the years, and the precise nature of the debate has morphed from differences of opinion about which methods are the most credible to how best to combine the strengths of each through mixed methods. The intensity of the debate has varied among disciplines and across institutional settings. It cuts across the domains of research and evaluation, so this may be a good time to revisit that distinction since this book is about both. Research has as its primary purpose contributing to knowledge, and evaluation has as its primary purpose informing action: using data to support decision making; informing judgments about merit, worth, and significance; and improving interventions (programs, policies, and products). (For more on what evaluation is, see American Evaluation Association, 2014.)

As we'll see, some elements of the debate endure and constitute part of the context for engaging in qualitative inquiry today for both research and evaluation purposes. That's why we need to take this short walk through history to understand how we got to where we are now, including some of the lingering effects and continuing divisions that sometimes surface today when methods decisions are being made. *Forewarned is forearmed.*

Understanding the Paradigms Debate: *Quants Versus Quals*

Intellectual Walls and Fences

> We build too many walls and not enough bridges.
>
> —Sir Isaac Newton (1642–1726)
> English physicist, mathematician,
> astronomer, natural philosopher
>
> Don't ever take a fence down until you know why it was put up.
>
> —Robert Frost (1874–1963)
> American poet

Once upon a time, not so very long ago, a group of statisticians (hereafter known as *quants*) and a party of case study aficionados (*quals*) found themselves together on a train traveling to the same professional meeting. The *quals*, all of whom had tickets, observed that the *quants* had only one ticket for their whole group.

"How can you all travel on one ticket?" asked a *qual*.

"We have our methods," replied a *quant*.

Later, when the conductor came to punch tickets, all the *quants* slipped quickly behind the door of the toilet. When the conductor knocked on the door, the head *quant* slipped their one ticket under the door, thoroughly fooling the conductor.

On their return from the conference, the two groups again found themselves on the same train. The qualitative researchers, having learned from the *quants*, had schemed to share a single ticket. They were chagrined, therefore, to learn that, this time, the statisticians had boarded with no tickets.

"We know how you traveled together with one ticket," said a *qual*, "but how can you possibly get away with no tickets?"

"We have ever more sophisticated methods," replied a *quant*.

Later, when the conductor approached, all the *quals* crowded into the toilet. The head statistician followed them and knocked authoritatively on the toilet door. The *quals* slipped their one and only ticket under the door. The head *quant* took the ticket and joined the other *quants* in a different toilet. The *quals* were subsequently discovered without tickets, publicly humiliated, and tossed off the train at its next stop.

Quants Versus Quals

Who are *quants*? They're numbers people who, in rabid mode, believe that if you can't measure something, it doesn't exist. They live by Galileo's admonition, "Measure what is measurable, and make measurable what is not so." Their mantra is "What gets measured gets done." And *quals*? They quote management expert W. Edwards Deming: "The most important things cannot be measured." *Quals* find meaning in words and stories and are ever ready to recite Albert Einstein's observation that "everything that can be counted does not necessarily count; everything that counts cannot necessarily be counted"—relatively speaking, of course.

Quants demand "hard" data: statistics, equations, charts, and formulae. *Quals*, in contrast, are "softies," enamored with narrative and case studies. *Quants* love experimental designs and believe that the only way to prove that an intervention caused an outcome is with a randomized controlled trial (RCT). *Quants* are control freaks, say the *quals*: simplistic, even simpleminded, in their naive belief that the world can be reduced to independent and dependent variables. The *qual's* world is complex, dynamic, interdependent, textured, nuanced, unpredictable, and understood through stories, more stories, and still more stories. *Quals* connect the causal dots through the unfolding patterns that emerge within and across these many stories and case studies. *Quants* aspire to operationalize key predictor variables and generalize across time and space—the holy grail of truth: If *x*, then *y*, and the more of *x*, the more of *y*. *Quals* distrust generalizations and are most comfortable immersed in the details of a specific time and place, understanding a story in the richness of context and the fullness of thick description. For *quals*, the patterns they extrapolate from cross-case analyses are possible principles to think about in new situations but are not the generalized, formulaic prescriptions that *quants* admire and aspire to. *Quants* produce *best practices* that assert, "Do this because it's been proven to work in rigorous studies." *Quals* produce themes and suggest, "Think about this and what it might mean in your own context and situation."

Are these overgeneralized stereotypes? In some contexts, yes; in others, not so much. Is there a middle ground? Absolutely. In this chapter, we will move beyond the extremes and find more than one place in what is sometimes called *the radical middle*. We will also look at core values shared by the full range of scientific researchers. But before doing all of that, we're going to linger for a moment where the differences are manifest. *Quals* and *quants* do hold contrasting, even opposite, worldviews at times—and the history of those differences provides an important context for appreciating current efforts at *détente* and integration.

So do opposites attract? Indeed, they do. They attract debate, derision, and dialectical differentiation—otherwise known as *the paradigms war*. The story of the *quals* and *quants* offers a window into how the paradigms debate has ebbed and flowed. This debate about the relative merits of quantitative/experimental methods versus qualitative/naturalistic methods has periodically run out of intellectual steam, but like mythical zombies, it seems never to die. Some, like Carol Weiss (2006), one of the distinguished founders of the field of evaluation research, refused to engage in the quantitative/qualitative debate because she thought it was "pretty silly" (p. 483), an elegantly phrased scholarly judgment. But one person's silliness is another person's battleground. I understand and appreciate the advice of one reviewer of this book, who wrote, "I would prefer comment that the Paradigm Wars are ridiculous and so he won't bother to give them space." But I've found that it's helpful to know a bit about the nature and history of the qualitative/quantitative debate because remnants remain and can surface unexpectedly at inopportune moments. Forewarned is forearmed, so some review seems prudent.

Competing Inquiry Paradigms

> Ideology is to research what Marx suggested the economic factor was to politics and what Freud took sex to be for psychology.
>
> —Michael Scriven (1972a, p. 94)
> Philosopher of science and evaluation pioneer

Controversy often accompanies and even engulfs qualitative methods. Students attempting to do qualitative dissertations can get caught up in and may have to defend philosophically as well as methodologically the use of qualitative inquiry to those who have little experience with or training in it. Evaluators may encounter policymakers and funders who dismiss qualitative data as mere anecdote. The statistically addicted may poke fun at what they call "the softness" of qualitative data. (In Western society, where anything can be and often is sexualized,

THE KINGDOM OF WISDOM

The king of the Kingdom of Wisdom was growing old. He had two sons to whom he commissioned the task of expanding the kingdom. One traveled south and built *Dictionopolis*, the city of words; the other went north and built *Digitopolis*, the city of numbers.

The two brothers and their two cities entered into a fierce rivalry. The people of Dictionopolis believed that words were wisdom and much more important than numbers. The people of Digitopolis believed that numbers were wisdom and much more important than words. The rivalry intensified.

To avoid war and settle the matter, they agreed to arbitration by two independent thinkers, Rhyme and Reason, princesses who lived in the capital city of Wisdom. After lengthy and thoughtful deliberation, they pronounced that words and numbers were both essential and equally valuable. Outraged, the brothers banished Rhyme and Reason from the Kingdom of Wisdom.

Eventually, after a great and dramatic rescue, Rhyme and Reason were returned to the Kingdom of Wisdom. To find out how, and why this story matters, you'll have to enter into the Kingdom of Wisdom through *The Phantom Tollbooth* (Juster, 1961).

the distinction between "hard" data and "soft" data has additional nuances of meaning and innuendo.) Such encounters derive from a long-standing methodological paradigms war. Not everyone has adopted a stance of methodological enlightenment and tolerance, namely, that methodological orthodoxy, superiority, and purity should yield to methodological appropriateness, pragmatism, and mutual respect. Therefore, as I noted in introducing this section, a brief review of the paradigms debate is in order. (Elsewhere I have provided a more extensive review of the methodological paradigms debate: Patton, 2008b, chap. 12; 2012a, chap. 11).

Philosophers of science and methodologists have been engaged in a long-standing epistemological debate about the nature of "reality" and knowledge. That philosophical debate finds its way into research and evaluation in differences of opinion about what constitutes "good" research and high-quality evidence. In its simplest and most strident formulation, this debate has centered on the relative value of two different and competing inquiry paradigms: (1) using quantitative and experimental methods to generate and test hypothetical-deductive generalizations *versus* (2) using qualitative and naturalistic approaches to inductively and holistically understand human experience in context-specific settings. To be sure, the variety of inquiry approaches has expanded well beyond the simplistic dichotomy between quantitative and qualitative paradigms. Still, the paradigms debate is part of our methodological heritage, and still all too alive in some places, so knowing a bit about it may deepen appreciation for the importance of a strategic approach to methodological decision making.

THE HIDDEN BIASES IN BIG DATA

The business and science worlds are focused on how large [quantitative] datasets can give insight on previously intractable challenges. The hype becomes problematic when it leads to ... "data fundamentalism," the notion that correlation always indicates causation, and that massive data sets and predictive analytics always reflect objective truth.

Data and data sets are not objective; they are creations of human design. We give numbers their voice, draw inferences from them, and define their meaning through our interpretations. Hidden biases in both the collection and analysis stages present considerable risks, and are as important to the big-data equation as the numbers themselves.

—Kate Crawford (2013)
Harvard Business Review

What Is an Inquiry Paradigm?

A paradigm is a worldview—a way of thinking about and making sense of the complexities of the real world. As such, paradigms are deeply embedded in the socialization of adherents and practitioners. Paradigms tell us what is important, legitimate, and reasonable. Paradigms are also normative, telling the practitioner what to do without the necessity of long existential or epistemological consideration. But it is this aspect of paradigms that constitutes both strength and weakness—a strength in that it makes decisions about what to do relatively easy, a weakness in that the very reason for a certain decision is hidden in the unquestioned assumptions of the paradigm. Thomas Kuhn, philosopher of science, explained how paradigms work in his influential classic *The Structure of Scientific Revolutions* (1970):

Scientists work from models acquired through education and through subsequent exposure to the literature often without quite knowing or needing to know what characteristics have given these models the status of community paradigms.... That scientists do not usually ask or debate what makes a particular problem or solution legitimate tempts us to suppose that, at least intuitively, they know the answer. But it may only indicate that neither the question nor the answer is felt to be relevant to their research. Paradigms may be prior to, more binding, and more complete than any set of rules for research that could be unequivocally abstracted from them. (p. 46)

So why does all this matter to the student interested in pursuing some research or evaluation question? It matters because paradigm-derived biases are the source of the derogatory distinctions between "hard data" and "soft data," empirical studies versus "mere anecdotes," and "objective" research versus "subjective" studies. These labels reveal value-laden prejudices about what constitutes credible and valuable contributions to knowledge. Such prejudices and paradigmatic blinders limit methodological choices, flexibility, and creativity. Adherence to a methodological paradigm can lock researchers into unconscious patterns of perception and behavior that disguise the biased, predetermined nature of their methods "decisions." Methods decisions tend to stem from disciplinary prescriptions, concerns about scientific status, old methodological habits, and comfort with what the researcher knows best. Training and academic socialization tend to make researchers biased in favor of and against certain approaches.

Methodological Pragmatism, Eclecticism, and Mixed Methods

Methodology depends primarily on the subject matter of what is investigated, and on certain background assumptions. Debates about methodology are productive only if the subject matter is considered first.

—Ernest House (1994, p. 13)
Evaluation theorist and methodologist

Many researchers and evaluators still conduct their studies with a predominantly quantitative or qualitative

"QUALITATIVE METHODS ARE RUBBISH."

What were you thinking?! Qualitative methods?

What kind of limp-wristed latte-sipping excuse for real science is this?

—World War II spoof depicting the vociferous emotional and dismissive reactions qualitative methods can provoke
YouTube (4 minutes): http://www.youtube.com/watch?v=FIdN8NKjqts

Now...Calm down a bit...I'm not criticizing your operation down here...I'm just asking...in your entry evaluation process, do you operate from a qualitative or quantitative paradigm?

Hellish Inquiry
Are There Paradigms After Death?

©2002 Michael Quinn Patton and Michael Cochran

paradigm, but many others have become eclectic and creative, mixing methods. Exhibit 3.1 summarizes the methods paradigms debate by presenting 10 contrasting emphases between the quantitative/experimental, the qualitative/naturalistic, and the mixed-methods paradigms.

In the quantitative/qualitative debate, the mixed-methods paradigm is sometimes called "the radical middle."

It is not enough for mixed methodology researchers to exist in an epistemological space that lies somewhere between the quantitative and qualitative epistemological spaces. Rather, mixed researchers should strive for what is the radical middle, which should not be a passive and comfortable middle space wherein the status quo among quantitative and qualitative epistemologies is maintained, but rather a new theoretical and methodological space in which a socially just and productive coexistence among all research traditions is actively promoted, and in which mixed research is

consciously local, dynamic, interactive, situated, contingent, fluid, strategic, and generative. (Onwuegbuzie, 2012, p. 192)

A central argument for mixing methods comes from "critical multiplism" (Shadish, 1993), which argues that any single method is biased and that only by thinking critically about the strengths and weaknesses of various methods, and using multiple methods to overcome single-methods biases, can rigorous results be achieved. (See the sidebar on critical multiplism, p. 92).

EXHIBIT 3.1   The Methods Paradigms Debate: Ten Contrasting Emphases

Both qualitative and quantitative inquiry share basic values around and commitments to systematic inquiry, matching methods to questions, conscientious data collection, appropriate analysis, and detailed reporting of the procedures followed, including acknowledging both the strengths and the weaknesses of the inquiry approach taken and the results attained. Exhibit 9.1 (pp. 659–660)

presents 10 general scientific research quality criteria that apply to all kinds of inquiries: qualitative, quantitative, and mixed methods. Still, beyond generally shared methodological standards, there are different emphases between qualitative, quantitative, and mixed-methods inquiry approaches. These are matters of *emphasis within ideal types*, not rigid orthodoxy.

	QUANTITATIVE/ EXPERIMENTAL PARADIGM	QUALITATIVE/ NATURALISTIC PARADIGM	MIXED-METHODS PARADIGM
1. *Aspiration*	Empirical generalizations across time and space	In-depth, holistic, contextually sensitive understandings of phenomena	Integrating in-depth qualitative understandings with broader generalizations
2. *Philosophy of science roots*	Positivism and logical positivism, scientific empiricism and realism	Social constructivism, phenomenology, hermeneutics, *Verstehen* tradition	Pragmatism
3. *Inquiry approach*	Specifying independent and dependent variables to test causal hypotheses: *hypothetical-deductive inquiry*	Entering real-world settings to observe, interact, and understand what emerges: *naturalistic inquiry*	Qualitative inquiry for exploration followed by quantitative inquiry for verification and generalization: *sequential inquiry*
4. *Data collection*	Quantitative data through valid and reliable surveys, tests, and statistical indicators	Qualitative data through fieldwork observations, participant observation, and in-depth interviewing	Combining quantitative and qualitative data
5. *Researcher/ evaluator stance*	Objective, independent, detached, and value-free	Engaged, subjectivity acknowledged, value laden, reflexive	Neutral, adaptable in accordance with the nature of the inquiry
6. *Conceptual approach*	Operationalize (quantify) variables; the concept of interest doesn't exist until it can be measured	Open-ended inquiry into *sensitizing concepts* to find out what people mean by and how they use concepts	Measure what can be measured statistically and explore additional meanings through interviews
7. *Approach to dealing with natural variation*	Control variation through experimentation with carefully designed treatment and control group comparisons	Study and document diversity and natural variation; minimize control to study how things unfold in the real world	Design flexibility depending on the nature of the research question
8. *Sampling strategy*	Random, probabilistic samples to achieve representativeness and high internal validity in RCTs	Strategic case selection and purposeful sampling of rich information for in-depth study to document diversity	Multiple and combined sampling approaches depending on the nature of the inquiry
9. *Analysis and comparisons*	Use standardized instruments to measure central tendencies and variation statistically; test hypotheses derived from theory: *deductive analysis*	Look for themes and patterns across case studies; theory emerges from cases (grounded theory): *inductive analysis*	Compare statistical results with qualitative patterns
10. *Criticisms of other paradigms*	Qualitative data are "soft," ambiguous, too susceptible to researcher bias; designs can't establish causality; samples too small for generalizations	Important things can't be reduced to standardized instruments; quantitative data only tap the surface of human experience; experimental designs are too rigid and controlling	Both quantitative and qualitative approaches have strengths and weaknesses; need to mix methods to overcome the weaknesses of each: *critical multiplism* (Shadish, 1993)

Paradigm Flexibility and Methodological Appropriateness

While a paradigm offers a coherent worldview, an anchor of stability and certainty in the real-world sea of chaos, operating narrowly within any singular paradigm, including a mixed-methods paradigm, can be quite limiting. My practical stance involves judging the quality of a study by its intended purposes, available resources, procedures followed, and results obtained, all within a particular context and for a specific audience. When a new drug is tested before being made available to the general population, a triple-blind randomized experiment to determine efficacy is the design of choice, with careful attention to controlled and carefully measured dosage and outcome interactions, including side effects. But if the concern is whether people are taking the new drug appropriately, and if one wants to know what a group of people think about the new drug (e.g., an antidepressant), how they make sense of taking or not taking it, what they believe about themselves as a result of experiencing the drug, and how those around them deal with it, then in-depth interviews and observations are the more appropriate method of inquiry. The importance of understanding alternative research paradigms is to sensitize researchers and evaluators to the ways in which their methodological prejudices, derived from their disciplinary socialization experiences, may reduce their methodological flexibility and adaptability.

Being practical and flexible allows one to eschew methodological orthodoxy in favor of *methodological appropriateness* as the primary criterion for judging methodological quality, recognizing that different methods are appropriate for different situations. Situational responsiveness means designing a study that is appropriate for a specific inquiry situation or interest. A major purpose of this book is to identify the kinds of research questions and program evaluation situations for which qualitative inquiry is an especially appropriate method of choice. A wide range of possibilities exist when selecting methods. The point is to do what makes sense and report fully on what was done, why it was done, and what the implications are for findings.

CRITICAL MULTIPLISM AS A RESEARCH STRATEGY

Scientists who rely on disciplinary training for scientific strategy are bound to produce research that is biased by paradigmatic errors of omission and commission. If so, then one of the more pressing needs in science is for the development of strategies that can uncover the biases of omission and commission that are inevitably present in all scientific methods and then ensure that they do not operate in the same direction in a study or in a research literature to yield a biased conclusion....

Critical multiplism is a research strategy aimed at doing just that: It advises scientists to put together packages of imperfect methods and theories in a manner that minimizes constant biases.... Briefly, multiplism refers to the fact that any task in science can usually be conducted in any one of several ways, but in many cases no single way is known to be uniformly best. Under such circumstances, a multiplist advocates making heterogeneous those aspects of research about which uncertainty exists, so that the task is conducted in several different ways, each of which is subject to different biases. *Critical* refers to rational, empirical, and social efforts to identify the assumptions and biases present in the options chosen. Putting the two concepts together, we can say that the central tenet of critical multiplism is this: When it is not clear which of several defensible options for a scientific task is least biased, we should select more than one, so that our options reflect different biases, avoid constant biases, and leave no plausible bias overlooked. That multiple options yield similar results across operationalizations with different biases increases our confidence in the resulting knowledge. If different results occur when we do the task in different ways, then we have an empirical and conceptual problem to solve if we are to explain why this happened, but we are saved from a premature conclusion that a particular piece of knowledge is plausible. (Shadish, 1993, p. 18)

Fools' Gold: The Widely Touted Methodological "Gold Standard" Is Neither Golden nor a Standard

I am offering one personal rumination per chapter. These are issues that have persistently engaged, sometimes annoyed, occasionally haunted, and often amused me over more than 40 years of research and evaluation practice. Here's where I state my case on the issue and make my peace.

The *Wikipedia* entry for randomized controlled trials (RCTs), reflecting common usage, designates such designs as the "gold standard" for research. News reports of research findings routinely repeat and reinforce the "gold standard" designation for RCTs (e.g., *New York Times*, 2014). Government agencies and scientific associations that review and rank studies for methodological quality acclaim RCTs as the gold standard. For example, researchers at the University of Saint Andrews reviewed "what counts as good evidence" and found that

> when the research question is "what works?", different designs are often placed in a hierarchy to determine the standard of evidence in support of a particular practice or programme. These hierarchies have much in common; randomised experiments with clearly defined controls (RCTs) are placed at or near the top of the hierarchy and case study reports are usually at the bottom. (Nutley, Powell, & Davies, 2013, p. 10)

The Gold Standard Versus Methodological Appropriateness

A consensus has emerged in evaluation research that evaluators need to know and use a variety of methods in order to address the priority questions of particular stakeholders in specific situations. But researchers and evaluators get caught in contradictory assertions: (a) select methods appropriate for a specific evaluation purpose and question, and use multiple methods—both quantitative and qualitative—to triangulate and increase the credibility and utility of findings, *but* (b) one question is more important than others (the causal attribution question), and one method (RCTs) is superior to all other methods in answering that question. This is what is known colloquially as talking out of both sides of your mouth. Thus, we have a problem. The ideal of researchers and evaluators being situationally responsive, methodologically flexible, and sophisticated in using a variety of methods runs headlong into the conflicting ideal that experiments are the gold standard and all other methods are, by comparison, inferior. Who wants to conduct (or fund) a second-rate study if there is an agreed-on gold standard?

The Rigidity of a Single, Fixed Standard

The gold standard allusion derives from international finance, in which the rates of exchange among national currencies were fixed to the value of gold. Economic historians share a "remarkable degree of consensus" about the gold standard as the primary cause of the Great Depression:

> the mechanism that turned an ordinary business downturn into the Great Depression. The constraints of the gold-standard system hamstrung countries as they struggled to adapt during the 1920s to changes in the world economy. And the ideology, mentality and rhetoric of the gold standard led policy makers to take actions that only accentuated economic distress in the 1930s. (Eichengreen & Temin, 1997, pp. 1–2)

The gold standard system collapsed in 1971 following the United States' suspension of convertibility from dollars to gold. *The system failed because of its rigidity*. And not just the rigidity of the standard itself but also the rigid ideology of the people who believed in it: Policymakers across Europe and North America clung to the gold standard despite the huge destruction it was causing. There was a clouded mind-set with a moral and epistemological tinge that kept them advocating the gold standard until political pressure emerging from the disaster became overwhelming.

Treating RCTs as the gold standard is no less rigid. Asserting a gold standard inevitably leads to demands for standardization and uniformity (Timmermans & Berg, 2003). Distinguished evaluation pioneer Eleanor Chelimsky (2007) has offered an illuminative analogy:

> It is as if the Department of Defense were to choose a weapon system without regard for the kind of war being fought; the character, history, and technological advancement of the enemy; or the strategic and tactical givens of the military campaign. (p. 14)

Indeed, while inquiries into the effectiveness of new pharmaceutical drugs or agricultural production techniques may benefit from RCTs, such designs are both inappropriate and misleading for inquiring into the complex, dynamic

(Continued)

(Continued)

phenomena that characterize much of the human condition. The problem is that treating experimental designs as the gold standard cuts off serious consideration of alternative methods and channels millions of dollars of research and evaluation funds into support for a method that has not only strengths but also significant weaknesses. The gold standard accolade means that funders and policymakers begin by asking, "How can we do an experimental design?" rather than asking, "Given the state of knowledge and the priority inquiry questions at this time, what is the appropriate design?" Here are examples of the consequences of this rigid mentality:

- At an African evaluation conference, a program director came up to me in tears. She directed an empowerment program with women in 30 rural villages. The funder, an international agency, had just told her that to have the funding renewed, she would have to stop working in half the villages (selected randomly by the funder) in order to create a control group going forward. The agency was under pressure for not having enough "gold standard evaluations." But, she explained, the villages and the women were networked together and were supporting each other. Even if they didn't get funding, they would continue to support each other. That was the empowerment message. Cutting half of them off made no sense to her. Or to me.
- At a World Bank conference on youth service learning, the director of a university program that placed student interns in rural villages in Cambodia offered her program for an exercise in evaluation design. She explained that she carefully selected 40 students each year and matched them to villages that needed the kind of assistance the students could offer. *Matching students and villages was key*, she explained. A senior World Bank economist told her and the group to forget matching. He advised an RCT in which she would randomly assign students to villages and then create a control group of qualified students and villages that did nothing to serve as a counterfactual. He said, "That's the only design we would pay any attention to here. You must have a counterfactual. Your case studies of students and villages are meaningless and useless." The participants were afterward aghast that he had completely dismissed the heart of the intervention: matching students and villages.
- I've encountered several organizations, domestic and international, that give bonuses to managers who commission RCTs for evaluation to enhance the organization's image as a place that emphasizes rigor. The incentives to do experimental designs are substantial and effective. Whether they are appropriate or not is a different question. In effect, it is *gold from those who enforce the gold standard to those who implement it*. One senior economist told me, "We're applying the basic economic principle of giving rewards for what we want. We pay for gold standard designs. We promote those who commission and conduct

gold standard designs. So we get gold standard designs. You get what you reward." This explanation of how the world works rolled off his tongue like an oft-told story, which I suspect it was. He was smiling broadly at his cleverness, awash in his wisdom and self-congratulatory certainty. What I heard was rigidity, narrow-mindedness, misguided and distorted incentives, and an utter incapacity to detect even a hint of a problem.

Those experiences, multiplied 100 times, are what have generated this rumination.

Heaven or Gold Standard

Professor David Storey (2006) of Warwick Business School, University of Warwick, has offered a competing metaphor to replace the gold standard. He has posited "seven steps to heaven" in conducting evaluations, where heaven is a randomized experiment. So materially oriented and worldly evaluators are admonished to aspire to the gold standard, while the more spiritually inclined can aspire to follow the path to heaven, where heaven is an RCT.

In contrast, the metaphors of naturalistic inquiry are more along the lines of *staying grounded*, looking at *the real world* as it unfolds, *going with the flow*, *being adaptable*, and *seeing what emerges*.

Evidence-Based Medicine and RCTs

Medicine is often held up as the bastion of RCT research in its commitment to evidence-based medicine (EBM). But here again, gold standard designation has a downside, as observed by the psychologist Gary Klein (2014):

Sure, scientific investigations have done us all a great service by weeding out ineffective remedies. For example, a recent placebo-controlled study found that arthroscopic surgery provided no greater benefit than sham surgery for patients with osteoarthritic knees. But we also are grateful for all the surgical advances of the past few decades (e.g., hip and knee replacements, cataract treatments) that were achieved without randomized controlled trials and placebo conditions. Controlled experiments are therefore not necessary for progress in new types of treatments and they are not sufficient for implementing treatments with individual patients who each have unique profiles.

Worse, reliance on EBM can impede scientific progress. If hospitals and insurance companies mandate EBM, backed up by the threat of lawsuits if adverse outcomes are accompanied by any departure from best practices, physicians will become reluctant to try alternative treatment strategies that have not yet been evaluated using randomized controlled trials. Scientific advancement can become stifled if front-line physicians, who blend medical expertise with respect for research, are prevented from exploration and are discouraged from making discoveries.

Parachutes and RCTs

Gordon C. S. Smith and Jill P. Pell (2003) published a clever and widely disseminated systematic review of RCTs on parachute use in the *British Medical Journal*. They arrived at the following conclusions:

- No RCTs of parachute use have been undertaken.
- The basis for parachute use is purely observational (qualitative).
- "Individuals who insist that all interventions need to be validated by a randomized controlled trial need to come down to earth with a bump" (p. 1460).

On the other hand, they added that RCT's parachute research could still be done: "We feel assured that those who advocate evidence-based medicine and criticize use of interventions that lack an evidence base will not hesitate to demonstrate their commitment by volunteering for a double blind, randomized, placebo controlled, crossover trial" (p. 1460).

RCTs and Bias

RCTs aim to control bias, but implementation problems turn out to be widespread:

> Even in the most stringent research designs, bias seems to be a major problem. For example, there is strong evidence that selective outcome reporting, with manipulation of the outcomes and analyses reported, is a common problem even for randomized trails. (Chan, Hrobjartsson, Haahr, Gotzsche, & Altman, 2004, p. 2457)

The result is that "a great many published research findings are false" (Ioannidis, 2005).

Methodological Appropriateness as the *Platinum Standard*

It may be too much to hope that the gold standard designation will disappear from popular usage. So perhaps we need to up the ante and aim to supplant the gold standard with a new platinum standard: *methodological pluralism and appropriateness.* To do so, I offer the following seven-point action plan:

1. Educate yourself about the strengths *and weaknesses* of RCTs. (See resources below.)

2. Never use the gold standard designation yourself. If it comes up, refer to the "so-called gold standard."

3. When you encounter someone referring to RCTs as the gold standard, don't be shy. Explain the negative consequences and even dangers of such a rigid pecking order of methods.

4. Understand and be able to articulate the case for methodological pluralism and appropriateness, to wit, adapting designs to the existing state of knowledge, the available resources, the intended uses of the inquiry results, and other relevant particulars of the inquiry situation. (See the resources listed below.)

5. Promote the *platinum standard* as higher on the hierarchy of research excellence (and even closer to heaven).

6. Don't be argumentative and aggressive in challenging gold standard narrow-mindedness. It's more likely a matter of ignorance than intolerance. Be kind, sensitive, understanding, and compassionate, and say, "Oh, you haven't heard. The old RCT gold standard has been supplanted by a new, more enlightened, Knowledge-Age platinum standard." (Beam wisely.)

7. Repeat Steps 1 to 6 over and over again.

Resources

- For 10 limitations of experimental designs, see Patton (2008b, pp. 447–450).
- For the case against a rigid methodological hierarchy, see Ornish (2014), Nutley, Powell, and Davies (2013), Chen, Donaldson, and Mark (2011), Donaldson, Christie, and Mark (2009), Patton (2008b), Scriven (2008), and Julnes and Rog (2007).
- For official statements advocating methodological pluralism, see American Evaluation Association (2003), "Scientifically Based Evaluation Methods," and European Evaluation Society (2007), "The Importance of a Methodologically Diverse Approach to Impact Evaluation."
- For alternative and mixed methods, see *Broadening the Range of Designs and Methods for Impact Evaluations* (Department for International Development, 2012), *Social Research Methods: Qualitative and Quantitative Approaches* (Bernard, 2013), *The SAGE International Handbook of Educational Evaluation* (Ryan & Cousins, 2009), and the alternative approaches for impact evaluation described in DFID (2012).
- In platinum standard references, the designation "platinum standard" has been used by Deborah Lowe Vandell, chair of the Department of Education at the University of California at Irvine, to describe an evaluation design that uses a range of data collection approaches (e.g., observations, interviews, surveys) to collect qualitative and quantitative data on program implementation and outcomes (*Evaluation Exchange*, 2010). See also "Toward a Platinum Standard for Evidence-Based Assessment by 2020," which incorporates case studies, comparative methods, triangulation, and alternative causal approaches in recognition of the wide array of goals and methodologies that are appropriate for assessing programs and policies in a dynamic and globalizing world (Khagram & Thomas, 2010); "Evaluating Outside the Box: Mixing Methods in Analyzing Social Protection Programmes" (Devereux & Roelen, 2013); and Scriven's (2008) article, which refers to a higher standard than gold for causal research, a platinum standard (p. 18).

From Core Strategies to Rich Diversity

> There's more than one way to skin a cat.
>
> —Old English proverb

Alternative proverbial meanings:

- More than one way to do almost anything
- Many ways to get something you want
- Many ways to make money
- More than one way to prepare a catfish (used in the southern American states, where catfish is often abbreviated to *cat*, a fish that is indeed usually skinned in preparing it for eating)
- A gymnastic feat that involves passing the feet and legs between the arms while hanging by the hands from a horizontal bar (as if turning an animal's skin inside out when separating the skin from the body)

> —Michael Quinion (1999)
> Lexicographer

Qualitative inquiry is not a single, monolithic approach to research and evaluation. The remainder of this chapter moves beyond the qualitative/quantitative debate to present a rich menu of alternative possibilities *within* qualitative research. While there are 12 core strategies that undergird any qualitative inquiry, as discussed in the last chapter (see Exhibit 2.1, pp. 46–47), there has emerged a proliferation of approaches *within* the qualitative paradigm. Frameworks for undertaking qualitative inquiry manifest great diversity. Different philosophical, epistemological, and theoretical perspectives and traditions have given rise to competing ways of engaging in qualitative inquiry. Those new to qualitative inquiry are often confused and even discombobulated by the diverse terminology and contested practices they encounter—phenomenology, hermeneutics, ethnomethodology, semiotics, heuristics, phenomenography. Such language! Exhibit 3.2

reproduces a letter of lamentation I received from a graduate student following publication of the first edition of this book, which did not include the current chapter. She felt lost in the arcane terminology. This chapter clarifies the major alternatives. The increasing

---

EXHIBIT 3.2 Plea for Help: Which Approach Is Right?

Dear Dr. Patton,

I desperately need your help. I am a graduate student in education planning to do my dissertation observing classrooms and teachers identified as innovative and effective. I want to see if they share any common approaches or wisdom that might be considered "best practices." I took this idea to one professor who asked me if I was proposing a phenomenological or Grounded Theory study. When I asked what the difference was, he said it was my job to find out. I've read about both but am still confused. Another professor told me I could do a qualitative study, but that asking about "best practices" meant that I was a positivist not a phenomenologist. Another grad student was told to "use a hermeneutic framing," but she's in a different department with a different topic. I'm a former school teacher and, I think, a pretty good observer and interviewer. I got very excited reading your book about the value of in-depth observations and interviewing, and that's where I got the idea for my dissertation, but now I'm being told I have to fit into one of these categories. Please tell me which one is right for my study. I don't care which one it is. I just want to get on with studying innovative classrooms. I feel lost and am on the verge of just doing a questionnaire where these philosophy questions don't seem to get asked. But if you can tell me which approach is right, I might still be able to do what I want to do. Help!!!!!!!!!!

* * * * *

This chapter is my full response to her poignant plea for help. I noted in my original personal response that her dilemma was all too common. Distinctions among frameworks are contentious; not everyone agrees about what the inquiry labels and traditions mean. What is certain is that there is no "right" approach any more than there is a "right" fruit—apples, oranges, passion fruit. What you eat is a matter of personal taste, availability, price, history, and preference. Each qualitative inquiry framework offers a different emphasis, framework, or focus.

diversity of qualitative inquiry frameworks has created both opportunity and confusion. This chapter aims to elucidate the opportunities and reduce the confusion.

In the remainder of this chapter, we shall look at how varying inquiry traditions emphasize different questions and how these particular emphases can affect the analytical framework that guides fieldwork and interpretation. Understanding the divergent theoretical and philosophical traditions that have influenced qualitative inquiry is especially important in the design stage, when the focus of fieldwork and interviewing is determined. Weaving together theory-based inquiry traditions and qualitative methods will reveal a rich tapestry with many threads of differing texture, color, length, and purpose.

Qualitative Inquiry Frameworks: Distinguishing Core Questions

There is no definitive way to categorize the various philosophical and theoretical perspectives that have influenced and that distinguish the types of qualitative inquiry. This chapter distinguishes theoretical perspectives by their foundational questions. A foundational or burning question, like the mythic burning bush of Moses, blazes with heat (controversy) and light (wisdom) but is not consumed (is never fully answered). Disciplines given birth by the mother of all disciplines, philosophy, can be distinguished by their core burning questions. For sociology, the burning question is the Hobbesian question of order: What holds society or social groups together? What keeps societal groups from falling apart? Psychology

asks why individuals think, feel, and act as they do. Political science asks, "What is the nature of power, how is it distributed, and with what consequences?" Economics studies how resources are produced and distributed.

Disciplines and subdisciplines reveal layers of questions. Biologists inquire into the nature and variety of life. Botanists ask how plants grow, while agriculturists investigate the production of food, and agronomists narrow their focus still further to field crops. To be sure, reducing any complex and multifaceted discipline to a single burning question oversimplifies the focus of that discipline. But what is gained is clarity about what distinguishes one lineage of inquiry from another. It is precisely that clarity and focus I shall strive for in identifying the burning questions that distinguish major lineages of qualitative inquiry. In doing so, I shall displease those who prefer to separate paradigms from philosophies and distinguish theoretical orientations from design strategies. For example, social constructivism may be viewed as a paradigm, ethnography may be considered a research strategy, and symbolic interactionism may be examined as a theoretical framework. However, distinctions between the paradigmatic, strategic, and theoretical dimensions within any particular approach are both arguable and arbitrary. Therefore, I have circumvented those distinctions by focusing on and distinguishing core inquiry questions as the basis for understanding and contrasting long-standing and emergent qualitative inquiry approaches. Exhibit 3.3 presents the core question for each inquiry framework reviewed here and serves as a guide for navigating through this chapter. We begin with ethnography's inquiry into culture.

| EXHIBIT 3.3 | Alternative Qualitative Inquiry Frameworks: Core Questions and Disciplinary Roots |

QUALITATIVE INQUIRY FRAMEWORK	DISCIPLINARY ROOTS	CORE QUESTIONS	PAGES IN THIS CHAPTER
1. *Ethnography*	Anthropology	What is the culture of this group of people? How does culture explain their perspectives and behaviors?	100–101
2. *Autoethnography*	Literary arts	How does my own experience of my culture offer insights about my culture, situation, event, and way of life?	101–104

(Continued)

(Continued)

QUALITATIVE INQUIRY FRAMEWORK	DISCIPLINARY ROOTS	CORE QUESTIONS	PAGES IN THIS CHAPTER
3. *Reality testing, foundationalist epistemologies* *Positivism* *Postpositivism* *Empiricism* *Objectivism*	Philosophy	What's the nature of the real world? What's true? How can one inquire objectively so that findings correspond to reality?	105–108
4. *Grounded theory*	Social sciences methodology	What theory, grounded in fieldwork, emerges from systematic comparative analysis so as to explain what has been observed?	109–111
5. *Realism*	Philosophy	What are the causal mechanisms that explain how and why reality unfolds as it does in a particular context?	111–114
6. *Phenomenology*	Philosophy	What is the meaning, structure, and essence of the lived experience of this phenomenon for this person or group of people?	115–118
7. *Heuristic inquiry*	Humanistic psychology	What is my experience of this phenomenon and the essential experience of others who also experience this phenomenon intensely?	118–120
8. *Social constructionism/ constructivism*	Sociology	How have the people in this setting constructed their reality? What is perceived as real? What are the consequences of what is perceived as real?	121–127
9. *Narrative inquiry*	Social sciences, literary criticism, literary nonfiction	How can this narrative (story) be interpreted to understand and illuminate the life and culture that created it? What does this narrative or story reveal about the person and world from which it came?	128–131
10. *Ethnomethodology*	Sociology	How do people make sense of their everyday activities so as to behave in socially acceptable ways?	132–133
11. *Symbolic interaction*	Social psychology	What common set of symbols and understandings have emerged to give meaning to people's interactions?	133–134
12. *Semiotics*	Linguistics	What signs (words and symbols) carry and convey special meaning in particular contexts? What are the implications of those signs for human beliefs, behaviors, and interactions?	134–136

QUALITATIVE INQUIRY FRAMEWORK	DISCIPLINARY ROOTS	CORE QUESTIONS	PAGES IN THIS CHAPTER
13. *Hermeneutics*	Linguistics, philosophy, literary criticism theology	What are the conditions and context under which a human act took place or a product was produced that makes it possible to interpret its meanings?	136–138
14. *Systems theory*	Interdisciplinary	How and why does this system function as it does? What are the system's boundaries and interrelationships, and how do these affect perspectives about how and why the system functions as it does?	139–144
15. *Complexity theory*	Theoretical physics, biological sciences, ecology	How can the emergent and nonlinear dynamics of complex adaptive systems be captured, illuminated, and understood?	145–151
16. *Pragmatism and generic qualitative inquiry*	Philosophy and program evaluation	What are the practical consequences and useful applications of what we can learn about this issue or problem?	152–156

Ethnography and Autoethnography

Culture as Central to Understanding Human Diversity

> The purpose of anthropology is to make the world safe for human differences.
>
> —Ruth Benedict (1887–1948)
> Ethnography pioneer
>
> Human culture exists in large measure to restrain the natural desires of the species.
>
> —David Brooks (2011, p. 15)
> *The Social Animal*

All 7 billion+ people alive on Earth today share a common female ancestor, dubbed *Mitochondrial Eve*, who lived in Africa about 140,000 years ago (Genomics, 2010; Oppenheimer, 2003). We all share her mitochondrial DNA. But we are not all the same. *Culture* explains the huge range of human diversity that cannot be attributed to genetic inheritance. Thus, I start this review of qualitative inquiry frameworks with ethnography, the study of culture.

Ethnography

> Core inquiry questions: *What is the culture of this group of people? How does culture explain their perspectives and behaviors?*

Ethnography, the primary method of anthropology, is the earliest distinct tradition of qualitative inquiry. The notion of culture is central to ethnography. *Ethnos* is the Greek word for "a people" or cultural group. The study of *ethnos,* then, or ethnography, is "devoted to describing ways of life of humankind . . . , a social scientific description of a people and the cultural basis of their peoplehood" (Vidich & Lyman, 2000, p. 38). Ethnographic inquiry takes as its central and guiding assumption that any human group of people interacting together for a period of time will evolve a culture. Culture is that collection of behavior patterns and beliefs that constitute "standards for deciding what is, standards for deciding what can be, standards for deciding how one feels about it, standards for deciding what to do about it, and standards for deciding how to go about

doing it" (Goodenough, 1971, pp. 21–22). The primary method of ethnographers is participant observation in the tradition of anthropology. This means intensive fieldwork in which the investigator is immersed in the culture under study. Ethnography becomes not just observation but *a way of seeing* (Wolcott, 2008).

Anthropologists have traditionally studied non-literate cultures in remote settings, what were often thought of as "primitive" or "exotic" cultures. As a result, anthropology and ethnographers became intertwined with Western colonialism, sometimes resisting imperialism in efforts to sustain native cultures and sometimes as handmaidens to conquering empires as their findings were used to overcome resistance to change and manage subjugated peoples.

Modern anthropologists apply ethnographic methods to the study of contemporary society to understand phenomena such as Information Age culture, the culture of poverty, school culture, the culture of addiction, intercultural marriages, youth culture, and the spread of disease (ethno-epidemiology), to give but a few of many examples. Ethnography is being used to study international relations and global diplomacy (Lie, 2013), corporate environments (Jordon, 2012), and consumer behavior (Sutherland & Denny, 2007). Understanding how group culture affects individual behavior can be a core ethnographic question; for example, what makes a 17-year-old girl decide to wrap a bomb around her body, walk into a supermarket, and detonate it, killing herself and an 18-year-old girl shopping there (Handwerker, 2009)? *Public ethnography* aims to help people speak to issues that affect them and their communities and to support change (Beaver, 2012).

The importance of understanding culture, especially in relation to change efforts of all kinds, is the cornerstone of "applied ethnography" (Chambers, 2000; Pelto, 2013). This can be seen in the ongoing reports of members of the Society for Applied Anthropology since its founding in 1941.

Since the 1980s, understanding culture has become central in organizational studies (Morgan, 1986, 1989; Pettigrew, 1983) and in much organizational development work, including major efforts to change the culture of an organization or program. Organizational ethnography has a distinguished history that can be traced back to the influential Hawthorne electric plant study that began in 1927 (Schwartzman, 1993). Programs develop cultures, just as organizations do. A program's culture can be thought of as part of the program's treatment. As such, the culture affects both program processes and outcomes. Improving a program, then, may include changing the program's culture (Patton, 2012a, pp. 144–147). An ethnographic evaluation would both facilitate and assess such change.

While traditionally ethnographers have used the methods of participant observation and intensive fieldwork to study everything from small groups to nation-states, what it means to "participate" or be in the "field" has changed with the advent of the Internet and social media. Nevertheless, whether doing ethnography in virtual space, a nonliterate community, a multinational corporation, or an inner-city school, what makes the approach distinct is the matter of interpreting and applying the findings from a *cultural perspective*. The *Ethnographer's Toolkit* (Schensul & LeCompte, 2010) offers an introductory overview of ethnographic methods, as does Fetterman's (2009) *Ethnography: Step by Step*, Walcott's (2010) primer *Ethnography Lessons*, Murchison's (2010) *Ethnography Essentials*, and O'Reilly's (2012a) *Ethnographic Methods*. Skinner (2013) focuses on an ethnographic approach to interviewing. As the myriad references available for guiding ethnography indicate, it is a well-established approach. Ethnography has also spawned a derivative but substantially different way of looking at and inquiring into culture: *autoethnography*.

Autoethnography

> Core inquiry question: *How does my own experience of my culture offer insights about this culture, situation, event, and way of life?*

We turn now from one of the earliest qualitative traditions, ethnography, to a recent and still emergent approach: autoethnography. Ethnography first emerged as a method for studying and understanding the *other*. It was their fascination with *exotic otherness* that attracted Europeans to study the peoples of Africa, Asia, the South Sea Islands, and the Americas. "The life world of the 'primitive' was thought to be the window through which the prehistoric past could be seen, described, and understood" (Vidich & Lyman, 2000, p. 46). In the United States, for educated, white, university-based Americans, the *others* were blacks, American Indians, recent immigrants, working-class families, and the inner-city poor (and, for that matter, anyone else not well educated, white, and university based). In recent times, when ethnography began to be used in program evaluations, the *other* became the program client, the student, the welfare recipient, the

Cross-cultural Perspective

©2002 Michael Quinn Patton and Michael Cochran

patient, the alcoholic, the homeless, the victim, the perpetrator, or the recidivist. In organizational studies, the *other* was the worker, the manager, the leader, the follower, and/or the board of directors. The *others* were observed, interviewed, and described, and their culture was conceptualized, analyzed, and interpreted. Capturing and being true to the perspective of those studied, what came to be called the *emic* perspective or the "insider's perspective," was contrasted with the ethnographer's perspective, the *etic*, or outsider's view. The etic viewpoint of the ethnographer implied some important degree of detachment or "higher" level of conceptual analysis and abstraction. To the extent that ethnographers reported on their own experiences as participant-observers, it was primarily methodological reporting related to how they collected data and how, or the extent to which, they maintained detachment. To "go native" was to lose perspective.

In the twenty-first-century postcolonial and postmodern world, the relationship between the observed and the observer has been called into question at every level. Postcolonial sensitivities raise questions about imbalances of power, wealth, and privilege between ethnographers and those they would study, including critical political questions about how findings will be used. Postmodern critiques and deconstruction of classic ethnographies have raised fundamental questions about how the values and cultural background of the observer affect what is observed, while also raising doubts about the desirability, indeed the possibility, of detachment. Then there is the basic question of how an ethnographer might study her or his own culture. What if there is no *other* as the focus of study but I want to study the culture of my own group, my own community, and my own organization and the way of life of people like me or people I regularly encounter, or my own cultural experiences?

These developments have contributed to the emergence of autoethnography—studying one's own culture and oneself as part of that culture—and its many variations. David Hayano (1979) is credited with originating the term *autoethnography* to describe studies by anthropologists of their own cultures. Goodall (2000) calls this *the new ethnography*: "creative narratives shaped out of a writer's personal experiences within a culture and addressed to academic and public audiences" (p. 9). In their extensive review of methods that focus on studying one's own culture and oneself as part of that culture to understand and illuminate a way of life, Ellis and Bochner (2000) cite a large number of phrases that have been used and conclude that "autoethnography has become the term of choice in describing studies and procedures that connect the personal to the cultural" (p. 740).

Carolyn Ellis, an autoethnography pioneer, describes it this way:

Autoethnography is an autobiographical genre of writing and research that displays multiple layers of consciousness, connecting the personal to the cultural. Back and forth autoethnographers gaze, first through an ethnographic wide-angle lens, focusing outward on social and the cultural aspects of their personal experience; then, they look inward, exposing a vulnerable self that is moved by and may move through, refract, and resist cultural interpretations. As they zoom backward and forward, inward and outward, distinctions between the personal and cultural become blurred, sometimes beyond distinct recognition. Usually written in first-person voice, autoethnographic texts appear in a variety of forms—short stories, poetry, fiction, novels, photographic essays, personal essays, journals, fragmented and layered writing, and social science prose. In these texts, concrete action, dialogue, emotion, embodiment, spirituality, and self-consciousness are featured, appearing as relational and institutional stories affected by history, social structure, and culture, which themselves are dialectically revealed through action, feeling, thought, and language. (Ellis & Bochner, 2000, p. 739)

In autoethnography, then, you use your own experiences to garner insights into the larger culture or subculture of which you are a part. Autoethnography

is both a perspective and a method (Chang, 2008; Muncey, 2010). As the *Handbook of Autoethnography* (Jones, Adams, & Ellis, 2013) demonstrates, great variability exists in the extent to which autoethnographers make themselves the focus of the analysis, how much they keep their role as social scientists in the foreground, the extent to which they use the sensitizing notion of culture, at least explicitly, to guide their analysis, and how personal their writing is. At the center, however, what distinguishes autoethnography from ethnography is self-awareness about and reporting of one's own experiences and introspections as a primary data source. Ellis describes this process as follows:

I start with my personal life. I pay attention to my physical feelings, thoughts, and emotions. I use what I call systematic sociological introspection and emotional recall to try to understand an experience I've lived through. Then I write my experience as a story. By exploring a particular life, I hope to understand a way of life. (Ellis & Bochner, 2000, p. 737)

HONORING ONE'S OWN EXPERIENCE

One writes out of one thing only—one's own experience. Everything depends on how relentlessly one forces from this experience the last drop, sweet or bitter, it can possibly give. This is the only real concern of the artist, to recreate out of the disorder of life that order which is art.

—James Baldwin (1990)
Introduction

Tony Adams (2011) used autoethnography to elucidate of the meaning of "the closet" through his lived experience in the gay community. Anthropologist Mary Catherine Bateson's (2000) autoethnographic description of teaching at a seminar at Spelman College in Atlanta includes detailed attention to the personal challenge she experienced in trying to decide how to engage with students of different ages, for example, calling older participants "elders." Aaron Turner (2000) has explored using one's own body as a source of data in ethnography—what he calls "embodied ethnography." Doing autoethnography in collaboration has emerged as an alternative to working entirely alone (Chang, Wambura, & Hernandez, 2013; Norris, Sawyer, & Lund, 2012). Performing autoethnographically offers new ways of communicating findings (Spry, 2011).

Such personal writing is controversial because of its "rampant subjectivism" (Crotty, 1998, p. 48). Some critics object to the way it blurs the lines between social science and literary writing. One sociologist told me angrily that those who want to write creative nonfiction or poetry should find their way to the English department of the university and leave sociology to sociologists. Laurel Richardson (2000b), in contrast, sees the integration of art, literature, and social science as precisely the point, bringing together the creative and critical aspects of inquiry. She suggests that what these various new approaches and emphases share is that "they are produced through *creative analytic practices*," which leads her to call "this class of ethnographies *creative analytic practice ethnography*" (p. 929). While the ethnographic aspect of this work is constructed on a foundation of careful research and fieldwork (p. 937), the creative element resides primarily in the writing, which she emphasizes is itself "a *method of inquiry*, a way of finding out about yourself and your topic" (p. 923). Exhibit 3.4 summarizes Richardson's criteria for judging the quality of an autoethnography.

Ellis warns that autoethnographic writing is hard to do:

It's amazingly difficult. It's certainly not something that most people can do well. Most social scientists don't write well enough carry it off. Or they're not sufficiently introspective about their feelings or motives, or the contradictions they experience. Ironically, many aren't observant enough of the world around them. The self-questioning autoethnography demands is extremely difficult. So is confronting things about yourself that are less than flattering. Believe me, honest autoethnography exploration generates a lot of fears and doubts—and emotional pain. Just when you think you can't stand the pain anymore, well, that's when the real work has only begun. Then there's the vulnerability of revealing yourself, not being able to take back what you've written or having any control over how readers interpret it. It's hard not to feel your life is being critiqued as well as your work. It can be humiliating. And the ethical issues. Just wait until you've written about family members and loved ones who are part of your story. (Ellis & Bochner, 2000, p. 738)

In my own major effort at autoethnographic inquiry (Patton, 1999), the struggle to find an authentic voice—authentic first to me, then to others who knew me, and finally to those who did not know

---

| **EXHIBIT 3.4** | **Criteria for Judging the Quality of an Autoethnography** |

1. *Substantive contribution.* Does this piece contribute to our understanding of social life? . . .

2. *Aesthetic merit:* . . . Does the use of creative analytic practices open up the text, invite interpretive responses? Is the text artistically shaped, satisfying, complex, and not boring?

3. *Reflexivity:* . . . How has the author's subjectivity been both a producer and a product of this text? Is there adequate self-awareness and self-exposure for the reader to make judgments about the point of view? . . .

4. *Impact:* Does this affect me? Emotionally? Intellectually? Does it generate new questions? Move me to write? Move me to try new research practices? Move me to action?

5. *Expression of a reality:* Does this text embody a fleshed out, embodied sense of lived experience? Does it seem a "true"—a credible account of a cultural, social, individual, or communal sense of the "real"?

SOURCE: Richardson (2000a, p. 254; 2000b, p. 937).

Confronting a Critic of Autoethnography

©2002 Michael Quinn Patton and Michael Cochran

me—turned what I thought would be a one-year effort into seven years of often painful, discouraging writing. And I was only writing about a 10-day period, a Grand Canyon hike with my son in which we explored what it means to *come of age*, or be initiated into adulthood, in modern society. My son started and graduated from college while I was learning how to tell the story of what we experienced together. To make the story work as a story and to make scattered interactions coherent, I had to rewrite the conversations that took place over several days into a single evening's dialogue, I had to reorder the sequence of some conversations to enhance the plotline, and I had to learn to follow the novelist's mantra to "show, don't tell," advice particularly difficult for those of us who make our living telling. More difficult still was revealing my emotions, foibles, doubts, weaknesses, and uncertainties. But once the story was told, the final chapter of the book, which contrasts alternative coming-of-age/initiation paradigms, emerged relatively painlessly.

Autoethnography integrates ethnography with personal story, a specifically autobiographical manifestation of the more general "turn to biographical methods in social science" that strive to "link macro and micro levels of analysis . . . [and] provide a sophisticated stock of interpretive procedures for relating the personal and the social" (Chamberlayne, Bornat, & Wengraf, 2000, pp. 2–3).

By opening this chapter with the contrast between ethnography and autoethnography, we have moved from the beginnings of qualitative methods in anthropological fieldwork more than a century ago, where the ethnographer was an outsider among exotically distinct nonliterate peoples, to the most recent manifestation of qualitative inquiry in the postmodern age of mass communications, where autoethnographers struggle to find a distinct voice by documenting their own experiences in an increasingly all-encompassing and commercialized global culture. Next, we look at how quantitatively oriented inquiry traditions have influenced qualitative inquiry.

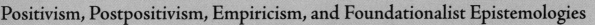

 Positivism, Postpositivism, Empiricism, and Foundationalist Epistemologies

Truth, Reality, and Objectivity

> A judgment is said to be true when it conforms to the external reality.
>
> —Thomas Aquinas (1225–1274)
> Philosopher and theologian

> Truth, as any dictionary will tell you, is a property of certain of our ideas. It means their "agreement," as falsity means their disagreement, with reality.
>
> —William James (1842–1910)
> Philosopher and psychologist

Reality–Testing Inquiry Frameworks: Foundationalist Epistemologies

Core inquiry questions: *What's the nature of the real world? What's true?*

How can one inquire objectively so that findings correspond *to reality?*

What these questions have in common is the presumption that there is a real world with verifiable patterns that can be observed and predicted—that reality exists and truth is worth striving for. Reality can be elusive, and truth can be difficult to determine, but describing reality and determining truth are the appropriate goals of scientific inquiry. Working from this perspective, researchers and evaluators seek methods that yield correspondence with the "real world"; thus, this is sometimes called a *correspondence* perspective.

> Correspondence theory: A statement is true if it describes reality accurately.
>
> —Epstein (2012, p. 20)

Reality-oriented inquiry and the search for truth have fallen on hard times in this skeptical postmodern age when honoring multiple perspectives and diverse points of view has gained ascendance in reaction against the oppressive authoritarianism and dogmatism that seemed so often to accompany claims of having found "truth." Yet many people, especially policymakers

and those who commission evaluation research, find it difficult to accept the notion that all explanations and points of view hold equal merit. Some people in programs seem to be helped more than others. Some students seem to learn more than others. Some claims of effectiveness are more plausible and have more merit than others. To test a claim of effectiveness by bringing data to bear on it, including qualitative data, is to be engaged in a form of reality testing that uses evidence to examine assertions and corroborate claims. How we can study, test, and come to know *reality* is a matter of considerable debate among philosophers of science and has led to distinctions among even those who share the belief that there is a fundamental reality to be discovered. I'll briefly review the reality-focused traditions in the philosophy of science and then examine the implications for qualitative methods.

Positivism and Postpositivism

Positivism, following August Comte (1798–1857), asserted that only verifiable claims based directly on experience could be considered genuine knowledge. Comte was especially interested in distinguishing the empirically based "positive knowledge" of experience from theology and metaphysics, which depended on fallible human reason and gullible belief. Logical positivism emerged in Austria and Germany in the early part of the twentieth century by combining Comte's emphasis on direct experience with a rigorous and systematic application of rational thought based in logic. Logical positivism subsequently came to be associated with philosophical efforts to specify the basic requirements for what could be considered truly *scientific* knowledge, which included (a) the search for universal laws through (b) empirical verification of logically deduced hypotheses with (c) key concepts and variables operationally defined and carefully formulated to (d) permit replication, falsification, and, ultimately, generalization across time and space. Thus, *real scientific knowledge* (as opposed to mere beliefs) is limited to what can be logically deduced from theory, operationally measured, and empirically validated. Though influential in the first half of the twentieth century, logical positivism has been "almost universally rejected" as a basis for social science inquiry because of its narrow definition of both science and knowledge (Campbell, 1999a, p. 132). The legacy of the fleeting influence of logical positivism is that the term lives on as an epithet hurled in paradigm debates and routinely used incorrectly though persistently. Shadish (1995b) argues that one would be hard-pressed to find any contemporary social scientist, philosopher, or evaluator who really adheres to the absolute and narrow tenets of logical positivism. Rather,

> the term has become the linguistic equivalent of "bad," a rhetorical device aimed at depriving one's opponent of credibility by name-calling. This is particularly true in the quantitative-qualitative debate where some qualitative theorists are fond of labeling all quantitative opponents as logical positivists, a fundamental but common "error." (p. 64)

Postpositivism, which takes into account the criticisms against and weaknesses of rigid positivism, now informs much of contemporary social science research, including truth- and reality-oriented qualitative inquiry. As articulated by the eminent methodologist Donald T. Campbell in his collected writings about and vision for an "Experimenting Society" (Campbell & Russo, 1999), postpositivism tempers positivism by positing that

- discretionary judgment is unavoidable in science;
- proving causality with certainty in human affairs is problematic;
- knowledge is inherently embedded in historically specific paradigms and is therefore relative rather than absolute; and
- all methods are imperfect, so multiple methods, both quantitative and qualitative, are needed to generate and test theory, improve understanding over time of how the world operates, and support informed policy making and social program decision making.

While they are modest in asserting what can be known with any certainty, postpositivists do assert that it is possible, using empirical evidence, to distinguish between more and less plausible claims, to test and choose between rival hypotheses, and to distinguish between "belief and *valid* [italics added] belief" (Campbell, 1999b, p. 151).

Logical Empiricism

Logical empiricism, another more moderate version of logical positivism, asserts that both natural and social sciences share the goal of describing, understanding, and explaining reality. Empiricists "maintain that objects exist regardless of whether we have any sense or consciousness of them—they are 'out there'—quite

apart from whether we know about them" (Epstein, 2012, p. 10). Schwandt (2001) helpfully explains the difference between empiricism as philosophy of science and empirical research as any inquiry that is data based: Empiricism is founded on

> the premise that knowledge begins with sense experience. . . . All forms of qualitative inquiry are empirical research to the extent that they deal in the data of experience. . . . To say that qualitative inquiry is empirical research, however, is not to say that it rests on the philosophy of empiricism. (pp. 67, 69)

Objectivism and Foundationalist Epistemologies

Objectivism is another philosophy that asserts "an independently existing world of objective reality that has a determinate nature that can be discovered" (Schwandt, 2001, p. 176).

Positivism, empiricism, and objectivism share a common foundation as "foundationalist epistemologies":

> These epistemologies assume the possibility and necessity of the ultimate grounding of knowledge claims. To be considered genuine, legitimate, and trustworthy (and not simply mere belief), knowledge must rest on foundations that require no further justification or interpretation. (Schwandt, 2001, p. 101)

Practice Implications for Qualitative Inquiry

What are the practical implications for qualitative inquiry of operating within a truth- and reality-oriented perspective? It means using the language and concepts of mainstream science to design naturalistic studies, inform data gathering in the field, analyze results, and judge the quality of qualitative findings. Thus, if you are a researcher or evaluator operating from a reality-oriented stance, you incorporate and use scientific terms like *validity, reliability, causality, generalizability,* and *objectivity.* You realize that completely value-free inquiry is impossible, but you worry about how your values and preconceptions may affect what you see, hear, and record in the field, so you wrestle with your values, try to make any biases explicit, take steps to mitigate their influence through rigorous field procedures, and discuss their possible influence in reporting findings. You may establish an "audit trail" to verify the rigor of your fieldwork and confirmability of the data collected because you want to minimize bias, maximize accuracy, and report impartially. In reporting, you emphasize the empirical findings—good, solid description and analysis—not your own personal perspective or voice, though you acknowledge that some subjectivity and judgment may enter in. You include triangulation of data sources and analytical perspectives to increase the accuracy and credibility of findings. Your criteria for quality include the "truth value" and plausibility of findings; credibility, impartiality, and independence of judgment; confirmability, consistency, and dependability of data; and explainable inconsistencies or instabilities. You may *generalize* case study findings, depending on the cases selected and studied, to generate or test theory, establish causality, or inform program improvement and policy decisions from the patterns established and lessons learned. In short, *you incorporate the language and principles of twentieth-century science into naturalistic inquiry and qualitative analysis* to convey a sense that you are dedicated to getting as close as possible to what is really going on in whatever setting you are studying. Realizing that absolute objectivity of the pure positivist variety is impossible to attain, you are prepared to admit and deal with imperfections in a complex, messy, and methodologically imperfect world, but you still believe that objectivity is worth striving for. As Kirk and Miller (1986) have asserted,

> Objectivity, though the term has been taken by some to suggest a naive and inhumane version of vulgar positivism, is the essential basis of all good research. Without it, the only reason the reader of the research might have for accepting the conclusions of the investigator would be an authoritarian respect for the person of the author. Objectivity is a simultaneous realization of as much reliability and validity as possible. Reliability is the degree to which the finding is independent of accidental circumstances of the research, and validity is the degree to which the finding is interpreted in a correct way. (p. 20)

The truth- and reality-testing framework informs and undergirds most research and evaluation supported by government agencies, international organizations, scientific institutes, philanthropic foundations, and university programs. It will not be called that or labeled thus because the assumptions and perspective of this framework are deeply embedded at the level of a paradigm. It is equated to good "science." Most quantitative research operates from this traditional scientific perspective and only adds counting and measuring as operational procedures. What is counted and measured, if valid and reliable, is real and true. The underlying philosophical premises are not made explicit, because they are basic assumptions, deeply shared, that don't need to be made explicit. They simply are. When this framework is transferred

from quantitative research into qualitative research, the methods of data collection are changed, but the inquiry framework is not.

SIDEBAR

SPEAKING TRUTH: A DIFFERENT VIEW OF REALITY

More than any other time in history, mankind faces a crossroads. One path leads to despair and utter hopelessness. The other, to total extinction. Let us pray we have the wisdom to choose correctly. I speak, by the way, not with any sense of futility, but with a panicky conviction of the absolute meaninglessness of existence, which could easily be interpreted as pessimism. It is not. It is merely a healthy concern for the predicament of modern man.

—Woody Allen (1981)
Filmmaker and humorist

What makes the underlying positivist framework hard to detect is that the qualitative literature, as represented by qualitative handbooks (e.g., Denzin & Lincoln, 2011), qualitative encyclopedias (e.g., Given, 2008), qualitative dictionaries (e.g., Schwandt, 2007), qualitative journals (e.g., *Qualitative Inquiry, Qualitative Research*), and major textbooks, presents a diverse array of inquiry frameworks, as represented by this chapter. Indeed, the diversity of inquiry frameworks is as impressive as it is confusing and contentious. But diversity of possibility is not diversity of practice. Most qualitative research and evaluation sponsored by official agencies and funders defaults to the truth- and reality-testing inquiry framework, expects objectivity to be valued and aimed for, and looks for findings that are presented as valid, reliable, generalizable, and, essentially, true in the common-sense meaning of *true*—that is, factual, credible, and supported by empirical evidence.

To emphasize this point, I close this review of truth- and reality-oriented qualitative inquiry with an excerpt from a medical journal in which health researchers are defending qualitative research to an audience known to be skeptical. Their approach is to associate qualitative research closely with accepted and credible forms of experimental research. Such a perspective epitomizes the truth- and reality-testing orientation:

> What, then, does the qualitative researcher do once he or she accomplishes a careful and trustworthy

understanding of the language and behavior of an individual human being? Here is where we rely on our positivist skills and methods. . . . Once we carefully examine and articulate that which we understand one human being to be doing, we attempt to collate the language and behavior of many human beings, so many that we might be able to test the relationships, for example, between setting and behavior or between age and hope. Included within the domains of qualitative science or narrative research, then, are efforts to generalize, to predict, and to relate initial states to outcomes. These efforts require the same evidence-based activities that are used in testing any hypotheses. (Charon, Greene, & Adelman, 1998, p. 68)

As Weiss and Bucuvalas (1980) found in their classic research, real-world decision makers as primary users of research and evaluation apply simple "truth tests" and "utility tests." Is it true? If so, can we do something with it? The truth- and reality-testing inquiry framework aims to pass the truth test first and foremost. As we'll see later, pragmatism picks up and focuses on the utility test. But, first, let's look at qualitative frameworks that focus on generating and testing theory to understand reality.

Our participants have been telling us valuable stories. We've learned so much about what we're doing right and where we can improve.

Did you get their emails? If so, we can survey them and get some real evidence.

freshspectrum.com

SOURCE: © Chris Lysy—freshspectrum.com

Grounded Theory and Realism

Theory Generation and Testing: Explaining How the World Works and Why It Works as It Does

> It is a capital mistake to theorize before one has data. Insensibly one begins to twist facts to suit theories, instead of theories to suit facts.
>
> —Arthur Conan Doyle
> *Sherlock Holmes*

> I am using the word *theory* as a scientist means it: a set of ideas so well established by observations and physical models that it is essentially indistinguishable from fact. That is different from the colloquial use that means "guess." To a scientist, you can bet your life on a theory. Remember, gravity is "just a theory" too.
>
> —Philip C. Plait
> Astronomer

Grounded Theory

Core inquiry question: *What theory, grounded in field-work, emerges from systematic comparative analysis so as to explain what has been observed?*

Influential qualitative theorist and methodologist Norman K. Denzin (1997a) observed some time ago that the "grounded theory approach is the most influential paradigm for qualitative research in the social sciences today" (p. 18). That assessment remains accurate, in my judgment, as it does in the judgment of those involved in further developing and applying this approach: "Grounded Theory is probably the most commonly used qualitative method, surpassing ethnography, and it is used internationally" (Morse et al., 2009). Why is it so influential? In part, it may be because it's an approach that comfortably incorporates and applies quantitative concepts like validity,

reliability, causality, and generalizability to qualitative inquiry, as just discussed in the previous section; in part because the procedures for grounded theory are systematized and prescriptive as well as contextually adaptable (Bryant & Charmaz, 2010); in part because Glaser and Strauss (1967), who developed the approach (before later taking their ideas in different directions), mentored a generation of students who have practiced widely and published prolifically; in part because grounded theory has won widespread acceptance as sufficiently rigorous to serve as an acceptable framework for academic dissertations precisely because of the emphasis on data-based theory; and, finally, in part because it unabashedly admonishes the researcher to strive for *objectivity*.

> It is important to maintain a balance between the qualities of objectivity and sensitivity when doing analysis. Objectivity enables the researcher to have confidence that his or her findings are a reasonable, impartial representation of a problem under investigation, whereas sensitivity enables creativity and the discovery of new theory from data. (Strauss & Corbin, 1998, p. 53)

Concern for theory development is often quite marked in the literature on qualitative methods. The classic writings of Glaser (1978, 2000), Strauss and Corbin (1998), Denzin (1978b), Lofland and Lofland (1984), Blumer (1969), Whyte (1984), and Becker (1970), to name but a few well-known qualitative inquiry advocates, take as a major contribution to knowledge the task of theory construction and verification. What distinguishes the discussion of theory in much of the qualitative methods literature is an emphasis on *inductive strategies* of theory development in contrast to theory generated by logical deduction from a priori assumptions.

> In contrasting Grounded Theory with logico-deductive theory and discussing and assessing their relative merits in ability to fit and work (predict, explain, and be relevant), we have taken the position that the adequacy of a theory for sociology today cannot be divorced from the process by which it is generated. Thus one canon for judging the usefulness of a theory is how it was generated—and we suggest that it is likely to be a better theory to the degree that it has been inductively developed from social research. . . . Generating a theory

from data means that most hypotheses and concepts not only come from the data, but are systematically worked out in relation to the data during the course of the research. *Generating a theory involves a process of research.* (Glaser & Strauss, 1967, pp. 5–6)

This theory–method linkage means that how you study the world determines what you learn about the world. Grounded theory depends on methods that take the researcher into and close to the real world, so that the results and findings are "grounded" in the empirical world. Herbert Blumer has offered a *lifting-the-veil* metaphor for explaining what it means to generate grounded theory by being immersed in the empirical world.

> The empirical social world consists of on-going group life and one has to get close to this life to know what is going on in it. The metaphor that I like is that of lifting the veils that obscure or hide what is going on. The task of scientific study is to lift the veils that cover the area of group life that one purposes to study. The veils are not lifted by substituting, in whatever degree, preformed images for first-hand knowledge. The veils are lifted by getting close to the area and by digging deep in it through careful study. Schemes of methodology that do not encourage or allow this betray the cardinal principle of respecting the nature of one's empirical world. . . . The merit of naturalistic study is that it respects and stays close to the empirical domain. (Blumer, 1978, p. 38)

All of the approaches to theory and research in this chapter use qualitative methods to stay grounded in the empirical world. Yet they vary considerably in their conceptualizations of what is important to ask and consider in elucidating and understanding the empirical world. While the phrase "grounded theory" is often used as a general reference to inductive, qualitative analysis, as an identifiable approach to qualitative inquiry, it consists of quite specific methods and systematic procedures (Glaser, 2000, 2001). In their book on techniques and procedures for developing grounded theory, Strauss and Corbin (1998, p. 13) emphasized that analysis is the interplay between researchers and data; so what grounded theory offers as a framework is a set of "coding procedures" to "help provide some standardization and rigor" to the analytical process.

> In all versions, grounded theory begins with very early close coding of collected data. The initial coding aims to ask what is happening in these data and invokes short analytic labels in the form of gerunds to identify specific processes and treat them

theoretically. . . . When researchers define a set of tentative codes, they use these codes to compare, sort, and synthesize large amounts of data. Throughout the process, grounded theorists write memos elaborating their codes by identifying their properties, the conditions under which the code arises, and comparisons with specific data and their codes. Memo writing (a) engages researchers with their data and emerging comparative analyses, (b) helps them to identify analytic gaps, (c) provides material for sections of papers and chapters, and (d) encourages researchers to record and develop their ideas at each stage of the research project. By writing successively more analytic memos, researchers raise the theoretical level of their work. (Charmaz & Bryant, 2008, p. 375)

Grounded theory emphasizes steps and procedures for connecting induction and deduction through the constant comparative method, comparing research sites, doing theoretical sampling (Morse, 2010), and testing emergent concepts with additional fieldwork. Grounded theory is meant to "build theory rather than test theory." It strives to "provide researchers with analytical tools for handling masses of raw data." It seeks to help qualitative analysts "consider alternative meanings of phenomena." It emphasizes being "systematic and creative simultaneously." Finally, it elucidates "the concepts that are the building blocks of theory." Glaser (1993) and Strauss and Corbin (1997) have collected together in edited volumes a range of grounded theory exemplars that include several studies of health (life after heart attacks, emphysema, chronic renal failure, chronically ill men, tuberculosis, and Alzheimer's disease), organizational head-hunting, abusive relationships, women alone in public places, selfhood in women, prison time, and characteristics of contemporary Japanese society.

While grounded theory has become widely thought of as an approach specific to qualitative inquiry, Glaser does not limit it in that way.

> Let me be clear. Grounded theory is a general method. It can be used on any data or combination of data. It was developed partially by me with quantitative data. It is expensive and somewhat hard to obtain quantitative data, especially in comparison to qualitative data. Qualitative data are inexpensive to collect, very rich in meaning and observation, and very rewarding to collect and analyze. So, by default, due to ease and growing use, grounded theory is being linked to qualitative data and is seen as a qualitative method, using symbolic interaction, by many. Qualitative grounded theory accounts for the global spread of its use. (Glaser, 2000, p. 7)

GROUNDED THEORY'S FOCUS ON ACTIONS AND PROCESSES

How grounded theorists use their methodological strategies differs from other qualitative researchers who study topics and structures instead of actions and processes. How we collect data and what we do with it matters. Research *actions* distinguish grounded theory from other types of qualitative inquiry. Grounded theorists . . . engage in the following actions:

1. Conduct data collection and analysis simultaneously in an iterative process.

2. Analyze actions and processes rather than themes and structure.

3. Use comparative methods.

4. Draw on data (e.g., narratives and descriptions) in service of developing new conceptual categories.

5. Develop inductive categories through systematic data analysis.

6. Emphasize theory construction rather than description or application of current theories.

7. Engage in theoretical sampling.

8. Search for variation in studied categories or process.

9. Pursue developing a category rather than covering a specific empirical topic. (Charmaz, 2011, p. 364)

The guidelines for grounded theory offered by Strauss and Corbin (1998, 1990), in the view of Charmaz (2000), "structure objectivist grounded theorists' work." These guidelines are didactic and prescriptive rather than emergent and interactive.

> Objectivist grounded theory accepts the positivistic assumption of an external world that can be described, analyzed, explained, and predicted: truth, but with a small *t*. . . . It assumes that different observers will discover this world and describe it in similar ways. (p. 524)

As a matter of philosophical distinctness, then, grounded theory is best understood as fundamentally objectivist in orientation, emphasizing disciplined and procedural ways of getting the researcher's biases out of the way but adding healthy doses of creativity to the analytic process. It is also important to note that grounded theory has evolved and has been subdivided into competing camps with different emphases that reflect "tussles, tensions, and resolutions" leading to a multiple-pathways genealogy of ground theory

(Morse, 2009, p. 17; see also Charmaz & Bryant, 2008; Clarke, 2005). We shall consider the analytic procedures of grounded theory in more detail in the chapter on analyzing qualitative data. As a general inquiry framework, I have included it in this chapter because of its emphasis on generating theory as the primary purpose of qualitative social science and its overt embrace of objectivity as a research stance. Janice Morse, moderator of the 2008 Banff Symposium on Grounded Theory, sponsored by the International Institute for Qualitative Methodology, summarized well the core of grounded theory's contributions in her opening remarks:

> To repeat the mantra one more time: Grounded theory is a way of thinking about data—processes of conceptualization—of theorizing from data, so that the end result is a theory that the scientist produces from data collected by interviewing and observing everyday life. (Morse, 2009, p. 18)

Realism

Core inquiry question: *What are the causal mechanisms that explain how and why reality unfolds as it does in a particular context?*

Realism emphasizes that truth is context dependent. Any given study "reveals its truths but in ways that are highly conditional and multiply contingent" (Pawson, 2013, p. 189). More specific to evaluation research, Ray Paulson argues in his *realist manifesto* "against absolute truths and against relative truths and for the idea that only partial truths emerge from evaluative inquiry. . . . Truth is thus accretive" (Pawson, 2013, p. 192).

Philosophically, realism begins with the premise that reality exists independent of perception. There is a real world. It is patterned (displays regularities) such that it can be studied, known, and explained. Joseph Maxwell (2012), in *A Realist Approach for Qualitative Research*, asserts,

> The idea that there is a real world with which we interact, and to which our concepts and theories refer, has proved to be a resilient and powerful one that has attracted increased philosophical attention following the demise of positivism. (p. 78; see also pp. 79–80, 88, 100)

Ray Pawson (2013) explains that from the perspective of realism,

> the world consists of three domains: i) the empirical, ii) the actual and iii) the real. As a first approximation, we

can say that the empirical domain consists of our experiences of the world, the observations and perceptions that clamour for our attention (fireworks exploding). The actual domain is realized at the level of pattern, when we see that our experiences are patterned in sequences of events (in the presence of flame, fireworks explode). The final and supreme level, the real, is discovered when we penetrate to an understanding of the subterranean mechanism that creates regular patterns of events (fireworks explode in the presence of a flame because of the chemical composition of gunpowder). (p. 68)

Theory about how the world works (or in evaluation, how a program or intervention works) informs realist inquiry, including especially a realist approach to choosing cases in qualitative research: Realist cases are purposefully chosen "to test and refine theory" (Emmel, 2013, p. 109). Realists aim to move beyond description to explanation. To do so requires theory. "Theory tells us where to but also what to look for. Theory provides explanations and so directs us to vital explanatory components within the world, their interrelationships and the things that bring about those interrelationships" (Pawson, 2013, p. 62). Realists view theoretical constructs as describing actual characteristics of a real world. For example, realists have no problem treating human thoughts, feelings, attitudes, intentions, and mental states as real even though they are not directly observable, a position denied by positivism.

WHAT IS REAL IN REALISM?

Concepts, meanings and intentions are as real as rocks; they are just not as accessible to direct observation and description as rocks. In this way they are like quarks, black holes, the meteor impact that supposedly killed the dinosaurs, or William Shakespeare are: we have no way of directly observing them, and our claims about them are based on a variety of sorts of indirect evidence.

—Joseph Maxwell (2012, p. 18)
A Realist Approach for Qualitative Research

A theoretical construct central to realism is the concept of *causality*. This is what Pawson is referring to in the quotation above as "the final and supreme level, the real, . . . when we penetrate the subterranean mechanism that creates regular patterns of events." *Mechanisms* means *causal mechanisms*. Causal mechanisms become "transfactual truths" that, by explaining how an effect is produced within a particular context,

can transcend the particular context to offer hypotheses for more general "law-producing mechanisms" (Pawson, 2013, p. 70).

Realism, in all of its brands and in relation to both the physical and social world, uses the notion of causal powers or underlying generative mechanisms as the seat of action and as the axis of explanation. (Pawson, 2013, p. 67)

Most realists see causality as a real phenomenon, an explanatory concept that is intrinsic to either the nature of the world or to our understanding of it. (Maxwell, 2012, p. 9)

A realist understanding of causality thus provides a philosophical justification for the ways in which qualitative researchers typically approach explanation. . . . There are three specific aspects of qualitative research that are supported by this understanding: the use of qualitative methods to identify causality in single cases, the essential role of context in causal explanation, and the legitimacy of seeing individuals' beliefs, values, motives, and meanings as causes. (Maxwell, 2012, p. 38)

The idea of social factors as causal mechanisms was introduced by eminent sociologist Robert K. Merton (1968a) and illustrated by the self-fulfilling prophecy. This occurs when people define a situation inaccurately by acting on their inaccurate understanding and then make what they perceived as true actually become true. Merton illustrated the self-fulfilling prophecy with an example, inspired by events of the Great Depression in the 1930s, of solvent banks becoming insolvent because of the spread of inaccurate rumors of their pending insolvency. As word spreads, wrongly, that a bank cannot meet the demand for cash withdrawals, people rush to withdraw their cash, and the ensuing panic actually makes the bank insolvent (Merton, 1968b).

This cycle of events—belief formation, acting on that belief, creating the situation that was believed to be true—is a very general and powerful mechanism. We can also see some of the conditions that trigger and maintain the mechanism: The definition of the situation must be public and communicable, it must be strong enough to lead to the actions through which the mechanism operates, and the results of these actions must change the situation in the direction of the—initially wrong—definition. (Gläser & Laudel, 2013, section 2)

The ascendance of realism as an inquiry framework for qualitative research is represented by the conceptual

FOUNDATIONS OF REALISM

Ontological realism—the belief that reality and its components exist independently of any consciousness.

Epistemological realism—the belief in the "knowability" of things, which presupposes that propositions about reality must be either true or false, regardless of which is which. (Epstein, 2012, p. 10)

journey of Matthew Miles and Michael Huberman. In the introduction to the first edition of their widely used and influential sourcebook *Qualitative Data Analysis* (1984), they stated modestly, "We think of ourselves as logical positivists who recognize and try to atone for the limitations of that approach. Soft-nosed logical positivists, maybe" (p. 19). They went on to explain what this means and, in so doing, provided a succinct summary of the reality-oriented approach to qualitative research:

> We believe that social phenomena exist not only in the mind but also in the objective world—and that there are some lawful and reasonably stable relationships to be found among them. . . . Given our belief in social regularities, there is a corollary: Our task is to express them as precisely as possible, attending to their range and generality and to the local and historical contingencies under which they occur.
>
> . . . We consider it important to evolve a set of valid and verifiable *methods* for capturing these social relationships and their causes. We want to interpret and explain these phenomena *and* have confidence that others, using the same tools, would arrive at analogous conclusions. (Miles & Huberman, 1984, pp. 19–20)

Ten years later, in their revised and expanded qualitative sourcebook, Miles and Huberman (1994) called themselves "realists" rather than logical positivists. Realism as a qualitative stance is clearly reality oriented, and much of the language quoted above remains in the revised edition. They acknowledge that knowledge is socially and historically constructed, but continue,

> Our aim is to register and "transcend" these processes by building theories to account for a real world that is both bounded and perceptually laden, and to test these theories in our various disciplines. . . .
>
> Our explanations flow from an account of how differing structures produced the events we observed.

We aim to account for events, rather than simply to document their sequence. We look for an individual or a social process, a mechanism, a structure at the core of events that can be captured to provide a *causal description* of the forces at work.

Transcendental realism calls both for causal explanation and for the evidence to show that each entity or event is an instance of that explanation. So we need not only an explanatory structure but also a grasp of the particular configuration at hand. That is one reason why we have tilted toward more inductive methods of study. (Miles & Huberman, 1994, p. 44)

Thus, realist philosophy (Baert, 1998, pp. 189–197; Bhaskar, 1975; Putnam, 1987, 1990) provides an inquiry framework for qualitative inquiry in both research and evaluation (Mark, Henry, & Julnes, 2000; Pawson & Tilley, 1997). It is beyond our purpose here to distinguish the various forms and schools of realism: agential realism, analytic realism, constructive realism, critical realism, depth realism, direct realism, emergent realism, ethnographic realism, experiential realism, innocent realism, metaphysical realism, naive realism, natural realism, perspectival realism, phenomenological realism, philosophic realism, scientific realism, subtle realism, transcendental realism, and universal realism (Altheide & Johnson, 2011; Cumpa & Tegtmeier, 2009; Madill, 2008; Mark et al., 2000; Maxwell, 2012; Miles & Huberman, 1994; Pawson, 2013; Schwandt, 2001). Rather than getting into the splitting of terminological hairs among realists themselves, for our purposes, it is their shared fundamental assumptions and inquiry framework that make realism a distinct perspective within which to undertake qualitative studies, namely, that entities and phenomena, both concrete and abstract, are real and constitute reality, whether we are aware of their existence or not. Because they exist and the world is patterned and knowable, reality can be theorized about, empirically studied, known, and explained. This includes what is sometimes called "commonsense realism":

> Realism presumes the existence of an external world in which events and experiences are triggered by underlying (and often unobservable) mechanisms and structures. Commonsense realism also gives standing to everyday experiences. It is antiformalist in the sense of not expecting logical, formal solutions to vexing problems such as the nature of truth. And it places a priority on practice and the lessons drawn from practice. . . . And as commonsense realists, we believe that although there is a world out there to be made sense of, the specific constructions and

construals that individuals make are critical and need to be considered. (Mark et al., 2000, pp. 15–16)

Qualitative methods are particularly useful in capturing and interpreting "the specific constructions and construals that individuals make" about how the world works. Realism places and interprets those constructions and construals within both a specific theoretical and a particular real-world context.

Realism covers a range of complex ontological and epistemological positions within which research can be conducted. The philosophy of science is in continual development, and new and modified realisms are under debate. Many of these are relevant to qualitative research. . . . Realist positions offer qualitative analysis grounding for research findings—sophisticated, complex, and compatible with the ethos of many qualitative methods. (Madill, 2008, p. 735)

Phenomenology and Heuristic Inquiry

Phenomenology

Core inquiry question: *What is the meaning, structure, and essence of the lived experience of this phenomenon for this person or group of people?*

> Pure phenomenology is the science of pure consciousness.
>
> —Edmund Husserl (1859–1938)
> German philosopher
>
> Phenomenology asks for the very nature of a phenomenon, for that which makes a some-"thing" what it is—and without which it could not be what it is.
>
> —Van Manen (1990, p. 10)

In Search of the Essence of Lived Experience

What the various phenomenological approaches share in common is a focus on exploring how human beings make sense of experience and transform experience into consciousness, both individually and as shared meaning. This requires methodologically, carefully, and thoroughly capturing and describing how people experience some phenomenon—how they perceive it, describe it, feel about it, judge it, remember it, make sense of it, and talk about it with others. To gather such data, one must undertake in-depth interviews with people who have directly experienced the phenomenon of interest; that is, they have "lived experience" as opposed to secondhand experience.

> Phenomenology aims at gaining a deeper understanding of the nature or meaning of our everyday experiences....

Anything that presents itself to consciousness is potentially of interest to phenomenology, whether the object is real or imagined, empirically measurable or subjectively felt. Consciousness is the only access human beings have to the world. Or rather, it is by virtue of being conscious that we are already related to the world. Thus all we can ever know must present itself to consciousness.

Whatever falls outside of consciousness therefore falls outside the bounds of our possible lived experience.... A person cannot reflect on lived experience

LIVED EXPERIENCE IN PHENOMENOLOGY

The term *lived experience* derives from the German *erlebnis—experience* as we live through it and recognize it as a particular type of experience. It could be argued that human experience is the main epistemological basis for many other qualitative research traditions, but the concept of lived experience possesses special methodological significance for phenomenology....

Our language can be seen as an immense linguistic map that names the possibilities of human lived experiences. The value of phenomenology is that it prioritizes and investigates how the human being experiences the world: how the patient experiences illness, how the teacher experiences the pedagogical encounter, how the student experiences a moment of success or failure, and how we experience novel ways of interacting with others and the world through computer mediated devices, social network technologies, new media, and so forth. Every lived experience (phenomenon) can become a topic for phenomenological inquiry. The phenomenological attitude keeps us reflectively attentive to the ways human beings live through experiences in the immediacy of the present that is only recoverable as an elusive past.

Phenomenology is interested in recovering the living moment of the now—even before we put language to it or describe it in words. Or to say this differently, phenomenology tries to show how our words, concepts, and theories always shape (distort) and give structure to our experiences as we live them. But the living moment of the present is always already absent in our effort to return to it.... We may identify and rate with empirical descriptors the nature and intensity of various forms of pain, but the actual moment of being struck by pain or the lingering discomfort of suffering pain somehow seem to be beyond words as we try to retrospectively appropriate the experience. These experiences can be described, but ultimately the meaning of the primal experience is beyond propositional discourse.

—Adams and Van Manen (2008, pp. 616–617)

while living through the experience. For example, if one tries to reflect on one's anger while being angry, one finds that the anger has already changed or dissipated. Thus, phenomenological reflection is not *introspective* but *retrospective*. Reflection on lived experience is always recollective; it is reflection on experience that is already passed or lived through. (Van Manen, 1990, pp. 9–10)

The phenomenon that is the focus of inquiry may be an emotion like loneliness, jealously, or anger; an experience like being pregnant, hiking the wilderness, or surviving cancer; or a relationship, a marriage, or a job. The phenomenon may be a program, an organization, or a culture. When used as a framework for program evaluation, phenomenology aims to capture the essence of program participants' experiences. For example, Breustedt and Puckering (2013) conducted in-depth interviews with vulnerable pregnant women in a prenatal parenting program to examine the effectiveness of psychological and practical techniques aimed at reducing anxiety and promoting well-being.

Phenomenology as a philosophical tradition was first applied to social science by the German philosopher Edmund H. Husserl (1913/1954) to study how people describe things and experience them through their senses. His most basic philosophical assumption was that we can only know what we experience by attending to perceptions and meanings that awaken our conscious awareness. Initially, all our understanding comes from sensory experience of phenomena, but that experience must be described, explicated, and interpreted. Yet descriptions of experience and interpretations are so intertwined that they often become one. Interpretation is essential to an understanding of experience, and the experience includes the interpretation. Thus, phenomenologists focus on how we put together the phenomena we experience in such a way as to make sense of the world and, in so doing, develop a worldview. There is no separate (or objective) reality for people. There is only what they know their experience is and means. The subjective experience incorporates the objective thing and becomes a person's reality, thus the focus on *meaning making* as the essence of human experience.

From a phenomenological point of view, we are less interested in the factual status of particular instances: whether something happened, how often it tends to happen, or how the occurrence of an experience is related to the prevalence of other conditions or events. For example, phenomenology does not ask, "how do these children learn this particular material?" but it asks, "What is the nature or essence of the experience of learning (so that I can now better understand what this particular learning experience is like for these children)?" (Van Manen, 1990, p. 10)

There are two implications of this perspective that are often confused in discussing qualitative methods. The first implication is that what is important to know is what people experience and how they interpret the world. This is the subject matter, the focus, of phenomenological inquiry. The second implication is methodological. The only way for us to really know what another person experiences is to experience the phenomenon as directly as possible for ourselves. This leads to the importance of participant observation and in-depth interviewing. In either case, in reporting phenomenological findings, "the essence or nature of an experience has been adequately described in language if the description reawakens or shows us the lived quality and significance of the experience in a fuller and deeper manner" (Van Manen, 1990, p. 10).

Phenomenological Reduction and Essence

There is one final dimension that differentiates a phenomenological approach: the assumption that *there is an essence or essences to shared experience*. These essences are the core meanings mutually understood through a phenomenon commonly experienced. The experiences of different people are bracketed, analyzed, and compared to identify the essences of the phenomenon, for example, the essence of loneliness, the essence of being a mother, or the essence of being a participant in a particular program. The assumption of essence, like the ethnographer's assumption that culture exists

Hi, I'm a graduate student, and I'm learning about a new type of research that focuses on the "lived experience" of different life events. I just have a few questions.

©2002 Michael Quinn Patton and Michael Cochran

Phenomenological Abduction

and is important, becomes the defining characteristic of a purely phenomenological study. *"Phenomenological research is the study of essences"* (Van Manen, 1990, p. 10). Phenomenologists are

> rigorous in their analysis of the experience, so that basic elements of the experience that are common to members of a specific society, or all human beings, can be identified. This last point is essential to understanding the philosophical basis of phenomenology, yet it is often misunderstood. On the other hand, each person has a unique set of experiences which are treated as truth and which determine that individual's behavior. In this sense, truth (and associate behavior) is totally unique to each individual. Some researchers are misled to think that they are using a phenomenological perspective when they study four teachers and describe their four unique views. A phenomenologist assumes a commonality in those human experiences and must use rigorously the method of bracketing to search for those commonalities. Results obtained from a phenomenological study can then be related to and integrated with those of other phenomenologists studying the same experience, or phenomenon. (Eichelberger, 1989, p. 6)

Getting at essence, methodologically, means undertaking phenomenological reduction systematically and rigorously in two ways:

> (1) The researcher has to bracket personal past knowledge and all other theoretical knowledge, not based on direct intuition, regardless of its source, so that full attention can be given to the instance of the phenomenon that is currently appearing to his or her consciousness, and

> (2) the researcher withholds the positing of the existence or reality of the object or state of affairs that he or she is beholding. The researcher takes the object or event to be something that is appearing or presenting itself to him or her but does not make the claim that the object or event really exists in the way that it is appearing. It is seen to be a phenomenon. . . .

> Husserl wants to limit our epistemological claim to the way that an event was experienced rather than to leaping to the claim that the event really was the way it was experienced. To make the latter claim is to make an existential or reality affirmation rather than staying within the confines of experience. To limit oneself to experiential claims is to stay within the phenomenal realm. (Giorgi, 2006, p. 355)

Through eidetic reduction and attention to imaginative variation, a phenomenologist aims to describe

> an essential finding that is intrinsically general. . . . I may observe a specific chair. But nothing prevents me from switching attitudes and taking a more general perspective toward the particular chair and seeing it as a cultural object designed to support the human body in the posture of sitting. That more general description is as true as the particular details of the chair that is taken as an example of a particular perception. There is no way to prevent one from assuming such a more general perspective. The switch results in eidetic findings which are intrinsically general. (Giorgi, 2006, p. 356)

In short, conducting a study with a phenomenological focus (i.e., getting at the essence of the experience of some phenomenon) is different from using phenomenology to philosophically justify the methods of qualitative inquiry as legitimate in social science research. Both contributions are important. But a phenomenological study (as opposed to a phenomenological perspective) is one that focuses on descriptions of what people experience and how it is that they experience what they experience. One can employ a general phenomenological perspective to elucidate the importance of using methods that capture people's experience of the world without conducting a phenomenological study that focuses on the essence of shared experience (at least that is my experience and interpretation of the phenomenon of phenomenology).

Qualitative researchers and evaluators using a phenomenological inquiry framework should immerse themselves in its historical evolution. Earlier, I noted the seminal work of Husserl (1913/1954; Husserl & Moran, 2012). Alfred Schutz's work (1977) helped establish phenomenology as a major social science perspective. Other important influences have been Merleau-Ponty (1962), Whitehead (1958), Giorgi (1971), and Zaner (1970). More recently, phenomenology has become a distinct approach to psychotherapy (Moustakas, 1988, 1995). But along the way and over time, the term *phenomenology* has become so popular and widely embraced that its meaning has become confused and diluted. It can refer to a philosophy (Husserl, 1967), an inquiry paradigm (Lincoln, 1990), an interpretive theory (Denzin & Lincoln, 2000a, p. 14), a social science analytical perspective (Harper, 2000, p. 727; Schutz, 1967, 1970), a major qualitative tradition (Creswell, 2012), or a research methods framework (Moustakas, 1994). Varying types of phenomenology complicate the picture even more: Transcendental, existential, and hermeneutic phenomenology offer different nuances of focus—the essential meanings of individual experience, the social construction of group reality, and the language and structure of communication, respectively (Schwandt, 2001, pp. 191–194). Phenomenological traditions in sociology and psychology vary in their unit of analysis, group or individual (Creswell, 1998, p. 53).

So, while there are certain basics to understand and master (Giorgi, 1985; Lewis & Staehler, 2010), applications are both diverse and contentious. See Exhibit 3.5 for advice to dissertation students about avoiding fundamental errors in phenomenological inquiries.

Heuristic Inquiry

Core inquiry question: *What is my experience of this phenomenon and the essential experience of others who also experience this phenomenon intensely?*

Heuristics is a form of phenomenological inquiry that brings to the fore the personal experience and insights of the researcher.

"Heuristic" research came into my life when I was searching for a word that would meaningfully encompass the processes that I believed to be essential in investigations of human experience. The root meaning of heuristic comes from the Greek word *heuriskein*, meaning to discover or to find. It refers to a process of internal search through which one discovers the nature and meaning of experience and develops methods and procedures for further investigation and analysis. The self of the researcher is present throughout the process and, while understanding the phenomenon

---

EXHIBIT 3.5   Six Common Errors in Phenomenological Dissertations: What Would Husserl Prescribe?

Distinguished phenomenological writer and teacher Amedeo Giorgi (1985) examined dissertations that "claimed to follow the phenomenological method" but did so inadequately "based upon phenomenology and the logic of research, and not personal biases" (Giorgi, 2006, p. 353). He applied what he considers classic and pure criteria based on Husserl's prescriptions as the founder of phenomenology. Below are six common errors he identified and critiqued:

1. Failure to ground the dissertation in the philosophy of phenomenology:

   Unlike certain other qualitative methods, such as grounded theory or certain narrative strategies, the phenomenological method requires a background in phenomenological philosophy which at certain times specifies criteria other than empirical ones. Phenomenology is not against empiricism, but it is broader than empirical philosophy. That is because its method interrogates phenomena which are not reducible to facts. (Giorgi, 2006, p. 354)

2. Citing different approaches to phenomenology without realizing that they contain conflicting perspectives that can be neither ignored nor simply combined:

   Some students seem to consider it a virtue to refer to as many phenomenologists as possible when discussing the logic and steps of the phenomenological method. However, . . . there are irreconcilable differences among them. Rather, the researcher has to choose one methodologist and stick with the

logic proposed by the methodologist. (Giorgi, 2006, p. 355)

3. Failure to mention *phenomenological reduction* at all, or if discussed, applying it incorrectly or inadequately: "If one is going to use a phenomenological method that is based upon the thought of Husserl, . . . then the phenomenological reduction has to be implemented. However, it seems that few practitioners get this part of phenomenology correct" (Giorgi, 2006, p. 355).

4. Failure to cite or use *imaginative variation* in order to discover essential characteristics (Giorgi, 2006, pp. 355–356).

5. Using inappropriate and "misguided" strategies to verify findings, like using independent judges and/ or presentation of the findings to the participants for them to verify (Giorgi, 2006, p. 357): Phenomenological findings can only be checked by phenomenological procedures, and the ordinary person is unlikely to know those procedures, "so the so-called verification by the participant has to remain dubious" (Giorgi, 2006, p. 358).

6. Failure to stay focused on contributing to generalizable knowledge:

   Should one want to study an aspect of a given individual's experience one has only to change the goal of the research and the method is equally applicable. The goal of the research is not then the phenomenon as such, but a specific phenomenon as experienced by John Doe. But such a strategy does not usually contribute to general psychological knowledge. (Giorgi, 2006, pp. 359–360) with increasing depth, the researcher also experiences growing self-awareness and self-knowledge. Heuristic processes incorporate creative self-processes and self discoveries. (Moustakas, 1990, p. 9)

There are two focusing or narrowing elements of heuristic inquiry within the larger framework of phenomenology. First, the researcher *must* have personal experience with and intense interest in the phenomenon under study. Second, others (coresearchers) who are part of the study must share *an intensity* of experience with the phenomenon. Heuristic inquiry focuses on intense human experiences, intense from the point of view of the investigator and coresearchers. It is the combination of personal experience and intensity that yields an understanding of the essence of the phenomenon. "Heuristics is concerned with meanings, not measurements; with essence, not appearance; with quality, not quantity; with experience, not behavior" (Douglass & Moustakas, 1985, p. 42).

The reports of heuristic researchers are filled with the discoveries, personal insights, and reflections of the researchers. Discovery comes from being wide open to the thing itself, a recognition that one must relinquish control and be tumbled about with the newness and drama of a searching focus, "asking questions about phenomena that disturb and challenge" (Douglass & Moustakas, 1985, p. 47).

The fundamental methods text on heuristic inquiry is by the primary developer of this approach, Clark Moustakas (1990). His classic works in this tradition include studies of loneliness (Moustakas, 1961, 1972, 1975) and humanistic therapy (Moustakas, 1995). Heuristic inquiry has strong roots in humanistic psychology (Maslow, 1956, 1966; Rogers, 1961, 1969, 1977) and in Polanyi's (1962) emphasis on personal knowledge, indwelling, and tacit knowledge (Polanyi, 1967). Polanyi (1967) explained tacit knowing as the inner essence of human understanding, what we know but can't articulate.

> Tacit knowing now appears as an act of indwelling by which we gain access to a new meaning. When exercising a skill we literally dwell in the innumerable muscular acts which contribute to its purpose, a purpose which constitutes their joint meaning. Therefore, since all understanding is tacit knowledge, all understanding is achieved by indwelling. (p. 160)

The rigor of heuristic inquiry comes from systematic observation of and dialogues with self and others, as well as in-depth interviewing of coresearchers. Heuristic inquiry is grounded in phenomenology but is different from it in four major ways:

1. Heuristics emphasizes connectedness and relationship, while phenomenology encourages more detachment in analyzing an experience.

2. Heuristics leads to reporting essential meanings and personal significance, while phenomenology emphasizes definitive descriptions of the structures of experience.

3. Heuristics concludes with a "creative synthesis" that includes the researcher's intuition and tacit understandings, while phenomenology presents a distillation of the structures of experience.

4. "Whereas phenomenology loses the persons in the process of descriptive analysis, in heuristics the research participants remain visible in the examination of the data and continue to be portrayed as whole persons. Phenomenology ends with the essence of experience; heuristics retains the essence of the person in experience" (Douglass & Moustakas, 1985, p. 43).

Identification of the *experiential essence* of the phenomenon that is the focus of heuristic inquiry emerges from systematic steps: initial engagement, immersion, incubation, illumination, explication, creative synthesis, and validation (see Exhibit 3.6).

What is important about heuristics for my purpose in this chapter—describing diverse inquiry frameworks for undertaking qualitative inquiry—is that heuristic research epitomizes the phenomenological emphasis on meanings and knowing through personal experience; it exemplifies and places at the fore the way in which the researcher is the primary instrument in qualitative inquiry; and it challenges traditional scientific concerns about researcher objectivity and detachment, much as autoethnography, described earlier in this chapter, does. The psychologist Dave Hiles, of De Montfort University, Leicester, has summarized heuristic inquiry through his own experiences of engaging in it to study phenomena such as near-death experiences, being a victim of serious crimes, and involvement in the helping professions, that is, counselors, nurses, social workers, and advocates.

1. Heuristic inquiry is a research process that is difficult to set any clear boundaries to, with respect to duration and scope. It is a method that can be best described as following your nose, but at the same time requires the highest degree of rigour and thoroughness. It is a method of inquiry that should not be undertaken lightly.

2. In heuristic inquiry, the research question chooses you, and invariably, the research question is deeply

EXHIBIT 3.6   Essential Elements and Stages of Heuristic Inquiry

Initial Engagement and Focus

Let an intense interest or matter of passion come into focus, something that compels the inquirer, that demands deeper understanding. The focus of inquiry becomes a burning question.

Immersion

The heuristic inquirer takes the question inside, dialogues with it through lived experience, being open to and enveloped by the inquiry, living it in mindful and alert reflection and introspection, and carrying the inquiry beyond awakened times into sleeping and even dreams. Anything and everything connected with the inquiry becomes raw material for reflection during immersion.

Incubation

Openness to tacit knowledge within and beyond consciousness facilitates illumination at a deeper, more subtle level of experience and self-knowledge; intuition comes to the fore, allowing seeing parts within a whole, integrating experience and, thereby, understanding.

Illumination

Indwelling, self-dialogue, and deep introspection eventually generate insight, what amounts to a breakthrough awakening and enhanced awareness that replaces old ways of knowing with a new understanding and experience of the phenomenon that is the focus of inquiry.

Explication

The bridge from illumination to explication is indwelling: a fully mindful, conscious, and deliberate process of going inward, gazing with unwavering attention and concentration into the lived experience of the focus of inquiry. This moves the inquiry into a period of systematic identification, organization, and elaboration of core themes that fully and comprehensively depict the essential nature of the phenomenon—the essence of lived experience deeply understood.

Creative Synthesis

Out of indwelling and deep reflection, out of solitude and meditation, out of illumination and explication of core themes emerges a sense of the whole that is communicated in the form of a creative synthesis, meanings and understandings expressed as a story, poem, narrative, drawing, painting, or other form of representation and communication. Ivana Djuraskovic wrote letters to her grandfather, her son, and herself as a form of creative synthesis (Djuraskovic & Arthur, 2010, pp. 1580–1583).

Validation

This involves moving back and forth between the data and insights generated during the inquiry, and the creative synthesis to check for meaning, accuracy, and validity in portraying the essence of the phenomenon of inquiry. This process moves outward by engaging with the experiences of others who have shared the experience of the phenomenon under inquiry and can provide additional data, insights, and feedback.

SOURCE: Moustakas (1990).

personal in origin. Indeed, it is my own experience that the research question has been a preoccupation of mine for at least thirty years, and probably much longer than that.

3. There is a very striking similarity between the methods of heuristic inquiry and the practice of counselling and psychotherapy, particularly with respect to the use of the "self." It is therefore a method of research that particularly resonates with inquiry into counselling and therapy-related issues.

4. Heuristic inquiry highlights the importance of working with the heuristic process of others, especially with the historical recordings of previous inquiry (especially spiritual texts). Indeed, it turns out that the works of writers, poets, artists, spiritual leaders, and scientists can all be usefully treated as the creative products of heuristic inquiry.

5. In the light of this last observation, it would seem that heuristic inquiry was probably the first research method adopted for psychological inquiry many, many centuries ago. It should really be regarded as the most ancient of methods, with a proven track record well before the advent of modern science and psychology. It is a method of inquiry that is desperately in need of being reinvented! (Hiles, 2001, section 3)

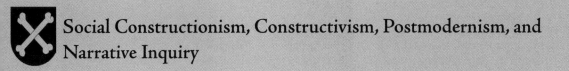

Social Constructionism, Constructivism, Postmodernism, and Narrative Inquiry

Understanding Reality as Perceived, Constructed, and Interpreted

> All things are subject to interpretation. Whichever interpretation prevails at a given time is a function of power and not truth.
>
> —Friedrich Nietzsche (1844–1900)
> German philosopher

> Lady, I do not make up things. That is lies. Lies are not true. But the truth could be made up if you know how. And that's the truth.
>
> —Lily Tomlin
> Comedienne and actress

Social Constructionism and Constructivism

> Core inquiry questions: *How have the people in this setting constructed their reality? What is perceived as real? What are the consequences of what is perceived as real?*

Social constructionism begins with the premise that the human world is different from the natural, physical world and therefore must be studied differently (Guba & Lincoln, 1990). Rocks don't think and feel. People do. Because human beings have evolved the capacity to interpret and *construct* reality—indeed, they cannot do otherwise—the world of human perception is not real in an absolute sense; for example, the sun is real but is "made up" and shaped by cultural and linguistic constructs, for example, the sun as a god.

Phenomenology seeks to discover and illuminate essence. In contrast, social constructionism asserts that things do not and cannot have essence because they are defined interpersonally and intersubjectively by people interacting in a network of relationships. A group of people can assign essence to a phenomenon and do so regularly, but essence does not then reside in the phenomenon but rather in the group that

EXHIBIT 3.7   Classic Social Construction Theorems

Thomas theorem: If men define situations as real, they are real in their consequences.

—William I. Thomas and Dorothy Swaine Thomas (1928, p. 572)
Sociologists

Historical note from the eminent sociologist Robert K. Merton (1995)

The designation *Thomas theorem* "does not, of course, adopt the term theorem in the strict mathematical sense (as, say, with the binomial theorem). It refers, rather, to an idea that is being proposed or accepted as sound, consequential, and empirically relevant." In proposing the Thomas eponym for both mnemonic and commemorative purposes (Merton, 1942), I had fastened on the term theorem rather than such less formidable terms as dictum, maxim, proposition, or aphorism in order to convey my sense that this was "probably the single most consequential sentence ever put in print by an American sociologist" (Merton, 1976, p. 174). (p. 380)

Mead theorem: If a thing is not recognized as true, then it does not function as true in the community.

—George H. Mead (1936, p. 29)
Sociologist

constructs and designates the phenomenon's essence. From a constructionist perspective, the notion of phenomenological essence is a social construction.

To say that the socially constructed world of humans is not physically real like the sun doesn't mean that it isn't perceived and experienced as real by people. Quite the contrary. A basic social psychological theorem is that *what is perceived as real is real in its consequences* (see Exhibit 3.7). So constructionists study the multiple realities constructed by different groups of people and the implications of those constructions for their lives and interactions with others. Any notion of "truth," then, becomes a matter of shared meanings and consensus among a group of people, not correspondence with some supposedly objective reality. Constructionist philosophy is built on the thesis of *ontological relativity*, which holds that all tenable statements about existence depend on a worldview and no worldview is uniquely determined by empirical or sense data about the world. Hence, two people can live in the same empirical world, even though one's world is haunted by demons, the other's by subatomic particles. Constructionist philosophy is *epistemologically subjectivist* in that the qualitative inquirer is also engaged in social construction as opposed to objectively depicting reality. (Exhibit 3.8 distinguishes between *constructionism* and *constructivism*, terms that are often used interchangeably because they share the same ontology and epistemology.)

How all-encompassing is the constructionist view? Michael Crotty (1998) asserts,

> It is not just our thoughts that are constructed for us. We have to reckon with the social construction of emotions. Moreover, constructionism embraces the whole gamut of meaningful reality. All reality, as

meaningful reality, is socially constructed. There is no exception. . . . The chair may exist as a phenomenal object regardless of whether any consciousness is aware of its existence. It exists *as a chair*, however, only if conscious beings construe it as a chair. As a chair, it too "is constructed, sustained and reproduced through social life." (pp. 54–55)

Elsewhere, Crotty (1998) uses the example of a tree:

> What the "commonsense" view commends to us is that the tree standing before us is a tree. It has all the meaning we ascribe to a tree. It would be a tree, with that same meaning, whether anyone knew of its existence or not. We need to remind ourselves here that it is human beings who have constructed it as a tree, given it the name, and attributed to it the associations we make with trees. It may help if we recall the extent to which those associations differ even within the same overall culture. "Tree" is likely to bear quite different connotations in a logging town, an artists' settlement and a treeless slum. (p. 43)

How, then, does operating from a constructionist perspective actually affect qualitative inquiry? While a realist would seek to both describe and explain a singular reality within some context by identifying and validating causal mechanisms, a constructionist would seek to capture diverse understandings and multiple realities about people's definitions and experiences of the situation. A singular or universal explanation would not be sought. In this way, constructionist qualitative inquiry honors *the idea of multiple realities*.

> One way in which the idea of multiple realities is honored is through the place of the researcher in the research. The researcher is not objective. It is expected that the researcher will bring biases to the research. Qualitative research addresses this by calling for the researcher to disclose his or her biases and explain how they may have affected the research. There also is acknowledgment that the researcher provides only one perspective.

> Those who participate in the study provide additional perspectives. Each person who participates in the study provides a different view on the topic being investigated. Each brings his or her own assumptions, beliefs, and perspective. This is commonly shown through the use of different quotes from participants. Quotes may show that participants do not agree on the topic and/or that they have had different experiences. Consistency is not necessarily the goal. Dissonant points of view are acceptable. Qualitative research frequently illustrates the complexity of multiple realities.

EXHIBIT 3.8   **Constructionism Versus Constructivism**

Michael Crotty (1998) makes a distinction between constructivism and constructionism, a distinction that illustrates how the process of social construction unfolds among scholars. It remains to be seen whether this distinction will gain widespread use since the two terms are so difficult to distinguish and easy to confuse.

> It would appear useful, then, to reserve the term *constructivism* for the epistemological considerations focusing exclusively on "the meaning-making activity of the individual mind" and to use *constructionism* where the focus includes "the collective generation [and transmission] of meaning." . . .

> Whatever the terminology, the distinction itself is an important one. Constructivism taken in this sense points out the unique experience of each of us. It suggests that each one's way of making sense of the world is as valid and worthy of respect as any other, thereby tending to scotch any hint of a critical spirit. On the other hand, social constructionism emphasizes the hold our culture has on us: it shapes the way in which we see things (even in the way in which we feel things!) and gives us a quite definite view of the world. (p. 58)

A third way in which multiple realities are honored is through flexible guidelines. Due to their emergent nature, qualitative methodologies tend to be malleable.... There is not a set procedure that must be followed. (Norum, 2008, p. 739)

Social Construction for Program Evaluation

Let's consider the implications of social construction for program evaluation. A constructionist evaluator would expect that different stakeholders involved in a welfare program (e.g., staff, clients, families of clients, administrators, and funders) would have different experiences and perceptions of the program, all of which deserve attention and all of which are experienced as real. The constructionist evaluator would capture these different perspectives through open-ended interviews and observations, and then examine the implications of the different perceptions (or multiple "realities"), but would not pronounce which set of perceptions was "right" or more "true" or more "real," as would a truth- and reality-oriented (postpositivist) evaluator. The constructionist evaluator would compare various program participants' perceptions with one another and with those of funders or program staff. The analysis could include interpreting the effects of differences in perceptions and experiences on attainment of the stated program goals. For example, those welfare recipients who perceive the program as an opportunity to get out of poverty may have different experiences and outcomes from those who view the program as merely meeting a requirement to keep receiving welfare payments. The inquiry could include looking at what shapes these different experiences and perceptions as reported by participants themselves.

In the evaluation of a diversity and inclusion project in the Saint Paul Schools, a major part of the design included capturing and reporting the experiences of people of color. Providing a way for African American, Native American, Chicano-Latino, and Hmong immigrant parents to tell their stories to mostly white, corporate funders was an intentional part of the design, one approved by those same white corporate funders. The final report was written as a "multi-vocal, multicultural" panorama that presented different experiences with and perceptions of the program's impacts rather than reaching singular conclusions. The medium of the report carried the message that multiple voices needed to be heard and valued as a manifestation of diversity (Stockdill, Duhon-Sells, Olson, & Patton, 1992). The parents and many of the staff were most interested in using the evaluation processes to make themselves heard by those in power. *Being heard was an end in itself*, quite separate from the use of findings to make decisions about the program.

CONSTRUCTED, EVER-CHANGING REALITY

Each moment of our lives, each thing we say, is equally true and false. It is true, because at the very moment we are saying it that is the only reality, and it is false because the next moment another reality will replace it.

—Charles Simic (2000, p. 11)
Poet

I go to the University of Colorado every year in order to [analyze] a film one shot at a time. We spend 10 hours going with a stop action projector or laser disc through an entire film. Every 10 years I do *La Dolce Vita*. And what I said the last time I did it was "When I saw this movie in 1962 for the first time, it represented everything that I dreamed would happen to me. When I saw it the second time, in 1972, it represented everything that I was stuck in. When I did it in 1982, it represented everything I had escaped from and the next time I do it, it will represent a pretty good movie that I remember from my youth." But the movie has not changed. It's a time capsule and as you see it, it evokes all of those feelings from all of those earlier viewings. The movie is the same. The emotions we bring to it are different. And if it's a good movie—because bad movies aren't worth seeing more than once, if once—but good movies are worth seeing over and over again because like good music or a good book, they read differently every time.

—Roger Ebert (1996)
Distinguished film critic

Primary Assumptions of Constructivism

Guba and Lincoln (1989) included among the primary assumptions of constructivism the following, whether for evaluation or research more generally:

- "Truth" is a matter of consensus among informed and sophisticated constructors, not of correspondence with objective reality.

- "Facts" have no meaning except within some value framework, hence there cannot be an "objective" assessment of any proposition.

- "Causes" and effects do not exist except by imputation....

- Phenomena can only be understood within the context in which they are studied; findings from one context cannot be generalized to another; neither problems nor solutions can be generalized from one setting to another....

• Data derived from constructivist inquiry have neither special status nor legitimation; they represent simply another construction to be taken into account in the move toward consensus. (pp. 44–45)

The idea that social groups like street gangs or religious adherents construct their own realities has a long history in sociology, especially the sociology of knowledge (e.g., Berger & Luckmann, 1967). Indeed, Restivo and Croissant (2008) proclaim that "social constructionism is a fundamental theorem of the sociological imagination" (p. 225). The ascendance of social constructionism as a major qualitative inquiry framework is evident throughout the comprehensive *Handbook of Constructionist Research* (Holstein & Gubrium, 2008).

But it wasn't until this idea of socially constructed knowledge was applied to scientists that constructionism became an influential methodological paradigm. No work has been more influential in that regard than Thomas Kuhn's classic *The Structure of Scientific Revolutions* (1970). Before Kuhn, most people thought that science progressed through heroic individual discoveries that contributed to an accumulating body of knowledge that got closer and closer to the way the world really worked. In contrast, Kuhn argued that tightly organized communities of specialists were the central forces in scientific development. Ideas that seemed to derive from brilliant individual scientific minds were actually shaped by and dependent on paradigms of knowledge that were socially constructed and enforced through group consensus. Rather than seeing scientific inquiry as progressing steadily toward truth about nature, he suggested that science is best seen as a series of power struggles between adherents of different scientific worldviews.

Kuhn emphasized the power of preconceived and socially constructed ideas to control the observations of scientists. He insisted that without the focusing effect of agreed-on constructs, investigators would not be able actually to engage in research. A fully "open" mind would not be able to focus on the details necessary to engage in "normal" science, that is, testing specific propositions derived from a theory or "scientific paradigm." What made his analysis so important and controversial was that it undermined the presumption that scientists were open-minded, value-free, and unencumbered by inherited ideas. Quite the contrary, he argued, scientists, like everyone else, operate within socially constructed preconceptions that, taken together, constitute a paradigm.

Socially constructed scientific paradigms determine what questions get priority, what methods of inquiry are preferred, what findings deserve attention, and which scientists get status and recognition. Kuhn applied to science the kind of language normally used to describe confrontations between opposing political and ideological communities, especially during revolutions. He argued (and showed with natural science examples) that communities of scientists, like ideological or religious communities, were organized by certain traditions that periodically came under strain when new problems arose that couldn't be explained by old beliefs. New explanations and ideas would then compete until the old ideas were discarded or revised, sometimes sweepingly. But the competition was not just intellectual. Power was involved. The leaders of scientific communities wielded power in support of their positions just as political leaders do. The assessment of Kuhn's contribution, three decades after his work first appeared, by Berkeley historian David Hollinger (2000) shows the importance of his analysis: "*The Structure of Scientific Revolutions* presented the strongest case ever made for the dependence of valid science on distinctly constituted, historically particular human communities" (p. 23).

Scientists constitute a critical case for social constructionism. If scientific knowledge is socially constructed and consensually validated, as opposed to consisting of empirical truths validated by nature, then surely all knowledge is socially constructed. "Accordingly, not only the social scientist but equally the natural scientist has to deal with realities that, as meaningful realities, are socially constructed. They are on equal footing in this respect" (Crotty, 1998, p. 55).

Social Construction and Postmodernism

Kuhn's analysis, though remaining controversial and heavily critiqued (e.g., Fuller, 2000), became a cornerstone of the postmodern skepticism about scientific truth. Postmodernism, Radavich (2001) asserts, has become "the most prevalent mode of thinking in our time.... Postmodernist discourse is precisely the discourse that denies the possibility of ontological grounding" (p. 6). In other words, no truth or "true meaning" about any aspect of existence is possible, at least not in any absolute sense; it can only be constructed. To understand constructionism and its implications for qualitative inquiry, a brief review of postmodernism may be helpful in that it has shaped contemporary intellectual discourse in both science and art.

The term "postmodernism" suggests something coming after modernism, which was the philosophical and aesthetic cultural dominant until approximately 1960, some postmodern theorists argue. Around 1960 there was a huge cultural shift, and the philosophical system we call postmodernism took control of life. The term has been described by a postmodern theorist, Jean-François Lyotard, as "incredulity toward meta-narratives," which means that the old overarching philosophical systems that shaped our beliefs, such as liberalism and a belief in progress, were jettisoned, and people created their own micro-narratives, which shaped their lives. The problem was that without anything to guide people, they became erratic, and there was a crisis of legitimation. What is the right thing to do if you have no philosophical or religious belief to guide you? (Berger, 2010, p. 100)

Belief in science as generating truth was one of the cornerstones of modernism inherited from the Enlightenment. Postmodernism attacked this faith in science by questioning its capacity to generate truth, in part because, like all human communications, it is dependent on language, which is socially constructed and, as such, distorts reality. Postmodernism asserts that no language, not even that of science, can provide a direct window through which one can view reality. Language inevitably and inherently is built on the assumptions and worldview of the social group that has constructed it and the culture of which it is a part. Thus, language does not and cannot fully capture or represent reality, a posture called the "crisis of representation" (Denzin & Lincoln, 2000a; Turner, 1998, p. 598). Translated into Kuhn's framework, scientific language and constructs are paradigm based and dependent.

It follows from this that the continuity of knowledge over time and across cultures is called into question. Modernism's faith in science included the assumption that knowledge increased over time and that such accumulation constituted continuous progress toward deeper and deeper truths. "Postmodernists argue that because there is not a truth that exists apart from the ideological interests of humans, discontinuity of

---

POSTMODERN EXISTENTIAL PHENOMENOLOGICAL INQUIRY

SIDEBAR

Should the starting point for the understanding of history be ideology or politics, or religion or economics? Should we try to understand a doctrine from overt content or from the psychological makeup and the biography of its author? We must seek an understanding from all these angles simultaneously; everything has meaning, and we shall find the same structure of being underlying relationships. All these views are true provided that they are not isolated, that we delve deeply into history and reach the unique core of existential meaning which emerges in each perspective.

—Merleau-Ponty (1908–1961)
French philosopher
The Phenomenology of Perception (1962, pp. xviii–xix)

What does this quote mean and what is its significance? . . . Perhaps this is a rhetorical question but, in case it is not, we may find that all of these aspects of historical underpinnings are not only helpful, but essential to the understanding of contemporary [postmodern] research. By the same token, when he asks if we should understand a doctrine from its overt content or from the psychological makeup and biography of its author, there is a recognition that both concrete and abstract conditions need to be met in order to gain clarity. Merleau-Ponty proceeds to suggest that we must seek understanding from a multiplicity of perspectives in order to gain a truer picture of the nature of anything that we are questioning.

The concept is elegant in its simplicity. Everything contains meaning, and we shall find that this is true of any question that we ask about anything. This is deceptively complex and serves to send us scurrying throughout the world gathering our information puzzle pieces in order to arrive at something that is asymptotic—always almost arriving but never quite achieving true or full understanding. Merleau-Ponty comes to our rescue by noting that all views are true provided that they are not isolated views. We must delve deeply into a multiplicity of perspectives in order to arrive at a core of meaning that each perspective provides.

Why is this important at all? It is important because, in essence, the meaning which emerges from each perspective speaks to some part of the primacy of our very existence. Thus, the search for meaning, for understanding, is therefore essentially an existential undertaking. Simply put, this quote (Merleau-Ponty, 1962) represents what [postmodern research] is all about.

—Cooper and White (2012, p. 2)
Qualitative Research in the Post-modern Era

knowledge is the norm, and a permanent pluralism of cultures is the only real truth that humans must continually face" (Turner, 1998, p. 599). Constructionism, then, consistent with postmodernism, is relativistic, meaning that knowledge is viewed as relative to time and place, never absolute across time and space, thus the reluctance to generalize and the suspicion of generalizations asserted by others.

Power comes into the picture here because, since views of reality are socially constructed and culturally embedded, those views that are dominant at any time and place will serve the interests and perspectives of those who exercise the most power in a particular culture. By exercising control over language, and therefore control over the very categories of reality that are opened to consciousness, those in power maintain their power and privilege (Ball, 2013).

Some postmodernists and constructionists question the possibility of ever finding and expressing true reality, even in the physical world, because language creates a screen between human beings and physical reality. "Vocabularies are useful or useless, good or bad, helpful or misleading, sensitive or coarse, and so on; but they are not 'more objective' or 'less objective' nor more or less 'scientific'" (Rorty, 1994, p. 57). This is because discovering the "true nature of reality" is not the real purpose of language; the purpose of language is to communicate the social construction of the dominant members of the group using the language.

Deconstruction

> There are things known and there are things unknown, and in between are the doors of perception.
>
> —Aldous Huxley (1894–1963)
> English novelist and essayist

The postmodern perspective, and its many variations—for postmodernism is not a unitary perspective (e.g., Constas, 1998; Pillow, 2000)—has given rise to an emphasis on *deconstruction*, which means to take apart the language of a text to expose its critical assumptions and the ideological interests being served. Perspective and power occur as hand in glove in postmodern critiques. Social constructions are presumed to serve someone's interests, usually those of the powerful. As Denzin (1991) has asserted with reference to deconstructing mass media messages, a critical analysis should "give a voice to the voiceless, as it deconstructs those popular culture texts which reproduce stereotypes about the powerless" (p. 153). Thus, deconstruction constitutes a core analytical tool of constructionists.

DECONSTRUCTION EXAMPLE: THE CONCEPT OF ENTREPRENEURSHIP

Ogbor (2000) has examined "mythicizing and reification in entrepreneurial discourse." The analysis discusses "the effects of ideological control in conventional entrepreneurial discourses and praxis." Ogbor argues that the concept of entrepreneurship is "discriminatory, gender-biased, ethnocentrically determined and ideologically controlled, sustaining not only prevailing societal biases, but serving as a tapestry for unexamined and contradictory assumptions and knowledge about the reality of entrepreneurs" (p. 605).

Deconstructing constructionism and constructivism, one finds a range of assumptions and positions from the radical "absolutely no reality ever" to a milder "Let's capture and honor different perspectives about reality." These positions share an interest in the subjective nature of human perceptions and skepticism about the possibility of objectivity. Reality-oriented researchers, in kind, are skeptical of the subjective knowledge of constructivism. How contentious is the debate? One gets some sense of the gulf that can separate these views from an assessment of postmodernism and constructivism by Rutgers mathematician Norman Levitt

TWO VIEWS OF OBJECTIVITY

Whether you are managing an organization or trying to manage your own life, your ability to cope successfully, to succeed—even survive—depends on *objectivity*. The Japanese word *sunao* translates roughly as "the untrapped mind." It describes the ability to see the world as it is, not as one wishes it to be.

—Arnold Brown and Edith Weiner
Management consultants
(Quoted in Safire & Safire, 1991, p. 158)

versus

If I were objective or if you were objective or if anyone was, he would have to be put away somewhere in an institution because he'd be some sort of vegetable.

—David Brinkley
Distinguished journalist
(Public television interview, December 2, 1968)

(1998) in an article titled "Why Professors Believe Weird Things": "Scientific evidence—which is to say the only meaningful evidence—cannot be neutralized by 'subjective knowledge,' which is to say bullshit" (p. 34). He goes on to comment on constructivism as a particular manifestation of postmodernism: "a particular technique for getting drunk on one's own words" (p. 35). Of course, that's Levitt's social construction of the scientific debate articulated from within his own paradigm assumptions about the nature of reality. The debate, especially about whether constructionism applies to both the social and the natural world, goes on (Hacking, 2000). What is not debatable is that social constructionism is an important inquiry framework for qualitative inquiry.

Further deconstructing the inquiry framework of social construction, one may find "inescapable connotations of manufacturing," as if people sat around and made things up. But

to say that people produce the world is not the same as saying that they are solipsists, that they are able to fashion the world according to their whims.... One cannot ordinarily produce an imaginary or nonsensical phenomenon and expect to be taken seriously. The mistake is to think of the process of production as one that is free of constraints when in fact it is a structure of constraints. (Watson & Goulet, 1998, p. 97)

Attending to the social construction of reality, then, points us not only to what is constructed but how it is constructed, and the very question of what it means to say it is constructed. Exhibit 3.9 summarizes the core contributions of social construction to qualitative inquiry. The premise that what is perceived as real is real in its consequences is at the heart of the social constructionist inquiry framework.

EXHIBIT 3.9 Ten Core Elements of the Social Construction Inquiry Framework

1. Inquire into the perceptions of reality shared by different groups of people. What is each group's perception and experience of reality?

2. Inquire into how what is perceived as real is real in its consequences. What are the behavioral and interaction consequences of what is perceived as real by people in a group?

3. Capture and honor multiple perspectives, and seek to understand multiple realities. What are the similarities and differences of perceptions about and experiences of reality within and between groups of people?

4. Inquire into the ways in which social groups come to share a worldview. How do shared social constructions emerge?

5. Be sensitive to the critical importance of context for understanding social constructions of reality. How does the social, political, cultural, economic, and environmental *context* illuminate social constructions?

6. Track the ways in which social constructions and worldviews change over time. How have the social constructions of a group (or comparative groups) evolved and changed?

7. Deconstruct language, and pay particular attention to the ways in which language as a social and cultural construction shapes, distorts, and structures perceptions of reality. What words are used to make sense of shared perceptions of reality?

8. Inquire into how power differentials affect and shape social constructions and perceptions of reality. Who exercises control over a group's shared meanings and benefits from particular perceptions of reality? Who is disadvantaged by those shared constructions?

9. Be reflective about how the subjective perspectives of researchers and evaluators affect their methods choices, shape findings, and inform their understandings of others and themselves. What social constructions and paradigm assumptions affect the inquirers' inquiry?

10. Inquire into how social constructions emerge out of and are affected by relationships, including the relationship between the inquirer and those being studied. How do the experiences of people being studied and their perceptions about the researcher or evaluator affect what is learned and how it is communicated (represented)?

Constructivist Credo

The Constructivist Credo is a set of foundational principles in the form of 150 propositional statements based on the work of Yvonna Lincoln and Egon Guba (2013), pioneers in articulating the constructivist paradigm for qualitative inquiry.

Narrative Inquiry

Core inquiry questions: *How can this narrative (story) be interpreted to understand and illuminate the life and culture that created it? What does this narrative or story reveal about the person and world from which it came?*

The narrative approach to qualitative inquiry focuses on stories. The beginning, middle, and end is the story. Narrative inquiry examines human lives through the lens of a narrative, honoring lived experience as a source of important knowledge and understanding (Clandinin, 2013). The more academic and scholarly term *narratology* was coined by Todorov in 1969 "in an effort to elevate the form 'to the status of an object of knowledge for a new science'" (Riessman, 1993, p. 1); but that term hasn't caught on. Narrative inquiry is the more general terminology and framework.

Personal narratives, family stories, suicide notes, graffiti, literary nonfiction, and life histories reveal cultural and social patterns through the lens of individual experiences. Stories organize and shape our experiences and also tell others about our lives, relationships, journeys, decisions, successes, and failures. Researchers and evaluators collect stories about formal education and planned program experiences and outcomes, as well as informal experiences of daily life, critical events, and life's many surprises as they unfold in particular situations, contexts, or circumstances. The stories are captured through interviews, transcribed, and analyzed for the patterns and themes they reveal, helping us learn about both specific individuals and society and culture more generally. The "narrative turn" in qualitative inquiry (Bochner, 2001) honors people's stories as data that can stand alone as pure description of experience, be worthy as a narrative documentary of experience (the core of phenomenology), or be analyzed for connections between the psychological, sociological, cultural, political, and dramatic dimensions of human experience.

Story and narrative are not the same thing, however. Narrative inquiry is more than just telling or capturing stories (Bell, 2002). One distinction is to treat the story as data and the narrative as analysis, which involves interpreting the story, placing it in context, and comparing it with other stories. Another distinction is that the story is what happened and the narrative is how the telling of what happened is structured and scripted within some context for some purpose. However, in data collection, the word *story* is more straightforward to ordinary people and carries a connotation different from that of narrative or another common label, case study. For example, in program evaluations, people may be invited to share their stories instead of being asked to participate in case studies or share narratives (Krueger, 2010). The central idea of narrative analysis remains, that stories offer especially translucent windows into cultural and social meanings when understood and analyzed as narratives.

In addition to stories collected through interviewing, written texts and rhetoric of all kinds can be fodder for narrative analysis, for example, the rhetoric of politicians or teachers (Graham, 1993). Policymakers construct narratives, for example, a "national strategic narrative" of U.S. military actions (Slaughter, 2011). Robert Coles, Harvard Professor of Psychiatry and Medical Humanities (his title offers interesting narratological fodder), has written about *The Call of Stories* (1989) as a basis for teaching, learning, and moral reflection. Michael White and David Epston's (1990) analysis *Narrative Means to Therapeutic Ends* looks at the power of stories in the lives of individuals and families and the connection between storytelling and therapy. They suggest that people have adjustment difficulties because the story of their life, as created by themselves or others, does not match their *lived experience.* They propose that therapists can help their patients by guiding them in rewriting their life stories. Another example of the applied potential of narrative inquiry is Columbia University Medical School's (2013) *Program in Narrative Medicine*, which involves practicing medicine with narrative understanding and competence, including awareness and appreciation of the complex, interactive narrative relationships among doctors, patients, colleagues, funders, policymakers, and the public.

In a quite different applied arena, Lisa Roberts (1997) used narratives to document, explore, and deepen understanding of people's experiences in museums. The title of her book, *From Knowledge to Narrative*, reverses the usual order. Narratives are usually analyzed to generate knowledge. Roberts looked at people's experiences with knowledge (museum exhibits) to see how that knowledge was translated into and communicated to others as narratives, as stories.

Analyzing Narrative Reality

The term *narrative reality* is used by Gubrium and Holstein (2009) "to flag the socially situated practice of storytelling, which would include accounts provided both within and outside of formal interviews." They continue,

> The term suggests that the contexts in which stories are told are as much a part of their reality as the

NARRATIVES: FACTS, MEMORIES, AND PERSPECTIVES

Social scientists have explored the impact of memory on people's stories. The evidence is that stories "are not simply recounted as a remembered set of events but are evaluated and interpreted, leading to developmental changes in understanding through the act of reminiscing" (Bold, 2012, p. 28).

The issue of memory is one that concerns those who seek facts about a situation, for example a road traffic incident in which identification of fault is necessary for insurance claims and prosecution processes. In professional workplace research, the identification of facts may be less important since the narrative data collected will most likely contribute to a collection of different perceptions and actions within the context. Each narrative adds to the others, creating a patchwork of information on the whole situation from which similarities and differences are identifiable. In research that supports the development of practice in the workplace, exploring the ways that participants understand each other's narratives is highly relevant since relationships between people rely on developing common understandings of events and situations. In addition, the impact of memory and previous experience on the telling or the understanding of a narrative serves to remind us that any narrative is a representation of actuality. Many researchers keep a research diary or an aide-memoire, which is a fluid narrative of their research process, maintaining chronological records and sustaining reflective thought. As they continue to reread and reflect further, the meaning they attach to the content will alter.

SOURCE: Bold (2012, p. 29).

and its circumstances designate as meaningful and important, together with the various purposes stories serve. (Gubrium & Holstein, 2009, p. 2)

Analyzing Narrative Reality . . . considers the circumstances, conditions, and goals of accounts. We are especially concerned with how storytellers work up, and what they do with, the accounts they present. We are particularly interested in the functions stories serve in different situations. Adapting familiar phrasing from philosopher Ludwig Wittgenstein (1953), storytellers not only tell stories, they *do* things with them. What speakers do with stories shapes their meaning for listeners, as well as the consequences of their communication. (Gubrium & Holstein, 2009, pp. xv–xvi)

Interpreting Narratives

Much of the methodological focus in narrative studies concerns the nature of interpretation, as in Norman Denzin's seminal qualitative works *Interpretive Biography* (1989a), *Interpretive Interactionism* (1989b), and *Interpretive Ethnography* (1997b). Interpretation of narrative poses the problem of how to analyze "talk and text" (Silverman, 2000). Tom Barone (2000) has entered into literary nonfiction to hone his interpretive aesthetic.

Budding Narratologist

©2002 Michael Quinn Patton and Michael Cochran

texts themselves. The narrative reality of the stories of WWII and Vietnam veterans, for instance, would not only include narrative elicited in research interviews, but extend to other circumstances in which such stories were incited and told. It could extend, for example, to occasions such as veterans' reunions, therapy groups, or the exaggerated "war stories" that some circumstances encourage. Stories not only are told in interviews, but they also wend their way through the lives of storytellers. They are boundless in that regard, are told and retold, with no definitive beginnings, middles, or ends in principle, even while a sense of narrative wholes and narrative organization is always in tow. This obliges us to analyze stories and storytelling in an extended framework, with the aim of documenting what the communication process

All great literature, I think, lures those who experience it away from the shores of literal truth and out into uncharted waters where meaning is more ambiguous. . . .

Ultimately, I erased the boundary between the realm of text which purports to give only the facts and that of the metaphor-laden story which dares (as Sartre once put it) to lie in order to tell the truth. But I did so haltingly, and not in a single confident stroke of understanding. Indeed, my insight came only gradually, after confronting a form of writing that aims to straddle the boundary between actual and virtual worlds, one foot firmly planted in each. These works are hybrids of textual species, essays/stories written in a literary style but shelved (curiously) in the nonfiction section of the library. (pp. 61–62)

Here, we have an example of personal narrative in the form of the narrative researcher's report of his journey into cross-genre exploration of the nature of textual interpretation. Later, he uses narrative as a method for exploring what it means to be a professional educational researcher, exploring the narratives researchers construct about themselves and the implications of those narratives for their relationships with nonresearchers (Barone, 2000, pp. 201–228).

W. Tierney (2000), in contrast, examines historical biographies and *testimonios* to explore interpretive challenges in using life histories in the postmodern age. His narrative analysis looks at the intersection of the interpreted purpose of a text, the constructed and interpreted "truth" of a text, and the *persona* of the author in text creation, all of which are called into interpretive question in the postmodern age.

Tedlock (2000) examines different genres of ethnography as constituting varying forms of narrative. She distinguishes life histories and memoirs from "narrative ethnography," a hybrid form that was created in an attempt to portray accurately the biographies of people in the culture studied and also to include ethnographers' own experiences in their texts. She assesses this as a "sea change in ethnographic representation" because it unsettled "the boundaries that had been central to the notion of a self studying an other" and replaced it with an "ethnographic interchange" between self and other within a single text (pp. 460–461).

Narrative analysis has also now emerged as a specific approach to studying organizations. As such, it takes at least four forms:

1. Organizational research that is written in story-like fashion (tales from the field)

2. Organizational research that collects organizational stories (tales of the field)

3. Organizational research that conceptualizes organizational life as story making and organizational theory as story reading (interpretative approaches)

4. A disciplinary reflection that takes the form of literary critique (Czarniawska, 1998, pp. 13–14)

The Story

The story is at the center of narrative analysis, whether they are stories of teaching (Preskill & Jacobwitz, 2000), stories of and by students (Barone, 2000, pp. 119–131), stories of participants in programs (Kushner, 2000), stories of fieldwork (Van Maanen, 1988), stories of relationships (Bochner, Ellis, & Tillman-Healy, 1997), or stories of illness (Frank, 1995, 2000). How to interpret stories and, more specifically, the texts that tell the stories is at the heart of narrative analysis.

SIDEBAR

NARRATIVE INQUIRY, SOCIAL MEDIA, AND GLOBAL INTERCONNECTEDNESS

Intentional use of narrative reinforces the world's trend toward networked connection.

Narrating experience integrates (1) the chaos of our conscious world and (2) subconscious tacit knowledge. . . . Although the linguistic and visual sound bites of text messages, Twitter, and Facebook verge on the point of banality or even meaninglessness for some people, others perceive them as condensed narrative platforms that afford the development of personalized and perpetually evolving narrative language; they create a multidimensional intertwine of social exchange. Over time, space, and context, these linguistic and visual condensations yield personalized iconic styles and become narrative tools by which to connect multiple generations and global participants.

—Nancy L. Pfahl (2011, p. 81)

Catherine Kohler Riessman (2008), one of the thought leaders in narrative analysis, cautions that high-quality data consist of detailed and lengthy accounts with many different crosscutting themes rather than short, succinct, and fragmented segments. Critical and illuminative contextual and structural elements of a story can be lost when it is coded narrowly and reduced to bite-size units. As we'll see in the chapter on interviewing, gathering such detailed and complex narrative accounts involves highly specific techniques and skillful interviewing (Andrews, Squire, & Taboukou, 2008; Wengraf, 2001).

Practice Stories as a Particular Narrative Niche

Practice stories explain something (e.g., how working-class children end up in working-class jobs) by describing how it develops over time as norms, rules, and organizational arrangements are acted on and adapted by people as part of their daily lives and in the context of their social lives (their communities, groups, networks, and families). One form of practice stories elucidates professional practice in a specific field, like program evaluation. More generally, practice stories help make sense of things as ongoing processes, both shaped by and shaping general patterns, arrangements, rules, norms, and other structures. Practice stories are a specific approach to capturing and reporting narratives developed by sociologist and ethnographer Karen O'Reilly (2012a, 2012b, 2013).

> Practice stories pay attention to people's feelings and emotions, their experiences and their free choices, but also to the wider constraints and opportunities within which they act. More than that, practice stories take account of how these different features of social life interact, and thereby how structures (like social classes, for example) get produced or reproduced. Theoretically, practice stories can draw from a wide range of social theory that comes under the description of structuration theory or practice theory. (O'Reilly, 2013, p. 1)

O'Reilly developed practice stories to address what she found at the heart of a great deal of sociological theory: an agency/structure dualism, that is, a tendency to perceive the agency of individual human actors as distinct and separate from social structures. In early sociology, she explains, the emphasis was on structures. In the work of Durkheim, for example, "social facts," such as laws, religion, education, and other more relational aspects such as norms, are depicted as having a force of their own on societies, independently of the individuals and their actions. In Marx's work, socioeconomic forces work independently to shape human societies.

> This approach was gradually challenged by a variety of schools of thought we might call "subjectivism,"

including symbolic interactionism, ethnomethodology, phenomenological sociology, social constructionism, and hermeneutics. These approaches emphasized the creative, reflexive and dynamic aspects of social life. They were especially influenced by a set of philosophical ideas known as interpretivism. For interpretivists, it is essential to see humans as actors in the social world rather than as re-acting like objects in the natural world. Sociology has now reached something of a consensus, in which it is impossible to ignore all that has been learned on either side and scholars are seeking ways to understand the ongoing interaction of structure and agency. These approaches tend to be called structuration or practice theories. (O'Reilly, 2013, p. 1)

Collecting Data to Tell Practice Stories

O'Reilly (2013) drew on these various theoretical perspectives to start thinking about how social life unravels in practice.

> Methodologically, this involves conceptualizing and learning about the wider structures that frame the practice of a given community or group. This can use both grand theorizing as well as learning practically about the smaller, local, relevant context. But abstract-level arguments should always be linked overtly to the analysis of the practice of daily life.

> Practice stories should reveal the complexity of people's daily lives, should try to understand cultural differences, and challenge stereotypes and typifications. But structures are both internal and external, so agents' perceptions can never be divorced from structural contexts.

> Furthermore, a researcher might understand aspects of the context not perceived by the agent. A methodology that enables a perspective beyond just that of the agent seems crucial. The gaze of the researcher cannot be restricted to the "present moment" or to "individual action." We have to study broader institutional systemic and structural frames and wider forces, but the focus is on how these are manifested in practice. (p. 1)

Narratives and stories reveal and communicate our human experiences, our social structures, and how we make sense of the world. The flow of a story—beginning, middle, end—is essentially a sense-making structure. As we interact with each other, we create and tell stories. Qualitative inquiry focused on capturing and analyzing those stories reveals our quintessentially social nature.

Ethnomethodology, Semiotics, Symbolic Interaction, and Hermeneutics

Human Beings as Quintessentially Social Animals

> Man is by nature a social animal; an individual who is unsocial naturally and not accidentally is either beneath our notice or more than human. Society is something that precedes the individual. Anyone who either cannot lead the common life or is so self-sufficient as not to need to, and therefore does not partake of society, is either a beast or a god.
>
> —Aristotle
> Philosopher, Ancient Greece
>
> We are primarily the products of thinking that happens below the level of awareness.
>
> —David Brooks (2011)
> *The Social Animal*

Ethnomethodology

Core inquiry question: *How do people make sense of their everyday activities so as to behave in socially acceptable ways?*

Ethnomethodology focuses on the ordinary, the routine, and the details of everyday life. Harold Garfinkel (1967) invented the term. While working with the Yale cross-cultural files, Garfinkel came across labels such as "ethnobotany," "ethnophysiology," and "ethnophysics." At the time, he was studying jurors. He decided that the deliberation methods of the jurors, or for that matter of any group, constituted an "ethnomethodology," wherein "ethno" refers to the "availability to a member of common-sense knowledge of his society as common-sense knowledge of the 'whatever'" (Turner, 1974, p. 16). For the jurors, this was their ordinary, everyday understanding of what it meant to deliberate as a juror. Such an understanding made jury duty possible.

Wallace and Wolf (1980) defined ethnomethodology as follows: "If we translated the 'ethno' part of the term as 'members' (of a group) or 'folk' or 'people,' then the term's meaning can be stated as: members' methods of making sense of their social world" (p. 263). Ethnomethodology gets at the norms, understandings, and assumptions that are taken for granted by people in a setting because they are so deeply understood that people don't even think about why they do what they do. It studies "the ordinary methods that ordinary people use to realize their ordinary actions" (Coulon, 1995, p. 2). Or put another way, ethnomethodologists study, reveal, and make explicit "*taken-for-granted methods of producing order that constitute sense*" (Rawls, 2008, p. 701). In this sense, revealing *common sense* is the sense-making focus of ethnomethodology.

Ethnomethodology has been particularly important in sociology.

> Ethnomethodology is, as the name suggests, a study of methods. It asks not why, but how. It asks how people get things done—how they transform situations or how they persevere situation "unchanged," step by step, and moment to moment. As its name also suggests, it is interested in ordinary methods, the methods of the people rather than their theorists. (Watson & Goulet, 1998, p. 97)

Ethnomethodologists elucidate what a complete stranger would have to learn to become a routinely functioning member of a group, a program, or a culture. To do this, ethnomethodologists conduct in-depth interviews and undertake participant observation. They stray from the nonmanipulative and unobtrusive strategies of most qualitative inquiry by employing "ethnomethodological experiments." During these experiments, the researcher "violates the scene" and disrupts ordinary activity by doing something out of the ordinary. A very simple and well-known example of such an experiment is turning to face the other people on an elevator instead of facing the doors. In such qualitative experiments, "the researchers are interested in what the subjects do and what they look to in order to give the situation an appearance of order, or to 'make sense' of the situation" (Wallace & Wolf, 1980, p. 278). Garfinkel (1967) offered a number of such experiments in his classic book, and those undertaking ethnomethodological studies have created a great many more (Rawls, 2002).

Ethnomethodologists also observe and document naturally occurring experiments where people are thrust into new or unexpected situations that require that they make sense of what is happening, "situations in which meaning is problematic" (Wallace & Wolf, 1980, p. 280). Such situations would include intake into a program, immigration clearance centers, the first few weeks in a new school or job, and major transition points or critical incidents in the lives of people, programs, and organizations.

In some respects, ethnomethodologists attempt to make explicit what might be called the group's tacit knowledge, to extend Polanyi's (1967) idea of tacit knowledge from the individual to the group. Ethnomethodologists get at a group's tacit knowledge by forcing it to the surface through disrupting violations of ordinary experience, since ordinary routines are what keep tacit knowledge at an unconscious level.

In short, ethnomethodologists "bracket or suspend their own belief in reality to study the reality of everyday life" (Taylor & Bogdan, 1984, p. 11). Elucidating the taken-for-granted realities of everyday life in a program or organization can become a force for understanding, change, and establishing a new reality based on the kind of everyday environment desired by people in the setting being studied. The findings of an ethnomethodological evaluation study would create a programmatic self-awareness that could facilitate program change and improvement. Ethnomethodology can also contribute to public policy by illuminating the everyday practices that emanate from policies, for example, what happens in courtrooms (Dupret, 2011; Jenkings, 2013).

Both ethnomethodology and symbolic interaction, the next inquiry framework we'll examine, focus on human beings as social animals. Sociologist Aaron Wildavsky summarizes this perspective succinctly:

> What matters to people is how they should live with other people. The great questions of social life are "Who am I?" (To what kind of a group do I belong?) and "What should I do?" (Are there many or few prescriptions I am expected to obey?). Groups are strong or weak according to whether they have boundaries separating them from others. Decisions are taken either for the group as a whole (strong boundaries) or for individuals or families (weak boundaries). Prescriptions are few or many indicating the individual internalizes a large or a small number of behavioral norms to which he or she is bound. (Wildavsky, quoted in Berger, 2010, p. 65)

Symbolic Interaction

Core inquiry question: *What common set of symbols and understandings have emerged to give meaning to people's interactions?*

Symbolic interaction, grounded in social psychology and most closely associated with George Herbert Mead (1934) and Herbert Blumer (1969), emphasizes the importance of meaning and interpretation as essential human processes. In this regard, symbolic interaction was a reaction against behaviorism and mechanical stimulus–response models in psychology. People create shared meanings through their interactions, and those meanings become their reality. Blumer articulated three major premises as fundamental to symbolic interactionism

> 1. People act toward things, including each other, on the basis of the meanings things have for them.
>
> 2. The meanings of things are derived out of social interaction with others.
>
> 3. The meanings of things are managed and transformed through an interpretive process that people use to make sense of and handle the objects that constitute their social worlds. (Snow, 2001, p. 367)

These premises led symbolic interactionists to qualitative inquiry as the only real way of understanding how people perceive, understand, and interpret the world. Only through close contact and direct interaction with people in open-mined, naturalistic inquiry and inductive analysis could the symbolic interactionist come to understand the symbolic world of the people being studied. Blumer was one of the first to use group discussion and interview methods with key informants. He considered a carefully selected group of naturally acute observers and well-informed people to be a real "panel of experts" about a setting or situation, experts who would take the researcher inside the phenomenon of interest, for example, drug use. As we shall see in the chapter on interviewing, group interviews and focus groups have now become highly valued and widely used qualitative methods.

Labeling theory—the proposition that what people are called has major consequences for social interaction—has been a primary focus of inquiry in symbolic interaction. For example, using a sample of 46 participants in a 12-step group, Debtors Anonymous, Hayes (2000) studied how people who are unable to manage their finances responsibly come to feel shame. In program evaluation, labeling theory can be applied to terms such as *drop-outs* and *at-risk* youth because language matters to staff and participants and can affect how they approach attaining desired outcomes (Hopson, 2000; Patton, 2000).

The Sociological Quarterly celebrated the 25th anniversary of symbolic interaction as an inquiry framework with a special issue in 1964. In that retrospective review, social psychologist Manford H. Kuhn (1964) characterized "the past twenty-five years [as] the age of inquiry in symbolic interactionism" (p. 63). As an example of such inquiry, he praised Howard Becker's (1953) classic study on becoming a marijuana user, asserting that "there is, in the literature, no more insightful account of the relation of the actor to a social object through the processes of communication and of self-definition, than Becker's account of becoming a marijuana user" (Kuhn, 1964, p. 73).

Though this perspective emerged in the 1930s, symbolic interactionists are showing that they can keep up with the times, for example, by applying their perspective to "cybersex" on the Internet. Computer-mediated human interaction constitutes

> a uniquely disembodied environment that potentially transforms the nature of self, body, and situation. Sex—fundamentally a bodily activity—provides an ideal situation for examining these kinds of potential transformations. In the disembodied context of on-line interaction both bodies and selves are fluid symbolic constructs emergent in communication and are defined by sociocultural standards. Situations such as these are suggestive of issues related to contemporary transgressions of the empirical shell of the body, potentially reshaping body-to-self-to-social-world relationships. (Waskul, Douglass, & Edgley, 2000, p. 375)

Perusing any issue of the journal *Symbolic Interaction* provides a sense of the wide range of contemporary issues being explored through application of symbolic interaction as an inquiry framework: understanding *sheng nu* ("leftover women"), the phenomenon of late marriage among Chinese professional women (To, 2013); "dark secrets and the performance of inflammatory bowel disease" (Thompson, 2013); and the social ecology of "normal adolescence"—insights from DiabetesCare (Allen, 2013).

For our purposes, the importance of symbolic interactionism to qualitative inquiry is its distinct emphasis on the importance of symbols and the interpretative processes that undergird interactions as fundamental to understanding human behavior. For program evaluation, organizational development, and other applied research, the study of the original meaning and influence of symbols and shared meanings can shed light on what is most important to people, what will be most resistant to change, and what will be most necessary to change if the program or organization is to move in new directions. The subject matter and methods of symbolic interactionism also emphasize the importance of paying attention to how particular interactions give rise to symbolic understandings when one is engaged in changing symbols as part of a program improvement or organizational development process.

NAKED VERSUS NUDE: A SYMBOLIC INTERACTION ILLUMINATION

To be naked is to be oneself. To be nude is to be seen as naked by others and yet not recognized for oneself. A naked body has to be seen as an object in order to become a nude. (The sight of it as an object stimulates the use of it as an object.) Nakedness reveals itself. Nudity is placed on display.

—John Berger (1972)
Ways of Seeing
(Quoted in Berger, 2010, p. 31)

Semiotics

Core inquiry questions: *What signs (words and symbols) carry and convey special meaning in particular contexts? What are the implications of those signs for human beliefs, behaviors, and interactions?*

Humans are distinctively symbol-generating and sign-using animals. As social animals, we communicate with each other through signs and symbols. Semiotics as an inquiry framework blends linguistics and social science by analyzing the creation, use, and meanings of signs. Umberto Eco (1976) provided the most basic classic definition: "Semiotics is concerned with everything that can be taken as a sign" (p. 7). And a sign is "anything that stands for something else" (Chandler, 2007, p. 2). Semiotics includes studying the rules and forms of language as well as the relationship between language and human behavior (Manning, 1987). Walker Percy (1990) has explained the insights generated by studying language as opposed to studying the physics of a solar eclipse or the biological basis of rat behavior, namely, that *language brings us face to face with the nature and essence of being human because humans are distinctively sign-using and symbol-generating animals* (pp. 150, 243). Semiotics offers an inquiry framework for "analyzing talk and text" (Silverman, 2000, p. 826) and studying "organizational symbolism" (Jones, 1996), or communications and media (Moore, 2011). It can include attention to nonverbal signals, for "ninety percent of emotional communication is nonverbal. Gestures are an unconscious language that we use to express not only our feelings but to constitute them" (Brooks, 2011, p. 12).

A building simultaneously provides shelter and sends a message through its architecture, imagery, materials, and location. What we wear not only serves straightforward utilitarian purposes (warmth, protection, hiding our nakedness) but also communicates something, often a great deal, about who we are. The location of an office—upper floor, corner with window versus middle of the room windowless cubicle—speaks volumes. An evaluation report contains findings, but its thickness, cover graphics, print quality, and binding communicate meanings beyond its content.

What we say and how we look (nonverbal communications, body language) can send different messages. Smells stimulate olfactory sensations and send messages: perfume, flatulence, curry, sweat—these are signs that carry different meanings. Political campaigns are rife with signs. Terrorists send messages through acts of terror. Mothers and fathers send signs of love. A practical problem of semiotics is the search for a universal symbol for danger. Exhibit 3.10 offers one candidate, but it has to be evaluated in many different contexts among many different peoples to determine its universal meaning. Or perhaps you'd be more interested in a universal sign of safety. Or love.

Qualitative methods are used in semiotic inquiries to observe how people create signs, how we behave in the presence of various signs, and how to interview people about the meanings they attach to signs. Exhibit 3.11 presents a summary of fundamental assumptions about qualitative semiotic inquiry.

EXHIBIT 3.10 **Semiotic Inquiry: Might This Be a Universal Sign of Danger?**

SOURCE: Wikimedia Commons (2012).

EXHIBIT 3.11 **Semiotic Inquiry Framework: Five Basic Assumptions**

It seems a strange thing, when one comes to ponder over it, that a sign should leave its interpreter to supply part of its meaning; but the explanation of the phenomenon lies in the fact that the entire universe—not merely the universe of existents, but all that wider universe, embracing the universe of existents, as a part, the universe which we are all accustomed to refer to as "the truth"—that all this universe is perfused with signs, if it is not composed exclusively of signs.

—Charles S. Peirce
(Quoted in Berger, 2010, p. 16)

Semiotic research is motivated by a set of fundamental assumptions about research and the world:

1. ***The world can be read just like a text:*** . . . We are interested in how the world can come together as a coherent and meaningful whole and in the implicit and explicit codes that allow this to happen.

2. ***The world is always talking to us:*** [Semiotist pioneer Charles] Peirce once insisted that the world was perfused with signs. Since we are always searching for meaning in order to avoid genuine doubt, we are always drawing guidance and affirmation from the world. Therefore, there is usually a meditative or even contemplative dimension to semiotic research where we allow the order of the world to come to us on its own terms.

3. ***The world is rich in meaning:*** Semioticians go beyond the notion that we make all the meaning we find, either singly or collectively. We share Peirce's sense that there is some form of logical order to the world, and that there are things in the world that are meaningful on their own terms, and not necessarily on ours. Stepping outside our own personal or cultural frames of meaning can be one of the most exciting forms of semiotic research.

(Continued)

(Continued)

4. *The presence or absence of specific things often serve as clues to the nature of reality:* One good way to think of semiotic researchers is to compare them to detectives. Signs are clues and symptoms and omens of things. Our job is to find them in intelligent and creative ways.

5. *Although semioticians assume that the world can be understood, we do not necessarily assume* *that it can always be understood easily in human terms:* . . . Simply put, semioticians try not to impose order on things, unless they are things we are supposed to impose order upon. Most of the time, we watch, learn, synthesize, and organize. If we are patient, we are often rewarded with startling, and even beautiful, insights.

SOURCE: Shank (2008, pp. 806–810).

Hermeneutics

Core inquiry question: *What are the conditions and context under which a human act took place or a product was produced that makes it possible to interpret its meanings?*

Understanding How and Why Context Matters So Much

> Without context, words and actions have no meaning at all. This is true not only of human communication in words but also of all communication whatsoever, of all mental process, of all mind, including that which tells the sea anemone how to grow and the amoeba what he should do next.
>
> —Gregory Bateson (1988, p. 15)
> Anthropologist
> *Mind and Nature*

Qualitative inquiry includes document analysis. Anything written can be a source of data for qualitative analysis: evaluation reports, meeting minutes, client case files, political speeches, news accounts of current events, editorials, blogs, Facebook posts, and on and on. The modern age is an age of texts. Everything and anything from historical writings to real-time Twitter posts can be material for qualitative study, seeking meanings, patterns, and themes that provide insights into how those who wrote what is being analyzed understood and interpreted the world. While observation, interviews, and fieldwork

dominate qualitative methods, analysis of documents and text is taking on increased importance in an information and communications age when *texting* has emerged as a verb. Hermeneutics examines "how we read, understand, and handle texts, especially those written in another time or in a context different from our own" (Thiselton, 2009, p. 1). In the real-time world of the Internet, texts "written in another time" can mean just minutes ago, and texts "written in a context different from our own" can mean anywhere else in the world, from down the street to the other side of the globe.

In this brief (or not so brief, depending on your perspective) excursion through the varieties of qualitative inquiry, hermeneutics is yet another approach that can inform qualitative inquiry. It can also help put all the other inquiry frameworks in this chapter in perspective, in that it reminds us that what something means depends on the cultural context in which it was originally created as well as the cultural context within which it is subsequently interpreted. This reminds us that each of the inquiry frameworks reviewed in this chapter emerged from a particular context to address specific concerns at that time—and each of these inquiry frameworks have generated deviations, adaptations, debates, new applications, and departures from the original, as time has moved on and new contexts have emerged with new issues and old issues addressed in new ways. As we consider the contributions of these diverse inquiry frameworks to the study of current interests and concerns, we do so in a different historical, scholarly, political, and cultural context.

Hermeneutic philosophy, first developed by Frederich Schleiermacher (1768–1834) and applied to human science research by Wilhelm Dilthey (1833–1911) and other German philosophers, focuses on the problem of *interpretation*. Hermeneutics provides a theoretical framework for interpretive understanding, or meaning, with special attention to context and original purpose. The term *hermeneutics* derives from the Greek word *hermeneuein*, meaning "to understand" or "to interpret."

> There is an obvious link between *hermeneuein* and the god Hermes. Hermes is the fleet-footed divine messenger (he has wings on his feet!). As a messenger, he is the bearer of knowledge and understanding. His task is to explain to humans the decisions of the gods. Whether *hermeneuein* derives from Hermes or the other way round is not certain. (Crotty, 1998, p. 88)

In modern usage, hermeneutics offers a perspective for interpreting legends, stories, and other texts, especially biblical and legal texts. Hermeneutics constitutes an interpretive theory (Porter & Robinson, 2011) that includes guidance for analyzing any kind of narrative data and text. To make sense of and interpret a text, it is important to know what the author wanted to communicate, to understand intended meanings, and to place documents in a historical and cultural context (Palmer, 1969). Following that principle, hermeneutics itself must be understood as part of a nineteenth- and twentieth-century

> broad movement away from an empiricist, logical atomistic, designative, representational account of meaning and knowledge.... Logical empiricism worked from a conception of knowledge as correct representation of an independent reality and was (is) almost exclusively interested in the issue of establishing the validity of scientific knowledge claims. (Schwandt, 2000, p. 196)

In other words, hermeneutics challenged the assertion that an interpretation can ever be absolutely correct or true. It must remain only and always an *interpretation*. The meaning of a text, then, is negotiated among a community of interpreters and to the extent that some agreement is reached about meaning at a particular time and place, that meaning can only be based on consensual community validation. Texts, then, must be "situated" within some literacy context (Barton, Hamilton, & Ivanič, 2000).

Kvale (1987) has suggested that "the attempts to develop a logic of validation within the hermeneutical tradition are relevant for clarifying the validity of interpretation in the qualitative research interview."

> The interpretation of meaning is characterized by a *hermeneutical circle*, or spiral. The understanding of a text takes place through a process where the meaning of the separate parts is determined by the global meaning of text. In principle, such a hermeneutical explication of the text is an infinite process while it ends in practice when a sensible meaning, a coherent understanding, free of inner contradictions has been reached. (p. 62)

SIDEBAR

A HERMENEUTICS OF VULNERABILITY

Clifford (1988) developed the construct of a "hermeneutics of vulnerability" to discuss the constitutive effects of relationships between researchers and research participants on research practice and research findings. A hermeneutics of vulnerability foregrounds the ruptures of fieldwork, the multiple and contradictory positionings of all participants, the imperfect control of the researcher, and the partial and perspectival nature of all knowledge. Among the primary tactics for achieving a hermeneutics of vulnerability . . . is the tactic of self-reflexivity, which may be understood in at least two senses. In the first sense, self-reflexivity involves making transparent the rhetorical and poetic work of the researcher in representing the object of her or his study. In the second (and, we think, more important) sense, self-reflexivity refers to the efforts of researchers and research participants to engage in acts of self-defamiliarization in relation to each other. (Kamberelis & Dimitriadis, 2011, p. 560)

Exhibit 3.12 offers four principles for hermeneutic inquiry and analysis that can be applied beyond the interpretation of legends, literature, and historical documents to any kind of qualitative data. This includes interviews, family conversations, works of art, photographs, or anything written or visual, for, from a hermeneutic perspective, the world is essentially narrative and text (Flick, 2009, p. 350). The *hermeneutic viewpoint* involves the belief that there is no such thing as a pure description; every communicative act involves interpretation, and therefore, when a social researcher writes about an experience, this is always an act of reconstruction (Seale, 2012, p. 448). "Social science understanding, therefore, is always essentially an understanding of understanding, an understanding 'of the second degree'" (Soeffner, 2010, p. 97).

Thus, hermeneutic researchers use qualitative methods to establish context and meaning for what people do. Hermeneutists

> are much clearer about the fact that they are *constructing* the "reality" on the basis of their interpretations of data with the help of the participants who provided the data in the study.... If other researchers had different backgrounds, used different methods, or had different purposes, they would likely develop different types of reactions, focus on different aspects of the setting, and develop somewhat different scenarios. (Eichelberger, 1989, p. 9)

EXHIBIT 3.12   Principles for Hermeneutic Inquiry

Below are four principles for hermeneutic inquiry and analysis that can be applied beyond the interpretation of legends, literature, and historical documents to guide interpretation of any qualitative data.

1. *Understanding a human act or product, and hence all learning, is like interpreting a text.*

2. *All interpretation occurs within a tradition.*

3. *Interpretation involves opening myself to a text (or any qualitative data) and questioning it.*

4. *I must interpret a text (or data of any kind) in the light of my situation.*

SOURCE: Adapted from Kneller (1984, p. 68).

Thus, one must know about the researcher as well as the researched to place any qualitative study in a proper, hermeneutic context. Hermeneutic theory argues that one can only interpret the meaning of something from some perspective, a certain standpoint, a praxis, or a situational context, whether one is reporting on one's own findings or reporting the perspectives of the people being studied (and thus reporting their standpoint or perspective). This means that "prior understandings and prejudices shape the interpretive process" (Denzin, 2010, p. 123). These ideas have become commonplace in much of contemporary social science and are now fundamental, even basic, in qualitative inquiry, but such was not always the case. Two centuries of philosophical dialogue provide our current foundation for understanding the centrality of interpretivism in qualitative research. As Michael Crotty concluded after his review of the historical development of hermeneutics and its influence on qualitative theory, "our debt to the hermeneutic tradition is large" (1998, p. 111). To follow the ongoing contributions of hermeneutics to scientific inquiry, you can track the conferences and publications of the International Society for Hermeneutics and Science, which, in 2013, held its 20th anniversary conference in Vienna, Austria, and Budapest, Hungary.

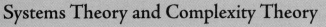

 Systems Theory and Complexity Theory

Understanding Systems Dynamics

> We contend that the primary challenge in research-practice integration is a failure to frame the effort from a systems perspective.
>
> —Jennifer Brown Urban and William Trochim (2009, p. 539)
> Evaluation researchers
>
> The core aspects of systems thinking are gaining a bigger picture (going up a level of abstraction) and appreciating other people's perspectives on an issue or situation. An individual's ability to grasp a bigger picture or a different perspective is not usually constrained by lack of information. The critical constraints are usually in the way that the individual thinks and the assumptions that they make—both of which are usually unknown to that individual.
>
> —Jake Chapman (2004, p. 14)
> *System failure:*
> *Why governments must learn to think differently*

Systems Theory

Core inquiry questions: *How and why does this system function as it does? What are the system's boundaries and interrelationships, and how do these affect perspectives about how and why the system functions as it does?*

In the eleventh century, King Canute, the sovereign who reigned simultaneously over Denmark, Norway, and England, stood at the seashore and ordered the tide to stop rolling in and out. His order had no effect. He repeated the order, shouting at the top of his lungs for the rolling tides to cease. Again, no effect. How are we to make sense of this story?

An ethnographer would look at what the story reveals about the culture of eleventh-century northern Europe. A positivist would say that he had positively demonstrated that human commands do not affect the tides. A realist would say that he had tested the causal hypothesis that kings can command the seas and determined that such a mechanism of causality does not exist. A grounded theorist would seek other examples of royal commands and look for patterns to generate a theory explaining royal commands using the constant comparative method. A phenomenologist would seek the essence of the lived experience of attempting to control nature. A narrative inquirer would examine the story for its larger narrative themes. A social constructionist would seek to understand the meanings of this story as interpreted by those involved. A semiotician would study the command to the tides as a sign or symbol of sovereign significance. A hermeneuticist would place the story in the context of the eleventh century to determine its meaning contextually. An ecological psychologist would revel in the story's rich material about the relationship between humans and nature. So what would systems thinking bring to the table?

Systems analysis would have us inquire into the perspective of the king and his subjects concerning the interrelationship between the sea and human beings and the interrelationship of the king with the people he ruled. In particular, the very fact that the story concerns a king directs our attention to a hierarchical system of power and privilege. The king's command to the tide would have to be understood as occurring within a system of power-based relationships that would determine its meaning. Indeed, following that line of inquiry, according to the legend, the cunning King Canute conducted this experiment to demonstrate to the hyperbolic flatterers who surrounded him, ever assuring him that his every command would always be obeyed, that he knew the limits of royal power. He did not expect the tides to cease on his command. He was demonstrating for all to see his knowledge of the systemic limitations on his power to command and control. The natural inclination to flatter those in power, however, has proved as persistent as the tides. Social scientists inquire into theories of human behavior, like flattery, as part of a social system. Natural scientists test theories of nature, like the ebb and flow of tides, as part of the ecological system. The story of King Canute looks at what happens when those two systems intersect.

Parallel to the philosophical and methodological paradigms debate between positivists and constructivists about the nature of reality, real or socially constructed, there has been another and corresponding paradigms debate about mechanistic/linear constructions of the world versus organic/systems constructions. Classic organizational theorists expounded systems thinking as providing both conceptual and methodological alternatives for studying and understanding how organizational systems function (Azumi & Hage, 1972; Burns & Stalker, 1972; Gharajedaghi, 1985; Lincoln, 1985; Morgan, 1986, 1989). Systems inquiry includes attention to the differences between closed systems and open systems and the implications of such boundary definitions for research, theory, and practice in understanding programs, organizations, entire societies, and even *the whole world* (Wallerstein, 2004, 2011).

It is important to note at the outset that the term *systems* has many and varied meanings. In the digital age, systems analysis often means looking at the interface between hardware and software, or the connectivity of various networks. The idea of "systems thinking" was popularized as the crucial "fifth discipline" of organizational learning in Peter Senge's (1990) best-selling book by that name. A number of management consultants have made systems thinking and analysis the centerpiece of their organizational development work (e.g., Ackoff, 1999a, 1999b, 1987; Anderson & Johnson, 1997; Kim, 1993, 1994, 1999). Indeed, since the publication of Ludwig Von Bertalanffy's classic *General System Theory* (1976), a vast literature about systems theory and applied systems research has developed (e.g., Checkland, 1999; Checkland & Poulter, 2006).

Some of it is highly quantitative and involves complex computer applications and simulations. Indeed, part of the challenge of using systems thinking as an inquiry framework is that there are so many different systems meanings, models, approaches, and methods, including system dynamics, soft systems methodology, cultural-historical activity theory, and critical systemic thinking, each of which has specific implications for research and evaluation (Williams, 2005; Williams & Hummelbrunner, 2011). At the same time, a literature on systems thinking and evaluation has emerged that offers distinct and important alternative ways to focus an evaluation (Fujita, 2010; Funnell & Rogers, 2011; Morell, 2010; Patton, 2011, 2012a; Williams & Iman, 2006).

Given this broad and multifaceted context, my purpose is quite modest. I want to call to the your attention three points: (1) a systems perspective is becoming increasingly important in dealing with and understanding real-world interconnections and interrelationships, viewing things as whole entities embedded in context and still larger wholes; (2) some approaches to systems research lead directly to and depend heavily on qualitative inquiry; and (3) a systems orientation

can be very helpful in framing questions and, later, making sense out of qualitative data.

Systems Thinking

Holistic thinking is central to a systems perspective. A system is a whole that is both greater than and different from its parts. The systems theorists Gharajedaghi and Ackoff (1985) are quite insistent that a system as a whole cannot be understood by analysis of its separate parts. They argue that "the essential properties of a system are lost when it is taken apart; for example, a disassembled automobile does not transport and a disassembled person does not live" (p. 23). Furthermore, the function and meaning of the parts are lost when separated from the whole. They insist that, instead of taking things apart, a systems approach requires "synthetic thinking":

> *Synthetic thinking* is required to explain system behavior. It differs significantly from analysis. In the first step of analysis the thing to be explained is taken apart: in synthetic thinking it is taken to be a part of a larger whole. In the second step of analysis, the contained parts are explained: in synthetic thinking, the containing whole is explained. In the final step of analysis, knowledge of the parts is aggregated into knowledge of the whole: in synthetic thinking understanding of the containing whole is disaggregated to explain the parts. It does so by revealing their *role* or *function* in that whole. Synthetic thinking reveals function rather than structure: it reveals why a system works the way it does, but not how it does so. Analysis and synthesis are complementary: neither replaces the other. Systems thinking incorporates both.

> Because the effects of the behavior of the parts of a system are interdependent, it can be shown that if each part taken separately is made to perform as efficiently as possible, the system as a whole will not function as effectively as possible. For example, if we select from all the automobiles available the best carburetor, the best distributor, and so on for each part required for an automobile, and then try to assemble them, we will not even obtain an automobile, let alone the best one, because *the parts will not fit together.* The performance of a system is not the sum of the independent effects of its parts; it is the product of their interactions. Therefore, effective management of a system requires managing the interactions of its parts, not the actions of its parts taken separately. (Gharajedaghi & Ackoff, 1985, pp. 23–24)

This kind of systems thinking has profound implications for program evaluation and policy analysis, where the parts are often evaluated in terms of strengths, weaknesses, and impacts, with little regard for how the parts are embedded in and interdependent with the whole program or policy. For example, Benko and Sarvimaki (2000) applied systems theory as a framework for patient-focused evaluation in nursing and other health care areas. Such a framework, they found, allowed interactive processes in health care to be captured by conducting simultaneous analyses of relationships on different levels. Their "systemic model" generated insights into system dynamics in both "downward" and "upward" directions—and into the interconnections of these systems dynamics in affecting patient care and outcomes. Surgeon Atul Gawande (2002) has examined how hospital errors (e.g., giving someone the wrong medication, amputating the wrong leg) must be understood as systemic problems and not just individual mistakes. In a similar vein, Harvard professor of medicine Jerome Groopman (2008) incorporated systems understandings in his analysis *How Doctors Think*. Bruce Perry and Maia Szalavitz (2006) used case studies to show the failures of the American child protection system and the long-lasting damage those failures cause in children.

Communities of Practice as Systems

In working to select the appropriate qualitative inquiry framework for studying a particular problem or phenomenon, a systems perspective is called for when the focus of the inquiry is a system. For example, Ettiene Wenger (1998) has pioneered inquiries into communities of practice. Communities of practice are people engaged in systematic reflective practice and learning together (Wenger, McDermott, & Snyder, 2002). Studying a community of practice through a systems framework is especially appropriate because the community of practice is a system of interconnected and interacting people who can constitute either a network or community or both. Notice the language and images of systems in how Wenger and his colleagues talk about their focus.

> We prefer to think of community and network as two aspects of social structures in which learning takes place.
>
> - The *network* aspect refers to the set of relationships, personal interactions, and connections among participants who have personal reasons to connect. It is viewed as a set of nodes and links with affordances for learning, such as information flows, helpful linkages, joint problem solving, and knowledge creation.
>
> - The *community* aspect refers to the development of a shared identity around a topic or set of challenges. It represents a collective intention—however tacit

DEVELOPMENTAL SYSTEMS THEORY AND METHODOLOGY

There exists a long tradition in theoretical psychology and theoretical biology in which developmental processes are explained as the result of self-organizing processes with emergent properties that have complex, dynamic interactions with environmental influences. The general denotation for this tradition is developmental systems theory....

Accordingly, the tenets of developmental systems theory are well established as the superordinate developmental frame in contemporary developmental science. In addition, there are strong conceptual links between these theories and other contemporary theoretical models, such as dynamical systems, biological systems theory, and artificial neural networks (e.g., connectionism). (Molenaar, Lerner, & Newell, 2014, pp. 3–4)

Science in general can be characterized as an inductive developmental system in which different scientific models constitute competing webs of beliefs. (Molenaar et al., 2014, p. 11)

Developmental systems theory and methodology employ theory-predicated methods to enhance understanding of

the mutually influential relations between individuals and the multiple levels of their context that constitute the developmental system.... A new era in the conduct of developmental science [is] one that captures the complexity of the developmental system and enhances the means to not only describe and explain intraindividual change and interindividual differences in intraindividual change but, as well, provides new means to generate evidence-based actions that optimize the course of health and positive functioning across the life span. (Molenaar et al., 2014, p. 12)

and distributed—to steward a domain of knowledge and to sustain learning about it. (Wenger, Trayner, & de Laat, 2011, p. 9)

Mapping Systems

Analyzing a system will often produce a visual map of that system. Such maps are a form of qualitative representation of findings. I became involved in a systems analysis study examining how a new light rail transit line would affect interdependent and interacting systems along the 11 miles of construction, not only the transportation systems (buses, cars, trains, airport connections) but also housing systems, neighborhood organizations and community systems, service delivery systems, small businesses and shopping centers, nonprofit agencies serving people along the rail corridor, government/political systems, land use systems, zoning systems, utility systems, school systems, and economic systems. Exhibit 3.13 shows how qualitative inquiry (key informant interviews and focus groups with different stakeholder and constituency groups) was used to create a baseline map of the existing interacting systems. That map was then the basis for envisioning, engaging in, and evaluating systems change.

In addition to its influence in organizational development, systems approaches are important in family research and therapy (Hoffman, 1981; Montgomery & Fewer, 1988; Schultz, 1984; Smith-Acuña, 2011). A systems approach has also become one of the central orientations to international development efforts. Specifically, the Farming Systems approach to development (Farming Systems Support Project [FSSP], 1986) illustrates some unique ways of engaging in qualitative inquiry to support development, intervention, and evaluation from a systems perspective. The Farming Systems approach to evaluation and research is worth examining in detail because it has developed as a theory-based yet practical solution to agricultural development problems.

EXHIBIT 3.13   Systems Map Example

The Central Corridor light rail line covers the 11 miles (18 kilometers) between downtown Minneapolis and downtown Saint Paul (the Twin Cities) in Minnesota. The new transit line runs along a major commercial street, with many small businesses and shopping areas, through residential areas, including diverse low-income neighborhoods and immigrant communities, and through the campus of the University of Minnesota. The potential was high for disrupting neighborhoods, displacing residents, and bankrupting small businesses. At the same time, new opportunities for community and business development were emerging. In 2010, to seize on the opportunities and to minimize impacts, a collaboration of funders and government applied to the Living Cities Integration Initiative (http://www.livingcities.org/integration/) to support work along the Central Corridor, among other activities. The Integration Initiative's initial goals were the following:

- Improve the lives of low-income people
- Create a new framework for solving complex problems
- Challenge obsolete conventional wisdom
- Drive the private market to work on behalf of low-income people
- Create a "new normal"/systems change

The Region & Communities Program of The McKnight Foundation, headquartered in Minneapolis, provided support to develop the local application for the initiative, including hosting focus groups and key informant interviews to identify how the existing system was functioning. The results provided information to create a baseline map of the existing system. That map, reproduced here, served as a foundation for those involved in the initiative to envision and undertake systems change together. Systems changes can be monitored and mapped against this baseline systems graphic.

The baseline systems map showed the following:

- Key subsystems engaged in developing the Corridors of Opportunity were (1) Land Use Planning and Development and (2) Transit Planning and Development. These two subsystems functioned mostly separately ("in silos") on "opposite sides of the tracks" (the new light rail system that is the centerpiece of the graphic).

- Low-income residents, businesses in the corridors of opportunity, and citizens' groups were having difficulty accessing, engaging with, being heard by, and influencing these subsystems (lower left-hand corner).

- Equity was not a priority for these planning and development subsystems at that time.

The "Baseline Systems" graphic depicts the situation at the beginning of light rail construction; the key system actors, structures, and processes; and the lack of integration among these subsystem elements. The purpose here is not to explain this initiative in detail but to provide an example of how qualitative inquiry can be used to depict a system and support the conceptualization and evaluation of systems change.

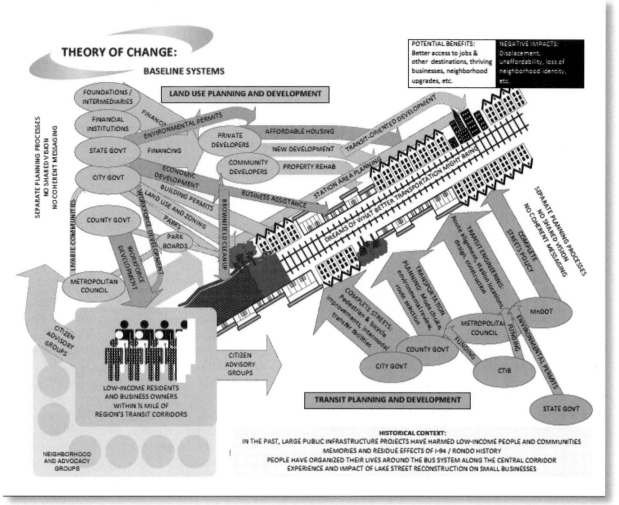

THEORY OF CHANGE:

BASELINE SYSTEMS

SOURCES: Regions & Communities, the McKnight Foundation: http://www.mcknight.org/grant-programs/region-and-communities/central-corridor-funders-collaborative; Corridors of Opportunity/Living Cities Integration Initiative: http://www.corridorsofopportunity.org/; Living Cities Integration Initiative: http://www.livingcities.org/integration/. Reprinted with permission.

NOTE: The ecosystem map was developed by Libby Starling, Metropolitan Council, and Mary Kay Bailey, The Saint Paul Foundation.

Farming Systems Research as a Systems Analysis Exemplar

In the first three decades following World War II, much of international development was conceived as a direct technology transfer from more developed to less developed countries. Scientists and change agents made technology transfer recommendations within their disciplinary areas of specialization, for example, crops, livestock, water, and so on. This approach to development epitomized a mechanistic orientation.

In reaction to the dismal failures of the mechanistic, specialized technology transfer approach to development, a farming systems approach emerged (Shaner, Philipp, & Schmehl, 1982b). Several elements are central to a farming systems perspective,

elements that lead directly to qualitative methods of research.

1. Farming systems research and development (FSRD) is *a team effort* (Shaner, Philipp, & Schmehl, 1982a).

2. FSRD is *interdisciplinary.* The team consists of representatives from a mix of both agricultural and social science disciplines (Cernea & Guggenheim, 1985).

3. FSRD takes place *in the field*, on real farms, not at a university or government experiment station (Simmons, 1985).

4. FSRD is *collaborative*—scientists and farmers work together on agricultural productivity within the goals, values, and situation of participating farmers (Galt & Mathema, 1987).

5. FSRD is *comprehensive*, including attention to all farm family members; all farming operations, both crops and livestock; all labor sources; all income sources; and all other factors that affect small-farm development (Harwood, 1979).

6. FSRD is *inductive and exploratory*, beginning by open-ended inquiry into the nature of the farming system from the perspective of those in the system (Holtzman, 1986).

7. FSRD begins with qualitative description. The first team task is fieldwork to qualitatively describe the system (Sands, 1986).

8. FSRD is sensitive to context, placing the farming system in the larger agroecological, cultural, political, economic, and policy environments of which it is a part.

9. FSRD is interactive, dynamic, and process oriented. The interdisciplinary team begins with inductive exploration and then moves to trying out system changes, observing the effects, and adapting to emergent findings. The work is ongoing and developmental (FSSP, 1986).

10. FSRD is situationally responsive and adaptive. There are many variations in FSRD projects depending on priority problems, available resources, team member preferences, and situation-specific possibilities (FSSP, 1987; Sands, 1986).

A farming systems approach uses mixed methods, incorporating both qualitative and quantitative forms of inquiry. It includes direct observations, informal interviews, naturalistic fieldwork, and inductive analysis, all within a systems framework. There may be no larger-scale example of efforts to integrate naturalistic inquiry, quantitative methods, and a systems perspective through interdisciplinary evaluation and research teamwork for the purpose of promoting long-term social and economic developments. As farming systems research has evolved, it has sometimes become framed in even more general terms as *agroecology*:

> a discipline in which agriculture can be conceptualized within the context of global change and studied as a coupled system involving a wide range of social and natural processes . . . [addressing] the key challenges of mitigating environmental impacts of agriculture while dramatically increasing global food production, improving livelihoods, and thereby reducing chronic hunger and malnutrition over the coming decades. (Tomich et al., 2011, p. 193)

Whole-Systems Frameworks

Farming and food systems agroecology is just one example of a systems approach to intervention, research, and evaluation. What this and other systems approaches offer is an inquiry framework based on the premise that the interconnected world of human beings cannot be fully captured and understood by simply adding up carefully measured and fully analyzed parts. At *the system level* (the *whole* program, the *whole* farm, the *whole* family, the *whole* organization, and the *whole* community), there is a qualitative difference in the kind of thinking that is required to make sense of what is happening. Qualitative inquiry facilitates that qualitative difference in understanding human or "purposeful systems" (Ackoff & Emery, 2005).

A final story will reinforce this point—the fable of the nine blind people and the elephant, which I used in the second chapter to illustrate the importance of context and which I repeat here because it illustrates so well the real challenge of systems thinking. Besides, good stories have layers of meaning, and this one has phenomenological, hermeneutic, constructionist, and even ethnographic implications, which you may want to reflect on, but I'll simply reintroduce it as a systems tale. Ironically, it is often offered as an example of systems thinking but is, in its usual Western telling, actually quite linear and mechanical.

As the story goes, nine blind people encounter an elephant. One touches the ear and proclaims that an elephant is like a fan. Another touches the trunk and says the elephant most surely resembles a snake. The third feels the elephant's massive side and insists that it is like a wall. Yet a fourth, feeling a solidly planted leg, counters that it resembles more a tree trunk. The fifth grabs hold of the tail and experiences the elephant as a rope. And so it goes, each blindly touching only a part and generalizing inappropriately to the whole. The usual moral of the story is that only by putting all the parts together in right relation to one another can one get a complete and whole picture of the elephant.

Yet, from a systems perspective, such a picture yields little real understanding of the elephant. To understand the elephant, it must be seen and understood in its natural ecosystem, whether in Africa or Asia, as one element in a complex system of flora and fauna. Only in viewing the movement of a herd of elephants across a real terrain, over time and across seasons, and in interaction with plants, trees, and other animals will one begin to understand the evolution and nature of elephants and the system of which elephants are a part. That understanding can never come at a zoo.

Complexity Theory

Core inquiry question: *How can the emergent and non-linear dynamics of complex adaptive systems be captured, illuminated, and understood?*

Science has explored the microcosms and the macrocosms; we have a good sense of the lay of the land. The great unexplored frontier is complexity.

—Heinz Pagels (1988)

American physicist

Complexity writings are filled with metaphors that try to make complex phenomena understandable to the human brain's hardwired need for order, meaning, patterns, sense making, and control, ever feeding our illusion that we know what's going on. So complexity theorists talk of flapping butterfly wings that change weather systems and spawn hurricanes, individual slime molds that remarkably self-organize into organic wholes, ant colonies whose frantic service to the queen mesmerize us with their collective intelligence, avalanches that reconfigure mountain ecologies, bacteria that *know* the systems of which they are a part without any capacity for self-knowledge, and black swans that appear suddenly and unpredictably to change the world. Complexity science offers insights into the billions of interactions in the global stock market, the spread of disease throughout the world, volatile weather systems, the evolution of species, large-scale ecological changes, and the flocking of migrating birds. Complexity theorists explain the rise and fall of civilizations and the rise and fall of romantic infatuation. That's a lot of territory. But the vision is vast: Complexity theorists

believe that they are forging the first rigorous alternative to the kind of linear, reductionist thinking that has dominated science since the time of Newton—and that [Newtonian thinking] has now gone about as far as it can go in addressing the problems of our modern world. They believe they are creating the sciences of the twenty-first century. (Waldrop, 1992, p. 13)

So what is a complex system? Professor Melanie Mitchell of the Santa Fe Institute, the world's leading think tank on complexity theory, offers this definition:

A system in which large networks of components with no central control and simple rules of operation give rise

to complex collective behavior, sophisticated information processing, and adaptation via learning or evolution. . . .

Systems in which organized behavior arises without an internal or external controller or leader are sometimes called *self-organizing*. Since simple rules produce complex behavior in hard-to-predict ways, the macroscopic behavior of such systems is sometimes called *emergent*. Here is an alternative definition of a *complex system*: a system that exhibits nontrivial emergent and self-organizing behaviors. The central question of the sciences of complexity is how this emergent self-organized behavior comes about. (Mitchell, 2009, p. 13)

Complexity theory has been viewed as a new paradigm of inquiry and explanation for natural sciences for some time (Gleick, 1987; Hall, 1993; Holte, 1993; Murali, 1995; Nadel & Stein, 1995; Waldrop, 1992). It is now established as a framework of inquiry in the social sciences (Cronbach & Gleick, 1988; Stacey, 2001, 2007). The openness, flexibility, and adaptability of qualitative methods make complexity theory an especially useful framework for qualitative inquiries into complex dynamic situations and phenomena. Complex dynamic systems are dynamic, unpredictable, and ever adapting. "It's like walking through a maze whose walls rearrange themselves with every step you take" (Gleick, 1987, p. 24). This metaphor fits a great deal of fieldwork in real-world settings, but the implications can be so threatening to our need for order that we ignore the rearranging walls and try to impose order on the morphing maze with a single, static diagram. Complexity theory challenges us to attend to the unpredictable, the disorderly, the messy, and the ever-changing and adapting—in short, to look at and inquire into the world as a complex entity.

Distinguished anthropologist Michael Agar used complexity theory to interpret fieldwork findings in his study of a heroin epidemic among suburban youth in Baltimore County. He concluded,

Complexity [theory] served, at least at the metaphorical level, to better define a research problem—explaining heroin trends—and it helps articulate why traditional social research has not answered this most basic question of drug research: How and why do trends occur? It also points at the kind of data we need to obtain and organize to do just that, however difficult that data might be to obtain. Furthermore, complexity handles some current anthropological research issues—like the inclusion of the researcher, broadening historical and political context, and the issue of prediction—as part of its central themes. With characteristics like holism, emergence,

COMPLEXITY IN NATURE: HONEYBEE COLONY COLLAPSE DISORDER

The mysterious and dramatic disappearance of U.S. honeybees since 2006, known as "colony collapse," manifests complex, dynamic, and intertwined factors: a parasitic mite, multiple viruses, bacteria, poor nutrition, genetics, habitat loss, and pesticides. Bees, especially honeybees, are crucial to pollination of crops. The multiple, interacting, and overlapping causes mean that there can be no single intervention strategy to reverse the decline that has seen one third of bees disappearing each winter since 2006.

"Consensus is building that a complex set of stressors and pathogens is associated with CCD (Colony Collapse Disorder), and researchers are increasingly using multifactorial approaches to studying causes of colony losses" (National Honey Bee Health Stakeholder Conference Steering Committee, 2012, p. 5).

and feedback that map onto anthropological assumptions more so than any previous formal models, complexity is clearly worth a closer look. (Agar, 1999, p. 119)

Complexity can inform qualitative inquiry by deepening attention to cultural diversity in both data gathering and analysis because other "theoretical models may often be unrealistic and/or ethnocentric: in other words, they fail to take account of the range of variation in orientation to be found even within a single society" (Hammersley, 2008a, p. 42).

A second kind of complexity emphasized by qualitative researchers arises from the *processual* character of social life. This is perhaps more fundamental. The argument is that, rather than the outcome of any sequence of events being a pre-determined or law-like result of some set of causal factors operating on it, the outcome is an emergent and contingent product, one which would not necessarily be produced in other similar cases.

One source of this kind of complexity in social life, to which some qualitative researchers have given great emphasis, is the ability of human beings to suspend a line of action in which they are engaged, and to redirect it. . . . All this suggests that social processes cannot be understood in terms of some immanent, fixed pattern of causal relations. Instead, the routes followed by social processes (and therefore the outcomes) are in an important sense contingent, though at the same time not simply inexplicable. (Hammersley, 2008a, p. 42)

Complexity theory has been used for qualitative, case-based inquiry across a wide range of phenomena

where a simple cause–effect conceptualization cannot capture and do justice to the nonlinear dynamics and emergent properties of the phenomenon of interest:

- Studying businesses that thrive under conditions of chaos and uncertainty (Collins & Hansen, 2011)
- Explaining the dynamics of financial markets, economic booms and busts, and political change (Ormerod, 2001; J. B. Stewart, 2009; Taleb, 2005, 2007, 2012)
- How cities function, how our brains operate, and how software is created (Johnson, 2001)
- Illuminating organizational change and adaptation (Allison, 2000; Eoyang & Holladay, 2013; Stacey, 2001, 2007)
- Making sense of how major social movements like Mothers Against Drunk Driving and microfinance have changed the world (Westley, Zimmerman, & Patton, 2006)
- Understanding particular groups that work in complex dynamic environments, for example, leaders (Wheatley, 1992) and social workers at midlife (Karpiak, 2006)
- Evaluating international humanitarian aid (Ramalingam, 2013; Ramalingam & Jones, 2008)
- Developmental evaluation based on complexity theory to evaluate a wide range of social innovations (Patton, 2011, 2012a)

Capturing complexity is a common rationale for undertaking qualitative research (Hammersley, 2008a), but complexity theory involves more than seeing the world as complex. Complexity theory and the concepts that are central to complexity science have important implications for how qualitative inquiries are undertaken. Exhibit 3.14 presents core complexity concepts and their implications for qualitative research and evaluation.

Distinguishing Between Simple, Complicated, and Complex

The challenge in any research or evaluation design is to match the inquiry framework to the nature of the phenomenon being studied and the questions being asked. In doing so, it can be helpful to distinguish between the simple, the complicated, and the complex. The simple, complicated, and complex distinctions do not constitute a taxonomy of operationally distinct, mutually exclusive, and exhaustive categories. The distinctions constitute a typological continuum. Phenomena are *relatively* simple, *relatively* complicated, and *relatively* complex. Exhibit 3.15 presents these distinctions and their implications for qualitative research and evaluation.

EXHIBIT 3.14	Complexity Theory Concepts and Qualitative Inquiry Implications

COMPLEXITY CONCEPTS	IMPLICATIONS FOR QUALITATIVE INQUIRY
1. *Nonlinearity.* A butterfly flapping its wings in Tokyo may affect the weather in New York—next month or next year. "The butterfly effect" has a technical name: sensitive dependence on initial conditions, which means that small actions can stimulate large reactions; thus, we have the *butterfly wings metaphor* (Gleick, 1987); *black swans* (Taleb, 2007), in which highly improbable, unpredictable, and unexpected events have huge impacts; and tipping points (Gladwell, 2002), when major shifts occur changing the whole landscape of action.	Watch for, sample, and study significant critical incidences when things shift. Assess and map tipping points and other changes in the inquiry setting and landscape. Use mixed methods to capture when cumulative quantitative changes in key indicators become substantively significant qualitative shifts. Small, minute events can make critical differences. Qualitative importance is not dependent on quantitative magnitude. "For want of a nail . . . , the war was lost." Look for contextual changes that shift local patterns, forks in the road that move the complex dynamic system in new directions, and sudden (or gradual) responses to unexpected developments.
2. *Emergence.* Patterns emerge from self-organization among interacting agents. Each agent or element pursues its own path, but as that path intersects with, and the agent interacts with, others, also pursuing their own paths, patterns of interaction emerge and the whole of the interactions cohere, becoming greater than the separate parts. What emerges can be beyond, outside of, and oblivious to any notion of shared intentionality (Johnson, 2001). There can be "wild disorder" among "islands of structure. . . . A complex system can give rise to turbulence and coherence at the same time," each of which is important (Gleick, 1987, p. 56)	Be especially alert to the formation of self-organizing subgroups that have different experiences and perspectives from each other. Anticipate and expect emergent issues, and take seriously the search for unanticipated consequences, tracking interactions among key players, both formal and informal, planned and unplanned. Map networks, system relationships, and subgroups. Track information flows, communications, and emergent issues. Emergence applies to both processes and outcomes. Watch for and assess not only what emerges but what declines or even disappears. Disappearance is the other side of the phenomenon of emergence. The unplanned emerges, the planned disappears. Both are important, as is what unfolds as planned. The inquiry design is also emergent.
3. *Adaptive.* Interacting elements and agents respond and adapt to each other and to their environment, so that what emerges is a function of ongoing adaptation among both interacting elements and the responsive relationships interacting agents have with their environment. Adaptation means that "the act of playing the game has a way of changing the rules" (Gleick, 1987, p. 24).	Regularly capture perspectives from key actors in different but interacting systems about what's going on. Put these perspectives into dialogue with each other to capture and track adaptations and their significance. *The inquiry itself must be adaptive.* An adaptive mind-set essentially involves learning by doing and observing. This parallels the process recommended by knowledge management consultant David Snowden when facing complexity: probe, sense, respond (Snowden & Boone, 2007). Probing is the doing. Sensing is the observing (where chance ever favors the prepared mind). And responding is the adaptation.
4. *Uncertainty.* Under conditions of complexity, interactions are unpredictable, uncontrollable, and unknowable in advance. Emergent and adaptive self-organization can create idiosyncratic	Identify and acknowledge sources of uncertainty, including inadequate knowledge about what to look for and how to make sense of what is observed, disagreements and value conflicts among key actors, and

(Continued)

(Continued)

COMPLEXITY CONCEPTS	IMPLICATIONS FOR QUALITATIVE INQUIRY
mountains or at other times erode into nothingness, and it's impossible to know ahead of time which pattern, if either, will prevail. Not acknowledging and dealing with uncertainty and unexpected events can lead to a spiral of disruption with things getting worse (Weick & Sutcliffe, 2001, p. 2) Uncertainty is a defining characteristic of complexity (Morell, 2010; Westley et al., 2006).	turbulence in the larger environment. Interview key informants on an ongoing basis to understand the implications of uncertainty. Nurture tolerance for ambiguity and messiness. This means resisting the temptation to address uncertainty by imposing order and control through preexisting constructs. In early stages of fieldwork, the unexpected may give off *weak signals*. "The overwhelming tendency is to respond to weak signals with a weak response." Understanding the potential significance of weak signals and responding strongly "holds the key to managing the unexpected" (Weick & Sutcliffe, 2001, p. 4). Much of qualitative analysis attempts to bring order from seeming chaos. We need to learn to observe, describe, and value disorder and turbulence without forcing patterns onto the dynamic uncertainties of complexity.
5. *Coevolutionary*. As interacting and adaptive agents self-organize, ongoing connections emerge that become *coevolutionary* as the agents evolve together (coevolve) within and as part of the whole system over time.	A qualitative inquiry will evolve in ways that affect both the findings and the people in the setting where the inquiry is taking place. This is the process of *cocreation*. The inquiry will not be independent and separate from the setting but will be interdependent with it and with those involved in it. Both participants and researchers are changed. The interactions between those being studied and those doing the studying are dynamic, interdependent, and coevolving. Documenting the nature and implications of these interactions is reported in the study's methodology.

EXHIBIT 3.15   Simple, Complicated, Complex

	NATURE OF THE PHENOMENON	QUALITATIVE RESEARCH IMPLICATIONS	EVALUATION IMPLICATIONS
Simple	Problem to be studied is well-defined; a great deal is already known about the phenomenon of interest. Clear, direct, linear, predictable, and controllable cause–effect patterns have already been documented from previous research.	Highly focused inquiry: Increase the depth of knowledge about the phenomenon. Examine how the phenomenon plays out in a new context or with different people. Validate existing knowledge. Document expertise: Interview knowledgeable key informants.	Evaluate how already identified *best practices* fit new and different contexts. Detect unanticipated consequences and context-specific implementation problems for an intervention or program.

	NATURE OF THE PHENOMENON	QUALITATIVE RESEARCH IMPLICATIONS	EVALUATION IMPLICATIONS
Complicated	There are disagreements about how to define the problem, conflicting or confusing findings from past research and/or evaluations. Phenomenon of interest manifests multiple dimensions that interact, thereby increasing uncertainty and reducing predictability. Cause–effect linkages are context contingent, discoverable with careful analysis, but neither obvious nor certain. Contingencies are discernible—there are known unknowns.	Document diversity: Seek multiple perspectives and points of view about the phenomenon. Pay special attention to effects of variations in context. Examine and document how multiple dimensions interact. Illuminate contingencies. Engage in systems inquiry.	Focus on understanding the system(s) and context(s) within which diverse actions unfold. Detect and measure both outcomes and contingencies. Design a reasonable test of the theory of change. Facilitate interpretation of less-than-certain findings.
Complex	There is high uncertainty about how to even define the nature of the phenomenon and great disagreement among diverse perspectives about the nature of the problem and how to study it. The phenomenon is dynamic, ever-changing, highly dependent on initial conditions and variations in context, nonlinear interactions within a dynamic system. No definitive findings are likely; key variables and their interactions are not known in advance. Each situation is unique and in flux. Causality is elusive.	The qualitative design must be open, emergent, and flexible. Focus on process dynamics, what is unfolding as interactions occur and the context evolves, keeping up with rapid change. High tolerance for ambiguity and uncertainty will be needed. Reduce time between data collection and analysis as these happen together so that what is learned as fieldwork unfolds can inform the next focus of data collection and real-time sense making. Learning and insight emerge from engaging with the complex phenomenon, not being outside and looking in. The qualitative inquirer is embedded in complexity.	The evaluation design will need to be flexible and emergent. The evaluation challenge will be keeping up with the rapid pace of change in turbulent and dynamic environments and documenting emergent developments. High level of ongoing interaction and communication will be needed. Provide timely ongoing feedback about changes observed. Facilitate interpretation of emergent findings for action.

SOURCES: For the distinction between simple, complicated, and complex, see Patton (2011, chap. 4); see also Westley et al. (2006), Snowden and Boone (2007), Gawande (2010), Funnell and Rogers (2011), and Williams and Hummelbrunner (2011).

Complexity Theory as a Worldview

Complexity theory offers much more than a framework for inquiry. It alters the way we think about the world and our relationship to it. Ilya Prigogine, a physicist and chemist, began by applying complexity theory to the physical world. He then came to see its implications of complexity (emergence, nonlinearity, and instability) for the social world—for how we study, understand, and ultimately engage the world.

> We have not chosen the world we describe, we are born into a certain world and we must take account of this world as it is, reducing as far as possible our *a priori*

feelings. This world is unstable [complex]—this is not a capitulation, but on the contrary an encouragement to combine new experimental and theoretical research which takes account of this unstable character. The world is not a victim offered up for us to dominate; we must respect it. The world of unstable phenomena is not a world which we can control, any more than we can control human society in the sense that extrapolation in classical physics led us to believe. . . .

We need to be aware that our knowledge is still a limited window on the universe; because of instability we must abandon the dream of total knowledge of the universe. From our window, we must extrapolate and guess what the mechanisms could be. An unstable world means that although we may know the initial conditions to an infinite number of decimal points, the future remains impossible to forecast.

There is a close analogy with a work of literature: in its first chapter a novel begins with a description of the situation in a finite number of words, but it is still open to numerous possible developments and this is ultimately the pleasure of reading: discovering which one of the possible developments will be used. Similarly, in a Bach fugue, once the theme has been given it allows a great number of developments out of which Bach has chosen. (Prigogine, 2009, pp. 235–236)

In-depth qualitative fieldwork and the case studies that result from such fieldwork share the kind of emergence Prigogine evokes in the complex imagery of an unfolding novel or a crescendoing orchestration of classical music: discovering from the initial conditions we observe what will evolve, emerge, adapt, and ultimately open to us a new view of an ever-changing, ever new world.

This [complex, unstable] world is very different from the classical world, and it extends to all of physics and cosmology. Instability leads to a new rationality, which puts an end to the idea of absolute control, and with it an end to any possible idealization of a society under absolute control. The real is not controllable in the sense which science claimed. (Prigogine, 2009, p. 236)

Focusing on the Simple: Entering Into Complexity or Avoiding Complexity?

Relationship of Systems Theory to Complexity Theory

Complexity theory is sometimes viewed as a subset of systems theory. In other framings, complexity theory and systems theory are sufficiently distinct to constitute separate and unique but overlapping approaches to understanding the world, like seeing and hearing. Seeing someone speak can enhance hearing and deepen understanding of what the person is saying. Listening to someone is given additional meaning by watching that person's expressions. Both are senses. They operate separately but can overlap to reinforce what we take in and make sense of in an interaction.

Exhibit 3.16 depicts these two contrasting perspectives. Figure 1 in Exhibit 3.16 shows complexity as one of many systems approaches. Figure 2 shows systems theory and complexity theory as parallel but overlapping frameworks, as presented in this chapter. For the purpose of identifying and presenting distinctly different theoretical orientations that can inform qualitative inquiry, I find it useful to treat systems theory and complexity theory as depicted in Figure 2: separate but overlapping approaches to understanding the world.

EXHIBIT 3.16   **Relationship of Systems Theory to Complexity Theory**

Complexity theory is sometimes viewed as a subset of systems theory. In other framings, complexity theory and systems theory are sufficiently distinct to constitute separate but overlapping approaches to understanding the world.

- Figure 1 shows complexity as one of many systems approaches.

- Figure 2 shows systems theory and complexity theory as parallel but overlapping frameworks, as presented in this chapter. For the purpose of identifying and presenting distinctly different theoretical orientations that can inform qualitative inquiry, I find it useful to treat them as depicted in Figure 2.

FIGURE 1   **Systems Theory as the Core Inquiry Framework: Complexity Theory as One of Many Systems Approaches**

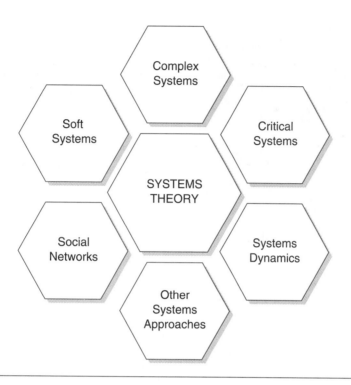

SOURCE: Williams and Hummelbrunner (2011).

FIGURE 2   **Systems Theory and Complexity Theory as Distinct but Overlapping Inquiry Frameworks (Patton, 2011)**

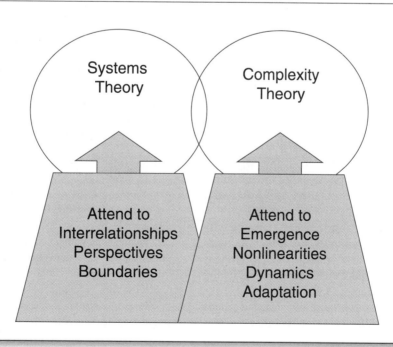

Pragmatism, Generic Qualitative Inquiry, and Utilization-Focused Evaluation

> Utility is its own demonstration of truth. Thus, what is useful is true.
>
> —Halcolm
>
> Essentially, all models are wrong, but some are useful.
>
> —George E. P. Box,
> (Box & Draper, 1987, p. 424) Mathematician

Pragmatism and Generic Qualitative Inquiry

Core inquiry question: *What are the practical consequences and useful applications of what we can learn about this issue or problem?*

The *pragmatic theory of truth* emerged at the turn of the twentieth century through the works of Charles Sanders Peirce, William James, and John Dewey. While important differences distinguish their contributions and perspectives, at the core they argued that truth is verified and confirmed by testing ideas and theories in practice. Pragmatic theory posits that "a statement is true if it works" (Seale, 2012, p. 20). As a qualitative inquiry framework, pragmatism directs us to seek practical and useful answers that can solve, or at least provide direction in addressing, concrete problems. Examples of pragmatic questions illustrate what this means.

- What have outstanding teachers learned about good teaching? (*Implication:* Interview and conduct case studies of outstanding teachers to seek useful and applicable patterns in their collective wisdom.)
- Why are participants dropping out of the program? (*Implication:* Interview dropouts to find out what can be done to reduce the dropout problem.)
- What lessons for living can we glean from the wisdom of older people? (*Implication:* Observe, interview, and do case studies of reflective senior citizens willing and able to reflect on life's lessons for living; see, for example, Pillemer, 2011.)
- What have successful immigrants learned about making the transition to a new country, society,

and culture? (*Implication:* Study the experiences and capture the lessons of successful immigrants compared with unsuccessful ones; see, for example, Blewett, 2009; Colic-Peisker, 2008.)

What these questions have in common is striving for practical understandings and wisdom about concrete, real-world issues. These questions aren't aimed at validating the nature of reality, getting at the essence of some phenomenon, generating grounded theory, or deconstructing social constructions. Rather, these are action research questions, seeking practical and useful insights to inform action. For pragmatists, findings that carry no practical value are meaningless precisely because they are useless. The pragmatist seeks what William James called the "cash-value" of knowledge: "From this it is only one further step to the pragmatist definition of truth as that which has fruitful consequences" (Russell, 1959, p. 279).

A Pragmatic Example

Here's a great example of a pragmatic qualitative inquiry from Michael Agar (2013), distinguished author of *The Lively Science: Remodeling Human Social Research.* His "lively science" is grounded in the understanding that human beings are different from minerals, plants, and other animals and need to be engaged interactively in the field to understand human situations and find solutions to human problems. He describes being asked to help solve a waiting room problem in an outpatient chemotherapy clinic. The clinic analyzed records, conducted surveys, did time-and-motion studies, and ran computer models but could only reduce the average of several hours of waiting time by a few minutes. So they sought his help.

> I hang around, listen to patients and front line staff, read a lot of things....
>
> It turns out that maybe reducing waiting time is impossible. But patients have different ways of looking at waiting time with a common thread. The thread is uncertainty—how much better things go if you know why you are waiting, both in terms of what's going on with the clinic and what's going on with your disease. So we plan a system to address that. Even though not much can be done about the number of minutes, a lot

PRAGMATISM AS A PHILOSOPHICAL SYSTEM

Pragmatism demonstrates that it is itself a unique philosophical worldview:

- In contrast to philosophies that emphasize the nature of reality, pragmatists emphasize the nature of experience.
- In place of questions about the nature of truth, pragmatists focus on the outcomes of action.
- Instead of concentrating on individuals as isolated sources of beliefs, pragmatists examine shared beliefs.

Thus, pragmatism as a philosophical stance is quite different from many other philosophical systems—and even more different from the crude summary of pragmatic behavior as "what works."

> As a philosopher, Dewey was especially interested in the concept of *inquiry* as a form of experience that helps to resolve uncertainty. Inquiry is thus a conscious response to situations in which how one should act is not immediately clear. When you are faced with such situations, pragmatism asks the key question: What difference would it make to act in one way rather than another? And the only way you can answer this question is by tracing out the likely consequences of different lines of action and ultimately deciding on a way of acting that is likely to resolve the original uncertainty in the situation....
>
> Thus, when you think about what difference it would make to use one method rather than another, you are thinking about the potential consequences of this choice. Moreover, the potential results from your choice can only be evaluated in terms of the goals and purposes behind your original research question.

—David L. Morgan (2014, pp. 28–29)
Integrating Qualitative and Quantitative Methods: A Pragmatic Approach

Ernie decided it was time to leave the university. He wanted to be pragmatic, not discuss pragmatism.

questions in search of useful and actionable answers. Second is making pragmatic decisions while conducting the inquiry based on real-world constraints of limited time and resources. This means making methods decisions based on the situation and opportunities that emerge rather than adherence to a pure paradigm, theoretical inquiry tradition, or fixed design. It can also mean mixing methods and adapting data collection as the fieldwork unfolds.

Bricolage: *Emergent Research Designs and Fieldwork Pragmatism*

The creative, practical, and adaptive qualitative inquirer draws on varied inquiry traditions and uses diverse techniques to fit the complexities of a particular fieldwork situation. It is in this regard that we may be thought of as *bricoleurs*. The term comes from anthropologist Levi-Strauss (1966), who defined a *bricoleur* as a "jack of all trades or a kind of professional do-it-yourself person" (p. 17). He brought into the world of research the tradition of the French *bricoleur*, who traveled the countryside using odds and ends, whatever materials were at hand, to perform fix-it work. The qualitative inquirer

> as bricoleur or maker of quilts uses the aesthetic and material tools of his or her craft, deploying whatever strategies, methods, or empirical materials are at hand. If new tools or techniques have to be invented, or pieced together, then the researcher will do this. (Denzin & Lincoln, 2011, p. 4)

can be done about reducing the uncertainty of those minutes. Even if it's bad news, which it usually isn't on a day-to-day basis, it's better to know one way or the other rather than to assume your cancer just jumped a stage when the real explanation is that a subway delay or another patient's morning test results messed up scheduling. (Agar, 2013, pp. 24–25)

There are two ways in which pragmatism informs qualitative inquiry. First is inquiring into practical

Drawing on creativity and pragmatism opens up new possibilities, the *bricolage* of combining old things in new ways, including alternative and emergent forms of data collection and combining inquiry traditions.

> By nature reluctant to take strong paradigmatic stands, pragmatic qualitative researchers move comfortably within and among various discourses whether postpositivist, interpretivist, feminist, or constructivist . . . allowing qualitative methods to showcase their strengths without ideological imperatives attached. (Padgett, 2004, p. 7)

Creswell (2009) adds that "for the mixed methods researcher, pragmatism opens the door to multiple methods, different worldviews, and different assumptions, as well as different forms of data collection and analysis" (p. 11). Thus, while it's quite feasible to pragmatically mix methods in principle, it remains quite challenging interpersonally where the mixing process involves quantitative and qualitative researchers working together; power dynamics and dysfunctional interactions do not disappear merely because researchers with different methodological expertise agree to collaborate (Lund, Heggen, & Strand, 2013).

Generic Qualitative Inquiry

Not all inquiries need to be formally conceptualized within one of the specific inquiry traditions reviewed in this chapter. The quite concrete and practical questions of people working to make the world a better place (and wondering if what they're doing is working) can be addressed without allegiance to a particular epistemological or philosophical tradition. While these intellectual, philosophical, and theoretical traditions have greatly influenced the focus, value, and legitimacy of particular types of qualitative inquiry, it is not necessary to swear vows of allegiance to any single epistemological perspective to use qualitative methods. Experienced pragmatic researcher David Morgan (2007) has characterized this as "paradigms lost and pragmatism regained," especially with regard to a pragmatic approach to "integrating qualitative and quantitative methods" (Morgan, 2014).

While students writing dissertations and academic scholars will necessarily be concerned with ontology and epistemology, there is a very practical side to qualitative methods that simply involves skillfully asking open-ended questions of people and observing matters of interest in real-world settings to solve problems, improve programs, or develop policies. This is sometimes called *generic qualitative inquiry* (Caelli, Ray, & Mill, 2003; McLeod, 2001; Merriam, 1997).

SIDEBAR

PRAGMATISM AS A BASIS FOR MIXED METHODS

Pragmatist philosophy has been especially influential as a justification for mixed methods. Evaluation pioneer and distinguished scholar Lois-ellin Datta (1997) made the case for pragmatism as the foundation for mixed methods in a classic article in *New Directions for Evaluation*:

> I propose for our field that "pragmatic" mean *the essential criteria for making design decisions are practical, contextually responsive, and consequential.* "Practical" implies a basis in one's experience of what does and does not work. "Contextually responsive" involves understanding the demands, opportunities, and constraints of the situation in which the evaluation will take place. "Consequential" in this discussion is defined by pragmatic theory . . . [which] holds that the truth of a statement consists of its practical consequences, particularly the statement's agreement with experience. These practical consequences form standards by which concepts are analyzed and their validity determined. . . .
>
> A Pragmatic Framework. Four questions about practical consequences of design decisions are:
>
> 1. Can salient evaluation questions be adequately answered?
>
> 2. Can the design be successfully carried out, taking into consideration such issues as access to information, time available, evaluators' skills, and money or other resources required for the evaluation?
>
> 3. Are design trade-offs (for example, between depth of understanding and generalizability) optimized?
>
> 4. Are the results usable? (pp. 34–35)

Pragmatism is a thread that runs through the writings of advocates for mixed methods (e.g., Creswell, 2009; Maxcy, 2003; Morgan, 2007, 2014; Patton, 1985b, 1988b; Rallis & Rossman, 2003; Rossman & Rallis, 2011; Tashakkori & Teddlie, 2003).

In short, in real-world practice, methods can be separated from the epistemology out of which they have emerged. One can use statistics in straightforward ways without doing a philosophical literature review of logical empiricism or realism. One can make an interpretation without studying hermeneutics. And one can conduct open-ended interviews or make observations without reading treatises on phenomenology. The methods of qualitative inquiry now stand

GENERIC QUALITATIVE INQUIRY

Generic qualitative inquiry uses qualitative methods—in-depth interviewing, fieldwork observations, and document analysis—to answer straightforward questions without framing the inquiry within an explicit theoretical, philosophical, epistemological, or ontological tradition. Say you have to understand what motivates people to volunteer their time to a nonprofit. You could frame this as an ethnographic inquiry (culture of volunteerism), a phenomenological study (the lived experience of volunteering), ethnomethodological research (everyday norms of volunteering), or a realist inquiry (identifying the mechanisms that support and explain volunteering within a particular context). Indeed, you could frame the study within any of the theoretical orientations reviewed in this chapter. Or you could just interview people about volunteering and observe them while they volunteer, without using an explicit theoretical framework. This is generic qualitative inquiry (Caelli et al., 2003; Cooper & Endacott, 2007; McLeod, 2001; Merriam, 1997).

Let me offer an analogy. You can give some money to someone in need because your religious faith directs you to be charitable or you feel an explicit ethical obligation. Or you can give money to a person in need because you see the need and respond as a fellow human being, without placing that charitable act within a religious or ethical tradition. You would be engaged in a generically charitable act instead of a religiously based or explicitly ethically oriented act. The behavior is the same in both cases. The interpretive framing of the act can vary.

Young lady, the court doesn't need to hear any more about difficulties with validity and reliability in qualitative research methods, discrepancies between different postmodern epistemologies, or this "Great Paradigm Debate." Please just tell us what you saw.

Generic Qualitative Request

©2002 Michael Quinn Patton and Michael Cochran

on their own as reasonable ways to find out what is happening in programs and other human settings.

Utilization–Focused Evaluation

Standards for high-quality evaluations emphasize utility and feasibility as well as propriety, accuracy, and accountability (Joint Committee on Standards, 2010). Utilization-focused evaluation (Patton, 2008b, 2012b) is explicitly pragmatic, eclectic, and situationally adaptive and responsive. Exhibit 3.17 presents examples of pragmatic principles for utilization-focused evaluations. In the same vein, *Real-World Evaluation* is a pragmatic approach to "working under budget, time, data, and political constraints" (Bamberger, Rugh, & Mabry, 2011).

Pragmatism, like all inquiry frameworks, has its critics. Attacking the practical and contextual focus of pragmatic inquiries, detractors criticize their failure to produce generalizable knowledge. With regard to the adaptive approach to fieldwork, Pawson (2013),

in his *Realist Manifesto*, takes aim at the allures of the pragmatic perspective, which he asserts "must be reckoned with . . . because it emanates from the labor of evaluation practitioners rather than an abstract, preconceived model" (p. 71). But he is concerned that emergent designs responding to unforeseen problems and opportunities, in-the-field adjustments driven by practicality, and a proliferation of novel methods and *bricoleur* creations lack the rigor, and therefore the credibility and validity, of systematic scientific inquiry. In practice, he believes, pragmatism rapidly becomes unmanageable and, ironically, impractical and unfeasible:

> Solutions are arrived at piecemeal. The field researcher is always best placed to spot the latest intricacy, the first to herald complexity's newest twist. They will draft up a methodological solution which, if it shows promise, will be imitated by other evaluators. One splendid idea after another is added to the evaluation toolbox—but with what result? Methodological pragmatism bursts evaluation at its seams. The toolbox requires a truck to transport it. No evaluation study can encompass all of its contents. Any single evaluation will always fall a crucial spanner or two short. (p. 72)

EXHIBIT 3.17   Utilization-Focused Evaluation: Pragmatism in Practice

Utilization-focused evaluation begins with the premise that evaluations should be judged by their utility and actual use; therefore, evaluators should facilitate the evaluation process and design any evaluation with careful consideration of how everything that is done, *from beginning to end*, will affect use. Use concerns how real people in the real world apply evaluation findings and experience the evaluation process.

Pragmatic Design Principles

Principles offer guidance. They are not recipes, laws, or absolute prescriptions. Principles help in dealing with practical trade-offs in the less than perfect real world of evaluation design. Below are three common evaluation challenges with corresponding pragmatic principles to use in the face of real-world constraints.

EVALUATION CHALLENGE	PRAGMATIC PRINCIPLE
1. Providing the best possible data in time to affect decisions	1. Providing less than perfect data that is available on time to affect decisions is better than using more perfect data that is available *after* decisions have been taken.
2. Providing methodologically rigorous data on important questions	2. The meaning of "rigor" depends on context. Rigor includes not just validity and accuracy but whether the findings are actionable and useful.
3. Providing comprehensive findings	3. Timeliness trumps comprehensiveness. Less is more when the evaluation can cut to the chase and focus on what is most useful.

SOURCE: Adapted from *Essentials of Utilization-Focused Evaluation* (Patton, 2012a).

His final judgment is that pragmatism produces "endless answers to never-ending problems" (p. 72). But perhaps we need some cultural context to understand the vehemence of his objections. Pragmatism is "a uniquely American philosophy" (Schwandt, 2001, p. 204). Pawson is quintessentially British. Americans tend to like thinking of themselves as pragmatic. But to people in other countries, American pragmatism comes with baggage, some of it negative. Thus, the 10th principle of pragmatic inquiry is as follows:

Be prepared to communicate to others how and why the philosophy of pragmatism has informed the inquiry. That people will understand the philosophy of *judging truth by utility*, or agree with it, cannot be assumed.

For the other 9 principles of pragmatic inquiry, see Exhibit 3.18.

EXHIBIT 3.18　Ten General Pragmatic Principles of Inquiry

1. Focus the inquiry on getting useful answers to practical questions.

2. Select and mix methods to get diverse perspectives on and triangulated insights into the problem being studied, recognizing that all methods have strengths and weaknesses (Creswell, 2009, pp. 10–11; Morgan, 2007, 2014).

3. Adapt the design to *real-world* constraints of limited time, resources, and access—and acknowledge those constraints and their implications for findings (Bamberger et al., 2011).

4. Integrate inquiry frameworks and epistemological traditions as appropriate, doing so thoughtfully, intentionally, and with attention to the implications of combining frameworks and the controversies that may ensue (Denzin & Lincoln, 2011, pp. 4–6).

5. Look for *actionable findings*: Analyze data with an eye toward informing action (Davidson, 2012).

6. Use multiple analytic reasoning processes: deduction (reasoning from the general to the specific), induction (reasoning from the specific to the general), and abduction (working back and forth between general and specific to solve a problem) (Patton, 2011, pp. 284–285; Reichertz, 2004).

7. Examine and report how the inquiry itself and the processes involved in data collection have affected what is learned (Flick, von Kardoff, & Steinke, 2004, p. 17).

8. Be explicit about the values that undergird and inform the inquiry. Pragmatism is not value-free. Utility, like beauty, is in the eye of the beholder. It involves making judgments: value judgments. The very definition of a problem to be studied involves values about how the world might be better—with human rights, social justice, gender equity, and shared governance as examples (Davidson, 2005, pp. 85–98; Julnes, 2012; Schwandt, 2002, pp. 137–155).

9. It is easier to establish what doesn't work than what does work, and what works in one context and at one time may not work at another place and in another time, so both utility and truth derived from utility are context dependent, subject to further inquiry, and constitute at best "partial truth" (Pawson, 2013, p. 192).

10. Be prepared to communicate to others how the philosophy of pragmatism has informed the inquiry. That people will understand the philosophy of *judging truth by utility*, or agree with it, cannot be assumed.

Patterns and Themes Across Inquiry Frameworks: Chapter Summary and Conclusions

> Research on humans in their social world by other humans is not a traditional science like the one created by Galileo and Newton. It's not that the creators were wrong. Far from it. The ones who were wrong were the historical figures who tried to imitate the way the creators worked, neglecting the fact that learning how people make it through the day is different from dropping balls from the Leaning Tower of Pisa or getting hit on the head by falling apples. Galileo didn't have to communicate with the balls. Besides, he didn't have to worry that the balls might look down 185 feet and refuse to jump and throw him over the parapet instead. (Agar, 2013, p. 6)

All the inquiry frameworks reviewed in this chapter grapple with the challenge of how to study human beings. All have turned to qualitative methods as the best way to do so. What distinguishes these inquiry frameworks is that they focus on and prioritize different questions and draw on different philosophical and epistemological traditions to inform their inquiry.

Six Distinguishing Questions

The variety of qualitative inquiry frameworks are distinguished by answers to six core questions (one for each day of the week plus a day left over to integrate your answers):

1. *What do we believe about the nature of reality?* Diverse answers derive from varying ontological premises and debates, for example, the possibility of a singular, verifiable reality and truth versus the inevitability of socially constructed multiple realities.

2. *How do we know what we know?* Diverse answers stem from divergent epistemological premises and debates about the possibility and desirability of objectivity, subjectivity, causality, validity, and generalizability.

3. *How should we study the world?* Different inquiry frameworks specify different methodological preferences, priorities, and procedures about what kinds of data and design to emphasize for what purposes and with what consequences.

4. *What is worth knowing?* Answers to this question are grounded in philosophical debates about what matters and why.

5. *What questions should we ask?* Different disciplines (e.g., anthropology, sociology) and interdisciplinary perspectives (e.g., ecological psychology, cultural studies) have emerged from focusing on different burning questions and different methodological preferences for answering those questions credibly (through peer review) within each discipline.

6. *How do I personally engage in inquiry?* This involves different perspectives and debates about injecting personal experiences and values into the inquiry, including issues of voice and using the inquiry to bring about change.

The same phenomenon, program, organization, or community studied by researchers from different perspectives will lead to quite different studies even though they might all undertake observations, interviews, and document analysis. Nor would it necessarily be possible to synthesize the descriptions and findings of such different studies even though they took place in the same setting. When researchers and evaluators operate from different inquiry frameworks, their results will not be readily interpretable by or meaningful to one another. While the frameworks provide guidance and a basis for interaction among researchers operating within the same framework, the different theoretical frameworks constitute barriers that impede interaction across and among different perspectives. In effect, each theoretical framework is a mini paradigm with its own internal logic and assumptions.

This means that one cannot reasonably ask which theoretical framework is "right," best, or most useful. It depends on what one wants to do and which assumptions one shares. Gareth Morgan stated the problem quite succinctly after presenting a variety of research perspectives:

> There was the question as to how the reader could come to some conclusion regarding the contrary nature, significance, and claims of the different perspectives. . . . I realized that there was a major problem here. . . . There is a fallacy in the idea that the propositions of a system can be proved, disproved, or evaluated on the basis of axioms within that system. . . . This means that it is not possible to judge the validity or contribution of different research perspectives in terms of the ground assumptions of any one set of perspectives, since the process is self-justifying. Hence the attempts in much social science debate to judge the utility of different research strategies in terms of universal criteria based on the importance of generalizability, predictability and control, explanation of variance, meaningful understanding, or whatever *are inevitably flawed: These criteria inevitably favor research strategies consistent with the assumptions that generate such criteria as meaningful guidelines for the evaluation of research.* It is simply inadequate to attempt to justify a particular style of research in terms of assumptions that give rise to that style of research. . . . Different research perspectives make different kinds of knowledge claims, and the criteria as to what counts as significant knowledge vary from one to another. (Morgan, 1983, pp. 14–15)

Essentially, then, you must make your own decision about the relative value of any given inquiry framework and perspective. Each has strengths. Each has limitations. No universal standard can be applied to choose among these different frameworks. Quite the contrary, the diversity itself is a good indicator of the complexity of human phenomena and the challenges involved in conducting research.

Themes That Cut Across the Inquiry Traditions and Frameworks

To conclude this chapter, I'm going to present the results of my own qualitative analysis of themes that emerge in looking across inquiry frameworks.

1. *Definitions of inquiry frameworks vary.* Charles Peirce, William James, and John Dewey defined pragmatism differently (Russell, 1959, pp. 276–279). In his *Dictionary of Qualitative Inquiry,* Thomas Schwandt (2001) introduces phenomenology with this caveat: "This complex, multifaceted philosophy defies characterization because it

is not a single unified philosophical standpoint" (p. 191). Of symbolic interaction, he warns, "Like all frameworks informing qualitative studies, this theory comes in a variety of forms and thus is difficult to summarize briefly" (p. 244). Systems theory definitions and inquiry frameworks abound and befuddle (Pawson, 2013, p. 53; Williams, 2005; Williams & Hummelbrunner, 2011). While ethnographers agree that ethnography is the study of culture, they disagree about how to define culture. The inevitable discombobulating conclusion is this: *Do not go in search of the definitive, agreed-on, "right" definition for any inquiry framework.* You will not find it. You will have to select that definition that resonates with you and your inquiry purpose and be prepared to explain, justify, and even defend whichever definition, or definitions, you select.

2. *Identification of founding theorists and important contributing figures within each tradition can be a matter of dispute.* Sometimes, the first person who labeled a distinct inquiry framework is recognized as the founder of that inquiry approach. Harold Garfinkel (1967), for example, coined the term *ethnomethodology.* Clark Moustakas (1990) originated heuristic inquiry. Glaser and Strauss (1967) are credited with developing grounded theory but with this caveat:

> Of course, the method did not arise solely from the work of Glaser and Strauss. If we look at previous work in the Chicago School [of Sociology] we see common elements—the basic social process, for instance, appears in the writings of that time. (Morse et al., 2009, p. 9)

But more general frameworks like ethnography, phenomenology, hermeneutics, narrative inquiry, and complexity theory have mixed genealogies, multiple contributors, and disagreements among theorists about who contributed what, who influenced whom, and what significance to attach to various contributions. We find the intersection of the definition challenge (Item 1 above) and the founders' challenge in the very definition of phenomenology offered by Denzin and Lincoln (2011): "A complex system of ideas associated with the works of Edmund Husserl, Martin Heidegger, Jean-Paul Sartre, Maurice Merleau-Ponty, and Alfred Schutz" (p. 16).

3. *The perspectives of pioneers are often misrepresented in secondary summaries of their views. Consult original sources to understand what major inquiry pioneers and theorists actually said.* Be wary of

relying entirely on secondary and tertiary sources, like this book, if you intend to work within a particular framework. This chapter is a menu, but to experience the full meal of any inquiry framework, you must go to the original cookbooks and feast on the pioneering writings of those in whose footsteps you intend to follow.

A wonderful example of why this is so can be found in a confessional essay by distinguished sociologist Robert K. Merton (1995) about how he and others misattributed the Thomas theorem, which is the foundation of the social constructionist inquiry framework: "If men define situations as real, they are real in their consequences" (see Exhibit 3.7). The book in which it first appeared was coauthored by William I. Thomas and Dorothy Swaine Thomas (1928). However, Merton confesses that in his own influential writings he attributed the theorem only to W. I. Thomas, omitting his wife's coauthorship and contribution. Others followed suit *and continue to do so to this day.* Merton openly reflected on whether this omission, which he deeply regretted, was a matter of sexism, scholarly sloppiness, innocent inattention to detail, or all of these in combination. The moral of the story is that secondary and tertiary sources are filled with inaccuracies, out-of-context quotations, distorting summaries, abysmal abstracts, misattributions, misrepresentations, mischaracterizations, and other mischief that can only be corrected by consulting original sources.

A parallel caveat concerns the labels attached to theorists. These can be misleading or can change over time. For example, Miles and Huberman described themselves as positivists in the first edition of *Qualitative Data Analysis* (1984) but came to think of themselves as realists in the second edition (1994). Or take the case of Paul-Michel Foucault (1926–1984), a French philosopher and social historian, who analyzed and addressed power and its functions, including how those in power control knowledge. That led to his being labeled variously as a postmodernist or poststructuralist, both of which he rejected, preferring to classify his thought as a critical history of modernity (Ball, 2013; Foucault, 2013). To know what Foucault wanted to be called, you have to consult his own writings, not rely just on secondary characterizations and labels.

4. *Fidelity disputes abound: Advocates, theorists, practitioners, and critics debate what is central, distinct, and essential in each inquiry framework.* Giorgi (2006) advocates what he considers the right, high-fidelity approach to phenomenological inquiry. Exhibit 3.5 presents his review of common errors and deviations he found in reviewing doctoral dissertations that claimed to be following phenomenology. Fidelity becomes an issue for every model or approach as followers adapt the original idea to new contexts. The question is this: How much variation and adaptation can there be before what is being done is so different from the original that it no longer falls within the original idea? Consider the case of *empowerment evaluation,* a well-defined participatory inquiry framework for program evaluation (Fetterman & Wandersman, 2005). Miller and Campbell (2006) examined 47 published studies labeled "empowerment evaluations." They found wide variation in what was done and which principles were followed, including a case of an evaluation that was designed and executed by an evaluator "with no input or involvement from stakeholders.... [But the evaluator thought] the project was an empowerment evaluation because by allowing a disenfranchised population to respond to a survey, the population was afforded a voice" (p. 306).

5. *Subdivisions and special interest subgroups have emerged within each inquiry tradition.* Phenomenology can scarcely be considered a coherent inquiry framework given the many subdivisions: existential phenomenology, transcendental phenomenology, hermeneutic phenomenology, humanistic phenomenology, sociological phenomenology, and psychological phenomenology. Realism follows suit: phenomenological versus scientific realism (Cumpa & Tegtmeier, 2009); ontological realism, philosophic realism, critical realism, experiential realism, perspectival realism, subtle realism, emergent realism, natural realism, innocent realism, and agential realism (Maxwell, 2012, pp. 3–4).

The "tussles, tensions, and resolutions" of different versions of grounded theory have been thoughtfully reviewed by Janice Morse (2009). Given the Internet age and increasing use of social media, communities of practice have emerged to facilitate communications dialogue and debate among those who feel affinity to a particular inquiry framework or one of its subdivisions. A quick search will take you to websites, listservs, and professional organizations convened around diverse inquiry frameworks and special interests within them.

6. *Combinations of inquiry frameworks have become commonplace—and controversial.* Some combinations involve just two inquiry traditions. For

example, Charmaz (2009) has created constructivist grounded theory. Hermeneutical phenomenology combines two major traditions. One can do a heuristic feminist study engaging in heuristic inquiry from a feminist perspective. Or one can do "critical ethnography" (Thomas, 1993), combining elements of critical theory and ethnography. Bentz and Shapiro (1998) have created what they call "mindful inquiry" as a synthesis of phenomenology, hermeneutics, critical theory, and Buddhism. From phenomenology, they take the focus on experience and consciousness. From hermeneutics, they take the focus on texts, on the process of understanding, and on letting new meanings emerge from the research process. From critical theory, they direct attention to the social and historical context of both the researcher and the research topic, including attention to domination, injustice, and oppression. From Buddhism, they take the focus on becoming aware of one's own "addictions" and attachments and on practicing compassion. In positing this synthesis, they aim to place the researcher, rather than research techniques, at the center of the research process. This adds something of a reflexive, autoethnographic orientation as another foundation of mindful inquiry because the mindful inquirer uses awareness of personal, social, and historical context and personal ways of knowing to shape the research.

Combining inquiry frameworks can spark severe criticisms. Combinations are controversial. The hope for such combinations is that they combine the strengths of different approaches, but they can just as easily undermine the strength of high-fidelity adherence to a particular framework and create a combination that weakens each of the frameworks brought together. Indeed, Blaikie (1991) has asserted that those who fail to acknowledge the ontological and epistemological differences built into different methods are guilty of "ignorance or misunderstanding" (pp. 126, 128). Giorgi (2006) is highly skeptical of dissertation students thinking they can willy-nilly add other inquiry frameworks to phenomenology. Thus, theorists and proponents of various qualitative epistemologies argue that they should be implemented as pure forms to fully realize the focus and purpose of each. Pragmatists argue that though there are times when some approaches can be mixed, the burden of proof that this is a good and fruitful idea rests with the researcher proposing to integrate approaches. It's not easy. It should not be undertaken cavalierly. Just because you consider yourself an effective multitasker doesn't mean you can effectively implement multiple

inquiry frameworks. It should be the exception rather than the rule, like mixing alcohol and prescription drugs: possible on special occasions but potentially dangerous; be sure you have a designated driver! A clear and convincing case must be presented, both methodologically and epistemologically, that combining multiple approaches actually serves to deepen the answer to the research question and that whatever approaches are combined can be implemented with quality. It shouldn't just be dallying around with different approaches and using more than one because you're having trouble deciding or focusing, which is what I see with some students who, unable to distinguish approaches enough to figure out which one is most appropriate, take what they hope will be the easy way out and design a combination. It might work. And, again, it might not.

7. ***All frameworks have attracted passionate criticism. Criticism comes with the territory.*** Having noted above the criticism that combining frameworks can generate, it's worth noting that any particular inquiry will also attract critics, skeptics, naysayers, and cynics. William Blake famously observed, "Without contraries is no progression." There's a reason why academic disciplines are characterized as "disputatious communities"; and not for nothing have debates about inquiry frameworks been called "paradigm wars" (Paulson, 1990). Pawson (2013) brings all his legendary wit to bear in a scathing attack on systems and complexity theories: "There are so many varieties of systems thinking that I hardly know where to start (an old complexity theory joke by the way)" (p. 53); he is no less acerbic about the inadequacies of pragmatism (pp. 71–80). He denigrates my own complexity-pragmatic approach, *developmental evaluation* (Patton, 2011), as "(a) evaluation-by-adjective, and (b) evaluation-by-personality" (Pawson, 2013, p. 77).

Phillips (2000) offers a comprehensive defense of "naturalistic social science"—that social science can attain the same rigor and certainty as natural science. He affirms traditional scientific values of truth and objectivity while critiquing what he calls the "beasts" of holism, postpositivistism, Kuhnian relativism, hermeneutics, narrative research, and constructivism.

Martyn Hammersley (2008a) has published a whole book of essays "questioning qualitative inquiry" that collects together in one volume critiques of particular frameworks as well as of qualitative inquiry generally.

Any given inquiry framework has its origins in criticism of some existing approach, critiquing it and offering what the critic considers a better alternative. Positivism spawned postpositivism and antipositivism; realism has given rise to antirealism. Naturalists can enjoy entertaining dinners with antinaturalists, as can modernists with postmodernists, empiricists with postempiricists, and structuralists with poststructuralists. Objectivists versus subjectivists constitute a macro-epistemological and ontological debate; phenomenological realism versus scientific realism is a more micro debate (Cumpa & Tegtmeier, 2009). Ethnomethodology was created by Harold Garfinkel as a critique of functionalist sociology; he indicted mainstream sociology for treating "human actors as 'judgmental dopes' who passively carried out prescribed actions on the basis of internalized norms" (Schwandt, 2001, p. 81).

You get the point. The dialectic of thesis giving rise to antithesis runs deeply through the history of inquiry frameworks. So what does this mean practically? I offer this: Whatever inquiry framework you choose to use, study the pioneering theorists and more recent practitioners, but study also the critics. There is much to learn, not least of which is where the pitfalls lie in any given approach.

8. *The extent to which an inquiry framework specifies precise methods of data collection and analysis varies.* Ethnography prescribes in-depth fieldwork, but ways of conducting fieldwork vary greatly (Bernard, 1998). By comparison, grounded theory prescribes some specific data collection processes and analysis techniques. Phenomenology is more philosophical than methodological, but heuristic inquiry, an offspring of phenomenology, is quite detailed methodologically and procedurally. Narrative inquiry can take a variety of forms but has generated a detailed and in-depth approach to biographic interviewing (Wengraf, 2001).

The extent to which philosophical and epistemological frameworks have or even can generate specific methods is a matter of debate.

> In relation to these arguments, however, we should ask whether it is true that different sources of data, or even different methods, do involve conflicting ontological or epistemological assumptions; and if they do, whether this implies incompatibility. A number of writers assert that empiricism, interpretivism, and realism are fundamentally different philosophical orientations that underpin various social research methods. However, these authors do not effectively establish that conflicting epistemological and ontological assumptions are *necessarily* built into the use of specific methods. (Hammersley, 2008b, pp. 28–29)

In later chapters, we will take an in-depth look at methods: interviewing techniques, observation methods, and qualitative data analysis approaches. Many of the methodological choices are more informed by pragmatism than philosophy. So this is just a cautionary note that if, as recommended earlier (Item 2), you turn to the original writings of pioneering inquiry framework theorists, be prepared to find a lot of philosophy about how to study the world but often not much methodological direction. The methods have developed through the experiences of practitioners trying to apply the philosophical guidance to real-world inquiries. The theory-to-practice translation is seldom smooth and often is another source of debate.

9. *The primary unit of analysis in different frameworks varies: Some focus more on understanding individuals, others on understanding social groups.* Ethnography focuses studies on groups of people who share a culture. Autoethnography begins with individual experience and reflection as the entry to culture. Phenomenology aspires to get at the essence of lived experiences for humans generally. Heuristic inquiry, like autoethnography, makes individual experience the focus of inquiry. Social constructionism aims to illuminate how groups of people come to share perspectives. Constructivism examines how individuals develop their perspectives (see Exhibit 3.8). Symbolic interaction, with a social-psychological sensibility, looks at interactions among individuals. Ethnomethodology focuses on people in specific situations and interactions. In essence, different frameworks put different boundaries around what aspects of the human experience are of priority concern. This is most directly reflected in different units of analysis.

10. *There is no definitive list of alternative inquiry frameworks and no agreed-on way of categorizing or grouping them.* Take a look again at Exhibit 3.3 (pp. 97–99), which summarizes the inquiry frameworks reviewed in this chapter. The purpose of this review has been to accentuate distinctions so as to facilitate choices. But as this summary of crosscutting and overarching themes makes abundantly and even painfully clear, the boundaries between perspectives remain fuzzy. Those of us who try to make sense of the options through some overall conceptual framework, as I have done in this chapter by reviewing contrasting inquiry frameworks and focusing on core questions, emphasize different approaches. No consensus exists about how to classify the varieties of qualitative research. Crotty (1998, p. 5) has elaborated five major theoretical perspectives as the foundations

of social research: positivism (and postpositivism), interpretivism (which includes phenomenology, hermeneutics, and symbolic interactionism), critical inquiry, feminism, and postmodernism (to which he adds an "etc." to suggest the open-ended nature of such a classification). Creswell (2012) also settled on five approaches, but a different five: (1) narrative research, (2) phenomenology, (3) grounded theory, (4) ethnography, and (5) case study. Jacob (1987) chose yet another set of five for a qualitative taxonomy: (1) ecological psychology, (2) holistic ethnography, (3) ethnography of communication, (4) cognitive anthropology, and (5) symbolic interactionism. Schwandt (2000) highlighted three "epistemological stances for qualitative inquiry": (1) interpretivism, (2) hermeneutics, and (3) social constructivism. Denzin and Lincoln (2000) organized their review of qualitative variety around seven historical periods and seven "paradigms/theories": (1) positivist/postpositivist, (2) constructivist, (3) feminist, (4) ethnic, (5) Marxist, (6) cultural studies, and (7) queer theory. Wolcott (1992) created a family tree of 20 distinct branches showing different "qualitative strategies." Tesch (1990) identified 27 varieties. Having examined some of the various attempts to classify qualitative approaches, Miles and Huberman (1994) concluded,

As comprehensive and clarifying as these catalogs and taxonomies may be, they turn out to be basically incommensurate, both in the way different qualitative strands are defined and in the criteria used to distinguish them. The mind boggles in trying to get from one to another. (p. 5)

Schwandt, who has studied these distinctions as much as anyone and is the lexicographer of the *Dictionary of Qualitative Inquiry* (2001), offers this reflection on theoretical distinctions:

It seems to be a uniquely American tendency to categorize and label complicated theoretical perspectives as either this or that. Such labeling is dangerous, for it blinds us to enduring issues, shared concerns, and points of tension that cut across the landscape of the movement, issues that each inquirer must come to terms with in developing an identity as a social inquirer. In wrestling with the ways in which these philosophies forestructure our efforts to understand what it means to "do" qualitative inquiry, what we face is not a choice of which label—interpretivist, constructivist, hermeneuticist, or something else—best suits us. Rather, we are confronted with choices about how each of us

QUEER THEORY AND QUALITATIVE INQUIRY QUESTIONS

Queer researchers are in good company with other scholars drawing on poststructuralist and postmodernist approaches such as some feminist, anti-racist and postcolonial scholars.... In research deemed "queer," the methods we use often let us speak to or interact with people, usually on the basis of sexual/gender identities and within anti-normative frameworks.... If, as queer thinking argues, subjects and subjectivities are fluid, unstable and perpetually becoming, how can we gather "data" from those tenuous and fleeting subjects using the standard methods of data collection such as interviews or questionnaires? What meanings can we draw from, and what use can we make of, such data when it is only momentarily fixed and certain?

And what does this mean for our thinking about ourselves as researchers? ... We found it disconcerting that we often do not apply our queer re-theorising, re-considering and re-conceptualising to our social science methodologies and choice of methods. As queer approaches to research proliferate across the social sciences, we argue there is a certain "sweeping under the carpet" of how we actually "do" research as "social scientists" given our attractions, and attachments, to queer theory....

Theory, data and method cannot be understood in isolation from each other and the relationship between theory and data is a methodological problem. When we think of this intersection, persistent and unresolved questions emerge: What impact, if any, could (or should) queer conceptualisations have on our methodological choices and in what ways? Can social science methods be "queered" or even made "queer enough"? How can social science methodologies feed into and question queer epistemological paradigms? How are social science methodologies, and the knowledges created by them, addressed in queer theories? If methodologies are meant to coherently link ontological and epistemological positions to our choice of methods, are methodologies automatically queer if queer conceptualisations are used? Can we have queer knowledges if our methodologies are not queer? Is there such a thing as queer method/methodology/research? These sorts of questions [are] often overlooked in the excitement of new revelations and insights gained by queer interventions into empirical data.

—Browne and Nash (2010, pp. 1–2)
Queer Methods and Methodologies

wants to live the life of a social inquirer. (Schwandt, 2000, p. 205)

Exhibit 3.19 presents an overview of these themes.

EXHIBIT 3.19 Distinguishing and Understanding Alternative Inquiry Frameworks: Crosscutting Themes

1. Definitions of inquiry frameworks vary.

2. Identification of founding theorists and important contributing figures within each tradition can be a matter of dispute.

3. The perspectives of pioneers are often misrepresented in secondary summaries of their views. Consult original sources to understand what major inquiry pioneers and theorists actually said.

4. Fidelity disputes abound: Advocates, theorists, practitioners, and critics debate what is central, distinct, and essential in each inquiry framework.

5. Subdivisions and special-interest subgroups have emerged within each inquiry tradition.

6. Combinations of inquiry frameworks have become commonplace—and controversial.

7. All frameworks have attracted passionate criticism. Criticism comes with the territory.

8. The extent to which an inquiry framework specifies precise methods of data collection and analysis varies.

9. The primary unit of analysis in different frameworks varies: Some focus more on understanding individuals, others on understanding social groups.

10. There is no definitive list of alternative inquiry frameworks and no agreed-on way of categorizing or grouping them.

Note: Each of these themes is discussed and elaborated on pages 159–164.

Chapter Summary and Conclusion

The chapter opened with a review of the quantitative/qualitative methods paradigms debate. Exhibit 3.1

(pp. 90–91) presented 10 contrasting emphases and contributions of quantitative, qualitative, and mixed methods approaches. Debate about these contrasting and competing perspectives has been an important part of the history of research and evaluation. While this chapter demonstrates that the variety of inquiry approaches has expanded well beyond the simplistic dichotomy between quantitative and qualitative paradigms, the debate is still alive in many places. The paradigms debate is part of our methodological heritage. Knowing about it and understanding the passions the debate has aroused, and still can arouse, is an important foundation for making methods decisions. At its core, the paradigms debate "addresses one of the most important and contentious issues challenging applied research and evaluation practice today: what constitutes credible evidence" (Donaldson et al., 2009, p. vii). Part of your responsibility as a researcher and/or evaluator is to develop and be able to articulate your own position with regard to what constitutes credible evidence and why. Chapter 9, this book's final chapter, will return to this issue as its focus.

Having reviewed the paradigms debate as context for methodological decision making, the bulk of this chapter has consisted of distinguishing theoretical perspectives by their foundational questions and discussing the implications for inquiry of positioning your study within a particular intellectual and theoretical tradition. Exhibit 3.3 (p. 97) summarizes the 16 contrasting and competing theoretical inquiry frameworks we've discussed: ethnography, autoethnography, reality-testing, foundationalist epistemologies (positivism, postpositivism, and empiricism), grounded theory, realism, phenomenology, heuristic inquiry, social constructionism and constructivism, narrative inquiry, ethnomethodology, semiotics, symbolic interaction, hermeneutics, systems theory, complexity theory, and pragmatism.

There is no definitive list of alternative inquiry frameworks and no agreed-on way of categorizing or grouping them. The 16 frameworks reviewed and discussed here are neither exhaustive of possibilities nor without controversy. While these diverse approaches reflect different inquiry traditions and priorities, they also provide windows into some central and enduring qualitative inquiry issues. Thus, I concluded the review with an analysis and synthesis of 10 themes that emerge in looking across inquiry frameworks (see Exhibit 3.19).

In concluding, I would reiterate that you must make your own decision about the relative value of any given inquiry framework and perspective based on your own interests, theoretical preferences, intellectual tradition, and inquiry context. Each approach has strengths.

Each has limitations. No universal standard can be applied to choose among these different frameworks and others not reviewed here. The diversity of possibilities reflects the complexity of human phenomena and the challenges involved in conducting research. Theories guide us through the world's maze. Prolific cultural theorist Arthur Asa Berger offered a concluding insight, a "coda," about how to use diverse theories productively.

> In the course of my career, a number of people have asked me how I was able to publish more than sixty books. I've actually written more books but not all of them have been published. I generally tell them that I know some theories and have learned how to apply the concepts found in these theories to any number of different topics. It is through theories and concepts that I make sense of things that interest me. When I have some topic that I want to write about, in my mind I think "round up the usual suspects," and then I see what ideas these "suspects" (theorists) have to offer that I can use. It also helps if you know how to type with ten fingers and are disciplined. (Berger, 2010, p. 149)

From Theory to Practice

The next chapter explores some of the ways in which qualitative inquiry can contribute to practical knowledge and pragmatic understandings. To help make that transition, this chapter ends with a practical, cautionary tale from "Halcolm the Gourmand."

APPLICATION EXERCISES

1. *Paradigms debate:* The chapter opened with a discussion of the quantitative/ qualitative paradigms debate. Exhibit 3.1 (pp. 90–91) summarizes the differences. Discuss how the debate shows up in your own discipline or field of interest. What is at the heart of the debate? Why does it matter? What is your own position on the paradigms debate? Explain and support your position. (If you are not sure, or undecided, explain why.)

2. *Advice to a graduate student:* Exhibit 3.2 (p. 96) is a letter from a graduate student asking for advice. Based on your reading and understanding of this chapter, write a letter of response.

3. *Examining a theoretical perspective in depth:* Exhibit 3.3 (pp. 97–99) provides an overview summary of the diverse and competing theoretical orientations discussed in this book. Select any two of these approaches, and apply them to a topic that you would be interested in researching. Demonstrate how the two approaches would lead you to frame your inquiry in different ways. Show that you understand how a theoretical inquiry framework influences the formulation of your research question.

4. *Pragmatic:* The review of different theoretical orientations ends with a discussion of pragmatism (see Module 18, including Exhibits 3.17 and 3.18). Select a research question of interest to you, and discuss how you would approach it from a pragmatic perspective. Demonstrate that you understand pragmatism as an inquiry framework.

5. *Generic qualitative inquiry:* What are the strengths and weaknesses of generic qualitative inquiry? (See pp. 154–155.) Demonstrate that you understand why some advocate generic qualitative inquiry and others criticize it.

6. *Halcolm (How come?):* Discuss the "Halcolm the Gourmand" cartoon at the end of the chapter (pp. 166–167). What does that story have to do with this chapter? What are its meanings and implications in relation to the module titles of this chapter and the practical focus of the next chapter?

Practical and Actionable Qualitative Applications

 The origin of the @ symbol is a mystery. The first documented use was in 1536, when a Florentine merchant, Francesco Lapi, used @ to denote units of wine called amphorae, which were shipped in large clay jars. The symbol's modern use emerged in 1971, when computer scientist Ray Tomlinson chose it to connect computer programmers with one another. He was searching for a symbol that wasn't used much, which eliminated an exclamation point, a comma, and even an equal sign, so he chose @. Thus did "the ancient @, once nearly obsolete, became the symbolic linchpin of a revolution in how humans connect" (Allman, 2012, p. 63). Simple yet elegant, it is the ultimate practical and useful symbol.

Apprenticeship in Pragmatism

A young carpenter, at the beginning of his career, came to Halcolm in distress. He had studied diligently to master carpentry. At the completion of his apprenticeship, the master carpenters said that his technical competence and skill were unmatched for one so young.

Halcolm knew all this, for word of the young man's mastery had reached even the great one. Yet Halcolm could also see that the young carpenter was in great distress. "What troubles you?" Halcolm asked gently.

"My parents, my townspeople, my master teachers have been most generous. Upon completion of my apprenticeship, they joined together to give me a fine set of tools. I have been trained by the best. I am told that my skills are—What can I say without being immodest?—my skills are adequate." The young man paused, his distress obvious and growing even as he spoke.

"Then what is the problem?" asked Halcolm. The young man looked down, embarrassed in the presence of the great one. It was a long time before he spoke, and then only in a whisper. "I have nothing to build."

"Ah, I see," said Halcolm.

"No one will give me any orders," continued the young man.

"Let me make some inquiries," offered Halcolm. "Return in a week and I'll tell you what I have learned."

The seven days were an agonizing eternity for the young man. At last, it was time to find out what Halcolm had learned through his discrete interviews with a few knowledgeable and well-connected people.

"I have confirmed all that you told me," Halcolm began. "Your skill is much respected. Your tools are the finest, given with much affection. Your competence is not in doubt. And yet, you have nothing to build."

The young man waited for Halcolm to continue.

"During your apprenticeship you did the latticework on the new cathedral. The craftsmanship you displayed is admired by all. You designed and constructed the intricate woodwork of the new town hall directrix—another work of great art admired even by your masters. You carved and installed the elaborate wine racks in the guardians' estate. In all these efforts you have distinguished yourself and pleased those for whom you built."

The young man was pleased but perplexed as he heard Halcolm affirm the quality of his work. Indeed, hearing the affirmations deepened his distress at finding himself now with nothing to build. Halcolm continued.

"You now have nothing to build because the towns-people believe your artistry and craftsmanship are far superior to their simple needs. They need simple chairs, tables, and doors. You work on cathedrals, town halls, and estates. You have designed objects of great beauty and complexity. You have not designed and built objects of great simplicity and practicality. To do the latter only looks easier, but takes no less skill.

"Build me a simple, functional and practical bookcase at reasonable cost, and let us see what the townspeople think. Apply your skills to the everyday needs of the people and you shall not lack for work."

Halcolm expected the young man to be delighted at the prospect of a solution to his problem—and regular employment. Instead, he saw the distress deepen into despair.

"I do not know how to build simple, practical, and functional things," lamented the young man. "I have never applied my skills and my tools to such things."

"Then your apprenticeship is not over," said Halcolm.

And they went down to Halcolm's workshop where the young man began to learn anew.

—From Halcolm's
Applied Arts and Sciences

Chapter Preview

Qualitative methods offer ways of finding out what people do, know, think, and feel by observing, inter-viewing, and analyzing documents. The previous chapter reviewed how qualitative methods draw on and contribute to generating and confirming social science theory. This chapter reviews how qualitative methods can contribute to *useful* evaluation, *practical* problem solving, real-world decision making, action research, policy analysis, and organizational, com-munity, and international development. This chap-ter offers examples of how qualitative methods can help answer concrete questions, support development, improve programs, and contribute to social change.

Module 20 opens the chapter by examining prac-tical purposes, concrete questions, and actionable answers that emanate from qualitative inquiry. Illumi-nating and enhancing *quality* provides a special focus: Quality of life, quality of programs, managerial qual-ity, and quality assurance encompass special concerns that lend themselves to qualitative methods. Module 21 then examines program evaluation applications, especially a focus on outcomes, often considered the domain for quantitative indicators. We'll look at the qualitative dimensions of program outcomes. Mod-ule 22 delves into specialized qualitative evaluation applications, including discovering and documenting unanticipated outcomes, understanding innovations and social change interventions, process evaluation, legislative auditing and monitoring, and my rumina-tion debunking so-called best practices. Module 23 will review program evaluation models and theories of change that are especially aligned with qualitative methods, like responsive evaluation, implementation evaluation, comparing diverse programs in differ-ent settings, preventing problems, and documenting development over time. Module 24 takes up interac-tive and participatory qualitative applications. Module 25 reviews democratic evaluation, indigenous research and evaluation, capacity building, and cultural com-petence. Module 26 examines special methodological applications: exploratory research; adding depth, detail, and meaning to quantitative analyses; rapid reconnais-sance; validity checks on the numbers; opening and looking into the experimental design black box; break-ing the routine to generate new insights; and futuring scenarios. Module 27 closes the chapter with a vision of the utility of qualitative methods in support of solv-ing problems and creating a better world.

It's a lot to cover, but even with all the examples and applications we will review, we'll just scratch the surface of the diverse and wide-ranging ways in which qualita-tive methods are being used. Thus, this chapter is meant to provide a window through which to get a closer look at practical and actionable qualitative applications. Much more lies beyond what we can see through the window of this chapter. My hope is that what you do see will encourage you to continue and expand the journey on your own, pursuing your own interests and applications.

Practical Purposes, Concrete Questions, and Actionable Answers: Illuminating and Enhancing Quality

> There is nothing so practical as a good theory.
>
> —Kurt Lewin (1890–1947)
> Pioneer of social and applied psychology
>
> Good theory is derived from observations of practice.
>
> —Halcolm

Practical Ethnography (Ladner, 2014), the *Phenomenology of Practice* (van Manen, 2014), and *Development Evaluation* (Rogers & Fraser, 2014) illustrate the integration of theory and practice. Practical qualitative inquiry harnesses research methods to inform and support social change (Leavy, 2014). This includes engaging in *Action Science* (Argyris, Putnam, & Smith, 1985), or what distinguished applied anthropologist Michael Agar (2013) has dubbed *The Lively Science.* Here are a few examples:

- Ethnographers work with human rights advocates to investigate allegations of the abuse, rape, and torture of indigenous peoples. Such inquiries require skilled and sensitive interviewing of victims; well-documented case studies of what has happened, sometimes to entire villages; and scrupulous collection and analysis of documents, photographic evidence, and observations from key informants (international aid workers, medical staff in clinics who have treated victims, local police, religious leaders, etc.). Credibility is of utmost importance, for at stake are claims that can affect compensation to victims, prosecution of crimes against humanity, and the legitimacy of governments.
- Qualitative field work is a core part of epidemiological investigations tracking outbreaks of diseases like polio, HIV/AIDS, SARS, H1N1 virus, and other contagious diseases. Interviews with people who have contracted diseases, or with the relatives of those who have died, help establish how diseases are transmitted.
- In-depth clinical case studies provide insights into the nature and impacts of mental health problems. Bruce Perry's case studies of children who

have suffered severe childhood trauma, like *The Boy Who Was Raised as a Dog* (Perry & Szalavitz, 2006), have opened up new ways of treating these children so that they can have healthy and productive lives.

- Comparative case studies of businesses and organizations identify the factors that separate the good from the great (Collins, 2001a); those that thrive under conditions of chaos from those that are overwhelmed by complexity and fail (Collins & Hansen, 2011); businesses that endure from those that do not (Collins & Porras, 2004); and organizations that are successfully strategic from those that flounder (Mintzberg, 2007). Such qualitative findings have been hugely influential in improving management practices.
- From 2006 to 2010, 55 train and engine service employees died during rail yard switching operations. A group representing varied interests and roles in the railroad industry, including government regulation of the industry, formed the Switching Operations Fatality Analysis (SOFA) Working Group to carry out forensic in-depth case study investigations of the accidents' causes. The group's analysis concluded that injuries and deaths resulting from railroad switching operations were not random, or simply unfortunate events, or just plain bad luck. SOFA's rigorous cross-case analysis found underlying patterns that yielded recommendations for safety approaches that would reduce deaths (SOFA, 2010).
- Danah Boyd (2014) interviewed teenagers about their online, virtual lives and network connections and interactions. She has documented the complicated adaptations teenagers make to protect their privacy, including creating forms of interpersonal encryption that disguise and control their meanings when communicating with each other. She has also documented the cyber risks they experience if private conversations and chat histories are exposed.

This chapter samples some of the research and evaluation questions for which qualitative inquiry strategies are especially appropriate and powerful. The actual and potential applications of qualitative methods are so diverse that this review, while including a great variety of applications, is far from exhaustive. My purpose is to expand the horizons of what is possible and appropriate for both practitioners and decision makers. Because the opportunities for qualitative inquiry are so vast, the examples offered here can be no more than the tip of the proverbial iceberg, merely hinting at the enormous array of qualitative applications that are possible. Think of this chapter as a continuation of the qualitative inquiry wine tasting begun in Chapter 1, this time sampling applied varieties of qualitative inquiry. The purpose remains enhancing your capacity to design and implement useful studies and stimulating your appetite to do so.

Being Pragmatic and Utilization Focused

Well-trained and thoughtful interviewers and observers in all arenas of action can get meaningful and actionable answers to practical questions through systematic fieldwork. Pragmatic and utilization-focused frameworks guide the applied qualitative inquiry perspective of this chapter. The emphasis will be on using qualitative inquiry not just to understand the world or test theory (the emphasis in the previous chapter) but to make the world a better place. Exhibit 4.1 highlights five critical elements of practical, problem-focused, solution-oriented qualitative inquiry.

EXHIBIT 4.1   Practical Qualitative Inquiry Principles to Get Actionable Answers

Evaluators, policy analysts, action and applied researchers, community change agents, and communications and organizational development specialists have studied extensively how to conduct inquiries that are useful and actually used. Below is a distillation of what has been learned.

PRACTICAL QUALITATIVE INQUIRY PRINCIPLES	APPLYING THE PRINCIPLES IN PRACTICE	POTENTIAL PITFALLS
1. Frame the inquiry's purpose as practical and action oriented.	Be clear that the purpose of the study is to improve the program or policy based on feedback from knowledgeable interviewees and key informants.	Taking a purely academic approach that is interesting and may even yield an academic publication but is not action oriented
2. Identify and work with people who can bring about change.	Actions and changes have to be done by people who can take action and make changes. Know who they are. Find out what they need to know to act.	Thinking in general audience terms without going through a process of identifying specific intended users and their intended uses (Patton, 2012a).
3. Ask action-oriented questions that will yield concrete answers.	Understand the problem at sufficient depth to generate solutions. Specific feedback from knowledgeable key informants may include answers to these questions: What doable improvements do well-informed people in the situation recommend? What practical solutions do they offer? What is the basis for their suggestions? What are the barriers to change? What are the factors supporting change?	Here are examples of interesting epistemological questions that are *not* directly actionable: How do people in a situation arrive at recommendations? How do recommendations emerge from lived experience? These questions make people's thinking process the focus rather than what to do to solve a problem.
4. Work with intended users to facilitate engagement with and interpretation of the findings.	Schedule data engagement sessions with key users as case studies are completed and qualitative analysis unfolds. Do so in a timely fashion that is meaningful and relevant to those who can take action.	Researchers often like to wait until their analysis is complete and ready for publication before sharing findings, attending only to their own timeline. This risks missing key decision points where findings could inform action and may miss opportunities to increase users' understanding and buy-in along the way.

PRACTICAL QUALITATIVE INQUIRY PRINCIPLES	APPLYING THE PRINCIPLES IN PRACTICE	POTENTIAL PITFALLS
5. Follow through and evaluate the extent to which recommended actions are implemented and, if they are, whether they've solved the problem in whole or at least in part (or, alternatively, made the situation worse).	For those committed to having their findings used, the work is not done when the findings are reported. To optimize use, following up with intended users to support action is part of the full process of using findings to make a difference.	Failing to inquire into the unintended consequences and/or negative ripple effects of any findings acted on or recommended actions.

SOURCE: Patton (2008b, 2011, 2012a).

The remainder of this chapter reviews some specific contributions of qualitative inquiry, including enhancing quality; program evaluation, especially individualized outcomes evaluation; organization and community development; understanding and improving interventions; identifying unanticipated consequences of interventions; developing practical theories of change; preventing problems; and future research to anticipate future challenges. Let's begin with illuminating and enhancing quality.

Illuminating and Enhancing Quality

> ### Come, give us a taste of your quality.
> —Shakespeare (Hamlet, Act II, Scene ii)

Quality care. Quality education. Quality parenting. Quality time. Total quality management. Continuous quality improvement. Quality control. Quality assurance. Malcolm Baldrige Quality Award. *Quality* is the watchword of our times. People in "knowledge-intensive societies . . . prefer 'better' to 'more'" (Cleveland, 1989, p. 157). *More* requires quantitative analysis; *better* evokes qualitative criteria and judgments of excellence.

Understanding and Improving Quality

> ### Quality is everyone's responsibility.
> —W. Edwards Deming (1900–1993)
> Quality management pioneer

Distinguished action scientist Kenneth E. Boulding (1985) devoted a book to the grand and grandiose topic of "Human Betterment." He defined development as "the learning of quality" (chap. 8). He struggled, ultimately in vain, to define "quality of life." He found the idea beyond determinative explication and certainly beyond numerical measurement. It is a subject particularly well suited for in-depth, holistic qualitative inquiry.

Concern for quality surrounds us in the postmodern age. Successful businesses are attentive to customers' demands for quality. This stems in part from the fact that we simply don't have time for things to break down. We hate wasting time waiting for repairs. We can't afford the lost productivity of not having things work (either products or programs). We have taken to heart the admonition that, in the long run, it is cheaper to do it right the first time—whatever "it" or "right" may refer to. It is within this context that we shall examine what qualitative inquiry brings to the challenge of studying and evaluating *quality*.

The "fathers" of the quality movement—W. Edwards Deming and Joseph M. Juran—were preaching quality in manufacturing before World War II. In the 1930s, for example, Juran was applying concepts of empowered worker teams and continuous quality improvement to reduce defects at Western Electric's Hawthorne Works in Chicago (Deutsch, 1998; Juran, 1951). Deming and his disciples have long viewed quality from the customer perspective, defining quality as meeting or exceeding customer expectations. Philip B. Crosby (1979) wrote a best-selling book, *The Art of Making Quality Certain*. His assertions have become widely repeated wisdom:

- "The cost of quality is the expense of doing things wrong" (p. 11).
- "Quality is ballet, not hockey" (p. 13).

EXHIBIT 4.2   Examples of Qualitative Inquiry Into Quality in Different Settings

UNIT OF ANALYSIS	QUALITATIVE INQUIRY QUESTION	PRACTICAL USES FOR FINDINGS
Individual elderly	What actionable factors affect quality of life for the elderly?	Improve quality of life for the elderly.
Hospital emergency room	What are the variations in quality in emergency conditions? What are the implications of those variations for patients?	Improve emergency health care.
Neighborhood or community	What qualities of this community do residents especially value? What are their concerns about quality of life?	Generate support for and organize residents to improve community quality.
Program	What constitutes program quality? What is excellence? What is the quality of the program's design, implementation, and results?	Increase quality and excellence. Enhance effectiveness. Generate and demonstrate a model of quality.
Organization	What quality of work life do people in this organization desire? How does the actual work life quality compare with their aspirations?	Identify steps to enhance work life quality in the organization.
University	What qualities of the university are especially valued? What would improve the quality of student learning?	Enhance university quality.

- "The problem of quality management is not what people don't know about. The problem is what they think they do know" (p. 13).
- "Quality management is a systematic way of guaranteeing that organized activities happen the way they are planned. . . . Quality management is needed because nothing is simple anymore, if indeed it ever was" (p. 19).

Efforts to implement these and other principles swelled crescendo-like into a national movement as management consultants everywhere sang of total quality management and continuous quality improvement. The Malcolm Baldrige National Quality Awards became, and remain, the pinnacle of recognition that the mountaintop of quality has been reached.

Understanding what people value and the meanings they attach to experiences, from their own personal and cultural perspectives, are major inquiry arenas for qualitative inquiry. This is especially true when making judgments about quality, or *valuing quality*. Exhibit 4.2 presents variations on the theme of qualitatively evaluating quality for different units of

analysis: individuals, communities, programs, organizations, and universities.

Quality Assurance and Program Evaluation

And on the ninth day, God looked down at the flourishing workshops and factories on the good Earth and said, "These enterprises need a caretaker, so God made a quality manager" (Micklewright, 2013, p. 1).

Program evaluation and quality assurance have developed as separate functions with distinct purposes, largely separate literatures, different practitioners, and varying jargon. Each began quite narrowly, but each has broadened its scope, purposes, methods, and applications to the point where there is a great deal of overlap and, most important, both functions can now be built on a single, comprehensive program information system.

Program evaluation traces its modern beginnings to the educational testing work of Thorndike and colleagues in the early 1900s. Program evaluation was originally focused on measuring attainment of goals and objectives—finding out if a program "works," that is, if it's effective. This came to be called "summative" evaluation, which originally relied heavily on experimental designs and quantitative measurement of outcomes. In recent years, program improvement (formative) evaluation and adapting programs to changes in complex dynamic environments (developmental evaluation; Patton, 2011) have become at least as important and pervasive as summative evaluation (Patton, 2008b).

Quality assurance in the United States had its official birth with the passage of the Community Mental Health Act Amendments of 1975 (Public Law 94-63). This law required federally funded community mental health centers to have quality assurance efforts with utilization and peer review systems. Its purpose was to assure funders, including insurers and consumers, that established standards of care were being provided. Quality assurance systems involve data collection and evaluation procedures to document and support the promise made by health and mental health care providers to funding sources, including third-party insurance carriers and consumers, "that certain standards of excellence are being met. It usually involves measuring the quality of care given to individual clients in order to improve the appropriateness, adequacy, and effectiveness of care" (Lalonde, 1982, pp. 352–353). As quality assurance systems have developed, special emphasis has been placed on detecting problems, correcting deficiencies, reducing errors, and protecting individual patients. In addition, "quality assurance" aims to control costs of health care by preventing overutilization of services and overbilling by providers.

Important methods of quality assurance include clinical case investigations and peer reviews. All cases that fail to meet certain standards are reviewed in depth and detail. For example, patients who remain hospitalized beyond an accepted or expected period may trigger a review. An original difference between program evaluation and quality assurance was that quality assurance focused on individuals on a case-by-case basis while program evaluation focused on the overall program. The traditional concerns with unique individual cases and quality in quality assurance systems continue to be quite consonant with qualitative methods. Moreover, quality assurance efforts have now moved beyond health care to the full spectrum of human service programs and government services.

In-depth reviews of the quality of care for participants in programs can draw heavily on clinical case files only if the files contain appropriate and valid information. When files are to be used for research and evaluation purposes, clinicians need special training and support in how to gather and report the highly descriptive qualitative data in clinical case files. Because there can be great variation in the quality of clinical case records, particularly the descriptive quality, a program of quality assurance must include evaluation of and attention to the quality of qualitative data available for quality assurance purposes.

Quality Control Versus Quality Enhancement

It is important and useful to distinguish quality control from quality enhancement. Quality control efforts identify and measure minimum acceptable results, for example, minimum competency testing in schools or maximum acceptable waiting times before seeing a physician in an emergency room. Quality enhancement, in contrast, focuses on *excellence*—that is, levels of attainment well beyond minimums. Quality control requires clear, specific, standardized, and measurable levels of acceptable results. Excellence, however, often involves individualization and professional judgment, which cannot and should not be standardized. Excellence is manifest in quality responses to special cases or especially challenging circumstances. Thus, while *quality control* relies on standardized statistical measures, comparisons, and benchmarks, *quality enhancement* relies on nuances of judgment that are often best captured qualitatively through case studies and cross-case comparisons.

The distinctions between quality assurance and program evaluation have lost much of their importance as both functions have expanded. Program evaluation has come to pay much more attention to program processes, implementation issues, and qualitative data. Quality assurance has come to pay much more attention to outcomes, aggregate data, and cumulative information over time. What has driven the developments in both program evaluation and quality assurance—and what now makes them more similar than different—is concern about program improvement and gathering really useful information. Both functions had their origins in demands for accountability. Quality assurance began with a heavy emphasis on quality control, but attention has shifted to concern for quality enhancement. Program evaluation began with an emphasis on summative judgments about whether a program was effective or not but has shifted to improving program effectiveness. In their shared concern for gathering useful information to support program improvement, program evaluation and quality assurance now overlap and find common ground. Accountability demands can be well served, in part, by evidence that programs are improving.

QUALITY OF LIFE FOR THE FRAIL ELDERLY

Quality of life depends on people involved and their situation. Quality of life for university students versus migrant workers involves different dimensions of experience and quite varied expectations. Qualitative inquiry documents and interprets these differences. For example, in rural northern Minnesota, a program offers support to frail elderly people to help them stay safely in their homes (avoid institutionalization) with a reasonable quality of life. Here are excerpts from interviews that provide a glimpse of quality-of-life challenges for the frail elderly and document important variations in their situations.

Volunteer Experiences

- I provide respite care for a woman whose husband has dementia. She really appreciates the times she can step away and relax. She comes back clearly refreshed. I do a variety of things with her husband when she's away. Sometimes we go to Subway for lunch. He really seems to enjoy that. Sometimes I listen to his stories. I bet I've heard his five stories 20 times. Each telling is worthy of my attention, and he is worthy of my respect, so I listen and engage.

- One of my clients is a couple. He's 90 and she's 88. They're doing wonderfully in their own home. I would go out to visit last summer and find him working in the vegetable garden. He still drives, but there was a period of time when they needed transportation. I'd take them where they needed to go until he could drive again. Ours is more of a friendship than a business arrangement, but I think my visits help them. I keep an eye on how they're doing, and we talk about things. They tell me how things are going and if they need anything. The number of my visits depends on the issues that arise.

- A woman I visit is 96 years old. She's been blind for seven years, and it's very depressing for her. Also, she's new to the area. Her son moved her here to be closer to him. She has no dementia, and she's interested in what goes on around her, so it's hard that everything is new. I go to visit every Wednesday at 10 a.m. She's always waiting for me. We talk and listen to music. One week we listened to the Academy Awards.

- I have a woman client who's 95. She's still at home, which is good, but her days can get long, and she gets lonely. She really appreciates me coming by to visit her. It's different from having your daughter stop by to check on you. I'm just a friend stopping by for a visit. Companionship really increases your quality of life. Sometimes I read to her, and other times I listen to the stories of her life. I feel blessed to be spending time with her—just hanging out.

Perspectives From Family Members Who Sought Services for Their Frail Elderly Parents

- My father died two years ago. My mother is 89 and still lives in her own home, which is a block away from my place. It seemed the right time to get systems in place so that when my mother begins to need more help, we're ready. About six months ago, we did a consultation with the program, and she's now a client, although she says she has no needs and needs help with nothing. I say to my mother—you don't need anything today, but things will change, and the systems are in place.

- A volunteer comes to visit mother once a week. She's really enjoying her. Sometimes they talk; sometimes they do things like bake cookies. She's still fairly active for 98. She has a cleaning person every two weeks. She's now getting the medical alert necklace through the senior center. It's an emotional aid to have someone to contact. It keeps me less frustrated. It improves my mother's quality of life—and mine.

- My husband has dementia, and it's hard. The visiting nurse makes subtle suggestions. As caregivers, we always feel that we're not doing as well as we should, so she was gentle. She'd observe and suggest things, like about safety—put the knives away. I know if there is anything about my husband that I'm concerned about I can call her and she'll help. That's a huge relief.

- Our first priority is to have mother stay in her place. She doesn't want to be dependent if she can avoid it, but we know it will happen. This support allows me to tolerate what we're doing longer. Her quality of life at home is so much better than it would be in an institution. Simply staying in her own home *is* what quality of life means to her.

Quantity and Quality in Evaluation

This may be a good place to revisit the different contributions of quantitative and qualitative data. Program outcomes can be examined in terms of both quantity of change and quality of change. Consider reading outcomes. Quantitative outcomes include the number of books students read and their scores on a standardized achievement test. But to find out *what it means* to a student to have read a certain number of books is an issue of quality. How did reading those books affect students personally and intellectually? The answer requires capturing students' perspectives and judgments such that the meaning of the experience for the students is elucidated.

WATER-QUALITY STUDIES

Water-quality studies include both quantitative measurements and qualitative observations. In describing and evaluating the physical characteristics and water quality of a stream-edge (riparian) habitat, quantitative measures include chemical assessment, stream flow velocity, and average stream discharge. Qualitative data come from direct observations of erosion, soil conditions along stream banks, and detailed descriptions and photographs of specific physical stream and habitat characteristics. Both quantitative and qualitative findings are used to create a stream map that can be used to draw conclusions and make judgments about local stream water quality (Center for Earth and Environmental Science, 2014).

Or consider programs that emphasize deinstitution-alization, for example, community mental health programs, community corrections, and community-based programs for the elderly. Evaluation typically includes counting the number of people placed in the community and measuring their satisfaction on various dimensions of quality of life. However, to fully grasp the meaning of a change in life for particular persons, an inquiry into quality of life would include questions like the following: What is your daily life like? Who do you interact with? What do you like about your life right now? What do you dislike? How would you describe your quality of life? These are areas of qualitative inquiry that support quality enhancement efforts and insights.

The failure to find statistically significant differences in comparing people on some outcome measure does not mean that there are no important differences among those people on those outcomes. The differences may simply be qualitative rather than quantitative. A carpenter is reported to have explained this point to William James. The carpenter, having worked for many different people, observed, "There is very little difference between one man and another; but what little there is, is very important." Those differences are qualitative.

Program evaluation is the systematic collection of information about the activities, characteristics, and results of programs to make judgments about the program, improve or further develop program effectiveness, inform decisions about future programming, and/or increase understanding. *Utilization-focused program evaluation* is evaluation done for and with specific intended primary users for specific, intended uses. (Patton, 2008b, p. 39)

The increased importance of qualitative evaluation has developed within a larger public policy context that includes increased calls for accountability, greater emphasis on evidence-based policy and "best practices," reduced resources for programs of all kinds given national and global financial crises, and the concomitant geometric growth of monitoring and evaluation internationally, especially in support of social and economic issues in poorer countries (Bamberger et al., 2011; Kusters, van Vugt, Wigboldus, Williams, & Woodhill, 2011). In such a high-stakes political environment, qualitative evaluation makes an important contribution by identifying and validating especially effective, evidence-based practices. Practical approaches to evaluation are especially important in the international development arena, where building evaluation capacity globally is a priority, as highlighted by the declaration of 2015 as the *International Year of Evaluation* (EvalPartners, 2014), with the following theme:

Evidence for the world we want:

Using evaluation to improve people's lives through better policy making

Qualitative evidence should be a part of evaluation's contribution to improving people's lives (Brandon & Ah Sam, 2014; Goodyear, Jewiss, Usinger, & Barela, 2014; Nutley et al., 2013; Schorr, 2012).

Outcomes Evaluation Distinguished From Outcomes Measurement

It is important to distinguish outcomes evaluation from outcomes measurement. *Outcomes evaluation* refers generally to determining the results and impacts of an intervention. Did the youth development program lead to increased youth strengths and assets? Did the HIV/AIDS prevention program lead to safer sex practices and reduce the incidence and prevalence of HIV/AIDS? Outcomes evaluation includes any data brought to bear on these questions, both qualitative and quantitative. *Outcomes measurement*, in contrast, refers only to quantitative indicators of outcomes and is, therefore, a narrower frame of reference.

SIDEBAR

EVALUATING HUMAN RIGHTS ADVOCACY OUTCOMES

What Is Measurable Is Potentially at Odds With What Is Right

The moral imperative of human rights work means that results can be amorphous, long-term, and potentially unattainable—the opposite of measurable. As one human rights advocate articulated,

> I spent eight years defending political prisoners. There was no hope of their release. I lost every case. What was my [observable] impact? Zero. Should I have done it? I haven't found one person who says no. So, that's an issue. How *do* you really measure your capacity for transformation when not much transformation happens in front of you.

—Schlangen (2014, p. 6)

In addition, planning for a specific result at a specific point in time risks oversimplifying human rights work. The drive for specific measures can mask complexity and can have unintended consequences, as illustrated by the following quote from another advocate.

> We used the number of political prisoners [as an indicator]. . . . The numbers went down but it might be because the government was just shooting prisoners instead of holding them.

—Schlangen (2014, p. 6)

It is also important and helpful to distinguish program evaluation from routine monitoring and management information system data. Routine monitoring data are used to manage programs, for example, track the flow of inputs (resources), participation rates, and participant completion of program activities. Management information systems data are descriptive. Such routine data report on the basic functioning of a program, often in comparison with targets. For example, an immunization program may aim to reach 100% of preschool children. The monitoring data report what percentage were actually vaccinated (e.g., 90). Program evaluation asks not only what has occurred and what was accomplished but *why*. What was the model being implemented? To what extent was the model implemented as designed (the fidelity question)? What are the variations in participation, and what explains those variations? What are the variations in outcomes, and what explains those variations? To what extent can documented outcomes be attributed to the intervention (the attribution question)? What, if any, unanticipated outcomes and impacts occurred? These are *evaluation* questions. Qualitative outcomes evaluation contributes to answering these kinds of evaluation questions. (For more on different kinds of program evaluation questions, see Mathison, 2005; Patton, 2008b, chap. 4.)

While funders and other users of evaluation findings value outcomes measurement and want to know outcome statistics, they also need case studies, stories, and the personal perspectives of program participants to make sense of the numbers and to place the results in a larger community context. Program staff members use case studies to reflect on their practices and improve programs. Policymakers and politicians understand that they need to know both the level of outcomes (quantitative data) and the *meaning* of those outcomes to the people whose lives have been affected. Qualitative data illuminate those meanings.

Qualitative Outcomes Evaluation

> For programs engaged in healing, transformation, and prevention, the best source and form of information are client stories. It is through these stories that we discover how program staff interact with clients, with other service providers, and with family and friends of their clients, to contribute to outcomes; and how the clients, themselves, grow and change in response to program inputs and other forces and factors in their lives. There is a richness here that numbers alone cannot capture. It is only for a story not worth telling, due to its inherent simplicity, that numbers will suffice.
>
> —Barry M. Kibel (1999, p. 13)
> *Success Stories as Hard Data*

Quantitative outcomes evaluation focuses on measuring increases in desired outcomes (e.g., higher student achievement test scores) and decreases in undesirable outcomes (reductions in rates of child abuse and neglect). Qualitative data show the human faces behind the numbers, providing critical context when interpreting statistical outcomes and ensuring that the numbers can be understood as constituting meaningful changes in the lives of real people.

Qualitative outcomes data can be, and increasingly are, presented with quantitative data (mixed methods) to present a fuller picture of both the level and the meaning of outcomes attained (or not attained). The following examples illustrate the kinds of qualitative outcome evaluations that have become more valued, and therefore more common, over the past decade.

- A study of child protection services documents the number and types of cases, adding qualitative case studies to illuminate the family situations that gave rise to the abuse, capture details about the services provided, and gather evidence from those affected about the effects on their lives over time, including how services did (or did not) support child safety and a more functional and healthy family situation; these findings are used to improve the services.
- An evaluation of a residential center to help chronic alcoholics get off the streets uses in-depth case studies to follow the effects on the health and well-being of residents and how they interact with each other, staff, and people in the community to help funders and policymakers decide whether to continue and/or expand the program.
- An in-depth, longitudinal case study of a neighborhood documents its community development efforts, examining the collaboration among human service and health organizations and capturing the perspectives and experiences of community leaders and residents to help government

and philanthropic funders assess both the extent to which targeted community outcomes have been attained and the meaning of those outcomes to people in the community.

- In an intervention aimed at reducing obesity, the outcome evaluation reported the number of people who attained the desired weight reduction targets (the quantitative outcome). What it meant to people to succeed in losing weight (or failing to meet targets in other cases) and how weight loss affected other aspects of their health and lives involved qualitative data. Integrating these data sources provided a complete picture of outcomes and their meanings. Mixed methods can provide a more balanced picture of outcomes than either quantitative or qualitative data alone.

- In an adult literacy program, the test results showed an average increase of 2.7 grade levels over a three-month period. The following people were included in the program:

  o A Puerto Rican man who was learning to read English so that he could help his young daughter with schoolwork

  o An 87-year-old African American grandmother who, having worked hard throughout her life to make sure that her children and grandchildren completed school, was now attending to her own education so that she could read the Bible herself

  o A manager in a local corporation who years earlier had lied on her job application about having a high school diploma and was now studying at night to attain a GED (Graduate Equivalency Degree)

Capturing such diversity is one of the major contributions of qualitative outcome studies. Statistics tend to focus on central tendencies and standardized outcome dimensions. Qualitative data capture and portray individual, family, and community differences. This facilitates attention to the significance and implications of diverse outcomes while also identifying common patterns of outcomes.

Case Example of a Qualitative Evaluation

Evaluation of a residence for chronic alcoholics and drug users illustrates how qualitative methods can be used to inform decision making and funding for an innovative human services model. The San Marco is a residence for chronic alcoholics, most of them homeless, aimed at providing a safe environment in which residents can achieve increased health, mental health, and quality-of-life outcomes. Ten of the 30 residents were

---

OUTCOME HARVESTING

Outcome harvesting is like forensic science in that it applies a broad spectrum of techniques to yield evidence-based answers to the following questions:

- What happened?
- Who did it (or contributed to it)?
- How do we know this? Is there corroborating evidence?
- Why is this important? What do we do with what we found out?

Answers to these questions provide important information about the contributions made by a specific program toward a given outcome or outcomes.

—Wilson-Grau and Britt (2012, p. 1)

---

selected for evaluation case studies using a *criterion-based maximum variation sample* in which participants were as different as possible from one another. This meant selecting both male and female cases, younger and older participants, people of different racial and ethnic backgrounds, and people with different health and mental health histories. This kind of qualitative sample permits documentation of diversity as well as identification of any common patterns in experiences and outcomes that cut across that diversity. Participants signed an authorization form that covered standard informed consent and voluntary participation issues.

Individuals were interviewed initially to establish a baseline and again after six months. A focus group was also conducted to clarify and validate the individual interviews. Participants were given a stipend for each interview and the focus group. Interviews were recorded, transcribed, and used to create participant case studies. Major themes and outcome patterns were identified through cross-case analysis of individual and focus group interviews. Sample quotations are included to illustrate the outcome themes.

San Marco Outcomes

1. *Participants reported more stable lives:* "I'm eating better, have medical care and a roof over my head. I have a family here that I didn't have out on the streets."

2. *Participants reported feeling safe at the San Marco:* "On the street you have to be real vigilant. You could get jumped in the middle of the night, or if you're coming back from the liquor store you could get jumped for money or liquor."

"I feel safe because I don't have to worry about my friends getting me evicted. I have a little contact with old friends by cell phone. A couple of them showed up here. I wasn't drinking and the staff just got rid of them."

3. *Participants reported being members of a community that feels like a family:* "The San Marco is like a little community. I know a lot of people in this building, so I feel like I'm with friends that are more like a family—a big family. We get along like families do. There are some differences but nothing that can't be handled. We have parties and celebrate holidays together."

4. *Participants feel supported and cared for:* "The staff here is excellent. They're helpful, courteous and treat residents like family. I get along with them well. They help me with my personal needs like appointments and shopping. Today they took me to pick up glasses. They let me know if I need to clean my room or mop floors. They worry about me. I went to see a friend for three or four days and they put out an APB [All Points Bulletin]. It showed they really care."

5. *Participants have a place to live that feels like home:* "Best of all, I have my own place. I've fixed it up the way I want it—fixed it up like a home. It's just a room, but, still, I can come home here, and they always wonder if I'm doing okay when I walk in. It's like coming home—that's what I call it."

6. *Most participants report drinking less:* "I drink less here. There is alcohol all around the building, but I still drink less. Out on the street I drank to keep warm and to drown out feelings. Around here I have people I can talk to. I don't drink every day. Some weeks I take two or three days off. I now drink maybe ten beers at a time and three screwdrivers. I'm not waking up sick every morning and not passing out at night. I'm remembering to take pills, remembering to eat right and I'm more sociable."

7. *Participants report looking beyond survival and short term goals:* "I'd like to stop drinking and move to the other side of the San Marco. Staff knows I can do it. When I'm sober, I do wood burning. I don't open my door or answer my phone. I'm very artistic, and I can sit all day long and listen to my favorite music. I would like more of it. People knock on my door and want this and want that. I say—just get away from me for awhile, please. I burn a little sage and I'm content."

8. *Participants have maintained or increased contact with family:* "I'm going to my dad's on weekends. He picks me up after work. We have a car project. I don't drink when I'm with my dad. He has invited me to his house for a month, but there's nothing to do there." (Patton & Gornick, 2011b, pp. 39–41)

Patterns of Program Effectiveness

In addition to documenting and illustrating outcomes, the qualitative evaluation findings provided feedback about patterns of program effectiveness. The feedback at the program level both affirmed key elements of the human services model being piloted and identified areas to be engaged for improvement.

- *Pattern 1:* Staff plays an important role in parenting/nurturing the San Marco residents.

- *Pattern 2:* The visitation rules are important to creating a sense of safety for residents.

- *Pattern 3:* People like being able to come and go. They don't want to be institutionalized. Partly, it is that they like being treated with respect and as adults, and partly because they don't like being cooped up. They like to be outdoors. There was feedback in the focus group about the inadequacy of the patio.

- *Pattern 4:* People need additional support to maintain sobriety. (Patton & Gornick, 2011b, p. 41)

These findings were used by program staff to improve the program and by funders, both government and philanthropic, to determine the merit and worth of the pilot project.

Detailed case studies can be especially useful for program improvement. To simply know that a targeted indicator has been met (or not met) provides little information for program improvement. Getting into case details better illuminates what worked and didn't work along the journey to outcomes—the kind of understanding a program needs to undertake improvement initiatives.

Exhibit 4.3 presents highlights of a case study from an employment training program. In addition to illuminating what the outcome of a "job placement" actually meant to a particular participant, the case documents attainment of hard-to-measure outcomes such as "understanding the American workplace culture" and "speaking up for oneself," which can be critical to long-term job success for an immigrant like the woman in the story. The next section looks at the special capacity of qualitative inquiry to document *individualized outcomes* in programs that especially value individualization.

EXHIBIT 4.3   The Qualitative Outcomes Story of Li: Behind the Outcome Numbers of an Employment Program

Outcome Statistics for Li

- She completed the WORK program.
- She stayed in the job she was placed in for one year (program outcome target).
- Her highest wage before the WORK program was $8.25 without benefits; her wage following WORK program completion is $11.75 with benefits, a 42% increase.
- She graduated from technical school with a 3.66 grade point average (out of 4.0).
- Her TABE test (math and language skills) results: reading, 7th-grade equivalent; language, 4th grade; spelling, 10.6; math computation, 12.9—highest score possible; applied math, 9.9. Average increase during the program: 5.4 grade levels.
- Her participation data: attended 89 classes at WORK, missed 6, and was late to 1.

Li's Outcome Story

Li entered the employment training program called WORK two years after she arrived in the United States from Vietnam. As a recent immigrant, Li faced language and cultural barriers at school, at work, and in her day-to-day living. She was originally from Saigon City, where she took the national test twice to study at the university but failed both times. She gained entrance to the Vietnamese Technical College, where she completed a degree in payroll and human resources when she was 23. She went to work for a Vietnamese company as an accountant.

Li married a man from North Vietnam despite her family's opposition. Neither Li nor her husband was accepted in their parents' homes, even though tradition calls for the married couple to live in the home of the husband's parents. As a result, the couple rented a small room, where they struggled to improve their lives. Shortly after the birth of their daughter, the couple discovered that they might be eligible to go to the United States because the U.S. government was granting visas to former officers of the Vietnamese army. After successfully negotiating the difficult application and emigration process, they arrived in Minnesota, where they knew no one.

Li's husband began treating her very badly, according to Li. He immediately got a job but would not give her any financial support. Within a year, he deserted her. She found a cleaning job at a bakery and later at a hotel. Li had always believed that through education she would get ahead, so she again took up the study of accounting, begun in Vietnam, by enrolling in a local technical college four months after arriving in the United States. After two years in the United States, she found a data entry job for a retail business, where she received $8.25 per hour without benefits. She lost that job when she had to stay home for a week to care for her very sick daughter. She then went on welfare.

Li first heard about the employment program WORK from some friends at the technical college. She entered and concentrated on improving her English, technical skills, and assertiveness with support from program staff. WORK provided tuition assistance, cost-of-living support, and bus passes. She also received tutoring in accounting (1.5 hours per week) from the program's accountant. The program secured for her a work experience placement at a local bank at the rate of $7.25 per hour. This work environment proved very stressful because her language skills were inadequate, and she was teased and harassed by other workers.

The WORK program supported her leaving the bank job so that she could more intensively study business English, refine her workplace communications skills, and enhance her accounting software skills. The program case records show that at times she expressed feelings of hopelessness. Her staff worker counseled her to focus on all the barriers that she had overcome by moving to this country and her hopes for a positive future for her daughter. The program supported her return to the technical college, where she graduated with dual diplomas in accounting and data entry. Shortly thereafter, she began attending interviews for jobs. She went through more than seven interviews with different companies before she received a final placement offer in an accounting position at a retail firm. She said that she had expected more help with placement from the program than she got. According to WORK staff, there was some confusion about their role in finding a placement for participants. At any rate, Li persevered and felt good about the end result:

> Even though I failed many interviews and I thought I might never get a job, all the staff encouraged me to keep trying. They talked to me about my good qualities. They were all positive, not negative. In the end, they really helped me get a good job even though I didn't understand the limits of their role at the time.

While her language and math tests showed great improvement (from elementary-level results to high school–level results over the two-year period in WORK), Li says that what was most important was what she learned at WORK about how to be professional in the workplace and how "to mesh with American culture." Li specifically mentioned that she found it important to learn interview skills, since interviews are not common in Vietnam. She emphasized that WORK played a critical role in helping her find a professional job in "American corporate culture," explaining, "I compare myself with my friends at the Tech College, and even now they don't have a job. If I didn't have WORK, I probably wouldn't have a job either." The program staff emphasized the challenges Li faced in gaining confidence and learning to speak up for herself, issues that the staff continued to work on with her after her job placement.

Li believes that WORK played a vital role in supporting her to overcome the challenges in her life. She said, "I can talk to staff about any problem—about my job, money, and my daughter. I talk to them to figure things out and solve problems. Even now that I have a job, I call and get help from staff." She recounted getting help dealing with a fellow worker who was making life difficult for her. Program staff talked her through the process of discussing the situation with her supervisor and getting help to resolve the situation. Li said,

> Even though I wasn't been successful at home with my husband, I feel like WORK is my family now. It makes people feel safe here. The staff encourages us to go forward. If there is a problem, they help us solve it. The important thing is that we treat each other like a family. I learned so many things here. Things I can't get in school. I learned interview skills, workplace skills, empowerment skills. I learned English that is more effective on a professional level for communication. WORK helps us not be afraid. They gave me a computer and some cost-of-living support, and they paid for a pronunciation class. They helped me so much.

A friend of Li's interviewed for this case study (with Li's permission) reported that WORK was "especially helpful to Li in making the adjustment from her culture to this culture. It was a touchstone for her. It helped her in her adapting. It was vital."

Li now works full time. She wakes at 5:30 a.m., catches the bus at 6:30, arrives at work at 7:30, and begins work at 8:00. She works until 4:30 p.m. and immediately takes the bus, arriving home at 5:30. Several times a week, Li swims after work at a local community pool. Each evening, she prepares dinner for her daughter and helps with her homework. After her daughter goes to bed, Li frequently works on her computer to improve her skills, or does other self-study.

Li says that she is still working on communicating better in the workplace. Her supervisor said in an interview, "We encouraged her to become more assertive. We wanted her to know that it's okay to stand up for herself. It's okay to be more assertive. It's the American way. She's learning." Her job supervisor continued,

> WORK has been here a couple of times to sit and talk with myself and the personnel manager when we were having difficulty with Li and one coworker. We wanted their input on what to do. So we all kind of worked together to try and solve the problem. They worked really well with us and are continuing to do so.

Li feels that her primary challenge is the fact that she is a "foreigner." She says, "An American learns one thing, and I have to learn double. I'm slower than another American worker, because it's all new for me."

According to Li, her "life has changed a lot because of WORK." She explained that before she received WORK support, it was hard to imagine herself in a professional position in an office environment in this country. Li noted, "I was so excited to get an office job. It is the luckiest thing in my life, and the biggest challenge." According to Li, one result of her effort with WORK is that she feels she can talk and interact with anyone at work or in the community. She's more secure in dealing with others and has more confidence in herself. In addition, Li is able to manage her own financial situation and make confident decisions about raising her daughter. Her interactions with her daughter have improved, and she says that she is no longer ashamed of her divorce.

Li's future includes short- and long-term goals. In the short term, she would like to save money to buy the software for Professional Payroll Accounting, complete the training, and move into payroll accounting. She doesn't see herself staying at her current company for more than several years. Eventually, she would like to go back to school to become a CPA (Certified Public Accountant) and increase her income to support herself and her daughter. One day, Li hopes to travel to Vietnam to visit her mother.

Documenting and Evaluating Individualized Outcomes

Individualization means matching program services and treatments to the needs of individual clients. Successful social and educational programs adapt their interventions to the needs and circumstances of specific individuals and families (Patton, 2011; Schorr, 1988, p. 257). Flexibility, adaptability, and individualization can be important to the effectiveness of educational and human service programs. Highly individualized programs operate under the assumption that outcomes will be different for different participants. Not only will outcomes vary along specific common dimensions, but outcomes will be qualitatively different and will involve qualitatively different dimensions for different clients. Under such conditions, program staff are justifiably reluctant to generate standardized criteria and scales against which all participants are compared. They argue that their evaluation needs are for documentation of the unique outcomes of individuals rather than for measures standardized across all participants.

Open education offers an example. Its emphasis on individualization rests on a pedagogical foundation that the outcomes of education for each child are unique. Open and experiential approaches to education offer diverse activities to achieve diverse and individualized outcomes. Moreover, the outcomes of having engaged in a single activity may be quite different for different students. For example, a group of primary school students may take a field trip to a fire station, followed by dictating stories to the teachers and volunteers about that field trip, and then learning to read their stories. For some students, such a process may involve learning about the mechanics of language: sentence structure, parts of speech, and verb conjugation, for example. For other students, the major outcome of such a process may be learning how to spell certain words. For some other students, the important outcome may be having generated an idea from a particular experience. For yet other students, the important outcome may have been something that was learned in the exercise or experience itself, such as knowledge about the firehouse or the farm that was visited. Other students may become more articulate as a result of the dictation exercise. Still other students may have learned to read better as a result of the reading part of the exercise. The critical point is that *a common activity for all students can result in drastically different outcomes for different students* depending on how they approach the experience, what their unique needs were, and which part of the activity they found most stimulating. Educators involved in individualized approaches, thus, need evaluation methods that permit documentation of a variety of outcomes, and they resist measuring the success of complex, individualized learning experiences by any limited set of standardized outcome measures (e.g., improved reading scores, better spelling, or more knowledge about some specific subject). Qualitative case studies and portfolios of each student's work offer methods for capturing and reporting individualized outcomes.

A similar case can be made with regard to the individualization of leadership development, criminal justice interventions, community mental health initiatives, job training, welfare, and health programs. Take, for example, the goal of increased independence among a group of clients with developmental disabilities. It is possible to construct a test or checklist that can be administered to a large group of people measuring their relative degrees of independence. Indeed, such tests exist. They typically involve checking off what kind of activities a person takes responsibility for, such as personal hygiene, transportation, initiatives in social interaction, food preparation, and so on. In many programs, measuring such criteria in a standardized fashion provides the information that program staff would like to have. However, in programs that emphasize individualization of treatment and outcomes, program staff may argue, quite justifiably, *that independence has a different meaning for different people under different life conditions.* Thus, for example, for one person, independence may have to do with changed relationships with parents. For another person, independence may have to do with nonfamily relationships—that is, interactions with persons of the opposite sex, social activities, and friendships. For still other clients, the dominant motif in independence may have to do with employment and economic factors. For still others, it may have to do with learning to live alone. While clients in each case may experience a similar psychotherapeutic intervention process, the meaning of the outcomes for their personal lives will be quite different. What program staff want to document under such conditions is the unique meaning of the outcomes for each client. They need descriptive information about what a client's life was like on entering treatment, the client's response to treatment, and what the client's life was like following treatment. They also want to report documented outcomes within the context of a client's life, for "*successful programs see the child in the context of the family and the family in the context of its surroundings*" (Schorr, 1988, p. 257). Such descriptive information results in a set of individual case studies. By analyzing patterns across these case histories, it is possible to construct an overview of the patterns of outcomes for a particular treatment facility or modality.

Evaluating Individualized Outcomes for Programs Serving Homeless Youth

An evaluation in Minnesota of programs serving homeless youth provides an excellent example of individualized outcomes using qualitative case studies for evaluation. Here's the introduction to the evaluation report that makes the case for an individualized, case study approach.

> All homeless young people have experienced serious adversity and trauma. The experience of homelessness is traumatic enough, but most also have faced poverty, abuse, neglect or rejection. They have been forced to grow up way too early. Most have serious physical or mental health issues. Some are barely teenagers; others may be in their late teens or early twenties. Some homeless youth have family connections, some do not; all crave connection and value family. They come from the big city, small towns and rural areas. Most are youth of color and have been failed by systems with institutionalized racism and programs that best serve the white majority. Homeless youth are straight, gay, lesbian, bisexual, transgender or questioning. Some use alcohol or drugs heavily. Some have been in and out of homelessness. Others are new to the streets.
>
> *The main point here is that, while all homeless youth have faced trauma, they are all unique. Each homeless youth has unique needs, experiences, abilities, and aspirations. Each is on a personal journey through homelessness and, hopefully, to a bright future.*
>
> Because of their uniqueness, how we approach and support each homeless young person also must be unique. No recipe exists for how to engage with and support homeless youth. As homeless youth workers and advocates, we cannot apply rigid rules or standard procedures. To do so would result in failure, at best, and reinforce trauma in the young person, at worst. Rules don't work. We can't dictate to any young person what is best. The young people know what is best for their future and need the opportunity to engage in self-determination. (Youth Homelessness Initiative Collaboration, 2014, p. 1)

The evaluation included 14 diverse, in-depth case studies to document and analyze how the programs serving homeless youth were engaging individual by individual based on guiding principles (Murphy, 2014). Exhibit 7.19 in Chapter 7 (pp. 508–511) presents one of these case studies.

The more a program moves beyond standardized procedures and the same outcomes for all participants to individualized interventions and outcomes, the more qualitative case studies will be needed to capture the range of program experiences and outcomes attained. Returning to the example of a leadership development program, a leadership program that focuses on basic concepts of planning, budgeting, and communications skills may be able to measure outcomes with a standardized, quantitative instrument. But a leadership program that engages in helping participants think in systems terms about how to find leverage points and intervention strategies to transform their own organizations will need case studies of the actual transformation efforts undertaken by participants, for their individual endeavors are likely to vary significantly. One may be the director of a small community-based nonprofit. Another may be a middle-level government manager. Still another may be part of a large national organization. "Transformation" will mean very different things in these different settings. Under such circumstances, qualitative case study methods and design strategies can be particularly useful for evaluation of individualized participant outcomes and organization-level impacts.

Exhibit 4.4 summarizes the premises and contributions of qualitatively evaluating individualized outcomes.

Other Units of Analysis for Outcomes Evaluation

Outcomes evaluation has traditionally focused on individual outcomes resulting from program participation. While individual outcomes remain important, other units of analysis have arisen as important for qualitative outcomes evaluation:

- *Family outcomes*, for example, documenting changes in family systems and among family members as the unit of inquiry and analysis
- *Organizational development outcomes*, in which increased organizational effectiveness, efficiency, and sustainability are the focus of qualitative and quantitative documentation
- *Community case studies* of community changes and outcomes, the qualitative equivalent of community indicators
- *Neighborhood outcomes*, for example, poverty programs, which often focus on neighborhood change and quality-of-life outcomes that include health and mental health factors and trends, children and families connected to and feeling a part of the neighborhood, and human services in those neighborhoods as an interconnected systems web, in which the health and functioning of that neighborhood web is a neighborhood-level outcome and organizational outcomes, for example, increased capacity and enhanced organizational effectiveness, are outcomes at the organization level connected to neighborhood outcomes.

EXHIBIT 4.4   Premises and Implications of Qualitatively Evaluating Individualized Outcomes

PREMISES	QUALITATIVE INQUIRY IMPLICATION
1. *Baseline.* Participants vary in what they bring to program experiences. Those variations will affect what happens to them during the program and what they take away from it.	Baseline data should include data on individualized background experiences. This can involve baseline data on standardized dimensions (educational status, income, health indicators) as well as qualitative descriptions of and reflections of participants' relevant experiences prior to program participation.
2. *Program experiences.* Participants with diverse backgrounds and life experiences will engage in program experiences differently in ways that are qualitatively important.	Program evaluation should include in-depth descriptive and reflective data about participants' unique personal, individual experiences throughout a program.
3. *Individualized outcomes.* Outcomes for diverse participants will vary based on what they bring to the program and what they individually experience during the program.	In addition to any standardized outcome measures used, evaluating individualized outcomes requires individual case studies of what participants get out of the program, including changes in attitudes, knowledge, and behavior.
4. *Long-term individualized and context-sensitive impacts.* The longer-term impacts of program experiences will depend on the interactions between what participants get out of a program and the individual context in which they live and work following a program experience.	Follow-up of participants will be needed to learn about long-term individualized outcomes and impacts, with particular attention to contextual factors that affect the manifestation and implications of short-term, end-of-program outcomes over time.

This module has focused on qualitative program evaluation, especially evaluation of individualized outcomes. Evaluation documents intended outcomes and emergent outcomes, both quantitative and qualitative, as the case example of Li illustrates (Exhibit 4.3).

Evaluation also inquires into, identifies, and documents unanticipated outcomes. Qualitative inquiry, being an open-ended, inductive approach, is especially well suited for evaluation of unanticipated consequences, side effects, and ripple effects.

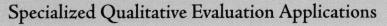

 Specialized Qualitative Evaluation Applications

Particular Problems and Appropriate Solutions Are a Matter of Perspective and Context

A family on vacation stopped at a small-town gas station in the Midwest. The gasoline pump failed to shut off automatically, and gas spurted out of the tank all over the teenage son, who was filling the tank. The mother complained to the station attendant, who pointed to a sign on the pump that said the automatic shutoff was broken.

On the return trip, they stopped at the same gas station because a good family restaurant was nearby. The sign warning about the broken automatic shutoff was gone. But again the pump failed to shut off and ruined another set of traveling clothes, this time the father's.

The parents and kids all filed into the service station office to complain. "There's no sign about the broken shutoff," the father shouted. "Look what's happened to my clothes."

"The sign blew away in last night's storm," the attendant explained.

"Why didn't you make a new one, then?" demanded the mother.

"No paper," said the attendant.

"How in the world can you run a business with no paper?" asked the father.

"'Cause we're in the gas business, not the paper business," said the attendant.

Module Preview

This module delves into special qualitative applications aimed at solving particular methodological challenges and problems. These challenges and corresponding solutions include discovering and documenting unanticipated outcomes, understanding innovations and social change interventions, process evaluation, and legislative auditing and monitoring. Also included, at no extra charge (well, there is the expense of your time), is my rumination debunking so-called best practices.

Discovering and Documenting Unanticipated Outcomes

> The law of unintended consequences pushes us ceaselessly through the years, permitting no pause for perspective.
>
> —Richard Schickel
> American journalist and documentary filmmaker

Distinguished sociologist Robert K. Merton, in a classic 1936 article credited with bringing modern scholarly attention to the phenomenon, identified five possible causes of unanticipated consequences.

1. *Ignorance:* We simply cannot anticipate everything no matter how hard we try.

2. *Error:* Our biases and selective perception lead to incorrect analysis of the problem.

3. *Immediate interest:* Short-term concerns and short attention spans trump long-term understandings.

4. *Basic values:* Strongly held beliefs and values limit our capacity to consider how certain actions may have negative results.

5. *Self-defeating prophecy:* Fear of what *might* occur drives people to find solutions before the problem manifests; thus, the nonoccurrence of the problem is not anticipated.

More recently, we have come to understand that the greater the complexity of the interventions being considered and the more those interventions occur in complex dynamic systems, the greater the likelihood of unanticipated consequences (Morell, 2010; Patton, 2011). For all these reasons, an important contribution of qualitative inquiry is to identify and understand unanticipated consequences, including how *unanticipated outcomes* intersect with targeted outcomes in program evaluation. This potential of qualitative fieldwork to uncover unanticipated outcomes deserves emphasis and is another reason why qualitative studies have become more valued. Statistical studies of outcomes can only measure what has been thought of, conceptualized, and operationalized in advance. The open-ended nature of qualitative inquiry, in contrast, is especially useful in capturing unanticipated outcomes. Indeed, one can only turn up side effects, ripple effects, emergent outcomes, and unanticipated impacts through open-ended fieldwork. For example, in studying the outcomes of a support program for people in poverty with severe social, emotional, and health needs, one of the authors found that intense relationships were formed with staff, which did not end when the program ended. The participants wanted and needed ongoing support and demanded such support from staff despite having completed the formal program. This created stress for staff and participants as they struggled with boundary management issues. The fact that the formal model treated the participant as having completed the program didn't mean that participants saw it that way, which had implications for the sustainability of outcomes and the program's resource capacity to meet the additional needs of the participants.

Here's another classic example of how an incentive in support of one goal can become a disincentive for a competing goal. New York City urgently needed more shelter space for the homeless. To expand housing for the homeless, the city began paying landlords more than they could get for the same apartments in the private market, so that they could use them to house the homeless. As a result, landlords began evicting paying tenants to make more money housing the homeless (Barrett & Greene, 2013).

Here are some other examples of unanticipated consequences from international development projects:

- *A food-for-work program in Central America:* Many women were forbidden by their husbands from participating in the project; some women were seriously beaten by their husbands for even attending meetings about the status of women.
- *A slum-upgrading project in Southeast Asia:* Prior to the official start of the project, many slum dwellers were forced (sometimes at gunpoint) to sell their houses at a very low price to people with political contacts.
- *A road construction project in East Africa:* Just prior to the start of the project, the government destroyed a number of houses to avoid paying compensation to the families once the project began, as specified in the loan agreement governing the project (Bamberger, 2012, p. 4).

Recognition of the importance of attending to unintended outcomes has increased over the past decade with greater attention to systems thinking and complexity theory in program evaluation (Patton, 2011; Williams & Imam, 2006). Outcomes evaluation has been dominated traditionally by linear logic models of program interventions. Reconceptualizing programs as complex nonlinear and dynamic systems in which unpredictable interactions affect targeted outcomes and give rise to unanticipated and emergent outcomes has increased the value of in-depth qualitative case studies that can document and portray such dynamic systems. For example, in the residential program for chronic alcoholics mentioned earlier, the original targeted outcomes focused on individual well-being and functioning patterns, but the in-depth individual interviews, case studies, and focus groups revealed that the interactions among participants was critical, that their relationships with staff went beyond the narrow mandates of the residence program, and that how residents interacted in the community was an important outcome.

Exhibit 4.5 provides a summary of the challenges of getting serious about investigating unanticipated consequences—and qualitative inquiry solutions to those challenges.

| EXHIBIT 4.5 | Getting Serious About Unanticipated Consequences: Eyes-Wide-Open Qualitative Research and Evaluation |

CHALLENGE	QUALITATIVE INQUIRY SOLUTION
1. Evaluators and researchers give lip service to the importance of investigating unanticipated consequences but don't actually include such inquiry in their proposals and designs.	Evaluators and researchers can only turn up unanticipated consequences, side effects, ripple effects, and unexpected, emergent outcomes through open-ended fieldwork. Build open-ended fieldwork into evaluation proposals and designs.
2. "Development programs almost never work out as planned" (Bamberger, 2012, p. 2). The outcomes of programs include both intended/targeted outcomes and unintended, unpredictable, and unexpected outcomes.	*Expect the unexpected.* Unanticipated outcomes are common, not rare. Look for them. Make unanticipated consequences, side effects, ripple effects, and unexpected, emergent outcomes a core part of evaluation.
3. Programs operate in complex dynamic systems with constantly changing economic, political, security, natural, and sociocultural environments. This means that program planners, implementation staff, and administrators have much less control than they often think. Their own narrow views of what is happening mean that they may not be aware of unanticipated consequences.	Gather data beyond the official views of the people in charge of a program. Develop diverse sources inside and outside the program to make sure you, the evaluator, know what people in informal networks know. Develop creative ways to get information on, about, and from nonbeneficiaries, dropouts, and groups that may be worse off.

CHALLENGE	QUALITATIVE INQUIRY SOLUTION
4. Evaluation and research designs typically specify all data collection at the beginning and are therefore inflexible and rigid. Traditional contracting and budgeting processes often require full design and data collection specification before the evaluation begins.	Emergent qualitative designs are especially well suited to investigation of unanticipated consequences and unexpected outcomes. Emergent designs support flexibility to pursue new leads as the inquiry unfolds.
5. Budgets don't include open-ended fieldwork for inquiry into potential unanticipated consequences.	Budget open-ended fieldwork. Include contingency funds in the evaluation budget to support pursuit of new leads that emerge during the evaluation. Negotiate some budget flexibility to fund serious inquiry into the unexpected and emergent.
6. Funders only want evaluators to look at whether their programs achieve their intended outcomes or goals. "Don't rock the boat" by looking beyond intended goal attainment.	Evaluators should be insistent that inquiry into unanticipated consequences is not an option but rather a standard for professional evaluation practice. Ethical and balanced evaluation practice includes attention to both intended and unintended effects.
7. Despite rhetoric about unanticipated consequences, actual evaluation practices indicate that inquiry into unanticipated consequences is seldom taken seriously. Even when undertaken, investigation of unanticipated consequences is an add-on, last-minute part of the evaluation, which undermines the depth and quality of inquiry.	Engage in *eyes-wide-open* evaluation: Find out what is really happening on the ground. Commit to serious inquiry into unanticipated consequences. Build in time, budgetary support, and an eyes-wide-open frame of mind from the beginning.

SOURCE: Adapted in part from Bamberger (2012).

Understanding Interventions

> *Fatalism:* The belief that the world will unfold as it is supposed to and we are powerless to affect the future.
>
> *Interventionism:* The belief that we can change the future by our actions.
>
> Fatalists experience interventionists as crazy. Interventionists experience fatalists as deluded. Both are right.
>
> —From Halcolm's *How Things Are*

The term *interventions* is a catchall phrase for a broad range of change efforts aimed at making the world a better place in some way. These take the form of projects, programs, initiatives, policies, strategies, organizational development, and community organizing. The previous sections examined the unique contributions of qualitative inquiry to studying quality, evaluating individualized outcomes, and discovering unanticipated consequences. We turn now to how interventions aimed at change occur—with emphasis on the *how*.

Evidence-Based Principles

> A model is not something to be replicated but rather it is a demonstration of the feasibility of a principle.
>
> —John Dewey (1859–1952)
> Philosopher

It's an intriguing idea that one should have evidence to support claims of effectiveness. Thus, the label "evidence based" has become widely used to give credence to a variety of models, programs, treatments, and interventions—many of which lack any actual evidence of effectiveness other than the fervent beliefs of their advocates. The question is "What counts as credible evidence in applied research and evaluation practice?" (Donaldson et al., 2009). What kinds of evidence are appropriate for different purposes, programs, and situations? Evidence about program effectiveness involves systematically gathering and carefully analyzing data about the extent to which observed outcomes can be attributed to a program's interventions. It is useful to distinguish three types of evidence-based conclusions:

1. *Single evidence-based program:* Rigorous and credible summative evaluation of a single program provides evidence for the effectiveness of that program, and only that program.

2. *Evidence-based model:* Systematic meta-analysis (statistical aggregation) is conducted of the results of several programs all implementing the same model in a high-fidelity, standardized, and replicable manner, and evaluated with randomized controlled trials (ideally), to determine overall effectiveness of the model. This is the basis for claims that a model is a "best practice."

3. *Evidence-based principles:* This is a synthesis of case studies, including both processes and outcomes, of a group of diverse programs all adhering to the same principles but each adapting those principles to its own particular target population within its own context. If the findings show that the principles have been implemented systematically, and analysis connects implementation of the principles with desired outcomes through detailed and in-depth contribution analysis, the conclusion can be drawn that the practitioners are following *effective evidence-based principles.* Exhibit 4.6

summarizes these three approaches to evidence and the role of qualitative data in each.

"Best" or "proven" practices" are like recipes in that they provide standardized directions that must be followed precisely to achieve the desired outcome. (See MQP Rumination # 4, page 191, about so-called "best practices.") In contrast, guiding principles provide direction but must be adapted to context and situation, like advice on how to be a good parent. Proven practices are specific and highly prescriptive: *Add one-quarter teaspoon of salt.* Principles provide guidance: *Season to taste.* A time management "proven practice" prescribes setting aside the last hour of the workday to respond to nonurgent e-mails. The principled approach suggests that you distinguish urgent from nonurgent e-mails and manage e-mail time accordingly.

Effective principles have to be interpreted and adapted to context. Best-selling management books like *In Search of Excellence* (Peters & Waterman, 1982), *Good to Great* (Collins, 2001a), and *The 7 Habits of Highly Effective People* (Covey, 2013), and "Lessons Learned From Peter Drucker" (Sources of Insight, 2009) offer effective principles, not so-called best or proven practices.

EXHIBIT 4.6   Types of Evidence-Based Interventions

TYPE OF EVIDENCE-BASED INTERVENTION	EXAMPLES	EVALUATION FOCUS, METHODS, AND FINDINGS
Single program, summative evaluation	A local job training program	Evidence of effectiveness is tested for one particular site. A single program with extensive, systematic, multiyear monitoring and evaluation data, including external summative evaluation of job placement and retention outcomes for participants. Compared with a control or comparison group. Positive findings yield an evidence-based program.
Statistical meta-analysis to validate a best-practice model	A standardized model of a job training program implemented in multiple sites, each with a rigorous evaluation	Evidence of effectiveness is tested across multiple sites through meta-analysis, statistically aggregating the separate program evaluation results to validate the model's overall effectiveness. The separate programs must be implemented as a standardized, prescribed model, applying the same measurement criteria to all. Positive results will yield an *evidence-based best-practices model.*
Principles-focused qualitative synthesis	A youth homelessness initiative engaging programs operated by six organizations that share common principles and values but operate independently (see Exhibit 4.8)	Evidence of effective principles: Each program is unique and provides different services, but all work from a common set of principles of engagement. A synthesis of findings from case studies of diverse programs' processes and outcomes tests for effective principles. Positive results will yield *evidence-based effective principles.*

MQP Rumination # 4

Best Practices Aren't

I am offering one personal rumination per chapter. These are issues that have persistently engaged, sometimes annoyed, occasionally haunted, and often amused me over more than 40 years of research and evaluation practice. Here's where I state my case on the issue and make my peace.

Designating something a "best practice" is a marketing ploy, not a scientific conclusion. Calling something "best" is a political and ideological assertion dressed up in research-sounding terminology. The claim may be innocent or ignorant. It doesn't matter. Such claims are dangerous and need to be debunked. So let's do some debunking.

The "Best Practices" Mania

Designating one's preferred model a "best practice" has swept like wildfire through all sectors of society. Do an Internet search and prepare to be astounded by how much *best-ness* there is in the world. Governments and international agencies publish best practices for education, health, highways, welfare reform, and on and on it goes. The "best practices" disseminating business is thriving worldwide. Philanthropic foundations are anxious to discover, fund, and disseminate best practices. Corporations advertise that they follow best practices. Management consultants teach best practices. Identifying, following, and promoting "best practices" has led to the creation of Best Practices databases (e.g., Conference of Mayors, 2014). Here's an example of what's being promulgated. Best Practices Benchmarking™ reports that by studying "world-class customer service practices," it has identified "best practice research findings" that are the key to "world-class excellence":

- Linking Best Practices to Strategy Fulfillment
- Best Practice Identification Systems
- Best Practice Recognition Systems
- Communicating Best Practices
- Ongoing Nurturing of Best Practices (Best Practices, LLC, 2014)

"Best practices" are top of the mountain. They are not just effective practices, decent practices, better practices, promising practices, or smart practices—but the *best*.

Why There Can't Be a Best Practice

The connotation embedded in the phrase *best practice* is that there is a single best way to do something. Karl Popper, whose contributions to the philosophy of science were monumental, argued that nothing could ultimately be proved, only disproved. Applied to best practices, this means that *we can never prove that a practice is the best* but we can distinguish practices that are poor and ineffective from those that are less poor and more effective, no small contribution. Popper identified a natural tendency for people to overgeneralize and speculate well beyond the data, thus the importance of carefully considering rival hypotheses and falsification (disproving claims). Science, he argued, was not so much a set of methods as a form of argumentation in which different interpretations are subjected to careful scrutiny and criticism.

Moreover, "best" is inevitably a matter of perspective and criteria. Like beauty, what is *best* will reside in the eye and mind of the beholder and the criteria, comparisons, and evidence that the beholder finds credible. In a world that manifests vast diversity, many paths exist for reaching a desired destination; some may be more difficult and some more costly, but such criteria reveal the importance of asking, *"'Best' from whose perspective using what criteria?"*

From a systems point of view, a major problem with many "best practices" is the way they are offered without attention to context. Suppose automobile engineers identified the best fuel injection system, the best transmission, the best engine cooling system, the best suspension system, and so forth. Let us further suppose, as is likely, that these best subsystems (fuel injection, etc.) come from different car models (Lexus, Infiniti, Audi, Mercedes, etc.). When one had assembled all the "best" systems from all the best cars, they would not constitute a working car. Each best part (subsystem) would have been designed to go together with other specifically designed parts for a specific model of car. They're not interchangeable. Yet a lot of best-practices rhetoric presumes context-free adoption.

Best-Practices Distortions in Practice

John Bare (2013), vice president of the Arthur M. Blank Family Foundation, Atlanta, tells of teaching an evaluation class for nonprofit executives.

(Continued)

(Continued)

> I had one woman explain that her organization had a foundation grant to deliver health services to low-income families. The funder required all grantees to adhere strictly to the preferred [best practice] implementation model. To keep their funding, grantees had to deliver detailed reports to the funder describing how they were following the model. For this executive, however, her staff had discovered that they could get better results by varying the model based on the needs of client families. Varying the model was not permitted. To keep the funder happy while also pursuing authentic social change for client families, the organization decided to keep two sets of books. The group used one set of reports to satisfy the evaluation requirements of the funder. The second set of books accurately tracked the work and the outcomes. (p. 86)

The assertion that a best practice has been discovered and must now be followed easily becomes a new orthodoxy, and then a mandate, when that alleged best practice aligns with the ideological preferences of the powerful. It is a dangerous, simple-minded, and authoritarian way to disguise preference as scientific knowledge.

Best Practices and Evidence-Based Medicine

Gary Klein is a psychologist who has studied evidence-based medicine (EBM). Klein (2014) warns against the notion that "best practices" can become the foundation of EBM.

> The concept behind EBM is certainly admirable: a set of best practices validated by rigorous experiments. EBM seeks to provide healthcare practitioners with treatments they can trust, treatments that have been evaluated by randomized controlled trials, preferably blinded. EBM seeks to transform medicine into a scientific discipline rather than an art form. What's not to like? We don't want to return to the days of quack fads and unverified anecdotes.
>
> But we should only trust EBM if the science behind best practices is *infallible* and *comprehensive*, and that's certainly not the case. Medical science is not infallible. Practitioners shouldn't believe a published study just because it meets the criteria of randomized controlled trial design. Too many of these studies cannot be replicated. Sometimes the researcher got lucky and the experiments that failed to replicate the finding never got published or even submitted to a journal (the so-called publication bias). In rare cases the researcher has faked the results. Even when the results can be replicated they shouldn't automatically be believed—conditions may have been set up in a way that misses the phenomenon of interest so a negative finding doesn't necessarily rule out an effect.
>
> And medical science is not comprehensive. Best practices often take the form of simple rules to follow, but practitioners work in complex situations. EBM relies on controlled studies that vary one thing at a time, rarely more than two or three. Many patients suffer from multiple medical problems, such as Type 2 diabetes compounded with asthma. The protocol that works with one problem may be inappropriate for the others. EBM formulates best practices for general populations but practitioners treat individuals, and need to take individual differences into account. A treatment that is generally ineffective might still be useful for a sub-set of patients. Further, physicians aren't finished once they select a treatment; they often have to adapt it. They need expertise to judge whether a patient is recovering at an appropriate rate. Physicians have to monitor the effectiveness of a treatment plan and then modify or replace it if it isn't working well. A patient's condition may naturally fluctuate and physicians have to judge the treatment effects on top of this noisy baseline. . . .
>
> Worse, reliance on EBM can impede scientific progress. If hospitals and insurance companies mandate EBM, backed up by the threat of lawsuits if adverse outcomes are accompanied by any departure from best practices, physicians will become reluctant to try alternative treatment strategies.

Embrace Humility and Acknowledge Uncertainty

Commenting on the proliferation of supposed "best practices" in evaluation reports, Harvard-based evaluation research pioneer Carol Weiss (2002) advised wisely that evaluators should exercise restraint and demonstrate "a little more humility" about what we conclude and report to our sponsors and clients. Humility acknowledges, even embraces, uncertainty. An assertion of certainty about what is *best* fosters ideological orthodoxy, intolerance, self-righteousness, rigidity, and pride. In asserting the seven deadly sins—anger, covetousness, envy, gluttony, lust, sloth, and pride—Saint Thomas Aquinas considered pride to be chief among the deadly sins, not because of its inherent gravity but because of its potential for leading to still other sins. There's pridefulness in proclaiming that one is practicing what is "best." Humility, in contrast, acknowledges respectfully that others, looking at the same facts, may arrive at different interpretations and draw different conclusions and that in a different context others might effectively pursue different actions. Of course, exercising and rewarding humility is admittedly hard to do in a world that feeds on hype.

> What works in a poverty neighborhood in Chicago may not stand a ghost of a chance in Appalachia. We need to understand the conditions under which programs succeed and the interior components that

actually constitute the program in operation, as well as the criteria of effectiveness applied. We need to look across . . . programs in different places under different conditions and with different features, in an attempt to tease out the factors that matter. . . . But even then, I have some skepticism about lessons learned, particularly in the presumptuous "best practices" mode. With the most elegant tools at our disposal, can we really confidently identify the elements that distinguished those program realizations that had good results from those that did not? (Weiss, 2002, pp. 229–230)

What to Do? Five Ideas for Your Consideration

1. Avoid either asking or entertaining the question "Which is best?" As is so often the case, the problem begins with the wrong question. Ask a more nuanced question to guide your inquiry, a question that in its very framing undermines the notion of a best. Ask what works for whom in what ways with what results under what conditions and in what contexts over what period of time.

2. Eschew the label "best practice." Don't use it even casually, much less professionally.

3. When you hear others use the term, inquire into the supporting evidence. It will usually turn out to be flimsy, opinion masquerading as research. Where the findings are substantially credible, the findings will still not rise to the standard of certainty and universality required by the designation "best." I then offer that *the only best practice in which I have complete confidence is avoiding the label "best practice."*

4. When there is credible evidence of effectiveness, use less hyperbolic terms like *better practices, effective practices,* or *promising practices,* which tend less toward overgeneralization.

5. Instead of supporting the search for *best-ness,* foster dialogue about and deliberation on multiple interpretations and perspectives. Qualitative data are especially useful in portraying and contextualizing diversity.

Why This Matters

The allure and seduction of best-practice thinking poisons genuine dialogue about both what we know and the limitations of what we know. What is at stake is the extent to which researchers and evaluators model the dialogic processes that support and nurture ongoing scientific discovery and generation of new knowledge. But even more is at stake: supporting democratic dialogue and peaceful exchanges about what we know and don't know, with what degrees of certainty, thereby helping create a context in which humility is possible and valued. As scholars, we contribute not just by the findings we generate but more crucially, and with longer effect, by the way we facilitate engagement with those findings—fostering mutual respect among those with different perspectives and interpretations. That modeling of and nurturing deliberative, inclusive, and, yes, humble dialogue may make a greater contribution to societal welfare than the search for generalizable, "best-practice" findings—conclusions that risk becoming the latest rigid orthodoxies even as they are becoming outdated anyway. At least that is the history of science so far.

Two Approaches to Diffusing and Scaling Interventions

Distinguishing "best" practices from effective principles leads to two fundamentally different ways of looking at the issue of achieving impact at scale—that is, spreading an evidence-based "best-practices" model versus spreading evidence-based effective principles. The best-practices approach involves replicating the validated, proven practices. This "universalist" approach entails high fidelity, controlled replication of a successful intervention, idea, or technology. In contrast, the "contextualist" approach for diffusing effective principles involves adaptation to different situations as the diffusion expands (Hancock, 2003; Patton, 2011, chap. 6).

Evaluation of the dissemination of best practices focuses on validating high-fidelity adherence to a proven model. Evaluation of the dissemination of effective principles focuses on capturing contextual interpretations and adaptations, assessing their effects and consequences, and feeding back the findings to inform ongoing principles-based adaptation. Such inquiries require interviews, observations, and case studies to document in depth and detail how principles are implemented and adapted in practice.

Evidence-based best practices involve stable, consistent, standardized, and controllable treatments and clearly quantifiable outcomes. The evidence-based principles approach, based on synthesizing qualitative and mixed-methods case studies, is especially appropriate for innovative and adaptive programs in complex dynamic environments. Innovative programs are often changed and contextualized as practitioners learn what works and what doesn't and as they experiment, learn, and adapt.

Earlier, in discussing case studies to capture individualized outcomes, I cited an example of programs serving homeless youth. Exhibit 4.7 presents the principles-focused evaluation of these programs.

EXHIBIT 4.7   **Principles-Focused Evaluation Research: Studying Principles for Working With Homeless Youth**

The Otto Bremer Foundation, a family foundation in Minnesota, funded six organizations serving homeless youth to examine the extent to which the principles they articulate as informing their practice actually show up in case studies with youth. Three of the programs are emergency shelters, two are youth opportunity centers, and one is a street outreach organization. All six organizations based their work on the following nine principles.

PRINCIPLES FOR WORKING WITH HOMELESS YOUTH
1. *Journey oriented.* Interact with youth to help them understand the interconnectedness of past, present, and future as they decide where they want to go and how to get there.
2. *Trauma-informed care.* Recognize that all homeless youth have experienced trauma; build relationships, responses, and services on that knowledge.
3. *Nonjudgmental engagement.* Interact with youth without labeling or judging them on the basis of their background, experiences, choices, or behaviors.
4. *Harm reduction.* Contain the effects of risky behavior in the short term, and seek to reduce its effects in the long term.
5. *Trusting youth–adult relationships.* Build relationships by interacting with youth in an honest, dependable, authentic, caring, and supportive way.
6. *Strengths-based approach.* Start with and build on the skills, strengths, and positive characteristics of each youth.
7. *Positive youth development:* Provide opportunities for youth to build a sense of competency, usefulness, belonging, and power.
8. *Holistic.* Engage youth in a manner that recognizes that mental, physical, spiritual, and social health are interconnected and interrelated.
9. *Collaboration.* Establish a principles-based, youth-focused system of support that integrates practices, procedures, and services within and across agencies, systems, and policies.

The purpose of the evaluation research was to understand how the nine shared principles were experienced by 14 unaccompanied homeless youth. Case studies were conducted with these youth, who were between the ages of 18 and 24, lived in the Minneapolis/Saint Paul metropolitan area, and had used services by two or more of the six agencies. Data for the case studies was generated through interviews with the youth themselves, street workers, agency staff, and other people important to understanding the youths' experiences. Youth case files were also reviewed. A reflective practice group comprising 18 individuals representing program leaders and foundation staff convened monthly to engage with the data and participate in the analysis. Individual case studies were written and coded to identify if, where, and how the youth experienced the principles in action. The case studies did find evidence of the principles present in the youths' stories. The study documented and described in depth and detail what the principles look like in practice and traced how they have affected the trajectory of the 14 youth interviewed.

Exhibit 7.20 in Chapter 7 (pp. 511–516) presents one of the case studies.

SOURCE: Murphy (2014).

Process Evaluation

> Process matters. How things are done matters. Outcomes are important, but how outcomes are achieved also matters. A lot. Especially to those who experience the process. So, pay attention to process. Learn from studying process. Not the least because, sometimes the process is the outcome.
>
> —From Halcolm's *How Things Are*

Focusing on process involves looking at *how* something happens. Process inquiries, especially for program evaluation, aim at elucidating and understanding the internal dynamics of how a program, organization, community, system, or relationship operates by asking the following kinds of questions: What are the things people experience that make this experience what it is? How are participants brought into the program, and how do they move through the program once they are participants? What activities do they undertake? What do they experience? How is what people do related to what they're trying to (or actually do) accomplish? What are the strengths and weaknesses of the program from the perspective of participants and staff? What is the nature of staff–participant interactions?

Qualitative inquiry is highly appropriate for studying process because (a) depicting process requires detailed descriptions of what happens and how people engage with each other; (b) people's experience of processes typically vary in important ways, so their experiences and perceptions of their experiences need to be captured in their own words; (c) process is fluid and dynamic, so it can't be fairly summarized on a single rating scale at one point in time; and (d) the process may be the outcome. Exhibit 4.8 (p. 196) shows how qualitative inquiry into these four questions illuminated a philanthropic foundation's grant-making process.

Process evaluations not only look at formal activities and anticipated outcomes but also investigate informal patterns and unanticipated interactions. A variety of perspectives may be sought from people with dissimilar relationships to the program—that is, inside and outside sources. Process descriptions

are also useful in permitting people not intimately involved in a program—for example, external funders, public officials, and external agencies—to understand how a program operates, enabling them to make better informed decisions about the program. Process evaluations have been especially widely used in documenting public health interventions (Steckler & Linnan, 2002), and the Centers for Disease Control and Prevention (2008) published a guide on process evaluation to help state and federal program managers and evaluation staff design and implement valid and reliable process evaluations. While their examples focus on tobacco use prevention and control programs, the general guidance is relevant for any process evaluation.

A good example of what can emerge from a process study comes from an evaluation of the efforts of outreach workers at a prenatal clinic in a low-income neighborhood. The outreach workers were going door to door identifying women, especially teenagers, in need of prenatal care, to get them into the community prenatal clinic. Instead of primarily doing recruiting, however, the process evaluation

PROCESS HAS IMPACT

Whoever fights monsters should see to it that in the process he does not become a monster. And if you gaze long enough into an abyss, the abyss will gaze back into you.

—Friedrich Nietzsche (1844–1900)
Philosopher

SIDEBAR

found that the outreach workers were spending a great deal of time responding to the immediate problems they were encountering, for example, helping with rat control, helping people locate and get into English-as-second-language classes, and protection from neglect, abuse, or violence (Philliber, 1989). What actually took place in the door-to-door contacts (helping people get help for a range of problems) turned out to be significantly different from the way the door-to-door process was designed and conceptualized (recruiting pregnant women into prenatal care). These findings, which emerged from interviews and observations, had important implications for staff recruitment and training and for how much time to allocate to cover a neighborhood.

EXHIBIT 4.8   Process as Outcome

The Committed Connections Foundation (pseudonym) aims to help build healthy rural communities. The foundation's strategic mission focuses on building and nurturing *committed connections*. Sustaining strong relationships with people in communities was identified as a core way of doing business. Examining case studies of funded projects, the senior leadership team participated in an in-depth inquiry into the nature, implications, and impacts of *committed connections*.

QUALITATIVE INQUIRY QUESTIONS	FINDINGS AND IMPLICATIONS
1. What kinds of relationships are built as part of the philanthropic grant-making process?	Comparing cases revealed a range of relationships that varied by context and type of grant. This understanding allowed staff to become more intentional about the nature of the relationship that needed to be built early in the grant-making process. Monitoring how the relationship was unfolding became more systematic, thereby supporting adaptations to deepen the committed connections as appropriate.
2. What are the experiences of the foundation's grantees in working with the foundation and with each other?	One important area of inquiry and learning concerned the inevitable power dynamics that emerge between those who are seeking funds and those who are granting funds. Rhetoric about "collaboration" and "partnership" may disguise these power dynamics but do not eliminate them. Becoming more reflective and open about the way money and power affect committed connections was necessary to deepen those relationships in more authentic ways.
3. To what extent and in what ways are committed connections relationships fluid and dynamic? What are the implications of these dynamics?	The foundation has done surveys getting feedback on grantees' satisfaction with their relationship with the foundation. But simply measuring satisfaction at some moment in time provides no insight into the ups and downs of relationships—periods of intense engagement and periods of relative disengagement. Understanding and monitoring these dynamics are essential to deepening committed connections over time.
4. How do the processes of building *committed connections* relate to the foundation's outcome of helping build healthy and vital communities? The process may be the outcome.	The process of building committed connections is also the outcome. Healthy and vital communities are those in which people can act together and feel connected. Where social capital is strong, communities are more resilient. Thus, process and outcome became understood as integrated, hand in glove. This insight informed staff job descriptions, program work plans, allocations of staff time, staff development, and evaluation priorities going forward.

Closely related to process evaluation is implementation evaluation. Both help avoid an evaluation version of Type III errors—choosing the wrong problem representation—which includes discovering that "the program you evaluated is not the program you *thought* you evaluated!" (Rain, 2012, p. 1).

Implementation Evaluation

> A good plan implemented today is better than a perfect plan implemented tomorrow.
>
> —General George S. Patton
> World War II commander of
> American forces in Europe

Implementation evaluation documents the extent to which an intervention has been implemented as planned and what departures from the plan, if any, have occurred—and why.

Decision makers can use implementation information to make sure that a policy is being put into operation according to design—or to test the very feasibility of the policy. Unless a program is operating as designed, there is little reason to expect it to produce the desired outcomes. When outcomes are evaluated without knowledge of implementation, the results seldom provide a direction for action because the decision maker lacks information about what produced the observed outcomes (or lack of outcomes). The classic study in the implementation genre is by Pressman and Wildavsky (1984), in which the book's title tells the essence of the story:

> *Implementation: How Great Expectations in Washington Are Dashed in Oakland: Or, Why It's Amazing That Federal Programs Work at All, This Being a Saga of the Economic Development Administration as Told by Two Sympathetic Observers Who Seek to Build Morals on a Foundation of Ruined Hopes.*

Studying program implementation involves gathering detailed, descriptive information about how an intervention or program operated, or is operating, by answering the following kinds of questions: What services are provided to participants, and what activities do they engage in? In what sequence? What are the staff's roles? How is the program organized? How does the way the program is actually implemented compare with how it was designed and supposed to be implemented? What explains the differences? Rigorous observations of the quality of program implementation can reveal insights into the program's feasibility (Brandon, Lawton, & Harrison, 2014).

Implementation evaluations routinely find that deviations from what was planned are quite common and natural. A classic study in this regard was the Change Agent Study, which examined 293 federal programs supporting educational change. The researchers found that national programs are implemented incrementally by adapting to local conditions, organizational dynamics, programmatic uncertainties, and contextual sensitivities.

> Where implementation was successful, and where significant change in participant attitudes, skills, and behavior occurred, implementation was characterized by a process of mutual adaptation in which project goals and methods were modified to suit the needs and interests of the local staff and in which the staff changed to meet the requirements of the project. This finding was true even for highly technological and initially well-specified projects; unless adaptations were made in the original plans or technologies, implementation tended to be superficial or symbolic, and significant change in participants did not occur. (McLaughlin, 1976, p. 169)

If a process of ongoing adaptation to local conditions characterizes program implementation, then the methods used to study implementation should correspondingly be open-ended, discovery oriented, and capable of describing developmental processes and program changes. Qualitative methods are ideally suited to the task of describing such program implementation. Here are some examples of implementation evaluation findings.

- A day care program that was supposed to serve low-income families was actually serving mostly middle-income families.
- An agricultural extension program for small farmers was ineffective because seeds and fertilizer for demonstration plots didn't arrive in time for the growing season.
- A polio vaccination campaign in Pakistan was canceled because of terrorist attacks against the young woman who was administering the vaccinations.

Failure to monitor and describe the nature of implementation, case by case and program by program, can render useless standardized, quantitative measures of program outcomes. The national evaluation of Follow Through was a prime example of this point. Follow Through was a planned variation "experiment" in compensatory education featuring 22 different models of education to be tested in 158 school districts on 70,000 children throughout the nation. The evaluation alone employed 3,000 people to collect data on program effectiveness. The multimillion-dollar evaluation focused almost entirely on standardized outcomes aimed at comparing the effectiveness of the 22 models. It was assumed in the evaluation plan that models could be and would be implemented in some systematic, uniform fashion. That was not the case. The lack of implementation details seriously undermined the evaluation. Because the evaluators did not know what was implemented in the various sites, they could not compare the effectiveness of the models (Tucker, 1977, pp. 11–12).

The Follow Through data analysis showed greater within-group variation than between-group variation; that is, the 22 models failed to show treatment effects as such. Most of the effects were null; some were negative; but "of all our findings, the most pervasive, consistent, and suggestive is probably this: *The effectiveness of each Follow Through model depended more on local circumstances than on the nature of the model*" (Anderson, 1977, p. 13). The evaluators, however, failed to study the local circumstances that affected variations in program implementation and outcomes (Elmore, 1976, p. 119).

The study of these important program implementation questions requires case data rich with the details of program content and context. Because it is impossible to anticipate in advance how programs will adapt to local conditions, needs, and interests, it is impossible to anticipate what standardized quantities could be used to capture the essence of each program's implementation. Under these evaluation conditions, a strategy of naturalistic inquiry is particularly appropriate.

Legislative Auditing and Monitoring

> We've funded nine of these so-called regional transportation coordinating boards around the state and we have no real idea who's involved, how well they're working, or even what they're really doing. We need an independent perspective on what the hell's going on. Think you can find out?
>
> —Phone inquiry from a powerful Minnesota state legislator

On occasion, some legislative body or board that has mandated and appropriated funds to a new program wants information about the extent to which the program is operating in accordance with legislative intent. Legislative intent may involve achieving certain outcomes or may focus on whether some mandated delivery specifications are being followed.

Sometimes the precise nature of the legislated delivery system is only vaguely articulated. For example, mandates such as deinstitutionalization, decentralization, services integration, and community-based programming involve varied conceptualizations of legislative intent that do not lend themselves easily to quantitative specification. Indeed, for the evaluator to unilaterally establish some quantitative measure of deinstitutionalization that provides a global, numerical summary of the nature of program operations may hide more than it reveals.

To monitor the complexities of program implementation in the delivery of government services, it can be particularly helpful to decision makers to have detailed case descriptions of how programs are operating and what they're accomplishing. Such legislative monitoring would include descriptions of program facilities, decision making, outreach efforts, staff selection procedures, and the nature of services offered, and descriptions from clients about the nature and results of their experiences. An exemplar of such an effort was an in-depth study of "transportation partnerships" throughout Minnesota that included case studies in each region and an extensive cross-case analysis of patterns and variations (DeCramer, 1997a, 1997b).

Busy legislators cannot be expected to read in detail a large number of such histories. Rather, legislators or funders are likely to be particularly interested in the case histories of those programs that are within their own jurisdiction or legislative district. More generally, legislative staff members who are particularly interested in the program can be expected to read such case histories with some care. From a political point of view, programs are more likely to be in trouble or cause trouble for legislators because they failed to follow legislative intent in implementation rather than because they failed to achieve the desired outcomes. In this case, *the purpose of legislative monitoring or auditing is to become the eyes and ears of the legislature or board.* This means providing descriptions of programs that are sufficiently detailed and evocative that the legislator or legislative staff can read such descriptions and have a good idea of what that program is really like. Having such descriptions enables legislators to decide whether their own interpretations of legislative intent are being met. The observation of a parent education program reported in the first chapter is an example of fieldwork done for the purpose of monitoring legislative intent. There are excellent program evaluation units within a number of state legislative audit commissions that use fieldwork to do policy research and evaluation for legislators. When done well, such fieldwork goes beyond simple compliance audits by using qualitative methods to get at program processes, implementation details, and variations in impacts.

Detailed case histories may also be of considerable service to the programs being monitored because they permit them to tell their own story in some detail. Thus, where they have deviated from legislative intent, such case histories would be expected to include information from program administrators and staff about the constraints under which the program operates and the decisions staff have made that give the program its character. At the same time, the collection of such case histories through site visits and program monitoring need not neglect the possibility of including more global statements about statewide patterns in programs, or even nationwide patterns. It is quite possible through content analysis to identify major patterns of program operations and outcomes for a number of separate cases. Thus, qualitative methods used for legislative monitoring allow one to document common patterns across programs as well as unique developments within specific programs.

Evaluating Program Models and Theories of Change, and Evaluation Models Especially Aligned With Qualitative Methods

Models and theories explain how and why the world works as it does. As a warm-up for this module, consider this insightful explanation of how context matters to understand how things work.

> Outside of a dog, a book is a man's best friend.
>
> Inside a dog it's too dark to read.
>
> —Groucho Marx (1890–1977)
> Pioneering film and television comedian

Qualitative Inquiry Into Program Logic Models, Program Theories, and Theories of Change

Implementation failure means an intervention was not sufficiently implemented to find out if the idea would achieve the desired outcomes. In contrast, theory failure (or idea failure) means the idea for the intervention was poorly conceived and didn't work in practice. It is critical to distinguish idea failure from implementation failure. For example, abstinence-only sex education programs don't reduce either teenage pregnancy rates or sexually transmitted diseases (Beh & Diamond, 2006). The idea of focusing sex education only on abstinence doesn't work. To find out that the theory didn't work required evaluating programs that fully implemented an abstinence-only sex education curriculum.

To find out if an intervention idea works in practice, it has to be conceptualized with specific detail that can be systematically evaluated. A logic model or theory of change depicts, usually in graphic form, the connections between program inputs, activities and processes (implementation), outputs, immediate outcomes, and long-term impacts. For example, the classic educational model of the popular DARE (Drug Abuse Resistance Education) program in schools followed the following simple logic model: (a) Recruit and train select police officers (inputs) to teach children the dangers of drug use; then (b) have the police, in uniform, teach children in special classes in school (implementation); as a result, (c) the children will find the teaching credible (process evaluation) and (d) learn facts about drugs (cognitive outcome) that

will (e) convince them not to use drugs (attitude change outcome), which will result in (f) students not using drugs (behavior change outcome), which will ultimately show up in community indicators showing less drug use (impact). At least that was the model, or theory. In practice, evaluations of DARE consistently showed that the theory didn't work in practice, even when carefully and fully implemented as designed.

Attention to program theory has become a major focus of program evaluation (Rogers, 2008). Qualitative inquiry can be used to identify a program's theory, evaluate the nature and extent of implementation of the theory, and study how much the theorized results are actually produced. The best book on how and why to do so is *Purposeful Program Theory* (Funnell & Rogers, 2011).

> A program theory is an explicit theory or model of how an intervention, such as a project, a program, a strategy, an initiative, or a policy, contributes to a chain of intermediate results and finally to the intended or observed outcomes. A program theory ideally has two components: a theory of change and a theory of action. The theory of change is about the central processes or drivers by which change comes about for individuals, groups, or communities—for example, psychological processes, social processes, physical processes, and economic processes. The theory of change could derive from a formal, research-based theory or an unstated, tacit understanding about how things work. For example, the theory of change underpinning some health promotion programs is that changes in perceived social norms lead to behavior changes. The theory of action explains how programs or other interventions are constructed to activate these theories of change. For example, health promotion programs might use peer mentors, advertisements with survey results, or some other strategy to change perceptions of social norms. (p. xix)

In considering logic models and program theories, there can be some confusion about terminology. I find it useful to distinguish a logic model from a theory of change. The only criterion for a logic model is that it be, well, *logical*, that is, it portrays a reasonable, defensible, and sequential order from inputs through activities to outputs, outcomes, and impacts. A theory of change, in contrast, bears the burden of specifying and explaining assumed, hypothesized, or tested causal linkages. Logic models are first and foremost *descriptive*, that is, describing the steps of a program from intake through completion. A well-conceptualized theory-of-change model is *explanatory* and *predictive*. For example, a group of national and local program staff, administrators, and evaluation researchers spent a day together working on a logic model for a program, Road to Recovery, that transported cancer patients to treatment. As a result of thorough contextual analysis and interviews with key informants, including patients and volunteer drivers, the program was reconceptualized as a *treatment compliance strategy* aimed at cancer control rather than "just" a transportation program. The reason for transporting patients was to ensure complete and consistent treatment (getting the full dose of chemo). This reconceptualization had significant implications for how the program was implemented, the outcomes that were measured, and the importance of the program in relation to American Cancer Society strategic priorities.

Espoused Theories Versus Theories-in-Use

> Inductive [program theory] development involves observing the program in action and deriving the theories that are implicit in people's actions when implementing the program. The theory in action may differ from the espoused theory: what people do is different from what they say they do, believe they are doing, or believe they should be doing according to policy or some other principle.
>
> —Sue Funnell and Patricia Rogers (2011, p. 111)
> *Purposeful Program Theory*

Organizational theorist Chris Argyris (1982) introduced what has become a classic distinction between "espoused theories" and "theories-in-use." The espoused theory is what people say they do; it's the official version of how the program or organization operates. The theory-in-use (or theory in action) is what really happens. (These distinctions are similar to Kaplan's [1973] classic differentiation between "reconstructed logic" and "logic-in-use" in decision-making analysis; see also Yanarella, 1975.) Interviewing supervisory or managerial staff and administrators, and analyzing official documents, reveals the *espoused theory*. Interviewing

participants and frontline staff, and directly observing the program, reveals the *theory-in-use*. The resulting analysis can include comparing the stated ideals (espoused theory) with the real priorities (theory-in-use) to help all concerned understand the reasons for and implications of discrepancies.

I was commissioned by a major philanthropic foundation to facilitate an organizational development process. The rhetoric of the foundation (espoused theory) was that the organization valued and supported innovation and risk taking. The staff was encouraged to speak openly about new ideas and take an experimental attitude—at least, that was the rhetoric. The board was "on board" with innovation as the foundation's "brand." I interviewed the staff and people in the community who knew the foundation well. The reality was that the foundation was highly risk averse. The message from the board was "Take risks and be innovative, just make sure that what you do works." Ensuring that things always work meant, to staff, *not* taking risks, *not* being innovative. They did what was safe, tried and true, and conservative—which was how people in the community saw the foundation in my interviews. The theory-in-use (reality of actual practice) did not come close to matching the rhetoric (espoused theory). Closing that gap became the focus of the organizational development process.

Comparing Programs: Focus on Diversity

Programs and organizations that carry the same title may not be doing the same thing. Mothers Against Drunk Drivers (MADD) campaigns to protect families from drunk driving and underage drinking and helps victims and survivors. MADD operates in every state and scores of countries around the world. But while sharing a common mission, MADD chapters are independent and diverse. Priorities vary by location and leadership. Programs differ from place to place because places differ and people in those places vary.

> Programs will be implemented very differently, in different sites and by different people. Therefore, there may be many significantly different interventions, all operating under the same program banner. Many of these differences may reflect quite appropriate adaptations to local contexts such as is often required by programs with complex aspects operating in complex situations. Understanding the theories that are operating in these different contexts can help to enrich the program theory by identifying what works for whom, under what circumstances, and why various adaptations are useful. (Funnell & Rogers, 2011, p. 112)

Local sites that are part of national or even international organizations show considerable variation in implementation and outcomes. These variations are not such that they can be fully captured and measured along standardized scales; they are differences in *kind*—differences in content, in process, in goals, in implementation, in politics, in context, in outcomes, and in program quality. To understand these differences, a holistic picture of each unique site is needed. Using only standardized measures to compare programs can seriously distort what is actually occurring in diverse sites. Consider data from a national educational program that measured staff–student ratios across the country. A few programs had student–staff ratios as high as 75:1 according to the uniform measures used; other programs had student–staff ratios as small as 15:1. What these data did not reveal was that some of the programs with large staff–student ratios made extensive use of volunteers. These regularly participating additional volunteer staff made the real adult–student ratios much smaller. The global and uniform reporting of the data, however, did not allow for that nuance to be recorded.

A good example of the diversity that can emerge from attention to the qualitative differences among programs is a classic study of national teacher center programs. Although funded as a single national program with common core goals and the shared label "teacher centers," qualitative inquiry revealed that three quite different types of center programs had emerged: "behavioral" centers, "humanistic" centers, and "developmental" centers. Differences among the three types are summarized in Exhibit 4.9.

EXHIBIT 4.9   **Evaluation Example Comparing Diverse Programs: Variations in Types of Teacher Centers**

The National Program for Teacher Centers originally discussed teacher centers in general terms as if all were the same. The implementation evaluation identified different types of teacher centers that constituted three distinct models for working with and supporting teachers.

TYPES OF CENTERS	PRIMARY PROCESS FOR WORKING WITH TEACHERS	PRIMARY OUTCOMES OF THE PROCESS
1. Behavioral centers	Curriculum specialists directly and formally instruct teachers.	Teachers adopt new curriculum content and methods.
2. Humanistic centers	Informal, undirected teacher exploration; teachers select their own resources based on their own needs.	Teachers feel supported and important; they pick up concrete and practical ideas and materials for immediate use in classes.
3. Developmental centers	Advisers establish warm, interpersonal, and trusting relationships with teachers over time.	Teachers think in new ways about what they do and why they do it, developing new insights and fundamental skills and capabilities.

SOURCE: Adapted from Feiman (1977).

This analysis highlights the ways in which different teacher centers were trying to accomplish different outcomes through distinct approaches to teacher center programming. Uniform, quantitative measures applied across all programs might capture some of these critical differences, and such measures have the advantage of facilitating direct comparisons. However, qualitative descriptions permit documentation of deeper and unanticipated program differences, idiosyncrasies, and uniquenesses. If decision makers want to understand variations in program implementation and outcomes, qualitative case studies of local programs can provide such detailed information.

Prevention Interventions and Evaluation

> While drinking with his friends late into the evening, a man boasted that he held the supernatural power to keep lions away. "But," protested his drinking companion, "we are not anywhere near a jungle. There have never been lions in this part of the world."
>
> "You see how effective my power is," replied his boastful colleague.
>
> —From Halcolm's *Fools' Tales*

There may be nothing more difficult to evaluate than prevention—the nonoccurrence of some problem. Yet there may be no more important direction for the long-term solution of health and social problems than prevention efforts. For decades, three stories have been endlessly repeated: (1) about the stream of ambulances at the bottom of the cliff instead of building fences at the top; (2) about the numerous dead bodies coming down the river, but all we do is build more impressive services for fishing them out; and (3) about giving someone a fish versus the value of teaching that person how to fish. In reviewing these archetypal cautionary tales, distinguished Australian action researcher and qualitative inquirer Yolande Wadsworth (2011a) has commented that they are reminders of our repeated tendency to go for the short-term quick fix rather than examine, come to understand, and take action to change how a system is functioning that creates the very problems being addressed. This is the essence of the prevention challenge. And this is why the National Resource Center for Community-Based Child Abuse Prevention (2009) published a guide on using qualitative data in program evaluation to tell the story of prevention.

The usual designs for evaluating prevention programs use experimental and control groups or time-series designs. In an experimental design, a sample targeted for prevention is compared with one that is not. There are often ethical problems with such designs (e.g., withholding needed services from the control group), and there can be problems in control (i.e., the control group may be subject to some other intervention). A 10-year study of heart disease prevention found no differences between the treatment and control groups because the whole society had moved toward healthier lifestyle practices during the 1980s, essentially wiping out the control aspect of the intended control group. Time-series designs examine changes in indicators of interest, for example, teenage suicide or alcoholism, but determining the causes of changes can be illusive.

An excellent example of a prevention-oriented qualitative inquiry is a study of heroin use among suburban youth in Baltimore County (Agar, 1999, 2013). The fieldwork looked at how young people began to experiment with heroin and the stories that were spread around based on initial experimentation. Those early stories tended to be positive: "Heroin puts one into a blissful, relaxed, dreamy state; the stresses, strains, and worries of everyday life fall away." Over time, the positive stories were followed by negative stories as some youth become addicted and a few overdosed. "The image of the addict, as opposed to the experimenter, was uniformly and strongly negative among the youth we interviewed, mostly based on their own observations or stories they had heard."As a result of his in-depth work on how heroin addiction spreads through the youth community, Agar (1999) learned how to be more effective in prevention education.

> I learned in the heroin lecture to emphasize the dangers of sliding into physical dependency—true addiction—and what post-addicted life was like. The theme contrasts with the normal approach, which conveys to youth that addiction and death may occur even with one experiment. Since youth rely on stories to evaluate drug effects and since so many stories contradict the normal premise, drug education loses its credibility. (pp. 115–116)

His fieldwork also included looking at the supply and demand of heroin, and the larger system of which drug use was a part, all using qualitative methods to enhance efforts to prevent heroin use.

Internationally, a good example is a qualitative evaluation of prevention training for health care providers in Mozambique working with prevention interventions in HIV care and treatment settings (Gutin et al., 2013).

I evaluated a prevention program aimed at helping elderly people remain living in their homes—that is, preventing institutionalization. In some cases, the delay in institutionalization was only six months to a year. In other words, through volunteer visits, home nursing care, and meals-on-wheels programs, elderly people could continue living in their homes, sometimes years beyond what would otherwise have been possible. While statistical data could document comparative

PREVENTING PROSTITUTION

Prostitution is not a monolithic phenomenon. Preventing prostitution requires contextually sensitive and situationally specific approaches. "Prostitution varies enormously across time, place, and sector—with important consequences for [sex] workers' health, safety, and job satisfaction" (Weitzer, 2007a, p. 33). Interviews with prostitutes show that many "indoor workers [e.g., brothels, escorts] made conscious decisions to enter the trade; they do not see themselves as oppressed victims and do not feel that their work is degrading. Consequently, they express greater job satisfaction than their street-level counterparts" (Weitzer, 2007a, p. 28). Some of the escorts interviewed

> took pride in their work and viewed themselves as morally superior to others: They consider women who are not "in the life" to be throwing away a woman's major source of power and control, while they as prostitutes are using it to their own advantage as well as for the benefit of society. (Weitzer, 2007a, p. 30)

Moral crusades against prostitution generally lack understanding of the variety of sex worker types and their diverse experiences and situations (Weitzer, 2007b). Both oppressive and romanticized "social constructions" of sex work fail to take into account data directly from those who engage in prostitution. Prevention efforts are therefore largely ineffective (Weitzer, 2007a, 2007b).

death rates, hospitalization rates, and costs, only direct interviews and observations could reveal quality-of-life differences—what it meant to these elderly people to stay in their own homes. I remember in particular the story of one very frail woman in her 80s who took the interviewer to her bathroom, opened her medicine cabinet, pointed to a bottle of capsules, and said, "If they come to take me away, I'll excuse myself for a moment, come in here, and take these pills. I'm never leaving here. Never." This scenario was all clearly worked out in her mind and routinely practiced mentally. Understanding prevention includes understanding what people think and do as a result of prevention efforts.

Documenting Development Over Time

> Everything happens to everybody sooner or later if there is time enough.
>
> —George Bernard Shaw (1856–1950)
> Irish playwright

Documenting development over time returns us to the value of qualitative inquiry for process studies, discussed earlier in this chapter, for development is best understood as a process. Organizational development, community development, human development, leadership development, and professional development—these are process-oriented approaches to facilitating change. Pre- and posttests do not do justice to dynamic development processes. Pre- and postmeasures imply a linear, ever upward, less to more image of growth and development. In reality, development usually occurs in fits and starts, some upward or forward progress, then backsliding or consolidation. Pre- and posttesting tells you where you started and where you ended up but not what happened along the way. Quantitative measures can parsimoniously capture snapshots of pre- and poststates, and even some interim steps, but qualitative methods are more appropriate for capturing evolutionary and transformational developmental dynamics. For example, I worked with a new and innovative employment training program that went through several bouts of reorganizing staff and participant teams, realigning courses, trying new mixes of activities, and restructuring as growth occurred. We found that even quarterly reflective practice sessions were insufficient to capture all the changes occurring. To study this highly dynamic program required ongoing fieldwork because the program development really was like walking through a maze whose walls rearrange themselves with every step you take—and only ongoing qualitative inquiry could capture those rearrangements.

Even routine monitoring using a standardized management information system can only provide an overview of up and down patterns. A break in a trend line or sudden blips in a time series can indicate that some qualitative difference may have occurred. Statistical data from monitoring and information systems can be used to trigger more in-depth qualitative studies that focus on finding out what the changes in the statistical indicators mean. Consider this analogy from the northern climes: The thermometer on a furnace thermostat shows the temperature throughout winter; if the temperature starts falling quickly in a house, someone had better have a look at the furnace. Just monitoring the temperature gauge won't solve the problem. So it is with management information systems. Blips or changes in indicators are a signal that fieldwork is needed to find out what's really going on where the action is. (For a full discussion of documenting development over time, see Patton, 2011).

From Evaluation Issues to Evaluation Models

Thus far, in this chapter, we have reviewed ways in which qualitative methods are especially appropriate for

UNDERSTANDING AND EVALUATING COMMUNITIES OF PRACTICE

Communities of practice have emerged around the globe and across a huge variety of endeavors to support professional development and knowledge generation. The *community of practice* (CoP) approach to creating and sharing knowledge originally developed as a framework for understanding how newcomers to a trade learn through informal and social processes. For example, the trade and artisan guilds of the Middle Ages controlled the flow of knowledge within their own groups. Learning happened by becoming a member and participating in mentored work. The concept has evolved into a knowledge management approach to cultivate learning and innovation among any group of people with shared interests—for example, early-childhood educators, environmental activists, leaders of nonprofit agencies, or qualitative evaluators. In a CoP, learning doesn't rely on formal training or detailed how-to manuals. Rather, learning happens interactively through the act of story sharing. Capturing these stories, extracting their lessons, and examining how they are used is the primary method for evaluating a CoP (Cox, 2005; Wenger, 1998).

- A CoP is the centerpiece of The McKnight Foundation's Collaborative Crop Research Program supporting agricultural development in Africa and South America. The CoP generates and shares lessons about agroecological system innovations and adaptations from different projects. The CoP supports cross-project and cross-continent communications, learning, and evaluation.

- In British Columbia, Canada, a Drug Treatment Funding Program aimed at strengthening substance abuse systems created a CoP to support professional development and learning across different treatment facilities, programs, and staff. The CoP used a Knowledge Exchange framework to support the cocreation of situations, conditions, and ecologies that support a culture of shared learning through knowledge creation, translation, dissemination, uptake, and evaluation.

CoPs aim to turn individual participants' tacit knowledge into explicit shared group knowledge—and, in so doing, build social capital: a series of social connections, a sense of trust, and deeper knowledge about a shared common interest or understanding (Lesser & Storck, 2001). It takes time for a CoP to generate significant social capital and deeper knowledge that translate into changed practice. Qualitative inquiry into communities of practice revealed five stages of development: potential, coalescing, maturing, stewardship, and transformation (Wenger et al., 2002). Systematic story collecting is the primary method for assessing the value of a CoP. Key elements of a story are how knowledge resources are developed and applied. This requires documenting (a) the knowledge development activity, (b) the knowledge resource that was generated, and (c) how the knowledge resource was applied to create value. Both positive and negative stories should be captured and analyzed systematically to illuminate the dynamics and contributions of CoPs.

---

dealing with some fairly common challenges: understanding and improving quality, including quality assurance and enhancement; documenting program outcomes through in-depth case studies; capturing individualized outcomes; discovering and documenting unanticipated outcomes; studying potentially evidence-based effective-intervention principles; process studies and implementation evaluation; clarifying a program's theory or logic model by linking processes and implementation to outcomes; comparing diverse programs; legislative auditing; evaluating prevention efforts; documenting development over time; and investigating sudden changes in management information system indicators. This is by no means an exhaustive list of practical qualitative applications, but each in some way illustrates a *practice-oriented application* of qualitative methods, the focus of this chapter. Qualitative findings can be used to enhance quality, improve programs, generate deeper insights into the root causes of significant problems, and

help prevent problems. The next section reviews some of the major *models* of evaluation research that are most closely associated with qualitative methods: goal-free evaluation, responsive evaluation, illuminative evaluation, values-driven evaluation, connoisseurship, developmental evaluation, and utilization-focused evaluation.

Evaluation Models Attuned to Qualitative Methods: Particularly Well-Aligned Approaches

Holopis kuntal baris: Indonesian phrase uttered to gain extra strength when carrying heavy objects.

—Rheingold (1988, p. 49)

Conducting an evaluation can be a heavy load. Evaluation models help with the heavy lifting. Models provide frameworks, like the metal frame on a backpack that gives support and shape to the load on a hiker's back. Models offer evaluators structure and support. They structure certain methodological decisions, offer guidance about the appropriate steps to follow in design, provide direction in the ways of dealing with stakeholders, and identify important issues to consider in undertaking an evaluation (Alkin, Vo, & Hanson, 2013; Patton, 2012a; Stufflebeam & Shinkfield, 2007). Models provide frameworks rather than recipes, helping evaluators and evaluation users identify and distinguish among alternative approaches. The models briefly presented here have in common that they have been closely associated with or rely heavily on qualitative methods.

Goal-Free Evaluation

The dominant model of evaluation focuses on and measures goal attainment. However, an evaluator can turn up some very interesting results by undertaking fieldwork in a program *without* knowing the goals of the program, or at least without designing the study with goal attainment as the primary focus. Essentially, goal-free evaluation means doing fieldwork and gathering data on a broad array of *actual* effects or outcomes and then comparing the observed outcomes with the actual *needs* of program participants. The evaluator makes a deliberate attempt to avoid all rhetoric related to program goals; no discussion about goals is held with staff; no program brochures or proposals are read; only the program's observable outcomes and documentable effects are studied in relation to participant needs. Philosopher-evaluator Michael Scriven, who first proposed the idea of goal-free evaluation, identified four primary advantages of this approach:

1. Avoid the risk of narrowly studying stated program objectives and thereby missing important unanticipated outcomes.

2. Remove the negative connotations attached to the discovery of unanticipated effects: "The whole language of 'side-effect' or 'secondary effect' or even 'unanticipated effect' tended to be a put-down of what might well be the crucial achievement, especially in terms of new priorities" (Scriven, 1972a, pp. 1–2).

3. Eliminate the perceptual biases introduced into an evaluation by knowledge of goals.

4. Maintain evaluator independence by avoiding dependence on goals—a staff or administrative creation and, often, fiction—which can limit the evaluator's range and freedom of inquiry.

Goal-free evaluation, in its search for "actual effects," employs an inductive and holistic strategy. It was a radical departure from virtually all traditional evaluation thinking and practice. For example, prominent evaluation researcher Peter Rossi asserted that "a social welfare program (or for that matter any program) which does not have clearly specified goals cannot be evaluated without specifying some measurable goals. This statement is obvious enough to be a truism" (Rossi & Williams, 1972, p. 18).

Goal-free evaluation, in contrast, opens up the option of gathering data directly on program effects and effectiveness without being constrained by a narrow focus on stated goals. Qualitative inquiry is especially compatible with goal-free evaluation because it requires capturing directly the actual experiences of program participants in their own terms. The evaluator can thus be open to whatever data emerge from the phenomena of the program itself and participants' experiences of the program. But, as Ernie House (1991) has reported, somewhat tongue in cheek, actually conducting a goal-free evaluation can be challenging for all concerned because during on-site fieldwork, staff keep dropping hints about program goals:

> Many are indignant that you do not want to know their objectives and incensed at the idea of your looking at what they are doing rather than what they are professing. As they chauffeur you around in their car, some will blurt out the goals as if accidentally, with sly apologies for their indiscretion. Others will write them on restroom walls—anonymously. (p. 112)

Scriven has proposed that goal-free evaluations might be conducted in parallel with goals-based evaluations but with separate evaluators using each approach, to maximize the strengths and minimize the weaknesses of each approach. (For a more detailed discussion of goal-free evaluation and critiques of this idea, see Alkin, 1972; Patton, 2008b.)

Responsive Evaluation

> Responsive evaluation enables an alert reporter to capture the deep-felt opinions of those most affected by the program, a product of grass-roots populism.
>
> —House (1991, p. 113)

Robert Stake's (1975) "responsive approach to evaluation" places particular emphasis on the importance of personalizing and humanizing the evaluation process. Being responsive requires having face-to-face contact with people in the program and learning firsthand about diverse stakeholders' perspectives, experiences, and concerns.

> Responsive evaluation is an alternative, an old alternative, based on what people do naturally to evaluate things, they observe and react. . . . To do a responsive evaluation, the evaluator conceives of a plan of observations and negotiations . . . , arranges for various persons to observe the program, and with their help prepares brief narrative portrayals, product displays, graphs, etc. . . . Of course, he checks the quality of his records: he gets program personnel to react to the accuracy of his portrayals; and audience members to react to the relevance of his findings. He does most of this informally—iterating and keeping a record of action and reaction. (p. 14)

Guba and Lincoln (1981) integrated naturalistic inquiry and responsive evaluation into an overall framework for improving the meaningfulness of evaluation results. The openness of naturalistic inquiry permits the evaluator to be especially sensitive to the differing perspectives of various stakeholders. This sensitivity allows the evaluator to collect data and report findings with those differing perspectives clearly in mind but with special attention to those whose perspectives are heard less often. Subsequently, Guba and Lincoln (1989) added an explicitly constructivist perspective to responsive evaluation in proposing "responsive constructivist evaluation" as the "fourth generation" of evaluation. The first generation focused on measurement, the second on description, the third on judgment, and the fourth on issues-derived, values-based perspectives.

Responsive evaluation includes the following primary emphases:

1. Identification of issues and concerns based on direct, face-to-face contact with people in and around the program

2. Use of program documents to further identify important issues

3. Direct, personal observations of program activities before formally designing the evaluation to increase the evaluator's understanding of what is important in the program and what can or should be evaluated

4. Designing the evaluation based on issues that emerged in the preceding three steps, with the design to include continuing direct qualitative observations in the naturalistic program setting

5. Reporting information in direct, personal contact through themes and portrayals that are easily understandable and rich with description

6. Matching information reports and reporting formats to specific audiences, with different reports and different formats for different audiences

Illuminative Evaluation

> Ignorance: a particular condition of knowledge: the absence of fact, understanding, insight, or clarity about something. . . .
>
> This is knowledgeable ignorance, perceptive ignorance, insightful ignorance.
>
> —Stuart Firestein (2012, p. 7)
> *Ignorance: How It Drives Science*

Illuminative evaluation aims to replace ignorance with illumination and understanding. To do so, it emphasizes context and interpretation.

> The aims of illuminative evaluation are to study the innovative program: how it operates; how it is influenced by the various school situations in which it is applied; what those directly concerned regard as its advantages and disadvantages; and how students' intellectual tasks and academic experiences are most affected. It aims to discover and document what it is like to be participating in the scheme, whether as teacher or pupil, and, in addition, to discern and discuss the innovation's most significant features, recurring, concomitant, and critical processes. In short, it seeks to address and to illuminate a complex array of questions. (Parlett & Hamilton, 1976, p. 144)

Both responsive and illuminative evaluations are based on the fundamental assumptions that undergird qualitative research: the importance of understanding people and programs in context, a commitment to study naturally occurring phenomena without introducing external controls or manipulation, and the assumption that understanding emerges most meaningfully from an inductive analysis of open-ended, detailed, descriptive data gathered through direct interactions and transactions with the program and its participants.

Values-Driven Evaluation: Harmonizing Program and Evaluation Values

> Happiness is when what you think, what you say, and what you do are in harmony.
>
> —Mahatma Gandhi (1869–1948)
> Indian political and spiritual leader

In the outcomes mania of our accountability-focused and results-driven culture, how things get done—processes, values, principles—has been relegated to secondary importance, if given any attention at all. But for values-driven social change activists and innovators, *how* outcomes are attained is at least as important as, if not more important than, the outcomes themselves. Process matters. The means to ends matter, not just the ends. Indeed, given the uncertainties of complex interventions and interactions, where the ends (outcomes, impacts, results) are uncontrollable, unpredictable, and emergent, values can become the anchor—the *only* knowable in an otherwise uncertain, unpredictable, uncontrollable, and complex world.

In *Getting to Maybe*, our inquiry into the motivations of social innovators revealed that those involved expressed *a sense of calling*. They saw things through their own personal and community lenses of strong values, and when they looked, they saw things that were unacceptable and problems that were outrageous, and they felt compelled to act (Westley et al., 2006). They were driven by a vision of how the world should be, and could be, a better place. These were not management-by-objectives folks. These were values-driven visionaries, and if they were to engage with evaluators, those evaluators would need to be able to engage with them around their values. Here are sample questions to guide a values-focused evaluation:

- What are the priority values that will guide how we engage the world?
- How will we track and judge whether we are true to those values, whether we are walking our talk?
- How do we get feedback from those with whom we engage about how they experience our values and the consequences of those values?
- Where does living out our values take us, to what actions and results, to what differences in the world?

I worked with a family foundation that faced the challenge of transitioning to the next generation when the founding parents both died within a short time interval. This led to a substantial infusion of new assets into the endowment, a much larger philanthropic operation, more formal processes, increased staff, and greatly expanded grant making. I was asked to facilitate a board retreat (adult children and grandchildren of the founders and a couple of trusted, long-time family friends) to begin a strategic planning process that would focus the foundation's mission going forward. After interviewing the board members individually and hearing over and over again about the values that the founders lived by, I suggested that we devote the retreat to articulation of those values. Over the course of a two-day retreat, they told stories about the founders that made explicit how they lived their values. Those values became the foundation's guiding strategic document, one that they returned to year after year in subsequent retreats, always asking the values-driven evaluation question: *Are we walking the talk?* Are we operating in a way that the founders would recognize as upholding their values? Are the things we're accomplishing with the endowment they created true to what they cared about? And one of the things the founders cared about was courageously taking risks and supporting innovation. So we evaluated the grants portfolio using the criteria of innovation and risk taking.

Values matter to values-driven social innovators because a deep sense of values undergirds their initiatives as their vision gets implemented through day-to-day operations and interactions. Strategies and tactics must be values based. Outcomes are thought of as manifestations of values. Innovation is what you promote, but *values are who you are*. Values inform decisions about which way to go at inevitable forks in the road. When a problem arises that challenges the innovator in ways not foreseen by strategy, then values provide guidance for reconciling tensions. Collecting and analyzing stories through in-depth qualitative inquiry offers an especially powerful approach for values-focused evaluation. (For more on values-driven evaluation, see Patton, 2011, pp. 246–250.)

Aligning Program and Evaluation Values

> In the middle of the word evaluation is *valu[e]*.

As the preceding section emphasizes, the values orientations of the intended users of a study are among the criteria to be considered when making methods decisions. One example of a framework for supporting harmony between a program philosophy and evaluation approach is presented in

Exhibit 4.10. This framework illustrates how the decision to use qualitative methods in evaluation can flow from the values of the people who will use the evaluation information.

EXHIBIT 4.10   **Matching Program Philosophy and Evaluation Approach: Marcy Open School Illustration**

The values of our school and our evaluation approach should be aligned. Three themes are core to our school: (1) a personalized curriculum, (2) learning as experiential, and (3) holistic learning. These are intricately intertwined. Their meanings for the school and evaluation are explained in the Marcy Open School evaluation framing document.

MARCY OPEN SCHOOL PHILOSOPHY	EVALUATION OF MARCY OPEN SCHOOL
1. *Personalized curriculum.* Curriculum will vary for each child as teaching extends from the interests, needs, and abilities of each child or group of children. The school personnel seek to be aware of each child as an individual and make that awareness the basis of learning activities and materials. The individual child, the teacher, and the parents make decisions on the personalized curriculum.	1. *Personalized evaluation.* The determination of the success of the school will vary depending on the values and perspectives of interested people. The evaluation will describe school activities and what was accomplished by children. Judgments about the success of the school and the validity of activities must be left to the individuals reading the evaluation report, according to their own perspectives.
2. *Experiential nature of learning.* The school seeks to have children experience language rather than only learning to read, to experience computation rather than only learning math, to be in and to learn from the community rather than only learning about social studies. Participants in the school believe that experience is the best transmitter of knowledge. Furthermore, students are expected to interact with their environment—to have an effect on it in the process of experiencing it—to change it, or to recognize ways in which they seek to move toward change.	2. *Experiential nature of evaluation.* This evaluation will provide an opportunity for the reader to experience the school and its children. It will provide not only charts and statistics but also photographs, drawings, and works of children and adults. Even so, any such report can only present an imperfect picture of the school and its processes. Furthermore, the evaluation report will not be presented as a final statement of the accomplishments of the school. It is, instead, a report-in-process, to be reacted to and sent back for new data, details, and descriptions for ongoing development.
3. *Holistic nature of learning.* Much emphasis has been placed on the interrelatedness of learning. Organizational structures, activities, and materials involve multidimensional goals. Goal statements and staff development activities give conscious attention to children's feelings about themselves and their world, their relationships with others and how these relate to their interests and abilities. Staff seek activities that allow students to experience the relationship between language, computation, and other knowledge rather than compartmentalizing them into separate content areas.	3. *Holistic evidence for evaluation.* Three of the school's goals for children have been chosen for special attention in this evaluation. The evidence presented attempts to observe the natural order of events as they happen in the school, with a minimum of distortion through compartmentalization. Both objective figures and subjective judgments are included and are considered valid. The activities and products of the children are viewed, as much as possible, in terms of their multidimensional effects.

Understanding, relevance, interest, and use are all increased when evaluators and users share values about methods. The final design of an evaluation depends on calculated trade-offs, including political/philosophic/ value considerations. The design also depends on resources, time constraints, and commitment. What is to be avoided is the routine selection of a design without consideration of its strengths and weaknesses in relation to this complex constellation of both values and technical factors in an effort to make methods and values relatively harmonious.

Connoisseurship Evaluation

While responsive evaluation places the program's stakeholders at the center of the evaluation process, *connoisseurship* evaluation places the evaluator's perceptions and expertise at the center of the evaluation process. The researcher as connoisseur or expert uses qualitative methods to study a program or organization but does so from a particular perspective, drawing heavily on his or her own judgments about what constitutes excellence, thus the term *connoisseur* (Eisner, 1985; Smith, 2005).

> A critic might be invited to a school or classroom without a prefigured focus and after several days or weeks perceive an aspect of the school or classroom that is of considerable significance but which could not have been anticipated. For example, one of my students received permission from a secondary school English teacher to observe and to write an educational criticism about her class. What emerged during my student's observation was the extraordinary way in which the teacher used satire in her teaching. It was not the case that the teacher was herself teaching satire; it was that she was satirical in her teaching. Such a process or an approach could hardly have been prefigured. The point here is that the focus of criticism can be either prefigured as a part of the research bargain between the critic and the teacher, or it can be emergent or it can be both. (Eisner, 1985, pp. 184–185)

The imagery of evaluators as "connoisseurs" making critical appraisals of programs is analogous to the traditional way in which literary and artistic connoisseurs and critics work. The approach involves naturalistic inquiry through direct observation and immersion in the setting under study. Educational criticism or connoisseurship is, according to Eisner, first and foremost descriptive, *both factually and artistically.* The factual component reports direct observations and interviews.

> The artistic aspect of description is literary and metaphorical; indeed, it can even be poetic in places. . . . In order to optimize communication, the potential of language is exploited so that the literary and the factual complement each other. (Eisner, 1985, p. 182)

The connoisseur, then, is explicitly and purposefully a qualitative researcher and an artistic critic of the phenomenon being studied.

Here's an example. A major philanthropic foundation wanted an independent review of its agricultural development programs in three countries in Latin America. The person selected to conduct the review had 40 years' experience working in agricultural development worldwide, including extensive work in the countries where the review would take place. He was also well-known as a professional evaluator and had written about evaluating development programs. He reviewed years of reports and annual evaluations, interviewed key informants by Skype, and did three weeks of fieldwork in the countries of interest. The credibility of the report's findings and recommendations rested on his credentials, knowledge, and experience, including knowledge of and experience with the countries in the study. This was a connoisseurship evaluation, an expert *meta-evaluation* (evaluation of evaluations) in which the evaluator was as much the instrument as the qualitative methods used. The inquiry methods, and therefore the credibility of findings, derived from and depended on the skill and knowledge of the inquirer, that is, to wit, the connoisseur. Sine qua non. Without this expertise, the evaluation has nothing.

Developmental Evaluation

> Every public action which is not customary, either is wrong, or, if it is right, is a dangerous precedent. It follows that nothing should ever be done for the first time.
>
> —Francis Mcdonald Cornford (1874–1943)
> English classical scholar and poet

Developmental evaluation (Patton, 2011) supports innovation *development* to guide adaptation to emergent and dynamic realities in complex environments. Innovations can take the form of new projects, programs, products, organizational changes, policy reforms, and system interventions. A complex system is characterized by a large number of interacting and interdependent elements in which there is no central control; self-organizing and emergent behaviors based on sophisticated information processing generate learning, evolution, and development. Complex environments for social interventions and innovations are those in which what to do to solve problems is uncertain, and key stakeholders are in conflict about how to proceed. Informed by systems thinking and sensitive to complex nonlinear dynamics, developmental evaluation supports social innovation and adaptive management.

Evaluation processes include asking evaluative questions, applying evaluation logic, and gathering timely data to inform ongoing decision making and adaptations. The developmental evaluator is often part of a development team whose members collaborate to conceptualize, design, and test new approaches in a long-term, ongoing process of continuous development, adaptation, and experimentation, keenly sensitive to unintended results and side effects. The evaluator's primary function in the team is to infuse team discussions with evaluative questions, thinking, and data and to facilitate systematic data-based reflection and decision making in the developmental process.

Developmental evaluation typically makes heavy use of qualitative inquiry because naturalistic inquiry and in-depth interviewing allow the evaluator to find out and track what is emerging in an innovation and the effects of what is emerging. A core complexity concept is *emergence*, which means that in complex dynamic situations, what will happen and the implications of what will happen cannot be known in advance. The openness of qualitative inquiry allows the evaluator to watch what emerges as it emerges. For example, an employment training program originally targeted chronically unemployed men of color. When public policy changed to emphasize getting women off welfare and into the job market, the program began taking in women. This led to the need for child care, new safety and security procedures at the agency, new connections with employees more inclined to hire women, and new rules against sexual harassment. Female staff had to be hired. The change in gender composition changed virtually everything about the program. Only open-ended, qualitative inquiry (interviews, observations, document analysis) could fully track and interpret these dramatic, interrelated developments—the core of developmental evaluation.

Utilization-Focused Evaluation

Utilization-focused evaluation (Patton, 2008b, 2012a) offers an evaluative process, a strategy, and a framework for making decisions about the content, focus, and methods of an evaluation.

Utilization-focused evaluation begins with the premise that evaluations should be judged by their utility and actual use; therefore, evaluators should facilitate the evaluation process and design any evaluation with careful consideration of how everything that is done, *from beginning to end*, will affect use. Use concerns how real people in the real world apply evaluation findings and experience the evaluation process. Therefore, the *focus* in utilization-focused evaluation is on *intended use by intended users*. Since no evaluation can be value-free,

EVALUATING "FAITH-BASED" INITIATIVES

First, some context. Shortly after his election as president of the United States in 2000, George W. Bush created a White House office to support funding religious organizations engaged in social and educational programs. Such religious efforts were called "faith-based" initiatives.

Following a national speech to nonprofit and philanthropic leaders on strategies for enhancing evaluation use, I was asked to comment on the effectiveness of "faith-based" programs.

I responded, "From an evaluation perspective, any program is faith based unless and until it has evaluation evidence of effectiveness. By that criterion, most programs have always been and remain essentially faith based."

utilization-focused evaluation answers the question of whose values will frame the evaluation by working with clearly identified, primary intended users who have the responsibility to apply evaluation findings and implement recommendations.

Utilization-focused evaluation is highly personal and situational. The evaluation facilitator develops a working relationship with intended users to help them determine what kind of evaluation they need. This requires negotiation in which the evaluator offers a menu of possibilities. Utilization-focused evaluation does not advocate any particular evaluation content, model, method, theory, or even use. Rather, it is a process for helping primary intended users select the most appropriate content, model, methods, theory, and uses for their particular situation. *Situational responsiveness* guides the interactive process between the evaluator and the primary intended users. This means that the interactions between the evaluator and the primary intended users focus on fitting the evaluation to the particular situation, with special sensitivity to context. A utilization-focused evaluation can include any evaluative purpose (formative, summative, developmental), any kind of data (quantitative, qualitative, mixed), any kind of design (e.g., naturalistic, experimental), and any kind of focus (processes, outcomes, impacts, costs, and cost–benefit, among many possibilities). *Utilization-focused evaluation is a process for making decisions about these issues in collaboration with an identified group of primary users, focusing on their intended uses of evaluation.*

A psychology of use undergirds and informs utilization-focused evaluation: Intended users are more likely to use evaluations if they understand

and feel ownership of the evaluation process and findings; they are more likely to understand and feel ownership if they've been actively involved; by actively involving primary intended users, the evaluator is training users in use, preparing the groundwork for use, and reinforcing the intended utility of the evaluation every step along the way. Qualitative inquiry strategies may emerge as appropriate in a particular utilization-focused evaluation as a result of defining the information needs of the specific intended evaluation needs.

Creative, practical evaluators need a full repertoire of methods to use in studying a variety of issues. This repertoire should include, but not be limited to, qualitative methods. By offering intended users methodological options, utilization-focused evaluators collaborate in making critical design and data collection decisions so as to increase the intended users' understanding and buy-in, thereby facilitating increased commitment to use findings. The evaluator's responsibility is to interact with decision makers about the strengths, weaknesses, and relative merits of various methods so that mutually agreed, informed methods decisions can be made. The evaluator may well challenge entrenched methodological biases while remaining ultimately respectful of the importance of users getting something they will believe in and use.

Measurement and methods decisions are not simply a matter of expertly selecting the best techniques. Researchers *and* decision makers operate within quite narrow methodological paradigms about what constitutes valid and reliable data, rigorous and scientific design, and personal or impersonal research methods. One of the tasks to be accomplished during the interactions between evaluators and intended evaluation users is to mutually explore design and data biases so that the evaluation generates information that is useful and believable to all concerned.

Utilization-focused evaluation was developed from a study of the factors that seemed to explain variations in the actual use of evaluations. That study used qualitative methods to study the uses of 20 federal health evaluations. Our study team interviewed evaluators, funders, and program managers to find out how evaluation findings were used (Patton, 1997). Many others have since confirmed and elaborated the major elements of utilization-focused evaluation, again using qualitative methods (Alkin, Daillak, & White, 1979; Campbell, 1983; Ferguson, 1989; Holley & Arboleda-Florez, 1988; King & Pechman, 1982). Utilization-focused evaluation has been further refined through experience and practice and by studying how well-known and effective evaluators actually consult with clients and conduct themselves in doing evaluation research (Patton, 2008a). Utilization-focused evaluation, then, is a process for creatively and flexibly interacting with intended evaluation users about their information needs and alternative methodological options, taking into account the decision context in which an evaluation is undertaken.

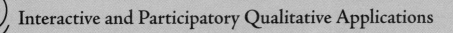

@ Interactive and Participatory Qualitative Applications

> If people don't want to come out to the park, nobody's going to stop them.
>
> —Yogi Berra (1925–)
> New York Yankees baseball player and manager

Thus far, this chapter has been describing various evaluation issues that are especially appropriate for qualitative inquiry and some specific evaluation models that are closely associated with and rely heavily on qualitative methods. The next sections in this chapter review interactive applications of qualitative methods—practical and pragmatic forms of inquiry in which the researcher is especially sensitive to the perspectives of others and interacts closely with them in designing and/or implementing the study. Such interactive approaches personalize and humanize research and evaluation and support action learning and reflective practice. Processes like appreciative inquiry and participatory research facilitate collaborative qualitative inquiry.

Learning and Problem-Solving Applications: Action Research, Action Learning, Reflective Practice, and Learning Organization Inquiries

> You learn something new every day. Actually, you learn something old every day. Just because you just learned it, doesn't mean it's new. Other people already knew it. Columbus is a good example.
>
> —George Carlin (1997, p. 135)
> Comedian

A variety of organizational, program, and staff development approaches emerged at the end of the past century that involved inquiry within organizations aimed at learning, improvement, and development (e.g., Argyris & Schon, 1978; Aubrey & Cohen, 1995; Senge, 1990; Torres, Preskill, & Piontek, 1996; Watkins & Marsick, 1993). These efforts are called various things:

- Action learning (McNamara, 1996; Pedler, 1991, 2008; Zuber-Skerritt, 2009)
- Action research (Burns, 2007; Dick, 2009a, 2009b; Elliott, 2005; Heer & Anderson, 2005; Stringer, 2007; Wadsworth, 2008a, 2008b)
- Community of practice inquiry (Wenger, 1998; Wenger et al., 2002)
- Developmental evaluation (Patton, 2011)
- Interactive evaluation practice (King & Stevahn, 2013)
- Participatory action research (McGarvey, 2007; McIntyre, 2007; Whyte, 1989)
- Reflective practice (Schon, 1983b, 1987; Tremmel, 1993)
- Team learning (Jarvis, 1999)

These problem-solving and learning-oriented processes often use qualitative inquiry and case study approaches to help a group of people reflect on ways of improving what they are doing or understand it in new ways.

For example, the teaching staff of an evening basic skills adult education program undertook an action learning inquiry into the experiences of students who had recently immigrated to the United States. Each staff person interviewed three students and wrote short case studies based on the students' reports about their experiences entering the program. In an ongoing "reflective practice group," the teachers shared their cases and interpreted the results. Based on these reflections and thematic cross-case analysis, the staff revised their emphasis on goal setting as the priority first step in the program. The interviews showed that the new immigrants lacked sufficient experience with and knowledge of American educational and employment opportunities to set meaningful and realistic goals. As a result, these students typically acquiesced to counselor suggestions just to get the mandated goal-setting forms completed, but they came out of the orientation without any substantial commitment to or ownership of the proposed program of study, supposedly based on their own goals. Here, we have an example of a short-term, rapid-turnaround qualitative inquiry for reflective practice or action learning aimed at program improvement. By participating in the process, the staff also deepened their sensitivity to the perspectives and needs of new immigrant students, thereby developing professionally.

The learning that occurs as a result of these processes is twofold: (1) The inquiry can yield specific insights and findings that can change practice, and (2) those who participate in the inquiry learn to think more systematically about what they are doing and their own relationship with those with whom they work. In many cases, the specific findings are secondary to the more general learnings that result from being involved in the process, what I have called "process use" as opposed to findings use (Patton, 2012a). Process use is greatly enhanced in "developmental evaluation" (Patton, 2011), in which the purpose of the evaluation is ongoing learning, internal improvement, and program development rather than generating reports and summative judgments for external audiences or accountability.

A lot of attention has been paid to action learning in recent years as a way of helping people in organizations cope with change. Mwaluko and Ryan (2000) offer a case study showing how action learning programs can succeed if they are carefully designed and implemented systemically to deal with organic, cultural, and power complexities. In qualitative terms, this means action learning inquiry that is holistic, grounded, and sensitive to context.

Harvey and Denton (1999) examined the twin and interrelated themes of "organizational learning" and "the learning organization" in the business sector. The qualitative research underpinning their study was conducted over a three-year period (1994–1997) and involved detailed examination of organizational learning aspirations and practices within the British operations of five major manufacturing companies. Sixty-six interviewees were classified into three groups—strategy, human resources, and research and development (R&D). They identified a set of six antecedents that together explain the rise to prominence of emphasis on organizational learning: (1) the shift in the relative importance of factors of production away from capital and toward labor, particularly intellectual labor; (2) the ever more rapid pace of change

REFLECTIVE PRACTICE: LEARNING FROM STORIES

Planned Lifetime Advocacy Network in Canada has been working on reframing how communities engage with people with disabilities. Early in their work, one of their guiding principles was "A home is a sanctuary, not a warehouse." Inquiring together into what this meant for engagement with people with disabilities led them to questions about related concepts: quality of life and relationships. As the work unfolded and they reflected systematically on what they were hearing, experiencing, and doing—sharing stories, reflecting on what they heard people saying—they came to a fundamental principle that has guided their work ever since:

Relationships do not lead to quality of life; they are quality of life.

in the business environment; (3) widespread acceptance of knowledge as a prime source of competitive advantage; (4) greater demands being placed on all businesses by customers; (5) increasing dissatisfaction, among managers and employees, with the traditional, command-and-control management paradigm; and (6) the intensely competitive nature of global business. Their analysis of qualitative interviews featured the interplay of thoughts and feelings between management practitioners and organizational theorists in getting inside the idea of organizational learning.

While the preceding example focused on business applications, the idea of being a learning organization has also become a prominent theme in the non-profit, government, and philanthropic sectors. Preskill and Mack (2013) interviewed key informants to assess the state of evaluation in the philanthropic and nonprofit sectors and concluded that a more strategic approach to evaluation was needed. One such strategic approach is to create a Learning Lab, a systematic process for group reflection that uses qualitative stories and cases as the basis for shared learning (Beynon & Fisher, 2013). Frances Westley and colleagues at the University of Waterloo Institute for Social Innovation and Resilience have created a Social Innovation Lab to guide groups and organizations toward identifying and implementing transformative change in complex problem domains. These and many similar initiatives use qualitative inquiry as practiced through the lenses of action learning and reflective practice as a foundation for shared learning. Exhibit 4.11 provides a detailed guide to engaging in story-based reflective practice.

EXHIBIT 4.11 Reflective Practice Guide

Reflective practice is a systematic process for group learning based on sharing stories and analyzing cross-cutting themes to inform future action.

1. The reflective practice facilitator helps the group identify a focus for inquiry and learning. An important concept, basic premise, or fundamental value often offers a useful focus: an idea that provides direction and vision to the desired change but the meaning of which is still emergent.

Example: A group of agency directors struggling with collaboration undertook reflective practice by sharing stories of good and poor collaboration experiences.

2. Turn the concept, idea, value, or vision into an experiential inquiry question. The question is not an abstract question for intellectual discussion. It is a question that evokes experience.

Example: Share an example of what you consider a positive, effective collaboration experience. What made it effective? Now share an example of a bad collaboration experience, whatever "bad" means to you. What made it "bad"?

3. Participants in the reflective practice group share their personal experiences (real-life anecdotes, lived experience stories).

Helpful facilitation guidelines:

   a. It has to be that person's own experience, not sharing an experience the person knows about from someone else or heard about secondhand or read about. It's *one's own* experience.

   b. It's not one's life story. The person has to be able to tell the story in three to five minutes.

   c. The person just tells the story but doesn't explain it. The group doesn't analyze it. Each person tells his or her story: the beginning, middle, and end (or where it is now if there is no ending).

   d. For larger reflective practice groups, break into smaller groups (four to six people) for the sharing and interacting. Later, the small groups report back to the full group.

4. Group members can ask short, clarifying questions. This is not a time for discussion or analysis. The focus is on understanding the story, on what happened and why.

Example: If it's not clear what made the collaboration a positive experience, ask the person telling the story to describe that connection. "Please say a little say more about what made this an effective collaboration. . . . I think I understand, but just say a bit more so that I don't misinterpret."

Facilitation guidelines:

   a. Establish norms of confidentiality; what is shared stays in the group.

   b. Be prepared for emotions; alert the group to expect that some stories may evoke strong feelings. That's okay. That's part of the process. Be sensitive. But it's also not a therapy group. *It's a reflective practice group.*

   c. Enforce timelines gently and sensitively. Don't be obnoxious and anal-compulsive about time. These are people's stories, their lives they are sharing. But time is short and precious. Nudge people along as needed.

5. After all stories have been shared, participants are asked to identify patterns and themes in the stories.

Facilitation guidelines:

   a. You may or may not distinguish patterns from themes, depending on the sophistication of the group. The first time I'm working with a group, I don't emphasize the distinction. Over time, part of capacity building with the group can be learning to make the distinction. Oh, and what is the distinction? I'm glad you asked. Actually, there's no hard-and-fast distinction. The term *pattern* usually refers to common, specific observable behaviors and reports. For example, I facilitated reflective practice with university administrators who participated in a wilderness program. One descriptive *pattern* was "Almost all participants reported feeling fear when they rappelled down the cliff." A

(Continued)

(Continued)

theme takes a more categorical, general, or topical form: FEAR. So putting the two together, the reflective practice analysis revealed a pattern of participants reporting being afraid when rappelling down cliffs, running river rapids, and hiking along ledges; many also initially experienced the group process of sharing personal feelings as evoking some fear. Those patterns make *dealing with fear* a major theme of the wilderness education program experience. Patterns are common, concrete observations grounded in the data. Themes cut across patterns and involve moving to an interpretation of meaning.

b. Not every story has to have the theme or pattern. In a group of six, commonalities among three people's stories is a pattern. The purpose is not generalization. The purpose is deepening understanding, generating shared insights and learning, often for further, ongoing inquiry.

c. Someone in the group needs to keep track of the stories, and either that person or someone else needs to be identified to report out the themes and patterns to the larger group (if more than one group is engaged in the reflective practice exercise).

d. The qualitative reflective practice facilitator needs to be sure to get the notes from each group, so alert them in advance that their summary notes will be collected; when they are collected, make sure the summary notes are legible, understandable, and interpretable.

6. If there is more than one small group engaged in the reflective practice exercise, each group reports its themes and patterns to the full group. The facilitator records the themes and patterns and combines those that appear to be similar or duplicative.

7. Once patterns and themes are identified, turn to implications. The group picks one or two themes that have important implications for the work at hand. The members discuss those implications. This often involves identifying important lessons.

Example: Positive collaboration experiences involved people who shared a strong commitment to an issue and took the time to build trust. Bad collaborative experiences involved people who were in competition with one another for resources and didn't trust each other.

8. Generate action agreements and next steps for future reflective practice.

Variations and additional guidance:

a. It helps the process if participants come prepared with their stories written in advance. This can increase thoughtfulness and anticipation and makes it far easier for the qualitative facilitator to capture the stories (data) from the reflective practice participants. *Potential downside:* Less spontaneity.

b. Tape the stories as a group record, and transcribe them so the qualitative reflective practice facilitator—*and the group*—has a record of the "data" from the stories.

c. If the reflective practice stories (as opposed to just the patterns and themes) are to be used for formal, external evaluation reporting, how the stories are reported (e.g., whether identities are disguised) has to be negotiated with the reflective practice group and informed consent procedures should be followed.

d. Instead of simply listing themes, arrange them into a generic story or a systems map showing how themes are interrelated and interconnected (a spider web of themes rather than a linear list).

e. A variation on the sharing process with large groups is to do a second round of thematic analysis after the first round. So let's say we have a group of 30. Five small groups of 5 people each engage in the initial reflective practice process. Then, instead of reporting their themes to the full group, each group numbers off from 1 to 5. Everyone has the list of themes from their group. The 1s assemble as a new cross-cutting group; likewise the 2s, and so on. Now we have five new groups that can synthesize the patterns and themes they bring from their first-round-of-analysis groups. These groups then report to the whole group, which as a whole synthesizes the final set of patterns and themes. This second round of small-group analysis deepens triangulation of findings.

Story-Focused Learning and Organizational Development

> Stories are how we learn. Stories tell what we know. The stories a group shares define the group. The stories the group's members tell to newcomers and retell to each other are windows into the essence of the group. Without shared stories, there is no group.
>
> —From Halcolm's *Stories and More Stories and Still More Stories*

In *The Springboard*, Stephen Denning (2001) explains "how storytelling ignites action in knowledge-era organizations." He teaches storytelling as a powerful and formal discipline for organizational change and knowledge management. What he calls "springboard" stories are those that communicate new or envisioned strategies, structures, identities, goals, and values to employees, partners, and even customers. He argues that storytelling has the power to transform individuals and organizations. He offers as an example his frustrated efforts, as director of knowledge management of the World Bank, to convince colleagues to share information throughout the organization. His reports and graphs were not proving effective in making his case, so he told the staff a story. In 1995, a health care worker in Zambia was searching for a treatment for malaria. He found what he needed at the website of the Centers for Disease Control. The story communicated what his memos had not: the potential life-and-death importance of having critical knowledge available and easily accessible to any World Bank worker in any remote place in the world.

In another management book, *Managing by Storying Around*, David Armstrong (1992) turns the noun *story* into a verb, "storying," to emphasize the direct and active impact of constructing stories to influence organizational values and culture. Shaw, Brown, and Bromiley (1998) have reported how capturing and using "strategic stories" helped the multinational 3M company improve its business planning both internally, for those involved, and externally, for those to whom strategic results were reported. "A good story (and a good strategic plan) defines relationships, a sequence of events, cause and effect, and a priority among items—*and those elements are likely to be remembered as a complex whole*" (p. 42).

STORYTELLING IS NOT A MAGIC BULLET

Sean Buvala coaches organizations on how to tell their story. He is a strong advocate of the power of stories to communicate. His website (Buvala, 2013) is a rich resource on storytelling. He also knows the limits of storytelling. He warns that stories must be true and should communicate the central focus of an organization's work. He adds that stories cannot and will not (a) solve problems created by inept leadership or financial mismanagement, (b) keep clients who receive bad service, or (c) keep employees who are mistreated.

> Story still contains the same power to change lives, connect people and build communities just as it has done throughout history. Having realistic expectations of this tool will help you use story to its full potential. (Buvala, quoted in Kaplan, 2013, p. 1)

Qualitative inquiry can be used to discover, capture, present, and preserve the stories of organizations, programs, communities, and families. Barry Kibel (1999) developed a process for capturing the "success stories" of clients in programs and aggregating them in a method he called "results mapping." His approach involves an arduous and rigorous coding process that can be challenging to manage, but the fundamental idea is that "for programs engaged in healing, transformation, and prevention, the best source and form of information are client stories" (p. 13). Richard Krueger (2010) has written insightfully about *using stories in evaluation*.

Most Significant Change Stories

Story collecting can be integrated into ongoing program evaluation, monitoring, and development (organizational learning) processes. For example, in Bangladesh, Rick Davies (1998a, 1998b) developed a story-based change-monitoring approach called Most Significant Changes, which was subsequently further developed and adapted by the Institute of Land and Food Resources at the University of Melbourne in Australia (Dart, Drysdale, Cole, & Saddington, 2000; Davies, 1996; Davies & Dart, 2005; Lennie, 2011). The process involves several steps:

1. Key program stakeholders and participants (e.g., farmers in an extension program) come to an agreement on which "domains of change" to monitor with stories.

2. Monthly stories of change written by farmers and field staff are collected.

3. Volunteer reviewers and evaluators using agreed-on criteria select the "most significant stories" during regional and statewide committee meetings.

4. At the end of the year, a document is produced containing all the "winning" stories.

5. This document forms the basis for a roundtable discussion with "key influentials" and funders of the project, who then also select the most significant stories according to their views.

This approach goes beyond merely capturing and documenting client stories; each story is accompanied by the storyteller's interpretation, and after review the stories are also accompanied by the reviewer's interpretation. One of the ideas behind the process is that it promotes a slow but extensive dialog up and down the project hierarchy each month. (Dart, 2000, personal communication, March 20, 2000)

See Exhibit 4.12 for an example of a Most Significant Change story. This story also illustrates the qualitative approach to outcomes documentation discussed earlier in this chapter.

EXHIBIT 4.12   Example of a "Most Significant Change" Story

Title: *"I'll Not Be Milking Cows When I'm 55"*

Name of person recording story: **Mark Saddington, dairy farmer**

Region: **Gippsland**

Who was involved: **Farmer and family**

What happened? *We did the pilot Dairy Business Focus Program in March; and for the first time, my wife came along. We were able to look at our farm as a business, not just as a farm. As a consequence of doing the program, we did a few sums and made a few decisions. We worked out that we can afford to have her on the farm, and she has left her job at the bank. We will generate enough income on the farm to make it more profitable for her to be here. The kids will benefit from seeing her a lot more, and they won't be in day care. So far this year, this has made the calving so much easier, we have a joint input, and it has been such a turnaround in my lifestyle. It has been so good.*

We actually went to the accountant yesterday to get some financial advice on how we should be investing off-farm. He was amazed that what we are doing is treating the farm as a business. I said: "Now everything that we earn on this farm is going to be put away so that I am not milking cows when I am 55 years old!"

We have got a debt-reduction program running for the next 12 months, but after that the money will be channeled to off-farm investment. I want to retire young enough to enjoy what we have been working towards for the last 20 or 30 years. My boss is 77 and is still working on the farm. If I am that fit when I am his age, I want to be touring around the world.

It has opened up our lives. We are now looking at off-farm investment, as capital investment on-farm is not that great.

We are not going to invest in new machinery but are going to invest in contractors to do any work we can't do. There is no point buying new machinery, as it depreciates. Instead, we will buy shares and invest off the farm. This proves that you can farm on 120 cows, you don't have to get big, and you don't have to milk a lot of cows. It just depends what you do with your money. If only we could educate the younger farmers to think ahead instead of buying the largest SS Commodore or the latest dual cab. I followed the same track for a few years until we sat down and worked out where we were going and where we could be. We made a few mistakes in the past, but the past is the past.

Feedback from the statewide committee:

- *This story generated lots of discussion. But is it really about profitability or quality of life or changes in farm practice?*
- *The general consensus was that there needed to be more detail in the story for it to be clearly about profitability.*
- *It is a really powerful story that shows considerable change.*

Feedback from the Round-table Meeting:

- *The story showed strong evidence of attitudinal change, leading to self-improvement and goal setting. These people will be high achievers and reap the rewards. They will be good role models for others who desire similar rewards.*
- *This approach is okay, but it isn't necessarily a prescription for others.*
- *It has some good messages, but it hasn't got all the answers.*
- *This is a very good example of achieving the goal of the DBF Program: i.e., getting strategic thinking/planning followed by farmer action.*
- *I liked this story as it highlights the diversity in personal goals and ways to get there.*

SOURCE: Dart, Drysdale, Cole, and Saddington (2000, pp. 52–53).

Cognitive scientists have found that stories are more memorable and better support learning and understanding than nonstory narratives (Shaw et al., 1998, p. 42). Language scholar Richard Mitchell (1979) has observed, "Our knowledge is made up of the stories we can tell, stories that must be told in the language that we know. . . . Where we can tell no story, we have no knowledge" (p. 34).

As I noted in the previous chapter in discussing narrative analysis, which focuses on stories as a particular form of qualitative inquiry, the language of "storytelling" is less intimidating to nonresearchers than the language of "case studies" or "ethnography," words that sound heavy and academic. It makes quite a difference when talking to people in a community to say, "We'd like to hear and record your stories" versus "We'd like to do a case study of you." Narratives illuminate the meanings of experiences and place those meanings in social and historical contexts to deepen our understanding (Bochner, 2014).

One of my favorite examples of narrative meaning was told by anthropologist Gregory Bateson—the story of attempts to create a computer that could think like human beings. Two scientists became particularly enthralled with this idea and devoted their lives to the quest. Each time they thought they had succeeded, their efforts failed critical tests, and they had to go back to the drawing board to search still deeper for the breakthrough that would lead to a computer that could truly think. They made revisions and adaptations based on all their prior failures until, finally, they felt more hopeful than ever that they had succeeded. The computer program passed all the preliminary tests. They became increasingly excited. The moment arrived for the penultimate test. They asked the computer, "Can you think like a human being?"

The computer processed the question and on the screen appeared the answer: "That question reminds me of a story." They knew then that they had succeeded (Bateson, 1977).

Appreciative Inquiry

> Appreciation is a wonderful thing: It makes what is excellent in others belong to us as well.
>
> —Voltaire (1694–1778)
> French philosopher

Appreciative inquiry has emerged as a popular organizational development approach that emphasizes building on an organization's assets rather than focusing on problems, or even problem solving. Conceived and described in the work of David Cooperrider and his colleagues at Case Western Reserve's school of Organization Behavior, appreciative inquiry is being offered by its advocates as

> a worldview, a paradigm of thought and understanding that holds organizations to be affirmative systems created by humankind as solutions to problems. It is a theory, a mindset, and an approach to analysis that leads to organizational learning and creativity. (Watkins & Cooperrider, 2000, p. 6)

What interests us here is that appreciative inquiry is grounded in qualitative understandings and prescribes a particular process of qualitative inquiry within an organization that includes a dialogue process among participants based on their interviewing each other. They ask each other questions that "elicit the creative and life-giving events experienced in the workplace."

> 1. Looking at your entire experience with the organization, remember a time when you felt most alive, most fulfilled, or most excited about your involvement in the organization. . . .
>
> 2. Let's talk for a moment about some things you value deeply; specifically, the things you value about yourself, about the nature of your work, and about this organization. . . .
>
> 3. What do you experience as the core factors that give life to this organization? Give some examples of how you experience those factors.
>
> 4. What three wishes would you make to heighten the vitality and health of this organization? (Watkins & Cooperrider, 2000, p. 9)

These questions aim at generating specific examples, stories, and metaphors about the positive aspects of organizational life. Participants in the process analyze the results in groups, looking for the themes and topics that can become the foundation for positive organizational development going forward.

> For example, if the original data suggests that COMMITMENT is an important factor in many of the stories about the best of times in the organization, then the workgroup might choose to ask more questions from others in the workplace about their experiences with commitment. This second round of interviews produces information about four to six topics that become the basis for building "Possibility

Propositions" that describe how the organization will be in the future. Each topic or theme can be fashioned into a future statement. And these statements become an integral part of the vision for the organization. Often, this process is completed with a future search conference that uses the Appreciative Inquiry data as a basis for imaging a positive and creative future for the organization. (Watkins & Cooperrider, 2000, p. 10)

Appreciative inquiry has been criticized for being unbalanced and uncritical in its emphasis (critics say *over*emphasis) on accentuating the positive. It may even, ironically, discourage inquiry by discouraging constructive criticism (Golembiewski, 2000). However, appreciative inquiry's questioning strategies can be incorporated into more balanced approaches, including for program evaluation (Preskill, 2005; Preskill & Catsambas, 2006; Preskill & Coghlan, 2003). Appreciative inquiry integrates inquiry and action within a particular developmental framework that guides analysis and processes of group interaction. The qualitative questioning and thematic analysis processes constitute a form of intervention by the very nature of the questions asked and the assets-oriented framework used to guide analysis. In this way, inquiry and action are completely integrated. Other forms of participatory inquiry also seek integration of inquiry and action.

Participatory Research and Evaluation: Valuing and Facilitating Collaboration

> In the long history of humankind (and animal kind, too) those who learned to collaborate and improvise most effectively have prevailed.
>
> —Charles Darwin (1809–1882)
> English naturalist

Let's start with three quite different examples of participatory qualitative inquiry.

In the film *Awakenings*, based on the real-life experience of Dr. Oliver Sacks (played by actor Robin Williams), the medical researcher and physician engages the assistance of nurses, orderlies, janitorial staff, and even patients in joint efforts to discover what might "reach" certain patients suffering from a rare form of catatonia. The film powerfully displays a form of collaborative research in which trained researchers and nonresearchers undertake an inquiry together. One orderly tries music. A nurse tries reading to patients.

A volunteer tries card games. Together they figure out what works.

In an African village, the women were skeptical of public health workers' admonitions to use well water rather than surface water for drinking during the rainy season. Going to the well was more work. Instead of simply trying to convince the women of the wisdom of this public health advice, the extension educators created an experiment in which half the village used well water and half used surface water. They kept track of illnesses with matchsticks. At the end of three months, the villagers could see for themselves that there were many more matchsticks in the surface water group. By participating in the study rather than just receiving results secondhand, the findings became more meaningful to them and more useful.

Early in my career, I was commissioned by a provincial deputy minister in Canada to undertake an evaluation in a school division he considered mediocre. I asked what he wanted the evaluation to focus on. "I don't care what the focus is," he replied. "I just want to get people engaged in some way in inquiring into what's going on. Education has no life there. Parents aren't involved. Teachers are just putting in time. Administrators aren't leading. Kids are bored. I'm hoping that having them participate in an evaluation will stir things up and get people involved again. See if you can get them participating in a meaningful evaluation process." The resulting collaborative evaluation led to major changes and an infusion of energy into the school division.

When conducting research in a collaborative mode, professionals and nonprofessionals become coresearchers. Participatory action research encourages collaboration within a mutually acceptable inquiry framework to understand and/or solve organizational or community problems. Participatory approaches are widely used in community research, health research, educational evaluation, and organizational development (Blumenthal, DiClemente, Braithwaite, & Smith, 2013; Chevalier & Buckles, 2013; Hacker, 2013; Israel, Eng, Schulz, & Parker, 2013; Minkler & Wallerstein, 2008). Participatory evaluation involves program staff and participants in all aspects of the evaluation process to increase evaluation understanding and use while also building capacity for future inquiries (Baker & Sabo, 2004; Cousins & Chouinard, 2012; King, Cousins, & Whitmore, 2007; King, 2005; King & Stevahn, 2013). Feminist methods are participatory in that "the researcher invites members of the setting to join her in creating the study" (Reinharz, 1992, p. 184). "Empowerment evaluation" aims to foster "self-determination" among those who participate in the inquiry process (Fetterman, 2000, 2013; Fetterman & Wandersman, 2005).

This can involve forming "empowerment partnerships" between researchers and participants (Weiss & Greene, 1992) and teaching participants to do research themselves (Wadsworth, 1984, 2011b). In-depth interviewing and description-oriented observations are especially useful methods for supporting collaborative inquiry because the methods are accessible to and understandable by people without much technical expertise.

Interest in participatory research has exploded in recent years, especially as an element of larger community change efforts. The principal researcher trains the participating coresearchers to observe, interview, reflect, and/or keep careful records or diaries. Those involved come together periodically to share in the data analysis process. The purpose of such shared inquiry is typically to elucidate and improve the nature of practice in some arena of action.

Qualitative, collaborative research efforts in educational settings have a distinguished history. In a classic study, John Elliott (1976) worked with classroom teachers as coresearchers doing action research, "developing hypotheses about classrooms from teachers' practical constructs." Bill Hull (1978) worked with teachers in a reflective, research-oriented process to study children's thinking. The Boston Women's Teachers' Group (1986) was organized as a research collaborative for studying the effects of teaching on teachers.

Genuinely collaborative approaches to research and evaluation require power sharing. One of the negative connotations often associated with evaluation is that it is something done *to* people. Participatory evaluation, in contrast, involves working *with* people. Instead of being research "subjects," the people in the research setting become "coinvestigators" or "co-inquirers." The process is facilitated by the researcher but is controlled by the people in the program or community. They undertake a formal, reflective process for their own development and empowerment.

Participatory evaluation has been used with great success as part of international and community development efforts by a number of nongovernmental groups and private voluntary organizations in the Third World. A collaborative group called Private Agencies Collaborating Together (1986) published one of the first guides to *Participatory Evaluation*, as well as a more general *Evaluation Sourcebook* (Pietro, 1983). Both remain relevant today. The guide includes techniques for actively involving nonliterate people as active participants in evaluating the development efforts they experience, often using qualitative methods. International studies of development have demonstrated how participatory research can be a means of understanding

HOW COLLABORATION IS LIKE TEENAGE SEX

SIDEBAR

At the end of a conference on youth participatory research, a parent-participant observed that after three days of hearing about the challenges of engaging in genuinely collaborative projects, she had come to understand how collaboration is a lot like teenage sex:

Everyone talks about it all the time;

everybody thinks everyone else is doing it;

those who are doing it aren't doing it very well;

despite that, they all talk about how wonderful it is.

and bridging diverse perspectives for responsive development (Better Evaluation, 2014; Mansuri & Vijayendra, 2012; Salmen & Kane, 2006).

The processes of participation and collaboration have an impact on participants and collaborators quite beyond whatever findings or report they may produce by working together. In the process of participating in research, participants are exposed to and have the opportunity to learn the logic of evidence-based inquiry and the discipline of evidentiary reasoning. Skills are acquired in problem identification, criteria specification, and data collection, analysis, and interpretation. Through acquisition of inquiry skills and ways of thinking, a collaborative inquiry process can have an impact beyond the findings generated from a particular study.

Moreover, people who participate in creating something tend to feel more ownership of what they have created and make more use of it. Active participants in research and evaluation, therefore, are more likely to feel ownership not only of *their* findings but also of the inquiry process itself. Properly, sensitively, and authentically done, it becomes *their* process. Participants and collaborators can be community members, villagers, organizational workers, program staff, and/or program participants (e.g., clients, students, farmers). Sometimes administrators, funders, and others also participate, but the usual connotation is that the primary participants are "lower down" in the hierarchy. *Participatory evaluation is bottom-up.* The trick is to make sure that participation is genuine and authentic, not just token or rhetorical, especially in participative evaluation, where differing political and stakeholder agendas often compete.

Uphoff (1991) reviewed a number of such grassroots community development projects that used participatory approaches for self-evaluation. He found that if the process of self-evaluation is carried out regularly and openly, with all group members participating, the answers they arrive at are in themselves not so important as what is learned from the discussion and from the process of reaching consensus on what questions should be used to evaluate group performance and capacity, and on what answers best describe their group's present status. (p. 272)

It was not a group's specific questions or answers that Uphoff found most affected the groups he observed. It was the process of reaching consensus about questions and engaging with each other in the meaning of the answers that turned up. The process of participatory self-evaluation, in and of itself, provided useful learning experiences for participants.

Viewing participatory inquiry as a means of creating an organizational culture committed to ongoing learning, as discussed in the previous section, has become an important theme in the literature linking program evaluation to "learning organizations" (e.g., Caister, Green, & Worth, 2011; King, 1995; King & Stevahn, 2013; Leeuw, Rist, & Sonnichsen, 1993). "The goal of a participatory evaluator is eventually to put him or herself out of work when the research capacity of the organization is self-sustaining" (King, 1995, p. 89). Indeed, "the self-evaluating organization" (Wildavsky, 1985) constitutes an important direction in the institutionalization of evaluation logic and processes.

I advise care in using labels like "participatory inquiry," "empowerment evaluation," or "collaborative research," for these terms mean different things to different people—and serve different purposes. Some use these terms interchangeably or as mutually reinforcing concepts. Wadsworth (1993b) distinguishes "research *on* people, *for* people or *with* people" (p. 1). Levin (1993) distinguished three purposes for collaborative research: (1) the pragmatic purpose of increasing use of findings by those involved, as emphasized by Cousins and Earl (1992); (2) the philosophical or methodological purpose of grounding data in participants' perspectives; and (3) the political purpose of mobilizing for social action, for example, empowerment evaluation, or what is sometimes called "emancipatory" research (Cousins & Earl, 1995, p. 10). A fourth purpose identified here is teaching inquiry logic and skills. Since no definitive definitions exist for "participatory" and "collaborative evaluation," these phrases must be defined and given meaning in each setting where they're used.

Exhibit 4.13 presents the primary principles for fully participatory and genuinely collaborative

EXHIBIT 4.13 Principles of Fully Participatory and Genuinely Collaborative Inquiry

1. The inquiry process involves participants in learning inquiry logic and skills, for example, the nature of evidence, establishing priorities, focusing questions, interpreting data, data-based decision making, and connecting processes to outcomes.

2. Participants in the process *own* the inquiry. They are involved authentically in making major focus and design decisions. They draw and apply conclusions. Participation is real, not token.

3. Participants work together as a group, and the inquiry facilitator supports group cohesion and collective inquiry.

4. All aspects of the inquiry, from research focus to data analysis, are undertaken in ways that are understandable and meaningful to participants.

5. The researcher or evaluator acts as a facilitator, collaborator, and learning resource to make the experience meaningful and mutually respectful.

6. The inquiry facilitator recognizes and values participants' perspectives and expertise and works to help participants recognize and value their own and each other's expertise.

7. So far as it's possible, practical, and authentic, status and power differences between the inquiry facilitator and the participants in the inquiry are minimized, without patronizing or game playing.

inquiry. This list can be a starting point for working with participants in a research or evaluation setting to decide what principles they want to adopt for their own process.

Regardless of the terminology—participatory, collaborative, cooperative, or empowerment—these approaches share a commitment to involving the people in the setting being studied as co-inquirers, at least to some important extent, though the degree and nature of involvement vary widely. Our purpose here has been to point out that these participatory approaches often employ qualitative methods because those methods are understandable, teachable, and usable by people without extensive research training.

Democratic Evaluation, Indigenous Research and Evaluation, Capacity Building, and Cultural Competence

Ancient Chinese philosopher Confucius was asked by one of his disciples what he would do if he were given his own territory to govern. He replied that he would "rectify the names," that is, make words correspond to reality. He explained,

> If the names are not correct, if they do not match realities, language has no object. If language is without an object, action becomes impossible and therefore, all human affairs disintegrate and their management becomes pointless.

Supporting Democratic Dialogue and Deliberation

Most of this chapter has examined relatively small-scale applications of qualitative methods in evaluating programs, developing organizations, supporting planning processes and needs assessments, and providing insights into communities. This section considers a much larger agenda—that of strengthening democracy. House and Howe (2000) have articulated three requirements for evaluation done in a way that supports democracy: *inclusion, dialogue, and deliberation.* They worry about the power that derives from access to evaluation and the implications for society if only the powerful have such access.

> We believe that the background conditions for evaluation should be explicitly democratic so that evaluation is tied to larger society by democratic principles argued, debated, and accepted by the evaluation community. Evaluation is too important to society to be purchased by the highest bidder or appropriated by the most powerful interest. Evaluators should be self-conscious and deliberate about such matters....
>
> If we look beyond the conduct of individual studies by individual evaluators, we can see the outlines of evaluation as an influential societal institution, one that can be vital to the realization of democratic societies. Amid the claims and counterclaims of the mass media, amid public relations and advertising, amid the legions of those in our society who represent particular interests for pay, evaluation can be an institution that stands

apart, reliable in the accuracy and integrity of its claims. But it needs a set of explicit democratic principles to guide its practices and test its intuitions. (House & Howe, 2000, p. 4)

Qualitative inquiry figures into this democratic approach to evaluation because, as discussed in the section on participatory research, qualitative methods are especially accessible to and understandable by non-researchers, and because case studies can be an excellent resource for supporting inclusion and dialogue. In Europe, the democratic evaluation model of Barry MacDonald (1987) illustrates these emphases. He argued that "the democratic evaluator" recognizes and supports value pluralism, with the consequence that the evaluator seeks to represent the full range of interests in the course of designing an evaluation. In this way, an evaluator can support an informed citizenry, the sine qua non of a strong democracy, by acting as an information broker between groups that want and need knowledge of one another. The democratic evaluator must make the methods and techniques of evaluation accessible to nonspecialists—that is, the general citizenry. MacDonald's democratic evaluator seeks to survey a range of interests by assuring confidentiality to sources, engaging in negotiation between interest groups, and making evaluation findings widely accessible. The guiding ethic is the public's right to know.

Saville Kushner (2000) has carried forward, deepened, and updated MacDonald's democratic evaluation model. He sees evaluation as a form of personal expression and political action, with a special obligation to be critics of those in power. He uses qualitative methods to place at the center of evaluation the experiences of people in programs. The experiences and perceptions of the people in programs, the supposed beneficiaries, is where, for Kushner, we will find the intersection of Politics (big *P*—Policy) and politics (small *p*—people). He uses qualitative case studies to capture the perspectives of real people—children and teachers and parents—and the realities of their lives in program settings as *they* experience those realities. He feels a special obligation to focus on, capture, report, and, therefore, honor the views of marginalized peoples. He calls this "personalizing evaluation," but the larger agenda is strengthening democracy. Consider these reflections on the need for evaluators and evaluations to address questions of social justice and the democratic contract:

> Where each social and educational program can be seen as a reaffirmation of the broad social contract (that is, a re-confirmation of the bases of power, authority, social structure, etc.), each program evaluation is an opportunity to review its assumptions and consequences. This is commonly what we do at some level or another. *All programs expose democracy and its failings; each program evaluation is an assessment of the effectiveness of democracy in tackling issues in the distribution of wealth and power and social goods* [italics added]. Within the terms of the evaluation agreement, taking this level of analysis into some account, that is, renewing part of the social contract, is to act more authentically; to set aside the opportunity is to act more inauthentically, that is, to accept the fictions. (pp. 32–33)

Writings on evaluation's role in supporting democratic processes reflect a significant shift in the nature of evaluation's real and potential contribution to strengthening democracy. A decade ago, the emphasis was all on increasing use of findings for enhanced decision making and program improvement and, therefore, making sure that findings reflected the diverse perspectives of multiple stakeholders, including the less powerful, through "inclusive evaluation" (Mertens, 1998, 1999) and genuinely engaging participants in programs, instead of just staff, administrators, and funders. While this thrust remains important, a parallel and reinforcing use of evaluation focuses on *helping people learn to think and reason evaluatively* and on how rendering such help can contribute to strengthening democracy over the long term. I turn now to elaborate that contribution.

Start with the premise that a healthy and strong democracy depends on an informed citizenry. A central contribution of policy research and evaluation, then, is to help ensure an informed electorate as well by disseminating findings, as well as to help the citizenry weigh evidence and think evaluatively. This involves thinking processes that must be learned. It is not enough to have trustworthy and accurate information (the informed part of the informed citizenry). People must also know how to use the information, that is, to weigh evidence, consider the inevitable contradictions and inconsistencies, articulate values, interpret findings, deal with complexity, and examine assumptions, to note but a few of the things meant by "thinking evaluatively." Moreover, in-depth democratic thinking includes political sophistication about the origins and implications of the categories, constructs, and concepts

that shape what we experience as information and "knowledge" (Minnich, 2004).

Philosopher Hannah Arendt was especially attuned to this foundation of democracy. Having experienced totalitarianism, then having fled it, she devoted much of her life to studying it and its opposite, democracy. She believed that thinking thoughtfully in public deliberations and acting democratically were intertwined. Totalitarianism is built on and sustained by deceit and thought control. To resist efforts by the powerful to deceive and control thinking, Arendt (1968) believed that people needed to practice thinking. Toward that end, she developed "eight exercises in political thought." She wrote that "experience in thinking . . . can be won, like all experience in doing something, only through practice, through exercises" (p. 4). From this point of view, might we consider every participatory research and evaluation inquiry as an opportunity for those involved to practice thinking? In this regard, we might aspire to have policy research, action research, participatory research, and collaborative evaluation do what Arendt hoped her exercises in political thought would do, namely, help us "to gain experience in *how* to think." Her exercises "do not contain prescriptions on what to think or which truths to hold" but, rather, on the act and process of thinking. For example, Arendt thought it important to help people think conceptually, to

> discover the real origins of original concepts in order to distill from them anew their original spirit which has so sadly evaporated from the very keywords of political language—such as freedom and justice, authority and reason, responsibility and virtue, power and glory—leaving behind empty shells. (pp. 14–15)

Might we add to her conceptual agenda for examination and public dialogue terms such as *outcomes* and *performance indicators*, *interpretation* and *judgment*, and *beneficiary* and *stakeholder*, among many evaluative possibilities?

Helping people learn to think evaluatively by participating in real evaluation exercises is, as I noted earlier, what I've come to call *process use*: changes in thinking and behaving that occur among those involved in evaluation as a result of the learning that occurs *during the evaluation process*. (Changes in program or organizational procedures and culture may also be manifestations of process impacts, but that is not our focus here.) This means that an evaluation can have dual tracks of impact: (1) a more informed electorate through use of findings and (2) a more thoughtful and deliberative citizenry though helping people learn to think and engage each other evaluatively.

Capacity Building Through Process Use

One way of thinking about *process use* is to recognize that research and evaluation constitute cultural perspectives. When we engage other people in systematic inquiry processes, we are providing them with a cross-cultural experience. This culture of inquiry, which we take for granted once trained in research methods, is quite alien to many people without such training. Examples of the values of evaluation include clarity, specificity, and focusing; being systematic and making assumptions explicit; operationalizing program concepts, ideas, and goals; distinguishing inputs and processes from outcomes; valuing empirical evidence; and separating statements of fact from interpretations and judgments. These values constitute ways of thinking that are not natural to people and, indeed, are quite alien to many. When we take people through a process of participatory research or evaluation, they are in fact learning how to think in these ways. Interviewing a variety of people and understanding in-depth case examples are especially effective ways of enhancing nonresearcher involvement in evaluation and research to help them increase their capacity to think about evidence and draw appropriate conclusions from data. Interactive and participatory forms of evaluation build capacity for critical thinking that can be applied to new challenges. Evaluation done in this way becomes a catalyst for social change (King & Stevahn, 2013; Lennie & Tacchi, 2013).

Helping people learn to think in these ways can make a study have a more enduring impact than use of specific findings generated in that same study. Findings have a very short "half-life"—to use a physical science metaphor. They deteriorate very quickly as the world changes rapidly. Specific findings typically have a small window of relevance. In contrast, learning to think and act evaluatively can have ongoing impact. The experience of being involved in an evaluation, then, for those stakeholders actually involved, can have a lasting impact on how they think, on their openness to reality testing, on how they view the things they do, and on their capacity to engage thoughtfully in democratic processes.

Participatory evaluations debunk the myth that methods and measurement decisions are purely technical. Nonresearchers then become savvier about both the technical and the nontechnical dimensions of evaluation. Moreover, we know that use is enhanced when practitioners, decision makers, and other users fully understand the strengths and weaknesses of evaluation data and that such understanding is increased by being involved in making methods decisions. We know that use is enhanced when intended users participate in making sure that, as trade-offs are considered, as they inevitably are because of limited resources and time, the path chosen is informed by relevance. We know that use is enhanced when users buy into the design and find it credible and valid within the scope of its intended purposes as determined by them. And we know that when evaluation findings are presented, the substance is less likely to be undercut by debates about methods if users have been involved in those decisions prior to data collection (Patton, 2008b, 2013).

At its roots, democratic evaluations are informed by a fundamental confidence in the wisdom of an informed citizenry and a willingness to engage ordinary citizens respectfully in all aspects of research and evaluation, including methodological discussions and decisions. This point is worth emphasizing because some—not all, to be sure, but some—resistance to participatory evaluation derives from the status associated with research expertise and an elitist or patronizing attitude toward nonresearchers (they are, after all, "subjects"). Qualitative evaluation pioneer Egon Guba (1978) has described in powerful language this archetype:

> It is my experience that evaluators sometimes adopt a very supercilious attitude with respect to their clients; their presumptuousness and arrogance are sometimes overwhelming. We treat the client as a "child-like" person who needs to be taken in hand; as an ignoramus who cannot possibly understand the tactics and strategies that we will bring to bear; as someone who doesn't appreciate the questions he *ought* to ask until we tell him—and what we tell him often reflects our own biases and interests rather than the problems with which the client is actually beset. . . . And if what we are looking for are ways to manipulate clients so that they will fall in with *our* wishes and cease to resist our blandishments, I for one will have none of it. (p. 1)

For others who will have none of it, one way to address the issue of methodological quality in democratic evaluations is to reframe the policy analyst's function from an emphasis on generating expert judgments to an emphasis on supporting informed dialogue, including methodological dialogue. The traditional, expert-based status of researchers, scholars, and evaluators has fueled the notion that we provide scientifically based answers and judgments to policymakers while, by our independence, assuring accountability to the general public. Playing such a role depends on a knowledge paradigm in which correct answers and independent judgments can be conceived of existing. However, postmodernism, deconstruction, critical theory, feminist theory, empowerment evaluation, and constructivism, among other perspectives, share skepticism about the traditional, truth-oriented knowledge paradigm. They offer, in contrast, an emphasis on interest-acknowledged interpretations articulated and discussed within an explicit context (political, social, historical, economic, and cultural). Constructivist orientations to qualitative inquiry have played a critical role in the emergence of dialogical forms of inquiry and analysis. Participatory methods have increased the access of nonresearchers to both research findings and processes. In combination, constructivist, dialogical, and participatory approaches offer a vision of research and evaluation that can support deliberative democracy in the postmodern knowledge age. Such a grandiose, even bombastic, vision derives from recognition that in this knowledge age, researchers have larger responsibilities than just publishing in academic journals.

Indigenous Research and Evaluation

While discussing participatory approaches to qualitative inquiry, one focus area deserves special attention: indigenous research and evaluation. Indigenous people have long been the object of study by anthropologists. The findings and reports of such study have not always been either accurate or well received by those same indigenous people when they learned the conclusions anthropologists had drawn about them. They have had particularly bad experiences with evaluation as something done *to* them rather than *with* them.

In beginning an evaluation training program with Native Americans, I started off by asking them, as part of introducing themselves, to mention any experiences with and perceptions of evaluation they cared to share. With 15 participants, I expected the process to take no more than a half-hour. But deep feelings surfaced, and a dialogue ensued that took more than two hours. Here is some of what they said.

- "I'm frustrated that what constitutes 'success' is always imposed on us by somebody who doesn't know us, doesn't know our ways, doesn't know me."
- "By white standards, I'm a failure because I'm poor, but spiritually I'm rich. Why doesn't that count?"
- "I have a hard time with evaluation. We need methods that are true to who we are."
- Said through tears by a female elder: "All my life I've worked with grant programs, and evaluation has been horrible for us—horribly traumatic. Painful. Made us look bad, feel bad. We've tried to give the funders what they want in numbers, but we know that those numbers don't capture what is happening. It's been demeaning. It's taken a toll. I didn't want to come here today."
- Spoken in his native language by a spiritual leader who had opened the session with a smudge ceremony and blessing, translated by his son: "Everything I do is connected to who I am as an Oglala Lakota elder, to our way as a people, to what you call our culture. Everything is connected. Evaluation will have to be connected if it is to have meaning. That's why I brought my son, and my neighbor, and my friend, and my granddaughter. They aren't signed up for this thing we're here to do. But they are connected, so they are here."

In her important book *Decolonizing Methodologies* (2012), Linda Tuhuwai Smith, a Maori researcher, provides the context that conditions indigenous experiences with research historically.

> From the vantage point of the colonized . . . the term "research" is inextricably linked to European imperialism and colonialism. The word itself, "research"; is probably one of the dirtiest words in the indigenous world's vocabulary. When mentioned in many indigenous contexts, it stirs up silence, it conjures up bad memories, it raises a smile that is knowing and distrustful. . . . The ways in which scientific research is implicated in the worst excesses of colonialism remains a powerful remembered history for many of the world's colonized peoples. It is a history that still offends the deepest sense of our humanity. . . . It galls us that Western researchers and intellectuals can assume to know all that it is possible to know of us, on the basis of their brief encounters with some of us. It appalls us that the West can desire, extract and claim ownership of our ways of knowing, our imagery, the things we create and produce, and then simultaneously reject the people who created and developed those ideas and seek to deny them further opportunities to be creators of their own culture and own nations. (p. 1)

Countering the history of colonial methodology and the stance of research *done on* indigenous people is the new stance of doing research *with* indigenous people. For example, Northern Plains tribal college students from four different communities participated as data gatherers and coresearchers, using reflective interviews in their communities for a participatory research project to study perceptions of service learning among the participating tribal college students (Cowden, McDonald, & Littlefield, 2012). This is one example among many articles published in the new *Journal of Indigenous Research* with the mission of coming "Full Circle: Returning Native Research to the People."

Qualitative inquiry is especially consistent with the oral story-sharing traditions of indigenous peoples around the world. Indigenous research certainly includes quantitative and mixed-methods approaches, but qualitative methods and data especially resonate because they are accessible and understandable. Participatory qualitative methods also conform readily to the principles for engaging in indigenous research. Exhibit 4.14 features "Guidelines for Ethical Research in Australian Indigenous Studies" from the Australian Institute of Aboriginal and Torres Strait Islander Studies (2011).

EXHIBIT 4.14 **Guidelines for Ethical Research in Australian Indigenous Studies**

1. Recognition of the diversity and uniqueness of peoples, as well as of individuals, is essential.

2. The rights of Indigenous peoples to self-determination must be recognised.

3. The rights of Indigenous peoples to their intangible heritage must be recognised.

4. Rights in the traditional knowledge and traditional cultural expressions of Indigenous peoples must be respected, protected and maintained.

5. Indigenous knowledge, practices and innovations must be respected, protected and maintained.

6. Consultation, negotiation and free, prior and informed consent are the foundations for research with or about Indigenous peoples.

7. Responsibility for consultation and negotiation is ongoing.

8. Consultation and negotiation should achieve mutual understanding about the proposed research.

9. Negotiation should result in a formal agreement for the conduct of a research project.

10. Indigenous people have the right to full participation appropriate to their skills and experiences in research projects and processes.

11. Indigenous people involved in research, or who may be affected by research, should benefit from, and not be disadvantaged by, the research project.

12. Research outcomes should include specific results that respond to the needs and interests of Indigenous people.

13. Plans should be agreed for managing use of, and access to, research results.

14. Research projects should include appropriate mechanisms and procedures for reporting on ethical aspects of the research and complying with these guidelines.

SOURCE: Australian Institute of Aboriginal and Torres Strait Islander Studies (2011).

A rapidly growing literature provides guidance for conducting culturally appropriate and sensitive indigenous research and evaluation (e.g., Chilisa, 2012; Denzin, Lincoln, & Smith, 2008; Jacob & Desautels, 2013; Kovach, 2009; Mertens, Cram, & Chilisa, 2013; Stewart, 2009; Wilson, 2009). The issues are not just methodological but include philosophical, political, ethical, cross-cultural, interpersonal, epistemological, and ontological concerns. Indeed, the issues are also professional, including the question of competence to undertake research and evaluation in culturally appropriate ways.

Cultural Competence

The American Evaluation Association (2013) has adopted a statement on *cultural competence* based on the principle that "to ensure recognition, accurate interpretation, and respect for diversity, evaluators should ensure that the members of the evaluation team collectively demonstrate cultural competence."

Cultural competence is a stance taken toward culture, not a discrete status or simple mastery of particular knowledge and skills. A culturally competent evaluator is prepared to engage with diverse segments of communities to include cultural and contextual dimensions important to the evaluation. Culturally competent evaluators respect the cultures represented in the evaluation.

The challenge is to engage in research and evaluation in ways that are respectful, ethical, and useful. Utility, in particular, cannot be taken for granted, as Maori researcher Linda Tuhuwai Smith (2012), quoted earlier, reminds us.

Most research is produced on the basis that it will contribute to something greater than itself, and that it adds value to society for the future. . . . The critical question for indigenous communities is that research has never really demonstrated that it can benefit communities—because the benefits never reach indigenous peoples or are used as a *ploy* or tactic to coerce indigenous communities into sacrificing their cultural values, leaving their homes, giving up their languages and surrendering control over basic decision making in their own lives.

In other words, research exists within a system of power. What this means for indigenous researchers as well as indigenous activists and their communities is that indigenous work has to "talk back to" or "talk up to" power. There are no neutral spaces for the kind of work required to ensure that traditional indigenous knowledge flourishes; that it remains connected intimately to indigenous people as a way of thinking, knowing and being; that it is sustained and actually grows over future generations. . . . Getting the story right and telling the story well are tasks that indigenous activists and researchers must both perform. (p. 226)

Getting the story right and telling the story well are essential qualitative methods commitments—*and skills*.

@ Special Methodological Applications

These final pages of this chapter examine some special methodological applications and challenges where qualitative strategies are especially appropriate: unobtrusive measures; exploratory research; adding depth, detail, and meaning to quantitative analyses; rapid reconnaissance; validity checks on the numbers; opening and looking into the experimental design black box; breaking the routine to generate new insights; and futuring scenarios.

Unobtrusive Measures

Qualitative observations can be particularly appropriate where the administration of standardized instruments, assigning people to comparison groups, and/or the collection of quantitative data would affect program operations by being overly intrusive. Examples of unobtrusive measures include the following:

- Observing and measuring wear and tear on carpets placed in front of different museum exhibits to evaluate visitor interest in different exhibits
- Reviewing disciplinary referrals to the principal's office as an unobtrusive measure of what is treated as student behavior problems
- Observing people in a public place without taking notes, to see what naturally unfolds
- Using documents or reports prepared for other purposes (e.g., clinical case notes)
- Studying racial interaction patterns in colleges by observing groups of students in dining halls and student centers
- Content analysis of graffiti on toilet walls as data about publicly taboo topics
- Studying hikers' compliance with the wilderness mandate to "take only photos and leave only footprints" by observing trash picked up from trails by National Park staff and volunteers
- Monitoring social media postings to obtain a great volume of data unobtrusively

Observations of program activities and informal interviews with participants can often be carried out in a less obtrusive fashion than having people complete a test or questionnaire. Indeed, administration of such instruments may produce artificial results because respondents are affected by the process, what measurement specialists call "reactivity," a source of invalidity in measurement. In their imaginative classic *Unobtrusive Measures*, distinguished researchers Eugene J. Webb, Donald T. Campbell, Richard Schwartz, and Lee Sechrest (1999) discussed at length the problems of "reactive measurement effects." They found that research subjects' awareness that they are part of a study (as they complete questionnaires or take tests) can distort and confound findings. It is likely that such problems are magnified in evaluation research (e.g., Holley & Arboleda-Florez, 1988). Thus, creative designs with unobtrusive methods should be part of the repertoire of those who study human behavior as "unobtrusive researchers" (Kellehear, 1993). While qualitative methods are also subject to certain reactivity problems (to be discussed in later chapters), use of unobtrusive observations can reduce or even eliminate distorting reactivity.

Exploratory Research and Lack of Proven Quantitative Instrumentation

Qualitative methods are especially appropriate for inquiries where no acceptable, valid, and reliable measures exist. The extent to which one believes that particular instruments, like personality tests, are useful, valid, and reliable can be a matter of debate and judgment. Moreover, for desired program outcomes where measures have not been developed and tested, it can be more appropriate to gather descriptive information about what happens as a result of program activities than to use some scale that has the merit of being quantitative but whose validity and reliability are suspect.

Creativity is a prime example. While there are some instruments that purport to measure creativity, the applicability of those instruments in diverse situations is at least open to question. Thus, a program that aims to support students or participants in being more creative might do better to document in detail the activities, products, behaviors, feelings, and actual creations of participants instead of administering some standardized instrument of marginal or dubious relevance. Qualitative documentation can be inspected and judged by those interested in the evaluation findings to make their own interpretations of the extent to which creativity was exhibited by the products produced.

Even hallowed concepts such as self-esteem are open to considerable controversy when it comes to specifying measurement criteria. In addition, for people whose self-esteem is already quite high, instruments that measure self-esteem will encounter ceiling effects, where pretest results are so high that no increase is possible on a posttest. Staff development or leadership training programs that include enhanced self-esteem as a desired outcome may find it more useful to do case studies to document changes actually experienced by participants rather than rely on a standardized measurement scale of problematic relevance and sensitivity.

The same point can be made with regard to the controversy surrounding even long-standing measurement instruments. The use of standardized achievement tests to measure student learning is a prime example. The way in which norm-referenced, standardized achievement tests are constructed reduces their relevance and validity for particular local programs, especially those that serve populations where scores are likely to cluster at the lower or higher extremes of the normal curve. For such programs, more accurate evaluation results can be produced through documentation of actual student portfolios—that is, developing case histories of what students can do and have done over time rather than relying on their responses to a standardized instrument administered under artificial conditions at one moment in time (Buxton, 1982; Carini, 1975, 1979).

A related state-of-the-art consideration is exploratory research. In new fields of study where little work has been done, few definitive hypotheses exist, and little is known about the nature of the phenomenon, qualitative inquiry is a reasonable beginning point for research. Excellent examples of such exploratory research are Angela Browne's (1987) classic study of *Battered Women Who Kill*, a qualitative study of female child sexual offenders in a Minnesota treatment program (Mathews, Matthews, & Speltz, 1989), follow-up interviews documenting the effects of reunification on sexually abusive families (Mathews, Raymaker, & Speltz, 1991), Jane Gilgun's (1994, 1995) work on the resilience of and intergenerational transmission of child sexual abuse, and related frontline, small-scale studies of family sexual abuse (Patton, 1991). These studies occurred at a time when family violence and child sexual abuse were just emerging into societal consciousness and as a focus

of scholarly inquiry. Exploratory work of this kind, using qualitative methods, is the way new fields of inquiry are developed, especially in the policy arena.

Confirmatory and Elucidating Research: Adding Depth, Detail, and Meaning to Quantitative Analyses

At the opposite end of the continuum from exploratory research is the use of qualitative methods to add depth and detail to statistical findings. For example, when a large-scale survey has revealed certain marked and significant patterns of responses, it can be illuminating to a sample of survey respondents to find out what they meant by their responses. Take the case of a program satisfaction survey. Program participants are asked how satisfied they were with various program components. But what does "somewhat satisfied" really mean? Follow-up interviews with a subsample of respondents can provide meaningful additional detail to help make sense out of and interpret survey results. Qualitative data can put flesh on the bones of quantitative results, bringing the results to life through in-depth elaboration.

Moreover, while the role of qualitative research in exploratory inquiry is relatively well understood, the confirmatory and elucidating roles of qualitative data are less well appreciated. Adding depth and detail to statistical findings is one aspect of confirmation and elucidation. Within major traditions of theory-oriented qualitative inquiry (Chapter 3), qualitative methods are also the methods of choice in extending and deepening the theoretical propositions and understandings that have emerged from previous field studies. In short, *Battered women who kill, a qualitative study just for exploratory purposes.*

Rapid Reconnaissance

Sometimes information is needed quickly. Indeed, this is increasingly the case in our rapidly changing world. There may be no time to search the literature, develop hypotheses, select probabilistic samples based on definitive population enumerations, and develop, pilot test, and administer new instruments. One major advantage of qualitative methods is that you can get into the field quickly.

Experimental, deductive, hypothesis-testing strategies can require a lot of front-end work. You've got to be quite certain about design and instrumentation *before* data collection because once the study is under way, changes in design and measurement will undermine both internal and external validity. Naturalistic inquiry, in contrast, permits the researcher to enter the field with relatively little advance conceptualization, allowing the inquirer to be open to whatever becomes salient to pursue. The design is emergent and flexible. The questions unfold as the researcher pursues what makes sense.

The terms *rapid reconnaissance*, *rapid review*, or *rapid assessment* connote doing fieldwork and analysis speedily, often to meet tight decision deadlines for policymakers (Beebe, 2001; Saul, Willis, Bitz, & Best, 2013). "Quick Ethnography" (Handwerker, 2001) is another such approach. In our highly dynamic world, it's important to stay close to the action. In crisis epidemiological work, as in the outbreak of highly contagious diseases (e.g., the Ebola virus in Africa) or the emergence of AIDS, rapid reconnaissance teams made up of medical personnel, public health researchers, and social scientists are deployed to investigate the crisis and determine the immediate interventions and longer-term actions needed. Rapid assessments are important to inform timely decision making in highly fluid natural disaster and political turmoil situations, where understanding and meeting the critical needs of refugees and displaced persons are urgent (Active Learning Network for Accountability and Performance in Humanitarian Action, 2014; Weiss, Bolton, & Shankar, 2000). Rapid assessments are especially aligned with interventions and strategies that seek rapid results, like the Rapid Results Institute.

Organizations involved in strategic planning may use rapid reconnaissance for *environmental scanning*, which can include a range of observations: content analysis of newspapers and periodicals, conducting focus groups with emerging new client groups, interviewing key informants well placed to identify cutting-edge issues, and making systematic observations of what is captured in monitoring reports and staff observations. Issues teams doing environmental scanning gather both quantitative and qualitative information to identify trends, scan the environment, and formulate new initiatives based on emergent issues. These teams sometimes undertake rapid reconnaissance fieldwork to get detailed, descriptive information about a developing situation, for example, an influx or out-migration of new people in an area, the impact of sudden economic changes, the sudden appearance of a crop disease or pest, or a rapid increase in some problem, like teenage pregnancies, crack cocaine–addicted infants, homelessness, or elderly in need of long-term care. Rapid reconnaissance research teams can gather data on emerging issues quickly, going where the action is, talking to people, and observing what is happening.

The farming systems approach to international development was developed using rapid reconnaissance teams (Shaner et al., 1982a, 1982b). Interdisciplinary teams conduct fieldwork and informal interviews to construct an initial, exploratory, and qualitative portrayal of the farming system in a particular agroecological area. Through the fieldwork, which may last from one to three weeks, the teams are able to identify system characteristics, farmers' needs, extension possibilities, and new applied agricultural research priorities. The Caribbean Agricultural Extension Project, for example, used 10-day rapid reconnaissance studies in each of 10 different islands to assess needs and develop interventions for the extension services in those countries (Alkin & Patton, 1987; Patton, 1988b).

In the farming systems literature, these rapid reconnaissance surveys are often called *sondeos*, after the Spanish word meaning "sounding." A sondeo is a qualitative "sounding out" of what is happening in an agroecological system (Butler, 1995). Informal interviews and observations are done on farms and in homes to document and understand variations within some defined geographical area. Once interventions begin, either research or extension interventions, the sondeos may be periodically repeated to monitor system changes and development, as well as to evaluate the interventions.

The point here is that the very nature of qualitative inquiry makes it possible to get into the field quickly to study emerging phenomena and assess quickly developing situations in a world of rapid change. (For an online course on Rapid Reconnaissance Methods, see U.S. Agency for International Development, 2011.)

Validity Checks on the Numbers

> Government agencies are very keen on amassing statistics. They collect them, add them, raise them to the nth power, take the cube root and prepare wonderful diagrams. But you must never forget that every one of these figures comes in the first instance from the village watchman, who just puts down anything he damn well pleases.
>
> —Stamp's Law, formulated by
> Sir Josiah Charles Stamp (1880–1941)
> British civil servant, economist, and statistician;
> director of the Bank of England; and chairman of
> the London, Midland and Scottish Railway

When I was a Peace Corps volunteer doing agricultural extension in Burkina Faso in 1968, we were asked to assist with research on the productivity of African famers. Without training, we were sent surveys to complete and told to pick five "typical" farmers in our area for the survey data. The survey asked for size of field in hectares. The fields of the Gourma farmers I worked with looked like amoebas. Calculating field size would have required calculus, not to mention some form of measuring tape, which we were without. Moreover, we didn't know whether to include fallow land and collective land or just individually tenured land. The situation was no better in trying to assess yields stored in variously sized straw granaries. We basically guessed at all the numbers asked for in the survey. Two years later, in graduate school, I came across the report based on those surveys. The methodology section simply said that local extension agents gathered the data in African countries. Let's just say that I was skeptical of the numbers—and the findings.

To find out how valid data collection procedures are, qualitative field audits have to be conducted as part of quantitative research and evaluation projects. Here's a sample of common problems:

- Time allocation data for a time study is supposed to be completed hourly, or at least daily, but is actually filled in only monthly.
- Youth participation data at an inner-city after-school program are guesses at best because no system exists to track who comes to the center, how frequently, or for what activities. The evaluation reports submitted turn out to be nothing more than guestimates to meet compliance reporting requirements.
- Police close unsolved cases to meet performance goals and get bonuses for closing cases (what gets measured gets done).

One of my favorite examples of qualitative fieldwork to verify the numbers and figure out what they mean comes from a program audit of a weatherization program in Kansas.

> Kansas auditors visited several homes that had been weatherized. At one home, workers had installed 14 storm windows to cut down on air filtration in the house. However, one could literally see through the house because some of the siding had rotted and either pulled away from or fallen off the house. The auditors also found that the agency had nearly 200 extra storm windows in stock. Part of the problem was that the supervisor responsible for measuring storm windows was afraid of heights; he would "eyeball" the size of second-story windows from the ground.... If these storm windows did not fit, he ordered new ones. (Hinton, 1988, p. 3)

SIDEBAR

INQUIRING INTO WHAT THE RATINGS MEAN

An Indianapolis charter school received a grade of A, the highest rating possible. How was the grade achieved? Here's the story behind the rating.

Tony Bennett, Florida's education commissioner in 2013, came to power and fame as a strong advocate for hard-core, numbers-based, standardized testing accountability. He was forced to resign when the story came to light that he increased the rating of an underperforming charter school from its original "C" grade to an "A."

> When it appeared an Indianapolis charter school run by a prominent Republican donor might receive a poor grade, Bennett's education team frantically overhauled his signature "A–F" school grading system to improve the school's marks.

—McGrory and Solochek (2013)
Miami Herald

Bottom line: How do you find out if statistics and performance indicators are accurate and meaningful? You inquire into how the data are collected. This means interviewing those who enter the data, observing how they collect and enter data, and examining the whole quality control process for data collection and analysis.

Opening and Looking Into the Experimental Black Box

In science and engineering, a black box is a device or system that receives inputs and generates outputs, but how inputs are transformed into outputs is hidden from external view. Knowledge of the internal workings of the input-to-output process takes place within a "black box," meaning the process can't be seen. The opposite of a black box is a system where the inner components or logic are available for inspection, which is sometimes known as a clear box, a glass box, or a white box. The problem, then, is that to know what has happened, you have to be able to study what has happened. You have to be able to open the black box. That sounds straightforward, but it seldom is.

The issue arises in research when conducting randomized controlled trials. A pharmaceutical company wants to test a drug and does so by randomly assigning people to a treatment group (people who receive the drug) and a control group (people who receive a placebo). The participants in the study are tested before and after some period of time (let's say three months) to see if those in the treatment group (taking the real drug) have better outcomes. The black box question concerns what else those in the study have been doing during those three months. First, are they actually taking the drug and placebo as prescribed? Are there any other changes they experience in diet, exercise, or taking of other drugs? What about other changes in their lives? To have confidence in the experimental design results, we need to know that other changes during the three months do not affect the findings. Here are examples of implementation problems in randomized controlled trials turned up by qualitative inquiry into the black box (the lives of people in the experiment during the experimental period).

- In a randomized controlled trial of worm medicine for treating diarrhea in a developing country, those in the treatment group broke their pills into smaller parts and shared them with people not in the study, thereby significantly reducing the dosage and making consumption of the medicine inconsistent and irregular.
- In an experimental study of computer usage and Internet search behavior, the participants let friends and family members use their computer despite being told that only they were supposed to use the computer during the study.
- In a diet study, many of those in the treatment group were found to be sneaking additional snack food into their diet, thereby undermining the study results.

Careful treatment implementation studies are an important component of high-quality experimental designs. Black box designs, where such implementation studies are not done, or not well done, yield problematic and less credible results.

Personalizing and Humanizing Research and Evaluation

> Your lack of interest in personal interaction makes you an ideal candidate for working with the dead.
> —Helen on *Daria* TV, "It Happened One Nut" MTV, July 7, 1999

Because qualitative inquiry is interactive and interpersonal, it is also personal. Program staff and participants in qualitative studies experience it as personal. This resonates especially powerfully with organizations based on humanistic concerns and principles that resist "turning people into numbers" because of the perception that numbers and standardized categorization are cold, hard, and impersonal. The issue here is not whether such objections are reasonable. The point is that such objectification *feels real* to those who hold such views.

The personal nature of qualitative inquiry derives from its openness, the evaluator's close contact with the program, and the procedures of observation and in-depth interviewing, particularly the latter, that communicate respect to respondents by making *their* ideas and opinions (stated in their own terms) the important data source for the evaluation. Saville Kushner's insightful book *Personalizing Evaluation* (2000) epitomizes this emphasis especially in advocating that the perspective of participants be given primacy.

> I will be arguing for evaluators approaching programs through the experience of individuals rather than through the rhetoric of program sponsors and managers. I want to emphasize what we can learn about programs from Lucy and Ann. This does not mean ignoring the rights of program managers and sponsors with access to evaluation. There is no case for using evaluation against any stakeholder group; though there is a case for asserting a compensatory principle in favor of those who start out with relatively lower levels of access to evaluation. I don't think there is a serious risk of evaluators losing touch with their contractual obligations to report on programs and to support program management and improvement; I don't think there is a danger that evaluators will ever lose their preoccupation with program effects. There is always a risk, however, that evaluators lose contact with people; and a danger that in our concern to report on programs and their effects we lose sight of the pluralism of programs. So my arguments will robustly assert the need to address "the person" in the program. (pp. 9–10)

Qualitative methods may also be experienced as more personal because of their inductive nature. This means that, again, rather than imposing on people or a program some predetermined model, hypotheses, or even survey scales, qualitative findings unfold in a way that takes into account idiosyncrasies, uniqueness, and complex dynamics.

Personalizing and humanizing evaluation are particularly important for education, therapy, and development efforts that are based on humanistic values (Patton, 1990). Humanistic values that undergird both qualitative inquiry and humanistic approaches to intervention and change include the core principles listed in Exhibit 4.15. Where people in a program, organization, or community hold these kinds of values, qualitative inquiry is likely to be experienced as particularly appropriate.

EXHIBIT 4.15 Ten Humanistic Principles Undergirding Qualitative Inquiry

1. Each person or community is unique.

2. Each person or community deserves respect.

3. Equity, fairness, mutual respect, and reciprocity should be the foundations of human interactions.

4. One expresses respect for and concern about others by learning about them, their perspective, and their world—and by being personally interactive.

5. Change processes should be person centered, attentive to the effects on real people as individuals with their own unique needs and interests.

6. Emotion, feeling, and affect are natural, healthy dimensions of human experience.

7. The change agent, therapist, researcher, or evaluator is nonjudgmental, accepting, and supportive in respecting others' right to make their own decisions and live as they choose. *Empathic neutrality* as a stance expresses deep interest in people's experiences, perspective, and stories but is neutral (nonjudgmental) about content.

8. People and communities should be understood in context and holistically.

9. The process (how things are done) is as important as the outcomes (what is achieved).

10. Information should be openly shared and communicated honestly as a matter of mutual respect and in support of transparency as a value.

Breaking the Routine to Generate New Insights

Both individuals and organizations can easily become trapped in routine ways of doing things, including routine ways of thinking about and conducting evaluations or engaging in research. Thus, I end this chapter with one final rationale for using qualitative methods:

breaking the routine or "making the familiar strange" (Erickson, 1973, p. 10). Programs and organizations that have established ongoing evaluation systems or management information approaches may have become lulled into a routine of producing statistical tables that are no longer studied with any care. Inertia and boredom can seriously reduce the usefulness of program evaluation results. After program staff or other decision makers have seen the same statistical results used in the same kinds of statistical tables year after year, those results can begin to have a numbing effect. Even though the implications of those results may vary somewhat from year to year, the very format used to report the data can reduce the impact of the reports.

Mao Tse-tung commented on the tendency of human beings to settle into numbing routines when he said that a revolution is needed in any system every 20 years. Revolutions in the collection of evaluation data may be needed much more often. One such revolution may be to introduce a totally new approach to evaluation simply for the purpose of attracting renewed attention to the evaluation process. At the time, changing the method may produce new insights or at least force people to deal with the old insights in a new way.

Of course, collection of qualitative data can also become routine. Programs based on a humanistic philosophy and/or programs with an emphasis on individualization may find that the collection of qualitative data has become a routine and that new insights can be gained through even the temporary use of some quantitative measures. This suggestion for periodically changing methods derives from a concern with enhancing the use of evaluation research. Given the ease with which human beings and social systems settle into inertia and routine, evaluators who want their results to make a difference need to find creative ways to get people to deal with the empirical world. Exploring methodological variations may be one such approach.

It is also worth noting that evaluators can also settle into routines and inertia. Evaluators who have been using the same methods over and over may have lost the cutting edge of their own creativity. Utilization-focused evaluators need to have at their disposal a large repertoire of possible data collection techniques and approaches. Evaluators can be more helpful to programs if they themselves are staying alert to the many possibilities available for looking at the world. Indeed, a change in methods may do as much or more to reenergize the evaluator as it does to renew the program evaluation process.

Futuring Applications: Anticipatory Research and Prospective Policy Analysis

> I am often surprised, but I am never taken by surprise.
>
> —Robert E. Lee (1807–1870)
> Civil War general, Southern Confederacy

Lee's statement captures, in essence, what it means to anticipate the future. Our rapidly changing world has increased interest in and the need for futures studies. Such work has moved beyond the supermarket tabloids to become a focus of scholarly inquiry. Much futures work involves statistical forecasting and computer simulations. But qualitative futuring research offers different insights.

One important futuring tool is scenario construction (Kahane, 2012; Lindgren & Bandhold, 2009). Scenarios are narrative portrayals of future systems. A scenario can be constructed for an organization, a community, a farming system, a society, or any other unit of interest. Useful scenarios are highly descriptive. A powerful technique for writing scenarios is to base the scenario on imagined future fieldwork. The scenario is written as if it were a qualitative study of that future system. As such, it would include interview results, observational findings, and detailed descriptions of the phenomena of interest.

Qualitative methods can also be used to gather the data for scenario development. The Minnesota Extension Service undertook a community development effort in rural areas called PROJECT FUTURE. Part of the development effort involved teams of community people interviewing people in their own communities about their visions, expectations, hopes, and fears for the future. The community teams then analyzed the results and used them to construct alternative future scenarios for their community. The community next reviewed these scenarios, changed them through discussion, and selected a future to begin creating. This is an example of what is sometimes called community-based ethnographic futures research (Domaingue, 1989).

Ethnographic futures research was pioneered by Robert B. Textor (1980), who described himself as an *anticipatory anthropologist*. As a method, ethnographic futures research integrates in-depth interviewing and scenario creation to study people's ideas about the future, their values, and their concepts of change (Avery, 2013).

EXPLORING THE FUTURE OF THE DIGITAL DIVIDE THROUGH ETHNOGRAPHIC FUTURES RESEARCH

Ethnographic futures research adapts ethnography to study perceptions of a culture's future. Matthew Mitchell interviewed 13 leaders who directed various efforts to bridge the digital divide in Washington State. Each interview posed three possible scenarios: optimistic, pessimistic, and most probable for Washington State's sociocultural system in 2016. The interviewees then provided recommendations for what action they believed was required to render the optimistic scenario more probable. The digital divide was discussed within the context of the future sociocultural systems described in the three scenarios and the interviewees' recommendations.

The study framed the digital divide as a social problem that is caused, in part, by inequities in the ability to access and use information communication technologies. Thus, the interviews inquired into how the digital divide affects opportunities for participation in important social and economic processes. A general theme that emerged was that significant optimism existed that Washington State would build and maintain a more just and equitable sociocultural system in the future (Mitchell, 2002).

The Evaluation Unit of the General Accounting Office developed methods for "prospective studies" to help policymakers anticipate the implications and consequences of proposed laws. Prospective studies can include interviewing key knowledgeables in a field to solicit the latest and best thinking about a proposal, sometimes feeding back the findings for a second round of interviews (a qualitative Delphi technique). Prospective methods can also include doing a synthesis of existing knowledge to pull together a research base that will help inform policy making. The General Accounting Office (1989) handbook on prospective methods includes attention to qualitative and quantitative synthesis techniques, with particular attention to the problem of drawing together diverse data sources, including case studies. Indeed, the *Prospective Methods* guidebook presents much of the material in a case study format.

Rapid reconnaissance fieldwork can also be used for anticipatory or futuring research. Being able to get into the field quickly to get a sense of emerging developments can be critical to futures-oriented needs assessment techniques and forward-looking planning processes. Constructing futures scenarios can also be an effective way to contextualize evaluation findings and recommendations to help decision makers think about the varying future conditions that could affect implementation of alternative recommendations (Patton, 2008b, pp. 505–506).

In summary, while most evaluation work involves looking at the effectiveness of past efforts to improve the future effectiveness of interventions, a futuring perspective involves anticipatory research and forward thinking to affect current actions toward creating desirable futures. Qualitative inquiry can play a role in both studies of the past and anticipatory research on the future.

A Vision of the Utility of Qualitative Methods: Chapter Summary and Conclusion

> Where there is no vision, the people perish.
>
> —Proverbs 29:18

A while back, my local coffee shop went out of business after only two years. This particular coffee shop was a bikers' hangout. That is, it catered to aficionados of large Harley Davidson motorcycles affectionately called "Hawgs." The owners, Scott and Connie, a young husband-and-wife team, were also bikers. Because the coffee shop was near my office, I had become a regular, despite lacking the appropriate leather attire and loud steel machine. One morning, Connie mentioned that she had decided to have LASIK surgery to correct her near-sighted vision. I left for a trip, wishing her a positive outcome. When I returned two weeks later, the coffee shop was closed, and I found a barely legible handwritten note on the door:

Closed indefinitely—vision problem

I made inquiries at a nearby gasoline station, but all I learned was that the shop had closed very suddenly without notice. Some three weeks later, I happened to see Scott riding his motorcycle on the street and waved him over to ask how Connie was doing. He said she was fine. What about the sign on the coffee shop door? That had nothing to do with Connie's surgery, he explained. "I just couldn't vision myself serving coffee the rest of my life."

This chapter has been aimed at helping you decide if you can envision yourself interviewing people, doing fieldwork, constructing case studies, and otherwise using qualitative methods in practical applications.

New applications of qualitative methods continue to emerge as people in a variety of endeavors discover the value of in-depth, open-ended inquiry into people's perspectives and experiences. For example, consumer product designers have begun using qualitative methods to better understand how consumers interact with products in their everyday lives. In developing innovative products and services, designers must be concerned with satisfying the needs of users of their products. Qualitative methods, including extensive use of videotape to capture people using products and services, offer designers insights into the cultural and environmental contexts within which real consumers use real products. Social media is another area where qualitative innovations are occurring.

The story of the invention of modern running shoes illustrates these principles and provides a helpful metaphor to close this chapter. The design of sneakers varied little until the 1960s, when competitive runners began to turn to lighter-weight shoes. Reducing the weight of shoes clearly improved performance, but problems of traction remained. A running coach, Bill Bowerman, went into the sneaker business in 1962. He paid close attention to the interest in lighter-weight shoes and the problems of traction.

One morning while he was making waffles, he had an idea. He heated a piece of rubber in the waffle iron to produce the first waffle-shaped sole pattern, which became the world standard for running shoes (Panati, 1987, pp. 298–299). Subsequently, engineers and computers would be used to design and test the best waffle patterns for different athletic purposes. But the initial discovery came from paying attention, being open, making connections, drawing on personal experience, getting a feel for what was possible, exploring, documenting initial results, and applying what he learned.

Chapter Summary and Conclusion

Exhibit 4.16 lists the applications offered in this chapter, the situations and questions for which qualitative methods are particularly appropriate. These are by no means exhaustive of the possibilities for applying qualitative approaches, but they suggest the wide range of arenas in which qualitative methods are being and can be used.

The emphasis in this chapter on practical and useful applications stands in contrast to the philosophical and theoretical focus of the previous chapter. Taken together, these two chapters demonstrate the importance of qualitative inquiry to both social science theory and practice-oriented inquiry, especially in evaluation and organizational development.

Practical applications of qualitative methods emerge from the power of observation, openness to what the world has to teach, and inductive analysis to make sense of the world's lessons. While there are elegant philosophical rationales and theoretical

EXHIBIT 4.16   Qualitative Inquiry Applications: Summary Checklist of Particularly Appropriate Uses of Qualitative Methods

Focus on Practical Purposes, Concrete Questions, and Actionable Answers

Illuminating Quality

- Understanding and improving quality
- Quality assurance, including quality control and quality enhancement

Program Evaluation Applications

- Outcomes evaluation
- In-depth outcome case studies
- Evaluating individualized outcomes
- Discovering and documenting unanticipated outcomes

Understanding Interventions

- Principles-focused evaluation research
- Process evaluation
- Implementation evaluation
- Evaluating logic models and theories of action
- Comparing programs: focus on diversity
- Legislative auditing and monitoring
- Prevention interventions and evaluation
- Documenting development over time

Evaluation Models and Qualitative Methods: Particularly Well-Aligned Approaches

- Goal-free evaluation
- Responsive evaluation
- Illuminative evaluation
- Values-driven evaluation
- Connoisseurship evaluation
- Developmental evaluation
- Utilization-focused evaluation

Interactive and Participatory Applications

- Learning and problem-solving applications: action research, action learning, reflective practice, and learning organization inquiries
- Story-focused learning and organizational development
- Appreciative inquiry
- Participatory research and evaluation
- Supporting democratic dialogue and deliberation
- Capacity building through process use
- Indigenous research and evaluation

Special Methodological Applications

- Unobtrusive measures
- Exploratory research and lack of proven quantitative instrumentation
- Confirmatory and elucidating research: adding depth, detail, and meaning to quantitative analyses
- Rapid reconnaissance
- Validity checks on the numbers
- Opening and looking into the experimental design black box
- Personalizing and humanizing research and evaluation
- Breaking the routine to generate new insights
- Futuring applications: anticipatory research and prospective policy analysis

underpinnings to qualitative inquiry, the practical applications come down to a few very basic and simple ideas: pay attention, listen and watch, be open, think about what you hear and see, document systematically (memory is selective and unreliable), and apply what you learn.

I close with the reminder that any inquiry begins with formulation of the questions that will guide your inquiry. The Halcolm graphic comic that closes this chapter takes us back to that beginning insight as a foundation for designing qualitative studies, the focus of the next chapter.

APPLICATION EXERCISES

1. *Qualitative outcomes analysis:* Exhibit 4.4 (p. 185) presents a case study of Li, a graduate of an employment training program. The case begins with her attainment of standardized program outcomes, and then tells her story to place those outcomes in a broader context.

   a. What qualitative outcomes does the case study document?

   b. Describe and comment on what the qualitative case study adds to an evaluation of the WORK program.

   c. What elements of Li's story stood out to you? Why?

   d. What does the case study illustrate about the contribution of qualitative data to an outcomes evaluation?

2. *Qualitative evaluation:* Search the Internet for an example of a qualitative evaluation of a program on an issue that concerns you personally (e.g., antipoverty programs, homelessness, living with HIV/AIDS, etc.). What specific qualitative methods were used in the evaluation? What was the particular contribution of qualitative methods to the evaluation's findings?

3. *Other qualitative applications:* This chapter illustrates various contributions of qualitative methods, but it is far from exhaustive. Identify an area of practical inquiry not discussed in this chapter that makes significant use of qualitative methods. *Give at least three examples of how qualitative methods contribute in practical ways to the field of inquiry you have chosen.* To stimulate your thinking about possibilities, here are examples of important qualitative applications not discussed in this chapter:

   • Oral histories of communities, families, organizations, eras, and significant events. (See the Oral History Association: http://www.oralhistory.org/.)
   • In-depth clinical case studies in medicine. (See the Center for Individualized Medicine, The Mayo Clinic: http://mayoresearch.mayo.edu/center-for-individualized-medicine/.)
   • Market research to understand consumers' reactions to products. (See Market Research Association: http://www. marketingresearch.org/node/12310.)
   • Look at articles in the journal *Qualitative Health Research* (http://qhr .sagepub.com/) and/or *Qualitative Nursing Research* (http://nursingplanet .com/research/qualitative_research.html.)

4. *Practical qualitative inquiry principles to get actionable answers:* Using Exhibit 4.1 (p. 172) as a guide, specify a research or evaluation question that interests you. Write a proposal for a qualitative inquiry project based on the principles in Exhibit 4.1. Show how you would address each principle in your study.

PART 2

Qualitative Designs and Data Collection

- Always be suspicious of data collection that goes according to plan.

- Research subjects have also been known to be people.

- A researcher's scientific observation is some person's real-life experience. Respect for the latter must precede respect for the former.

- Total trust and complete skepticism are twin losers in the field. All things in moderation, especially trust and skepticism.

- Evaluators are presumed guilty until proven innocent.

- Make sure when you yield to temptation in the field that it appears to have something to do with what you are studying.

- A fieldworker should be able to sweep the floor, carry out the garbage, carry in the laundry, cook for large groups, go without food and sleep, read and write by candlelight, see in the dark, see in the light, cooperate without offending, suppress sarcastic remarks, smile to express both pain and hurt, experience both pain and hurt, spend time alone, respond to orders, take sides, stay neutral, take risks, avoid harm, be confused, seem confused, care terribly, become attached to nothing. . . . The nine-to-five set need not apply. (Inspired by Halcolm's mentor, Lazarus Long)

- Always carry extra batteries and getaway money.

—From Halcolm's *Fieldwork Laws*

5 Designing Qualitative Studies

Symbol of continuous quest for knowledge

NEA ONNIM NO SUA A, OHU

One who does not know can learn.

—Arthur (2001)
Cloth as Metaphor: (Re)reading the Adinkra Cloth Symbols of the Akan of Ghana

A Design Is a Plan

There's a reason you're undertaking the study you're going to do. You have a purpose. You presumably have a question or questions you want to answer in keeping with that purpose. The design sets forth how you will fulfill your purpose and answer the questions you've identified. A design is a plan. As John Steinbeck has his character assert at a moment of desperation in the Great Depression novel *Of Mice and Men*, "A plan is a real thing." It sets direction. It gets you moving into the field. But it is also a flexible and emergent thing. The tension between following the plan you create at the beginning and adapting it as you learn along the way means that design is not a contained, bounded, do-it-and-be-done step. Yes, you do it. Then, you implement it. But it's not a mechanical, linear, set-in-stone plan. Rather, design is a process and a way of thinking.

You don't just do a design. You *think design*. You *engage design*. You follow the *Rule of the Loop* (Schell, 2008, p. 80): *The more times you test and improve your design, the better it will be.* Better, but not perfect. As you think through design options and their implications, improve your design, and think through the improvements and their implications, you come to know deeply, even intimately, your design's strengths and limitations. Designs are inevitably constrained by limited resources, time, and the complexities of

the real world that do not yield easily to our design parameters. But thoughtful designs are also laden with the energy of potential. Feed on that energy. You are creating something. Something real, palpable. A plan is a real thing. A design is a real thing. Be guided by the Thomas theorem: *What is perceived as real is real in its consequences.* Designs have consequences. Your design will determine what you learn. Your design will accompany you as a fellow traveler throughout your inquiry. You'll think you understand it well at the beginning, but as the design unfolds, becomes data, guides analysis, and turns into findings, you'll find that you've come a long way together, and that you've changed, and developed and learned together. So prepare for design immersion. *Think design. Engage design.* Create a worthy fellow traveler to both guide and accompany you throughout your inquiry.

Chapter Preview

Module 28 takes a deep dive into design thinking. Inquiry questions derive from the purpose, and design answers questions. Module 29 examines data collection options, parameters, and decisions, which leads to Module 30 on purposeful sampling and case selection. This is a critical design discussion because what you sample is what you have something to say about at the end. Strategic and purposeful case selection is hugely important, hugely misunderstood, and therefore often hugely controversial. This module sets the stage for the most comprehensive presentation of qualitative case selection options ever assembled, a recognition of and tribute to the flowering of qualitative applications and approaches in the past decade.

The final three modules discuss sample size for qualitative designs and mixed-methods designs, with a concluding review of methods choices and decisions.

This chapter is the pivot chapter of both the book and qualitative inquiry. Chapters 1 through 4 presented examples of qualitative studies, the 12 core strategies of qualitative inquiry, diverse theoretical traditions that inform alternative frameworks for qualitative methodology, and a panorama of practical applications. Chapters 6 thorough 9 will discuss data collection and analysis. Chapter 5 is the pivot chapter on design; it builds on the foundation of the previous chapters and anticipates the fieldwork and analysis of the remaining chapters. Designs must be theoretically and conceptually strong; methodologically feasible, rigorous, and credible; and appropriate to the inquiry question, the available resources, the context for the study, and your own interests, capacities, and capabilities. Here's where the rubber hits the road, where it gets real, where trade-offs are negotiated, where competing ideas funnel into a design, and where the nature of the eventual results is determined.

So prepare for design immersion. *Think design. Engage design. Game on.*

The First Evaluation, the First Design

The young people gathered around Halcolm. "Tell us again, Teacher of Many Things, about the first evaluation."

"The first evaluation," he began, "was conducted a long, long time ago, in Ancient Babylon when Nebuchadnezzar was King. Nebuchadnezzar had just conquered Jerusalem in the third year of the reign of Jehoiakim, King of Judah. Now Nebuchadnezzar was a shrewd ruler. He decided to bring carefully selected children of Israel into the palace for special training so that they might be more easily integrated into Chaldean culture. This special program was the forerunner of the gifted-and-talented education programs that would become so popular in the twentieth century. The three-year program was royally funded with special allocations and scholarships provided by Nebuchadnezzar. The ancient text from the Great Book records that

> the king spake unto Ashpenaz the master of his eunuchs that he should bring certain of the children of Israel, and of the King's seed, and of the princes; Children in whom was no blemish, but well-favored and skillful in all wisdom, and cunning in knowledge, and understanding science, and such as had ability in them to stand in the king's palace, and whom they might teach the learning and the tongue of the Chaldeans.
>
> And the king appointed them a daily provision of the king's meat, and of the wine which he drank; so nourishing them three years, that at the end thereof they might stand before the king. (Daniel 1:3–5)

"Now this program had scarcely been established when the program director, Ashpenaz, who happened also to be prince of the eunuchs, found himself faced with a student rebellion led by a radical named Daniel, who decided for religious reasons that he would not consume the king's meat and wine. This created a serious problem for the director. If Daniel and his coconspirators did not eat their dormitory food, they might fare poorly in the program and endanger not only future program funding but also the program director's head! The Great Book says:

> But Daniel purposed in his heart that he would not defile himself with the portion of the king's meat, nor with the wine which he drank; therefore he requested of the prince of the eunuchs that he might not defile himself.
>
> And the prince of the eunuchs said unto Daniel, I fear my lord the king, who hath appointed your meat and your drink; for why should he see your faces worse liking than the children which are of your sort? Then shall ye make me endanger my head to the king. (Daniel 1:8, 10)

"At this point, Daniel proposed history's first educational experiment and program evaluation. He and three friends (Hananiah, Mishael, and Azariah) asked to be placed on a strict grain legume (pulse) and water diet for ten days, while other students continued on the king's rich diet of meat and wine. At the end of ten days the program director would inspect the treatment group for any signs of physical deterioration and judge the value of Daniel's alternative diet plan. Daniel proposed the experiment thusly:

> Prove thy servants, I beseech thee, ten days; and let them give us pulse to eat, and water to drink. Then let our countenances be looked upon before thee, and the countenance of the children that eat of the portion of the king's meat: and as thou seest, deal with thy servants.
>
> So he consented to them in this matter, and proved them ten days. (Daniel 1:12–14)

"During the ten days of waiting Ashpenaz had a terrible time. He couldn't sleep, he had no appetite, and he had trouble working because he was preoccupied worrying about how the evaluation would turn out. He had a lot at stake. Besides, in those days they hadn't quite worked out the proper division of labor so he had to play the roles of both program director and evaluator. You see. . . ."

The young listeners interrupted Halcolm. They sensed that he was about to launch into a sermon on the origins of the division of labor when they still wanted to hear the end of the story about the origins of evaluation. "How did it turn out?" they asked. "Did Daniel end up looking better or worse from the new diet? Did Ashpenaz lose his head?"

"Patience, patience," Halcolm pleaded. "Ashpenaz had no reason to worry. The results were quite amazing. The Great Book says that

> at the end of ten days their countenances appeared fairer and fatter in flesh than all the children which did eat the portion of the king's meat.

> Thus Melzar took away the portion of their meat, and the wine that they should drink; and gave them pulse.

> As for these four children, God gave them knowledge and skill in all learning and wisdom; and Daniel had understanding in all visions and dreams. Now at the end of the days that the king had said he should bring them in, then the prince of the eunuchs brought them in before Nebuchadnezzar. And in all matters of wisdom and understanding, that the king inquired of them, he found them ten times better than all the magicians and astrologers that were in all his realm. (Daniel 1:15–18, 20)

"And that, my children, is the story of the first evaluation. Those were the good old days when evaluations really got used. Made quite a difference to Ashpenaz and Daniel. Now off with you—and see if you can do as well."

—From *Halcolm's Evaluation Histories*

A Meta-evaluation

A *meta-evaluation* is an evaluation of an evaluation. A great deal can be learned about qualitative designs by conducting a meta-evaluation of history's first program evaluation. Let us imagine a panel of experts conducting a rigorous critique of this evaluation of Babylon's compensatory education program for Israeli students:

1. Small sample size ($n = 4$)

2. Selectivity bias, because recruitment into the program was done by "creaming," that is, only the best prospects among the children of Israel were brought into the program

3. Sampling bias, because students were self-selected into the treatment group (diet of pulse and water)

4. Failure to clearly specify and control the nature of the treatment, thus allowing for the possibility of treatment contamination because we don't know what other things, apart from a change in diet, either group was involved in that might have explained the outcomes observed

Observing Countenance

5. Possibility of interaction effects between the diet and the students' belief system (i.e., potential Hawthorne and halo effects)

6. Outcome criteria vague: just what is "countenance"

7. Outcome measurement poorly operationalized and nonstandardized

8. Single observer with deep personal involvement in the program, introducing the possibility of selective perception and bias in the observations

9. Validity and reliability data not reported for the instruments used to measure the final outcome ("He found them ten times better than all the magicians and astrologers")

10. Possible reactive effects from the students' knowledge that they were being evaluated (Hawthorne and halo effects)

Despite all of these threats to internal validity, not to mention external validity, the information generated by the evaluation appears to have been used. The 10-day evaluation was used to make a major decision about the program, namely, to change the diet for Daniel and his friends. The end-of-program evaluation conducted by the king was used to judge the program a success. (Daniel was placed first in his class.) Indeed, it would be difficult to find a more exemplary model for the uses of evaluation in making educational policy decisions than this "first evaluation" conducted under the auspices of Nebuchadnezzar so many years ago. This case study is an exemplar of evaluation research having an immediate, decisive, and lasting impact on an educational program. Modern evaluation researchers, flailing away in seemingly futile efforts to affect contemporary governmental decisions, can be forgiven a certain nostalgia for the "good old days" in Babylon when evaluation really made a difference.

But should the results have been used? Given the apparent weakness of the evaluation design, was it appropriate to make a major program decision on the basis of data generated by such a seemingly weak research design?

I would argue that not only was use impressive in this case, it was also appropriate because the research design was exemplary. Yes, *exemplary*, because the study was set up in such a way as to provide precisely the information needed by the program director to make the decision he needed to make. Certainly, it is a poor research design to study the relationship between nutrition and educational achievement. It is even a poor design to decide if *all* students should be placed on a vegetarian diet. But those were not the issues. The question the director faced was whether to place four specific students on a special diet at their request. The information he needed concerned the consequences of that specific change and *only* that specific change. He showed no interest in generalizing the results beyond those four students, and he showed no interest in convincing others that the measures he made were valid and reliable. Only he and Daniel had to trust the measures used, and so data collection (observation of countenance) was done in such a way as to be meaningful and credible to the primary intended evaluation users, namely, Ashpenaz and Daniel. If any bias existed in his observations, given what he had at stake, the bias would have operated against a demonstration of positive outcomes rather than in favor of such outcomes.

While there are hints of whimsy in the suggestion that this first evaluation was exemplary, I do not mean to be completely facetious. I am serious in suggesting that the Babylonian example is an exemplar of utilization-focused evaluation. It contains and illustrates all the factors modern evaluation researchers have verified as critical from studies of utilization (Patton, 2008b, 2012a). The decision makers who were to use the findings generated by the evaluation were clearly identified and were deeply involved in every stage of the evaluation process. The evaluation question was carefully focused on needed information that could be used in the making of a specific decision. The evaluation methods and design were appropriately matched to the evaluation question. The results were understandable, credible, and relevant. Feedback was immediate, and utilization was decisive. Few modern evaluations can meet the high standards for evaluation set by Ashpenaz and Daniel more than 3,000 years ago.

This chapter discusses some ways in which evaluation and research designs can be appropriately matched to inquiry questions, in an attempt to emulate the exemplary match between evaluation problem and research design achieved in the Babylonian evaluation. As in the previous chapters, I shall emphasize the importance of being both strategic and practical in creating evaluation and research designs. Being strategic begins with being clear about the purpose of the intended research or evaluation.

Clarity About Purpose: A Typology

Purpose is the controlling force in research. Decisions about design, measurement, analysis, and reporting all flow from purpose. Therefore, the first step in a research process is getting clear about purpose. The centrality of purpose in making methods decisions becomes evident from examining alternative purposes along a continuum from theory to action.

1. *Basic research:* contribute to fundamental knowledge and theory

2. *Applied research:* illuminate a societal concern or problem in the search for solutions

3. *Summative evaluation:* determine if a solution (policy or program) works

4. *Formative evaluation:* improve a policy or program as it is being implemented

5. *Action research:* understand and solve a specific problem as quickly as possible

Basic and applied researchers publish in scholarly journals, where their audience is other researchers who will judge their contributions using disciplinary standards of rigor, validity, and theoretical import. In contrast, evaluators and action researchers publish reports for specific stakeholders who will use the results to make decisions, improve programs, and solve problems.

Standards for judging quality vary among these five different types of research. Expectations and audiences are different. Reporting and dissemination approaches are different. Because of these differences, the researcher must be clear at the beginning about which purpose has priority. *No single study can serve all of these different purposes and audiences equally well.* With clarity about purpose and primary audience, the researcher can go on to make specific design, data-gathering, and analysis decisions to meet the priority purpose and to address the intended audience.

In the Babylonian example, the purpose was simply to find out if a vegetarian diet would negatively affect the healthy appearances (countenances) of four participants—not *why* their countenances appeared healthy or not (a causal question) but *whether* the dietary change would affect countenance (a descriptive question). The design, therefore, was appropriately simple to yield descriptive data for the purpose of making a minor program adjustment. No contribution to general knowledge. No testing or development of theory. No generalizations. No scholarly publication. No elaborate report on methods. Just find out what would happen to inform a single decision about a possible program change. The participants in the program were involved in the study; indeed, the idea of putting the diet to an empirical test originated with Daniel. In short, we have a very nice example of simple *formative evaluation*.

The king's examination of the program participants at the end of three years was quite different. We might infer that the king was judging the overall value of the program. Did it accomplish his objectives? Should it be continued? Could the outcomes he observed be attributed to the program? This is the kind of study we have come to call *summative evaluation*—summing up judgments about a program to make a major decision about its value, whether it should be continued, and whether the demonstrated model can or should be generalized to and replicated for other participants or in other places.

Now imagine that researchers from the University of Babylon wanted to study the diet as a manifestation of culture in order to develop a theory about the role of diet in transmitting culture. Their sample, their data collection, their questions, the duration of fieldwork, and their presentation of results would all be quite different from the formative evaluation undertaken by Ashpenaz and Daniel. The university study would have taken much longer than 10 days and might have yielded empirical generalizations and contributions to theory, yet it would not have helped Ashpenaz make his simple decision. On the other hand, we might surmise that University of Babylon scholars would have scoffed at a study done in just 10 days with such problematic (from their perspective) measures. Different purposes. Different criteria for judging the research contribution. Different methods. Different audiences. Different kinds of inquiry.

These are examples of how purpose can vary. This is not the only typology of purpose distinctions, but it serves to illustrate and emphasize the critical importance of matching design to purpose and audience. Previous chapters have presented the nature and strategies of qualitative inquiry, its philosophical and theoretical foundations and practical applications. In

effect, you have been presented with a large array of options, alternatives, and variations. *How do you sort it all out to decide what to do in a specific study? The answer is to get clear about purpose.*

Exhibit 5.1 summarizes some of the major differences among the different kinds of research and evaluation. The framework provided in Exhibit 5.1 is meant to facilitate clarity about purpose and the implications for design, quality criteria, use of findings, and publication. The framework also illustrates how one can organize a mass of observations into a coherent typology—a major analytical tool of qualitative inquiry.

Are they doing research or evaluation?

The Purpose of Purpose Distinctions

Different purposes typically lead to different ways of conceptualizing problems, different designs, different types of data gathering, and different ways of publicizing and disseminating findings. Researchers engaged in inquiry at various points along the continuum can have very strong opinions and feelings about researchers at other points along the continuum, sometimes generating opposing opinions and strong emotions. Basic and applied researchers, for example, would often dispute calling formative and action research by the name "research." The standards that basic

EXHIBIT 5.1   A Typology of Research Purposes

TYPES OF RESEARCH	PURPOSE	FOCUS OF RESEARCH	DESIRED RESULTS	DESIRED LEVEL OF GENERALIZATION	KEY ASSUMPTIONS	PUBLICATION MODE	STANDARD FOR JUDGING
Basic research	Knowledge as an end in itself; discover truth	Questions deemed important by one's discipline or personal intellectual interest	Contribution to theory	Across time and space (ideal)	The world is patterned; those patterns are knowable and explainable.	Major refereed scholarly journals in one's discipline; scholarly books	Rigor of research; universality and verifiability of theory
Applied research	Understand the nature and sources of human and societal problems	Questions deemed important by society	Contributions to theories that can be used to formulate problem-solving programs and interventions	Within as general a time and space as possible, but clearly limited application context	Human and societal problems can be understood and solved with knowledge.	Specialized academic journals, applied research journals within disciplines, interdisciplinary problem-focused journals	Rigor and theoretical insight into the problem
Summative evaluation	Determine the effectiveness of human interventions and actions (programs, policies, personnel, products)	Goals of the intervention	Judgments and generalizations about effective types of interventions and the conditions under which those efforts are effective	All interventions with similar goals	What works in one place under specified conditions should work elsewhere.	Evaluation reports for program funders and policymakers, specialized journals	Generalizability to future efforts and to other programs and policy issues
Formative evaluation	Improve an intervention: a program, policy, organization, or product	Strengths and weaknesses of the specific program, policy, product, or personnel being studied	Recommendations for improvements	Limited to the specific setting studied	People can and will use information to improve what they're doing.	Oral briefings; conferences; internal reports; limited circulation to similar programs, other evaluators	Usefulness to and actual use by intended users in the setting studied
Action research	Solve problems in a program, organization, or community	Organization and community problems	Immediate action; solving problems as quickly as possible	Here and now	People in a setting can solve problems by studying themselves.	Interpersonal interactions among research participants; informal unpublished material	Feelings about the process among research participants; feasibility of the solution generated

---

EXHIBIT 5.2 **Family Research Example: Research Questions Matched to Research Category**

Basic research	What are the variations in types of families, and what functions do those variations serve?
Applied research	What is the divorce rate among different kinds of families in the United States, and what explains the different rates of divorce among different groups?
Summative evaluation	What is the overall effectiveness of a publicly funded educational program that teaches family members communication skills, where the desired program outcomes are enhanced communications among family members, a greater sense of satisfaction with family life, effective parenting practices, and reduced risk of divorce?
Formative evaluation	How can a program teaching family communication skills be improved? What are the program's strengths and weaknesses? What do the participants like and dislike?
Action research	A self-study by people in a particular organization (e.g., church, neighborhood center) to figure out what activities would be attractive to families with children of different ages to solve the problem of low participation in family-oriented activities

---

researchers apply to what they would consider "good" research exclude even some applied research because it may not manifest the conceptual clarity and theoretical rigor in real-world situations that basic researchers value. Formative and action researchers, on the other hand, may attack basic research for being esoteric, academic, and irrelevant.

Debates about the meaningfulness, rigor, significance, and relevance of various approaches to research are regular features of university life. On the whole, within universities and among scholars, the status hierarchy in science attributes the highest status to basic research, secondary status to applied research, little status to summative evaluation research, and virtually no status to formative and action research. The status hierarchy is reversed in real-world settings, where people with problems attribute the greatest significance to action and formative research that can help them solve their problems in a timely way and attach the least importance to basic research, which they consider remote and largely irrelevant to what they are doing on a day-to-day basis.

Examples of Types of Research Questions: A Family Research Example

To further clarify these distinctions, it may be helpful to take a particular issue and look at how it would be approached for each type of research. For illustrative purposes, let's examine the different kinds of questions that can be asked about families for different inquiry purposes. All of the inquiry questions in Exhibit 5.2 focus on families, but the purpose and focus of each

type of inquiry are quite different. With clarity about purpose, it is possible to turn to formulating the overarching inquiry questions.

With the overarching inquiry questions formulated, the design can be constructed. The design specifies what data will be collected to answer the inquiry questions. This will immediately raise challenges of choosing among alternative strategies and methods, including decisions about critical trade-offs in design, our next topic.

Framing Qualitative Inquiry Questions

Judge a man by his questions rather than his answers.

—Voltaire (1694–1778)
French Enlightenment philosopher

Every year, I review scores of research and evaluation proposals with a range of purposes on a wide variety of topics. A common weakness, I find, is how the inquiry question is framed. Here are the seven most common forms of feedback I find myself giving about formulating the qualitative inquiry question.

1. ***First the question, then the methods, then back to the question:*** Question formulation is not a simple sequential process. You begin with a problem statement or question, but as you move to how to answer the question, the methods deliberations are likely to reshape and sharpen the question.

SIDEBAR

FAMILY STORIES SUPPORT RESILIENCE

Marshall Duke, a psychologist at Emory University, and Robyn Fivush, director of Emory's Family Narratives Lab, ask school-children 20 questions about their families. They have found that kids who know the most about their families tend to be the most resilient when they face adversity. Here are some examples of the questions they ask:

Do you know where your grandparents grew up?

Do you know where your parents went to school and how they met?

Do you know the story of your birth?

Duke and Fivush have found that children who know their family history have a strong sense of their "intergenerational selves" and feel that they belong to something bigger than themselves. Family storytelling, they hypothesize, may be a key to resilience in children (Kurylo, 2013). Happy families have stories that bind them together (Feiler, 2013a, 2013b).

Robert Stake (2010) has described the iterative and evolving nature of question formulation in qualitative inquiry:

Better, first, to ask what you need to know; then how to go about finding it. (p. 72)

For most of us, most of the time, the research problem should have first priority—but a question cannot be conceptualized without some thought of method and place of study. One cannot think deeply about the content of research without thinking of its meanings as studied one way or another. And the reality of studying it one place rather than others quickly forms in our minds. In other words, first conceptualization of the study happens pretty much all together, the focus shifting from question to method to place and back to question, each time hopefully refining the idea. And the refining will continue well into the time you are gathering data and writing up patches for the report. (p. 74)

2. *Frame a specific study's inquiry questions within a larger inquiry context and make the linkage explicit:* Chapter 3 reviewed a variety of paradigmatic, philosophical, and theoretical frameworks like phenomenology, ethnography, hermeneutics, or systems and complexity theory. These distinct frameworks serve to provide an inquiry context that should guide the formulation of specific study questions. Yet I regularly see proposals and reports that

discuss both the overarching theoretical framework and a study's specific focus *but don't connect the two.* Each inquiry framework in Chapter 3 is distinguished by a core question. The question asked by a specific study working within that general framework should be derived from and informed by that overarching question. Likewise, the practical and actionable inquiry frameworks in Chapter 4 also provide overarching perspectives and inquiry traditions that can and should inform specific study questions that serve a pragmatic and actionable purpose. The degree of alignment between an overarching inquiry framework or purpose and a specific study focus constitutes a test of the relevance of the general framework to guide the specific inquiry you are proposing as well as the likelihood that your specific study will contribute to knowledge within a more general and established inquiry tradition.

3. *Formulate one or more questions to guide your overall inquiry:* I see lots of proposals that muse about a journey of inquiry without ever bringing the study into focus with actual questions. Here's a recent example:

I want to study leadership by engaging with some leaders about leadership and inviting them to reflect on and share their leaderships perceptions and experiences. It's important to capture the voices of leaders about leadership and the best way to do that is to interview them about leadership so that they can express how they view leadership and what they've learned.

And the overarching question is what? We can infer the question. It's implicit. But reread the proposed inquiry. What would you say is the overarching inquiry question? I've used this example in qualitative methods workshops and asked participants to formulate an overall inquiry question based on the musings above. They come up with different questions. What questions you ask matters. Questions determine the realm where you'll be traveling on your inquiry journey—and the nature and range of the answers you'll find. Ask carefully. Ask thoughtfully. But most of all, *ask.*

4. *Ask open-ended questions:* Qualitative inquiry begins with descriptive questions. Avoid dichotomous (yes/no) questions. This is true in qualitative interviewing (Chapter 7), and it is equally true in framing the overall inquiry. Surveys that grammatically pose yes/no questions are inherently overly simplistic and reflect narrow categorical thinking that is inconsistent with the complexity

and richness of qualitative inquiry, not to mention the complex and diverse nature of the world. It's the difference between asking,

"Do university students rely on social media to nurture their primary relationships?"

versus

"How do university students use social media with their primary relationships?"

Linger for a moment on the huge differences between these two questions. Think about what kind of data would most appropriately answer each question. Think about what the results would look like. Think about the thinking process involved in engaging these two questions.

Asking dichotomous questions invites Type III errors: *getting the right answer to the wrong question*. A Type I error is concluding that there is a difference between groups when there is not. A Type II error is concluding that there is no difference between groups when there is. These errors are associated with statistical analysis based on significance tests, though conceptually, these errors can occur in qualitative analysis. A Type III error, however, is more fundamental than drawing an incorrect inference. A Type III error is introduced right at the beginning of an inquiry by asking the wrong question. Asking dichotomous questions constitutes one such error in qualitative inquiry: the right answer for the wrong question. Exhibit 5.3 gives examples of open-ended questions compared with dichotomous questions.

5. ***Avoid a laundry list of questions:*** I recently reviewed a proposal that listed 30 questions. This would have been too much for a detailed interview protocol much less for an overall inquiry proposal. This researcher had fallen into an all-too-common trap of believing that asking lots and lots of questions showed an inquisitive mind. What it actually showed was a lack of overall direction. It is about such people that the observation is made that he or she "can't see the forest for the trees." When I encounter a proposal with a laundry list of questions, my feedback is "Less is more." And I often share the guidance of American writer Richard Bach, which is both instructional and inspirational: "You don't want a million answers as much as you want a few forever questions. The questions are diamonds you hold in the light."

EXHIBIT 5.3 | **Asking Open-Ended Questions**

Qualitative Inquiry Begins With Descriptive Questions.
Avoid Dichotomous (Yes/No) Questions.

FOCUS OF THE INQUIRY	OPEN-ENDED INQUIRY QUESTION	CLOSED QUESTION INQUIRY FRAMING (*TO BE AVOIDED*)
Immigration experiences	What are the processes that immigrants experience during immigration? What are the implications of these processes for how they engage where they have immigrated?	Do immigrants' experiences during immigration affect how they engage in the community after immigration?
Program evaluation	What works for whom in what ways with what results and in what contexts?	Does the program work?
Homeless youth	What are the experiences of homeless youth? How do they perceive and talk about their experience of homelessness?	Are there patterns in the experiences of homeless youth?
Ecology and climate change	How, if at all, is the ecological system of the Great Lakes changing? What factors are contributing to those changes? What are the implications of those changes for the future health of the ecosystem?	Is the social ecological system of the Great Lakes changing? Is climate change causing the ecological system to change? Can the implications for the future be identified?

6. ***Distinguish questions from hypotheses:*** The language of hypothesis testing is quantitative language. Statistical analysis is built around hypothesis testing. Statistical significance tests reveal whether the null hypothesis is confirmed or rejected and, in either case, with what degree of confidence. There are precise rules for statistical hypothesis testing. Qualitative analysis lacks both those rules and that degree of precision. Certainly, there are qualitative analysts working comfortably within the positivist tradition who unabashedly use the quantitative language of independent and dependent variables and hypothesis testing (e.g., George & Bennett, 2005; Gerring, 2007; King, Keohane, & Verba, 1994). But my preference (and advice) is to leave the language of hypothesis testing to quantitative/experimental designs, where it has specific, agreed-on meaning. Qualitative inquiry explores questions. We do not test hypotheses in the way that phrase is normally understood, so to use the language of hypothesis testing invites both confusion and criticism. An alternative phrase preferred by some is *foreshadowed problems:* "a qualitative version of hypotheses . . . in which the researcher enters a setting with topics to explore" (Hays & Singh, 2012, pp. 41, 423). I think it's simplest to articulate questions.

7. ***Questions can evolve:*** Questions are beginning points. Don't treat them as written in cement. All aspects of qualitative inquiry can be emergent, including inquiry questions. Part of the inquiry journey can be discovering new questions. In evaluating a Minnesota program for high school dropouts, we began with the staff's concern about how to better serve a sudden influx of recent immigrants. The program was individualized and portfolio based so that students could pursue their own interests. The purpose of the inquiry was to adapt the program to this new, diverse population. The developmental evaluation question was "What do immigrants want to learn?"

In interviews and focus groups, we found that immigrants were confused by the question. Why were they being asked what they wanted to learn? They had enrolled in the program to get a high school degree. Didn't the American teachers know what they were supposed to learn? They came from countries where educational authorities and teachers dictated the curriculum. It was unsettling to them that they, new to the United States, were being asked what they should learn.

We asked them what questions they thought program staff and the evaluation should ask. The following questions emerged: In what ways, with what impacts, can the program support immigrants to share and build on their experiences and knowledge? How can Minnesotans learn and benefit from interactions with recent immigrants? The program staff had not imagined asking those questions, but once they emerged during the inquiry, the staff embraced them, and the evaluation shifted focus.

Inquiry as Asking and Pursuing Answers to Questions

> The art and science of asking questions is the source of all knowledge.
>
> —Thomas Berger
> American writer
>
> There's a lot of ways to take a lot of data, mangle what you're doing with it, not ask good questions, and get yourself in trouble. . . . People blame the data, when they should be asking better questions.
>
> —Nate Silver (2012)
> *The Signal and the Noise: Why So Many Predictions Fail—but Some Don't*

Questions matter. The quality of questions matters. The thoughtfulness of questions matters. It's not true that there are no dumb questions. I see them in proposals and reports all the time. Over more than 40 years of reviewing designs, I have observed a huge and varied panorama of sentences that have one thing in common: They end with a question mark. As I've provided feedback on revising questions, and received both appreciation and pushback, I've concluded that dumb questions derive from ignorance and thoughtlessness, not stupidity. After reading this section, you can no longer plead ignorance and have no excuse for thoughtlessness. So, by way of review, here's my short, nonexhaustive list of dumb questions.

Dichotomous questions are dumb in that they frame the complexity of the world as reducible to yes/no simplicities. Laundry lists of questions are dumb in that they confuse and complicate the inquiry rather than guiding it. Treating questions as rigid and unalterable is dumb because qualitative inquiry is a journey of discovery, and that includes learning what deeper questions to ask as the inquiry unfolds. Exhibit 5.3 (p. 253) compares appropriately worded, open-ended questions with inappropriate, closed-ended questions.

Designs are built around the questions we ask. Then, understanding, insight, and knowledge emerge from inquiry into the questions we ask. That means determining what data to collect and what cases to study.

Data Collection Decisions

> Constructing a research design successfully means to define who or what shall be studied (and who or what shall not).
>
> —Flick (2007a, p. 44)

Qualitative inquiry collects data from in-depth interviews, focus groups, open-ended questions on surveys, postings in social media, direct observations in the field, and analysis of documents. A mixed-methods design can add any of the broad panorama of quantitative data to be collected in conjunction with qualitative data. Subsequent chapters will examine these specific data collection approaches in depth. We'll look at different ways of making observations and engaging in fieldwork. We'll examine options for interviewing and the implications of different approaches. First, however, we're going to look at the strategic design decisions

that must be made about the depth of inquiry given the resource and time constraints.

Nature of Data Collection: One Point in Time Versus Longitudinal Inquiry

Two fundamentally different approaches to data collection involve how much contact the inquirer will have with the people and places from which data will be collected. Exhibit 5.4 compares these approaches. The one-point-in-time approach involves one interview per person or one site visit per place. For example, it is quite common in a program evaluation to interview a sample of program participants at the end of the program to document their experiences, reactions, and outcomes. The end-of-program interview is the only point of contact with these participants.

A longitudinal study, in contrast, involves multiple points of contact over some period of time. Again, using a program evaluation example, a longitudinal study would interview participants several times: (a) at

EXHIBIT 5.4 **Contrasting Designs: One-Point-in-Time Data Collection Versus Longitudinal Data Collection**

CONTRASTS	ONE-POINT-IN-TIME DATA COLLECTION	LONGITUDINAL DATA COLLECTION
1. Inquiry question example	How do a sample of homeless youth describe how they became homeless and the experiences of being homeless?	What happens to a sample of homeless youth over the period of a year after they first enter a program aimed at ending homelessness?
2. Nature of contact	One interview per person or one site visit per place	Multiple interviews per person over time; multiple field observations over time
3. Design strength	Gets all the data needed at one time; makes comparisons across interviewees or sites easier on common questions; whole study can take place in a relatively narrow time frame	Captures changes over time to document how things unfold, evolve, and emerge; more in-depth data over time; opportunity to build a relationship that enhances trust and openness
4. Design weakness	Changes must be captured retrospectively (recall) rather than as they happen	May have missing data if some interviews or site visits don't happen
5. Resource implications for a fixed budget	Larger sample size because more people can be interviewed or more places observed for a fixed amount of resources compared with a longitudinal design	Smaller number of cases can be followed over time to get multiple data points compared with all data collection occurring at one point in time, but more in-depth data results

the beginning of the program to document their backgrounds and expectations; (b) once or more during the program to find out how the experience is unfolding, the reactions to the experience, and what the participant is getting from the program; (c) at the end of the program to capture perceptions and outcomes as the program experience concludes; and (d) one or more follow-up interviews to get at lasting impacts and digested perceptions.

These two quite different inquiry strategies have major time and resource implications. Each has strengths and weaknesses. Deciding which design to implement involves trade-offs between depth and breadth.

THE ULTIMATE LONGITUDINAL STUDY

In 1938, Harvard University began following 268 male undergraduate students in what became the longest-running longitudinal study of human development in history. The study aimed to determine what factors contribute most strongly to "human flourishing." The men were followed into their 90s, documenting life from college through the years until long after retirement and studying a wide range of experiences, including relationships, politics, religion, coping strategies, and health, both physical and mental.

Highlights of the Findings:

- Those who do well in old age did not necessarily prosper in midlife, and vice versa.
- Recovery from an unhappy childhood is possible, but a happy childhood is a source of strength throughout life.
- Marriages bring much more contentment after age 70.
- Physical aging after 80 is determined less by heredity than by habits formed prior to age 50.

—George E. Vaillant (2013)
Triumphs of Experience

Critical Trade-Offs in Design

Purposes, strategies, data collection options, and trade-offs—these themes go together. A discussion of design strategies and trade-offs is necessitated by the fact that *there are no perfect research designs*. There are always trade-offs. Limited resources, limited time, and limits on the human ability to grasp the complex nature of social reality necessitate trade-offs.

The very first trade-offs come in framing the research or evaluation questions to be studied. The problem here is to determine the extent to which it

is desirable to study one or a few questions in great depth or to study more questions (though not a laundry list) but in less depth—the "boundary problem" in naturalistic inquiry (Guba, 1978). Once a potential set of inquiry questions has been generated, it is necessary to begin the process of prioritizing those questions in order to decide which of them ought to be pursued. For example, for an evaluation, should all parts of the program be studied or only certain parts? Should all clients be interviewed or only some subset of clients? Should the evaluator aim to describe all program processes or only certain selected processes in depth? Should all outcomes be examined or only certain outcomes of particular interest to inform a pending decision? These are questions that are discussed and negotiated with intended users of the evaluation. In basic research, these kinds of questions are resolved by the nature of the theoretical contribution to be made. In dissertation research, the doctoral committee provides guidance on focusing. And always there are fundamental constraints of time and resources.

Converging on focused priorities typically proves more difficult than the challenge of generating potential questions at the beginning of a study or evaluation. Doctoral students can be especially adept at avoiding focus, conceiving instead to propose sweeping, comprehensive studies that make the whole world their fieldwork oyster. In evaluations, once the involved users begin to take seriously the notion that they can learn from finding out whether what they think is being accomplished by a program is what is really being accomplished, they soon generate a long list of things they'd like to find out. The evaluation facilitator's role is to help them move from a rather extensive list of potential questions to a much shorter list of realistically possible questions and finally to a focused list of *essential and necessary questions*.

Review of the relevant literature can also bring focus to a study. What is already known? Unknown? What are the cutting-edge theoretical issues? Yet reviewing the literature can present a quandary in qualitative inquiry because it may bias the researcher's thinking and reduce openness to whatever emerges in the field. Thus, sometimes a literature review may not take place until after data collection. Alternatively, the literature review may go on simultaneously with fieldwork, permitting a creative interplay among the processes of data collection, literature review, and researcher introspection (Marshall & Rossman, 2011, pp. 38–40). As with other qualitative design issues, trade-offs appear at every turn, for there are decided advantages *and* disadvantages to reviewing the literature before, during, or after fieldwork—or on a continual basis throughout the study.

A specific example of possible variations in focus will illustrate the kinds of trade-offs involved in designing a study. Suppose some educators are interested in studying how a school program affects the social development of school-age children. They want to know how the interactions of children with others in the school setting contribute to the development of social skills. They believe that those social skills will be different for different children, and they are not sure of the range of social interactions that may occur, so they are interested in a qualitative inquiry that will capture variations in program experience and relate those experiences to individualized outcomes. What, then, are the trade-offs in determining the final focus?

We begin with the fact that any given child has social interactions with a great many people. The first problem in focusing, then, is to determine how much of the social reality experienced by children we should attempt to study. In a narrowly focused study, we might select one particular set of interactions and limit our study to those—for example, the social interactions between teachers and children. Broadening the scope somewhat, we might decide to look at only those interactions that occur in the classroom, thereby increasing the scope of the study to include interactions not only between teacher and child but also among peers in the classroom and between any volunteers and visitors to the classroom and the children. Broadening the scope of the study still more, we might decide to look at all of the social relationships that children experience in schools; in this case, we would move beyond the classroom to look at interactions with other personnel in the school—for example, the librarian, school counselors, special-subject teachers, the custodian, and/or school administrative staff. Broadening the scope of the study still further, the educators might decide that it is important to look at the social relationships children experience at home as well as at school so as to better understand how children experience and are affected by both settings, so we would include in our design interactions with parents, siblings, and others in the home. Finally, one might look at the social relationships experienced throughout the full range of societal contacts that children have, including church, clubs, and even mass media contacts.

A case could be made for the importance and value of any of these approaches, from the narrowest focus, looking at only student–teacher interactions, to the broadest focus, looking at students' full, complex social world. Now let's add the real-world constraint of limited resources—say, $50,000 and three months—to conduct the study. At some level, any of these research endeavors could be undertaken for $50,000. But it becomes clear, immediately, that *there are trade-offs between breadth and depth*. A highly focused study of student–teacher interactions could consume our entire budget but allow us to investigate the issue in great depth. On the other hand, we might attempt to look at all social relationships that children experience, but to look at each of them in a relatively cursory way in order, perhaps, to explore which of those relationships is primary. (If school relationships have very little impact on social development in comparison with relationships outside the school, policymakers could use that information to decide whether the school program ought to be redesigned to have greater impact on social development or, alternatively, if the school should forget about trying to directly affect social development at all.) The trade-offs involved are the classic ones between breadth and depth.

Breadth Versus Depth

In some ways, the differences between quantitative and qualitative methods involve trade-offs between breadth and depth. Qualitative methods permit inquiry into selected issues in great depth with careful attention to detail, context, and nuance; that data collection need not be constrained by predetermined analytical categories contributes to the potential breadth of qualitative inquiry. Quantitative instruments, on the other hand, ask standardized questions that limit responses to predetermined categories (less breadth and depth). This has the advantage of making it possible to measure the reactions of many respondents to a limited set of questions, thus facilitating comparison and statistical aggregation of the data. In contrast, qualitative methods typically produce a wealth of detailed data about a much smaller number of people and cases.

However, the breadth-versus-depth trade-off also applies within qualitative design options. Human relations specialists tell us that we can never *fully* understand the experience of another person. The design issue is how much time and effort we are willing to invest in trying to increase our understanding about any single person's experiences. So, for example, we could look at a narrow range of experiences for a larger number of people or a broader range of experiences for a smaller number of people. Take the case of interviews. Interviewing with an instrument that provides respondents with largely open-ended stimuli typically takes a great deal of time. In an education study, we developed an open-ended interview for elementary students consisting of 20 questions that included items like "What do you like most about school?" and "What don't you like about school?" These interviews took between half an hour and two hours depending on the students' ages and how articulate they were. It would certainly have been possible to have longer interviews. Indeed, I have conducted in-depth interviews with people that ran from 6 to 16 hours over a period of a couple of days. On the other hand, it would have been possible to ask fewer questions, make the interviews shorter, and probe in less depth.

Or consider another example with a fuller range of possibilities. It is possible to study a single individual over an extended period of time—for example, the study, in depth, of one week in the life of one child. This involves gathering detailed information about every occurrence in that child's life and every interaction involving that child during the week of the study. With more focus, we might study several children

during that week but capture fewer events. With a still more limited approach, say a daily half-hour interview, we could interview a yet larger number of children on a smaller number of issues. The extreme case would be to spend all of our resources and time asking a single question of as many children as we could interview given the time and resource constraints.

No rule of thumb exists to tell a researcher precisely how to focus a study. The extent to which a research or evaluation study is broad or narrow depends on the purpose, the resources available, the time available, and the interests of those involved. In brief, these are not choices between good and bad but choices among alternatives, all of which have merit.

Exhibit 5.5 summarizes with examples some of the primary trade-offs involved between depth and

EXHIBIT 5.5   Design Trade-Off Example: Depth Versus Breadth

Inquiry questions: *What is the nature and variety of social interactions and relationships that secondary school students experience? What are the functions and implications of students' various interactions and relationships?*

Design Constraints: Limited Time and Resources That Necessitate Trade-Offs and Choices

DESIGN ISSUE	DEPTH	MIDDLE GROUND	BREADTH
Focus	*Narrow focus.* Nature of interactions between students and teachers	Focus on two or three primary student interactions, e.g., with teachers and peers in the classroom	*Broad inquiry.* Nature of student interactions with a range of people, e.g., teachers, peers, administrators, parents, community people
Inquiry boundary	*Tight.* Interactions in the classroom	Interactions in the school	*Open.* Interactions in the classroom, in school, at home, and in community activities
Methods	*Single method.* In-depth interviews with teachers and students	Interviews and records but no direct classroom observations	*Multiple methods.* Interviews, observations, focus groups, analysis of student records and teachers' lesson plans
Sample	*In-depth case study.* One classroom studied for a full school year	Four classrooms studied during a three-month time period	*Larger sample.* Six teachers and 30 students from two different schools interviewed once for an hour each
Openness of inquiry	*Open to what emerges.* Overarching questions and initial sample as a starting point, with depth of inquiry to unfold flexibly as initial findings provide opportunities to go deeper and pursue whatever emerges, including adding to the sample	Most of the design is predetermined, but some limited time and resources are set aside for emergent possibilities	*Inquiry questions, procedures, sample, and design fixed.* What can be studied (breadth of inquiry) is determined at the design stage for prefieldwork approval, and only the approved design can be implemented

breadth. Especially important in focusing a study and navigating trade-offs is clarity about what case or cases will be studied and how they will be selected. We turn now to these critical design issues.

Case Study Designs: What Is a Case?

First, the bad news: Social scientists and methodologists do not agree about what constitutes a case. Eminent case study methodologist Robert Stake (2006) states the problem succinctly and bluntly: "The terms 'case' and 'study' defy clear definition. . . . Here and there, researchers will call anything they please a case study" (p. 8).

In an intriguing and provocative book titled *What Is a Case?* Ragin and Becker (1992) document the variety of perspectives and definitions.

> To the question "What is a case?" most social scientists would have to give multiple answers. A case may be theoretical or empirical or both; it may be a relatively bounded object or a process; and it may be generic and universal or specific in some way. Asking "What is a case?" questions many different aspects of empirical social science. (Ragin, 1992, p. 3)

Cases can be "empirical units" (individuals, families, organizations) or "theoretical constructs" (resilience, excellence, living with HIV). They can be quite specific, for example, the U.S. war in Vietnam, or general, such as military campaigns (Engberg, 2013). Cases can be physically real (deaths of children in foster care), socially constructed (excellence, medical "mistakes"), or historical/political (case study of the Congressional Budget Office, Joyce, 2011). Ragin (1992) expands the definition of what is a case to treat any study of any kind on any topic using any methods as a case:

> While it is tempting to see the case study as a type of qualitative analysis, and perhaps even to equate the two, virtually every social scientific study is a case study or can be conceived as a case study, often from a variety of viewpoints. At a minimum, every study is a case study because it is an analysis of social phenomena specific to time and place. . . . The tendency to conflate *qualitative study* and *case study* should be resisted. (pp. 2–4)

Different views on case studies are also tied to varying theoretical traditions (Chapter 3). A phenomenological case is different from an ethnographic case. A symbolic interaction case is different from a systems case. A positivist case is different from a constructivist case, which is in turn different from case study research in complexity science (Anderson, Crabtree, Steele, & McDaniel, 2005). Given the varied assumptions and approaches of these diverse theoretical orientations, "it is increasingly obvious that there are quite different understandings of case study research" (Blatter, 2008, p. 71).

Now, the good news: The variety of approaches to defining a case gives you an opportunity (and responsibility) to define what a case is within the context of your own field and focus of inquiry. To help you do so, let's review some of the issues and options.

Case studies are often talked about as a product. The case study stands on its own as a detailed and rich story about a person, organization, event, campaign, or program—whatever the focus of study (unit of analysis). From this perspective, the prime meaning of a case study is the case, not the methods by which the case is created. "The first objective of a case study is to understand the case. . . . The prime referent in case study is the case, not the methods by which the case operates (Stake, 2006, p. 2).

With a different emphasis, Merriam (1995) focuses on the case study as *a method of inquiry* in which the researcher examines in depth a program, event, activity, process, or one or more individuals, using a variety of data collection procedures over a sustained period of time.

Creswell (1998) elevates the case study to a distinct "qualitative inquiry and research design tradition" that is both an object of study and a methodology in which the inquirer bounds the case by time and place.

King et al. (1994), in their highly influential book in political science *Designing Social Inquiry*, and George and Bennett (2005), in *Case Studies and Theory Development in the Social Sciences*, focus on case studies as a method for developing and testing theory.

Flyvbjerg (2011), describing himself as commonsensical, finds the dictionary definition of a case study adequate: "An intensive analysis of an individual unit (as a person or community) stressing development factors in relation to environment" (p. 301).

Despite differences in emphasis, a common thread in defining a case for study is the necessity of placing a boundary around some phenomenon of interest—and where the boundary is placed is both inevitably arbitrary and fundamentally critical because that boundary-setting process determines what the case is and therefore the focus of inquiry.

- A case study is an exploration of a "bounded system" or a case (or multiple cases) over time through detailed, in-depth data collection involving multiple sources of information rich in context. This *bounded system* is bounded by time and place, and it is the *case* being studied—a program, an event, an activity, or individuals (Creswell, 1998, p. 61).

- A "case" is a bounded entity (person, organization, behavioral condition, event, or other social phenomenon), but the boundary between the case and its contextual condition—in both spatial and temporal dimension—may be blurred (Yin, 2012, p. 6; see also Yin, 2009, 2011).

The inevitable arbitrary nature of bounding cases means that what constitutes a case study varies broadly:

> When presenting their results, investigators manipulate both empirical cases and theoretical cases, and these different cases may vary by level, as when they are nested or hierarchically arrayed, and they may vary in specificity. (Ragin, 1992, p. 2)

An excellent way to experience the diversity of types of and approaches to case studies is to consult the *Encyclopedia of Case Study Research* (Mills, Durepos, & Wiebe, 2010) or to read Yin's (2004) *The Case Study Anthology*. The 19 cases include a public health scare (the feared 1976 "swine flu" epidemic), the 1962 Cuban Missile Crisis, communities ("Middletown" and "Yankee City"), an implementation process, a Korean technological development, two organization case studies (the U.S. Department of Defense and the New York City Police Department), two educational innovations, a labor union, a school reform, a technology company, a riot, a public health intervention program (methadone), military base closures, and an urban school district. Yin characterizes these as "some of the best case studies that may ever have been done" (p. xi).

What this diversity of types of case studies means is that, as I said earlier and here reiterate:

> *The variety of approaches to defining a case gives you an opportunity (and responsibility) to define what a case is within the context of your own field and focus of inquiry.*

To help you do this, the next section looks at cases through the concept of unit of analysis. I'll then present an in-depth discussion of alternative strategies for selecting cases to study: purposeful sampling.

Units of Analysis

A design specifies the unit or units of analysis to be studied. Decisions about what cases to study—issues

"I know your research is important. I'm just not ready to commit to being your case."

of both sampling strategies and sample size—depend on prior decisions about the appropriate units of analysis to study. Exhibit 5.6 presents a range of unit of analysis options.

Often, individual people, participants in programs, or students are the unit of analysis. This means that the primary focus of data collection will be on what is happening to individuals in a setting and how individuals are affected by the setting. Individual case studies and variation across individuals would focus the analysis.

Comparing groups of people in a program or across programs involves a different unit of analysis. One may be interested in comparing demographic groups (males compared with females, whites compared with African Americans) or programmatic groups (dropouts versus people who complete the program, people who do well versus people who do poorly, people who experience group therapy versus people who experience individual therapy). One or more groups are selected as the unit of analysis when there is some important characteristic that separates people into groups and when that characteristic has important implications for the program.

A different unit of analysis involves focusing on different parts of a program. Different classrooms within a school might be studied, making the classroom the unit of analysis. Outpatient and inpatient programs in a medical facility could be studied. The intake part of a program might be studied separately from the service delivery part of a program as separate units of analysis. Entire programs can become the unit of analysis. In state and national programs where there are a number of local sites, the appropriate unit of analysis may be local projects. The analytical focus in such multisite studies is on variations among project sites more than on variations among individuals within projects.

EXHIBIT 5.6   Examples of Units of Analysis for Case Studies, Comparisons, and Response Analysis

PEOPLE FOCUSED	STRUCTURE FOCUSED
Individuals	Projects
Small, informal groups (friends, gangs)	Programs
Families	Organizations
Couples, partners	Units in organizations
	Collaborations
PERSPECTIVE/WORLDVIEW FOCUSED	
People who share a culture	
People who share a common experience or perspective, e.g., dropouts, graduates, leaders, parents, Internet listserv participants, survivors, etc.	
PLACE BASED	
Neighborhoods	Villages
Cities	Farms
States	Regions
Countries	Markets
ACTIVITY FOCUSED	
Critical incidents	Time periods
Celebrations	Crises
Quality assurance violations	Events
TIME BASED	
Particular days, weeks, months, or years	Vacations
A holiday season	Rainy season
Ramadan	Dry season
Full moons	School term
A political term of office	An election period
A historical period	A generation
ANALYSIS FOCUSED (CROSS-SECTIONAL ANALYSIS)	
Specific interview questions	Focus group responses
Open-ended questionnaire responses	Site visit observations
DOCUMENTS	
Diary entries	Media items (e.g., news clippings)
Crime reports	Medical records
School portfolios	Clinical files
E-mails	Blogs
Social media postings	Letters

Note: These are not mutually exclusive categories.

Different units of analysis are not mutually exclusive. However, each unit of analysis implies a different kind of data collection, a different focus for the analysis of data, and a different level at which statements about findings and conclusions would be made. Neighborhoods can be units of analysis, or communities, cities, states, cultures, and even nations in the case of international programs.

One of the strengths of qualitative analysis is looking at program units holistically. This means doing more than aggregating data from individuals to get overall program results. When a program, group, organization, or community is the unit of analysis, qualitative methods involve observations and descriptions focused directly on that unit: The program, organization, or community becomes the case study focus, not just the individual people in those settings.

Particular events, occurrences, or incidents may also be the focus of study (unit of analysis). For example, a quality assurance effort in a health or mental health program might focus only on those critical incidents in which a patient fails to receive quality treatment according to established standards of quality. A criminal justice evaluation could focus on violent events or instances in which juveniles run away from treatment. A cultural study may focus on celebrations.

Time-Bounded Units of Analysis

A time period can also be the unit of analysis, for example, studying farming practices during spring seeding or harvest practices at the end of the growing season. Studying the orientation period for new employees can reveal a great deal about organizational culture.

SIDEBAR

INTIMATE PARTNERS AS THE UNIT OF ANALYSIS:

An HIV Example

Me, my intimate partner, and HIV:

Fijian self-assessments of transmission risks.

—Hammer (2011a)

The study's aim was to strengthen Fiji's response to HIV and AIDS by collecting and analyzing qualitative data about Fijian perceptions of their risks of HIV transmission and of other sexually transmitted diseases. The project interviewed 74 couples, with the respondents being interviewed separately and by different researchers so as to protect their confidentiality and anonymity. They belonged to one of six "target groups": people in sex work, gays and lesbians, Christian pastors, university students, taxicab drivers, and health care workers.

Both halves of each couple were recruited, enrolled, and interviewed separately but simultaneously using multiple research methods and instruments. Not all of the intimate partners of the Christian pastors were wives, some of the health care workers were also gays and lesbians, and not all of the people in sex work were heterosexual. All of the taxicab drivers were married males, and many of the intimate partners of the university students were themselves students.

The research team also interviewed 20 key informants, including expatriates; conducted 14 audiotaped focus groups; led 74 face-to-face interviews; collected 148 drawings by research participants that depicted "Me, My Intimate Partner, and HIV" and "How I Try to Prevent HIV Transmission"; and collected another 222 drawings depicting "Risky Behaviour," "Risky Person," and "Risky Setting."

Highlights of the Findings:

- The respondents talked easily and openly about sex.
- Of those who reported being HIV positive, some had not yet disclosed their status to their intimate partner. (One partner found out during a focus group.)
- General awareness of HIV and AIDS was high.
- Beliefs in the abilities of "deep seawater," "faith healing," "prayer," and "Fiji medicine" to cure if not also prevent HIV infection and/or transmission were common, and they were just about as commonly shared by health care workers as by sex workers.
- Specific knowledge of the signs, symptoms, names for, and causes of various STDs was minimal, including among health care workers.
- Rates of condom use were low, especially when omitting university students, who didn't tend to use them with intimate partners anyway.
- Low-likelihood sources of HIV transmission (e.g., rugby matches, car crashes, blood donation, and "renegade, syringe-wielders on the dance floor") were exaggerated.
- Only four (all females) perceived HIV or STD transmission risks to emanate from their intimate partner; only one (a female) drew herself as being in a risky situation; and not one respondent perceived HIV or STD transmission risk to emanate from self (Hammer, 2011b).

Studying new parents during the first month after their first child is born would examine how couples adapt to a child; couples might be compared with single parents during the same one-month (postpartum) period.

In observational studies, continuous and ongoing observation in a setting contrasts with fixed-interval sampling, in which one treats units of time (e.g., 15-minute segments) as the unit of observation. "The advantages of fixed interval sampling over continuous monitoring are that field workers experience less fatigue and can collect more information at each sampling interval than they could on a continuous observation routine" (Johnson & Sackett, 1998, p. 318). However, the decision about whether and how to sample using units of time should be based primarily on the nature of the phenomenon being observed. Sensitivity to time (sampling periods or units of time) can be especially important in evaluation because programs, organizations, and communities may function in different ways at different times during the year. Of course, in some programs, there never seems to be a good time to collect data. In doing school evaluations in the United States, I've been told by educators to avoid collecting data before Halloween because the school year is just getting started and the kids and teachers need time to get settled in. But the period between Halloween and Thanksgiving is really too short to do very much, and then, of course, after Thanksgiving, everybody's getting ready for the holidays, so that's not a typical or convenient period. It then takes students a few weeks after the new year to get their attention focused back on school, and then the winter malaise sets in and both teachers and students become deeply depressed with the endlessness of winter (at least in northern climes). Then, of course, once spring hits, attention turns to the close of school, and the kids want to be outside, so that's not an effective time to gather data either. In African villages, I was given similar scenarios about the difficulties of data collection for every month in the annual cycle of the agricultural season. A particular period of time, then, is both an important context for a study and a sampling issue.

Cross-Case Item Analysis

Case studies make the whole case the unit of analysis, but for some inquiries, the specific interview questions are the unit of interest for data collection and analysis. In such inquiries, the analysis focuses on analyzing patterns across interview questions, focus group responses, open-ended questionnaire items, or site visit observations. For example, a program evaluation could do case studies of participants to capture their holistic experiences of a program and the resulting outcomes. In such a study, the program participants are the unit of analysis. But it is also quite common to interview participants and focus the analysis on the patterns that are found question by question: (a) patterns in how participants learned about the program, (b) patterns in what activities they found most valuable, and (c) patterns in outcomes. Responses to these specific questions become the unit of analysis.

Documents

The unit of analysis can also be documents: diary entries, letters, media items (e.g., news clippings), crime reports, medical records, school portfolios, clinical files, e-mails, blog entries, and social media postings.

Unit of Analysis Decisions

Choosing among all of the possible options for units of analysis and cases for study can become overwhelming, whether the decision is about which time periods to sample, which activities to observe, or which people to interview. *The trick is to keep coming back to the criterion of usefulness.* What data collected during what time period describing what activities will most likely illuminate the inquiry? What focus will be most useful given the purpose of the inquiry? *There are no perfect inquiry designs, only more and less useful ones.*

The key to making decisions about the appropriate unit of analysis and cases to study is in determining what you want to be able to say something about at the end of the study. Do you want to have findings about individuals, families, groups, or some other unit of analysis? For scholarly inquiries, disciplinary traditions provide guidance about relevant units of analysis. For evaluations, one has to determine what decision makers and primary intended users really need information about. Do they want findings about the different experiences of individuals in programs, or do they want to know about variations in program processes at different sites? Or both?

Clarity about the unit of analysis is needed to select a study sample. In Chapter 2, I identified purposeful sampling as one of the core distinguishing strategic themes of qualitative inquiry. The next section presents variations in, rationales for, and details of how to design a study based on a purposeful sample.

Sampling Purposefully

> It is necessary to locate *excellent* participants to obtain excellent data.
>
> —Janice Morse (2010, p. 231)

Purposeful sampling: *Selecting information-rich cases to study, cases that by their nature and substance will illuminate the inquiry question being investigated.*

Perhaps nothing better illustrates the difference between quantitative and qualitative methods than the different logics that undergird their sampling approaches. Qualitative inquiry typically focuses in depth on relatively small samples, even single cases ($n = 1$), selected for a quite specific purpose. Quantitative methods typically depend on larger samples selected randomly. Not only are the techniques for sampling different, but the very logic of each approach is unique because the purpose of each strategy is different. While qualitative methodologists prefer the term *purposeful sampling*, quantitative methodologists are more likely to label these strategies "nonprobability sampling," making explicit the contrast to probability sampling (e.g., American Association for Public Opinion Research, 2013).

The logic and power of random sampling derives from statistical probability theory. A random and statistically representative sample permits confident generalization from a sample to a larger population. Random sampling also controls for selection bias. The purpose of probability-based random sampling is generalization from the sample to a population and control of selectivity errors.

What would be "bias" in statistical sampling, and therefore a weakness, becomes intended focus in qualitative sampling, and therefore a strength. The logic and power of purposeful sampling lies in selecting *information-rich cases* for in-depth study. Information-rich cases are those from which one can learn a great deal about issues of central importance to the purpose of the inquiry, thus the term *purposeful* sampling. Studying information-rich cases yields insights and in-depth understanding rather than empirical generalizations. For example, if the purpose of an evaluation is to increase the effectiveness of a program in reaching lower-socioeconomic groups, one may learn a great deal by studying in depth a small number of carefully selected poor families. Purposeful sampling focuses on selecting information-rich cases whose study will illuminate the questions under study.

Alternative Purposeful Sampling Strategies

Case selection is the foundation of qualitative inquiry. What you find from your inquiry will be determined by the cases you study. The type of sample you select should follow from and support inquiry into the questions you are asking. The purpose of a purposeful sample is to focus case selection strategically in alignment with the inquiry's purpose, primary questions, and data being collected. Exhibit 5.7 illustrates this relationship.

The previous edition of this book presented 15 purposeful sampling strategies. This edition expands the options to 40. This reflects the emergence of more distinct and nuanced strategic options over the past decade. Because of the larger number of options, I've organized them into eight categories:

1. Single significant case

2. Comparison-focused sampling

3. Group characteristics sampling

4. Concept or theoretical sampling

5. Instrumental-use multiple-case sampling

6. Sequential and emergence-driven sampling strategies during fieldwork

7. Analytically focused sampling

8. Mixed, stratified, and combination sampling strategies

Exhibit 5.8 presents all 40 sampling options distributed within these eight categories. Each type of sampling is discussed following Exhibit 5.8. The importance of understanding sampling options is that they constitute design options, in essence, different ways of thinking and strategizing about what to study. In the opening of this chapter, I wrote, "Prepare for design immersion. *Think design. Engage design.*" Now, we focus more narrowly on one core qualitative design issue. So prepare for sampling immersion. *Think of sampling as a core design issue. Engage sampling as purposeful and strategic thinking.*

PURPOSEFUL VERSUS PURPOSIVE SAMPLING VERSUS NONPROBABILITY SAMPLING

Purposeful sampling: *Strategically selecting information-rich cases to study, cases that by their nature and substance will illuminate the inquiry question being investigated*

Purposeful sampling is also called purposive sampling. There is generally no difference in meaning. It's a matter of which term one prefers. In the first edition of this book (Patton, 1980), I substituted "purposeful" sampling for "purposive" sampling. I did so for three reasons:

1. My evaluation and applied research work involved close collaboration with nonresearchers who told me that they found the term *purposive* too academic, off-putting, jargon-ish, and unclear. One said, "Who talks like that?" In contrast, they could understand "purposeful." It meant for a purpose. No one I knew, including myself, had ever used "purposive" in common conversation. Ordinary people did know, understand, and sometimes used "purposeful." My aim was to communicate clearly and be user-friendly, so I introduced "purposeful sampling" and ceased using "purposive." However, readers should know that purposive remains the term of choice among academic qualitative researchers.

2. In deciding whether to abandon "purposive" for "purposeful," I investigated the origin of the term and found that it originated as a type of statistical sampling. The 1925 meeting in Rome of the International Statistics Institute included vigorous debate on various sampling approaches. The delegates ended by adopting a formal resolution that distinguished two principal kinds of representative sampling: random and purposive. Purposive sampling involved sampling population elements such that the chosen groups should have average values approximately equal to the population averages for the characteristics already known

of the population. By the 1930s, however, random sampling was ascendant, and "purposive sampling lost its appeal" (Kruska & Mosteller, 1980, p. 188). Subsequently, *quota sampling* emerged as

a kind of purposive sampling that attempts to mimic the population in particular respects. Like all purposive sampling, quota sampling procedures suffer from potential bias because the similarity to the population in some respects by no means implies similarity in others. (Kruska & Mosteller, 1980, p. 190)

This origin of purposive sampling can still be found in some qualitative definitions of the approach. For example, in Barbour's (2001) checklist for improving rigor in qualitative research, purposive sampling is defined as aiming to capture the diversity within a population. Given the historical origins of purposive sampling as an attempt to get a statistically representative sample of a population in order to generalize, I thought it appropriate to abandon the term altogether. So I introduced *purposeful sampling* as a specifically qualitative approach to case selection. My mistake, I now realize (and confess), was in not explaining the substitution when I originally made it in 1980. I simply started using the new term, and have ever since. Now it's your choice.

3. While qualitative methodologists use the terms *purposeful* or *purposive sampling*, quantitative methodologists are more likely to label these strategies "nonprobability sampling," making explicit the contrast to probability sampling (e.g., American Association for Public Opinion Research, 2013). This defines qualitative sampling by what it is not (nonprobability) rather than by what it is (strategically purposeful).

EXHIBIT 5.7 Steps for Design Alignment

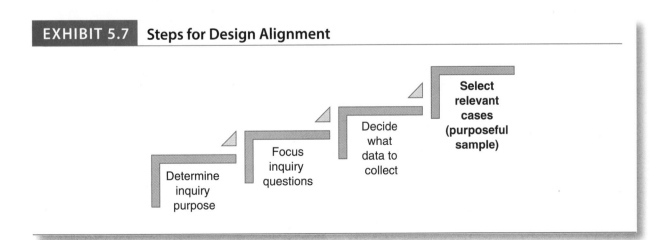

Determine inquiry purpose → Focus inquiry questions → Decide what data to collect → **Select relevant cases (purposeful sample)**

EXHIBIT 5.8   Purposeful Sampling Strategies

Selecting Information-Rich, Illuminative Cases for Qualitative Inquiry

(For detailed discussion and references, see the accompanying text following this exhibit.)

PURPOSEFUL SAMPLING STRATEGY	PURPOSE: DEFINITION/EXPLANATION	PURPOSEFUL EXAMPLES
A. Single significant case	One in-depth case ($n = 1$) that provides rich and deep understanding of the subject and breakthrough insights, and/or has distinct, stand-out importance.	Single-case examples.
1. Index case	The first documented case to manifest a phenomenon; in epidemiology, the first person exhibiting a condition, syndrome, or cure; an index case often becomes the "classic" case in the literature on the phenomenon.	Understand the discovery and impact of *HeLa cells*, the first grown in culture that do not die, which enabled major medical breakthroughs on a number of frontiers. Document and understand the impacts of the discovery on the family of Henrietta Lacks, the source of the original cells.
2. Exemplar of a phenomenon of interest	Examining an issue in depth and over time through a single case that manifests the important major dimensions of the issue and that is accessible for intense longitudinal study.	Understand and illuminate how an elderly man and his family coped with and adapted to his severe memory loss over time, and the various medical, behavioral, and psychological interventions attempted over a span of 15 years.
3. Self-study: making oneself the case	Examining one's own experience of a phenomenon of interest.	Carl Jung recorded and analyzed his own dreams as a basis for research on the nature and meaning of dreams.
4. High-impact case	A case studied and documented in depth because of the impacts it illuminates and its significance to a field, problem, or society; a high-visibility case.	Extract lessons from a case study of the successful campaign to overturn the juvenile death penalty leading up to the Supreme Court's *Roper v. Simmons* case.
5. Teaching case	An in-depth case study that offers such deep insights into a phenomenon that it serves as a source of substantial illumination about the issues documented in the case and is written to use in teaching based on the case method.	Generate a high-quality case study to teach how the Robert Wood Johnson Foundation's 20-year investment in end-of-life grant making illustrates strategic philanthropy and leads to the development of a new field.
6. Critical case (or crucial case)	The weight of evidence from a single critical case permits logical generalization and maximum application of information to other, highly similar cases, because if it's true of this one case, it's likely to be true of all other cases in that category: If it works here, it will work anywhere, or if it doesn't work here, it won't work anywhere.	Document the generalizing value of one critical case when Australian medical researcher Barry Marshall injected a bacterium, *Helicobacter pylori*, into himself to demonstrate the physiological cause of peptic ulcers and to discredit the psychosomatic theory of ulcers dominant at the time. His carefully documented study of his own body's reactions constituted a critical medical case.

PURPOSEFUL SAMPLING STRATEGY	PURPOSE: DEFINITION/EXPLANATION	PURPOSEFUL EXAMPLES
B. Comparison-focused sampling	Select cases to compare and contrast to learn about the factors that explain similarities and differences.	Comparison-focused examples.
7. Outlier sampling	Cases on the tails of a distribution that would have little or no visibility in a statistical analysis; but outlier cases can reveal a great deal about intense manifestations of the phenomenon of interest.	In program evaluation, comparing successes (high outcomes) with failures (low outcomes and dropouts); studying high achievers, excellent organizations, and/or outstanding communities; studying disasters manifesting great ineffectiveness.
8. Intensity sampling	Information-rich cases that manifest the phenomenon intensely but not extremely.	Good students/poor students; above average/below average.
9. Positive-deviance comparisons	Comparing individuals or communities that have discovered solutions to problems (positive deviants) with those peer individuals or communities where the problem endures.	In the rural area of a developing country, most children were malnourished, but a few communities had healthy children: "positive deviates." Understanding the differences led to scalable solutions.
10. Matched comparisons	Studying and comparing cases that differ significantly on some dimension of interest to understand what factors explain the difference.	Comparing "great" companies with "good" companies"; comparing select public, private, and charter schools; comparing resilient with nonresilient youth.
11. Criterion-based case selection	Based on an important criterion, all cases that meet the criterion are studied, implicitly (or explicitly) comparing the criterion cases with those that do not manifest the criterion. This includes *critical incident sampling*.	Quality assurance programs: Deaths in foster care, hospitals, or prisons, as examples, trigger an in-depth investigation into cause of death. Accidents or errors in hospitals trigger case studies. All airline crashes are investigated.
12. Continuum or dosage sampling	Select cases along a continuum of interest to deepen understanding of the nature and implications of different levels or positions along the continuum.	Continuum: five stages of women's ways of knowing. Dosage comparisons for levels of program participation: high dosage—participated in all sessions; medium dosage—participated in more than half the sessions; low dosage—participated in fewer than half the sessions.
C. Group Characteristics Sampling	Select cases to create a specific information-rich group that can reveal and illuminate important group patterns.	Group characteristics sampling examples.
13. Maximum variation (heterogeneity) sampling	Purposefully picking a wide range of cases to get variation on dimensions of interest; two purposes: (1) to document diversity and (2) to identify important common patterns that are common across the diversity (cut through the noise of variation) on dimensions of interest.	Case studies of 20 leaders from different kinds of organizations with different background characteristics: First, document the diversity, then identify any common leadership traits or patterns.

(Continued)

(Continued)

PURPOSEFUL SAMPLING STRATEGY	PURPOSE: DEFINITION/EXPLANATION	PURPOSEFUL EXAMPLES
14. Homogeneous sampling	Select cases that are very similar to study the characteristics they have in common.	Study the subpopulation of female immigrants in a job training program to evaluate the program from their unique perspective as a subgroup. Focus groups typically bring similar people together.
15. Typical cases	Select and study several cases that are average to understand, illustrate, and/or highlight what is typical, normal, and average.	Sample and study average participants who complete a program for students who dropped out of secondary school. "Average" students are those who competed in an "average" amount of time with "average" educational outcomes.
16. Key informants, key knowledge-ables, and reputational sampling	Identify people with great knowledge and/or influence (by reputation) who can shed light on the inquiry issues.	Interview long-time employees about the history of an organization, or long-time residents about the history of a community.
17. Complete target population	Interview and/or observe everyone within a unique group of interest.	Following up all 19 marathon runners who lost limbs in the 2013 Boston Marathon terrorist bombing.
18. Quota sampling	A predetermined number of cases are selected to fill important categories of cases in a larger population. Quota sampling ensures that certain categories are included in a study regardless of their size and distribution in the population. Quota sampling can be flexible, a beginning point to frame the initial design and to facilitate entering the field rather than a fixed and rigid sample size; the size and composition of the sample can be adjusted based on what is learned as the inquiry deepens.	A statewide evaluation contract specifies that the study will conduct 12 community case studies with three cases from each of four regions. Having equal cases from each region is a political consideration. Janice Morse (2000) advocates quota sampling as the foundation for "organic" qualitative inquiry that grows as the inquiry unfolds and deepens.
19. Purposeful random sample	Adds credibility to a qualitative study when those who will use the findings have a strong preference for random selection, even for small samples; it can be perceived to reduce bias; purposeful random sampling is especially appropriate when the potential number of cases within a purposeful category is more than what can be studied with the available time and resources.	Resources are available to do in-depth case studies of 20 village programs in an area of hundreds of villages. Even though the sample size is too small to be considered representative or generalizable, random selection will avoid controversy about potential selection bias.
20. Time–location sample	Interview everyone present at a particular location during a particular time period.	Study everyone who works out at a fitness center every day after work (or every day before work).

PURPOSEFUL SAMPLING STRATEGY	PURPOSE: DEFINITION/EXPLANATION	PURPOSEFUL EXAMPLES
D. Theory-focused and concept sampling	Select cases for study that are exemplars of the concept or construct that is the focus of inquiry to illuminate the theoretical ideas of interest.	Concept or theory-focused sampling examples.
21. Deductive theoretical sampling; operational construct sampling	Find case manifestations of a theoretical construct of interest so as to examine and elaborate the construct and its variations and implications. Theoretical constructs are based in, are derived from, and contribute to scholarly literature. This involves deepening or verifying theory.	Cases of resilience, posttraumatic stress, or other psychological constructs; case studies of organizational culture, or social capital in communities as sociological constructs; studying ecological diversity in natural ecosystems.
22. Inductive grounded and emergent theory sampling	Open-ended fieldwork reveals concepts that become the basis for subsequent sampling. "Participants are selected according to the descriptive needs of the emerging concepts and theory" (Morse, 2010, p. 235). Grounded theoretical sampling becomes more selective as the emerging theory focuses the inquiry. Additional cases are added to support constant comparison as a theory-sharpening analysis process.	In *Awareness of Dying* (Glaser & Strauss, 1965), the inquiry began by open-ended observations between nurses and patients in diverse settings; emergent concepts directed further sampling and comparisons. Grounded theory inquiries into social justice, which "often focuses on people who experience horrendous coercion and oppression" (Charmaz, 2011, p. 371).
23. Realist sampling	"Theory always precedes data collection in a scientific realist sampling strategy" (Emmel, 2013, p. 95). Explanation and interpretation in a realist sampling strategy tests and refines theory. Sampling choices seek out examples of mechanisms in action, or inaction toward being able to say something explanatory about their causal powers. Sampling . . . is both pre-specified and emergent, it is driven forward through an engagement with what is already known about that which is being investigated and ideas catalyzed through engagement with empirical accounts. (Emmel, 2013, p. 85).	A realist researcher evaluating a program aimed at helping poor families transition out of poverty would begin by identifying ideas about how the program was intended to help these families, essentially identifying hypothesized causal mechanisms within the context of the program. This conceptualization would direct the sampling. "The realist sample is always bootstrapped with theory" (Emmel, 2013, p. 83). See an example of studying perceptions of health in a Mumbai slum (Emmel, 2013, pp. 103–105).
24. Causal pathway case sampling	A *pathway case* provides "uniquely penetrating insight into causal mechanisms" (Gerring, 2007, p. 122). "Maximize leverage over the causal hypothesis" (George & Bennett, 2005, p. 172).	Case study of how leadership-initiated reform leads to increased democratization, to trace the causal pathway from leadership to reform to democratization (Gerring, 2007, p. 123).
25. Sensitizing concept exemplars sampling	Select information-rich cases that illuminate sensitizing concepts: terms, labels, and phrases that are given meaning by the people who use them *in a particular context*.	Empowerment, leadership, inclusivity, sustainable development, complexity, and innovation are sensitizing concepts.

(Continued)

(Continued)

PURPOSEFUL SAMPLING STRATEGY	PURPOSE: DEFINITION/EXPLANATION	PURPOSEFUL EXAMPLES
26. Principles-focused sampling	Principles provide guidance and direction for working with people in need or trying to bring about change. Principles, unlike rules, involve judgment and have to be adapted to the context and situation. Principles-focused sampling identifies cases that illuminate the nature, implementation, outcomes, and implications of the principles.	Several agencies serving homeless youth posited a shared set of principles. Case studies of youth and staff focused on generating and analyzing evidence of the meaningfulness and effectiveness of these principles.
27. Complex, dynamic systems case selection: ripple effect sampling	Selecting cases where complex dynamic processes can be tracked, studied, and documented over time; such studies are inherently highly emergent. The unit of analysis is a complex phenomenon.	Tracking and documenting complex dynamic phenomena, like vicious and virtuous circles, ripple effects, adoption and adaption of innovations, viral-like spreading phenomena, crowd-sourcing dynamics, network dynamics, etc.
E. Instrumental-use multiple-case sampling	Select multiple cases of a phenomenon for the purpose of generating generalizable findings that can be used to inform changes in practices, programs, and policies.	Instrumental-use multiple-case sampling examples.
28. Utilization-focused sampling	Select a set of cases concerning a problem or issue where sufficient depth and detail in specific cases will support rigorously identifying key factors that can credibly inform future decision making.	Detailed, systematic case studies of railroad switching operation fatalities identified five changed practices and 10 hazards to be corrected that would save lives.
29. Systematic qualitative evaluation reviews	Selecting diverse evaluation studies already completed on a program or policy and synthesizing findings across those separate and diverse evaluations to reach conclusions about *what is effective*; systematic reviews serve a meta-evaluation function (Purposeful Sampling Strategy 37 is a synthesis of qualitative *research* studies).	The Cochrane Collaboration Qualitative & Implementation Methods Group supports the synthesis of qualitative evidence and the integration of qualitative evidence with other evaluation evidence (mixed methods) in Cochrane intervention reviews on the effectiveness of health interventions.
F. Sequential and emergence-driven sampling strategies during fieldwork	Build the sample during fieldwork. One case leads to another, in sequence, as the inquiry unfolds. Follow leads and new directions that emerge during the study.	Sequential and emergence sampling examples.
30. Snowball or chain sampling	Start with one or a few relevant and information-rich interviewees and then ask them for additional relevant contacts, others who can provide different and/or confirming perspectives. Create a chain of interviewees based on people who know people who know people who would be good sources given the focus of inquiry. Researcher does the recruiting.	Seek to understand the effective relationships between agricultural extension agents and small farmers in Burkina Faso. Begin with a few farmers referred by extension agents, and then build the sample by getting referrals from the first farmers interviewed.
31. Respondent-driven sampling, network sampling, and link-tracing sampling	A network-based strategy in which a small number of initial participants from the target population ("seeds") are studied and asked to recruit up to three new contacts in their network; initial interviewees do the recruiting, usually for compensation; used to find hard-to-access research participants because of the rarity of the phenomenon of interest and/or the sensitivity of the topic being studied.	Identify sex workers living with HIV to document and understand their coping and mutual support strategies. Trust and confidentiality are critical to obtaining a sample among rare and hard-to-reach people involved in or affected by illegal or stigmatized behaviors.

PURPOSEFUL SAMPLING STRATEGY	PURPOSE: DEFINITION/EXPLANATION	PURPOSEFUL EXAMPLES
32. Emergent phenomenon or emergent subgroup sampling	Selecting a sample after the study is under way when important subgroups self-organize, or critical issues arise that affect some and not others, so those affected become the target sample.	Understand the different experiences and outcomes for two subgroups that naturally emerge in an early-childhood parent education program. Women participants with and without spouses connect and cohere together. These two subgroups become an emergent comparison sample not anticipated in advance.
33. Opportunity sampling	During fieldwork, the opportunity arises to interview someone or observe an activity, neither of which could have been planned in advance.	During a site visit to a youth homeless shelter, a former staff member who helped open the shelter happens to have come for lunch and a visit with current staff friends. This is an opportunity to learn about the history of the shelter.
34. Saturation or redundancy sampling	Analyzing patterns as fieldwork proceeds and continuing to add to the sample until nothing new is being learned (especially with snowball or response-driven sampling).	After 18 interviews with female sex workers about their knowledge of and approaches to sexually transmitted diseases, the same things were being said by each new interviewee, so no new cases were added.
G. Analytically focused sampling	Cases are selected to support and deepen qualitative analysis and interpretation of patterns and themes. This is a form of emergent sampling at the analysis stage.	Analytically focused examples.
35. Confirming and disconfirming cases	Elaborate and deepen initial analysis by seeking additional cases, e.g., variations on or exceptions to the patterns identified in the original sample.	Case studies of nonprofit leaders showed that all those in the initial sample of 15 had spent their whole careers in the not-for-profit sector. Diversify the sample by seeking nonprofit leaders who had worked in the private sector to explore the significance of background experience on leadership approach.
36. Illumination and elaboration additions to the original sample	Deepen understanding of a finding that emerges in analysis of the original sample by adding cases that can specifically illuminate that finding.	Case studies of homeless youth ($n = 14$) turned up evidence that gay, lesbian, and transgender youth faced special problems. Three such cases were added to the sample, when that pattern emerged, to strengthen the findings.
37. Qualitative research synthesis	Selecting qualitative research studies to analyze for crosscutting findings; quality criteria for which studies to include in the synthesis is a sampling issue (systematic evaluation reviews, # 29 in this table, are a form of qualitative synthesis that focuses on evaluation findings).	A substantial number of separate and independent ethnographic studies of coming-of-age and initiation ceremonies have been conducted across cultures. A synthesis involves selecting studies to include in the cross-study analysis..
38. Sampling politically important cases	Attract attention to a study (or avoid attracting controversy) by adding or omitting politically important cases.	A study of adult literacy programs across the state purposefully includes a program located in the district of the chair of the legislative committee that oversees those programs.

(Continued)

(Continued)

PURPOSEFUL SAMPLING STRATEGY	PURPOSE: DEFINITION/EXPLANATION	PURPOSEFUL EXAMPLES
H. Mixed, stratified, and nested sampling strategies	Meet multiple inquiry interests and needs; deepen focus, triangulation for increased relevance and credibility.	Mixed, stratified, and nested examples.
39. Combined or stratified purposeful sampling strategies	Begin with one sampling strategy, and then add a second to further focus the sample, e.g., (a) begin with outliers, then do snowball sampling or (b) begin with key informants to construct a maximum-variation sample.	Evaluating the implementation of the Paris Declaration Principles for Development Aid, key informants were sampled and interviewed to then identify exemplary cases of implementation for in-depth study (outliers).
40. Mixed probability and purposeful samples	Five examples of mixed sampling strategies: 1. *Stratified mixed methods.* Use statistical distribution to stratify for purposeful sampling, e.g., identify outliers, typical cases, or subgroups of interest. 2. *Sequential mixed methods.* Select cases from a probability sample for greater in-depth inquiry to illuminate and validate what the numbers mean. 3. *Parallel mixed methods.* Simultaneously do a survey of a probability sample for representativeness and generalizability and, at the same time, in-depth case studies purposefully chosen to provide depth of interpretation of what the survey results mean. 4. *Triangulated mixed methods.* Compare probability and purposeful samples, studied independently, with the triangulate and examine the consistency of findings with different methods and sampling strategies. 5. *Validity-focused mixed methods.* Do fieldwork (observations and interviews) to determine the validity of select statistical data, e.g., whether the procedures for gathering statistics have been followed rigorously, or find out if the control and treatment groups in an experimental design have complied with the design specifications.	A random, representative sample of people with disabilities was surveyed to determine their priority needs. A small number of respondents in different categories of need were then interviewed to understand in depth their situations and priorities and make the statistical findings more personal through stories. The qualitative sample was also used to validate the accuracy and meaningfulness of the survey responses

Discussion of the 40 purposeful sampling options in Exhibit 5.8 follows in Modules 31 through 38.

Single-Significant-Case Sampling as a Design Strategy

Sample purpose: *One in-depth case* (n = 1) *that provides rich and deep understanding of the subject and breakthrough insights, and/or has distinct, stand-out importance*

What can possibly be learned from a single case? We live in a world of big data and large samples. An *n* of 1? Worthless. More is better. Lots more is lots better. Can less be more? Can one make a difference? It turns out that, yes, single cases can provide quite powerful breakthroughs and insights. On March 3, 2013, headlines around the world proclaimed,

Scientists Report First Cure of HIV in a Child, Say It's a Game-Changer

The baby was the first child in the world known to have been cured since the virus touched off a global pandemic three decades earlier. Until this case, children born with HIV were considered permanently infected, their only hope being lifelong treatment with antiviral drugs to prevent HIV from becoming AIDS, destroying their immune system, and leading to death. World health statistics estimate that 330,000 children around the world get infected with HIV at or around birth every year. Part of what made this one case significant and credible was careful documentation of the circumstances of the mother and child, and their health status during pregnancy and immediately after the child's birth, and thorough documentation of an innovative treatment approach (Knox, 2013; Pollack & McNeil, 2013).

Index Case

This is what is called an *index case*: the first documented case to manifest a phenomenon. In epidemiology, an index case is the first person exhibiting a condition, syndrome, or cure. An index case often becomes *the classic case* in the literature on the phenomenon. I discussed the story of Henrietta Lacks and the discovery and impact of her *HeLa cells* in Chapter 1 as an example of the power and importance of an in-depth case study. She constitutes an index case, the person whose cells could be grown and sustained alive in a laboratory culture, which enabled major medical breakthroughs on a number of frontiers (Skloot, 2010).

Examples of index cases from several fields:

- The first human landing on the moon, Apollo 11 spacecraft, July 20, 1969.

- The first electronic *general-purpose* computer, ENIAC (Electronic Numerical Integrator and Computer), 1949.
- Sirimavo Ratwatte Dias Bandaranaike, the modern world's first female head of government, served as prime Minister of Ceylon and Sri Lanka three times, 1960 to 1965, 1970 to 1977, and 1994 to 2000.
- Lorenzo Odone suffered from adrenoleucodystrophy, a genetic disease that progressively destroys the brain of young boys. He was the first to be treated with oleic acid, which lowered his fatty acids more effectively than any other medical approach that had been tried. His treatment led to clinical trials and the discovery of a preventative protocol for boys genetically at risk of the life-threatening disease (BBC, 2004).

TYPHOID MARY: A RENOWNED INDEX CASE

Mary Mallon (1869–1938), an Irish immigrant cook, was the first person in the United States identified as an asymptomatic carrier of the pathogen associated with typhoid fever. As public health officials investigated contacts of people who contracted typhoid fever, she was the one common contact, though she never showed any symptoms herself. She was estimated to have infected more than 50 people, at least three of whom died; thus she was dubbed *Typhoid Mary*. She refused to cooperate with health authorities, withheld information or lied about her past, and made up names when she moved around trying to escape the investigators. She was quarantined twice by public health authorities and eventually spent nearly three decades in isolation. How she was able to infect others without succumbing to the illness herself is still a matter of scientific inquiry 75 years after her death (*Huffington Post*, 2013).

Sampling an Exemplar of a Phenomenon of Interest

But a single case doesn't have to be the first of its kind to be significant and to merit in-depth study and analysis. Any exemplar of a phenomenon of interest can be a worthy single-case study. In his classic *The Art of Case Study Research*, Robert Stake (1995) emphasized the value of what he called *intrinsic cases* in which the case offers insights that stand alone as important. Consider the case study of Eugene Pauly, who provided breakthroughs in our understanding of habit formation and behavior change in the face of memory loss and reduced cognitive functioning. The case study extended over 15 years and documented in day-by-day detail the life of an elderly man and his family coping with and adapting to severe memory loss, including the various medical, behavioral, social, and psychological interventions attempted over the years. The Pauly case provided significant advances in understanding the nature and power of habit and "revolutionized the scientific community's understanding of how the brain works by proving, once and for all, that it's possible to learn and make unconscious choices without remembering anything about the lesson or decision making" (Duhigg, 2013, pp. 24–25). Breakthroughs like the Pauly case study findings show why the "science of habit formation has exploded into a major field of study" (Duhigg, 2013, p. 25).

Individuals can be single-case studies, but so can events (e.g., the French Revolution), companies (Intel, Microsoft), and policies (e.g., a country's immigration policy). Here are some other examples of single cases that are exemplars of a phenomenon of interest where a community or program is the single case:

- Minnesota has the second largest Hmong immigrant population from Southeast Asia living in the United States. Treating the Hmong community as the unit of analysis, a case study of the Hmong people settling in Minnesota would provide insights into immigration and resettlement issues.
- Ujamaa Place is a program that serves African American men typically between 17 and 28 years of age who are economically disadvantaged and have experienced repeated cycles of failure. An evaluation of that program constitutes a single-case study where the program is the unit of analysis.
- Street Corner Society is a classic and groundbreaking single-case study done by William Foote Whyte (1943). For three and a half years, he lived in and observed an Italian immigrant slum district of Boston, a neighborhood considered dangerous. Among other findings, he documented how the local gangs operated.

Self-Study: Making Oneself the Case

Sometimes, the phenomenon of interest is your own experience, which makes you a single-case sample. Autoethnography examines how the inquirer's own experience of a culture offers insights about the culture, situation, event, and way of life. Heuristic inquiry,

derived from phenomenology, also makes the inquirer's own experience the primary focus of inquiry. (See Chapter 3 for an explanation and discussion of autoethnography and heuristic inquiry.) Psychoanalyst Carl Jung (1902) famously recorded and analyzed his own dreams. Humanistic phenomenologist Clark Moustakas (1961, 1972, 1975) studied loneliness through his own experience. He told me once, "I'm the best case I know and the case to which I have the deepest access. Why wouldn't I be an ideal case study?" He subsequently wrote extensively about how to bring rigor using oneself as the case (Moustakas, 1990, 1994).

High-Impact Cases

Another category of single case studies is high-impact cases. These are cases studied and documented in depth because of the impacts illuminated and the significance of the case to a field, problem, or society. High-impact cases have high visibility. Atlantic Philanthropies supported a campaign to overturn the juvenile death penalty leading up to the Supreme Court's 2005 Roper v. Simmons case, which did rule against the death penalty. Given the high visibility and high impact of the campaign, the case study identified critical factors that made the campaign exemplary and examined the question of the likelihood of the Court's decision having been influenced by the campaign (Patton, 2008a; Patton & Sherwood, 2007).

Rosa Parks, civil rights pioneer, and the Birmingham, Alabama, bus boycott exemplify a high-impact, high-visibility case, as does the 1963 march on Washington, the largest in U.S. history at the time, which inspired the "I have a dream" speech by Martin Luther King Jr., one of the most famous speeches in history.

A more recent example of a high-impact, high-visibility case is the story of Columbia University economics professor Jeffrey Sachs, author of the influential and visionary best-selling treatise The End of Poverty (2005). With $100 million from philanthropic donors, he launched the Millennium Villages Project in 10 African countries, with the promise to eliminate poverty in those villages with technical assistance and low-cost innovation in agriculture, health, water, energy, nutrition, and education. The successes and failures that ensued involve case studies of the village projects, but Jeffrey Sachs and his approach constitutes a single, high-visibility case (Munk, 2013).

A quite different example of a high-impact single case is Bruce Perry and Maia Szalavitz's (2006) work with and study of children from the Branch Davidian cult in Waco, Texas, after federal authorities raided the compound in 1993 and took custody of 21 children of ages 5 months to 12 years old. The ensuing treatment of the group over many months revealed the insidious effects of trauma and the challenges and possibilities of supporting recovery.

Teaching Cases

A high-impact case may become a teaching case: an in-depth case study that offers such deep insights into a phenomenon that it serves as a source of substantial illumination about the issues documented in the case and is written for use in teaching based on the case method (Barnes, Christensen, & Hansen, 1994; Patton & Patrizi, 2005). In business schools, law, medicine, and other professions, once the basics have been covered and learned, more advanced issues are best taught through the case method. Cases give students the opportunity to understand the context within which real decisions are made in the real world, and the inevitable ambiguities and uncertainties involved in making judgments in an uncertain, complex, and dynamic world. Here are three examples of high-quality evaluation teaching cases:

- A case study of the Robert Wood Johnson Foundation's 20-year investment in end-of-life grant making: It illustrates issues in strategic philanthropy and documents the creation of the new field of end-of-life medicine and care (Patrizi, Thompson, & Spector, 2008).
- Making Connections was an ambitious multisite, decade-long community change effort by the Annie E. Casey Foundation, started in 1999, which aimed at improving outcomes for the most vulnerable children by transforming their neighborhoods and helping their parents achieve economic stability, connect with better services and supports, and forge strong social networks. The evaluation of Making Connections spanned eight years and cost almost $60 million. The complex, multidimensional case study focuses on measurement choices and challenges.
- The case study of the Central Valley (California) Partnership of the James Irvine Foundation describes the evolution of the evaluator's role as the program evolved and developed and as the needs of the client and intended users changed over time. The initiative aimed to assist immigrants in California's Central Valley. The case illustrates important tensions among the accountability, learning, and capacity-building purposes of evaluation (Campbell, Patton, & Patrizi, 2003).

The preceding examples are part of a set of 10 philanthropy teaching cases of major evaluation studies available from the Evaluation Roundtable (2014).

TEACHING CASES TEACH THEORY

The roots of the "case method" in the teaching of law in this country, certainly the best-known approach to employing cases as vehicles for professional education, lie in their value for teaching theory, not practice. Christopher Columbus Langdell, who became dean of the Harvard University Law School in 1870, was responsible for advancing the case method of legal education. His rationale for employing this method was not its value as a way of teaching methods or approaches to practice. He believed that if practice were the essence of law, it had no place in a university. Instead, he advocated the case method of legal education because of its effectiveness in teaching law as science—in teaching legal *theory* through cases.

A case, properly understood, is not simply the report of an event or incident. To call something a case is to make a theoretical claim—to argue that it is a "case of something" or to argue that it is an instance of a larger class. A red rash on the face is not a case of something until the observer has invoked theoretical knowledge of disease. A case of direct instruction or of higher-order questioning is similarly a theoretical assertion. I am therefore not arguing that the preparation of teachers be reduced to the most practical and concrete; rather, using the power of a case literature to illuminate both the practical and the theoretical, I argue for development of a case literature whose organization and use will be profoundly and self-consciously theoretical.

Case Knowledge

Case knowledge is knowledge of specific, well-documented, and richly described events. Whereas cases themselves are reports of events or sequences of events, the knowledge they represent is what makes them cases. The cases may be examples of specific instances of practice—detailed descriptions of how an instructional event occurred—complete with particulars of contexts, thoughts, and feelings. On the other hand, they may be exemplars of principles, exemplifying in their detail a more abstract proposition or theoretical claim.

—Lee Shulman (1986, p. 11)
Stanford University
Presidential address, American Educational Research Association

Critical Case (or Crucial Case)

The final single-case example of purposeful sampling is *the critical case*, also sometimes called the crucial case. The weight of evidence from a single critical case permits logical generalization and maximum application of information to other highly similar cases because if it's true of this one case, it's likely to be true of all other cases in that category. Australian medical researcher Barry Marshall injected a bacterium, *H. pylori*, into himself to demonstrate the physiological cause of peptic ulcers and discredit the psychosomatic theory of ulcers dominant at the time (Academy of Achievement, 2005). His carefully documented study of his own body's reactions constituted a critical medical case.

Critical cases are those that can make a point quite dramatically or are, for some reason, particularly important in the scheme of things. A clue to the existence of a critical case is a statement to the effect that "if it happens there, it will happen anywhere" or, vice versa, "if it doesn't happen there, it won't happen anywhere." Another clue to the existence of a critical case is a key informant's observation to the effect that "if that group is having problems, then we can be sure all the groups are having problems." Or you know you likely have a critical case for evaluation when knowledgeable people say, "If it doesn't work here, it won't work anywhere"; or "If it works here, it will work anywhere." This latter statement has been dubbed the Frank Sinatra inference, after the line in the song he popularized, *New York, New York*: "If you can make it here, you can make it anywhere" (Levy, quoted by Gerring, 2007, p. 119).

Looking for the critical case is particularly important where resource constraints may limit the evaluation to the study of only a single site. Under such conditions, it makes strategic sense to pick *the site that would yield the most information and have the greatest impact on the development of knowledge*. While studying one or a few critical cases does not technically permit broad generalizations to all possible cases, *logical generalizations* can often be made from the weight of evidence produced in studying a single, critical case.

Physics provides a good example of such a critical case. In Galileo's study of gravity, he wanted to find out if the weight of an object affected the rate of speed at which it would fall. Rather than randomly sampling objects of different weights to generalize to all objects in the world, he selected a critical case—the feather. If in a vacuum, as he demonstrated, a feather fell at the same rate as some heavier object (a coin), then he could logically generalize from this one critical comparison to all objects. His finding was both useful *and* credible because the feather was a convincing critical case.

Critical cases can be found in social science and evaluation research if one is creative in looking for them. For example, suppose national policymakers want to get local communities involved in making decisions about how their local program will be run, but they aren't sure

that the communities will understand the complex regulations governing their involvement. The first critical case is to evaluate the regulations in a community of well-educated citizens. If they can't understand the regulations, then less-educated folks are sure to find the regulations incomprehensible. Or, conversely, one might consider the critical case to be a community consisting of people with quite low levels of education: "If they can understand the regulations, anyone can."

Identification of critical cases depends on recognition of the key dimensions that make for a critical case. For example, a critical case might come from a particularly difficult program location. If the funders of a new program are worried about recruiting clients or participants into a program, it may make sense to study the site where resistance to the program is expected to be greatest to provide the most rigorous test of program recruitment. If the program works in that site, it could work anywhere. That makes the critical case an especially information-rich exemplar and therefore worthy of study as the centerpiece in a small or "*n* of 1" sample.

World-renowned medical hypnotist Milton H. Erickson became a critical case in the field of hypnosis. Erickson was so skillful that he became widely known for "his ability to succeed with 'impossibles'—people who have exhausted the traditional medical, dental, psychotherapeutic, hypnotic and religious avenues for assisting them in their need, and have not been able to make the changes they desire" (Grinder, DeLozier, & Bandler, 1977, p. 109). If Milton Erickson couldn't hypnotize a person, no one could. He was able to demonstrate that, under his definition of hypnosis, anyone could be hypnotized.

Summary of Single Significant Case

I have discussed single-case purposeful sampling at some length and offered many examples because single cases constitute a powerful exemplar of what can be learned from in-depth study. Qualitative studies are often ignored, or even dismissed, because of their small sample size. But every discipline, profession, and field of endeavor has benefitted from breakthrough insights generated by in-depth single cases. Locate and learn about such cases in your own area of expertise. Then, determine which kind of single case it is:

1. An index case

2. An exemplar of a phenomenon of interest

3. Self-study: making yourself the case

4. A high-impact, high-visibility case

5. A teaching case

6. A critical case

Comparison-Focused Sampling Options

Sample purpose: Select cases to compare and contrast to learn about the factors that explain similarities and differences.

> There is always a level of generality at which any two things can be said to be essentially the same, and always a level of particularity at which they can be distinguished.
>
> —Michael Scriven (1970, p. 189)
> Philosopher and evaluation pioneer

EXHIBIT 5.9 Outliers

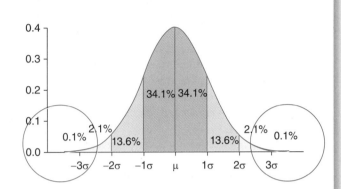

The circles show outliers on a normal distribution.

We've been looking at the value of in-depth study of single cases. Now we turn to multiple cases, which intrinsically invite comparisons. Comparison-focused sampling looks in depth at the significant similarities and differences between cases and the factors that explain those differences. We are hardwired in our brains to make comparisons. We do so all the time. We compare the new with the old, the known with the unknown, the best with the worst, the true with the false, the beautiful with the ugly, and what works with what doesn't. In all these comparisons, judgments have to be made: What is new? What is known? What is best? What is true? What is beautiful? What works?

Serious qualitative inquiry ensures that comparisons are systematic, rigorous, and meaningful. Different kinds of comparisons serve different purposes. That's where purposeful sampling comes in. Given the overall purpose of the inquiry, the specific question you're investigating, and the data you propose to collect, what kind of comparative sample would be most appropriate? This section reviews some major options for comparison-focused sampling.

Outlier Sampling

Outlier sampling, also called extreme or deviant case sampling, involves selecting cases that are information-rich, because they are unusual or special in some way, such as outstanding successes or notable failures. In statistical terms, extreme case sampling focuses on the endpoints of the bell-shaped curve normal distribution (outliers), which are often ignored (or even dropped) in aggregate data reporting. (See Exhibit 5.9.)

The influential study of "America's best run companies," published as *In Search of Excellence*, exemplifies

the logic of purposeful, outlier sampling. This classic study was based on a sample of 62 companies "never intended to be perfectly representative of U.S. industry as a whole . . . but a list of companies considered to be innovative and excellent by an informed group of observers of the business scene" (Peters & Waterman, 1982, p. 19). Lisbeth Schorr (1988) used a similar strategy in studying especially effective programs for families in poverty, published as the influential book *Within Our Reach*. Stephen Covey's (1989) best-selling book, *The 7 habits of Highly Effective People*, is based on a purposeful, extreme group sampling strategy. Studies of leadership have long focused on identifying the characteristics of highly successful leaders, as in Jim Collins's (2001b) case studies of 11 corporate executives in whom "extreme personal humility blends paradoxically with intense professional will" (p. 67), what he calls "Level 5 leaders," the highest level in his model. *Super achievers* are outliers (Sweeney & Gosfield, 2013).

Malcolm Gladwell's study *Outliers* (2008) exemplifies this sampling strategy. He posed the intriguing question "Why do some people succeed, living remarkably productive and impactful lives, while so many more never reach their potential?" He did case studies of the lives of hugely successful outliers like Mozart, Bill Gates (founder of Microsoft), and the Beatles. He found that success involves more than intelligence and talent; these outliers benefitted from favorable timing, connections, unusual opportunities, and simple luck.

Bill Gates attributes part of his success to learning from failure. Indeed, sometimes, cases of dramatic failure offer powerful lessons. Though "the wisdom of learning from failure is incontrovertible," asserts organizational learning expert Amy Edmondson (2011, p. 1), she has found that few organizations actually know how to engage in the kind of in-depth case analysis and interpretation that yields meaningful and actionable lessons. The legendary UCLA basketball coach John Wooden won 10 national championships from 1964 through 1975, an unparalleled sports achievement. But the game he remembered the most and said he learned the most from was UCLA's 1974 overtime loss to North Carolina State in the semifinals (quoted in the *Los Angeles Times*, December 21, 2000). Wooden's focus on that game—that extreme case—illustrates the learning psychology of extreme group purposeful sampling.

Examples of outlier sampling in applied research:

- In her book *When Battered Women Kill* (1987), Angela Browne describes the in-depth studies she conducted of the most extreme cases of domestic violence to elucidate the phenomenon of battering and abuse.
- In the early days of AIDS research, infections almost always resulted in death. However, a small number of people infected with HIV who did not develop AIDS became crucial outlier cases that provided important insights into directions researchers should take in combating AIDS.
- Ethnomethodologists (see Chapter 3) do field experiments to expose assumptions and norms on which everyday life is based by creating disturbances that deviate greatly from the norm. Observing the reactions to someone eating like a pig in a restaurant and then interviewing people about what they saw and how they felt illuminate the ordinary through outlier cases.

In evaluation, the logic of extreme case sampling is that lessons may be learned about unusual conditions or extreme outcomes that are relevant to improving more typical programs. Let's suppose that we are interested in studying a national program with hundreds of local sites. We know that many programs are operating reasonably well, even quite well, and that other programs verge on being disasters. We also know that most programs are doing "okay." This information comes from knowledgeable sources who have made site visits to enough programs to have a basic idea about what the variation is. With limited time and resources, an evaluator might study one or more examples of really poor programs and one or more examples of really excellent programs. The evaluation focuses on understanding under what conditions programs get into trouble and under what conditions programs exemplify excellence.

Brinkerhoff's (2003) success case method for program evaluation provides a systematic way to identify

---

EROTIC OUTLIERS

Erotic outliers are those whose sexual preferences depart from whatever a given society at a given time considers "normal." Erotic outliers are labeled "perverts" and other epithets, or by more clinical names:

- *Podophilia*, intense sexual interest in feet to the point of fetishism
- *Climacophilia*, the erotic compulsion to tumble down stairs
- *Melissophilia*, lust for bees
- *Titillagnia*, arousal from tickling
- *Zoophilia*, sexual attraction to certain animals
- *Sadism*, deriving pleasure and gratification from inflicting physical pain and/or humiliation
- *Masochism*, deriving pleasure from physical pain and/or humiliation
- *Partialism*, sexual interest with an exclusive focus on a specific part of the body

—Bering (2013)

---

successful programs, compare them with less effective programs, and figure out what makes the difference. In a single program, the same strategy may apply. Instead of studying some representative sample of people in the setting, the evaluator may focus on studying and understanding selected cases of special interest, for example, dropouts versus outstanding successes. In an evaluation of the Caribbean Agricultural Extension Project, we did case studies of the "outstanding extension agent" selected by peers in each of eight Caribbean countries to help the program develop curriculum and standards for improving extension practice. The sample was purposefully "biased," not to make the program look good but rather to learn from those who were exemplars of good practice. In a similar vein, Hogle and Moberg (2013) used success case studies to evaluate the impact of an awards program of the National Institutes of Health on the scientific achievements and career advancement of researchers. Barry Kibel (1999) developed a process for capturing the "success stories" of clients in programs and aggregating them in a method he called "results mapping."

In essence, the logic of outlier sampling is that extreme cases may be information-rich cases precisely because by being unusual they can illuminate both the unusual and the typical. In proposing an outlier sample, as in all purposeful sampling designs, the researcher has an obligation to present the rationale and expected benefits of this strategy as well as to note its weakness (limited generalizability).

SAMPLING THE BEST

SIDEBAR

Evaluating the Caribbean Agricultural Extension Project posed a challenge: how to provide funders with decision-relevant information about the long-term potential of agricultural extension in eight English-speaking Caribbean countries. The first phase of the project involved needs assessment, planning, and capacity building. These processes laid the foundation for training agricultural extension agents who would, it was hoped, help improve the productivity and profitability of small farms. However, a critical funding decision about the potential of the project had to be made at the end of the first phase, before actual training of extension agents had begun and prior to any impact on farmers. No data existed on the effectiveness of extension that could be used to calculate potential effectiveness. Funders were asking for concrete estimates of future impact, not just some hoped-for productivity increase pulled out of the air. The solution was to study a purposeful sample of "the best." Each agricultural extension service established a process for identifying and recognizing its own "outstanding agricultural extension agent." Those agents were each asked to identify 5 farm families whose farm productivity the agent believed he or she had increased. Independent case studies were done of these 40 farm families (5 families in each of the eight countries).

This sample was purposefully biased to establish case-based goals for increased extension effectiveness while also showing the diversity of small-farm situations and the variety of extension agent practices. It could be expected that the typical extension agent would have somewhat less impact on farmers than these outstanding agents. However, by gathering data about the impacts of the "best," it was possible to provide project funders with concrete examples of what might be accomplished over time if more extension agents were trained in the practices of the "best." These data allowed potential second-phase funders to engage the question "Given the impacts of those extension agents identified as the best, is it worth funding a training program aimed at creating more extension agents following those 'best' practices?" Without these concrete cases to examine, the funding decision had been based on abstract discussion and speculative guesses about extension agent activities and impacts. Having real cases to examine made the resulting discussions concrete, focused, and data based. The second phase was funded.

freshspectrum.com

SOURCE: © Chris Lysy—freshspectrum.com

that manifest the phenomenon of interest intensely (but not extremely). Extreme or deviant cases may be so unusual as to distort the manifestation of the phenomenon of interest. Using the logic of intensity sampling, one seeks excellent or rich examples of the phenomenon of interest but not highly unusual cases.

Heuristic inquiry (see Chapter 3; Moustakas, 1990) uses intensity sampling. Heuristic research draws explicitly on the intense personal experiences of the researcher, for example, experiences with loneliness or jealousy. Coresearchers who have experienced these phenomena intensely also participate in the study. The heuristic researcher is not typically seeking pathological or extreme manifestations of loneliness, jealousy, or whatever phenomenon is of interest. Such extreme cases might not lend themselves to the reflective process of heuristic inquiry. On the other hand, if the experience of the heuristic researcher and his or her coresearchers is quite mild, there won't be much to study. Thus, the researcher seeks a sample of sufficient intensity to elucidate the phenomenon of interest.

The same strategy can be applied in a program evaluation. Extreme successes or unusual failures may be discredited as being too extreme or unusual to yield useful information. Therefore, the evaluator may select cases that manifest sufficient intensity to illuminate the nature of success or failure but not at the extreme.

Intensity sampling involves some prior information and considerable judgment. The researcher or evaluator must do some exploratory work to determine the nature of the variation in the situation under study and then sample intense examples of the phenomenon of interest.

Intensity Sampling

Intensity sampling involves the same logic as extreme case sampling but with less emphasis on the extremes. An intensity sample consists of information-rich cases

Positive Deviance

Positive deviance is a special application of outlier sampling that focuses on finding people or communities that have solved a problem where the norm in the area is for the problem to remain unsolved. Those who have solved the problem are called "positive deviants." This approach emerged in the 1970s in nutrition when researchers discovered that some families in poverty had well-nourished children. The thing that distinguishes positive deviance from more general outlier sampling is that the search for positive deviants is specifically aimed at finding solutions; it is action oriented (Pant & Odame, 2009; Pascale, Sternin, & Sternin, 2010). Advocates and practitioners have formed the Positive Deviance Alliance to promote using this method. It is a highly participatory approach where people in the community help identify and learn from positive deviants (Better Evaluation, 2013).

Matched Comparisons

In contrast to outlier sampling, which explicitly or implicitly compares one end of a dimension of interest with the other end (successes with failures, excellence vs. ineffectiveness), a different form of comparison-focused purposeful sampling is matched comparisons. Jim Collins (2001a) and his team of case study experts have used this form of purposeful sampling to yield influential insights about factors that differentiate great companies from good companies, enduringly successful companies from those that decline (Collins, 2009; Collins & Porras, 2004), and companies that thrive in turbulent times from those that become overwhelmed by chaos and complexity (Collins & Hansen, 2011). As an example of this sampling method, for this last comparison, the team screened 20,400 companies in a search for those that met three selection criteria: (1) 15 or more years of "spectacular results" (2) achieved under conditions of high environmental turbulence and uncertainty with (3) a rise from vulnerability (young, small company) to a position of "greatness." They ended up with a sample of 7 companies that met all three criteria. They then selected comparison companies to achieve "the power of contrast":

> Our research rests upon having a comparison set. The critical question is not "What did the great companies have in common?" The critical question is "What did the great companies share in common that *distinguished* them from their direct comparisons?" Comparisons are companies that were in the same industry with the same or very similar opportunities during the same era . . . , yet did not produce great performance. (Collins & Hansen, 2011, p. 7)

With the seven matched comparisons selected, the team conducted in-depth historical case studies. The importance and credibility of their findings, despite

POSITIVE DEVIANCE EXAMPLE: DELAYING CHILDBIRTH IN INDIA

In 2013, Arvind Singhal led a positive deviance inquiry in urban slums of New Delhi, India, to design a mass media health campaign to (a) promote small family size (delay of first child and spacing between children), (b) enhance the health of new mothers and their infants, (c) counter the preference for male children, and (d) encourage adoption of contraceptive methods. Through analysis of archival data and key informant interviews, the team identified several positive deviants.

> One of our married woman respondents noted that she had significantly reduced her risk of getting pregnant by keeping a close track of her menstrual cycle, avoiding sex during the days she was at highest-risk for conception. During these *na na din* ("no, no days"), she employed a variety of *bahanas* (excuses) to avoid intercourse with her husband. She would, for instance, tell her husband that "I am keeping a *vrat* (fast) for a few days for your health," or "I am not feeling well these days." On her "yes, yes days," when she was not at high risk for conceiving, she noted: "I go out of my way to please him." (Singhal, 2013)

While the researchers noted that most married women in India would not be in a position to negotiate sex with their husbands, this "positive deviant" had found "creative, culturally-appropriate strategies to reduce her risk for conception. After all, how could an Indian husband overrule his wife's sacred *vrat*? And that, too, to preserve his health!"

The researchers also found a positive deviant health worker who achieved high rates of male vasectomy. To overcome normal fear and resistance, the health worker

> arranged for a few men, who were eager to undergo the vasectomy, to stride up—in open view of other men—and demand that they be the first to be snipped. And after they had undergone their vasectomy, usually a quick and painless procedure, they would stride out with *musteid chaal* (a stallion's stride), boasting about how easy the whole thing was. Such creative orchestration of theatrical elements helped reduce the anxiety of other men, significantly boosting rates of adoption of vasectomy. (Singhal, 2013)

the relatively small sample size, led to a best-selling management and organizational development book.

Here are examples of actual or potential purposefully matched comparison studies for different units of analysis.

- **Individual level:** Compare women who chose to have children with those who chose to be childless

(Hird & Abshoff, 2000; Kirkman, 2013; Mantel, 2013; Zhang, 2009).

- *Community level:* Compare social capital in matched urban and rural communities (Beaudoin & Thorson, 2004).
- *Ecological systems:* Compare ecologies with increasing and decreasing biodiversity (Minnesota Biological Survey, 2013).

Matched comparisons often begin with quantitative data and categorical distinctions as the basis for matching and then move to in-depth case studies to understand what explains the differences behind the numbers.

Criterion-Based Case Selection

The logic of criterion sampling is to review and study all cases that meet some predetermined criterion of importance, thereby explicitly (or implicitly) comparing the criterion cases with those that do not manifest the criterion. This is a strategy common in quality assurance efforts. For example, the expected range of participation in a mental health outpatient program might be 4 to 26 weeks. All cases that exceed 28 weeks are reviewed to find out why the expected range was exceeded and to make sure the case was being appropriately handled. Or a quality assurance standard may be that all patients entering a hospital emergency room who are not in a life-threatening situation receive care within one hour. Cases that exceed this standard are reviewed.

Quality management in manufacturing, as developed by W. Edwards Deming (2000), included the importance of understanding *knowledge of variation,* which requires distinguishing normal and acceptable variation from problematic variations that cause defects. When a production process shows an increased defect rate, the manager must go to the assembly line and investigate the nature of the problem with workers. The key is to establish a criterion for acceptable and unacceptable defect rates and to investigate firsthand (through observation, interviewing, and interaction with workers) whenever the criterion of acceptability is exceeded. This kind of quality assurance, though developed for manufacturing, can be applied to any service delivery or change process, including human services, education, and organizational development.

Critical incidents can be a focus of criterion sampling. For example, all incidents of client abuse in a program will generate an in-depth evaluation in a quality assurance effort. All former mental health clients who commit suicide within three months of release may constitute a sample for in-depth, qualitative study. In a school setting, all students who are absent 25% or more of the time may merit the in-depth attention of a case study. The point of criterion sampling is to be sure to understand cases that are likely to be information-rich because they may reveal major system weaknesses that become targets of opportunity for program or system improvement.

Criterion sampling can add an important qualitative component to a management information system or an ongoing program monitoring system. All cases in the data system that exhibit certain predetermined criterion characteristics are routinely identified for in-depth, qualitative analysis. Criterion sampling also can be used to identify cases from standardized questionnaires for in-depth follow-up, for example, all respondents who report having experienced ongoing workplace discrimination. (This strategy can only be used where respondents have willingly supplied contact information.)

W. Edwards Deming (2000) famously asserted, "The most important things cannot be measured." By this, he meant that monitoring statistical variation through a quality control process only tells you if something has changed, if some problem (outside acceptable variation) has emerged. To find out why, you have to go where the problem has emerged and then observe and study the situation firsthand. That involves qualitative fieldwork. Criterion-based sampling is a basis for identifying cases for further investigation in the field.

Continuum or Dosage Sampling

Continuum sampling derives from some kind of conceptual framework that distinguishes people, programs, organizations, or communities along a continuum. An example is Women's Ways of Knowing, which distinguishes five levels of "gaining voice" along a continuum: Silence, Received knowledge, Subjective knowledge, Procedural knowledge, and Constructed knowledge (Belenky et al., 1986; see Exhibit 1.5 in Chapter 1 for a description of each level). This framework derives from qualitative inquiry. To test the meaningfulness and relevance of the framework in new settings and contexts, the qualitative inquirer would purposefully select cases along the continuum. Continuum sampling is comparison based because it studies, describes, interprets, and compares cases along some dimension of interest, like "gaining voice." Exhibit 5.10 offers some well-known examples of continua that can and have served as a conceptual framework for case selection to illustrate different levels of a phenomenon of interest.

EXHIBIT 5.10 **EXHIBIT 5.10**   Examples of Well-Known Continua That Illustrate the Basis for Continuum Sampling

CONTINUUM	DIMENSION THAT DISTINGUISHES LEVELS	DISCIPLINARY ORIGIN	REFERENCE
1. Women's ways of knowing	Gaining voice	Epistemology	Belenky et al. (1986)
2. Maslow's hierarchy	Human needs	Psychology	Maslow (1954)
3. Folk–urban	Societal development	Anthropology	Redfield (1953)
4. Continuum of care	Service intensity (e.g., mental health, homelessness, aging)	Social work, public health	Evashwick (1989)
5. Autism spectrum	A range of neurodevelopmental disorders	Psychiatry	American Psychiatric Association (2013)
6. Approach and avoidance sexual goals	Motivations for having sex: self-focused and partner focused	Social psychology	Muise, Impett, and Desmarais (2013)
7. Adaptive cycle	Stages of biodiversity	Ecology	Gunderson and Holling (2002)

Dosage sampling is a type of continuum sampling used in evaluation to distinguish different degrees of engagement in a program linked to different levels of outcomes. Consider a youth after-school drop-in center that reports serving about 500 teenagers on a quarterly basis. The center helps with homework, offers recreational opportunities (basketball, volleyball), provides nutritious snacks, and offers staff support as needed. If we disaggregate the number 500, however, we find that about 400 only come to the center three or four times a quarter, usually for the monthly dance or pizza night. Another 70 come about once a week. Then, there are 30 participants who come to the center almost every day. This latter group is the high-dosage one (high participation, high potential impact). Average results for the whole population of 500 will be mediocre at best because the dosage level (degree of engagement) is so low. The high-dosage group, in contrast, would merit in-depth case studies to understand what they are getting from the center experiences.

Dosage matters. Qualitative inquiry elucidates how it matters. Dosage levels along a continuum from none to high can make all the difference, a fact observed in the Renaissance by German-Swiss physician, botanist, and alchemist Paracelsus (1493–1541):

Poison is in everything, and no thing is without poison. The dosage makes it either a poison or a remedy.

Summary of Comparison-Focused Sampling

We began this listing of purposeful sampling strategies with single significant case options ($n = 1$). In contrast, the distinguishing strategy of comparison-focused sampling is selecting multiple cases to study similarities and differences through comparative analysis. Comparisons are information-rich in that they provide insights into both the unique attributes and the common characteristics of the selected cases. The options we've reviewed are as follows (numbers correspond to those in the summary in Exhibit 5.8, p. 267):

7. Outlier sampling

8. Intensity sampling

9. Positive deviance comparisons

10. Matched comparisons

11. Criterion-based case selection

12. Continuum or dosage sampling

Group Characteristics Sampling Strategies and Options

Sample purpose: *Select cases to create a specific informa-tion-rich group that can reveal and illuminate important group patterns.*

One important purposeful sampling strategy is to create a group of cases that provide information-rich data-gathering and analysis possibilities. For example, if you're interested in documenting diversity of people, organizations, or places, you can create a diverse sample. In contrast, if you're interested in a particular type of person, organization, or place, you would create a homogeneous sample. This section reviews six approaches to sampling based on group characteristics.

Maximum Variation (Heterogeneity) Sampling

This strategy aims at capturing and describing the central themes that cut across a great deal of variation. For small samples, a great deal of heterogeneity can be a problem because individual cases are so different from each other. The maximum variation sampling strategy turns that apparent weakness into a strength by applying the following logic: Any common patterns that emerge from great variation are of particular interest and value in capturing the core experiences and central, shared dimensions of a setting or phenomenon.

How does one maximize variation in a small sample? You begin by identifying diverse characteristics or criteria for constructing the sample. Suppose a statewide program has project sites spread around the state, some in rural areas, some in urban areas, and some in suburban areas. The evaluation lacks sufficient resources to randomly select enough project sites to generalize across the state. The evaluator can study a few sites from each area and at least be sure that the geographical variation among sites is represented in the study. While the evaluation would describe the uniqueness of each site, it would also look for common themes across sites. Any such themes take on added importance precisely because they emerge out of great variation. For example, in studying community-based energy conservation efforts statewide using a maximum heterogeneity sampling strategy, we constructed a matrix sample of 10 communities in which each community was as different as possible from every other community on characteristics such as size, form of local government (e.g., strong mayor/weak mayor), ethnic diversity, strength of the economy, demographics, and region. In the analysis, what stood out across these diverse cases was the importance of the presence or absence of a local, committed cadre of people who made things happen.

In a study of the MacArthur Foundation Fellowship Program (popularly known as Genius Awards), the design focused on case studies of individual fellowship recipients. Over the 20 years of awards, more than 600 people had received fellowships. We only had sufficient resources to do 40 in-depth case studies. We maximized sample variation by creating a matrix in which each person in the sample was as different as possible from the others, using dimensions like nature of work, stage in career, public visibility, institutional affiliation, age, gender, ethnicity, geographical location, and field of endeavor. The thematic patterns of achievement that emerged from this diversity allowed us to construct a model to illuminate the primary dimensions of and factors in the award's impact. A signal emerged from all the static of heterogeneity. A theme song emerged from all the scattered noise. That's the power of maximum variation (heterogeneity) sampling.

Thus, when selecting a small sample of great diversity, the data collection and analysis will yield two kinds of findings: (1) high-quality, detailed descriptions of each case, which are useful for documenting uniquenesses and diversity, and (2) important shared patterns that cut across cases and derive their significance from having emerged out of heterogeneity.

Homogeneous Sampling

In direct contrast to maximum variation sampling is the strategy of picking a small homogeneous sample, the purpose of which is to describe some particular subgroup in depth. A program that has many different kinds of participants may need in-depth information about a particular subgroup. For example, a parent education program that involves many different kinds of parents may focus a qualitative evaluation on the experiences of single-parent female heads of households because that is a particularly difficult group to reach and keep in the program.

Focus group interviews are typically homogeneous. Focus groups involve open-ended interviews with groups of five to eight people on specially targeted or focused issues. The use of focus groups will be discussed at greater length in the chapter on interviewing. The point here is that sampling for focus groups typically involves bringing together people of similar backgrounds and experiences to participate in a group interview about major issues that affect them.

Typical Case Sampling

> Many highly intelligent people are poor thinkers. Many people of average intelligence are skilled thinkers. The power of a car is separate from the way the car is driven.
>
> —Edward de Bono
> Author of *Six Thinking Hats* (1985)
> and *Think! Before It's Too Late* (2009)

This strategy involves selecting and studying several cases that are average to understand, illustrate, and highlight what is typical and normal. Sometimes we're interested in outliers, but sometimes we're interested in the typical. These are different purposes, interests, and strategies. Each is valuable, but each serves a different purpose. For example, in describing a culture, community, or program to people not familiar with the setting studied, it can be helpful to provide a qualitative profile of "typical" cases. These cases are selected using statistical data about what is average, or with the cooperation of key informants, such as program staff or knowledgeable community members, who can help identify who and what are typical. Typical cases can be selected using survey data, a demographic analysis of averages, or other statistical data that provide a normal distribution of characteristics from which to identify "average-like" cases. Keep in mind that the purpose of a qualitative profile of one or more typical cases is to describe and illustrate what is typical to those unfamiliar with the setting—not to make generalized statements about the experiences of all participants. The sample is illustrative, not definitive.

When entire programs or communities are the unit of analysis, the processes and effects described for the typical program may be used to provide a frame of reference for case studies of "poor" or "excellent" sites. When the typical site sampling strategy is used, the site is specifically selected because it is not in any major way atypical, extreme, deviant, or intensely unusual. This strategy is often appropriate in sampling villages for community development studies in Third World countries. A study of a typical village illuminates key issues that must be considered in any development project aimed at that kind of village.

In evaluation and policy research, the interests of decision makers will shape the sampling strategy. I remember an evaluation in which the key decision makers had made their peace with the fact that there will always be some poor programs and some excellent programs, but the programs they really wanted more information about were what they called "those run-of-the-mill programs that are so hard to get a handle on precisely because they are so ordinary and don't stand out in any definitive way." Given that framing, we employed typical case sampling. It is important when using this strategy to attempt to get a broad consensus about which cases are "typical"—and what criteria are being used to define typicality.

Key Informants, Key Knowledgeables, and Reputational Sampling

Key informants are a prized group. These are people who are especially knowledgeable about a topic and are willing to share their knowledge. Key informant interviews were developed by ethnographers to help understand cultures other than their own. Because being an "informant" can have negative connotations, I prefer the term *key knowledgeables*. They inform our inquiry when we tap into their knowledge, experience, and expertise. Key informants are especially important sources on specialized issues. The evaluation of the implementation of the Paris Declaration Principles on development aid relied heavily on key informants in governments and international agencies because the topic is highly specialized. A study of New Zealand's Prostitution Reform Act identified and interviewed as key informants "those who have had direct experience of working with or as sex workers" (Mossman & Mayhew, 2007, p. 21).

Key knowledgeables may stand alone as a purposeful sampling strategy or be used in combination with other approaches. For example, identifying outliers or typical cases may begin with key informant interviews. Preskill and Beer (2013) used key informant interviews to identify exemplars of developmental evaluation and determine critical, core dimensions of the approach. Key knowledgeables provide valuable expertise on and insights into the root of problems, as in a study of the barriers philanthropic foundations experience that limit their capacity to plan and act strategically (Patrizi, Thompson, Coffman, & Beer, 2013). Key informant interviews are widely used to identify trends and future directions, as when Gopalakrishnan, Preskill, and Lu (2013) interviewed more than a dozen foundation leaders, evaluation practitioners, and social sector thought leaders to identify "Next Generation Evaluation" characteristics and approaches. I contacted qualitative evaluation methodologists for input into a presentation at the 2013 annual conference of the American Evaluation Association (AEA) on "Top Ten Trends in Qualitative Evaluation Over the Last Decade and Future Challenges." In essence, key knowledgeable interviews are among the most common sampling strategies for qualitative inquiry. The trick, of course, is identifying and gaining the cooperation of genuinely knowledgeable experts. So let me reiterate: As with all sampling, what you end up having something to say about depends on whom you sample.

Complete Target Population

Sometimes even a small number is everyone. A complete target population involves interviewing and/or observing everyone within a group of interest. For a small, local early-childhood parent education program with 20 participants, this would mean including everyone in the study. Here are other examples of complete target population samples:

- Following up all 19 marathon runners who lost limbs in the 2013 Boston Marathon terrorist bombing
- Studying the recovery of the 21 children (ages 5 months to 12 years) who survived the U.S. government raid on the 1993 Branch Davidian religious cult compound in Waco, Texas (Perry & Szalavitz, 2006, chap. 3)
- Doing case studies of all participants in one cohort of the Blandin Community Leadership program ($n = 24$)
- Long-term, in-depth case studies of the Suleman octuplets born in 2009: the world record survival rate for a complete set of octuplets surviving infancy

Each of these examples concerns an information-rich group that is unique and worthy of study in its own right. That's what makes for a complete population target group as the focus of qualitative inquiry.

Quota Sampling

In quota sampling, a predetermined number of cases are selected to fill important categories of cases in a larger population. Quota sampling ensures that certain categories are included in a study regardless of their size and distribution in the population. For example, a statewide evaluation contract specifies that the study will conduct 12 community case studies, with 3 cases from each of four regions. Having equal cases from each region is a political consideration.

> This sampling is a nonprobability technique because it requires only that the quota for each category be met without any further attention to how those sample members are actually located. For example, a study of social service organizations might set quotas for both the number of public and private providers and the number of larger and smaller agencies in the sample. Similarly, an interview study might select participants using a two-by-two table for gender and age so that one quarter of the informants were younger women, one quarter were older men, and so on. (Morgan, 2008b, p. 722)

Quota sampling is also used to ensure that certain important categories, whether demographic, geographical, or theoretical, are included in the study. Quota sampling also helps with budgeting and logistical calculations because the design specifies exactly how many cases (the quota) will be included in each category.

Quota sampling can be flexible, a beginning point to frame the initial design and facilitate entering the field rather than a fixed and rigid sample size. Janice Morse (2000) advocates quota sampling as the foundation for "organic" qualitative inquiry that grows as the inquiry unfolds and deepens. Quotas are a starting point, a baseline for launching an inquiry, but the size and composition of the sample can be adjusted based on what is learned as the inquiry deepens. This use of quota sampling has an emergent sensitivity. It is used to support theoretical sampling and is implemented in support of theory development and subsequent confirmation.

> Crucial to this organic sampling practice are the ways in which it supports theoretical generalisation and the evocative and illustrative accounts from the research. In seeking to undertake a particular kind of analysis researchers shape and form their sample reflexively.

> This . . . characterises researchers as making sampling decisions in the research from the very earliest stages of planning a study through to its completion. It recognises that all research is faced with resource and ethical constraints. The role of researchers in shaping the research they do through the ways in which they frame their study and make decisions about the sampling strategy are central to this approach. The emphasis here is not on a reflective responsiveness to the unfolding empirical findings in a study, but researchers as reflexive agents who make decisions about who or what to sample and the universe this relates to. (Emmel, 2013, p. 63)

Purposeful Random Sampling

A purposeful sampling strategy does not automatically eliminate any possibility for random selection of cases. For many audiences, random sampling, even of small samples, will substantially increase the credibility of the results. I recently worked with a program that annually appears before the state legislature and tells "war stories" about client successes and struggles, sometimes even including a few stories about failures to provide balance. To enhance the credibility of their reports, the director and staff decided to begin collecting evaluation information more systematically. Because they were striving for individualized outcomes, they rejected the notion of basing the evaluation entirely on a standardized pre–post instrument. They wanted to collect case histories and do in-depth case studies of clients, but they had very limited resources and time to devote to such data collection. In effect, staff at each program site, many of whom serve 200 to 300 families a year, felt that they could only do 10 or 15 detailed, in-depth clinical case histories each year. We systematized the kind of information that would be going into the case histories at each program site and then set up a random procedure for selecting those clients whose case histories would be recorded in depth, thereby systematizing and randomizing their collection of "war stories." While they cannot generalize to the entire client population on the basis of 10 cases from each program site, they will be able to tell legislators that the stories they are reporting were randomly selected *in advance of knowledge of how the outcomes would appear* and that the information collected was comprehensive. The credibility of systematic and randomly selected case examples is considerably greater than the personal, ad hoc selection of cases selected and reported *after* the fact—that is, after outcomes are known.

Another example of a purposeful random sampling involved selecting 20 village programs in an area of hundreds of villages. Even though the sample size was too small to be considered representative or generalizable, random selection avoided controversy about potential selection bias. Exhibit 5.11 presents another example, a mixed-methods sampling strategy in which a random sample was selected to narrow the huge number of potential cases (several thousand) to a manageable few to be considered for a small purposeful sample.

It is critical to understand that we are talking about a *purposeful random sample*, not a representative random sample. *The purpose of a small random sample is credibility and manageability, not representativeness.* A small, purposeful random sample aims to reduce suspicion about why certain cases were selected for study, but such a sample still does not permit statistical generalizations. To reiterate, the defining group characteristic of a purposeful random sample is its randomness.

Time–Location Sampling

The final strategy in the Group Characteristics category is time–location sampling, in which you interview everyone present at a particular location during a particular time period. For example, you may be interested in studying everyone who works out at a fitness center every day after work (or every day before work). The time (before work) and location (fitness center) combine to generate an information-rich sample. This strategy can be used in sampling social media. For example, more than 3,500 people registered for the 2013 annual meeting of the AEA in Washington, D.C. The conference offered a Twitter site (#eval13). All those who posted multiple tweets (Twitter being a virtual location) during the week of the conference (time boundary) would constitute a time–location sample. Finding how being an active part of the Evaluation Twitter Community during the conference could be an inquiry of interest and importance for the leadership of the AEA.

Summary of Group Characteristics Sampling

We began this listing of purposeful sampling strategies with single significant case options ($n = 1$) followed by comparison-focused sampling. The distinguishing strategy of Group Characteristics just discussed is selecting cases to create a specific information-rich group that can reveal and illuminate important group patterns. The options we've reviewed are as follows (numbers correspond to those in the summary in Exhibit 5.8 pp. 267–268):

13. Maximum variation (heterogeneity) sampling

14. Homogeneous sampling

15. Typical cases

16. Key informant, key knowledgeable, and reputational sampling

17. Complete target group

18. Quota sampling

19. Purposeful random sample

20. Time–location sampling

EXHIBIT 5.11	Mixed-Methods Sampling Example

Population survey as the basis for outlier sampling (success case method) with diversity criteria (urban–rural mix and size of program)

This study explored the factors that supported or hindered compliance of licensed practical nurses (LPNs) in Alberta, Canada, with the full scope of practice assigned to them by legislation.

> Due to the politically charged environment of this study, particular attention was given to devising an objective method for the selection of case study sites.... The success case method (Brinkerhoff, 2003), a province-wide survey, and statistical modeling were used to produce an objective and defensible platform for site selection as well as to enhance study rigor. (Barrington, Shimoni, & Legaspi, 2014, p. 1)

To identify high-scope sites (success cases) and low-scope sites (weak compliance cases) in an unbiased way, the researchers surveyed all LPNs in the province. They sent out 8,549 surveys to all practicing LPNs, providing both an online and a mail-in option; 2,313 LPNs responded, for a response rate of 27%. While they had hoped for a higher return rate, they found that the respondents could be tracked proportionately to staff deployment across the province. The absolute number of returns also gave them confidence that they could use the results to guide purposeful sample selection.

Based on the survey responses, the researchers distinguished four categories of sites that varied on compliance with practice ideals:

- Acute care sites in which LPNs work to high scope (success cases)
- Acute care sites in which LPNs work to low scope (weak-compliance cases)
- Long-term care sites in which LPNs work to high scope (success cases)
- Long-term care sites in which LPNs work to low scope (weak-compliance cases)

Geographic location and size of program were added as criteria in final site selection. Six case study sites were identified representing high- or low-scope workplaces for LPNs, environments in which they were either supported or hindered in their ability to use the competencies for which they were trained. The final case study sample included three acute care sites, one mixed site (providing both acute and long-term care), and two long-term care sites. Three sites were high scope, and three were low; three were urban, and three were rural.

> The resulting case studies provided a rich and detailed description of six particular healthcare sites in Alberta and as such have been able to inform the broader discussion about LPNs' scope of practice.... Use of a survey greatly enhanced the information obtained by the study and the two-phased approach allowed us to incorporate early findings into later research activities. (Barrington et al., 2014, p. 22)

The researchers concluded that "study findings were well received by the diverse stakeholders who represented key sectors in the health care system" (Barrington et al., 2014, p. 1).

> In the end, the decision to add a survey to our design strengthened our study immeasurably. The mix of quantitative and qualitative methods added depth and credibility to our findings but it also allowed us to explore a number of issues more fully. By staging the research over two phases, we had time to refine our focus as we went, so that by the time we actually visited the sites, we knew a lot more about LPN characteristics and salient workplace issues than we would have if we had gone there directly as initially planned. We were able to refine our case study tools based on survey findings, focusing quickly on key topics. For us, administering a survey first followed by in-depth case studies was a winning strategy. (Barrington et al., 2014, pp. 14–15)

Sample purpose: *Select cases for study that are exemplars of the concept or construct that is the focus of inquiry, to illuminate the theoretical ideas of interest.*

Let's do a quick review. The first set of purposeful sampling strategies involved looking at the value of in-depth study of single cases ($n = 1$). Then, we turned to comparison-focused sampling, which selects cases to compare and contrast significant similarities and differences between them and the factors that explain those differences. The third set of purposeful sampling strategies, just reviewed, involves creating a group of cases that provide information-rich data-gathering and analysis possibilities. Now we turn to a distinctly different purposeful sampling strategy: selecting cases to illuminate concepts and theories.

Theoretical Orientations and Sampling

Chapter 3 presented a variety of qualitative inquiry frameworks based on paradigmatic, philosophical, and theoretical orientations. These provide frameworks for determining inquiry priorities and core questions, but they provide little generic guidance on sampling. An ethnographic inquiry, for example, asks, "What is the culture of this group of people?" But the group studied can be one or more ethnic groups, tribes, villages, neighborhoods, schools, organizations, programs, gangs, clubs, and so forth and so on—any collection of people who may be said to share a cultural perspective. Case selection depends on what culture or cultures the ethnographer wants to study. That's the sampling challenge. For example, pioneering anthropologist Ruth Fulton Benedict's (1934) influential book *Patterns of Culture* compared and contrasted Pueblo cultures of the American Southwest with Native American cultures of the Great Plains. Hers was a matched comparison sample.

Phenomenology directs qualitative inquirers to ask, "What is the meaning, structure, and essence of the lived experience of this phenomenon for this person or group of people?" But the cases you choose to study will depend on what phenomenon interests you among what group of people. Barnett (2005) studied the experience of people living with chronic obstructive pulmonary disease. This was a homogeneous group sample recruited through clinical referrals.

So choosing to work within a major theoretical tradition doesn't tell you how to sample. The options we're reviewing are still open to you. That includes the

options of concept or theoretical sampling. Let's begin with deductive theoretical construct sampling.

Deductive Theoretical Sampling for Deepening or Verifying Theory-Derived Constructs; Operational Construct Sampling

> The same ideas that others had before you are waiting for you to bring them back to life in a new way. The part of who you are that is left behind within these old ideas is what makes them original all over again.
>
> —Ashly Lorenzana
> Author and diarist

Deductive theoretical construct sampling involves finding case manifestations of a theoretical construct of interest so as to examine and elaborate the construct, its variations and implications. Theoretical constructs are based in, derived from, and contribute to scholarly literature. In deductive theoretical construct sampling, the researcher samples incidents, slices of life, time periods, or people on the basis of their potential manifestation or representation of important theoretical constructs. Here are some examples of theoretical constructs studied qualitatively.

- *Culture:* Geertz (1988) used ethnographic inquiry and "thick description" to interpret the "theory of culture."
- *Resilience:* Buckholt (2001) studied *resilience* among adult abuse survivors; studying resilience among traumatized youth aims to integrate theory and practice (Perry & Szalavitz, 2006).
- *Trauma:* Perry and Szalavitz (2006) used case studies to study *trauma* and *posttraumatic stress* among children.
- *Respect:* Lawrence-Lightfoot (2000) used case-based qualitative "portraitures" to inquire into the meanings and nature of *respect*.
- *Incest:* Gilgun (1995) examined the concepts of *justice*, *care*, and *incest* among incest perpetrators.
- *Race:* Hilliard (1989) examined the concept of race through historical documents, images, and texts in ancient Egypt.

- *Total institution:* Erving Goffman originated this concept in studying asylums, and it has subsequently served as the conceptual focus for a variety of institutional studies (Davies, 1989; Goffman & Helmreich, 1961).
- *Epiphanies:* These are life-changing moments and events, "those life experiences that radically alter and shape the meanings persons give to themselves and their experiences" (Denzin, 2001, p. 1).

Deductive theoretical sampling involves deepening or verifying theory in new contexts, new time periods, or new situations. "Operational construct" sampling involves selecting for study real-world examples (i.e., *operational* examples) of the constructs in which one is interested. Studying a number of such examples is called "multiple operationalism" (Webb, Campbell, Schwartz, & Sechrest, 1966). For example, classic diffusion of innovations theory (Rogers, 1962) predicts that early adopters of some innovation will be different in significant ways from later adopters. Doing case studies on early and late adopters, then, would be an example of theory-based sampling. Such samples are often necessarily purposefully selected because the population of all early and late adopters may not be known, so random sampling is not an option.

When you are studying people, programs, organizations, or communities, the population of interest can be fairly readily determined. Constructs, however, do not have as clear a frame of reference:

> For sampling operational instances of constructs, there is no concrete target population. . . . Mostly, therefore, we are forced to select on a purposive basis those particular instances of a construct that past validity studies, conventional practice, individual intuition, or consultation with critically minded persons suggest offer the closest correspondence to the construct of interest. (Cook, Leviton, & Shadish, 1985, pp. 163–164)

What distinguishes deductive theoretical construct sampling is that the inquiry begins with selection of a concept of interest and importance within a discipline or field and then identifies a unit of analysis for studying the concept. That leads to sample selection. The sample becomes, by definition and selection, illuminative of the theoretical concept of interest.

Inductive Grounded and Emergent Theory Sampling

Inductive theoretical sampling is what grounded theorists define as "sampling on the basis of the emerging concepts, with the aim being to explore the dimensional range or varied conditions along which the properties of concepts vary" (Strauss & Corbin, 1998, p. 73). In grounded theory inquiries, open-ended fieldwork focuses on observing real-world social interactions that reveal concepts that become the basis for subsequent sampling. "Participants are selected according to the descriptive needs of the emerging concepts and theory" (Morse, 2010, p. 235). In essence, in an inductive strategy, emerging theory guides sample selection. The sample is built as the theory emerges moving from exploration to deepening to verification. Thus, grounded theoretical sampling becomes more selective as the emerging theory focuses the inquiry. Additional cases are added to support constant comparison as a theory-sharpening analysis process. Inductive theoretical sampling is challenging because the sample must be created as the inquiry unfolds. Such theoretical sampling

> is far more difficult than collecting data with preplanned groups; individuals and groups selected theoretically require that decisions are made informed by thought, analysis, and search. The sample's ongoing inclusion in the study is for a strategic reason, to test emerging theory. (Emmel, 2013, p. 15)

Evidence is not collected because it will accurately describe or verify some preconceived theoretical position. Indeed, deductively verifying existing theory is "the nemesis" of grounded theory:

> The logic is not one of: "I plan to sample this group because they use this service, and this group because they do not use this service." These rules of evidence hinder the discovery of theory. Groups are chosen because the data they produce relates to a particular category in the research. The search is for groups that display the category under investigation in different situations. (Emmel, 2013, p. 15)

In grounded theory, theoretical sampling supports the *constant comparative method of analysis*. That is, one does theoretical sampling in grounded theory to use the constant comparative method of analysis. The two go hand in glove, connecting design and analysis. Theoretical sampling permits elucidation and refinement of the variations in, manifestations of, and meanings of a concept as it is found in the data gathered during fieldwork. The constant comparative method involves systematically examining and refining the variations in emergent and grounded concepts. Variations in the concept must be sampled to rigorously compare and contrast them. (See Chapters 3 and 8 for more detailed discussions of grounded theory.)

Realist Sampling

> Theory always precedes data collection in a scientific realist sampling strategy.
>
> —Nick Emmel (2013, p. 95)

I reviewed realism as a distinct theoretical foundation for guiding qualitative inquiry in Chapter 3 (see pp. 111–114). Realism has become an influential perspective in qualitative inquiry generally (Maxwell, 2012) and program evaluation particularly (Astbury, 2013; Pawson, 2013; Rogers, 2014; Westhorp, 2013). Emmel (2013) has written the most comprehensive account of realist sampling. Theories are the basis for realist sampling "because, while constructed, they are nonetheless real in the sense of being and explaining reality" (p. 95). Sampling decisions can draw from a wide range of theories but must be theory based in some direct way. Here is a purposeful sample of Emmel's explanations of realist sampling:

- For a realist there is a relationship between mental processes and that which is directly observable and recordable. Both must be accounted for in any description of the research we do. By extension, sampling in realist qualitative research works out the relation between theory and evidence of the samples we select. (p. 71)

- For realists, generative mechanisms govern, mediate, impede, and facilitate sampling choices. These are the powers that describe sampling. (p. 74)

- In a realist account of the research, theory, like the generative mechanism of the way in which housing is allocated to abused women who have left their partners, is ontologically real. Explanation and interpretation in a realist sampling strategy tests and refines theory. Sampling choices seek out examples of mechanisms in action, or inaction towards being able to say something explanatory about their causal powers. Sampling . . . is both pre-specified and emergent, it is driven forward through an engagement with what is already known about that which is being investigated and ideas catalyzed through engagement with empirical accounts. (p. 85)

- [Realist researchers] will spend a considerable amount of time thinking about where the phenomenon they are interested in investigating is most likely to occur. . . . Researchers are often faced with opportunities in their fieldwork to choose a person, organization, or object to be sampled. These choices are made . . . because it is thought that the inclusion of this unit in the research will allow for the refining and elaboration of theory. (p. 80)

A realist approach to sampling for program evaluation is illuminative. Suppose a realist evaluator wants to evaluate a program aimed at helping poor families transition out of poverty. A realist sampling strategy would begin by identifying ideas about how the program was intended to help these families, essentially identifying hypothesized causal mechanisms within the context of the program. This conceptualization would direct sampling. "The realist sample is always bootstrapped with theory. . . . Ideas drive forward sampling choices" (Emmel, 2013, p. 83).

Realist sampling is guided by theory-based design decisions (deduction) while being open to what emerges (induction). This requires reflexivity.

> Considerations of reflexivity reinforce the observation that sampling is not procedural. These processes cannot be predicted in advance, nor do they conform to rules. They are matters of judgment made by researchers. The uncertainties and complexities of doing research are emphasized. We may accept, reject, or most likely refine theory through the theoretical work of sampling that happens in the research. This sifting, winnowing, and subsequent refining of theory advocated in a scientific realist sampling strategy cannot be described by rules of engagement with data and evidential account. Reflexive practice in sampling can only be shaped by guidelines of practice towards the task of knowledge production. (Emmel, 2013, p. 87)

Causal Pathway Case Sampling

If the primary purpose of a qualitative inquiry is to study causal mechanisms and elucidate or confirm causal relationships, the information-rich case will be the one that permits study of causality. This means designing the study and selecting cases to "maximize leverage over the causal hypothesis" (George & Bennett, 2005, p. 172). John Gerring (2007) has called this a *pathway case*:

One of the most important functions of case study research is the elucidation of causal mechanisms. . . . But what sort of case is most useful for this purpose? Although all case studies presumably shed light on causal mechanisms, not all cases are equally transparent. In situations where a causal hypothesis is clear and has already been confirmed by cross-case analysis, researchers are well advised to focus on a case where the causal effect of one factor can be isolated from other potentially confounding factors. I shall call this a *pathway case* to indicate its uniquely penetrating insight into causal mechanisms.

To clarify, the pathway case exists only in circumstances where cross-case covariational patterns are well studied but where the mechanism linking X_1 and Y remains dim. Because the pathway case builds on prior cross-case analysis, the problem of case selection must be situated within that sample. There is no stand-alone pathway case. (p. 122)

Gerring's (2007) examples are drawn from political science. For example, a case study of how leadership-initiated reform leads to increased democratization would trace the causal pathway from leadership to reform to democratization (p. 123). He concludes,

When researchers refer to a particular case as an "example" of a broader phenomenon, they are often referring to a pathway case. This sort of case illustrates the causal relationship of interest in a particularly vivid manner, and therefore may be regarded as a common trope among case study researchers. (p. 131)

Sensitizing Concept Exemplars Sampling

Sensitizing concept sampling involves finding information-rich cases that can illuminate the use and meaning of particular concepts within particular settings. The difference from deductive theoretical sampling (or operational construct sampling) is that the focus is on contextually specific uses and meanings rather than contribution to theory more generally. Sensitizing concepts could, theoretically and conceptually, be subsumed within theoretical sampling, but I consider the approach sufficiently distinct to merit its own attention. The reason, in part, is the ethnographic distinction between "emic" and "etic," to distinguish classification systems reported by anthropologists based on (a) the language and concepts used by the people in the culture studied, an *emic* approach, from (b) concepts and categories created by anthropologists based on their analysis of important cultural distinctions, an *etic*

approach (Pike, 1954). Sampling sensitizing concepts is an *emic* approach to conceptual sampling.

Qualitative sociologist and symbolic interactionist Herbert Blumer (1954) is credited with originating the idea of the "sensitizing concept" as a guide to fieldwork with special attention to the words and meanings that are prevalent among the people being studied. Thus, the distinction between theoretical construct sampling and sensitizing concept sampling is that the former concepts originate in scholarly literature while the latter are grounded in the language of ordinary people. Sensitizing concepts provide some initial direction to a study as the field-worker inquires into how the concept is given meaning in a particular place or set of circumstances being studied (Schwandt, 2001). A sensitizing concept makes the researcher or evaluator *sensitive* to a concept of importance to some group of people. Being sensitive means being attuned to how the concept is used, what it means, and how it provides insight into the perspectives and behaviors of the people using the concept. A sample must be identified and selected that will illuminate the sensitizing concept. Here are some examples:

- *Bullying:* Bullying in school is a matter of great concern. In extreme cases, bullying can lead to physical and psychological injures, and even death. A sample of students in a school might differentiate (a) those who are victims of bullying, (b) those who are bullies, and (c) those who observe the bullying but are neither victims nor perpetrators.

- *As Māori:* A sport and recreation program in New Zealand has been built around the idea of participating *as Māori*. This contrasts with the traditional government conceptual framework of increasing participation *by Māori*. Participation *as Māori* was a conceptual shift that became the focus of inquiry for a developmental evaluation of the program that received the 2013 Best Evaluation Policy and Systems Award from the Australasian Evaluation Society (McKegg, Wehipeihana Pipi, & Thompson, 2013).

- *System:* Seventeen organizations working collaboratively on early-childhood home-visiting programs talk about changing "the system." What do they mean by "the system"? Who and what is in their system? What does systems change involve and look like for them? (Hargreaves & Paulsell, 2009).

- *Awareness of dying:* How do people within a particular context communicate with and about people who are terminally ill? (Glaser & Strauss, 1965).

- *Health:* How do people in health institutions think about concepts like health, public heath, and health promotion? What does it mean to be healthy? Unhealthy? (Flick, 2000, 2007a, pp. 18–20)

- *Middle age:* Linguistic pundit William Safire (2007) devoted a *New York Times* column to pondering "middle age." He considered several operational definitions, judging each to be inadequate. Indeed, the more precise the definition (e.g., age 45–60 years), the more problematic is its general utility. He concluded that the inherent ambiguity of the term *middle life*, and the resulting implication that each of us must define it *in context*, made it not a euphemism but rather a "usefulism." For our purpose, it is also a *sensitizing concept*.

Sampling based on sensitizing concept exemplars essentially involves two dimensions: (1) a term or label used by a group of people (the sensitizing concept) and (2) identification of the people or situations where that sensitizing concept will be manifest. The purpose of the sample is to illuminate the sensitizing concept's uses and meanings within some specific context (the case or cases sampled).

And just in case you are looking for a simple but profound sensitizing concept to study, consider this request from best-selling author Gail Sheehy (2006):

> Would that there were an award for people who come to understand the concept of *enough*. Good *enough*. Successful *enough*. Thin *enough*. Rich *enough*. Socially responsible *enough*. When you have self-respect, you have *enough*. (p. 23)

Principles-Focused Sampling

> It is often easier to fight for a principle than to live up to it.
>
> —Adlai Stevenson (1900–1965)
> U.S. Ambassador to the
> United Nations (1961–1965)

Principles provide guidance and direction for working with people in need or trying to bring about change. Principles, unlike rules, involve judgment and have to be adapted to the context and situation. *Principles-focused evaluation*, discussed in Chapter 4, necessarily employs principles-focused sampling. This involves identifying and studying cases that illuminate the nature, implementation, outcomes, and implications of principles. Studying the implementation and outcomes of effective, *evidence-based principles* is a major new direction in developmental evaluation (Patton, 2011, pp. 167–168, 194–195).

A principles-based approach is appropriate when a group of diverse programs are all adhering to the same principles but each adapting those principles to its own particular target population within its own context. A principle is defined as a fundamental proposition that serves as the foundation for a system of belief or behavior or for a chain of reasoning. An evidence-based effective principles approach assumes that while the principles remain the same, in implementing them, there will necessarily and appropriately be adaptation within and across contexts. Taken together, a set of principles provides a framework that guides cohesive approaches and solutions. Evidence for the effectiveness of principles is derived from in-depth case studies of their implementation and implications. The results of the case studies are then synthesized across the diverse programs, all adhering to the same principles but each adapting those principles to its own particular target population within its own context. See Exhibit 4.7 in Chapter 4 for an example and discussion of principles-focused evaluation.

Theoretical constructs as a basis for sampling are derived from disciplinary studies and scholarly literature. Sensitizing concepts as a focus for case selection are focused on terms that have special meaning to people who use the concept of interest within a particular context. Principles-focused sampling aims to elucidate the nature and implications of prescriptive admonitions like "Plan your work. Work your plan." What does this mean in practice for a specific group of practitioners? To what extent and in what ways, if at all, is this principle followed? And if followed, how does adhering to this principle affect behavior? The sample for inquiry would consist of people in organizations or professions that profess adherence to this principle. In my view, principles have become so widely touted as a basis for action and, as a result, such an important focus of inquiry that principles-focused sampling deserves its own category among purposeful sampling options.

Complex, Dynamic Systems Sampling

> We bring together the case study method and complexity science. . . . The case study provides a method for studying systems. Complexity theory suggests that keys to understanding the system are contained in patterns of relationships and interactions among the system's agents. We propose some of the "objects" of study that are implicated by complexity theory and discuss how studying these using case methods may provide useful maps of the system. We offer complexity theory, partnered with case study method, as a place to begin the daunting task of studying a system as an integrated whole. (Anderson et al., 2005, p. 669)

The objects of study (units of analysis) to be sampled in complexity-focused case studies include interdependencies, nonlinear processes, emergent phenomena, and unexpected events.

Complex, dynamic systems sampling involves selecting cases where complex dynamic processes can be tracked, studied, and documented over time. Such studies are inherently highly emergent. This type of purposeful sampling derives from the increased attention being directed to systems thinking and complexity as overarching theoretical constructs for understanding the world (Hieronymi, 2013). Studying complex dynamic systems poses special challenges because they are characterized by turbulence, uncertainty, unpredictability, nonlinearity, and rapid change. This means tracking and documenting complex dynamic phenomena such as vicious and virtuous circles, ripple effects, adoption and adaption of innovations, viral-like spreading phenomena, crowd-sourcing dynamics, network interactions, and "black swan" events, which are highly improbable and have high impact (Taleb, 2007). The unit of analysis for this type of inquiry will be phenomena such as an innovation, a system, a critical incident, a major change initiative, a social movement, a disaster, and other highly complex, dynamic phenomena. Often,

such studies will be retrospective (after the fact), but some, for instance, major new innovative initiatives, can be identified prospectively. Here are some examples:

- Mapping the ripple effects of a major agricultural extension innovation initiative (Chazdon & Alviz, 2013)
- Studying dynamic social movements over time, such as Mothers Against Drunk Driving and Muhammad Yunus's microcredit innovation movement (Westley et al., 2006)
- Evaluating the impact and influence of emergency assistance during and after the 1994 Rwanda genocide (Dallaire & Beardsley, 2004; Danida, 2005; United Nations, 1996)
- Studying the 2008 global financial crisis (J. B. Stewart, 2009)
- Extracting lessons from the humanitarian response to the 2004 Indian Ocean earthquake and tsunami (Christoplos, 2006; UNICEF, 2008)

Each of these examples manifests complexity. I have singled out complex dynamic systems as a specific form of theory-focused purposeful sampling because it involves unique sampling challenges. Complex phenomena require emergent designs and flexible fieldwork to follow the action wherever it unfolds. Sampling is multifaceted: observing those engaged in the system, interviewing key actors, tracking decisions, locating and analyzing documents, and immersing oneself in the chaos as it proceeds on its unknown and unpredictable course. Complexity will be the dominant theme of the twenty-first century. Qualitative inquiry is especially attuned to capturing the emergent and dynamic nature of complex systems change. As with any inquiry, what is sampled is what is studied. So we would do well, it seems to me, to acknowledge complex phenomena as a distinct type of purposeful sample.

Summary of Theoretical and Concept Sampling

> Science has explored the microcosms and the macrocosms;
>
> We have a good sense of the lay of the land.
>
> The great unexplored frontier is complexity.
>
> —Heinz Pagels (1988)
> American physicist

The distinctive contribution of this form of purposeful sampling is elucidation and illumination of the meanings, variations, implications, and applications of ideas. Concepts and constructs are mental models. They wall off some part of human experience and say this is a category of experience that is important, different from other things, and worthy of inquiry as a phenomenon of interest and importance. Qualitative inquiry, where words are the coin of the realm and the object of our fascination, is especially adept at investigating what concepts and constructs mean. Concepts are the fundamental building blocks of theory, the bedrock of any professional practice, and essential in making distinctions in ordinary, day-to-day conversations. Thus, it is altogether critical to understand concept and theoretical construct sampling options. We've examined seven alternatives (numbers correspond to those in the summary in Exhibit 5.8, pp. 269–270):

21. Deductive theoretical sampling; operational construct sampling

22. Inductive grounded and emergent theory sampling

23. Realist sampling

24. Causal pathway case sampling

25. Sensitizing concept exemplars

26. Principles-focused sampling

27. Complex dynamic systems sampling

MISUNDERSTANDING ABOUT AND UNDERUSE OF PURPOSEFUL SAMPLING

The Evaluation Co-Operation Group (2012) of the Multilateral Development Banks has produced the *Big Book on Evaluation Good Practice Standards*. It stipulates that purposeful samples can only be used for learning, not for accountability or public reporting on evaluation of public sector operations (p. 23). Only randomly chosen representative samples can be used for accountability and public reporting. This narrow view of purposeful sampling, codified in official methodological guidance documentation, limits the potential contributions of strategically selected purposeful samples. This is but one example of many such narrow methodological policy and procedure guidebooks that show a misunderstanding about and underuse of purposeful case selection.

Instrumental-Use Multiple-Case Sampling

The purpose of instrumental-use multicase purposeful sampling is to select multiple cases of a phenomenon so as to understand the phenomenon and, in applied multicase studies, generate generalizable findings that can be used to inform changes in practices, programs, and policies.

> There are many purposes for case research, running from the most theoretical to the most practical. When the purpose of case study is to go beyond the case, we call it "instrumental" case study. When the main and enduring interest is in the case itself, we call it "intrinsic" case study. (Stake, 2006, p. 8)

Selecting cases for instrumental use is a purposeful sampling challenge. I have taken Stake's distinction between intrinsic and instrumental case studies and added the meaning of instrumental use in the evaluation utilization literature, namely, studies aimed at directly informing decisions. What Stake means by instrumental studies going "beyond the case" includes selecting multiple cases of a phenomenon for the purpose of generating findings that can be used to inform changes in practices, programs, and policies. Instrumental case studies can be used by program and policy decision makers, or by practitioners and funders, in what has come to be called evidence-based decision making. Instrumental case sampling makes qualitative findings part of the evidence for evidence-based decisions and actions. "Evidence-based research should enable people to attain a deeper conviction of how the thing works and what to do about it" (Stake, 2010, p. 123). To have influence in evidence-based discussions concerning "what to do about it" (where *it* is any program, policy, or practice under review), the cases must be selected for relevance and permit such depth of analysis that issues of generalizability, causality, and scaling can be addressed. These terms have particular interpretations in qualitative inquiry, interpretations we'll discuss in later chapters. For now, the critical point is that "with multicase study and its strong interest in the *quintain* [group of cases illuminating a phenomenon], the interest in the cases will be primarily instrumental" (Stake, 2006, p. 8). For instrumental multicase inquiries, Stake (2006) identifies three main criteria for selecting cases:

1. Relevance of each case to the multicase phenomenon that is the focus of inquiry

2. Selecting cases that provide diversity across contexts

3. Selecting cases that provide "good opportunities to learn about complexity and contexts" (p. 23).

In this section, I am highlighting two particular kinds of instrumental samples: utilization-focused sampling and systematic qualitative reviews.

Utilization-Focused Sampling

Utilization-focused evaluation is an approach to evaluation aimed at intended use by intended users (Patton, 2008b, 2012a). Within that overall framework, utilization-focused sampling involves selecting cases that will be relevant to the issues and decisions of concern to an identifiable group of stakeholders and intended users. Such sampling can employ any of the purposeful strategies already identified but *adds a requirement that cases selected for study will have credibility, relevance, and utility for primary intended users.* The ideal way to meet these criteria is to actively engage primary intended users in design and methods decisions, especially sampling. Focusing on generating useful and actionable findings, a common result of utilization-focused qualitative inquiry is identification of factors that explain the differences between what works and what doesn't work. Those factors and explanations can be used to inform decision-making and support programs, practices, and/or policy improvements. To accomplish this action-oriented result, cases must be selected with that kind of finding and use in mind. This means not only attending to the relevance, contextual diversity, and complexity but also having sufficient access to in-depth information about the cases to be able to decipher individual case dynamics and cross-case patterns that will support conclusions about causal factors and recommendations about actions to be taken more generally beyond the cases studied.

Exhibit 5.12 provides an example of a utilization-focused evaluation with a utilization-focused sampling approach. The example also illustrates criterion-based sampling and complete target population case selection: *all* railroad switching yard deaths. Purposeful sampling strategies can overlap. What makes this also a utilization-focused sampling example is that the decision to include all deaths was explicitly made by the primary intended users to ensure the relevance, credibility, and utility of the findings.

EXHIBIT 5.12   Utilization-Focused Sampling and Evaluation Example

One of the most rigorous case study evaluation processes I've worked with was a *Switching Operations Fatality Analysis* (SOFA, 2010). Fifty-five railroad employees died in switching yard accidents from 2005 to 2010. To analyze the causes of these deaths, SOFA Working Group, a multistakeholder, made up of representatives from the railroad industry, labor unions, locomotive engineers, and federal regulators, was established. These stakeholder groups have traditionally been suspicious of each other and often in conflict over regulations and procedures. However, they put aside those conflicts to rigorously investigate the causes of the switching operations fatalities. They decided that for credibility and utility, and to honor those who had lost their lives, every fatality would have to be analyzed. They procured time and resources to carefully review every case, coding a variety of variables related to conditions, contributing factors, and kinds of operations involved. They then looked for patterns in the qualitative case data and correlations in the quantitative cross-case data. They spent three to four hours coding each case and hours analyzing the patterns across cases.

They found that fatalities happen for a reason. Accidents are not random occurrences, unfortunate events, or just plain bad luck. The risks to employees engaged in switching operations are real, ever present, and preventable. The data showed patterns in why switching fatalities occur. Knowledge about the causes of railroad accidents has accumulated through this rigorous analysis over time and across cases. The first SOFA Working Group report was released in 1999, based on analysis of 76 railroad fatal accident case files. The 2004 report examined the 48 switching fatalities that had occurred since the first report. These analyses identified the *five lifesavers* and 10 special

switching hazards that contribute to fatal accidents. The 2010 report presented the latest findings on the causes of deadly accidents.

On February 25, 2010, more than 50 senior industry executives, labor leaders, and Federal Railroad Administration staff convened to discuss the draft findings and their implications. The meeting opened with a moment of silence in memory of the 179 lives lost in switching operations. The group then reviewed how the SOFA Working Group arrived at its findings and determined that the results were credible, accurate, and important. In making that determination, understanding how the cases were selected and analyzed was critical. These key stakeholders needed the opportunity to discuss together how to make sense of the data. In small groups, they discussed patterns in the findings and identified potential preventive initiatives. The final SOFA report incorporated the reactions at this summit meeting and invited those in the railroad industry, at all levels, to engage with the findings and take actions to prevent fatalities.

The goal remains zero deaths. The SOFA Working Group analysis process was evaluated by a team of external, independent evaluation professionals (Bonnet, Ranney, Snow, Coplen, & Patton, 2009). The evaluation included examining the quality of the analysis process, the validity of the findings, what lessons have been learned, and what has changed in the industry over the past decade that affects interpretation and use of new findings. The independent evaluation concluded that the SOFA analysis process was systematic, rigorous, comprehensive, and objective. The findings were deemed valid and significant. At the core of the meta-evaluation was the conclusion that the commitment to examine every case was the foundation for credibility and utility.

Qualitative Evaluation Systematic Reviews

A systematic evaluation review seeks to identify, appraise, select, and synthesize *all* high-quality evaluation research evidence relevant to a particular arena of knowledge that is the basis for interventions. Systematic evaluation reviews of randomized controlled trial studies have been the basis for *evidence-based medicine.* The cases being synthesized are completed, usually published, studies. The process involves identifying all high-quality, peer-reviewed studies on a problem and synthesizing findings across

those separate and diverse studies to reach conclusions about what is effective in dealing with the problem of concern, for example, female hormone replacement therapy or effective treatments for prostate cancer. Informing and setting guidelines for treatment is the instrumental use. Even though, on the face of it, the strategy is to include all high-quality evaluation studies, it becomes a sampling issue because of the necessity of determining and assessing what constitutes a high-quality study.

The important new direction in systematic reviews is including qualitative evaluation studies (Gough,

Oliver, & Thomas, 2012; Wright, 2013). For example, the internationally prestigious Cochrane Collaboration Qualitative & Implementation Methods Group supports the synthesis of qualitative evidence and the integration of qualitative evidence with other evidence (mixed methods) in Cochrane intervention reviews on the effectiveness of health interventions. (See sidebar.) Later, we will examine how qualitative syntheses generally involve the challenge of selecting high-quality cases for inclusion in the synthesis. Here, however, the focus is on the instrumental-use purpose of systematic qualitative reviews, which is grounded in selecting and synthesizing high-quality qualitative studies with actionable findings and implications. In effect, systematic evaluation reviews serve a meta-evaluation function. Qualitative syntheses more generally seek cross-study patterns but are not necessarily focused on patterns of effectiveness.

are exemplars of the concept or construct that is the focus of inquiry, with the purpose of illuminating theoretical ideas of interest. In contrast, *instrumental-use multiple-case sampling* is applied and evaluative in focus. The purpose is to inform professional practice and program decision making. We've examined two strategies (numbers correspond to those in the summary in Exhibit 5.8, p. 270):

28. Utilization-focused sampling

29. Systematic qualitative evaluation reviews

> In both of these strategies, the findings are judged by whether they are used. In that regard, instrumental sampling is subject to alternative and competing perceptions of both quality and utility. As evaluation pioneer Eleanor Chelimsky (1983) observed, "The concept of usefulness . . . depends upon the perspective and values of the observer. This means that one person's usefulness may be another person's waste" (p. 155).

COCHRANE COLLABORATION QUALITATIVE AND IMPLEMENTATION METHODS GROUP

The Cochrane Collaboration is an international network of health researchers, healthcare practitioners, policymakers, patients, and their advocates and caregivers. The Collaboration aims to inform decision making about effective health care by preparing, updating, and promoting *Systematic Reviews*. These reviews are recognized as setting the standard for high-quality information about the effectiveness of health care.

One working group within The Cochrane Collaboration is the Qualitative & Implementation Methods Group. The group supports the synthesis of qualitative evidence and the integration of qualitative evidence with other evidence (mixed methods) in Cochrane intervention reviews of health effects. The qualitative working group advises the Cochrane Collaboration and its network on policy and practice related to qualitative evidence synthesis, develops and maintains methodological guidance on qualitative syntheses, and provides training.

For further information and examples of qualitative syntheses, see http://cqim.cochrane.org/

Summary of Instrumental-Use Multiple-Case Sampling

The previous section on *theory-focused and concept sampling* looked at strategies for selecting cases that

ON OPENNESS IN INQUIRY: REFLECTIONS FROM CARL SAGAN (1934–1996)

(Eminent Astrophysicist, Cosmologist, and Science Communicator)

It seems to me what is called for is an exquisite balance between two conflicting needs: the most skeptical scrutiny of all hypotheses that are served up to us and at the same time a great openness to new ideas. . . . If you are only skeptical, then no new ideas make it through to you. You never learn anything new. You become a crotchety old person, convinced that nonsense is ruling the world. (There is, of course, much data to support you.) But every now and then, maybe once in a hundred cases, a new idea turns out to be on the mark, valid and wonderful. If you are too much in the habit of being skeptical about everything, you are going to miss or resent it, and either way you will be standing in the way of understanding and progress. On the other hand, if you are open to the point of gullibility and have not an ounce of skeptical sense in you, then you cannot distinguish the useful ideas from the worthless ones. (Sagan, 1987, para. 17)

The open-ended naturalistic nature of exploratory qualitative inquiry means that you sometimes have to build the sample during fieldwork. One case leads to another, in sequence, as the inquiry unfolds. The sample is expanded as you follow leads and new directions that emerge during the study.

Snowball or Chain Sampling

This is an approach for locating information-rich key informants or critical cases. The process begins by asking well-situated people, "Who knows a lot about _____? Whom should I talk to?" By asking a number of people who else to talk with, the snowball gets bigger and bigger as you accumulate new information-rich cases. In most programs or systems, a few key names or incidents are mentioned repeatedly. Those people or events, recommended as valuable by a number of different informants, take on special importance. The chain of recommended informants would typically diverge initially as many possible sources are recommended and then converge as a few key names get mentioned over and over again.

The Peters and Waterman (1982) classic study *In Search of Excellence* began with snowball sampling, asking a broad group of knowledgeable people to identify well-run companies. Rosabeth Moss Kanter's (1983) influential study of innovation, *The Change Masters*, focused on 10 core case studies of the "most innovative" companies. She began by asking corporate experts for candidate companies to study. Nominations snowballed as she broadened her inquiry, and then converged into a small number of core cases nominated by a number of different expert informants.

You can also build the sample as you interview by asking each interviewee for suggestions about people who have a similar or different perspective. This generates a chain of interviewees based on people who know people who know people who would be good sources given the focus of inquiry. Once identified, the researcher typically does the recruiting.

The "snowball" effect occurs as referrals multiply at each step. For example, if you got two referrals from each person, then starting from two people get four more, then eight, sixteen, and so on.

Snowball sampling uses a method beloved by sales people, where customer referrals to new prospects have

"So which part of extreme snowball sampling are you having ethical doubts about? Not knowing where it will stop or only interviewing the survivors?"

particular value as the relationship of trust and obligation between the identified person and the referrer makes it more likely that the new person will make a purchase.

The way that the sample is chosen by target people makes it liable to various forms of bias. People tend to associate not only with people with the same study selection characteristic but also with other characteristics. This increases the chance of correlations being found in the study that do not apply to the generalized wider population.

The need to get the person to give you a referral also means that the researcher has to form a relationship with the person and be nice to them. This can change the study results as affective biases in both the researcher and the target person change how they think and behave.

With care in selection (you do not have to use every referral) and avoiding personal bias, snowball sampling can still be a useful method, particularly if you have no other way of reaching the target population. (Changing Minds, 2013, p. 1)

Snowball sampling can be an effective and efficient way to generate a sample through the Internet or social media. McNabb (2013) sought teachers who were using the Internet in their English language arts curriculum on a regular basis. She interviewed national leaders in this area and from the interviews generated a list of 20 classroom teachers who were active/known leaders in the field of technology and reading/English language arts teaching. She sent the teachers an online survey with participation criteria and asked them to forward the survey request to up to 10 teachers they knew who met the criteria. She ended up with a snowball sample of about 200 teachers.

Respondent-Driven Sampling (Network Sampling)

Respondent-driven sampling, also called network sampling (American Association for Public Opinion Research, 2013), is a network-based strategy in which a small number of initial participants from the total target population, called "seeds," are studied and asked to recruit up to three new contacts in their network. The initial interviewees do the recruiting (rather than the researcher), usually for compensation. This approach is especially useful to locate hard-to-access research participants, like sex workers, gang members, or people with rare diseases, and thus confidentiality is critical because of the sensitive topic being studied. An example of seeking access to an inner circle of network connections would be identifying sex workers with HIV to document and understand their coping and mutual support strategies. Trust and confidentiality are critical to obtaining a sample among such rare and hard-to-reach people involved in or affected by illegal or stigmatized behaviors (Abdul-Quader, Heckathorn, Sabin, & Saidel, 2006; White et al., 2012).

Respondent-driven sampling originated as snowball sampling combined with a mathematical model that weighted the sample to compensate for the nonrandom purposeful design (Heckathorn, 1997). But in its core recruitment elements, it can be a completely qualitative strategy.

A major difference between snowball sampling and RDS [respondent-driven sampling] is that seeds recruit their peers (rather than identifying them to an investigator) using a set number of uniquely coded coupons which are redeemed at a fixed interview location within a set period of time (e.g., 10 days). RDS peer-to-peer recruitment removes selection bias of the survey staff and the coupon quota minimizes biases associated with the over-representation of those participants with large networks. . . . In addition, RDS requires that recruitment continue far beyond the seed and his or her recruits. The recruits of seeds (Wave 1) are also expected to recruit their peers (Wave 2), who in turn enroll in the survey and receive their own set of recruitment coupons to use in recruiting their peers (Wave 3). This process is encouraged until the final sample comprises long recruitment chains made up of several waves of participants (sometimes as long as 20 waves). Long recruitment chains allow for deeper penetration into the target population networks. (Johnston & Sabin, 2010, p. 39)

Also called "link-tracing sample design" or "link-tracing network sampling," the important core characteristic is the participation of the respondent in reaching other persons to whom he or she is linked, which is what makes it *respondent driven*. "The defining feature of this approach is that subsequent sample members are selected from among the network contacts of previous sample members" (American Association for Public Opinion Research, 2013, p. 49).

Two special populations are distinguished when using link-tracing designs: rare versus hidden populations. Rare populations are hard to reach because of low prevalence in society, for instance, people suffering from a rare disease. Hidden populations are hard to reach because of the difficulty in getting cooperation, for example, illegal immigrants. In both cases, trustworthy network connections and compensation are the keys to sample recruitment.

Emergent Phenomenon or Emergent Subgroup Sampling

Fieldwork often involves on-the-spot decisions about sampling to take advantage of what emerges during actual data collection. Being open to following wherever the data lead is a primary strength of qualitative fieldwork strategies. Moreover, during fieldwork, it is impossible to observe everything. Decisions must be made about what activities to observe, which people to observe and interview, and when to collect data. These decisions cannot all be made in advance. Emergent sampling takes advantage of whatever unfolds as it unfolds.

For example, in the course of conducting fieldwork, the researcher or evaluator may become aware of subgroups that were not known at the time of the original design. Indeed, in complex dynamic situations, subgroups may self-organize and emerge after fieldwork has begun. These newly discovered or emergent subgroups, after the study is under way, become new information-rich samples, especially where their emergence provides special insight into the phenomenon of interest. In evaluating a wilderness leadership program through participant observation, I had to reorient the design and data collection plan when, on the first day of hiking, the group of 20 participants was subdivided into a slow group, self-designated "the turtles," and a faster hiking group, dubbed "the truckers." The turtles and truckers stayed together through the 10 days of the program. They had different experiences and outcomes. What began as a single group sample became a comparison group sample.

Similarly, in *evaluating* an early-childhood parent education program, we had to change the design to understand the different experiences and outcomes for two subgroups that emerged naturally: women participants with and without spouses. These two groups formed and cohered together and led to an emergent comparison sample not anticipated in advance.

An important element of qualitative fieldwork flexibility is adjusting the sampling strategy in the field, as what you learn in fieldwork provides new options and insights into who to include in the sample and how to conceptualize the sample composition.

Opportunity Sampling

During fieldwork, the opportunity may arise to interview someone or observe an activity, neither of which could have been planned in advance. For example, during a site visit to a youth homeless shelter, a former staff member who helped open the shelter happens to have come for lunch and a visit with current staff friends. This is an opportunity to learn about the history of the shelter. This differs from emergent sampling in that something new has not emerged but an unanticipated opportunity presents itself.

In Chapter 2, I identified emergent flexible designs as one of the core strategic themes of qualitative inquiry and cited as an exemplar anthropologist Brackette Williams and her fieldwork on how Americans view violence in America:

> I do impromptu interviews. I don't have some target number of interviews in mind or predetermined questions. It depends on the person and the situation. Airports, for example, are a good place for impromptu interviews with people. So sometimes, instead of using airport time to write, I interview people about the death penalty or about killing or about death in their life. It's called *opportunity sampling*. . . . I'm following where the data take me, where my questions take me. (Personal interview)

Few qualitative studies are as fully emergent and open-ended as the fieldwork of Brackette Williams. Her approach exemplifies opportunity sampling.

Saturation or Redundancy Sampling

Saturation sampling is a purposeful strategy for dealing with the problem of small sample size. Lincoln

THE NATURE OF THE CASE EMERGES THROUGH FIELDWORK AND ANALYSIS

Distinguished qualitative sociologist Howard Becker invites researchers to be open to the emergence of the very nature of what they are studying. He cautioned against beginning research with a fixed notion of what the case is.

> Strong preconceptions are likely to hamper conceptual development. Researchers probably will not know what their cases are until the research, including the task of writing up the results, is virtually completed. What *it* is a case *of* will coalesce gradually, sometimes catalytically, and the final realization of the case's nature may be the most important part of the interaction between ideas and evidence.

> In short, Becker wanted to make researchers continually ask the question "What is this a case of?" The less sure that researchers are of their answers, the better their research may be. From this perspective, no definitive answer to the question "What is a case?" can or should be given, especially not at the outset, because *it depends*. The question should be asked again and again, and researchers should treat any answer to the question as tentative and specific to the evidence and issues at hand. Working through the relation of ideas to evidence answers the question "What is this a case of?" (Ragin, 1992, p. 6)

and Guba (1985) recommended sample selection to the point of redundancy.

> In purposeful sampling the size of the sample is determined by informational considerations. If the purpose is to maximize information, the sampling is terminated when no new information is forthcoming from new sampled units; thus *redundancy* is the primary criterion. (p. 202)

This strategy leaves the question of sample size open, another example of the emergent nature of qualitative inquiry. There remains, however, the practical problem of how to negotiate an evaluation budget or get a dissertation committee to approve a design if you don't have some idea of sample size. Sampling to the point of redundancy is an ideal, one that works best for basic research, unlimited timelines, and unconstrained resources.

To know if informational redundancy or saturation is reached implies and is founded on the assumption that data collection and analysis are going hand in hand. In other words, data is collected and analyzed, at least in a preliminary fashion, and this analysis informs subsequent data collection decisions.

It is important to keep in mind that saturation or informational redundancy can be reached prematurely if:

- one's sampling frame is too narrow

- one's analytical perspective is skewed or limited

- the method employed is not resulting in rich, in depth information

- the researcher is unable to get beyond the surface or "status quo" with respondents. (Robert Wood Johnson Foundation, 2008)

Another caution in judging saturation is whether early interviewees may be similar in ways that generate common responses but would be different from what a more diverse sample would yield. A medical researcher interviewed 30 physicians, and the early interviewing was dominated by older primary care physicians—saturation was reached at about five interviews, he concluded. But as he interviewed younger primary care physicians, along with pediatricians, orthopedic surgeons, neurologists, plastic surgeons, and emergency room physicians, it became clear that the physicians' responses varied in important ways by clinical context, specific role, and even personality. It

would have been a mistake to think that saturation had occurred with the first five interviews. Moreover, he found that saturation was achieved quickly with respect to some content areas (e.g., none of them liked insurance paperwork) but not others (e.g., attitudes about health care reform).

Finally, and most important, one must decide whether the areas of saturation achieved meet the purpose of the inquiry.

Summary of Sequential and Emergence-Driven Sampling

What the four sampling strategies just reviewed have in common is that they involve building the sample during fieldwork as the inquiry unfolds. This means that neither the sample size nor the nature of the sample will be fully known in advance. For these purposeful sampling approaches, the sample emerges as the inquiry deepens (numbers correspond to those in the sampling summary in Exhibit 5.8 pp. 270–271).

30. Snowball or chain sampling

31. Respondent-driven sampling

32. Emergent phenomenon or emergent subgroup sampling

33. Opportunity sampling

34. Saturation or redundancy sampling

Standard models of research describe distinct stages that proceed one after another in logical sequence: problem definition or question formulation, design, data collection, analysis, and reporting. The number of steps and labels for each step vary, but the image is one of finishing one stage before moving on to the next. That's largely true for quantitative studies. It's not necessarily true at all for qualitative inquiry (though dissertation students may have to pretend it is). Many qualitative research and evaluation studies are highly structured and fully planned in advance to implement a carefully prescribed design approved by those funding the study or by a supervising authority like a doctoral committee or institutional review board (IRB). In such cases, the sample is determined before fieldwork begins, and the sample as designed is the sample that is studied, provided recruitment goes as planned (which is not always the case).

But some qualitative studies are highly emergent, as we've just seen in reviewing sequential and emergence-driven sampling. Moreover, in more open-ended and emergent inquiries, some analysis is going on as fieldwork is conducted. That initial analysis can be deepened through analytically focused sampling. These purposeful sampling strategies are aimed at adding information-rich cases that answer questions that have emerged during fieldwork. These sampling additions are essential for bringing closure to the fieldwork.

Confirming and Disconfirming Cases

In the early part of open-ended qualitative fieldwork, the inquirer is exploring—gathering data and watching for patterns to emerge. Over time, the exploratory process gives way to confirmatory fieldwork. This involves testing ideas, confirming the importance and meaning of possible patterns, and checking out the viability of emergent findings with new data and additional cases. This stage of fieldwork requires considerable rigor and integrity on the part of the inquirer in looking for and sampling confirming as well as *disconfirming* cases. Howard Becker (1998) advises, "The trick, then, is to *identify the case that is likely to upset your thinking and look for it*" (p. 87).

Confirmatory cases are additional examples that fit already emergent patterns; these cases confirm and elaborate the findings, adding richness, depth, and credibility. Disconfirming cases are no less important

at this point. These are the examples that don't fit. They are a source of rival interpretations as well as a way of placing boundaries around confirmed findings. They may be "exceptions that prove the rule" or exceptions that disconfirm and alter what appeared to be primary patterns.

In a principles-focused study of homeless youth, confirming cases were those where interactions were clearly guided by a principle, like harm reduction or building trust. Disconfirming cases were those where the basis for the interactions appeared ambiguous or ad hoc, as when one staff person reported,

I was just making up what to do as we went along. It's not like I stopped and tried to figure out some principle to apply. I had to act in real time, in the moment. It was more instinct than like following what somebody thinks is—what do they call it?—"best practice." In a crisis, it's all instinct. And you hope your instincts are good. Working with kids on the streets is totally in the moment, reacting from instinct and experience.

While this appeared to be contrary to a principles-based approach, the subsequent interpretation of this example with staff who serve homeless youth led to in-depth reflection about the basis of "instinctual" action in real time. Principles-based interactions that are talked about and practiced over and over can become "instinctual" at some level; or at least that was the shared conclusion of those engaged in making sense of the case studies. The discussion also led to gathering more examples for further study and reflection.

The source of questions or ideas to be confirmed or disconfirmed may be from stakeholders or previous scholarly literature, in addition to the researcher's fieldwork.

Thinking about the challenge of finding confirming and disconfirming cases emphasizes the relationship between sampling and research conclusions. As I've reiterated throughout this discussion on sampling, *the sample determines what the inquirer will have something to say about—thus the importance of sampling carefully and thoughtfully.*

Illumination and Elaboration Additions to the Original Sample

Another analytically driven strategy is adding cases to more fully illuminate or elaborate emergent findings.

Case studies of homeless youth ($n = 14$) turned up evidence that gay, lesbian, and transgender youth faced special problems. Three such cases were added to the sample, when that pattern emerged, to strengthen a maximum homogeneity sample and to further diversify the findings.

Qualitative Research Synthesis

Meta-analysis is a method of aggregating findings from a number of different quantitative studies to establish the reliability and validity of generalizable patterns across the findings.

> Meta-analysis can be understood as a form of survey research in which research reports, rather than people, are surveyed. . . . It applies only to studies that produce quantitative findings. . . . This rules out qualitative forms of research like case studies, ethnographies, and "naturalistic" inquiry. (Lipsey & Wilson, 2001, pp. 1–2)

Methodologists and statisticians engage in lively debate about what criteria of rigor must be met for a quantitative study to be included in a meta-analysis (Gough et al., 2012).

Qualitative research synthesis, in contrast, involves seeking patterns across and integrating different qualitative studies (Finlayson & Dixon, 2008; Saini & Shlonsky, 2012). Any qualitative synthesis immediately faces the challenge of sampling: locating studies and deciding which are of sufficient quality and relevance to include in the synthesis. The purposeful sampling strategies we've already reviewed can be used in selecting cases for qualitative research synthesis (Brunton, Stansfield, & Thomas, 2012, p. 114). To include a qualitative research study in a synthesis requires determining what quality criteria must be met and how those criteria will be evaluated. Earlier, I presented and discussed *systematic qualitative evaluation reviews* as a type of instrumental-use multicase sampling. Systematic evaluation reviews are aimed at informing practice and decision making; the strategy also faces the qualitative synthesis sampling challenges but with the narrower focus on targeting evaluation studies that are relevant and actionable. (Chapter 9 will examine different frameworks and criteria for judging qualitative quality and credibility.)

This problem of what to include in the final qualitative synthesis even arises in a specific inquiry when, as is not unusual, case data vary in quality. Some interviewees may be more forthcoming than others. Some interviews may get cut short, interruptions may occur, or some unanticipated person may be present who undercuts the authenticity of the responses. Deciding whether to include such weaker cases in the final analysis is, in part, a qualitative synthesis sampling problem because it involves including cases in the final cross-case analysis and synthesis on the basis of explicit quality criteria. Here are eight diverse examples of qualitative research syntheses and systematic evaluation reviews for each of which the researchers had to decide what case studies to include in the synthesis.

- Qualitative findings from multiple studies on how to support breastfeeding mothers (McInnes & Chambers, 2008)
- Using qualitative research synthesis to build an actionable knowledge base (Denyer & Tranfield, 2006)
- Effective interventions to reduce hepatitis C risk among injecting drug users (Rhodes & Treloar, 2008)
- Review of 66 peer-reviewed research studies published between 1999 and 2010 that empirically evaluated the outcomes of environmental education programs for youth (ages 18 years and younger) to determine what works (or does not) and to uncover lessons for future initiatives and their evaluation (Stern, Powell, & Hill, 2013)
- Identifying factors that affect adherence to tuberculosis treatment (Stewart & Oliver, 2012, p. 231)
- Lessons learned and insights gained from evaluation across 61 Clinical and Translational Science Institutes (funded by the National Institutes of Health) with the aim of reducing translation time from a bench discovery to when the application of knowledge affects patients (Pincus, Abedin, Blank, & Mazmanian, 2013)
- Synthesis of 133 documents on evaluability assessments, approximately half of them by international development agencies (Davies, 2013)
- Realist synthesis of primary research on healthcare systems (Rogers, 2014)

What studies to include in a qualitative research synthesis involves both sampling and analysis since those cases in the pool of potential studies must be analyzed to determine whether they meet the criteria for inclusion. Moreover, which studies are included in the final synthesis will depend on what emerges as relevant and meaningful during the synthesis. "Assessing quality is also about examining how study findings fit (or do not fit) with the findings of other studies. How study findings fit with the findings of other studies cannot be assessed until the synthesis is completed" (Harden & Gough, 2012, p. 160). In essence, quality criteria in a qualitative research synthesis can be emergent. In being open to and engaging that emergence, the qualitative synthesizer exercises creativity, judgment, and even artistry: "combinations of scientific and artistic approaches [as] the synthesizer 'puzzles together' an interpretive account of qualitative studies" (Kinn, Holgersen, Ekeland, & Davidson, 2013, p. 1258).

Politically Sensitive Sampling

This final analytically driven strategy involves a different kind of analysis: political analysis. Evaluation is inherently and inevitably political. A variation on the critical case sampling strategy involves selecting (or sometimes avoiding) a politically sensitive site or unit of analysis. For example, a statewide program may have a local site in the district of a state legislator who is particularly influential. By studying carefully the program in that district, evaluation data may be more likely to attract attention and be used. This does not mean that the evaluator then undertakes to make that site look either good or bad depending on the politics of the moment. That would clearly be unethical. Rather, sampling politically important cases is simply a strategy for trying to increase the usefulness and relevance of information where resources permit the study of only a limited number of cases.

The same political perspective (broadly speaking) may inform case sampling in applied or even basic research studies. A political scientist or historian might select the election year 2000 Florida vote-counting case, the Clinton impeachment effort, Nixon's Watergate crisis, or Reagan's Iran-Contra scandal for study, not only because of the insights they provide about the American system of government but also because of the likely attention such a study would attract. A sociologist's study of a riot or a psychologist's study of a famous suicide would likely involve some attention during sampling to the public and political importance of the case. Such political calculations may enter into initial case selection, but sometimes, the political importance of cases to include (or avoid) only becomes apparent during fieldwork. Thus, sampling politically important cases, interviews, observation sites, or documents may be an emergent sampling strategy. In any event, the analysis and interpretation of politically sensitive cases will have an explicitly political overlay from beginning to end, and the political attention garnered may or may not work out as intended. For, as political satirist and science fiction author George Orwell observed, "In our age there is no such thing as 'keeping out of politics.' All issues are political issues, and politics itself is a mass of lies, evasions, folly, hatred and schizophrenia."

Summary of Analytically Focused Sampling

Saturation or redundancy sampling could have been included in this group. I chose to categorize it as an emergence-driven strategy, but it shares the defining characteristic that case selection and analysis can and do go on simultaneously in naturalistic, open-ended inquiry, as is true for these four purposeful strategies (numbers correspond to those in the summary in Exhibit 5.8, pp. 271–272):

35. Confirming and disconfirming cases

36. Illumination and elaboration additions to the original sample

37. Qualitative research synthesis

38. Politically sensitive sampling

Mixed, Stratified, and Nested Purposeful Sampling Strategies

Stratified or nested samples are samples within samples, a strategy within a strategy. Mixed strategies combine approaches. These combinations can serve to meet multiple inquiry interests and needs. They can deepen and narrow the focus of inquiry, like a funnel that channels the flow of a liquid more precisely, to increase relevance and credibility.

Stratified or Nested Samples

Purposeful samples can be stratified or nested by combining types of purposeful sampling. For example, you might combine typical case sampling with maximum heterogeneity sampling by taking *a stratified purposeful sample* of above-average, average, and below-average cases. This represents less than a full maximum variation sample but more than simple typical case sampling. The purpose of this nested purposeful sample is to capture major variations rather than to identify a common core, although the latter may also emerge in the analysis. Each of the strata would constitute a fairly homogeneous sample. This strategy differs from stratified random sampling in that the sample sizes are likely to be too small for generalization or statistical representativeness. Exhibit 5.13 illustrates an example of this stratifying or funneling process.

An outlier sample or maximum heterogeneity approach may yield an initial potential sample size that is still larger than the study can handle. The final selection, then, may be made randomly—a combination approach. Thus, purposeful strategies are not mutually exclusive. Each approach serves a somewhat different purpose. Because research and evaluations

---

EXHIBIT 5.13 **Example of Nesting Sampling Strategies**

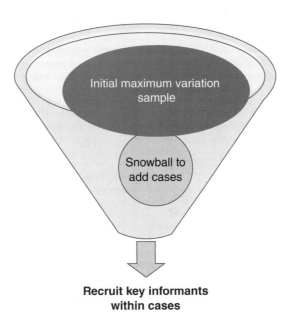

1. Begin with a maximum variation sampling based on known variations in the target population of interest, let's say early-childhood Head Start sites in Minnesota, to document program diversity and analyze core themes.

2. Once in the field, add sites to the sample through snowball sampling, seeking additional diversity.

3. At the site level, identify and recruit key informants to provide in-depth understanding of the important characteristics and attributes of each site.

often serve multiple purposes, more than one qualitative sampling strategy may be necessary. In long-term fieldwork, several sampling strategies may be used at some point.

Mixing Probability and Purposeful Sampling

Mixed-methods designs can include combining probability and purposeful sampling strategies. Indeed, mixed sampling is at the core of mixed methods. The very first issue of the *Journal of Mixed Methods Research* featured a typology of mixed-methods sampling (Teddlie & Yu, 2007). Here are five examples of such mixed-sampling strategies:

1. *Stratified mixed methods:* Begin with a statistical distribution to stratify for purposeful sampling; for example, identify outliers, typical cases, or subgroups of interest. Exhibit 5.13 illustrates this combination. In selecting companies to compare using in-depth case studies, Collins (2001a) began with extensive statistical analysis of the financial performance of companies within sectors to select cases that his team could study to explain the difference between "great" and "good" companies.

2. *Sequential mixed methods:* Select cases from a probability sample for greater in-depth inquiry to illuminate and validate what the numbers mean. A random, representative sample of people with disabilities was surveyed to determine their priority needs. A small number of respondents in different categories of need were then interviewed to understand in depth their situations and priorities and to make the statistical findings more personal through stories. The qualitative sample was also used to validate the accuracy and meaningfulness of the survey responses.

3. *Parallel mixed methods:* Simultaneously conduct a survey using a probability sample for representativeness and generalizability, and at the same time conduct in-depth case studies purposefully chosen to provide depth of interpretation to elucidate what the survey results mean.

4. *Triangulated mixed methods:* Compare probability and purposeful samples, studied independently, to triangulate and examine consistency of findings with different methods and sampling strategies. An evaluation of an early-childhood home visitation program used both quantitative (pre–post measures) and qualitative (case studies) methods. The findings provided quite different—and conflicting—perspectives on the program, which supported the need for further inquiry (Sherwood, 2005).

5. *Validity-focused mixed methods:* Do observations and interviews to determine the validity of select statistical data. For example, have the procedures for gathering statistical data been rigorously followed? In 1851, French political theorist Pierre-Joseph Proudhon observed that "to be governed is to be noted, registered, enumerated, accounted for, stamped, measured, classified audited, patented, licensed, authorized, endorsed, reprimanded, prevented, reformed, rectified, and corrected in every operation, every transaction, every movement" (quoted by Schulz, 2014, p. 34). A hundred and sixty–plus years later, the ways in which we are enumerated, measured, and classified have expanded geometrically. One form of qualitative inquiry is to investigate where the numbers come from, how they are entered, and what they mean.

Consider death certificates in the United States. Death certificates list cause of death and are a critically important source of statistics depicting mortality and disease trends. But studies of how death certificates are completed have revealed substantial errors. Medical residents are often assigned to fill out death certificates in hospitals without adequate (or sometimes any) training in how to do so; they often rely on secondhand or thirdhand reports of cause of death, and they run into a variety of administrative and procedural problems in completing death certificates. A 2010 survey of 521 doctors in 38 residency programs across New York City found that "only a third believed death certificates to be accurate. Nearly half reported knowingly listing an inaccurate cause of death, and that number rose to sixty percent among residents with the most experience" (Schulz, 2014, p. 36). Qualitative fieldwork and interviews are necessary to get at the factors that lead to significant inaccuracies in this important data source. Chapter 4 (pp. 231–232) discussed conducting fieldwork to do such validity checks. I gave the example of entering guesses on a survey of subsistence farmers' yields when I was working in agricultural extension in Burkina Faso in the 1960s, because we had no way of actually measuring yields; fieldwork on the data-gathering procedures would have revealed the numbers to be highly problematic.

Exhibit 5.11 (pp. 286–287) provides an example of combining probability and purposeful sampling, beginning with a population survey as the basis for outlier sampling (success case method) and then adding diversity criteria (urban–rural mix and size of program). This mixed-sampling approach increased the study's validity and credibility significantly. Such mixed-methods sampling strategies have become more widely used as researchers and evaluators have gained experience and knowledge about how to combine approaches and as mixed-methods designs have become more common, a trend likely to accelerate with the historic launch in 2014 of the Mixed Methods International Research Association, "a momentous development in mixed methods research" (Mertens, 2014, p. 3). Especially common in mixed-methods sampling is the sequence of (a) a small purposeful sample to explore issues and generate hypotheses, followed by (b) a probability sample to answer questions of representativeness in a population of interest, followed by (c) a new purposeful sample to enhance interpretation of the quantitative (probability sample) findings; this sequence was depicted in Exhibit 2.3 in Chapter 2 (p. 65). Another common mixed-methods sequence involves probability sampling first, for breadth of coverage, and then purposeful sampling as follow-up to add depth and enhance understanding and interpretation (Kemper, Stringfield, & Teddlie, 2003, pp. 284–285).

Summary of Mixed, Stratified, and Nested Sampling Strategies

Sampling is a means to an end. The end, or purpose, is generating knowledge and deepening understanding. In service of those purposes, we combine strategies and use multiple and mixed methods as much as is possible and appropriate. We've concluded this extensive and comprehensive discussion of purposeful sampling strategies with two flexible and integrating approaches (numbers correspond to those in the summary in Exhibit 5.8, p. 272):

39. Combined or stratified purposeful sampling strategies

40. Mixed probability and purposeful samples

> Sampling decisions are inherently practical. . . . It is in sampling, perhaps more than anywhere else in research, that theory meets the hard realities of time and resources
>
> —Kemper et al. (2003, p. 273)

Exhibit 5.8 on page 266 provided an overview of the 40 purposeful sampling strategies we've now discussed. The underlying principle that is common to all these strategies is selecting *information-rich cases*—cases from which one can learn a great deal about the focus of inquiry and which therefore are worthy of in-depth study.

Maxwell (2012) provides an illuminative example of the considerations that go into identifying an information-rich and accessible sample. A graduate student proposed to study classroom discourse norms in a college department. She could only do in-depth interviews with a small sample of students, so she needed criteria for selecting interviewees. Her dissertation committee recommended that she interview sophomores and seniors to get diverse experiences and perspectives. When she proposed this to the department where the study would occur, faculty in the department told her that

sophomores were too new to the department to fully understand the norms of discourse, while seniors were too deeply involved in their theses and in planning for graduation to be good informants. Juniors turned out to be the choice that would best meet the criteria of having the desired information and being most likely to provide this in interviews. (p. 96)

As this example nicely illustrates, reasons for site selections or individual case sampling need to be thoughtfully deliberated, carefully articulated, and made explicit in the methods report. Credibility concerns will have to be taken into consideration. In the process of developing the research design, the evaluator or researcher is trying to consider and anticipate the kinds of arguments that will lend credibility to the study as well as the kinds of arguments that might be used to attack the findings. Moreover, it is important to be open and clear about a study's limitations, that is, to anticipate and address criticisms that may be made of a particular sampling strategy, especially by people who think that the only high-quality samples are random ones.

Having weighed the evidence and considered the alternatives, evaluators and primary stakeholders make their sampling decisions, sometimes painfully but always with the recognition that there are no perfect designs. The sampling strategy must be selected to fit the purpose of the study, the resources available, the questions being asked, and the constraints being faced. This holds true for sampling strategy as well as sample size.

MQP Rumination # 5

Convenience Sampling Is *Not* Purposeful Sampling

I am offering one personal rumination per chapter. These are issues that have persistently engaged, sometimes annoyed, occasionally haunted, and often amused me over more than 40 years of research and evaluation practice. Here's where I state my case on the issue and make my peace.

Convenience sampling is "defined as a sample in which research participants are selected based on their ease of availability" (Saumure & Given, 2008, p. 124). This means interviewing whoever happens to be at a place during a site visit, stopping people on the street and asking them a few questions, or studying a village because it is near the main road and easy to get to.

In the previous edition of this book, I wrote,

> Sampling by convenience: doing what's fast and convenient. This is probably the most common sampling strategy—and the least desirable. Too often evaluators using qualitative methods think that, because the sample size they can study will be too small to permit generalizations, it doesn't matter how cases are picked, so they might as well pick ones that are easy to access and inexpensive to study. *While convenience and cost are real considerations, they should be the last factors to be taken into account* after strategically deliberating on how to get the most information of greatest utility from the limited number of cases to be sampled. Purposeful, strategic sampling can yield crucial information about critical cases. *Convenience sampling is neither purposeful nor strategic.* (Patton, 2002, pp. 241–242)

Having thus denigrated convenience sampling, or so I thought, I still made the mistake of including it in the summary table of sampling strategies as Item 15, where I wrote, "Do what's easy to save time, money and effort. Poorest rationale; lowest credibility. Yields information-poor cases" (Patton, 2002, p. 244). Why was this a mistake? I have reviewed a number of research publications, evaluation proposals, and qualitative reports that state in the methods section that they have used "Patton's purposeful sampling approach # 15." My bad. Sigh!

So let me emphatically reiterate, *convenience sampling is neither strategic nor purposeful.* It is lazy and largely useless. And it is omitted from the summary table of purposeful sampling strategies in this book, Exhibit 5.8 (p. 266).

Five Problems With Convenience Sampling

1. ***Information-poor:*** Yin (2011) has succinctly stated the problem with convenience sampling: "It is likely to produce an unknown degree of incompleteness because the most readily available sources of data are not likely to be the most informative sources. Similarly, convenience samples are likely to produce an unwanted degree of bias" (p. 88).

2. ***Dangerous:*** Maxwell (2012) cautions that to make convenience the primary or sole criterion for sampling decisions is "dangerous" both because it diminishes or ignores the primary sampling criterion of being purposeful about finding the best information and understandings you seek "and because it exposes your conclusions to serious validity threats." Unfortunately, he goes on to temper this warning by emphasizing that

> the realities of access, cost, time, and difficulty necessarily influence *every* decision about what settings and participants to include in a study, and to dismiss these considerations as "unrigorous" is to ignore the real conditions that will influence how data can be collected and the ability of these data to answer your research questions. (p. 95)

I would have preferred that he had ended with the conclusion that convenience sampling is "dangerous" and not offered practicality as a loophole and cop-out. The critical distinction, however, is that convenience should be, at best, a secondary or tertiary practical consideration and never the sole or primary criterion.

3. ***Limited utility:*** Morse (2010) also gets entangled in the convenience sample quagmire by arguing that convenience sampling allows researchers to go to accessible places where they are likely to see the social interactions that are the focus of their inquiry *as a way of getting started.* In effect, she is advocating convenience sampling as an open-ended, inductive way of early sampling to begin exploring the phenomenon of interest. Subsequent sampling will be purposefully theoretical to generate and deepen grounded theory. There are two points to be noted. (1) In this framing, convenience sampling does not stand alone as a purposeful strategy, but is simply an easy way to enter the field; inductive theoretical sampling is her real strategy. (2) I would much prefer to designate this approach as *exploratory sampling* or *entry-into-the-field sampling* or *dipping-your-toe-in-the-inquiry-water*

(Continued)

(Continued)

sampling. But whatever the alternative labeling, she is not advocating basing the whole inquiry on finding easily accessible cases, which is what convenience sampling connotes.

4. ***Lazy, not opportunistic:*** Opportunity sampling is sometimes treated as the same as convenience sampling, and the terms are used interchangeably. Let me distinguish the two. Opportunity sampling is one tactic in a larger strategy of in-depth fieldwork. As I noted in defining and discussing opportunity sampling (p. 300), during fieldwork an opportunity may arise to interview someone or observe an activity, neither of which could have been planned in advance. For example, during an evaluation of an after-school youth drop-in center, a college student who used to participate in the center's programs dropped by during a semester break to reconnect with former staff members. This offered an opportunity to learn about the history of the center and to see how a former participant viewed her experience. This differs from convenience sampling in that an unanticipated opportunity presents itself and is worth taking advantage of. But it is not the whole sampling strategy or even a major part of it.

5. ***Low credibility, easy to attack:*** Finally, those who are looking to attack the credibility of small qualitative samples love to highlight the worthlessness and laziness of convenience samples. Let's stop giving them ammunition. Stop doing convenience sampling, and stop treating it as a viable purposeful option. As I have shouted in titling this rumination: Convenience sampling is *not* purposeful sampling. *Make your qualitative sampling strategic and purposeful.* That's the criterion of qualitative excellence.

Following the success of our convenience sample, we mixed together old readily available questions to create a convenience interview protocol.

Now, thinking back to your third trimester...

freshspectrum.com

Sample Size for Qualitative Designs

Qualitative inquiry is rife with ambiguities. There are purposeful strategies instead of methodological rules. There are inquiry approaches instead of statistical formulas. Qualitative inquiry seems to work best for people with a high tolerance for ambiguity. (And we're still only discussing design. It gets worse when we get to analysis.)

Nowhere is this ambiguity clearer than in the matter of sample size. I get letters. I get calls. I get e-mails.

"Is 10 a large enough sample to achieve maximum variation?"

"I started out to interview 20 people for two hours each, but I've lost 2 people. Is 18 large enough, or do I have to find 2 more?"

"I want to study just one organization, but interview 20 people in the organization. Is my sample size 1 or 20 or both?"

My universal, certain, and confident reply to these questions is this: "It depends."

There are no rules for sample size in qualitative inquiry. Sample size depends on what you want to know, the purpose of the inquiry, what's at stake, what will be useful, what will have credibility, and what can be done with the available time and resources.

Earlier in this chapter, I discussed the trade-offs between breadth and depth. With the same fixed resources and limited time, a researcher could study a specific set of experiences for a larger number of people (seeking breadth) or a more open range of experiences for a smaller number of people (seeking depth). In-depth information from a small number of people can be very valuable, especially if the cases are information-rich. Less depth from a larger number of people can be especially helpful in exploring a phenomenon and trying to document diversity or understand variation. I repeat, the size of the sample depends on what you want to find out, why you want to find it out, how the findings will be used, and what resources (including time) you have for the study.

Janice Morse (2000), former editor of *Qualitative Health Research*, has explained insightfully how the nature of the data being collected and the theoretical tradition within which the study is positioned (see Chapter 3) affect sample size. She begins by articulating a principle to guide trade-offs in depth versus breadth:

The quality of the data and the number of interviews per participant determine the amount of useable data obtained. There is an inverse relationship between the amount of useable data obtained from each participant and the number of participants. The greater the amount of useable data obtained from each person (as number of interviews and so forth), the fewer the number of participants.

This principle links the number of participants with the research method used. If, when using semistructured interviews, one obtains a small amount of data per interview question (i.e., relatively shallow data), then to obtain the richness of data required for qualitative analysis, one needs a large number of participants (at least 30 to 60). If, on the other hand, one is doing a phenomenological study and interviewing each person many times, one has a large amount of data for each participant and therefore needs fewer participants in the study (perhaps only 6 to 10). Grounded theory, with two to three unstructured interviews per person, may need 20 to 30 participants, adjusted according to the factors discussed above. (pp. 4–5)

To understand the credibility problem of small samples in qualitative inquiry, it's necessary to place these small samples in the context of probability sampling. A qualitative inquiry sample *only seems small* in comparison with the sample size needed for representativeness when the purpose is generalizing from a sample to the population of which it is a part. Suppose there are 100 people in a program to be evaluated. It would be necessary to randomly sample 80 of those people (80%) to make a generalization at the 95% confidence level. If there are 1,000 people, 278 people must be sampled (28%); and if there are 5,000 people in the population of interest, 357 must be sampled (7%) to achieve a 95% confidence level in the generalization of findings. (See Fitz-Gibbon & Morris, 1987, p. 163, for a table on determining a representative sample size from a given population.)

The logic of purposeful sampling, seeking information-rich cases, is quite different. The problem is, however, that the utility and credibility of small purposeful samples are often judged on the basis of the logic, purpose, and recommended sample sizes of probability sampling. Instead, purposeful samples should be judged according to the purpose and rationale of the study: Does the sampling strategy support the study's purpose? "Determining your final

TO SAMPLE OR TO SELECT CASES

(With apologies to William Shakespeare's *Hamlet*)

To sample or not to sample, but to select cases, that is the question:

Whether 'tis Nobler in the mind to suffer

The Slings and Arrows of outraged statisticians,

For whom sampling is only and ever will be

Random sampling to generalize to a Population,

Or to take Arms against the Purists,

And by opposing them: *to sample purposefully*.

Yea, to Dare use the word *sample*.

Or, to sample no more and by a phrase,

To say we end debate, and talk only of

Case selection, purposeful case selection,

Yea, selecting cases for a specified purpose,

But not to sample, foreswearing that word and the debates

That ensue, The Heart-ache, and the thousand Natural shocks

That Flesh is heir to? 'Tis a resolution

Devoutly to be wished. To sample, to select cases,

To get on with data collection; Aye, there's the rub,

For in whatever inquiry may come,

When we have shuffled off this language contention,

Must give us pause, and focus: There's the respect,

For the people interviewed and observed,

Not subjects, nor cases, nor a sample,

But people with stories and lives,

That makes Calamity of so long life:

For who would bear the Whips and Scorns of time,

The Oppressor's wrong, the *proud* man's Contumely,

The pangs of *ineffective* programs, the Policy implementation's delay,

The insolence of numbers only, devoid of storied life.

So call it a sample, or case selection,

It matters not to those whose worlds we seek to enter,

But make it purposeful, whatever it be called.

Serve that specified purpose with intention and forethought,

Fulfill that purpose with rigor and resolve,

To grunt and sweat under a worthy quest,

To discover what is not known,

To understand with depth and illumination,

The undiscovered Country, that Puzzles the will,

And makes us rather bear those ills we have,

For enterprises of great *pitch* and moment,

With this regard our debates turn *awry*,

And lose the name of Truth in lieu of Action.

Soft you now, into the fray of inquiry,

And name the thing you do, Sampling or case selection,

And make this choice the least of thy sins remembered.

---

sample size is a matter of intellectual judgment based on the logic of making meaningful comparisons, developing and testing your explanations" (Mason, 2010, p. 139).

The sample size, like all other aspects of qualitative inquiry, must be judged in context—the same principle that undergirds analysis and presentation of qualitative data. Random probability samples cannot accomplish what in-depth, purposeful samples accomplish, and vice versa.

Here is wise advice on sample size from eminent case study methodologist Robert Stake (2006):

The benefits of multicase study will be limited if fewer than, say, 4 cases are chosen, or more than 10. Two or three cases do not show enough of the interactivity between programs and their situations, whereas 15 or 30 cases provide more uniqueness of interactivity than the research team and readers can come to understand. But for good reason, many multicase studies have fewer than 4 or more than 15 cases. (p. 22)

Small samples that are *truly in-depth* have provided many of the most important breakthroughs in our understanding of the phenomenon under study. Piaget contributed a major breakthrough to our understanding of how children think by observing his own two children at length and in great depth. Freud established the field of psychoanalysis based originally on fewer than 10 client cases. Bandler and Grinder (1975) founded neurolinguistic programming by studying three renowned therapists: Milton Erickson, Fritz Perls, and Virginia Satir. Peters and Waterman (1982) formulated their widely followed eight principles for organizational excellence by studying 62 companies, a relatively small sample from the thousands of companies one might study. Sands (2000) did a fine dissertation studying a single school principal, describing the leadership of a female leader who entered a challenging school situation and brought about constructive change.

Clair Claiborne Park's (2001) single-case study of her daughter's autism reports 40 years of data on every stage of her development, language use, emotions, capacities, barriers, obsessions, communication patterns, emergent artistry, and challenges overcome and not overcome. Park and her husband made systematic observations throughout the years. Eminent medical anthropologist Oliver Sacks reviewed the data and determined in his preface to the book that more data are available on the woman in this extraordinary case study than on any other autistic human being who has ever lived. Here, then, is the epitome of $n = 1$ in-depth inquiry.

The validity, meaningfulness, and insights generated from qualitative inquiry have more to do with the information richness of the cases selected and the observational/analytical capabilities of the researcher than with sample size.

Large Qualitative Samples

Advances in qualitative software for data management and analysis have made larger sample sizes much more manageable and common. Increasingly, large-scale qualitative studies include samples of 60 to 100 (Mason, 2010). In *Far From the Tree*, Andrew Solomon (2012) reports interviewing more than 300 families in his inquiry into family experiences of deafness, dwarfism, autism, schizophrenia, disability, prodigies, transgender, crime, and children born of rape. Strong (1979) studied 1,120 pediatric consultations. GlobalGiving help nonprofits in 144 countries raise funds and communicate their impacts. In an effort to allow people in local communities to tell their own stories, they created

a process for collecting perspectives at the grass-roots level from program staff and participants. In four years, they collected more than 57,000 stories for analysis and have used them for a variety of purposes, from needs assessment to program evaluation and trend analysis (Maxson, 2014). Such large samples capture vignettes and anecdotes, so they are not in-depth case studies, but they can and have been used to detect overall patterns in recipients' experiences of development assistance.

Sample Size as Emergent and Flexible

As noted in the earlier sections on emergent sampling and quota sampling, the actual size of a sample can be flexible. In the beginning, when the initial design is formulated (and approved by others, e.g., funders or an IRB, if necessary), a desired or targeted sample size may be specified. That sample size can be a starting point or minimum, but it may not be the final number. The size and composition of the sample can be adjusted based on what is learned as fieldwork is conducted and the inquiry deepens. The emergent nature of qualitative inquiry applies especially powerfully to sample size. The sample can grow, or if saturation is achieved sooner than expected, the size can be reduced.

The final sample size might also involve a trade-off between greater depth versus more breadth. Suppose that the original design, given time and resources, called for conducting 20 two-hour interviews. Once the interviews began, however, it emerged that to do justice to the inquiry, interviews were taking four hours instead of two. This could mean reducing the sample size from the original 20 to 5 or 6. Or the opposite scenario might occur. The interviewees only needed 45 minutes to tell their stories instead of the planned two hours. This could mean increasing the sample size from 20 to 30.

Thus, the challenge of determining sample size becomes even more complicated when emergent strategies are used, like snowball sampling, RDS, and sampling to the point of redundancy. These purposeful strategies leave the question of sample size open, a prime example of the emergent nature of qualitative inquiry. There remains, however, the practical problem of how to negotiate an evaluation budget or get a dissertation committee to approve a design if you don't have some idea of sample size. Sampling to the point of redundancy is an ideal, one that works best for basic research with unlimited timelines and unconstrained resources.

Bottom line: *The strategic principle of design emergence rules* (see Chapter 2, Exhibit 2.1, p. 46).

This issue of sample size is a lot like the problem students have when they are assigned an essay to write.

Student: How long does the paper have to be?

Instructor: Long enough to cover the assignment.

Student: But how many pages?

Instructor: Enough pages to do justice to the subject—no more, no less.

PURPOSEFUL SAMPLING SIZE IN DISSERTATIONS

Mason (2010) identified and analyzed 560 qualitative dissertations. The smallest sample was a single participant used in a life history study. The largest sample was 95. The median size was 28 (and mean was 31). However, the most common sample sizes he found were 20 and 30 (followed by 40, 10, and 25).

The significantly high proportion of studies utilizing multiples of 10 as their sample is the most important finding from this analysis. There is no logical (or theory driven) reason why samples ending in any one integer would be any more prevalent than any other in qualitative PhD studies using interviews. If saturation is the guiding principle of qualitative studies, it is likely to be achieved at any point, and is certainly no more likely to be achieved with a sample ending in a zero than with any other number.

He interpreted the finding that the most common sample size was a multiple of 10 as constituting a "pre-meditated approach," in which the sample size is fixed in advance rather than emergent or saturation determined.

Practical Purposeful Sampling

The solution to determining a purposeful sample size is judgment and negotiation. I recommend that qualitative sampling designs specify *minimum samples* based on expected reasonable coverage of the phenomenon given the purpose of the study and stakeholder interests. You may add to the sample as fieldwork unfolds. You may change the sample if information emerges that indicates the value of a change. The design should be understood to be flexible and emergent. Yet, at the beginning, for planning and budgetary purposes, one specifies a minimum expected sample size and builds a rationale for that minimum, as well as criteria that would alert the researcher to inadequacies in the original sampling approach and/or size.

In the end, sample size adequacy, like all aspects of research, is subject to peer review, consensual validation, and judgment. What is crucial is that the sampling procedures and decisions be fully described, explained, and justified, so that information users and peer reviewers have the appropriate context for judging the sample. The researcher or evaluator is obligated to discuss how the sample affected the findings, the strengths and weaknesses of the sampling procedures, and any other design decisions that are relevant for interpreting and understanding the reported results. Exercising care not to overgeneralize from purposeful samples, while maximizing to the full the advantages of in-depth, purposeful sampling, will do much to alleviate concerns about small sample size.

Protection of Human Subjects, Sampling Issues, and Emergent Designs

Committees for the protection of human subjects (commonly known as IRBs in the United States) may not include members with expertise in qualitative methods. With or without such expertise, conflicts can arise between qualitative ideals and interpretation of procedures for protecting participants in research. For example, I was recently contacted about how to respond to a committee that rejected purposeful sampling and insisted on random sampling, even for a small sample, to comply with the Belmont Principle of Justice that every individual in a program have equal access to participate in the study. The committee required that each participating agency collect the names of everyone who met the study criteria and then draw people at random so as to be fair.

The design solution involved a combination of purposeful and random sampling. First, agency records were used to establish a possible maximum variation pool of potential participants.

Those meeting the criteria were contacted and asked if they were interested in and willing to participate in an interview (with a $100 compensation) laying out the nature of the study. All those meeting the purposeful sampling criteria who responded positively within two weeks were considered in the potential pool. The target number for the study was 20. The first issue, then, was whether 20 were eligible and interested. At that stage, it would be possible to assess whether random sampling was the way to go or whether additional recruitment would be necessary to meet the target.

Let's say 30 respond positively. You can then sample randomly, with stratification to ensure diversity, to get to 20. Then, you sample both randomly and purposefully (stratification) to get the final 12. If anyone doesn't complete the interview among the sampled 12, you go to the next person on the random stratified list. This procedure meets the Belmont Justice standard of fairness because it permits any potential participants in the program meeting the minimum criteria to be considered and placed in the potential research pool.

Emergent sampling designs also pose special problems for IRBs. Such boards typically want to know, in advance of fieldwork, who will be interviewed and the precise questions that will be asked. If the topic is fairly innocuous and the general line of questioning relatively unobtrusive, an IRB may be willing to approve the framework of an emergent design with sample questions included but without full sample specification and a formal interview instrument.

Another approach is to ask for approval in stages. This means initially asking for approval for the general framework of the inquiry and specifically for the first exploratory stage of fieldwork, including procedures for assuring confidentiality and informed consent, and then returning periodically (e.g., quarterly or annually) to update the design and its approval. This is cumbersome for both the researcher and the IRB, but it is a way of meeting IRB mandates and still implementing an emergent design. This staged-approval approach can also be used when the evaluator is developing the design jointly with program staff and/or participants and therefore cannot specify the full design at the beginning of the participatory process.

These are just a couple of examples of the challenges involved in ensuring that qualitative designs meet ethical standards while maintaining as much as possible the creative and emergent elements of open, naturalistic inquiry.

I invited you all to join our focus group because you were the easiest to track down and therefore should provide us with the best feedback.

freshspectrum.com

How NOT to sample

SOURCE: © Chris Lysy—freshspectrum.com

A study may employ more than one sampling strategy. It may also include multiple types of data. The chapters on interviewing, observation, and analysis will include information that will help in making design decisions. Before turning to those chapters, however, I want to discuss briefly the value of using multiple methods in research and evaluation.

Triangulation

> The method must follow the question. Campbell, many decades ago, promoted the concept of triangulation—that every method has its limitations, and multiple methods are usually needed.
>
> (Gene V. Glass eulogizing pioneering methodologist Donald T. Campbell, quoted in Tashakkori & Teddlie, 1998, p. 22)

Triangulation strengthens a study by combining methods. This can mean using several kinds of methods or data, including using both quantitative and qualitative approaches. Denzin (1978b) has identified four basic types of triangulation: (1) *data triangulation*—the use of a variety of data sources in a study, (2) *investigator triangulation*—the use of several different researchers or evaluators, (3) *theory triangulation*—the use of multiple perspectives to interpret a single set of data, and (4) *methodological triangulation*—the use of multiple methods to study a single problem or program.

The term *triangulation* is taken from land surveying. Knowing a single landmark only locates you somewhere along a line in a direction from the landmark, whereas with two landmarks (and your own position being the third point of the triangle), you can take bearings in two directions and locate yourself at their intersection. Triangulation also works metaphorically to call to mind the world's strongest geometric shape—the triangle (e.g., the form used to construct geodesic domes). The logic of triangulation is based on the premise that

no single method ever adequately solves the problem of rival causal factors. Because each method reveals

different aspects of empirical reality, multiple methods of observations must be employed. This is termed triangulation. I now offer as a final methodological rule the principle that multiple methods should be used in every investigation. (Denzin, 1978b, p. 28)

Triangulation is ideal. It can also be expensive. A study's limited budget and time frame will affect the amount of triangulation that is practical, as will political constraints (stakeholder values) in an evaluation. Certainly, one important strategy for inquiry is to employ multiple methods, measures, researchers, and perspectives—but to do so reasonably and practically.

Most good researchers prefer addressing their research questions with any methodological tool available, using the pragmatist credo of "what works." For most researchers committed to the thorough study of a research problem, method is secondary to the research question itself, and the underlying worldview hardy enters the picture, except in the most abstract sense. (Tashakkori & Teddlie, 1998, p. 22)

A rich variety of methodological combinations can be employed to illuminate an inquiry question. Some studies mix interviewing, observation, and document analysis. Others rely more on interviews than observations, and vice versa. Studies that use only one method are more vulnerable to errors linked to that particular method (e.g., loaded interview questions, biased or untrue responses), unlike studies that use multiple methods, in which different types of data provide cross-data validity checks. Using multiple methods allows inquiry into a research question with "an arsenal of methods that have non-overlapping weaknesses in addition to their complementary strengths" (Brewer

Triangulation

& Hunter, 1989, p. 17). Mixed methods strengthen the credibility of evidence in evaluation (Mertens & Hesse-Biber, 2013).

However, a common misunderstanding about triangulation is that the point is to demonstrate that different data sources or inquiry approaches yield essentially the same result. But the point is really to *test for* such consistency. Different kinds of data may yield somewhat different results because different types of inquiry are sensitive to different real-world nuances. Thus, understanding the inconsistencies in findings across different kinds of data can be illuminative. Finding such inconsistencies ought not be viewed as weakening the credibility of results but, rather, as offering opportunities for deeper insight into the relationship between inquiry approach and the phenomenon under study.

Triangulation within a qualitative inquiry strategy can be attained by combining both interviewing and observations, mixing different types of purposeful samples (e.g., both intensity and opportunity sampling), or examining how competing theoretical perspectives inform a particular analysis (e.g., the transcendental phenomenology of Husserl versus the hermeneutic phenomenology of Heidegger).

A study can also be designed to cut across inquiry approaches and achieve triangulation by combining qualitative and quantitative methods. Doing so well, in a truly integrated manner, involves, indeed may require, a "mixed methods way of thinking" (Greene, 2007; see sidebar). Mixed methods have become particularly important, even preferred, in program evaluation, including impact evaluation (Bamberger, 2013; Rogers, 2012).

Mixing Data, Design, and Analysis Approaches

Borrowing and combining distinct elements from pure or coherent methodological strategies can generate creative mixed inquiry strategies that illustrate variations on the theme of triangulation. We begin by distinguishing measurement, design, and analysis components of the hypothetico-deductive (quantitative/experimental) and holistic-inductive (qualitative/naturalistic) paradigms. The ideal-typical qualitative methods strategy consists of three core elements: (1) qualitative data, (2) a holistic-inductive design of naturalistic inquiry, and (3) content or case analysis. In the traditional hypothetico-deductive approach to research, the ideal study would include (a) quantitative data from (b) experimental (or quasi-experimental) designs and (c) statistical analysis.

MIXED-METHODS WAY OF THINKING

The core meaning of mixed methods social inquiry is to invite multiple mental models into the same inquiry space for purposes of respectful conversation, dialogue, and learning one from the other, toward a collective generation of better understanding of the phenomena being studied. By definition, then, mixed methods social inquiry involves a plurality of philosophical paradigms, theoretical assumptions, methodological traditions, data gathering and analysis techniques, and personalized understandings and value commitments—because these are the stuff of mental models. . . .

A mixed methods way of thinking involves an openness to multiple ways of seeing and hearing, multiple ways of making sense of the social world, and multiple standpoints on what is important and to be valued and cherished. A mixed methods way of thinking rests on assumptions that there are multiple legitimate approaches to social inquiry, that any given approach to social inquiry is inevitably partial, and that thereby multiple approaches can generate more complete and meaningful understanding of complex human phenomena. A mixed methods way of thinking means genuine acceptance of other ways of seeing and knowing as legitimate. A mixed methods way of thinking involves an active engagement with difference and diversity.

—Jennifer C. Greene (2007, p. xii)

Measurement, design, and analysis alternatives can be mixed to create eclectic designs, like customizing an architectural plan to tastefully integrate modern, postmodern, and traditional elements or preparing an elegant dinner with a French appetizer, a Chinese entrée, and an American dessert—not to everyone's taste, to be sure, but the possibilities are endless. At least, that's the concept. To make the idea of mixed elements more concrete and to illustrate the creative possibilities that can emerge out of a flexible approach to research, it will be helpful to examine alternative design possibilities for a single program evaluation. The examples that follow have been constructed under the artificial constraint that only one kind of measurement, design, and analysis could be used in each case. In practice, of course, the possible mixes are much more varied, because any given study could include several measurement approaches, varying design approaches, and different analytical approaches to achieve triangulation.

The Case of Operation Reach-Out: Variations in Program Evaluation Design

Let's consider design alternatives for a comprehensive program aimed at high school students at high risk educationally (poor grades, poor attendance, poor attitudes toward school), with highly vulnerable health (poor nutrition, sedentary lifestyle, high drug use), and who are likely candidates for delinquency (alienation from dominant societal values, running with a "bad" crowd, angry). The program consists of experiential education internships through which these high-risk students get individual tutoring in basic skills, part-time job placements that permit them to earn income while gaining work exposure, and an opportunity to participate in peer-group discussions aimed at changing health values, establishing a positive peer culture, and increasing social integration. Several evaluation approaches are possible.

Pure Hypothetical-Deductive Approach to Evaluation: Experimental Design, Quantitative Data, and Statistical Analysis

The program does not have sufficient resources to serve all targeted youth in the population. A pool of eligible youth is established, with admission into the program on a random basis and the remaining group receiving no immediate treatment intervention. Before the program begins and one year later, all the youth, both those in the program and those in the control group, are administered standardized instruments measuring school achievement, self-esteem, anomie, alienation, and locus of control. Rates of school attendance, illness, drug use, and delinquency are obtained for each group. When all data have been collected by the end of the year, comparisons between the control and experimental groups are made using inferential statistics.

Pure Qualitative Strategy: Naturalistic Inquiry, Qualitative Data, and Content Analysis

Procedures for recruiting and selecting participants for the program are determined entirely by the staff. The evaluator finds a convenient time to conduct an in-depth interview with new participants as soon as they are admitted into the program, asking students to describe what school is like for them, what they do in school, how they typically spend their time, what their family life is like, how they approach academic tasks, their views about health, and their behaviors and attitudes with regard to delinquent and criminal activity.

In brief, participants are asked to describe themselves and their social world. The evaluator observes the program activities, collecting detailed descriptive data about staff–participant interactions and conversations, staff intervention efforts, and youth reactions. The evaluator finds opportunities for additional in-depth interviews with the participants to find out how they view the program, what kinds of experiences they are having, and what they're doing. Near the end of the program, in-depth interviews are conducted with the participants to learn what behaviors have changed, how they view things, and what their expectations are for the future. Interviews are also conducted with program staff and some parents. These data are content analyzed to identify the patterns of experiences participants bring to the program, what patterns characterize their participation in the program, and what patterns of change are reported by and observed in the participants.

Mixed Form: Experimental Design, Qualitative Data, and Content Analysis

As in the pure experimental design, potential participants are randomly assigned to treatment and control groups. In-depth interviews are conducted with all the youth, both those in the treatment group and those in the control group, both before the program begins and again at the end of the program. Content and thematic analyses are performed so that the control and experimental group patterns can be compared and contrasted.

Mixed Methods: Experimental Design, Qualitative Data, and Statistical Analysis

Participants are randomly assigned to treatment and control groups, and in-depth interviews are conducted both before the program and at the end. These interview data, in raw form, are then given to a panel of judges, who rate each interview along several outcome dimensions operationalized as a 10-point scale. For both the "pre" interview and the "post" interview, the judges assign ratings on dimensions such as likelihood of success in school (low = 1, high = 10), likelihood of committing criminal offenses (low = 1, high = 10), commitment to education, commitment to engaging in productive work, self-esteem, and manifestation of desired nutritional and health habits. Inferential statistics are then used to compare these two groups. Judges make the ratings without knowledge of which participants were in which group. Outcomes on the rated scales are also statistically related to background characteristics of the participants.

Mixed Methods: Naturalistic Inquiry, Qualitative Data, and Statistical Analysis

As in the pure qualitative form, students are selected for the program on the basis of whatever criteria staff members choose to apply. In-depth interviews are conducted with all students before and at the end of the program. These data are then submitted to a panel of judges, who rate them on a series of dimensions similar to those listed in the previous example. Change scores are computed for each individual, and changes are statistically related to background characteristics of the students to determine in a regression format which characteristics are likely to predict success in the program. In addition, observations of program activities are rated on a set of scales developed to quantify the climate attributes of activities: for example, the extent to which the activity involved active or passive participation, the extent to which student–teacher interaction was high or low, the extent to which interactions were formal or informal, and the extent to which participants had input into program activities. Quantitative ratings of activities based on qualitative descriptions are then aggregated to provide an overview of the treatment environment of the program.

Mixed Methods: Naturalistic Inquiry, Quantitative Data, and Statistical Analysis

Students are selected for the program according to staff criteria. The evaluator enters the program setting without any predetermined categories of analysis or presuppositions about important variables or variable relationships. The evaluator observes important activities and events in the program, looking for the types of behaviors and interactions that will emerge. For each significant type of behavior or interaction observed, the evaluator creates a category and then uses a time and space sampling design to count the frequency with which those categories of behavior and interaction are exhibited. The frequency of the manifestation of the observed behaviors and interactions is then statistically related to characteristics such as group size, duration of the activity, staff–student ratios, and social/physical density.

Pure and Mixed Strategies

Exhibit 5.14 summarizes the six alternative design scenarios we've just reviewed for evaluation of "Operation Reach-Out." As these alternative designs illustrate, purity of approach is only one option. Inquiry strategies, measurement approaches, and analysis procedures can be mixed and matched in the search for relevant and useful information. That said, it is worth considering the case for maintaining the integrity and purity of qualitative and quantitative paradigms. The 12 themes of qualitative inquiry described in the second chapter (Exhibit 2.1) do fit together as a coherent strategy. The openness and personal involvement of naturalistic inquiry mesh well with the openness and depth of qualitative data. Genuine openness flows naturally from an inductive approach to analysis, particularly an analysis grounded in the immediacy of direct fieldwork and sensitized to the desirability of holistic understanding of unique human settings.

Likewise, there is an internal consistency and logic to experimental designs that test deductive hypotheses derived from theoretical premises. These premises identify the key variables to consider in testing theory or measuring, controlling, and analyzing hypothesized relationships between program treatments and outcomes. The rules and procedures of the quantitative/experimental paradigm are aimed at producing internally valid, reliable, replicable, and generalizable findings.

Guba and Lincoln (1988) have argued that the internal consistency and logic of each approach, or paradigm, mitigates against methodological mixing of different inquiry modes and data collection strategies. Their cautions are not to be dismissed lightly. Mixing parts of different approaches is a matter of philosophical and methodological controversy. Yet the practical mandate in evaluation (Patton, 1981, 2012a) to gather the most relevant possible information for evaluation users outweighs concerns about methodological purity based on epistemological and philosophical arguments. The intellectual mandate to be open to what the world has to offer surely includes methodological openness. In practice, it is altogether possible, as we have seen, to combine approaches and to do so creatively (Patton, 1987); just as machines that were originally created for separate functions like printing, faxing, scanning, and copying have now been combined into a single integrated technological unit, so too methods that were originally created as distinct, stand-alone approaches can now be combined into more sophisticated and multifunctional designs.

Advocates of methodological purity argue that a single evaluator cannot be both deductive and inductive at the same time, or cannot be testing predetermined hypotheses and still remain open to whatever emerges from open-ended, phenomenological observation. Yet, in practice, human reasoning is sufficiently complex and flexible that it is possible to research predetermined questions and test hypotheses about certain aspects of a program while being quite open and

EXHIBIT 5.14   Data Collection, Design, and Analysis Combinations: Pure and Mixed Design Strategies

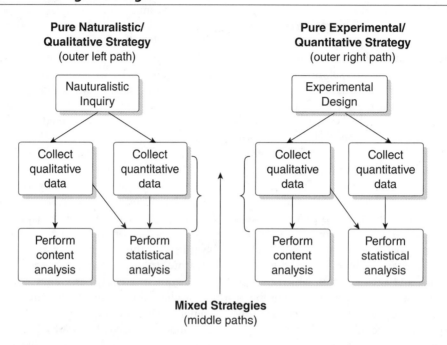

Pure Naturalistic/
Qualitative Strategy
(outer left path)

Pure Experimental/
Quantitative Strategy
(outer right path)

Mixed Strategies
(middle paths)

naturalistic in pursuing other aspects of the program. In principle, this is not greatly different from a questionnaire that includes both fixed-choice and open-ended questions. The extent to which a qualitative approach is inductive or deductive varies along a continuum. As evaluation fieldwork begins, the evaluator may be open to whatever emerges from the data—a discovery or inductive approach. Then, as the inquiry reveals patterns and major dimensions of interest, the evaluator will begin to focus on verifying and elucidating what appears to be emerging—a more deductively oriented approach to data collection and analysis.

The extent to which a study is naturalistic in design is also a matter of degree. This applies particularly to the extent to which the investigator places conceptual constraints on or makes presuppositions about the program or phenomenon under study. In practice, the naturalistic approach may often involve moving back and forth from inductive, open-ended encounters to more hypothetical-deductive attempts to verify hypotheses or solidify ideas that emerged from those more open-ended experiences, sometimes even manipulating something to see what happens.

These examples of variations in qualitative approaches are somewhat like the differences between experimental and quasi-experimental designs. Pure experiments are the ideal; quasi-experimental designs often represent

Sophisticated Emergent Design Strategy

what is possible and practical. Likewise, full participant observation over an extended period of time is the qualitative ideal. In practice, many acceptable and meaningful variations to qualitative inquiry can be designed.

This spirit of adaptability and creativity in designing studies is aimed at being pragmatic and responsive to real-world conditions and, when doing evaluations, at meeting stakeholder information needs. Mixed methods and strategies allow creative research adaptations to particular settings and questions, though certain designs pose constraints that exclude other possibilities. It is not possible, for example, to operate a program as an experiment by assigning participants to treatment and control groups while at the same time operating the program under naturalistic inquiry conditions, in which all eligible participants enter the program (and thus there is no control group and no random assignment). Another incompatibility: Qualitative descriptions can be converted into quantitative scales for purposes of statistical analysis, but it is not possible to work the other way around and convert purely quantitative measures into detailed, qualitative descriptions.

Qualitative Design Chapter Summary and Conclusion: Methods Choices and Decisions

> *Every path we take leads to fantasies about the paths not taken.*
>
> —Halcolm

This chapter opened with a case study example (the story of Daniel in Babylon) to set the context for qualitative design and methods decision making. I then presented a typology of different research purposes (basic research, applied research, summative evaluation, formative evaluation, and action research) to emphasize that, just as form follows function in architecture, design follows purpose in research and evaluation. Next came seven points of guidance for framing qualitative inquiry questions. After dealing with formulation of questions, we turned to design and data collection options, emphasizing critical trade-offs between depth and breadth. This led to consideration of different perspectives on what constitutes a case study, or even a case, and a variety of possible units of analysis. That set the stage for an extended presentation of purposeful sampling (case selection) strategies in the search for information-rich cases, and discussion of sample size. Finally, we examined a variety of mixed-methods designs based on different combinations of design, measurement, and analysis options. Exhibit 5.15 summarizes the issues discussed in this chapter that must be addressed in designing a study, which can also serve as an outline for a qualitative inquiry design proposal.

Which research design is best? Which strategy will provide the most useful information to decision makers? No simple and universal answer to these questions is possible. The answer in each case will depend on the purpose of the study; the scholarly or evaluation audience for the study (what intended users want to know); the funds available; the political, organizational, and cultural context; and the interests/abilities/perspectives of the researchers.

The word "design" is both a noun and a verb, denoting both the product and the process of our planning. Issues of context and culture are especially important considerations in both the product and process of designing an evaluation.

—Nick Smith (2013)
Evaluation scholar

In qualitative inquiry, the problem of design poses a paradox. The term *design* suggests a very specific blueprint, but "design in the naturalistic sense . . . means planning for certain broad contingencies without, however, indicating exactly what will be done in relation to each" (Lincoln & Guba, 1985, p. 226). Ideally, a qualitative design can remain sufficiently open and flexible to pursue whatever turns up during early interviewing and fieldwork. The degree of flexibility and openness is, however, a matter of considerable variation among designs.

What is certain is that different methods can produce quite different findings. The challenge is to figure out which design and methods are most appropriate, productive, and useful in a given situation. It's worth distinguishing arguments about which methods are most appropriate versus arguments about the intrinsic and universal superiority of one method over another.

Every cobbler thinks leather is the only thing. Most social scientists, including the present writer, have their favorite research methods with which they are familiar and have some skill in using. And I suspect we mostly choose to investigate problems that seem vulnerable to attack through these methods. But we should at least try to be less parochial than cobblers. Let us be done with the arguments of participant observation *versus* interviewing—as we have largely dispensed with the arguments for psychology *versus* sociology—and get on with the business of attacking our problems with the widest array of conceptual and methodological tools that we possess and they demand. This does not preclude discussion and debate regarding the relative usefulness of different methods for the study of specific problems or types of problems. But that is very different from the assertion of the general and inherent superiority of one method over another on the basis of some intrinsic qualities it presumably possesses. (Trow, 1970, p. 149)

Two overarching themes are at the core of this chapter and, indeed, this book. First, there are many design and data collection options—and varying perspectives about the relative strengths and weaknesses of each, though each has distinct strengths and weaknesses. Second, the variety of approaches gives you an opportunity—*and responsibility*—to design your study in support of your inquiry questions and within the context of your own field. In so doing, and in wrestling

EXHIBIT 5.15   Outline for a Qualitative Inquiry Design Proposal

DESIGN ISSUE TO ADDRESS	DESIGN OPTION EXAMPLES	RELEVANT EXHIBITS IN THIS BOOK
1. Primary overall purpose of the inquiry and nature of the contribution to be made	Basic research, applied research, summative evaluation, formative evaluation, action research	Exhibits 5.1 and 5.2: Typology of Research Purposes
2. Theoretical or philosophical tradition(s) within which the inquiry will be framed	Ethnography, grounded theory, hermeneutics, phenomenology, realism, systems, pragmatism	Exhibit 3.3: Alterative Inquiry Frameworks
3. Primary inquiry questions. Why are these questions important? Why are they worth answering?	Discipline-centered questions, interdisciplinary inquiry, evaluation questions	Exhibit 5.3: Asking Open-Ended Questions
4. Focus of inquiry	Broad vs. narrow, exploratory vs. confirmatory, single focus vs. multiple areas of inquiry	Exhibit 5.5: Design Trade-Offs
5. Time frame for the inquiry	One point in time vs. longitudinal	Exhibit 5.4: Contrasting Time Boundaries
6. Units of analysis	Individuals, families, programs, organizations, events, communities	Exhibit 5.6: Alternative and Diverse Units of Analysis
7. Purposeful sampling strategy	Outliers, typical cases, maximum variation, homogeneous, snowball, saturation, mixed methods	Exhibit 5.8: Comprehensive Menu of Purposeful Sampling Strategies; Exhibit 5.11: Nested Sampling
8. Types of data to be collected and data collection strategy	Interviews, observations, document analysis, mixed methods	Exhibit 1.1: Types of Qualitative Data; Exhibit 2.1: Themes of Qualitative Inquiry; Exhibit 5.12: Mixed Methods; Exhibit 5.13: Pure and Mixed Designs
9. Degree to which the design is open and emergent	Fixed design (predetermined sample and data collection) vs. design unfolding during fieldwork	Exhibit 5.5: Inquiry Openness; Exhibit 5.8: Fixed vs. Emergent Sampling Strategies
10. Analytical approach	Inductive/deductive; content, and thematic vs. statistical; combinations	Exhibit 5.13: Pure and Mixed Approaches to Analysis
11. Approach to ensuring credibility of findings	Establish quality criteria, nature and degree of triangulation	Exhibit 9.1: Analysis Strategies to Enhance Credibility and Utility
12. Ethical issues	Ensuring confidentiality, informed consent, protection of human subjects, reactivity	Exhibit 7.18: Ethical Issues Checklist
13. Practical issues	Time and resource constraints, logistics	Exhibit 4.1: Practical Principles to Get Actionable Answers
14. Strengths and weaknesses of the design	No perfect designs; there are always constraints and trade-offs	Exhibit 3.1: Paradigm Contrasts and Criticisms; Exhibit 5.5: Trade-Offs

with methodological options and competing perspectives, it is worth keeping in mind the admonition of Nobel Prize–winning physicist Percy Bridgman: "There is no scientific method as such, but the vital feature of the scientist's procedures has been merely to do his utmost with his mind, *no holds barred*" (quoted in Waller, 2004, p. 106).

But it all begins with what you choose to study. How do you decide that? Halcolm provides an answer in the cartoon that closes this chapter.

APPLICATION EXERCISES

1. *Distinguishing research purposes:* (a) Using Exhibits 5.1 and 5.2, identify an area of inquiry of interest to you. Generate inquiry questions to match the different kinds of research (as in Exhibit 5.2). (b) Discuss the connection between the five different *purposes of inquiry* (Exhibit 5.1) and *purposeful sampling.* How are these purposes connected conceptually and methodologically?

2. *Depth versus breadth trade-offs:* This chapter includes an extensive discussion of design trade-offs. Select a topic of interest to you, and illustrate the trade-offs between depth and breadth that you might face in designing a qualitative inquiry on that topic. Be as concrete and specific as possible about what a broad design would entail versus what a deep inquiry design would involve, with the same amount of money to support either one. Use Exhibit 5.5 to guide your analysis.

3. *Purposeful sampling:* Select a topic of interest to you. Among the 40 purposeful sampling strategies in Exhibit 5.8, choose 3 different strategies (in three different categories, A–H), and describe how you would use that purposeful sampling strategy to undertake your inquiry.

4. *Mixed-methods inquiry:* In your field of interest, find two studies that describe their approach as "mixed methods" or "multiple methods." Describe the studies, comparing and contrasting them. What did each mean by mixed methods? What was the nature of the "mix"? What are the strengths and weaknesses of the different kinds of data included in the mix?

6 Fieldwork Strategies and Observation Methods

The *intersection of lines of equal length* is an extremely old ideogram found in every part of the world, in prehistoric caves and engraved on rocks. In pre-Columbian America, the sign seems to have been associated with the *four points of the compass*. Semiotically, the vertical beam sometimes stands for the human arena whereas the transverse beam represents the physical world, so the intersection is where human culture and ecological context interact.

The cross with arms of equal length was common in the Neolithic Age. The so-called Ice Man, who died on the top of an Alpine pass at 3,000 meters about 3200 BCE, had this sign tattooed on one of his legs.

The *law of the polarity of meanings of elementary graphs* is well illustrated by ┼, which is both a *sign that unites*, as in logic, mathematics, and chemistry, and a *sign that separates*, as in numismatics and cartography; and is positive in many systems, but negative in the French hobo or gypsy system: *here they give nothing.* (Symbols, 2006)

The ┼ symbol also represents exploration, being open to going in any direction. Here, it is our symbol for fieldwork.

Enter Into the World; Observe and Wonder; Experience and Reflect

And the children said unto Halcolm, "We want to understand the world. Tell us, O Sage, what must we do to know the world?"

"Have you read the works of our great thinkers?"

"Yes, Master, every one of them as we were instructed."

"And have you practiced diligently your meditations so as to become One with the infinity of the universe?"

"We have, Master, with devotion and discipline."

"Have you studied the experiments, the surveys, and the mathematical models of the Sciences?"

"Beyond even the examinations, Master, we have studied in the innermost chambers where the experiments and surveys are analyzed, and where the mathematical models are developed and tested."

"Still you are not satisfied? You would know more?"

"Yes, Master. We want to understand the world."

"Then, my children, you must go out into the world. Live among the peoples of the world as they live. Learn their language. Participate in their rituals and routines. Taste the world. Smell it. Watch and listen. Touch and be touched. Write down what you see and hear, how they think and how you feel.

"Enter into the world. Observe and wonder. Experience and reflect. To understand a world you must become part of that world while at the same time remaining separate, a part of and apart from.

"Go, then, and return to tell me what you see and hear, what you learn, and what you come to understand."

—From Halcolm's *Methodological Chronicle*

Chapter Preview

You know that you've decided to work within theoretical tradition. You know what, if any, applied contribution you want to make. So your theory is informing your practice, and your practice is linked to your theory. You have a design. You know how and why you are purposefully sampling. You know where and from whom you will gather data. You are ready to enter the "field" to collect data—the "work" of qualitative inquiry—thus the designation *fieldwork*. When you undertake qualitative inquiry, you become a *fieldworker*. Add it to your resume. Announce your status to friends, family, lovers, and Facebook.

This chapter provides guidance for fieldwork and observation. The next chapter covers in-depth interviewing. Every interview is also an observation. Observation gives rise to interviewing opportunities. The two kinds of data collection involve different, distinct, but complementary skills. There's a lot to learn to become adept at either observation or interviewing, thus the length of each of these chapters.

Module 43 opens this chapter with reflections on and examples of the power of direct observation.

Module 44 presents an overview of variations in observational methods. Subsequent modules deal with specific issues of scope and method.

Module 45 Variations in Duration of Observations and Site Visits: From Rapid Reconnaissance to Longitudinal Studies Over Years

Module 46 Variations in Observational Focus and Summary of Dimensions Along Which Fieldwork Varies

Module 47 What to Observe: Sensitizing Concepts

Module 48 Integrating What to Observe With How to Observe

Module 49 Unobtrusive Observations and Indicators, and Documents and Archival Fieldwork

The chapter then moves into the nature and stages of fieldwork, beginning with observing yourself and documenting your personal and interpersonal experiences (**Module 50**) and the data-gathering process (**Module 51**). That leads to three stages:

Module 52 Stages of Fieldwork: Entry Into the Field

Module 53 Routinization of Fieldwork: The Dynamics of the Second Stage

Module 54 Bringing Fieldwork to a Close

The chapter concludes with reflections on the observer and what is observed (**Module 55**) and summary guidelines for conducting fieldwork (**Module 56**). Let's move into the field.

The Power of Direct Observation

The Power of Direct Observation

Some things must be observed directly. Bridges must be inspected. The sick must be examined. Births must be attended. Children on playgrounds must be watched. Melting ice caps must be seen and seen again and seen again. Weddings must be witnessed. Testimony at trial must be seen as well as heard. Astronomers observe the cosmos, biologists nature, and social scientists people. Our ancestors observed fire and figured out how to control it. Copernicus observed the movement of the planets and placed the sun at the center of the solar system. Newton observed falling objects and figured out gravity. Darwin observed species variation and theorized about evolution. *All scientific knowledge is rooted in observation.*

Timely and Important Contemporary Observations

Our knowledge of the modern world depends on trained, skilled, and dedicated observers paying attention to what's going on, systematically documenting what they see, and reporting what they learn for our common benefit—and scrutiny by other observers. Consider these diverse findings from observational studies and fieldwork.

Source of Escherichia coli outbreak found: German authorities concluded that contaminated sprouts from an organic farm in the country's north were the most likely cause of one of the world's worst outbreaks of *E. coli*. To reach their conclusion, health officials said they relied on an epidemiological study of the pattern of infection among patients, tracking the outbreak along the food chain from hospital beds to restaurants and back to the farm, southeast of Hamburg, at Bienenbüttel. The breakthrough came with the discovery of infected sprouts in a garbage can at the home of two infected patients in Cologne. At least 30 people in Germany died, "unsettling the nation" and throwing European agriculture into "disarray." "As the outbreak spread, German authorities blamed cucumbers, tomatoes or lettuce imported from Spain and urged people, particularly in northern Germany . . . to avoid the products." By observing the pattern of the infection as it spread, observers narrowed down the cause epidemiologically to sprouts. To do so, investigators examined 112 people, 19 of whom had been infected with *E. coli* during a group visit to a single restaurant, examined recipes for the food they had eaten, interviewed chefs, and examined photographs diners had taken of one another with their choices of food on the table (Cowell, 2011, p. A5).

Human testicles subtly but effectively regulate sperm temperature: Observations of temperature variation and control explain why one testicle is usually slightly lower than the other, why the skin of the scrotum sometimes becomes wrinkled, why the testicles retract during sexual arousal, and why blows to the testicles are so excruciatingly painful. Until recent systematic observations, these subtle temperature-regulating features had gone largely unnoticed by doctors and researchers, much less by ordinary people. The cremasteric muscle retracts the testicles to draw them up closer to the body in cold and lowers them when it gets warm. This up-and-down action occurs on a moment-to-moment basis so that the male body continually optimizes the gonadal climate for creating sperm, sperm storage, and, ultimately, ejaculation (Bering, 2012).

People with power tune out the less powerful: A growing body of recent research shows that people with the most social power pay scant attention to those with little such power. This tuning out has been observed, for instance, with strangers in a mere five-minute get-acquainted session, where the more powerful person shows fewer signals of paying attention, like nodding or laughing. Higher-status people are also more likely to express disregard, through facial expressions, and are more likely to take over the conversation and interrupt or look past the other speaker (Goleman, 2013, p. SR12).

Understanding toddler tantrums: Pediatric researchers dressed toddlers in a special outfit with a high-quality microphone inside. They recorded more than 100 naturally occurring tantrums and analyzed the acoustic patterns. They found that tantrums have two distinct phases. First comes anger, which involves yelling, screaming, and physical aggression, such as throwing things and/or writhing on the floor. This is followed by sadness, including tears, whimpering, and the desire for comfort. The researchers advise parents to end a tantrum by getting the child past the peak of anger as quickly as possible to sadness because sad children seek comfort. The quickest way past the anger is to do nothing (Green, Whitney, & Potegal, 2011).

Climate change in the Great North Woods: University of Minnesota forest researchers have established 100 observation sites in the Boundary Waters Canoe Area, a 1,090,000-acres (4,400 square kilometers) wilderness area within the Superior National Forest in northeastern Minnesota of the United States along the Canadian border. One major finding is the spread of earthworms, invasive creatures that "are re-engineering the forest floor as they go. . . . Each year, the worms can eat a season's worth of basswood leaves, depriving the forest floor of 'duff,' the carpet-like layer of decaying matter that is a critical component of northern American forests. In a healthy forest, the duff keeps tree roots cool, germinates tree seeds and mushrooms, and provides a home for ovenbirds, salamanders and other small creatures." But below a worm-infested basswood tree, the earth is bare, a circle of hard-packed dirt 30 feet in diameter. Hardwood trees that might fare better as the climate warms—red maple and basswood—can't take root in the packed dirt. Instead, the worms create ideal conditions for invasive, brushlike buckthorn and garlic mustard. University of Minnesota forester Lee Frelich calls it "global worming," because they are amplifying the effects of climate change (Marcotty, 2013, pp. A10–A11).

Folk Wisdom About Human Observation

> In the fields of observation, chance favors the prepared mind.
>
> —Louis Pasteur (1822–1895)
> French chemist who
> discovered the principles of vaccination

Every student who takes an introductory psychology or sociology course learns that human perception is highly selective. When looking at the same scene or object, different people will see different things. What people "see" is highly dependent on their interests, biases, and backgrounds. Our culture shapes what we see; our early-childhood socialization forms how we look at the world; and our value systems tell us how to interpret what passes before our eyes. How, then, can one trust observational data?

In their classic book *Evaluating Information: A Guide for Users of Social Science Research*, Katzer, Cook,

and Crouch (1998) titled their chapter on observation "Seeing Is Not Believing." They open with an oft-repeated story meant to demonstrate the problem with observational data.

> Once at a scientific meeting, a man suddenly rushed into the midst of one of the sessions. Another man with a revolver was chasing him. They scuffled in plain view of the assembled researchers, a shot was fired, and they rushed out. About twenty seconds had elapsed. The chairperson of the session immediately asked all present to write down an account of what they had seen. The observers did not know that the ruckus had been planned, rehearsed, and photographed. Of the forty reports turned in, only one was less than 20-percent mistaken about the principal facts, and most were more than 40-percent mistaken. The event surely drew the undivided attention of the observers, was in full view at close range, and lasted only twenty seconds. But the observers could not observe all that happened. Some readers chuckled because the observers were researchers, but similar experiments have been reported numerous times. They are alike for all kinds of people. (pp. 21–22)

Using this story to cast doubt on all varieties of observational research manifests two fundamental fallacies: (1) these researchers were not trained as social science observers and (2) they were not prepared to make observations at that particular moment. *Scientific inquiry using observational methods requires disciplined training, systematic preparation, and readiness.*

The fact that a person is equipped with functioning senses does not make that person a skilled observer. The fact that ordinary persons experiencing any particular incident will report different things does not mean that *trained and prepared observers* cannot report with accuracy, authenticity, and reliability that same incident. Exhibit 6.1 lists the training and attributes necessary for skilled observation.

Training observers can be particularly challenging because so many people think that they are "natural" observers and therefore have little to learn. Training to become a skilled observer is a no less rigorous process than the training necessary to become a skilled survey researcher or statistician. People don't "naturally" know how to write good survey items or analyze statistics—and people do not "naturally" know how to do systematic research observations. All forms of scientific inquiry require training and practice.

Careful preparation for entering into fieldwork is as important as disciplined training. Though I have considerable experience doing observational fieldwork, had I been present at the scientific meeting where the shooting scene occurred, my recorded observations might

Becoming a Skilled Observer

Training to become a skilled observer includes the following:

1. Learning to pay attention: Seeing what there is to see, and hearing what there is to hear

2. Writing descriptively writing descriptively

3. Acquiring expertise and discipline in recording field notes

4. Knowing how to separate detail from trivia in order to achieve the former without being overwhelmed by the latter

5. Using systematic methods to validate and triangulate observations

6. Reporting the strengths and limitations of one's own perspective, which requires both self-knowledge and self-disclosure

not have been significantly more accurate than those of my less trained colleagues because *I would not have been prepared to observe what occurred* and, lacking that preparation, would have been seeing things through my ordinary eyes rather than my scientific observer's eyes.

Preparation has mental, physical, intellectual, and psychological dimensions. Pasteur said, "In the fields of observation, chance favors the prepared mind." Part of preparing the mind is learning how to concentrate during the observation. Observation, for me, involves enormous energy and concentration. I have to "turn on" that concentration—turn on my scientific eyes and ears, my observational senses. A scientific observer cannot be expected to engage in systematic observation on the spur of the moment any more than a world class boxer can be expected to defend his title spontaneously on a street corner or an Olympic runner can be asked to dash off at record speed because someone suddenly thinks it would be nice to test the runner's time. Athletes, artists, musicians, dancers, engineers, and scientists require training and mental preparation to do their best. Experiments and simulations that document the inaccuracy of spontaneous observations made by untrained and unprepared observers are no more indicative of the potential quality of observational methods than an amateur community talent show is indicative of what professional performers can do.

Two points are critical, then, in this introductory section. First, the folk wisdom about observation

being nothing more than selective perception is true in the ordinary course of participating in day-to-day events. Second, the skilled observer is able to improve the accuracy, authenticity, and reliability of observations through intensive training and systematic preparation. The remainder of this chapter is devoted to helping evaluators and researchers move their observations from the level of ordinary looking to the rigor of systematic seeing.

The Value of Direct Observations

I'm often asked by students,

> Isn't interviewing just as good as observation. Do you really have to go see a program or spend time in a community to understand it? Can't you find out all you need to know by talking to people without going out and seeing things firsthand.

I reply by relating my experience of evaluating a leadership development program with two colleagues. As part of a formative evaluation aimed at helping staff and funders clarify and improve the program's design before undertaking a comprehensive follow-up study for a summative evaluation, we went through the program as participant-observers. After completing the six-day leadership retreat, we met to compare experiences. Our very first conclusion was that we would never have understood the program without personally experiencing it. It bore little resemblance to our expectations, what people had told us, or the official program description. Had we designed the follow-up study without having participated in the program, we would have completely missed the mark and asked inappropriate questions. To absorb the program's language, understand nuances of meaning, appreciate variations in participants' experiences, capture the importance of what happened outside formal activities (during breaks, over meals, in late-night gatherings and parties), and feel the intensity of the retreat environment—nothing could have substituted for direct experience with the program. Indeed, what we observed and experienced was that participants were changed as much or more by what happened outside the formal program structure and activities as anything that happened through the planned curriculum and exercises.

Indeed, a major purpose of observation is to see firsthand what is going on rather than simply assume we know. We go into a setting, observe, and describe what we observe.

> What does all that description do for us? Perhaps not the only thing, but a very important one, is that it

helps us get around conventional thinking. A major obstacle to proper description and analysis of social phenomena is that we think we know most of the answers already. We take a lot for granted, because we are, after all, competent adult members of our society and know what any competent adult knows. We have, as we say, "common sense." We know, for example, that schools educate children and hospitals cure the sick. "Everyone" knows that. We don't question what everyone knows; it would be silly. But, since what everyone knows is the object of our study, we must question it or at least suspend judgment about it, go look for ourselves to find out what schools and hospitals do, rather than accepting conventional answers. (Becker, 1998, p. 83)

The Characteristics and Strengths of High-Quality Observations

The first-order purpose of observational data is to *describe in depth and detail* the setting that was observed, the activities that took place in that setting, the people who participated in those activities, and the meanings of what was observed from the perspectives of those observed. The descriptions should be factual, accurate, and thorough without being cluttered by irrelevant minutiae and trivia. You can judge the quality of observational reports by the extent to which the observation takes you into the situation described and deepens your understanding. In this way, evaluation users, for example, can come to understand program activities and impacts through detailed descriptive information about what has occurred in a program and how the people in the program have reacted to what has occurred.

Naturalistic observations take place *in the field*. For ethnographers, the field is a cultural setting. For organizational researchers, the field will be an organization. For evaluators, the field is the program being studied. Many phrases are used for talking about field-based observations, including participant observation, fieldwork, qualitative observation, direct observation, or field research. "All these terms refer to the circumstance of being in or around an ongoing social setting for the purpose of making a qualitative analysis of that setting" (Lofland, 1971, p. 93).

In addition to the value of direct, personal contact with and observations of a setting through fieldwork, the inquirer is better able to understand and capture the context within which people interact—for understanding context is essential to a holistic perspective. In a site visit to an employment program, it became clear that the location of the program in a run-down,

WHAT IS INVOLVED IN PARTICIPANT OBSERVATION: AN EXEMPLAR

The Study of the Boys in White: Student Culture in Medical School

Organizational sociologists Howard Becker, Blanche Geer, Everett Hughes, and Anselm Strauss set the standard for organizational participant observation in their classic and revered study of student experience at medical school. Here's an overview of their methodology that provides insight into the commitment and rigor involved in participant observation.

They participated in the daily lives of the students. They did this by attending medical school with the students, "following them from class to laboratory to hospital ward." In studying the clinical years, they "attached" themselves to a group and "followed the members through the entire day's activities." They "went with students to lectures and to the laboratories in which they studied the basic sciences, watched their activities, and engaged in casual conversation with them." They "followed students to their fraternity houses and sat with them while they discussed their school experiences." They accompanied students in the hospital, watching them examine patients. They sat in on discussion groups and oral exams, had meals with the students, and took night calls with them. In short, they "went with the students wherever they went in the course of the day."

> Two aspects of our participant observation are important: it was *continued* and it was *total*. When we observed a particular group of students, we observed them day after day, more or less continuously, for periods ranging from a week to two months; and we observed the total day's activities of such a group (so far as possible) rather than simply some segment of them. (Becker, Geer, Hughes, & Strauss, 1961, p. 27)

They did not record every event they observed, commenting that "encyclopedic recording is neither possible nor particularly useful." They "made it a rule" to dictate a complete account of each day's observations as soon as possible. The recorded conversations they overheard were a part of "verbatim." Thus, they not only engaged in informal, conversational interviewing but also conducted formal interviews with a purposeful sample of students and faculty.

Their field notes and interviews yielded some 5,000 single-spaced typed pages. They indexed the field notes and labeled each entry with topical codes. Their conclusions were emergent: "What we ended with in the way of a theoretical position is somewhat different from what we started with" (Becker et al., 1961, pp. 25–32).

unsafe area was affecting how participants who came from elsewhere in the city felt about the program and contributed to the high dropout rate.

Moreover, firsthand experience with a setting and the people in the setting allows an inquirer to be open and discovery oriented and inductive because, by being on-site, the observer has less need to rely on prior conceptualizations of the setting, whether those prior conceptualizations are from written documents or verbal reports. In a visit to a rural African school that was supposed to serve both boys and girls, the first thing that stood out was that there were no girls. Routine monitoring reports submitted to headquarters showed roughly equal numbers of boys and girls. So where were the girls? They were on a waiting list. The school lacked space to serve everyone, so they enrolled everyone, boys and girls, and reported enrollment data to headquarters and the government, but only boys were allowed to actually attend.

Another strength of observational fieldwork is that the inquirer has the opportunity to see things that may routinely escape awareness among the people in the setting. For someone to provide information in an interview, he or she must be aware enough to report the desired information. Because all social systems involve routines, participants in those routines may take them so much for granted that they cease to be aware of important nuances that are apparent only to an observer who has not become fully immersed in those routines. In an evaluation of early-childhood programs, I was struck immediately by the bare walls in one center. All other centers had walls covered with children's art, posters of animals, and scenes of children playing in places around the world. I asked the program director about the bare walls. She looked quizzically, like she didn't understand my question, and then exclaimed, "Oh no, oh no. The walls are bare." She grabbed a child's drawing from her desk and taped it on the wall while she explained, "We formed a committee to decorate the walls. They got into a personality conflict and couldn't agree. Nothing happened. I guess we got busy and just got used to the bare walls. This is so embarrassing." (*Note:* Observing and interviewing in the field can affect what is going on, a matter we'll discuss at some length later in this chapter.)

The participant-observer can also discover things no one else has ever really paid attention to. One of the highlights of the leadership training program we experienced was the final evening banquet at which the staff were "roasted." For three nights after the training ended, the participants worked to put together a program of jokes, songs, and skits for the banquet. Staff were never around for these preparations, which lasted late into the night, but they had come to count on this culminating event. Month after month for two years each completely new training group had organized a final banquet event to both honor and make fun of staff. Staff assumed that either prior participants passed on this tradition or it was a natural result of the bonding among participants. We learned that neither explanation was true. What actually occurred was that, unknown to program staff, the dining hostess for the hotel where participants stayed initiated the roast. After the second evening's meal, when the staff routinely departed for a meeting, the hostess would tell the participants what was expected. She even brought out a photo album of past banquets and offered to supply joke books, costumes, music, or whatever. This 60-year-old woman had begun playing what amounted to a major staff role for one of the most important processes in the program—and the staff didn't know about it. We learned about it by being there.

Yet another value of direct observation is the chance to learn things that people would be unwilling to talk about in an interview. Interviewees may be unwilling to provide information on sensitive topics, especially to strangers. During participant observation of the wilderness leadership program, observing the emergence of romantic encounters raised issues that we would not have been likely to ask about if just doing interviews or surveys for the inquiry.

Ongoing fieldwork offers an opportunity to move beyond the selective perceptions of others. Interviews present the understandings of the people being interviewed. Those understandings constitute important, indeed critical, information. However, it is necessary for the inquirer to keep in mind that interviewees are always reporting perceptions—selective perceptions. A program director told me that one of the strengths of her program was its welcoming environment. And, indeed, whenever she was around, the reception staff were unfailingly friendly and attentive. When she wasn't around, they were rude, inattentive, and condescending toward participants.

At the same time, an important point to keep in mind is that field observers will also have selective perceptions. By reflecting on the basis for their perceptions and making their own perceptions part of the data—a matter of training, discipline, and self-awareness—observers can arrive at a more comprehensive view of the setting being studied and move beyond their preconceptions.

Finally, getting close to the people in a setting through firsthand experience permits the inquirer to draw on personal knowledge during the interpretation stage of

OBSERVATION SOCIETY

The recent sudden increase of ethnography can be explained with the hypothesis that we are entering an "observation society," a society in which observing has become a fundamental activity, and watching and scrutinizing are becoming important cognitive modes....

The clues that we are living in an "observation society" are many. Wherever we go, there is always a television camera ready to film our actions (unbeknownst to us): camera phones and the current fashion for making video recordings of even the most personal and intimate situations and posting them on the Internet; or logging on to webcams pointed at city streets, monuments, landscapes, plants, bird nests, coffee pots, etc., to observe movements, developments, and changes. Then, there is the trend of webcams worn by people so that they can lead us virtually through their everyday lives. These are not minor eccentricities but websites visited by millions of people around the world.

Observing and being observed are two important features of contemporary Western societies. Consequently, there is an increasing demand in various sectors of society—from marketing to security, television to the fashion industry!—for observation and ethnography. All of which suggests that ours is becoming an observation society.

—Giampietro Gobo (2011, p. 25)
Professor of Methodology
and Social Research and Evaluation
University of Milan

use of any particular datum. This wealth of information and impression sensitizes him to subtleties which might pass unnoticed in an interview and forces him to raise continually new and different questions, which he brings to and tries to answer in succeeding observations. (Becker & Geer, 1970, p. 133)

In the course of fieldwork, the inquirer develops an empathetic understanding: coming to understand a setting and its people not just intellectually but emotionally, what it feels like to be there, in that place, doing those things, with those people. In all of my participant observation experiences, I could look back through my notes and reflect on my experiences and identify when I had begun to feel the experience and understand the feelings of the people I was interacting with. Participant observation is a holistic encounter. All senses come into play. That is the great strength of being present, truly present, in the field.

Exhibit 6.2 summarizes the 10 strengths of direct observational fieldwork just presented.

Going Into the Real World

In a moment, we'll offer guidance on how to do fieldwork, but to inform that transition and reinforce the importance of direct observation in the real world, let me offer a perspective from the world of children's stories. Some of the most delightful, entertaining, and suspenseful fables concern tales of kings who discard their royal robes to take on the apparel of peasants so that they can move freely among their people to really understand what is happening in their kingdom. Our modern-day kings and political figures are more likely to take television crews with them when they make excursions among the people. They are unlikely to go out secretly disguised, moving through the streets anonymously, unless they're up to mischief. It is left, then, to applied researchers and evaluators to play out the fable, to take on the appropriate appearance and mannerisms that will permit easy movement among the people, sometimes secretly, sometimes openly, but always with the purpose of better understanding what the world is really like. They are then able to report those understandings to our modern-day version of kings so that policy wisdom can be enhanced and programmatic decisions enlightened. At least that's the fantasy. Turning that fantasy into reality involves a number of important decisions about what kind of fieldwork to do. We turn now to those decisions.

analysis. Reflection and introspection are important parts of field research. The impressions and feelings of the observer become part of the data to be used in attempting to understand a setting and the people who inhabit it. The observer takes in information and forms impressions that go beyond what can be fully recorded in even the most detailed field notes.

> Because he sees and hears the people he studies in many situations of the kind that normally occur for them, rather than just in an isolated and formal interview, he builds an ever-growing fund of impressions, many of them at the subliminal level, which give him an extensive base for the interpretation and analytic

EXHIBIT 6.2   Ten Strengths of High-Quality Observations

1. **Rich description:** Detailed, rich descriptions take readers into the setting observed, providing a vicarious experience and deepened understanding.

2. **Contextual sensitivity:** Being in a setting allows observation of the context and environment that is likely to affect what happens in the setting.

3. **Being open to what emerges:** Direct observation supports being open, discovery oriented, and inductive, taking in whatever is there in addition to and beyond any predetermined observational protocol.

4. **Seeing the unseen:** In the field, the inquirer has the opportunity to see things that may routinely escape awareness among the people in the setting. (The fish doesn't know it's swimming in water.)

5. **Testing old assumptions and generating new insights:** The participant-observer can discover things no one else has ever really paid attention to.

6. **Opening up new areas of inquiry:** Observations generate questions that can be pursued in interviews to help understand and interpret what has been observed.

7. **Delving into sensitive issues:** Fieldwork provides an opportunity to learn things that people may be unwilling to bring up in an interview.

8. **Getting beyond selective perceptions of others:** Direct observation provides an opportunity to move beyond the selective perceptions of others and see for oneself. Interview data alone do not allow the comparison between what is said to a fieldworker and what is directly observed.

9. **Getting beyond one's own selective perceptions:** By reflecting on the basis for their perceptions and making their own perceptions part of the data—a matter of training, discipline, and self-awareness— observers can arrive at a more comprehensive view of the setting being studied and move beyond their own preconceptions.

10. **Experiencing empathy:** Observers come to understand a setting and its people not just intellectually but emotionally, what it feels like to be there, in that place, doing those things, with those people.

Fieldwork Variations

©2002 Michael Quinn Patton and Michael Cochran

Observational inquiry explores the world in many ways for varied purposes. Observation for evaluation or action research is different from basic social science research because the purposes, questions, and timelines are typically different. While fieldwork methods originated in anthropology and qualitative sociology, using those methods for evaluation often requires adaptation. The sections that follow will discuss both the similarities and differences between evaluation field methods and research field methods and the options that exist for both.

Variations in Observer Involvement: Participant, Onlooker, or Both?

The first and most fundamental distinction that differentiates observational strategies concerns the extent to which the observer will be a *participant* in the setting being studied. This involves more than a simple choice between participation and nonparticipation. The extent of participation is a continuum that varies from complete immersion in the setting as full participant to complete separation from the setting as spectator, with a great deal of variation along the continuum between these two end points.

Nor is it simply a matter of deciding at the beginning how much the observer will participate. The extent of participation can change over time. In some cases, the observer may begin the study as an onlooker and gradually become a participant as fieldwork progresses. The opposite can also occur. An evaluator might begin as a complete participant to experience what it is like to be initially immersed in the program and then gradually withdraw participation over the period of the study until finally taking the role of occasional observer from an onlooker stance.

Full participant observation constitutes an omnibus field strategy in that it "simultaneously combines document analysis, interviewing of respondents and informants, direct participation and observation, and introspection" (Denzin, 1978b, p. 183). If, on the other hand, an evaluator observes a program as an onlooker, the processes of observation can be separated from interviewing. In participant observation, however, no such separation exists. Typically, anthropological fieldworkers combine in their field notes data from personal, eye witness observation with information

gained from informal, natural interviews and informants' descriptions (Pelto & Pelto, 1978, p. 5). Thus, the participant-observer employs multiple and overlapping data collection strategies: being fully engaged in experiencing the setting (participation) while at the same time observing and talking with others about whatever is happening.

In leadership programs I have evaluated through participant observation, I have been a full participant in all exercises and program activities, using the field of evaluation as my leadership arena (since all participants had to have an arena of leadership as their focus). As did the other participants, I developed close relationships with some people as the week progressed, sharing meals and conversing late into the night. I sometimes took detailed notes during activities if the activity permitted (e.g., group discussion), while at other times, I waited until later to record notes (e.g., after meals). If a situation suddenly became emotional, for example, during a small group encounter, I would cease to take notes so as to be fully present as well as to keep my note taking from becoming a distraction. Unlike other participants, I sat in on staff meetings and knew how staff viewed what was going on. Much of the time, I was fully immersed in the program experience as a participant, but I was also always aware of my additional role as evaluation observer.

The extent to which it is possible for an observer to become a participant in a setting will depend partly on the nature of the setting. In human service and education programs that serve children, an observer cannot participate as a child but may be able to participate as a volunteer, parent, or staff member in such a way as to develop the perspective of an insider in one of those adult roles. Gender can create barriers to participant observation. Males can't be participants in female-only settings (e.g., a battered women's shelter). Females doing fieldwork in nonliterate cultures may not be permitted access to male-only councils and ceremonies. Programs that serve special populations may also involve natural limitations on the extent to which the observer can become a full participant. For example, a researcher who is not chemically dependent will not be able to become a *full* participant, physically and psychologically, in a chemical dependency program, even though it may be possible to participate in the program as a client. Such participation in a treatment program can lead to important insights and understanding about what it is like to be in the program; however, the observer must avoid the delusion that participation has been complete. This point is illustrated by an exchange between an inmate and an observer who was doing participant observation in a prison.

Inmate: What are you in here for, man?

Observer: I'm here for a while to find out what it's like to be in prison.

Inmate: What do you mean—"find out what it's like"?

Observer: I'm here so that I can experience prison from the inside instead of just studying what it's like from out there.

Inmate: You got to be jerkin' me off, man. "Experience from the inside"...? Shit, man, you can go home when you decide you've had enough can't you?

Observer: Yeah.

Inmate: Then you ain't never gonna know what it's like from the inside.

Social, cultural, political, and interpersonal factors can limit the nature and degree of *participation* in participant observation. For example, if people in a setting all know each other intimately, they may object to an outsider trying to become part of their close circle. Where marked social class differences exist between a sociologist and people in a neighborhood, access will be more difficult; likewise, when, as is often the case, an evaluator is well educated and middle class while welfare program clients are economically disadvantaged and poorly educated, the participants in the program may object to any ruse of "full" participant observation. Program staff will sometimes object to the additional burden of including an evaluator in a program where resources are limited and an additional participant would unbalance staff–client ratios. Thus, in evaluation, the extent to which full participation is possible and desirable will depend on the precise nature of the program, the political context, and the nature of the evaluation questions being asked. Adult training programs, for example, may permit fairly easy access for full participation by evaluators. Offender treatment programs are much less likely to be open to participant observation as an evaluation method. Evaluators must therefore be flexible, sensitive, and adaptive in negotiating the precise degree of participation that is appropriate in any particular observational study, especially where reporting timelines are constrained such that entry into the setting must be accomplished relatively quickly. Social scientists who can take a long time to become integrated into the setting under study have more options for fuller participant observation.

As these examples illustrate, full and complete participation in a setting, what is sometimes called "going native," is fairly rare, especially in program evaluation. The degree of participation and nature of observations vary along a wide continuum of possibilities. *The ideal is to design and negotiate that degree of participation that will yield the most meaningful findings given the characteristics of the setting, the people in the setting, and the sociopolitical context of the program.* In summary, the purpose, scope, length, and setting of the study will dictate the range and types of participant observation that are possible.

One final caution: The observer's plans and intentions regarding the degree of involvement to be experienced may not be the way things actually turn out. Lang and Lang (1960) reported that two scientific participant-observers who were studying audience behavior at a Billy Graham evangelical crusade made their "decision for Christ" and left their observer posts to walk down the aisle and join Reverend Graham's campaign. Such are the occupational hazards (or benefits, depending on your perspective) of real-world fieldwork.

Insider and Outsider Perspectives: Emic Versus Etic Approaches

> People who are insiders to a setting being studied often have a view of the setting and any findings about it quite different from that of the outside researchers who are conducting the study.
>
> —Bartunek and Louis (1996, p. 1)
> *Insider/Outsider Team Research*

Ethnosemanticist Kenneth Pike (1954) coined the terms *emic* and *etic* to distinguish classification systems reported by anthropologists based on (a) the language and categories used by the people in the culture studied, an emic approach, in contrast to (b) categories created by anthropologists based on their analysis of important cultural distinctions, an etic approach. Leading anthropologists like Franz

Boas and Edward Sapir argued that the only meaningful distinctions were those made by people within a culture, that is, the emic perspective. However, as anthropologists turned to more comparative studies, engaging in cross-cultural analyses, distinctions that cut across cultures were based on the anthropologist's analytical perspective, that is, an etic perspective. The etic approach involved "standing far enough away from or outside of a particular culture to see its separate events, primarily in relation to their similarities and their differences, as compared to events in other cultures" (Pike, 1954, p. 10). For some years, a debate raged in anthropology about the relative merits of emic versus etic perspectives (Headland, Pike, & Harris, 1990; Pelto & Pelto, 1978, pp. 55–60), but as often happens over time, both approaches came to be understood as valuable, though each contributes something different. Nevertheless, tension between these perspectives remains:

> Today, despite or perhaps because of the new recognition of cultural diversity, the tension between universalistic and relativistic values remains an unresolved conundrum for the Western ethnographer. In practice, it becomes this question: By which values are observations to be guided? The choices seem to be either the values of the ethnographer or the values of the observed—that is, in modern parlance, either the *etic* or the *emic*.... Herein lies a deeper and more fundamental problem: How it is possible to understand the other when the other's values are not one's own? This problem arises to plague ethnography at a time when Western Christian values are no longer a surety of truth and, hence, no longer the benchmark from which self-confidently valid observations can be made. (Vidich & Lyman, 2000, p. 41)

Methodologically, the challenge is to do justice to both perspectives during and after fieldwork and to be clear with one's self and one's audience how this tension is managed.

A participant-observer shares as intimately as possible in the life and activities of the setting under study to develop *an insider's view* of what is happening, the emic perspective. This means that the participant-observer not only sees what is happening but also feels what it is like to be a part of the setting or program. Anthropologist Hortense Powdermaker (1966) has described the basic assumption undergirding participant observation as follows:

> To understand a society, the anthropologist has traditionally immersed himself in it, learning, as far as possible, to think, see, feel and sometimes act as

CAPTURING THE INSIDERS' PERSPECTIVE IN THEIR OWN WORDS

When social scientists study something—a community, an organization, an ethnic group—they are never the first people to have arrived on the scene, never newcomers to an unpeopled landscape who can name its features as they like. Every topic they write about makes up part of the experience of many other kinds of people, all of whom have their own ways of talking about it, their own distinctive words for the objects and events and people involved in that area of social life. Those words are never neutral, objective signifiers. Rather, they express the perspective and situation of the people who use them. The natives are already there, were there all along, and everything in that terrain has a name, more likely many names.

If we choose to name what we study with words the people involved already use, we acquire, with the words, the attitudes and perspectives the words imply. Since many kinds of people are involved in any social activity, choosing words from any of their vocabularies commits us to one or another of the perspectives in use by one or another of the groups already on the scene. Those perspectives invariably take much for granted, making assumptions about what social scientists might better treat as problematic....

What we call the things we study has consequences [italics added].

—Howard S. Beck (2007, p. 224)
Telling About Society

a member of its culture and at the same time as a trained anthropologist from another culture. (p. 9)

Experiencing the setting or program as an insider accentuates the participant part of participant observation. At the same time, the inquirer remains aware of being an outsider. The challenge is to combine participation and observation so as to become capable of understanding the setting as an insider while describing it to and for outsiders.

Obtaining something of the understanding of an insider is, for most researchers, only a first step. They expect, in time, to become capable of thinking and acting within the perspective of two quite different groups, the one in which they were reared and—to some degree—the one they are studying. They will also, at times, be able to assume a mental position peripheral to both, a position from which they will be able to perceive and, hopefully, describe those relationships, systems, and patterns of which an inextricably involved insider is not likely to be consciously aware. For what the social scientist realizes is that while the outsider simply does not know the meanings or the patterns, the insider is so immersed that he may be oblivious to the fact that patterns exist.... What field workers eventually produce out of the tension developed by this ability to shift their point of view depends upon their sophistication, ability, and training. Their task, in any case, is to realize what they have experienced and learned and to communicate this in terms that will illumine. (Wax, 1971, p. 3)

Who Conducts the Inquiry? Solo and Team Versus Participatory and Collaborative Approaches

The ultimate "insider" perspective comes from involving the insiders as coresearchers through collaborative or participatory research. Collaborative forms of fieldwork, participatory action research, indigenous peoples' ethical research principles, feminist research, and empowerment evaluation have become sufficiently important and widespread to make *degree of collaboration* a dimension of design choice in qualitative fieldwork. Participatory action research has a long and distinguished history (Kemmis & McTagery, 2000; Lincoln, Lynham, & Guba, 2011; Whyte, 1989). Collaborative principles of feminist inquiry include participatory processes that support equity and mutuality (Olesen, 2000; Podems, 2010, 2013). Guidelines for ethical research in indigenous studies call for collaboration between researchers and the indigenous people involved in a setting under observation: "Indigenous people have the right to full participation appropriate to their skills and experiences in research projects and processes" (Australian Institute of Aboriginal and Torres Strait Islander Studies, 2011, p. 10). In evaluation, participatory and collaborative approaches are widely used (Cousins & Chouinard, 2012; Cousins & Earl, 1995; Cousins & Whitmore, 2007; King & Stevahn, 2013). Empowerment evaluation, often using qualitative methods (Fetterman, 2000; Fetterman & Wandersman, 2005), involves the use of evaluation concepts and techniques to foster self-determination and help people help themselves by learning to study and report on their own issues and concerns.

What these approaches have in common is a style of inquiry in which the researcher or evaluator becomes a facilitator, collaborator, coach, and teacher in support of those engaging in the inquiry. While the findings from such participatory processes may be useful, a supplementary agenda is often to increase participants' sense of being in control of, deliberative about, and reflective on their own lives and situations. Chapter 4 discussed these approaches as examples of how qualitative inquiry can be applied in support of organizational or program development and community change.

FULLY PARTICIPATORY EVALUATION

Participant-directed evaluation holds the potential to engage individuals in meaningful studies of issues in their organization or community while simultaneously teaching them evaluative knowledge and skills. Consider, for example, a group of American Indian teenagers who, with the support of a native evaluator, conducted an evaluation of the perceived outcomes of their reservation's health programming. The evaluator was responsible for teaching the teenagers how to develop an interview protocol, how to conduct interviews, and how to record and analyze qualitative data. The teenagers then interviewed every one of the community's elders. After analyzing the data, the young people made a formal presentation to the tribal council. The council received good information about the elders' perspectives; the youth learned the skills of an evaluator (King & Stevahn, 2013, p. 32).

Degrees of collaboration and coresearcher participation vary along a continuum from the traditional solo fieldworker operating alone, or teams of professional researchers working together, sometimes in the same site, often in a multisite design, in contrast to completely shared processes in which "coresearchers," that is, people in the setting studied, own the inquiry from design through data collection to analysis and reporting. Along the middle of the continuum are various degrees of partial and periodic (as opposed to continuous) collaboration.

Overt Versus Covert Observations

A traditional concern about the validity and reliability of observational data has been the effects of the observer on what is observed. People may behave quite differently when they know they are being observed versus how they behave naturally when they don't think they're being observed. Thus, the argument goes, covert observations are more likely to capture what is really happening than are overt observations where the people in the setting are aware they are being studied.

Researchers have expressed a range of opinions concerning the ethics and morality of conducting covert research, what Mitchell (1993) called "the debate over secrecy" (p. 23). Historically, one end of the continuum is represented by Edward Shils (1959), who absolutely opposed all forms of covert research including "any observations of private behavior, however technically feasible, without the explicit and fully informed permission of the person to be observed." He argued that there should be full disclosure of the purpose of any research project and that even participant observation is "morally obnoxious ... manipulation" unless the observer makes explicit his or her research questions at the very beginning of the observation (Shils, 1959; quoted in Webb et al., 1966, p. vi).

At the other end of the continuum is the "investigative social research" of Jack Douglas (1976). Douglas argued that conventional anthropological field methods have been based on a consensus view of society that views people as basically cooperative, helpful, and willing to have their points of view understood and shared with the rest of the world. In contrast, Douglas adopted a conflict paradigm of society that led him to believe that any and all covert methods of research should be considered acceptable options in a search for truth.

> The investigative paradigm is based on the assumption that profound conflicts of interest, values, feelings, and actions pervade social life. It is taken for granted that many of the people one deals with, perhaps all people to some extent, have good reason to hide from others what they are doing and even to lie to them. Instead of trusting people and expecting trust in return, one suspects others and expects others to suspect him. Conflict is the reality of life; suspicion is the guiding principle.... It's a war of all and no one gives anyone anything for nothing, especially truth....
>
> All competent adults are assumed to know that there are at least four major problems lying in the way of getting at social reality by asking people what is going on and that these problems must be dealt with if one is to avoid being taken in, duped, deceived, used, put on, fooled, suckered, made the patsy, left holding the bag, fronted out, and so on. These four problems are (1) misinformation, (2) evasions, (3) lies, and (4) fronts. (Douglas, 1976, pp. 55, 57)

Just as degree of participation in fieldwork turns out to be a continuum of variations rather than an all or none proposition, so too is the question of how explicit to be about the purpose of fieldwork. The extent to which participants in a program under study are informed that they are being observed and are told the purpose of the research has varied historically from full disclosure to no disclosure, with a great deal of variation along the middle of this continuum (Junker, 1960). Discipline-based ethics statements (e.g., American Psychological Association, American Sociological Association) now generally condemn deceitful and covert research. Likewise, institutional review board (IRB) procedures for the protection of human subjects have severely constrained such methods. They now refuse to approve protocols in which research participants are deceived about the purpose of a study, as was commonly done in early psychological research. One of the more infamous examples was Stanley Milgram's New Haven experiments aimed at studying whether ordinary people would follow the orders of someone in authority by having these ordinary citizens administer what they were told were behavior modification electric shocks to help students learn, shocks that appeared to the unsuspecting citizens to go as high as 450 volts despite the screams and protests heard from supposed students on the other side of a wall. The real purpose of the study, participants later learned, was to replicate Nazi prison guard behavior among ordinary American citizens (Milgram, 1974).

IRBs also refuse to approve research in which people are observed and studied without their knowledge or consent, as in the infamous Tuskegee Experiment. For 40 years, physicians and medical researchers, under the auspices of the U.S. Public Health Service, studied untreated syphilis among black men in and around the county seat of Tuskegee, Alabama, without the informed consent of the men studied, men whose syphilis went untreated so that the progress of the disease could be documented (Jones, 1993). Other stories of abuse and neglect by researchers doing covert studies abound. In the late 1940s and early 1950s, schoolboys at the Walter E. Fernald State School in Massachusetts were routinely served breakfast cereal doused with radioactive isotopes, without permission of the boys or their guardians, for the dissertation of a doctoral student in nutritional biochemistry. In the 1960s, the U.S. Army secretly sprayed a potentially hazardous chemical from downtown Minneapolis rooftops onto unsuspecting citizens to find out how toxic materials might disperse during biological warfare. Native American children on the Standing Rock Sioux Reservation in the Dakotas were used to test an unapproved and experimental hepatitis A vaccine without the knowledge or approval of their parents. In the 1960s and 1970s, scientists tested skin treatments and drugs on prisoners in a Philadelphia County jail without informing them of potential dangers. Unfortunately, the stories of abuses and deceits make for a mountain of cases of treating those researched as subjects to be manipulated in service to the good (or might one say god) of science.

Doctoral students frustrated by having their fieldwork delayed while they await IRB approval need to remember that they are paying for the sins of their research forebears, for whom deception and covert observations were standard ways of doing their work. Those most subject to abuse were often the most vulnerable in society—children, the poor, people of color, the sick, people with little education, women and men incarcerated in prisons and asylums, and children in orphanages or state correctional schools. Anthropological research was commissioned and used by colonial administrators to maintain control over indigenous peoples. Protection of human subjects' procedures is now an affirmation of our commitment to treat all people with respect. And that is as it should be. But the necessity for such procedures comes out of a past littered with scientific horrors, for which those of us engaging in research today may still owe penance. At any rate, we need to lean over backward to be sure that such history is truly behind us—and that means being ever vigilant in fully informing and protecting the people who honor us by agreeing to participate in our research, whether they be homeless mothers (Connolly, 2000) or corporate executives (Collins, 2001a).

However, not all research and evaluation falls under IRB review, so the issue of what type and how much disclosure to make remains a matter of debate, especially where the inquiry seeks to expose the inner workings of cults and extremist groups or those whose power affects the public welfare, for example, corporations, labor union boards, political parties, and other groups with wealth and/or power. For example, Maurice Punch (1985, 1989, 1997), formerly of the Nijenrode Business School in the Netherlands, has written about the challenges of doing ethnographic studies of corruption in both private and public sector organizations, notably the police.

One classic form of deception in fieldwork involves pretending to share values and beliefs in order to become part of the group being studied. Sociologist Richard Leo carefully disguised his liberal political and social views, instead feigning conservative beliefs, to build trust with police and thereby gain admission to interrogation rooms (Allen, 1997, p. 32). Sociologist Leon Festinger and colleagues (1956) infiltrated a doomsday cult by lying about his profession and pretending to believe in the cult's prophecies. Sociologist

Laud Humphreys (1970) pretended to be gay to gather data for his dissertation on homosexual encounters in public parks. Anthropologist Carolyn Ellis (1986) pretended to be just visiting friends when she studied a Chesapeake Bay fishing culture. Her negative portrayals made their way back to the local people, many of whom were infuriated. She later expressed remorse about her deceptions (Allen, 1997).

Some covert research is conducted by engaging with a system to find out how it operates in practice (compared with how those in control say it operates). This is especially possible in public settings open to anyone (as opposed to inside organizations or groups). Here are some classic examples:

- To study the availability of condoms, trained observers can document how easy it is for teenagers to purchase condoms from pharmacies. Observers would also record where condoms are located and displayed in the store.
- To study compliance with laws that prohibit discrimination in housing, interracial couples can attempt to rent apartments advertised as available.
- To study hospital emergency room procedures, observers can document what medical issues and what types of patients get priority treatment.
- To study hand-washing behaviors in a public restroom, those being observed are not aware they are being observed and are not identified in the covert research (Hays & Singh, 2012, p. 226).

Evaluation Issues and Applications

In traditional scholarly fieldwork, the decision about the extent to which observations would be covert was made by researchers balancing the search for truth against their sense of professional ethics. In program evaluation research, the information users for whom the evaluation is done have a stake in what kind of methods are used, so the evaluator alone cannot decide the extent to which observations and evaluation purposes will be fully disclosed. Rather, the complexities of program evaluation mean that there are several levels at which decisions about the covert–overt nature of evaluation observations must be made. Sometimes only the funders of the program or of the evaluation know the full extent and purpose of observations. On occasion, program staff may be informed that evaluators will be participating in the program, but clients will not be so informed. In other cases, an evaluator may reveal the purpose and nature of program participation to fellow program participants and ask for their cooperation in keeping the evaluation secret from program staff. On still other occasions, a variety of people intimately associated with the program may be informed of the evaluation, but public officials who are less closely associated with the program may be kept "in the dark" about the fact that observations are under way. Sometimes the situation becomes so complex that the evaluator may lose track of who knows and who doesn't know, and of course, there are the classic situations where everyone involved knows that a study is being done and who the evaluator is—but the evaluator doesn't know that everyone else knows.

A Canadian evaluation studied the performance of immigration officers whose job was to identify individuals who should not be admitted into Canada without special documentation. The evaluation included sending actors to border posts and airports to play the roles of American travelers whose situation should call for an immigration interview.

Everything went fine until word got out. We [found] that immigration officers were two to three times more effective after knowing that they were observed than before. That allowed the evaluation to identify some of the weaknesses in the system—which were due not so much to information systems as they were to human resources. (Gauthier, 2013, p. 1)

In undertaking participant observation of the community leadership program mentioned earlier, my two evaluation colleagues and I agreed with the staff to downplay our evaluation roles and describe ourselves as "educational researchers" interested in studying the program. We didn't want the participants to think that *they* were being evaluated and therefore worry about our judgments. Our focus was on evaluating the program, not the participants, but to avoid increasing participant stress, we simply attempted to finesse our evaluation role by calling ourselves "educational researchers."

Our careful agreement on and rehearsal of this point with the staff fell apart during introductions (at the start of the six-day retreat) when the program director proceeded to tell the participants—for 10 minutes—that we were *just* participants and they didn't have to worry about our evaluating them. The longer he went on reassuring the group that they didn't have to worry about us, the more worried they got. Sensing that they were worried, he increased the intensity of his reassurances. While we continued to refer to ourselves as "educational researchers," the participants thereafter referred to us as evaluators. It took a day and a half to recover our full participating roles as the participants got to know us on a personal level as individuals.

Trying to protect the participants (and the evaluation) had backfired and made our entry into the group even more difficult than it otherwise would have been. However, this experience sensitized us to what we subsequently observed to be a pattern in many program situations and activities throughout the week and became a major finding of the evaluation: staff overprotection of and condescending attitudes toward participants.

Based on this and other evaluation experiences, I recommend full and complete disclosure. People are seldom really deceived or reassured by false or partial explanations—at least not for long. Trying to run a ruse or scam is simply too risky and adds to evaluator stress while holding the possibility of undermining the evaluation if (and usually when) the ruse becomes known. Program participants, over time, will tend to judge evaluators first and foremost as people, not as evaluators.

ETHICAL CHALLENGES IN QUALITATIVE RESEARCH

The relationship between informed consent and covert research serves as the ultimate illustration of ethical dilemmas in data collection. On the one hand, the distinction between overt and covert research is often unclear.... On the other hand, to make written informed consent mandatory would mean the end of much "street-style" ethnography....

As much as we appreciate proper practice on consent, it does raise issues about how and if we may be able to study phenomena such as crime, elites and private groups. If we accept that what we find out is a result of how we find it, then this is a good illustration of the close but at times not unproblematic relationship between research ethics and knowledge production. A great moral responsibility rests with decision makers here.

There are no standard answers to these dilemmas, and this was the very argument why they have been found too important to be left to researchers alone. We need to be prepared for all these challenges, which demand that we put them on the agenda from the very start of our projects and ask ourselves how they relate to our own particular project.

But, what should we do? A good piece of advice is always to invite experienced researchers with particular knowledge in research ethics and in your field to discuss matters with you.

—Anne Ryen (2011, pp. 418–419)
Ethics and Qualitative Research

The nature of the questions being studied in any particular evaluation will have a primary effect on the decision about who will be told that an evaluation is under way. In formative evaluations where staff members and/or program participants are anxious to have information that will help them improve their program, the quality of the data gathered may be enhanced by overtly soliciting the cooperation of everyone associated with the program. Indeed, the ultimate acceptance and usefulness of formative information may depend on such prior disclosure and agreement that a formative evaluation is appropriate. On the other hand, where program funders have reason to believe that a program is corrupt, abusive, incompetently administered, and/or highly negative in impact on clients, it may be decided that an external, covert evaluation is necessary to find out what is really happening in the program. Under such conditions, my preference for full disclosure may be neither prudent not practical. On the other hand, Whyte has argued that "in a community setting, maintaining a covert role is generally out of the question" (Whyte, 1984, p. 31).

Confidentiality

Finally, there is the related issue of confidentiality. Those who advocate covert research usually do so with the condition that reports conceal names, locations, and other identifying information so that the people who have been observed will be protected from harm or punitive action. Because the basic researcher is interested in truth rather than action, it is easier to protect the identity of informants or study settings when doing scholarly research. In evaluation research, however, while the identity of who said what may be possible to keep secret, it is seldom possible to conceal the identity of a program, and doing so may undermine the utility of the findings.

Evaluators and decision makers will have to resolve these issues in each case in accordance with their own consciences, evaluation purposes, political realities, and ethical sensitivities. We'll return to discussion of ethics in qualitative research in the final chapter. Before moving on, though, let's consider the special ethical issues raised by research and evaluation involving Internet communities.

Ethical Issues in Qualitative Research on Internet Communities

Research on the Internet raises new opportunities to engage in covert research—and new ethical challenges. It is easy to join online groups without disclosing one's true identity or research purpose. That doesn't make it ethical; it just makes it easy. Exhibit 6.3 presents ethical guidance for qualitative research on Internet communities.

Researchers lurking on Internet communities may be regarded with suspicion, and they may interfere with the functioning of the community. Consider this

---

EXHIBIT 6.3 **Seven Ethical Issues in Qualitative Research on Internet Communities**

1. *Intrusiveness.* Different ethical issues arise in passive analysis of Internet postings versus active participant observation by researcher involvement in the community through posting communications and interacting. Under what circumstances and for what purposes, if any, is covert participation in an Internet community ethical?

2. *Respect for privacy.* What is the perceived level of privacy of the Internet community being studied? Is it a closed group requiring registration or open to anyone? What are the group's norms and expectations regarding degree of privacy?

3. *Sensitivity to vulnerability.* How vulnerable is the Internet community? Internet support groups for victims of sexual abuse, survivors of cancer, or people living with AIDS patients may feel highly vulnerable, so extra sensitivity and precautions are necessary.

4. *Potential harm.* Consider how research or evaluation, and subsequent publication of findings, might harm individuals or the Internet community as a whole.

5. *Internet informed consent.* Consider whether informed consent is required or can be waived and, if required, how it will be obtained.

6. *Confidentiality.* How can the identity of participants be protected, especially when verbatim quotes are used in the inquiry?

7. *Intellectual property rights.* Some participants in an Internet community may prefer publicity to anonymity or, for other reasons, want to maintain control over their postings and communications.

—Adapted from Eysenbach
and Till (2001) and Berry (2004)

reaction by a participant in a health support group who quit when she learned that a researcher was monitoring the group:

> When I joined this, I thought it would be a support group, not a fishbowl for a bunch of guinea pigs. I certainly don't feel at this point that it is a safe environment, as a support group is supposed to be, and I will not open myself up to be dissected by students or scientists. (King, 1996, p. 120)

Online, virtual, and digital research and evaluation are rapidly evolving and will continue to evolve as new technologies and applications emerge. If you are studying virtual communities or networks, it will be important to stay up-to-date on innovative data gathering and reporting possibilities as well as corresponding ethical challenges (see, e.g., Boellstorff, Nardi, & Pearce, 2012; Ess, 2014; Eynon, Fry, & Schroeder, 2008; Fielding, Lee, & Blank, 2008; Gosling & Johnson, 2013).

Variations in Duration of Observations and Site Visits: From Rapid
Reconnaissance to Longitudinal Studies Over Years

Another important dimension along which observational studies vary is the length of time devoted to data gathering. In the anthropological tradition of field research, a participant-observer would expect to spend six months at a minimum, and often years, living in the culture being observed. The fieldwork of Napoleon Chagnon (1992) among the Yanomami Indians in the rain forest at the borders of Venezuela and Brazil spanned a quarter-century. To develop a holistic view of an entire culture or subculture takes a great deal of time, especially when, as in the case of Chagnon, he was documenting changes in tribal life and threats to the continued existence of these once-isolated people. The effects of his long-term involvement on the people he studied became controversial (Geertz, 2001; Tierney, 2000a, 2000b), a matter we shall take up later. The point here is that fieldwork in basic and applied social science aims to unveil the interwoven complexities and fundamental patterns of social life—actual, perceived, constructed, and analyzed. Such studies take a long time.

Educational researcher Alan Peshkin constitutes a stellar example of a committed fieldworker who lived for periods of time in varied settings to study the intersections between schools and communities. He did fieldwork in a Native American community; in a high school in a stable, multiethnic, midsized city in California; in rural, east-central Illinois; in a fundamentalist Christian school; and in a private, residential school for elites (Peshkin, 1986, 1997, 2000b). To collect data, he and his wife, Maryann, lived for at least a year in and with the community that he was studying. They shopped locally, attended religious services, and developed close relationships with civic leaders as well as teachers and students.

In contrast, evaluation and action research typically involve much shorter durations in keeping with their more modest aims: generating useful information for action. To be useful, evaluation findings must be timely. Decision makers cannot wait for years while fieldworkers sift through mountains of field notes. Many evaluations are conducted under enormous pressures of time and limited resources. Thus, the duration of observations will depend to a considerable extent on the time and resources available in relation to the information needs and decision deadlines of primary evaluation users. Later in this chapter, we'll include reflections from an evaluator about what it

was like being a part-time, in-and-out observer of a program for eight months but only present 6 hours a week out of the program's 40-hour weeks.

On the other hand, sustained and ongoing evaluation research may provide annual findings while, over years of study, accumulating an archive of data that serves as a source of more basic research into human and organizational development. Such has been the case with the extraordinary work of Patricia Carini at the Prospect School in North Bennington, Vermont. Working with the staff of the school to collect detailed case records on students of the school, she established an archive with as much as 12 years of detailed documentation about the learning histories of individual students and the nature of the school programs they experienced. Her data included copies of the students' work (completed assignments, drawings, papers, projects), classroom observations, teacher and parent observations, and photographs. Any organization with an internal evaluation information system can look beyond quarterly and annual reporting to building a knowledge archive of data to document development and change over years instead of just months. Participant observations by those who manage such systems can and should be an integral part of this kind of knowledge-building organizational data system that spans years, even decades.

Rapid Appraisal Studies: Qualitative Inquiries Built for Speed

On the other end of the time continuum from long-term ethnographic studies are rapid appraisal site visits and short-term studies that involve observations of a single segment of a program, sometimes for only an hour or two. Evaluations that include brief site visits to a number of program locations may serve the purpose of simply establishing the existence of certain levels of program operations at different sites. Chapter 1 presented just such an observation of a single two-hour session of an Early Childhood Parent Education program in which mothers discussed their child-rearing practices and fears. The site visit observations of some 20 such program sessions throughout Minnesota were part of an implementation evaluation that reported to the state legislature how these innovative (at the time) programs were operating in practice. Each site visit lasted no more than a day, often only a half-day.

Sometimes an entire segment of a program may be of sufficiently short duration that the evaluator can participate in the complete program. The leadership retreat we observed lasted six days, plus 3 one-day follow-up sessions during the subsequent year.

Rapid appraisal studies, also called rapid reconnaissance and rapid assessment, are aimed at quick, low-cost data collection to answer specific, urgent questions. The purpose is to obtain a narrow, but in-depth understanding of the conditions and needs of the targeted group in a specific area to inform immediate decisions. Qualitative methods are especially appropriate because, being open, naturalistic, and emergent, they make it possible to get into the field quickly. Applications include the following.

1. ***Update a needs assessment, situation analysis, or project proposal, and establish an up-to-date baseline:*** Often months, sometimes years, pass between the time a project is originally formulated and the time when the implementation actually begins. Approval and funding allocation processes can take a long time. The original needs assessment and/or situation analysis on which the proposal was based can become badly out of date. Conducting a rapid appraisal can be used to update the situation based on field visits, purposeful interviews, and retrieving any new documentation.

2. ***Launch a new initiative:*** Quick site visits to engage local people at the start of a project can serve to officially launch an initiative while gathering data about baseline perceptions and conditions and short-term priorities, building relationships, and documenting immediate concerns.

   Example: *When I became codirector in the field for the Caribbean Agricultural Extension Project, our purpose was to develop extension improvement plans in eight Caribbean countries in the Windward and Leeward Islands. By the time the project was funded and personnel recruited, the original needs assessment and project proposal was three years out of date. We also needed to begin engaging local extension and agriculture personnel in the participating countries. With the Caribbean Agricultural Research Development Institute and University of the West Indies, we formed multidisciplinary teams to conduct a rapid reconnaissance of each country's extension and agricultural situation. Each five-person team included two agricultural scientists (e.g., agronomist, soil scientists, plant specialist, etc.), a social scientist, an extension specialist, and an evaluator. The teams spent one week in each country visiting small farmers' fields, interviewing key knowledgeables, examining documents and reports, and assessing needs and capacities. Reports were produced*

by the teams before they left each country so that a briefing could be held with relevant officials and project partners prior to departure. These rapid appraisals served to update the project baseline, renew the situation analysis, and launch the project. Four separate teams (two countries per team) worked simultaneously so that all updated appraisals were completed in two weeks.

3. ***Crisis management:*** A rapid appraisal may be needed when a project falls into crisis. Such crises are precipitated by project staff turnover, changes in the government, or emergencies (floods, droughts, civil unrest).

   Example: *In 2013, Al-Qaeda began assassinating young women engaged in the national polio vaccination campaign in Pakistan. This was in retaliation for the U.S. Central Intelligence Agency having used a fake polio vaccination campaign as a way to locate Osama bin Laden. Pakistani and international humanitarian and health agencies had to halt the vaccination campaign and conduct a rapid appraisal of the changed situation to strategize about how to reconfigure the polio vaccination campaign (Khan, 2013).*

4. ***Solve problems:*** Expert observers are skilled at rapid appraisal. They bring specialized knowledge to bear quickly to solve problems. But to do so, they have to get into the field where they can see what is going on.

   Example: *Sri Lanka was experiencing a failing irrigation system that threatened their food security. Irrigation experts sponsored by the International Irrigation Management Institute conducted a rapid assessment and found inadequate drainage, lack of maintenance of systems, improper water management, excessive seepage, lack of demonstration projects, poor irrigation plots, improper operation, and lack of communication in headquarters (Chambers & Carruthers, 1986).*

5. ***Add independent assessment to controversial or highly visible findings:*** When the findings of a study enter public dialogue, those who are unhappy with the findings will commonly attack the study's methods. Since no study is perfect, flaws will be detected and highlighted. A return to the field to augment a study's findings with a rapid appraisal can help sort out the weight of evidence and validity of conclusions.

   Example: *Controversy arose on the H1N1 virus vaccination campaign in 2009 in Martha's Vineyard, Massachusetts, where towns registered to receive the vaccine independently of the local hospital and physicians' offices. When the planned islandwide vaccination clinic was postponed twice due to delays in vaccine delivery, finger-pointing ensued and the cause of the problem became a matter of contention and visibility. An independent qualitative rapid assessment was able to provide credible and timely data on root causes (Stoto, Nelson, & Klaiman, 2013).*

Quality and Utility of Rapid Appraisal Methods and Applications

Rapid appraisals respond to increasing demands for "real-time" data. Where decisions have to be taken with imperfect data, some data are better than none, and timely findings are better than reports that come in after a decision has been taken. While speed can contribute to utility, the quality and selectivity of *rapid recon* data can raise credibility questions. Rapid appraisal teams need to be carefully selected to be credible, represent diverse perspectives and skills, be politically and culturally sensitive, have the capacity to gather and analyze data quickly, communicate effectively, and work together smoothly under pressure. (For resources and applications, see Bamberger et al., 2011, pp. 67–72; Beebe, 2008; Hargreaves, 2013; Social Impact, 2013; U.S. Agency for International Development, 2010, 2011.)

The critical point of this module is that the length of time during which observations take place depends on the purpose of the study and the questions being asked, not some ideal about what a typical participant observation must necessarily involve. Field studies may be massive efforts with a team of people participating in multiple settings to do comparisons over several years. At times, then, and for certain studies, long-term fieldwork is essential. At other times and for other purposes, as in the case of short-term formative evaluations, it can be helpful for program staff to have an evaluator provide feedback based on just one hour of onlooker observation at a staff meeting, as I have also done.

My response to students who ask me how long they have to observe a program to do a good evaluation follows the line of thought developed by Abraham Lincoln during one of the Douglas–Lincoln debates. In an obvious reference to the difference in stature between Douglas and Lincoln, a heckler asked, "Tell us, Mr. Lincoln, how long do you think a man's legs ought to be?"

Lincoln replied, "Long enough to reach the ground."

Fieldwork should last long enough to get the job done—to answer the research or evaluation questions being asked and fulfill the purpose of the study. That said, short site visits pose special challenges. I've devoted this chapter's rumination (MQP Rumination # 6) to the problems associated with short-term fieldwork (also sometimes known as "quick-and-dirty" site visits).

MQP Rumination # 6 (A Long One)

Evaluation's Site Visit Scandal: Poor-Quality Fieldwork, Worse Ethics

I am offering one personal rumination per chapter. These are issues that have persistently engaged, sometimes annoyed, occasionally haunted, and often amused me over more than 40 years of research and evaluation practice. Here's where I state my case on the issue and make my peace.

Over the years I've heard more complaints from program staff and evaluators about wasteful and useless site visits than any other aspect of evaluation. What kinds of complaints? Little things like that site visits are ineptly conceived, planned, staffed, timed, and conducted. The stories I hear lead me to believe that site visit quality is too often poor in more developed countries and too often outrageous in the developing world.

The problem strikes me as huge, so be prepared; this is a long rumination.

What Are Site Visits?

A "site" is the location of an organization, program, project, community activity, meeting, or event. In short, a site is where something is going on that is of interest and importance to a funding or oversight agency (e.g., a philanthropic foundation, an international agency, or a government administrative unit) or is relevant for the inquiry of a researcher or student

(Continued)

(Continued)

doing fieldwork. The "visit" involves a person or team going to the site to observe what is happening, interview key people, review relevant documents, and write a report. Here are examples, both international and domestic. The World Bank is funding a five-year mother-and-infant nutrition project in Bangladesh, a well-digging project in Burkina Faso, or an agricultural development project in the Andes. Halfway through the project, a three-person team is selected by the World Bank to visit the project for a few days to assess progress. A philanthropic foundation is funding an adult literacy program in a rural community and sends a program officer out for a day to observe some classes and talk with participants. The U.S. Department of Transportation and several state agencies are funding driver education programs aimed at reducing use of cellular phones while driving; a team of national and state staff go for two-day site visits to observe four such programs in different states. In essence, evaluation site visits involve fieldwork *on location* to report independently on how a project is being implemented and what it is accomplishing.

Midterm and end-of-project evaluation teams may well be the most common form of formal evaluation activity in the world. There are thousands of such reviews commissioned by scores of funding agencies every year. Many are well done by competent site visitors (evaluators, researchers, program experts) who produce credible reports that are helpful both to people at the site and to the agency that has commissioned the site visit. So what's the problem? Many site visits are organized poorly, staffed inadequately, and conducted badly (poor-quality fieldwork), leading to useless reports. How and why does this occur? Here's the tip of the iceberg.

Site Visit Problems: Some Examples

- Contracting processes to hire independent international site visitors are often mired in bureaucratic solicitation procedures that create delays in getting teams assembled and into the field. Once funds are finally released, the procurement timelines are unmanageably short. I regularly see the solicitations seeking evaluators to conduct site visits, many such requests every month. I receive phone calls and e-mails asking if I'm available. But who's available to pick up and leave on two weeks' notice to spend a month doing site visits in remote areas of Eritrea? Or India? Or Ecuador? Except in unusually fortunate cases, the least qualified evaluators are the ones who are available on short notice, including novices and those who don't have much work because they simply aren't very good. International evaluation colleague Ricardo Wilson-Grau has dubbed the contract solicitation process "tendering madness," and he has his own ruminations about it, which includes this astute observation, with which I heartily agree:

Tendered evaluations commit the mortal sin of focusing on the formal written scope of work rather than on the flesh and blood. Even the most sophisticated checklists for selecting the best tenders, and which include, of course, the evaluators' qualifications, do not emphasize what matters most to the success of an evaluation: the evaluator.

I posed the question of why this problem is so pervasive to an international development aid site visit contractor with 25 years' experience. He told me, "Truthfully, at the end of the day the main people available are retired, alcoholic, burned out, cynical-to-the-core former aid workers and foreign service officers who want to escape their spouses and pick up a few bucks while drinking their way through the assignment." Exaggeration? Hyperbole? Rant? As I probed his perspective, based on his years of administering, organizing, managing, and supervising a large number and variety of site visits, it became clear that he knew firsthand that of which he spoke—for he was, himself, a near-retirement, alcoholic, burned-out, cynical-to-the-core aid administrator. He confessed that he didn't really care about the quality of the work of people hired as site visitors. "No one pays any attention to site visit reports anyway," he explained. "It's all part of the game, just window-dressing to give the appearance of some kind of accountability." So he hired people like himself who followed regulations, did the paperwork right, didn't make waves, and made good drinking buddies.

- I asked a highly experienced international colleague if he thought drunks conducting site visits was a widespread problem. He launched into his own war stories, including an especially embarrassing incident in an indigenous community that struggled with alcoholism where he was the last to recognize that a person assigned to his team, feigning illness, was actually drunk. The local program people had to point out the problem to him. Then he added, "And don't forget to mention the sexual predators and harassers."
- From program directors and staff I hear complaints about site visits being planned and imposed with no input from them as to timing, purpose, or process. The logistics are a nightmare. The site visit teams are ill prepared, haven't done their homework, take up lots of staff time, expect VIP treatment, and are condescending, arrogant, rude, and, too often, incompetent, corrupt, or both. Junior or novice evaluators are often assigned tasks that should be done by senior evaluators (who may be off carousing instead of working).
- Domestically, philanthropic foundations and governmental agencies too often send program officers and administrative staff into the field to conduct site

visits with little or no training, a vague scope of work, and little advance notice to sites and programs being visited.

Those complaints are somewhat abstract, so let me illustrate with some concrete examples.

Examples of the Underbelly of Site Visits

A session at the 2005 joint AEA/CES (American Evaluation Association/Canadian Evaluation Society) International Evaluation Conference in Toronto on evaluation site visits featured many horror stories documenting that external site visit evaluators hired by international donors often lack minimum evaluation competence and expertise to carry out their assignments. Alexey Kuzmin (2005) did his doctoral dissertation on evaluation capacity building. In the course of his fieldwork, he turned up a number of instances of unscrupulous evaluators engaged in unethical practices while conducting site visits. In one case, an American evaluation site visitor to a small program arrived late one day, left early the next, never visited the program, and refused to take the documents offered by the program. Sometime later, the program director received an urgent demand from the funder for required documentation (the very documentation that the evaluator was supposed to have taken, analyzed, and presented to the funder), which it took considerable expense to get to the funder by express courier. Subsequently, the program was denied additional funding for lack of an evaluation having been completed.

In another instance, an external evaluator arrived, one without any substantive expertise in the program's area of focus. He spent a couple of days with the program and then disappeared, leaving an unpaid hotel bill and expenses that the program had to cover because it had made arrangements for his visit. A couple of months later, the program director received a draft evaluation report for her comments, but the deadline for sending comments had already passed. The evaluation contained numerous errors and biased conclusions based on the evaluator's own prejudices. The program director spent a considerable amount of time writing a response to the evaluation, but the evaluation, with its errors and negative judgments, had already been posted on a prominent agency website. The program had to resort to expensive legal action to have the evaluation removed.

A third example concerns a very affable evaluator of a program undergoing an external midterm review. The American evaluation team leader was a charming man who gave a great deal of attention to the program director. In a private conversation with her, he talked in depth about an organizational development model he had recently implemented in another country. It gradually became clear that the evaluator was offering positive evaluation findings if the director hired him as a consultant to return and implement his model in her agency. He said, "I really want to say good things about your program. I want to support it, but I need to hear some things from you before I do that." The program director, feeling quite vulnerable, contacted a lawyer for advice about how to protect herself and her agency, advice she heeded, but found the whole experience traumatic.

A more recent example involved an evaluator who refused the full services of a translator during a three-day site visit. His scope of work included interviewing program participants. He said that he would use the translator to ask questions in the participants' African language, but he didn't want the responses translated. He explained, "I'm an expert at reading body language. I've been all over the world. I can tell by watching people whether they are having good program experiences. Translations can't be trusted. I trust what I see." The program director was flabbergasted but afraid of alienating him, so she deferred.

In a similar instance, the program director contacted the donor agency that had sent the site visit evaluator and got an appropriate response. The site visit was terminated. But the program director felt great trepidation about complaining to the donor agency about an inept evaluator, fearing that it would be interpreted as resistance to being accountable. The power dynamics are palpable.

Student Site Visits

Site visit problems are not always (or even usually) so dramatic. Nor are site visits just done by professionals. Students often conduct site visits to local agencies as part of a course, an academic paper assignment, or thesis work. Seldom do the professors who assign such site visits prepare or train students. They just send them off to get some experience in the real world. I've heard story after story from program directors about students who were attentive, engaged, sensitive, and a delight to involve—and I've heard stories about students who were insensitive, inappropriate, and clueless.

An evaluation colleague told me about observing a graduate student do a site visit while she was also on site for an evaluation she was conducting. "I remember the student sitting in one of the classrooms for quite a long time, more than a couple of hours. I circled back to this room and asked her how it was going. I was especially interested in what she was learning about the culture of the school, which was the focus of her observations. I asked, so what are you learning? She yawned, looked bored, and said, "Not much," and added that she would probably leave soon. I noticed that she had only a half-page of notes. The next time I looked for her, she was gone. Imagine my surprise when I found out that she had written up her observations as a chapter in a book on school innovation. The methods description in that chapter did not correspond to the degree and nature of fieldwork I knew had actually occurred."

(Continued)

(Continued)

Anecdotal Evidence

These anecdotes are sufficient, I hope, to illustrate the nature and variety of site visit problems. I have many more I could share, including many positive ones. MQP Rumination # 1 in Chapter 1 (pp. 31–33) was about valuing anecdotes as data. They are illuminative and illustrative, not representative or generalizable. What I've reported here are outlier (hopefully) anecdotes to illustrate the nature of the problem, not the scope of the problem, which is hard to determine. In discussing these examples with a former graduate student, Donna Podems, who is a full-time independent evaluation consultant working globally, she commented,

> I have been doing site or field visits for 20 years. It makes up a bulk of the work I do. I have had a range of experiences with site visits, some great, some awful, but the majority of my site visits are good. Not always great. Rarely bad. But good.

That strikes me as a fair and balanced picture.

But the patterns of ineptness and ethical lackadaisicalness in the negative anecdotes are sufficient, in my opinion, to raise concern. Certainly, I know of agencies that work hard to ensure quality site visits and evaluators who bring knowledge, skill, and competence to the work of site visits. In hopes of further enhancing sensitivity to the nature of the problem, let me invite you to a short stroll down memory lane.

Site Visit Ethics: Tainted History

Potemkin Village Site Visits

The phrase *Potemkin village* refers to a fake village, built only to deceive and falsely impress. According to legend, in 1787, Grigory Potemkin, governor of Crimea, erected fake settlements along the banks of the Dnieper River to fool Russian empress Catherine II and her entourage, including several foreign ambassadors, during their visit to the region. Potemkin wanted to show that the region, which had been devastated by war, had been rebuilt and was prospering under his governance. He constructed portable village facades to give the appearance of development. As the queen's barge passed by, Potemkin had his men dressed up as peasants and looked like they lived and worked in the flourishing village. Once the barge floated down river, the fake peasants would take down the phony facades and rebuild at a new site during the night, repeating the deception at various spots along the route.

The phrase *Potemkin village* has since come to be used generally to designate and denigrate any fake creation built to deceive people into thinking that conditions are better than they really are. For observers engaged in fieldwork, the story raises the question of whether what is being observed is real and typical. For program evaluation site visits in

particular, which are often short in duration and planned by the people whose program is being evaluated, the potential for making things appear rosier than they are is a serious validity concern. Observers need to be able to exercise some control over where they go, what they see, and who they talk to in order to ensure that they are not being manipulated into observing only what those in authority want them to observe.

WELCOME TO
Potemkin
NEW YORK

"Wait a minute! Isn't that the same village idiot we interviewed yesterday in Jersey?"

True or False

From a research and evaluation perspective, the *Potemkin village* story also symbolizes the problem of finding out what the truth is. Historians debate whether the *Potemkin village* deception actually happened, and if it did, whether the empress was in on the plot as a way of impressing the foreign ambassadors accompanying her. This last possibility is boosted by the fact that Potemkin and the empress were lovers, though it must be acknowledged, it is not unprecedented for lovers to deceive each other. Whether the original story is true or false, the story has endured and serves as a cautionary tale for observers to look behind the facades offered by those in positions of authority who want to make a positive impression. Cultivate multiple sources. Examine a situation from diverse perspectives. Dig beneath the surface to find out what's really going on. *Triangulate.*

The Infamous Theresienstadt Site Visit

Near the end of World War II, rumors began to circulate about the Nazi extermination of Jews. The Nazis wanted to refute those rumors, so they took one camp, called Theresienstadt, in what was then Czechoslovakia, and turned it, if ever so briefly, into a model town. They shot a movie there to prove how well they treated the Jews and invited the Red Cross to inspect it. This was 1944. In preparation for the Red Cross inspection, a beautification plan was implemented by the Nazi administration. They painted buildings, planted flowers, opened stores, and put up a bandstand in the town square. They built a kindergarten in a small park, near a children's home. They opened a coffee shop. They turned the concentration camp into a pleasant little town. And they made it a lot less crowded. Just before the Red Cross delegation arrived, the Nazis shipped 7,500 people off to the Auschwitz death camp, reducing crowding and creating more open spaces.

On June 23, 1944, the Red Cross delegates came for a one-day inspection. The Red Cross delegates went exactly where the Nazis took them—and only where they took them. They didn't question a single prisoner. The Swiss head of the delegation took pictures showing how happy the children looked—and included them in his report. The final report of the Red Cross delegation reported that Theresienstadt looked like a normal provincial town where the "elegantly dressed women all had silk stockings scarves and stylish handbags." The delegates also wrote that Theresienstadt was a final destination camp and that people who come there were not sent elsewhere—because that's what the inspection team was told. In fact, by the time of the visit, some 68,000 people had already been shipped from there to the death camps (CBS News, 2007).

Back to the Present

Think Theresienstadt is just yesterday's news? I recently had a lengthy exchange with an evaluator who was disturbed by having just come from a midterm site visit of a large and important but controversial project funded by an international agency in an Asian developing country. All sites visited were selected by the host government. Only project participants selected by the host government were interviewed and always with project and/or government people present. Most requested documents and data were "not available." The team had to produce and hand in their report in the country before they departed. It is not clear what happened to the report. On the midterm review team of four people, only one person had any evaluation background, knowledge, or training—the person who was telling me about the experience—and that person was the only one who saw any problem with the way the review was organized and the only one who complained, to no avail. I can't reveal details of the project without breaking confidentiality, but I can tell you that the project involved issues of life and death for the intended beneficiaries and huge potential embarrassment to the host country. Every aspect of the site visit was managed and manipulated to produce findings that would portray the government's policies in a positive light. The evaluator complained to me about the corruption of the process but admitted having signed off on the report

because "otherwise I wouldn't have been paid and my expenses wouldn't have been reimbursed." (At the end of this rumination, I'll suggest some guidelines to help avoid or at least ameliorate this kind of situation.)

The Perspective From the Site

Programs on the receiving end of being "visited" (what a euphemism!) for evaluation recognize the high stakes and act accordingly. A colleague summed up the situation succinctly:

> The site does not have direct power over a site visit as the visit is almost always externally mandated and contracted. The information collected will potentially be used to make decisions about the site or intervention. So a site visit is usually like a high stakes evaluation without any real accountability, affecting continuation, expansion, contraction, or even termination. So sites try to protect themselves, which is the absolutely rational thing to do. The site prepares for the visit by sweeping up messes, cleaning, priming community storytellers, arranging with counterparts or counterpart proxies to drop by for support, filing data in surprisingly obscure places, and even going so far as to arrange a preemptive invitation by a well-known, high-status expert on the pretext of providing advice to the project, but making sure the site visitor knows. The truth is that everyone is gaming the system on a site visit.

> But sometimes, *sometimes*, an external visitor is competent, attentive, and sincere, and appears to have something to contribute and wants to do so. Learning can occur. Mutual understanding can emerge. Good things can happen. One such visitor came to a long-term project where I was M&E [Monitoring and Evaluation] team leader and got the keys to the kingdom, listened well, offered insightful feedback, and left it a better place. He is still remembered.

Reflections From a Veteran Site Visitor

In discussing site visits with veteran evaluators, I collected many stories and reflections, more than I can include in this rumination. Most reported that, on balance, their experiences were positive. But all had run into some situations that gave rise to concerns. Doug Horton, an international evaluator with more than 40 years' experience in the field, told me he'd never witnessed the worst excesses reported here but he had his own list of problematic encounters. Here are his reflections:

To Party or to Evaluate? That Is the Question

> Sometimes it isn't the evaluators who want to drink, but the program team being evaluated or local collaborators or beneficiaries. A common strategy of program teams that organize site visits is to plan a

(Continued)

(Continued)

"show-and-tell" event that combines a presentation of the program's activities and a celebration with local partners or beneficiaries. In such cases, the evaluators may find that instead of presenting an opportunity for data gathering, the site visit becomes a party!

In one case that an international evaluator relayed to me, the evaluation team drove several hours to a remote community where 30 heads of household (mainly men) were lined up in the town square, ready to great the team. After formal greetings, the evaluators were informed that they had about an hour before lunch, to observe the program's activities, meet with people, and conduct interviews. During this hour, the gender specialist found it impossible to meet with women, who were all involved in preparing the lunch! After an hour, an array of local dishes was spread out on tables in the local school, and the lunch was accompanied by speeches (generally extolling the virtues of the program being evaluated), music, and drink.

After lunch, the festivities continued, severely limiting the possibility of "serious fieldwork," until the climax, when the evaluation team leader was presented with a live rooster, which he was expected to take back to the capital city—presumably to prepare a chicken soup to recover from the drinking.

In this case, the evaluators decided that it would be too rude to stop the party and focus on the evaluation, so they participated in the festivities and did their best to gather information on the sidelines. The following morning, while some of the program team members were nursing hangovers, the evaluators called a meeting with the program leaders to clarify the purposes and appropriate activities for future site visits and emphasize the importance of using the limited time available for gathering information that would be of use for program improvement and providing accountability to sponsors.

How to Stop the Speeding Train?

When confronted with an evaluation, program teams often try to give the evaluators so much detailed information that they will never be able to "see the forest for the trees." This is sometimes referred to as the *mushroom strategy*, since the program team is attempting to cultivate the evaluation the way one grows mushrooms: by burying them in bullshit and keeping them in the dark.

A variant of the mushroom strategy commonly used by those who plan site visits is to pack so much in and keep the evaluators moving around so much that they don't have time to reflect on their preliminary findings or adjust their work plan. A common feature of international site visits is that the organization or program being evaluated organizes such a "dense" series of site visits—often in several parts of the country and sometimes even in different countries— that there is precious little time left for the evaluators to review their notes (if they were able to take any), meet for joint discussions and reflections on their observations, or begin working on their report. In cases like this—which are very common—an important job for the evaluators is to renegotiate the schedule of site visits to focus on the collection of the most essential information from the most essential sites, and to carve out time for evaluation team members to discuss and reflect on their findings as they evolve and adjust the evaluation plan as needed. (Horton, personal e-mail communication, July 18, 2014. Used with permission.)

Reframing Site Visits

A highly experienced evaluation colleague has been "wrestling with the site visit ball-and-chain for some time." While the term *site visit* can be understood to mean something very formal, "a very planned dog-and-pony show by program staff to demonstrate the greatness of a project, we're trying very hard in a large, multisite evaluation to avoid those situations by reframing site visits as *case study data collection* opportunities." The evaluation team was working to avoid site visits becoming "program song and dance routines, but we're running up against a program culture of expectations that are hard to change." She went on to describe arriving at a site to conduct a planned two-hour interview with the leadership of a project, only to be confronted with a formal PowerPoint presentation and told that they could ask questions for a half-hour after the presentation. They were able to negotiate more time for an interview, but the program staff still felt a strong need to make a formal presentation, an illustration of the high-stakes politics and pressures that can surround site visits.

Site Visit Standards

Most site visits may well be conducted appropriately, professionally, and effectively. Certainly many are. Some of my most memorable experiences have been conducting site visits with engaged and insightful colleagues. Going into the field to see what is happening is essential. On the other hand, some proportion of site visits is conducted insensitively, incompetently, inappropriately, and unproductively. How many is not and cannot be known. My sense is that the numbers are sufficiently large and the poor practices are sufficiently widespread to constitute a significant problem. As a starting point toward reducing the problem, then, I propose 10 standards for the conduct of site visits.

Draft Standards for Site Visits

1. *Competence:* Ensure that site visit team members have skills and experience in qualitative observation and interviewing. Subject matter expertise, like an economics doctorate (to pick a random example), does not suffice. Nor does simply being available for the assignment.

2. *Knowledge:* Where the purpose of the site visit is evaluative, ensure that at least one team member, preferably the team leader, has evaluation knowledge and credentials. I apologize for being redundant, but this deserves emphasis: *Subject matter expertise does not suffice. Nor does simply being available for the assignment.* Evaluation is a profession grounded in a body of knowledge, standards, specific ethical principles, and skills, including cultural competence. It's a radical suggestion, I realize, but teams conducting evaluation site visits should include a professional evaluator.

3. *Preparation:* Site visitors should know something about the site being visited. Whether this knowledge is procured through background materials, briefings, prior experience, or some combination thereof, one of the most common complaints from people at the site is that site visitors display a blinding ignorance of place, context, and what goes on at the site. Those who commission, fund, and otherwise engage site visitors have an obligation to ensure some reasonable level of preparation.

4. *Site participation:* People at the site should be engaged in planning and preparation for the site visit to minimize disruption to program activities and services to program beneficiaries. Logistics, costs, scheduling, access to documents, and the nature of the inquiry should be planned and transparent, with reasonable consideration of how the inquiry may be intrusive.

5. *Do no harm:* Site visits are not benign. The stakes can be high. People can be put at risk. Programs can be put at risk. Good intentions, naïveté, and general cluelessness are not excuses. Apologies after the fact are insufficient. Be alert to what can go wrong. Take responsibility. Be accountable. Commit as a team that you will, first and foremost, do no harm. (This goes for those who commission evaluations as well as those who conduct them.)

6. *Credible fieldwork:* While those at the site should be involved and informed, they should not control the information collection in ways that undermine, significantly limit, or corrupt the inquiry. The site visitors, for example, should determine the sample of activities observed and people interviewed, insofar as possible, and should be able to arrange confidential interviews to enhance data quality.

7. *Neutrality:* A site visitor should not have a preformed position on the intervention or the intervention model. Openness to what the data reveal is essential but cannot be assumed.

8. *Debriefing and feedback:* Before departing from the site visit, some appropriate form of debriefing and feedback should be provided to key people at the site. In addition to providing highlights of the findings, insofar as possible and appropriate, those at the site should know when they will receive a report, either oral or written or both. If no reporting will be shared, why not? What follow-up will occur?

9. *Site review:* Those at the site should have an opportunity to respond in a timely way to site visitors' reports, to correct errors and provide an alternative perspective on findings and judgments, if necessary. Site visit reports, often done in haste, may be infused with site visitors' biases and preconceptions. The fact that they're independent and external doesn't make them neutral or accurate. Triangulation and a balance of perspectives should be the rule.

10. *Follow-up:* The agency commissioning the site visit should do some minimal follow-up (a brief survey or phone call) to assess the quality of the site visit from the perspective of the locals on-site. And the site visit team should know in advance that such a follow-up assessment will take place. A philanthropic foundation in Minnesota routinely has its own program officers conduct due diligence site visits. The executive director sends out a short follow-up survey that comes directly to her—*and that she reads and responds to*. The results have turned up a number of problems that have informed staff development on how to conduct a site visit appropriately.

Rethinking Site Visits

These proposed standards accept site visits as an ongoing method for conducting external evaluations. But the notion that a valid, meaningful, and useful evaluation can result from a short visit by an external team with limited knowledge of context and little time to collect data is hugely problematic. The widespread use of site visits as a central form of accountability evaluation needs to be rethought. But that's a different rumination, for another forum. So while my preferred solution is eliminating ritualistic site visits because the method is inherently flawed, the proposed standards are aimed at some modest improvements and preventing the worst abuses.

But 10 standards may be too many. In that regard, one principle may well suffice, the standard offered by Leslie Goodyear et al. (2014) as the final sentence in their important book *Qualitative Inquiry in the Practice of Evaluation*:

> The responsibility of a qualitative evaluator is to leave the setting enriched for having conducted the evaluation.

Variations in Observational Focus and Summary of Dimensions Along Which Fieldwork Varies

> Why so many options? Why so many variations?
>
> Why so many choices? Why so many decisions?
>
> Because it's a messy world. One size doesn't fit.
>
> —Halcolm

Module 44 discussed how observations vary in the extent to which the observer participates in the setting being studied, the tension between *insider* and *outsider* perspectives, and the extent to which the purpose of the study is made explicit. Module 45 discussed variations in the duration of the observations. A major factor affecting each of these other dimensions is the scope or focus of the study or evaluation. The scope can be broad, encompassing virtually all aspects of the setting, or it can be narrow, involving a look at only some small part of what is happening.

Parameswaran (2001) wanted to interview young women in India who read Western romance novels. Thus, her fieldwork had a very narrow focus. But to contextualize what she learned from interviews, she sought "active involvement in my informants' lives beyond their romance reading." How did she do this?

> I ate snacks and lunch at cafes with groups of women, went to the movies, dined with them at their homes, and accompanied them on shopping trips. I joined women's routine conversations during break times and interviewed informants at a range of everyday sites, such as college grounds, homes, and restaurants. I visited used-book vendors, bookstores, and lending libraries with several readers and observed social interactions between library owners and young women. To gain insight into the multidimensional relationship between women's romance reading and their experiences with everyday social discourse about romance readers, I interviewed young women's parents, siblings, teachers, bookstore managers, and owners of the lending libraries they frequented. (p. 75)

The tradition of ethnographic fieldwork has emphasized the importance of understanding whole cultural systems. The various subsystems of a society are seen as interdependent parts, so that the economic system, the cultural system, the political system, the kinship system, and other specialized subsystems could only be understood in relation to each other. In reality, fieldwork and observations have tended to focus on a particular part of the society or culture because of specific investigator interests and the need to allocate the most time to those things that the researcher considered most important. Thus, a particular study might present an overview of a particular culture but then go on to report in greatest detail about the religious system, kinship system, or technology system of that culture.

In evaluating programs, a broad range of possible foci makes choosing a specific focus challenging. One way of thinking about focus options involves distinguishing various program processes sequentially: (a) processes by which participants enter a program (the outreach, recruitment, and intake components), (b) processes of orientation to and socialization into the program (the initiation period), (c) the basic activities that make up program implementation over the course of the program (the service delivery system), and (d) the activities that go on around program completion, including follow-up activities and client outcomes over time. It would be possible to observe only one of these program components, some combination of components, or all of the components together. Which parts of the program and how many are studied will clearly affect issues such as the extent to which the observer is a participant, who will know about the evaluation's purpose, and the duration of observations.

Chapter 5 discussed how decisions about the focus and scope of a study involve trade-offs between breadth and depth. The very first trade-off comes in framing the research questions to be studied. The problem is to determine the extent to which it is desirable and useful to study one or a few questions in great depth or to study more questions but each in less depth. Moreover, in emergent designs, the focus can change over time.

Sensitivity of the Inquiry: Potential for Controversy

The degree to which fieldwork is inquiring into matters that are sensitive or potentially controversial will affect how the inquiry is conducted. Occasionally, I have been asked to help prepare university students for cross-cultural experiences in which they will be doing fieldwork on some topic of interest. Most have become accustomed to Western media norms where no questions are off-limits. They propose research on politics, sexual practices, gender relationships, sexual orientation, family power dynamics, income distribution, and religion as if these were straightforward topics for open study anywhere in the world. Quite the contrary. Most people in most cultures in the world do not openly and easily discuss these matters with each other, much less with strangers. Such obliviousness is potentially insulting and sometimes dangerous. Ultimately, engaging in fieldwork requires building trusting relationships with the people in the setting being studied. Inquiry into sensitive and potentially controversial issues requires delicacy, time, and judgment. Key informants who understand the local context will be crucial to learning how to approach sensitive issues.

Dimensions Along Which Fieldwork Varies: An Overview

We've examined 10 dimensions that can be used to describe some of the primary variations in fieldwork (Modules 44–45). Those dimensions are summarized in Exhibit 6.4. Each is a continuum. These dimensions can be used to help design observational studies and make decisions about the parameters of fieldwork. They can also be used to organize the methods section of a report or dissertation in order to document how research or evaluation fieldwork actually unfolded.

EXHIBIT 6.4   Dimensions of Fieldwork: Variations and Options Along Continua

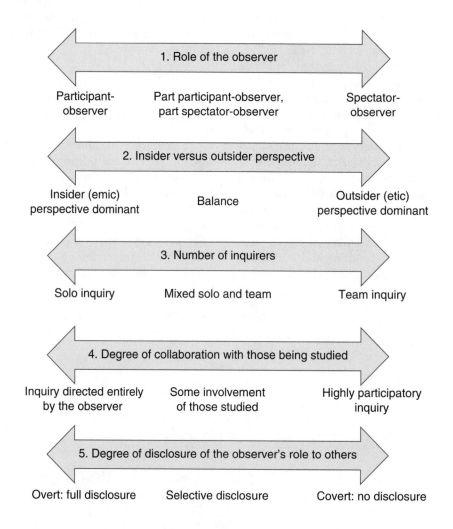

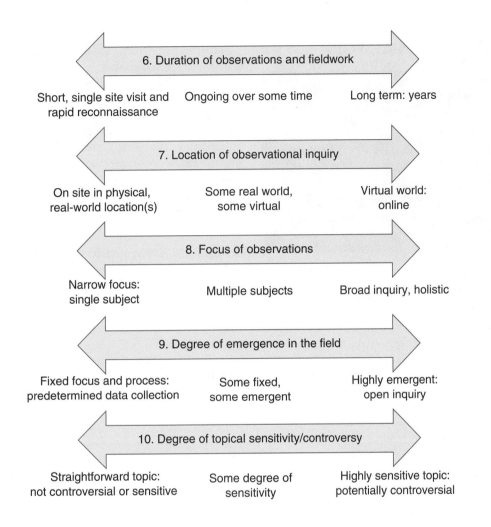

6. Duration of observations and fieldwork

Short, single site visit and   Ongoing over some time   Long term: years
rapid reconnaissance

7. Location of observational inquiry

On site in physical,   Some real world,   Virtual world:
real-world location(s)   some virtual   online

8. Focus of observations

Narrow focus: Multiple subjects   Broad inquiry, holistic
single subject

9. Degree of emergence in the field

Fixed focus and process: Some fixed,   Highly emergent:
predetermined data collection   some emergent   open inquiry

10. Degree of topical sensitivity/controversy

Straightforward topic: Some degree of   Highly sensitive topic:
not controversial or sensitive   sensitivity   potentially controversial

A SENSITIZING FRAMEWORK

I keep six honest serving men.

They taught me all I knew:

Their names are What and Why and When

And How and Where and Who.

—Rudyard Kipling

NASA (National Aeronautics and Space Administration) space scientist David Morrison (1999) once commented that, in astronomy, geology, and planetary science, observation precedes theory generation "and the journals in these fields never require the authors to state a 'hypothesis' in order to publish their results" (p. 8).

A recent example is the famous Hubble Space Telescope Deep Field in which the telescope obtained a single exposure of many days duration of one small field in an unremarkable part of the sky. The objective was to see fainter and farther than ever before, and thus to find out what the universe was like early in its history. No hypothesis was required—just the unique opportunity to look where no one had ever looked before and see what nature herself had to tell us.

In many other sciences the culture demands that funding proposals and published papers be written in terms of formulating and testing a hypothesis. But I wonder if this is really the way the scientific process works, or is this just an artificial structure imposed for the sake of tradition. (p. 8)

Part of the value of open-ended naturalistic observations is the opportunity to see what there is to see without the blinders of hypotheses and other preconceptions. Pure observation. As Morrison put it so elegantly, just the unique opportunity to look where no one has ever looked before and see what the world has to show us.

That's the most open and emergent approach to fieldwork. A different approach is to enter the field to test theory and, especially, causal hypotheses (George & Bennett, 2005; Gerring, 2007; King et al., 1994).

How to focus fieldwork, then, is a matter of differing perspectives. Certainly it's not possible to observe everything. The human observer is not a movie camera, and even a movie camera has to be pointed in the right direction to capture what is happening. For both the human observer and the camera, there must be focus. In fieldwork, this focus is provided by the study design and the nature of the questions being asked (Chapter 5). Once in the field, however, the observer must somehow organize the complex stimuli experienced so that observing becomes and remains manageable.

Sensitizing Concepts to Guide Fieldwork

Experienced observers often use "sensitizing concepts" to orient fieldwork. Chapter 5 discussed sensitizing concepts as an approach to sampling, essentially selecting cases that are exemplars of the concept of interest. As noted in the purposeful sampling discussion, qualitative sociologist and symbolic interactionist Herbert Blumer (1954) advocated using sensitizing concepts as a guide to fieldwork with special attention to the words and meanings that are prevalent among the people being studied. More generally, "a sensitizing concept is a starting point in thinking about the class of data of which the social researcher has no definite idea and provides an initial guide to her research" (van den Hoonaard, 1997, p. 2). Sensitizing concepts in the social sciences include loosely defined notions like victim, stress, stigma, and learning organization, which can provide some initial direction to a study as a fieldworker inquires into how the concept is given meaning in a particular place or set of circumstances being studied (Schwandt, 2001).

Rudyard Kipling's poem about his "six honest serving men," quoted above, constitutes a fundamental and insightful sensitizing framework identifying the central elements of good description. In social science, "group process" is a general sensitizing concept as is the focus on "outcomes" in evaluation. Kinship, leadership, socialization, power, and similar notions are sensitizing in that they alert us to ways of organizing observations and making decisions about what to record. Qualitative methodologist Norman Denzin (1978a) has captured the essence of how sensitizing concepts guide fieldwork:

The observer moves from sensitizing concepts to the immediate world of social experience and permits that world to shape and modify his conceptual framework. In this way he moves continually between the realm of more general social theory and the worlds of native people. Such an approach recognizes that social phenomena, while displaying regularities, vary by time, space, and circumstance. The observer, then, looks for repeatable regularities. He uses ritual patterns of dress and body-spacing as indicators of self-image. He takes special languages, codes and dialects as indicators of group boundaries. He studies his subject's prized social objects as indicators of prestige, dignity and esteem hierarchies. He studies moments of interrogation and derogation as indicators of socialization strategies. He attempts to enter his subject's closed world of interaction so as to examine the character of private versus public acts and attitudes. (p. 9)

The notion of "sensitizing concepts" reminds us that observers do not enter the field with a completely blank slate. While the inductive nature of qualitative inquiry emphasizes the importance of being open to whatever one can learn, some way of organizing the complexity of experience is virtually a prerequisite for perception itself. Exhibit 6.5 presents examples of common sensitizing concepts for various areas on inquiry. These concepts constitute ways of breaking the complexities of planned human interventions into distinguishable, manageable, and observable elements; they are by no means exhaustive of sensitizing concept possibilities but rather illustrate oft-used ways of organizing an agenda for inquiry. These concepts serve to guide initial observations as the inquirer watches for incidents, interactions, and conversations that illuminate these sensitizing concepts in a particular setting.

Overused Sensitizing Concepts Can Become Desensitizing

An Example of a Fieldwork Observation: Capturing the Interplay and Tension Between Two Evaluation Sensitizing Concepts

The Paris Declaration on Development Aid (OECD, 2005) posits five principles aimed at changing the relationships between donor countries and recipient countries. An evaluation was commissioned to assess implementation and impact. The credibility of the evaluation would depend, in part, on *evaluator independence*, a central principle of external evaluation and thus a sensitizing concept for observing how the evaluation was conducted. On the other hand, the relevance and utility of the evaluation would depend on the involvement of key stakeholders. To that end, an international advisory group was assembled to have input into the evaluation. *Stakeholder involvement*, then, is another central principle of evaluation and also a sensitizing concept for observing how the evaluation was conducted. These two principles, evaluator independence and stakeholder involvement, can come into conflict. Too much stakeholder involvement can undermine evaluator independence, and vice versa.

These two concepts are relational in that evaluator independence can only be meaningful in relation to some kind of real or potential effort to interfere with, undermine, intrude on, or otherwise threaten independence. Likewise, stakeholder involvement is a matter of degree, varies hugely in type and amount, and only has meaning in relationship to the evaluation and evaluator. Howard Becker (1998) has asserted, convincingly, I think, that

> all terms describing people are relational—that is, they only have meaning when they are considered as part of a system of terms. . . . That seems obvious enough. But it's one of those obvious things that people acknowledge and then ignore. (pp. 132–133)

So let's look at the relationship between stakeholder involvement and evaluator independence in the evaluation of the Paris Declaration as an example of using sensitizing concepts to guide fieldwork.

Observation

The international advisory group for the evaluation met with the core evaluation team for two days in Copenhagen in 2011 to review and provide comments on the draft final evaluation report shortly before it was due to be publically issued. (There had been two prior meetings of the advisory group at earlier stages of the evaluation.) A critical and revealing exchange arose near the end of the meeting when, near the end of two days of intense interaction, two advisory members pushed for one more round of review before the report was finalized. At this point, the pressure on the Core Evaluation Team to complete the report on time was quite intense. To write yet another draft and send it out for review would involve delay and threaten delivering the final report on time. The leader of the Core Evaluation Team expressed concern about such a delay and added, "The final evaluation report is *our* report. We have listened to your feedback and will incorporate it as much as we think is appropriate, but our independence requires that we, and we alone, produce the final report."

The tone of the demand from the advisory members who wanted to see one more draft became challenging. One said, in a loud voice, with finger pointing at the evaluation leader, "You tell us that you'll take our feedback into account, but now you tell us that we won't get a chance to review a new draft and react to whether you have accurately heard and incorporated our feedback."

Silence enveloped the room after this exchange; the tension was palpable. The chair of the meeting called for a short break. Negotiations ensued in the hallway among key actors in the process from both the evaluation and the advisory group sides. Arguments from both sides were vehemently repeated. Options were offered and rejected. But after 20 minutes, a resolution emerged. The Core Evaluation Team agreed to circulate one part of the report for final commentary and feedback, the Conclusions and Recommendations section. This was the part of the report that would likely receive the greatest attention when it was publicly issued. However, instead of producing the revised section in three languages (English, French, and Spanish, as was

---

EXHIBIT 6.5   Examples of Sensitizing Concepts in Various Contexts

Sensitizing concepts are terms, phrases, labels, and constructs that invite inquiry into what they mean to people in the setting(s) being studied. Quantitative inquiry begins by operationalizing concepts (determining measurement). Qualitative inquiry using sensitizing concepts leaves terms purposefully undefined to find out what they mean to people in a setting. *Sensitizing concepts are windows into a group's worldview.*

The inquiry questions are as follows: What does this concept mean in this context to these people? What are the variations in meaning and the implications of those variations?

FIELD OR CONTEXT	SENSITIZING CONCEPTS
Health	• Wellness • Quality of life • Continuum of care
Education	• Special needs • At risk • Gifted and talented • Racial integration
Environment	• Sustainability • Ecosystem health • Ecological footprint
Mental health	• Mental illness • Trauma • Stress
Government	• Accountability • Policy • Public interest
Philanthropy	• Collaboration • Impact • Portfolio
Planning	• Vision • Strategy • Goals

FIELD OR CONTEXT	SENSITIZING CONCEPTS
Program evaluation	• Stakeholder involvement • Use • Evaluator independence
Organization	• Learning organization • Leadership • Power
Community	• Diversity • Social capital • Healthy community
International agency	• Global system • Development • Equity
Journalism	• Fairness • Balance • Impartiality
University	• Interdisciplinary • Knowledge • Theory

Here is a note of caution about sensitizing concepts. When they become part of popular culture, they can lose much of their original meaning. Philip Tuwaletstiwa, a Hopi geographer, relates the story of a tourist cruising through Native American areas of the Southwest. He overheard the tourist, "all agog at half-heard tales about Hopi land," ask his wife, "Where are the power places?"

"Tell her that's where we plug-in TV," he said (quoted by Milius, 1998, p. 92).

the custom), the revised conclusion would only be available in English, and there would be only *one week for responses*. The evaluation team agreed to take the responses seriously but reiterated that the final report would be informed by feedback, especially about clarity and understandability, but the substantive conclusions would be those of the evaluation team. All agreed that this solution was pragmatic and reasonable and that it protected evaluator independence while ensuring stakeholder involvement. That is the process that was followed. That critical incident was an opportunity to observe directly the intersection of two sensitizing concepts that are central to evaluation theory and practice (Patton & Gornick, 2011b, pp. 42–43).

WISDOM AS A SENSITIZING CONCEPT

In 1950, renowned psychoanalyst Erik Erikson conceptualized the phases of life, identifying *wisdom* as a likely, but not inevitable, by-product of aging. As an old man writing the 4th edition of this book, I find myself strangely resonating to this developmental theory. Wisdom becomes ascendant during the eighth and final stage of psychosocial development, a time of "ego integrity versus despair." Ego integrity counters the potential despair of increasing infirmity and approaching death, yielding mellowness-inducing wisdom. Erikson, however, never operationalized *wisdom*, and a half-century later, psychologists still don't agree on what it is or how to measure it (Hall, 2007). This comforts me. The meaning and manifestation of wisdom no doubt vary by context. I experience *wisdom* as a *usefulism*—a sensitizing concept—something to ponder, look for, inquire into, and dialogue about. I confess that I find the possibility of at least this one positive outcome of aging reassuring (MQP).

Sensitizing Concepts Versus Operationalized Concepts

> All terms describing people are relational—that is, they only have meaning when they are considered as part of a system of terms.
>
> —Becker (1998, p. 132)

Chapter 5, on designing qualitative studies, contrasted purposeful sampling with probability sampling, making the case that these serve fundamentally different purposes that involve different logics, essentially leading to different types of inference. Social scientists trained only in random probability sampling have trouble grasping the value of small, information-rich purposeful samples. A fundamental difference between quantitative and qualitative inquiry strategies centers on these contrasting sampling strategies.

An equally great divide concerns how to inquire into concepts. Quantitative inquiry requires converting concepts into operationalized (measureable) variables. Qualitative inquiry leaves concepts open for exploration and illumination in the field. These are fundamentally different ways of undertaking an inquiry. Because these differences are a major source of misunderstanding and conflict between quantitative and qualitative researchers, it is worth taking a moment to look more deeply at this divide and its implications for fieldwork.

Operationalization involves translating an abstract construct into concrete measures for the purpose of gathering data on the construct. Through operationalization, a concept becomes a variable, and if the operationalization gains wide acceptance and use, the operational definition becomes the standard measurement of the concept, for example, intelligence tests to measure intelligence. This constitutes the dominant approach to science. The *Encyclopedia of Social Science Research Methods* (2004), in an entry on operationalization, affirms the scientific goal of standardizing definitions of key concepts. It notes that concepts vary in their degree of abstractness, using as an illustration the concepts human capital versus education versus number of years of schooling as moving from high abstraction to operationalization. The entry then observes,

> Social science theories that are more abstract are usually viewed as being the most useful for advancing knowledge. However, as concepts become more abstract, reaching agreement on appropriate measurement strategies becomes more difficult. (Mueller, 2004, p. 162)

Note here that a certain degree of abstraction is *useful* for advancing knowledge and building theory. The *Encyclopedia* entry goes on to discuss the controversy surrounding the relationship between the concept of intelligence and the operationalization of intelligence through intelligence tests, including the classic critique that the splendidly abstract and sensitizing concept of *intelligence* has been reduced by psychometricians to what intelligence tests measure.

> Operationalization as a value has been criticized because it reduces the concept to the operations used to measure it, what is sometimes called "raw empiricism." As a consequence, few researchers define their

concepts by how they are operationalized. Instead, nominal definitions are used . . . and measurement of the concepts is viewed as a distinct and different activity. Researchers realized that measures do not perfectly capture concepts, although . . . the goal is to obtain measures that validly and reliably capture the concepts. (Mueller, 2004, p. 162)

It appears that there is something of a conundrum here, some tension between social science theorizing and empirical research. A second entry in the *Encyclopedia of Social Science Research Methods* sheds more light on this issue.

Operationalism began life in the natural sciences . . . and is a variant of positivism. It specifies that scientific concepts must be linked to instrumental procedures in order to determine their values. . . . In the social sciences, operationalism enjoyed a brief spell of acclaim. . . . [O]perationalism remained fairly uncontroversial while the natural and social sciences were dominated by POSITIVISM but was an apparent casualty of the latter's fall from grace. (Williams, 2004, pp. 768–769)

The entry elaborates three problems with operationalization. First, "underdetermination" is the problem of determining "if testable propositions fully operationalize a theory" (p. 769). Examples include concepts like homelessness, poverty, and alienation, which have variable meanings in different social contexts. What "homeless" means varies historically and sociologically. A second problem is that objective scholarly definitions may not capture the subjective definition of those who experience something. *Poverty* offers an example: What one person considers poverty another may view as a pretty decent life. The Northwest Area Foundation, which has as its mission *poverty alleviation*, has struggled trying to operationalize poverty for outcomes evaluation; moreover, their researchers found that many quite poor people in states like Iowa and Montana, who fit every official definition of being in poverty, do not even see themselves as poor, much less *in poverty*. Third is the problem of disagreements among social scientists about how to define and operationalize key concepts. The second and third problems are related in that one researcher may use a local and context-specific definition to solve the second problem but that context-specific definition is likely to be different from and in conflict with the definition used by other researchers inquiring in other contexts.

One way to address problems of operationalization is to treat a concept of interest as a sensitizing concept and abandon the search for a standardized and universal operational definition. This means that any specific empirical study of a concept would generate a definition that fit the specific context for and purpose of the study, but operational definitions would be expected to vary. In mixed-methods inquiries, understandings that emerge from qualitative inquiry into a sensitizing concept could lead to a context-specific measurement of the concept for quantitative inquiry.

Context as a Sensitizing Concept

Throughout this book, I have noted that sensitivity to context is a strength of qualitative inquiry. Any particular qualitative inquiry is designed within some *context*, and researchers are admonished to take context into account, be sensitive to context, and watch out for changes in context. But what is context? It turns out that it works pretty well as a sensitizing concept but defies operationalization. This was especially evident at the 2009 annual conference of the American Evaluation Association (AEA; Rog, 2009), which had as its theme "Context and Evaluation." Systems thinkers posited that system boundaries are inherently arbitrary, so defining what is within the immediate scope of an evaluation versus what is within its surrounding context will inevitably be arbitrary, but the distinction is still useful. Indeed, being intentional about deciding what is in the immediate realm of action of an evaluation and what is in the enveloping context can be an illuminating exercise—and different stakeholders might well provide different perspectives. In that sense, the idea of *context* is a sensitizing concept. Those conference participants seeking an operational definition of context ranted in some frustration about the ambiguity, vagueness, and diverse meanings of what they, ultimately, decided was a useless and vacuous concept. Why? Because it had not been (and could not be) operationally defined—and they displayed a low tolerance for the ambiguity that is inherent in such sensitizing concepts.

A sensitizing concept raises consciousness about something and alerts us to watch out for it within a specific context as we undertake fieldwork. That's what the concept of *context* does. It says things are happening to people and changes are taking place as people in families, programs, organizations, and communities interact with people and processes in the surrounding environment, in their context(s). Watch out for those interactions and their effects. Pay attention. Something important may be happening. Units of observation (again, families, programs, organizations, communities, etc.) are not closed systems. Observe what's going on both within the setting of interest and across the boundaries of that setting.

The bottom line: Don't judge the maturity and utility of the concept by whether it has "achieved" a standardized and universally accepted operational definition.

Judge it instead by its utility in sensitizing us to how people experience the worlds in which they live and work. This means that specific qualitative studies of important concepts may lead to operational definitions as appropriate. Over time, many empirical studies may use the same or similar operational definitions. Periodically, syntheses and comparisons will be undertaken. We can learn a great deal from how different researchers define a concept, whether operationally (deductively and quantitatively), nominally (as a sensitizing concept), or inductively (exploring emergent meanings and manifestations). What I am arguing against is the notion that arriving at some standard operational definition is the desired target, some kind of "achievement" indicating maturity, consensus, shared understanding, and professional acceptance.

The *what* and *how* of qualitative inquiry are closely linked. Sources of data are derived from inquiry questions. Knowing what we want to illuminate helps us determine sources of data for that illumination. But there are also some general areas of observation that provide principles for engaging in qualitative fieldwork. In the sections that follow, we'll examine the following:

- Describing the context: the observation setting
- Attending to historical perspectives as part of fieldwork
- Observing the human, social environment
- Observing planned, formal interactions
- Observing informal, spontaneous interactions
- Capturing the language used and its meanings
- Making sense of nonverbal communication
- Observing unobtrusively
- Using documents as sources of data in fieldwork
- Observing what does not happen
- Observing oneself

Describing the Context: The Observation Setting

> Pay attention. It's all about paying attention. Attention is vitality.
>
> —Susan Sontag (1933–2004)
> American writer, filmmaker, and activist

Describing a setting, like a program setting, begins with paying attention to the physical environment within which the program takes place. The description of the program setting should be sufficiently detailed to permit the reader to visualize that setting. In writing a program description, the observer, unlike the novelist, should avoid interpretive adjectives except as they appear in quotes from participants about their reactions to and perceptions of that environment. Adjectives such as "comfortable," "beautiful," "drab," and "stimulating" interpret rather than describe—and interpret vaguely at that. More purely descriptive adjectives include

colors—"a room painted blue with a blackboard at one end";

space—"a 40-foot by 20-foot classroom with windows on one side"; and

purpose—"a library, the walls lined with books and tables in the center."

TO DESCRIBE WELL, MINIMIZE INTERPRETATION

Everyone knows that there is no "pure" description, that all description, requiring acts of selection and therefore reflecting a point of view, is what Thomas Kuhn said it was, "theory laden." That it is not possible to do away entirely with the necessity of selection, and the point of view it implies, does not mean that there aren't degrees of interpretation, that some descriptions can't be less interpretive (or perhaps we should say less conventionally interpretive) than others. We might even say that some descriptions require less inference than others. To say that someone looks like he is hurrying home with his shopping requires an inference about motivation that saying that he is walking rapidly doesn't.

—Howard S. Becker (1998, p. 79)
Tricks of the Trade

If you are just beginning to learn to do systematic observations, you can practice writing descriptively by sharing a description of a setting you've observed with a couple of people and asking them if they can visualize the setting described. Another helpful exercise involves two people observing the same environment and exchanging their descriptions, watching in particular for the use of interpretive adjectives instead of descriptive ones. Vivid description provides sufficient information so that the reader does not have to speculate about what is meant. For example, simply reporting "a crowded room" requires interpretation. Contrast with this:

The meeting room had a three-person couch across one side, six chairs along the adjoining walls next to the couch, and three chairs along the wall facing the couch, which included the door. With twenty people in the room, all standing, there was very little space between people. Several participants were overheard to say, "This room is really crowded."

SIDEBAR

Such descriptive writing requires attention to detail and discipline to avoid vague, interpretive phrases. But such writing can also be dull. Metaphors and analogies can enliven and enrich descriptions, helping readers connect through shared understandings and giving them a better *feel* for the environment being described. I once evaluated a wilderness education program that included time at the Grand Canyon. Exhibit 6.6 presents my feeble attempt to capture in words our first view of the Grand Canyon. Notice the metaphors that run through the description. Of course, this is one of those instances where a picture would be worth a mountain of words, which is why qualitative fieldwork increasingly includes photography and videography (Gubrium & Harper, 2013; Marion & Crowder, 2013). This excerpt aims at offering *a sense of the physical environment* more than it offers a literal description because, unless one has been there or seen pictures, the landscape is outside ordinary experience.

The physical environment of a setting can be important to what happens in that environment. In observing a program, what the reception area is like, the way the walls look in rooms, the amount of space available, how the space is used, the nature of the lighting, how people are organized in the space, and the reactions of program participants to the physical setting can be important information about both program implementation and the effects of the program on participants.

In neighborhoods and communities, the physical environment provides an essential context for understanding life in that setting. Chapter 2 featured Rebecca Skloot's case study of *The Immortal Life of Henrietta Lacks* (2010) to illustrate qualitative strategies in practice. Here is her description of entering the town of Clover in southern Virginia in search of Henrietta's family and burial place.

Downtown Clover started at a boarded-up gas station with **RIP** spray-painted across its front, and ended at an empty lot that once held the depot where Henrietta caught her train to Baltimore. The roof of the old movie theater on Main Street had caved in years ago, its screen landing flat in a field of weeds. The other businesses looked like someone left for lunch decades earlier and never bothered coming back: one wall of Abbott's clothing store was lined with boxes of new Red Wing work boots stacked to the ceiling and covered in thick dust; inside its long glass counter, beneath an antique cash register, lay rows and rows of men's dress shirts, still folded starch-stiff in their plastic. The lounge at Rosie's restaurant was filled with overstuffed chairs, couches, and shag carpet, all in dust-covered browns, oranges, and yellows. A sign in the front window said OPEN 7 DAYS, just above one that said CLOSED. At

| EXHIBIT 6.6 | Integrating Description and Metaphor to Provide a Sense of Place: A Participant Observation Example |

The excerpt that follows was aimed at providing context for a Wilderness Leadership Education Program. This was the group's first view of the Grand Canyon from Bright Angel Point, where the program was being launched. I was serving as participant-observer-evaluator.

We followed an asphalt path from the lodge a quarter-mile to Bright Angel Point, perhaps the most popular tourist site at the Grand Canyon because of its relatively easy access. With cameras aimed in all directions at the spectacular panorama, in a sea of domestic accents and foreign tongues, we waited our turn at the edge to behold the magnificent rock temples of Ottoman Amphitheater: Deva, Brahma, Zoroaster, and, in the distance, Thor. Each rises a half-mile above the undulating grayness of the stark Tonto Platform defining the eight-mile descent of Bright Angel Canyon, a narrow slit hiding the inner gorge that looks like it had been drawn in black ink to outline the base of the temples. Each begins as sheer Redwall that forms a massive foundation supporting a series of sloping sedimentary rock terraces, the Supai. These sweeping terraces, spotted green with sparse desert vegetation, point upward like arrow feathers to a white sandstone pedestal, the Coconino. A dark red pinnacle of Hermit shale uniquely crowns each temple. Eons of erosion have sculpted dramatic variations in every aspect save one: their common geologic history. I studied each separately, wanting to fix in my mind the differences between them, but the shared symmetry of strata melded them into a single, massive formation, a half-mile high and many miles around. Behind me I heard a participant say softly to no one in particular, almost under her breath, "It's too awesome. I feel overwhelmed." (Adapted from Patton, 1999)

Gregory and Martin Super Market, half-full shopping carts rested in the aisles next to decades-old canned foods, and the wall clock hadn't moved past 6:34 since Martin closed up shop to become an undertaker sometime in the eighties. (pp. 77–78)

A common mistake among observers is to take the physical environment for granted. Thus, an evaluator may report that the program took place in "a school." The evaluator may have a mental image of "school" that matches what was observed, but schools vary considerably in size, appearance, and neighborhood setting. Even more so, the interiors of schools vary considerably. The same can be said for criminal justice settings, health settings, community mental health programs, neighborhoods, and, indeed, many places where humans congregate.

During site visits to early-childhood education programs, we found a close association between the attractiveness of the facility (child-made decorations and colorful posters on the walls, well-organized learning materials, orderly teacher area) and other program attributes (parent involvement, staff morale, clarity of the program's goals, and theory of action). An attractive, well-ordered environment corresponded to an engaging, well-ordered program. In observing, as well as conducting workshops, I have noted how the arrangement of chairs affects participation. It is typically much easier to generate discussion with chairs in a circle than in lecture style. The dim lighting of many hotel conference rooms seems to palpably drain energy from people sitting in those rooms for long periods of time. Physical environments clearly affect people.

Variations in the settings for a wilderness training program for which I served as participant-observer provide an interesting example of how physical environments affect a program. The explicit purpose of holding the "field conferences" in the wilderness was to remove people from their everyday settings in largely urban environments surrounded by human-made buildings and the paraphernalia of modern industrial society. Yet wilderness environments are no more uniform than the environments of human service programs. During the yearlong program, participants were exposed to four different wilderness environments: (1) the autumn forest in the Gila Wilderness of New Mexico; (2) the rough terrain of Arizona's Kofa Mountains in winter; (3) the muddy, flooding San Juan River in the canyon lands of Utah during spring; and (4) among the magnificent rock formations of the Grand Canyon in summer, a desert environment. One focus of the evaluation, then, was to observe how participants responded to the opportunities and constraints presented by these different environments: forest, mountains, canyon-lined river, and Grand Canyon desert.

In addition, weather and seasonal differences accentuated the variations among these environments. Program activities were clearly affected by the extent to which there was rain, cold, wind, and shelter. In the program's theory, weather uncertainties were expected to be a natural part of the program, offering unpredictable challenges for the group to deal with. But the program theory also called for participants to engage deeply with each other during evening group discussions. During one 10-day winter "field conference" that was unusually cold and wet, participants were miserable, and it became increasingly difficult to carry on group discussions, thus reducing considerably the amount of group process time available and rushing the interactions that did occur because of participants' discomfort. Program staff learned that they needed to anticipate more the possible variations in physical environments, plan for those variations, and include the participants in that planning so as to increase their commitment to continuing the process under difficult physical conditions.

History as Context

> I never forget things. I just don't observe them in the first place.
>
> —Bauvard
> Avant-garde artist

Historical information can shed important light on a setting. The history of a place, program, community, or organization is an important part of the context for research and evaluation. Distinguished qualitative sociologist William Foote Whyte (1984), sometimes called the father of sociological field research, has reflected on how he came to value historical research as a critical part of his fieldwork.

> When we began our Peruvian research program, I viewed history as having little value for understanding the current scene. I thought I was only being sympathetic to the interests of our Peruvian researchers in suggesting that they gather historical data on each village for the last 50 years.
>
> Fortunately, the Peruvians refused to accept the 50-year limit and in some cases probed up to 500 years in the history of villages or areas. Much of these data on rural communities would be of interest only to historians. However, understanding the paradox of the Mantaro Valley required us to go back to the conquest of Peru, and, in the Chancay Valley, we traced the beginnings of the differentiation of Huayopampa from Pacaros back more than a century. (p. 153)

Documenting and understanding the context of a program as part of an evaluation will require delving into its history. How was the program created and initially funded? Who were the original people targeted for program services and how have target populations changed over time? To what extent and in what ways have goals and intended outcomes changed over time? What have staffing patterns been over time? How has the program's governance (board) been involved at various stages in the program's history? What crises has the program endured? If the program is embedded within a larger organizational context, what is the history of that organization in relation to the program? How has the larger political and economic environment changed over time, and how have those changes affected program development? What are the stories people tell about the program's history? These kinds of questions frame inquiry into the program's history to illuminate context.

Damiano is a community-based poverty alleviation program in Duluth, Minnesota. It began as a temporary soup kitchen providing emergency meals during an economic crisis in 1982. Over time, it added a variety of services and is still operating today. Understanding that story constitutes a "retrospective developmental evaluation" (Patton, 2011, pp. 297–303) without which observing current operations would be devoid of meaningful context.

In the 1990s, I evaluated a "free high school" that had been created during the struggles and turmoil of the 1960s. Little about the program's current programming could be understood outside the context of its historical emergence. The school's image of itself, its basic approach, including curriculum and policies, had been handed down and adapted from that intense period of early development. Doing fieldwork in the 1990s could only be done by traversing the memories and legends of the school's historical emergence in the 1960s.

The Human, Social Environment

> To acquire knowledge, one must study; but to acquire wisdom, one must observe.
>
> —Marilyn Vos Savant
> American author, formerly listed in the
> *Guinness Book of World Records* under "Highest IQ"

Just as physical environments vary, so too do social environments. The ways in which human beings interact create social ecological constellations that affect how participants behave toward each other in those environments. Rudolf Moos (1975) pioneered the social ecological view of human environments:

> The social climate perspective assumes that environments have unique "personalities," just like people. Personality tests assess personality traits or needs and provide information about the characteristic ways in which people behave. Social environments can be similarly portrayed with a great deal of accuracy and detail. Some people are more supportive than others. Likewise, some social environments are more supportive than others. Some people feel a strong need to control others. Similarly, some social environments are extremely rigid, autocratic, and controlling. Order, clarity, and structure are important to many people. Correspondingly, many social environments strongly emphasize order, clarity and control. (p. 4)

In describing the social environment, the observer looks for the ways in which people organize themselves into groups and subgroups. Patterns and frequency of interactions, the direction of communication patterns (Who controls? Who follows? Who isn't heard?), and changes in these patterns tell us things about the social environment. How people group together can be illuminative and important. All-male versus all-female groupings, male–female interactions, and interactions among people with different background characteristics, racial identities, and/or ages alert the observer to patterns in the social ecology of a setting.

Decision-making patterns can be a particularly important part of a social environment. Who makes decisions about what gets done? To what extent are decisions made openly, so that others are aware of the decision-making process? How are decisions communicated? Answers to these questions are an important part of the description of the power dynamics of a social setting. These issues are particularly important in program evaluation to understand and illuminate the program's decision-making patterns and environment.

An observer's descriptions of a social environment will not necessarily be the same as the perceptions of that environment expressed by participants. Nor is it likely that all participants will perceive the setting's human climate in the same way. At all times, it is critical that the observer records participants' comments in quotation marks, indicating the source—who said what?—so as to keep perceptions of participants separate from the observer's or evaluator's own descriptions and interpretations.

Formal Interactions and Planned Activities

> Never trust to general impressions, my boy, but concentrate yourself upon details.
>
> —Sir Arthur Conan Doyle (1859–1930)
> *The Adventures of Sherlock Holmes*

In this section and the next, I distinguish observing formal, planned activities and occasions from informal, serendipitous interactions and events. Early fieldwork observations tend to focus on what is known to be occurring in advance. For example, program evaluations focus at least some significant observations on planned program activities. What goes on in the program? What do participants and staff do? What is it like to be a participant? These are the kinds of questions evaluators bring to the program setting to document program implementation. Likewise, in organizations and communities, there will be formal, planned activities that are important to see and understand. For example, in studying organizations, I find it valuable to observe staff meetings and orientation sessions for new staff; if a community is the focus, I'd want to observe any formal neighborhood council meetings and community celebrations like *Cinco de Mayo* in a Hispanic neighborhood or *Carnival* in a Caribbean community.

Formal, planned activities have a kind of unity about them: a beginning, some middle point, and a closure point—things like a class session, a counseling session, mealtime in a residential facility, a meeting of some kind, a home visit in an outreach program, a consultation, or a registration procedure. Attending to such sequences illustrates how fieldwork progresses over the course of an observation. Initially, the observer will focus on how the activity is introduced or begun. Who is present at the beginning? What exactly was said? How did participants respond or react to what was said?

These kinds of basic descriptive questions guide the evaluator throughout the full sequence of observation. *Who* is involved? *What* is being done and said by staff and participants? *How* do they go about what they do? *Where* do activities occur? *When* do things happen? What are the variations in how participants engage in planned activities? How does it feel to be engaged in this activity? (The observer records his or her own feelings as part of the observation.) How do behaviors and feelings change over the course of the activity?

Finally, the observer looks for closure points. What are the signals that a particular activity is being ended? Who is present at that time? What is said? How do participants react to the ending of the activity? How is the completion of this unit of activity related to other program activities and future plans?

Each unit of activity is observed and treated as a self-contained event for the purpose of managing field notes. The observation of a single session of the Early Childhood Parent Education program presented in Chapter 1 is an example (pp. 29–31). Each observed event or activity can be thought of as a mini case study of a discrete incident, activity, interaction, or event. Choosing what to observe involves sampling decisions, and the options for sampling within a case are those discussed at length in Chapter 5. You can observe typical events, diverse interactions, or outlier occasions. Sensitizing concepts can guide what you select to observe. For example, if you are interested in power as a sensitizing concept, you will choose to focus observations on events and interactions that are organized by those with power but include those with less power. What is the nature of interactions on such occasions, for example, a political event or a ribbon-cutting ceremony with people in power on display?

Distinguish Cases From Observational Units Within Cases: Choosing What to Observe

> Distinguish cases from observations. "Cases" are understood as the broader units, that is, the broader research settings or sites within which analysis is conducted; "observations" are pieces of data, drawn from those research sites, that form the direct basis for descriptive and causal inference. (Collier, Seawright, & Munck, 2010, p. 51)

During analysis, one looks across discrete observations for patterns and themes, but during the initial stages of fieldwork, the observer will be kept busy just trying to capture self-contained units of activity (observational units) without worrying yet about looking for patterns across activities.

How do you decide what to observe? Observe those things that relate most directly to the focus of your inquiry. This is called maximizing leverage, where you maximize your fieldwork time and resources to get the most relevant data to your inquiry. At the same time, you're watching for opportunities to expand your understanding of your inquiry by being open to observations that might at first appear tangential but, in fact, take you in an important, emergent direction. So you watch for observations that have similar characteristics (unit homogeneity) and those that are diverse (unit heterogeneity). Part of the cross-unit analysis will involve examining "unit homogeneity" versus "unit heterogeneity" in comparing observational units like discrete events, formal interactions, and planned activities (George & Bennett, 2005, pp. 171–172).

Observing and documenting formal activities will constitute a central element in fieldwork. In evaluation, the formal, planned activities are the core of evaluating planned program implementation (evaluating the "model"), but to fully understand a program and its effects on participants, observations should not be restricted to formal, planned activities. The next section discusses observation of the things that go on between and around formal, planned program activities.

REALITY CHECK APPROACH: "LIGHT TOUCH" PARTICIPANT OBSERVATION

The Reality Check Approach (RCA) seeks to understand the views, perspectives, and opinions of people living in poverty by spending several nights living with poor families and engaging in informal conversations with them, their neighbors, and local service providers. It may be viewed as a "light touch" participant observation. The RCA is considered to be a useful contribution to the overall study in that it provides contextual and candid information directly from people regarding their particular experience of change. (GRM International, 2013, p. 6)

For an example of RCA in practice, see the study of the long-term impact of development interventions in the Koshi Hills of Nepal.

The main focus of the study was to establish how people experienced change over the past 30+ years, what changes they considered were significant, and what their ideas were regarding the drivers of these changes. To this end, the host households included in the study were purposely selected to comprise different generations so that discussions and debate could be facilitated across generations.

The RCA was a small-scale study involving twenty seven host families, their neighbours, and local service providers (in total about 600 people). (GRM International, 2013, p. 6)

Informal Interactions and Unplanned Activities

> I am turned into a sort of machine for observing . . .
>
> —Charles Darwin (1809–1882)
> Naturalist and pioneer in the study of evolution

If you stop seeing and observing as soon as a planned, formal activity ends, you will miss a great deal of data. For example, most programs build in free or unstructured time between activities, with the clear recognition that such periods provide opportunities for participants to assimilate what has occurred during formal programmatic activities as well as to provide participants with necessary breathing space. Rarely, if ever, can a program or institution plan every moment of participants' time.

During periods of informal interaction and unplanned activity, it can be particularly difficult to organize observations because people are likely to be milling around, coming and going, moving in and out of small groups, with some sitting alone, some writing, some seeking refreshments, and otherwise engaging in a full range of what may appear to be random behaviors. How, then, can the observer collect data during such a time?

This scenario illustrates beautifully the importance of staying open to the data and doing opportunity sampling. One can't anticipate all the things that might emerge during unplanned program time, so the observer watches, listens, and looks for opportunities to deepen observations, recording what people do, the nature of informal interactions (e.g., what subgroups are in evidence) and, in particular, what people are saying to each other. This last point is particularly important. During periods of unplanned activity, participants have the greatest opportunity to exchange views and to talk with each other about what they are experiencing in the program. In some cases, the evaluator will simply listen in on conversations, or there may be opportunities to conduct informal interviews, either with a single participant in natural conversation or with some small group of people, asking normal, conversational questions:

> "So what did you think of what went on this morning?"

> "What did you think of the session today?"

> "How do you think what went on today fits with the rest of what you've been experiencing?"

Such questioning should be done in an easy, conversational manner so as not to be intrusive or so predictable that every time someone sees you coming, they know what questions you're going to ask. "Get ready, here comes the evaluator with another endless set of questions." Also, when doing informal, conversational interviewing, be sure that you are acting in accordance with ethical guidelines regarding informed consent and confidentiality. (See the earlier discussion in this chapter about overt vs. covert fieldwork.)

How something is said should be recorded along with what is said. At a morning break in the second day of a two-day workshop, I joined the other men in the restroom. As the men lined up to use the facilities, the first man to urinate said loudly, "Here's what I think of this program." As each man finished, he turned to the man behind him and said, "Your turn to piss on the program." This spontaneous group reaction spoke volumes more than answers to formal interview questions and provided much greater depth of expression than checking "Very dissatisfied" on an evaluation questionnaire.

Everything that goes on in or around the program or other setting being observed is data. The fact that none of the participants talk about a session when it is over is data. The fact that people immediately split in different directions when a session is over is data. The fact that people talk about personal interests and share gossip that has nothing to do with the program is data. In many programs, the most significant participant learning occurs during unstructured time as a result of interactions with other participants. To capture a holistic view of the program, the evaluator-observer must stay alert to what happens during these informal periods. While others are "on break," the observer is still working. No breaks for the dedicated fieldworker! Well, not really. You've got to pace yourself and take care of yourself, or your observations will deteriorate into mush. Hopefully, you get the idea. You may be better off taking a break during part of a formal session, so you can work (collect data) while others are on break.

As happens in many programs, the participants in the wilderness education project I was observing/evaluating began asking for more free, unstructured time. When we weren't hiking or doing camp chores, a lot of time was spent in formal discussions and group activities. Participants wanted more free time to journal. Some simply wanted more time to reflect. Most of all, they wanted more time for informal interactions with other participants. I respected the privacy of one-to-one interactions when I observed them and would never attempt to eavesdrop. I would, however, watch for such interactions, and judging body language and facial expressions, I would speculate when serious interpersonal exchanges were taking place. I would then look for natural opportunities to engage each of those participants in conversational interviews, telling them I had noticed the intensity of their interaction and inquiring whether they were willing to share what had happened and what significance they attached to the interaction. Most of the participants appreciated my role in documenting the program's unfolding and its effects on participants and were open to sharing.

It was on the basis of those informal interviews and observations that I provided formative feedback to staff about the importance of free time and helped alleviate the feeling among some staff that they had responsibility to plan and account for every moment during the program.

Participant observation necessarily combines observing and informal interviewing. Observers need to be disciplined about not assuming that they know the meaning to participants of what they observe without checking with those participants. During one period of unstructured time in the wilderness program, following a fairly intensive group activity in which a great deal of interpersonal sharing had taken place, I decided to pay particular attention to one of the older men in the group who had resisted involvement. Throughout the week, he had taken every available opportunity to make it known that he was unimpressed with the program and its potential for having an impact on him. When the session ended, he immediately walked over to his backpack, pulled out his writing materials, and went off to a quiet spot where he could write. He continued writing, completely absorbed, until dinnertime an hour later. No one interrupted him. With legs folded, notebook in lap, and his head and shoulders bent over the notebook, he gave off clear signals that he was involved, concentrating, and working on something to which he was giving a great deal of effort.

I suspected as I watched that he was venting his rage and dissatisfaction with the program. I tried to figure out how I might read what he had written. I was so intrigued that I momentarily even considered covert means of getting my hands on his notebook, but I quickly dismissed such unethical invasion of his privacy. Instead, I looked for a natural opportunity to initiate a conversation about his writing. During the evening meal around the campfire, I moved over next to him, made some small talk about the weather, and then began the following conversation:

"You know, in documenting experiences people are having, I'm trying to track some of the different things folks are doing. The staff has encouraged people to keep journals and do writing, and I noticed that you were writing fairly intensely before dinner. If you're willing to share, it would be helpful for me to know how you see the writing fitting into your whole experience with the program."

He hesitated, moved his food about in his bowl a little bit, and then said, "I'm not sure about the program or how it fits in or any of that, but I will tell you what I was writing. I was writing . . . ," and he hesitated because his voice cracked, "a letter to my teenage son trying to tell him how I feel about him and make contact with him about some things. I don't know if I'll give the letter to him. The letter may have been more for me than for him. But the most important thing that's been happening for me during this week is the time to think about my family and how important it is to me and I haven't been having a very good relationship with my son. In fact, it's been pretty shitty, and so I wrote him a letter. That's all."

This short conversation revealed a very different side of this man and an important impact of the program on his personal and family life. We had several more conversations along these lines, and he agreed to be a case example of the family impacts of the program. Until that time, impacts on family had not even been among the expected or intended outcomes of the program. It turned out to be a major area of impact for a number of participants.

Capturing People's Language and Meanings

> The lunatic, the lover, and the poet
>
> Are of imagination all compact.
>
> One sees more devils than vast hell can hold;
>
> That is, the madman. The lover, all as frantic,
>
> Sees Helen's beauty in a brow of Egypt.
>
> The poet's eye, in a fine frenzy rolling,
>
> Doth glance from heaven to earth, from earth to heaven;
>
> And as imagination bodies forth the forms of things unknown,
>
> The poet's pen turns them to shapes
>
> And gives to airy nothing
>
> A local habitation and a name.
>
> —William Shakespeare
> *A Midsummer Night's Dream*, Act V, Scene 1

As noted in Chapter 2, the Whorf hypothesis alerts us to the power of language to shape our perceptions and experiences. It's worth repeating the story of Benjamin Whorf's discovery. As an insurance investigator, he was assigned to look into explosions in warehouses. He discovered that truck drivers were entering "empty" warehouses smoking cigarettes and cigars. The warehouses often contained invisible but highly flammable gases. He interviewed truckers and found that they associated the word "empty" with "harmless" and acted accordingly. Whorf's job, in Shakespeare's terms, was to turn the perception of "airy nothing" into the shape of possible danger.

An anthropological axiom insists that one cannot understand another culture without understanding the language of the people in that culture. Language organizes our world for us by shaping what we see, perceive, and pay attention to. The things for which people have special words tell others what is important to that culture. Thus, as students learn in introductory anthropology, indigenous Arctic peoples have many words for *snow* and desert-dwelling peoples may have many words for *camel*. Likewise, the artist has many words for *red* and to make space for the other changes kinds of brushes.

Roderick Nash (1986), in his classic study *Wilderness and the American Mind*, traces how changing European American perceptions of "wilderness" have affected at the deepest levels our cultural, economic, and political perspectives on deserts, forests, canyons, and rivers. He traced the very idea of *wilderness* to the eighth-century epic hero Beowulf, whose bravery was defined by his courage in entering the "*wildeor*"—a place of wild and dangerous beasts, dark and foreboding forests, and untamed, primordial spirits. In the Judeo-Christian tradition, wilderness came to connote a place of uncontrolled evil that needed to be tamed and civilized, while Eastern cultures and religions fostered love of wilderness rather than fear. Nash credits the Enlightenment with offering new ways of thinking about wilderness—and new language to shape that changed thinking.

Moving from the wilderness to the interior territory of communities, organizations, families, agencies, and programs, language still shapes experience and is therefore an important focus during fieldwork. People develop their own language to describe their concerns and the problems they deal with in their work. Educators who work with learning-disabled students have a complex system of language to distinguish different degrees and types of special needs and developmental disabilities, a language that changes as cultural and political sensitivities change. People in criminal justice generate language for distinguishing types of offenders or "perps" (perpetrators). Fieldwork involves learning the "native language" of the setting or program being studied and attending to variations in connotations and situational use. The field notes and reports of the observer should include the exact language used by participants to communicate the flavor and meaning of "native" language.

LANGUAGE INTENSITY IN THE SUPREME COURT

In decisions of the U.S. Supreme Court, justices writing dissents (on the losing side of the case) use more strident and intense language to express their disagreement than do those in the majority. Words like "clearly," "absolutely," and "undoubtedly" are used more often in dissents. These language "intensifiers" are accompanied by longer words.

Legal readers do not like intensifiers, long sentences, and long words. Nevertheless, when the legal reader becomes the legal writer and feels threatened with losing an appeal, or being on the dissenting side of a judicial opinion, the threatened legal writer will subconsciously resort to using more intensifiers, and maybe longer words and longer sentences, in an irrational attempt to "attack" the winning side, or to defend the losing argument. This response is explained by the . . . theory of argumentative threat . . . ; alienated from the majority, a Supreme Court Justice subconsciously (and irrationally) resorts to the universally censured intensifier in an attempt to bolster the losing argument. The theory of argumentative threat is consistent with social psychology theories suggesting that language use changes in response to a perceived threat. (Long & Christensen, 2013, pp. 958–959)

Language was especially important in the wilderness education project I evaluated. These were highly verbal people, well educated, reflective, and articulate, who spent a lot of program time in group discussions. Program staff understood how words can shape experiences. They wanted participants to view the time in the wilderness as a professional development learning experience, not a vacation, so staff called each week in the wilderness a "field conference"—hoping participants would see the program as a "conference" held in the "field." Despite the determined efforts of staff, however, the participants never adopted this language. Almost universally, they referred to the weeks in the wilderness as "trips." During the second "field conference," the staff capitulated. Interestingly enough, that capitulation coincided with negative reactions by participants to some logistical inadequacies, unsuccessful program activities, and bad weather, all of which undercut the "conference" emphasis. Staff language reflected that change.

Other language emerged that illuminated participants' experiences. One of the participants expressed the hope of "detoxifying" in the wilderness. He viewed his return to his everyday world as "poisonous retoxification." The group immediately adopted this language of "detoxification" and "retoxification" to refer to *wilderness time* versus ordinary *urban civilization time*, ultimately shortening the words to "detox" and "retox." This language came to permeate the program's culture.

The discussions in the wilderness often reflected the physical environment in which program activities took place. Participants became skilled at creating analogies and metaphors to contrast their urban work lives with their wilderness experiences. After backpacking all day, participants could be heard talking about learning how to "pace myself in my work" or "shifting the burdens of responsibilities that I carry so that the load is more evenly balanced" (a reference to the experience of adjusting the weight of the backpack). In the mountains, after rock climbing, participants referred to "the danger taking risks at work without support" (a reference to the *belay* system of climbing, where someone supports the climber with a safety rope below). One discussion focused on how to "find toeholds and handholds" to bring about change back home, "to get on top of the steep wall of resistance in my institution." They even assigned numbers to degrees of back-home institutional resistance corresponding to the numbers used to describe the degree of difficulty of various rock climbs. On the river, participant language was filled with phrases like "going with the flow," "learning to monitor professional development like you read and monitor the current," and "trying to find my way out of the eddies of life."

Because of the power of language to shape our perceptions and experiences, most participants wanted to know the names of the rock formations, winding canyons, and river rapids we encountered, while others, following *Desert Solitaire* author Eddy Abbey (1968), set for themselves the goal of suppressing the human tendency to personify natural forms. Thus began a sustained personal interest in how names in the wilderness shaped our experiences there (Patton & Patton, 2001).

When I took my son into the Grand Canyon for a "coming-of-age" initiation experience (Patton, 1999), he reacted to the problem of finding words for that awesome environment by making up words. For example, on seeing the Canyon for the very first time, he whispered, "*Buedüden*," which became our way of describing things too beautiful and awesome for ordinary words.

JARGON

SIDEBAR

Don Miller, in *The Book of Jargon*, tells of beginning the book with the intent of attacking and making fun of the jargon in various professions. To his surprise, he came to respect it. Whether within families, communities, peer groups, or professional associations, specialized language can add nuance and precision to communications and support connections.

I began to feel that there was something to be learned from the jargons, and that in fact it was much more interesting to have these different intellectual dialects in existence than to live in a perfectly homogeneous linguistic universe.... It is not only the major professions and subcultures that have their own styles and vocabularies; many smaller social units do, as well. Often within a single family, between brothers and sisters, or between lovers or old friends there exists a special, secret language, unknown to the rest of the world, which carries an intimate set of meanings and associations. In my own experience I have found these to be among the most beautiful forms in which language can be used. (Miller, 1981, p. xiii)

Capturing the precise language of participants honors the "emic" tradition in anthropology: recording participants' own understandings of their experiences. Observers must learn the language of participants in the setting or program they are observing to faithfully represent participants in their own terms and be true to their worldview.

Nonverbal Communications

Verbal and nonverbal activity is a unified whole, and theory and methodology should be organized or created to treat it as such.

—Kenneth L. Pike (1912–2000)
Linguist and anthropologist

Social and behavioral scientists have reported at length the importance of both verbal and nonverbal communication in human groups. While recording the language of participants, the observer should also attend to nonverbal forms of communication. For example, in educational settings, nonverbal communications would include how students get the attention of or otherwise approach instructors, like waving their hands in the air. In group settings, a great deal of fidgeting and moving about may reveal things about attention and involvement. How participants dress, express affection, and sit together or apart are examples of nonverbal cues about social norms and patterns.

Again, the wilderness program provides informative examples. Hugging emerged as a nonverbal way of providing support at times of emotional distress or celebration, for example, a way to recognize when someone had overcome some particularly difficult challenge, like making it up across a ledge along a cliff face. But subgroups differed in the amount of and comfort with hugging, and different "field conferences" manifested different amounts of hugging. When the group felt disparate, separated, with people on their own "trips" and isolated from each other, little hugging occurred either in pairs or around the group campfire. When the depth of connection was deeper, shoulder-to-shoulder contact around the campfire was common and group singing was more likely. Over time, it became possible to read the tenor of the group by observing the amount and nature of physical contact participants were having with each other—and participants in groups with a lot of hugging and connectedness reported noticeably greater personal change and deeper reflection.

In evaluating an international development project, I observed that the three host country nationals ("locals") had developed a subtle set of hand signals and gestures that the American staff never noticed. In meetings, the host country nationals regularly communicated with each other and operated as a team using these nonverbal signals. Later, having gained their confidence, I asked the local staff about the gestures. They told me that the Americans had insisted that each person participate as an individual on an equal footing in staff meetings, and to support an atmosphere of openness, the Americans asked them not to use their own language during staff meetings. But the locals wanted to operate as a unit to counter the power of the Americans, so they developed subtle gestures to communicate with each other since they were denied use of their own language.

Parameswaran (2001) has described how she relied on reading nonverbal cues to tell how potential interviewees reacted to her subject matter—the study of young middle-class women in India who read Western romance novels. She had to depend on observing and reading body language to pick up hostility, disapproval, support, or openness because the verbal formalities of some interactions offered fewer cues than nonverbal reactions. Among the young women, giggles, winks, animated interactions, lowered eyes, direct gaze—these became cues on how the fieldwork was progressing.

A caution is in order here. Nonverbal behaviors are easily misinterpreted, especially cross-culturally. Therefore, whenever possible and appropriate, having observed what appear to be significant nonverbal behaviors, some effort should be made to follow up with those involved to find out directly from them what the nonverbals really meant. I confirmed with other participants in the wilderness project the importance of hugging as a mechanism that they themselves used to sense the tenor of the group.

Unobtrusive Observations and Indicators; and Documents and Archival Fieldwork

Being observed can make people self-conscious and generate anxiety, especially when the observations are part of an evaluation. Regardless of how sensitively observations are made, the possibility always exists that people will behave differently under conditions where an observation or evaluation is taking place than they would if the observer were not present.

> Even when he is well-intentioned and cooperative, the research subject's knowledge that he is participating in a scholarly search may confound the investigator's data. . . . It is important to note early that the awareness of testing need not, by itself, contaminate responses. It is a question of probabilities, but the probability of bias is high in any study in which a respondent is aware of his subject status. (Webb, Campbell, Schwartz, & Sechrest, 1966, p. 13)

Concern about reactions to being observed has led some social scientists to recommend covert observations as discussed earlier in this chapter. An alternative strategy involves searching for opportunities to collect "unobtrusive measures" (Webb et al., 1966) or, qualitatively, to make unobtrusive observations. Unobtrusive measures and observations are those made without the knowledge of the people being observed and without affecting what is observed.

Robert Wolf and Barbara Tymitz (1978) included unobtrusive measures in their naturalistic inquiry evaluation of the National Museum of Natural History at the Smithsonian Institution, a study that has become a classic. They looked for "wear spots" as indicators of use of particular exhibit areas. They decided that worn rugs would indicate the popularity of particular areas in the museum. The creative observer can learn a number of things about patterns of behavior by looking for physical clues. Dusty equipment or files may indicate things that are not used. Areas that are used a great deal by children in a school will look different—that is, more worn—than areas that are little used.

In a weeklong staff training program for 300 people, I asked the kitchen to systematically record how much coffee was consumed in the morning, afternoon, and evening each day. Those sessions that I judged to be particularly boring had a correspondingly higher level of coffee consumption. Active and involving sessions showed less coffee consumption, regardless of time of day. (Participants could get up and get coffee whenever they wanted.)

In the wilderness program, the thickness of notebooks called "learning logs" became an unobtrusive indicator of how engaged participants were in self-reflective journaling. All participants were provided with "learning logs" at the beginning of the first "field conference" and encouraged to use them for private reflections and journaling. These three-ring binders contained almost no paper when first given to participants. Participants brought "learning logs" back each time they returned to the wilderness. (The program involved four different 10-day trips over the course of a year.) Despite differences in handwriting and spacing, the extent to which paper had been added to the notebooks was one rough indicator of the extent to which the logs were being used. I subsequently interviewed each participant about his or her use of the notebooks.

The personnel of the National Forest Service and the Bureau of Land Management have a kind of unobtrusive measure they use in "evaluating" the wilderness habits of groups that go through an area such as the San Juan River in Utah. The canyons along the San Juan River are a very fragile environment. The regulations for use of that land state the following: "Take only photographs, leave only footprints." This means that all garbage, including human waste and feces, are to be carried out. It takes several days to go down the river. By observing the amount and types of garbage groups carry out, park rangers could learn a great deal about the wilderness habits of various groups and their compliance with river regulations.

The creative observer, aware of the variety of things to be learned from studying physical and social settings, will look for opportunities to incorporate unobtrusive measures into fieldwork, thereby manifesting "sympathy toward multi-method inquiry, triangulation, playfulness in data collection, outcroppings as measures, and alternatives to self report" (Webb & Weick, 1983, p. 210). An example of such creative fieldwork is Jason De León's Undocumented Migrant Project, which combines ethnographic research (collecting oral histories) and unobtrusive archeological research: collecting desert garbage left by illegal immigrants entering the United States from Mexico. He treats the garbage as cultural artifacts worthy of study in what he calls the "archaeology of undocumented migration": viewing illegal immigration through desert debris (Wagner, 2011).

A particularly powerful example of unobtrusive fieldwork is Laura Palmer's (1988) study of letters and remembrances left at the Vietnam Veterans Memorial in Washington, D.C., a work she called *Shrapnel in the Heart*. For the unobtrusive part of her fieldwork, Palmer sampled items left at the memorial, all of which are saved and warehoused by the U.S. government. She categorized and analyzed types of items and the content of messages. In some cases, because of identifying information contained in letters or included with objects (photographs, baby shoes, artwork) she was able, through intensive investigative work, to locate the people who left the materials and interview them. Their stories, the intrusive part of her study, combined with vivid descriptions of the objects that led her to them, offer dramatic and powerful insights into the effects of the Vietnam War on the lives of survivors. In one sense, her analysis of letters, journals, photos, and messages can be thought of as a nontraditional and creative form of document analysis, another important fieldwork strategy.

Documents and Documentation

> ### If it isn't documented, it didn't happen.
> —Sohail Sangi (2009)

Fieldwork creates documentation. Systematic documentation of what is observed, heard, and experienced is what fieldwork is all about. Fieldwork also involves finding documents and documentation. Records, documents, artifacts, and archives, what has traditionally been called "material culture" in anthropology, constitute a particularly rich source of information about many organizations and programs. In contemporary society, all kinds of entities leave a trail of paper and artifacts, a kind of spoor that can be mined as part of fieldwork. Families keep photographs, children's schoolwork, letters, old Bibles with detailed genealogies, bronzed baby shoes, and other sentimental objects that can inform and enrich family case studies. People who commit suicide leave behind suicide notes that can reveal patterns of despair in a society (Wilkinson, 1999). Gangs and others inscribe public places with graffiti. Organizations of all kinds produce mountains of records, both public and private, on paper, digitally,

and online. Indeed, an oft-intriguing form of analysis involves comparing official statements found in public documents (brochures, board minutes, annual reports) with private memos and what the observer actually hears or sees occurring in direct observation.

SIDEBAR

ARCHIVAL INQUIRY INTO TIMING

University of Minnesota business professor Stuart Albert (2013) studies how timing affects results. He describes himself as *an archival hunter-gatherer*. He collected case examples from *The New York Times, The Wall Street Journal,* and *The Economist* magazine by reading from cover to cover for more than 20 years and clipping every article that had anything to do with timing. That yielded more than 2,000 articles. In essence, *timing* was his sensitizing concept.

He sought examples of mistakes that could have been prevented based on what could have been known at the time. His qualitative analysis then involved looking for underlying temporal elements and structures. He found six elements that are part of every situation people face in making decisions: (1) sequences—what follows what; (2) punctuation—where things begin, end, stop, and start; (3) intervals—how much time there is between events; (4) rate—how fast things are going; (5) the shape or rhythm by which things unfold; and (6) polyphony—a lot of things happening at the same time and how they interact. He used these six dimensions to generate a seven-step process for analyzing why things happen when they do that provides guidance on when to effectively act.

No interviews. No fieldwork. Just immersion in his 20-year archive of news clippings.

Client files in programs are another rich source of case data to supplement field observations and interviews. At the very beginning of fieldwork, access to potentially important documents and records should be negotiated. The ideal situation would include access to all routine records on clients, all correspondence from and to program staff, financial and budget records, and organizational rules, regulations, memoranda, charts, and any other official or unofficial documents generated by or for the program. These kinds of documents provide the inquirer with information about many things that cannot be observed. They may reveal things that have taken place before the study began. They may include private interchanges to which the inquirer would not otherwise be privy. They can reveal aspirations, arrangements, tensions, relationships, and decisions that might be otherwise unknown through direct observation.

In evaluating the mission fulfillment of a major philanthropic foundation, I examined 10 years of annual reports. Each report was professionally designed, elegantly printed, and widely disseminated—and each report stated a slightly different mission for the foundation. It turned out that the president of the foundation wrote an annual introduction and simply stated the mission from memory. The publication designer routinely lifted this "mission statement" from the president's letter and highlighted it in bold font at the beginning of the report, often on the cover page. From year to year, the focus changed until, over the course of 10 years, the stated mission had changed dramatically without official board action, approval, or even awareness. Further investigation through years of board minutes revealed that, in fact, the board had never adopted a mission statement at all, a matter of considerable surprise to all involved.

As this example shows, documents prove valuable not only because of what can be learned directly from them but also as a stimulus for paths of inquiry that can only be pursued through direct observation and interviewing. As with all information to which an inquirer has access during observations, the confidentiality of program records, particularly client records, must be respected. The extent to which actual references to and quotations from program records and documents are included in a final report depends on whether the documents are considered part of the public record and therefore able to be publicized without breach of confidentiality. In some cases, with permission and proper safeguards to protect confidentiality, some information from private documents can be quoted directly and cited.

Program Records

Program records can provide a behind-the-scenes look at program processes and how they came into being. In the wilderness program evaluation, program staff made their files available to me. I discovered a great deal of information not available to other program participants: letters detailing both conceptual and financial debates between the technical staff (who led the wilderness trips) and the project directors (who had responsibility for the overall management of the program). Without knowledge of those arguments, it would have been impossible to fully understand the nature of the interactions between field staff and executive staff in the project. Disagreements about program finances constituted but one arena of communication difficulties during the program, including time in the wilderness. Interviews with those involved revealed quite different perceptions of the nature of the conflicts,

ARCHIVAL FIELDWORK: AN AFRICAN EXAMPLE

The progressive government of Tanzania under President Julius Nyerere (1961–1985) sought to settle the *Wagogo* people, a nomadic herding tribe in the Dodoma Region of Central Tanzania. In the late 1960s, conflicts between nomadic herders and settled agricultural farmers were increasing as both farms and herds grew larger. The former British colonial administration had tried unsuccessfully to settle the *Wagogo*. Before launching a new settlement scheme, the Tanzanian Ministry of Agriculture commissioned a study of the extensive colonial archives to inventory and assess what prior attempts had been tried. The Tanzanian government wanted to distinguish any new scheme from former colonial policies as well as learn from past failures. I spent three months going through years of detailed (and dusty) colonial reports to create an inventory of 35 different settlement schemes that had been tried, including a variety of financial incentives; harsh penalties; confiscation of cattle; disruption of herding routes; relocation of water sources, fences, and armed guards; and removing children to residential schools. It was clear that the Tanzanian government would need a great deal of creativity, collaboration, and trust building to come up with a scheme that was not tainted by colonialism (Patton, 1970).

their intensity, and their potential for resolution. While participants became aware of some arguments among staff, for the most part, they were unaware of the origins of those conflicts and the extent to which program implementation was hampered by them.

My review of files also revealed the enormous complexity of the logistics for the wilderness education program. Participants (college deans, program directors, administrators) were picked up at the airport in vans and driven to the wilderness location where the field conference would take place. Participants were supplied with all the gear necessary for surviving in the wilderness. Prior to each field trip, staff had many telephone and written exchanges with individual participants about particular needs and fears. Letters from participants, especially those new to the wilderness, showed how little they understood about what they were getting into. One seasoned administrator and hard-core smoker inquired, with reference to the first 10-day hike in the heart of the Gila Wilderness, "Will there be a place to buy cigarettes along the way?" Talk about being clueless! But by the end of the year of field trips, he had given up smoking. His letter of inquiry alerted me to the importance of this pre–post observation.

Without having looked over this correspondence, I would have missed the extent to which preparation for the one-week experiences in the wilderness consumed the time and energy of program staff. The intensity of work involved before the field conferences helped explain the behavior of staff once the field trips got under way. So much had gone into the preparations, virtually none of which was appreciated by or known to program participants, that program staff would sometimes experience a psychological letdown effect and have difficulty energizing themselves for the actual wilderness experience.

Learning to use, study, and understand documents and files is part of the repertoire of skills needed for qualitative inquiry. Exhibit 6.7 provides an inventory of documentation that can deepen fieldwork and qualitative analysis. For an extended discussion of "the interpretation of documents and material culture," see Hodder (2000). For an exemplar of in-depth, carefully triangulated document analysis, you can't do much better than the examination of the legend of President Abraham Lincoln's last photograph before his assassination in 1865. Errol Morris (2013) examines multiple documents and photographs to separate the legend from the reality. It's a fascinating story, readily available online and illustrating what qualitative documentary and investigative scholarship involve.

EXHIBIT 6.7 **Types of Documentation and Artifacts to Support Qualitative Inquiry: A Suggestive Inventory**

INDIVIDUALS/FAMILIES	NONPROFIT ORGANIZATIONS
o Diaries and journals	o Mission, vision documents
o Photographs	o Strategic plan
o Schoolwork	o Annual reports
o Letters	o Budget documents
o Collectibles/art	o Board minutes
o Family heirlooms	o Staff meeting minutes
o Scrapbooks	o Public relations documents
COMMUNITIES	**PROGRAMS**
o Local newspapers	o Client files
o Local library historical records	o Program funding proposals
o Public legal documents (deeds, lawsuits, construction permits)	o Critical incident reports
o Government records (births, deaths, crime reports, public health records)	o Quarterly and annual reports
o Public notices	o Staff meeting minutes
o Documentation of special events and celebrations	o Websites (plural)
o Historical photos and documents	o Program implementation documents
o Social media	o Evaluation reports
	o Participants' outputs (projects done)
INTERNET GROUPS	**GOVERNMENT UNITS**
o Blog posts	o Enabling legislation
o Chat room transcripts	o Budget documents
o Listserv archives	o Public hearing and testimony transcripts
o Social media postings	o Reports
o Special groups of shared concerns	o Planning documents

Observing What Does Not Happen

> You can observe a lot by just watching.
>
> —Yogi Berra
> Baseball player and pundit
>
> Watch what's there. And watch what's not there.
>
> —Halcolm's *Yin and Yang of Observation*

The preceding sections have described the things one can observe in a setting or program. Observing context, activities, interactions, what people do, and what they say is important in a comprehensive approach to fieldwork. But what about observing what does *not* happen?

The potential absurdity of speculating about what does not occur is illustrated by a Sufi story. During a plague of locusts, the wise fool Mulla Nasrudin, always looking on the bright side, went from village to village encouraging people by observing how fortunate they were that elephants had no wings. "You people don't realize how lucky you are. Imagine what life would be like with elephants flying overhead. These locusts are nothing."

To observe that elephants have no wings is indeed data. Moreover, elephants have no fins, claws, feathers, or branches. Clearly, once one ventures into the area of observing what does not happen, there are a near-infinite number of things one could point out. The "absence of occurrence" list could become huge. It is therefore with some caution that I include among the tasks of the observer that of noting what does not occur.

If social science theory, program goals, implementation designs, and/or proposals suggest that certain things ought to happen or are expected to happen, then it is appropriate for the observer or evaluator to note that those things did not happen. If a community where water is scarce shows no evidence of conflict over water rights, an anthropologist could be expected to report and explain this absence of expected tension. If a school program is supposed to, according to its funding mandate and goals, provide children with opportunities to explore the community but no such explorations occur, it is altogether appropriate for the evaluator to note the said implementation failure. If the evaluator reported only what occurred, a question might be left in the mind of the reader about whether the other activities had occurred but had simply not been observed. Likewise, if a criminal justice program is supposed to provide one-to-one counseling to juveniles but no such counseling takes place, it is entirely appropriate for the evaluator to note the absence of counseling.

Earlier, for a different purpose, I recounted observing early-childhood programs and finding in one program the absence of children's art on the walls. Indeed, the absence of any colorful posters or art of any kind stood out because, in all other centers we observed, the walls were covered with colorful displays.

Thus, it can be appropriate to note that something did not occur when the observer's basic knowledge of and experience with the phenomenon suggests that the absence of some particular activity or factor is noteworthy. This clearly calls for judgment, common sense, and experience. As eminent qualitative methodologist Bob Stake (1995) has asserted,

> One of the principal qualifications of qualitative researchers is experience. Added to the experience of ordinary looking and thinking, the experience of the qualitative researcher is one of knowing what leads to significant understanding, recognizing good sources of data, and consciously and unconsciously testing out the veracity of their eyes and robustness of their interpretations. It requires sensitivity and skepticism. Much of this methodological knowledge and personality come from hard work under the critical examination of colleagues and mentors. (pp. 49–50)

Making informed judgments about the significance of nonoccurrences can be among the most important contributions an evaluator can make because such feedback can provide program staff or other evaluation users with information that they may not have thought to request. Moreover, they may lack the requisite experience or awareness to have noticed the absence of that which the evaluator observes. For example, the absence of staff conflict is typically noteworthy because staff conflict is common. Similarly, absence of conflict between administrative levels (local, state, and federal) would be noteworthy because such conflict is, in my experience, virtually universal.

In many such cases, the observation about what did not occur is simply a restatement, in the opposite, of what did occur. That restatement, however, will attract attention in a way that the initial observation might not. For example, if one were observing a program being conducted in a multiracial community, it is possible that program goals would include statements about the necessity of staff being sensitive to the particular needs, interests, and cultural patterns of minorities, but there may not be specific mention of the desired racial composition of program staff. If, then, the evaluator knows that the staff of the program consist entirely of self-identified Caucasians, it is appropriate to report that the staff are all white; that is, no people of color are among the program staff, the importance of which derives from the location and nature of the program.

Observations of staff interaction and decision-making processes also provide opportunities for evaluators to note things that do not happen. If, over time, the observer notes that program-planning processes never include participants' input in any systematic or direct way, it may well be appropriate for the evaluator to point out the absence of such input based on experiences indicating the significance of participant input in the planning processes of other programs.

My evaluation of the wilderness education program included observations about a number of things that did not occur. No serious injuries occurred on any of the six field conferences in the wilderness—important information for someone thinking about the possible risks involved in such a program. No participant refused to shoulder his or her share of the work that had to be done for the group to live and work together in the wilderness. This observation emerged from

discussions with technical field staff who often worked with juveniles in wilderness settings where uneven sharing of cooking, cleaning, and related responsibilities often led to major group conflicts. The fact that the groups I observed never had to deal with one or two people not helping out was worth noting.

Perhaps the most important observation about what did not happen came from observing staff meetings. Over time, I noticed a pattern in which staff held meetings to make decisions about important issues, but no such decisions were made. Staff sometimes thought that a decision had been made, but closure was not brought to the decision-making process, and no responsibility for follow-up was assigned. Many subsequent implementation failures and staff conflicts could be traced to ambiguities and differences of opinion that were left unresolved at staff meetings. By hearing me describe what both was and was not occurring, staff became more explicit and effective in making decisions. Reporting what did happen in staff meetings was important; but it was also extremely important to observe what did not happen.

Observing Oneself: Reflexivity and Creativity, and Review of Fieldwork Dimensions

> Physician, heal thyself.
>
> Observer, observe thyself.
>
> —Halcolm
>
> Observations often tell you more about the observer than the observed.
>
> —Chris Geiger
> Journalist and cancer survivor

In Chapter 2, I identified voice and perspective, or *reflexivity*, as one of the central strategic themes of qualitative inquiry. The term *reflexivity* has entered the lexicon as a way of emphasizing the importance of self-awareness, political/cultural consciousness, and ownership of one's perspective. Reflexivity reminds the qualitative inquirer to observe herself or himself so as to be attentive to and conscious of the cultural, political, social, linguistic, and ideological origins of one's own perspective and voice as well as, and often in contrast to, the perspectives and voices of those one observes and talks to during fieldwork. Reflexivity calls for self-reflection, indeed, critical self-reflection and self-knowledge, and a willingness to consider *how* who one is affects what one is able to observe, hear, and understand in the field as an observer and analyst. The observer, therefore, during fieldwork, must observe self as well as others, and interactions of self with others.

Exhibit 2.5 (p. 72) poses three interconnected, tri-angulated reflexive questions from three perspectives:

1. *Myself, as inquirer:* How do I know what I know? What shapes and has shaped my perspective?

2. *People in the setting being studied:* How do they know what they know? What shapes and has shaped their worldview? How do they perceive me? Why? How do I know? How do I perceive them?

3. *Audiences for the study:* How do they make sense of what I give them? What perspectives do they bring to the findings I offer? How do they perceive me? How do I perceive them?

These are questions to address explicitly in the methods section of a qualitative report. Reflecting on these questions informs the credibility of the conclusions you report from fieldwork.

SIDEBAR

REFLEXIVE/REFLECTIVE INQUIRY

Reflexivity involves "paying much attention to how one thinks about thinking . . . , a 'reflexivity' that constantly assesses the relationship between 'knowledge' and 'the ways of doing knowledge.'"

Reflective [reflexive] research, as we define it, has two basic characteristics: careful interpretation and reflection. The first implies that all references—trivial and nontrivial—to empirical data are the *results of interpretation. . . .* Interpretation comes to the forefront of the research work. This calls for the utmost awareness of the theoretical assumptions, the importance of language and pre-understanding, all of which constitute major determinants of the interpretation. The second element, reflection, turns attention "inwards" towards the person of the researcher, the relevant research community, society as a whole, intellectual and cultural traditions, and the central importance, as well as the problematic nature, of language and narrative (the form of presentation) in the research context. . . . Reflection can, in the context of empirical research, be defined as the *interpretation of interpretation* and the launching of a critical self-exploration of one's own interpretations of empirical material (including its construction).

—Alvesson and Sköldbery (2009, pp. 8–9)
Reflexive Methodology

Once again, for continuity, I would cite Parameswaran (2001), who has written a wonderfully self-reflective account of her experience returning to her native India to do fieldwork as a feminist scholar after being educated in the United States.

Because my parents were fairly liberal compared to many of my friends' parents, I grew up with a little more awareness than many middle- and upper-class Indians of the differences between my life and that of the vast majority of Indians. Although I questioned some restrictions that were specific to women of my class, I did not have the language to engage in a systematic feminist critique of patriarchy or nationalism. Feminism for me had been unfortunately constructed as an illness that struck highly Westernized intellectual Indian women who were out of touch with reality. . . . It was my dislocation from India to the relatively radicalized context of United States that prompted my political development as a feminist and a woman of color. (p. 76)

Given this background and the controversial focus of her fieldwork (reading of Western romance novels by young Indian women), she identified reflective questions to guide her reflexive inquiry during and after fieldwork:

> How do kinship roles assigned to native scholars shape social interactions in the field? How can commitments to sisterhood make it difficult for feminist ethnographers to achieve critical distance and discuss female informants' prejudiced views? (p. 76)

Her personal inquiry into these questions, reflecting on her own fieldwork experiences (Parameswaran, 2001), is a model of reflexivity.

Many years ago, Indian philosopher J. Krishnamurti (1964) commented on the challenges of self-knowledge. Although his reflections were directed to the importance of lifelong learning rather than to being reflexive in fieldwork, his ruminations offer a larger context for thinking about how to observe oneself, a context beyond concern about methodological authenticity, though his advice applies to that as well.

> Self-knowledge comes when you observe yourself in your relationship with your fellow-students and your teachers, with all the people around you; it comes when you observe the manner of another, his gestures, the way he wears his clothes, the way he talks, his contempt or flattery *and your response* [italics added]; it comes when you watch everything in you and about you and see yourself as you see your face in the mirror. . . . Now, if you can look into the mirror of relationship exactly as you look into the ordinary mirror, then there is no end to self-knowledge. It is like entering a fathomless ocean which has no shore . . . ; if you can just observe what you are and move with it, then you will find that it is possible to go infinitely far. Then there is no end to the journey, and that is the mystery, the beauty of it. (pp. 50–51)

I realize that Krishnamurti's phrase "There is no end to the journey" may strike terror in the hearts of graduate students reading this in preparation for dissertation fieldwork or evaluators facing a report deadline. But remember, he's talking about lifelong learning, of which the dissertation or a specific evaluation report is but one phase. Just as most dissertations and evaluations are reasonably expected to contribute incremental knowledge rather than make major breakthroughs, so too the self-knowledge of reflexive fieldwork is but one phase in a lifelong journey toward self-knowledge—but it's an important phase and a commitment of growing significance as reflexivity has emerged as a central theme in qualitative inquiry.

The point here, which we shall take up in greater depth in the chapters on analysis and credibility, is that the observer must ultimately deal with issues of authenticity, reactivity, and how the observational process may have affected what was observed as well as how the background and predispositions of the observer may have constrained what was observed and understood. Each of these areas of methodological inquiry depends on some degree of critical reflexivity.

"*Limiting your peripheral vision will not enhance your observational powers. You see fine. You're just a lousy observer.*"

Fieldwork Menu Summary

This lengthy review of options for what and how to observe during fieldwork constitutes a *sensitizing framework*. It is *not* a prescriptive guide—you must do this and all of this! Nor is it a formal checklist of the kind that airline pilots go through before takeoff. Rather, it is a menu of possibilities. You have to decide which items to incorporate into your own inquiry, adapt them to your own research or evaluation purpose and questions, and fill in the details within your own context. Exhibit 6.8 provides a summary graphic.

EXHIBIT 6.8   Dimensions of Fieldwork

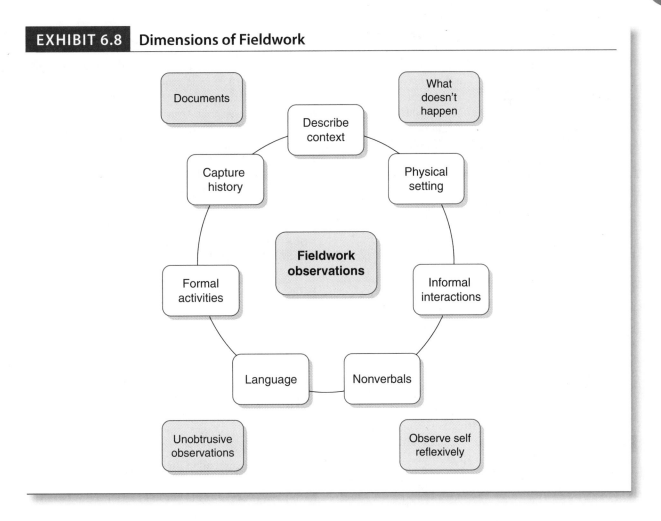

Nested and Layered Case Studies During Fieldwork

A case study is expected to catch the complexity of a single case. The single leaf, even a single toothpick, has unique complexities—but rarely will we care enough to submit it to case study. We study a case when it itself is of very special interest. We look for the detail of inter-action with its context. Case study is the study of the particularity and complexity of a single case, coming to understand its activity within important circumstances. (Stake, 1995, p. xi)

Months of fieldwork may result in a single case study that describes a village, community, neighbor-hood, organization, or program. However, that single case study is likely to be made up of many smaller cases, or observational units—the stories of specific individuals, families, organizational units, and other groups. Critical incidents and case studies of specific bounded activities, like a celebration, may also be pre-sented within the larger case. The qualitative analysis process typically centers on presentation of specific cases and thematic analysis across cases. Knowing this, fieldwork can be organized around nested and layered case studies, which means that some form of nested case sampling must occur.

Let me briefly review the centrality of case studies as a qualitative inquiry strategy. Chapter 1 opened by citing a number of well-known and influential books based on case studies, for example, Peters and Water-man's *In Search of Excellence: Lessons From Ameri-cas' Best-Run Companies* (1982), Angela Browne's important book *When Battered Women Kill* (1987), and Sara Lawrence-Lightfoot's six detailed case studies of *Respect* (2000). Chapter 2 presented the construction of *unique case studies* as a major strategic theme of qualitative inquiry. Chapter 3 reviewed the-oretical perspectives that are inductively case based. Chapter 4 reviewed at some length the importance in qualitative evaluation of *capturing and reporting individualized outcomes* based on case studies of how participants in programs change during a program and whether they maintain those changes afterward.

Chapter 5 discussed alternative views of what a case is, examined what it means to do a case study (both an inquiry method and a product of such an inquiry), and provided an extended discussion of purposeful sampling strategies for selecting cases. Now we come to the point of distinguishing and integrating various observational units and data sources in fieldwork. William Foote Whyte's (1943) classic study of *Street Corner Society* has long been recognized as an exemplar of the single (*n* = 1) community case study, even though his study of "Cornerville" includes the stories (case studies) of several individual lower-income youths, some of whom were striving to escape the neighborhood.

Integrating Cases: The Wilderness Education Program as an Example

Fieldwork can include a variety of kinds of data: formal interviews, conversational interviews, group interviews, spectator observations, participant observation, and document reviews. To illustrate this point, in the wilderness education program I've been using as an example throughout this chapter, our evaluation team generated case studies of participants (higher education leaders and administrators) using multiple sources of data from fieldwork: (1) background data gathered through interviews about participants' situations and perspectives on entering the year of field conferences; (2) observations of their experiences during field conferences; (3) informal and conversational interviews with them during the four wilderness trips, each for 10 days, over the course of a year; (4) quotations from formal group interviews (focus groups) held at various times during the trips; (5) excerpts from their journals and other personal writings when they were willing to share those with us, as they often were; and (6) follow-up telephone interviews with participants after each field trip and after the entire program was completed to track the impact of the program on individuals over time.

The overall case being studied was *the wilderness program*, a single case. But it operated with two different cohorts of participants over two years, so each year was also a case, as was each trip within a year. And for the individual case studies, each year (*n* = 36), the various data sources noted above were combined to construct an integrated, holistic case study of individual personal and professional outcomes. Thus, when more than one object of study or unit of analysis is included in fieldwork, case studies may be layered and nested within the overall, primary case approach.

Let's look at how the various qualitative pieces fit together. As noted, the three-year wilderness program constituted the overall, one might say *macro*, case study. Therefore, the final evaluation report presented conclusions about the processes and outcomes of the overall program, a case study of the three-year wilderness education initiative. As Exhibit 6.9 shows, however, within that overall evaluation case study were nested individual participant's case studies documenting individual experiences and outcomes; case studies of each yearlong group cohort; and case studies of each separate field conference, for example, the 10 days in the Gila Wilderness of New Mexico and the 10 days in the Kofa Mountains of Arizona. Slicing through the fieldwork and analysis in other ways were case studies of particular incidents, for example, the emotional catharsis experienced by one participant when she finally managed to overcome her terror and rappel down a cliff face, the whole group watching and urging her on, a process that took some 45 tense minutes. Other mini cases consisted of different *observational units*. A full day's hike could be an observational unit, as could running a specific dangerous rapid on the San Juan River. Each evening discussion constituted an observational unit, such that over the three years, we had notes on more than 80 discussions of various kinds. Staff meetings made for unique observational units, as could a specific decision made in a staff meeting.

In essence, extended fieldwork can and typically does involve many mini or micro case studies of various observational units (individuals, groups, specific activities, specific time periods, critical incidents), all of which *together* make up the overall case study, in this example, the final evaluation of the wilderness education program. Fieldwork, then, can be thought of as engaging in a series of multilayered and nested case studies, often with intersecting and overlapping units of analysis or observational units.

Creativity in Fieldwork

No checklist can be relied on to guide all aspects of fieldwork. A participant-observer must constantly make judgments about what is worth noting. Because it is impossible to observe everything, some process of selection is necessary. Plans made during design should be revised as appropriate when important new opportunities and sources of data become available. That's where flexibility and creativity help. Creativity can be learned and practiced (Patton, 1987). Creativity comes into play in a major way in integrating and synthesizing multiple observations and sources of data into a coherent whole—the case study story. Creative fieldwork means using every part of oneself to experience and understand what

EXHIBIT 6.9 Nested, Layered, and Overlapping Mini Case Studies During Fieldwork

The wilderness program evaluation for higher education leaders and administrators illustrates how case studies can be layered and nested. Evaluation of the three-year wilderness program constituted the overall *macro* case study, the primary evaluation case study. Nested and layered within that overall evaluation were various mini cases of overlapping and intersecting observation units that helped organize and frame fieldwork and note taking.

Nested, layered, and overlapping mini case studies (observational units)

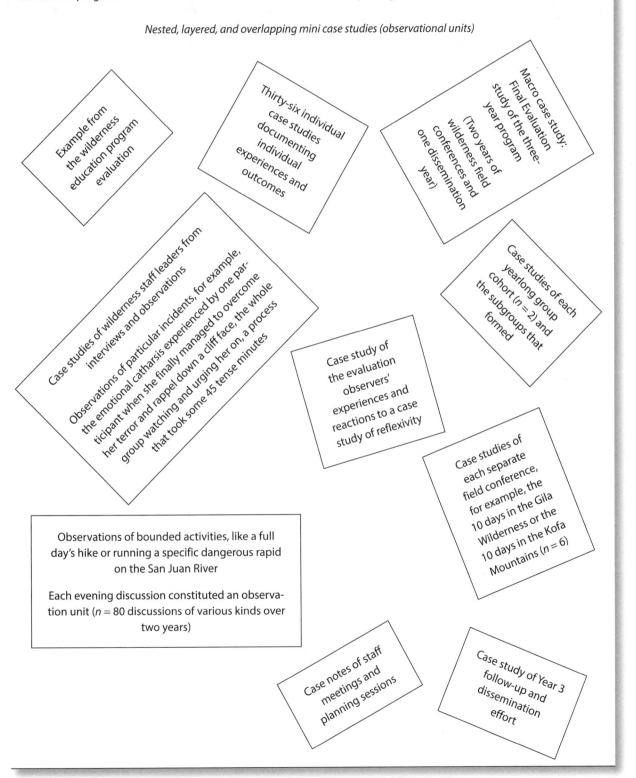

Example from the wilderness education program evaluation

Thirty-six individual case studies documenting individual experiences and outcomes

Macro case study: Final Evaluation study of the three-year program (Two years of wilderness field conferences and one dissemination year)

Case studies of wilderness staff leaders from interviews and observations

Observations of particular incidents, for example, the emotional catharsis experienced by one participant when she finally managed to overcome her terror and rappel down a cliff face, the whole group watching and urging her on, a process that took some 45 tense minutes

Case studies of each yearlong group cohort (n = 2) and the subgroups that formed

Case study of the evaluation observers' experiences and reactions to a case study of reflexivity

Case studies of each separate field conference, for example, the 10 days in the Gila Wilderness or the 10 days in the Kofa Mountains (n = 6)

Observations of bounded activities, like a full day's hike or running a specific dangerous rapid on the San Juan River

Each evening discussion constituted an observation unit (n = 80 discussions of various kinds over two years)

Case notes of staff meetings and planning sessions

Case study of Year 3 follow-up and dissemination effort

is happening. Creative insights come from being directly involved in the setting being studied.

I shall return to the issue of creativity in considering the interpretation of field notes later in this chapter and again in the analysis chapter (Chapter 8). For the moment, it is sufficient to acknowledge the centrality of creativity in naturalistic inquiry and to concur with Virginia Woolf:

> Odd how the creative power at once brings the whole universe to order. . . . I mark Henry James' sentence: observe perpetually. Observe the oncome of age. Observe greed. Observe my own despondency. By that means it becomes serviceable. (Quoted by Partnow, 1978, p. 185)

BOXED IN

Those who aspire to be creative are admonished to "think outside the box." This presumes that one has exhausted the possibilities of learning within the box. Before moving outside the box, make sure you know the box. Observe it. Look deep within. Find out the history of the box, how it came to be the box. What has it held? What has been taken from it? Examine the corners. Look underneath, on top, on all sides. Know the box. Understand the box. Learn what the box has to teach. Think inside the box. Only then will you truly be ready to "think outside the box."

—From Halcolm's *Boxing Guide*

SIDEBAR

Doing Fieldwork: The Data-Gathering Process

The purpose of the research has been clarified. The primary research questions have been focused. Qualitative methods using observations have been selected as one of the appropriate methods for gathering data. It is time to enter the field. Now begins the essential and arduous task of taking field notes.

Field Notes

> *Nota bene* (Latin): "Note well, observe carefully, pay close attention."

Many options exist for taking field notes. Variations include the writing materials used, the time and place for recording field notes, the symbols developed by observers as their own method of shorthand, and how field notes are stored. No universal prescriptions about the mechanics of and procedures for taking field notes are possible because different settings lend themselves to different ways of proceeding, and the precise organization of fieldwork is very much a matter of personal style and individual work habits. *What is not optional is the taking of field notes!*

> Aside from getting along in the setting, the fundamental work of the observer is the taking of field notes. Field notes are "the most important determinant of later bringing off a qualitative analysis. Field notes provide the observer's *raison d'etre*. If he is not doing them, he might as well not be in the setting." (Lofland, 1971, p. 102)

Field notes contain the description of what has been observed. They should contain everything that the observer believes to be worth noting. Don't trust anything to future recall. At the moment one is writing, it is very tempting, because the situation is still fresh, to believe that the details or particular elements of the situation can be recalled later. If it's important as part of your consciousness as an observer, if it's information that has helped you understand the context, the setting, and what went on, then as soon as possible that information should be captured in the field notes.

At the same time, you can't and won't capture everything. Reflecting his years of experience both doing qualitative studies and working with students

on their dissertations, Bob Stake (2010) offers this comforting observation:

> One of the largest worries of a new researcher is making an accurate record of what is happening. I sometimes think he or she worries too much about the accuracy. Yes, it has to be right, but there is more than one chance to get it right. The first responsibility of the observer is to know what is happening, to see it, to hear it, to try to make sense of it. That is more important than getting the perfect note or quote. Much of what we put down is an approximation that we can improve upon later—if we have a good idea what happened. (p. 94)

Description First

First and foremost, field notes are descriptive. They should be dated and should record basic information such as where the observation took place, who was present, what the physical setting was like, what social interactions occurred, and what activities took place. Field notes contain the descriptive information that will permit you to return to an observation later during analysis and, eventually, permit the reader of the study's findings to experience the activity observed through your report.

Exhibit 6.10 presents different kinds of descriptive field notes. On the left side are vague and overgeneralized field notes. On the right side are more detailed and concrete field notes from the same observation. These examples illustrate the problem of using general terms to describe specific actions and conditions. Words like *poor*, *anger*, and *uneasy* are insufficiently descriptive. Such interpretive words conceal what actually went on rather than reveal the details of the situation. Such terms have little meaning for a person present for the observation. Moreover, the use of such terms in field notes, without the accompanying detailed description, means that the fieldworker has fallen into the bad habit of primarily recording interpretations rather than description. Particularly revealing are terms that can only make sense in comparison with something else. The phrase "poorly dressed" requires some frame of reference about what constitutes "good" dress. No skill is more critical in fieldwork than learning to be descriptive, concrete, and detailed.

EXHIBIT 6.10   Observations From Field Notes: Poorly Done Compared With Well Done

VAGUE, INTERPRETATIVE, AND OVERGENERALIZED OBSERVATIONS	DETAILED, CONCRETE, AND DESCRIPTIVE OBSERVATIONS
1. The new client was uneasy waiting for her intake interview.	At first, the new client sat very stiffly on the chair next to the receptionist's desk. She picked up a magazine and let the pages flutter through her fingers very quickly without really looking at any of the pages. She set the magazine down, looked at her watch, pulled her skirt down, picked up the magazine again, set it back down, took out a cigarette, put it to her lips but didn't light it. She watched the receptionist out of the corner of her eye and glanced at the two or three other people waiting in the room. Her eyes moved from people to the magazine to the cigarette to the people to the magazine in rapid succession, but avoided eye contact. When her name was finally called, she jumped like she was startled.
2. The client was quite hostile toward the staff person.	When Judy, the senior staff member, told the client that she could not just do whatever she wanted to do, the client began to yell, screaming that Judy couldn't control her life, accused Judy of being on a "power trip," and said that she'd "like to beat the shit out of her," then told her to "go to hell." The client shook her fist in Judy's face and stomped out of the room, leaving Judy standing there with her mouth open, looking dumbfounded.
3. The next student who came in to take the test was very poorly dressed.	The next student who came into the room wore clothes quite different from the three previous students. The other students had hair carefully combed and clothes clean, pressed, and in good condition with colors coordinated. This new student wore soiled pants with a tear in one knee and a threadbare seat. His flannel shirt was wrinkled with one tail tucked into the pants and the other tail hanging out. His hair was disheveled and his hands looked like he'd been playing in the oily engine of a car.

Field notes also contain what people say. Direct quotations, or as near as possible recall of direct quotations, should be captured during fieldwork, recording what was said during observed activities as well as responses garnered during interviews, both formal and conversational. Quotations provide the "emic perspective" discussed earlier—the insider's perspective—which "is at the heart of most ethnographic research" (Fetterman, 1989, p. 30).

Field notes also contain the observer's own feelings, reactions to the experience, and reflections about the personal meaning and significance of what has been observed. Don't deceive yourself into thinking that such feelings can be conjured up again simply by reading the descriptions of what took place. Feelings and reactions should be recorded at the time they are experienced, while you are in the field. Both the nature and intensity of feelings should be recorded. In qualitative inquiry, the observer's own experiences are part of the data. Part of the purpose of being in a

setting and getting close to the people in the setting is to permit you to experience what it is like to be in that setting. If what it is like for you, the observer or participant-observer, is not recorded in your field notes, then much of the purpose of being there is lost.

Finally, field notes include your insights, interpretations, beginning analyses, and working hypotheses about what is happening in the setting and what it means. While you should approach fieldwork with a disciplined intention not to impose preconceptions and early judgments on the phenomenon being experienced and observed, nevertheless, as an observer, you don't become a mechanical recording machine on entering the field. Insights, ideas, inspirations—and yes, judgments too!—will occur while making observations and recording field notes. It's not that you sit down early on and begin the analysis and, if you're an evaluator, make judgments. Rather, it's in the nature of our intellects that ideas about the meaning, causes, and significance of what we experience find their way into our minds. These insights and inspirations become part of the data of fieldwork and should be recorded in context in field notes. I like to set off field interpretations with brackets. Others use parentheses, asterisks, or some other symbol to distinguish interpretations from description. The point is that interpretations should be understood to be just that, interpretations, and labeled as such. Field-based insights are sufficiently precious that you need not ignore them in the hope that, if really important, they will return later.

Field notes, then, contain the ongoing data that are being collected. They consist of descriptions of what is being experienced and observed, quotations from the people observed, the observer's feelings and reactions to what is observed, and field-generated insights and interpretations. Field notes are the fundamental database for constructing case studies and carrying out thematic cross-case analysis in qualitative research.

Procedurally Speaking

When field notes are written will depend on the kind of observations being done and the nature of your participation in the setting being studied. In an evaluation of a parent education program, I was introduced to the parents by the staff facilitator, who explained the purpose of the evaluation and assured the parents that no one would be identified. I then openly took extensive notes without participating in the discussions. Immediately following those sessions, I would go back over my notes to fill in details and be sure what I had recorded made sense. By way of contrast, in the wilderness education program, I was a full participant engaged in full days of hiking, rock climbing, and rafting/kayaking. I was so exhausted by the end of each day that I seldom stayed awake while the others slept, making field notes by flashlight. Rather, each night, I jotted down basic notes that I could expand during the time when the others were writing in their journals, but some of the expansion had to be completed after the weeklong field conference. In evaluating a leadership training program as a participant-observer, the staff facilitator privately asked me not to take notes during group discussions because it made him nervous, even though most of the other participants were taking notes.

The extent to which notes are openly recorded during the activities being observed is a function of the observer's role and purpose, as well as the stage of participant observation. If the observer or evaluator is openly identified as a short-term, external, nonparticipant observer, participants may expect him or her to write down what is going on. If, on the other hand, one is engaged in longer-term participant observation, the

early part of the process may be devoted to establishing the participant-observer role, with emphasis on participation, so that open taking of notes is deferred until the fieldworker's role has been firmly established within the group. At that point, it is often possible to openly take field notes since, hopefully, the observer is better known to the group and has established some degree of trust and rapport.

The wilderness program evaluation involved three 10-day trips ("field conferences") with participants at different times during the year. During the first field conference, I never took notes openly. The only time I wrote was when the others were also writing. During the second field conference, I began to openly record observations when discussions were going on if taking notes did not interfere with my participation. By the third field conference, I felt I could take notes whenever I wanted to, and I had no indication from anyone that they even paid attention to the fact that I was taking notes. By that time, I had established myself as a participant, and my participant role was more primary than my evaluator role.

The point here is that evaluator-observers must be strategic about taking field notes, timing their writing and recording in such a way that they are able to get their work done without unduly affecting either their participation or their observations. Given those constraints, *the basic rule of thumb is to write promptly*, to complete field notes as soon and as often as physically and programmatically possible.

Writing field notes is rigorous and demanding work. Lofland (1971) has described this rigor quite forcefully:

> Let me not deceive the reader. The writing of field notes takes personal discipline and time. It is all too easy to put off actually writing notes for a given day and to skip one or more days. For the actual writing of the notes may take as long or longer than did the observation! Indeed, a reasonable rule of thumb here is to expect and plan to spend as much time writing notes as one spent in observing. This is, of course, not invariant . . . but one point is inescapable. All the fun of actually being out and about monkeying around in some setting must also be met by cloistered rigor in committing to paper—and therefore to future usefulness—what has taken place. (p. 104)

Integrating Observations, Interviews, and Documentation: Bringing Together Multiple Sources for Triangulation

Fieldwork is more than a single method or technique. For example, evaluation fieldwork means that the evaluator is on-site (where the program is happening) observing, talking with people, and going through program records. Multiple sources of information are sought and used because no single source of information can be trusted to provide a comprehensive perspective on the program. By using a combination of observations, interviewing, and document analysis, the fieldworker is able to use different data sources to validate and cross-check findings. Each type and source of data has strengths and weaknesses. Using a combination of data types—triangulation, a recurring theme in this book—increases validity as the strengths of one approach can compensate for the weaknesses of another approach.

Limitations of observations include the possibility that the observer may affect the situation being observed in unknown ways, people may behave in some atypical fashion when they know they are being observed, and the selective perception of the observer may distort the data. Observations are also limited in focusing only on external behaviors—the observer cannot see what is happening inside people, what they're thinking and feeling. Moreover, observational data are often constrained by the limited sample of activities actually observed. Researchers and evaluators need other data sources to find out the extent to which observed activities are typical or atypical.

Interview data limitations include possibly distorted responses due to personal bias, anger, anxiety, politics, and simple lack of awareness since interviews can be greatly affected by the emotional state of the interviewee at the time of the interview. Interview data are also subject to recall error, reactions of the interviewee to the interviewer, and self-serving responses.

Observations provide a check on what is reported in interviews; interviews, on the other hand, permit the observer to go beyond external behavior to explore feelings and thoughts.

Documents and records also have limitations. They may be incomplete or inaccurate. Client files maintained by programs are notoriously variable in quality and completeness, with great detail in some cases and virtually nothing in others. Document analysis, however, provides a behind-the-scenes look at the program that may not be directly observable and about which the interviewer might not ask appropriate questions without the leads provided through documents.

By using a variety of sources and resources, the qualitative inquirer can build on the strengths of each type of data collection while minimizing the weaknesses of any single approach. This mixed-methods, triangulated approach to fieldwork is based on pragmatism and is illustrated by my attempt to understand some of the problems involved in staff communication during the wilderness education and leadership program evaluation. I mentioned this example earlier, but I'd like to expand it here. As noted, two kinds of staff worked in the program: (1) three people who had overall management and administrative responsibility (the program was their idea, and they procured the funding) and (2) the two technical staff who had responsibility for wilderness skills training, field logistics, and safety. The technical staff had extensive experience leading wilderness trips, but they also were skilled at facilitating group processes. During the trips, the lines of responsibility between technical staff and administrative staff were often blurred, and on occasion, these ambiguities gave rise to conflicts. I observed the emergence of conflict early on the first trip but lacked context for knowing what was behind these differences. Through interviews and casual conversations during fieldwork, I learned that all of the staff, both administrative and technical, had known each other prior to the program. Indeed, the program administrative directors had been the college professors of the technical staff, while the latter were still undergraduate students. However, the technical staff had introduced the directors to the wilderness as an environment for experiential education. Each of the staff members described in interviews his or her perception of how these former relationships affected the field operations of the program, including difficulties in communication that had emerged during planning sessions prior to the actual field conferences. Some of those conflicts were documented in letters and memos. Reading their files and correspondence gave me a deeper understanding of the different assumptions and values of various staff members. *But the documentation would not have made sense without the interviews, and the focus of the interviews came from the field observations. Taken together, these diverse sources of information and data gave me a complete picture of staff relationships.* Working back and forth among individual staff members and group staff meetings, I was able to use this information to assist the staff in their efforts to improve their communications during the final field conference. All three sources of information proved critical to my understanding of the situation and that understanding enhanced my effectiveness as a formative evaluator.

The Technology of Fieldwork and Observation

The classic image of the anthropological fieldworker is of someone huddled in an African hut writing voluminously by light of a lantern. Contemporary researchers, however, have available to them a number of technological innovations, which, when used judiciously, can make fieldwork more efficient and comprehensive. First and foremost is the digital recorder. For some people, including myself, dictating field notes saves a great deal of time while increasing the comprehensiveness of the report. Learning to dictate takes practice, effort, and critical review of early attempts. Voice recorders must be used judiciously so as not to become obtrusive and inhibit program processes or participant responses. A voice recorder is much more useful for recording field notes in private than as an instrument to be carried about at all times, available to put a quick end to any conversation into which the observer enters.

Technology for taking photographs has become standard in fieldwork. Photographs can help in recalling things that have happened as well as vividly capturing the setting for others. Digital photography, as well as advances in printing and photocopying, now makes it possible to economically reproduce photographs in research and evaluation reports. In the wilderness education program evaluation, I officially became the group photographer, making photographs available to all of the participants. This helped legitimize taking photographs and reduced the extent to which other people felt it necessary to carry their own cameras at all times, particularly at times when it was possible that the equipment might be damaged. Looking at photographs during analysis helped me recall the details of certain activities that I had not fully recorded in my written notes. I relied heavily on photographs to add details to descriptions of places where critical events occurred in the Grand Canyon initiation story I wrote about coming of age in modern society (Patton, 1999). We'll discuss the use of photographs at greater length in the analysis chapter (Chapter 8).

Video photography is another technological innovation that has become readily accessible and common enough that it can sometimes be used unobtrusively. For example, in a formative evaluation of a staff training program, I used video to provide visual feedback to staff. Videos of classrooms, training sessions, therapeutic interactions, and a host of other observational targets can sometimes be less intrusive than a note-taking observer. We had great success taking videos of mothers and children playing together in early-childhood education centers. Of course, use of such technology must be negotiated with program staff and participants, but the creative and judicious use of technology can greatly increase the quality of field observations and the utility of the observational record to others. Moreover, comfort with voice recorders and video has made it increasingly possible to use such technology without undue intrusion when observing programs

where professionals are the participants. In addition, sometimes videos originally done for research or evaluation can subsequently be used for future training, program development, and communications to policymakers and the public, making the costs more manageable because of added uses and benefits.

> Why should a social researcher wish to incorporate the analysis of images—paintings, photographs, film, videotape, drawings, diagrams and a host of other images—into their research? There are two good reasons. . . .
>
> The first good reason is that images are ubiquitous in society, and because of this some consideration of visual representation can potentially be included in all studies of society. No matter how tightly or narrowly focused a research project is, at some level all social research says something about society in general, and given the ubiquity of images, their consideration must at some level form part of the analysis.
>
> The second good reason why the social researcher might wish to incorporate the analysis of images is that a study of images or one that incorporates images in the creation or collection of data might be able to reveal some sociological insight that is not accessible by any other means. (Banks, 2007, pp. 3–4)

Here's the caveat: Visual technology can add an important dimension to fieldwork *if the observer knows how to use such technology and uses it well*—for there is much to learn beyond how to point the camera or turn on the video recorder, especially about integrating and analyzing visual data within a larger fieldwork context. Moreover, a downside to visual technology has emerged since it is now possible to not only capture images on film and video but also change and edit those images in ways that distort. In his extensive review of "visual methods" in qualitative inquiry, Douglas Harper (2000) concluded that "now that images can be created and/or changed digitally, the connection between image and 'truth' has been forever severed" (p. 721). This means that issues of credibility apply to using and reporting visual data as they do to other kinds of data. Still, visual qualitative data will only become more important over time, and visual technology has already become a specialized method, with its own handbooks to offer guidance about how to make the best use of visual data (Margolis & Pauwels, 2011).

Whether using technology is more intrusive than taking notes by hand will depend on the setting, the people being observed, and the inquirer's approach.

VISUAL QUALITATIVE DATA

Visual media, including photography, film, and more recently video, provide unprecedented opportunities for social science research. Consider video, for example: here is a cheap and reliable technology that enables us to record naturally occurring activities as they arise in ordinary habitats, such as the home, the workplace or the classroom. These records can be subject to detailed scrutiny. They can be repeatedly analyzed and they enable access to the fine details of conduct and interaction that are unavailable to more traditional social science methods. These records can be shown and shared with others not only fellow researchers, but participants themselves, or those with a more practical or applied interest in the activities and their organization. Unlike many forms of qualitative data, video can form an archive, a corpus of data that can be subject to a range of analytic interests and theoretical commitments, providing flexible resources for future research and collaboration. Video can also enable us to reconsider the ways in which we present the findings of social science research, not only to academic colleagues but more generally to the wider public.

—Heath, Hindmarsh, and Luff (2010)
Video in Qualitative Research:
Analyzing Social Interaction in Everyday Life

Creative Visual Documentation

Distinguished evaluator Michael Bamberger is also a wood sculptor. He created a work of art to depict the daily activities of women in Morocco. He explains,

This (wood carving) was inspired by a series of planning workshops with Moroccan village women. They were asked to draw a story illustrating their typical day. The figures in the carving illustrate these activities including collecting firewood, drawing water from the well, taking the goats to pasture, preparing meals, taking care of the children and then all going to sleep on one of the beautiful carpets that the women make. The numbers indicate the time of day when different activities take place. The gnarled branches in the center are the olive trees. (Bamberger, 2014)

SOURCE: Bamberger (2014). Used with permission.

Indeed, taking field notes can be nearly as intrusive as using visual technology, as illustrated in the fieldwork of anthropologist Carlos Castaneda. In the passage below, Castaneda (1972) reports on his negotiations with Don Juan to become his Native Indian key informant on sorcery and indigenous drugs. The young anthropologist records that Don Juan "looked at me piercingly."

"What are you doing in your pocket?" he asked, frowning. "Are you playing with your whanger?"

He was referring to my taking notes on a minute pad inside the enormous pockets of my windbreaker.

When I told him what I was doing he laughed heartily.

I said that I did not want to disturb him by writing in front of him.

"If you want to write, write," he said. "You don't disturb me." (pp. 21–22)

Whether one uses modern technology to support fieldwork or simply writes down what is occurring, some method of keeping track of what is observed must be established. In addition, the nature of the recording system must be worked out in accordance with the observer's role, the purpose of the study, and consideration of how the data-gathering process will affect the activities and persons being observed. Many of these issues and procedures must be worked out during the initial phase (entry period) of fieldwork, to which we now turn.

NEW DIRECTIONS IN TECHNOLOGY-BASED FIELDWORK

The Human Speechome Project

The Human Speechome Project studies how children learn the meaning of words through "analysis of observational recordings of child–caregiver interactions in natural contexts." Professor Deb Roy is recording his son's development at home by collecting approximately 10 hours of high-fidelity audio and video on a daily basis from birth to age three. The resulting data amount to more than 100,000 hours of multitrack recordings. They constitute "the most comprehensive record of a child's development made to date" (MIT Media Lab, 2014; Roy, 2009).

Google Glass

Google Glass is a pair of glasses that takes photographs by voice command and then transmits them by e-mail to a data collection site. It can be used in observations and interviews. It can also be intrusive, so extra care must be taken to respect people's privacy and get permission when using any visual data collection technique as part of observational fieldwork.

> All truth passes through three stages. First, it is ridiculed. Second, it is violently opposed. Third, it is accepted as being self-evident.
>
> —Arthur Schopenhauer (1788–1860)
> German philosopher

Thus far, fieldwork has been described as if it was a single, integrated experience. Certainly, when fieldwork goes well, it flows with certain continuity, but it is useful to look at the evolution of fieldwork through identifiable stages. Three stages are most often discussed in the participant observation literature: (1) the entry stage, (2) the routinization of the data-gathering period, and (3) the closing stage. The following sections explore each of these stages.

Entry Into the Field

> Beginnings matter. Where you begin a journey and the route you take will determine what you see and the experience you have. It has always been so. Pay attention, then, to where and how you begin.
>
> —Halcolm

The writings of anthropologists sometimes present a picture of the early period of fieldwork that reminds me of the character in Franz Kafka's haunting novel *The Castle*. Kafka's character is a wandering stranger, K., with no more identity than that initial. He doesn't belong anywhere, but when he arrives at the castle, he wants to become part of that world. His efforts to make contact with the faceless authorities who run the castle lead to frustration and anxiety. He can't quite figure out what is going on and can't break through their vagueness and impersonal nature. He doubts himself; then he gets angry at the way he is treated; then he feels guilty, blaming himself for his inability to break through the ambiguous procedures for entry. Yet he remains determined to make sense out of the incomprehensible regulations of the castle. He is convinced that, after all, where there are rules—and he does find that there are rules—they must fit together somehow, have some meaning, and manifest some underlying logic. There must be some way to make contact, to satisfy the needs of the authorities, to find some pattern of behavior that will permit him to be accepted. If only he could figure out what to do, if only he could understand the rules, then he would happily do what he was supposed to do. Such are the trials of entry into the field.

Entry into the field for research or evaluation involves two separate parts: (1) negotiation with gatekeepers, whoever they may be, about the nature of the fieldwork to be done and (2) actual physical entry into the field setting to begin collecting data. These two parts are closely related, for the negotiations with gatekeepers will establish the rules and conditions for how one goes about playing the role of observer and how that role is defined for the people being observed. In traditional scholarly fieldwork for the purpose of basic or applied research, the investigator unilaterally decides how best to conduct the fieldwork. In evaluation studies, the evaluator will need to take into account the perspectives and interests of the primary intended users of the evaluation. In either case, interactions with those who control entry into the field are primarily strategic, figuring out how to gain entry while preserving the integrity of the study and the inquirer's interests. The degree of difficulty involved varies depending on the purpose of the fieldwork, the extent to which it may involve controversial or sensitive issues, and, accordingly, the expected degree of resistance to the study. Where the field researcher expects cooperation, gaining entry may be largely a matter of establishing trust and rapport. At the other end of the continuum are those research settings where considerable resistance, even hostility, is expected, in which case gaining entry becomes a matter of "infiltrating the setting" (Douglas, 1976, p. 167). And sometimes, entry is simply denied. A doctoral student had negotiations for entry end abruptly in a school district where she had developed good relationships with school personnel and negotiations appeared to be going well. She later learned that she was denied entry far into the negotiation process because of community opposition. The local community had had a very bad experience with a university researcher more than 20 years earlier and still viewed all research with great suspicion.

Explaining Why You Are Undertaking the Inquiry

A major difference between the entry process in scholarly research in contrast to entry for evaluation research is how the purpose of the study is explained. In cross-cultural scholarly research, the usual explanation is some variation of "I'm here because I would like to understand you better and learn about your way of life because the people from my culture would like to know more about you." While anthropologists admit that such an explanation almost never makes sense to indigenous peoples in other cultures, it remains a mainstay initial explanation until mutual reciprocities can be established with enough local people for the observation process to become established and accepted in its own right.

Evaluators and action researchers, however, are not just doing fieldwork out of personal or professional interest. They are doing the fieldwork for some decision makers and information users who may be either known or unknown to the people being studied. It becomes critical, then, that evaluators, their funders, and evaluation users give careful thought to how the fieldwork is going to be presented. Because the word *evaluation* has such negative connotations for many people, having had negative experiences being evaluated, for example, at school or work, it may be appropriate to consider some other term to describe the fieldwork. In our onlooker, nonparticipatory observations for an implementation study of early-childhood programs in Minnesota, we described our role to local program participants and staff as follows:

> We're here to be the eyes and ears for state legislators. They can't get around and visit all the programs throughout the state, so they've asked us to come out and describe for them what you're doing. That way they can better understand the programs they have funded. We're not here to make any judgments about whether your particular program is good or bad. We are just here to be the eyes and ears for the legislature so that they can see how the legislation they've passed has turned into real programs. This is your chance to inform them about your work and give them your point of view.

Other settings lend themselves to other terms that are less threatening than *evaluator*. Sometimes, a fieldwork project can be described as "documentation." Another term I've heard used by community-based evaluators is *process historian*. In the wilderness education program, I was a full participant-observer, and staff described my role to participants as "keeper of the community record," making it clear that I was not there to evaluate individual participants. The staff of the project explained that they had asked me to join

the project because they wanted someone who did not have direct ego involvement in the success or outcomes of the program to observe and describe what went on, both because they were too busy running the program to keep detailed notes about what occurred and because they were too involved with what happened to be able to look at things dispassionately. We had agreed from the beginning that the community record I produced would be accessible to participants as well as staff.

In none of these cases did changing the language automatically make the entry process smooth and easy. Earlier in this chapter, I described our attempt to be viewed as "educational researchers" in evaluating a community leadership program. Everyone figured out almost immediately that we were really "evaluators"— and that's what participants called us. Regardless of the story told or the terms used, the entry period of fieldwork is likely to remain "the first and most uncomfortable stage of field work" (Wax, 1971, p. 15). It is a time when the observer is getting used to the new setting, and the people in that setting are getting used to the observer. Johnson (1975) has suggested that there are two reasons why the entry stage is both so important and so difficult:

> First, the achievement of successful entree is a precondition for doing the research. Put simply, no entree, no research. . . . Published reports of researcher's entree experiences describe seemingly unlimited contingencies which may be encountered, ranging from being gleefully accepted to being thrown out on one's ear. But there is a more subtle reason why the matter of one's entrance to a research setting is seen as so important. This concerns the relationship between the initial entree to the setting and the validity of the data that is subsequently collected. The conditions under which an initial entree is negotiated may have important consequences for how the research is socially defined by the members of the setting. These social definitions will have a bearing on the extent to which the members trust a social researcher, and the existence of relations of trust between an observer and the members of a setting is essential to the production of an objective report, one which retains the integrity of the actor's perspective and its social context. (pp. 50-51)

Reciprocity as a Guiding Principle

While the observer must learn how to behave in the new setting, the people in that setting are deciding how to behave toward the observer. Mutual trust, respect, and cooperation are dependent on the emergence of an exchange relationship, or reciprocity, in which the observer obtains data and the people being observed find something that makes their cooperation worthwhile, whether that something is a feeling of importance from being observed, useful feedback, pleasure from interactions with the observer, assistance in some task, or compensation in some form. This *reciprocity model* of gaining entry assumes that some reason can be found for participants to cooperate in the research and that some kind of mutual exchange can occur, not only initially but as the inquiry unfolds.

SIDEBAR

RECIPROCITY

Qualitative studies intrude into settings as people adjust to the researcher's presence. People may be giving their time to be interviewed or to help the researcher understand group norms; the researcher should plan to reciprocate. When people adjust their priorities and routines to help the researcher, or even just tolerate the researcher's presence, they are giving of themselves. The researcher is indebted and should be sensitive to this. Reciprocity may entail giving time to help out, providing informal feedback, making coffee, being a good listener, or tutoring. Of course, reciprocity should fit within the constraints of research and personal ethics and of maintaining one's role as a researcher. (Marshall & Rossman, 2011, p. 121; see also Crow, 2008)

Infiltration

Infiltration lies at the opposite end of the continuum from a negotiated, reciprocity model of entry. Many field settings are not open to observation based on cooperation. Douglas (1976, pp. 167–171) has described a number of infiltration strategies, including "worming one's way in," "using the crowbar to pry them open for our observations," showing enough "saintly submissiveness" to make members guilty enough to provide help, or playing the role of a "spineless boob" who could never possibly hurt the people being observed. He has also suggested using various ploys of misdirection where the researcher diverts people's attention away from the real purpose of the study. There is also the "phased-entree tactic" by which the researcher who is refused entry to one group begins by studying another group until it becomes possible to get into the group that is the real focus of the researcher's attention; for example, begin by observing children in a school when who you really want to observe are teachers or administrators.

Known Sponsor Approach

Often the best approach for gaining entree is the *known sponsor approach*. When employing this tactic, observers use the legitimacy and credibility of another person to establish their own legitimacy and credibility, for example, the director of an organization for an organizational study, a local leader, an elected official, or the village chieftain for a community study. Of course, it's important to make sure that the "known sponsor" is indeed a source of legitimacy and credibility. Some prior assessment must be made of the extent to which that person can provide halo feelings that will be positive and helpful. For example, in an evaluation, using a program administrator or funders as a known sponsor may increase suspicion and distrust among program participants and staff.

Entry Anxiety Syndrome

The initial period of fieldwork can be frustrating and can give rise to self-doubt. The fieldworker may lie awake at night worrying about some mistake, some faux pas, made during the day. There may be times of feeling embarrassed, feeling foolish, questioning the whole purpose of the project, and even experiencing feelings of paranoia. The fact that one is trained in social science does not mean that one is immune to all the normal pains of learning in new situations. On the other hand, the initial period of fieldwork can also be an exhilarating time; a period of rapid new learning, when the senses are heightened by exposure to new stimuli; and a time of testing one's social, intellectual, emotional, and physical capabilities. *The entry stage of fieldwork magnifies both the joys and the pains of doing fieldwork.*

Evaluators can reduce the *stick-out-like-a-sore-thumb* syndrome by beginning their observations and participation in a program at the same time when participants are beginning the program. In traditional fieldwork, anthropologists cannot become children again and experience the same socialization into the culture that children experience. Evaluators, however, can often experience the same socialization process that regular participants experience by becoming part of the initiation process and timing their observations to coincide with the beginning of a program. Such timing makes the evaluator one among a number of novices and substantially reduces the disparity between the evaluator's knowledge and the knowledge of other participants.

The Observer Is Also the Observed

Beginning the program with other participants, however, does not assure the evaluator of equal status.

Some participants may be suspicious that real difficulties experienced by the evaluator as a novice participant are phony—that the evaluator is playacting, only pretending to have difficulty. On the first day of my participation in the wilderness education program, we had our first backpacking experience. The staff leader began by explaining that "your backpack is your friend." I managed to both pack and adjust my "friend" incorrectly. As a result, as soon as we hit the trail, I found that the belt around my waist holding the backpack on my hips was so tight that my "friend" was making my legs fall asleep. I had to stop several times to adjust the pack. Because of these delays and other difficulties I was having with the weight and carriage of the pack, I ended up as the last participant along the trail. The next morning when the group was deciding who should carry the map and walk at the front of the group to learn map reading, one of the participants immediately volunteered my name. "Let Patton do it. That way he can't hang back at the end of the group to observe the rest of us." No amount of protest from me seemed to convince the participants that I had ended up behind them all because I was having trouble hiking (working out my "friendship" with my backpack). They were convinced that I had taken that position as a strategic place from which to evaluate what was happening. It is well to remember, then, that regardless of the nature of the fieldwork, during the entry stage more than at any other time, the observer is also the observed.

What You Say and What You Do

Fieldworkers' actions speak louder than their words. Researchers necessarily plan strategies to present themselves and their function, but participant reactions to statements about the researcher's role are quickly superseded by judgments based on how the person actually behaves.

The relative importance of words versus deeds in establishing credibility is partly a function of the length of time the observer expects to be in a setting. For some direct onlooker observations, the fieldworker may be present in a particular program for only a few hours or a day. The entry problem in such cases is quite different from the situation where the observer expects to be participating in the program over some longer period of time.

All field workers are concerned about explaining their presence and their work to a host of people. "How shall I introduce myself?" they wonder, or, "what shall I say I am doing?" If the field worker plans to do a very rapid and efficient survey, questions like these are extremely important. The manner in which an interviewer introduces himself, the precise words he uses, may mean the difference between a first-rate job and a failure. . . . But if the field worker expects to engage in some variety of participant observation, to develop and maintain long-term relationships, to do a study that involves the enlargement of his own understanding, the best thing he can do is relax and remember that most sensible people do not believe what a stranger tells them. In the long run, his host will judge and trust him, not because of what he says about himself or about his research, but by the style in which he lives and acts, by the way in which he treats them. In a somewhat shorter run, they will accept or tolerate him because some relative, friend, or person they respect has recommended him to them. (Wax, 1971, p. 365)

Qualitative sociology pioneer William Foote Whyte (1984) extracted and summarized entry strategies used in a number of groundbreaking sociological studies, including the Lynds' study of *Middletown*, W. Lloyd Warner's study of *Yankee City*, Burleigh Gardner's fieldwork in the Deep South, Elliot Liebow's hanging around *Tally's Corner*, Elijah Anderson's fieldwork in a black neighborhood, Ruth Horowitz's study of a Chicano neighborhood, Robert Cole's work in Japan, and Whyte's own experiences in *Cornerville* (pp. 37–63). All of these researchers, who produced classic qualitative studies from in-depth participant observation, had to adapt their entry strategy to the local setting, and they all ended up changing what they had planned to do as they learned from the initial responses to their efforts to gain acceptance. Exhibit 6.11 presents reflections of a part-time observer, a role that is quite common in evaluation, but rarely discussed in the qualitative literature.

EXHIBIT 6.11   Entry-to-the-Field Reflections From a Part-Time Observer

Introductory note: *In the reflections that follow, a senior staff member of the University of Minnesota Center for Social Research describes her entry into fieldwork as a part-time observer for a program evaluation. Because limitations of time and resources are common in evaluation, many situations call for a part-time observer. Her reflections capture some of the special entry problems associated with this "Now you're here, now you're gone" role. This is a common situation in evaluation studies but one that is seldom documented.*

One word can describe my role, at least initially, in a recent evaluation assignment: *ambiguous.* I was to be neither a participant-observer nor an outsider coming in for a brief but intensive stint. I was to allocate approximately six hours a week for seven months to observing the team development of a group of 23 professionals in an educational setting. At first, the ambiguity was solely on my side: What, really, was I to do? The team members, too busy in the beginning with defining their own roles, had little time to consider mine. Later on, as I became accustomed to my task, the team's curiosity about my function began to grow.

When I met with the group for the first time, I directed most of my energies to matching names and faces. I would be taking notes at most of the sessions, and it was essential that I could record not only what was said but who said it. At the first session, everyone, including me, wore a name tag. But within a few days, they were all well acquainted and had discarded their name tags; I was the only one still fumbling for names. While being able to greet each member by name was important, so was knowing something about each one's background. Coffee breaks allowed me to circulate among the group and carry on short conversations with as many as possible to try to fix in my mind who they were and where they came from, which provided insights into why they behaved in the group as they did.

But I could tell that in their eyes, I served no useful purpose that they could see. I was in the way a great deal of the time, inhibiting their private conversations. On the other hand, they appeared to be concerned about what I was thinking. Some of them—most of them—began to be friendly, to greet me as I came in, to comment when I missed a team meeting. They came to see me as I saw myself: neither really part of the group nor a separate, removed force.

Observing their interaction perhaps 6 hours a week out of their 40-hour work week obviously meant that I missed a great deal. I needed to develop a sense of when to be present, to choose among group meetings, subgroup meetings, and activities when all the members were to come together. At the same time, I was working on other contracts which limited the amount of adjustable time available. "Flexible" was the way I came to define my weekly schedule; others, not as charitable, would probably have defined it as "shifty."

A hazard that I encountered as I filled my ambiguous, flexible role was that I soon discovered I was not high on the priority list to be notified in the event of schedule changes. I would have firmly in mind that a subgroup was to meet on Tuesday at 10:00 a.m. in a certain place. I would arrive to find no one there. Later, I would discover that on Monday the meeting had been changed to Wednesday afternoon, and no one had been delegated to tell me. At no time did I seriously feel that the changes were planned to exclude me; on the contrary, the members' contrition about their oversight seemed quite genuine. They had simply forgotten me.

Another area of sudden change that caused me difficulty was in policy and procedure. What had seemed to be firm commitments on ways to proceed or tasks to be tackled were being ignored. I came to realize that while a certain amount of this instability was inherent in the program itself, other shifts in direction were outgrowths of planning sessions I had not attended or had not heard the results from after they had occurred. Therefore, keeping current became for me a high-priority activity.

With my observer role to continue over many months, I realized that I must maintain the difficult position of being impartial. I could not be thought of by the team members as being closely aligned with their leaders, nor could I expect the leaders to talk candidly and openly with me if they believed that I would repeat their confidences to the group members. Reluctantly, for I discovered several team members with whom friendships could easily have developed, I declined invitations to social activities outside of working hours.

Team members at first expressed a certain amount of enthusiasm for minutes to be taken of their meetings. This enthusiasm was short-lived, for willing volunteers to serve as secretary did not emerge. I noted (and ignored) a few passing suggestions that since I was obviously taking notes maybe I could. . . .

I took copious notes before I began to develop a sense of what was or was not important to record. Over time, my understanding of the group increased. I had to realize that, as a part-time observer, it was impossible for me to understand all of what was said. Often, I would have to jot down a reminder to myself to ask clarifying questions later.

Side-stepping sensitive questions from both leaders and team members became a fine art. I was frequently queried as to my perceptions of a particular individual or situation. On one occasion, I found a team member jumping into an elevator to ride two floors with me in a direction he didn't want to go so that he could ask me privately what I thought of another team member. My response was "I think she's a very interesting person," or something equally innocuous, and received from him a highly raised eyebrow, since the woman in question had just behaved in a very peculiar manner at the meeting we had both just attended.

In-depth interviews with each team member began in the fourth month of my observations. These filled in many of the gaps in my understanding. The timing was perfect: I had gained enough familiarity with both personnel and project by that time so that

I was knowledgeable, they had come to trust me, and they still cared deeply about the project. (This caring diminished for some as the project year drew to a close without any real hopes of refunding for a second year.) My interview design was intentionally simple and open-ended to balance my observations with their observations.

The amount of new information diminished throughout the six weeks or so that was required to interview all team members. My own performance unquestionably diminished too as the weeks went on. It was difficult to be animated and interesting as I asked the same questions over and over, devised strategies with which to probe, and recorded perceptions and incidents which I had heard many times before. Bit by bit, team members filled in holes in my information, and their repetitious descriptions of particular incidents helped me understand what was important and why. The interviews also helped me become aware of misconceptions on my part caused by seeing only part of the picture due to time constraints.

The experience was a new one for me, that of part-time observer. Quite frankly, this mode of evaluation probably will never be a favorite one. On the other hand, it provided a picture that no "snap-shot" evaluation method could have accomplished as interactions changed over time and in a situation where the full participant-observer role was clearly not appropriate.

> "What did you learn in your readings today?" asked Master Halcolm.
>
> "We learned that a journey of a thousand miles begins with the first step," replied the learners.
>
> "Ah, yes, the importance of beginnings," smiled Halcolm.
>
> "Yet I am puzzled," said a learner. "Yesterday I read that there are a thousand beginnings for every ending."
>
> "Ah, yes, the importance of seeing a thing through to the end," affirmed Halcolm.
>
> "But which is more important, to begin or end?"
>
> "Two great self-deceptions are asserted by the world's self-congratulators: that the hardest and most important step is the first, or that the greatest and most resplendent step is the last."
>
> "While every journey must have a first and last step, what ultimately determines the nature and enduring value of the journey are the steps in-between. Each step has its own value and importance. Be present for the whole journey, learners that you are. Be present for the whole journey."
>
> —From Halcolm's *Journey Into the Present*

During the second stage, after the fieldworker has established a role and purpose, made initial contacts, and entered into the field, the focus moves to high-quality data gathering and opportunistic investigation following emergent possibilities and building on what is observed and learned each step along the way. The observer, no longer caught up in adjustments to the newness of the field setting, begins to really *see* what is going on instead of just looking around. As Florence Nightingale said, "Merely looking at the sick is not observing."

Describing the second stage as "routinization of fieldwork" probably overstates the case. In emergent designs and ever-deepening inquiry, the human tendency toward routines yields to the ups and downs of new discoveries, fresh insights, sudden doubts, and ever-present questioning of others—and often of self. Discipline is needed to maintain high-quality, up-to-date field notes. Openness and perseverance are needed to keep exploring, looking deeper, diverging broader, and focusing narrower, always going where the inquiry and data take you. Fieldwork is intellectually challenging at times, mind-numbingly dull at times, and, for many, an emotional roller coaster.

Managing Relationships and Feelings During Fieldwork

One of the things that can happen in the course of fieldwork is the emergence of a strong feeling of connection with the people being studied. As you come to understand the behaviors, ideals, anxieties, and feelings of other people, you may find yourself identifying with their lives, their hopes, and their pain. This sense of identification and connection can be a natural and logical consequence of having established relationships of rapport, trust, and mutuality. For me, that awakening identification involves some realization of how much I have in common with these people whose world I have been permitted to enter. At times during fieldwork, I feel a great separation from the people I'm observing, then at other times, I feel a strong sense of our common humanity. For a fieldworker to identify, however briefly, with the people in a setting or for an evaluator to identify with the clients in a program can be a startling experience because social science observers are often quite separated from those they study by education, experience, confidence, and income. Such differences sometimes make the world of programs as exotic to evaluators as nonliterate cultures are exotic to anthropologists.

There come times, then, when a fieldworker must deal with his or her own feelings about and perspectives on the people being observed. Part of the sorting-out process of fieldwork is establishing an understanding of the relationship between the observed and the observer. When that happens, and as it happens, the person involved in fieldwork may be no less startled than Joseph Conrad's infamous character Marlowe in *Heart of Darkness*. Marlowe had followed Kurtz, the European ivory trader, up the deep river into the Congo, where Kurtz established himself as a man-god to the tribal people there. He used his position to acquire ivory. To maintain his position, he had to perform the indigenous rituals of human sacrifice and cannibalism. Marlowe, deeply enmeshed in the racism of his culture and time, was initially horrified by the darkness of the jungle and its peoples, but as he watched the rituals of those seeming savages, he found an emergent identification with them and even entertained the suspicion that they were *not* inhuman. He became aware of a linkage between himself and them:

> They howled and leaped and spun, and made horrid faces; but what thrilled you was the thought of their humanity—like ours—the thought of your remote kinship with this wild and passionate uproar. Ugly. Yes, it was ugly enough; but if you were man enough you would admit to yourself that there was in you just the faintest trace of a response to the terrible frankness of that noise, a dim suspicion of there being a meaning in it which you—you so remote from the night of the first ages—comprehend. And why not? (Conrad, 1960, p. 70)

In this passage, Conrad chronicles the possibility of awakening to unexpected realizations and intense emotions in the course of encounters with the unknown and those who are different from us. In many ways, it is our common humanity, whether we are fully aware of it at any given moment or not, that makes fieldwork possible. As human beings, we have the amazing capability to become part of other people's experiences, and through watching and reflecting, we can come to understand something about those experiences.

Finding Balance

As fieldwork progresses, the intricate web of human relationships can entangle the participant-observer in ways that will create tension between the desire to become more enmeshed in the setting so as to learn more versus the need to preserve some distance and perspective. Participant-observers carry no immunity to the political dynamics of the settings being observed. Virtually any setting is likely to include subgroups of people who may be in conflict with other subgroups. These factions or cliques may either woo or reject the participant-observer, but they are seldom neutral. During her fieldwork interviewing young women in India, Parameswaran (2001) reports efforts by parents and teachers to get her to inform on the women she interviewed or to influence them in a desired direction. She found herself in the middle of deep generational divisions between mothers and their daughters, teachers and students, bookstore owners and their clients. She could not risk either deeply alienating or completely acquiescing to any of these important and competing groups, for they all affected her access and the ultimate success of her fieldwork.

In evaluations, the evaluator can be caught between competing groups and conflicting perspectives. For example, where divisions exist among the staff and/or the participants in a program, and such divisions are common, the evaluator will be invited, often subtly, to align with one subgroup or the other. Indeed, the evaluator may want to become part of a particular subgroup in order to gain further insight into and understanding of that subgroup. How such an alliance occurs, and how it is interpreted by others, can greatly affect the course of the evaluation.

My experience suggests that it is impractical to expect to have the same kind of relationship—close or distant—with every group or faction. Fieldworkers—human beings with their own personalities and interests—will be naturally attracted to some people more than others. Indeed, to resist those attractions may hinder the observer from acting naturally and becoming more thoroughly integrated into the setting or program. Recognizing this, the observer will be faced with ongoing decisions about personal relationships, group involvement, and how to manage differential associations without losing perspective on what the experience is like for those with whom the fieldworker is less directly involved.

Capturing Varying Perspectives

Perhaps the most basic division that will always be experienced in program evaluation is the separation of staff and participants. While the rhetoric of many programs attempts to reduce the distinction between staff and participants, there is almost always a distinction between those who are paid for their responsibilities in the program (staff) and those who are primarily recipients of what the program has to offer (participants). Sociologically, it makes sense that staff and participants would be differentiated, creating a distance that can evolve into conflict or distrust. Participants will often view the evaluator as no different from the staff or administration, or even the funding sources—virtually any group except the participants. If the evaluator-observer is attempting to experience the program as a participant, special effort will be required to make participation real and meaningful, and to become accepted, even trusted, by the other participants. On the other hand, staff and administrators may be suspicious of the evaluator's relationships with funders or board members.

The point is this: Do not be naive about the tangled web of relationships the participant-observer will experience and the diverse perceptions that will be encountered. Be prepared. And be thoughtful about how fieldwork, data quality, and the overall inquiry are affected by interpersonal interactions and relationships, all of which have to be negotiated and attended to on an ongoing basis.

Alignments

Lofland (1971) has suggested that participant-observers can reduce suspicion and fear about a study by becoming openly aligned with a single broad grouping within a setting while remaining aloof from that grouping's own internal disputes.

> Thus, known observers of medical schools have aligned themselves only with the medical students, rather than attempting to participate extensively with both faculty and students. In mental hospitals, known observers have confined themselves largely to mental patients and restricted their participation with staff. To attempt to participate with both, extensively and simultaneously, would probably have generated suspicion about the observers among people on both sides of those fences. (pp. 96–97)

In contrast to Lofland's advice, in evaluating the wilderness education program, I found myself moving back and forth between a full participant role, where I was identified primarily as a participant, and a full staff role, where I was identified primarily with those who carried responsibility for directing the program. During the first field conference, I took on the role of full participant and made as visible as possible my allegiance to fellow participants while maintaining distance from the staff. Over time, however, as my personal relationships with the staff increased, I became more and more aligned with the staff. This coincided with a change of

emphasis in the evaluation itself, with the earlier part of the fieldwork being directed at describing the participant experience and the latter part of the fieldwork being aimed at describing the workings of the staff and providing formative feedback.

However, I was always aware of a tension, both within myself and within the group at large, about the extent to which I was a participant or a staff member. I found that as my observational skills became increasingly valued by the program staff, I had to more consciously and actively resist their desire to have me take on a more active and explicit staff role. They also made occasional attempts to use me as an informer, trying to seduce me into conversations about particular participants. The ambiguities of my role were never fully resolved. I suspect that such ambiguities were inherent in the situation and are to be expected in many evaluation fieldwork experiences.

Participatory Inquiry Relationships

Managing field relationships involves a different set of dynamics when the inquiry is collaborative or participatory. Under such designs, where the researcher involves others in the setting in fieldwork, a great deal of the work consists of facilitating the interactions with co-inquirers, supporting their data collection efforts, ongoing training in observation and interviewing, managing integration of field notes among different participant researchers, and monitoring data quality and consistency. These collaborative management responsibilities will reduce the primary researcher's own time for fieldwork and will affect how others in the settings, those who aren't participatory or collaborative researchers, view the inquiry and the fieldwork director, if that is the role taken on. In some cases, management of the collaborative inquiry effort is done by one of the participants and the trained field-worker serves primarily as a skills and process trainer and consultant to the group. Clarity about these roles and divisions of labor can make or break collaborative forms of inquiry. Having shared values about collaboration does not guarantee actually pulling it off. Collaborative inquiry is challenging work, often frustrating, but when it works, the findings will carry the additional credibility of collaborative triangulation, and the results tend to be rewarding for all involved, with enduring insights and new inquiry skills for those involved.

Key Informants and Key Knowledgeables

One of the mainstays of much fieldwork is the use of key informants or key knowledgeables as sources of information about what the observer has not or cannot experience, as well as a source of explanation for events the observer has actually witnessed. Key informants are people who are particularly knowledgeable about the inquiry setting and articulate about their knowledge—people whose insights can prove especially useful in helping an observer understand what is happening and why. Because the term *informants* is often associated with spies, key knowledgeables may be a more appropriate designation. The term *informants* emphasizes their role in the inquiry. The term *knowledgeables* emphasizes why they are being engaged in the inquiry.

Selecting key informants must be done carefully to avoid arousing hostility or antagonisms among those who may resent or distrust the special relationships between the fieldworker and the key informant. Indeed, how—and how much—to make visible this relationship involves considering how others will react and how their reactions will affect the inquiry. There's no formal announcement that the "position" of key informant is open or that it's been filled; the key informant is simply that person or those persons with whom the researcher or evaluator is likely to spend considerable time.

Key informants/key knowledgeables (I'll use the terms interchangeably) must be trained or developed in their role, not in a formal sense, but they will be more valuable if they understand the purpose and focus of the inquiry, the issues and questions under investigation, and the kinds of information that are needed and most valuable. Anthropologists Pelto and Pelto (1978) made this point in reflecting on their own fieldwork.

> We noticed that humans differ in their willingness as well as their capabilities for verbally expressing cultural information. Consequently, the anthropologist usually finds that only a small number of individuals in any community are good key informants. Some of the capabilities of key informants are systematically developed by the field workers, as they train the informants to conceptualize cultural data in the frame of reference employed by anthropologists. . . . The key informant gradually learns the rules of behavior in a role vis-à-vis the interviewer-anthropologist. (p. 72)

The danger in cultivating and using key knowledgeables is that the researcher comes to rely on them too much and loses sight of the fact that their perspectives are necessarily limited, selective, and biased. Data from informants represent perceptions, not truths. Information obtained from key informants should be clearly specified as such in the field notes so that the researcher's observations and those of the informants do not become confounded. This may seem like an obvious point, and it is, but over weeks and months of fieldwork,

KEY INFORMANTS/KEY KNOWLEDGEABLES

Key informants or key knowledgeables are people with in-depth understanding of the setting being studied and willing and able to share what they know as part of the inquiry. Because the term *informants* is often associated with spies, *knowledgeables* may be a more appropriate designation. The term *informants* emphasizes their role in the inquiry. The term *knowledgeables* emphasizes why they are being engaged in the inquiry.

Here is an excerpt on key informants from the *Encyclopedia of Qualitative Research Methods* (Given, 2008), which adds the designation "key actors" to the lexicon:

> Key informants, or key actors, are individuals who are articulate and knowledgeable about their community. They are often cultural brokers straddling two cultures. This role gives them a special vantage point in describing their culture. Key actors play a pivotal role in the theater of qualitative research, providing an understanding of cultural norms and responsibilities. Key informants represent an efficient source of invaluable cultural information. It is impossible to interview everyone and observe everything in a community and, logistically, it is easier to work with one or two reliable key informants than it is to assemble a series of focus groups.
>
> Key informants help to establish a link between the researcher and the community. They may provide detailed historical data, photographs, manuscripts, knowledge about interpersonal relationships, a contextual framework in which to observe and interpret behavior, and a wealth of information about the nuances of everyday life. . . . In research on communities and large organizational studies where there is a paucity of relevant archival documentation, particularly concerning vested interests and power dynamics, key informants are especially valuable. . . .
>
> Key informants provide not only personal feelings or opinions, but reflect on larger social patterns as well. They are considered "teachers" by some ethnographic researchers because they impart information, insight, and understanding. However, their insights are rarely accepted blindly. Their views are compared and combined with interviews, observations, and survey data in order to make a complete study. More to the point, key informant and qualitative researchers are collaborators, using questions, answers, and probes to better understand how and why things work. (Fetterman, 2008b, p. 477)

it can be become difficult to decipher what information came from what sources unless the fieldworker has a routine system for documenting sources and uses that system with great discipline, thoroughness and care.

Key informants can be particularly helpful in learning about subgroups to which the observer does not or cannot have direct access. During the first year of the wilderness education program evaluation, as noted previously, two groups emerged on the first day's hike and gave themselves names—turtles and truckers—to denote their differing styles of hiking. During the second year, with new participants, this history carried forward but took on a different connotation. One informal group, mostly women, adopted the name "turtles" to set themselves apart from participants, mostly men, who wanted to hike at a fast pace, climb the highest peaks, or otherwise demonstrate their prowess—a group they called somewhat disparagingly the "truckers" (trucks being unwelcome in the wilderness). Having had a full year of wilderness experiences the first year of the program and being male, I wasn't particularly welcome to become an intimate member of the "turtles." I therefore established an informant relationship with one of the "turtles," who willingly

kept me informed about the details of what went on in that group with the knowledge and permission of the group. Without that key informant relationship, I would have missed some very important information about the kinds of experiences the "turtle" participants were having and the significance of the project to them.

While being part of any setting necessarily involves personal choices about social relationships and political choices about group alliances, the emphasis on making strategic decisions in the field should not be interpreted as suggesting that the conduct of qualitative research in naturalistic settings is an ever-exciting game of chess in which players and pieces are manipulated to accomplish some ultimate goal. Fieldwork certainly involves times of both exhilaration and frustration, but the dominant motifs in fieldwork are hard work, long hours to do both observations and keep up-to-date with field notes, enormous discipline, attention to details, and concentration on mundane and day-to-day work. The routinization of fieldwork is a time of concentrated effort and immersion in gathering data. Alas, let the truth be told: *The gathering of field data involves very little glory and an abundance of nose-to-the-grindstone drudgery.*

Constructivist *Rashomon* Heaven: Multiple and Diverse Perspectives

Bringing Fieldwork to a Close

> Before you came to live with us, our lives were as always, and we were happy.
>
> We worked, we ate, and then we slept. When you came we were glad, for you brought us many fine gifts. And every night, instead of going to sleep, we sat with you, drank coffee, smoked your tobacco, and listened to your radio.
>
> But now you go, and we are sorry, for all of these things go with you.
>
> We now know pleasures to which we were unaccustomed, and we shall be unhappy.
>
> —The headman of Cabrua in 1953 to anthropologists Robert and Yolanda Murphy Placard on display at Harvard's Peabody Museum (2010)

In traditional scholarly fieldwork within anthropology and sociology, it can be difficult to predict how long fieldwork will last. The major determinant of the length of the fieldwork is the investigator's own resources, interests, and needs. Evaluation and action research typically have quite specific reporting deadlines, stated in a contract, that affect the duration of and resources available for fieldwork and the intended uses of evaluative findings.

In the previous section, we looked at the many complex relationships that get formed during fieldwork, relationships with key informants, hosts, and "sponsors" in the setting who helped with entry and may have supported ongoing fieldwork, helping solve problems and smoothing over difficulties. In collaborative research, relationships with coresearchers will have deepened. In any extended involvement within a setting, friendships and alliances are formed. As fieldwork comes to an end, an exit or disengagement strategy is needed. While a great deal of attention has traditionally been paid to entering the field, much less attention has been given to the disengagement process, what Snow (1980) has called the "neglected problem in participant observation research."

One side of the coin is disengagement. The other side is reentry back to one's life after extended fieldwork or an all-consuming project. When I went to do graduate research in Tanzania, our team received a lot of support and preparation for entry, much of it aimed at avoiding "culture shock." But when we returned home, we were given no preparation for what it would be like to return to America's highly commercial, materialistic, and fast-moving culture after months in an agrarian, community-oriented, slower-moving environment. The culture shock hit coming home, not going to Africa.

ETHICAL DIMENSIONS OF LEAVING THE FIELD

It is an issue of research ethics how field relationships are terminated....

Power relations between field researcher and those researched may become evident in the expectations of these parties when fieldwork ends. Researchers may have greater access to resources (physical, intellectual, cultural) than those they research. Also, during fieldwork, a number of formal or informal research bargains may have been struck between researcher and participants, such as an agreement not to undermine hierarchies or other commitments concerning behavior or confidentiality. Such research bargains with informants may meet psychosocial needs; for example, to have an interested, unengaged, or neutral ear to listen. Participants may be willing to talk or be observed, so long as their anonymity is respected. Sometimes there will be a more tangible bargain; for instance, to assist in some project or struggle or to provide feedback to support participants' objectives. When leaving the field, the researcher's commitment to these bargains needs to be confirmed explicitly so that participants are reassured that their trust and participation in the research are respected and rewarded.

Informal relations may also have developed during fieldwork, including emotional or psychosocial engagements with researchers. These affective relations must be addressed before a researcher leaves a field, and culturally appropriate forms of valediction must be undertaken. Researchers may need to use informants to check what kinds of expectations are held by participants concerning departure. (Fox, 2008, p. 483)

Interpersonal, cross-cultural, disengagement, and reentry issues all deserve attention as fieldwork comes to a close. Relationships with people change and evolve from entry, through the middle days, and up to the end of fieldwork. So does the fieldworker's relationship with the data and engagement in the inquiry process. That changed engagement in the inquiry process is what I want to focus on here.

Capturing Interpretive Insights While Still in the Field

As you near completion of data gathering, having hopefully become fairly knowledgeable about the setting being observed, more and more attention can be shifted to fine-tuning and confirming the observed patterns. Possible interpretations of and explanations for what was observed show up more in the field notes. Some of these explanations have been offered by others; some occur directly to the observer. In short, analysis and interpretation will have begun even before the observer has left the field.

Chapter 8 discusses analysis strategies at length. At this point, I simply want to recognize the fact that data gathering and analysis flow together in fieldwork, for there is usually no definite, fully anticipated point at which data collection stops and analysis begins. One process flows into the other. As the observer gains confidence in the quality and meaningfulness of the data, becoming more sophisticated about the setting under study and aware that the end draws near, additional data collection becomes increasingly selective and strategic.

Concluding Data Collection Rigorously

As fieldwork draws to a close, the researcher is increasingly concerned with *verification* of already collected data and less concerned with generating new inquiry leads. While in naturalistic inquiry one avoids imposing preconceived analytical categories on the data, as fieldwork comes to an end, experience with the setting will usually have led to thinking about prominent themes and dimensions that organize what has been experienced and observed. These emergent ideas, themes, concepts, and dimensions—generated inductively through fieldwork—can also now be deepened, further examined, and verified during the closure period in the field. Bringing closure to fieldwork includes, as a matter of rigor, the search for disconfirming cases as well as confirming cases. This is also when saturation sampling comes into play (see Chapter 5, pp. 300–301), the judgment that no new information is emerging or is likely to emerge from additional data collection.

Guba (1978) described fieldwork as moving back and forth between the discovery mode and the verification mode like a "wave." The ebb and flow of research involves moving in and out of periods when the investigator is open to new inputs, generative data, and opportunistic sampling and periods when the investigator is testing out hunches, fine-tuning conceptualization, sifting ideas, and verifying explanations.

The Psychology of Ending Fieldwork

When fieldwork has gone well, the observer grows increasingly confident that things make sense and begins to believe in the data. Glaser and Strauss (1967), commenting on grounded theory as an outcome of fieldwork, have described the feelings that the traditional field observer has as fieldwork moves to a close, data-based patterns have emerged, and the whole takes shape.

> The continual intermeshing of data collection and analysis has direct bearing on how the research is brought to a close. When the researcher is convinced that his conceptual framework forms a systematic theory, that it is a reasonably accurate statement of the matter studied, that it is couched in a form possible for others to use in studying a similar area, and that he can publish his results with confidence, then he has neared the end of his research....
>
> Why does the researcher trust what he knows? ... They are his perceptions, his personal experiences, and his own hard-won analyses. A field worker knows that he knows, not only because he has been in the field and because he has carefully discovered and generated hypotheses, but also because "in his bones" he feels the worth of his final analysis. He has been living with partial analyses for many months, testing them each step of the way, until he has built this theory. What is more, if he has participated in the social life of his subject, then he has been living by his analyses, testing them not only by observation and interview but also by daily living. (pp. 224–225)

This representation of bringing a grounded theory inquiry to a close represents the scholarly inquiry ideal. In the *contracted deliverables* world of program evaluation, with limited time and resources, and reporting schedules that may not permit as much fieldwork as is desirable, the evaluator may have to bring the fieldwork to a close before that state of real confidence has fully emerged. Nevertheless, I find that there is a kind of Parkinson's law in fieldwork: As time runs out, the investigator feels more and more the pressure of making sense out of things, and some form of order does indeed begin to emerge from the observations. This is a time to celebrate emergent understandings even while retaining the critical eye of the skeptic, especially useful in questioning one's own confident conclusions.

Feedback to Program Staff as Evaluation Fieldwork and Site Visits End

In doing fieldwork for program evaluation, in contrast to theory-oriented scholarly field research, the evaluator-observer must be concerned about providing feedback, making judgments, and generating recommendations. Thus, as the fieldwork draws to a close, the evaluator must consider what feedback is to be given to whom and how.

Giving feedback to program staff and/or participants can be part of the verification process in fieldwork. My own preference is to provide the participants and staff with descriptions and analysis, verbally and informally, and to include their reactions as part of the data. Part of the reciprocity of fieldwork can be an agreement to provide participants with descriptive information about what has been observed. I find that participants and staff are hungry for such information and fascinated by it. I also find that I learn a great deal from their reactions to my descriptions and analyses. Of course, it's neither possible nor wise to report everything one has observed. Moreover, the informal feedback that occurs at or near the end of fieldwork will be different from the findings that are reported formally based on the more systematic and rigorous analysis that must go on once the evaluator leaves the field. But that formal, systematic analysis will take more time, so while one is still in the field, it is possible to share at least some findings and to learn from the reactions of those who hear those findings.

Timing feedback in formative evaluations can be challenging. When the purpose is to offer recommendations to improve the program, the program staff will usually be anxious to get that information "ASAP" (as soon as possible). The evaluator-observer may even feel pressured to report findings prematurely before having confidence in the patterns that seem to be emerging. I experienced this problem throughout the evaluation of the wilderness education program evaluation. During the first year, we met with the staff at the end of each field conference program (the three 10-day field conferences were spread out over a year) to discuss what we had observed and to share interpretations about those observations. At the very first feedback session, the staff reaction was "I wish you'd

told us that in the middle of the week, when we could have done something about it. Why'd you hold back? We could have used what you've learned to change the program right then and there."

I tried to explain that the implications of what I observed had only become clear to me an hour or two before our meeting, when my co-evaluator and I had sat down with our field notes, looked them over, and discussed their significance together. Despite this explanation, which struck me as altogether reasonable and persuasive and struck the staff as altogether disingenuous, from that moment forth, a lingering distrust hung over the evaluation as staff periodically joked about when we'd get around to telling them what we'd learned next time. Throughout the three years of the project, the issue of timing feedback surfaced several times a year. As they came increasingly to value our feedback, they wanted it to come earlier and earlier during each field conference. During the second field conference in the second year, when a number of factors had combined to make the program quite different from what the staff had hoped for (major logistical and weather problems, personality conflicts among participants), the end-of-the-conference evaluation feedback session generated an unusual amount of frustration from the staff because my analyses of what had happened and why had not been shared earlier. Again, I found some distrust of my insistence that those interpretations had emerged later rather than sooner as the patterns became clear to me.

Evaluators who provide formative feedback on an ongoing basis need to be conscientious in resisting pressures to share findings and interpretations before they have confidence about what they have observed and sorted out important patterns—not certainty but at least some degree of confidence. The evaluator is caught in a dilemma: Reporting patterns before they are clearly established may lead program staff to intervene inappropriately; withholding feedback too long may mean that dysfunctional patterns become so entrenched that they are difficult, if not impossible, to change.

No ideal balance has ever emerged for me between continuing observations versus providing feedback. Timing feedback is a matter of judgment and strategy, and it depends on the nature of the evaluator's relationship with program staff and the nature of the feedback, especially the balance between what staff will perceive as negative and positive feedback. When in doubt, and where the relationship between the evaluator and program staff has not stabilized into one of long-term trust, I counsel evaluator-observers to err on the side of less feedback rather than more. As often happens in social relationships, negative feedback that was wrong is long remembered and often recounted. On the other hand, it may be a measure of the success of the feedback that program staff so fully adopt it that they make it their own and cease to credit the insights of the evaluator.

Once feedback is given, the role of the evaluator changes. Those to whom the feedback was presented are likely to become much more conscious of how their behavior and language is being observed. Thus, added to the usual effect of the fieldworker on the setting being observed, this feedback dimension of fieldwork increases the impact of the evaluator-observer on the setting in which he or she is involved.

Summary of Stages of Fieldwork

Fieldwork is dynamic and fluid. Each inquiry is unique. There are, however, patterns. One prominent pattern is that fieldwork must have a beginning and an ending, with ongoing fieldwork in the middle. Different issues surface at various stages of fieldwork. Exhibit 6.12 summarizes the issues we've been discussing at these three stages.

EXHIBIT 6.12   Overview of Stages of Fieldwork

FIELDWORK STAGE	SUBSTANTIVE PRIORITIES	INTERPERSONAL PROCESS PRIORITIES
1. Entry	a. Observe and record baseline descriptions of the setting and context.	a. Explain why you are there; adapt explanation based on feedback and questions.
	b. Identify and recruit key informants/key knowledgeables.	b. Make the acquaintance of key people with the help of a "champion" or "gatekeeper"; learn names, roles, who is connected to whom in what ways.
	c. Establish procedures for observations and disciplined note taking.	c. Find your way around the setting; get comfortable, settled in.
	d. Learn about important events, activities, or occasions that should be observed.	d. In collaborative inquiries, orient and train your coresearchers.
	e. Begin implementing the inquiry design, making adjustments as needed (and recording reasons for adaptations).	e. Get the logistics set up to facilitate your fieldwork.
		f. Watch for and start learning the specialized language people use.
2. Routinization	a. Stay disciplined about observing and taking field notes; don't get behind in documenting.	a. Observe and document reactions to your presence and role (watch for reactivity).
	b. Implement the inquiry design while being attentive to emergent possibilities for adaptation.	b. Observe and document your own experiences, feelings, and reactions (reflexivity).

FIELDWORK STAGE	SUBSTANTIVE PRIORITIES	INTERPERSONAL PROCESS PRIORITIES
	c. Make use of key informants/key knowledgeables.	c. Manage key informant/key knowledgeable relationships.
	d. Seek multiple perspectives and sources of data, including documents and artifacts: *triangulate*.	d. Deal sooner rather than later with challenging interpersonal and ethical dynamics (cross-cultural blunders, groups competing for your attention, getting entangled in political or interpersonal conflicts, etc.).
	e. Start making sense of the data: Note emergent patterns and themes; interpret what you're observing.	e. Continue to explain what you're doing and why.
	f. Deepen your understanding of local terminology and sensitizing concepts.	f. In collaborative inquiries, manage interpersonal team processes and tasks sensitively.
3. Closure	a. Complete the inquiry design; note and explain changes and incomplete aspects; if sampling to saturation, document the nature of and evidence for saturation.	a. Keep key people in the setting advised of your departure schedule.
	b. Look for and gather data on confirming and disconfirming cases.	b. Meet any obligations incurred, promises made, and reciprocities committed to.
	c. Examine field notes for gaps and fill in gaps.	c. Figure out what kind of feedback to provide about your findings and arrange to provide that feedback.
	d. Review baseline descriptions of context and setting and document any changes.	d. Observe and document reactions of people you've engaged with, including reactions to closure.
	e. Examine initial interpretations and gather additional data as appropriate and possible to deepen and confirm interpretations.	e. Observe and document your own reactions to the experience and imminent close (reflexivity).
		f. Prepare for entry into life after fieldwork.

The question of how the observer affects what is observed has natural as well as social science dimensions. The Heisenberg Uncertainty Principle in physics states that the instruments used to measure velocity and position of an electron alter the accuracy of measurement. When the scientist measures the position of an electron, its velocity is changed; and when velocity is measured, it becomes difficult to capture precisely the electron's position. The process of observing affects what is observed. These are real effects, not just errors of perception or measurement. The physical world can be altered by the intrusion of the observer. How much more, then, are social worlds changed by the intrusion of fieldworkers?

The effects of observation vary depending on the nature of the observation, the type of setting being studied, the personality and procedures of the observer, and a host of unanticipated conditions. Nor is it simply in fieldwork involving naturalistic inquiry that scientific observers affect what is observed. Experimentalists, survey researchers, cost–benefit analysts, and psychologists who administer standardized tests all affect the situations into which they introduce data collection procedures. The issue is not whether or not such effects occur; rather, the issue is how to monitor those effects and take them into consideration when interpreting data.

A strength of naturalistic inquiry is that the observer is sufficiently a part of the situation to be able to understand personally what is happening. Fieldworkers are called on to inquire into and be reflective about how their inquiry intrudes and how those intrusions affect findings. But that's not always easy. Consider the case of anthropologist Napoleon Chagnon, who did fieldwork for a quarter-century among the isolated and primitive Yanomami Indians, who lived deep in the rain forest at the borders of Venezuela and Brazil. He studied mortality rates by dispensing steel goods, including axes, as a way of persuading people to give him the names of their dead relatives in violation of tribal taboos. Brian Ferguson, another anthropologist knowledgeable about the Yanomami, believes that Chagnon's fieldwork destabilized relationships among villages, promoted warfare, and introduced disease. Chagon denies these charges but acknowledges extracting tribal secrets by giving informants gifts like beads and fishhooks, capitalizing on animosities between individuals, and bribing children for information when their elders were not around. He gave away machetes in exchange for blood samples for his genealogical studies. The long-term effects of his fieldwork have become a matter of

REACTIVITY: OBSERVER EFFECTS, OBSERVER AFFECTED

Reactivity, also known as the observer effect, takes place when the act of doing research changes the behavior of participants. . . .

A number of factors have been seen to influence the degree of reactivity. Conspicuous observers, or those who place themselves in the middle of the activities, are more intrusive than those who stand to the side. Characteristics of the observer (e.g., age, race, gender, dress) that differ substantially from those of participants are likely to cause more reactivity. Characteristics of participants may also influence behavior. For example, children usually return to naturally occurring behavior more quickly than do adults. Reactive behavior usually decreases as time passes, a process known as habituation. It is postulated that this return to normality arises from the development of rapport and trust between participants and the researcher and the fact that it is difficult to sustain unnatural behavior for a long period. Participants' understanding of the purpose of the research may cause reactivity. For example, if participants believe that the researcher is trying to document socially unacceptable or deviant activities, they may hide these behaviors.

In qualitative research, reactivity is usually seen as being inclusive of the researcher as well as the participants. The researcher keeps reflexive notes that document how his or her own behavior and understandings may have been affected by the research process. Reactivity is regarded as being inevitable in any research process that involves interaction among participants, the researcher, and a setting of interest. Reflexive analysis helps to uncover and respond to reactivity in appropriate ways. (McKechnie, 2008, pp. 729–730)

spirited debate and controversy within anthropology (Geertz, 2001; Tierney, 2000a, 2000b).

Reactivity in Collaborative Research and Evaluation

At the other end of the intrusion continuum, we find those qualitative designs where "intrusions" are intentionally designed because the qualitative inquiry is framed as an intended form of desired intervention. This is the case, for example, with collaborative and participatory forms of inquiry in which those people in the setting who become coresearchers are expected to learn from the inquiry. The processes of participation and collaboration can be designed and facilitated to have an impact on participants and collaborators quite beyond whatever findings they may generate by working together. In the process of participating, participants have the opportunity to learn the logic of research and data-based reasoning. Skills are acquired in problem identification, criteria specification, data collection, analysis, and interpretation. Acquisition of such research skills and ways of thinking can have a longer-term impact than the use of findings from a particular evaluation study. This *learning from the process* as an outcome of participatory and collaborative inquiry is called "process use," in contrast to findings use (Patton, 2008, chap. 5; Patton, 2012a, pp. 140–167).

While it is not possible to know precisely how collaboration will affect coresearchers or to fully anticipate how an observer will affect the setting being observed, both cases illustrate the need to be thoughtful about the interconnections between the observers and the observed. When designing a study and making decisions about the observer's degree of participation in the setting, the visibility and openness of fieldwork, and the duration of fieldwork (see Exhibit 6.4 earlier in this chapter), it is important to anticipate and be prepared for observation effects—and know how to handle them. For example, I have been involved as a participant-observer-evaluator in a number of professional development programs where participants were expected to exercise increasing control over the curriculum as the program evolved. Had I fully engaged in such participatory decision making, I could have influenced the direction of the program. Anticipating that problem and reviewing the implications with program staff, in each case, I decided not to participate actively in participant-led decision making, at least not to the full extent I might have had I not been involved in the role of evaluator-observer. I aimed my involvement at a level where I would not appear withdrawn from the process, yet at the same time, I attempted to minimize my influence, especially where the group was divided on priorities.

Another example comes from evaluation of a community leadership program mentioned previously in this chapter. As a three-person team of participant-observers, we participated fully in small-group leadership exercises. When the groups in which we participated were using concepts inappropriately or doing the exercise wrong, we went along with what

participants said and did so without offering corrections. Had we really been *only* participants—and not participant-evaluators—we would have offered corrections and solutions. Thus, our roles made us more passive than we tended naturally to be, so as not to dominate the small groups. We had anticipated this possibility in the design stage prior to fieldwork and had agreed on this strategy at that time.

The Dynamics of Participation

The role and impact of the evaluator-observer can change over the course of fieldwork. Early in the wilderness project, I kept a low profile during participant-led planning discussions. Later in the program, particularly during the final field conference of the second year, I became more engaged in discussions about the future direction of the project.

Methodological Reporting

Reporting on the relationship between the observer and the observed, then, and the ways in which the observer may have affected the phenomenon observed becomes part of the methodological discussion in published fieldwork reports and evaluation studies. In that methodological discussion (or the methods chapter of a dissertation), the observer presents both data about the effects of fieldwork on the setting and people therein as well as the observer's perspective on what has occurred. As Patricia Carini (1975) has explained, such a discussion acknowledges that findings inevitably are influenced by the observer's point of view during naturalistic inquiry:

> The observer has a point of view that is central to the datum and it is in the articulation—in the revelation of his point of view—that the datum of inquiry is assumed to emerge. In effect the observer is here construed as one moment of the datum and as such the fabric of his thought is inextricably woven into the datum as he is assumed to be constituent of its meaning. From this assumption it is possible to consider the relationship of the observer to the phenomenon under inquiry. Relatedness can be stated in many ways: opposition, identity, proximity, interpenetration, isolation, to name only a few. All imply that the way in which a person construes his relationship to the phenomenal world is a function of his *point of view* about it. That is, relationship is not a given nor an absolute, but depends upon a personal perspective. It is also true that perspective can shift, the only necessity of a person's humanity being that he takes some stance in relationship to the events about him. (pp. 8–9)

INSIDER VERSUS OUTSIDER PERSPECTIVES: "NOTHING ABOUT US, WITHOUT US"

As an example of reflexive methodological reporting, what follows are ruminations on a unique way of looking at insider (emic) versus outsider (etic) perspectives. In this case, the insider and outsider is the same person but at different times under different conditions. Understanding different perspectives from inside and outside a phenomenon goes to the core of qualitative inquiry. Experience affects perspective. Perspective shapes experience. In the next chapter, on interviewing, we shall continue to explore ways of capturing the experiences and getting deeply into the perspectives of those who have encountered whatever phenomenon interests us.

The reflections that follow look at the experience of mental illness from the inside and outside, and how being on the inside can dramatically change the view from the outside. Janice James (not her real name) moved from the outside (PhD researcher and mental health professional) to the inside as an involuntary participant-observer (a patient in a locked mental health facility) and back again to the outside (as a professional program evaluator). She recounts those transitions from outside to inside, and inside to outside, and how they have shaped her perspective on research and evaluation.

Outside to Inside, Inside to Outside: Shifting Perspectives

I was a high school and junior high science teacher, a school psychologist and guidance counselor, and finally an educational researcher, before I completed my PhD. Since I had been a scientist of some sort for all of my career life, I was a "natural" for the field of evaluation. I thought like a scientist. I was familiar with the practice of research in biology, bacteriology, field botany and had a special interest in medicine. But through some quirks of fate and personality, I found myself working as a clinician, providing therapy and case management to people with severe mental disorders. Then, after having been gainfully employed my entire adult life, and successfully raising two children, who had now produced one grandchild each, I was forced into the locked, psychiatric ward of a hospital for the third time in my life.

Back at work, after nearly two months in the hospital, I found myself, for the first time, looking at my professional work and reading professional literature with the eyes of one from the other side of the locked doors and medical charts. The irony of my situation was obvious: I treated people like me! Thus began a shift of viewpoint that has radically altered my practice of evaluation in the field of mental health.

First, I had to throw out some grand assumptions. As a scientist, I trusted scientific method and worshipped at the same shrine of true experimental design and random assignment as everyone else. But now I was much more conscious that the "lab rats," the subjects of our research, literally have minds of their own. Some of the most treasured assumptions of mental health research were looking awfully different from inside the maze. Probably, the most serious is that what I had assumed to be

"treatment" from the viewpoint of the researcher now looked like mostly futile efforts that were most often experienced as punishment and threat from the viewpoint of "patient." I was not asked if I would go into the hospital. I was told I had to. If I got angry at something, someone gave me powerful medications that made me feel like a zombie.

While some of what happened in my hospitalization helped, much of what I experienced made me feel much worse. I was incarcerated, and my jailors looked at me kindly, certain that I was being locked up, strapped to a bed, and injected with medications for my own good. Even though I actually entered the hospital "voluntarily," the threat of involuntary treatment and permanent damage to my ability to earn a living was the driving force that got me there, and kept me there, and taught me to "make nice" for the staff, lest they refuse to certify me sane and let me free again.

Providers are taught, in all sincerity and good intentions, to act in a kind of parental role toward clients, a benign dictatorship. But their "subjects" are people who have already had their dignity as adults medically removed, their privacy invaded, and the job and relationship underpinnings of American self-esteem destroyed; who are told that they will be sick for life; and who are medicated so they cannot perform sexually or sometimes even read a good book. Is it any wonder that many of them (us) will accept survival as an adequate "quality of life"?

Evaluation designs that test the effects of treatment programs on the individual don't address the problem of living in a neighborhood where life is stressful and taxis won't take you home after dark and where the threatening voices might just as likely be real as hallucinated. The provider who is confined to the office and distance of a professional relationship will not know when there is abuse in the home the clients never speak of, because the abuser is someone they love or who controls their money as payee. Mental health workers are put in the role of defenders of the public purse, and then we wonder why clients feel their safety net of services threatened with every dollar they are given or earn and fail to trust their "providers."

In short, I may apply many of the same ideas, theories, methods, and interpretations as I always did as a program evaluator. But now I always question not just the validity, reliability, and generalizability of the evaluation work itself but also the hidden assumptions that surround it. I will always be seeking to empower those disenfranchised by custom, poverty, and stigma. And I now know that if I want to be part of the solution, rather than part of the problem, I better be sure I know what the experience of different stakeholders really is, not constrained by the limited questions I may think to ask or guided too narrowly by work done in the past. Thus, I have adopted the motto of the people who do not claim the title of "consumer," because they were not given true choice about treatment when they found themselves pinned with psychiatric labels: *Nothing about us, without us."*

Carini (1975) is here articulating the interdependence between the observer and what is observed. Prior to data collection, the fieldworker plans and strategizes about the hoped-for and expected nature of that interdependence. But things don't always unfold as planned, so observers must make an effort to observe themselves observing and record the effects of their observations on the people observed and, no less important, reflect on changes they've experienced from having been in the setting. This means being able to balance observation with reflection and manage the tension between engagement and detachment.

Balancing Engagement and Detachment

Bruyn (1966), in his classic work on participant observation, articulated a basic premise of participant observation: that the "role of the participant observer requires both detachment and personal involvement" (p. 14). To be sure, there is both tension and ambiguity in this premise. How it plays out in any given situation will depend on both the observer and the phenomenon being observed.

> Thus, we may observe at the outset that while the traditional role of the scientist is that of a neutral observer who remains unmoved, uncharged, and untouched in his examination of phenomena, the role of the participant observer requires sharing the sentiments of people in social situations; as a consequence he himself is changed as well as changing to some degree the situation in which he is a participant. . . . The effects are reciprocal for observer and observed. The participant observer seeks, on the one hand, to take advantage of the changes due to his presence in the group by recording these changes as part of his study, and on the other hand, to reduce the changes to a minimum by the manner in which he enters into the life of the group. (Bruyn, 1966, p. 14)

Whether one is engaged in participant observation or onlooker observation, what happens in the setting being observed will, to some extent, be dependent on the role assumed by the observer. Likewise, the nature of the data collected will, to some extent, be dependent on the role and perspective of the observer. And just as the presence of the observer can affect the people observed, so too can the observer be affected.

The Personal Experience of Fieldwork

The intersection of social science procedures with individual capabilities and situational variation is what makes fieldwork a highly personal experience. At the end of her book on *Doing Field Work*, Wax (1971) reflected on how fieldwork had changed her.

CHANCE FAVORS THE PREPARED MIND AND OBSERVANT EYE

In 1949, an obscure Australian psychiatrist, John F. J. Cade, noticed that the urine of his manic patients was highly toxic to guinea pigs, and he began looking for the toxic chemical, which he suspected was uric acid. He began experimenting with lithium urate, not because of any psychiatric properties of lithium but because lithium urate was the most soluble salt of uric acid. To Cade's surprise, far from being toxic, the salt protected the guinea pigs against the urine of manics, and it also sedated the animals, effects Cade found were due to the lithium. He immediately tried other lithium salts on himself and, when they proved safe, on 10 hospitalized manic patients, all of whom recovered, some almost miraculously.

Cade's discovery is sometimes characterized as serendipitous, but the discovery of lithium as an antimanic agent resulted from one man's deep commitment, diligent observation, and reasoned inference (Ironside, 1993).

A colleague has suggested that I reflect on the extent to which I was changed as a person by doing field work. I reflected and the result astonished me. For what I realized was that I had not been greatly changed by the things I suffered, enjoyed or endured; nor was I greatly changed by the things I did (though they strengthened my confidence in myself). What changed me irrevocably and beyond repair were the things *learned*. More specifically, these irrevocable changes involved replacing mythical or ideological assumptions with the correct (though often painful) facts of the situation. (p. 363)

Fieldwork is not for everyone. Some, like Henry James, will find that "innocent and infinite are the pleasures of observation." Others will find observational research anything but pleasurable. Some students have described their experiences to me as tedious, frightening, boring, and "a waste of time," while others have experienced challenge, exhilaration, personal learning, and intellectual insight. More than once, the same student has experienced both the tedium and the exhilaration, the fright and the growth, the boredom and the insight. Whatever the adjectives used to describe any particular individual's fieldwork, of this much we are assured: The experience of observing provides the observer with experiences and observations, the interconnection being cemented by reflection. No less an authority than William Shakespeare gives us this assurance.

Armado. How hast thou purchased this experience?

Moth. By my penny of observation.

—William Shakespeare
Love's Labor Lost

Apart of and Apart From the World Observed

The personal, perspective-dependent nature of observations can be understood as both a strength and weakness, a strength in that personal involvement permits firsthand experience and understanding, and a weakness in that personal involvement introduces selective perception. In the deep engagement of naturalistic inquiry lies both its risks and its benefits. Reflection on that engagement, from inside and outside the phenomenon of interest, crowns fieldwork with reflexivity and makes the observer the observed—even if only by oneself. So we repeat Halcolm's refrain that opened this chapter:

Go out into the world. Live among the peoples of the world as they live. Learn their language. Participate in their rituals and routines. Taste the world. Smell it. Watch and listen. Touch and be touched. Write down what you see and hear, how they think and how you feel.

Enter into the world. Observe and wonder. Experience and reflect. To understand a world you must become part of that world while at the same time remaining separate, a part of and apart from.

Go then, and return to tell what you see and hear, what you learn, and what you come to understand.

Chapter Summary and Conclusion: Guidelines for Fieldwork

A reader who came to this chapter looking for specific fieldwork rules and clear procedures would surely be disappointed. Looking back, the major theme seems to be "What you do depends on the situation, the nature of the inquiry, the characteristics of the setting, and the skills, interests, needs, and point of view that you, as observer, bring to your engagement." Yet the conduct of observational research is not without direction. Exhibit 6.13 offers a modest list of 10 guidelines for fieldwork (not, please note, commandments, just guidelines) by way of reviewing some of the major issues discussed in this chapter. Beyond these seemingly simple but deceptively complex prescriptions, the point remains that what you do depends on a great number of situational variables, your own capabilities, and careful judgment informed by the strategic themes for qualitative inquiry presented in Chapter 2 (Exhibit 2.1, pp. 46–47).

The Future of Observational Inquiry

Angrosino and Rosenberg (2011), in a thoughtful review of "Observations on Observations," conclude that the proven value and therefore likely future contribution of observational inquiries will be illuminating the particular.

> Rather than attempting to describe the composite culture of a group or to analyze the full range of institutions that supposedly constitute the society, the observation-based researcher will be able to provide a rounded account of the lives of particular people, focusing on the lived experience of specific people and their ever-changing relationships. (p. 476)

I resonate with that forecast. In a world of increasingly short attention spans and marked tendencies to overgeneralize, understanding a particular time, place, and people feels anchoring, grounding, and important.

So having considered the guidelines and strategic themes for naturalistic field-based research, and after the situational constraints on and variations in the conduct of fieldwork have been properly recognized and taken into account in the design, there remains only the core commitment of qualitative inquiry to reaffirm. That core commitment was articulated by Nobel laureate Nicholas Tinbergen in his 1975 acceptance speech for the Nobel Prize in physiology and medicine: "watching and wondering." Tinbergen explained that it was by "watching and wondering" that he had, despite being neither a physiologist nor a medical doctor, discovered what turned out to be a major breakthrough in our understanding of autism. His observations revealed that the major clinical research on autism did not hold up outside clinical settings. His "watching and wondering" allowed him to see that normal individuals, those not clinically labeled as autistic, exhibited under a variety of circumstances all of the behaviors described as autistic in clinical research. He also noted that children diagnosed as autistic responded in nonautistic ways outside the clinical setting. By observing people in a variety of settings and watching a full range of behaviors, he was able to make a major medical and scientific contribution. His research methodology: "watching and wondering."

A Final Cautionary Tale About Observation

> People only see what they are prepared to see.
>
> —Ralph Waldo Emerson (1803–1882)
> American poet and philosopher

I opened this chapter, many, many pages ago, noting that a scientific observer cannot be expected to engage in systematic observation on the spur of the moment any more than a world-class boxer can be expected to defend his title spontaneously on a street corner or an Olympic runner can be asked to dash off at record speed because someone suddenly thinks it would be nice to test the runner's time. Athletes, artists, musicians, dancers, engineers, *and* scientists require training and mental preparation to do their best. Observers have to be prepared and ready to move their observations from the level of ordinary looking to the rigor of systematic seeing. Here is further evidence in support of that premise.

We have a family cabin in the Colorado Rocky Mountains, situated on a lake. Access is along a dirt road below the busy highway that leads into the town of Grand Lake, entrance to Rocky Mountain National Park. One spring morning in 2010, I looked out the upstairs guest room, which has a view of the brush-covered hill that rises about 40 yards up to the

EXHIBIT 6.13   Summary Guidelines for Fieldwork

1. *Design the fieldwork to anticipate, be clear about, and deal with classic tensions and trade-offs.*

   a. The role of the observer (degree of participation)

   b. The tension between insider (emic) and outsider (etic) perspectives

   c. The degree and nature of collaboration with coresearchers

   d. Disclosure and explanation of the observer's role to others

   e. Duration of observations (short vs. long)

   f. Focus of observation (narrow vs. broad) (see Exhibit 6.4)

2. *Take detailed, descriptive field notes.*

   a. Strive for thick, deep, and rich description.

   b. Be disciplined and conscientious in taking detailed field notes at all stages of fieldwork.

   c. Separate interpretation from description.

3. *Stay open and observant.*

   a. Gather a variety of information from different perspectives.

   b. Be opportunistic in following leads and sampling purposefully to deepen understanding.

   c. Allow the design to emerge flexibly as new understandings open up new paths of inquiry.

4. *Cross-validate and triangulate by gathering different kinds of data.*

   a. Collect and integrate observations, interviews, documents, artifacts, recordings, and photographs.

   b. Use multiple and mixed methods as appropriate and possible.

5. *Capture, use, and report direct quotations.*

   a. Represent people in their own terms. Capture participants' views of their experiences in their own words.

   b. Distinguish direct quote from summaries of what was said.

6. *Select key informants and key knowledgeables wisely, and use them carefully.*

   a. Draw on the wisdom of their informed perspectives, but keep in mind that their perspectives are selective.

---

highway. I saw sunlight shining off a couple of objects in the brush. I figured it was trash someone had thrown down from the highway. I climbed the hill toward the highway, found car parts that were the source of the reflection (a broken mirror, glass from broken windows, and chrome from the vehicle bumper), and found a dead body behind a bush: a woman, neck obviously broken, body stiff and contorted.

The previous Sunday, three days earlier, I had noticed a wrecked sports utility vehicle (SUV) in the neighbor's driveway across the dirt access road, sitting perfectly upright and centered in her driveway (see photo). This house isn't occupied. The next-door neighbor was outside, and I talked with him about it. He has lived there for 40 years, a year-round resident, knows everyone, and speculated that the owner was letting someone park the junker on her property, which she had done before. He said that if it was still there in a week, he'd call the sheriff and get an order to have it towed away. We stood talking for a half-hour in a light drizzle just 10 feet in back of the wreck.

It turned out that sometime in the wee hours of the previous Sunday morning, the driver, going very, very fast, drove off the highway, was ejected, and probably instantly killed. The SUV bounced off a large boulder, rolled down the hill, and came to rest perfectly upright in the driveway, as if it had been towed there and left.

All Wednesday afternoon, after I discovered the body, the authorities combed the hill collecting evidence to determine what had happened. Called to the scene were emergency medical personnel, the sheriff and his deputies, state troopers, a state crime scene investigator, fire fighters, and the county coroner.

The body had been there in the underbrush behind a boulder for three days before I discovered it. What was most disturbing to me was not discovering the body but that I had not been a more astute observer on the Sunday morning when the wreck appeared overnight in the driveway. I have spent most of my professional life working with people to help them question their assumptions, examine their mental maps, and become astute, systematic, careful, and thoughtful observers. Yet at a critical moment, I failed to do so myself. I've written more than 100 pages in this qualitative book on observation, yet I failed to thoughtfully observe and inquire. I deferred to local knowledge (the neighbors'

b. Observe how others in the setting react to the relationship with key informants/key knowledgeables.

7. *Be aware of and strategic about the different stages of fieldwork.*

   a. Build trust and rapport at the entry stage. Remember that the observer is also being observed and evaluated.

   b. Attend to relationships throughout fieldwork and the ways in which relationships change over the course of fieldwork, including relationships with hosts, sponsors within the setting, and coresearchers in collaborative and participatory research.

   c. Stay alert and disciplined during the more routine, middle phase of fieldwork.

   d. As fieldwork draws to a close, focus on pulling together a useful synthesis. Move from generating possibilities to verifying emergent patterns and confirming themes. Seek both confirming and disconfirming cases to complete data collection.

   e. In evaluations and action research, provide formative feedback as part of the verification process of fieldwork. Time that feedback carefully. Observe its impact.

8. *Be reflective and reflexive.*

   a. Be as involved as possible in experiencing the setting as fully as is appropriate and manageable while maintaining an analytical perspective grounded in the purpose of the fieldwork.

   b. Include in your field notes and reports your own experiences, thoughts, and feelings.

   c. Ponder and report the nature and implications of your own perspective in your methods section.

9. *Document reactivity.*

   Consider and report how your observations may have affected the observed as well as how you may have been affected by what and how you've participated and observed.

10. *Engage ethically.*

   At each stage of fieldwork, be sensitive and attuned to ethical challenges that arise from interpersonal relationships, power dynamics, and potential conflicts of interest.

---

explanation), which provided a reasonable story line for what we saw on Sunday morning but was completely inaccurate. Merely walking another 15 paces up to the side of wrecked SUV on Sunday morning and looking up the hill from that vantage point would have revealed what had happened and exposed the body. But it was someone else's property, so we stayed back and speculated, operating on false but reasonable assumptions. And the bottom line is as follows: I was not in mental observation mode, so *I looked but didn't see*.

Making Sense of What You Observe

Observing something doesn't necessarily mean we can interpret it. In particular, we can't know what it means to those involved in an event, interaction, or activity without talking to them or hearing from them in some way. The next chapter takes up interviewing. Every interview is also an observation. Interviewing adds to observation of the perceptions and sense making of the people being studied. The Halcolm graphic comic that ends this chapter is a cautionary tale about interpreting an observation without interviewing.

NOT WHAT IT SEEMS

Wrecked vehicle perfectly aligned and positioned in the driveway despite having tumbled and rolled 40 yards down a steep incline from the highway above.

APPLICATION EXERCISES

1. Locate an observation study in your field of interest or on a subject that interests you. Use Exhibit 6.4, "Dimensions of Fieldwork: Variations and Options Along Continua" (pp. 356–357) to classify and describe the nature of the study. Discuss how the 10 dimensions in Exhibit 6.4 are interconnected and interrelated. How does the location of the study on some dimensions affect the location on other dimensions?

2. Identify and discuss a *sensitizing concept* that you would like to study. Explain why the sensitizing concept is important and of interest to you. Design an inquiry into the sensitizing concept specifying (a) the setting where you would conduct the study, (b) what activities or interactions you would plan to observe, and (c) who you would plan to observe and interview. Explain the value of approaching the concept as a sensitizing concept instead of operationalizing it. What's the difference (sensitizing vs. operational concept) and why does it matter? Use your example and proposed study to illustrate and explain the differences. (See Exhibit 6.5 for examples.)

3. Conduct an observation. (1) Select a place you'd like to observe, like an art museum, a restaurant, a sports event, a public park, or a store. (2) Pick an area within the location for your observation, for example, an area serving food, the area around the public restrooms, the doors where people enter, or an activity area. (3) Observe for a half-hour, taking notes. Capture interactions, movements, patterns of being in that location, and anything that strikes you as important. (4) Write a detailed description of what you observed. (5) Share your written observation with someone. Ask them if they get a vicarious experience of the place and what happened through your description. Does your description take them there? Do they experience what you experienced? What are their questions? What doesn't your description tell them? (6) Interpret your description? What sense can you make of what you observed?

4. Write an entry script for entering a site to undertake fieldwork. Sketch out your qualitative design: Topic, field location, and inquiry question. Now write two entry scripts: (1) Write an explanation you would use with people in the setting to explain your study, why you want to observe them, and what you're going to do with the observations you make. (2) Write a script to recruit key informants/ key knowledgeables to assist you in your inquiry. Explain how and why you want to work with them. What will their role be in our inquiry? What will you offer them in return for their assistance? (3) Try out your scripts on two friends, family members, or class mates. How do they react?

For more practice exercises in observation, see Janesick (2011, chaps. 2 and 3).

Qualitative Interviewing

 The combination of a question mark and an exclamation point is called an *interrobang*. A sentence ending with an *interrobang* asks a question in an emotionally intense manner, expressing excitement, wonder, disbelief, doubt, or concern:

You did what ‽
You lost everything ‽

 The symbol on the left means
NO interrobanging.

Qualitative interviewing asks questions (?) without the exclamatory bang (!). You may want to insert this symbol on your interview guide to remind you:
NO interrobanging.

Beyond Silent Observation

After much cloistered study, three youth came before Halcolm to ask how they might further increase their knowledge and wisdom. Halcolm sensed that they lacked experience in the real world, but he wanted to have them make the transition from seclusion in stages.

During the first stage, he sent them forth under a six-month vow of silence. They wore the identifying garments of the muted truth seekers so that people would know they were forbidden to speak. Each day, according to their instructions, they sat at the market in whatever village they entered, watching but never speaking. After six months in this fashion, they returned to Halcolm.

"So," Halcolm began, "you have returned. Your period of silence is over. Your transition to the world beyond our walls of study has begun. What have you learned so far?"

The first youth answered, "In every village the patterns are the same. People come to the market. They buy the goods they need, talk with friends, and leave. I have learned that all markets are alike and the people in markets always the same."

Then the second youth reported, "I too watched the people come and go in the markets. I have learned that all life is coming and going, people forever moving to and fro in search of food and basic material things. I understand now the simplicity of human life."

Halcolm looked at the third youth: "And what have you learned?"

"I saw the same markets and the same people as my fellow travelers, yet I know not what they know. My mind is filled with questions. Where did the people come from? What were they thinking and feeling as they came and went? How did they happen to be at this market on this day? Who did they leave behind? How was today the same or different for them? I have failed, Master, for I am filled with questions rather than answers, questions for the people I saw. I do not know what I have learned."

Halcolm smiled. "You have learned most of all. You have learned the importance of finding out what people have to say about their experiences. You are ready now to return to the world, this time without the vow of silence.

"Go forth now and question. Ask and listen. The world is just beginning to open up to you. Each person you question can take you into a new part of the world. The skilled questioner and attentive listener know how to enter into another's experience. If you ask and listen, the world will always be new."

—From Halcolm's *Epistemological Parables*

Chapter Preview

This chapter opens with reflections on what it means to be living and undertaking research in what has been called "The Interview Society." From there, we will move into the knowledge and skills that are essential for high-quality interviewing. It may seem straightforward to assert that interviewing is a skill. Like any skill, it can be done well or poorly. More to the point, like any skill, it has to be learned and practiced. *And there's the rub.* A lot of people engaged in interviewing lack fundamental skills, have never been trained, and are actually lousy interviewers—and don't seem to know it. Does that sound like the start of a rant? It is—and the focus of MQP Rumination # 7 in this chapter: *Interviewing as an Unnatural Act: Overcoming the Overconfidence of Incompetence.*

This chapter begins by discussing different approaches to interviewing derived from diverse theoretical and methodological traditions, like ethnography versus phenomenology, and different uses of interviews, helping people (the counseling interview) versus finding wrongdoing (the investigative interview). We then move to different types of interview formats: standardized versus free flowing. Later sections consider the content of interviews and skills of interviewing: what questions to ask and how to phrase questions. The chapter ends with a discussion of how to record the responses obtained during interviews. This chapter will emphasize skill and technique as ways of enhancing the quality of interview data, but no less important is a genuine interest in and caring about the perspectives of other people. If what people have to say about their world is generally boring to you, then you will never be a great interviewer. Unless you are fascinated by the rich variation in human experience, qualitative interviewing will become drudgery. On the other hand, a deep and genuine interest in learning about people is insufficient without disciplined and rigorous inquiry based on skill and technique. Here's an overview of this chapter's modules:

Module 57 The Interview Society: Diversity of Applications

Module 58 Distinguishing Interview Approaches and Types of Interviews

Module 59 Question Options and Skilled Question Formulation

Module 60 Rapport, Neutrality, and the Interview Relationship

Module 61 Interviewing Groups and Cross-Cultural Interviewing

Module 62 Creative Modes of Qualitative Inquiry

Module 63 Ethical Issues and Challenges in Qualitative Interviewing

The chapter then closes with my personal reflections on interviewing:

Module 64 Personal Reflections on Interviewing, and Chapter Summary and Conclusion

The Interview Society: Diversity of Applications

The word interview has roots in Old French and meant something like "to see one another." (Narayan & George, 2012, p. 515)

The word interview has origins associated with the French entre voir, *meaning "to be in the sight of" and referring to a meeting of people face to face. It also has Latin origins with the prefix: "inter" meaning among and between and "view" referring to seeing, looking or inspection.* (Skinner, 2013, p. 16)

We live in an "interview society" (Fontana & Frey, 2000, p. 646), where "interviewing has become a fundamental activity and interviews seem to have become crucial for people to make sense of their lives" (Gobo, 2011, pp. 24–25). But the very popularity of interviewing may be its undoing as an inquiry method because so much interviewing is being done so badly that its credibility may be undermined. Television, radio, magazines, newsletters, and websites feature interviews. In their ubiquity, interviews done by social scientists become indistinguishable in the popular mind from interviews done by talk show hosts. The motivations of social scientists have become suspect, as have our methods. Popular business magazine *Forbes* (self-proclaimed "The Capitalist Tool") has opined, "People become sociologists because they hate society, and they become psychologists because they hate themselves" (quoted by Geertz, 2001, p. 19). Such glib sarcasm, anti-intellectual at the core, can serve to remind us that we bear the burden of demonstrating that our methods involve rigor and skill. Qualitative research interviewing, seemingly straightforward, easy, and universal, can be done well or poorly. This chapter is about doing it well. But as context for the challenges of doing it well, Exhibit 7.1 presents an overview of interviewing used for purposes other than research and evaluation. These diverse uses of interviewing compete with qualitative inquiry for attention, credibility, and utility. Exhibit 7.1 reminds us that the generic word *interview* covers a huge variety of applications, purposes, and uses, such as the following, each with its own criteria for quality, or lack thereof:

1. Journalism interviews
2. Celebrity television talk show interviews
3. Personnel evaluation and human resource interviews
4. Clinical and diagnostic interviews
5. Motivational interviewing
6. Audit and compliance interviews
7. Interrogation interviews
8. Cognitive interviewing for survey research
9. Cognitive interviewing for eyewitness enhancement
10. Religion-based interviewing

EXHIBIT 7.1 Ten Diverse Purposes and Uses of Interviews in the Interview Society

Qualitative inquiry for research and evaluation is but one of many uses of interviewing in the Information Age, also dubbed the "Interview Society" (Fontana & Frey, 2000, p. 646). Below are 10 types of interviews conducted to achieve a specific outcome other than research or evaluation. These diverse uses of interviewing compete with qualitative inquiry for attention, credibility, and utility. This reminds us that the generic word *interview* covers a huge variety of applications, purposes, and uses.

DIVERSE INTERVIEW TYPES AND USES	PRIMARY INTERVIEW PURPOSE	SPECIAL INTERVIEW CHALLENGES
1. Journalism interviews	Getting a story that will attract readers (print and Internet media) and viewers (television and cable news)	Tight deadlines; fierce competition for stories, space (print media), and time (TV and cable news). Errors can lead to loss of credibility and reputation (Farhi, 2013)

(Continued)

(Continued)

DIVERSE INTERVIEW TYPES AND USES	PRIMARY INTERVIEW PURPOSE	SPECIAL INTERVIEW CHALLENGES
2. Celebrity television talk show interviews	Entertainment (Bushkin, 2013)	Making the interviews interesting, titillating, and revealing (Public Radio Program Directors Association, 2008)
3. Personnel evaluation and human resource interviews	Hiring effective employees (Hoevemeyer, 2005); exit interviews to learn from departing employees (Work Institute, 2013); "stay interviews" to keep valued personnel from leaving and increase retention (Scott, 2011; Sullivan, 2013)	Getting beyond rehearsed and self-serving answers to real workplace behaviors and actual performance (Turner, 2004)
4. Clinical and diagnostic interviews	Diagnosing psychological needs and problems (Morrison, 2014); interviewing children and adolescents with severe emotional problems (McConaughy, 2013)	Establishing trust; appropriately interpreting responses to make a diagnosis and to determine an intervention
5. Motivational interviewing	Involves "attention to natural language about change" and then creating conversational exchanges in which people talk themselves into change, based on their own values and interests (Miller & Rollnick, 2012, p. 4). See also Hohman (2011) and Westra (2012)	Moving from talk to behavior; sustaining change
6. Audit and compliance interviews	"Interviews are a useful audit tool to gather information about internal controls and fraud risks. . . . First, employees involved in the day-to-day operations of a functional area . . . are in an excellent position to identify weak internal controls and fraud risks. Second, . . . they may have knowledge of suspected, or actual, frauds that interviews can bring to light. Third, . . . when interviewed, employees are often willing, even relieved, to talk about these issues" (Leinicke, Ostrosky, Rexroad, Baker, & Beckman, 2005, p. 1)	Uncovering illegal behavior when stakes are high for those being interviewed; verifying allegations
7. Interrogation interviews	Getting a confession to a crime or terrorist activity (Starr, 2013; Zulawski & Wicklander, 1998)	False confessions under the stress and manipulation of intense interrogation (Schwartz, 2010; Starr, 2013)
8. Cognitive interviewing for survey research	For survey questions that are nontrivial, the question-answering process involves a number of cognitive steps, some conscious, some not. Cognitive interviewing aims to decipher the thought processes involved in answering survey questions to increase validity and reliability (Willis, 1999)	Getting at and making explicit internal thought processes and unconscious reactions to language. Identifying and resolving data collection problems (Clarke & Long, 2013)
9. Cognitive interviewing for eyewitness enhancement	A nonhypnotic investigatory approach for enhancing and deepening eyewitness testimony to increase accuracy and completeness (Culver, 2013; Memon, Meissner, & Fraser, 2010)	Separating strong perceptions and beliefs from the reality of what is actually happening
10. Religion-based interviewing	Determining if a follower is true to the faith or assessing readiness for conversion from one religion to another (Richardson, 1998)	Distinguishing true beliefs from insincere expressions of belief done for convenience (e.g., marriage, access to resources, status, conformity)

IN-DEPTH INTERVIEWING

In-depth qualitative interviews are *long*, ranging from a couple of hours to full days, and in some cases of longitudinal interviews, they are for several days over an extended period of time. Why do people participate in such lengthy interviews?

I am persuaded that the long interview offers the respondent benefits as well as risks. When I proposed long interviews with individuals between the ages of 65 and 75, funding agencies expressed concern that these interviews might prove fatiguing. I, too, was alarmed that my respondents might be dangerously taxed by the experience of answering intimate questions over a long period. Our fears proved unfounded. Almost without exception, respondents proved more durable and energetic than their interviewer. Again and again, I was left clinging to consciousness and my tape recorder as the interview was propelled forward by respondent enthusiasm. Something in the interview process proved so interesting and gratifying that it kept replenishing respondent energy and involvement.

—Grant McCracken (1988, p. 27)
The Long Interview

Distinguishing Qualitative Inquiry for Research and Evaluation From Other Types of Interviewing

Qualitative inquiry can take place in any of the settings where other kinds of interviewing are occurring. So, for example, to study journalism is different from engaging in journalism. Moreover, qualitative research can be integrated into journalism in what Iorio (2009) has called *Taking It to the Streets*, where "It" refers to qualitative research and "Streets" references where journalists find their stories. In a similar vein, there's a critical and important difference between doing qualitative inquiry in a clinical setting (Hays & Singh, 2012) versus doing diagnostic interviews in a clinical setting (McConaughy, 2013; Morrison, 2014). Yet to the people being interviewed in a clinical setting, the differences in purpose and approach may not be at all obvious. To them, an interview is an interview is an interview. But the purpose of a diagnostic interview is to arrive at a diagnosis of a patient's problem and determine a treatment plan. In contrast, a qualitative research interview aims to understand the patient's experience of the clinical setting.

QUALITATIVE INTERVIEWING VERSUS THERAPEUTIC INTERVIEWING

Here are four ways by which qualitative interviewing differs from therapeutic interviewing.

1. *Aims and practices differ.* "In therapeutic interviewing the functioning of the patient is the object of concern. Whatever the therapist does is, or should be, motivated by the aim of helping the patient.... In research interviewing, on the other hand, the interviewer's questioning is motivated by the aim of eliciting information useful to a study. The interviewer is without license to produce change in the respondent's functioning and has no right to give interpretations or advice."

2. *Focus and substance of research interviewing and therapeutic interviewing are different.* "Therapists are likely to encourage patients not only to talk about their internal states but also to find sources in earlier life for current images and feelings.... The research interviewer is much more likely to want to hear about scenes, situations, and events the respondent has witnessed. The interviewer in a qualitative research study will want respondents to talk about their internal states only if this would be useful for the study...."

3. *Interview relationship is different.* "Therapists are responsible to patients for helping them improve in functioning. Because the patient looks to the therapist for help, the therapist will almost surely become an authoritative figure in the patient's life and thoughts. In contrast, the research interviewer is a partner in information development. The interviewing relationship is defined as one of equals, although interviewer and respondent have different responsibilities. And while therapists remain for some time important figures in the lives of patients, interviewers are ordinarily recognized by respondents as transient figures in their lives."

4. *Compensation differs.* "The patient pays the therapist for the therapist's help. The interviewer is paid not by the respondent but by the study. Indeed, the respondent may also be compensated by the study; at the very least, the respondent is likely to be thanked by the interviewer for the interview."

—Robert S. Weiss (1994, pp. 134–135)
Learning From Strangers

To emphasize the distinctive niche of qualitative inquiry interviewing, let's briefly review some diverse interview-based studies to get a sense of what in-depth qualitative interviews can yield.

Examples of Qualitative Interview Findings

- *Interviews with homeless youth:* Formerly homeless youth told their stories through in-depth interviews and reflected on the factors and principles they experienced in Minnesota shelters and programs that helped them move forward on their life's journey. The findings have been used to improve programs for homeless youth (Murphy, 2014). See Exhibit 7.20, pp. 511–516, at the end of this chapter for one of the case studies that came out of those interviews.

- *Interviews with Native Alaskans:* Colorectal cancer is the leading type of cancer among Native Alaskan people and the second leading cause of cancer mortality. Screening has the potential to reduce both colorectal cancer incidence and mortality. Culturally sensitive interviews captured reactions to an innovative colorectal cancer educational program (Cueva, Dignan, Lanier, & Kuhnley, 2013).

- *Interviews on the sensitizing concept "respect":* Harvard sociologist Sara Lawrence-Lightfoot (2000) has created powerful portraits of what *respect* means in modern society based on in-depth interviews. She has also studied "exits": how relationships come to an end (Lawrence-Lightfoot, 2012).

- *Interviews with immigrants:* Immigrants to the United States tell stories of how they've adapted, including answers to the intriguing question "What's the smartest thing you did in your home country that you maintain here and that Americans would benefit from?" (Kolkey, 2011).

- *Interviews with new mothers:* Mothers report how they learn to cope with their newborn baby's crying. Experienced mothers knew that the infant's frequent crying would diminish after a while, whereas first-time mothers had to learn how to respond to the crying. With growing experience, mothers could decipher the reason and urgency of different kinds of crying and used more successful soothing techniques. At the same time, they learned to assess and mitigate their own stress reactions by self-soothing and adopting realistic expectations of normal infant behavior (Kurth et al., 2013).

- *Interviews with incest perpetrators:* Getting inside the heads of child sexual abuse perpetrators to uncover their complete incapacity for empathy and how they think about and justify their actions (Gilgun, 1994, 1995; Gilgun & Connor, 1989).

- *Interviews with elderly people:* Five years of interviews with more than 1,000 Americans past the age of 65 yielded *30 Lessons for Living* (Pillemer, 2011).

Anywhere there are people there is the potential for interviewing them to capture their experiences, beliefs, fears, triumphs—any and all aspects of their stories. An interview, when done well, takes us inside another person's life and worldview. The results help us make sense of the diversity of human experience.

Inner Perspectives

> Interviewing is rather like a marriage: everybody knows what it is, an awful lot of people do it, and yet behind each closed door there is a world of secrets. (Oakley, 1981, p. 41)

We interview people to find out from them those things we cannot directly observe and to understand what we've observed. The issue is not whether observational data are more desirable, valid, or meaningful than self-reported data. The fact of the matter is that we cannot observe everything. We cannot observe feelings, thoughts, and intentions. We cannot observe behaviors that took place at some previous point in time. We cannot observe situations that preclude the presence of an observer. We cannot observe how people have organized the world and the meanings they attach to what goes on in the world. We have to ask people questions about those things.

The purpose of interviewing, then, is to allow us to enter into the other person's perspective. Qualitative interviewing begins with the assumption that the perspective of others is meaningful and knowable and can be made explicit. We interview to find out what is in and on someone else's mind to gather their stories.

Program evaluation interviews, for example, aim to capture the perspectives of program participants, staff, and others associated with the program. What does the program look and feel like to the people involved? What are their experiences in the program? What thoughts do people knowledgeable about the program have concerning program operations, processes, and outcomes? What are their expectations? What changes do participants perceive in themselves as a result of their involvement in the program? It is

THE ART OF ASKING QUESTIONS

One of the first books ever written on asking high-quality questions was by Stanley L. Payne in 1951. Drawing on his experience as president of Interview Research Institute, a pioneering market research organization, he wrote a short and insightful volume titled *The Art of Asking Questions*. He opened the book by describing why he thought it was needed. His rationale remains relevant today as a justification and context for this chapter.

What is the need for this first book on question wording? If it all boils down to the familiar platitudes about using simple, understandable, bias-free, nonirritating wordings, all of us recognize these obvious requirements anyway. Why say more?

Oblivious of the Obvious

One reason for elaborating on the subject is that all of us, from time to time, forget these requirements. Like some churchgoers, we appear to worship the great truths only one day a week and to ignore them on working days. Or we remember a certain example, but fail to see how it applies to other situations. In combatting our very human frailty, a more provocative set of examples and a detailed list of points to consider may be more helpful than the isolated examples and the broad generalities which we now so often disregard. (Payne, 1951, p. 3)

A Lecture on Taking too Much for Granted

If all the problems of question wording could be traced to a single source, their common origin would probably prove to be in taking too much for granted. We questioners assume that people know what we are talking about. We assume that they have some basis for testimony. We assume that they understand our questions. We assume that their answers are in the frame of reference we intend. Frequently, our assumptions are not warranted. (Payne, 1951, p. 16)

the interviewer into his or her world. We enter the interviewee's world through what he or she tells us. As eminent anthropologist Clifford Geertz (1986) insightfully observed, "Whatever sense we have of how things stand with someone else's inner life, we gain it through their expressions, not through some magical intrusion into their consciousness" (p. 373).

Rigorous and Skillful Interviewing

> I know how to listen when clever men are talking. That is the secret of what you call my influence.
>
> —Hermann Sudermann (1857–1928)
> German dramatist and novelist

The premise here is that *the quality of the information obtained during an interview is largely dependent on the interviewer*. This chapter discusses ways of obtaining high-quality information by talking with people who have that information. Skilled interviewing is about asking questions well so that interviewees want to share their stories. An interview is an interaction, a relationship. Every interview is also an observation—*a two-way observation*. You, as the interviewer, are being watched and assessed, even as you are observing the person you're interviewing and assessing the responses you're hearing. Establishing rapport matters. Being nonjudgmental matters. Being authentic and trustworthy matter. Interview skills include asking genuinely open-ended questions; being clear so that the person being interviewed understands what is being asked; asking follow-up questions and probing, as appropriate, for greater depth and detail; and making smooth transitions between sections of the interview or topics. Skilled interviewing requires distinguishing different kinds of questions—descriptive questions versus questions that ask for interpretations or judgments. It means distinguishing both questions and answers that are behavioral, attitudinal, or knowledge focused. And skilled interviewing involves the art of listening, and *really hearing*. These and other interviewing skills can and do affect the quality and meaningfulness of responses. Exhibit 7.2 sets the stage for this chapter by providing 10 principles of and skills for high-quality interviewing. We'll examine these in more depth later in the chapter.

the responsibility of the evaluator to provide a framework within which people can respond comfortably, accurately, and honestly to these kinds of questions. Evaluators enhance the utility of qualitative data by generating relevant and high-quality findings.

Any interviewer faces the challenge of making it possible for the person being interviewed to bring

EXHIBIT 7.2   Ten Interview Principles and Skills

An interview is an interaction, a relationship. The interviewer's skills and experience can and do affect the quality of responses. Here are 10 skills and competencies to cultivate.

INTERVIEW PRINCIPLES/SKILLS	ILLUSTRATIVE EXAMPLES
1. *Ask open-ended questions.* Ask relevant and meaningful open-ended questions that invite thoughtful, in-depth responses that elicit whatever is salient to the interviewee.	*What is a strong memory you have of your first year of high school?* not *Do you have any strong memories from high school?*
2. *Be clear.* Ask questions that are clear, focused, understandable, and answerable.	*What was most important to you about your experience?* not *What was important that you'll remember and can use and will tell people about and that made the program effective at least as you think about it now?*
3. *Listen.* Attend carefully to responses. Let the interviewees know that they've been heard. Respond appropriately to what you hear.	*That's very helpful. You've explained very well why that was important to you.*
4. *Probe as appropriate.* Follow up in complete responses with clarifying probes. Interviewees will only then learn what degree of depth and detail you seek through probes.	*It would be helpful to hear more about that. Tell me more about what happened and how you were involved.*
5. *Observe.* Watch the interviewee to guide the interactive process. Acknowledge what is going up. Adapt the interview as appropriate to fit the reactions of the person interviewed. Every interview is also an observation.	*I can see that the question evoked strong emotions. Take your time, or if you'd like, we can change topics for the moment and come back to this later.*
6. *Be both empathic and neutral.* Show interest and offer encouragement *nonjudgmentally: empathic neutrality.*	*I appreciate your willingness to share your story. Every story is unique and we've heard all kinds of things. There's no right or wrong answer to any of these questions. What matters is that it's your story.*
7. *Make transitions.* Help and guide the interviewee through the interview process.	*You've been describing how you got into the program. Now the next set of questions is about what you experienced in the program.*
8. *Distinguish types of questions.* Separate purely descriptive questions from questions about interpretations and judgments. Distinguish behavior, attitude, knowledge, and feeling questions.	Descriptive behavior question: *What did you do in your art class?* Interpretive opinion question: *What were the strengths and weaknesses of the class in your opinion?*
9. *Be prepared for the unexpected.* The world can intrude during an interview. Be flexible and responsive.	Despite a commitment to a two-hour interview, only a half-hour may be available. Make the most of it. Interruptions occur. Things may emerge that need more time.
10. *Be present throughout.* Interviewees can tell when the interviewer is distracted, inattentive, or uninterested.	Checking the time regularly, glancing at your text messages, looking around instead of staying engaged with the person talking, these things are noticed.

MQP Rumination # 7

Interviewing as an Unnatural Act: Overcoming the Overconfidence of Incompetence

I am offering one personal rumination per chapter. These are issues that have persistently engaged, sometimes annoyed, occasionally haunted, and often amused me over more than 40 years of research and evaluation practice. Here's where I state my case on the issue and make my peace.

Study after study shows that we tend to be overconfident about our competence. In experiments, people who perform poorly on a variety of tasks are typically quite confident about their competence. In fact, they're often more confident than those who actually perform well. This phenomenon has been dubbed "illusory superiority" (Hoorens, 1993). Certain overconfident people are "unskilled and unaware of it" (Dunning & Kruger, 1999). Their difficulties in recognizing their own incompetence lead to inflated self-assessments—"an absence of self-insight among the incompetent" (Ehrlinger, Johnson, Banner, Dunning, & Kruger, 2008). Incompetent people also lack the skill to recognize competence in others.

Illusory superiority is rampant in interviewing. Many people who think of themselves as good interviewers are unskilled and unaware of it. They display an absence of self-insight about their interviewing incompetence and lack the skill to recognize interviewing competence in others. They don't know what they don't know. Who are these people, and what evidence do I have for these assertions? Read on.

Evidentiary Ruminating

By definition, ruminations express opinions. A particularly strong opinion may become a rant. People who rant often display overconfidence about their overgeneralizations. The irony that I may be displaying my own illusory superiority is not lost on me. So, given that this is a research text, let me at least report the basis for my assertions and make this an evidentiary rumination—or rant. You decide which label is most appropriate.

In the 1970s, I directed the Minnesota Center for Social Research at the University of Minnesota. That's when I began training staff and graduate students in qualitative methods, especially interviewing. I noticed from the beginning that some displayed an affinity for interviewing and were receptive to training, while others thought they were naturally gifted as interviewers and were inattentive during training. I also became convinced, as the title of this

rumination asserts, that interviewing is an *unnatural act*. That's why training is necessary. Normal interpersonal and social interactions don't involve one person systematically asking questions, the other answering, and the first person listening, probing, taking the responses deeper, but not sharing his or her own experiences. Interviewing is an unnaturally unbalanced interaction.

Every interview is also an observation, so the skills required include appropriate questioning, focused listening, astute observation, and sensitive responding, all while taking notes (even when the interview is being recorded). There's a lot to learn and then practice, for it's not enough to grasp the basic concepts. Putting them into practice, actually becoming a skilled interviewer, requires practicing, listening to, and studying transcripts of your own interviews and getting feedback about how well you're doing—the usual things that accompany professional development and learning.

Having developed training processes for my own research and evaluation staff, I began offering qualitative workshops to others, first through the University of Minnesota and then through a variety of organizations that provide training. For many years, I have taught qualitative methods workshops for the American Evaluation Association, the Evaluators' Institute, and the International Program in Development Evaluation Training. I also take on special commissions from other organizations to train their staff or prepare people for fieldwork projects. All in all, I do at least 30 days of qualitative methods training a year with a heavy emphasis on interviewing skills as well as the other skills and topics covered in this book.

Finally, as a full-time independent organizational development and evaluation consultant, I engage in a lot of interviewing. When I undertake a new assignment, I routinely begin by interviewing key people. Part of what I ask is how they get information. I find out from junior staff how they experience the communication and interviewing skills of their managers and senior personnel in the organization. I also have opportunities to review a lot of research and evaluation projects, read interview transcripts, and enquire into the skills of colleagues and others engaged in qualitative fieldwork. Many are accomplished, highly skilled, and exemplary interviewers. Many others manifest remarkable patterns of *illusory superiority*, proclaiming their skill even as

(Continued)

(Continued)

they demonstrate fundamental incompetence. Here are some types to watch out for.

A Typology of Clueless Incompetent Interviewers

Myopic Managers

I was once involved in helping a major international organization set up an institutional learning system that featured interviewing as a core element to extract lessons in real time as projects unfolded. Junior staff interviewed senior staff as projects began to be implemented. Midway through projects, midlevel staff interviewed other midlevel staff in specialized program areas where they didn't ordinarily have contact with one another. Senior staff interviewed midlevel staff as projects came to an end. To implement this organization-wide learning system, interview protocols were developed, and everyone committed to interview training. Well, almost everyone. Senior managers were quite happy to commit their underlings to interview training, but they were insistent that they didn't need such basic training themselves. Having this reported to me by the senior vice president in charge of the learning system, I wrote the following memo, which he agreed to send to all senior staff:

> I understand that you are cooperating with the institutional learning process by committing your staff to participate in interview training but that you, yourselves, have opted out of training. I would ask you to reconsider. First, even if you are already a skilled interviewer, the project would benefit from each of you serving as a role model demonstrating support for this effort, not just by sending your staff but by committing your own time and leadership to this organization-wide initiative.
>
> Second, and more importantly, I doubt that most of you are skilled interviewers for this kind of open-ended, in-depth qualitative inquiry interviewing. Most managerial interviewing is directed at solving problems. Someone comes into your office, and you ask just enough questions to tell them what to do to solve the problem and get them out of your office. You may be good at that kind of problem-solving interviewing, but that's not in-depth, get-the-full-story, lessons-focused interviewing. Indeed, without naming names, I know from interviewing your staff about your interviewing skills that most of you ask bad questions, listen poorly, interrupt as the person is trying to explain what's going on, and give off all kinds of signals that you aren't paying attention, are annoyed that you're having to deal with the problem before you, and that you're not interested in understanding the situation in depth and detail.

> Let me be blunt. In my many years of experience, I've found managers in a wide variety of organizations to be at the top of the class in deceiving themselves about their competence as interviewers, even for problem solving. To do a good job and make your own contribution to this important initiative, I invite you to participate in the customized senior manager interview training we've designed for you. I believe it will not only increase your skills for this learning initiative, but you'll pick up techniques that will enhance your day-to-day problem solving and strategic interviews with your staff. If you come and don't find it worthwhile, I'll apologize for wasting your time and return my training fee.
>
> And one last thing: I've heard that the junior staff is betting that there's nothing I can say to convince you that you need training because you regularly and consistently avoid any professional development training. I invite you to prove them wrong in this case.

The senior managers did decide to participate in interview training though only for a half-day instead of the full day I proposed. The feedback was generally positive. No one asked for an apology or return of my fee. They affirmed that they learned some things that they could apply to their work generally. Their feedback also affirmed a finding from the research on illusory superiority: Those who begin a learning experience by overestimating their baseline competence do improve when they are given focused training and feedback. The trick is to get them into training and provide meaningful feedback.

DSM-Enthralled Therapists

A second group especially susceptible to illusory superiority in interviewing is therapists, especially those who are well trained in and focused on making a diagnosis based on the *Diagnostic and Statistical Manual of Mental Disorders* (*DSM*) or any other diagnostic framework. Therapists engage in in-depth interviewing, but it is for the purpose of diagnosing a problem and prescribing a treatment or intervention. That is a crucial medical task, but it is not open-ended qualitative interviewing. I've had a lot of therapists in my classes and workshops over the years. They pose special challenges in reorienting them to qualitative inquiry interviewing because they have developed deeply ingrained habits of diagnostic interviewing. They have as much to unlearn as they have to learn, and I find that they are generally surprised, and initially discouraged, by the degree of difficulty in making the transition. Those who prevail, however, find that the world opens up to them in quite a different way when they are in discovery and leaning mode rather than mired in a diagnostic mind-set and set of protocols.

Endemic Academics

Illusory superiority might well be considered endemic to academia. Somehow having attained a doctorate in whatever field makes high-achieving scholars suddenly and miraculously qualified to teach, administer, advise, evaluate—the list goes on and on and includes interviewing. I review all kinds of projects, site visits, evaluations, and fieldwork that require interviewing as a primary or at least major form of data collection. When I am asked to review such qualitative studies, I enquire into the training and experience of the people doing the data collection. Too often I find that the scholars, researchers, and evaluators conducting these studies have never had any interview training. They describe themselves, when challenged, as being self-trained, which I treat as a euphemism for "untrained but don't know it." I know of prestigious researchers and evaluators who undertake a great deal of fieldwork and think they're great interviewers—and tell their colleagues, students, and friends as much. They're very smart, have impressive disciplinary knowledge, are held in high esteem for their peer-reviewed publications, but are lousy interviewers. Lousy how? They ask leading questions, interrupt respondents' responses, talk as much or more than they listen, make instantaneous judgments, and commit the whole series of novice errors warned against in this chapter. Worst of all, they are oblivious to doing so and feel insulted and defensive when their competence is challenged.

Illusory Superiority: The Good News

The three illusorily superior types I've been ruminating about—(1) myopic mangers, (2) *DSM*-enthralled therapists, and (3) endemic academics—do not exhaust the possibilities. Not even close. Exhibit 7.1 (pp. 423–424) identifies 10 types of interviewing done for purposes other than research and evaluation. Those well schooled in these other interviewing approaches and purposes will find it difficult to become skilled qualitative interviewers. *The skill set and mind-set are different.*

There is, however, some good news in this otherwise bleak rumination. As I noted earlier, those who begin a learning experience by overestimating their baseline competence do improve when they are given focused training and feedback. The trick is to get them training and feedback. Reading this book can be a starting point. But you must find ways of practicing, getting feedback, and continuing to learn. Interviewing is a skill. Acquire it. Then, like many other skills, once acquired, *use it or lose it.*

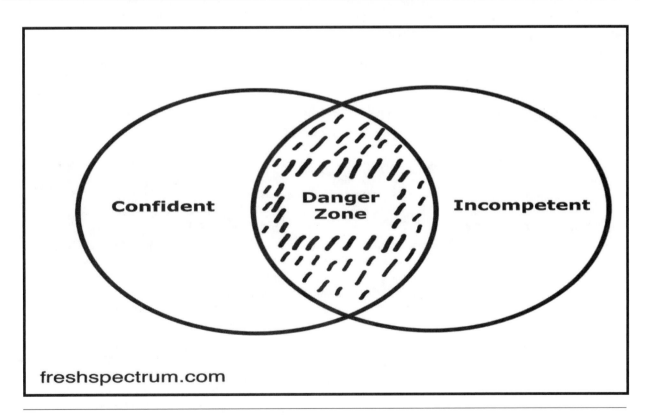

freshspectrum.com

> Learn from yesterday, live for today, hope for tomorrow. The important thing is to not stop questioning.
>
> —Nobel Prize–winning physicist
> Albert Einstein (1879–1955)
> *Relativity: The Special and the General Theory* (1916)

Different inquiry traditions emphasize different questions and fieldwork methods. Interviewing varies in important ways, then, within different traditions and inquiry approaches. Traditional social science interviewing emphasizes standardized questions and consistency across interviewers and interviewees, while social constructionist interviewing places priority on individualized interactions and adapting the interview as appropriate to the emergent relationship that is formed between the interviewer and the interviewee in the course of an interview. Ethnographic interviewing involves conversational interactions that are part of long-term, in-depth fieldwork to support and deepen direct observations in the field about cultural patterns. Phenomenological interviewing, in contrast, aims to elicit a personal description of a *lived experience* so as to describe a phenomenon as much as

possible in concrete and lived-through terms. Exhibit 7.3 contrasts the focus and methods of interviewing across 12 different qualitative inquiry traditions. These distinctions and comparisons build on the discussion in Chapter 3 of the variety of philosophical, theoretical, and methodological frameworks for undertaking qualitative inquiry. Exhibit 7.3 highlights the following interviewing approaches:

1. The ethnographic interview
2. The traditional social science research interview
3. The phenomenological interview
4. Social constructionist interviewing
5. The hermeneutic interview
6. Narrative inquiry interviewing
7. The life story interview
8. Interpretive interactionism
9. Oral history interviewing
10. Postmodern interviewing
11. Investigative interviewing
12. Pragmatic interviews

EXHIBIT 7.3   Twelve Contrasting Interview Approaches Grounded in Different Qualitative Inquiry Traditions and Frameworks

QUALITATIVE INQUIRY INTERVIEW APPROACHES	INTERVIEWING TO GATHER QUALITATIVE DATA WITHIN A SPECIFIC INQUIRY TRADITION	DIFFERENT THEORETICAL, PHILOSOPHICAL, AND METHODOLOGICAL APPROACHES AFFECT HOW INTERVIEWS ARE CONDUCTED AND ANALYZED
TYPE OF INQUIRY	PURPOSE AND FOCUS	EXPLANATION AND ELABORATION
1. *The ethnographic interview*	Conversational interactions that are part of long-term, in-depth fieldwork. Interviews support direct observations in the field (Skinner, 2013).	Ethnographic interviews "serve comparative and representative purposes—comparing responses and putting them in the context of common group beliefs and themes . . ., most valuable when the fieldworker comprehends the fundamentals of a community from the insider's perspective. . . . They seem to be casual conversations . . . but have a specific but implicit research agenda. The researcher uses informal approaches to discover how the people conceptualize their culture and organize it into meaningful categories" (Fetterman, 2008a, pp. 290–291).

TYPE OF INQUIRY	PURPOSE AND FOCUS	EXPLANATION AND ELABORATION
2. *The traditional social science research interview*	Standardize questions so that each person gets the same stimulus and interviewer effects are minimized. Interviewers must be well trained to maintain objectivity and control bias (Hyman, 1954). The interviewer "asks questions and records answers. At this point, the researcher's concern shifts to measurement error, or inaccuracies in the responses to questions" (Singleton & Straits, 2012, p. 86). Open-ended questions have "the potential to provide an outcome that is not representative of a respondent's perception of events . . ., distorted by the interaction of the interviewer and respondent" (Frey, 2004, p. 768).	Interviews follow "a closely prescribed form" (Bornat, 2004, p. 771). The interviewer, according to Hyman (1954), "has a kind of professional task orientation which enables him to preserve objectivity; that interviewers themselves regard over-involvement in the interview socially to be a fault to be avoided" (p. 283). Experienced interviewers are superior in competence and in "the avoidance of bias" (p. 300). They make "continuous efforts to eliminate or reduce bias in interviewing by intensive instruction and training, by means of manuals, specifications for particular surveys, and by continuing supervision and inspection of the interviewer's work" (p. 305). Quality, reliability, and validity are attained through standardization of questions and procedures, and conscientious efforts to maintain objectivity and control bias. "Rules of standardized interviewing: 1. Read the questions exactly as written. 2. . . . Use nondirective follow-up probes to elicit a better answer. 3. Record answers to questions without interpretation or editing. . . . 4. Maintain a professional, neutral relationship with the respondent" (Singleton & Straits, 2012, p. 86).
3. *The phenomenological interview*	The interview focuses on capturing *lived experience* (Van Manen, 1990). "The phenomenological interview involves an informal, interactive process . . . aimed at evoking a comprehensive account of the person's experience of the phenomenon" (Moustakas, 1994, p. 114; see also King & Horrocks, 2012).	The interview evokes "descriptions of lived-through moments, experiential anecdotal accounts, remembered stories of particular experiences, narrative fragments, and fictional experiences. . . . By capturing a personal description of a lived experience, the researcher aims to describe a phenomenon as much as possible in concrete and lived-through terms. In other words, the focus is on the direct description of a particular situation or event as it is lived through without offering causal explanations or interpretive generalizations" (Adams & Van Manen, 2008, p. 618).
4. *Social constructionist interviewing*	Social constructionist interviews are dialogical performances, social meaning-making acts, and co-facilitated knowledge exchanges (Koro-Ljungberg, 2008, p. 430). The constructionist approach involves "interpretive practice" that "requires flexibility and dexterity that cannot be	According to Koro-Ljungberg (2008)," In order for researchers to understand the meaning-making activities that take place during an interview, they must focus on the actions of individuals that influence the immediate social process and context of the interview, as well as those actions that have been influenced by other sociopolitical contexts or discourses" (p. 430). Constructionist

(Continued)

(Continued)

TYPE OF INQUIRY	PURPOSE AND FOCUS	EXPLANATION AND ELABORATION
	captured in mechanical scriptures or formulas" (Holstein & Gubrium, 2011, p. 347; see also Gergen & Gergen, 2008).	interviewing "should shift the focus from mining individual minds to the coconstruction of (temporarily) shared discourses. . . . Rather than the researcher studying what participants know about a particular topic or what kind of experiences they have had, they instead engage in dialogue with participants and thus actively contribute to the knowledge production. The goal of the interview is to examine how knowing subjects . . . experience or have experienced particular aspects of life as they are coconstructed through dialogue" (p. 431).
5. *The hermeneutic interview*	"The research interview is a conversation about the human life world, with the oral discourse transformed into texts to be interpreted" (Kvale, 1996, p. 46). An interview generates text through which we "enter the hermeneutic circle and build an interpretation of the text" (Brown, Tappan, Gilligan, Miller, & Argyris, 1989, p. 147).	"Hermeneutics is then doubly relevant to interview research, first by elucidating the dialogue producing the interview texts to be interpreted, and then by clarifying the subsequent process of interpreting the interview texts produced, which may again be conceived as a dialogue or a conversation with the text" (Kvale, 1996, p. 46). "Even the 'facts' of lived experience need to be captured in language . . . and this is inevitably an interpretive process" (Van Manen, 1990, p. 181).
6. *Narrative inquiry interviewing*	Narrative inquiry revolves around an interest in life experiences as narrated by those who live them (Chase, 2008, p. 421; see also King & Horrocks, 2012).	Narrative analysis takes as its object the story itself—the interviewees' own life experience. "The approach is slow and painstaking, requiring attention to subtlety; nuances of speech, the organization of a response, relations between researcher and subject, social and historical contexts" (Riessman, 2003, p. 342).
7. *The life story interview*	What distinguishes the life story interview is that it keeps the presentation of the life story in the words of the person telling the story. The finished product is entirely a first-person narrative, with the researcher removed as much as possible from the text. The life story interview reveals "how a specific human life is constructed and reconstructed in representing that life as a story" (Atkinson, 2012, p. 116).	"A life story is the story a person chooses to tell about the life he or she has lived, told as completely and honestly as possible, what the person remembers of it and what he or she wants others to know of it, usually as a result of a guided interview by another. The resulting life story is the narrative essence of what has happened to the person. It can cover the time from birth to the present or before and beyond. It includes the important events, experiences, and feelings of a lifetime" (Atkinson, 2002, p. 125; see also Atkinson, 1998, 2012). Wengraf's *Biographic-Narrative Interpretive Method* (2013) involves several lengthy sessions beginning with the opening question "Please tell me the story of your life, all the events and experiences that have been important to you personally; begin wherever you like, I won't interrupt, I'll just take some notes for afterwards," and sticks to the promises given in the question (p. 163).

TYPE OF INQUIRY	PURPOSE AND FOCUS	EXPLANATION AND ELABORATION
8. *Interpretive interactionism*	Interpretive interactionism interviewing brings symbolic interactionist and sociological perspectives to biography: "the intersection between the public and private lives of individuals, groups, and human collectivities" (Denzin, 1983, pp. 129–130).	"Interpretive interactionism takes as its fundamental subject matter the everyday life world. . . . An interpretive, phenomenological, symbolic interactionism that may be synthesized or combined with a structuralist approach to the study and analysis of power, knowledge, and control in everyday life. . . . A fundamental thrust is on the hermeneutic interpretation of ongoing, lived social history. Interpretive interactionism attempts to study biographies as these articulate a particular historical moment in the life world. . . . Individuals are placed back in history; without living people, there is no experience" (Denzin, 1983, pp. 129–130). The focus is on the epiphany: life-changing moments and events, "radical experiences that radically alter and shape the meanings people give to themselves and their life experiences" (Denzin, 2001, p. 34).
9. *Oral history interviewing*	Oral history is "pursuing the past through the spoken word" (Henige, 1988, p. 3). The synthesis of people's stories and reminiscences providing firsthand knowledge of a particular place, event, and/or period of time, "new knowledge about and insights into the past through an individual biography" (Shopes, 2011, p. 451).	Oral history "is the act of recording the speech of people with something interesting to say and then analyzing their memories of the past" (Abrams, 2010, p. 1). A collection of stories creates a sense of history, which (a) empowers us, (b) may serve to illuminate the present situation, (c) forces us to make sense of who we are, (d) requires us to document a life, and (e) inspires respect and awareness that other persons' stories are as valid as our own (Janesick, 2010, p. 15). According to Shopes (2011), "An oral history interview is an inquiry in depth . . ., not a casual or serendipitous conversation but a planned and scheduled, serious and searching exchange, one that seeks a detailed, expansive, and reflective account of the past" (p. 452). Community oral history is a participatory approach. "An oral history interview is an inquiry in depth . . ., not a casual or serendipitous conversation but a planned and scheduled, serious and searching exchange, one that seeks a detailed, expansive, and reflective account of the past" (p. 452). Community oral history is a participatory community-based approach (Mackay, Quinlan, & Sommer, 2013).
10. *Postmodern interviewing*	"The interview conversation is constructed as much within the interview as it stems from predetermined questions. Interview roles are less clear than they once were;	Standardized representation gives way to "representational invention, where the dividing line between fact and fiction is blurred to encourage richer understanding . . . with radical postmodernism completely displacing reality with

(Continued)

(Continued)

TYPE OF INQUIRY	PURPOSE AND FOCUS	EXPLANATION AND ELABORATION
	in some cases they are even exchanged to promote new opportunities for understanding the shape and evolution of selves and experience" (Gubrium & Holstein, 2003, p. 3).	representation. Here, questions shift from the substance, process, and indigenous constitution of experience, to the representational devices used by society and researchers to convey the image of objective or subjective reality" (Gubrium & Holstein, 2003, pp. 4–5). "Postmodern sensibilities" inform postmodern interviewing. "Although there is no such thing as postmodern interviewing per se, postmodern epistemologies have profoundly influenced our understanding of the interview as both a product and a process" (Borer & Fontana, 2012, p. 46.).
11. *Investigative interviewing*	Based on a conflict paradigm of society, Douglas (1976) says, "It's a war of all against all and no one gives anyone anything for nothing, especially the truth" (p. 55). "The problems are considerable in getting the truth from others in our society. . . . Researchers have to use more in-depth and investigative methods to get at these private regions of life than they would to study the public realms which are open to almost anyone" (p. 9).	According to Douglas (1976), "People are extremely adept at constructing complex and convoluted forms of falsehoods and deceptions to front out others, such as researchers, and sometimes even themselves, from the most important parts of their lives" (p. 9). All competent adults are assumed to know that there are at least four major problems lying in the way of getting at social reality by asking people what is going on and that these problems must be dealt with if one is to avoid being taken in, duped, deceived, used, put on, fooled, suckered, made the patsy, left holding the bag, fronted out, and so on. These four problems are (1) misinformation, (2) evasions, (3) lies, and (4) fronts. Three major problems that are less apparent to most people but often of more importance to a social researcher seeking to understand what is going on and why are (1) taken-for-granted meanings, (2) problematic meanings, and (3) self-deceptions (p. 57). Investigative interviewing involves "the sociologist as detective" (Sanders, 1976), as in Cressey's (1976) classic study of violations of financial trust that involved interviewing convicted felons.
12. *Pragmatic interviews*	Straightforward questions about real-world issues aimed at getting straightforward answers that can yield practical and useful insights. "Pragmatism results in a problem-solving, action-oriented inquiry process" (Teddlie & Tashakkori, 2011, p. 290). "Pragmatism [involves] . . . intense reliance on personalized seeing, hearing, experiencing in specific social settings" (Van Maanen, 2011, p. 156).	The "pragmatic maxim" calls for focusing on the practical effects of beliefs and actions (Peirce, 1878). This is the basis for interviewing program participants from a *utilization-focused evaluation* perspective (Patton, 2008, 2012a) or undertaking action research interviews with people in organizations and communities (Argyris, 1982; Argyris et al., 1985; Argyris & Schön, 1978; Schon, 1983a; Stringer, 2007). "Action-oriented" qualitative research seeks solutions to problems (Eder & Fingerson, 2002, p. 187). These tend to be relatively short, focused interviews, often lasting an hour or less.

Variations in Qualitative Interview Question Formats: Alternative Protocols and Instruments

On her deathbed, the writer Gertrude Stein asked her beloved companion, Alice B. Toklas, "What is the answer?"

When Alice, unable to speak, remained silent, Gertrude asked, "In that case, what is the question?"

The question in this section is how to format questions. There are three basic approaches to collecting qualitative data through open-ended interviews. They involve different types of preparation, conceptualization, and instrumentation. Each approach has strengths and weaknesses, and each serves a different purpose. The three alternatives are as follows:

1. The informal conversational interview

2. The interview guide

3. The standardized open-ended interview

These three approaches to the design of the interview differ in the extent to which interview questions are determined and standardized *before* the interview occurs. Exhibit 7.4 presents an overview of variations in interview instrumentation. Let's look at each approach in greater depth for each serves a different purpose and poses quite varying interviewer challenges.

The Informal Conversational Interview

The informal conversational interview is the most open-ended approach to interviewing. It relies entirely on the spontaneous generation of questions in the natural flow of an interaction, often as part of ongoing participant observation fieldwork. Thus, the conversational interview is sometimes referred to as "ethnographic interviewing." It is also called "unstructured interviewing" (Fontana & Frey, 2000, p. 652). The conversational interview offers maximum flexibility to pursue information in whichever direction appears to be appropriate, depending on what emerges from observing a particular setting or from talking to one or more individuals in that setting. Most of the questions will flow from the immediate context. No predetermined set of questions would be appropriate under many emergent field circumstances where the fieldworker doesn't know beforehand what is going to happen, who will be present, or what will be important to ask during an event, incident, or experience.

Data gathered from informal conversational interviews will be different for each person interviewed. The same person may be interviewed on different occasions with questions specific to the interaction or event at hand. Previous responses can be revisited and deepened. This approach works particularly well where the researcher can stay in the setting for some period of time so as not to be dependent on a single interview opportunity. Interview questions will change over time, and each new interview builds on those already done, expanding information that was picked up previously, moving in new directions, and seeking elucidations and elaborations from various participants.

Being unstructured doesn't mean that conversational interviews are unfocused. Sensitizing concepts and the overall purpose of the inquiry inform the interviewing. But within that overall guiding purpose, the interviewer is free to go where the data and respondents lead.

The conversational interviewer must "go with the flow." Depending on how the interviewer's role has been defined, the people being interviewed may not know during any particular conversation that "data" are being collected. In many cases, participant-observers do not take notes during such conversational interviews, instead writing down what they learned later. In other cases, it can be both appropriate and comfortable to take notes or even use a tape recorder.

The strength of the informal conversational method resides in the opportunities it offers for flexibility, spontaneity, and responsiveness to individual differences and situational changes. Questions can be personalized to deepen communication with the person being interviewed and to make use of the immediate surroundings and situation to increase the concreteness and immediacy of the interview questions.

A weakness of the informal conversational interview is that it may require a greater amount of time to collect systematic information because it may take several conversations with different people before a similar set of questions has been posed to each participant in the setting. Because this approach depends on the conversational skills of the interviewer to a greater extent than do more formal, standardized formats, this go-with-the-flow style of interviewing may be susceptible to interviewer effects, leading questions, and biases, especially with novices. The conversational interviewer must be able to interact easily with people in a variety of settings, generate rapid insights, formulate questions quickly and smoothly, and guard against asking questions that impose interpretations on the situation by the structure of the questions.

Data obtained from informal conversational interviews can be difficult to pull together and analyze.

EXHIBIT 7.4   Variations in Interview Instrumentation

TYPE OF INTERVIEW	CHARACTERISTICS	STRENGTHS	WEAKNESSES
Informal conversational interview	Questions emerge from the immediate context and are asked in the natural course of things; there is no predetermination of question topics or wording.	It increases the salience and relevance of questions; interviews are built on and emerge from observations; the interview can be matched to individuals and circumstances.	Different information is collected from different people with different questions. It is less systematic and comprehensive if certain questions do not arise naturally. Data organization and analysis can be quite difficult.
Interview guide approach	Topics and issues to be covered are specified in advance, in outline form; interviewer decides the sequence and wording of questions in the course of the interview.	The outline increases the comprehensiveness of the data and makes data collection somewhat systematic for each respondent. Logical gaps in data can be anticipated and closed. Interviews remain fairly conversational and situational.	Important and salient topics may be inadvertently omitted. Interviewer flexibility in sequencing and wording questions can result in substantially different responses from different perspectives, thus reducing the comparability of responses.
Standardized open-ended interview	The exact wording and sequence of questions are determined in advance. All interviewees are asked the same basic questions in the same order. Questions are worded in a completely open-ended format.	Respondents answer the same questions, thus increasing comparability of responses; data are complete for each person on the topics addressed in the interview. Interviewer effects and bias are reduced when several interviewers are used. It permits evaluation users to see and review the instrumentation used in the evaluation and facilitates organization and analysis of data.	There is little flexibility in relating the interview to particular individuals and circumstances; standardized wording of questions may constrain and limit naturalness and relevance of questions and answers.
Closed, fixed-response interview	Questions and response categories are determined in advance. Responses are fixed; respondent chooses from among these fixed responses.	Data analysis is simple; responses can be directly compared and easily aggregated; many questions can be asked in a short time.	Respondents must fit their experiences and feelings into the researcher's categories; the interview may be perceived as impersonal, irrelevant, and mechanistic and can distort what respondents really meant or experienced by so completely limiting their response choice.

Because different questions will generate different responses, the researcher has to spend a great deal of time sifting through responses to find patterns that have emerged at different points in different interviews with different people. By contrast, interviews that are more systematized and standardized facilitate analysis but provide less flexibility and are less sensitive to individual and situational differences.

The Interview Guide

An interview guide lists the questions or issues that are to be explored in the course of an interview. An interview guide is prepared to ensure that the same basic lines of inquiry are pursued with each person interviewed. The guide provides topics or subject areas within which the interviewer is free to explore, probe, and ask questions that will elucidate and illuminate that particular subject. Thus, the interviewer remains free to build a conversation within a particular subject area, to word questions spontaneously, and to establish a conversational style but with the focus on a particular subject that has been predetermined. The guide serves as a checklist during the interview to make sure that all relevant topics are covered.

The advantage of an interview guide is that it makes sure that the interviewer/evaluator has carefully decided how best to use the limited time available in an interview situation. The guide helps make interviewing a number of different people more systematic and comprehensive by delimiting in advance the issues to be explored. A guide is essential in conducting focus group interviews for it keeps the interactions *focused* while allowing individual perspectives and experiences to emerge. With an interview guide in hand,

the investigator has a rough travel itinerary with which to negotiate the interview. It does not specify precisely what will happen at every stage of the journey, how long each lay-over will last, or where the investigator will be at any given moment, but it does establish a clear sense of the direction of the journey and the ground it will eventually cover. (McCracken, 1988, p. 37)

Interview guides can be developed in more or less detail, depending on the extent to which the interviewer is able to specify important issues in advance and the extent to which it is important to ask questions in the same order to all respondents. Exhibit 7.5 provides an example of an interview guide used with participants in an employment training program. This guide provides a framework within which the interviewer could develop questions, sequence those questions, and make decisions about which information to pursue in greater depth. Usually, the interviewer would not be expected to go into totally new subjects that are not covered within the framework of the guide. The interviewer does not ask questions, for example, about previous employment or education, how the person got into the program, how this program compares with other programs the trainee

has experienced, or the trainee's health. Other topics might still emerge during the interview—topics of importance to the respondent that are not listed explicitly on the guide and therefore would not normally be explored with each person interviewed. For example, trainees might comment on family support (or lack thereof) or personal crises. Comments on such concerns might emerge when, in accordance with the interview guide, the trainee is asked for reactions to program strengths, weaknesses, and so on, but if family is not mentioned by the respondent, the interviewer would not raise the issue.

The Standardized Open-Ended Interview

This approach requires carefully and fully wording each question before the interview. For example, the interview guide for the employment training program above simply lists "work experiences" as a topic for inquiry. In a fully structured interview instrument, the question would be completely specified:

You've told me about the courses you've taken in the program. Now I'd like to ask you about any work experiences you've had. Let's go back to when you first entered the program and go through each work experience up to the present. Okay? So what was your first work experience?

Possible probes

Who did you work for?

What did you do?

What do you feel you learned doing that?

What did you especially like about the experience, if anything?

What did you dislike, if anything?

Transition

Okay, tell me about your next work experience.

Why so much detail? To be sure that each interviewee gets asked the same questions—the same stimuli—in the same way and in the same order, including standard probes. The *standardized open-ended interview* consists of a set of questions carefully worded and arranged with the intention of taking each respondent through the same sequence and asking each respondent the same questions with essentially the same words. Flexibility in probing is more or less limited, depending on the nature of the interview and the skills of interviewers.

EXHIBIT 7.5   Evaluation Interview Guide for Participants in an Employment Training Program

What has the trainee done in the program?

- ✓ Activities?
- ✓ Courses?
- ✓ Groups?
- ✓ Work experiences?

What are his achievements?

- ✓ Skills attained?
- ✓ Products produced?
- ✓ Outcomes achieved?
- ✓ Knowledge gained?
- ✓ Things completed?
- ✓ What can the trainee do that is *marketable*?

How has the trainee been affected in areas other than job skills?

- ✓ Feelings about self?
- ✓ Attitudes toward work?
- ✓ Aspirations?
- ✓ Interpersonal skills?

What aspects of the program have had the greatest impacts?

- ✓ Formal courses

- ✓ Relationships with staff
- ✓ Peer relationships
- ✓ The way he or she was treated in the program
- ✓ Contacts
- ✓ Work experiences

What problems has the trainee experienced?

- ✓ Work related
- ✓ Program related
- ✓ Personal
- ✓ Family, friends, world outside the program

What are the trainee's plans for the future?

- ✓ Work plans?
- ✓ Income expectations?
- ✓ Lifestyle expectations/plans?

What does the trainee think of the program?

- ✓ Strengths? Weaknesses?
- ✓ Things liked? Things disliked?
- ✓ Best components? Poorest components?
- ✓ Things that should be changed?

The standardized open-ended interview is used when it is important to minimize variation in the questions posed to interviewees. A doctoral committee may want to see the full interview protocol before approving a dissertation proposal. The institutional review board for protection of human subjects may insist on approving a structured interview, especially if the topic is controversial or intrusive. In evaluations, funders and other key stakeholders may want to be sure that they know what program participants will be asked. In team research, standardized interviews ensure consistency across interviewers. In multisite studies, structured interviews provide comparability across sites.

In participatory or collaborative studies, inexperienced and nonresearcher interviewers may be involved in the process so that the standardized questions can compensate for variability in skills. Some evaluations rely on volunteers to do the interviewing; at other times, program staff may be involved in doing some interviewing; and in still other instances, interviewers may be novices, students, or others who are not social scientists or professional evaluators. When a number of different interviewers are used, variations in data created by differences among interviewers will become particularly apparent if an informal conversational approach to data gathering is used or even if each interviewer uses a basic guide. The best way to guard against variations among interviewers is to carefully word questions in advance and train the interviewers not to deviate from the prescribed forms. The data collected are still open ended, in the sense that the respondent supplies his or her own words, thoughts, and insights in answering the questions, but the precise wording of the questions is determined ahead of time.

When doing action research or conducting a program evaluation, it may only be possible to interview participants once for a short, fixed time, like a half-hour, so highly focused questions serve to establish priorities for the interview. At other times, it is possible and desirable to interview participants before they enter the program, when they leave the program, and again after some period of time (e.g., six months) after they have left the program. For example, a chemical dependency program would ask participants about sobriety issues before, during, at the end of, and after the program. To compare answers across these time periods, the same questions need to be asked in the same way each time. Such interview questions are written out in advance *exactly* the way in which they are to be asked during the interview. Careful consideration is given to the wording of each question before the interview. Any clarifications or elaborations that are to be used are written into the interview itself. Probes are placed in the interview at appropriate places to minimize interviewer effects by asking the same question of each respondent, thereby reducing the need for interviewer judgment during the interview. The standardized open-ended interview also makes data analysis easier because it is possible to locate each respondent's answer to the same question rather quickly and to organize questions and answers that are similar.

Exhibit 7.19 (pp. 508–511) at the end of this chapter, provides an example of a standardized open-ended evaluation interview sequence for participants in an Outward Bound wilderness program for disabled persons. The first interview was conducted at the beginning of the program, the second interview was used at the end of the 10-day experience, and the third interview took place six months after the program. Questions are specified for able-bodied and disabled participants.

In summary, the four major reasons for using standardized open-ended interviews are as follows:

1. The exact instrument used in the study is available for inspection by those who will use the findings of the study.

2. Variation among interviewers can be minimized where a number of different interviewers must be used.

3. The interview is highly focused so that interviewee time is used efficiently.

4. Analysis is facilitated by making responses easy to find and compare.

In program evaluations, potential problems of legitimacy and credibility for qualitative data can make it politically wise to produce an exact interview form that the evaluator can show to primary decision makers and evaluation users. Moreover, when generating a standardized form, evaluation users can participate more completely in writing the interview instrument. They will not only know precisely what is going to be asked, but, no less important, they will also understand what is *not* going to be asked. This reduces the likelihood of the data being attacked later because certain questions were missed or asked in the wrong way. By making it clear, in advance of data collection, exactly what questions will be asked, the limitations of the data can be known and discussed before evaluation data are gathered.

While the conversational and interview guide approaches permit greater flexibility and individualization, these approaches also open up the possibility, indeed, the likelihood, that more information will be collected from some program participants than from others. Those using the findings may worry about how conclusions have been influenced by qualitative differences in the depth and breadth of information received from different people.

In contrast, in fieldwork done for basic and applied research, the researcher will be attempting to understand the holistic worldview of a group of people. Collecting the same information from each person poses no credibility problem when each person is understood as a unique informant with a unique perspective. The political credibility of consistent interview findings across respondents is less an issue under basic research conditions.

The weakness of the standardized approach is that it does not permit the interviewer to pursue topics or issues that were not anticipated when the interview was written. Moreover, a structured interview reduces the extent to which individual differences and circumstances can be queried.

Combining Approaches

These contrasting interview strategies are by no means mutually exclusive.

A conversational strategy can be used within an interview guide approach, or you can combine a guide approach with a standardized format by specifying certain key questions exactly as they must be asked while leaving other items as topics to be explored at the interviewer's discretion. This combined strategy offers the interviewer flexibility in probing and in determining when it is appropriate to explore certain subjects in greater depth, or even to pose questions about new areas of inquiry that were not originally anticipated in the interview instrument's development. A common combination strategy involves using a standardized interview format in the early part of an interview and then leaving the interviewer free to pursue any subjects of interest during the latter parts of the interview. Another combination would include using the informal conversational interview early in an evaluation project, followed midway through by an interview guide, and then closing the program evaluation with a standardized open-ended interview to get systematic information from a sample of participants at the end of the program or when conducting follow-up studies of participants.

A *sensitizing concept* can provide the bridge across different types of interviews. In doing follow-up interviews with recipients of MacArthur Foundation fellowships, the sensitizing concept "enabling," a concept central to the Fellowship's purpose, allowed us to focus interviews on any ways in which receiving the fellowship had *enabled* recipients. *Enabling*, or *being enabled*, broadly defined and open-ended, gave interviewees room to share a variety of experiences and outcomes while also letting us identify some carefully worded, standardized questions for all interviewees, some interview guide topics that might or might not be pursued, and a theme for staying centered during completely open-ended conversations at the end of the interviews.

Summary of Interviewing Formats

All three qualitative formats for interviewing share the commitment to ask genuinely open-ended questions that offer the persons being interviewed the opportunity to respond in their own words and to express their own personal perspectives. While the three strategies vary in the extent to which the wording and sequencing of questions are predetermined, no variation exists in the principle that the response format should be open-ended. The interviewer never supplies or predetermines the phrases or categories that must be used by respondents to express themselves, as is the case in fixed-response questionnaires. The purpose of qualitative interviewing is to capture how those being interviewed view their world, to learn *their* terminology and judgments, and to capture the complexities of *their* individual perceptions and experiences. This openness distinguishes qualitative interviewing from the closed questionnaire or test used in quantitative studies. Such closed instruments force respondents to fit their knowledge, experiences, and feelings into the researcher's categories. The fundamental principle of qualitative interviewing is to provide a framework within which respondents can express *their own* understandings in their own terms.

CULTURE AS A SENSITIZING CONCEPT

"Culture" is certainly one of the more contentious and complex words in our lexicon. Like the term "force" to a physicist or "*life*" to a biologist, or even "god" to a theologian, "culture" to the ethnographer is multivocal, highly ambiguous, shape-shifting, and difficult if not impossible to pin down. When put into use, contradictions abound. Culture is taken by some of its most distinguished students as cause and consequence, as material and immaterial, as coherent and fragmented, as grand and humble, as visible (to some) and invisible (to many). . . .

One of the charming but endlessly frustrating things about culture is that everybody uses the term, albeit in vastly different ways. The notion of culture as used by ethnographers today is more a loose, sensitizing concept than a strict, theoretical one. It signals a conviction that agency and action (be it word or deed) rest on social meanings that range from the rather bounded and particularistic to more or less institutionalized and broad. . . . Certainly the view of culture as an integrated, shared system of interlocking ideas, routines, signs, and values passed on more or less seamlessly from generation to generation has withered away (thankfully). . . .

But as long as meanings are taken to be central to accounts of human activity and meanings are seen as coming forth—somehow, someway—from human interaction, it is most unclear what conceptual framework might step up to replace culture as a way to imagine and think about such matters as "how things work." . . . Culture and the meaning making and remaking processes associated with the concept, however trimmed down and inevitably flawed, still seem to me indispensable.

—John Van Maanen (2011, p. 154)
Tales of the Field: On Writing Ethnography

Anticipating Analysis and Reporting in Developing an Interview

Look to the end at the beginning to increase the likelihood of ending up where you want to be at the end.

—Halcolm

The overarching purpose of an inquiry frames analysis, and anticipating analysis can inform the kind of interview protocol and instrumentation developed.

For example, a program evaluation might interview 20 participants, about an hour each, covering a set of discrete topics, like the following:

1. What influenced your decision to participate in the program?

   (Program recruitment)

2. Describe the program activities in which you participated.

   (Program processes)

3. What did you get out of participating in the program?

   (Program outcomes)

The evaluation report will analyze and present the responses from the 20 participants *question by question*, or section by section, what is called cross-sectional analysis. Organizing and sequencing the interview in anticipation of this format for the final report will greatly facilitate analysis.

A different kind of analysis and reporting involves constructing case studies. A case study integrates all the responses in an interview into a coherent story of that person or place (the case). If 20 participants are interviewed for individual case studies, the analysis and report will be presented as holistic cases and patterns across cases rather than question by question.

Yet a third kind of analysis is illumination of one or more sensitizing concepts. Responses across participants and across questions will be synthesized to present varying perspectives on and meanings of the sensitizing concept, for example, *empowerment*, or *power* (Sheehan, De Cieri, Cooper, & Brooks, 2014).

Thus, the interview protocol and format is part of the overall design of the study and flows from prior decisions. The overarching purpose of the study and the primary inquiry questions lead to a purposeful sampling strategy. Those inquiry and design decisions anticipate the kind of analysis that will be needed to fulfill the purpose of the study, whether research or evaluation. The interview format (degree of standardization vs. degree of flexibility) flows from this combination of design considerations, the background and experience of the interviewer(s), and the expectations of the audience for the study (What approach will be most credible?). Exhibit 7.6 summarizes these analysis alternatives and their implications for formatting and constructing qualitative interviews.

EXHIBIT 7.6 **Anticipating Analysis and Reporting to Organize, Sequence, and Format Interviews**

Three examples of how anticipating the analysis and report can influence the qualitative interview format.

ANTICIPATED REPORT FORMAT	UNIT OF ANALYSIS	NATURE OF THE ANALYSIS	INTERVIEW FORMAT ALIGNMENT
1. Cross-sectional analysis	Responses to specific interview questions	Report findings question by question.	Standardized questions asked in a standardized sequence facilitates analysis question by question.
2. Case studies	Each interviewee	Integrate and synthesize interview responses from throughout the interview into a coherent story.	Interview guide enhances flexibility to pursue various topics in greater or lesser depth depending on relevance to each person (case).
3. Sensitizing concept illumination	The sensitizing concept	Extract quotations and perspectives from across different interviewees while also placing those quotes in the context of each interviewee as a case.	A combination of interview approaches (some standardized questions, some guide topics, and pursuing some conversational leads) will permit focusing on the sensitizing concept in a variety of ways as the interview unfolds.

Know What You Want to Find Out, and Listen Attentively to Whether Your Question Was Answered

> If you ask me, I'm gonna tell you.
>
> —Comedienne Roseanne Barr

Six kinds of questions can be asked of people. On any given topic, it is possible to ask any of these questions. Distinguishing types of questions forces you, the interviewer, to be clear about what is being asked and helps the interviewee respond appropriately.

Experience and Behavior Questions

Questions about what a person does or has done aim to elicit behaviors, experiences, actions, and activities that would have been observable had the observer been present: "If I followed you through a typical day, what would I see you doing? What experiences would I observe you having?" "If I had been in the program with you, what would I have seen you doing?"

Opinion and Values Questions

Questions aimed at understanding the cognitive and interpretive processes of people ask about opinions, judgments and values—"head stuff" as opposed to actions and behaviors: Answers to these questions tell us what people *think* about some experience or issue. They tell us about people's goals, intentions, desires, and expectations. "What do you believe?" "What do you think about _____?" "What would you like to see happen?" "What is your opinion of_____?"

Feeling Questions

Emotional centers in the brain can be distinguished from cognitive areas. Feeling questions aim at eliciting emotions—*feeling* responses of people to their experiences and thoughts. Feelings tap the affective dimension of human life. In asking feeling questions—for example, "How do you feel about that?"—the interviewer is looking for adjective responses: anxious, happy, afraid, intimidated, confident, and so on.

Opinions and feelings are often confused. It is critical that interviewers understand the distinction between the two in order to know when they have the kind of answer they want to the question they are asking. Suppose an interviewer asks, "How do you feel about that?" If the response is "I think it's probably the best that we can do under the circumstances," the question about feelings has not really been answered. Analytical, interpretive, and opinion statements are not answers to questions about feelings.

This confusion sometimes occurs because interviewers give the wrong cues when asking questions—for example, by asking opinion questions using the format "How do you feel about that?" instead of "What is your opinion about that?" or "What do you think about it?" When you want to understand the respondents' emotional reactions, you have to ask about *and listen for* feeling-level responses. When you want to understand what someone thinks about something, the question should explicitly tell the interviewee that you're searching for opinions, beliefs, and considered judgments—not feelings.

Knowledge Questions

Knowledge questions inquire about the respondent's factual information—what the respondent knows. Certain things are facts, like whether it is against the law to drive while drunk and how the law defines drunkenness. These things are not opinions or feelings. Knowledge about a program may include knowing what services are available, who is eligible, what the rules and regulations of the program are, how one enrolls in the program, and so on.

Sensory Questions

Sensory questions ask about what is seen, heard, touched, tasted, and smelled. Responses to these questions allow the interviewer to enter into the sensory experience of the respondent. "When you walk through the doors of the program, what do you see? Tell me what I would see if I walked through the doors with you." Or again, "What does the counselor ask you when you meet with him? What does he actually say?" Sensory questions attempt to have interviewees describe the stimuli that they experience. Technically, sensory data are a type of behavioral or experiential data—they capture the experience of the senses. However, the types of questions asked to gather sensory data are sufficiently distinct to merit a separate category.

Background/Demographic Questions

Age, education, occupation, and the like are standard background questions that identify characteristics of the person being interviewed. Answers to these questions help the interviewer locate the respondent in relation to other people. Asking these questions in an open-ended rather than closed manner elicits the respondent's own categorical worldview. Asked about age, a person aged 55 might respond "I'm 55" or "I'm middle-aged" or "I'm at the cusp of old age" or "I'm still young at heart" or "I'm in my mid-50s" or "I'm 10 years from retirement" or "I'm between 40 and 60 (smiling broadly)," and so forth. Responses to open-ended, qualitative background inquiries tell us about how people categorize themselves in today's endlessly categorizing world. Perhaps nowhere is such openness more important and illuminative than in asking about race and ethnicity. For example, professional golf star Tiger Woods has African, Thai, Chinese, American Indian, and European ancestry—and resists being "assigned" to any single ethnic category. He came up with the name "Cablinasian" to describe his mixed heritage. In an increasingly diverse world with people of mixed ethnicity and ever-evolving labels (e.g., Negro, colored, black, African American, person of African descent), qualitative inquiry is a particularly appropriate way of finding out how people perceive and talk about their backgrounds.

Distinguishing Question Types

Behaviors, opinions, feelings, knowledge, sensory data, and background demographics—these are the universe of kinds of questions it is possible to ask in an interview. Any kind of question one might want to ask can be subsumed in one of these categories. Keeping these distinctions in mind can be particularly helpful in planning an interview, designing the inquiry strategy, focusing on priorities for inquiry, and ordering the questions in some sequence. Before considering the sequence of questions, however, let's look at how the *dimension of time* intersects with the different kinds of questions.

The Time Frame of the Questions

Questions can be asked in the present, past, or future tense. For example, you can ask people what they're doing now, what they have done in the past, and what they plan to do in the future. Likewise, you can inquire about present attitudes, past attitudes, or future attitudes. By combining the time frame of questions with the different type of questions, we can construct a

matrix that generates 18 different types of questions. Exhibit 7.7 shows that matrix.

EXHIBIT 7.7   A Matrix of Questions Options

QUESTION FOCUS	PAST	PRESENT	FUTURE
Behaviors/ experiences			
Opinions/ values			
Feelings/ emotions			
Knowledge			
Sensory			
Background			

Asking all 18 questions about any particular situation, event, or programmatic activity may become somewhat tedious, especially if the sequence is repeated over and over throughout the interview. The matrix constitutes a set of options to help you think about what information is most important to obtain.

Sequencing Questions

No recipe for sequencing questions can or should exist, but the matrix of questions suggests some possibilities. The challenges of sequencing vary, of course, for different strategies of interviewing. Informal conversational interviewing is flexible and responsive so that a predetermined sequence is seldom possible or desirable. In contrast, standardized open-ended interviews must establish a fixed sequence of questions to fit the structured format. I offer, then, some suggestions about sequencing.

I prefer to begin an interview with questions about noncontroversial present behaviors, activities, and experiences like "What are you currently working on in school?" Such questions ask for relatively straightforward descriptions; they require minimal recall and interpretation. Such questions are hopefully fairly easy to answer. They encourage the respondent to talk *descriptively*. Probes should focus on eliciting greater detail—filling out the descriptive picture.

Once some experience or activity has been described, then opinions and feelings can be solicited, building on and probing for interpretations of the experience. Opinions and feelings are likely to be more grounded and meaningful once the respondent has verbally "relived" the experience. Knowledge and skill questions also need a context. Such questions can be quite threatening if asked too abruptly. The interviewer doesn't want to come across as a TV game show host quizzing a contestant. So, for example, in evaluation interviewing, it can be helpful to ask knowledge questions (What are the eligibility requirements for this program?) as follow-up questions about program activities and experiences that have a bearing on knowledge and skills (How did you become part of the program?). Finding out from people what they know works best once some rapport and trust have been established in the interview.

Questions about the present tend to be easier for respondents than questions about the past. Future-oriented questions involve considerable speculation, and responses to questions about future actions or attitudes are typically less reliable than questions about the present or past. I generally prefer to begin by asking questions about the present; then, using the present as a baseline, I ask questions about the same activity or attitude in the past. Only then do I broach questions about the future.

Background and demographic questions are basically boring; they epitomize what people hate about interviews. They can also be somewhat uncomfortable for the respondent, depending on how personal they are. I keep such questions to a minimum and prefer to space them strategically and unobtrusively throughout the interview. *I advise never beginning an interview with a long list of routine demographic questions.* In qualitative interviewing, the interviewee needs to become actively involved in providing descriptive information as soon as possible instead of becoming conditioned to providing short-answer, routine responses to uninteresting categorical questions. Some background information may be necessary at the beginning to make sense out of the rest of the interview, but such questions should be tied to descriptive information about present life experience as much as possible. Otherwise, save the socio-demographic inquiries (age, socioeconomic status, birth order, etc.) for the end.

Practical Guidance on Wording Questions

An interview question is a stimulus aimed at eliciting a response from the person being interviewed. How a question is worded and asked affects how the interviewee responds. As Payne (1951) observed in his classic book on questioning, *asking questions is an art.*

PRAGMATIC INTERVIEWING: SKILLFULLY WORDING QUESTIONS

How qualitative interview questions are worded depends on a number of factors, including but not limited to these: the theoretical tradition that informs the inquiry, the nature and focus of the study, the relationships and interactions between the interviewer and the interviewee, the length of the interview and setting where the interview talks place, the interviewer's experience as an interviewer, and the interview format being used (conversational, interview guide, standardized questions). Thus, general guidance on how to word questions must be adapted to purpose and context. What can be said with confidence is that skillful wording is a core competence that enhances the quality of responses. In the sections that follow, I offer some practical guidance about how to be strategic, intentional, thoughtful, and skillful in wording questions.

In qualitative inquiry, "good" questions should, at a minimum, be open-ended, neutral, singular, and clear. Let's look at each of these criteria.

Ask Truly Open-Ended Questions

Qualitative inquiry—strategically, philosophically, and methodologically—aims to minimize the imposition of predetermined responses like fixed survey items ("strongly agree"). Rather, questions should be asked in a truly open-ended fashion so people can respond in their own words. Those open-ended responses are the heart of qualitative data, and they emerge from asking open-ended questions.

The standard fixed-response item in a questionnaire provides a limited and predetermined list of possibilities: "How satisfied are you with the program? (a) very satisfied, (b) somewhat satisfied, (c) not too satisfied, (d) not at all satisfied." The closed and limiting nature of such a question is obvious to both questioner and respondent. Many researchers seem to think that the way to make a question open-ended is simply to leave out the structured response categories. But doing so does not make a question truly open-ended. It merely disguises what still amounts to a predetermined and implicit constraint on likely responses.

Consider the question "How satisfied are you with this program?" Asked without fixed-response choices, this can appear to be an open-ended question. On closer inspection, however, we see that the dimension along which the respondent can answer has already been identified—*degree of satisfaction.* The interviewee can use a variety of modifiers for the word *satisfaction*—"pretty satisfied," "kind of satisfied," "mostly satisfied," and so on. But, in effect, the possible response set has been narrowly limited by the wording of the question. The typical range of answers will vary only slightly more than what would have been obtained had the categories been made explicit from the start while making the analysis more complicated.

A truly open-ended question does not presuppose which dimension of feeling or thought will be salient for the interviewee. The truly open-ended question allows the person being interviewed to select from among that person's full repertoire of possible responses those that are most *salient*. Indeed, in qualitative inquiry, one of the things the inquiry is trying to determine is what dimensions, themes, and images/words people use among themselves to describe their feelings, thoughts, and experiences. Examples, then, of truly open-ended questions would take the following format:

What's your reaction to _____?

How do you feel about _____?

What do you think of _____?

The truly open-ended question permits those being interviewed to take whatever direction and use whatever words they want to express what they have to say. Moreover, to be truly open-ended, a question cannot be phrased as a dichotomy.

The Horns of a Dichotomy

Dichotomous response questions provide the interviewee with a grammatical structure suggesting a "yes" or "no" answer. Are you satisfied with the program? Have you changed as a result of your participation in this program? Was this an important experience for you? Do you know the procedures for enrolling in the program? Have you interacted much with the staff in the program? By their grammatical form, all of these questions invite a yes/no reply.

In contrast, in-depth interviewing strives to get the person being interviewed to talk—to talk about experiences, feelings, opinions, and knowledge. Far from encouraging the respondent to talk, dichotomous response questions limit expression. They can even create a dilemma for respondents who may not be sure whether they are being asked a simple yes/no question or if, indeed, the interviewer expects a more elaborate response. Often, in teaching interviewers and reviewing their fieldwork, I've found that those who report having difficulty getting respondents to talk are posing a string of dichotomous questions that program the respondent to be largely reactive and binary.

Consider this classic exchange between a parent and teenager. (Teenager returns home from a date.)

"Do you know that you're late?"

"Yeah."

"Did you have a good time?"

"Yeah."

"Did you go to a movie?"

"Yeah."

"Was it a good movie?"

"Yeah, it was okay."

"So, it was worth seeing?"

"Yeah, it was worth seeing."

"I've heard a lot about it. Do you think I would like it?"

"I don't know. Maybe."

"Anything else happen?"

"No. That's about it."

Teenager then goes off to bed. One parent turns to the other and says, "Sure is hard to get him to talk to us. I guess he's at the age where kids just don't want to tell their parents anything."

Dichotomous questions can turn an interview into an interrogation or quiz rather than an in-depth conversation. In everyday conversation, our interactions with each other are filled with dichotomous questions that we unconsciously ignore and treat as if they were open-ended. If a friend asks, "Did you have a good time?" you're likely to offer more than a yes/no answer. In a more formal interview setting, however, the interviewee will be more conscious of the grammatical structure of questions and is less likely to elaborate beyond "yes" or "no" replies when hit with dichotomous queries. Indeed, the more intense the interview situation, the more likely the respondent will react to the "deep-structure" stimulus of questions—which includes their grammatical framing—and to take questions literally (Bandler & Grinder, 1975).

In training interviewers, I like to play a game where I only respond literally to the questions asked without volunteering any information that is not clearly demanded in the question. I do this before explaining the difficulties involved in asking dichotomous questions. I have played this game hundreds of times, and the interaction seldom varies. On getting dichotomous responses to general questions, the interviewer will begin to rely on more and more specific dichotomous response questions, thereby digging a deeper and deeper hole, which makes it difficult to pull the interview out of the dichotomous-response pattern. Exhibit 7.8 provides a transcription of an actual interview from a training workshop. In the left column, I have recorded the interview that took place; the right column records truly open-ended alternatives to the dichotomous questions that were asked.

The questions on the left in Exhibit 7.8 illustrate a fairly extreme example of posing dichotomous questions in an interview. Notice that the open-ended questions on the right side generate richer answers and quite different information than was elicited from the dichotomous questions. In addition, dichotomous

---

EXHIBIT 7.8 **Interview Training Demonstration: Closed Versus Open-Ended Questions**

Instruction to workshop participants: Okay, now we're going to play an interviewing game. I want you to ask me questions about an evaluation I just completed. The program evaluated was a leadership development initiative that involved taking higher education professionals into a wilderness setting for a week. That's all I'm going to tell you at this point. I'll answer your questions as precisely as I can, but *I'll only answer what you ask.* I won't volunteer any information that isn't directly asked for by your questions.

In the left column below, I have recorded the interview questions actually asked and my answers; the right column records truly open-ended alternatives to the dichotomous questions that were asked and the answers I would have given to more open-ended questions.

ACTUAL QUESTIONS ASKED AND ANSWERS GIVEN	GENUINELY OPEN-ENDED ALTERNATIVES WITH RICHER RESPONSES
Q1. Were you doing a formative evaluation? A. Mostly.	Q1a. What were the purposes of the evaluation? A. First, to document what happened, then, to provide feedback to staff and help them identify their "model," and finally to report to funders.
Q2. Were you trying to find out if the people changed from being in the wilderness? A. That was part of it.	Q2a. What were you trying to find out through the evaluation? A. Several things. How participants experienced the wilderness, how they talked about the experience, what meanings they attached to what they experienced, what they did with the experience when they returned home, and any ways in which it affected them.
Q3. Did they change? A. Some of them did.	Q3a. What did you find out? How did participation in the program affect participants? A. Many participants reported "transformative" experiences—their term—by which they meant something life changing. Others became more engaged in experiential education themselves. A few reported just having a good time. You'd need to read the full case studies to see the depth of variation and impacts.
Q4. Did you interview people both before and after the program? A. Yes.	Q4a. What kinds of information did you collect from the evaluation? A. We interviewed participants before, during, and after the program; we did focus groups; we engaged in participant observation with conversational interviews; and we read their journals when they were willing. They also completed open-ended evaluation forms that asked about aspects of the program.

ACTUAL QUESTIONS ASKED AND ANSWERS GIVEN	GENUINELY OPEN-ENDED ALTERNATIVES WITH RICHER RESPONSES
Q5. Did you find that being in the program affected what happened? A. Yes.	Q5a. How do you think your participation in the program affected what happened? A. We've reflected a lot on that and we talked with staff and participants about it. Most agreed that the evaluation process made everyone more intentional and reflective—and that increased the impact in many cases.
Q6. Did you have a good time? A. Yes.	Q6a: What was the wilderness experience like for you? A: First, I learned a great deal about participant observation and evaluation. Second, I came to love the wilderness and have become an avid hiker. Third, I began what I expect will be a deep and lifelong friendship with one staff member.

questions can easily become leading questions. Once the interviewer begins to cope with what appears to be a reluctant or timid interviewee by asking ever more detailed dichotomous questions, guessing at possible responses, the interviewer may actually impose those responses on the interviewee.

One sure sign that an interview is going poorly is when the interviewer is doing more talking than the person being interviewed. Consider the excerpt from an actual interview in Exhibit 7.9. The interviewee was a teenager who was participating in a chemical dependency program. The person conducting this interview said that she wanted to find out two things in this portion of the interview: (1) What experiences were most salient for John? (2) How personally involved was John becoming in his treatment? As you'll see in the transcript, she learns that the "hot-seat" therapeutic experience was highly salient for John, but she really gets very little information about the reasons for that salience. With regard to the question of his personal involvement, the only data she has come from his acquiescence to leading questions. To see how few qualitative data are actually generated in her interview, I've listed below the actual *data* from the interview—his verbatim responses:

Okay.

Yeah, . . . the hot seat.

Right.

One person does it every day.

Yeah, it depends.

Okay, let's see, hmmm . . . there was this guy yesterday who really got nailed. I mean he really caught a lot of crap from the group. It was really heavy.

No, it was them others.

Yeah, right, and it really got to him.

He started crying and got mad and one guy really came down on him and afterwards they were talking, and it seemed to be okay for him.

Yeah, it really was.

It was pretty heavy.

The lack of a coherent story line in these responses reveals how little we've actually learned about John's perspective and experiences. Study the transcript, and you'll find that the interviewer was talking more than the interviewee. The questions put the interviewee in a passive stance, able to confirm or deny the substance provided by the interviewer but not really given the opportunity to provide in-depth, descriptive detail in his own words.

EXHIBIT 7.9   Interview Transcript With Commentary

The interview below took place with a teenager during his residency in a chemical dependency treatment program.

The interview transcript is on the left. My commentary is in the column on the right.

INTERVIEW TRANSCRIPT	COMMENTARY
Q1. Hello, John. It's nice to see you again. I'm anxious to find out what's been happening with you. Can I ask you some questions about your experience?	The opening is dominated by the interviewer. No informal give and take. The interviewee is set up to take a passive/reactive role.
A. Okay.	
Q2. I'd like you to think about some of the really important experiences you've had here. Can you think of something that stands out in your mind?	An introductory cue sentence is immediately followed by a dichotomous response question.
A. Yeah, . . . the hot seat.	John goes beyond the dichotomous response.
Q3. The hot seat is when one person is the focus of attention for the whole group, right?	The interviewer has provided the definition rather than getting John's own definition of the hot seat.
A. Right.	
Q4. So, what was it like . . . ? Was this the first time you've seen the "hot seat" used?	Began open ended, then changed the question and posed a dichotomous question. The question is no longer singular or open.
A. One person does it everyday.	Not really an answer to the question.
Q5. Is it different with different people?	Question follows the previous answer, but still a dichotomous format.
A. Yeah, it depends.	
Q6. Can you tell me about one that really stands out in your mind?	Can you? Is this an inquiry about willingness, memory, capacity, or trust?
A. Okay, let's see, hmm. . . . There was this guy yesterday who really got nailed. I mean, he really caught a lot of crap from the group. It was really heavy.	Before responding to the open request, John reacts to the dichotomous format.
Q7. Did you say anything?	Dichotomous question.
A. No, it was them others.	
Q8. So what was it like for you? Did you get caught up in it? You said it was really heavy. Was it heavy for you or just the group?	Multiple questions. Unclear connections. Ambiguous, multiple-choice format at the end.
A. Yeah, right, and it really got to him.	John's positive answer ("Yeah, right") is actually uninterpretable, given the questions asked.
Q9. Did you think it was good for him? Did it help him?	Dichotomous questions.
A. He started crying and got mad, and one guy really came down on him, and afterwards, they were talking, and it seemed to be okay for him.	The question asks for a judgment. John wants to describe what happened. The narrowness of the interview questions are limiting his responses.
Q10. So it was really intense?	Leading question, setting up an easy acquiescence response.

INTERVIEW TRANSCRIPT	COMMENTARY
A. Yeah, it really was.	Acquiesces to leading question. Accepts interviewer's term, *intense*, so we don't learn what word he would have chosen.
Q11. And you got really involved.	Another leading question.
A. It was pretty heavy.	John doesn't actually respond to the question. Ambiguous response.
Q12. Okay, I want to ask you something about the lecture part of the program. Anything else you want to say about the hot seat?	Transition. John is cued that the hot-seat questions are over. No response really expected.
(John doesn't answer verbally. He sits and waits for the next questions.)	

Asking Singular Questions

One of the basic rules of questionnaire writing is that each item must be singular—that is, no more than one idea should be contained in any given question. Consider this example:

How well do you know and like the staff in this program?

a. A lot

b. Pretty much

c. Not too much

d. Not at all

This item is impossible to interpret in analysis because it asks two questions:

1. How well do you know the staff?

2. How much do you like the staff?

When one turns to open-ended interviewing, however, many people seem to think that singular questions are no longer called for. Precision gives way to vagueness and confused multiplicity, as in the illustration in this section. I've seen transcripts of interviews conducted by experienced and well-known field researchers in which several questions have been thrown together, which they might think are related but which are likely to confuse the person being interviewed about what is really being asked.

To help the staff improve the program, we'd like to ask you to talk about your opinion of the program: What you think are the strengths and weaknesses of the program? What you like? What you don't like? What you think could be improved or should stay the same? Those kinds of things—and any other comments you have.

The evaluator who used this question regularly in interviewing argued that by asking a series of questions he could find out which was most salient to the person being interviewed because the interviewee was forced to choose what he or she most cared about in order to respond to the question. The evaluator would then probe more specifically in those areas that were not answered in the initial question.

It's necessary to distinguish, then, between giving an overview of a series of questions at the beginning of a sequence, and then asking each one singularly, versus laying out a whole series of questions at once, and then seeing which one strikes a respondent's fancy. In my experience, multiple questions create tension and confusion because the person being interviewed doesn't really know what is being asked. An analysis of the strengths and weaknesses of a program is not the same as reporting what one likes and dislikes. Likewise, recommendations for change may be unrelated to strengths, weaknesses, likes, and dislikes. The following is an excerpt from an interview with a parent participating in a family education program aimed at helping parents become more effective as parents.

Q: Based on your experience, what would you say are the strengths of this program?

A: The other parents. Different parents can get together and talk about what being a parent is like for them. The program is really parents with parents. Parents really need to talk to other parents about what they do, and what works and doesn't work. It's the parents, it really is.

Q: What about weaknesses?

A: I don't know. . . . I guess I'm not always sure that the program is really getting to the parents who need it the most. I don't know how you do that, but I just think there are probably a lot of parents out there who need the program and . . . especially maybe single-parent families. And fathers. It's really hard to get fathers into something like this. It should just get to everybody and that's real hard.

Q: Let me turn now to your personal likes and dislikes about the program. What are some of the things that you have really liked about the program?

A: I'd put the staff right at the top of that. I really like the program director. She's really well educated and knows a lot, but she never makes us feel dumb. We can say anything or ask anything. She treats us like people, like equals even. I like the other parents. And I like being able to bring my daughter along. They take her into the child's part of the program, but we also have some activities together. But it's also good for her to have her activities with other kids, and I get some time with other parents.

Q: What about dislikes? What are some things you don't like so much about the program?

A: I don't like the schedule much. We meet in the afternoons after lunch, and it kind of breaks into the day at a bad time for me, but there isn't any really good time for all the parents, and I know they've tried different times. Time is always going to be a hassle for people. Maybe they could just offer different things at different times. The room we meet in isn't too great, but that's no big deal.

Q: Okay, you've given us a lot of information about your experiences in the program, strengths and weaknesses you've observed, and some of the things you've liked and haven't liked so much. Now I'd like to ask you about your recommendations for the program. If you had the power to change things about the program, what would you make different?

A: Well, I guess the first thing is money. It's always money. I just think they should put, you know, the legislature should put more money into programs like this. I don't know how much the director gets paid, but I hear that she's not even getting paid as much as school teachers. She should get paid like a professional. I think there should be more of these programs and more money in them.

Oh, I know what I'd recommend. We talked about it one time in our group. It would be neat to have some parents who have already been through the program come back and talk with new groups about what they've done with their kids since they've been in the program, you know, like problems that they didn't expect or things that didn't work out, or just getting the benefit of the experiences of parents who've already been through the program to help new parents. We talked about that one day and thought that would be a neat thing to do. I don't know if it would work, but it would be a neat thing. I wouldn't mind doing it, I guess.

Notice that each of these questions elicited a different response. Strengths, weaknesses, likes, dislikes, and recommendations—each question meant something different and deserved to be asked separately. Qualitative interviewing can be deepened through thoughtful, focused, and distinct questions.

A consistent theme runs through this discussion of question formulation: *The wording used in asking questions can make a significant difference in the quality of responses elicited.* The interviewer who throws out a bunch of questions all at once to see which one takes hold puts an unnecessary burden on the interviewee to decipher what is being asked. Moreover, multiple questions asked at the same time suggests that the interviewer hasn't figured out what question should be asked at that juncture in the interview. Taking the easy way out by asking several questions at once transfers the burden of clarity from the interviewer to the interviewee.

Asking several questions at once can also waste precious interview time. In evaluation interviews, for example, both interviewers and respondents typically have only so much time to give to an interview. To make the best use of that time, it is helpful to think through priority questions that will elicit relevant responses. This means that the interviewer must know what issues are important enough to ask questions about and to ask those questions in a way that the person being interviewed can clearly identify what he or she is being asked—that is, to ask clear questions.

Clarity of Questions

> If names are not correct, language will not be in accordance with the truth of things.
>
> —Confucius

The interviewer bears the responsibility to pose questions that make it clear to the interviewee what is being asked. Asking understandable questions facilitates establishing rapport. Unclear questions can make the person being interviewed feel uncomfortable, ignorant, confused, or hostile. Asking singular questions helps a great deal to make things clear. Other factors also contribute to clarity.

First, in preparing for an interview, find out what special terms are commonly used by people in the setting. For example, state and national programs often have different titles and language at the local level. CETA (Comprehensive Employment and Training Act Programs) was designed as a national program in which local contractors were funded to establish and implement services in their area. We found that participants only knew these programs by the name of the local contractor, such as "Youth Employment Services," "Work for Youth," and "Working Opportunities for Women." Many participants in these programs did not even know that they were in CETA programs. Conducting an interview with these participants where they were asked about their "CETA experience" would have been confusing and disruptive to the interview.

When I was doing fieldwork in Burkina Faso, the national government was run by the military after a coup d'etat. Local officials carried the title "commandant" (commander). However, no one referred to the government as a military government. To do so was not only politically incorrect but risky too. The appropriate official phrase mandated by the rulers in the capitol, Ouagadougou, was "the people's government."

Second, clarity can be sharpened by understanding what language participants use among themselves in talking about a setting, activities, or other aspects of life. When we interviewed juveniles who had been placed in foster group homes by juvenile courts, we had to spend a good deal of preparatory time trying to find out how the juveniles typically referred to the group home parents, to their natural parents, to probation officers, and to each other in order to ask questions clearly about each of those sets of people. For example, when asking about relationships with peers, should we use the word *juveniles, adolescents, youth, teenagers,* or what? In preparation for the interviews, we checked with a number of juveniles, group home parents, and court authorities about the proper language to use. We were advised to refer to "the other kids in the group home." However, we found no consensus about how "kids in the group home" referred to group home parents. Thus, one of the questions we had to ask in each interview was "What do you usually call Mr. and Mrs. _____?" We then used the language given to us by that youth throughout the rest of the interview to refer to group home parents.

Third, providing clarity in interview questions may mean avoiding using labels altogether. This means that when asking about a particular phenomenon or program component, it may be better to first find out what the interviewee believes that phenomenon to be and then ask questions about the descriptions provided by the person being interviewed. In studying officially designated "open classrooms" in North Dakota, I interviewed parents who had children in those classrooms. (*Open classrooms* were designed to be more informal, integrated, community based, project oriented, and experiential than traditional classrooms.) However, many of the teachers and local school officials did not use the term *open* to refer to these classrooms because they wanted to avoid political conflicts and stereotypes that were sometimes associated with the notion of "open education." Thus, when interviewing parents, we could not ask their opinions about "open education." Rather, we had to pursue a sequence of questions like the following:

> What kinds of differences, if any, have you noticed between your child's classroom in the past year and the classroom this year? (Parent responds.)
>
> Ok, you've mentioned several differences. Let me ask you your opinion about each of the things you've mentioned. What do you think about _____?

This strategy avoids the problem of collecting responses that later turn out to be uninterpretable because you can't be sure what respondents meant by what they said. Their opinions and judgments are grounded in descriptions, in their own words, of what they've experienced and what they're assessing.

A related problem emerged in interviewing children about their classrooms. We wanted to find out how basic skills were taught in "open" classrooms. In preparing for the interviews, we learned that many teachers avoided terms like *math time* or *reading time* because they wanted to integrate math and reading into other activities. In some cases, we learned during parent interviews, children reported to parents that they didn't do any "math" in school. These same children would be working on projects, such as the construction of a model of their town using milk cartons, that required geometry, fractions, and reductions to scale, but they did not perceive of these activities as "math" because they associated math with worksheets and workbooks. Thus, to find out the kind of math activities children were doing, it was necessary to talk with them in detail about specific projects and work they were engaged in without asking them the simple question, "What kind of math do you do in the classroom?"

Another example of problems in clarity comes from follow-up interviews with mothers whose children were victims of sexual abuse. A major part of the interview focused on experiences with and reactions to the child protection agency, the police, welfare workers, the court system, the school counselor, probation officers, and other parts of the enormously complex system constructed to deal with child sexual abuse. We learned quickly that mothers could seldom differentiate the parts of the system. They didn't know when they were dealing with the courts, the child protection people, the welfare system, or some treatment program. It was all "the system." They had strong feelings and opinions about "the system," so our questions had to remain general, about the system, rather than specifically asking about the separate parts of the system (Patton, 1991).

The theme running through these suggestions for increasing the clarity of questions centers on the importance of using language that is understandable and part of the frame of reference of the person being interviewed. It means taking special care to find out what language the interviewee uses. Questions that use the respondent's own language are most likely to be clear. This means being sensitive to "languaculture" by attending to "meanings that lead the researcher beyond the words into the nature of the speaker's world" (Agar, 2000, pp. 93–94). This sensitivity to local language, the "emic perspective" in anthropology, is usually discussed in relation to data analysis in which a major focus is illuminating a setting or culture through its language. Here, however, we're discussing languaculture not as an analytical framework but as a way of enhancing data collection during interviewing by increasing clarity, communicating respect, and facilitating rapport.

Using words that make sense to the interviewee, words that reflect the respondent's worldview, will improve the quality of data obtained during the interview. Without sensitivity to the impact of particular words on the person being interviewed, an answer may make no sense at all—or there may be no answer. A Sufi story makes this point quite nicely.

> A man had fallen between the rails in a subway station. People were all crowding around trying to get him out before the train ran him over. They were all shouting. "Give me your hand!" but the man would not reach up.
>
> Mulla Nasrudin elbowed his way through the crowd and leaned over the man. "Friend," he asked, "what is your profession?"
>
> "I am an income tax inspector," gasped the man.
>
> "In that case," said Nasrudin, "take my hand!"
>
> The man immediately grasped the Mulla's hand and was hauled to safety. Nasrudin turned to the amazed by-standers. "Never ask a tax man to give you anything, you fools," he said. (Shah, 1973, p. 68)

Before leaving the issue of clarity, let me offer one other suggestion: Be especially careful asking "why" questions.

Why to Take Care Asking "Why?"

> Three Zen masters were discussing a flapping flag on a pole. The first observed dryly: "The flag moves."
>
> "No," said the second. "Wind is moving."
>
> "No," said the third. "It is not flag. It is not wind. It is mind moving."

"Why" questions presuppose that things happen for a reason and that those reasons are knowable. "Why" questions presume cause–effect relationships, an ordered world, and rationality. "Why" questions move beyond what has happened, what one has experienced, how one feels, what one opines, and what one knows to the making of analytical and deductive inferences.

The problems in deducing causal inferences have been thoroughly explored by philosophers of science (Bunge, 1959; Nagel, 1961). On a more practical level and more illuminative of interviewing challenges, reports from parents about "why" conversations with their children document the difficulty of providing causal explanations about the world. The infinite regression quality of "why" questions is part of the difficulty engendered by using them as part of an interview. Consider this parent–child exchange:

Dad, why does it get dark at night?

Because our side of the earth turns away from the sun.

Dad, why does our side of the earth turn away from the sun?

Because that's the way the world was made.

Dad, why was the world made that way?

So that there would be light and dark.

Dad, why should there be dark? Why can't it just be light all the time?

Because then we would get too hot.

Why would we get too hot?

Because the sun would be shining on us all the time.

Why can't the sun be cooler sometimes?

It is, that's why we have night.

But why can't we just have a cooler sun?

Because that's the way the world is.

Why is the world like that?

It just is. Because . . . .

Because why?

Just because.

Oh.

Daddy?

Yes.

Why don't you know why it gets dark?

In a program evaluation interview, it might seem that the context for asking a "why" question would be clearer. However, if a precise reason for a particular activity is what is wanted, it is usually possible to ask that question in a way that does not involve using the word *why*. Let's look first at the difficulty posed for the respondent by the "why" question and then look at some alternative phrases.

"Why did you join this program?" The actual reasons for joining the program probably consist of some constellation of factors, including the influences of other people, the nature of the program, the nature of the person being interviewed, the interviewee's expectations, and practical considerations. It is unlikely that an interviewee can sort through all of these levels of possibility at once, so the person to whom the question is posed must pick out some level at which to respond.

- "Because it takes place at a convenient time." (*programmatic* reason)
- "Because I'm a joiner." (*personality* reason)
- "Because a friend told me about the program." (*information* reason)

- "Because my priest told me about the program and said he thought it would be good for me." (*social influence* reason)
- "Because it was inexpensive." (*economic* reason)
- "Because I wanted to learn about the things they're teaching in the program." (*outcomes* reason)
- "Because God directed me to join the program." (*personal* motivation reason)
- "Because it was there." (*philosophical* reason)

Anyone being interviewed could respond at any or all of these levels. The interviewer must decide before conducting the interview which of these levels carries sufficient importance to make it worth asking a question. If the primary evaluation question concerns characteristics of the program that attracted participants, then instead of asking "Why did you join?" the interviewer should ask something like the following: "What was it about the program that attracted you to it?" If the evaluator is interested in learning about the social influences that led to participation in a program, either voluntary or involuntary participation, a question like the following could be used:

> Other people sometimes influence what we do. What other people, if any, played a role in your joining this program?

In some cases, the evaluator may be particularly interested in the characteristics of participants, so the question might be phrased in the following fashion:

> I'm interested in learning more about you as a person and your personal involvement in this program. What is it about you—your situation, your personality, your desires, whatever—what is it about you that you think led you to become part of this program?

When used as a probe, "why" questions can imply that a person's response was somehow inappropriate. "Why did you do that?" may sound like doubt that an action (or feeling) was justified. A simple "Tell me more, if you will, about your thinking on that" may be more inviting.

The point is that by thinking carefully about what you want to know and being sensitive to what the interviewee will hear in your question, there is a greater likelihood that respondents will supply answers that make sense—and are relevant, usable, and interpretable. My cautions about the difficulties raised with "why" questions come from trying to analyze such questions when responses covered such a multitude of dimensions that it was clear different people were responding to different things. This makes analysis unwieldy.

Perhaps my reservations about the use of "why" questions come from having appeared the fool when asking such questions during interviews with children. In our open classroom interviews, several teachers had mentioned that children often became so involved in what they were doing that they chose not to go outside for recess. We decided to check this out with the children.

"What's your favorite time in school?" I asked a first grader.

"Recess," she answered quickly.

"Why do you like recess?"

"Because we go outside and play on the swings."

"Why do you go outside?" I asked.

"Because that's where the swings are!"

She replied with a look of incredulity that adults could ask such stupid questions, then explained helpfully, "If you want to swing on the swings, you have to go outside where the swings are."

Children take interview questions quite literally, and so it becomes clear quickly when a question is not well thought out. It was during those days of interviewing children in North Dakota that I learned about the problems with "why" questions.

Rapport, Neutrality, and the Interview Relationship

Rapport Through Neutrality

As an interviewer, I want to establish rapport with the person I am interviewing, but that rapport must be established in a way that it does not undermine my neutrality concerning what the person tells me. I must be nonjudgmental. Neutrality means that the person being interviewed can tell me anything without engendering either my favor or disfavor. I cannot be shocked; I cannot be angered; I cannot be embarrassed; I cannot be saddened. Nothing the person tells me will make me think more or less of her or him. Openness and trust flow from nonjudgmental rapport.

At the same time that I am neutral with regard to the content of what is being said to me, I care very much that that person is willing to share with me what he or she is saying. *Rapport is a stance vis-à-vis the person being interviewed. Neutrality is a stance vis-à-vis the content of what that person says.* Rapport means that I respect the people being interviewed, so what they say is important because of who is saying it. I want to convey to them that their knowledge, experiences, attitudes, and feelings are important. Yet I will not judge them for the content of what they say to me.

EMPATHIC NEUTRALITY

Empathic neutrality is one of the 12 core strategies of qualitative inquiry discussed in Chapter 2. (See Exhibit 2.1 and pages 46–47.)

Naturalistic inquiry involves fieldwork that puts one in close contact with people and their problems. What is to be the researcher's cognitive and emotional stance toward those people and problems? No universal prescription can capture the range of possibilities, for the answer will depend on the situation, the nature of the inquiry, and the perspective of the researcher. But thinking strategically, I offer the phrase "empathic neutrality" as a point of departure. It connotes a middle ground between becoming too involved, which can cloud judgment, and remaining too distant, which can reduce understanding. What is *empathic neutrality*? In essence, it is understanding a person's situation and perspective without judging the person—and communicating that understanding with authenticity to build rapport, trust, and openness.

Reflections on Empathic Neutrality From an Experienced International Program Evaluator

For me there is a big difference between being empathetic when you interview (like when I work with full-blown AIDS patients, abused women, or orphans—seriously how can you not be touched?) and taking on evaluations because you are empathetic (as I often do) and therefore care about getting good empirical data. Let me explain.

Being empathetic, and really caring, should make an evaluator do an even better EMPIRICAL job. For me, when I am involved emotionally (which I usually am), I need to try even harder because I realize that it is UP TO ME (I take that all on. . . .) to make sure that when the critics come around, I have ensured that the program has a strong Theory of Change and Theory of Action and data to evaluate the program, and, data to support making changes in order to make it a stronger program. That holds true for impact evaluations as well. If I don't find a program effective then I also think how that money could be spent on a MORE EFFECTIVE program that WILL change people's lives. It's not that I want the program to be successful in itself (thus my neutrality), I want people to have better lives (the basis for empathy). If I was neutral about influencing people's lives, I wouldn't try so hard. I care, so I work very hard to do a good job.

I am sometimes told, "But you are [an] inside evaluator so you are biased." AGHH! I am an inside evaluator so I know all the dirty laundry, which makes me better able to identify critical points of data to inform improvements. If anything being inside enables me to [be] MORE critical, because you can't hide anything from me.

—Donna Podems, PhD
Director, Otherwise Research and Evaluation
Cape Town, South Africa

Rapport is built by conveying empathy and understanding without judgment. In this section, I want to focus on ways of wording questions that are particularly aimed at conveying that important sense of neutrality.

Using Illustrative Examples in Questions

One kind of question wording that can help establish neutrality is *the illustrative examples format.* When phrasing questions in this way, I want to let the person I'm interviewing know that I've pretty much heard it all—the bad things and the good—and so I'm not interested in something that is particularly sensational, particularly negative, or especially positive. I'm really only interested in what that person's genuine experience has been. I want to elicit open and honest judgments from them without making them worry about my judging what they say. Consider this example of the illustrative examples format from interviews we conducted with juvenile delinquents who had been placed in foster group homes. One section of the interview was aimed at finding out how the juveniles were treated by group home parents.

> Okay, now I'd like to ask you to tell me how you were treated in the group home by the parents. Some kids have told us they were treated like one of the family; some kids have told us that they got knocked around and beat up by the group home parents; some kids have told us about sexual things that were done to them; some of the kids have told us about fun things and trips they did with the parents; some kids have felt they were treated really well and some have said they were treated pretty bad. What about you—how have you been treated in the group home?

A closely related approach is the *illustrative extremes format*—giving examples only of extreme responses. This question is from a follow-up study of award recipients who received a substantial fellowship with no strings attached.

> How much of the award, if any, did you spend on entirely personal things to treat yourself well? Some fellows have told us they spent a sizable portion of the award on things like a new car, a hot tub, fixing up their house, personal trips, and family. Others spent almost nothing on themselves and put it all into their work. How about you?

In both the illustrative examples format and the illustrative extremes format, it is critical to avoid asking leading questions. Leading questions are the opposite of neutral questions; they give the interviewee hints about what would be a desirable or appropriate kind of answer. Leading questions "lead" the respondent in a certain direction. Below are questions I found on transcripts during review of an evaluation project carried out by a reputable university center.

> "We've been hearing a lot of really positive comments about this program. So what's your assessment?"

or

> "We've already heard that this place has lots of troubles, so feel free to tell us about the troubles you've seen."

or

> "I imagine it must be horrible to have a child abused and have to deal with the system, so you can be honest with me. How bad was it?"

Each of these questions builds in a response bias that communicates the interviewer's belief about the situation prior to hearing the respondent's assessment. The questions are "leading" in the sense that the interviewee can be led into acquiescence with the interviewer's point of view.

In contrast, the questions offered below to demonstrate the *illustrative examples format* included several dimensions to provide a balance between what might be construed as positive and negative kinds of responses. I prefer to use the illustrative example format primarily as a clarifying strategy after having begun with a simple, straightforward, and truly open-ended question: "What do you think about this program?" or "What has been your experience with the system?" Only if this initial question fails to elicit a thoughtful response, or if the interviewee seems to be struggling, will I offer illustrative examples to facilitate a deeper response.

Role-Playing and Simulation Questions

Providing context for a series of questions can help the interviewee hone in on relevant responses. A helpful context provides cues about the level at which a response is expected. One way of providing such a context is to role-play with persons being interviewed, asking them to respond to the interviewer as if he or she were someone else.

> *Suppose I was a new person who just came into this program, and I asked you what I should do to succeed here, what would you tell me?*

or

> *Suppose I was a new kid in this group home, and I didn't know anything about what goes on around here, what*

would you tell me about the rules that I have to be sure to follow?

These questions provide a context for what would otherwise be quite difficult questions, for example, "How does one get the most out of this program?" or "What are the rules of this group home?" The role-playing format emphasizes the interviewees' expertise—that is, it puts him or her in the role of expert because he or she knows something of value to someone else. The interviewee is *the insider with inside information.* The interviewer, in contrast, as an outsider, takes on the role of novice or apprentice. The "expert" is being asked to share his or her expertise with the novice. I've often observed interviewees become more animated and engaged when asked role-playing questions. They get into the role.

A variation on the role-playing format involves the interviewer dissociating somewhat from the question to make it feel less personal and probing. Consider these two difficult questions for a study of a tough subject: teenage suicide.

"What advice would you give someone your age who was contemplating suicide?"

versus

"Think of someone you know and like who is moody. Suppose that person told you that he or she was contemplating suicide. What would you tell that person?"

The first question comes across as abrupt and demanding, almost like an examination to see if he or she knows the right answer. The second question, with the interviewee allowed to create a personal context, is softened and, hopefully, more inviting. While this technique can be overused and can sound phony if asked insensitively, with the right intonation communicating genuine interest and used sparingly, with subtlety, the role-playing format can ease the asking of difficult questions to deepen answers and enhance the quality of responses.

Simulation questions provide context in a different way, by asking the person being interviewed to imagine himself or herself in the situation in which the interviewer is interested.

Suppose I was present with you during a staff meeting. What would I see going on? Take me there.

or

Suppose I was in your classroom at the beginning of the day when the students first come in. What would

I see happening as the students came in? Take me to your classroom, and let me see what happens during the first 10 to 15 minutes as the students arrive—what you'd be doing, what they'd be doing, what those first 15 minutes are like.

In effect, these questions ask the interviewee to become an observer. In most cases, a response to this question will require the interviewee to visualize the situation to be described. I frequently find that the richest and most detailed descriptions come from a series of questions that ask a respondent to reexperience and/or simulate some aspect of an experience.

Presupposition Questions

Presupposition questions involve a twist on the theme of empathic neutrality. Presuppositions have been identified by linguists as a grammatical structure that creates rapport by assuming shared knowledge and assumptions (Bandler & Grinder, 1975; Kartunnen, 1973). Natural language is filled with presuppositions. In the course of our day-to-day communications, we often employ presuppositions without knowing we're doing so. By becoming aware of the effects of presupposition questions, we can use them strategically in interviewing. The skillful interviewer uses presuppositions to increase the richness and depth of responses.

What then are presuppositions? Linguists Bandler & Grinder (1975) define presuppositions as follows:

When each of us uses a natural language system to communicate, we assume that the listener can decode complex sound structures into meanings, i.e., the listener has the ability to derive the Deep-Structure meaning from the Surface-Structure we present to him auditorily. . . . We also assume the complex skill of listeners to derive extra meaning from some Surface-Structures by the nature of their form. Even though neither the speaker nor the listener may be aware of this process, it goes on all the time. For example, if someone says: I want to watch Kung Fu tonight on TV we must understand that Kung Fu is on TV tonight in order to process the sentence "I want to watch . . ." to make any sense. These processes are called presuppositions of natural language. (p. 241)

Used in interviewing, presuppositions communicate that the respondent has something to say, thereby increasing the likelihood that the person being interviewed will, indeed, have something to say. Consider the following question: "What is the most important experience you have had in the program?" This question presupposes that the respondent has had an

EXHIBIT 7.10   Illustrative Dichotomous Versus Presupposition Questions

Listed below, on the left, are typical dichotomous questions used to introduce a longer series of questions. On the right are presupposition questions that bypass the dichotomous lead-in query and, in some cases, show how adding "if any" retains a neutral framing.

DICHOTOMOUS LEAD-IN QUESTION	PRESUPPOSITION LEAD-IN QUESTION
1. Do you feel you know enough about the program to assess its effectiveness?	1. How effective do you think the program is? (Presupposes that a judgment can be made)
2. Have you learned anything from this program?	2. What, if anything, have you learned from this program? (Presupposes that some learning is likely)
3. Do you do anything now in your work that you didn't do before the program began?	3. What, if anything, do you do now that you didn't do before the program began? (Presupposes change)
4. Are there any conflicts among the staff?	4. What kinds of staff conflicts have you observed here? (Presupposes conflicts)

important experience. Of course, the response may be "I haven't had any important experiences." However, it is more likely that the interviewee will internally access an experience to report as important rather than dealing first with the question of whether or not an important experience has occurred.

Contrast the presupposition format—"What is the most important experience you have had in the program?"—to the following dichotomous question: "Have you had any experiences in the program that you would call really important?" This dichotomous framing of the question requires the person to make a decision about what an important experience is and whether one has occurred. The presupposition format bypasses this initial step by asking directly for description rather than asking for an affirmation of the existence of the phenomenon in question. Exhibit 7.10 contrasts typical dichotomous response questions with presuppositions that bypass the dichotomous lead-in query and, in some cases, show how adding "if any" retains a neutral framing. Compare the two question formats, and think about how you would likely respond to each.

A naturalness of inquiry flows from presuppositions making more comfortable what might otherwise be embarrassing or intrusive questions. The presupposition includes the implication that what is presupposed is the natural way things occur. It is natural for there to be conflict in programs. The presupposition provides a stimulus that asks the respondent to mentally access the answer to the question directly without making a decision about whether or not something has actually occurred.

I first learned about interview presuppositions from a friend who worked with the agency in New York City that had responsibility for interviewing carriers of venereal disease. His job was to find out about the carrier's previous sexual contracts so that those persons could be informed that they might have venereal disease. He had learned to avoid asking men "Have you had any sexual relationships with other men?" Instead, he asked, "How many sexual contacts with other men have you had?" The dichotomous question carried the burden for the respondent of making a decision about some admission of homosexuality and/or promiscuity. The presupposition form of the open-ended question implied that some sexual contacts with other men might be quite natural and focused on the frequency of occurrence rather than on whether or not such sexual contacts have occurred at all. The venereal disease interviewers found that they were much more likely to generate open responses with the presupposition format than with the dichotomous response format.

The purpose of in-depth interviews is to find out what someone has to say. By presupposing that the person being interviewed does, indeed, has something to say, the quality of the descriptions received may be enhanced. However, a note of warning: Presuppositions, like any single form of questioning, can be overused. Presuppositions are *one* option. There are many times when it is more comfortable and appropriate to check out the relevance of a question with a dichotomous inquiry (Did you go to the lecture?) before asking further questions (What did you think of the lecture?)

EXHIBIT 7.11	Summary of Question Formats to Facilitate Communicating Interviewer Neutrality

TYPE OF QUESTION	EXAMPLES FROM AN EVALUATION INTERVIEW FOR A PROFESSIONAL LEADERSHIP DEVELOPMENT PROGRAM
1. Illustrative examples format	Now, I'd like to ask you to tell me how you were treated by staff in the program. Some participants have told us that they were treated with respect; some have said that the staff was arrogant and condescending. Some have said staff played favorites; others have said that staff gave the same attention to everyone. Some participants have reported forming close relationships with certain staff; others have said that they didn't get close to any staff. So we've heard lots of diversity in how participants felt treated by staff. What about you—what was your experience with the program staff?
2. Illustrative extremes format	Next, we'd like to ask you about the impact of participating in the program. Some participants have described the experience as transformative; others have said it didn't have much impact on them. What was the impact on you?
3. Role-playing question	You've now completed this program and know a lot about it. If I was a new participant just coming into this program, what advice would you give me about how to get the most out of it?
4. Simulation question	You said the group trust-building exercise gave you important new insights into yourself. Take me into that exercise. Take me through what happened from the beginning to the end.
5. Presupposition questions	What significant relationships did you form with other participants, if any? (Presumes that some significant relationships were probably formed)

Summary of Skillful Questioning to Communicate Neutrality

We've reviewed five question formats to communicate neutrality and help establish rapport. Exhibit 7.11 reviews and summarizes these question types: (1) illustrative examples format, (2) illustrative extremes format, (3) role-playing questions, (4) simulation questions, and (5) presupposition questions. But as important as skillful questioning is, and I believe it's quite important, it is even more important to be attentive to and mindful about the overall patterns of interaction that emerge during an interview and ultimately determine the kind of relationship that gets built between interviewer and interviewee. Rapport and empathy reside in that relationship.

Beyond Skillful Questioning: Relationship-Focused, Interactive Interview Approaches

The interview is a specific form of conversation where knowledge is produced through the interaction between an interviewer and an interviewee. . . .

If you want to know how people understand their world and their lives, why not talk with them? Conversation is a basic mode of human interaction. Human beings talk with each other, they interact, pose questions and answer questions. Through conversations we get to know other people, get to learn about their experiences, feelings and hopes and the world they live in. In an interview conversation, the researcher asks about, and listens to, what people themselves tell about their lived world, about their dreams, fears and hopes, hears their views and opinions in their own words, and learns about their school and work situation—their family and social life. The research interview is an interview where knowledge is constructed in the inter-action between the interviewer and the interviewee. (Kvale, 2007, pp. xvii, 1)

As we discussed earlier in this chapter, some approaches to interviewing focus on standardizing the interview protocol so that each person interviewed is asked the same questions in the same way. That standardization is considered the foundation of validity and reliability in traditional social science interviewing (see Exhibit 7.1, pp. 423–424). In contrast, other qualitative methodologists emphasize that the depth, honesty, and quality of responses in an interview depend on the *relationship* that develops between the interviewee and the interviewer (Josselson, 2013). Exhibit 7.12 presents six relationship-focused, interactive interview approaches aimed at establishing rapport, empathy, mutual respect, and mutual trust. It is important not to get so caught up in trying to word questions perfectly that you miss the dynamics of the unfolding relationship that is at the heart of interactive interviewing.

EXHIBIT 7.12   **Six Relationship-Focused, Interactive Interview Approaches**

Below are six relationship-focused, interactive interview approaches aimed at establishing rapport, empathy, mutual respect, and mutual trust.

INTERVIEW APPROACH	INTERVIEW EMPHASIS	ELABORATION OF THE APPROACH AND RESOURCES
1. *The interview conversation* (Kvale, 1996; Kvale & Brinkmann, 2008)	"An interview is literally an *inter view*, an inter change of views between two persons conversing about a theme of mutual interest" (Kvale, 1996, p. 2).	"In an interview conversation, the researcher listens to what people themselves tell about their lived world, hears them express their views and opinions in their own words, learns about their views on their work situation and family life, their dreams and hopes" (Kvale, 1996, p. 1). "Conversation is a basic mode of human interaction. Human beings talk with each other—they interact, pose questions, and answer questions. Through conversations we get to know other people, get to learn about their experiences, feelings, and hopes and the world they live in.... The research interview is based on the conversations of daily life and is a professional conversation" (Kvale, 1996, p. 5).
2. *Responsive interviewing* (Rubin & Rubin, 2012)	Responsive interviewing "emphasizes flexibility of design and expects the interviewer to change questions in response to what he or she is learning" (Rubin & Rubin, 2012, p. 7).	"Responsive interviewing accepts and adjusts to the personalities of both conversational partners.... Responsive interviewing brings out new information, often of startling candor, and often suggests unanticipated interpretations. The freshness and depth of the interviews makes them exciting to do and, later on, to read.... Responsive interviewing is generally gentle and cooperative, feels respectful, and is ethical" (Rubin & Rubin, 2012, p. 7).
3. *The active interview* (Holstein & Gubrium,1995, 2003)	An interview is a social interaction with the interviewer and the interviewee sharing in constructing a story and its meanings; both are participants in the meaning-making process. The interviewer facilitates the interviewees in "subjectively creating" their story (Holstein & Gubrium, 1995, p. 8).	"Construed as active, the subject behind the respondent not only holds facts and details of experience but, in the very process of offering them up for response, constructively adds to, takes away from, and transforms the facts and details.... This activated subject pieces experiences together, before, during, and after occupying the respondent role. As a member of society, he or she mediates and alters the knowledge that the respondent conveys to the interviewer; he or she is 'always already' an active maker of meaning" (Holstein & Gubrium, 1995, pp. 8–9).
4. *Creative interviewing* (Douglas, 1985)	Situationally adaptive interviewing that varies interview questions and interviewing processes to	"*Creative interviewing is purposefully situated interviewing.* Rather than denying or failing to see the situation of the interview as a determinant of what goes on in the questioning and answering processes, creative interviewing

INTERVIEW APPROACH	INTERVIEW EMPHASIS	ELABORATION OF THE APPROACH AND RESOURCES
	fit the particular situation and context in which the interview interaction occurs.	embraces the immediate, concrete situation; tries to understand how it is affecting what is communicated; and, by understanding these effects, changes the interviewer's communication processes to increase the discovery of the truth about human beings" (Douglas, 1985, p. 22).
5. *Reflective interviewing* (Roulston, 2010)	Integrates the theoretical conception of the interview, the researcher's relationship to the inquiry and participants, and the methodological review of the interview interaction to inform design (Roulston, 2010, p. 1).	*"The reflective interviewer understands*[italics added]: research subjectivities; subject positions occupied by the researcher in relation to research participants . . . ; theoretical perspectives and assumptions that relate to interviewing . . . ; and how to analyze interview interaction to inform data analysis and interview practice" (Roulston, 2010, p. 4). "Qualitative interviewers and social researchers learn by doing, and reflection on doing" (Roulston, 2010, p. 6).
6. *Portraiture interviewing* (Lawrence-Lightfoot & Davis, 1997)	Portraiture is a negotiated co-creation between the social scientist and the person being depicted. "The relationship between the two is rich with meaning and resonance and becomes the arena for navigating the empirical, aesthetic, and ethical dimensions of authentic and compelling narrative" (Lawrence-Lightfoot & Davis, 1997, p. xv).	Portraiture interviewing "blurs the boundaries of aesthetics and empiricism in an effort to capture the complexity, dynamics, and subtlety of human experience and organizational life. Portraitists seek to record and interpret the perspectives and experience of the people they are studying, documenting their voices and their visions—their authority, knowledge, and wisdom. The drawing of the portrait is placed in social and cultural context and shaped through dialogue between the portraitist and the subject, each one negotiating the discourse and shaping the evolving image" (Lawrence-Lightfoot & Davis, 1997, p. xv).

Pacing and Transitions in Interviewing

The longer an interview, the more important it is to be aware of pacing and transitions. The popular admonition to "go with the flow" alerts us to the fact that flow varies. Sometimes there are long stretches of calm water in a river trip; then around a corner, there are rapids. A river narrows and widens as it flows. To "go with the flow," you must be aware of the flow. Interviews have their own flow. Here are some ways to manage the flow.

Prefatory Statements and Announcements

The interaction between interviewer and interviewee is greatly facilitated when the interviewer alerts the interviewee to what is about to be asked before it is actually asked. Think of it as warming up the respondent, or ringing the interviewee's mental doorbell. This is done with prefatory statements that introduce a question. These can serve two functions. First, a preface alerts the interviewees to the nature of the question that is coming, directs their awareness, and focuses their attention. Second, an introduction to a question gives respondents a few seconds to organize their thoughts before responding. Prefaces, transition announcements, and introductory statements help smooth the flow of the interview. Any of several formats can be used.

The *transition format* announces that one section or topic of the interview has been completed and a new section or topic is about to begin.

We've been talking about the goals and objectives of the program. Now I'd like to ask you some questions about actual program activities. What are the major activities offered to participants in this program?

or

We've been talking about your childhood family experiences. Now I'd like to ask you about your memories of and experiences in school.

The transition format essentially says to the interviewee, "This is where we've been . . . , and this is where we're going." Questions prefaced by transition statements help maintain the smooth flow of an interview.

An alternative format is the *summarizing transition*. This involves bringing closure to a section of the interview by summarizing what has been said and asking if the interviewee has anything to add or clarify before moving on to a new subject.

Before we move on to the next set of questions, let me make sure I've got everything you said about the program goals and objectives. You said the program had five goals. First, . . . Second, . . .

Before I ask you some questions about program activities related to these goals, are there any additional goals or objectives that you want to add?

The *summarizing transition* lets the person being interviewed know that the interviewer is actively listening to and recording what is being said. The summary invites the interviewee to make clarifications, corrections, or additions before moving on to a new topic.

The *direct announcement format* simply states what will be asked next. A preface to a question that announces its content can soften the harshness or abruptness of the question itself. Direct prefatory statements help make an interview more conversational and easily flowing. The transcriptions below show two interview sequences, one without prefatory statements and the other with prefatory statements.

- *Question without preface:* How have you changed as a result of the program?
- *Question with preface:* Now, let me ask you to think about any changes you see in yourself as a result of participating in this program. (Pause). How, if at all, have you been changed by your experiences in this program?

The *attention-getting preface* goes beyond just announcing the next question to making a comment about the question. The comment may concern the importance of the question, the difficulty of the question, the openness of the question, or any other characteristic of the question that would help set the stage. Consider these examples:

- This next question is particularly important to the program staff. How do you think the program could be improved?

CONTEXTUALIZING A QUESTION MAKES A DIFFERENCE

How you introduce a question can have a significant impact on how interviewees respond. Kim Manturuk, Senior Research Associate at the University of North Carolina Center for Community Capital, interviewed low-income people about whether they had enough food to eat. Manturuk (2013) found that asking the question straight out could be offensive.

We added an introduction along the lines of "In these difficult economic times, more people are finding it difficult to always have the kinds of foods they like." . . . We found that, by setting it up so that the cause of food insecurity was the economy (and not an individual failing), people seemed more comfortable answering. (p. 1)

- This next question is purposefully vague so that you can respond in any way that makes sense to you. What difference has this program made to the larger community?
- This next question may be particularly difficult to answer with certainty, but I'd like to get your thoughts on it. In thinking about how you've changed during the past year, how much has this program caused those changes compared with other influences on your life at this time?
- This next question is aimed directly at getting your unique perspective. What's it like to be a participant in this program?
- As you may know, this next issue has been both controversial and worrisome. What kinds of staff are needed to run a program like this?

The common element in each of these examples is that some prefatory comment is made about the question to alert the interviewee to the nature of the question. The attention-getting format communicates that the question about to be asked has some unique quality that makes it particularly worthy of being answered.

Making statements about the questions being asked is a way for the interviewer to engage in some conversation during the interview without commenting judgmentally on the answers being provided by the interviewee. What is said concerns the questions and not the respondent's answers. In this fashion, the interview can be made more interesting and interactive. However, all of these formats must be used selectively and strategically. Constant repetition of the same format or mechanical use of a particular approach will make the interview more awkward rather than less so. Exhibit 7.13 summarizes the pacing and transition formats we've just discussed.

EXHIBIT 7.13 Summary of Pacing and Transition Formats

PACING AND TRANSITION FORMATS	EXAMPLES FOR A WORKPLACE CULTURE INTERVIEW
1. Straight transition statements	We've been talking about what it was like when you first started working here. Now, I'd like to ask about the people you work with most closely.
2. Summarizing transition	So you've told me about how senior management roles are different from middle management roles. Your descriptions of the differences are very helpful. Before we move on to how work teams operate, is there anything that you want to add to what you've said about management roles?
3. Preparatory preface	Now, let me ask you to think about the training you've received since you came to work here. (Pause). Let's start by listing to the different kinds of training you've received; then, I'd like to ask you to tell me about each one.
4. Attention-getting preface	This next set of questions is very important to understanding what it's like to work here. I'm going to ask you to tell me about conflicts among people. I want to reiterate that your answers are confidential and will not be identified as coming from you. We find that all organizations have at least some conflicts among people. So let me ask you to think about an example of a conflict you've experienced and tell me about it.

Probes and Follow-Up Questions

Probes are used to deepen the response to a question, increase the richness and depth of responses, and give cues to the interviewee about the level of response that is desired. The word *probe* is usually best avoided in interviews—a little too proctological. The expression "Let me probe that further" can sound as if you're conducting an investigation of something illicit or illegal. Quite simply, a probe is a follow-up question used to go deeper into the interviewee's responses. As such, probes should be conversational, offered in a natural style and voice, and used to follow up initial responses.

One natural set of conversational probes consists of *detail-oriented questions*. These are the basic questions that fill in the blank spaces of a response.

When did that happen?

Who else was involved?

Where were you during that time?

What was your involvement in that situation?

How did that come about?

Where did that happen?

These *detail-oriented probes* are the basic "who," "where," "what," "when," and "how" questions that are used to obtain a complete and detailed picture of some activity or experience.

At other times, an interviewer may want to keep a respondent talking about a subject by using *elaboration probes*. The best cue to encourage continued talking is nonverbal—gently nodding your head as positive reinforcement. However, overenthusiastic head nodding may be perceived as endorsement of the content of a response or as wanting the person to stop talking because the interviewer has already understood what the respondent has to say. Gentle and strategic head nodding is aimed at communicating that you are listening and want to go on listening.

The verbal corollary of head nodding is the quiet "uh-huh." A combination may be necessary; when the respondent seems about to stop talking and the interviewer would like to encourage more comment, a combined "uh-huh" with a gentle rocking of the whole upper body can communicate interest in having the interviewee elaborate.

Elaboration probes also have direct verbal forms:

- Would you elaborate on that?
- That's helpful. I'd appreciate a bit more detail.
- I'm beginning to get the picture. (The implication is that I don't have the full picture yet, so please keep talking.)

If something has been said that is ambiguous or an apparent non sequitur, a *clarification probe* may be useful. Clarification probes tell the interviewee that you need more information, a restatement of the answer, or more context.

> You said the program is a "success." What do you mean by "success"? I'm not sure I understand what you meant by that. Would you elaborate, please?

> I want to make sure I understand what you're saying. I think it would help me if you could say some more about that.

A *clarification probe* should be used naturally and gently. It is best for the interviewer to convey the notion that the failure to understand is the fault of the interviewer and not a failure of the person being interviewed. The interviewer does not want to make the respondent feel inarticulate, stupid, or muddled. After one or two attempts at achieving clarification, it is usually best to leave the topic that is causing confusion and move on to other questions, perhaps returning to that topic at a later point.

Another kind of clarifying follow-up question is the *contrast probe* (McCracken, 1988, p. 35). The purpose of a contrast probe is to "give respondents something to push off against" by asking, "How does *x* compare with *y*?" This is used to help define the boundaries of a response. How does this experience/feeling/action/term compare with some other experience/feeling/action/term?

A major characteristic that separates probes from general interview questions is that probes are seldom written out in an interview. *Probing is a skill that comes from knowing what to look for in the interview, listening carefully to what is said and what is not said, and being sensitive to the feedback needs of the person being interviewed.* Probes are always a combination of verbal and nonverbal cues. Silence at the end of a response can indicate as effectively as anything else that the interviewer would like the person to continue. Probes are used to communicate what the interviewer wants. More detail? Elaboration? Clarity?

ENTERING INTO THE WORLDS OF OTHERS THROUGH INTERVIEWING IN THE FIELD

In *Far From the Tree*, Andrew Solomon (2012) reports his inquiry into family experiences of deafness, dwarfism, autism, schizophrenia, disability, prodigies, transgender, crime, and children born of rape. Over 10 years, he interviewed more than 300 families. He reflects,

> I had to learn a great deal to be able to hear these men and women and children. On my first day at my first dwarf convention, I went over to help an adolescent girl who was sobbing. "This is what I look like," she blurted between gasps, and it seemed she was half laughing. "These people look like me." Her mother, who was standing nearby, said, "You don't know what this means to my daughter. But it also means a *lot* to me, to meet these other parents who will know what I'm talking about." She assumed I, too, must be a parent of a child with dwarfism; when she learned that I was not, she chuckled, "For a few days, now, you can be the freakish one" (p. 41).

Follow-Up Questions: Listening for Markers

Follow-up questions pick up on cues offered by the interviewee. While probes are used to get at information the interviewer knows is important, follow-up questions are more exploratory. The interviewee says something almost in passing, maybe as an afterthought or a side comment, and you note that there may be something there worth following up on. Weiss (1994) calls these "markers."

> I define a marker as a passing reference made by a respondent to an important event or feeling state. . . . Because markers occur in the course of talking about something else, you may have to remember them and then return to them when you can, saying, "A few minutes ago you mentioned . . . ." But it is a good idea to pick up a marker as soon as you conveniently can if the material it hints at could in any way be relevant for your study. Letting the marker go will demonstrate to the respondent that the area is not of importance for you. It can also demonstrate that you are only interested in answers to your questions, not in the respondent's full experience. (p. 77)

In interviewing participants in the wilderness-based leadership program, one section asked about any concerns leading up to participation. In the midst of talking about having diligently followed the suggested training routine (hiking at least a half-hour every day leading up to the program), a participant mentioned that she knew a friend whose boyfriend had died in a climbing accident on a wilderness trip; then, she went on talking about her own preparation. When she had finished describing how she had prepared for the program, I returned to her comment about the friend's boyfriend. Quite unexpectedly, pursuing this casual offhand comment (marker) opened up strong feelings, not about possible injuries or death but about the possibility of forming romantic relationships on the 10-day wilderness experiences. It turned out to mark a decisive shift in the interview into a series of important and revealing interpersonal issues that provided a crucial context for understanding her subsequent program experiences.

Probes and follow-up questions, then, provide guidance to the person interviewed. They also provide the interviewer with a way to facilitate the flow of the interview, a subject we now go to.

Process Feedback During the Interview

Previous sections have emphasized the importance of thoughtful wording so that interview questions are clear, understandable, and answerable. Skillful question formulation and probing concern the content of the interview. This section emphasizes feedback about how the interview *process* is going.

A good interview feels like a connection has been established in which communication is flowing two ways. Qualitative interviews differ from interrogations or detective-style investigations. The qualitative interviewer has a responsibility to communicate clearly what information is desired and why that information is important, and to let the interviewee know how the interview is progressing.

An interview is an interaction in which an interchange occurs and a temporary interdependence is created. The interviewer provides stimuli to generate a reaction. That reaction from the interviewee, however, is also a stimulus to which the interviewer responds. You, as the interviewer, must maintain awareness of how the interview is flowing, how the interviewee is reacting to questions, and what kinds of feedback are appropriate and helpful to maintain the flow of communication.

Support and Recognition Responses

A common mistake among novices is failing to provide *reinforcement and feedback*. This means letting the interviewee know from time to time that the purpose of the interview is being fulfilled. Words of thanks,

support, and even praise will help make the interviewee feel that the interview process is worthwhile and support ongoing rapport. Here are some examples:

- Your comments about program weaknesses are particularly helpful, I think, because identification of the kind of weaknesses you describe can really help in considering possible changes in the program.
- It's really helpful to get such a clear statement of what it feels like to be an "outsider" as you've called yourself. Your reflections are just the kind of thing we're trying to get at.
- We're about halfway through the interview now, and from my point of view, it's going very well. You've been telling me some really important things. How's it going for you?
- I really appreciate your willingness to express your feelings about that. You're helping me understand— and that's exactly why I wanted to interview you.

You can get clues about what kind of reinforcement is appropriate by watching the interviewee. When verbal and nonverbal behaviors indicate that someone is really struggling with a question, going mentally deep within, working hard trying to form an answer, after the response it can be helpful to say something like the following: "I know that was a difficult question and I really appreciate your working with it because what you said was very meaningful and came out very clearly. Thank you."

At other times, you may perceive that only a surface or shallow answer has been provided. It may then be appropriate to say something like the following:

> I don't want to let that question go by without asking you to think about it just a little bit more, because I feel you've really given some important detail and insights on the other questions and I'd like to get more of your reflections about this question.

In essence, the interviewer, through feedback, is "training" the interviewee to provide high-quality and relevant responses.

Maintaining Control and Enhancing the Quality of Responses

Time is precious in an interview. Long-winded responses, irrelevant remarks, and digressions reduce the amount of time available to focus on critical questions. These problems exacerbate when the interviewer fails to maintain a reasonable degree of control over the process. Control is facilitated by (a) knowing what you want to find out, (b) asking focused questions to get relevant answers, (c) listening attentively

CREATIVE BUMBLING

If doing nothing can produce a useful reaction, so can the appearance of being dumb. You can develop a distinct advantage by waxing slow of wit. Evidently, you need help. Who is there to help you but the person who is answering your questions? The result is the opposite of the total shutdown that might have occurred if you had come on glib and omniscient. If you don't seem to get something, the subject will probably help you get it. If you are listening to speech and at the same time envisioning it in print, you can ask your question again, and again, until the repeated reply will be clear in print. Who is going to care if you seem dumber than a cardboard box? Reporters call that *creative bumbling*.

—John McPhee (2014, p. 50)
Journalist

describe what the interviewee thinks *ought* to happen (opinions). Since the interviewer wants behavioral data, it is necessary to first recognize that the responses are not providing the kind of data desired, and then to ask appropriate follow-up questions that will lead to behavioral responses, something like this:

Interviewer: Okay, you try to establish contact with each person and generate enthusiasm at the beginning. What would help me now is to have you actually take me into a training session. Describe for me what the room looks like, where the trainees are, where you are, and tell me what I would see and hear if I were right there in that session. What would I see you doing? What would I hear you saying? What would I see the trainees doing? What would I hear the trainees saying? *Take me into a session so that I can actually experience it, if you will.*

It is the interviewer's responsibility to work with the person being interviewed to facilitate the desired kind of responses. At times, it may be necessary to give very direct feedback about the difference between the response given and what is desired.

Interviewer: I understand what you try to do during a training session—what you hope to accomplish and stimulate. Now I'd like you to describe to me what you actually do, not what you expect, but what I would actually see happening if I were present at the session.

It's not enough to simply ask a well-formed and carefully focused initial question. Neither is it enough to have a well-planned interview with appropriate basic questions. The interviewer must listen actively and carefully to responses to make sure that the interview is working. I've seen many well-written interviews that have resulted in largely useless data because the interviewer did not listen carefully and thus did not recognize that the responses were not providing the information needed. The first responsibility, then, in facilitating the interview interaction is knowing what kind of data you are looking for and managing the interview so as to get quality responses.

Verbal and Nonverbal Feedback

Giving appropriate feedback to the interviewee is essential in pacing an interview and maintaining control of

to assess the quality and relevance of responses, and (d) giving appropriate verbal and nonverbal feedback to the person being interviewed.

Knowing what you want to find out means being able to recognize and distinguish relevant from irrelevant responses. It is not enough just to ask the right questions. You, the interviewer, must listen carefully to make sure that the responses you receive provide meaningful answers to the questions you ask. Consider the following exchange:

Q: What happens in a typical interviewer training session that you lead?

A: I try to be sensitive to where each person is at with interviewing. I try to make sure that I am able to touch base with each person so that I can find out how she or he is responding to the training, to get some notion of how each person is doing.

Q: How do you begin a session, a training session?

A: I believe it's important to begin with enthusiasm, to generate some excitement about interviewing.

Earlier in this chapter, I discussed the importance of distinguishing different kinds of questions and answers: behavior, opinion, knowledge, feelings (see Exhibit 7.7, p. 445). In the interaction above, the interviewer is asking descriptive, behavioral questions. The responses, however, are about beliefs and hopes. The answers do not actually describe what happened. Rather, they

SKILLFUL AND EFFECTIVE INTERVIEWING

SIDEBAR

Anyone who uses interviewing in their evaluation work—and, let's face it, most of us do—learns over time the importance of following a fruitful line of questioning:

Focused but not narrow; flexible but not directionless.

We need to make sure that our interview (1) goes after what's most important; (2) doesn't go off on tangents; (3) keeps the interviewee engaged; and (4) doesn't get prematurely evaluative by throwing in conclusions while still gathering the evidence.

—E. Jane Davidson and Patricia Rogers
Genuine Evaluation Blog (January 4, 2013)

the interview process. Head nodding, taking notes, "uh-huhs," and silent probes (remaining quiet when a person stops talking to let him or her know that you're waiting for more) are all signals to the person being interviewed that the responses are on the right track. These techniques encourage greater depth in responses, but you also need skill and techniques to stop a highly verbal respondent who gets off the track. The first step in stopping the long-winded respondent is to cease giving the usual cues that encourage talking: stop nodding the head, interject a new question as soon as the respondent pauses for breath, stop taking notes, or call attention to the fact that you've stopped taking notes by flipping the page of the writing pad and sitting back, waiting. When these nonverbal cues don't work, you simply have to interrupt the long-winded respondent.

> Let me stop you here for a moment. I want to make sure I fully understand something you said earlier. (Then ask a question aimed at getting the response more targeted.)

or

> Let me ask you to stop for a moment because some of what you're talking about now I want to get later in the interview. First, I need to find out from you. . . .

Interviewers are sometimes concerned that it is impolite to interrupt an interviewee. It certainly can be awkward, but when done with respect and sensitivity, the interruption can actually help the interview. It is both patronizing and disrespectful to let the respondent run on when no attention is being paid to what is said. It is respectful of both the person being interviewed and the interviewer to make good use of the short time available to talk. It is the responsibility of the interviewer to help the interviewee understand what kind of information is being requested and to establish a framework and context that makes it possible to collect the right kind of information.

Reiterating the Purpose of the Interview as a Basis for Focus and Feedback

Helping the interviewee understand the purpose of the overall interview and the relationship of particular questions to that overall purpose are important pieces of information that go beyond simply asking questions. While the reason for asking a particular question may be absolutely clear to the interviewer, don't assume it's clear to the respondent. You communicate respect for the persons being interviewed by giving them the courtesy of explaining why the questions are being asked. Understanding the purpose of a question will increase the motivation of the interviewee to respond openly and in detail.

The overall purpose of the interview is conveyed in an opening statement. Specific questions within the interview should have a connection to that overall purpose. (We'll deal later with issues of informed consent and protection of human subjects in relation to opening statements of purpose. The focus here is on communicating purpose to improve responses. Later, we'll review the ethical issues related to informing interviewees about the study's purpose.) While the opening statement at the beginning of an interview provides an overview about the purpose of the interview, it will still be appropriate and important to explain the purpose of particular questions at strategic points throughout the interview. Here are two examples from evaluation interviews.

- This next set of questions is about the program staff. The staff has told us that they don't really get a chance to find out how people in the program feel about what they do, so this part of the interview is aimed at giving them some direct feedback. But as we agreed at the beginning, the staff won't know who said what. Your responses will remain confidential.
- This next set of questions asks about your background and experiences before you entered this program. The purpose of these questions is to help us find out how people with varying backgrounds have reacted to the program.

The One-Shot Question

Informal, conversational interviewing typically takes place as a natural part of fieldwork. It is opportunistic and often unscheduled. A chance arises to talk with someone when the interview is under way. More structured and scheduled interviewing takes place by way of formal appointments and site visits. Yet the best-laid plans for scheduled interviews can go awry. You arrive at the appointed time and place only to find that the person to be interviewed is unwilling to cooperate or needs to run off to take care of some unexpected problem. When faced with such a situation, it is helpful to have a single, one-shot question in mind to salvage at least something. This *one-shot question* is the one you ask if you are only going to get a few minutes with the interviewee.

For an agricultural extension needs assessment, I was interviewing farmers in rural Minnesota. The farmers in the area were economically distressed, and many of them felt alienated from politicians and professionals. I arrived at a farm for a scheduled interview, but the farmer refused to cooperate. At first, he refused to even come out of the barn to call off the dogs surrounding my truck. Finally, he appeared and said,

> I don't want to talk to you tonight. I know I said I would, but the wife and I had a tiff and I'm tired. I've always helped with your government surveys. I fill out all the forms the government sends. But I'm tired of it. No more. I don't want to talk.

I had driven a long way to get this interview. The fieldwork was tightly scheduled, and I knew that I would not get another shot at this farmer, even if he later had a change of heart. And I didn't figure it would help much to explain that I wasn't from the government. Instead, to try and salvage the situation, I took my one-shot question, a question stimulated by his demeanor and overt hostility.

> I'm sorry I caught you at a bad time. But as long as I'm here, let me ask you just one quick question; then, I'll be on my way. Is there anything you want to tell the bastards in Saint Paul?

He hesitated for just a moment, grinned, and then launched into a tirade that turned into a full, two-hour interview. I never got out of the truck, but I was able to cover the entire interview (though without ever referring to or taking out the written interview schedule). At the end of this conversational interview, which had fully satisfied my data collection needs, he said, "Well, I've enjoyed talkin' with you, and I'm sorry about refusin' to fill out your form. I just don't want to do a survey tonight."

I told him I understood and asked him if I could use what he had told me as long as he wasn't identified. He readily agreed, having already signed the consent form when we set up the appointment. I thanked him for the conversation. My scheduled, structured interview had become an informal, conversational interview developed from a last ditch, one-shot question.

Here's a different example. The story is told of a young ethnographer studying a village that had previously been categorized in anthropological studies as aggressive and war oriented. He sat outside the school at the end of the day and asked each boy who came out his one-shot, stupid, European question: "What do men do?" The responses he obtained overwhelmingly referred to farming and fishing and almost none to warfare. In one hour, he had a totally different view of the society from that portrayed by previous researchers.

The Final or Closing Question

In the spirit of open-ended interviewing, it's important in qualitative interviewing to provide an opportunity for the interviewee to have the final say: "That covers the things I wanted to ask. Anything you care to add?" I've gotten some of my richest data from this question with interviewees taking me in directions that had never occurred to me to pursue.

In a routine evaluation of an adult literacy program, we were focused on what learning to read meant to the participants. At the end of the interview, I asked, "What should I have asked you that I didn't think to ask?" Without hesitation one young Hispanic woman replied, "About sexual harassment." The program had a major problem that the evaluation ended up exposing.

Experienced qualitative methodologist David Morgan (2012) offers this advice about bringing closure to an interview:

> I think there are two basic points [for] closure: first [is] to avoid ending with the sense "OK, I've got my data, so long." The second is to give the participant[s] a chance to express their thoughts in their own voice.
>
> The question I use most often is: "Out of all the things we've talked about today—or maybe some topics we've missed—what should I pay most attention to? What should I think about when I read your interview?" (p. 1)

It can also be helpful to leave interviewees with a way to contact you in case they want to add something that they forgot to mention, or clarify some point.

> This way, you don't "close the door," but leave it ajar— and it is up to them whether they want to open it again.

I think it also gives the participants some power in the process—to decide when the process is over, rather than the researcher doing so. (Gustason, 2012, p. 1)

Beyond Technique

We've been looking with some care at different kinds of questions in an effort to polish interviewing technique and increase question precision. Below I'll offer suggestions about the mechanics of managing data collection—things like recording the data and taking notes. Before moving on, though, it may be helpful to step back and remember the larger purpose of qualitative inquiry so that we don't become overly technique oriented. You're trying to understand a person's world and worldview. That's why you ask focused questions in a sensitive manner. You're hoping to elicit relevant answers that are meaningful and useful in understanding the interviewee's perspective. That's basically what interviewing is all about.

HOW MUCH TECHNIQUE?

Sociologist Peter Berger is said to have told his students, "In science as in love, a preoccupation with technique may lead to impotence."

To which Halcolm adds, "In love as in science, ignoring technique reduces the likelihood of attaining the desired results. The path of wisdom joins that of effectiveness somewhere between the outer boundaries of ignoring technique and being preoccupied with it."

This chapter has offered ideas about how to do quality interviews, but ultimately, no recipe can prescribe the single right way of interviewing. No single correct format exists that is appropriate for all situations, and no particular way of wording questions will always work. The specific interview situation, the needs of the interviewee, and the personal style of the interviewer all come together to create a unique situation for each interview. Therein lies the challenge of qualitative interviewing.

Maintaining focus on gathering information that is useful, relevant, and appropriate requires concentration, practice, *and the ability to separate that which is foolish from that which is important*. In his great novel *Don Quixote*, Cervantes (1964) describes a scene in which his uneducated sidekick Sancho is rebuked by the knight errant Don Quixote for trying to impress his cousin by repeating deeply philosophical questions and answers that he has heard from others, all the

while trying to make the cousin think that these were Sancho's own insights.

> "That question and answer," said Don Quixote, "are not yours, Sancho. You have heard them from someone else."
>
> "Whist, sir," answered Sancho, "if I start questioning and answering, I shan't be done till tomorrow morning. Yes, for if it's just a matter of asking idiotic questions and giving silly replies, I needn't go begging help from the neighbors."
>
> "You have said more than you know, Sancho," said Don Quixote, "for there are some people who tire themselves out learning and proving things that, once learned and proved, don't matter a straw as far as the mind or memory is concerned." (p. 682)

Regardless of which interview strategy is used—the informal conversational interview, the interview guide approach, or a standardized open-ended interview—the wording of questions will affect the nature and quality of the responses received. So will careful management of the interview process. Constant attention to *both content and process*, with both informed by the purpose of the interview, will reduce the extent to which, in Cervantes's words, researchers and evaluators "tire themselves out learning and proving things that, once learned and proved, don't matter a straw as far as the mind or memory is concerned."

Mechanics of Gathering Interview Data

Recording the Data

No matter what style of interviewing you use and no matter how carefully you word questions, it all comes to naught if you fail to capture the actual words of the person being interviewed. The raw data of interviews are the actual quotations spoken by interviewees. Nothing can substitute for these data—the actual things said by real people. That's the prize sought by the qualitative inquirer.

Data interpretation and analysis involve making sense out of what people have said, looking for patterns, putting together what is said in one place with what is said in another place, and integrating what different people have said. These processes occur primarily during the analysis phase, after the data have been collected. During the interviewing process itself—that is, during the data collection phase—the purpose of each interview is to record as fully and fairly as possible that particular interviewee's perspective. Some method for recording the verbatim responses of people being interviewed is therefore essential.

As a good hammer is essential to fine carpentry, a good recorder is indispensable to fine fieldwork. Recorders do not "tune out" conversations, change what has been said because of interpretation (either conscious or unconscious), or record words more slowly than they are spoken. (Recorders, do, however, malfunction.) Obviously a researcher doing conversational interviews as part of covert fieldwork does not walk around with a recorder. However, most interviews are arranged in such a way that recorders are appropriate if properly explained to the interviewee:

> I'd like to record what you say so I don't miss any of it. I don't want to take the chance of relying on my notes and maybe missing something that you say or inadvertently changing your words somehow. So, if you don't mind, I'd very much like to use the recorder. If at any time during the interview you would like to stop the recorder, juts let me know.

When it is not possible to use a recorder because of some sensitive situation, interviewee request, or recorder malfunction, notes must become much more thorough and comprehensive. It becomes critical to gather actual quotations. When the interviewee has said something that seems particularly important or insightful, it may be necessary to say,

> I'm afraid I need to stop you at this point so that I can get down exactly what you said because I don't want to lose that particular quote. Let me read back to you what I have and make sure it is exactly what you said.

This point emphasizes again the importance of capturing what people say in their own words.

But such verbatim note taking has become the exception now that most people are familiar and comfortable with recorders. More than just increasing the accuracy of data collection, using a recorder permits the interviewer to be more attentive to the interviewee. If you tried to write down every word said, you'd have a difficult time responding appropriately to interviewee needs and cues. Ironically, *verbatim* note taking can interfere with listening attentively. The interviewer can get so focused on note taking that the person speaking gets only secondary attention. Every interview is also an observation, and having one's eyes fixed on a note pad is hardly conducive to careful observation. In short, the interactive nature of in-depth interviewing can be seriously affected by an attempt to take verbatim notes. Lofland (1971) has made this point forcefully:

One's full attention must be focused upon the interviewee. One must be thinking about probing for further explication or clarification of what he is now saying; formulating probes linking up current talk with what he has already said; thinking ahead to putting in a new question that has now arisen and was not taken account of in the standing guide (plus making a note at that moment so one will not forget the question); and attending to the interviewee in a manner that communicates to him that you are indeed listening. All of this is hard enough simply in itself. Add to that the problem of writing it down—even if one takes shorthand in an expert fashion—and one can see that the process of note-taking in the interview decreases one's interviewing capacity. Therefore, if conceivably possible, *record*; then one can interview. (p. 89)

ADVICE FROM AN EXPERIENCED JOURNALIST ON RECORDING AN INTERVIEW

Whatever you do, don't rely on memory. Don't even imagine that you will be able to remember verbatim in the evening what people said during the day. And don't squirrel notes in a bathroom—that is, runoff to the john and write surreptitiously what someone said back there with the cocktails. From the start, make clear what you are doing and who will publish what you write. Display your notebook as if it were a fishing license. While the interview continues, the notebook may serve other purposes, surpassing the talents of a taperecorder. As you scribble away, the interviewee is, of course, watching you. Now, unaccountably, you slowdown, and even stop writing, while the interviewee goes on talking. The interviewee becomes nervous, tries harder, and spills out the secrets of a secret life, or may be just a clearer and more quotable version of what was said before. Conversely, if the interviewee is saying nothing of interest, you can pretend to be writing, just to keep the enterprise moving forward.

—John McPhee (2014, p. 50)
Distinguished journalist

So if verbatim note taking is neither desirable nor really possible, what kinds of notes are taken during a recorded interview?

Taking Notes During Interviews

The use of the recorder does not eliminate the need for taking notes, but you take strategic and focused notes, not verbatim notes. Notes can serve at least four purposes:

1. Notes taken during the interview can help the interviewer formulate new questions as the interview moves along, particularly where it may be appropriate to check out something said earlier.

2. Looking over field notes before transcripts are done helps make sure the inquiry is unfolding in the hoped-for direction and can stimulate early insights that may be relevant to pursue in subsequent interviews while still in the field—the emergent nature of qualitative inquiry.

3. Taking notes will facilitate later analysis, including locating important quotations, from the recording.

4. Notes are a backup in the event the recorder has malfunctioned or, as I've had happen, a recording is erased inadvertently during transcription.

When a recorder is being used during the interview, notes will consist primarily of key phrases, lists of major points made by the respondent, and key terms or words shown in quotation marks that capture the interviewee's own language. It is useful to develop some system of abbreviations and informal shorthand to facilitate taking notes, for example, in an interview on leadership writing "L" instead of the full word. Some important conventions along this line include (a) using quotation marks *only* to indicate full and actual quotations; (b) developing some mechanism for indicating interpretations, thoughts, or ideas that may come to mind during the interview, for example, the use of brackets to set off one's own ideas from those of the interviewee; and (c) keeping track of questions asked as well as answers received. Questions provide the context for interpreting answers.

Note taking serves functions beyond the obvious one of taking notes. Note taking helps pace the interview by providing nonverbal cues about what's important, providing feedback to the interviewee about what kinds of things are especially "noteworthy"— literally. Conversely, the failure to take notes may indicate to the respondent that nothing of particular importance is being said. And don't start making out your work "to do" list while someone is droning on endlessly. The person might think that you're taking notes. Enchanted, he or she will keep on talking. The point is that taking notes affects the interview process. Be mindful of those effects.

After the Interview: Quality Control

The period after an interview or observation is critical to the rigor and validity of qualitative inquiry. This is a time for guaranteeing the quality of the data.

Immediately after a recorded interview, check the recording to make sure it worked. If, for some reason, a malfunction occurred, you should immediately make extensive notes of everything that you can remember. Even if the recorder functioned properly, you should go over the interview notes to make certain that they make sense and to uncover areas of ambiguity or uncertainty. If you find things that don't quite make sense, as soon possible check back with the interviewee for clarification. This can often be done over the telephone. In my experience, people who are interviewed appreciate such a follow-up because it indicates the seriousness with which the interviewer is taking their responses. Guessing the meaning of a response is unacceptable; if there is no way of following up the comments with the respondent, then those areas of vagueness and uncertainty simply become missing data.

The immediate postinterview review is a time to record details about the setting and your observations about the interview. Where did the interview occur? Under what conditions? How did the interviewee react to questions? How well do you think you did asking questions? How was the rapport?

Answers to these questions establish a context for interpreting and making sense of the interview later. Reflect on the quality of information received. To what extent did you find out what you really wanted to find out in the interview? Note weaknesses and problems: poorly worded questions, wrong topics, poor rapport. Reflect on these issues, and make notes on the interview process while the experience is still fresh in your mind. These process notes will inform the methodological section of your research report, evaluation, or dissertation.

Reflection as Qualitative Data

This period after an interview or observation is a critical time for reflection and elaboration. It is a time of quality control to guarantee that the data obtained will be useful. This kind of postinterview ritual requires discipline. Interviewing and observing can be exhausting, so much so that it is easy to forego this time of reflection and elaboration, put it off, or neglect it altogether. To do so is to seriously undermine the rigor of qualitative inquiry. Interviews and observations should be scheduled so that sufficient time is available for data clarification, elaboration, and evaluation afterward. Where a team is working together, the whole team needs to meet regularly to share observations and debrief together. This is the beginning of analysis, because, while the situation and data are fresh, insights can emerge that might otherwise be lost. Ideas and interpretations that emerge following an interview or observation should be written down and clearly marked as emergent, field-based insights to be further reviewed later.

I think about the time after an interview as a period for postpartum reflection, a time to consider what has been revealed and what has been birthed. In eighteenth-century Europe, the quaint phrase "in an interesting condition" became the genteel way of referring to an expectant mother in "polite company." The coming together of an interviewer and an interviewee makes for "an interesting condition." The interviewer is certainly expectant, as may be the interviewee. What

emerged? What was created? Did it go okay? Is some form of triage necessary? As soon as a child is born, a few basic observations are made, and tests are performed to make sure that everything is alright. That's what you're doing right after an interview—making sure that everything came out okay.

Such an analogy may be a stretch for thinking about a postinterview debrief, but interviews are precious to those who hope to turn them into dissertations, contributions to knowledge, and evaluation findings. It's worth managing the interview process to allow time to make observations about, reflect on, and learn from each interview.

Up to this point, we've been focusing on techniques to enhance the quality of the standard one-on-one interview. We turn now to some important variations in interviewing and specialized approaches. We begin with interviewing groups.

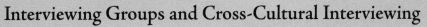

> Human beings are social creatures. We are social not just in the trivial sense that we like company, and not just in the obvious sense that we each depend on others. We are social in a more elemental way: simply to exist as a normal human being requires interaction with other people.
>
> —Atul Gawande
> Surgeon and journalist

Group interviews take a variety of forms and serve diverse purposes. Focus groups, interviews with naturally occurring groups, family interviews, and documenting consciousness-raising groups offer a variety of opportunities to engage in qualitative inquiry. What these group interviewing approaches have in common is recognition that, *as social beings*, qualitative inquiry with groups makes data collection a social experience. That social experience is presumed to increase the meaningfulness and validity of findings because our perspectives are formed and sustained in social groups. Our interactions with each other are how we come to more deeply understand our own views, test our knowledge, get in touch with our feelings, and make sense of our behaviors.

The purpose of the inquiry drives the use of group interviews. We may use groups to get diverse perspectives or to facilitate consensus. We may interview naturally occurring groups to compare and contrast their activities and views. We may conduct group interviews in certain cross-cultural settings because people are most comfortable in groups and are not accustomed to one-on-one inquiries. In the sections that follow, we'll examine and compare several group-interviewing approaches. Exhibit 7.14 provides an overview of 12 varieties of group interviews with different purposes and priorities.

Focus Group Interviews

A focus group is an interview with a small group of people on a specific topic. Groups are typically of 6 to 10 people with similar backgrounds who

participate in the interview for one to two hours. In a given study, a series of different focus groups will be conducted to get a variety of perspectives and increase confidence in whatever patterns emerge. Focus group interviewing was developed in recognition that many consumer decisions are made in a social context, often growing out of discussions with other people. Thus, market researchers began using focus groups in the 1950s as a way of stimulating the consumer group process of decision making to gather more accurate information about consumer product preferences (Higginbotham & Cox, 1979). On the academic side, distinguished sociologist Robert K. Merton pioneered the seminal work on research-oriented focus group interviews in 1956: *The Focused Interview* (Merton, Riske, & Kendall).

The focus group interview is, first and foremost, an *interview*. It is not a problem-solving session. It is not a decision-making group. It is not primarily a discussion, though direct interactions among participants often occur. It is an *interview*. The twist is that, unlike a series of one-on-one interviews, in a focus group, participants get to hear each other's responses and to make additional comments beyond their own original responses as they hear what other people have to say. However, participants need not agree with each other or reach any kind of consensus. Nor is it necessary for people to disagree. The object is to get high-quality data in a social context where people can consider their own views in the context of the views of others.

Focus group experts Richard Krueger and Mary Anne Casey (2008) explain that a focused interview should be carefully planned to create a "permissive, nonthreatening environment," a setting that "encourages participants to share perceptions and points of view without pressuring participants to vote or reach consensus" (p. 4). A focus group is conducted in a way that is comfortable and even enjoyable for participants as they share their ideas and perceptions. Group members influence each other by responding to the ideas and comments they hear.

The term focus group *moderator* may be used instead of *interviewer* because

this term [*moderator*] highlights a specific function of the interviewer—that of moderating or guiding the discussion. The term *interviewer* tends to convey a more limited impression of two-way communication

EXHIBIT 7.14   Twelve Varieties of Group Interviews

TYPE OF GROUP INTERVIEW	PURPOSE AND NICHE
1. Research focus group	Interview with a small group of relatively similar people (homogeneity sampling) on a specific topic of research interest (Carey, Asbury, & Tolich, 2012; Morgan, 1997a, 1997b)
2. Evaluation focus group	Interview with a group of program participants to get their perspectives on and experiences with a specific program (Krueger & Casey, 2008; Marczak & Sewell, 2013)
3. Marketing focus group	Learn consumer preferences to inform marketing campaigns (Beall, 2010) and business decisions (Eriksson & Kovalainen, 2008). "Extract information . . . to manipulate people more effectively" (Denzin & Lincoln, 2011, p. 419)
4. Diversity-focused group	Interview people with diverse perspectives and experiences regarding some issue to compare and contrast their perspectives as they interact
5. Convergence-focused group	Interview people with relatively homogeneous experiences to identify commonalities and shared patterns
6. Group interviews with naturally occurring or already existing groups	Interview with a group of teenage girls who are friends in India (Parameswaran, 2001). Interviews with members of a church group, a gang, a political organization, or any already existing group of any kind
7. Family interviews	Interviews with married couples or families (Gale & Dolbin-MacNab, 2014; Miller & Johnson, 2014)
8. Dyadic interviews	Two interviewees interact together in response to open-ended research questions. For example, examine in pairs experiences of people with early-stage dementia, explore the experiences of staff who work together to provide services to elderly housing residents, and examine barriers and facilitators to substance abuse treatment (Morgan, Ataie, Carder, & Hoffman, 2013)
9. Consciousness-raising inquiry groups	"Focus groups in the service of radical political work designed within social justice agendas . . . to mobilize empowerment agendas and to enact social change," especially in the feminist movement (Kamberelis & Dimitriadis, 2011, pp. 550–551)
10. Generative, multifunctional constructionist group conversations	Collective conversations where researchers and participants interact "in a performative turn" to integrate pedagogy, politics, and empirical reflection through the dynamics of shared inquiry and "synergistic articulations" (Kamberelis & Dimitriadis, 2011, pp. 547, 556)
11. Internet focus groups	Convening invited participants to engage in either synchronous or asynchronous focus group interview using an Internet platform (Krueger & Casey, 2012; Lee & O'Brien, 2012)
12. Social media group interviews	General outreach through social media to identify a group of people to engage in an online interactive interview process (Pew, 2013; Poynter, 2010)

between an interviewer and an interviewee. By contrast, the focus group affords the opportunity for multiple interactions not only between the interviewer and respondent but among all participants in the group. The focus group is not a collection of simultaneous individual interviews but rather a group discussion where the conversation flows because of the nurturing of the moderator. (Krueger, 1994, p. 100)

GIVING VOICE TO MARGINALIZED PEOPLE THROUGH FOCUS GROUPS

By bringing together people who share a similar background, focus groups create the opportunity for participants to engage in meaningful conversations about the topics that researchers wish to understand. This ability to learn about participants' perspectives by listening to their conversations makes focus groups especially useful for hearing from groups whose voices are often marginalized within the larger society. Focus groups are thus widely used in studies of ethnic and cultural minority groups, along with studies of sexuality and substance use. . . .

For evaluation research, focus groups are used in both preliminary phases, such as needs assessment or program development, and in follow-up or summative evaluation, to hear about the participants' experiences with a program. (Morgan, 2008a, p. 352)

THE QUESTIONING ROUTE AND GROUP INTERACTIONS

The series of questions used in a focused interview—the questioning route—looks deceptively simple. Typically, a focused interview will include about a dozen questions for a two-hour group. If you asked these questions in an individual interview, the respondent could probably tell you everything he or she could think of related to the questions in just a few minutes. But when these questions are asked in a group environment, the discussion can last for several hours. Part of the reason is in the nature of the questions and the cognitive processes of humans. As participants answer questions, their responses spark ideas from other participants. Comments provide mental cues that trigger memories or thoughts of other participants—cues that help explore the range of perceptions. (Krueger & Casey, 2008, p. 35)

The combination of moderating and interviewing is sufficiently complex that Krueger recommends that teams of two conduct the groups so that one person can focus on facilitating the group while the other takes detailed notes and deals with mechanics like recorders, cameras, and any special needs that arise, for example, someone needing to leave early or becoming overwrought. Even when the interview is recorded, good notes help in sorting out who said what when the recording is transcribed.

Postmodern *Unfocused* Group Interview

Strengths of Focus Groups

Focus group interviews have several advantages for qualitative inquiry.

- *Focus groups offer cost effective data collection:* In one hour, you can gather information from eight people instead of only one, significantly increasing sample size. "Focus group interviews are widely accepted within marketing research because they produce believable results at a reasonable cost" (Krueger, 1994, p. 8).
- *Focus groups highlight diverse perspectives:* Focus groups should be homogeneous in terms of background and not attitudes. Differences of opinion "lend 'bite' to focus group discussions. Provided that we are not cavalier about mixing together people who are known to have violently differing perspectives on emotive issues, a little bit of argument can go a long way towards teasing out what lies beneath 'opinions' and can allow both focus group facilitators and participants to clarify their own and others' perspectives. Perhaps, in some contexts, this can even facilitate greater mutual understanding. In terms of generating discussion, a focus group consisting of people in agreement about everything would make for very dull conversation and data lacking in richness" (Barbour, 2007, p. 59).

- *Interactions among participants enhance data quality:* Participants tend to provide checks and balances on each other, which weeds out false or extreme views (Krueger & Casey, 2000, 2008). Moreover, how people talk about a topic is important, not just what they say about it. The researcher attends "not only the content of the conversation, but also what the conversation situation is like in terms of emotions, tensions, interruptions, conflicts and body language . . . ; how people tell and retell different narratives or how they draw on and reproduce discourses in interaction" (Eriksson & Kovalainen, 2008, p. 175).
- *Silences and topics avoided are revealing:* What is not said in focus groups and what precipitates periods of silence can generate fruitful insights.
- *Analysis unfolds as the interview unfolds:* The extent to which there is a relatively consistent, shared view or great diversity of views can be quickly assessed.
- *Focus groups tend to be enjoyable to participants:* They draw on our human tendencies as social animals to enjoy interacting with others.

Limitations of Focus Groups

Focus groups, like all forms of data collection, also have *limitations*.

- The number of questions that can be asked is greatly restricted in the group setting.
- The available response time for any particular individual is restrained to hear from everyone. *A rule of thumb:* With eight people and an hour for the group, plan to ask no more than 10 major questions.
- Facilitating and conducting a focus group interview requires considerable group process skill beyond simply asking questions. The moderator must manage the interview so that it's not dominated by one or two people, and so that those participants who tend not to be highly verbal are able to share their views.
- Those who realize that their viewpoint is a minority perspective may not be inclined to speak up and risk negative reactions.
- Focus groups appear to work best when people in the group, though sharing similar backgrounds, are strangers to one another. The dynamics are quite different and more complex when participants have prior established relationships.
- Controversial and highly personal issues are poor topics for focus groups (Kaplowitz, 2000).
- Confidentiality cannot be assured in focus groups. Indeed, in market research, focus groups are often videotaped so that marketers can view them and see for themselves the emotional intensity of people's responses.
- "The focus group is beneficial for identification of major themes but not so much for the micro-analysis of subtle differences" (Krueger, 1994, p. x).
- Compared with most qualitative fieldwork approaches, focus groups typically have the disadvantage of taking place outside the natural settings where social interactions normally occur (Madriz, 2000, p. 836).

LEARNING IN FOCUS GROUPS

Focus group participants sometimes engage in problem solving as they respond to questions. Sharing what they think and know, participants generate new knowledge as a group that can affect individual knowledge and beliefs, and even subsequent behavior. Expressing disagreement can also stimulate learning as participants challenge each other, defend their own views, and sometimes modify their viewpoints. Thus, while the quotations from focus groups constitute evaluation findings, the interactions and learnings in the group can constitute learning among the participants (Wiebeck & Dahlgren, 2007).

The *Focus* in Focus Groups

As these strengths and limitations suggest, *the power of focus groups resides in their being focused*. The focused questions typically seek reactions to something (a product, a program, an idea, or a shared experience) rather than exploring complex life issues in depth and detail. The groups are focused by being formed homogeneously. The facilitation is focused, keeping responses on target. Interactions among participants are focused, staying on topic. Use of time must be focused, because the time passes quickly. Despite some of the limitations introduced by the necessity of sharp focus, applications of focus groups are widespread and growing (Fontana & Frey, 2000; Krueger & Casey, 2000, 2008; Madriz, 2000; Morgan, 1988, 2008a).

Focus groups have entered into the repertoire of techniques for qualitative researchers and evaluators involved in participatory studies with coresearchers. For community research, collaborative action research, and participatory evaluations, local people who are not professional researchers are being successfully trained and supported to do focus groups (King & Stevahn, 2013; Krueger & King, 1997). Rossman and Rallis (1998) have done focus groups effectively with children in elementary schools.

Because the focus group "is a collectivistic rather than an individualistic research method," focus groups have also emerged as a collaborative and empowering approach in feminist research (Madriz, 2000, p. 836). Sociologist and feminist researcher Esther Madriz explains,

> Focus groups allow access to research participants who may find one-on-one, face-to-face interaction "scary" or "intimidating." By creating multiple lines of communication, the group interview offers participants . . . a safe environment where they can share ideas, beliefs, and attitudes in the company of people from the same socioeconomic, ethnic, and gender backgrounds. . . .

> For years, the voices of women of color have been silenced in most research projects. Focus groups may facilitate women of color "writing culture together" by exposing not only the layers of oppression that have suppressed these women's expressions, but the forms of resistance that they use every day to deal with such oppressions. In this regard, I argue that focus groups can be an important element in the advancement of an agenda of social justice for women, because they can serve to expose and validate women's everyday experiences of subjugation and their individual and collective survival and resistance strategies. (Madriz, 2000, pp. 835–836)

I experienced firsthand the potential of focus groups to provide safety in numbers for people in vulnerable situations. I conducted focus groups among low-income recipients of legal aid as one technique in an evaluation of services provided to people in a large public housing project. As the interview opened, the participants, who came from different sections of the project and did not know each other, were reserved and cautious about commenting on the problems they were experiencing. As one woman shared in vague terms a history of problems she had had in getting needed repairs, another woman jumped in and supported her saying, "I know exactly what you're talking about and *who* you're talking about. It's really bad. Bad. Bad. Bad." She then shared her story. Soon others were telling similar stories, often commenting that they had no idea so many other people were having the same kind of problems. Several also commented at the end of the interview that they would have been unlikely to share their stories with me in a one-on-one interview because they would have felt individually vulnerable, but they drew confidence and a sense of safety and camaraderie from being part of the interview group.

On the other hand, Kaplowitz (2000) studied whether sensitive topics were more or less likely to be discussed in focus groups versus individual interviews.

Ninety-seven year-round residents from the Chelem Lagoon region in Yucatan, Mexico, participated in one of 12 focus groups or 19 individual in-depth interviews. A professional moderator used the same interview guide to get reactions to a shared mangrove ecosystem. The 31 sessions generated more than 500 pages of transcripts that were coded for the incidence of discussions of sensitive topics. The findings showed that the individual interviews were 18 times more likely to raise socially sensitive discussion topics than the focus groups. Additionally, the study found the two qualitative methods, focus groups and individual interviews, to be complementary to each other, each yielding somewhat different information.

Internet Focus Groups

Internet and social media platforms have created new opportunities for focus groups (Krueger & Casey, 2012; Lee & O'Brien, 2012). Walston and Lissitz (2000) compared the reactions of Internet-based versus face-to-face focus groups discussing academic dishonesty and found that the Internet environment appeared to reduce participants' anxiety about what the moderator thought of them, making it easier for them to share embarrassing information.

Interviews With Existing Groups

Not all group interviews are of the focus group variety. During fieldwork, unstructured conversational interviews may occur in groups that are not at all focused. In evaluating a community leadership program, much of the most important information came from talking with groups of people informally during breaks from the formal training. During the fieldwork for the wilderness education program described extensively in the previous chapter, informal group interviews became a mainstay of data collection, sometimes with just two or three people, and sometimes in dinner groups as large as 10.

Parameswaran (2001) found important differences in the data she could gather in group versus individual interviews during her fieldwork in India. Moreover, she found that she had to do group interviews with the young women *before* she could interview them one-on-one. She was studying the reading of Western romance novels among female college students in India. She reports,

> To my surprise, several young women did not seem happy or willing to spend time with me alone right away. When I requested the first group of women to

meet me on an individual basis and asked if they could meet me during their breaks from classes, I was surprised and uncomfortable with the loud silence that ensued. . . .

When I faced similar questions from another group of women who also appeared to resist my appeals to meet with them alone, I realized that I had arrogantly encroached into their intimate, everyday rituals of friendship. . . . Knowing well that without collecting data from these possibly recalcitrant subjects, I had no project, I reluctantly changed my plans and agreed to accept their demands. I began talking to them in groups first, and gradually, more than 30 women agreed to meet me in individual sessions. Later, I discovered that they preferred to respond to me as a group first because they were wary about the kinds of questions I planned to ask about their sexuality and romance reading. The more public nature of group discussions meant that it was a safe space where I might hesitate to ask intrusive and personal questions. . . .

Group interviews in which women spoke about love, courtship, and heterosexual relations in Western romance fiction became opportunities to debate, contradict, and affirm their opinions about a range of gendered social issues in India such as sexual harassment of women in public places, stigmas associated with single women, expectations of women to be domestic, pressures on married women to obey elders in husbands' families, and the merits of arranged versus choice/love marriages. . . . In contrast to these collective sessions where young women's discussions primarily revolved around gender discrimination toward women as a group, in individual interviews, many women were much more talkative about restrictions on their sexuality, and several women shared their frustrations with immediate, everyday problems pertaining to family members' control over their movements. (pp. 84–86)

Parameswaran's (2001) work illustrates well the different kinds of data that can be collected from groups versus individuals, with both kinds of data being important in long-term or intensive fieldwork. I had similar experiences in Burkina Faso and Tanzania, where villagers much preferred group interviews for formal and structured interactions and the only way to get individual interviews was informally when walking somewhere or doing something with an individual. On several occasions, I tried scheduling and conducting individual interviews only to have many other people present when I arrived. Such cross-cultural differences in valuing individual versus group interactions provide a segue to the next section on cross-cultural interviewing.

Cross-Cultural Interviewing

> Culture and place demand our attention not because our concepts of them are definitive or authoritative, but because they are fragile and fraught with dispute.
>
> —Jody Berland (1997, p. 9)
> Cultural studies scholar, York
> University, Toronto, Canada

Cross-cultural inquiries add layers of complexity to the already-complex interactions of an interview. The possibility for misunderstandings is increased significantly. Ironically, economic and cultural globalization, far from reducing the likelihood of misunderstandings, may simply make miscommunication more nuanced and harder to detect because of false assumptions about shared meanings. Whiting (1990) tellingly explored cross-cultural differences in his book, *You Gotta Have Wa*, on how Americans and Japanese play seemingly the same game, baseball, quite differently—and then uses those differences as entrées into the two cultures.

Ethnographic interviewing has always been inherently cross-cultural but has the advantage of being grounded in long-term relationships and in-depth participant observations. More problematic, and the focus of this section, are short-term studies for theses or student exchange projects and brief evaluation site visits sponsored by international development agencies and philanthropic foundations. In the latter case, teams of a few Westerners are flown into a developing country for a week to a month to assess a project, often with counterparts from the local culture. These rapid appraisals revolve around cross-cultural interviewing and are more vulnerable to misinterpretation and miscommunication than traditional, long-term anthropological fieldwork. Examples of the potential problems, presented in the sections that follow, will hopefully help sensitize students, short-term site visitors, and evaluators to the precariousness of cross-cultural interviewing. As Rubin and Rubin (1995) have noted,

> You don't have to be a woman to interview women, or a sumo wrestler to interview sumo wrestlers. But if you are going to cross social gaps and go where you are ignorant, you have to recognize and deal with cultural barriers to communication. And you have to accept that how you are seen by the person being interviewed will affect what is said. (p. 39)

Language Differences

The data from interviews are words. It is tricky enough to be sure what a person means when using a common language, but words can take on a very different meaning in other cultures. In Sweden, I participated in an international conference discussing policy evaluations. The conference was conducted in English, but I was there two days, much of the time confused, before I came to understand that their use of the term *policy* corresponded to my American use of the term *program.* I interpreted policies from an American context, to be fairly general directives, often very difficult to evaluate because of their vagueness. In Sweden, however, policies were articulated and even legislated at such a level of specificity that they resembled programmatic prescriptions more than the vague policies that typically emanate from the legislative process in the United States.

The situation becomes more precarious when a translator or interpreter must be used. Special and very precise training of translators is critical. Translators need to understand what, precisely, you want them to ask and that you will need full and complete translation of responses as verbatim as possible. Interpreters often want to be helpful by summarizing and explaining responses. This contaminates the interviewee's actual response with the interpreter's explanation to such an extent that you can no longer be sure whose perceptions you have—the interpreter's or the interviewee's.

Some words and ideas simply can't be translated directly. People who regularly use the language come to know the unique cultural meaning of special terms. One of my favorites from the Caribbean is *liming*, meaning something like hanging out, just being, doing nothing—*guilt free.* In interviews for a Caribbean program evaluation, a number of participants said that they were just "liming" in the program. That was not meant as criticism, however, for liming is a highly desirable state of being, at least to the participants. Funders viewed the situation somewhat differently.

Rheingold (2000) published a whole book on "untranslatable words and phrases" with special meanings in other cultures. Below are four examples that are especially relevant to researchers and evaluators.

1. *Schlimmbesserung* (German): A so-called improvement that makes things worse
2. *bigapeula* (Kiriwina, New Guinea): Potentially disruptive, unredeemable true statements
3. *animater* (French): A word of respect for a person who can communicate difficult concepts to general audiences
4. *ta* (Chinese): To understand things and thus take them lightly

In addition to the possibility of misunderstandings, there may be the danger of contracting a culturally specific disease, including for some what the Chinese call *koro*—"the hysterical belief that one's penis is shrinking" (Rheingold, 1988, p. 59).

Attention to language differences cross-nationally can, hopefully, make us more sensitive to barriers to understanding that can arise even among those who speak the same language. Joyce Walker undertook a collaborative study with 18 women who had written to each other annually for 25 years, from 1968 to 1993. She involved them actively in the study, including having them confirm the authenticity of her findings. In reviewing the study's findings, one participant reacted to the research language used: "Why call us a cohort? There must be something better—a group, maybe?" (Walker, 1996, p. 10).

Differing Norms and Values

The high esteem in which science is held has made it culturally acceptable in Western countries to conduct interviews on virtually any subject in the name of scholarly inquiry and the common good. Such is not the case worldwide. Researchers cannot simply presume that they have the right to ask intrusive questions. Many topics discussed freely in Western societies are taboo in other parts of the world. I have experienced cultures where it was simply inappropriate to ask questions to a subordinate about a superordinate. Any number of topics may be insensitive to ask or indelicate if brought up by strangers, for example, family matters, political views, who owns what, how people came to be in certain positions, and sources of income.

Interviewing farmers for an agricultural extension in Central America became nearly impossible to do because for many their primary source of income came from growing illegal crops. In an African dictatorship, we found that we could not ask about "local" leadership because the country could have *only* one leader. Anyone taking on or being given the designation "local leader" would have been endangered. Interviewees can be endangered by insensitive and inappropriate questions; so can naive interviewers. I know of a case where an American female student was raped following an evening interview in a foreign country because the young man interpreted her questions about local sexual customs and his own dating experiences as an invitation to have sex.

EXHIBIT 7.15 Ten Examples of Variations in Cross-Cultural Norms That Can Affect Interviewing and Qualitative Fieldwork

CULTURAL NORM	CROSS-CULTURAL VARIATIONS
1. Attention to time and being "on time"	Some cultures value being on time as a matter of respect. In others, people routinely show up late, sometimes by a considerable amount.
2. Engaging in pre-interview "small talk"	Some cultures value getting right down to business. In others, it would be impolite to begin an interview without first discussing the weather, exchanging family information, or having coffee or tea together before getting on with the task at hand.
3. Eye contact	Some cultures value direct eye contact; others consider direct eye contact invasive.
4. Physical proximity	Norms vary greatly about how near to sit or stand when interacting.
5. Gender issues	Cultures vary greatly in appropriate gender interactions, whether males can interview females, and vice versa.
6. Nonverbal gestures	Behaviors such as shaking hands, crossing one's legs while sitting, head nodding, and other nonverbal behaviors can be welcome or offensive depending on the cultural norms.
7. Who grants permission for interviews	Who controls access to potential interviewees has cultural, political, social, and even economic implications that require understanding and respecting power and social status dynamics in different settings.
8. Appropriate topics for inquiry	Cultures vary greatly in what issues can be addressed in interviews. Issues related to religion, politics, sexual norms, and money may be off-limits. In many cultures, people with little education are reluctant to express opinions.
9. Gift giving and food sharing	Norms about how to welcome visitors can come into play when interviewers are treated as honored guests. Sharing food or exchanging small gifts may be expected.
10. What words can be used	In many languages, there is no word for "interview," "research," or "evaluation." Describing and explaining the process of gathering data, for what purposes with what uses, can be among the most challenging aspects of cross-cultural communications.

As noted in the previous section on group interviews, different norms govern cross-cultural interactions. I remember going to an African village to interview the chief and finding the whole village assembled. Following a brief welcoming ceremony, I asked if we could begin the interview. I expected a private, one-on-one interview. He expected to perform in front of and involve the whole village. It took me a while to understand this, during which time I kept asking to go somewhere else so we could begin the interview. He did not share my concern about and preference for privacy. What I expected to be an individual interview soon became a whole village group dialogue.

In many cultures, it is a breach of etiquette for an unknown man to ask to meet alone with a woman. Even a female interviewer may need the permission of a husband, brother, or parent to interview a village woman. A female colleague created a great commotion, and placed a woman in jeopardy, by pursuing a personal interview without permission from the male headman. Exhibit 7.15 lists some of the variations in cultural behaviors that can affect cross-cultural interviewing and qualitative fieldwork.

The value of Cross-Cultural Interviewing

As difficult as cross-cultural interviewing may be, it is still far superior to standardized questionnaires for collecting data from nonliterate villagers. Salmen (1987) described a major water project undertaken by The World Bank based on a needs assessment survey. The project was a failure because the local people ended up opposing the approach used. His reflection on the project's failure includes a comparison of survey and qualitative methods.

Although it is difficult to reconstruct the events and motivation that led to the rejection there is little question that a failure of adequate communication between project officials and potential beneficiaries was at least partly responsible. The municipality's project preparation team had conducted a house-to-house survey in Guasmo Norte before the outset of the project, primarily to gather basic socioeconomic data such as family size, employment and income. The project itself, however, was not mentioned at this early stage. On the basis of this survey, World Bank and local officials had decided that standpipes would be more affordable to the people than household connections. It now appears, from hindsight, that the questionnaire survey method failed to elicit the people's negative attitude toward standpipes, their own criterion of affordability, or the opposition of their leaders who may have played on the negative feelings of the people to undermine acceptance of the project. Qualitative interviews and open discussions would very likely have revealed people's preferences and the political climate far better than did the preconstructed questionnaire. (Salmen, 1987, pp. 37–38)

Appropriately, Salmen's book is called *Listen to the People* and advocates qualitative methods for international project evaluation of development efforts.

Interviewers are not in the field to judge or change values and norms. Researchers are there to understand the perspectives of others. Getting valid, reliable, meaningful, and usable information in cross-cultural environments requires special sensitivity to and respect for differences, including concerns about decolonizing research in cross-cultural contexts (Denzin & Lincoln, 2011, pp. 92–93; L. T. Smith, 2012; Mutua & Swadener, 2011) and cultural competence (American Evaluation Association, 2013). Moreover, specific challenges emerge in specific countries, as Bubaker, Balakrishnan, and Bernadine (2013) demonstrated when conducting quality studies in Libya and Malaysia. (For additional discussion and examples of cross-cultural research and evaluation, see Ember & Ember, 2009; Lonner & Berry, 1986; Patton, 1985a.)

One final observation on international and cross-cultural evaluations may help emphasize the value of such experiences. Connor (1985) found that doing international evaluations made him more sensitive and effective in his domestic evaluation work. The heightened sensitivity we expect to need in exotic, cross-cultural settings can serve us well in our own cultures. Sensitivity to and respect for other people's values, norms, and worldviews is as much needed at home as abroad.

CROSS-CULTURAL PERSPECTIVES ON DEVELOPMENT

The titles of these books communicate the challenges and importance of generating cross-cultural understanding and appropriate action around the world. These books also illustrate the global scale at which mixed-methods inquiries are now being done.

- *Can Anyone Hear Us? Voices of the Poor*
  The voices of more than 40,000 poor women and men in 50 countries from the World Bank's participatory poverty assessments (Narayan, Patel, Schafft, Rademacher, & Koch-Schulte, 2000)

- *Crying Out for Change: Voices of the Poor*
  Fieldwork on poverty conducted in 23 countries (Narayan, Chambers, Shah, & Petesch, 2000)

- *Voices of the Poor: From Many Lands*
  Regional patterns of poverty and country case studies (Narayan & Petesch, 2002)

> Creativity is just connecting things. When you ask creative people how they did something, they feel a little guilty because they didn't really do it, they just saw something. It seemed obvious to them after a while. That's because they were able to connect experiences they've had and synthesize new things.
>
> —Steve Jobs (1955–2011)
> Apple Computer founder

We have looked in some depth at various kinds of standard qualitative interviewing approaches: conversational interviews during fieldwork; in-depth, open-ended, one-on-one formal interviews; interactive, relationship-based interviews; focus group interviews; and cross-cultural interviews. These represent mainstream qualitative interviewing approaches. Now, we move beyond these well-known and widely used interviewing approaches to consider innovative, creative, and pioneering approaches to qualitative interviewing.

Photo Elicitation Interviewing

Photo elicitation involves using photographs to stimulate reflections, support memory recall, and elicit stories as part of interviewing. Photo elicitation methods are proving powerful supplements in in-depth interviewing. For example, to study the meaning of change in dairy farming in northern New York, sociologist Douglas Harper (2008) showed elderly farmers photographs from the 1940s (a period when they had been teenage or young adult farmers) and asked them to remember events, stories, or commonplace activities that the photos brought to mind. He used archived documentary photographs from the era that the elderly farmers had experienced at the beginnings of their careers. The photographs inspired "detailed and often deep memories."

> The farmers described the mundane aspects of farming, including the social life of shared work. But more important, they explained what it meant to have participated in agriculture that had been neighbor based,

environmentally friendly, and oriented toward animals more as partners than as exploitable resources.

> In this and other photo elicitation studies, photographs proved to be able to stimulate memories that word-based interviewing did not. The result was discussions that went beyond "what happened when and how" to themes such as "this was what this had meant to us as farmers." (pp. 198–199)

Visual data generally are becoming increasingly important in qualitative research and evaluation (Azzam & Evergreen, 2013a, 2013b; Banks, 2007; Prosser, 2011). A classic example by Hedy Bach (1998) involved documenting the daily life of four schoolgirls who, in addition to regular schoolwork, engaged in art, drama, ballet, and music programs in and out of school. Using disposable cameras, the students visually documented their lives both inside and outside classrooms. Their photos served as the basis for one-on-one interviews with them about their images. The analysis of the images and interviews revealed an "evaded curriculum" within adolescent life, exposing pain, pleasure, and the intensity of joy in making and creating schoolgirl culture.

Using photos as part of interviewing can reduce

> the awkwardness that an interviewee may feel from being put on the spot and grilled by the interviewer . . . ; direct eye contact need not be maintained, but instead interviewee and interviewer can both turn to the photographs as a kind of neutral third party. Awkward silences can be covered as both look at the photographs, and in situations where the status difference between interviewer and interviewee is great (such as between an adult and a child) or where the interviewee feels they are involved in some kind of test, the photographic content always provides something to talk about. (Banks, 2007, pp. 65–66)

Lorenz (2011) used photo elicitation as one way to generate empathy in research with acquired brain injury survivors. Brain injury patients can experience

> a lack of empathy that leads to feelings of being disrespected and powerless. . . . Practicing empathy by using photos to create discursive spaces in research relationships may help us to learn *about* ourselves as we learn *with* patients. (p. 259)

As powerful as photography is, a picture being legendarily worth a thousand words, visual methods have limitations, as do all methods. Qualitative methodologists debate "the myth of photographic truth," recognizing that "photographs represent a highly selected sample of the 'real' world" (Prosser, 2011, p. 480). Seeing is not necessarily believing, as photos can be manipulated, distorted, and removed from context (Morris, 2011). In this regard, visual methods do not stand alone but are most credible and useful when used in conjunction with other data: interviews, direct observations, and supporting documents.

Video Elicitation Interviewing

Teachnological developments have increased access to video and lowered costs making video elicitation interviewing and storytelling practical as well as innovative—the frontier of visual qualitatiuve inquiry. During video elicitation interviews, researchers interview using video to stimulate responses. An example is using video elicitation interviews for investigating physician–patient interactions (Henry & Fetters, 2012). Patients or physicians were interviewed about a recent clinical interaction using a video recording of that interaction as an elicitation tool.

Video elicitation is useful because it allows researchers to integrate data about the content of physician–patient interactions gained from video recordings with data about participants' associated thoughts, beliefs, and emotions gained from elicitation interviews. This method also facilitates investigation of specific events or moments during interactions. Video elicitation interviews are logistically demanding and time-consuming, and they should be reserved for research questions that cannot be fully addressed using either standard interviews or video recordings in isolation. As many components of primary care fall into this category, high-quality video elicitation interviews can be an important method for understanding and improving physician–patient interactions in primary care (Henry & Fetters, 2012, p. 118).

Nick Agafanoff (2014) calls the use of video *real ethnography*—using documentary film techniques to capture and communicate stories and narratives. An example of the kind of experimentation going on in video interviewing is the approach of award-winning documentary filmmaker Errol Morris (*The Thin Blue Line*). He invented the "interrotron" to increase rapport and deepen eye contact when videotaping interviews. Morris believes that Americans are so comfortable with television sets that doing his interviews through a television enhances rapport. Morris asks his questions through a specially designed video camera in the same room with the interview subject. The interviewee sees and hears him on a television and responds by talking into the television. Morris, in turn, watches the interview live on TV. The whole interaction takes place face-to-face through televisions placed at right angles to each other in the same room. For the results of this approach, see Morris (2000, 2003, 2012).

Writing as Preparation for or as Part of the Interview

The subject–object interview methodology illustrates another basis for interviewing—including writing as part of the interview. Prior to interviewing the research participants about the 10 ideas, they are given 15 to 20 minutes to jot down things on the index cards. They subsequently choose which cards to talk about and can use their jottings to facilitate their verbal responses. Such an approach gives interviewees a chance to think through some things before responding verbally.

In reflective practice group interviews (Patton, 2011, pp. 266–269), participants are asked to come with stories written that address the focused question for the session:

- Tell about an effective collaboration you have experienced.
- Tell about a time you felt excluded from a group.
- Tell about a time when you overcame a huge obstacle.

Writing the stories in advance facilitates documenting the participants' experiences and helps the group move more quickly to identifying patterns and themes across their stories. The process of asking and clarifying questions about one another's stories deepens and enriches the stories.

Critical incidents are a revealing focus for reflective practice with groups. Interviewees can be invited to identify and write about incidents they consider "critical" to their own development, their family's history, or the work of their organization. Dialogue ensues to understand what made the incident "critical."

Projection Elicitation Techniques

Projection techniques are widely used in psychological assessment to gather information from people. The best-known projective test is probably the Rorschach Test. The general principle involved is to have people react to something other than a question—an inkblot, a picture, a drawing, a photo, an abstract painting, a film, a story, a cartoon, or whatever is relevant. This approach is especially effective in interviewing children, but it can be helpful with people of any age. I found, for example, when doing follow-up interviews two years after completion of a program, that some photographs of the program site and a few group activities greatly enhanced recall.

Students can be interviewed about work they have produced. In the wilderness program evaluation, we interviewed participants about the entries they shared from their journals. Walker (1996) used letters exchanged between friends as the basis for her study of a generation of American women. Holbrook (1996) contrasted official welfare case records with a welfare recipient's journals to display two completely different constructions of reality. Hamon (1996) used proverbs, stories, and tales as a starting point for her inquiry into Bahamian family life. Rettig, Tam, and Magistad (1996) extracted quotes from transcripts of public hearings on child support guidelines as a basis for their fieldwork. Laura Palmer (1988) used objects left in memory of loved ones and friends at the Vietnam War Memorial in Washington, D.C., as the basis for her inquiry and later interviews. An ethnomusicologist will interview people as they listen and react to recorded music. The options for creative interviewing stretch out before us like an ocean teeming with myriad possibilities, some already known, many more waiting to be discovered or created.

Robert Kegan (1982) and colleagues (Helsing, Howell, Kegan, & Lahey, 2008) have had success basing interviews on reactions to 10 words in what they call the subject–object interview. To understand how the interviewee organizes interpersonal and intrapersonal experience, real-life situations are elicited from a series of 10 uniform probes. The interviewee responds to 10 index cards, each listing an idea, concept, or emotion: Angry, Success, Sad, Moved/Touched, Change, Anxious/Nervous, Strong Stand/Conviction, Torn, Lost Something, and Important To Me.

Reactions to these words provide data for the interviewer to explore the interviewee's underlying epistemology or "principle of meaning-coherence" based on Kegan's work *The Evolving Self* (1982). The subject–object interview is a complex and sophisticated method that requires extensive training for proper application and theoretical interpretation. For my purposes, the point is that a lengthy and comprehensive interview interaction can be based on reaction to 10 deceptively simple ideas presented on index cards rather than fully framed questions.

Think-Aloud Protocol Interviewing

"Protocol analysis" or, more literally, the "*think-aloud protocol*" approach, aims to elicit the inner thoughts or cognitive processes that illuminate what's going on in a person's head during the performance of a task, for example, painting or solving a problem. The point is to undertake interviewing as close to the action as possible. While someone engages in an activity, the interviewer asks questions and probes to get the person to talk about what he or she is thinking as he or she does the task. In "teaching rounds" at hospitals, senior physicians do a version of this when they talk aloud about how they're engaging in a diagnosis while medical students listen, presumably learning the experts' thinking processes by hearing them in action. (For details of the think-aloud protocol method, see Ericsson & Simon, 1993; Krahmer & Ummelen, 2004; Pressley & Afflerbach, 1995.)

Wilson (2000) used a protocol research design in a doctoral dissertation that investigated student understanding and problem solving in college physics. Twenty students in individual 45-minute sessions were videotaped and asked to talk aloud as they tried to solve three introductory physics problems of moderate difficulty involving Newton's second law. Wilson was able to pinpoint the cognitive challenges that confronted the students as they tried to derive the acceleration of a particle moving in various directions and angles with respect to a particular reference frame.

Real-Time Qualitative Data Collection

The *experience sampling method* was developed to collect information on people's reported feelings in real time in natural settings during selected moments of the day. Participants in experience sampling studies carry a handheld computer that prompts them several times during the course of the day (or days) to answer a set of questions immediately. They indicate their physical location, the activities in which they were engaged just before they were prompted, and the people with whom they were interacting. They also report their current subjective experience by indicating their feelings (Kahneman & Krueger, 2006, p. 9). While the experience sampling method as originally developed had participants respond on quantitative scales, the pervasive use of handheld devices in contemporary society means that qualitative data (e.g., texts, tweets) can be collected with this technique.

Creative Interviewing

As the preceding examples illustrate, qualitative inquiry need not be confined to traditional written interview protocols and taking field notes. Researchers and evaluators have considerable freedom to creatively adapt qualitative methods to specific situations and

PERFORMING THE INTERVIEW

Interviews are performance texts. A performative social science uses the reflexive, active interview as a vehicle for producing moments of performance theater, a theater that is sensitive to the moral and ethical issues of our time. This interview form is gendered and dialogic. In it, gendered subjects are created through their speech acts. Speech is performative. It is action. The act of speech, the act of being interviewed, becomes a performance itself. The reflexive interview, as a dialogic conversation, is the site and occasion for such performances; that is, the interview is turned into a dramatic, poetic text. In turn, these texts are performed, given dramatic readings.

—Norman K. Denzin (2003, p. 84)
Performance Ethnography

purposes using anything that comes to mind—and works—as a way to enter into the world and worldview of others. Not only are there many variations in what stimuli to use and how to elicit responses, but creative variations also exist for who conducts interviews. We turn now to those options.

Collaborative and Participatory Interviewing

Chapter 4, "Practical and Actionable Qualitative Applications," discussed conducting research and evaluation in a collaborative mode in which professionals and nonprofessionals become co-inquirers. Participatory approaches are widely used in community research, health research, educational evaluation, and organizational development. For example, participatory action research encourages collaboration within a mutually respectful inquiry relationship to understand and/or solve organizational or community problems. Participatory evaluation involves program staff and participants in all aspects of the evaluation process to increase evaluation understanding and use while also building capacity for future inquiries. Empowerment evaluation aims to foster self-determination among those who participate in the inquiry process. This can involve forming empowerment collaborations between researchers and participants and teaching participants to do research themselves. Feminist methods are often highly participatory (Ackerly & True, 2010; Bamberger & Podems, 2002; Brisolara, Seigart, & SenGupta, 2014, Hesse-Biber, 2011; Hesse-Biber & Leavy, 2006b; Hughes, 2002; Mertens, 2005; Olesen, 2011; Podems, 2014b).

In-depth interviewing is especially useful for supporting collaborative inquiry because the methods are accessible to and understandable by nonresearchers and community-based people without much technical expertise. Exhibit 4.13 (p. 222) presents *Principles of Fully Participatory and Genuinely Collaborative Inquiry.* The following sections provide some specific examples of collaborative and participatory interviewing approaches.

PARTICIPATORY ACTION RESEARCH

Participatory action research is the sum of its individual terms, which have had and continue to have multiple combinations and meanings, as well as a particular set of assumptions and processes. . . . Participation is a major characteristic of this work, not only in the sense of collaboration, but in the claim that all people in a particular context (for both epistemological and, with it, political reasons) need to be involved in the whole of the project undertaken. Action is interwoven into the process because change, from a situation of injustice toward envisioning and enacting a "better" life (as understood from those in the situation) is a primary goal of the work. Research as a social process of gathering knowledge and asserting wisdom belongs to all people, and has always been part of the struggle toward greater social and economic justice locally and globally.

—Brydon-Miller, Kral, Maguire, Noffke, and Sabhlok (2011, p. 388)
Jazz and the Banyan Tree: Roots and Riffs on Participatory Action Research

Participant Interview Chain

As a participant-observer in the wilderness training program for adult educators, I was involved in (a) documenting the kinds of experiences program participants were having and (b) collecting information about the effects of those experiences on the participants and their work situations. In short, the purpose of the evaluation was to provide formative insights that could be used to help understand the personal, professional, and institutional outcomes of intense wilderness experiences for these adult educators. But the two of us doing the evaluation didn't have sufficient time and resources to track everyone—40 people—in depth. Therefore, we began discussing with the program staff ways in which the participants might become involved in the data collection effort to meet both program and evaluation needs. The staff liked the idea of involving participants, thereby introducing them to observation and interviewing as ways of expanding their own horizons and deepening their perceptions.

The participants' backpacking field experience was organized in two groups. We used this fact to design a data collection approach that would fit with the programmatic needs of sharing information between the two groups. Participants were paired to interview each other. At the very beginning of the first trip, before people knew each other, all of the participants were given a short, open-ended interview of 10 questions. They were told that each of them, as part of their project participation, was to have responsibility for documenting the experiences of their pairmate throughout the year. They were given a little bit of interview training and a lot of encouragement about probing and were told to record responses fully, thereby taking responsibility for helping build this community record of individual experiences. They were then sent off in pairs and given two hours to complete the interviews with each other, recording the responses by hand.

At the end of the 10-day experience, when the separate groups came back together, the same pairs of participants, consisting of one person from each group, were again given an interview outline and sent off to interview each other about their respective experiences. This served the program need of sharing of information and an evaluation need of collection of information. The trade-off, of course, was that with the minimal interview training given to the participants and the impossibility of carefully supervising, controlling, and standardizing the data collection, the results were of variable quality. This mode of data collection also meant that confidentiality was minimal and certain kinds of information might not be shared. But we gathered a great deal more data than we could have obtained if we had had to do all the interviews ourselves.

Exhibit 7.16 presents an example of participatory inquiry in a study of a program aimed at helping women engaged in prostitution transition into a new life. The program participants, women who had themselves been prostitutes, were trained to conduct the evaluation interviews.

Limitations certainly exist as to how far one can push participant involvement in data collection and analysis without interfering in the program or burdening participants. But before those limits are reached, a considerable amount of useful information can be collected by involving program participants in the actual data collection process. I have since used similar participant interview pairs in a number of

| EXHIBIT 7.16 | Training Nonresearchers as Focus Group Interviewers: Women Leaving Prostitution |

Rainbow Research studied the feasibility of developing a transitional housing program for prostituted women. To assist us we recruited 5 women who had been prostituted, trained them in focus group facilitation and had them do our interviews with women leaving prostitution. For them the experience was empowering and transformational. They were excited about learning a new skill, pleased to be paid for this work and thought it rewarding that it might benefit prostituted women. Especially thrilling for them during the interviews was the validation and encouragement they received from their peers for the work they were doing. Our work together also had its light moments. During a group simulation, our interviewers loudly and provocatively bantered with one another as they might have on the street.

Our interviewers were proud of their contribution. At project's end they requested certificates acknowledging the training they had received and the interviews successfully performed. And, because they had performed well, we were pleased to oblige. In the simulations they critiqued our interview guide, leading us to edit the language, content, order and length, introduce new questions and drop others. Clearly they had rapport with their peers based on shared discourse and experience, allowing them to gather information others without the experience of prostitution would have been hard pressed to secure. This was apparent in the reliability of our data. Comparing across the interviews, responses to the same items were highly consistent. For all concerned it was a positive experience, with findings that most definitely shaped our final recommendations.

—Barry B. Cohen
Executive Director, Rainbow Research
Minneapolis, Minnesota

program evaluations with good results. The trick is to integrate the data collection into the program. Such cooperative and interactive interviewing can deepen the program experience for those involved by making qualitative data collection and reflection part of the change process (Cousins, Whitmore, & Shulha,

2014; Fetterman, Rodríguez-Campos, Wandersman, & O'Sullivan, 2014; King & Stevahn, 2013; Patton, 2011, 2012a).

Photovoice

Earlier, I discussed photo elicitation approaches to enhance interviewing. Incorporating photography into data collection can also be a highly participatory process. The best-known and most widely used participatory photo elicitation approach is *Photovoice*, which combines photography and qualitative inquiry with grassroots social action (see sidebar). The Minnesota Department of Health (2013) sponsored a Photovoice project in which youth from 13 communities were trained in Photovoice and given cameras to answer questions about their everyday lives through photography.

SIDEBAR

PHOTOVOICE

Photovoice combines photography and qualitative inquiry with grassroots social action. Community participants present their experiences and perspectives by taking photographs, developing narratives to go with their photos, analyzing them together, and using the results to advocate for change with policymakers and philanthropic funders. Photovoice, as a form of qualitative participatory photography, was developed by Caroline C. Wang and Mary Ann Burris while working in Beijing, China, in 1992. They sought a way for rural women of Yunnan Province, China, to tell their stories and influence the policies and programs that affected them (Wang & Burris, 1994). It has evolved into a global movement to combine visual evidence, qualitative narratives, and social change (http://www.photovoice.org/).

Data Collection by Program Staff

Program staff constitute another resource for data collection that is often overlooked. Involving program staff in data collection raises objections about staff subjectivity, data contamination, loss of confidentiality, the vested interests of staff in particular kinds of outcomes, and the threat that staff can pose to clients or students from whom they are collecting the data. Balancing these objections are the things that can be gained from staff involvement in data collection: (a) greater staff commitment to the evaluation, (b) increased staff reflectivity, (c) enhanced understanding of the data collection process that comes from training staff in data collection procedures, (d) increased understanding by staff of program

participants' perceptions, (e) increased data validity because of staff rapport with participants, and (f) cost savings in data collection.

One of my first evaluation experiences involved studying a program to train teachers in open education at the University of North Dakota. The faculty were interested in evaluating that program, but there were almost no resources available for a formal evaluation. Certainly not enough funds existed to bring in an external evaluation team to design the study, collect data, and analyze the results. The main means of data collection consisted of in-depth interviews with student teachers in 24 different schools and classrooms throughout North Dakota and structured interviews with 300 parents who had children in those classrooms. The only evaluation monies available would barely pay for the transportation and the actual mechanical costs of data collection. Staff and students at the university agreed to do the interviews as an educational experience. I developed structured interview forms for both teacher and parent interviews and trained all of the interviewers in a full-day session. Interviewers were assigned to geographical areas, making sure that no staff collected data from their own student teachers. The interviews were recorded and transcribed. I did follow-up interviews with a 5% sample as a check on the validity and reliability of the student and staff data.

After data collection, seminars were organized for staff and students to share their personal perceptions based on their interview experiences. Their stories had considerable impact on both staff and students. One outcome was the increased respect both staff and students had for the parents. They found the parents to be perceptive, knowledgeable, caring, and deeply interested in the education of their children. Prior to the interviewing, many of the interviewers had held fairly negative and derogatory images of North Dakota parents. The systematic interviewing had put them in a situation where they were forced to listen to what parents had to say, rather than telling parents what they (as educators) thought about things, and in learning to listen, they had learned a great deal. The formal analysis of the data yielded some interesting findings that were used to make some changes in the program, and the data provided a source of case materials that were adapted for us in training future program participants, but it is likely that the major and most lasting impact of the evaluation came from the learnings of students and staff who participated in the data collection. That experiential impact was more powerful than the formal findings of the study—an example of "evaluation process use" (Patton, 2008, 2012a), using an evaluation process for participant and organizational learning.

By the way, had the interviewers been paid at the going commercial rate, the data collection could have cost at least $30,000 in personnel expenses. As it was, there were no personnel costs, and a considerable human contribution was made to the university program by both students and staff.

Such participatory action research remains controversial. As Kemmis and McTaggart (2000) noted in their extensive review of participatory approaches,

> In most action research, including participatory action research, the researchers make sacrifices in methodological and technical rigor in exchange for more immediate gains in face validity: whether the evidence they collect makes sense to them in their context. For this reason, we sometimes characterize participatory action research as "low-tech" research: it sacrifices in methodological sophistication in order to generate timely evidence that can be used and further developed in a real-time process of transformation (of practices, practitioners, and practice settings). (p. 591)

Whether some loss of methodological sophistication is merited depends on the primary purpose of the inquiry and the primary intended users of the results. Participatory research will have lower credibility among external audiences, especially among scholars who make rigor their primary criterion for judging quality. Participants involved in improving their work or lives, however, lean toward pragmatism where *what is useful determines what is true.* As Kemmis and McTaggart (2000) conclude,

> The inevitability—for participants—of having to live with the consequences of transformation provides a very concrete "reality check" on the quality of their transformative work, in terms of whether their practices are more efficacious, their understandings are clearer, and the settings in which they practice are more rational, just, and productive of the kinds of consequences they are intended to achieve. For participants, the point of collecting compelling evidence is to achieve these goals, or, more precisely, to avoid subverting them intentionally or unintentionally by their action. Evidence sufficient for this kind of "reality checking" can often be low-tech (in terms of research methods and techniques) or impressionistic (from the perspective of an outsider who lacks the contextual knowledge that the insider draws on in interpreting this evidence). But it may still be "high-fidelity" evidence from the perspective of understanding the nature and consequences of particular interventions in transformations made by participants, in their context— where they are privileged observers. (p. 592)

Interactive Group Interviewing and Dialogues

The involvement of program staff or clients as colleagues or coresearchers in action research and program evaluation changes the relationship between evaluators and staff, making it interactive and cooperative rather than one-sided and antagonistic. William Tikunoff (1980), a pioneer in "interactive research" in education projects, found that putting teachers, researchers and trainer/developers together as a team increased both the meaningfulness and the validity of the findings because teacher cooperation with and understanding of the research made the research less intrusive, thus reducing rather than increasing reactivity. Their discussions were a form of group interviews in which *they all asked each other questions.*

The problem of how research subjects or program clients will react to staff involvement in an evaluation, particularly involvement in data collection, needs careful scrutiny and consideration in each situation in which it is attempted. Reactivity is a potential problem in both conventional and nonconventional designs. Breaches of confidence and/or reactivity-biased data cannot be justified in the name of creativity. On the other hand, as Tikunoff's experiences indicate, interactive designs may increase the validity of data and reduce reactivity by making evaluation more visible and open, thereby making participants or clients less resistant or suspicious (King & Stevahn, 2013).

These approaches can reframe inquiry from a duality (interviewer–interviewee) to a dialogue in which all are co-inquirers. Miller and Crabtree (2000) advocate such a collaborative approach even in the usually closed and hierarchical world of medical and clinical research:

> We propose that clinical researchers investigate questions emerging from the clinical experience with the clinical participants, pay attention to and reveal any underlying values and assumptions, and direct results toward clinical participants and policy makers. This refocuses the gaze of clinical research onto the clinical experience and redefines its boundaries so as to answer three questions: Whose question is it? Are hidden assumptions of the clinical world revealed? For whom are the research results intended? . . . Patients and clinicians are invited to explore their own and/or each other's questions and concerns with whatever methods are necessary. Clinical researchers share ownership of the research with clinical participants, thus undermining the patriarchal bias of the dominant paradigm and opening its assumptions to investigation. This is the situated knowledge . . . where space is created to find a larger, more inclusive vision of clinical research. (p. 616)

MEDIATED CONVERSATIONS

SIDEBAR

Mediated conversations are an innovative method for the generation of rich qualitative data on complex issues such as those that researchers often face in education contexts. The method draws from the sociocultural paradigm (Wertsch, 1991), which highlights the role of *artifacts* and *audience* in mediating participants' actions and thoughts. Mediated conversations were developed during the Curriculum Implementation Exploratory Studies (in New Zealand), when we invited school leaders and teachers to a series of one-day data-generating "workshops." The conversations that took place were mediated by two kinds of artifacts: one that participants had been asked to bring and discuss and the other the 10 contract research questions. The artifacts served as a practical situated exemplification of the participants' descriptions of their work in relation to curriculum implementation. They served as a locus and prompt for questions and discussion. We convened the mediated conversations for our *research* purposes, but the participants experienced them as rich *professional learning*. Mediated conversations appear to provide a participatory method for data generation that has immediate benefits for researchers and research participants.

SOURCE: Cowie et al. (2009) and Hipkins, Cowie, Boyd, Keown, and McGee (2011).

Creativity and Data Quality: Qualitative Bricolage

> I realized quite early in this adventure that interviews, conventionally conducted, were meaningless. Conditioned cliches were certain to come. The question-and-answer technique may be of some value in determining favored detergents, toothpaste and deodorants, but not in the discovery of men and women.
>
> —Studs Terkel (Quoted by Douglas, 1985, p. 7)

No definitive list of creative interviewing or inquiry approaches can or should be constructed. Such a list would be a contradiction in terms. Creative approaches are those that are situationally responsive and appropriate, credible to primary intended users,

and effective in opening up new understandings. The approaches just reviewed are deviations from traditional research practice. Each idea is subject to misuse and abuse if applied without regard for the ways in which the quality of the data collected can be affected. I have not discussed such threats and possible errors in depth because I believe it is impossible to identify in the abstract and in advance all the trade-offs involved in balancing concerns for accuracy, utility, feasibility, and propriety. For example, having program staff do client interviews in an outcomes evaluation could (a) seriously reduce the validity and reliability of the data, (b) substantially increase the validity and reliability of the data, or (c) have no measurable effect on data quality. The nature and degree of effect would depend on staff relationships with clients, how staff were assigned to clients for interviewing, the kinds of questions asked, the training of the interviewers, the attitudes of clients toward the program, the purposes to be served by the evaluation, the environmental turbulence of the program, and so on and so forth. Program staff might make better or worse interviewers than external evaluation researchers depending on these and other factors. An evaluator must grapple with these kinds of data quality questions for all designs, particularly nontraditional approaches.

Practical, but creative, data collection consists of using whatever resources are available to do the best job possible. Constraints always exist and do what constraints do—constrain. Our ability to think of alternatives is limited. Resources are always limited. This means data collection will be imperfect, so dissenters from research and evaluation findings who want to attack a study's methods can always find some grounds for doing so. A major reason for actively involving intended evaluation users in methods decisions is to deal with weaknesses and consider trade-off threats to data quality *before* data are collected. By strategically calculating threats to utility, as well as threats to validity and authenticity, it is possible to make practical decisions about the strengths of creative and nonconventional data collection procedures (Patton, 2008, 2012a).

The creative, adaptive inquirer using diverse techniques may be thought of as a "bricoleur." The term comes from Levi-Strauss (1966), who defined a bricoleur as a "jack of all trades or a kind of professional do-it-yourself person" (p. 17).

> The qualitative researcher as bricoleur or maker of quilts uses the aesthetic and material tools of his or her craft, deploying whatever strategies, methods, or empirical materials are at hand. If new tools or techniques have to be invented, or pieced together, then the researcher will do this. (Denzin & Lincoln, 2000, p. 4)

Creativity begins with being open to new possibilities, the bricolage of combining old things in new ways, including alternative and emergent forms of data collection, transformed observer–observed relations, and reframed interviewer–interviewee interconnections. Naturalistic inquiry calls for ongoing openness to whatever emerges in the field and during interviews. This openness means avoiding forcing new possibilities into old molds. The admonition to remain open and creative applies throughout naturalistic inquiry, from design through data collection and into analysis. Failure to remain open and creative can lead to the error made by a traveler who came across a peacock for the first time, a story told by Halcolm.

A traveler to a new land came across a peacock. Having never seen this kind of bird before, he took it for a genetic freak. Taking pity on the poor bird, which he was sure could not survive for long in such deviant form, he set about to correct nature's error. He trimmed the long, colorful feathers; cut back the beak; and dyed the bird black. "There now," he said, with pride in a job well done, "you now look more like a standard guinea hen."

Adapting Interviewing for Particular Interviewees

This chapter has reviewed general principles of and approaches to in-depth qualitative interviewing. However, most researchers, evaluators, and practitioners specialize in working with and studying specific target populations using finely honed interviewing techniques. Interviewing "elites" or "experts" often requires an interactive style.

> Elites respond well to inquiries about broad topics and to intelligent, provocative, open-ended questions that allow them the freedom to use their knowledge and imagination. In working with elites, great demands are placed on the ability of the interviewer, who must establish competence by displaying a thorough knowledge of the topic or, lacking such knowledge, by projecting an accurate conceptualization of the problem through shrewd questioning. (Rossman & Rallis, 1998, p. 134)

Robert Coles (1990) became adept at interviewing children, as have Guerrero-Manalo (1999), Graue and Walsh (1998), Holmes (1998), and Greig and Taylor (1998). Rita Arditti (1999) developed special culturally and politically sensitive approaches for gaining access to and interviewing grandmothers whose children were among "the disappeared" during

---

ON INTERVIEWING POLITICIANS

Political people could always be more frank. . . . There is a truth that can't be spoken. You can come to a shorthand understanding of certain realities—I know what we can't talk about.

—Mark Leibovich (2013)
New York Times journalist

SIDEBAR

Argentina's military regime. Guerrero (1999a, 1999b) developed special participatory approaches for interviewing women in developing countries. Judith Arcana (1981, 1983) drew on her own experiences as a mother to become expert at interviewing 180 mothers for two books about the experiences of mothering (one about mothers and daughters, one about mothers and sons).

At the other end of the continuum from Cole's delightful stories of childhood innocence are Jane Gilgun's haunting interviews with male sexual offenders and Angela Browne's intensive interviews with women incarcerated in a maximum security prison. Gilgun (1991, 1994, 1995, 1996, 1999) conducted hundreds of hours of interviews with men who have perpetrated violent sex offenses against women and children, most of them having multiple victims. She learned to establish relationships with these men through repeated life history interviews but did so without pretending to condone their actions, and sometimes challenging their portrayals. Two examples offer a sense of the challenges for someone undertaking such work through long hours and horrific details.

One man, engaged to be married at the time of his arrest, confessed to seven rapes; Gilgun interviewed him for a total of 14 hours over 12 different interviews, including detailed descriptions of his sexual violence. Another man molested more than 20 boys; at the time of his arrest, he was married, sexually active with his wife, and a stepfather of two boys; she obtained 20 hours of tape over 11 interviews. These cases were particularly intriguing as purposeful samples because both men were white college graduates in their early 30s who were employed as managers with major supervisory responsibilities and who came from upper middle-class, two-parent, never divorced families where the fathers held executive positions and mothers were professionals. Gilgun worked with an associate in transcribing and interpreting the interviews.

The data were so emotionally evocative that we spent a great deal of time working through our personal responses. Almost two years went by before we found we had any facility in articulating the meanings of the discourses we identified in the informants' accounts. Their way of thinking was for the most part outside our frames of reference. As we struggled through these interpretive processes, we made notes of our responses. . . . Most compelling to us [was the men believing they were] entitled to take what they wanted and of defining persons and situations as they wished . . . , to suit themselves. Overall, the discourses they invoked served oppressive hegemonic ends. We also found that the men experienced chills, thrills, and intense emotional gratification as they imposed their wills on smaller, physically weaker persons. (Gilgun & McLeod, 1999, p. 175)

The life work of Angela Browne illustrates a similar commitment to in-depth, life history, interviewing with people who are isolated from the mainstream and whose experiences are little understood by the general culture. After her groundbreaking study of women who kill violent partners in self-defense (*When Battered Women Kill*, 1987), Browne began gathering life history narratives from women incarcerated in a maximum security prison. These interviews, conducted in a small room off a tunnel in the middle of the facility and six hours in length, included women with lifetime histories of trauma, much of it at the hands of family members in early childhood. Some had witnessed brutal homicides; others were serving time for crimes of violence they had committed. Their stories were painful to tell and to hear. Interviews often were so emotionally draining that Browne came away exhausted, sometimes needing to debrief on both the impact of what she had heard and the dynamics of the interviewing process. On several occasions, we had

lengthy phone conversations immediately following the interviews, while she was still within the prison walls. This kind of extreme interviewing takes unusual skill, dedication, and self-knowledge, coupled with a keen interest in the dynamics of human interaction. Although Browne returned home drained after each full week of conducting these day-long interviews, her enthusiasm for the task and her appreciation of the respondents' strength and lucidity never dimmed.

The works of Gilgun and Browne illustrate the intensity, commitment, long hours, and hard work involved with certain in-depth and life history approaches to interviewing.

Adapting Interviewing to Particular Target Populations

Applications and methods of qualitative inquiry, especially interviewing techniques for specially targeted populations and specialized disciplinary approaches, continue to evolve as interest in qualitative methods grows exponentially (a metaphoric rather than a statistical estimation). General interview skills include asking open-ended questions, listening carefully to ask follow-up questions, effective and sensitive probing, distinguishing different kinds of questions, and pacing the interview. But additional challenges arise in interviewing particular target populations. General interview skills are necessary as a foundation for in-depth qualitative interviewing, but those skills must then be adapted to particular target groups and situations. Exhibit 7.17 on the next page highlights some specialized interviewing approaches and accompanying challenges, including online interviewing. As applications and techniques have proliferated, so have concerns about the ethical challenges of qualitative inquiry, our next topic.

EXHIBIT 7.17   Special Interviewing Challenges for Particular Target Populations: Five Examples

This table highlights some of the specific challenges that arise in interviewing particular target populations. This is by no means an exhaustive list of target populations or challenges. It is meant to illustrate how general interview skills are necessary as a foundation for in-depth qualitative interviewing but those skills must then be adapted to particular target groups and situations.

SPECIAL TARGET POPULATIONS	PARTICULAR CHALLENGES	RESOURCES
1. *Online:* Interviews with online and social media individuals and groups. (Important to distinguish real time vs. asynchronous interviews; they involve different processes.)	Face-to-face interviews are also observations; nonverbal communication is part of the interaction. Online interviews lose the nonverbal communications and connection	Online qualitative interviewing guides (James & Busher, 2012; Salmons, 2010, 2012)
2. *Children and youth:* Interviewing children of different ages (5–8, 9–12, and teenagers) poses different challenges.	Gearing questions to their level of understanding, keeping their interest and attention, getting informed consent from parents and guardians, and distinguishing research and evaluation interviewing from clinical, diagnostic, and forensic interviewing	Special considerations and resources for interviewing children and adolescents (Alderson, 2001; Coles, 1989, 1990; Eder & Fingerson, 2003; Graue & Walsh, 1998; Greig & Taylor, 1998; Holmes, 1998; Kavanagh, 2013; Lambert, Glacken, & McCarron, 2013)
3. *Marginalized and fringe groups:* People who perceive themselves as outside the mainstream and may be especially suspicious of inquiring strangers.	Gaining trust, getting "insider" information that interviewees fear may be misused, and distinguishing what is true from what is constructed to protect or keep safe the truth	*Examples:* (a) John E. Mack (2007, 2010) of Harvard Medical School conducted more than 200 interviews with people who believed that they had been abducted by aliens from outer space. (b) Miriam Boeri (2002) interviewed women who were religious cult members.
4. *Interviewing elites:* People in positions of power, wealth, and prestige.	Gaining access, distinguishing what is true from scripted responses aimed at creating a favorable impression, and getting beyond the surface to deeper issues and insights	Rice (2009), reflections on interviewing elites, and Morris (2009), the truth about interviewing elites
5. *Interviewing older people, the elderly (over the age of 65):* As the population ages, different ages (65–80, 81–95, and over age 95) may pose different challenges.	May require a lot of time to establish a trusting relationship and hear lengthy stories; staying focused; remembering accurately; may evoke strong memories, sometimes painful	Robertson and Hale (2011), reflections on interviewing older people, and Erber (2010) and Wenger (2003)

Ethical Issues and Challenges in Qualitative Interviewing

Interviews are interventions. They affect people. A good interview evokes thoughts, feelings, knowledge, and experience not only to the interviewer but also to the interviewee. The process of being taken through a directed, reflective process affects the persons being interviewed and leaves them knowing things about themselves that they didn't know—or least were not fully aware of—before the interview. Two hours or more of thoughtfully reflecting on an experience, a program, or one's life can be change inducing; 10, 15, or 20 hours of life history interviewing can be transformative—or not. Therein lies the rub. Neither you nor the interviewee can know, in advance, and sometimes even after the fact, what impact an interviewing experience will have or has had.

The purpose of a research interview is first and foremost to gather data, not change people. Earlier, in the section on neutrality, I asserted that an interviewer is not a judge. Neither is a research interviewer a therapist. Staying focused on the purpose of the interview is critical to gathering high-quality data. Still, there will be many temptations to stray from that purpose. It is common for interviewees to ask for advice, approval, or confirmation. Yielding to these temptations, the interviewer may become the interviewee—answering more questions than are asked.

On the other hand, the interviewer, in establishing rapport, is not a cold slab of granite—unresponsive to learning about great suffering and pain that may be reported and even reexperienced during an interview. In a major farming systems needs assessment project to develop agricultural extension programs for distressed farm families, I was part of a team of 10 interviewers (working in pairs) who interviewed 50 farm families. Many of these families were in great pain. They were losing their farms. Their children had left for the city. Their marriages were under stress. The two-hour interviews traced their family history, their farm situation, their community relationships, and their hopes for the future. Sometimes questions would lead to husband–wife conflict. The interviews would open old wounds, lead to second-guessing decisions made long ago, or bring forth painful memories of dreams never fulfilled. People often asked for advice— what to do about their finances, their children, government subsidy programs, even their marriages. But we were not there to give advice. Our task was to get information about needs that might, or might not, lead to new programs of assistance. Could we do more

than just ask our questions and leave? Yet, as researchers, could we justify in any way intervening? Yet again, our interviews were already an intervention. Such are the ethical dilemmas that derive from the power of interviews.

What we decided to do in the farm family interviews was leave each family a packet of information about resources and programs of assistance, everything from agricultural referrals to financial and family counseling. To avoid having to decide which families really needed such assistance, we left the information with all families—separate and identical packages for both husband and wife. When interviewees asked for advice during the interview, we could tell them that we would leave them referral information at the end of the interview.

While interviews may be intrusive in reopening old wounds, they can also be healing. In doing follow-up interviews with families who had experienced child sexual abuse, we found that most of the mothers appreciated the opportunity to tell their stories, vent their rage against the system, and share their feelings with a neutral, but interested, listener. Our interviews with elderly residents participating in a program to help them stay in their homes and avoid nursing home institutionalization typically lasted much longer than planned because the elderly interviewees longed to have company and talk. When interviewees are open and willing to talk, the power of interviewing poses new risks. People will tell you things they never intended to tell you. This can be true even with reluctant or hostile interviewees—a fact depended on by journalists. Indeed, it seems at times that the very thing someone is determined *not* to say is the first thing he or she tells, just to release the psychological pressure of secrecy or deceit.

I repeat, people in interviews will tell you things they never intended to tell. Interviews can become confessions, particularly under the promise of confidentiality. But beware that promise. Social scientists can be summoned to testify in court. We do not have the legal protection of clergy and lawyers. In addition, some information must be reported to the police, for example, evidence of child abuse. Thus, the power of interviewing can put the interviewees at risk. *The interviewer needs to have an ethical framework for dealing with such issues.*

There are also direct impacts on interviewers. The previous section described the wrenching interviews conducted by Jane Gilgun with male sex offenders and Angela Browne with incarcerated women, and the physical and emotional toll of those interviews on them as interviewers, exposed for hours on end to horrendous details of violence and abuse. In a family sex abuse project (Patton, 1991), we found that the interviewers needed to be extensively debriefed, sometimes in support groups together, to help them process and deal with the things they heard. They could only take in so much without having some release, some safety valve for their own building anger and grief. Middle-class interviewers going into poor areas may be shocked and depressed by what they hear and see. It is not enough to do preparatory training before such interviewing. Interviewers may need debriefing—and their observations and feelings can become part of the data on team projects.

These examples are meant to illustrate the power of interviewing and why it is important to anticipate and deal with the ethical dimensions of qualitative inquiry. Because qualitative methods are highly personal and interpersonal, because naturalistic inquiry takes the researcher into the real world where people live and work, and because in-depth interviewing opens up what is inside people, qualitative inquiry may be more intrusive and involve greater reactivity than surveys, tests, and other quantitative approaches.

Exhibit 7.18 presents a checklist of 12 common ethical issues as a starting point in thinking through ethical issues in design, data collection, analysis, and reporting. The next sections elaborate some common issues of special concern.

- Informed consent and confidentiality
- Confidentiality versus people owning their own stories
- How much of an interview must be approved in advance?
- Reciprocity: Should interviewees be compensated? If so, how?
- How hard to push for sensitive information?
- Being careful in the face of danger in the field

EXHIBIT 7.18   Ethical Issues Checklist

1. *Explaining purpose.* How will you explain the purpose of the inquiry and the methods to be used in ways that are accurate and understandable?

   - What language will make sense to the participants in the study?
   - What details are critical to share? What can be left out?
   - What's the expected value of your work to society and to the greater good?

Guiding principle: Be clear, honest, and transparent about purpose.

2. *Reciprocity.* What's in it for the interviewee?

   - Why should the interviewee participate in the interview?
   - What is appropriate compensation?

Guiding principle: Honor the gift of an interviewee's time in a meaningful and tangible way.

3. *Promises.* Don't make promises lightly, for example, promising a copy of the recording or the report.

Guiding principle: If you make promises, keep them.

4. *Risk assessment.* In what ways, if any, will conducting the interview put people at risk?

   - Psychological stress
   - Legal liabilities

- In evaluation studies, continued program participation (if certain things become known)
- Ostracism by peers, program staff, or others for talking
- Political repercussions

How will you describe these potential risks to interviewees?

How will you handle them if they arise?

Guiding principle: First, do no harm.

5. *Confidentiality.* What are reasonable promises of confidentiality that can be fully honored? Know the difference between confidentiality and anonymity. (Confidentiality means you know but won't tell. Anonymity means you don't know, as in a survey returned anonymously.)

   - What things can you *not* promise confidentiality about, for example, illegal activities, evidence of child abuse, or neglect?
   - Will names, locations, and other details be changed? Or do participants have the option of being identified? (See discussion of this in the text.)
   - Where will data be stored?
   - How long will data be maintained?

Guiding principle: Know the ethical and legal dimensions of confidentiality.

6. *Informed consent.* What kind of informed consent, if any, is necessary for mutual protection?

- What are your local institutional review board (IRB) guidelines and/or requirements, or those of an equivalent committee for protection of human subjects in research?
- What has to be submitted, under what timelines, for IRB approval, if applicable?

Guiding principle: Know and follow the standards of your discipline or field.

7. *Data access and ownership.* Who will have access to the data? For what purposes?

- Who owns the data in an evaluation? (Be clear about this in the contract.)
- Who has right of review before publication? (For example, of case studies, by the person or organization depicted in the case, or of the whole report, by a funding or sponsoring organization)

Guiding principle: Don't wait until publication to deal with data ownership issues; anticipate data access and ownership issues from the beginning.

8. *Interviewer mental health.* How will you and other interviewers likely be affected by conducting the interviews?

- What might be heard, seen, or learned that may merit debriefing and processing?
- Who can you talk to about what you experience without breaching confidentiality?

Guiding principle: Fieldwork is engaging, intellectually and emotionally. Take care of yourself and your coresearchers.

9. *Ethical advice.* Who will be the researcher's confidant and counselor on matters of ethics during a study? Not all issues can be anticipated in advance. Knowing who you will go to in the event of difficulties can save precious time in a crisis and bring much needed comfort.

Guiding principle: Plan ahead, and know who you will consult on emergent ethical issues.

10. *Data collection boundaries.* How hard will you push for responses from interviewees?

- What lengths will you go to in trying to gain access to data you want? What won't you do?
- How hard will you push interviewees to respond to questions about which they show some discomfort?

Guiding principles: Know yourself. Err on the side of caution. Don't let the ends justify the means in overstepping boundaries.

11. *Intersection of ethical and methodological choices.*

- Methods and ethics are intertwined. Understand the intersection.
- As a qualitative methodologist, based on your own work, contribute to the field by reporting the ethical challenges you face.

Guiding principle: Include ethical dilemmas faced and handled in your methods discussion.

12. *Ethical versus legal.* What ethical framework and philosophy informs your work and ensures respect and sensitivity for those you study beyond whatever may be required by law?

- What disciplinary or professional code of ethical conduct will guide you?

Guiding principles: Don't make up ethical responses along the way. Know your profession's ethical standards. Know what the law in your jurisdiction requires.

For additional guidance on ethics in qualitative inquiry, see Christians (2000), Rubin and Rubin (2012, pp. 85–93), and Silverman and Marvasti (2008).

Informed Consent and Confidentiality

Informed consent protocols and opening statements in interviews typically cover the following issues:

What is the purpose of collecting the information?

Who is the information for? How will it be used?

What will be asked in the interview?

How will responses be handled, including confidentiality?

What risks and/or benefits are involved for the person being interviewed?

The interviewer often provides this information in advance of the interview and then again at the beginning of the interview. Providing such information does not, however, require making long and elaborate speeches. Statements of purpose should be simple, straightforward, transparent, and understandable. Long statements about what the interview is going to be like, and how it will be used, when such statements are made at the beginning of the interview, are usually either boring or anxiety producing. The interviewee will find out soon enough what kinds of questions are going to be asked and, from the nature of the questions, will make judgments about the likely use of such information. The basic messages to be communicated in the opening statement are (a) that the information is important, (b) the reasons for that importance, and (c) the willingness of the interviewer to explain the purpose of the interview out of respect for the interviewee. Here's an example of an opening interview statement from an evaluation study.

> I'm a program evaluator brought in to help improve this program. As someone who has been in the program, you are in a unique position to describe what the program does and how it affects people. And that's what the interview is about: your experiences with the program and your thoughts about your experiences.
>
> The answers from all the people we interview, and we're interviewing about 25 people, will be combined for our report. Nothing you say will ever be identified with you personally. As we go through the interview, if you have any questions about why I'm asking something, please feel free to ask. Or if there's anything you don't want to answer, just say so. The purpose of the interview is to get your insights into how the program operates and how it affects people.
>
> Any questions before we begin?

This may seem straightforward enough, but dealing with real people in the real world, all kinds of complications can arise. Moreover, some genuine ethical quandaries have arisen in recent years around the ethics of research in general and qualitative inquiry in particular.

How Much of an Interview Must Be Approved in Advance?

In the chapter on observation and fieldwork, I discussed the problems posed by approval protocols aimed at protecting human subjects given the emergent and flexible designs of naturalistic inquiry. IRBs for the protection of human subjects often insist on approving actual interview questions, which can be done when using the standardized interview format discussed in this chapter, but it works less well when using an interview guide and doesn't work at all for conversational interviewing, where the questions emerge at the moment in context. A compromise is to specify those questions one can anticipate and list other possible topics while treating the conversational

SIDEBAR

WHAT DO YOU DO WITH DANGEROUS REVELATIONS? THE ETHICAL CHALLENGE OF CONFIDENTIALITY

Robert Weiss (1994) reported the problem posed by a woman respondent who was HIV positive.

> She said that all her life, from the time she was a child, she had been treated brutally by men. Contracting HIV from a boyfriend was only the most recent instance. Now she wanted to get even with the whole male sex. She visited barrooms every evening to pick up men with whom she could have intercourse, in the hope that she would infect them. The woman's sister had already reported her to a public health agency, mostly because she wanted the woman stopped before she was hurt by some man she had tried to infect. The public health agency did nothing.
>
> In our final interview I learned the woman was no longer seeking revenge through sex. She had met a man who had become her steady boyfriend and who remained with her even after he was told—by that same sister—that she was HIV positive. (His first reaction was to yell at the woman and, I think, push her around.) If in our final interview the woman had reported continuing her campaign to spread HIV among men, I would have told her to stop. I can't believe that would have done much good, but I would have told her anyway. I also would have discussed her report with the head of the clinic where she was being treated, with the thought of devising some way to interrupt her behavior. (p. 132)

component as probes, which are not typically specified in advance. Adopt a strategy of "planned flexibility" insofar as possible, and be prepared to educate your IRB about why this is appropriate both methodologically and ethically (Silverman & Marvasti, 2008, p. 48). In a fully naturalistic inquiry design, interview questions can and should change as the interviewer understands the situation more fully and discovers new pathways for questioning. The tension between specifying questions for approval in advance versus allowing questions to emerge in context in the field led Elliott Eisner (1991) to ask, "Can qualitative studies be informed ... [since] we have such a hard time predicting what we need to get consent about?" (p. 215). An alternative to specifying precise questions for approval in advance is to specify areas of inquiry that will be avoided—that is, to anticipate ways in which respondents might be put at risk and affirm that the interviewer will avoid such areas.

Uninformed Consent Seeker

Conversational Interviewing Poses Special Informed Consent Problems:

Whipping out an informed consent statement and asking for a signature can be awkward at best. To the extent that interviews are an extension of a conversation and part of a relationship, the legality and formality of a consent form may be puzzling to your conversational partner or disruptive to the research. On the one hand, you may be offering conversational partners anonymity and confidentiality, and on the other asking them to sign a legal form saying they are participating in the study. How can they later deny they spoke to you—which they may need to do to protect themselves—if you possess a signed form saying they were willing to participate in the study? (Rubin & Rubin, 1995, p. 95)

Qualitative methodologists find that many IRBs are not competent to judge qualitative research. When engaging with an IRB, distinguish between legal compliance with human subjects protection requirements versus conscientious ethical behavior:

You cannot achieve ethical research by following a set of preestablished procedures that will always be correct. *Yet, the requirement to behave ethically is just as strong in qualitative interviewing as in other types of research on*

humans—*maybe even stronger*. You must build ethical routines into your work. You should carefully study code of ethics and cases of unethical behavior to sensitize yourself to situations in which ethical commitments become particularly salient. Throughout your research, keep thinking and judging what are your ethical obligations. (Rubin & Rubin, 1995, p. 96)

New Directions in Informed Consent: Confidentiality Versus People Owning Their Own Stories

Confidentiality norms are also being challenged by new directions in qualitative inquiry. Traditionally, researchers have been advised to disguise the locations of their fieldwork and change the names of respondents, usually giving them pseudonyms, as a way of protecting their identities. The presumption has been that the privacy of research subjects should always be protected. This remains the dominant presumption, as well it should. It is being challenged, however, by participants in research who insist on "owning their own stories." Some politically active groups take pride in their identities and refuse to be involved in research that disguises who they are. Some programs that aim at empowering participants emphasize that participants "own" their stories and should insist on using their real names. In a program helping them overcome a history of violence and abuse, I encountered women who were combating the stigma of their past by telling their stories and attaching their real names to their stories as part of healing, empowerment, and pride. Does the researcher, in such cases, have the right to impose confidentiality against the wishes of those involved? Is it patronizing and disempowering for a university-based human subjects committee to insist that these women are incapable of understanding the risks involved if they choose to turn down an offer of confidentiality? On the other hand, by identifying themselves, they give up not only their own privacy but also perhaps that of their children, other family members, and current or former partners.

A doctoral student studying a local church worked out an elaborate consent form in which the entire congregation decided whether to let itself be identified in his dissertation. Individual church members also had the option of using their real names or choosing pseudonyms. Another student studying alternative health practitioners offered them the option of confidentiality and pseudonyms or using their real identities in their case studies. Some chose to be identified, while some didn't. A study of organizational leaders offered the same option. In all of these cases, the research participants also had the right to review and approve the final versions of their case studies and transcripts before they were made public. In cases of collaborative inquiry, where the researcher works with "coresearchers" and data collection involves more of a dialogue than an interview, the coresearchers may also become coauthors as they choose to identify themselves and share in publication.

These are examples of how the norms about confidentiality are changing and being challenged as tension has emerged between the important ethic of protecting people's privacy and, in some cases, their desire to own their stories. Informed consent, in this regard, does not automatically mean confidentiality. Informed consent can mean that participants understand the risks and benefits of having their real names reported and choose to do so. Protection of human subjects properly insists on informed consent. That does not now automatically mean confidentiality, as these examples illustrate.

Reciprocity: Should Interviewees Be Compensated? If So, How?

> *Quid pro quo*: "something for something"
> (Ancient Roman principle)

The issues of whether and how to compensate interviewees involve questions of both ethics and data quality. Will payment, even of small amounts, affect people's responses, increasing acquiescence or, alternatively, enhancing the incentive to respond thoughtfully and honestly? Is it somehow better to appeal to people on the basis of the contribution they can make to knowledge or, in the case of evaluation, improving the program, instead of appealing to their pecuniary interest? Can modest payments in surveys increase response rates to ensure an adequate sample size? Does the same apply to indepth interviewing and focus groups? The interviewer is usually getting paid. Shouldn't the time of interviewees be respected, especially the time of low-income people, by offering compensation? What alternatives are there to cash for compensating interviewees? In Western capitalist societies, issues of compensation are arising more and more often both because people in economically disadvantaged communities are reacting to being overstudied and undervalued and because private sector marketing firms routinely compensate focus group participants; so this practice has spread to the public and nonprofit sectors.

Professionals in various fields differ about compensation for interviewees. Below are some comments from a lively discussion of these issues that took place on *EvalTalk*, the American Evaluation Association Internet Listserv.

- I believe in paying people, particularly in areas of human services. I am thinking of parenting and teen programs where it can be very difficult to get participation in interviews. If their input is valuable, I believe you should put your money where your mouth is. However, I would always make it very clear to the respondents that, although they are being paid for their time, they are *not* being paid for their responses and should be as candid and forthright as possible.

- One inner-city project offered vouchers to parent participants in a focus group to buy books for their kids, which for some low-income parents proved to be their first experience owning books rather than always borrowing them.

- Cash payment for participation in interviews is considered income and is therefore taxable. This can create problems if the payment comes from a public agency, as our county attorney has pointed out in the past. Consequently, when we "pay" for participation, we use incentives other than cash, for example, vouchers or gift certificates donated by local commercial vendors, such as a discount store. These seem to be as effective.

- If you are detaining a person with a face-to-face interview, and it isn't a friendly conversation, rather it is a business exercise, it is only appropriate to offer to pay the prospective respondent for his or her time and effort. This should not preclude, and it would certainly help to explain, the importance of his or her contribution.

- We have paid and not paid incentives for focus groups for low-income folks as well as professionals and corporate CEOs. The bottom line is that in most cases the incentive doesn't make a lot of difference in terms of participation rates, especially if you have well-trained interviewers and well-designed data collection procedures.

One of my concerns is that we are moving in a direction in which it is assumed (with very little substantive foundation) that people will only respond if given incentives. My plea here is that colleagues not fall into the trap of using incentives as a crutch but that they constantly examine and reexamine the whole issue of incentives and not simply assume that they are either needed or effective.

Alternatives to cash can instill a deeper sense of reciprocity. In doing family history interviews, I found that giving families a copy of the interview was much appreciated and increased the depth of responses because they were speaking not just to me, the interviewer, but to their grandchildren and great-grandchildren in telling the family's story. In one project in rural areas, we carried a tape duplicator in the truck and made copies for them instantly at the end of the interview. Providing complete transcripts of interviews can also be attractive to participants. In an early-childhood parenting program where data collection included videotaping parents playing with their children, copies of the videotapes were prized by the parents. The basic principle informing these exchanges is reciprocity. Participants in research provide us with something of great value—their stories and their perspectives on their world. We show that we value what they give us by offering something in exchange.

How Hard to Push for Sensitive Information?

Skillful interviewers can get people to talk about things they may later regret having revealed. Or sharing revelations in an interview may unburden a person, letting one get something off one's chest. Since one can't know for sure, interviewers are often faced with an ethical challenge concerning how hard to push for sensitive information, a matter in which the interviewer has a conflict of interest since the interviewer's predilection is likely to be to push for as much as possible. Herb and Irene Rubin tell of interviewing an administrator in Thailand and learning that two months after their fieldwork, he committed suicide, "leaving us wondering if our encouraging him to talk about his problems may have made them more salient to him" (Rubin & Rubin, 1995, p. 98).

In deciding how hard to push for information, the interviewer must balance the value of a potential response against the potential distress for the respondent. This requires sensitivity, but it is not a burden the interviewer need to take on alone. When I see that someone is struggling for an answer, seems hesitant or unsure, or I simply know that an area of inquiry may be painful or uncomfortable, I prefer to make the interviewee a partner in the decision about how deeply to pursue the matter. I say something like this:

I realize this is a difficult thing to talk about. Sometimes people feel better talking about something like this and, of course, sometimes they don't. You decide how much is comfortable for you to share. If you do tell me what happened and how you feel and later you wish you hadn't, I promise to delete it from the interview. Okay?

Obviously, I'm very interested in what happened, so please tell me what you're comfortable telling me.

Be Careful. It's Dangerous Out There.

> In our teaching and publications we tend to sell students a smooth, almost idealized, model of the research process as neat, tidy, and unproblematic. . . . Perhaps we should be more open and honest about the actual pains and perils of conducting research in order to prepare and forewarn aspiring researchers.
>
> —Maurice Punch (1986, pp. 13–14)

In an old police investigation television show, *Hill Street Blues*, the precinct duty sergeant ended each daily briefing of police officers by saying, "Let's be careful out there." The same warning applies to qualitative researchers doing fieldwork and interviewing: "Be careful. It's dangerous out there." It's important to protect those who honor us with their stories by participating in our studies. It's also important to protect yourself.

I was once interviewing a young man at a coffee shop for a recidivism study when another man showed up, an exchange took place, and I realized I had been used as a cover for a drug purchase. In doing straightforward outcomes evaluation studies, I have discovered illegal and unethical activities that I would have preferred not to have stumbled across. When we did the needs assessment of distressed farm families in rural Minnesota, we took the precaution of alerting the sheriffs' offices in the counties where we would be interviewing in case any problems arose. One sheriff called back and said that a scam had been detected in the county that involved a couple in a pickup truck soliciting home improvement work and then absconding with the down payment. Since we were interviewing in couple teams and driving pickup trucks, the sheriff, after assuring himself of the legitimacy of our work, offered to provide us with a letter of introduction, an offer we gratefully accepted.

I supervised a dissertation that involved interviews with young male prostitutes. We made sure to clear that study with the local police and public prosecutors and to get their agreement that promises of confidentiality would be respected given the potential contribution of the findings to reducing both prostitution and the spread of AIDS. This, by the way, was a clear case where it would have been inappropriate to pay the interviewees. Instead of cash, the reciprocity incentive the student offered was the result of a personality instrument he administered.

One of the more famous cases of what seemed like straightforward fieldwork that became dangerous involved dissertation research on the culture of a bistro in New York City. Through in-depth interviews, graduate student Mario Brajuha gathered detailed information from people who worked and ate at the restaurant—information about their lives and their views about others involved with the restaurant. He made the usual promise of confidentiality. In the midst of his fieldwork, the restaurant was burned, and the police suspected arson. Learning of his fieldwork, they subpoenaed his interview notes. He decided to honor his promises of confidentiality and ended up going to jail rather than turning over his notes. This case, which dragged on for years, disrupting his graduate studies and his life, reaffirmed that researchers lack the protection of clergy and lawyers when subpoenas are involved, promises of confidentiality notwithstanding. (For details, see Brajuha & Hallowell, 1986.)

It helps to think about potential risks and dangers prior to gathering data, but Brajuha could not have anticipated the arson. Anticipation, planning, and ethical reflection in advance only take you so far. As Maurice Punch (1986) has observed, sounding very much like he is talking from experience,

How to cope with a loaded revolver dropped in your lap is something you have to resolve on the spot, however much you may have anticipated it in prior training. (p. 13)

So be careful. It's dangerous out there.

> The word *interview* has roots in Old French and meant something like "to see one another."
>
> —Narayan and George (2003, p. 449)

Personal Reflections on Interviewing

Though there are dangers, there are also rewards.

I find interviewing people invigorating and stimulating—the opportunity for a short period of time to enter another person's world. If participant observation means "walk a mile in my shoes," in-depth interviewing means "walk a mile in my head." New worlds are opened up to the interviewer on these journeys.

I'm personally convinced that to be a good interviewer you must like doing it. This means being interested in what people have to say. You must yourself believe that the thoughts and experiences of the people being interviewed are worth knowing. In short, you must have the utmost respect for these persons who are willing to share with you some of their time to help you understand their world. There is a Sufi story that describes what happens when the interviewer loses this basic sensitivity to and respect for the person being interviewed.

An Interview With the King of the Monkeys

A man once spent years learning the language of monkeys so that he could personally interview the king of monkeys. Having completed his studies, he set out on his interviewing adventure. In the course of searching for the king, he talked with a number of monkey underlings. He found that the monkeys he spoke to were generally, to his mind, neither very interesting nor very clever. He began to doubt whether he could learn very much from the king of the monkeys after all.

Finally, he located the king and arranged for an interview. Because of his doubts, however, he decided to begin with a few basic questions before moving on to the deeper meaning-of-life questions that had become his obsession.

"What is a tree?" he asked.

"It is what it is," replied the king of the monkeys. "We swing through trees to move through the jungle."

"And what is the purpose of the banana?"

THE PERSONAL EXPERIENCE OF QUALITATIVE INQUIRY

Question. What's the most memorable or meaningful evaluation that you have been a part of—and why?

I worked briefly on an evaluation of a rail safety intervention program aimed at reducing safety-related incident rates through peer-to-peer observation (Zuschlag, Ranney, Coplen, & Harnar, 2012). To collect some qualitative insights, I was sent to San Antonio where I got a full tour, a ride aboard a locomotive, and the opportunity to talk with some engineers and conductors about changes they saw in safety related to the program. One of their stories, in particular, touches me every time I tell it.

A trained safety observer noticed an older engineer not wearing any hearing protection, even though he was right alongside a locomotive. The observer said something to the order of "Don't you want to hear your granddaughter's voice when she talks to you?" This relatively casual but purposeful effort had a tremendous impact. After that, every time the "old-timer" saw our observer he would point to his hearing protection to say, "Look, I'm wearing them!"

In my youth, I spent long days playing in the dead yards of the Reading Railroad (yes, the one from the Monopoly game, pronounced Redding), and I never dreamed of someday riding the tracks as a researcher and learning about that world through the eyes of its practitioners. Every time I recall the story it takes me back to that day in the locomotive and the hours I spent talking with those engineers and conductors, and it reminds me that one conversation can change a person's world . . . and maybe let him hear his granddaughter's voice in his old age.

—Michael A. Harnar, Mosaic Network, Inc.
Senior Associate, Research and Evaluation
American Evaluation Association Newsletter,
November, 2012

"Purpose? Why, to eat."

"How do animals find pleasure?"

"By doing things they enjoy."

At this point, the man decided that the king's responses were rather shallow and uninteresting, and he went on his way, crushed and cynical. Soon afterward, an owl flew into the tree next to the king of the monkeys. "What was that man doing here?" the owl asked.

"Oh, he was only another silly human," said the king of the monkeys. "He asked a bunch of simple and meaningless questions, so I gave him simple and meaningless answers."

Not all interviews are interesting, and not all interviews go well. Certainly, there are uncooperative respondents, people who are paranoid, respondents who seem overly sensitive and easily embarrassed, aggressive and hostile interviewees, timid people, and the endlessly verbose, who go on at great length about very little. When an interview is going badly, it is easy to call forth one of these stereotypes to explain how the interviewee is ruining the interview. Such blaming of the victim (the interviewee), however, does little to improve the quality of the data. Nor does it improve interviewing skills.

I prefer to believe that there is a way to unlock the internal perspective of every interviewee. My challenge and responsibility as an interviewer involve finding the appropriate and effective interviewing style and question format for a particular respondent. It is my responsibility as the interviewer to establish an interview climate that facilitates open responses. When an interview goes badly, as it sometimes does, even after all these years, I look first at my own shortcomings and miscalculations, not the shortcomings of the interviewee. That's how, over the years, I've gotten better and come to value reflexivity, not just as an intellectual concept but also as a personal and professional commitment to learning and engaging people with respect.

Overview and Conclusion

Interviewing has become central in our modern knowledge age *Interview Society*. This chapter began by discussing different approaches to interviewing for different purposes: journalism, therapeutic interviews, forensic investigations, celebrity interviews, and personnel interviewing, among others (see Exhibit 7.1, pp. 423–424). We examined differences in how interviews are used to guide diverse theoretical and methodological traditions: ethnography, phenomenology, social constructionism, hermeneutics, and other frameworks for inquiry (see Exhibit 7.3, pp. 432–436).

A major focus of this chapter has been that obtaining high-quality data from interviews requires *skilled interviewing*. Exhibit 7.2 (p. 428) presented 10 core interviewing skills. We then moved to different types of interview formats: standardized questions, interview guides, and conversational interviewing (see Exhibit 7.4, pp. 437–438). We looked at how to phrase questions and probe responses. A point of emphasis was the importance of anticipating analysis and reporting to organize, sequence, and format interviews (see Exhibit 7.6, p. 443).

Every interview is also an observation, a two-way interaction, and, therefore, a relationship. Exhibit 7.12 (pp. 462–463) summarized six distinct approaches to undertaking interviewing as a relationship. We also compared interviews with individuals with group interviews, like focus groups and 11 other types of group interviews (see Exhibit 7.14, pp. 475–476).

Cross-cultural interviewing and qualitative fieldwork presents special challenges (see Exhibit 7.15, p. 482). Special target populations pose particular challenges in qualitative inquiries: interviewing children, older people, elites, and marginalized people or conducting online interviews (see Exhibit 7.17, pp. 493–494).

The interpersonal nature of in-depth qualitative interviewing raises special ethical dilemmas and concerns (see Exhibit 7.18, pp. 496–497). The people interviewed can be affected by the inquiry, but so can the person conducting the interviews. Nora Murphy (2014) interviewed homeless youth for her dissertation.

> This also took a personal toll. At times, the harsh reality of the experiences the youths have gone through had me in tears at home. In fact, in some interviews, the youth and I cried together. The stories are so powerful because they are real and true, and sharing these truths means opening wounds. I felt guilty that I could not help them more and feared exploiting them. Sometimes I emailed advisors just to vent to him about how difficult it was; playing the role of an objective researcher who could not step in to help felt unnatural and uncomfortable. It helped when the youth thanked me for listening and glowed when I read their stories back to them. They often said things like, "You made connections about me that I didn't realize . . . No one else in my life knows all of this . . . I want to share this with my brother. It will help me explain things to him that I've never been able to say before . . . I want to make this into a book." I left each interview with the feeling that the youths appreciated being listened to, that they felt gratitude toward me and the process. I still harbor guilt that I could not do more for each of them, but I also believe the process, overall, has done more good than harm. I still carry some of the sadness I experienced as I sat with the youth and listened to their tales of their difficult journeys, but I also carry a hope inspired by their optimism and their strength. (p. 428)

While this chapter has emphasized skill and technique as ways of enhancing the quality of interview data, no less important is a genuine interest in and caring about the perspectives of other people. If what people have to say about their world is generally boring to you, then you will never be a great interviewer. Unless you are fascinated by the rich variation in human experience, qualitative interviewing will become drudgery. On the other hand, a deep and genuine interest in learning about people is insufficient without disciplined and rigorous inquiry based on skill, technique, and a deep capacity for empathic understanding.

Conclusion: Halcolm on Interviewing

Ask.

Listen and record.

Ask.

Listen and record.

Asking involves a grave responsibility.

Listening is a privilege.

Researchers, listen and observe. Remember that your questions will be studied by those you study. Evaluators, listen and observe. Remember that you shall be evaluated by your questions.

To ask is to seek entry into another's world. Therefore, ask respectfully and with sincerity. Do not waste questions on trivia and tricks, for the value of the answering gift you receive will be a reflection of the value of your question.

> Blessed are the skilled questioners, for they shall be given mountains of words to ascend.
>
> Blessed are the wise questioners, for they shall unlock hidden corridors of knowledge.
>
> Blessed are the listening questioners, for they shall gain perspective.
>
> —From Halcolm's *Beatitudes*

Looking Forward

Part I of this book presented the niche of qualitative methods in studying the world (Chapter 1), presented 12 fundamental qualitative strategies (Chapter 2), reviewed major theoretical traditions that inform diverse approaches to qualitative inquiry (Chapter 3), and examined practical applications for qualitative methods (Chapter 4). Part II presented qualitative design options (Chapter 5), fieldwork approaches (Chapter 6), and interview methods (Chapter 7). We now turn to Part III: analyzing and reporting qualitative findings (Chapter 8) and ways of enhancing the credibility and utility of qualitative results (Chapter 9). As we make the transition from Part II to Part III, from ways of gathering data to ways of making sense of the data gathered, the Halcolm story that ends this chapter reminds us that *things are not always what they seem*. The story also reminds us of the centrality of Thomas's theorem for interpretation of qualitative data: *What is perceived as real is real in its consequences*.

EXHIBIT 7.19   Examples of Standardized Open-Ended Interviews

The edited interviews below were used in evaluation of an Outward Bound program for the disabled. Outward Bound is an organization that uses the wilderness as an experiential education medium. This particular program consisted of a 10-day experience in the Boundary Waters Canoe Area of Minnesota. The group consisted of half able-bodied participants and half disabled participants, including paraplegics; persons with cerebral palsy, epilepsy, or other developmental disabilities; blind and deaf participants; and, on one occasion, a quadriplegic. The first interview was conducted at the beginning of the program; the second interview was used at the end of the 10-day experience; and the third interview took place six months later. To save space, many of the probes and elaboration questions have been deleted, and space for writing notes has been eliminated. The overall thrust and format of the interviews have, however, been retained.

Precourse Interview: Minnesota Outward Bound School Course for the Able-Bodied and the Disabled

This interview is being conducted before the course as part of an evaluation process to help us plan future courses. You have received a consent form to sign, which indicates your consent to this interview. The interview will be recorded.

1. First, we'd be interested in knowing how you became involved in the course. How did you find out about it?

   a. What about the course appealed to you?

   b. What previous experiences have you had in the outdoors?

2. Some people have difficulty deciding to participate in an Outward Bound course, and others decide fairly easily. What kind of decision process did you go through in thinking about whether or not to participate?

   a. What particular things were you concerned about?

   b. What is happening in your life right now that stimulated your decision to take the course?

3. Now that you've made the decision to go on the course, how do you feel about it?

   a. How would you describe your feelings right now?

   b. What lingering doubts or concerns do you have?

4. What are your expectations about how the course will affect you personally?

   a. What changes in yourself do you hope will result from the experience?

   b. What do you hope to get out of the experience?

5. During the course you'll be with the same group of people for an extended period of time. What feelings do you have about being part of a group like that for nine full days?

   a. Based on your past experience with groups, how do you see yourself fitting into your group at Outward Bound?

For the Disabled

6. One of the things we're interested in understanding better as a result of these courses is the everyday experience of disabled people. Some of the things we are interested in are as follows:

   a. How does your disability affect the types of activities you engage in?

   b. What are the things that you don't do that you wish you could do?

   c. How does your disability affect the kinds of people you associate with? *Clarification:* Some people find that their disability means that they associate mainly with other disabled persons. Others find that their disability does not affect their contacts with people. What has your experience been along these lines?

   d. Sometimes people with disabilities find that their participation in groups is limited. What has been your experience in this regard?

For the Able-Bodied

7. One of the things we're interested in understanding better as a result of these courses is feelings that able-bodied people have about being with disabled folks. What kinds of experiences with disabled people have you had in the past?

   a. What do you personally feel you get out of working with disabled people?

   b. In what ways do you find yourself being different from your usual self when you're with disabled people?

   c. What role do you expect to play with disabled people on the Outward Bound course? *Clarification:* Are there any particular things you expect to have to do?

d. As you think about your participation in this course, what particular feelings do you have about being part of an outdoor course with disabled people?

8. About half of the participants on the course are disabled people, and about half are people without disabilities. How would you expect your relationship with the disabled people to be different from your relationship with course participants who are not disabled?

9. We'd like to know something about how you typically face new situations. Some people kind of like to jump into new situations, whether or not some risk is be involved. Other people are more cautious about entering situations until they know more about them. Between these two, how would you describe yourself?

10. Okay, you've been very helpful. Are there other thoughts or feelings you'd like to share with us to help us understand how you're seeing the course right now. Anything at all you'd like to add?

Postcourse Interview

We're conducting this interview right at the end of your course with Minnesota Outward Bound. We hope this will help us better understand what you've experienced so that we can improve future courses. You have signed a form giving your consent for material from this interview to be used in a written evaluation of the course. This interview is being tape-recorded.

1. To what extent was the course what you expected it to be?

   a. How was it different from what you expected?

   b. To what extent did the things you were concerned about before the course come true?

   b-1. Which things came true?

   b-2. Which didn't come true?

2. How did the course affect you personally?

   a. What changes in yourself do you see or feel as a result of the course?

   b. What would you say you got out of the experience?

3. During the past nine days, you've been with the same group of people constantly. What kind of feelings do you have about having been a part of the same group for that time?

a. What feelings do you have about the group?

b. What role do you feel you played in the group?

c. How was your experience with this group different from your experiences with other groups?

d. How did the group affect you?

e. How did you affect the group?

f. In what ways did you relate differently to the able-bodied and disabled people in your group?

4. What is it about the course that makes it have the effects it has? What happens on the course that makes a difference?

   a. What do you see as the important parts of the course that make an Outward Bound course what it is?

   b. What was the high point of the course for you?

   c. What was the low point?

5. How do you think this course will affect you when you return to your home?

   a. Which of the things you experienced this week will carry over to your normal life?

   b. What plans do you have to change anything or do anything differently as a result of this course?

For the Disabled

6. We asked you before the course about your experience of being disabled. What are your feelings about what it's like to be disabled now?

   a. How did your disability affect the type of activities you engaged in on the course? *Clarification:* What things didn't you do because of your disability?

   b. How was your participation in the group affected by your disability?

For the Able-Bodied

7. We asked you before the course your feelings about being with disabled people. As a result of the experiences of the past nine days, how have your feelings about disabled people changed?

   a. How have your feelings about yourself in relation to disabled persons changed?

   b. What did you personally get out of being/working with disabled people on this course?

(Continued)

(Continued)

  c. What role did you play with the disabled people?

  d. How was this role different from the role you usually play with disabled people?

8. Before the course, we asked you how you typically faced a variety of new situations. During the past nine days, you have faced a variety of new situations. How would you describe yourself in terms of how you approached these new experiences?

  a. How was this different from the way you usually approach things?

  b. How do you think this experience will affect how you approach new situations in the future?

9. Suppose you were being asked by a government agency whether or not they should sponsor a course like this. What would you say?

  a. What arguments would you give to support your opinion?

10. Okay, you've been very helpful. We'd be very interested in any other feelings and thoughts you'd like to share with us to help us understand your experience of the course and how it affected you.

Six-Month Follow-Up Interview

This interview is being conducted about six months after your Outward Bound course to help us better understand what participants experience so that we can improve future courses.

1. Looking back on your Outward Bound experience, I'd like to ask you to begin by describing for me what you see as the main components of the course? What makes an Outward Bound course what it is?

  a. What do you remember as the highlight of the course for you?

  b. What was the low point?

2. How did the course affect you personally?

  a. What kinds of changes in yourself do you see or feel as a result of your participation in the course?

  b. What would you say you got out of the experience?

3. For nine days, you were with the same group of people, how has your experience with the Outward Bound group affected your involvement with groups since then?

For the Disabled

(*Check previous responses before the interview. If the person's attitude appears to have changed, ask if he or she perceives a change in attitude.)

4. We asked you before the course to tell us what it's like to be disabled. What are your feelings about what it's like to be disabled now?

  a. How does your disability affect the types of activities you engage in? *Clarification:* What are some of the things you don't do because you're disabled?

  b. How does your disability affect the kinds of people you associate with? *Clarification:* Some people find that their disability means that they associate mainly with other disabled persons. Other people with disabilities find that their disability in no way limits their contacts with people. What has been your experience?

  c. As a result of your participation in Outward Bound, how do you believe you've changed the way you handle your disability?

For the Able-Bodied

5. We asked you before the course to tell us what it's like to work with the disabled. What are your feelings about what it's like to work with the disabled now?

  a. What do you personally feel you get out of working with disabled persons?

  b. In what ways do you find yourself being different from your usual self when you are with disabled people?

  c. As you think about your participation in the course, what particular feelings do you have about having been part of a course with disabled people?

6. About half of the people on the course were disabled people, and about half were people without disabilities. To what extent did you find yourself acting differently with disabled people compared with the way you acted with able-bodied participants?

7. Before this course, we asked you how you typically face new situations. For example, some people kind of like to jump into new situations even if some risks are involved. Other people are more cautious, and so on. How would you describe yourself along these lines right now?

a. To what extent, if at all, has the way you have approached new situations since the course been a result of your Outward Bound experience?

8. Have there been any ways in which the Outward Bound course affected you that we haven't discussed? If yes, how? Would you elaborate on that?

   a. What things that you experienced during that week carried over to your life since the course?

   b. What plans have you made, if any, to change anything or do anything differently as a result of the course?

9. Suppose you were being asked by a government agency whether or not they should support a course like this. What would you say?

   a. Who shouldn't take a course like this?

10. Okay, you've been very helpful. Any other thoughts or feelings you might share with us to help us understand your reactions to the course and how it affected you?

   a. Anything at all you'd like to add?

EXHIBIT 7.20   Interview Case Study Example

**The Experience of Youth Homelessness:
Thmaris Tells His Story**

The following case study is one of 14 done as part of a study of youth homelessness in Minnesota (Murphy, 2014).

Thmaris (the name he created for himself for this case study)

Thmaris was born in Chicago, Illinois, in 1990. He has an older and a younger brother, and was raised by his mother. His family moved to Minnesota in 1996 but didn't have housing when they moved here. As a result, they moved around a lot between extended family members' homes and family shelters, never staying anywhere for more than a year. His mother was addicted to alcohol and drugs, and it often fell on Thmaris and his older brother to take care of their younger brother. Consequently, Thmaris has been earning money for the family since he turned 12. Sometimes this was through a job—like the construction apprenticeship he had at age 12—and sometimes this was through stealing or dealing weed. Even though he stole and dealt when he had to, he recognized early on that he was happiest when he was working with his hands and providing for his family through a job.

When Thmaris was 13, his mother entered an alcoholic treatment program, and he and his brothers went to stay with his uncle. Thmaris remembers this as one of the lowest points in his life. There were six kids in a three-bedroom house, with people coming and going. There was never enough food or clothing, and it didn't feel safe. When his mother returned from the treatment center to get them from the uncle's home, Thmaris thought things would get better, but they got much worse. His mother accused Thmaris's older brother of molesting Thmaris and their little brother, and these accusations tore his family apart. To this day, he doesn't understand why she did this.

She really went above and beyond to try to prove it and try to accuse him. I know that made him feel like nothing. I know it made him feel like the worse kinds of person. It wasn't true. There was no truth to it. My bigger brother, he's super protective and he's not that type of dude. It just hurt me for her to do something like that and to accuse her own son of something like that. She just accused him of the worst crime type ever. That really stuck with us.

Not only was it painful to watch his mother accuse their brother of something he didn't do, but also as a result, they lost their older brother, their protector, and the closest thing to a father figure that they had. This placed Thmaris in a tough position, where he felt that he had to align his loyalties either with this older brother or with his mother. If he placed his loyalties with his older brother, it would mean that he couldn't have a relationship with his younger brother.

He had to place us to a distance. He stepped back from being in our lives a lot. When he went to Chicago to stay with my dad, it was like we never saw him, we never talked to him. It was confusing for me because I'm the middle child. So I have a little brother and I have an older brother. When she sent my brother away, I always felt like she abandoned him, just threw him out of our lives. . . . I would go up to Chicago to see my brother, I'd

(Continued)

(Continued)

stay there for a couple of months, just to get that big brother/small brother bond again. Then, I would come back because I didn't want my little brother growing up like, "Dang. Both of my brothers left me." It was just always really hard and I just felt like [my mother] just put me in a position to choose, to choose whether I want to be around my little brother or my big brother.

Once again, Thmaris found himself having to grow up quickly. With his older brother gone, he felt that he had become the sole provider for and protector of his younger brother.

His mother continued to cycle in and out of treatment programs, and Thmaris and his younger brother likewise cycled in and out of foster care. Foster care was a time of relative peace and stability for Thmaris and his brother as they developed a relationship with their foster mother that they maintain to this day. He was always drawn back to family though and ended up staying with his mother when he was 16. He remembers this year as the longest year of his life. At this time, she wasn't able to have the boys stay with her because she had Section 8 housing and they weren't named on her lease. He was angry at the world and fighting with anyone and everyone. He started running with a gang and was getting into a lot of trouble. There were many nights that his mother would lock him out as a consequence of this behavior.

After being locked out several times, Thmaris decided that he'd had it and told his mother that he was moving out. This was the night Thmaris became homeless. When he first moved out, he stayed with his girlfriend. When he couldn't stay with her, he would couch hop or sleep outside in public places. Thmaris's girlfriend at the time introduced him to the drop-in centers in Minneapolis and St. Paul. She had a baby and used to visit the drop-in centers to get pampers, wipes, and other baby supplies. Thmaris gravitated toward the drop-in center, which he felt had a smaller, more intimate feel to it. First and foremost, he used it as a place to get off of the streets and be safe. He also appreciated the support they provided for doing things such as writing resumes, applying for jobs, and locating apartments, but he only took advantage of them sporadically.

When asked to recall his first impressions of the drop-in center, Thmaris describes a place that had some clear rules and expectations. While at the drop-in center, Thmaris knew he couldn't use profanity, fight, smoke, bring in drugs, or be intoxicated. The expectation was that you try your hardest to succeed in whatever you want to do.

They help you with a lot of stuff. I just feel like there's nothing that you really need that you can't really get from here. If you really need underwear and socks and t-shirts, they have closets full of stuff like that. If you really need toothpaste and toothbrushes and deodorant and all that stuff, all the hygiene stuff, they have that here. If you have a whole lot of stuff but nowhere to go, they have lockers here where you can leave your stuff here and can't nobody really get in them. They have showers here that you can use if you really need it. I just feel like they have so many resources for you it's ridiculous.

During his first few years utilizing the drop-in center, Thmaris continued to be in a gang and was frequently in and out of jail. Being a gang member was his primary source of income, and he was loyal to his fellow gang members even when it got him in trouble. He recalls a time when he got arrested for auto theft.

I was in jail a lot. I remember back in 2009 I was running with this group of guys and they was doing breaking and entering and stuff like that. I knew it could come with jail time or whatever, but I just seen these guys with a lot of money and I was very broke at the time. So I'm like, well, I'm doing it with these guys.

One time we go to this one house, and there's a car there. The guy I'm with gets the car and he was driving it around. We get to the house, and he's like, "Well, I forgot to go to the store to get some something." He's like, "Could you go to the store?"

I'm like, "Yeah." So I get the car and I get the keys and I go to the store. And right away I was just surrounded by cops, and I got thrown in jail for auto theft. I could have easily just gave those guys up, but I had always been taught to be a man of responsibility. If I did something, then I should take responsibility for it. So I got thrown in jail; I got a year and a day of jail time over my head with five years' probation.

After this conviction, Thmaris was starting to use the resources at the drop-in center more consistently and secured a job delivering newspapers. Because he had a job and was making progress, he was accepted into a Transitional Living Program (TLP). Things were moving in a direction that Thmaris felt good about, but he violated the probation related to the auto theft and lost his place in the TLP program. He describes violating his probation because his probation officer was in Washington County and Thmaris was staying in St. Paul. It was hard to report to a parole officer in St. Cloud when Thmaris didn't have consistent access to transportation, so Thmaris decided to serve his year and a day instead. He spent time in two correctional facilities before being released after eight months. By the time he was released, he had lost his job and his spot in the TLP.

Turning Point

Going to jail and losing his TLP was a turning point for Thmaris. When he got out of jail, he recalls saying to himself,

> *I need to stop with the crimes and committing crimes and just try to find something different to do with my life. . . . So I got back in school. I went to school for welding technology, and that's been my main focus ever since I got out at that time. It's been rare that I would go back to jail. If I did, it would be for arguing with my girlfriend or not leaving the skyway when they wanted me to, so it was like trespassing and stuff like that. After I got out of Moose Lake, it just really dawned on me that I needed to change my life and that the life that I was living wasn't the right path for me.*

But he didn't know how to do this on his own, so when he got out of prison, he went straight to the drop-in center to ask his case manager for help.

Thmaris's case manager is supported by a Healthy Transitions grant through the Minnesota Department of Health. He works specifically with homeless, unaccompanied youth who spent at least 30 days in foster care between the ages of 16 and 18. Each time a young person visits the drop-in center for the first time, he or she is asked about his or her previous experience with the foster care system. If they report being in foster care between the ages of 16 and 18 for at least 30 days, then they will typically be up on the case load of one of the workers supported by the Healthy Transitions grant. Thmaris had two case workers previous to Rahim but was transferred to Rahim when Rahim began working with youth under Healthy Transitions. This was a lucky move for Thmaris because this relationship developed into one that has been deeply meaningful to him.

> *I just feel that ever since I turned 20 I realized that I'm an adult and that I have to make better choices, not just for me but the people around me. Didn't nobody help me with that but Rahim. . . . the things that he was able to do, he made sure that he did them. I remember days that I'd come down to Safe Zone, and I'd be like,*

> *Rahim, I haven't eaten in two days or, Rahim, I haven't changed my underwear in like a week or whatever. He would give me bus cards to get to and from interviews. He would give me Target cards to go take care of my personal hygiene. He would give me Cub cards to go eat. It was like every problem or every obstacle I threw in front of him, he made sure that I would overcome it with him. He was like the greatest mentor I ever had. I've never had nobody like that.*

> *Out of all the other caseworkers I had, nobody ever really sat me down and tried to work out a resolution for my problems. They always just gave me pamphlets like, "Well, go call these people and see if they can do something for you." Or, "You should go to this building because this building has that, what you want." It was like every time I come to him I don't have to worry about anything. He's not going to send me to the next man, put me on to the next person's caseload. He just always took care of me. If I would have never met Rahim, I would have been in a totally different situation, I would have went a totally different route.*

Without the support of Rahim and the resources at the drop-in center, Thmaris is sure he would still be in a gang and dealing weed and would eventually end up in jail again.

And it wasn't just that Rahim was there for him, it's also that Rahim has been with the drop-in center for more than five years. This consistency has meant a lot to Thmaris, who shared, "I just seen a lot of case managers come and go. Rahim is the only one that has never went anywhere. So many years have gone past, and Rahim is here." After his turning point, Rahim recalls Thmaris coming in to the drop-in center with an intense focus on changing his life.

> *He was here, pretty close to everyday, and he'd never been here that frequently before. He was here all the time, working on job searches. He started to take school really, really seriously, which has been a really positive and strong thing for him. And we kind of sat together and figured out a path that hopefully would help provide for him over time. He got his high school diploma, which was really huge. I think just being successful at something was helpful. He's a smart kid. I think getting his high school diploma helped convince him of that.*

Thmaris recalls this time similarly. "Everything I was doing [with Rahim] was productive. When you get that feeling like you're accomplishing something and you're doing good, it's like a feeling that you can't describe."

(Continued)

(Continued)

Going Back to School

When Thmaris first thought about going back to school, he was planning to go just to get a student loan check like many of the homeless youth around him. But Rahim helped him see a different path. Knowing of Thmaris's past positive experiences with construction and building, Rahim urged Thmaris to consider careers that would allow him to work with his hands and to focus on the big picture. He also made sure that Thmaris had the support of a learning specialist from Saint Paul Public Schools, who helped Thmaris figure out what steps he would need to take to get from where he was then to his dream career of welding.

The first time I ever talked to him about school. I was like, "Yeah, man. I just feel like I should go to school and get a loan." And he was like, "Well, it's bigger than that. That loan money is going to be gone like that."

He didn't tell me, no, you shouldn't do that. He just said just think about the future. And I just thought about it. I was thinking and I was thinking. And then I was like, "What if I went to school for something that I want to get a job in?" He was, "Yeah, that's the best way to go."

I talked to him about construction, and he told me you already basically have experience in that because he knows all the jobs I had. So he was like you should go into something that's totally different but that pays a lot of money. Then I researched it and then just knew that I liked working with my hands. So I just put two and two together and was like, well, I want to go to school for welding. Ever since I learned how to weld it's like . . . I love it!

Thmaris loved welding but was intimidated by the engineering part of the learning, which he called "book work." Again, it was Rahim's belief in him that helped him persevere.

I started doing the book work, and I was getting overwhelmed. It was like every time I came down here—even if I just came down here to use the phone—Rahim would be like, "How's that welding class going?"

He just kept me interested in it. I don't know how to explain it. I was just thirsty—not even to get a job in it—but to show Rahim what I'd learned and what I accomplished.

He would tell me, "I'm proud of you." Then I'd show him my book work, and he'd be like, "Man, I don't even know how to read this stuff." And it just made me feel like I

was actually on the right path. It made me feel like I was doing what I was intended to do with my life. That's just how he makes me feel. He makes me feel like when I'm doing the right thing, and he makes sure that I know it.

Thmaris finished his welding certificate with a 4.0 grade point average and the high positive regard of his teachers, and he did this against great odds. He was homeless while completing the program, meaning that he had no consistent place to sleep, to do homework, or to even keep his books. Getting to campus was challenging because tokens were hard to come by. During this time, he was also dealing with one of his greatest challenges, unhealthy relationships with older women. In part, these relationships are a survival technique. Any time living with these women is time off of the streets. But these relationships are also, in part, a symptom of Thmaris's desire to be loved, to have a family, and to take care of others.

I just super easily fall in love. I always look for the ones that have been through the most, the ones that have always had a rough time, and I try to make their life better. That's the biggest thing for me, is trying to stay away from love. I don't know. I'm just so into love. It's probably because I wanted my mom and dad to be together so bad it killed me. Every time she brung a new man home it killed me inside. So I just want a family.

The relationship he was in while attending his first semester of welding classes ended badly. She threw away or burned everything Thmaris owned and paid people to jump him and "beat the hell out of him." They broke his hand and wrist so badly that surgery was required, and the healing time was long and painful. Despite all of this, Thmaris persevered. He completed his welding certificate and is now completing his general education requirements.

Leaving the Gang

To choose this different path, Thmaris had to leave his gang. His case manager helped him with this too by talking through what it was that the gang offered him. Certainly, it offered money and friends, but as Thmaris got older, it also offered more opportunities to serve extended jail time. While Thmaris was in jail for auto theft, no one from the gang came to see him, put money in the book for him, or checked on his little brother. Thmaris identified this as one of the greatest challenges he's had to overcome.

I think the first biggest thing I had to do was leave a lot of friends that weren't on the same level or the same type of mentality that I was on. I had to leave them alone, regardless of if I knew them my whole life or not.

Thmaris's case manager thinks his commitment to school was critical in helping Thmaris leave his gang. He described leaving a gang as follows.

Leaving a gang is almost like drug addiction. You have to replace it with something. And in Thmaris's instance, he actually can replace it with this really furious effort towards getting his education, looking for jobs, and trying to do something different with himself. If he had just tried to quit the gang and replaced it with nothing and did nothing all day, it would've been a lot harder. But he replaced it with this welding certificate, and that was really good. He got very, very excited about welding. It's nice because he is naturally good at it.

Becoming a Father

Thmaris has also recently become a father. He currently has a three-month-old son, named after him, who is his pride and joy. But he and the mother have a rocky relationship, and it's frustrating to Thmaris, who would like to raise his child with his child's mother. He didn't grow up with his own father in the picture and would like something different for his own son. He feels that his son would be happier in life waking up each day seeing both of his parents.

Man, I just have so many plans for him. I just . . . I don't want to put him on a pedestal or anything because I don't want him to go through having a dad that thinks so highly of him and then it's so hard for him to meet my goals. I want him to make his goals. I want him to be happy. That's all I care about. Regardless if he wants to work at Subway instead of going to school . . . if that's your choice, that's your choice.

Becoming a father has also helped him give up the gang life and weed. He realizes that if he doesn't stop selling drugs, then he might not live long enough to see his son grow up.

I don't want to sell drugs all my life and then when I die and my son be like, "Wow. Dad didn't leave me nothing or he didn't teach anything but how to sell drugs." No, that's not what you would want for your son. You want your son to know what it is to work for his money. You want your son to know how it feels to come from a long day's work, tired, like, "Damn. I earned my money though."

Couch Hopping

Throughout his time visiting drop-in centers, Thmaris has never stayed at a youth or adult shelter. He typically couch hops or stays with women he is dating. Some nights, he walks the skyway all night or sleeps in a stairwell. After staying in family shelters when he was younger, he has vowed to himself that he would never stay in another shelter again.

When we first came to Minnesota that's all we did. We stayed in shelter as a family. It was like traumatizing to me because [in shelters] you see like humans at their weakest point. You see them hungry, dirty. I didn't like that. I don't like being around a whole bunch of people that was . . . I'm not saying that I feel like I'm better than anybody because I definitely don't. But I just felt like it was too much to take in. It was too stressful, and it just made me want to cry. It was crazy. I don't want to go back to a shelter ever again.

Where He Is Now

Currently, Thmaris isn't stably housed. He spends some nights at his baby's mother's house, some nights with friends, other nights outside, and some nights at a hotel room. But despite this, he feels very positive and hopeful about his life.

The fact that I have my certificate for welding and I'm certified for welding, that just blows me away. I would have never thought in a million years that I would have that. Even though I've still got to look for a job and I still got a long ways to go, I just feel proud of myself. . . . I know a lot of people that don't even have high school diplomas or a GED and they're struggling to get into college and people going to college just to go get a loan and stuff like that.

I just feel like I'm bettering myself. I've learned a lot over these past seven years. I've matured a great deal. I honestly feel that I'm bettering myself. I don't feel like I'm taking any steps back, regardless or not if I have employment or if I have my own house. I just feel like each day I live more and I learn more and I just feel . . . I'm just grateful to be alive, grateful to even go through the things I'm going through.

Thmaris credits having someone believe in him as critically important in helping him learn how to believe in himself. He says the following about his case manager:

(Continued)

(Continued)

He just saw more in me. I didn't even see it at the time. He saw great potential, and he told me that all the time. "Man, I see great potential in you. I see that. You can just be way much more than what you are." Just to keep coming down here and having somebody have that much faith in you and believe in you that much, it's a life changer.

My mom always used to tell me that I wasn't shit, you know what I'm saying? She was a super alcoholic, and when she gets drunk she always say that. "You ain't shit, your daddy ain't shit, you ain't going to be shit." She was just always down on me. Just to hear somebody really have an interest in you or want you to better yourself, it just changed my life.

I honestly feel like if I didn't have Rahim in my corner, I would have been doing a whole bunch of dumb shit. I would have been right back at square one. I probably would have spent more time in jail than I did. I just felt like if it wasn't for him I probably wouldn't be here right now talking to you.

Through this relationship, Thmaris was able to learn things that others may take for granted, such as how to create an e-mail account, write a resume, apply for a job online, or use time productively.

To be honest, I never knew what a resume was, I never knew how to create an e-mail account, I never knew how to send a resume online, I never knew how to do an application for a job online. He taught me everything there was about that. He taught me how to look for an apartment, he taught me how to look for a job, he taught me how to dress, he taught me how to talk to a boss, how to talk to a manager, how to get a job. There's just a lot of stuff like that.

I also learned that there's always something to do productively instead of wasting your time. So I just thought about making my resume better. And I just thought about sending e-mails to companies that I knew were hiring, and just doing productive stuff. I never knew what "productive" was until Rahim. I just didn't think about time that was being wasted.

Thmaris hopes to open his own shop someday doing car modifications. He wants to make his son proud, and despite their difficult history and relationship, he wants to make his mother proud. He knows it will take a lot of work to do this but feels motivated and determined to do so.

That's what I basically learned from being at [the drop-in center]. I understand now the value of doing what you need to do versus what you want. A lot of people say, "Men do what they want and boys do what they can." But that's not it. It's "Men do what they need to do and boys do what they want." I'm so glad I learned that for real. Because I was always just doing what I wanted to do.

He's proud that he's been able to overcome the challenges in his life in order to get a high school diploma and graduate from his welding program. Feeling successful has just fueled Thmaris's ambition to experience more success.

You don't know how it felt when I graduated high school. I was like, "Wow, I did this on my own?" And it just felt so good. I'm thirsty again to get another certificate or diploma or whatever just because it's just the best feeling in the world. It's better than any drug. It's like, man, I don't even know how to explain it. It just felt like you just climbed up to the top of the mountain and just like you made it.

Thmaris feels that without the drop-in center and his case manager he might be in jail right now. He sees other young people "wasting their time" at the drop-in center and wishes he could tell them what he now knows.

If you're still stuck in that stage where you don't know what you want to do with your life, then come here and sit down with a case manager. Try to talk to somebody, and they'll help you better your situation.

For Thmaris, the drop-in center and his case manager were key to helping him quit his addiction, leave his gang, get a high school diploma and welding certificate, and start to build a life for himself that he's proud of.

APPLICATION EXERCISES

1. Construct an interview that includes (a) a section of standardized open-ended questions and (b) interview guide topics. (See Exhibit 7.4, pp. 437–438.) Interview at least three people. As part of the interview, look for opportunities to add in emergent, conversational questions not planned in advance. After doing the interviews, discuss the differences you experienced among these three interview format approaches. What are the strengths and weaknesses of each?

2. Discuss what in-depth interviewing can and cannot do. What are the strengths and weaknesses of in-depth, open-ended interviewing? As part of your discussion, comment on the following quotation by financial investment advisor Fred Schwed (2014):

   > Like all of life's rich emotional experiences, the full flavor of losing important money cannot be conveyed by literature. Art cannot convey to an inexperienced girl what it is truly like to be a wife and mother. There are certain things that cannot be adequately explained to a virgin either by words or pictures. Nor can any description I might offer here even approximate what it feels like to lose a real chunk of money that you used to own.

3. Exhibit 7.7 (p. 445) presents a matrix of question options. To understand how these options are applied in an actual study, review a real interview. The Outward Bound standardized interview, Exhibit 7.19 (pp. 508–511), at the end of this chapter, can be used for this purpose. Identify questions that fit in as many cells as you can. That is, which cell in the matrix (Exhibit 7.7) is represented by each question in the Outward Bound interview protocol?

4. Exhibit 7.12 (pp. 462–463) presents six approaches to *interactive, relationship-based interviewing*. What do these approaches have in common? How are they different from each other? Why are these approaches controversial from the perspective of traditional social science interviewing? (see Exhibit 7.3, Item 2, p. 433)

5. Exhibit 7.20 (pp. 511–516), at the end of this chapter, presents a case study of a homeless youth, Thmaris, based on an in-depth interview with him. What is your reaction to the case study? What purposes does it serve? How would you expect it might be used? What makes the case study effective?

6. Examine an oral history data set. Select and compare three oral histories from a collection. Here are examples:

   - The Voices of Feminism Oral History Project, with transcripts available through Smith College: http://www.smith.edu/libraries/libs/ssc/vof/vof-intro.html
   - Oral history project transcripts made available by University of South Florida: http://guides.lib.usf.edu/ohp

7. Locate a qualitative study in your area of interest, discipline, or profession that made extensive use of interviewing as the primary method of data collection. What approach to interviewing was used? Why? How was the study designed and conducted to ensure high-quality data? What challenges, if any, are reported? How were they handled? What ethical issues, if any, are reported and discussed? Overall, assess how well the study reports on the interviewing methods used to allow you to make a judgment about the quality of the findings. What questions about the interviewing approach and its implications are left unanswered?

PART 3

Analysis, Interpretation, and Reporting

Halcolm will tell you this:

> *Because you can name something does not mean you understand it.*
>
> *Because you understand it does not mean it can be named.*

And this:

> *What you do not see you cannot describe.*
>
> *What you cannot describe you cannot interpret.*
>
> *But because you can describe something*
>
> *does not mean you can interpret it.*

And yet this:

> *The riddle about the sound of one hand clapping arose from watching the first decision maker reading the first evaluation report.*

And finally this:

> *Where the sun shines, there too is shadow.*
>
> *Be illumined by the light of knowledge*
>
> *no less than by its shadow.*

Complete Analysis Isn't

The moment you cease observing, pack your bags and leave the field, you will get a remarkably clear insight about that one critical activity you should have observed . . . but didn't.

The moment you turn off the tape recorder, say goodbye, and leave the interview, it will become immediately clear to you what perfect question you should have asked to tie the whole thing together . . . but didn't.

The moment you begin analysis it will become perfectly clear to you that you're missing the most important pieces of information, and that without those pieces of information there is absolutely no hope of making any sense out of what you have.

Know, then, this:

The complete analysis isn't.

Analysis finally makes clear what would have been most important to study, if only we had known beforehand.

Evaluation reports finally make clear to decision makers what they had really wanted to know, but couldn't articulate at the time.

Analysis brings moments of terror that nothing sensible will emerge and times of exhilaration from the certainty of having discovered ultimate truth. In between are long periods of hard work, deep thinking and weight-lifting volumes of material.

—From Halcolm's *Iron Laws of Evaluation Research*

8 Qualitative Analysis and Interpretation

In set theory, an empty set is denoted by a pair of empty braces: { }. A *set* is defined as a collection of *things* that are brought together because they have something in common. In mathematical set theory, the items in a set must obey a clear, definitive *rule*, for example, whole numbers that are multiples of 2 (2, 4, 6, 8, 10, . . .) or words that begin with the letters *Qu* and have only one syllable (Queen, Queer, Quinn, Quit, . . .). The rule must be without ambiguity, so that it is clear and uncontested that an item does or does not belong in the set.

In qualitative analysis, in contrast, what items belong in a set (e.g., a grouping, a category, a pattern, a theme) is a matter of judgment. Judgments can vary depending on who is doing the judging, with what criteria, and for what purpose. Thus, unlike rules, judgments can be ambiguous. Love and hate may be considered in the same set (strong emotions) or may be judged to belong in different sets (things that bring humans together vs. things that divide us). What constitutes a set in the game of tennis is clear, defined by a rule. What constitutes a set in qualitative analysis must be defined and created anew each time the game, qualitative analysis, is played. Play on.

Chapter Preview

Part 1 of the book provided an overview of qualitative inquiry, with chapters on the nature, niche, and value of qualitative inquiry; strategic themes in qualitative inquiry; a variety of qualitative inquiry frameworks (paradigmatic, philosophical, and theoretical orientations); and practical and actionable qualitative applications. Part 2 covered qualitative designs and data collection, with chapters covering purposeful sampling and design options, fieldwork strategies and observation methods, and qualitative interviewing. Part 3 presents the two final chapters: Chapter 8, "Qualitative Analysis and Interpretation," followed by Chapter 9, "Enhancing the Quality and Credibility of Qualitative Studies."

Module 65 in this chapter opens by covering the basics of analysis, with a focus on establishing a strong foundation for qualitative analysis. Module 66 presents the importance of thick description and constructing case studies. Module 67 turns to pattern, theme, and content analysis. Module 68 looks in depth at the intellectual and operational work of analysis. Module 69 presents logical and matrix analyses and explains how to synthesize qualitative studies. Module 70 takes on the critical processes of interpreting findings and determining substantive significance, with special attention to phenomenological and hermeneutic examples. Module 71 examines causal explanation thorough qualitative analysis. Module 72 opens the window on new analysis directions: contribution analysis, participatory analysis, and qualitative counterfactuals. Module 73 provides advice and examples on writing up and reporting findings, including using visuals. Module 74 addresses special analysis and reporting issues—mixed methods and focused communications—and provides a principles-focused report exemplar. Finally, Module 75 summarizes and concludes the chapter, plus providing case study exhibits. We begin with basic analysis.

SOURCE: Brazilian cartoonist Claudius Ceccon. Used with permission.

Establishing a Strong Foundation for Qualitative Analysis: Covering the Basics

> Good field methods are necessary, but not sufficient, for good research. You may be a skilled and diligent observer and interviewer and gather "rich data," but, unless you have good ideas about how to focus the study and analyze those data, your project will yield little of value.
>
> —William Foot Whyte (1984, p. 225)
> *Learning From the Field*

The Challenge

Qualitative analysis transforms data into findings. No formula exists for that transformation. Guidance yes, but no recipe. Direction can and will be offered, but the final destination remains unique for each inquirer, known only when—and if—arrived at.

Medieval alchemy aimed to transmute base metals into gold. Modern alchemy aims to transform raw data into knowledge, the coin of the Information Age. Rarity increases value. Fine qualitative analysis remains rare and difficult—and therefore valuable.

Metaphors abound. Analysis begins during a larval stage that, if fully developed, metamorphoses from a caterpillar-like beginning into the splendor of the mature butterfly. Or this: The inquirer acts as a catalyst on raw data, generating an interaction that synthesizes new substance born anew of the catalytic conversion. Or this: Findings emerge like an artistic mural created from collage-like pieces that make sense in new ways when seen and understood as part of a greater whole.

Consider the patterns and themes running through these metaphors: transformation, transmutation, conversion, synthesis, whole from parts, and sense making. Such motifs run through qualitative analysis like golden threads in a royal garment. They decorate the garment and enhance its quality, but they may also distract attention from the basic cloth that gives the garment its strength and shape—the skill, knowledge, experience, creativity, diligence, and work of the garment maker. No abstract processes of analysis, no matter how eloquently named and finely described, can substitute for the skill, knowledge, experience, creativity, diligence, and work of the qualitative analyst. Thus, Stake (1995) writes classically of the *art* of case study research. Van Maanen (1988) emphasizes the storytelling motifs of qualitative writing in his ethnographic book on *telling tales*. Golden-Biddle and Locke (2007) make *story* the central theme in their book *Composing Qualitative Research*. Corrine Glesne (2010), a researcher and a poet, begins with the story analogy, describing qualitative analysis as "finding your story," then later represents the process as "improvising a song of the world." Lawrence-Lightfoot and Davis (1997) evoke "portraits" in naming their form of qualitative analysis *The Art and Science of Portraiture*. Brady (2000) explores "anthropological poetics." Janesick (2000) uses the metaphor of dance in "the choreography of qualitative research design," which suggests that, for warming up, we may need "stretching exercises" (Janesick, 2011). Hunt and Benford (1997) call to mind theatre as they use "dramaturgy" to examine qualitative inquiry. Denzin (2003) and Hamera (2011) call for ethnography to be "performative." Richardson (2000b) reminds us that qualitative analysis and writing involve us not just in making sense of the world but also in making sense of our relationship to the world and therefore in discovering things about ourselves even as we discover things about some phenomenon of interest. In this complex and multifaceted analytical integration of disciplined science, creative artistry, skillful crafting, rigorous sense making, and personal reflexivity, we mold interviews, observations, documents, and field notes into *findings*.

The challenge of qualitative analysis lies in making sense of massive amounts of data. This involves reducing the volume of raw information, sifting the trivial from the significant, identifying significant patterns, and constructing a framework for communicating the essence of what the data reveal. In analyzing qualitative data, guidelines exist but no recipes; principles provide direction, but there is no significance test to run that determines whether a finding is worthy of attention. No ways exist of perfectly replicating the researcher's analytical thought processes. No straightforward tests can be applied for reliability and validity.

DISTINGUISHING SIGNAL FROM NOISE

When you try to locate a clear radio station signal through the static noise that fills the airways between signals, you are engaged in the process of distinguishing *signal from noise*. Nate Silver (2012) used that metaphor as the title of his best-selling book on "why so many predictions fail—but some don't." Silver found that the best predictions—whether of election outcomes, economic patterns, social trends, spread of disease, winning sports teams, stock market indicators, or any of the many arenas in which humans attempt predictions—are those that use both quantitative and qualitative data and use theory to turn data into a feasible, meaningful, and compelling story. Data are noise, lots and lots of noise. The more data, the more noise. *Big data: loud noise.* The story that detects, makes sense of, interprets, and explains meaningful patterns in the data is *the signal.*

But signals are not constant or static. They vary with context and change over time. So the quest to distinguish signal from noise is ongoing.

In short, no absolute rules exist, except perhaps this: Do your very best with your full intellect to fairly represent the data and communicate what the data reveal given the purpose of the study.

Frameworks for analyzing qualitative data can be found in abundance (Saldaña, 2011), and studying examples of qualitative analysis can be especially helpful, as in the Miles, Huberman, and Saldaña (2014) qualitative analysis sourcebook. But guidelines, procedural suggestions, and exemplars are not rules. Applying guidelines requires judgment and creativity. Because each qualitative study is unique, the analytical approach used will be unique. Because qualitative inquiry depends, at every stage, on the skills, training, insights, and capabilities of the inquirer, qualitative analysis ultimately depends on the analytical intellect and style of the analyst. The human factor is the great strength and the fundamental weakness of qualitative inquiry and analysis—a scientific two-edged sword.

That said, let's get on with it. Exhibit 8.1 offers 12 tips for laying a strong foundation for qualitative analysis. These tips are neither exhaustive nor universal. They won't all apply to everyone, but perhaps they will stimulate you to think of other things you might do to get yourself ready for the challenges of qualitative analysis. I'll elaborate several of these tips and then delve into alternative ways of conducting qualitative analysis.

Elaboration of Some Tips for Ensuring That a Strong Foundation for Qualitative Analysis Begins During Fieldwork

> Field methods have the advantage of flexibility, allowing us to explore the field, to refine or change the initial problem focus, and to adapt the data gathering process to ideas that occur to us even in late stages of our exploration.
>
> —Whyte, (1984, p. 225)

Research texts typically make a hard-and-fast distinction between data collection and analysis. For data collection based on surveys, standardized tests, and experimental designs, the lines between data collection and analysis are clear. But the fluid and emergent nature of naturalistic inquiry makes the distinction between data gathering and analysis far less absolute. In the course of fieldwork, ideas about directions for analysis will occur. Patterns take shape. Signals start to emerge from the noise. Possible themes spring to mind. Thinking about implications and explanations deepens the final stage of fieldwork. While earlier stages of fieldwork tend to be generative and emergent, following wherever the data lead, later stages bring closure by moving toward confirmatory data collection—deepening insights into and confirming (or disconfirming) patterns that seem to have appeared. Indeed, sampling, confirming, and disconfirming cases require a sense of what there is to be confirmed or disconfirmed.

Ideas for making sense of the data that emerge while still in the field constitute the beginning of analysis; they are part of the record of field notes. Sometimes insights emerge almost serendipitously. When I was interviewing recipients of MacArthur Foundation fellowships (popularly dubbed "Genius Awards"), I happened to interview several people in major professional and personal transitions, followed by several in quite stable situations. This happenstance of how interviews were scheduled suggested a major distinction that became important in the final analysis—distinguishing the impact of the fellowships on recipients in transition from those in stable situations.

Recording and tracking analytical insights that occur during data collection is part of fieldwork and the beginning of qualitative analysis. I've heard graduate students being instructed to repress all analytical thoughts while in the field and to concentrate on data

| EXHIBIT 8.1 | Twelve Tips for Ensuring a Strong Foundation for Qualitative Analysis |

1. ***Begin analysis during fieldwork:*** Note and record emergent patterns and possible themes while still in the field. Add confirming cases to deepen analysis and possible disconfirming cases to test thematic ideas while still in the field.

2. ***Inventory and organize the data:*** Make sure you have all the interviews, observations, and documents that constitute the raw data of your qualitative inquiry. Check that the data elements and sources are labeled, dated, and complete.

3. ***Fill in gaps in the data:*** As soon as possible, fill in the gaps in the data while connections in the field are fresh. If later interviews turn up issues that need to be checked out with earlier interviewees, do so quickly. If documents are missing, take steps to get them.

4. ***Protect the data:*** Back them up. Make sure the data are secure.

5. ***Express appreciation:*** Thank those who have provided you with data. Fieldwork creates relationships. Once out of the field, analysis and writing can take a lot of time. Don't wait until it's all done to show your appreciation to those who have provided you with data. Follow up with appropriate expressions of appreciation sooner rather than later. Procrastination can too easily lead to never getting it done. Do it!

6. ***Reaffirm the purpose of your inquiry:*** Restate the purpose of your enquiry, and therefore the purpose of your analysis. Purpose drives analysis. Design frames and sets the stage for analysis. Be clear about why you're doing this work, revisit and reengage with the questions and the purposeful sampling strategy that have guided your inquiry, and get clear about the primary product that you are producing that will fulfill your study's purpose and answer priority questions in accordance with your design.

7. ***Review exemplars for inspiration and guidance:*** Reexamining classic works in your field can be a source of inspiration. They are classics for a reason. Keep those exemplars nearby to reinvigorate and motivate you when the drudgery of analysis sets in or doubts about what you're finding emerge. Those who wrote the classics experienced analysis fatigue and doubts as well. They persevered. So will you.

8. ***Make qualitative analysis software decisions:*** If you're using qualitative data management software, learn how to use it effectively. Locate technical support. Practice data entry and some simple analysis. All qualitative analysis software has a steep learning curve. Leave time and mental space to learn if you're new to the software. If you're an old hand, check out new versions and features that might be useful.

9. ***Schedule intense, dedicated time for analysis:*** Qualitative analysis requires immersion in the data. It takes time. Make time. Set a realistic schedule. Enlist the support of family, friends, and colleagues to help you stay focused, and give the analysis the dedicated time it deserves.

10. ***Clarify and determine your initial analysis strategy:*** Inductive qualitative analysis can follow a number of pathways. Various theoretical traditions (ethnography, phenomenology, constructivism, realism, etc.) provide frameworks and guidance. Data can be organized and reported in different ways: case studies, question-by-question interview analysis, storytelling, elucidating sensitizing concepts or principles, and thematic analysis, among others. Grounded theory provides a highly prescriptive framework for analysis. Decide what strategy fits your purpose, write it down with your rationale, and get started analyzing. This process involves reconnecting with the theoretical and strategic framework that presumably guided design decisions and the formulation of your inquiry questions.

11. ***Be reflective and reflexive:*** Monitor your thought processes and decision-making criteria. Be in touch with predispositions, biases, fears, hopes, constraints, blinders, and pressures you're under. Qualitative analysis is ultimately highly personal and judgmental. You are the analyst. Observe yourself. Learn about yourself and your analysis processes, both cognitively and emotionally.

12. ***Start and keep an analysis journal:*** Document analysis decisions, emergent ideas, forks in the road, false starts, dead ends, breakthroughs, eureka moments, what you learn about analysis, what you learn about the focus of the inquiry, and what you learn about yourself. You may think you'll remember these things. You won't. Document the analytical process—in depth, systematically, and regularly. That documentation is the foundation of rigor. Qualitative analysis is a new stage of fieldwork in which you must observe and document your own processes even as you are doing the analysis.

See elaboration of these tips in the next sections.

collection. Such advice ignores the emergent nature of qualitative designs and the power of field-based analytical insights. Certainly, this can be overdone. Too much focus on analysis while fieldwork is still going on can interfere with the openness of naturalistic inquiry, which is its strength. Rushing to premature conclusions should be avoided. But repressing analytical insights may mean losing them forever, for there's no guarantee they'll return. And repressing in-the-field insights removes the opportunity to adapt data collection to test the authenticity of those insights while still in the field and fails to acknowledge the confirmatory possibilities of the closing stages of fieldwork. In the MacArthur Fellowship study, I added transitional cases to the sample near the end of the study to better understand the varieties of transitions the fellows were experiencing—an in-the-field form of emergent, purposeful sampling driven by field-based analysis. Such overlapping of data collection and analysis improves both the quality of the data collected *and* the quality of the analysis, so long as the fieldworker takes care not to allow these initial interpretations to overly confine analytical possibilities. Indeed, instead of focusing additional data collection entirely on confirming emergent patterns while still in the field, the inquiry should become particularly sensitive to looking for alternative explanations and patterns that would invalidate the initial insights.

In essence, when data collection has ended and it is time to begin the formal and focused analysis, the qualitative inquirer has two primary sources to draw from in organizing the analysis: (1) the questions that were generated during the conceptual and design phases of the study, prior to fieldwork, and (2) the analytic insights and interpretations that emerged during data collection.

Even when analysis and writing are under way, fieldwork may not be over. On occasion, gaps or ambiguities found during analysis cry out for more data collection—so, where possible, interviewees may be recontacted to clarify or deepen responses, or new observations are made to enrich descriptions. This is called *member checking*—verifying data, findings, and interpretations with the participants in the study, especially key informants. While writing a Grand Canyon–based book that describes modern male coming-of-age issues (Patton, 1999), I conducted several follow-up and clarifying interviews with my two key informants and returned to the Grand Canyon four times to deepen my understanding of Canyon geology and add descriptive depth. Each time I thought that, at last, fieldwork was over and I could just concentrate on writing, I came to a point where I simply could not continue without more data

collection. Such can be the integrative, iterative, and synergistic processes of data collection and analysis in qualitative inquiry. A final caveat, however: *Perfectionism breeds imperfections*. Often, additional fieldwork isn't possible, so gaps and unresolved ambiguities are noted as part of the final report. Dissertation and publication deadlines may also obviate additional confirmatory fieldwork. And no amount of additional fieldwork can, or should, be used to force the vagaries of the real world into hard-and-fast conclusions or categories. Such perfectionist and forced analysis ultimately undermines the authenticity of inductive, qualitative analysis. Finding patterns is one result of analysis. Finding vagaries, uncertainties, and ambiguities is another.

Inventory and Organize the Raw Data for Analysis

> It wasn't curiosity that killed the cat.
>
> It was trying to make sense of all the data curiosity generated.
>
> —Halcolm

The data generated by qualitative methods are voluminous. I have found no way of preparing students for the sheer mass of information they will find themselves confronted with when data collection has ended. Sitting down to make sense out of pages of interviews and whole files of field notes can be overwhelming. Organizing and analyzing a mountain of narrative can seem like an impossible task.

How big a mountain? Consider a study of community and scientist perceptions of HIV vaccine trials in the United States done by the Centers for Disease Control. A large, complex, multisite effort called Project LinCS: Linking Communities and Scientists, the study's 313 interviews generated more than 10,000 pages of transcribed text from 238 participants on a range of topics (MacQueen & Milstein, 1999). Now that's an extreme case, but on average, a one-hour interview will yield 10 to 15 single-spaced pages of text; 10 two-hour interviews will yield roughly 200 to 300 pages of transcripts.

Getting organized for analysis begins with an inventory of what you have. Are the field notes complete? Are there any parts that you put off to write later but never got to doing that need to be finished, even at this late date, before beginning analysis? Are there any glaring holes in the data that can still be data management and preparation. Doing all or some of your own interview transcriptions (instead of having them done by a transcriber), for example, provides an opportunity to get immersed in the data, an experience that usually generates important insights. Typing and organizing handwritten field notes offer another opportunity to immerse yourself in the data, a chance to get a feel of the cumulative data as a whole. Doing your own transcriptions, or at least checking them by listening to the tapes as you read them, can be quite different from just working off transcripts done by someone else.

Depth and Detail in Qualitative Analysis

Transcriptions

filled by collecting additional data before the analysis begins? Are all the data properly labeled with a notation system that will make retrieval manageable (dates, places, interviewee-identifying information, etc.)? Are interview transcriptions complete? Assess the quality of the information you have collected. *Get a sense of the whole.*

Fill in Gaps in the Data

The problem of incomplete data is illustrated by the experience of a student who had conducted 30 in-depth pre- and post interviews with participants in a special program. The transcription process took several weeks. She made copies of three transcripts and brought them to our seminar for assistance in doing the analysis. As I read the interviews, I got a terrible sinking feeling in my stomach. While other students were going over the transcriptions, I pulled her aside and asked her what instructions she had given the typist. It was clear from reading just a few pages that she did not have *verbatim* transcriptions—the essential raw data for qualitative analysis. The language in each interview was the same. The sentence structures were the same. The answers were grammatically correct. People in natural conversations simply do not talk that way. The grammar in natural conversations comes out atrocious when transcribed. Sentences hang incomplete, interrupted by new thoughts before the first sentence is completed. Without the knowledge of this student, and certainly without her permission, the typist had decided to summarize the participants' responses because "so much of what they said was just rambling on and on about nothing," the transcriber later explained. All of the interviews had to be transcribed again before analysis could begin.

Earlier, I discussed the transition from fieldwork to analysis. Transcribing offers another point of transition between data collection and analysis as part of

Protect Your Data

Thomas Carlyle lent the only copy of his handwritten manuscript on the history of the French Revolution, his masterwork, to philosopher John Stuart Mill, who lent it to a Mrs. Taylor. Mrs. Taylor's illiterate housekeeper thought it was waste paper and burned it. Carlyle reacted with nobility and stoicism and immediately set about rewriting the book. It was published in 1837 to critical acclaim and consolidated Carlyle's reputation as one of the foremost men of letters of his day. We'll never know how the acclaimed version compared with the original, or what else Carlyle might have written in the year lost after the fireplace calamity.

So it is prudent to make backup copies of all your data, putting one master copy away someplace secure. Indeed, if data collection has gone on over any long period, it is wise to make copies of the data as they are collected, being certain to put one copy in a safe place where it will not be disturbed and cannot be lost, stolen, or burned. One of my graduate students kept all of his field notes and transcripts in the truck of his car. The car was vandalized, and he lost everything, with no backup copies. The data you've collected are unique and precious. The exact observations you've made, the exact words people have spoken in interviews—these can never be recaptured in precisely the same way, even if new observations are undertaken and new interviews are conducted. Moreover, you've likely made promises about protecting confidentiality, so you have an obligation to take care of the data. Field notes and interviews should be treated as the valuable material they are. Protect them.

Beyond Thomas Carlyle's cautionary tale, my advice in this regard comes from two more recent disasters. I was at the University of Wisconsin when antiwar protestors bombed a physics building, destroying the life work of several professors. I also had a psychology doctoral student who carried her dissertation work, including all the raw data, in the back seat of her car.

An angry patient from a mental health clinic with whom she was working firebombed her car, destroying all of her work. Tragic stories of lost research, while rare, occur just often enough to remind us about the wisdom of an ounce of prevention.

Once a copy is put away for safekeeping, I like to have one hard copy handy throughout the analysis, one copy for writing on, and one or more copies for cutting and pasting. A great deal of the work of qualitative analysis involves creative cutting and pasting of the data, even if done on a computer, as is now common, rather than by hand. Under no circumstances should one yield to the temptation to begin cutting and pasting the master copy. The master copy or computer file remains a key resource for locating materials and maintaining the context for the raw data.

Qualitative data analysis software (QDAS) facilitates saving data in multiple digital locations, such as external hard drive, server, DVD copy, and flash drive. However, the researcher must be disciplined enough to update backups frequently and not destroy all old copies, especially preserving the original master copy.

Purpose Drives Analysis

> "Data" linked to real human social worlds is where human social science, whatever else it does, has to start and has to finish.
>
> —Michael Agar (2013, p. 19)
> *The Lively Science*

Purpose drives analysis: This follows from the theme of Chapter 5, that purpose drives design. Design gives a study direction and focus. Chapter 5 presented a typology of inquiry purposes: basic research, applied research, summative evaluation research, formative evaluation, and action research. (See Exhibit 5.1, p. 250.) These distinct purposes inform design decisions and inquiry focus and subsequently undergird analysis because they involve different norms and expectations for generating, validating, presenting, and using findings.

Basic qualitative research is typically reported through a scholarly monograph or published article, with primary attention to the contribution of the research to social science theory. The theoretical framework within which the study is conducted will heavily shape the analysis. As Chapter 3 made clear,

the theoretical framework for an ethnographic study will differ from that for ethnomethodology, heuristics, or hermeneutics.

Applied qualitative research may have a more or less scholarly orientation depending on primary audience. If the primary audience is scholars, then applied research will be judged by the standards of basic research, namely, research rigor and contribution to theory. If the primary audience is policymakers, the relevance, clarity, utility, and applicability of the findings will become most important.

For *scholarly qualitative research*, the published literature on the topic being studied focuses the contribution of a particular study. Scholarship involves an ongoing dialogue with colleagues about particular questions of interest within the scholarly community. The analytical focus, therefore, derives in part from what one has learned that will make a contribution to the literature in a field of inquiry. That literature will likely have contributed to the initial design of the study (implicitly or explicitly), so it is appropriate to revisit that literature to help focus the analysis.

Focus in *evaluation research* should derive from questions generated at the very beginning of the evaluation process, ideally through interactions with primary intended users of the findings. Too many times, evaluators go through painstaking care, even agony, in the process of working with primary stakeholders to clearly conceptualize and focus evaluation questions before data collection begins. But then, once the data are collected and analysis begins, they never look back over their notes to review and renew their clarity on the central issues in the evaluation. It is not enough to count on remembering what the evaluation questions were. The early negotiations around the purpose of an evaluation usually involve important nuances. To reestablish those nuances for the purpose of helping focus the analysis, it is important to review the notes on decisions that were made during the conceptual part of the evaluation. (This assumes, of course, that the evaluator has treated the conceptual phase of the evaluation as a field experience and has kept detailed notes about the negotiations that went on and the decisions that were made.)

In addition, it may be worth reopening discussions with intended evaluation users to make sure that the original focus of the evaluation remains relevant. This accomplishes two things. First, it allows the evaluator to make sure that the analysis will focus on needed information. Second, it prepares evaluation users for the results. At the point of beginning formal analysis, the evaluator will have a much better perspective on what kinds of questions can be answered with the data that have been collected. It pays to check out which questions should take priority in the final report and to suggest new possibilities that may have emerged during fieldwork.

Summative evaluations will be judged by the extent to which they contribute to making decisions about a program or intervention, usually decisions about overall effectiveness, continuation, expansion, and/or replication in other sites. A full report presenting data, interpretations, and recommendations is required. In contrast, *formative evaluations*, conducted for program improvement, may not even generate a written report. Findings may be reported primarily orally. Summary observations may be listed in outline form, or an executive summary may be written, but the timelines for formative feedback and the high costs of formal report writing may make a full, written report impractical. Staff and funders often want the insights of an experienced outsider who can interview program participants effectively, observe what goes on in the program, and provide helpful feedback. The methods are qualitative, the purpose is practical, and the analysis is done throughout fieldwork; no written report is expected beyond a final outline of observations and implications. Academic theory takes second place to understanding the program's theory of action as actually practiced and implemented. In addition, formative feedback to program staff may be ongoing rather than simply at the end of the study. However, in some situations, funders may request a carefully documented, fully developed, and formally written formative report. The nature of formative reporting, then, is dictated by user needs rather than scholarly norms. For qualitative evaluators, a primary purpose of inquiry, analysis, and interaction around findings is to "foster learning"; a qualitative evaluator "serves as an educator, helping program staff and participants understand the evaluation process and ways in which they can use that process for their own learning" (Goodyear et al., in press).

Action research reporting also varies a great deal. In some action research, *the process is the product*, so no report will be produced for outside consumption. On the other hand, some action research efforts are undertaken to test organizational or community development theory, and therefore, they require fairly scholarly reports and publications. Action research undertaken by a group of people to solve a specific problem may involve the group sharing the analysis process to generate a mutually understood and acceptable solution, with no permanent, written report of findings.

Students writing *dissertations* will typically be expected to follow very formal and explicit analytical procedures to produce a scholarly monograph with careful attention to methodological rigor. Graduate students will be expected to report in detail on all aspects of methodology, usually in a separate chapter, including a thorough discussion of analytical procedures, problems, and limitations.

The point here is that the process, duration, and procedures of analysis will vary depending on the study's purpose and audience. Likewise, the reporting format will vary. First and foremost, then, analysis depends on clarity about purpose (as do all other aspects of the study). Knowing what kind (or kinds) of findings are needed and how those results will be reported constitutes the foundation for analysis.

Design Frames Analysis

Purpose drives analysis through purposeful sampling. Design reflects purpose and therefore frames analysis.

You don't wait until you've collected data to figure out your analysis approach. Design decisions (Chapter 5) anticipate what kind of analysis will be done. In particular, the purposeful sampling strategy you've followed is based on what kind of results you want to produce. The data you have to analyze are based on your design. What you have sampled determines what you will address in analysis. Exhibit 8.2 shows the connection between purposeful sampling strategy and analysis approach. The purposeful sampling strategies are taken from Exhibit 5.8 in Chapter 5 (pp. 266–272).

Take Guidance and Inspiration From Examples and Exemplars

> Since we learn from examples, it pays to carefully select good examples to learn from.
>
> —Halcolm

Reexamining classic works in your field can be a source of inspiration. They are classics for a reason. Keep those exemplars nearby to reinvigorate and inspire you when the drudgery of analysis sets in or doubts about what you're finding emerge. Those who wrote the classics experienced analysis fatigue and doubts as well. They persevered. So will you.

The first chapter presented several examples of important qualitative studies from different fields and disciplines:

- Patterns in women's ways of knowing (Belenky et al., 1986)

EXHIBIT 8.2 Connecting Design and Analysis: Purposeful Sampling and Purpose-Driven Analysis

PURPOSEFUL SAMPLING STRATEGY	PURPOSE: DEFINITION/EXPLANATION	ANALYSIS STRATEGY
a. *Single significant case* (see pp. 273–276)	One case (*n* = 1) is studied in depth that provides rich and deep understanding of the subject and breakthrough insights, and/or has distinct, stand-out importance.	Construct an in-depth, detailed, holistic case study. Tell the story that illuminates why the sampled case is significant.
b. *Comparison-focused sampling* (see pp. 277–282)	Select cases to compare and contrast to learn about the factors that explain the similarities and differences.	Construct case studies. Identify dimensions of similarity and difference, and plug in data from the cases to illustrate those similarities and differences. Interpret their implications.
c. *Group characteristics sampling* (see pp. 283–287)	Select cases to create a specific, information-rich group that can reveal and illuminate important group patterns.	Identify patterns and themes for the set of cases in the group. For example, with a maximum variation sample, first document the diversity of cases, and then look for patterns that cut across that diversity.
d. *Theory-focused and concept sampling* (see pp. 288–294)	Select cases for study that are exemplars of the concept or construct that is the focus of inquiry, to illuminate the theoretical ideas of interest.	Analyze what the cases illuminate about the sensitizing concept or principles being studied. Inductively generate and document the emergent grounded theory, or deductively examine whether and how the patterns identified illuminate the theory that frames the inquiry.
e. *Instrumental-use multiple-case sampling* (see pp. 295–297)	Select multiple cases of a phenomenon for the purpose of generating generalizable findings that can be used to inform changes in practices, programs, and policies.	Analyze the cases so that intended users can act on the findings to make decisions, improve programs, and engage in policy making. The emphasis is on generating actionable, useful findings.
f. *Sequential and emergence-driven sampling strategies during fieldwork* (see pp. 298–301)	Build the sample during fieldwork. One case leads to another, in sequence, as the inquiry unfolds. Follow leads and new directions that emerge during the study.	Analyze the data, and report the analysis to answer the questions that guided the inquiry and tell the story of what emerged.
g. *Analytically focused sampling* (see pp. 302–304)	Cases are selected to support and deepen qualitative analysis and interpretation of patterns and themes. This is a form of emergent sampling at the analysis stage.	Deepen and enhance credibility of the initial analysis by adding information-rich illuminative cases and/or confirming and disconfirming cases as the analysis unfolds.
h. *Mixed, stratified, and nested sampling strategies* (see pp. 305–307)	Meet multiple-inquiry interests and needs; deepen focus and triangulation for increased relevance and credibility.	Begin with separate analyses of different kinds of samples; for example, in a mixed-methods study that includes a probability sample and a purposeful sample, analyze each, then integrate and synthesize the findings from each analysis.

NOTE: The first and second columns correspond to purposeful sampling strategies presented in Exhibit 5.8, pp. 266–272.

The methods section of a qualitative report can use this same connecting framework (Exhibit 8.2), showing how design informs analysis. An excellent example of such explicit connecting design and analysis is Kaczynski, Salmona, and Smith, (2014).

- Eight characteristics of organizational excellence (Peters & Waterman, 1982)
- Seven habits of highly effective people (Covey, 2013)
- Case studies and cross-case analysis illuminating why battered women kill (Browne, 1987)
- Three primary processes that contribute to the development of an interpersonal relationship (Moustakas, 1995)
- Case examples illustrating the diversity of experiences and outcomes in an adult literacy program (Patton & Stockdill, 1987)
- Teachers' reactions to an oppressive school accountability system (Perrone & Patton, 1976)

Reviewing these examples of qualitative findings from the first chapter will ground this discussion of analytical processes in samples of the real fruit of qualitative inquiry. And this chapter will add many more examples.

Chapter 3 presented theoretical orientations associated with qualitative inquiry (ethnography, phenomenology, constructivism, etc.). These theoretical frameworks have implications for analysis in that the fundamental premises articulated in a theoretical framework or philosophy are meant to inform how one makes sense of the world. If you have positioned your inquiry within one of those traditions, exemplars will help immerse you in the way in which that tradition guides analysis. Later in this chapter, I'll examine in more depth two of the major theory-oriented analytical approaches, phenomenology and grounded theory, as examples of how theory informs analysis.

Using Qualitative Analysis Software

Computers and software are tools that *assist* analysis. Software doesn't really analyze qualitative data. Qualitative software programs facilitate data storage, coding, retrieval, comparing, and linking— but human beings do the analysis. That said, one reviewer of this book added the following elaboration of how software analysis facilitates engagement with the data.

It is correct to emphasize that Qualitative Data Analysis Software (QDAS) is just a tool which the researcher as instrument must remain in control of. However, the tool is both a data management tool and a qualitative analysis tool. When working with QDAS the researcher is building relationships [with the data] which are a process that is more than just content analysis. The tool helps the researcher build connections which promote further development of complex insights. This ongoing process of meaning

construction is analysis. It should be also noted that QDAS can be used with any number of theoretical approaches.

Software has eased significantly the old drudgery of manually locating a particular coded paragraph. Analysis programs speed up the processes of searching for certain words, phases, and themes; labeling interview passages and field notes for easy retrieval and comparative analysis; locating coded themes; grouping data together in categories; and comparing passages in transcripts or incidents from field notes. But the qualitative analyst doing content analysis must still decide what things go together to form a pattern, what constitutes a theme, what to name it, and what meanings to extract from case studies. The human being, not the software, must decide how to frame a case study, how much and what to include, and how to tell the story. Still, software can play a useful role in managing the volume of qualitative data to facilitate analysis, just as quantitative software does.

Quantitative programs revolutionized that research by making it possible to crunch numbers, more accurately, more quickly, and in more ways.... Much of the tedious, boring, mistake-prone data manipulation has been removed. This makes it possible to spend more time investigating the meaning of their data.

In a similar way, QDA [qualitative data analysis] programs improve our work by removing drudgery in managing qualitative data. Copying, highlighting, cross-referencing, cutting and pasting transcripts and field notes, covering floors with index cards, making multiple copies, sorting and resorting card piles, and finding misplaced cards have never been the highlights of qualitative research. It makes at least as much sense for us to use qualitative programs for tedious tasks as it does for those people down the hall to stop hand-calculating gammas. (Durkin, 1997, p. 93)

The analysis of qualitative data involves creativity, intellectual discipline, analytical rigor, and a great deal of hard work. Computer programs can facilitate the work of analysis, but they can't provide the creativity and intelligence that make each qualitative analysis unique. Moreover, since new software is being constantly developed and upgraded, this book can do no more than provide some general guidance. Most of this chapter will focus on the human thinking processes involved in analysis rather than the mechanical data management challenges that computers help solve. Reviews of computer software in relation to various theoretical and practical issues in qualitative analysis can help you decide what software fits your needs.

What began as distinct software approaches have become more standardized as the various software packages have converged to offer similar functions, though sometimes with different names for the same functions. They all facilitate marking text, building codebooks, indexing, categorizing, creating memos, and displaying multiple text entries side by side. Import and export capabilities vary. Some support team work and multiple users more than others. Graphics and matrix capabilities vary, but they are becoming increasingly sophisticated. All take time to learn to use effectively. The greater the volume of data to be analyzed, the more helpful these software programs are. Moreover, knowing which software program you will use *before* data collection will help you collect and enter data in the way that works best for that particular program.

Qualitative discussion groups on the Internet regularly discuss, rate, compare, and debate the strengths and weaknesses of different software programs. While preferences vary, these discussions usually end with the consensus that any of the major programs will satisfy the needs of most qualitative researchers. Increasingly, distinctions depend on "feel," "style," and "ease of use"—matters of individual taste—more than differences in function. Still, differences exist, and new developments can be expected to solve existing limitations. Exhibit 8.3 lists resources for comparing and using qualitative software programs.

In considering whether to use software to assist in analysis, keep in mind that this is partly a matter of individual style, comfort with computers, amount of data to be analyzed, and personal preference. Computer analysis is not necessary and can interfere with the analytic process for those who aren't comfortable spending long hours in front of a screen. Some self-described "concrete" types like to get a physical feel for the data, which isn't possible with a computer. Participants on a qualitative *listserv* posted these responses to a thread on software analysis:

- The best advice I ever received about coding was to read the data I collected over and over and over. The more I interacted with the data, the more patterns and categories began to "jump out" at me. I never even bothered to use the software program I installed on the computer because I found it much easier to code it by hand.
- I found that hand coding was easier and more productive than using a computer program. For me, actually seeing the data in concrete form was vital in recognizing emerging themes. I actually printed multiple copies of data and cut it into individual "chunks," color coding as I went along, and actually physically manipulating the data by grouping chunks by apparent themes, filing in colored folders, and so on. This technique was especially useful when data seemed to fit more than one theme and facilitated merging my initial and later impressions as themes solidified. Messy, but vital for us concrete people.

So, though software analysis has become common and many swear by it because it can offer leaps in

QUALITATIVE ANALYSIS TECHNOLOGY REVOLUTION: PAST AND FUTURE

In the early 1980s, as qualitative researchers began to grapple with the promise and challenges of computers, a handful of innovative researchers brought forth the first generation of what would come to be known as CAQDAS (Computer Assisted Qualitative Data Analysis Software), or QDAS (Qualitative Data Analysis Software).... These stand-alone software packages were developed, initially, to bring the power of computing to the often labor-intensive work of qualitative research. While limited in scope at the beginning to text retrieval tasks, for instance, these tools quickly expanded to become comprehensive all-in-one packages....

Close to 30 years later, QDAS packages are comprehensive, feature-laden tools of immense value to many in the qualitative research world. However, with the advent of the Internet and the emergence of web-based tools known as Web 2.0, QDAS is now challenged on many fronts as researchers seek out easier-to-learn, more widely available and less expensive, increasingly multimodal, visually attractive, and more socially connected technologies....

Truly, qualitative research and technology is in the midst of a revolution.

QDAS 2.0 offers spectacular possibilities to qualitative researchers.... What is emerging: Web 2.0 tools with various capacities.... As we move more deeply into the digital age, their use, which was once a private choice, will become a necessity. (Davidson & di Gregorio, 2011, pp. 627, 639)

productivity for those adept at it, using software is not a requisite for qualitative analysis. Whether you do or do not use software, the real analytical work takes place in your head.

Being Reflective and Reflexive

> Distinguishing signal from noise requires both scientific knowledge and self-knowledge.
>
> —Nate Silver (2012, p. 453)

The strategies, guidelines, and ideas for analysis offered in this book are meant to be suggestive and facilitating rather than confining or exhaustive. In actually doing analysis, you will have to adapt what is presented here to fit your specific situation and study. However analysis is done, *analysts have an obligation to monitor and report their own analytical procedures and processes as fully and truthfully as possible.* This means that qualitative analysis is a new stage of fieldwork in which analysts must observe their own processes even as they are doing the analysis. The final obligation of analysis is to analyze and report on the analytical process as part of the report of actual findings. The extent of such reporting will depend on the purpose of the study. Module 73, later in this chapter, will discuss reflexivity and voice in depth.

EXHIBIT 8.3   **Examples of Resources for Computer-Assisted Qualitative Data Analysis Software Decisions, Training, and Technical Assistance**

Reviews of qualitative analysis software and web-based programs in relation to significant theoretical and practical issues in qualitative analysis can help you decide what software, if any, fits your needs. Here are some examples of resources that provide information and training.

- ***The Computer-Assisted Qualitative Data Analysis Software (CAQDAS):***
  The software provides practical support, training, and information in the use of a range of software programs designed to assist qualitative data analysis; platforms for debate concerning the methodological and epistemological issues arising from the use of such software packages; and research into methodological applications of CAQDAS (CAQDAS Networking Project, 2014).

- ***International Institute for Qualitative Methodology, University of Alberta, Canada:***
  The institute facilitates the development of qualitative research methods across a wide variety of academic disciplines, offering training and networking opportunities through annual conferences and online workshops on qualitative software analysis programs (http://www.iiqm.ualberta.ca/AboutUs.aspx).

- ***Mobile and Cloud Qualitative Research Apps, The Qualitative Report:***
  http://www.nova.edu/ssss/QR/apps.html

- ***Qualitative Research, Software & Support Services, University of Massachusetts, Amherst:***
  http://www.umass.edu/qdap/

- ***Coding Analysis Toolkit (CAT), a free service of the Qualitative Data Analysis Program (QDAP) hosted by the University Center for Social and Urban Research at the University of Pittsburgh:***
  http://cat.ucsur.pitt.edu/

- ***CAQDAS list of options and links:***
  (a) Open source and free, (b) proprietary, and (c) web based (http://en.wikipedia.org/wiki/Computer-assisted_qualitative_data_analysis_software)

- ***Learning qualitative data analysis on the web:***
  Comparative reviews of software (http://onlineqda.hud.ac.uk/Intro_CAQDAS/reviews-of-sw.php)

- ***Make inquiries to the qualitative research listserv, QUALRS-L:***
  QUALRS-L-request@listserv.uga.edu/?

SIDEBAR

OBSERVATIONS AND ADVICE FROM A QUALITATIVE SOFTWARE CONSULTANT

Asher E. Beckwitt has built a consulting business advising graduate students and novice researchers on how to engage in qualitative research using software for analysis (www .qualitativeresearch.org). I asked her to share her experiences as a consultant.

Question: What are the most common challenges you encounter in using qualitative methods with clients and in advising graduate students on their qualitative dissertations?

Answer: The most common challenges are as follows:

1. Students do not understand the differences between qualitative and quantitative methods.

2. They do not understand that there are different types of qualitative analysis (e.g., grounded theory, phenomenology, narrative analysis, etc.).

3. They do not understand that there are different methodologists' approaches within these areas (e.g., Glasser and Strauss approach grounded theory differently than Strauss and Corbin).

4. They do not understand how to code and analyze their data according to their chosen approach.

Question: You work a lot with qualitative software. What do you tell clients and students that qualitative software does well—and doesn't do?

Answer: Software is an excellent tool for organizing and coding information. It allows you to store all of the collected data and the codes in one place, as well as code multiple sources (papers/interviews/focus group transcripts/field notes). The organizational capacity of software permits you to code in more detail (e.g., you may have hundreds of codes).

Question: What are the most common misunderstandings about qualitative software?

Answer: Most people assume software codes the data for you. This is incorrect, because the software only stores the information. The researcher is still responsible for coding the text and making decisions about what (and how) to code the data. People also assume software is a method. Software is not a method, it is a software program. Clients and students also erroneously assume using software will make their project more rigorous or valid. Software simply stores the information inputted by the researcher and provides a more efficient way to retrieve and query that information.

Question: What are the greatest advantages of qualitative software?

Answer: It stores the information in one place and allows you to code in more detail.

Question: Advice about using software? Cautions? Like with any software, it is important to save your work often. Your wisdom about qualitative software?

Answer: Most of the people I encounter think the software will do the work for them. As mentioned, software does not code the data for you. The researchers must understand the type of qualitative method and methodology they are using and understand how to implement and translate this approach to code their data in software.

Thick Description and Case Studies: The Bedrock of Qualitative Analysis

> *Bedrock* is a solid, firm, strong, and stable foundation.
>
> We need a bedrock of story and legend in order to live our lives coherently.
>
> —Alan Moore
> English writer and graphic comic artist

Thick, rich description provides the foundation for qualitative analysis and reporting. Good description takes the reader into the setting being described. In his classic *Street Corner Society*, William Foote Whyte (1943) took us to the "slum" neighborhood where he did his fieldwork and introduced us to the characters there, as did Elliot Liebow in *Tally's Corner* (1967), a description of the lives of unemployed black men in Washington, D.C., during the 1960s. In Constance Curry's (1995) oral history of school integration in Drew, Mississippi, in the 1960s, she tells the story of an African American mother Mae Bertha Carter and her seven children as they faced day-to-day and night-to-night threats and terror from resistant, angry whites. Through in-depth case study descriptions, Angela Browne (1987) helps us experience and understand the isolation and fear of a battered woman whose life is controlled by a rage-filled, violent man. Through detailed description and rich quotations, Alan Peshkin (1986) showed readers *The*

SIDEBAR

ANALYZING AND REPORTING HOW PROGRAM DESCRIPTIONS VARY BY PERSPECTIVE

In describing and evaluating different models of youth service programs, Roholt, Hildreth, and Baizerman (2009) began by capturing and reporting different perspectives: (a) official programmatic descriptions, (b) youth programmatic descriptions, and (c) adult descriptions of the programs.

Evaluation is necessary for youth civic engagement (YCE) programs if they are to receive funding and if they want to improve their work.... We sought to understand the program from the points of view and experiences of the multiple participants.... We wanted to know what they did, how they made sense of it, and what consequences this had for them and others. We did this by asking young people, teachers, youth workers, volunteers, coordinators, principals, parents, and other non-involved young people to teach us as much as possible about the program and about their participation experiences.

Five questions guided the evaluation:

1. What does this project say it is about?

2. How is it organized and carried out?

3. What are young people doing in the program?

4. What meaning do they give to their work?

5. What consequences does this work have for them, others, the larger community?

What was different about our study was that we worked as if we did not understand what anyone was telling us. For example, when a young person said the word "citizenship," we assumed that we did not understand what he or she meant. This strategy opened the door to a deeper interrogation of the insider's experience and brought us to the many ways these projects are experienced and understood by different types of participants—young people, adult coaches, school principals, community leaders, teachers, group leaders, and by different individuals. This gave us a look at what was meaningful to them, as well as to what was done in the program and why. (Roholt et al., 2009, pp. 72, 73, 75)

The detailed descriptions were used to compare different program models and the different perspectives on those models of people in diverse roles and in varying relationships with the programs. This comparative analysis was possible because the descriptions were "thick" and rich. The bedrock of the analysis was description from different perspectives.

Total World of a Fundamentalist Christian School, as Erving Goffman (1961) had done earlier for other "total institutions," closed worlds like prisons, army camps, boarding schools, nursing homes, and mental hospitals. Howard Becker (1953, 1985) described how one learns to become a marijuana user in such detail that you almost get the scent of the smoke from his writing.

These classic qualitative studies share the capacity to open up a world to the reader through rich, detailed, and concrete descriptions of people and places—"thick description" (Denzin, 1989c; Geertz, 1973)—in such a way that we can understand the phenomenon studied and draw our own interpretations about meanings and significance.

Description forms the bedrock of all qualitative reporting, whether for scholarly inquiry, as in the examples above, or for program evaluation. For evaluation studies, basic descriptive questions include the following: How do people get into the program? What is the program setting like? What are the primary activities of the program? What happens to people in the program? What are the effects of the program on participants? Thick evaluation descriptions take those who need to use the evaluation findings *into* the experience and outcomes of the program.

Description Before Interpretation

A basic tenet of research is careful separation of description from interpretation. Interpretation involves explaining the findings, answering "why" questions, attaching significance to particular results, and putting patterns into an analytic framework. It is tempting to rush into the creative work of interpreting the data before doing the detailed, hard work of putting together coherent answers to major descriptive questions. But description comes first.

Alternative Ways of Organizing and Reporting Descriptions

Several options exist for organizing and reporting descriptive findings. Exhibit 8.4 presents various options depending on whether the primary organizing motif centers on telling the story of what occurred, presenting case studies, or illuminating an analytical framework.

These are not mutually exclusive or exhaustive ways of organizing and reporting qualitative data. Different parts of a report may use different reporting approaches. The point is that one must have some initial framework for organizing and managing the voluminous data collected during fieldwork.

Where variations in the experiences of individuals are the primary focus of the study, it is appropriate to begin by writing a case study using all the data for each person. Only then are cross-case analysis and comparative analysis done. For example, if one has studied 10 juvenile delinquents, the analysis would begin by doing a case description of each juvenile before doing cross-case analysis. On the other hand, if the focus is on a criminal justice program serving juveniles, the analysis might begin with description of variations in answers to common questions, for example, what were the patterns of major program experiences, what did they like, what did they dislike, how did they think they had changed, and so forth.

Likewise, in analyzing interviews, the analyst has the option of beginning with case analysis or cross-case analysis. Beginning with case analysis means writing a case study for each person interviewed or each unit studied (e.g., each critical event, each group, or each program location). Beginning with cross-case analysis means grouping together answers from different people to common questions or analyzing different perspectives on central issues. If a standardized open-ended interview has been used, it is fairly easy to do cross-case or cross-interview analysis for each question in the interview. With an interview guide approach, answers from different people can be grouped by topics from the guide, but the relevant data won't be found in the same place in each interview. An interview guide, if it has been carefully conceived, actually constitutes a descriptive analytical framework for analysis.

A qualitative study will often include both kinds of analysis—individual cases and cross-case analyses—but one has to begin somewhere. Trying to do both individual case studies and cross-case analysis at the same time will likely lead to confusion.

Case Studies

> Case study is not a methodological choice but a choice of what is to be studied.... We could study it analytically or holistically, entirely by repeated measures or hermeneutically, organically or culturally, and by mixed methods—but we concentrate, at least for the time being, on the case.
>
> —Robert E. Stake
> "Case Studies" (2000, p. 435)

Case analysis involves organizing the data by specific cases for in-depth study and comparison.

Well-constructed case studies are *holistic* and *context sensitive*, two of the primary strategic themes of qualitative inquiry discussed in Chapter 2. Cases can be individuals, groups, neighborhoods, programs, organizations, cultures, regions, or nation-states.

"In an ethnographic case study, there is exactly one unit of analysis—the community or village or tribe" (Bernard, 1994, pp. 35–36). Cases can also be critical incidents, stages in the life of a person or of a program, or anything that can be defined as a

---

EXHIBIT 8.4   Options for Organizing and Reporting Qualitative Data

STORYTELLING APPROACHES	
Chronology, historical sequence	Describe what happened chronologically, over time, telling the story from beginning to end. This focuses on some development over time to portray the life of a person, the history of an organization or community, or the story of a family.
Flashback, retrospective approach	Start at the end (outcome), then tell the story of how the ending emerged. For example, in an evaluation study, a participant's story might begin with the outcome realized (or unrealized), and then the story is presented that illuminates how that outcome came about.
CASE STUDY APPROACHES	
People	If individuals or groups are the primary unit of analysis, then case studies of people or groups may be the focus for analysis. Exhibit 7.20 (pp. 511–516) presents a case study of a homeless youth.
Critical incidents	Critical incidents or major events can constitute self-contained descriptive units of analysis, often presented in order of importance rather than in sequence of occurrence. McClure (1989) reported a case study of a university through the critical incidents that shaped it.
Various settings	Describe various places, sites, settings, or locations (doing case studies of each) before doing cross-setting pattern analysis. In an evaluation of multinational efforts to preserve ancient buildings, we reported on cases in Japan, England, and Indonesia before drawing on cross-cultural conclusions.
ANALYTICAL FRAMEWORK APPROACHES	
Processes	Qualitative data may be organized to describe important processes. For example, an evaluation of a program may describe recruitment processes, socialization processes, decision-making and communication processes, and so on. Distinguishing important processes becomes the analytical framework for organizing qualitative descriptions.
Issues	An analysis can be organized to illuminate key issues, often the equivalent of the primary evaluation questions, for example, variations in how participants changed as a result of the program. In a study of leadership training, we organized the qualitative report around key issues such as conflict management, negotiation skills, enhancing creativity, and effective communications—all important training issues.
Questions	Responses to interviews can be organized question by question, especially where a standardized interviewing format was used. For example, if an evaluation includes questions about perceived strengths and perceived weaknesses, responses to these questions would be grouped together.
Sensitizing concepts	Where sensitizing concepts like "leadership" versus "followership" have played an important role in guiding fieldwork, the data can be organized and described through those sensitizing concepts.

"specific, unique, bounded system" (Stake, 2000, p. 436). Cases are units of analysis. What constitutes a case, or unit of analysis, is usually determined during the design stage and becomes the basis for purposeful sampling in qualitative inquiry (see Chapter 5 for a discussion of case study designs and purposeful sampling). Sometimes, however, new units of analysis, or cases, emerge during field-work or from the analysis after data collection. For example, one might have sampled schools as the unit of analysis, expecting to do case studies of three schools, and then, reviewing the fieldwork, one might decide that classrooms are a more mean-ingful unit of analysis and shift to case studies of classrooms instead of schools, or add case studies of particular teachers or students. Contrariwise, one could begin by sampling classrooms and end up doing case studies on schools. This illustrates the critical importance of thinking carefully about the question "What is a case?" (Ragin & Becker, 1992).

The case study approach to qualitative analysis constitutes a specific way of collecting, organizing, and analyzing data; in that sense, it represents an analysis *process*. The purpose is to gather comprehensive, sys-tematic, and in-depth information about each case of interest. The analysis process results in a *product*: a case study. Thus, the term *case study* can refer to either the process of analysis or the product of analysis, or to both.

Analyzing patterns and identifying themes across multiple case studies has become a significant way of conducting qualitative analysis (Stake, 2006). Case studies may be layered or nested. For example, in evaluation, a single program may be a case study. However, within that single program case ($n = 1$), one may do case studies of several participants. In such an approach, the analysis would begin with the individ-ual case studies; then, the cross-case pattern analysis of the individual cases might be part of the data for the program case study. Likewise, if a national or state program consists of several project sites, the analysis may consist of three layers of case studies: (1) individ-ual participant case studies at project sites combined to make up project site case studies, (2) project site case studies combined to make up state program case studies, and (3) state programs combined to make up a national program case study. Exhibit 8.5 shows this layered case study approach.

This kind of layering recognizes that you can always build larger case units out of smaller ones—that is, you can always combine studies of individu-als into studies of a program—but if you only have program-level data, you can't disaggregate it to con-struct individual cases.

Case Study Rule

Remember this rule: No matter what you are studying, always collect data on the lowest level unit of analysis possible.

> Collect data about individuals, for example, rather than about households. If you are interested in issues of pro-duction and consumption (things that make sense at the household level), you can always package your data about individuals into data about households during analysis. . . . You can always aggregate data collected on individuals, but you can never disaggregate data col-lected on groups. (Bernard, 1994, p. 37)

Though a scholarly or evaluation project may con-sist of several cases and include cross-case compari-sons, the analyst's first and foremost responsibility consists of doing justice to each individual case. All else depends on that.

> Ultimately, we may be interested in a general phe-nomenon or a population of cases more than in the individual case. And we cannot understand this case without knowing about other cases. But while we are studying it, our meager resources are concentrated on trying to understand its complexities. For the while, we probably will not study comparison cases. We may simultaneously carry on more than one case study, but each case study is a concentrated inquiry into a single case. (Stake, 2000, p. 436)

Case data consist of all the information one has about each case: (a) interview data, (b) observations, (c) the documentary data (e.g., program records or files, newspaper clippings), (d) impressions and statements of others about the case, and (e) contex-tual information—in effect, all the information one has accumulated about each particular case goes into that case study. These diverse sources make up the raw data for case analysis and can amount to a large accumulation of material. For individual people, case data can include (a) interviews with the person and those who know her or him, (b) clinical records and background and statistical information about the per-son, (c) a life history profile, (d) things the person has produced (diaries, photos, writings, paintings, etc.), and (e) personality or other test results (yes, quanti-tative data can be part of a qualitative case study). At the program level, case data can include (a) program documents, (b) statistical profiles, (c) program reports and proposals, (d) interviews with program partici-pants and staff, (e) observations of the program, and (f) program histories.

EXHIBIT 8.5 **Case Study: Layers of Possible Analysis**

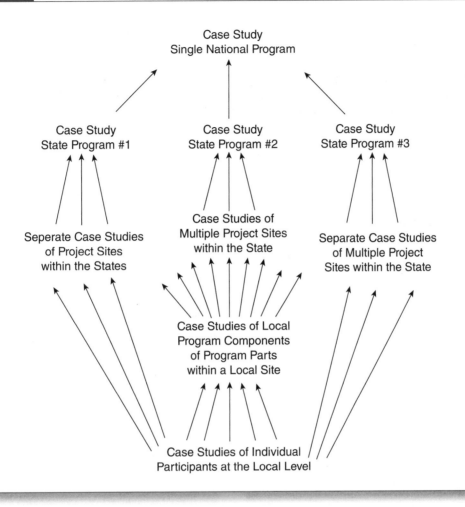

Case Study
Single National Program

Case Study
State Program #1

Case Study
State Program #2

Case Study
State Program #3

Seperate Case Studies
of Project Sites
within the States

Case Studies of
Multiple Project Sites
within the State

Separate Case Studies
of Multiple Project
Sites within the State

Case Studies of Local
Program Components
of Program Parts
within a Local Site

Case Studies of Individual
Participants at the Local Level

From Data to Case Study

Once the raw case data have been accumulated, the researcher may write a case record. The case record pulls together and organizes the voluminous case data into a comprehensive, primary resource package. The case record includes all the major information that will be used in doing the final case analysis and writing the case study. Information is edited, redundancies are sorted out, parts are fitted together, and the case record is organized for ready access either chronologically or topically. The case record must be complete but manageable; it should include all the information needed for subsequent analysis, but it is organized at a level beyond that of the raw case data.

A case record should make no concessions to the reader in terms of interest or communication. It is a condensation of the case data aspiring to the condition that no interpreter requires to appeal behind it to the raw data to sustain an interpretation. Of course, this criterion cannot be fully met: some case records will be better than others. The case record of a school attempts a portrayal through the organization of data alone, and a portrayal without theoretical aspirations. (Stenhouse, 1977, p. 19)

The case record is used to construct a case study appropriate for sharing with an intended audience, for example, scholars, policymakers, program decision makers, or practitioners. The tone, length, form, structure, and format of the final case presentation depend on audience and study purpose. The final case study is what will be communicated in a publication or report. The full report may include several case studies that are then compared and contrasted, but the basic unit of analysis of such a comparative study remains the distinct cases, and the credibility of the overall findings will depend on the quality of the individual case studies. Exhibit 8.6 shows this sequence of moving from raw case data to the written case study. The second step—converting the raw data to a case record before writing the actual case study—is optional. A case record is only constructed when a great deal of unedited raw data from interviews, observations, and documents must be edited and organized before writing the final case study. In many studies, the analyst will work directly and selectively from raw data to write the final case study.

The case study should take the reader into the case situation and experience—a person's life, a group's life, or a program's life. Each case study in a report stands alone, allowing the reader to understand the case as a unique, holistic entity. At a later point in analysis, it is possible to compare and contrast cases, but initially, each case must be represented and understood as an idiosyncratic manifestation of the phenomenon of interest. A case study should be sufficiently detailed and comprehensive to illuminate the focus of inquiry without becoming boring and laden with trivia. A skillfully crafted case feels like a fine weaving. And that, of course, is the trick. How to do the weaving? How to tell the story? How to decide what stays in the final case presentation and what gets deleted along the way. Elmore Leonard (2001), the author of *Glitz* and other popular detective thrillers, was once asked how he managed to keep the action in his books moving so quickly. He said, "I leave out the parts that people skip" (p. 7). Not bad advice for writing an engaging case study.

In doing biographical or life history case studies, Denzin (1989a) has found particular value in identifying what he calls "epiphanies"—"existentially problematic moments in the lives of individuals" (p. 129).

EXHIBIT 8.6 **The Process of Constructing Case Studies**

Step 1. *Assemble the raw case data*

These data consist of all the information collected about the person, program, organization, or setting for which a case study is to be written.

Step 2. (optional) *Construct a case record*

This is a condensation of the raw case data, organized, classified, and edited into a manageable and accessible file.

Step 3. *Write a final case study narrative*

The case study is a readable, descriptive picture of or story about a person, program, organization, or other unit of analysis, making accessible to the reader all the information necessary to understand the case in all its uniqueness. The case story can be told chronologically or presented thematically (sometimes both).

The case study offers a holistic portrayal, presented with any context necessary for understanding the case.

HOLISTIC CASE STUDIES

Neuroscientists have long used case studies of victims of traumatic brain injuries to understand how the brain works.

Depending on what part of the brain suffered, strange things might happen. Parents couldn't recognize their children. Normal people became pathological liars. Some people lost the ability to speak—but could sing just fine. These incidents have become classic case studies, fodder for innumerable textbooks and bull sessions around the lab. The names of these patients—H. M. Tan, Phineas Gage—are deeply woven into the lore of neuroscience. (Kean, 2014, p. SR8)

Science journalist Sam Kean (2014) has reflected on such outlier case studies and concluded that "in the quest for scientific understanding, we end up magnifying patients' deficits until deficits are all we see. The actual person fades away" (p. SR8). He has concluded that more holistic case studies are needed and are even critical to a fuller understanding.

When we read the full stories of people's lives . . . , we have to put ourselves into the minds of the characters, even if those minds are damaged. Only then can we see that they want the same things, and endure the same disappointments, as the rest of us. They feel the same joys, and suffer the same bewilderment that life got away from them. Like an optical illusion, we can flip our focus. Tales about bizarre deficits become tales of resiliency and courage. (p. SR8)

SIDEBAR

It is possible to identify four major structures, or types of existentially problematic moments, or epiphanies, in the lives of individuals. First, there are those moments that are major and touch every fabric of a person's life. Their effects are immediate and long term. Second, there are those epiphanies that represent eruptions, or reactions, to events that have been going on for a long period of time. Third are those events that are minor yet symbolically representative of major problematic moments in a relationship. Fourth, and finally, are those episodes whose effects are immediate, but their meanings are only given later, in retrospection, and in the reliving of the event. I give the following names to these four structures of problematic experience: (1) the major epiphany, (2) the cumulative epiphany, (3) the illuminative, minor epiphany, and (4) the relived epiphany. (Of course, any epiphany can be relived and given new retrospective meaning.) These four types may, of course, build upon one another. A given event may, at different phases in a person's or relationship's life, be first, major, then minor, and then later relived. A cumulative epiphany will, of course, erupt into a major event in a person's life. (p.129)

Programs, organizations, and communities have parallel types of epiphanies, though they're usually called critical incidents, crises, transitions, or organizational lessons learned. For a classic example of an organizational development case study in the business school tradition, see the analysis of the Nut Island sewage treatment plant in Quincy, Massachusetts—the complex story of how an outstanding team, highly competent, deeply committed to excellence, focused on the organizational mission, and working hard still ended up in a "catastrophic failure" (Levy, 2001).

Studying such examples is one of the best ways to learn how to write case studies. The section titled "Thick Description," earlier in this chapter, cited a number of case studies that have become classics in the genre. Chapter 1 presented case vignettes of individuals in an adult literacy program. An example of a full individual case study is presented as Exhibit 8.33, at the end of this chapter (pp. 638–642). Originally prepared for an evaluation report that included several participant case studies, it tells the story of one person's experiences in a career education program. This case represents an exemplar of how multiple sources of information can be brought together to offer a comprehensive picture of a person's experience, in this instance, a student's changing involvement in the program and changing attitudes and behaviors over time. The case data for each student in the evaluation study included the following:

1. Observations of selected students at employer sites three times during the year

2. Interviews three times per year with the students' employer-instructors at the time of observation

3. Parent interviews once a year

4. In-depth student interviews four times a year

5. Informal discussions with program staff

6. A review of student projects and other documents

7. Twenty-three records from the files of each student (including employer evaluations of students, student products, test scores, and staff progress evaluations of students)

Initial interview guide questions provided a framework for analyzing and reviewing each source. Information from all of these sources was integrated to produce a highly readable narrative that could be used by decision makers and funders to better understand what it was like to be in the program (Owens, Haenn, & Fehrenbacher, 1976). The evaluation staff of the Northwest Regional Educational Laboratory went to great pains to carefully validate the information in the case studies. Different sources of information were used to cross-validate the findings, patterns, and conclusions. Two evaluators reviewed the material in each case study to independently make judgments and interpretations about its content and meaning. In addition, an external evaluator reviewed the raw data to check for biases or unwarranted conclusions. Students were asked to read their own case studies and comment on the accuracy of fact and interpretation in the study. Finally, to guarantee the readability of the case studies, a newspaper journalist was employed to help organize and edit the final versions. Such a rigorous case study approach increases the confidence of readers that the cases are accurate and comprehensive. Both in its content and in the process by which it was constructed, the Northwest Lab case study presented at the end of this chapter (Exhibit 8.33) exemplifies how an individual case study can be prepared and presented.

How one compares and contrasts cases will depend on the purpose of the study and how the cases were sampled. As discussed in Chapter 5, critical cases, extreme cases, typical cases, and heterogeneous cases serve different purposes. Once case studies have been written, the analytic strategies described in the remainder of this chapter can be used to further analyze, compare, and interpret the cases to generate cross-case themes, patterns, and findings. Exhibit 8.7 summarizes the central points I've discussed for constructing case studies.

DIVERSE CASE STUDY EXEMPLARS

Case Studies in This Book

- *The story of Henrietta Lacks*. This in-depth case study tells the story of a poor African American tobacco farmer who grew up in the South. In 1951, when she died of cervical cancer, the cells from her tumor were taken for research, without her knowledge or permission, by an oncologist. The cells manifest unique characteristics: They could be cultured, sustained, reproduced, and distributed to other researchers, the first tissue cells discovered with these extraordinary characteristics. How this affected her family and the medical world shows how layers of case studies can be interwoven (Skloot, 2010). (See Chapter 2, Exhibit 2.7, pp. 78–80.)

- *Story of Li*. This case study presents highlights of a participant case study used to illuminate a Vietnamese woman's experience in an employment training program; in addition to describing what a job placement meant to her, the case was constructed to illuminate hard to measure outcomes such as "understanding the American workplace culture" and "speaking up for oneself," learnings that can be critical to long-term job success for an emigrant. (See Chapter 4, Exhibit 4.3, pp. 182–183.)

- *Thmaris*. A case study of a homeless youth and his journey to a more stable life, this case illuminates the challenges and long-term effects of dealing with childhood trauma and failed relationships. His experience in homeless shelters is central to the case study. (See Chapter 7, Exhibit 7.20, pp. 511–516.)

- *Mike's story*. This case study tells the story of one person's experiences in a career education program. This case represents an exemplar of how multiple sources of information can be brought together to offer a comprehensive picture of a person's experience, in this instance, a student's changing involvement in the program and changing attitudes and behaviors over time. (See Chapter 8, Exhibit 8.33, pp. 638–642.)

Examples of Excellent Published Case Studies

- *Education case studies.* Brizuela, Stewart, Carrillo, and Berger (2000), Stake, Bresler, and Mabry (1991), Perrone (1985), and Alkin, Daillak, and White (1979)
- *Family case studies.* Sussman and Gilgun (1996)
- *International development cases.* Wood et al. (2011), Salmen (1987), and Searle (1985)
- *Government accountability case study.* Joyce (2011)
- *Case studies of effective antipoverty programs.* Schorr (1988)
- *Case studies of research influencing policy in developing countries.* (Carden, 2009)
- *Philanthropy case studies.* Evaluation Roundtable (2014) and Sherwood (2005)
- *Public health cases.* White (2014)
- *Business cases.* Collins (2001a, 2009), Collins and Porras (2004), and Collins and Hansen (2011)

EXHIBIT 8.7   Guidelines for Constructing Case Studies

1. *Focus first on capturing the uniqueness of each case.* The qualitative analyst's first and foremost responsibility consists of doing justice to each individual case. Don't jump ahead to formal cross-case analysis until the individual cases are fully constructed.

2. *Construct cases for smaller units of analysis first.* You can always aggregate data collected on individuals into groups for analysis, but you can never disaggregate data collected only on groups to construct individual case studies.

3. *Use multiple sources of data.* A case study includes and integrates all the information one has about each case—interview data, observations, and documents.

4. *Write the case to tell a core story.* Structure the case with a beginning, middle, and end.

5. *Make the case coherent for the reader.* The case study should take the reader into the case situation and experience—a person's life experience, a group's cohesion, a program's coherence as a program, or a community's sense of community.

6. *Balance detail with relevance.* A case study should be sufficiently detailed and comprehensive to illuminate the focus of inquiry without becoming boring and laden with trivia.

7. *Readability and coherence check.* Have someone read the case and give you feedback about its coherence and readability, and any gaps or ambiguities that need attention.

8. *Accuracy check.* For individual case studies, have the person whose story you've written review the case for accuracy. For other units of analysis (programs, communities, organizations) have a key informant review the case.

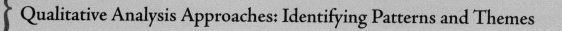

Qualitative Analysis Approaches: Identifying Patterns and Themes

> The ability to use thematic analysis appears to involve a number of underlying abilities, or competencies. One competency can be called *pattern recognition*. It is the ability to see patterns in seemingly random information.
>
> —Boyatzis (1998, p. 7)

This module will present the kinds of findings that result from qualitative analysis. I'll examine what is meant by content analysis and distinguish patterns from themes. I'll contrast inductive and deductive analytical approaches and introduce some specific analytical approaches like grounded theory and analytic induction. I'll discuss sensitizing concepts as a focus for analysis and differentiate indigenous concepts and typologies from analyst-created concepts and typologies. Exhibit 8.10, at the end of this module (pp. 551–552), will summarize the 10 analytical approaches reviewed in this module.

The next module will go into detail about the actual coding and analytical procedures for making sense of qualitative data. I could have started with those procedural processes for analysis, but I think it's helpful to understand first what kinds of findings can be generated from qualitative analysis before delving very deeply into the mechanics and operational processes. Thus, this module will provide examples of patterns, themes, indigenous concepts and typologies, and analyst-constructed concepts and typologies—*the fruit of qualitative analysis*, a metaphor that harks back to Chapter 1. The next module will explain how you harvest qualitative fruit once you know more about the variety of fruit that can be harvested.

Content Analysis

No consensus exists about the terminology to apply in differentiating varieties and processes of qualitative analysis. *Content analysis* sometimes refers to searching text for and counting recurring words or themes. For example, a speech by a politician might be analyzed to see what phrases or concepts predominate, or speeches of two politicians might be compared to

see how many times and in what contexts they used a phrase like "global economy" or "family values." More generally, *content analysis* usually refers to analyzing text (interview transcripts, diaries, or documents) rather than observation-based field notes. Even more generally, *content analysis* refers to any qualitative data reduction and sense-making effort that takes a volume of qualitative material and attempts to identify core consistencies and meanings. Case studies, for example, can be content analyzed.

Patterns Are the Basis for Themes

The core meanings found through content analysis are patterns and themes. The processes of searching for patterns and themes may be distinguished as pattern analysis and theme analysis, respectively. I'm asked frequently about the difference between a pattern and a theme. The term *pattern* refers to a descriptive finding, for example, "Almost all participants reported feeling fear when they rappelled down the cliff," while a theme takes a more categorical or topical form, interpreting the meaning of the pattern: FEAR. Putting these terms together, a report on a wilderness education study might state,

> The *content analysis* revealed a *pattern* of participants reporting being afraid when rappelling down cliffs and running river rapids; many also initially experienced the group process of sharing personal feelings as evoking some fear. Those patterns make *dealing with fear* a major *theme* of the wilderness education program experience.

Inductive and Deductive Qualitative Analyses

Qualitative deductive analysis: Determining the extent to which qualitative data in a particular study support existing general conceptualizations, explanations, results, and/or theories

Qualitative inductive analysis: Generating new concepts, explanations, results, and/or theories from the specific data of a qualitative study

> Francis Bacon is known for his emphasis on *induction, the use of direct observation to confirm ideas and the linking together of observed facts to form theories or explanations*

of how natural phenomenon work. Bacon correctly never told us how to get ideas or how to accomplish the linkage of empirical facts. Those activities remain essentially humanistic—you think hard. (Bernard, 2000, p. 12)

SIDEBAR

HUMAN PATTERN RECOGNITION

Pattern detection is an evolutionary capacity developed in and passed on from our Stone Age ancestors.

> Human beings do not have very many natural defenses. We are not all that fast, and we are not all that strong. We do not have claws or fangs or body armor. We cannot spit venom. We cannot camouflage ourselves. And we cannot fly. Instead, we survive by means of our wits. Our minds are quick. We are wired to detect patterns and respond to opportunities and threats without much hesitation. (Silver, 2012, p. 12)

Both the capacity and the drive to find patterns is much more developed in humans than in other animals, explains Tomaso Poggio, a neuroscientist who studies how human brains process information and make sense of the world. The problem is that these evolutionary instincts sometimes lead us to see patterns when there are none. People do that all the time, Poggio has found—"finding patterns in random noise." Thus, unless we work *actively* to become aware of our biases and avoid overconfidence when identifying patterns, we can fail to accurately distinguish signal (pattern) from noise (random occurrences and relationships) (Silver, 2012, p. 12).

So beware of the allure and dangers of *apophenia*: identifying meaningful patterns in meaningless randomness.

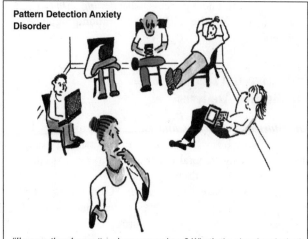

Pattern Detection Anxiety Disorder

"I'm sure there's a pattern here somewhere? What's the signal and what's the noise? And what does that even mean? If only I could see the pattern. If only I could be sure. And is the pattern really there, or am I imposing it? Or is it a theme and not a pattern? Or both a pattern and a theme? Oh I so wish I'd just done a survey."

Bacon (1561–1626) is recognized as one of the founders of scientific thinking, but he also has been awarded "the dubious honor of being the first martyr of empiricism." Still pondering the universe at the age of 65, he got an idea one day while driving his carriage in the snow in a farming area north of London. It occurred to him that cold might delay the biological process of putrefaction, so he stopped, purchased a hen from a farmer, killed it on the spot, and stuffed it in the snow. His idea worked. The snow did delay the rotting process, but he subsequently contracted bronchitis and died a month later (Bernard, 2000, p. 12). As I noted in Chapter 6, fieldwork can be risky. Engaging in analysis, on the other hand, is seldom life threatening, though you do risk being disputed and sometimes ridiculed by those who arrive at contrary conclusions.

Inductive analysis involves *discovering* patterns, themes, and categories in one's data. Findings emerge out of the data, through the analyst's interactions with the data. In contrast, when engaging in *deductive analysis*, the data are analyzed according to an existing framework. Qualitative analysis is typically inductive in the early stages, especially when developing a codebook for content analysis or figuring out possible categories, patterns, and themes. This is often called "open coding" (Strauss & Corbin, 1998, p. 223), to emphasize the importance of being open to the data. "Grounded theory" (Glaser & Strauss, 1967) emphasizes becoming immersed in the data—being *grounded*—so that embedded meanings and relationships can emerge. The French would say of such an immersion process, *Je m'enracine* ("I root myself"). The analyst becomes implanted in the data. The resulting analysis grows out of that groundedness.

From Inductive to Deductive

Once patterns, themes, and/or categories have been established through inductive analysis, the final, confirmatory stage of qualitative analysis may be deductive in testing and affirming the authenticity and appropriateness of the inductive content analysis, including carefully examining deviate cases or data that don't fit the categories developed. Generating theoretical propositions or formal hypotheses after inductively identifying categories is considered deductive analysis by grounded theorists Strauss and Corbin (1998): "Anytime that a researcher derives hypotheses from data, because it involves interpretation, we consider that to be a deductive process" (p. 22). Grounded theorizing, then, involves both inductive and deductive processes: "At the heart of theorizing lies the interplay of making inductions (deriving concepts, their

SIDEBAR

INDUCTIVE PROGRAM THEORY DEVELOPMENT

Inductive [program theory] development involves observing the program in action and deriving the theories that are implicit in people's actions when implementing the program. The theory in action may differ from the espoused theory: what people do is different from what they say they do, believe they are doing, or believe they should be doing according to policy or some other principle.

The program in action could be observed at the point of service delivery in the field that is closest to the clients of the program. It could include observation of the program in action, including through participant observation. Interviews can be conducted with staff about how they implement the program and about why they undertake some activities that may appear to be at variance with the program design or omit parts of the program design. Program participants can be interviewed about how they experience the program (or have experienced it), if data are gathered through exit interviews; and how they would like to experience it. . . .

Theory in action could also be identified from looking at how program managers have interpreted the program, as indicated by the types of practices they adopt. However, it is important to confirm that inferences drawn are correct. For example, what they consider to be important about the program and how they interpret the program's intent might be inferred from their choice of particular performance indicators and how they use them. This inference would need to be confirmed with program managers, since the selection of indicators may have been imposed on management and staff as, for example, part of national nonprogram specific monitoring requirements. Or the indicators may have been selected simply because they were available and easy to measure and report, but not necessarily considered by staff to be meaningful.

—Funnell and Rogers (2011, pp. 111–112)

DEDUCTIVE ANALYSIS EXAMPLE: AGENCY RESISTANCE TO OUTCOME MEASUREMENT

Identifying factors that support evaluation use and overcoming resistance to evaluation have been two of the central concerns of the evaluation profession for 40 years (Alkin, 1975; Patton, 1978b). A great volume of research, much of it qualitative case studies of use and nonuse, point to the importance of high-quality stakeholder involvement to enhance use (Brandon & Fukunaga, 2014; Patton, 2008, 2012a). Strickhouser and Wright (2014) contributed to this arena of inquiry by interviewing the directors and staff of eight human service nonprofit agencies and their one common funder in a large southeastern metropolitan area. They found that agencies continue to resist, and in some cases sabotage, evaluation reporting requirements. They found that, as shown in previous studies, program evaluators often find it difficult to conceptualize and evaluate outcomes. Tensions around outcome measurement make communication between agencies and their funders difficult and frustrating on both sides. This is an example of a primarily deductive qualitative analysis because the questions asked and the analysis conducted draw on issues and concepts that are already well established. The qualitative inquiry tests whether already identified factors continue to be manifest in a specific group not previously studied. The findings are confirmatory and illuminating, but they do not generate any new concepts or factors.

---

properties, and dimensions from data) and deductions (hypothesizing about the relationships between concepts)" (Strauss & Corbin, 1998, p. 22).

From Deduction to Induction: Analytic Induction

Analytic induction as a distinct qualitative analysis approach begins with an analyst's deduced propositions or theory-derived hypotheses and "is a procedure for verifying theories and propositions based on qualitative data" (Taylor & Bogdan, 1984, p. 127). Sometimes, as with analytic induction, qualitative analysis is first deductive or quasi-deductive and then inductive, as when, for example, the analyst begins by examining the data in terms of theory-derived sensitizing concepts or applying a theoretical framework developed by someone else (e.g., testing Piaget's developmental theory on case studies of children). After or alongside this deductive phase of analysis, the researcher strives to look at the data afresh for undiscovered patterns and emergent understandings (inductive analysis). I'll discuss both grounded theory and analytic deduction at greater length later in this chapter.

Because, as identified and discussed in Chapter 2, inductive analysis is one of the primary characteristics of qualitative inquiry, we'll focus on strategies for thinking and working inductively. There are two distinct ways of analyzing qualitative data inductively. First, the analyst can identify, define, and elucidate the categories developed and articulated by the people studied to focus analysis. Second, the analyst may also become aware of categories or patterns for which the people studied did not have labels or terms, and the analyst develops terms to describe these inductively generated categories. Each of these approaches is described below.

Inductive Approaches: Indigenous and Analyst-Constructed Patterns and Themes

Indigenous Concepts and Practices

A good place to begin inductive analysis is to inventory and define key phrases, terms, and practices that are special to the people in the setting studied. What are the indigenous categories that the people interviewed have created to make sense of their world? What are the practices they engage in that can only be understood within their worldview? Anthropologists call this *emic* analysis and distinguish it from *etic* analysis, which refers to labels imposed by the researcher. (For more on this distinction and its origins, see Chapter 6, which discusses emic and etic perspectives in fieldwork.) "Identifying the categories and terms used by informants themselves is also called *in vivo* coding" (Bernard 1998, p. 608).

Consider the practice among traditional Dani women of amputating a finger joint when a relative dies. The Dani people live in the lush Baliem Valley of Irian Java, Indonesia's most remote province, in the western half of New Guinea. The joint is removed to honor and placate ancestral ghosts. Missionaries have fought against the practice as sinful, and the government has banned it as barbaric, but many traditional women still practice it.

> Some women in Dani villages have only four stubs and a thumb on each hand. In tribute to her dead mother and brothers, Soroba, 38, has had the tops of six of her fingers amputated. "The first time was the worst," she said. "The pain was so bad, I thought I would die. But it's worth it to honor my family." (Sims, 2001, p. 6)

Analyzing such an indigenous practice begins with understanding it from the perspective of its practitioners, within the indigenous context, in the words of the local people, in their language, within their worldview.

> According to this view, cultural behavior should always be studied and categorized in terms of the inside view—the actors' definition—of human events. That is, the units of conceptualization in anthropological theories should be "discovered" by analyzing the cognitive processes of the people studied, rather than "imposed" from cross-cultural (hence, ethnocentric) classifications of behavior. (Pelto & Pelto, 1978, p. 54)

Anthropologists, working cross-culturally, have long emphasized the importance of preserving and reporting the indigenous categories of the people studied. Franz Boas (1943) was a major influence in this direction: "If it is our serious purpose to understand the thoughts of a people, the whole analysis of experience must be based on their concepts, not ours" (p. 314).

In an intervention program, certain terms may emerge or be created by participants to capture some essence of the program. In the wilderness education program I evaluated, the idea of "detoxification" became a powerful way for participants to share meaning about what being in the wilderness together meant (Patton, 1999, pp. 49–52). In the Caribbean Extension Project evaluation, the term *liming* had special meaning from the participants. Not really translatable, it essentially means passing time, hanging out, doing nothing, shooting the breeze—but doing so agreeably, without guilt, stress, or a sense that one ought to be doing something more productive with one's time. *Liming* has positive, desirable connotations because of its social group meaning—people just enjoying being together without having to accomplish anything. Given that uniquely Caribbean term, what does it mean when participants describe what happened in a training session or instructional field trip as primarily "liming"? How much "liming" could acceptably be built into training for participant satisfaction and still get something done? How much programmatic liming was acceptable? These became key formative evaluation issues.

In evaluating a leadership training program, we gathered extensive data on what participants and staff meant by the term *leadership*. Pretraining and post-training exercises involved having participants write a paragraph on leadership; the writing was part of the program curriculum, not designed for evaluation, but the results provided useful qualitative evaluation data. There were small-group discussions on leadership. The training included lectures and group discussions on leadership, which we observed. We participated in and took notes on informal discussions about leadership. Because the very idea of *leadership* was central to the program, it was essential to capture variations in what participants meant when they talked about "leadership." The results showed that the ongoing confusion about what leadership meant was one of the problematic issues in the program. Leadership was an indigenous concept in that staff and participants throughout the training experience used it extensively, but it was also a *sensitizing concept* since we knew going into the fieldwork that it would be an important notion to study.

Sensitizing Concepts

In contrast to purely indigenous concepts, sensitizing concepts refer to categories that the analyst brings to the data. Experienced observers often use sensitizing concepts to orient fieldwork, an approach discussed in Chapter 6 (pp. 357–363). These sensitizing concepts have their origins in social science theory, the research literature, or evaluation issues identified at the beginning of a study. Sensitizing concepts give the analyst "a general sense of reference" and provide "directions along which to look" (Blumer, 1969, p. 148). Using sensitizing concepts involves examining how the concept is manifest and given meaning in a particular setting or among a particular group of people.

Conroy (1987) used the sensitizing concept "victimization" to study police officers. Innocent citizens are frequently thought of as the victims of police brutality or indifference. Conroy turned the idea of victim around and looked at what it would mean to study police officers as victims of the experiences of law enforcement. He found the sensitizing concept of victimization helpful in understanding the isolation, lack of interpersonal affect, cynicism, repressed anger, and sadness observed among police officers. He used the idea of victimization to tie together the following quotes from police officers:

- As a police officer and as an individual I think I have lost the ability to feel and to empathize with people. I had a little girl that was run over by a bus and her mother was there and she had her little book bag. It was really sad at the time but I remember feeling absolutely nothing. It was like a mannequin on the street instead of some little girl. I really wanted to be able to cry about it and I really wanted to have some feelings about it, but I couldn't. It's a little frightening for me to be so callous and I have been unable to relax.

- I am paying a price by always being on edge and by being alone. I have become isolated from old friends. We are different. I feel separate from people, different, out of step. It becomes easier to just be with other police officers because they have the same basic understanding of my environment, we speak the same language. The terminology is crude. When I started I didn't want to get into any words like *scumbags* and *scrotes*, but it so aptly describes these people.

- I have become isolated from who I was because I have seen many things I wish I had not seen. It's frustrating to see things that other people don't see, won't see, can't see. I wish sometimes, I didn't see the things. I need to be assertive, but don't like it. I have to put on my police mask to do that. But now it is getting harder and harder to take that mask off. I take my work home with me. I don't want my work to invade my personal life but I'm finding I need to be alone more and more. I need time to recharge my batteries. I don't like to be alone, but must. (p. 52)

Two additional points are worth making about these quotations. First, by presenting the actual data on which the analysis is based, the readers are able to make their own determination of whether the concept "victimization" helps in making sense of the data. By presenting respondents in their own words and reporting the actual data that were the basis of his interpretation, Conroy invites readers to make their own analysis and interpretation. The analyst's constructs should not dominate the analysis, but rather, they should facilitate the reader's understanding of the world under study.

Second, these three quotations illustrate the power of qualitative data. The point of analysis is not simply to find a concept or label to neatly tie together the data. What is important is understanding the people studied. Concepts are never a substitute for direct experience with the descriptive data. What people actually say and the descriptions of events observed remain the essence of qualitative inquiry. The analytical process is meant to organize and elucidate telling the story of the data. Indeed, the skilled analyst is able to get out of the way of the data to let the data tell their own story. The analyst uses concepts to help make sense of and present the data, but not to the point of straining or forcing the analysis. The reader can usually tell when the analyst is more interested in proving the applicability and validity of a concept than in letting the data reveal the perspectives of the people interviewed and the intricacies of the world studied.

Analyst-Created Concepts

Sensitizing concepts are used during fieldwork to guide the inquiry and subsequent analysis. The analysis puts flesh on the bare bones of a sensitizing concept, deepening its meaning and revealing its implications. Concepts can also emerge during analysis that were not yet imagined or conceptualized during fieldwork. At a conference for fathers of teenagers aimed at illuminating strategies for dealing with the challenges of guiding one's child through adolescence, the term *reverse incest anxiety* emerged in our analysis to describe some fathers' fear of expressing physical affection for teenage daughters lest it be perceived as inappropriate.

SIDEBAR

TEMPLATE ANALYSIS

Template analysis is an approach being used in organizational research to organize and make sense of rich, unstructured qualitative data. The analytical framework provides guidance for defining codes, hierarchical coding, and parallel coding. Template analysis involves identifying conceptual themes, clustering them into broader groupings, and, subsequently, identifying "master themes" and subsidiary constituent themes across cases. In organizational research and program evaluation, template analysis "works particularly well when the aim is to compare the perspectives of different groups of staff within a specific context" (King, 2004, p. 257).

SOURCES: King (2012); King and Horrocks (2012); Waring and Wainwright (2008).

British social scientist Guy Standing (2011) created the term *precariat* to describe a new class of people in industrialized societies who are in the precarious position of only getting occasional short-term and part-time work and whose quality of life and living standards are made precarious. They have no career path and stability, and they experience multiple forms of economic and social insecurity.

Having suggested how singular concepts can bring focus to inductive analysis, the next level of analysis, constructing typologies, moves us into a somewhat more complex analytical strategy.

Typologies and Continua: Indigenous and Analyst-Constructed Frameworks

> There are two kinds of people in the world: those who think there are two kinds of people in the world and those who don't.
>
> —Humorist Robert Benchley (1889–1945)
> Law of Distinction

Indigenous Typologies

Typologies are classification systems made up of categories that divide some aspect of the world into parts along a continuum. They differ from taxonomies, which completely classify a phenomenon through mutually exclusive and exhaustive categories, like the biological system for classifying species. Typologies, in contrast, are built on ideal types or illustrative end points rather than a complete and discrete set of categories. Well-known and widely used sociological typologies include Redfield's folk–urban continuum (*gemeinschaft–gesellschaft*) and Von Wiese's and Becker's sacred–secular continuum (for details, see Vidich & Lyman, 2000, p. 52). Sociologists classically distinguish ascribed from achieved characteristics. Psychologists distinguish degrees of mental illness (neuroses to psychoses). Political scientists classify governmental systems along a democratic–authoritarian continuum. Economists distinguish laissez-faire from centrally planned economic systems. Systems analysts distinguish open from closed systems. In all of these cases, however, the distinctions involve matters of degree and interpretation rather than absolute distinctions. All of these examples have emerged from social science theory and represent theory-based typologies constructed by analysts. We'll examine that approach in greater depth in a moment. First, however, let's look at identifying indigenous typologies as a form of qualitative analysis.

Illuminating indigenous typologies requires an analysis of the continua and distinctions used by people in a setting to break up the complexity of reality into distinguishable parts. The language of a group of people reveals what is important to them in that they name something to separate and distinguish it from other things with other names. Once these labels have been identified from an analysis of what people have said during fieldwork, the next step is to identify the attributes or characteristics that distinguish one thing from another. In describing this kind of analysis, Charles Frake (1962) used the example of a hamburger. Hamburgers can vary a great deal in how they are cooked (rare to well done) or what is added to them (pickles, mustard, ketchup, lettuce), and they are still called hamburgers. However, when a piece of cheese is added to the meat, it becomes a cheeseburger. The task for the analyst is to discover what it is that separates a "hamburger" from a "cheeseburger"—that is, to discern and report "how people construe their world of experience from the way they talk about it" (Frake, 1962, p. 74).

An analysis example of this kind comes from a formative evaluation aimed at reducing the dropout rate among high school students. In observations and interviews at the targeted high school, it became important to understand the ways in which teachers categorized students. With regard to problems of truancy, absenteeism, tardiness, and skipping class, the teachers had come to label students as either

THE SIGNAL AND THE NOISE: LIVING WITH, LEARNING FROM, AND HONORING THE *NOISE*

The metaphor of distinguishing signal from noise is a powerful way to talk about pattern detection in qualitative analysis. In reporting findings, qualitative analysts typically focus on and highlight the patterns and themes found, what they mean, and their implications for theory and/or practice. The signal versus noise distinction can appear to connote that what is valuable is the signal. Of course, one person's noise can be another person's signal, so the distinction depends on perspective. But however distinguished, once the signal (pattern, theme, or meaning) is detected, the noise fades into the background. Still, much can be learned from dwelling with and understanding the noise. Describing, characterizing, making sense of, portraying, and understanding the noise can be, in and of itself, a qualitative analysis contribution.

- Evaluation professional Nora Murphy, cofounder of the *TerraLuna Collaborative*, described to me observing dinner meetings of teachers participating in an innovative initiative and just immersing herself in the predinner chatter and other activities going on—small groups forming and disbanding, people moving around and milling around (taking in the head nodding, head shaking, furled brows, and animated hands; reconnecting hugs and handshakes; and hearing the laughter)—literally *experiencing the noise* of the interactions to get a sense of how these teachers were coming together with each other.
- Seasoned educator Eleanor Coleman, of the Minnesota Humanities Center, told me how she and her team of district leaders went through an exercise of listing all the new initiatives that had been introduced in the school district over the past five years. The walls were soon covered with a list of more than 100 initiatives that had been

introduced, demanding their attention and participation, plus ongoing demands from needy students, concerned parents, paper-pushing administrators, union leaders, and elected officials, while they tried to lead their lives, take care of their families, maintain relationships with friends and neighbors, and feed their spiritual needs. Messy lives. Noisy lives. How, through all that noise, would yet another initiative become an innovative and valued signal, catching the attention of and engendering commitment from teachers? Answering that question—indeed, even beginning to answer that question—meant dwelling more deeply with and understanding the noise.

In *The Signal and the Noise*, Nate Silver (2012) recounts a conversation with an international terrorism expert in which the expert distinguishes the challenge of finding a proverbial needle in a haystack from the even more daunting challenge of finding one particular needle in a large stack of needles. In both cases, the focus is on finding the needle—the thing you're looking for, the signal, the pattern, the thing that stands out. And to find that one particular needle you're looking for, you have to take apart the haystack or the stack of needles. But before doing so, imagine first inquiring into the stack, whether of hay or needles (How did the stack come to be there? What's the context within which the stack has been stacked? What are the characteristics of the stack? What can be learned about and from the stack) before destroying it in search of the needle.

A comprehensive, holistic qualitative inquiry will describe, analyze, attend to, and attempt to understand *both* the signal and the noise. And sometimes, *the noise is the signal.*

"chronics" or "borderlines." One teacher described the chronics as "the ones who are out of school all the time, and everything you do to get them in doesn't work." Another teacher said, "You can always pick them out, the chronics. They're usually the same kids." The borderlines, on the other hand,

skip a few classes, waiting for a response, and when it comes they shape up. They're not so different from your typical junior high student, but when they see the chronics getting away with it, they get more brazen in their actions.

Another teacher said, "Borderlines are gone a lot but not constantly like the chronics."

Not all teachers used precisely the same criteria to distinguish "chronics" from "borderlines," but all teachers used these labels in talking about students. To understand the program activities directed at reducing high school dropouts and the differential impact of the program on students, it became important to observe differences in how "borderlines" and "chronics" were treated. Many teachers, for example, refused even to attempt to deal with chronics. They considered it a waste of their time. Students, it turned out, knew what labels were applied to them and how to manipulate these labels to get more or less attention from teachers. Students who wanted to be left alone called themselves "chronics" and reinforced their "chronic image with teachers. Students who wanted to graduate, even if only barely and with minimal school attendance, cultivated an image as 'borderline.'"

Another example of an indigenous typology emerged in the wilderness education program I evaluated. As I explained earlier, when I used this example to discuss participant observation, one subgroup started calling themselves the "turtles." They contrasted themselves to the "truckers." On the surface, these labels were aimed at distinguishing different styles of hiking and backpacking, one slow and one fast. Beneath the surface, however, the terms came to represent different approaches to the wilderness and different styles of experience in relation to the wilderness and the program.

Groups, cultures, organizations, and families develop their own language systems to emphasize distinctions they consider important. Every program gives rise to special vocabulary that staff and participants use to differentiate types of activities, kinds of participants, styles of participation, and variously valued outcomes. These indigenous typologies provide clues to analysts that the phenomena to which the labels refer are important to the people in the setting and that to fully understand the setting it is necessary to understand those terms and their implications.

Analyst-Constructed Typologies

Once indigenous concepts, typologies, and themes have been surfaced, understood, and analyzed, the qualitative analysis may move to a different inductive task to further elucidate findings—constructing non-indigenous typologies based on *analyst-generated* patterns, themes, and concepts. Such constructions must be done with considerable care to avoid creating things that are not really in the data. The advice of biological theorist John Maynard Smith (2000) is informative in this regard: Seek models of the world that make sense and whose consequences can be worked out, for "to replace a world you do not understand by a model of a world you do not understand is no advance" (p. 46).

Constructing ideal types or alternative paradigms is one simple form of presenting qualitative comparisons. Exhibit 8.8 presents my ideal-typical comparison of "coming-of-age paradigms," which contrasts tribal initiation themes with contemporary coming-of-age themes (Patton, 1999). A series of patterns are distilled into contrasting themes that create alternative ideal types. The notion of "ideal types" makes it explicit that the analyst has constructed and interpreted something that supersedes purely descriptive analysis.

In creating analyst-constructed typologies through inductive analysis, you take on the task of identifying and making explicit patterns that appear to exist but remain unperceived by the people studied. The danger is that analyst-constructed typologies impose a world of meaning on the participants that better reflects the observer's world than the world under study. One way of testing analyst-constructed typologies is to present them to the people whose world is being analyzed to find out if the constructions make sense to them.

> The best and most stringent test of observer constructions is their recognizability to the participants themselves. When participants themselves say, "yes, that is there, I'd simply never noticed it before," the observer can be reasonably confident that he has tapped into extant patterns of participation. (Lofland, 1971, p. 34)

Exhibit 8.9, using the problem of classifying people's ancestry, shows what can happen when indigenous and official constructions conflict, a matter of some consequence to those affected.

A good example of an analyst-generated typology comes from an evaluation of the National Museum of Natural History, Smithsonian Institution, done by Robert L. Wolf and Barbara L. Tymitz (1978). This

TAO QUALITIES

When Beauty is recognized in the World
Ugliness has been learned;

When Good is recognized in the World

Evil has been learned.

In this way:

Alive and dead are abstracted from growth;

Difficult and easy are abstracted from progress;

Far and near are abstracted from position;

Strong and weak are abstracted from control,

Song and speech are abstracted from harmony;

After and before are abstracted from sequence.

Comparative Analysis

A newborn is soft and tender,

A crone, hard and stiff.

Plants and animals, in life, are supple and juicy;

In death, brittle and dry.

So softness and tenderness are attributes of life,

And hardness and stiffness, attributes of death.

—*Tao Te Ching* of Lao Tzu

EXHIBIT 8.8 Coming-of-Age Paradigms

Ideal-typical comparison of "coming-of-age paradigms," which contrasts indigenous tribal initiation themes with contemporary, analyst-constructed coming-of-age themes

DIMENSIONS OF COMPARISON	TRIBAL INITIATION FROM AN INDIGENOUS PERSPECTIVE	MODERN COMING OF AGE: ANALYST-CONSTRUCTED DIMENSIONS
1. View of life passages	One-time transition from child to adult	Multiple passages over a lifetime journey
2. Territory	Tribal territory	Earth: global community
3. Ancestry	Creation myth	Evolutional story of humankind
4. Identity	Becoming a man or woman	Becoming a complete person
5. Approach	Standardized	Individualized
6. Outcome	Tribe-based identity	Personal identity, sense of self
7. Message	You are first and foremost a member of the tribe	You are first and foremost a person in your own right

EXHIBIT 8.9 Qualitative Analysis of Ancestry at the U.S. Census

To count different kinds of people—the job of the Census Bureau—you need categories to count them in. The long form of the 2000 census, given to one in six households, asked an open-ended, fill-in-the-blank question about "ancestry." Analysts then coded the responses into 604 categories, up from 467 in 1980. The government doesn't ask about religion, so if people respond that they are "Jewish," they don't get their ancestry counted. However, those who write in that they are Amish or Mennonite do get counted because those are considered cultural categories.

Ethnic minorities that cross national boundaries, such as French and Spanish Basques, and groups affected by geopolitical change, like Czechs and Slovaks or groups within the former Yugoslavia, are counted in distinct categories. The Census Bureau, following advice from the U.S. State Department, differentiates Taiwanese Americans from Chinese Americans, a matter of political sensitivity.

Can Assyrians and Chaldeans be lumped together? When the Census Bureau announced that they would combine the two in the same "ancestry code," an Assyrian group sued over the issue but lost the lawsuit. Assyrian Americans trace their roots to a biblical-era empire covering much of what is now Iraq and believe that Chaldeans are a separate religious subgroup. A fieldworker for the Census Bureau did fieldwork on the issue.

"I went into places where there were young people playing games, went into restaurants, and places where older people gathered," says Ms. McKenney.... She paid a visit to Assyrian neighborhoods in Chicago, where a large concentration of Assyrian-Americans lives. At a local community center and later that day at the Assyrian restaurant next door, community leaders presented their case for keeping the ancestry code the same. Over the same period, she visited Detroit to look into the Chaldean matter ....

"I found that many of the people, especially the younger people, viewed it as an ethnic group, not a religion," says Ms. McKenney. She and Mr. Reed (Census Bureau Ancestry research expert) concurred that enough differences existed that the Chaldeans could potentially qualify as a separate ancestry group.

In a conference call between interested parties, a compromise was struck. Assyrians and Chaldeans would remain under a single ancestry code, but the name would no longer be Assyrian, it would be Assyrian/Chaldean/Syriac—Syriac being the name of the Aramaic dialect that Assyrians and Chaldeans speak. "There was a meeting of the minds between all the representatives, and basically it was a unified decision to say that we're going to go under the same name," says the Chaldean Federation's Mr. Yono. (Kulish, 2001, p. 1) has become a classic in the museum studies field. They conducted a naturalistic inquiry of viewers' reactions to an exhibit on "Ice Age Mammals and Emergence of Man." From their observations, they identified four different kinds of visitors to the exhibit.

These descriptions are progressive in that each new category identifies a person more serious about the exhibit hall.

- *The commuter:* A person who merely uses the hall as a vehicle to get from the entry point to the exit point. . . .

- *The nomad:* A casual visitor, a person who is wandering through the hall, apparently open to become interested in something. The Nomad is not really sure why he or she is in the hall and not really sure that s/he is going to find anything interesting in this particular exhibit hall. Occasionally the Nomad stops, but it does not appear that the nomadic visitor finds any one thing in the hall more interesting than any other thing.

- *The cafeteria type:* This is the interested visitor who wants to get interested in something, and so the entire museum and the hall itself are treated as a cafeteria. Thus, the person walks along, hoping to find something of interest, hoping to "put something on his or her tray" and stopping from time to time in the hall. While it appears that there is something in the hall that spontaneously sparks the person's interest, we perceive this visitor has a predilection to becoming interested, and the exhibit provides the many things from which to choose.

- *The V.I.P.—very interested person:* This visitor comes into the hall with some prior interest in the content area. This person may not have come specifically to the hall, but once there, the hall serves to remind the V.I.P.'s that they were, in fact, interested in something in that hall beforehand. The V.I.P. goes through the hall much more carefully, much slower, much more critically—that is, they move from point to point, they stop, they examine aspects of the hall with a greater degree of scrutiny and care. (pp. 10–11)

This typology of types of visitors became important in the full evaluation because it permitted analysis of different kinds of museum experiences. Moreover, the evaluators recommended that when conducting interviews to get museum visitors' reactions to exhibits, the interview results should be differentially valued depending on the type of person being interviewed—commuter, nomad, cafeteria type, or VIP.

A different typology was developed to distinguish how visitors learn in a museum, "Museum Encounters of the First, Second, and Third Kind," a takeoff on the popular science fiction movie *Close Encounters of the Third Kind*, which referred to direct human contact with visitors from outer space.

- *Museum encounters of the first kind:* This encounter occurs in halls that use display cases as the primary approach to specimen presentation. Essentially, the visitor is a passive observer to the "objects of interest." Interaction is visual and may occur only at the awareness level. The visitor is probably not provoked to think or consider ideas beyond the visual display.

- *Museum encounters of the second kind:* This encounter occurs in halls that employ a variety of approaches to engage the visitor's attention and/or learning. The visitor has several choices to become active in his/her participation. . . . The visitor is likely to perceive, question, compare, hypothesize, etc.

- *Museum encounters of the third kind:* This encounter occurs in halls that invite high levels of visitor participation. Such an encounter invites the visitor to observe phenomena in process, to create, to question the experts, to contribute, etc. Interaction is personalized and within the control of the visitor. (Wolf & Tymitz, 1978, p. 39)

Here's a sample of a quite different classification scheme, this one developed from fieldwork by sociologist Rob Rosenthal (1994) as "a map of the terrain" of the homeless.

- *Skidders:* Most often women, typically in their 30s, who grew up middle or upper class but "skidded" into homelessness as divorced or separated parents

- *Street people:* Mostly men, often veterans, rarely married; highly visible and know how to use the resources of the street

- *Wingnuts:* People with severe mental problems, occasionally due to long-term alcoholism, a visible subgroup (*Note to readers:* Including this example and the label "wingnuts" is not an endorsement of its insensitivity. The label is offensive. Labeling is treacherous and will appear especially inappropriate when removed from the context in which it was generated; in this case, the term was sometimes used among homeless people themselves.)

- *Transitory workers:* People with job skills and a history of full-time work who travel from town to town, staying months or years in a place and then heading off to greener pastures

EXHIBIT 8.10 Ten Types of Qualitative Analysis

These varying types of qualitative analysis are distinct, not mutually exclusive. An analysis can include, and typically does include, several approaches. It is worth distinguishing them because they involve different ways of approaching the challenge of making sense of qualitative data.

TYPE OF ANALYSIS	DEFINITION	EXAMPLES USING CASE STUDIES WITH HOMELESS YOUTH
1. Content analysis	General term for identifying, organizing, and categorizing the content of narrative text	Analyzing and grouping the reactions of homeless youth to agency programs based on their responses to the open-ended question "What did you get out of this program?" • New ideas • New relationships • New skills • Changed attitudes • Changed behaviors • New goals
2. Case study	Qualitative data organized to coherently tell the story of the case (person, organization, community, etc.) that has been purposefully sampled	Case studies of homeless youth: Focus on capturing and communicating the unique and particular nature of each young person's story
3. Cross-case pattern analysis	Descriptions of actions, perceptions, experiences, relationships, and behaviors that are similar enough to be considered a manifestation of the same thing	"Difficulty sustaining positive relationships": homeless youth reporting failed relationships with family members, coworkers, friends, lovers, counselors, etc.
4. Cross-case thematic analysis	Interpreting and assigning meaning to a documented pattern by giving it a thematic name, a term that connotes and interprets the implications of the pattern	Relationship doubt and distrust: The pattern of failed relationships leads homeless youth to approach relationships with doubt and distrust
5. Inductive analysis	Searching the qualitative data for patterns and themes without entering the analysis with preconceived analytical categories; begin with specific cases, generate general patterns, and discover common themes through cross-case analysis	Discovering in interviews with homeless youth that their interactions with program staff often involve treating homelessness as a period in a lifelong journey rather than as an identity-defining status ("being a homeless person")
6. Deductive analysis	Examining the data for illumination of predetermined sensitizing concepts or theoretical relationships	Programs serving homeless youth claim to follow the principle of "harm reduction"; examining the case studies of homeless youth for manifestations of "harm reduction" is a deductive-analytical approach
7. Indigenous concept analysis (emic analysis)	Identifying and reporting indigenous terminology and concepts: documenting their meanings and interpreting their implications *from the perspective of those interviewed*	"Doin' it DE MAN'S way": phrase used to describe conforming with societal norms and power structure expectations
8. Analyst-generated concepts (etic analysis)	Concepts, labels, and terms created by the analyst to describe an observed phenomenon	Homeless youth as members of the *precariat*—a class of people in industrialized societies who live on the margin in precarious situations and experience multiple forms of economic and social insecurity (Standing, 2011)
9. Indigenous typologies	A continuum made up of contrasting end points that people being studied use to divide some aspect of the world into distinct categories or ideal types	"Jails" versus "parks": A *jail* is a homeless shelter with lots of rules, zero tolerance, heavy controls, and mean staff; a *park* is a homeless shelter that offers space, lets young people hang out, is an easy place to be, and has welcoming staff
10. Analyst-constructed typologies	A continuum or classification system *made up by analysts* to divide some aspect of the world into distinct categories or ideal types	Categories of how homeless youth spend their time: • Hanging out • Getting by • Getting ahead (Rosenthal, 1994)

COMPLEMENTARY PAIRS AS CONCEPTUALLY SENSITIZING CONTINUA

Nobel laureate Niels Bohr's maxim is as follows:

Contraria sunt complementa ("Contraries are complementary")

Contraries are not contradictory:...We replace all related but slightly different terms like contraries, polar opposites, duals, opposing tensions, binary oppositions, dichotomies, and the like with the all-encompassing term "complementary pairs." (Kelso & Engstrom, 2006, p. 7)

Sampling of Complementary Pairs From Various Field of Endeavor

Anatomy: organ/organism; form/function

Art: foreground/background; original/reproduction

Culture: permissible/taboo; public/private

Economics: boom/bust; equilibrium/disequilibrium

Education: knowledge/ignorance; student/teacher

Entertainment: amateur/professional; comedy/tragedy

Mathematics: problem/solution; finite/infinite

Medicine: curative/palliative; invasive/noninvasive; prevention/cure

Military: all/enemy; defensive/offensive; peace/war

Mythology: hero/villain; beauty/ugliness

Philosophy: faith/reason; physical/spiritual; truth/falsehood

Politics: conservative/liberal; rights/responsibilities

Psychology: abnormal/normal; extraversion/introversion

Sociology: folk/urban; general/particular; task oriented/process oriented; social/antisocial

SOURCE: Kelso and Engstrom (2006, pp. 257–262).

Categories of How Homeless People Spend Their Time:

- Hanging out
- Getting by
- Getting ahead

As these examples illustrate, the first purpose of typologies is to distinguish aspects of an observed pattern or phenomenon *descriptively*. Once identified and distinguished, these types can later be used to make interpretations, and they can be related to other observations to draw conclusions, but the first purpose is description based on an inductive analysis of patterns that appear in the data. Kenneth Bailey's (1994) classic on typologies and taxonomies in qualitative analysis remains an excellent resource.

Summary of Pattern, Theme, and Content Analysis

Purpose drives analysis. Design frames analysis. Purposeful sampling strategies determine the unit of analysis. Different analytical approaches will yield different kinds of findings based on distinct analysis procedures and priorities. There is no single right way to engage in qualitative analysis. Distinguishing signal from noise (detecting patterns and identifying themes) results from immersion in the data, systematic engagement with what the data reveal, and judgment about what is meaningful and useful. The next module gets into some of the nitty-gritty operational processes involved in analysis. Exhibit 8.10 presents the 10 analytical approaches reviewed in this module.

> Classification is Ariadne's clue through the labyrinth of nature.
>
> —George Sand (1869)
> *Nouvelles Lettres d'un Voyageur*

Coding Data, Finding Patterns, Labeling Themes, and Developing Category Systems

Thus far, I've provided lots of examples of the fruit of qualitative inquiry: patterns, themes, categories, and typologies. Let's back up now to consider how you recognize patterns in qualitative data and turn those patterns into meaningful categories and themes. This chapter could have started with this section, but I think it's helpful to understand what kinds of findings can be generated from qualitative analysis before delving very deeply into the mechanics and operational processes, especially because the mechanics vary greatly and are undertaken differently by analysts in different disciplines working from divergent frameworks. That said, some guidance can be offered.

Raw field notes and verbatim transcripts constitute the undigested complexity of reality. Simplifying and making sense out of that complexity constitutes the challenge of content analysis. Developing some manageable classification or coding scheme is the first step of analysis. Without classification, there is chaos and confusion. Content analysis, then, involves identifying, coding, categorizing, classifying, and labeling the primary patterns in the data. This essentially means analyzing the core *content* of interviews and observations to determine what's significant. In explaining the process, I'll describe it as done traditionally, which is without software, to highlight the thinking involved. Software programs provide different tools and formats for coding, but the principles of the analytical process are the same whether doing it manually or with the assistance of a computer program.

I begin by reading through all of my field notes or interviews and making comments in the margins or even attaching post-it notes that contain my notions about what I can do with the different parts of the data. This constitutes the first cut at organizing the data into topics and files. Coming up with topics is like constructing an index for a book or labels for a file

system: You look at what is there and give it a name, a label. The copy on which these topics and labels are written becomes the indexed copy of the field notes or interviews. Exhibit 8.11 shows examples of codes from the field note margins of the evaluation of a wilderness education program. You create your own codes, or in a team situation, you create codes together.

EXHIBIT 8.11 First-Cut Coding Examples

P is for participants, S is for staff.

Sample codes from the field note margins of the evaluation of a wilderness education program

CODE ABBREVIATIONS	ABBREVIATION MEANINGS
Ps Re Prog	Participants' reactions to the program
Ps Re Ps	Participants' reactions to other participants
Ob PP	Observations of participants' interactions
Ob SS	Observations of staff interactions
Ob SP	Observations of staff–participant interactions
S-G	Subgroup formations
Phil	Quotes about program philosophy
Prc	Program processes
P/outs	Effects of program on participants: outcomes
C!	Conflicts

The shorthand codes (abbreviations) are written in the margins directly on the relevant data passages or quotations. The full labels in the second column of the above table are the designations for separate files that contain all similarly coded passages.

The shorthand codes (abbreviations) are written directly on the relevant data passages, either in the margins or with an attached tab on the relevant page. Many passages will illustrate more than one theme or pattern. The first reading through the data is aimed at developing the coding categories or classification system. Then a new reading is done to actually start the formal coding in a systematic way. Several readings of the data may be necessary before field notes or interviews can be completely indexed and coded. Some people find it helpful to use colored highlighting pens—color coding different idea or concepts. Using self-adhesive colored dots or post-it notes is another option.

> If sensing a pattern or "occurrence" can be called *seeing*, then the encoding of it can be called *seeing as*. That is, you first make the observation that something important or notable is occurring, and then you classify or describe it. . . . The *seeing as* provides us with a link between a new or emergent pattern and any and all patterns that we have observed and considered previously. It also provides a link to any and all patterns that others have observed and considered previously through reading. (Boyatzis, 1998, p. 4)

Where more than one person is working on the analysis, it is helpful to have each person (or small teams for large projects) develop the coding scheme independently, then compare and discuss similarities and differences. Important insights can emerge from the different ways in which two people look at the same set of data—a form of analytical triangulation.

Often an elaborate classification system emerges during coding, particularly in large projects where a formal scheme must be developed that can be used by several trained coders. In our study of evaluation use, which is the basis for *Utilization-Focused Evaluation* (Patton, 2008), graduate students in the evaluation program at the University of Minnesota conducted lengthy interviews with 60 project officers, evaluators, and federal decision makers. We developed a comprehensive classification system that would provide easy access to the data by any of the student or faculty researchers. Had only one investigator been intending to use the data, such an elaborate classification scheme would not have been necessary. However, to provide access to several students for different purposes, every paragraph in every interview was coded using a systematic and comprehensive coding scheme made up of 15 general categories with subcategories. Portions of the codebook used to code the utilization of evaluation data appear as Exhibit 8.34 at the end of this chapter (pp. 642–643), as an example of one kind of qualitative analysis codebook. This codebook was developed from four sources: (1) the standardized

open-ended questions used in interviewing; (2) review of the utilization literature for ideas to be examined and hypotheses to be reviewed; (3) our initial inventory review of the interviews, in which two of us read all the data and added categories for coding; and (4) a few additional categories added during coding when passages didn't fit well into the available categories.

Every interview was coded twice by two independent coders. Each individual code, including redundancies, was entered into our qualitative analysis database so that we could retrieve all passages (data) on any subject included in the classification scheme, with brief descriptions of the content of those passages. The analyst could then go directly to the full passages and complete interviews from which the passages were extracted to keep quotations in context. In addition, the computer analysis permitted easy cross-classification and cross-comparison of passages for more complex analyses across interviews.

Some such elaborate coding system is routine for very rigorous analysis of a large amount of data. Complex coding systems with multiple coders categorizing every paragraph in every interview constitutes a labor-intensive form of coding, one that would not be used for most small-scale formative evaluation or action research projects. However, where data are going to be used by several people or where data are going to be used over a long period of time, including additions to the data set over time, such a comprehensive and computerized system can be well worth the time and effort required. This is the case, for example, where an action research project involves a number of people working together in an organizational or community context where the stakes are high.

Classifying and coding qualitative data produces a framework for organizing and describing what has been collected during fieldwork. (For published examples of coding schemes, see Bernard, 2000, pp. 447–450; Bernard & Ryan, 2010, pp. 325–328, 387–389, 491–492, 624; Boyatzis, 1998; Miles, Huberman, & Saldaña, 2014; Strauss & Corbin, 1998.) This descriptive phase of analysis builds a foundation for the interpretative phase, when meanings are extracted from the data, comparisons are made, creative frameworks for interpretation are constructed, conclusions are drawn, significance is determined, and, in some cases, theory is generated.

Convergence and Divergence in Coding and Classifying

In developing codes and categories, a qualitative analyst must first deal with the challenge of "convergence" (Guba, 1978)—figuring out what things fit together.

ERRORS TO AVOID
OR FIX WHEN FOUND

In a clever graphic comic titled *The Good, the Bad, and the Data: Shane the Lone Ethnographer's Basic Guide to Qualitative Data Analysis*, Sally Galman (2013) identifies an alphabet soup of qualitative errors to be avoided or fixed. Here's a sample:

A is for Anemia. As you code, you find that your data are terribly thin. Go back to the field to beef up your data.

J is for Jaws (the shark). Something is lurking under the surface—you are in denial of the disconfirming evidence, so you avoid it.

O is for "Oh no! I didn't OBSERVE." All you have are notes filled with interpretations rather than actual observations.

P is for Procrastination. Don't let too much time pass before you analyze.

X is for Extra Stuff. You've completed your analysis, and you have all this extra data that do not seem to fit. Maybe it's time to revisit your questions and design.

Z is for Zealotry. Did you make room for discovery or did you only confirm your own ideas? (pp. 82–85)

large number of unassignable or overlapping data items is good evidence of some basic fault in the category system" (Guba, 1978, p. 53). The analyst then works back and forth between the data and the classification system to verify the meaningfulness and accuracy of the categories and the placement of data in categories. If several different possible classification systems emerge or are developed, some priorities must be established to determine which are more important and illuminative. Prioritizing is done according to the utility, salience, credibility, uniqueness, heuristic value, and feasibility of the classification schemes. Finally, the category system or set of categories is tested for completeness.

1. The set should have internal and external plausibility, a property that might be termed "integratability." Viewed internally, the individual categories should appear to be consistent; viewed externally, the set of categories should seem to comprise a whole picture. . . .

2. The set should be reasonably inclusive of the data and information that do exist. This feature is partly tested by the absence of unassignable cases, but can be further tested by reference to the problem that the inquirer is investigating or by the mandate given the evaluator by his client/sponsor. If the set of categories did not appear to be sufficient, on logical grounds, to cover the facets of the problem or mandate, the set is probably incomplete.

3. The set should be reproducible by another competent judge. . . . The second observer ought to be able to verify that a) the categories make sense in view of the data which are available, and b) the data have been appropriately arranged in the category system. . . . The category system auditor may be called upon to attest that the category system "fits" the data and that the data have been properly "fitted into" it.

4. The set should be credible to the persons who provided the information which the set is presumed to assimilate. . . . Who is in a better position to judge whether the categories appropriately reflect their issues and concerns than the people themselves? (Guba, 1978, pp. 56–57)

Begin by looking for *recurring regularities* in the data. These regularities reveal patterns that can be sorted into categories. Categories should then be judged by two criteria: (1) internal homogeneity and (2) external heterogeneity. The first criterion concerns the extent to which the data that belong in a certain category hold together or "dovetail" in a meaningful way. The second criterion concerns the extent to which differences among categories are bold and clear. "The existence of a

After analyzing for convergence, the mirror analytical strategy involves examining divergence. By this, Guba means that the analyst must "flesh out" the patterns or categories. This is done by the processes of *extension* (building on and going deeper into the patterns and themes already identified), *bridging* (making connections among different patterns and themes), and *surfacing* (proposing new

©2002 Michael Quinn Patton and Michael Cochran

categories that ought to fit and then verifying their existence in the data). The analyst brings closure to the process when sources of information have been exhausted, when sets of categories have been saturated so that new sources lead to redundancy, when clear regularities have emerged that feel integrated, and when the analysis begins to "overextend" beyond the boundaries of the issues and concerns guiding the it. Divergence also includes careful and thoughtful examination of data that do not seem to fit, including *deviant cases* that don't fit the dominant identified patterns.

This sequence, convergence then divergence, should not be followed mechanically, linearly, or rigidly. The processes of qualitative analysis involve both technical and creative dimensions. As noted early in this chapter, no abstract processes of analysis, no matter how eloquently named and finely described, can substitute for the skill, knowledge, experience, creativity, diligence, and work of the qualitative analyst. "The task of converting field notes and observations about issues and concerns into systematic categories is a difficult one. No infallible procedure exists for performing it" (Guba, 1978, p. 53).

(For in-depth guidance on qualitative coding and analysis see Bernard & Ryan, 2010, *Analyzing Qualitative Data: Systematic Approaches*; Boeije, 2010, *Analysis in Qualitative Research*; Guest, MacQueen, & Namey, 2012, *Applied Thematic Analysis*; Northcutt & McCoy, 2004, *Interactive Qualitative Analysis: A Systems Method for Qualitative Research*; Saldaña, 2009, *The Coding Manual for Qualitative Researchers*.)

SIDEBAR

FEAR OF FINDING NOTHING

Students beginning dissertations often ask me, their anxiety palpable and understandable, "What if I don't find out anything?" Bob Stake, of "responsive evaluation" and "case study" fame, said at his retirement, "Paraphrasing Milton: They also serve who leave the null hypothesis tenable.... It is a sophisticated researcher who beams with pride having, with thoroughness and diligence, found nothing there" (Stake, 1998, p. 364, with a nod to Michael Scriven for inspiration).

True enough. But in another sense, it's not possible to find nothing there, at least not in qualitative inquiry. The case study is there. It may not have led to new insights or confirmed one's predilections, but the description of that case at that time and in that place is there. That is much more than nothing. The interview responses and observations are there. They, too, may not have led to headline-grabbing insights or confirmed someone's eminent theory, but the thoughts and reflections from those people at that time and in that place are there, recorded and reported. That is much more than "nothing."

Halcolm will tell you this:

> You can only find nothing if you stare at a vacuum.
>
> You can only find nothing if you immerse yourself in nothing.
>
> You can only find nothing if you go nowhere.
>
> Go to real places.
>
> Talk to real people.
>
> Observe real things.
>
> You will find something.
>
> Indeed, you will find much, for much is there.
>
> You will find the world.

MQP Rumination # 8

Make Qualitative Analysis First and Foremost Qualitative

I am offering one personal rumination per chapter. These are issues that have persistently engaged, sometimes annoyed, occasionally haunted, and often amused me over more than 40 years of research and evaluation practice. Here's where I state my case on the issue and make my peace.

Here's the scenario. I've conducted 15 key informant interviews with executive directors of nonprofit agencies that receive funds from the same philanthropic foundation. I'm presenting the results to the foundation's senior staff and trustees. I report as follows:

Most of those I interviewed report being quite frustrated with your evaluation reporting requirements. They don't think you're asking the most important questions and they are dubious that anyone here is reading or using their reports. Most said that they get no feedback after submitting the required reports.

I then share three examples of direct quotes supporting this overall conclusion:

- "I do the reports because we're required to, and we take them seriously and answer seriously. But there are important things we'd like to report and think they'd like to know that aren't asked, and there's no space for. That feels like a lost opportunity."
- "Look, I've been at this for years. It's very frustrating. We know it's just a compliance thing. No one reads our reports. We do them because they're required. That's it. End of story."
- "Truth be told, it's a waste of time, a frustrating waste of time."

I then invite questions, comments, and reactions.

The board chair asks, "How many said it was a waste of time?"

I take a deep breath, and bite my tongue (metaphorically) to stop myself from saying, "You have a problem here. Does it really matter whether it's 7 people or 9 or 12? You have a problem! It's not about the number. It's about the substance. YOU HAVE A PROBLEM!"

The Allure of Precision

This scenario occurs over and over again. It's the knee-jerk response to the ambiguities of qualitative findings: "Many said," "some said," "a few said," and so on. When presenting findings at a major international evaluation that involved 20

key informant interviews, the response from the conference chair was to dismiss the report as "evaluation by adjective." He wanted to know how many said what? "What are the percentages?" he demanded.

I refused. I invite you to refuse. Here are 5 reasons why. (Count them. There are exactly 5 reasons. Now I could have generated 10 reasons or just offered my top 3. But I decided on 5. Elsewhere, I've offered lists of 10, 12, or 3, but 5 struck me as about right for a rumination. So that's what you get: 5.)

1. *Open-ended interviews generate diverse responses.* That's the purpose of an open-ended question, to find out what's salient in the interviewees' own words. We then group together those responses that manifest a common theme. The three quotes above all fall into a category of *Feeling Frustrated*. Only one person used the phrase "waste of time." Another said, "I put it off as long as I can and do it just in time to meet the deadline for submission, because I have a lot of more important things to do and it's not a great use of my time. But I do it." Not quite "waste of time," but pretty close. What responses go together is a matter of interpretation and judgment. Coding, categorizing, and theme analysis are not precise. The result is qualitative. Stay qualitative.

2. *The adjectives "most," "many," "some," or "a few" are actually more accurate than a precise number.* It's common to have a couple of responses that could be included in the category or left out, thus changing the number. I don't want to add a quote to a category just to increase the number. I want to add it because, in my judgment, it fits. So when I code 12 of 20 saying some version of "feeling frustrated," I'm confident in reporting that "many" felt frustrated. It could have been 10, or it could have been 14, depending on the coding. But it definitely was *many*.

3. *Percentages may be misleading.* With a key informant sample of 20, each response is 5%. Thus, going from 12 of 20 to 14 of 20 is a jump from 60% to 70%. In a survey of 300 respondents, a 10% difference is significant. In a small purposeful sample, it's not. Going from 12 to 14 is still "many."

4. *The "how many" question can distract from dealing with substantive significance.* I regularly conduct workshops with 20 to 40 participants. The workshop sponsors

(Continued)

(Continued)

usually have some standardized evaluation form that solicits ratings and then invites an overall open-ended response. Over the years, a single, particularly insightful and specific response has proved more valuable to me than a large number of general comments (e.g., "I learned a lot"). The point of qualitative analysis is not to determine how many said something. The point is to generate substantive insight into the phenomenon. One or two very insightful and substantive responses can easily trump 15 general responses. Here's an example. I interviewed 15 participants in an employment training program. Two female participants said they were on the verge of dropping out because of sexual harassment by a staff member. That's "only" 13%. That's just 2 of 15. But *any* sexual harassment is unacceptable. The program has a problem, a potentially quite serious problem.

5. *Small purposeful samples pose confidentiality challenges.* When I'm reporting qualitative findings, I say in the methods section that I will not report that "all" or "no one" responded in a certain way because that would potentially break the confidentiality pledge. In the example that opened this rumination, all the 15 agency directors I interviewed complained about the foundation's evaluation reporting process, especially the lack of feedback. But I reported that "many" complained, and refused attempts to get me to provide a number (which would have been 20 of 20), so as not to put any of the directors at risk.

Reasons Galore

So there you have 5 reasons to keep qualitative analysis qualitative. But maybe that doesn't seem like enough. Maybe you'd be more persuaded and feel more confident if I gave you 10 reasons. No sooner asked, than done. Here are 5 more rumination-inspired reasons to keep qualitative analysis first and foremost qualitative:

6. Doing so demonstrates integrity.

7. It reinforces the message that the inquiry is qualitative.

8. It requires people to think about meanings.

9. Numbers are easily manipulated and analysis is corrupted under pressure to increase the number. (Hmmm, is that 2 reasons or just 1?)

10. Meaning is essentially qualitative and about qualities.

And a bonus item: *Generating numbers is not the purpose of qualitative inquiry.* If someone wants precise numbers, tell them to do a survey and ask closed questions and count the responses. That's what quantitative methods are for!

Pragmatism

Readers of this book know by now that I'm fundamentally a pragmatist. Thus, the prior points notwithstanding, sometimes numbers are appropriate, sometimes they are illuminative, and sometimes they are simply demanded by those who commission evaluations. My point is not to be rigid but to place the burden of proof on justifying *quantitizing*. Do not go gently down that primrose path. Use numbers when appropriate, and then in moderation.

Here's an example where numbers are appropriate. Psychologist Marvin Eisenstadt studied the link between career achievement and loss of a parent in childhood by identifying famous people from ancient Greece through to modern times whose lives merited significant entries in encyclopedias. He generated a list of 573 eminent people and did extensive research on their childhoods, an inquiry that took 10 years. "A quarter had lost at least one parent before the age of ten. By age 15, 34.5 percent had at least one parent die, and by the age of twenty, 45 percent" (Gladwell, 2013, p. 141). This conversion of qualitative codes to quantitative distributions is appropriate because the sample size is large, the numbers are accurate, and the focus of the inquiry is on a single variable. When there is something meaningful to be counted, then count. As sample sizes increase, especially in mixed-methods studies, quantizing is likely to become even more pervasive. Now let me offer an example where quantitizing strikes me as considerably less appropriate and meaningful.

Feeding the Quantitative Beast

The opening scenario in this rumination involved a board chair reacting to my qualitative presentation by asking how many said what. But those involved in qualitative studies exacerbate the problem by turning their reports into numbers even before being asked to do so. As I was completing this chapter, I received an analysis from a graduate student who had taken interviews I had given and counted how many times I used various words, a form of so-called content analysis that actually diminishes the meaning of both "content" and "analysis." Having counted my use of various words, he then correlated them. He was seeking my interpretation of a couple of statistical correlations that he couldn't explain. My response was that the entire analytical approach struck me as meaningless since I adapt my language in an interview to context, audience, and whatever I'm working on at the time. To lose the contextual meaning of words by counting them as isolated data points strikes me as highly problematic—and certainly not qualitative meaning making.

I receive a substantial number of qualitative evaluations to review each year. The most common pattern I see, and criticize in my review, is a qualitative study filled with numbers. Here's an example that just came to me the very week I was writing this rumination. I'm afraid my response was rather intemperate.

Qualitative Report Excerpts

- Of the 20 students interviewed, 14 mentioned gaining leadership skills; 8 of 21 staff said leadership skills were important; 7 out of 13 field personnel said this, as did 4 out of 6 community leaders.
- Eighteen of the 20 students said they were more committed to scholarly publication; 2 said they didn't want to be university scholars.
- Two out of the seven program directors at different universities felt that the purpose of the professional development program was mainly to train advanced students how to write for academic publication; the other five emphasized writing for policymakers.

- Out of the 14 university researchers interviewed, 7 had no opinion about students becoming better teachers because they were not sure what the program was doing to train students as teachers. Four other interviewees claimed that combining teaching skills with research skills caused confusion. Three said combining the two made sense and was valuable.

The 20-page report was filled with this kind of quantitative gibberish—I'm sorry, analytical reporting. Qualitative software easily generates such numbers, so that may feed this trend and give it the appearance of being appropriate and expected. *It is not appropriate and should not be expected.* Indeed, I urge those involved in qualitative evaluations to make it clear at the outset to those who will be receiving the findings that numbers will generally not be reported. The focus will be on substantive significance. The point is not to be anti-numbers. The point is to be *pro-meaningfulness. Keep qualitative analysis first and foremost qualitative.*

Welcome to Qualitative Data Visualization 101. Let's get started...

Step 1. Quantify your data

> Logic: The art of thinking and reasoning in strict accordance with the limitations and incapacities of the human misunderstanding.
>
> —Satirist Ambrose Bierce (1842–1914)
>
> Contrariwise, if it was so, it might be; and if it were so, it would be; but as it isn't, it ain't. That's logic.
>
> —Author Lewis Carroll (1832–1898)

Logical Analysis

While working inductively, the analyst is looking for emergent patterns in the data. These patterns, as noted in the preceding sections, can be represented as dimensions, categories, classification schemes, and themes. Once some dimensions have been constructed, using either participant-generated constructions or analyst-generated constructions, it is sometimes useful to cross-classify different dimensions to generate new insights about how the data can be organized and to look for patterns that may not have been immediately obvious in the initial, inductive analysis. Creating cross-classification matrices is an exercise in logic.

The logical process involves creating potential categories by crossing one dimension or typology with another and then working back and forth between the data and one's logical constructions, filling in the resulting matrix. This logical system will create a new typology all parts of which may or may not actually be represented in the data. Thus, the analyst moves back and forth between the logical construction and the actual data in search of meaningful patterns.

In the high school dropout program described earlier, the focus of the program was reducing absenteeism, skipping classes, and tardiness. An external team of change agents worked with teachers in the school to help them develop approaches to the dropout problem. Observations of the program and interviews with the teachers gave rise to two dimensions. The first dimension distinguished *teachers' beliefs about what kind of programmatic intervention was effective* with dropouts—that is, whether they primarily favored maintenance (i.e., caretaking or warehousing

of kids to just keep the schools running), rehabilitation efforts (helping kids with their problems), or punishment (no longer letting them get away with the infractions they had been committing in the past). Teachers' behaviors toward dropouts could be conceptualized along a continuum from taking direct responsibility for doing something about the problem at one end to shifting responsibility to others at the opposite end. Exhibit 8.12 shows what happens when these two dimensions are crossed. Six cells are created, each of which represents a different kind of teacher role in response to the program.

The qualitative analyst working with these data had been struggling in the inductive analysis to find the patterns that would express the different kinds of teacher roles manifested in the program. He had tried several constructions, but none of them quite seemed to work. The labels he came up with were not true to the data. When he described to me the other dimensions he had generated, I suggested that he cross them, as shown in Exhibit 8.12. When he did, he said that "the whole thing immediately fell into place." Working back and forth between the matrix and the data, he generated a full descriptive analysis of diverse and conflicting teacher roles.

The description of teacher roles served several purposes. First, it gave teachers a mirror image of their own behaviors and attitudes. It could thus be used to help teachers make more explicit their own understanding of roles. Second, it could be used by the external team of consultants to more carefully gear their programmatic efforts toward different kinds of teachers who were acting out the different roles. The matrix makes it clear that an omnibus strategy for helping teachers establish a program that would reduce dropouts would not work in this school; teachers manifesting different roles would need to be approached and worked with in different ways. Third, the description of teacher roles provided insights into the nature of the dropout problem. Having identified the various roles, the evaluator–analyst had a responsibility to report on the distribution of roles in this school and the observed consequences of that distribution.

Abductive Analysis

One must be careful about purely logical analysis. It is tempting for an analyst using a logical matrix to force data into the categories created by the cross-classification to fill out the matrix and make it work. Logical analysis

EXHIBIT 8.12 An Empirical Typology of Teacher Roles in Dealing With High School Dropouts

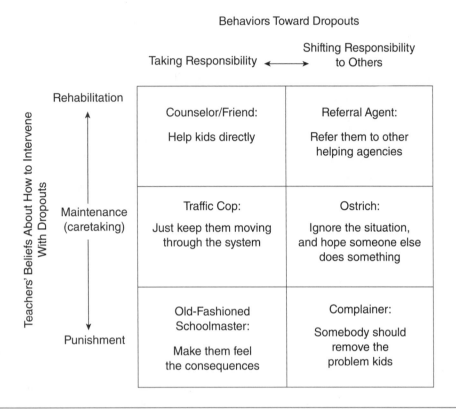

to generate new sensitizing concepts must be tested out and confirmed by the actual data. *Such logically derived sensitizing concepts provide conceptual possibilities to test.* Levin-Rozalis (2000), following American philosopher Charles Sanders Pierce of the pragmatic school of thought, suggests labeling the logical generation and discovery of hypotheses and findings *abduction* to distinguish such logical analysis from data-based inductive analysis and theory-derived deductive analysis.

Denzin (1978b) has explained abduction in qualitative analysis as a combination of inductive and deductive thinking with logical underpinnings:

Naturalists inspect and organize behavior specimens in ways which they hope will permit them to progressively reveal and better understand the underlying problematic features of the social world under study. They seek to ask the question or set of questions which will make that world or social organization understandable. They do not approach that world with a rigid set of preconceived hypotheses. They are initially directed toward an interest in the routine and taken-for-granted features of that world. They ask how it is that the persons in

question know about producing orderly patterns of interaction and meaning. . . . They do not use a full-fledged deductive hypothetical scheme in thinking and developing propositions. Nor are they fully inductive, letting the so-called facts speak for themselves. Facts do not speak for themselves. They must be interpreted. Previously developed deductive models seldom conform with empirical data that are gathered. The method of abduction combines the deductive and inductive models of proposition development and theory construction. It can be defined as *working from consequence back to cause or antecedent.* The observer records the occurrence of a particular event, and then works back in time in an effort to reconstruct the events (causes) that produced the event (consequence) in question. (pp. 109–110)

The famous fictional detective Sherlock Holmes relied on abduction more than deduction or induction, at least according to William Sanders's (1976) review of Holmes's analytical thinking in *The Sociologist as Detective.* We've already suggested that the qualitative analyst is part scientist and part artist. Why not add the qualitative analyst as detective? Here's an example.

An Example of Abductive Qualitative Analysis

In the evaluation of the rural community leadership program, we did follow-up interviews with participants to find out how they were using their training when they returned to their home communities. We found ourselves with a case not unlike the "Silver Blaze" story, in which Sherlock Holmes made much of the fact that the dog at the scene of the crime had not barked during the night while the crime was being committed. He inferred that the criminal was someone known to the dog. In our case, we discovered that graduates of the leadership program were not leading. In fact, they weren't doing much of anything. Were their skills inadequate after only 1 week of intensive training? Did they lack confidence? Were they discouraged? Disinterested? Intimidated? Incompetent? Unmotivated?

So we had a finding. We had an outcome—or more precisely, the lack of an outcome. We worked backward from the experience of the training, examined what had happened during the training right up to the final session, and tried to connect the dots between what had happened and this unexpected result. We also returned to the participants for further reflections that might explain this general lack of follow-up action.

The participants expressed great interest in and commitment to exercising community leadership and engaging in community development. They expressed confidence in their abilities and felt they were competent to use the skills they had learned. But at perhaps the most teachable moment of all, in the final session of training, as the participants enthusiastically prepared to return to their communities and begin to use their learnings, the director of the program had offered a closing word of caution:

> Take your time when you return. Don't go back like a cadre of activists invading your community. You've had an intense experience together. Let things settle. It can be pretty overwhelming to the people back home when they get a sense that you've been through what for many of you has been a transformative experience. So go easy. Take your time. Resettle.

And so they did—more than he imagined. What he had neglected was any guidance about how to know when it was time to begin engaging after the reentry period of resettling. So they waited. And waited. And waited. Not wanting to get it wrong.

Of all the explanations we considered, that one fit the evidence the best. Its accuracy was further borne out when the director changed his parting advice at the end of subsequent programs and we found a different result in the communities. A qualitative inquirer is a detective, using both data and reasoning to solve the mystery of how a story has unfolded and why it has unfolded in the way documented.

Abductive Analysis Caution

Abduction is not widely known, understood, or appreciated. I think it provides a distinct and important alternative to deductive and inductive reasoning. But one external reviewer of this book worried that including abduction might confuse novice researchers and weaken the emphasis on induction as the core of qualitative reasoning. I disagree but think the caution is worth acknowledging, so here it is:

> Although interesting, the discussion of abductive analysis will likely only confuse the novice researcher. Perhaps this section should start with a stronger disclaimer that advises novice qualitative researchers to give their undivided attention to the skills needed for inductive inquiry and clearly identify potential logical positivist pitfalls.

Abductive Matrix Analysis

The empty cell of a logically derived matrix (the cell created by crossing two dimensions for which no name or label immediately occurs) creates an intersection of a possible consequence and antecedent that begs for abductive exploration and explanation. Each such intersection of consequence and antecedent sensitizes the analyst to the possibility of a category of activity or behavior that has either been overlooked in the data or that is logically a possibility in the setting but has not yet been documented. The latter cases are important to note for their importance derives from the fact that they did not occur. The next section will look in detail at a process–outcomes matrix ripe with abductive possibilities. First, Exhibit 8.13 shows a matrix for mapping stakeholders' stakes in a program or policy. This matrix can be used to guide data collection as well as analysis.

A Process–Outcomes Matrix

The linkage between processes and outcomes constitutes such a fundamental issue in many program evaluations that it provides a particularly good focus for illustrating qualitative matrix analysis. As discussed in Chapter 4, qualitative methods can be particularly appropriate for evaluation where program processes,

EXHIBIT 8.13 Power Versus Interest Grid for Analyzing Diverse Stakeholders' Engagement With a Program, a Policy, or an Evaluation

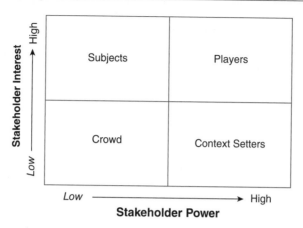

Players. High interest, high power: They make things happen.

Context setters. High power, low interest: They watch what unfolds and can become players if they get interested.

Subjects. High interest, low power: They do not act unless organized and empowered.

Crowd. Low interest, low power: They are disengaged until they become aware that they have a stake in what is unfolding.

This matrix can be used to map the stakeholder environment for any initiative by gathering data about the perspectives, interests, and nature of power of people in diverse relationships to the initiative (Bryson & Patton, 2010, p. 42; Patton, 2008, p. 80).

impacts, or both are largely unspecified or difficult to measure. This can be the case because the outcomes are meant to be individualized; sometimes one is simply uncertain as to what a program's outcomes will be; and in many programs, neither processes nor impacts have been carefully articulated. Under such conditions, one purpose of the evaluation may be to illuminate program processes, program impacts, and the linkages between the two. This task can be facilitated by constructing a process–outcomes matrix to organize the data.

Exhibit 8.14 shows how such a matrix can be constructed. Major program processes or identified implementation components are listed along the left side. Types or levels of outcomes are listed across the top. The category systems for program processes and outcomes are developed from the data in the same way that other typologies are constructed (see previous sections). The cross-classification of any process with any outcome produces a cell in the matrix—for example, the first cell in Exhibit 8.14 is created by the intersection of Process 1 with Outcome a. The information that goes in Cell 1-a (or any other cell in the matrix) describes linkages, patterns, themes, experiences, content, or actual activities that help us understand the relationships between processes and

outcomes. Such relationships may have been identified by participants themselves during interviews or discovered by the evaluator in analyzing the data. In either case, the process–outcomes matrix becomes a way of organizing, thinking about, and presenting the qualitative connections between program implementation dimensions and program results.

Application of the Process–Impact Matrix: An Example

An example will help make the notion of the process–outcomes matrix (Exhibit 8.14) more concrete and, hopefully, useful. Suppose we have been evaluating a juvenile justice program that places delinquent youth in foster homes. We have visited several foster homes, observed what the home environments are like, and interviewed the juveniles, the foster home parents, and the probation officers. A *regularly recurring process theme* concerns the importance of "letting kids learn to make their own decisions." A *regularly recurring outcomes theme* involves "keeping the kids straight" (reduced recidivism). Crossing the program process ("kids making their own decisions") with the program outcome ("keeping kids straight") creates a data analysis

EXHIBIT 8.14   Process–Impact Matrix

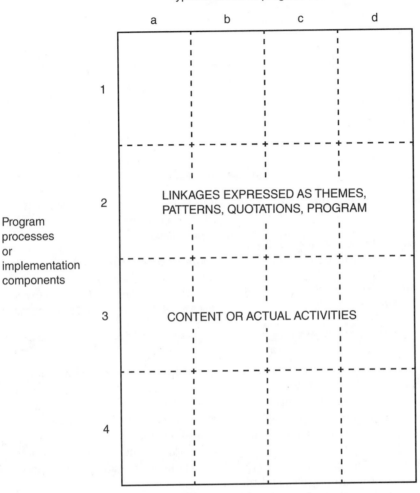

Matrix of linkages between program processes and impacts

Types or levels of program outcomes

a   b   c   d

Program
processes
or
implementation
components

1

2   LINKAGES EXPRESSED AS THEMES,
PATTERNS, QUOTATIONS, PROGRAM

3   CONTENT OR ACTUAL ACTIVITIES

4

question: What actual decisions do juveniles make that are supposed to lead to reduced recidivism? We then carefully review our field notes and interview quotations, looking for data that help us understand how people in the program have answered this question based on their actual behaviors and practices. When we describe what decisions juveniles actually make in the program, the decision makers to whom our findings are reported can make their own judgments about the strength or weakness of the linkage between this program process and the desired outcome. Moreover, once the process–outcomes descriptive analysis of linkages has been completed, the evaluator is at liberty to offer interpretations and judgments about the nature and quality of this process–outcome connection.

An In-Depth Analysis Example: Recognizing Processes, Outcomes, and Linkages in Qualitative Data

Because of the centrality of the sensitizing concepts "program process" and "program outcome" in evaluation research, it may be helpful to provide a more detailed description of how these concepts can be used in qualitative analysis. How does one recognize a program process? Learning to identify and label program processes is a critical evaluation skill. This sensitizing notion of "process" is a way of talking about the common action that cuts across program activities, observed interactions, and program content. The example I shall use involves data from the wilderness education program I evaluated and discussed throughout the observations chapter (Chapter 6). That program, titled the Southwest Field Training Project, used the wilderness as a training arena for professional educators in the philosophy and methods of experiential education by engaging those educators in their own experiential learning process. Participants went from their normal urban environments into the wilderness for 10 days at a time, spending at least 1 day and night completely alone in some wilderness spot "on solo." At times, while backpacking, the group was asked to walk silently so as not to be distracted from the wilderness sounds and images by conversation. In group discussions, participants were asked to talk about what they had observed about the wilderness and how they felt about being in the wilderness. Participants were also asked to write about the wilderness environment in journals. *What do these different activities have in common, and how can that commonality be expressed?*

We begin with several different ways of abstracting and labeling the underlying process:

- Experiencing the wilderness
- Learning about the wilderness
- Appreciating the wilderness
- Immersion in the environment
- Developing awareness of the environment
- Becoming conscious of the wilderness
- Developing sensitivity to the environment

Any of these phrases, each of which consists of some verb form (experiencing, learning, developing, etc.) and some noun form (wilderness, environment, etc.), captures some nuance of the process. The qualitative analyst works back and forth between the data (field notes and interviews) and his or her conception of what it is that needs to be expressed to find the most fitting language to describe the process. What language do people in the program use to describe what those activities and experiences have in common? What language comes closest to capturing the essence of this particular process? What level of generality or specificity will be most useful in separating out this particular set of things from other things? How do program participants and staff react to the different terms that could be used to describe the process?

It's not unusual during analysis to go through several different phrases before finally settling on the exact language that will go into a final report. In the Southwest Field Training Project, we began with the concept label "Experiencing the Wilderness." However, after several revisions, we finally described the process as "Developing Sensitivity to the Environment" because this broader label permitted us to include discussions and activities that were aimed at helping participants understand how they were affected by and acted in

their normal institutional environments. "Experiencing the wilderness" became a specific subprocess that was part of the more global process of "developing sensitivity to the environment." Program participants and staff played a major role in determining the final phrasing and description of this process.

Other processes identified as important in the implementation of the program were as follows:

- Encountering and managing stress
- Sharing in group settings
- Examining professional activities, needs, and commitments
- Assuming responsibility for articulating personal needs
- Exchanging professional ideas and resources
- Formally monitoring experiences, processes, changes, and impacts

As you struggle with finding the right language to communicate themes, patterns, and processes, keep in mind that there is no absolutely "right" way of stating what emerges from the analysis. There are only more and less useful ways of expressing what the data reveal.

Identifying and conceptualizing program outcomes and impacts can involve induction, deduction, abduction, and/or logical analysis. *Inductively*, the qualitative analyst looks for changes in participants, expressions of change, program ideology about outcomes and impacts, and ways that people in the program make distinctions between "those who are getting *it*" and "those who aren't getting *it*" (where *it* is the desired outcome). In highly individualized programs, the statements about change that emerge from program participants and staff may be global. Outcomes such as "personal growth," increased "awareness," and "insight into self" are difficult to operationalize and standardize. That is precisely the reason why qualitative methods are particularly appropriate for capturing and evaluating such outcomes. The task for the qualitative analyst, then, is to describe what actually happens to people in the program and what they say about what happens to them.

Logically (or abductively), constructing a process–outcomes matrix can suggest additional possibilities. That is, where data on both program processes and participant outcomes have been sorted, analysis can be deepened by organizing the data through a logical scheme that links program processes to participant outcomes. Such a logically derived scheme was used to organize the data in the Southwest Field Training Project. First, a classification scheme that described different types of outcomes was conceptualized:

a. Changes in knowledge

b. Changes in attitudes

c. Changes in feelings

d. Changes in behaviors

e. Changes in skills

These general themes provided the reader of the report with examples of and insights into the kinds of changes that were occurring and how those changes were perceived by participants to be related to specific program processes. I emphasize that the process–outcomes matrix is merely an organizing tool; the data from participants themselves and from field observations provide the actual linkages between processes and outcomes.

What was the relationship between the program process of "developing sensitivity to the environment" and these individual-level outcomes? Space permits only a few examples from the data.

> *Skills:* "Are you kidding? I learned how to survive without the comforts of civilization. I learned how to read the terrain ahead and pace myself. I learned how to carry a heavy load. I learned how to stay dry when it's raining. I learned how to tie a knot so that it doesn't come apart when pressure is applied. You think those are metaphors for skills I need in my work? You're damn right they are."

> *Attitudes:* "I think it's important to pay attention to the space you're in. I don't want to just keep going through my life oblivious to what's around me and how it affects me and how I affect it."

> *Feelings:* "Being out here, especially on solo, has given me confidence. I know I can handle a lot of things I didn't think I could handle."

> *Behaviors:* "I use my senses in a different way out here. In the city you get so you don't pay much attention to the noise and the sounds. But listening out here, I've also begun to listen more back there. I touch more things too, just to experience the different textures."

> *Knowledge:* "I know about how this place was formed, its history, the rock formations, the effects of the fires on the vegetation, where the river comes from, and where it goes."

A different way of thinking about organizing data around outcomes was to think of the different levels of impact: (a) effects at the individual level, (b) effects on the group, and (c) effects on the institutions from which participants came into the program. The staff hoped to have impacts at all of these levels. Thus, it also was possible to organize the data by looking at what themes emerged when program processes were crossed with levels of impact. How did "developing sensitivity to the environment" affect individuals? How did the process of "developing sensitivity to the environment" affect the group? What was the effect of "developing sensitivity to the environment" on the institutions to which participants returned after their wilderness experiences? The process–outcomes matrix thus becomes a way of asking questions of the data, an additional source of focus in looking for themes and patterns in the hundreds of pages of field notes and interview transcriptions. Exhibit 8.35, at the end of this chapter (pp. 643–649), presents an extended excerpt from the qualitative evaluation report.

A Three-Dimensional Qualitative Analysis Matrix

To study how schools used planning and evaluation processes, Campbell (1983) developed a 500-cell matrix (Exhibit 8.15) that begins (but just begins) to reach the outer limits of what one can do in a three-dimensional space. Campbell used this matrix to guide data collection and analysis in studying how the mandated, statewide educational planning, evaluation, and reporting system in Minnesota was being used. She examined five levels of use (high school, . . . , community, district), 10 components of the statewide project (planning, goal setting, . . . , student involvement), and 10 factors affecting utilization (personal factor, political factors, . . .). Exhibit 8.15 again illustrates matrix thinking for both data organization and analytical/conceptual purposes.

Miles et al. (2014) have provided a rich source of ideas and illustrations on how to use matrices in qualitative analysis. Their *Sourcebook* provides a variety of ideas for analytical approaches to qualitative data, including a variety of mapping and visual display techniques.

Synthesizing Qualitative Studies

We turn now to analysis across a different unit of analysis: synthesizing patterns, themes, and findings across qualitative studies where a completed study is the unit of analysis. In Chapter 5, Exhibit 5.8 (pp. 266–272), I introduced two purposeful sampling strategies that involve sampling completed studies as the unit of analysis: (1) *qualitative research synthesis* and (2) *systematic qualitative evaluation reviews*. One is research focused, the other evaluation focused. I distinguished these as different purposeful sampling strategies because they serve different purposes and select different kinds of qualitative studies for synthesis. (1) A qualitative

EXHIBIT 8.15   Conceptual Guide for Data Collection and Analysis: Utilization of Planning, Evaluation, and Reporting

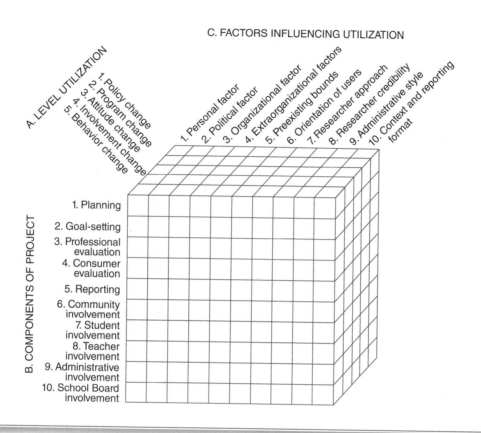

research synthesis selects qualitative studies to analyze for cross-cutting research findings and contributions to theory. For example, there have been a substantial number of separate and independent ethnographic studies of coming-of-age and initiation ceremonies across cultures. A synthesis involves analyzing and interpreting findings across those myriad studies. (2) In contrast, a systematic qualitative evaluation review seeks patterns across diverse qualitative evaluations to reach conclusions about *what is effective.*

Qualitative Research Synthesis

A qualitative research synthesis involves seeking patterns across and integrating different qualitative studies (Finlayson & Dixon, 2008; Hannes & Lockwood, 2012; Saini & Shlonsky, 2012). Which studies are included in the final synthesis will depend on what emerges as relevant and meaningful during the synthesis. "Assessing quality is also about examining how study findings fit (or do not fit) with the findings of other studies. How study findings fit with the

findings of other studies cannot be assessed until the synthesis is completed" (Harden & Gough, 2012, p. 160). In essence, quality criteria in a qualitative synthesis can be emergent.

Hannes and Lockwood (2012) have examined diverse approaches and frameworks for synthesizing qualitative research. The diversity of synthesis approaches reflects the diversity of qualitative methods and the variety of theoretical perspectives that inform qualitative inquiries (see Chapter 3). What all syntheses share in common is having to address at a minimum three separate processes: (1) identifying and aggregating studies to include; (2) analyzing patterns, themes, and findings across the studies; and (3) interpreting the results.

In one sense, each qualitative study is a case. Synthesis of different qualitative studies on the same subject is a form of cross-case analysis. Such a synthesis is much more than a literature review. Noblit and Hare (1988) describe synthesizing qualitative studies as "meta-ethnography," in which the challenge is to "retain the uniqueness and holism of accounts even as we synthesize them in the translations" (p. 7).

Systematic Qualitative Evaluation Reviews

Systematic qualitative evaluation serves a meta-evaluation function and is aimed at increasing confidence in actions to be taken to solve particular problems by synthesizing patterns of effectiveness across separate and independent evaluation studies. Such a systematic review seeks to identify, appraise, select, and synthesize *all* high-quality evaluation research evidence relevant to a particular arena of knowledge that is the basis for interventions (Funnell & Rogers, 2011, pp. 508–514).

Systematic reviews of randomized controlled trial (RCT) studies have been the basis for *evidence-based medicine*. The cases being synthesized are completed, usually published, studies. The process involves identifying all high-quality, peer-reviewed studies on a problem and synthesizing findings across those separate and diverse studies to reach conclusions about what is effective in dealing with the problem of concern, for example, female hormone replacement therapy or effective treatments for prostate cancer. Informing and setting guidelines for treatment is the instrumental use. The important new direction in systematic reviews is including qualitative evaluation studies (Gough et al., 2012; Wright, 2013). For example, the internationally prestigious Cochrane Collaboration Qualitative & Implementation Methods Group supports the synthesis of qualitative evidence and the integration of qualitative evidence with other evidence (mixed methods) in Cochrane intervention reviews on the effectiveness of health interventions. In effect, systematic reviews serve a meta-evaluation function.

Systematic Evaluation Reviews of Lessons Learned

Evaluators can synthesize lessons from a number of case studies to generate generic factors that contribute to program effectiveness—as, for example, Lisbeth Schorr (1988) did for poverty programs in her review and synthesis *Within Our Reach: Breaking the Cycle of Disadvantage*. Three decades ago, the U.S. Agency for International Development began commissioning lessons-learned synthesis studies on subjects such as irrigation (Steinberg, 1983), rural electrification (Wasserman & Davenport, 1983), food for peace (Rogers & Wallerstein, 1985), education development efforts (Warren, 1984), contraceptive social marketing (Binnendijk, 1986), agricultural policy analysis and planning (Tilney & Riordan, 1988), and agroforestry (Chew, 1989). In synthesizing separate evaluations to identify lessons learned, evaluators build a store of knowledge for future program development, more effective program implementation, and enlightened policy making.

QUALITATIVE RESEARCH SYNTHESES VERSUS SYSTEMATIC QUALITATIVE EVALUATION REVIEWS

I distinguish *qualitative research syntheses* from *systematic qualitative evaluation reviews* because they involve different purposeful sampling strategies that serve different purposes. You won't find this distinction elsewhere. I make the distinction, and believe it is worth making, because this book addresses both research and evaluation methods. Qualitative research synthesis serves research purposes—selecting qualitative studies to analyze for cross-cutting findings and contributions to theory. Systematic qualitative evaluation reviews serve evaluation purposes—analyzing diverse qualitative evaluations to reach conclusions about *patterns of effectiveness*. The criteria for what studies to include in each case will be different, as will the criteria for judging the final result: contribution to theory (qualitative research syntheses) versus contributions to practice (systematic evaluation reviews).

For scholarly inquiry, the qualitative research synthesis is a way to build theory through induction, deduction, interpretation, and integration. For evaluators, a qualitative systematic review can identify and extrapolate lessons learned to inform future program designs, identify effective intervention approaches, and build program theory.

The sample for synthesis studies usually consists of case studies with a common focus, for example, elementary education, health care for the elderly, and so on. However, one can also learn lessons about effective human intervention processes more generically by synthesizing case studies on quite different subjects. I synthesized three quite different qualitative evaluations conducted for The McKnight Foundation: (1) a major family housing effort, (2) a downtown development endeavor, and (3) a graduate fellowship program for minorities. Before undertaking the synthesis, I knew nothing about these programs, nor did I approach them with any particular preconceptions. I was not looking for any specific similarities, and none were suggested to me by either McKnight or program staff. The results were intended to provide insights into The McKnight Foundation's operating philosophy and strategies *as exemplified in practice by real operating programs*. Independent evaluations of each program had already been conducted and presented to The McKnight Foundation, showing that these programs had successfully attained and exceeded the intended outcomes. But why were they successful? That was the intriguing and complex question on which the synthesis study focused.

The synthesis design included fieldwork (interviews with key players and site visits to each project) as well as an extensive review of their independent evaluations. I identified common success factors that were manifest in all three projects. Those were illuminating but not surprising. The real contribution of the synthesis was in how the success factors fit together, an unanticipated pattern that deepened the implications for understanding effective philanthropy.

The 10 success factors common to all three programs were as follows:

1. *Strong leadership*, developed, engaged, and supported throughout the initiative

2. A *sizeable amount of money* ($15 million), able to attract attention and generate support

3. Effective use of *leverage* at every level of program operation (McKnight insisted on sizable matching funds and use of local in-kind resources from participating universities.)

4. A *long-term perspective* on and commitment to a sustainable program with cumulative impact over time—in perpetuity (Support for the programs was converted to an endowment.)

5. A carefully melded *public–private partnership*

6. A program based on a *vision* made real through a carefully designed model that was true to the vision

7. Taking the time and effort to carefully *plan* in a process that generated *broad-based community and political support* throughout the state

8. The careful structuring of *local board control* so that responsibility and ownership resided among key influentials

9. Taking advantage of the right *timing* and climate for this kind of program

10. Clear *accountability and evaluation* so that problems could be corrected and accomplishments could be recognized

While each of these factors provided insight into an important element of effective philanthropic programming, the unanticipated pattern was how these factors fit together to form a *constellation of excellence*. I found that I couldn't prioritize these factors because they worked together in such a way that no one factor was primary or sufficient; rather, each made a critical contribution to an integrated, effectively functioning whole. The lesson that emerged for effective philanthropy was not a series of steps to follow but rather a mosaic to create; that is, effective

philanthropy appears to be a process of matching and integrating elements so that the pieces fit together in a meaningful and comprehensive way as a solution to complex problems. This means matching people with resources; bringing vision and values to bear on problems; and nurturing partnerships through leverage, careful planning, community involvement, and shared commitments—*and doing all these things in mutually reinforcing ways*. The challenge for effective philanthropy, then, is putting all the pieces and factors together to support integrated, holistic, and high-impact efforts and results—and to do so creatively (Storm & Vitt, 2000, pp. 115–116).

Another example of a major synthesis focused on evaluation of the 2005 Paris Declaration on Aid Effectiveness, endorsed by more than 150 countries and international organizations. An independent evaluation examined what difference, if any, the Paris Declaration made to development processes and results (Wood et al., 2011). The final report was a synthesis of case studies done in 22 developing countries and 18 donor agencies. The synthesis identified the factors that contributed to international aid reform and barriers to more effective aid. (For the lessons learned, see Dabelstein & Patton, 2013a.)

Qualitative synthesis has become a major and important approach to making sense of multiple and diverse qualitative studies. It is possible only because there are now many qualitative studies in diverse fields of interest available for synthesis. Synthesis findings are elevating the contribution of qualitative results to both research and evaluation.

REALIST SYNTHESIS

SIDEBAR

"Realist synthesis might best be thought of as a way of assembling rocks, or nonstandardized pieces of knowledge" (Funnell & Rogers, 2011, p. 514). Developed by British sociologist and evaluator Ray Pawson (2013), realist synthesis selects and integrates any quality evidence on a topic, including experiments, quasi-experimental studies, and case studies. "Quality is not assessed with reference to a hierarchy of research designs for the entire study but by assessing whether threats to validity have been adequately addressed in terms of the specific piece of evidence being used" (p. 515). Realist synthesis uses purposeful sampling of diverse kinds of evidence available, both quantitative and qualitative, to develop, refine, and test theories about how, for whom, and in what contexts policies will be effective . . . [and to] identify causal mechanisms that operate only in particular contexts. Realist synthesis is inherently a process of building, testing, and refining program theory. (p. 515)

MODULE

70

Interpreting Findings, Determining Substantive Significance, Elucidating Phenomenological Essence, and Hermeneutic Interpretation

> Simply observing and interviewing do not ensure that the research is qualitative; the qualitative researcher must also interpret the beliefs and behaviors of participants.
>
> —Valerie J. Janesick (2000, p. 387)

Interpreting for Meaning

Qualitative interpretation begins with elucidating meanings. The analyst examines a story, a case study, a set of interviews, or a collection of field notes and asks, "What does this mean? What does this tell me about the nature of the phenomenon of interest?" In asking these questions, the analyst works back and forth between the data or story (the evidence) and his or her own perspective and understandings to make sense of the evidence. Both the evidence and the perspective brought to bear on the evidence need to be elucidated in this choreography in the search for meaning. Alternative interpretations are tried and tested against the data.

Interpretation, by definition, involves going beyond the descriptive data. Interpretation means attaching significance to what was found, making sense of findings, offering explanations, drawing conclusions, extrapolating lessons, making inferences, considering meanings, and otherwise imposing order on an unruly but surely patterned world. The rigors of interpretation and bringing data to bear on explanations include dealing with rival explanations, accounting for disconfirming cases, and accounting for data irregularities as part of testing the viability of an interpretation. All of this is expected—and appropriate—as long as the researcher owns the interpretation and makes clear the difference between description and interpretation. A good example is Reid Zimmerman's (2014) description and interpretation of the *seven deadly sayings of a nonprofit leader.*

Schlechty and Noblit (1982) concluded that an interpretation may take one of three forms:

1. Making the obvious obvious

2. Making the obvious dubious

3. Making the hidden obvious

This captures rather succinctly what research colleagues, policymakers, and evaluation stakeholders expect: (1) confirm what we know that is supported by data, (2) disabuse us of misconceptions, and (3) illuminate important things that we didn't know but should know. Accomplish these three things, and those interested in the findings can take it from there.

Explaining findings is an interpretive process. For example, when we analyzed follow-up interviews with participants who had gone through intensive community leadership training, we found a variety of expressions of uncertainty about what they should do with their training. In the final day of a six-day retreat, after learning how to assess community needs, work with diverse groups, communicate clearly, empower people to action, and plan for change, they were cautioned to go easy in transitioning back to their communities and to take their time in building community connections before taking action. What program staff meant as a last-day warning about not returning to the community as a bull in a china shop and charging ahead destructively had, in fact, paralyzed the participants and made them afraid to take any action at all. The program, which intended to position participants for action, had inadvertently left graduates in "action paralysis" for fear of making mistakes. The *meaning-laden phrase* "action paralysis" emerged from the data analysis through interpretation. No one used that specific phase. Rather, we interpreted *action paralysis* as the essence of what the interviewees were reporting through a haze of uncertainties, ambiguities, worried musings, and wait-and-see-before-acting reflections.

Interpreting Findings at Different Levels of Analysis

In the early 1990s, The McKnight Foundation in Minnesota invested $13 million in an innovative initiative titled The Aid to Families in Poverty Program. The initiative funded 34 programs using a variety of strategies. Our evaluation team conducted evaluations

GENERAL THEORY OF INTERPRETATION

Interpretation is dependent on value, and judgments in all domains requiring interpretation are necessarily value judgments of some kind or other—though they start out from the descriptive facts.

Whenever we are faced with a normative system like law, whose point is to govern conduct, we cannot understand it simply as a pattern of behavior. We have to see it as an internalized set of standards and principles that the participants take to justify their behavior. But we need not share that point of view in order to understand it, even though we must rely on our own capacity for value judgments when we interpret how others see things as right that we believe to be wrong, and vice versa. We will not understand a bad system unless we see how its participants see it as good.

—Dworkin's General Theory of Interpretation
From *Ronald Dworkin: The Moral Quest* (Nagel, 2013, p. 56)

at the program level, the overall initiative level (synthesis of findings across all 34 programs), and the level of the broader policy environment that set the context for antipoverty interventions in Minnesota and the nation. The qualitative synthesis team interviewed program staff, conducted site visits, reviewed individual program evaluation reports, met with foundation program officers, and reviewed the national literature about poverty issues and programs. Here are examples of major findings and how we interpreted them. I invite you to pay special attention to how the findings and interpretations change at different levels of analysis for different units of analysis.

1. *Outcomes for Families in Poverty*

   Finding: Program participants and staff emphasized the importance of helping families move out of crisis. Taking first steps and arresting decline were consistently reported as important outcomes. A common early outcome was *increased intentionality*—helping families in poverty come up with a plan, a sense of direction, and a commitment to making progress.

   Interpretation: Funders, policymakers, and evaluators place heavy emphasis on achieving long-term outcomes—getting families out of poverty. In doing so, they undervalue the huge amount of work it takes to establish trust with families in

need, help stabilize families, and begin the process toward long-term outcomes. Early indicators of progress foreshadow longer-term outcomes and should be reported and valued.

2. *Characteristics of Effective Program Staff*

   Finding: Effective staff approach program participants on a case-by-case basis. Their approach is *respectful and individualized.* They recognize that progress occurs in varying ways and at different rates, and what may be limited progress for one family may represent enormous strides for another. Moreover, effective staff are highly *responsive* to individual participants' situations, needs, capabilities, interests, and family context. In being responsive and respectful, they work to raise hopes and empower participants by helping them make concrete, intentional changes.

   Interpretation: Effective staff are the key to effective programs and achieving desired program outcomes for families in poverty. It wasn't the program model or approach that made the difference. More important than the conceptual model or theory of change being implemented was how staff interacted with program participants. This has implications for staff recruitment, training, support, and performance evaluation.

3. *Characteristics of Effective Programs*

   Finding: Effective programs support staff responsiveness by being flexible and giving staff discretion to take whatever actions assist participants to climb out of poverty. Flexible, responsive programs affect the larger systems of which they are a part by pushing against boundaries, arrangements, rules, procedures, and attitudes that hinder their capability to work flexibly and responsively—and therefore effectively—with participants.

   Interpretation: Patterns of effectiveness cut across different levels of operation and impact and showed up in what we would call the *program culture.* How people are treated affects how they treat others. How staff are treated affects how they treat program participants. Responsiveness reinforces responsiveness, flexibility supports individualization, and empowerment breeds empowerment. Program directors, professional staff, and organizational administrators will often need training and technical assistance in setting up and working with flexible, responsive approaches.

4. *Philanthropic Foundation Lessons*

   Finding: The foundation chose the initiative's name without consultation and review in the community. Our interviews found that many program staff and participants reacted negatively to the initiative's name: Aid to Families in Poverty Program.

Interpretation: Language matters to people. What an initiative is called sends messages. A name for the initiative that conveyed hope and strength rather than deficiency would have been received more positively. Failure to consult with people outside the foundation about the name increased the risk that the initiative's title would inadvertently carry negative connotations (Patton, 1993).

Substantive Significance

In lieu of statistical significance, qualitative findings are judged by their substantive significance. The analyst makes an argument for substantive significance in presenting findings and conclusions, but readers and users of the analysis will make their own value judgments about significance. In determining substantive significance, the analyst addresses these kinds of questions:

- To what extent and in what ways do the findings increase and deepen understanding of the phenomenon studied (*verstehen*)?
- To what extent are the findings useful for their intended purpose, for example, contributing to theory, informing policy, improving a program, informing decision making about some action, or problem solving in action research?
- To what extent are the findings consistent with other knowledge? A finding supported by and supportive of other work has confirmatory significance. A finding that breaks new ground has discovery or innovative significance.
- How solid, coherent, and consistent is the evidence in support of the findings? Triangulation, for example, can be used in determining the strength of evidence in support of a finding.

Examples of Substantive Significance

Here are three examples of findings that, when interpreted, can be judged as substantively significant:

1. Case studies of 30 postdoctoral fellowship recipients found only 2 whose lives had been disrupted during the fellowship. Those 2 permanently lost the fellowship funds and were deemed failures by the fellowship's sponsor and the fellows' institutions. The case studies showed that the failures were due to the inflexibility of the funders in having no process for allowing a temporary sabbatical from the fellowship under conditions of sudden hardship. The consequences for the fellows were dramatic and long term, while a solution was readily at hand that could alleviate such dire consequences, which were likely to occur again in the future.

2. During a polio immunization campaign in India, a community of Muslim mothers heard a rumor that the immunization was a Hindu plot to sterilize Muslim children. So they hid their children from the vaccinators. A year later, those children were the source of an outbreak of polio. The numbers who resisted were small, but the consequences were great. The implication was that immunization campaigns need to be inquiring into community perceptions in real time so as to intervene and correct misperceptions in real time.

3. An international funder spent a large sum to build typhoon shelters in Bangladesh in low-lying areas near the ocean, where thousands of poor people were especially vulnerable to storms. When a typhoon hit, the shelters went largely unused because animals were prohibited and the resources of these poor people were their animals, which they would not abandon. The funding agency had been told of this potential problem when it was identified from a few key informant interviews, but the agency dismissed the findings because of the small sample size.

Determining substantive significance requires critical thinking about the broader consequences of findings. Exhibit 8.16 provides another example of substantive significance.

Interpretation Requires Both Critical Thinking and Creativity

Identifying patterns, themes, and categories involves using both creative and critical faculties in making carefully considered judgments about what is meaningful and substantively significant in the data. Since as a qualitative analyst you do not have a statistical test to help tell you when an observation or pattern is significant, you must rely first on your own sense making, understandings, intelligence, experience, and judgment; second, you should take seriously the responses of those who were studied or who participated in the inquiry about what they have reported to you as meaningful and significant; and third, you should consider the responses and reactions of those who read and review the results. Where all three—the qualitative analyst, those studied,

EXHIBIT 8.16   Substantive Significance Example: *Minimally Disruptive Medicine*

Chronic disease requires ongoing, lifetime management. This means that a patient must find a way to fit medical and lab appointments, exercise, medications, and dietary changes into a life already busy with family and work. Victor Montori, a professor of medicine at Mayo Clinic, Carl May, a professor of medical sociology at Newcastle University, and Francis Mair, a professor of primary care research at University of Glasgow, decided to study the burden of treatment by interviewing patients with multiple chronic comorbidities or cognitive impairment—two groups that are often excluded from studies examining compliance.

Case Study Examples

- A man who in the previous two years had visited specialist clinics for appointments, tests, and treatment 54 times (the equivalent of a full day every two weeks)
- A woman whose doctors had prescribed medications to be taken at 11 separate times during the day and was having trouble managing this (Montori, 2014)

Though initially from a small purposeful sample, the findings were sufficiently substantive to inspire conceptualization of a new category of health intervention: *Minimally Disruptive Medicine.* Minimally disruptive medicine is minimally disruptive not because it is minimal but because it is designed to fit naturally into a patient's life and be manageable on an ongoing basis.

and reviewers—agree, one has *consensual validation* of the substantive significance of the findings. Where disagreements emerge, which is more usual, you get a more interesting life and the joys of debate.

Determining substantive significance involves distinguishing signal from noise, which involves risking two kinds of errors. First, the analyst may decide that something is not a signal—that is, is not significant—when in fact it is, or second, and conversely, the analyst may attribute significance to something that is meaningless (just noise). The Halcolm story presented as a graphic comic at the end of Chapter 6 (pp. 418–419) is worth repeating here to illustrate this challenge of making judgments about what is really significant.

Halcolm was approached by a woman who handed him something. Without hesitation, Halcolm returned the object to the woman. The many young disciples

who followed Halcolm to learn his wisdom began arguing among themselves about the special meaning of this interchange. A variety of interpretations were offered.

When Halcolm heard of the argument among his young followers, he called them together and asked each one to report on the significance of what they had observed. They offered a variety of interpretations. When they had finished, he said, "The real purpose of the exchange was to enable me to show you that you are not yet sufficiently masters of observation to know when you have witnessed a meaningless interaction."

INTEROCULAR SIGNIFICANCE

SIDEBAR

If we are interested in real significance, we ignore little differences . . . . We ignore them because, although they are very likely real, they are very unlikely to hold up in replications. Fred Mosteller, the great applied statistician, was fond of saying that he did not care much for statistically significant differences, he was more interested in *interocular differences*, the differences that hit us between the eyes. (Scriven, 1993, p. 71)

Phenomenology as an Interpretative Framework: Elucidating Essence

> Phenomenology asks for the very nature of a phenomenon, for that which makes a some-"thing" what it is—and without which it could not be what it is.
>
> —Van Manen (1990, p. 10)

Phenomenology as a qualitative theoretical framework was discussed at length in Chapter 3 (pp. 115–118). In this module, we're going to focus on phenomenological analysis as an interpretative framework. Phenomenological analysis seeks to grasp and elucidate the meaning, structure, and essence of the lived experience of a phenomenon for a person or group of people. Before I present the steps of one particular approach to phenomenological analysis, it is important to note that phenomenology has taken on a number of meanings, has a number of forms, and encompasses varying traditions, including transcendental phenomenology, existential phenomenology, and hermeneutic phenomenology (Schwandt, 2007). Moustakas (1994) further distinguishes empirical phenomenology from transcendental phenomenology. Gubrium and Holstein (2000) add the label social phenomenology. Van Manen (1990) prefers "hermeneutical phenomenological reflection." Sonnemann (1954) introduced the term *phenomenography* to label phenomenological investigation aimed at "a descriptive recording of immediate subjective experience as reported" (p. 344). Harper (2000) talks of looking at images through "the phenomenological mode"—that is, from the perspective of the self: "From the phenomenological perspective, photographs express the artistic, emotional, or experiential intent of the photographer" (p. 727). To this confusion of terminology is added the difficulty of distinguishing phenomenological philosophy from phenomenological methods and phenomenological analysis, all of which increases the tensions and contradictions in qualitative inquiry (Gergen & Gergen, 2000).

> The use of the term phenomenology in contemporary versions of qualitative inquiry in North America tends to reflect a subjectivist, existentialist, and non-critical emphasis not present in the Continental tradition represented in the work of Husserl and Heidegger. The latter viewed the phenomenological project, so to speak, as an effort to get beneath or behind subjective experience to reveal the genuine, objective nature of things, and as a critique of both taken-for-granted meanings and subjectivism. Phenomenology, as it is commonly discussed in accounts of qualitative research, emphasizes just the opposite: It aims to identify and describe the subjective experiences of respondents. It is a matter of studying everyday experience from the point of view of the subject, and it shuns critical evaluation of forms of social life. (Schwandt, 2001, p. 192)

Phenomenological analysis involves and emphasizes different elements depending on which type of phenomenology you are using as a framework. I have chosen to focus on the phenomenological approach to analysis taken by Clark Moustakas, founder of The Center for Humanistic Studies (Detroit, Michigan). More than most, he has focused on the analytical process itself (Douglass & Moustakas, 1985; Moustakas, 1961, 1988, 1990, 1994, 1995). As we go deeper into the perspective and language of phenomenological analysis, let me warn you that the terminology and distinctions can be hard to grasp at first. But don't skip over them lightly. These distinctions constitute windows into the world of phenomenological analysis.

They matter. See if you can figure out why they matter. If you can, you will have grasped phenomenological interpretation.

Consciousness, Intentionality, Nomea, and Noesis

Husserl's transcendental phenomenology is intimately bound up in the concept of intentionality. In Aristotelian philosophy the term *intention* indicates the orientation of the mind to its object; the object exists in the mind in an intentional way....

Intentionality refers to consciousness, to the internal experience of being conscious of something; thus the act of consciousness and the object of consciousness are intentionally related. Included in understanding of consciousness are important background factors such as stirrings of pleasure, shapings of judgment, or incipient wishes. Knowledge of intentionality requires that we be present to ourselves and things in the world, that we recognize that self and world are inseparable components of meaning.

Consider the experience of joy on witnessing a beautiful landscape. The landscape is the *matter*. The landscape is also the object of the intentional act, for example, its perception in consciousness. The matter enables the landscape to become manifest as an object rather than merely exist in consciousness.

The *interpretive form* is the perception that enables the landscape to appear; thus the landscape is self-given; my perception creates it and enables it to exist in my consciousness. The *objectifying quality* is the actuality of the landscape's existence, as such, while the *non-objectifying quality* is a joyful feeling evoked in me by the landscape.

Every intentionality is composed of a *nomea* and *noesis*. The *nomea* is not the real object but the phenomenon, not the tree but the appearance of the tree. The object that appears in perception varies in terms of when it is perceived, from what angle, with what background of experience, with what orientation of wishing, willing, or judging, always from the vantage point of the perceiving individual.... The tree is out there present in time and space while the perception of the tree is in consciousness....

Every intentional experience is also noetic....

In considering the *nomea-noesis* correlate, ... the "perceived as such" is the *nomea*; the "perfect self-evidence" is the *noesis*. Their relationship constitutes the intentionality of consciousness. For every nomea, there is a noesis; for every noesis, there is a nomea. On the noematic side is the uncovering and explication, the unfolding and becoming distinct, the clearing of what is actually presented in consciousness. On the noetic side is an explication of the intentional processes themselves. . . .

Summarizing the challenges of intentionality, the following processes stand out:

1. explicating the sense in which our experiences are directed;

2. discerning the features of consciousness that are essential for the individuation of objects (real or imaginary) that are before us in consciousness (Noema);

3. explicating how beliefs about such objects (real or imaginary) may be acquired, how it is that we are experiencing what we are experiencing (Noesis); and

4. integrating the noematic and noetic correlates of intentionality into meanings and essences of experience. (Moustakas, 1994, pp. 28–32)

Epoche

If those are the challenges, what are the steps for meeting them? The first step in phenomenological analysis is called *epoche*.

> Epoche is a Greek word meaning to refrain from judgment, to abstain from or stay away from the everyday, ordinary way of perceiving things. In a natural attitude we hold knowledge judgmentally; we presuppose that what we perceive in nature is actually there and remains there as we perceive it. In contrast, Epoche requires a new way of looking at things, a way that requires that we learn *to see* what stands.

> In the Epoche, the everyday understandings, judgments, and knowings are set aside, and the phenomena are revisited, visually, naively, in a wide-open sense, from the vantage point of a pure or transcendental ego. (Moustakas, 1994, p. 33)

In taking on the perspective of *epoche*, the researcher looks inside to become aware of personal bias, eliminate personal involvement with the subject material—that is, eliminate, or at least gain clarity about, preconceptions. Rigor is reinforced by a "phenomenological attitude shift" accomplished through *epoche*.

The researcher examines the phenomenon by attaining an attitudinal shift. This shift is known as the phenomenological attitude. This attitude consists of a different way of looking at the investigated experience. By moving beyond the natural attitude or the more prosaic way phenomena are imbued with meaning, experience gains a deeper meaning. This takes place by gaining access to the constituent elements of the phenomenon and leads to a description of the unique qualities and components that make this phenomenon what it is. In attaining this shift to the phenomenological attitude, *Epoche* is a primary and necessary phenomenological procedure.

> *Epoche* is a process that the researcher engages in to remove, or at least become aware of prejudices, viewpoints or assumptions regarding the phenomenon under investigation. *Epoche* helps enable the researcher to investigate the phenomenon from a fresh and open view point without prejudgment or imposing meaning too soon. This suspension of judgment is critical in phenomenological investigation and requires the setting aside of the researcher's personal viewpoint in order to see the experience for itself. (Katz, 1987, pp. 36–37)

According to Ihde (1977), "*Epoche* requires that looking precede judgment and that judgment of what is 'real' or 'most real' be suspended until all the evidence (or at least sufficient evidence) is in" (p. 36). As such, *epoche* is an ongoing analytical process rather than a single fixed event. The process of *epoche* epitomizes the data-based, evidential, and empirical (vs. empiricist) research orientation of phenomenology.

Phenomenological Reduction, Bracketing, and Theme Analysis

Following *epoche*, the second step is phenomenological reduction. In this analytical process, the researcher brackets out the world and presuppositions to identify the data in pure form, uncontaminated by extraneous intrusions.

> Bracketing is Husserl's (1913) term. In bracketing, the researcher holds the phenomenon up for serious inspection. It is taken out of the world where it occurs. It is taken apart and dissected. Its elements and essential structures are uncovered, defined, and analyzed. It is treated as a text or a document; that is, as an instance of the phenomenon that is being studied. It is not interpreted in terms of the standard meanings given to it by the existing literature. Those preconceptions, which were isolated in the deconstruction phase, are suspended and put aside during bracketing. In

bracketing, the subject matter is confronted, as much as possible, on its own terms. Bracketing involves the following steps:

(1) Locate within the personal experience, or self-story, key phrases and statements that speak directly to the phenomenon in question.

(2) Interpret the meanings of these phrases, as an informed reader.

(3) Obtain the subject's interpretations of these phrases, if possible.

(4) Inspect these meanings for what they reveal about the essential, recurring features of the phenomenon being studied.

(5) Offer a tentative statement, or definition, of the phenomenon in terms of the essential recurring features identified in step 4. (Denzin, 1989b, pp. 55–56)

Imaginative Variation and Textural Portrayal

Once the data are bracketed, all aspects of the data are treated with equal value—that is, the data are "horizontalized." The data are spread out for examination, with all elements and perspectives having equal weight. The data are then organized into meaningful clusters. Then, the analyst undertakes a delimitation process whereby irrelevant, repetitive, or overlapping data are eliminated. The researcher then identifies the invariant themes within the data to perform an *imaginative variation* on each theme. This can be likened to moving around a statue to see it from differing views. Through imaginative variation, the researcher develops enhanced or expanded versions of the invariant themes.

Using these enhanced or expanded versions of the invariant themes, the researcher moves to the textural portrayal of each theme—a description of an experience that doesn't contain that experience (i.e., the feelings of vulnerability expressed by rape victims). The textural portrayal is an abstraction of the experience that provides content and illustration but not yet essence.

Phenomenological analysis then involves a "structural description" that contains the "bones" of the experience for the whole group of people studied, "a way of understanding *how* the co-researchers as a group experience *what* they experience" (Moustakas, 1994, p. 142). In the structural synthesis, the phenomenologist looks beneath the affect inherent in the experience to deeper meanings for the individuals who, together, make up the group.

Synthesis and Essence

The final step requires "an integration of the composite textual and composite structural descriptions, providing a synthesis of the meanings and essences of the experience" (Moustakas, 1994, p. 144). In summary, the primary steps of the Moustakas transcendental phenomenological model are as follows: (a) *Epoche*, (b) phenomenological reduction, (c) imaginative variation, and (d) synthesis of texture and structure. Other detailed analytical techniques are used within each of these stages (see Moustakas, 1994, pp. 180–181).

Heuristic Inquiry

According to Moustakas (1990), heuristic inquiry applies phenomenological analysis to one's own experience. As such, it involves a somewhat different, highly personal analytical process. Moustakas describes five basic phases in the heuristic process of phenomenological analysis: (1) immersion, (2) incubation, (3) illumination, (4) explication, and (5) creative synthesis.

Immersion is the stage of steeping oneself in all that is—of contacting the texture, tone, mood, range, and content of the experience. This state "requires my full presence, to savor, appreciate, smell, touch, taste, feel, know without concrete goal or purpose" (Moustakas, 1988, p. 56). The researcher's total life and being are centered on the experience. He or she becomes totally involved in the world of the experience—questioning, mediating, dialoging, daydreaming, and indwelling.

The second state, *incubation*, is a time of "quiet contemplation" where the researcher waits, allowing space for awareness, intuitive or tacit insights, and understanding. In the incubation stage, the researcher deliberately withdraws, permitting meaning and awareness to awaken in their own time. One "must permit the glimmerings and awakenings to form, allow the birth of understanding to take place in its own readiness and completeness" (Moustakas, 1988, p. 50). This stage leads the way toward a clear and profound awareness of the experience and its meanings.

In the phase of *illumination*, expanding awareness and deepening meaning bring new clarity of knowing. Critical textures and structures are revealed so that the experience is known in all of its essential parameters. The experience takes on a vividness, and understanding grows. Themes and patterns emerge, forming clusters and parallels. New life and new visions appear along with new discoveries.

In the *explication* phase, other dimensions of meanings are added. This phase involves a full unfolding of the experience. Through focusing, self-dialogue, and reflection, the experience is depicted and further delineated. New connections are made through further explorations into universal elements and primary themes of the experience. The heuristic analyst refines emergent patterns and discovered relationships.

> It is an organization of the data for oneself, a clarification of patterns for oneself, a conceptualization of concrete subjective experience for oneself, and integration of generic meanings for oneself, and a refinement of all these results for *oneself*. (Craig, 1978, p. 52)

What emerges is a depiction of the experience and a portrayal of the individuals who participated in the study. The researcher is ready now to communicate findings in a creative and meaningful way. *Creative synthesis* is the bringing together of the pieces that have emerged into a total experience, showing patterns and relationships. This phase points the way for new perspectives and meanings, a new vision of the experience. The fundamental richness of the experience and of the experiencing participants is captured and communicated in a personal and creative way. In heuristic analysis, the insights and experiences of the analyst are primary, including drawing on "tacit" knowledge that is deeply internal (Polanyi, 1967).

These brief outlines of phenomenological and heuristic analysis can do no more than hint at the in-depth *living with the data* that is intended. The purpose of this kind of disciplined analysis is to elucidate the essence of the experience of a phenomenon for an individual or a group. The analytical vocabulary of phenomenological analysis is initially alien, and potentially alienating, until the researcher becomes immersed in the holistic perspective, rigorous discipline, and paradigmatic parameters of phenomenology. As much as anything, this outline reveals the difficulty of defining and sequencing the internal intellectual processes involved in qualitative analysis more generally.

Phenomenology seeks to describe, elucidate, and interpret human experience. The product is a deep understanding of the nature and essence of the phenomenon studied. We turn now to a quite different analytical priority: not just describing and interpreting but also *explaining the world*.

The Hermeneutic Circle and Interpretation

> Hermes was messenger to the Greek gods.... Himself the god of travel, commerce, invention, eloquence, cunning, and thievery, he acquired very early in his life a reputation for being a precocious trickster. (On the day he was born he stole Apollo's cattle, invented the

Heuristic Inquiry Reactivity

lyre, and made fire.) His duties as messenger included conducting the souls of the dead to Hades, warning Aeneas to go to Italy, where he founded the Roman race, and commanding the nymph Calypso to send Odysseus away on a raft, despite her love for him. With good reason his name is celebrated in the term "hermeneutics," which refers to the business of interpreting.... Since we don't have a godly messenger available to us, we have to interpret things for ourselves. (Packer & Addison, 1989, p. 1)

Hermeneutics focuses on interpreting something of interest, traditionally a text or work of art; but in the larger context of qualitative inquiry, it has also come to include interpreting interviews and observed actions. The emphasis throughout concerns the nature of interpretation, and various philosophers have approached the matter differently, some arguing that there is no method of interpretation per se because everything involves interpretation (Schwandt, 2000, 2001). For our purposes here, *the hermeneutic circle*, as an analytical process aimed at enhancing understanding, offers a particular emphasis in qualitative analysis, namely, relating parts to wholes and wholes to parts.

> Construing the meaning of the whole meant making sense of the parts, and grasping the meaning of the parts depended on having some sense of the whole.... The hermeneutic circle indicates a necessary condition of interpretation, but the circularity of the process is only temporary—eventually the interpreter can come to something approximating a complete and correct understanding of the meaning of a text in which whole and parts are related in perfect harmony. Said somewhat differently, the interpreter can, in time, get outside of or escape the hermeneutic circle in discovering the "true" meaning of the text. (Schwandt, 2001, p. 112)

The *method* involves playing the strange and unfamiliar parts of an action, text, or utterance off against the integrity of the action, narrative, or utterance as a whole until the meaning of the strange passages and the meaning of the whole are worked out or accounted for. (Thus, for example, to understand the meaning of the first few lines of a poem, I must have a grasp of the overall meaning of the poem, and vice versa.) In this process of applying the hermeneutic method, the interpreter's self-understanding and socio-historical location neither affects nor is affected by the effort to interpret the meaning of the text or utterance. In fact, in applying the method, the interpreter abides by a set of procedural rules that help insure that the interpreter's historical situation does not distort the bid to uncover the actual meaning embedded in the text, act, or utterance, thereby helping to insure the objectivity of the interpretation (Schwandt, 2001, p. 114).

The circularity and universality of hermeneutics (every interpretation is layered in and dependent on other interpretations, like a series of dolls that fit one inside the other, and then another and another) pose for the qualitative analyst the problem of where to begin. How and where do you break into the hermeneutic circle of interpretation? Packer and Addison (1989), in adapting the hermeneutic circle as an inquiry approach for psychology, suggest beginning with "practical understanding":

> Practical understanding is not an origin for knowledge in the sense of a foundation; it is, instead, the starting place for interpretation. Interpretive inquiry begins not from an absolute origin of unquestioned data or totally consistent logic, but at a place delineated by our everyday participatory understanding of people and events. We begin there in full awareness that this understanding is corrigible, and that it is partial in the twin senses of being incomplete and perspectival. Understanding is always moving forward. Practical activity projects itself forward into the world from its starting place, and shows us the entities we are home among. This means that neither commonsense nor scientific knowledge can be traced back to an origin, a foundation. . . . (p. 23)

> The circularity of understanding, then, is that we understand in terms of what we already know. But the circularity is not, Heidigger argues, a "vicious" one where we simply confirm our prejudices, it is an "essential" one without which there would be no understanding at all. And the circle is complete; there is accommodation as well as assimilation. If we are persevering and open, our attention will be drawn to the projective character of our understanding

and—in the backward arc, the movements of return—we gain an increased appreciation of what the forestructure involves, and where it might best be changed. . . . (p. 34)

> Hermeneutic inquiry is not oriented toward a grand design. Any final construction that would be a resting point for scientific inquiry represents an illusion that must be resisted. If all knowledge were to be at last collected in some gigantic encyclopedia this would mark not the triumph of science so much as the loss of our human ability to encounter new concerns and uncover fresh puzzles. So although hermeneutic inquiry proceeds from a starting place, a self-consciously interpretive approach to scientific investigation does not come to an end at some final resting place, but works instead to keep discussion open and alive, to keep inquiry under way. (p. 35)

At a general level and in a global way, hermeneutics reminds us of the interpretive core of qualitative inquiry, the importance of context and the dynamic whole–part interrelations of a holistic perspective. At a specific level and in a particularistic way, the hermeneutic circle offers a process for formally engaging in interpretation.

Theory-Driven Qualitative Findings

This module has looked in some depth at how two theory-based inquiry perspectives, phenomenology and hermeneutics, prescribe different analytical processes and produce different kinds of findings. To further emphasize this point, Exhibit 8.17 highlights how 10 different theoretical perspectives yield *different kinds of findings* due to the distinct focus of inquiry embedded in each theoretical perspective. Ethnography directs the inquiry to elucidate the nature of culture, whether for tribes, organizations, or programs. Social constructionism captures mental models and worldviews. Realism documents contextually operative causal mechanisms. Theories of change differentiate patterns of change and change trajectories. Systems theory illuminates interrelationships. Complexity theory invites studies of adaptation patterns, emergence, and simple rules. Phenomenology aims to elucidate the essence of the phenomenon studied. Grounded theory inquiries yield theoretical propositions and hypotheses. Hermeneutics interprets the meanings of texts. Pragmatism enquires into how things work and how they are used. Thus, operating within a particular theoretical orientation (see Chapter 3) provides focus, inquiry processes, and analytical procedures and yields certain kinds of findings that are a matter of core interest and priority for the community of inquiry engaged in studying the world through the lenses offered by that shared theoretical perspective. Exhibit 8.17 cites research and evaluation examples illuminating the differences in types of findings among these theoretical orientations.

EXHIBIT 8.17 Findings Yielded by Various Theoretical Perspectives With Research and Evaluation Examples

Different theoretical perspectives yield *different kinds of findings* due to the distinct focus of inquiry embedded in each theoretical perspective. Conducting research or evaluation within a particular theoretical orientation provides focus, inquiry processes, and analytical procedures and yields certain kinds of findings that are a matter of core interest and priority for the community of inquiry engaged in studying the world through the lenses offered by that shared theoretical perspective. This exhibit cites research and evaluation examples illuminating the differences in types of findings among 10 theoretical orientations.

FINDINGS YIELDED BY VARIOUS THEORETICAL PERSPECTIVES	RESEARCH EXAMPLES	PROGRAM EVALUATION EXAMPLES
1. *Ethnography.* Cultural patterns	Dionysian versus Apollonian cultures (Benedict, 1934)	Program cultural patterns: supportive foster group homes versus jail-like foster group homes (Patton, 2008, pp. 487–492)
2. *Social constructionism.* Mental models and worldviews	Women's ways of knowing (Belenky et al., 1986)	Personalized evaluation: how participants make sense of program experiences (Kushner, 2000); how police view and therefore respond to domestic violence (Funnell & Rogers, 2011, p. 107)
3. *Realism.* Contextually operative causal mechanisms	Effects of social class on people and society (Pawson, 1989); realist synthesis of the "naming-and-shaming theory of change identified different outcomes when this was used for a car theft index naming automobile manufacturers, tables of school educational results, registers of sex offenders, and naming of poll tax protesters in local newspapers and explained them with reference to Merton's typology of aspirations to group membership" (Funnell & Rogers, 2011, p. 516)	Realist matrix linking context, mechanism, and outcome for a computer-assisted learning model (Funnell & Rogers, 2011, p. 249)
4. *Theories of change.* Differentiating patterns of change; change trajectories	Qualitative comparative analysis—e.g., social forces leading to the emergence of welfare states in Western Europe (Rihoux & Ragin, 2009)	Purposeful program theory archetypes: education, incentives, case management, capacity building, and direct service (Funnell & Rogers, 2011, pp. 351–385); theory-driven evaluation (Chen, 2014)

(Continued)

(Continued)

FINDINGS YIELDED BY VARIOUS THEORETICAL PERSPECTIVES	RESEARCH EXAMPLES	PROGRAM EVALUATION EXAMPLES
5. *Systems theory.* Interrelationships, perspectives, tensions	Linking social and ecological systems (Berkes & Folke, 2000)	Program attributes of success and failure different for different subgroups (Funnell & Rogers, 2011, p. 215); mental health system dynamics model (Funnell & Rogers, 2011, p. 274); identifying causal factors and their interrelationships that are contributing to a problem (Funnell & Rogers, 2011, p. 157)
6. *Complexity theory.* Adaptation patterns, emergence, simple rules	Understanding complexity dynamics and transformations in human and natural systems (Gunderson & Holling, 2002)	Program principles adapted situationally in complex systems (Patton, 2011)—e.g., principles adapted across different homeless programs serving homeless youth (Murphy, 2014)
7. *Phenomenology.* Essence	Essence of loneliness (Moustakas, 1961, 1972, 1975)	Program evaluation through phenomenological inquiry (Mertens & Wilson, 2012)
8. *Grounded theory.* Theoretical propositions, hypotheses	Grounded theory in a study of poverty in Greenland to explain how to transition clients from a state of damaging dependence to a less dependent state (Christiansen, Scott, & Sørensen, 2013)	Using a grounded theory approach to evaluate professional development to explain immediate outcomes and short-term impacts on participants' learning (Robinson, 2013)
9. *Hermeneutics.* Meanings of texts	The meanings of permissive and prohibitive sexuality in sacred Christian texts and the U.S. Constitution (Kahn, 1990)	Meanings of academic language used in diverse classrooms and the effects of terminology used (Gottlieb & Ernst-Slavit, 2013)
10. *Pragmatism.* How things work in practice	Experiential learning by doing among school children (Dewey, 1956)	Practical solution to a program's drop-out problem; practitioner wisdom; lessons learned (Patton, 2011, 2012a)

Summary: Interpretation

This module has focused on interpreting findings, determining substantive significance, elucidating phenomenological essence, understanding hermeneutic interpretation, and distinguishing how different theory-based inquiries yield certain kinds of findings. The core theme has been *interpretation*. Interpretation is how we make sense of data. Interpretation is both an inevitability and a necessity. But as philosopher Friedrich Nietzsche added, "Necessity is not an established fact, but an interpretation." The same can be said of causal inference, our next subject.

SOURCE: Brazilian cartoonist Claudius Ceccon. Used with permission.

A scientist puts a frog on a table and yells, "Jump!" The frog jumps. He then surgically removes a leg from the frog, puts the frog back on the table, and yells, "Jump!" The frog jumps again. The scientist surgically removes another limb and repeats the experiment. The frog jumps when commanded, just like before. He does it a third time. "Jump!" he exclaims after removing another limb, and the frog jumps. He removes the last limb on the frog and tries again. "Jump!" he yells, but the frog remains still. "Jump!" he repeats, but no response.

The scientist writes his findings in his notebook: "Upon removing all four limbs, the frog becomes deaf."

This rather grotesque frog story is meant to facilitate the transition from Module 70, which focused on interpreting what qualitative data mean, to this module, which focuses on causal explanation. In essence, we are moving from interpreting findings to explaining them. Causal inference is a particularly treacherous and important form of qualitative analysis and interpretation.

Causality as Explanation

> We construct reality, not exclusively but importantly, in terms of cause and effect. We must—else our individual worlds would be chaotic jumbles of actions unrelated to consequences.
>
> —Northcutt and McCoy (2004, p. 169)
> *Interactive Qualitative Analysis*

Description tells what happened and how the story unfolded. Interpretation elucidates what the description means and judges what makes it significant. Then comes the "why?" question. Why did things unfold as they did? The answer: Because.

"Causation is intimately related to *explanation*; asking for explanation of an event is often to ask *why* it happened" (Schwandt, 2001, p. 23).

- The homeless youth moved into housing and enrolled in school or got a job *because* of what they experienced, how they were treated, and what they learned in the program serving homeless youth.

- The crime rate in the community is low *because* people look out for and support each other.
- Climate change is occurring *because* of human activity.
- The gap between rich and poor is increasing *because* of public policies favoring wealth accumulation by those already rich.

To use the verb "because" is to posit an explanation and assert a causal connection. We make such assertions all the time in casual conversation. "You caused me to be late for my meeting because you didn't wake me." "You caused the car accident because you were texting on your phone and not paying attention to the road." When researchers and evaluators make causal claims, however, evidence is required to support the assertion. Yet what counts as credible evidence is a matter of vociferous debate (Donaldson, Christie, & Mark, 2008). "Exactly how to define a causal relationship is one of the most difficult topics in epistemology and the philosophy of science. There is little agreement on how to establish causation" (Schwandt, 2001, p. 23).

Stake (1995) has emphasized that

> explanations are intended to promote understanding and understanding is sometimes expressed in terms of explanation—but the two aims are epistemologically quite different . . . , a difference important to us, the difference between case studies seeking to identify cause and effect relationships and those seeking understanding of human experience. (p. 38)

Appreciating and respecting this distinction, once case studies have been written and descriptive typologies have been developed and supported, the tasks of organization and description are largely complete. Whether to take the next step in analysis and posit causality depends on the purpose of the inquiry and whether it has been designed to answer causal questions. As emphasized at the beginning of this chapter, analysis is driven by purpose. Purpose determines design. Design frames and focuses analysis.

In program evaluation, for example, explanations about which things appear to lead to other things, which aspects of a program produce certain effects, and how processes lead to outcomes are natural areas for analysis. When careful study of the data gives rise to ideas about causal linkages, there is no reason to deny those interested in the study's results the benefit of those insights, with presentation of the supporting evidence from interviews, case studies, and field observations.

WHY WE'RE SO OBSESSED WITH CAUSALITY

We are pattern seekers, believers in a coherent world, in which regularities appear not by accident but as a result of mechanical causality or of someone's intention.

—Daniel Kahneman (cognitive scientist)
(Quoted in Pomeroy, 2013, p. 1)

Humans are creatures of causality. We like effects to have causes, and we detest incoherent randomness. Why else would the quintessential question of existence give rise to so many sleepless nights, endear billions to religion, or single-handedly fuel philosophy?

This predisposition for causation seems to be innate. In the 1940s, psychologist Albert Michotte theorized that "we see causality, just as directly as we see color," as if it is omnipresent. To make his case, he devised presentations in which paper shapes moved around and came into contact with each other. When subjects—who could only see the shapes moving against a solid-colored background—were asked to describe what they saw, they concocted quite imaginative causal stories. . . .

Humanity's need for concrete causation likely stems from our unceasing desire to maintain some iota of control over our lives. That we are simply victims of luck and randomness may be exhilarating to a madcap few, but it is altogether discomforting to most. By seeking straightforward explanations at every turn, we preserve the notion that we can always affect our condition in some meaningful way. Unfortunately, that idea is a facade. Some things don't have clear answers. Some things are just random. Some things simply can't be controlled. (Pomeroy, 2013, p. 1)

Reporting the Causal Assertions of Others Versus Inferring Causality Yourself

To the extent that you describe and report the causal linkages suggested by and believed in by those you've interviewed, you haven't crossed the line from description into causal interpretation. You are simply reporting their beliefs and assertions.

Much qualitative inquiry stops at reporting the explanations of the people studied. This can take the form of presenting indigenous explanations through quotations, presenting case data, and/or identifying cross-case patterns and themes that describe how people explain key phenomena of interest, but without the analyst adding any additional explanation of "why" the indigenous causal assertions take the form they do. People and groups construct causal explanations, for example, the gods are angry when it thunders. How people explain things can suffice. Indeed, the study of the attributions people make (how they explain things) offers a vast panorama for qualitative inquiry.

> Attributions refer to people's understandings of themselves and their environment. . . . Someone who has just passed a driving test at the first attempt may attribute his or her success to good fortune; another might attribute it to a happy choice of driving school; a third might attribute it to their own natural driving talent; and so on. Attributions such as these are often made about matters of moment in our lives. . . .
>
> Interest in attribution stems from its crucial role as a mediator of perceptions, emotions, motivations and behaviours. Indeed, how people see cause and effect has implications for or may influence their interpersonal relations, their psychopathology, their response to psychotherapy, their decision making, and their adjustment to illness. . . . Attributions may serve many other functions, including enhancement of one's perception of control, preservation of self-esteem, presentation of a particular picture of the self, and emotional release. (Harvey, Turnquist, & Agostinell, 1998, pp. 32–33)

Causal Explanation Grounded in Fieldwork

A researcher who has lived in a community for an extensive period of time will likely have insights into why things happen as they do there. A qualitative analyst who has spent hours interviewing people will likely come away from the analysis with possible explanations for how the phenomenon of interest takes the forms and has the effects it does. An evaluator who has studied a program, lived with the data from the field, and reflected at length about the patterns and themes that run through the data is in as good a position as anyone else at that point to interpret meanings, make judgments about significance, and offer explanations about relationships. Moreover, if decision makers and evaluation users have asked for such information—and in my experience they virtually always welcome causal analyses—there is no reason not to share insights with them to help them think about their own causal presuppositions and hypotheses, and to explore what the data do and do not support in the way of interconnections and potential causal relationships. But doing so remains controversial among those who lack training in qualitative causal analysis. I take the position in this module that qualitative causal analysis has now advanced in rigor and credibility to a point where it should be valued on its merits—and this module will demonstrate those merits. In this module, we'll examine different perspectives on qualitative causal explanations and ways of increasing the credibility and utility of such analyses.

Historical Context:
Approaching Causation Cautiously

The qualitative/quantitative debate of the past century positioned qualitative methods as primarily descriptive and quantitative methods as explanatory. Internal validity in experimental designs concerns the extent to which causal claims are warranted and can be substantiated. Randomized controlled trials (RCTs) involve manipulating a variable (the independent variable) to measure what effect it has on a second variable of interest (the dependent variable). This quantitative/experimental method, first used in agricultural and pharmaceutical studies and then applied in psychology and social science, became advocated as the "gold standard" for establishing causality. My MQP Rumination opposing this gold standard designation is in Chapter 3 (pp. 93–95). For our purpose here, the historically important point is that the experimental method became synonymous with causal research. RCTs had high internal validity. Case studies had low to no internal validity, meaning that they could describe but not explain (Campbell & Stanley, 1963; Shadish, Cook, & Campbell, 2001).

This remains the dominant perspective to this day, but developments in qualitative methods and analysis over the past two decades have demonstrated that *qualitative analysis can yield causal explanations rigorously and credibly.* This assertion is controversial and remains controversial largely because the advocates of experimental designs have been so successful in positioning experiments as the gold standard and the only way to establish causality. This module will present and explain how qualitative analysis can be used to generate causal explanations. Exhibit 8.19, in Module 72 (pp. 600–601), summarizes approaches to causal explanation in qualitative analysis.

SIDEBAR

THE CAUSAL WARS ARE RAGING

The causal wars are still raging, and the amount of collateral damage is increasing.

The causal wars are about what is to count as scientifically impeccable evidence of a causal connection, usually in the context of the evaluation of interventions into human affairs.

The collateral damage comes from the policy that the RCT camp has been supporting with considerable success, here referred to as "the exclusionary policy," which recommends that no (or almost no) programs be funded whose claims of good effects cannot be supported by randomized controlled trials (RCT)-based evidence. This means terminating many demonstrably excellent programs currently saving huge numbers of life-years.

After reviewing the causal wars and alternatives for making causal inferences, Scriven (2008) concludes, "In sum, there is absolutely nothing imperative, and nothing in general superior, about . . . RCT designs" (p. 23).

SOURCE: From the introduction to "A Summative Evaluation of RCT Methodology and an Alternative Approach to Causal Research," by philosopher of science and evaluation research pioneer Michael Scriven (2008, p. 11).

Qualitative Causal Analysis as Speculative: Classic Advice to Proceed With Caution

> Qualitative researchers . . . have generally denied that they were seeking causal explanations, arguing that their goal was the interpretive understanding of meanings rather than the identification of causes.
>
> —Maxwell and Mittapalli (2008, pp. 322–323)

Lofland's (1971) advice in his classic and influential book *Analyzing Social Settings* offers a cautionary perspective on the role of causal "speculation" in qualitative analysis. He argued that the strong suit of the qualitative researcher is the ability "to provide an orderly description of rich, descriptive detail" (p. 59); the consideration of causes and consequences using qualitative data should be a "tentative, qualified, and subsidiary task" (p. 62).

It is perfectly appropriate that one be curious about causes, so long as one recognizes that whatever account or explanations he develops is conjecture. In more legitimacy-conferring terms, such conjectures are called hypotheses or theories. It is proper to devote a portion of one's report to conjectured causes of variations so long as one clearly labels his conjectures, hypotheses or theories as being that. (p. 62)

Neuendorf (2002) follows in this cautious tradition by setting such a high standard for determining causality, especially from content analysis of qualitative data, that it "is generally impossible to fully achieve . . .; true causality is essentially an unattainable goal" (pp. 47–48).

In contrast to this cautious approach to drawing causal inferences from qualitative data, other qualitative analytical frameworks focus on causal explanation as a primary purpose, an attainable outcome, and even the strength of case studies.

Rigorous Qualitative Causal Analysis

The Case for Valuing Direct Observation

The most straightforward form of causal attribution involves direct observation in a short time frame where you can link the cause and effect in a straightforward manner. For example, you go out to a restaurant with friends, and all eat raw oysters. After the meal, all of you experience stomach sickness. It's not a wild speculation to conclude that the oysters caused the sickness.

During a site visit to an employment training program, I witnessed a staff member yelling at a participant. The participant immediately left the building. I followed him and asked if I could talk with him for a moment. He said, "No, I'm finished with that program. Done. I won't stand to be yelled at like that. I'm so out of there." He turned and went his way. I checked subsequent attendance. He didn't return. It seems reasonable to conclude that being yelled at was at least a contributing cause of his dropping out.

The idea of *establishing causality* has taken on such heavy meaning philosophically, methodologically, and epistemologically that the very idea can be daunting and intimidating (Maxwell, 2004).

Thus, qualitative analysts have often been advised to eschew using causal language and avoid making assertions. Reclaiming causal analysis as a reasonable and valid form of qualitative analysis begins with a recognition that direct, critical observation yields causal understandings.

The real "gold standard" for causal claims is the same ultimate standard as for all scientific claims; it is critical observation. Causation can be directly observed, in [a] lab or home or field, usually as one of many contextually embedded observations, such as lead being melted by heating a crucible, eggs being fried in a pan, or a hawk taking a pigeon. And causation can also be inferred from non-causal direct observations with no experimentation, as by the forensic pathologist performing an autopsy to determine the cause of death. (Scriven, 2008, p. 18)

One end of the observational continuum is seeing an action and a reaction that allows a direct, immediate conclusion that the action caused the reaction. At the other end of the continuum is long-term observation, including participant observation. Such in-depth qualitative studies yield detailed, comprehensive data about what happened in the observed settings, both how and why what happened *happened*. Spending a long time in a field setting engaging in ongoing observations, interviewing people, and tracking the details of what occurred makes it possible to *connect the dots* to explain what has occurred. When I read a classic high-quality, in-depth, comprehensive, field-based case study (e.g., Becker, Geer, Hughes, & Strauss, 1961; Goffman, 1961), the causal explanations offered strike me as valid, warranted, well supported, and consistent with the great volume of evidence presented.

What Constitutes Credible Evidence?

The fact that I find in-depth case studies persuasive does not mean that others do. Valuing different kinds of evidence is what the causal attribution war is about. Consider the challenge of eradicating intestinal worms in children, a widespread problem in many developing countries. Suppose we want to evaluate an intervention in which school-age children with diarrhea are given deworming medicine to increase their school attendance and performance. To attribute the intervention to the desired outcome, advocates of RCTs would insist on an evaluation design in which students suffering from diarrhea are randomly divided into a treatment group (those who receive worm medicine) and a control group (those who do not receive the medicine). The school attendance and test performance of the two groups would then be compared. If, after a month on the medicine, those receiving the intervention show higher attendance and school performance at a statistically significant level compared with the control group (the counterfactual), then the increased outcomes can be attributed to the intervention (the worm medicine).

Advocates of qualitative inquiry would question the value of the control group in this case. Suppose that students, parents, teachers, and local health professionals are interviewed about the reasons why students miss school and perform poorly on tests. Independently, each of these groups assert that diarrhea is a major cause of the poor school attendance and performance. Gathering data separately from different informant groups (students, parents, teachers, and health professionals) is a form of *triangulation*, a way of checking the consistency of findings from different data sources. Following the baseline interviews, students are given a regimen of worm medicine. Those taking the medicine show increased school attendance and performance, and in follow-up interviews, the students, parents, teachers, and health professionals independently affirm their belief that the changes can be attributed to taking the worm medicine and being relieved of the symptoms of diarrhea. Is this credible, convincing evidence?

Those who find such a design sufficient would argue that the results are both reasonable and empirical and that the high cost of adding a control group is not needed to establish causality. Nor, they would assert, is it ethical to withhold medicine from students with diarrhea when relieving their symptoms has merit in and of itself. The advocates of RCTs would respond that without the control group, other unknown factors may have intervened to affect the outcomes and that *only the existence of a counterfactual* (control group) will establish with certainty the impact of the intervention.

As this example illustrates, those evaluators and methodologists on opposite sides of this debate have different worldviews about what constitutes sufficient evidence for attribution and action in the real world. This is not simply an academic debate. At stake are millions of dollars of evaluation funds and the credibility of different kinds of and approaches to evaluations around the world.

Thus far, I've been presenting the case for using high-quality, detailed, context-specific qualitative data to make causal inferences. When causal findings from fieldwork are explained through relevant theory, the explanations move to a higher level of generalizability. We turn now to theory-based approaches to causal inference.

Theory-Based Causal Analysis

Realist Analysis to Identify Causal Mechanisms

Qualitative inquiry grounded in realist philosophy and methods makes causal analysis the central focus. The overarching question for realist inquiry is "What are the causal mechanisms that explain how and why reality unfolds as it does in a particular context?" (See Chapter 3, pp. 111–114.) Causal explanation comes

from carefully documenting and analyzing *the actual mechanisms and processes that are involved in particular events and situations.*

> These mechanisms and processes can include mental phenomena as well as physical phenomena and can be identified in unique events as well as through regularities. This position's emphasis on understanding processes, rather than on simply showing an association between variables, provides an alternative approach to causal explanation that is particularly suited to qualitative research. It incorporates qualitative researchers' emphasis on meaning for actors and on unique contextual circumstances, and by treating causal processes as real events, it implies that these may be observed directly rather than only inferred. Thus, it removes the restriction that causal inference requires the comparison of situations in which the presumed cause is present or absent. (Maxwell & Mittapalli, 2008, p. 323)

Emmel (2013) does not hesitate to make drawing causal inferences a priority purpose of realist inquiries because a "scientific realist sampling strategy" (p. 95) is always based in theory.

> Explanation and interpretation in a realist sampling strategy tests and refines theory. Sampling choices seek out examples of mechanisms in action, or inaction towards being able to say something explanatory about their causal powers. Sampling . . . is both pre-specified and emergent, it is driven forward through an engagement with what is already known about that which is being investigated and ideas catalyzed through engagement with empirical accounts. (p. 85)

This illustrates how purpose and design drive analysis. Cases sampled are chosen purposefully *because* they will illuminate causal mechanisms. The inquiry involves detailed description and analysis of the real connection between events in such a way that they can be validly explained as "causally connected" (Emmel, 2013, p. 99). Emmel (2013) uses the image of splitting open a chicken ("spatchcocking") to study its inner parts as analogous to realist qualitative analysis:

> Splitting a chicken down its breastbone and opening it up to reveal the details of its thoracic and abdominal cavities. In a similar way, in research we will split these things, these variables . . . open and lay bare their anatomy for scrutiny and explanation through theorisation and empirical investigation. In the process of which we will be able to better describe, interpret, and, ultimately, explain the sample. (p. 100)

SIDEBAR

WHY THE GERMAN BOMBING OF LONDON IN WORLD WAR II DID NOT DEMORALIZE THE BRITISH

Both British and German leaders expected the intense aerial bombing of London in World War II to create panic. London's residents were expected to flee to the countryside, leaving the city abandoned. Over an eight-month period of incessant bombing in 1940 and 1941, more than 40,000 people were killed, 46,000 injured, and a million buildings damaged or destroyed (Gladwell, 2013, p. 129). But the panic never came. Why?

Canadian psychiatrist J. T. MacCurdy (1943) offered an explanation in his book *The Structure of Morale*. He categorized three groups of people affected by the bombing: (1) those killed, (2) near misses, and (3) remote misses. He interviewed and conducted case studies of near misses and remote misses. His qualitative comparative analysis led to a causal explanation.

So why were Londoners so unfazed by the Blitz? Because forty thousand deaths and forty-six thousand injuries—spread across a metropolitan area of more than eight million people—means that there were many more remote misses who were emboldened by the experience of being bombed than there were near misses who were traumatized by it. (Gladwell, 2013, p. 133)

"We are all of us not merely liable to fear," MacCurdy went on,

we are also prone to be afraid of being afraid, and the conquering of fear produces exhilaration.... When we have been afraid that we may panic in an air-raid, and, when it has happened, we have exhibited to others nothing but a calm exterior and we are now safe, the contrast between the previous apprehension and the present relief and feeling of security promotes a self-confidence that is the very father and mother of courage. (MacCurdy, quoted by Gladwell, 2013, p. 133)

In the midst of the Blitz, a middle-aged laborer in a button-factory was asked if he wanted to be evacuated to the countryside. He had been bombed out of his house twice. But each time he and his wife had been fine. He refused. "What, and miss all this?" he exclaimed. "Not for all the gold in China! There's never been nothing like it! Ever! And never will be again." (Gladwell, 2013, pp. 132–133)

Grounded Theory and Causal Analysis

Theory denotes a set of well-developed categories (e.g., themes, concepts) that are systematically interrelated through statements of relationship to form a theoretical framework that explains some relevant social, psychological, educational, nursing, or other phenomenon. The statements of relationship explain who, what, when, where, why, how, and with what consequences an event occurs. Once concepts are related through statements of relationship into an explanatory theoretical framework, the research findings move beyond conceptual ordering to theory.... A theory usually is more than a set of findings; it offers an explanation about phenomena. (Strauss & Corbin, 1998, p. 22)

Chapter 3 provided an overview of grounded theory in the context of other theoretical perspectives like ethnography, constructivism, phenomenology, and hermeneutics. As I noted in Chapter 3, grounded theory has opened the door to qualitative inquiry in many traditional academic social science and education departments, especially as a basis for doctoral dissertations. I believe this is, in part, because of its overt emphasis on the importance of and specific procedures for generating explanatory theory. Being systematic gets particular emphasis.

By systematic, I still mean systematic every step of the way; every stage done systematically so the reader knows exactly the process by which the published theory was generated. The bounty of adhering to the whole grounded theory method from data collection through the stages to writing, using the constant comparative method, show how well grounded theory fits, works, and is relevant. Grounded theory produces a core category and continually resolves a main concern, and through sorting the core category organizes the integration of the theory.... Grounded theory is a package, a lock-step method that starts the researcher from a "know nothing" to later become a theorist with a publication and with a theory that accounts for most of the action in a substantive area. The researcher becomes an expert in the Substantive area.... And if an incident comes his way that is new he can humbly through constant comparisons modify his theory to integrate a new property of a category.

Grounded theory methodology leaves nothing to chance by giving you rules for every stage on what to do and what to do next. If the reader skips any of these steps and rules, the theory will not be as worthy as it could be. The typical falling out of the package is to yield to the thrill of developing a few new, capturing categories and then yielding to use them in unending conceptual description and incident tripping rather that analysis by constant comparisons. (Glaser, 2001, pp. 1–2)

NO PRECONCEPTIONS: THE GROUNDED THEORY DICTUM

Preconceived questions, problems, and codes all block emergent coding, thereby undermining the foundation of Grounded Theory. Entering the field without preconceptions is the fundamental grounded theory dictum. The grounded theory researcher begins an inquiry without knowing the participant's issues, worldview, basic concepts, or sense-making framework. These emerge through the course of the inquiry.

—Barney G. Glaser (2014b)
No Preconceptions: The Grounded Theory Dictum

Other grounded theory resources are as follows:

STOP, WRITE!: Writing Grounded Theory (Glaser, 2012)

Memoing: A Vital Grounded Theory Procedure (Glaser, 2014a)

In their book on techniques and procedures for developing grounded theory, Strauss and Corbin (1998) emphasize that analysis is the interplay between the researcher and data, so what grounded theory offers as a framework is a set of "coding procedures" to "help provide some standardization and rigor" to the analytical process. Grounded theory is meant to "build theory rather than test theory." It strives to "provide researchers with analytical tools for handling masses of raw data." It seeks to help qualitative analysts "consider alternative meanings of phenomenon." It emphasizes being "systematic and creative simultaneously." Finally, it elucidates "the concepts that are the building blocks of theory." Grounded theory operates from a *correspondence perspective* in that it aims to generate explanatory propositions that correspond to real-world phenomena. The characteristics of a grounded theorist, they posit, are these:

1. The ability to step back and critically analyze situations

2. The ability to recognize the tendency toward bias

3. The ability to think abstractly

4. The ability to be flexible and open to helpful criticism

5. Sensitivity to the words and actions of respondents

6. A sense of absorption and devotion to the work process (p. 7)

According to Strauss and Corbin (1998), grounded theory begins with *basic description*, then moves to *conceptual ordering* (organizing data into discrete categories "according to their properties and dimensions and then using description to elucidate those categories" [p. 19]) and then *theorizing*: "conceiving or intuiting ideas—concepts—then also formulating them into a logical, systematic, and explanatory scheme" (p. 21).

In doing our analyses, we *conceptualize and classify* events, acts, and outcomes. The categories that emerge, along with their relationships, are the foundations for our developing theory. This abstracting, reducing, and relating is what makes the difference between *theoretical and descriptive coding* (or theory building and doing description). Doing line-by-line coding through which categories, their properties, and relationships emerge automatically takes us beyond description and puts us into a *conceptual mode of analysis*. (p. 66)

Strauss and Corbin (1998) have defined terms and processes in ways that are quite specific to grounded theory. It is informative to compare the language of grounded theory with the language of phenomenological analysis presented in the previous module. Here's a sampling of important terminology.

Microanalysis: "The detailed line-by-line analysis necessary at the beginning of a study to generate initial categories (with their properties and dimensions) and to suggest relationships among categories; a combination of open and axial coding" (p. 57).

Theoretical sampling: "Sampling on the basis of the emerging concepts, with the aim being to explore the dimensional range or varied conditions along which the properties of concepts vary" (p. 73).

Theoretical saturation: "The point in category development at which no new properties, dimensions, or relationships emerge during analysis" (p. 143).

Range of variability: "The degree to which a concept varies dimensionally along its properties, with variation being built into the theory by sampling for diversity and range of properties" (p. 143).

Open coding: "The analytic process through which concepts are identified and their properties and dimensions are discovered in data" (p. 101).

Axial coding: "The process of relating categories to their subcategories, termed 'axial' because coding occurs around the axis of the category, linking categories of the level of properties and dimensions" (p. 123).

Relational statements: "We call these initial hunches about how concepts relate 'hypotheses' because they make two or more concepts, explaining the what, why, where, and how of phenomenon" (p. 135).

Comparative Analysis

According to Strauss and Corbin (1998), comparative analysis is a core technique of grounded theory development. Making *theoretical comparisons*—systematically and creatively—engages the analyst in "raising questions and discovering properties and dimensions that might be in the data by increasing researcher sensitivity" (p. 67). Theoretical comparisons are one of the techniques used when doing microscopic analysis. Such comparisons enable

identification of *variations* in the patterns to be found in the data. It is not just one form of a category or pattern in which we are interested but also how that pattern varies dimensionally, which is discerned through a comparison of properties and dimensions under different conditions. (p. 67)

Strauss and Corbin (1998) offer specific techniques to increase the systematic and rigorous processes of comparison, for example, "the flip-flop technique":

This indicates that a concept is turned "inside out" or "upside down" to obtain a different perspective on the event, object, or actions/interaction. In other words, we look at opposites or extremes to bring out significant properties (p. 94).

In the course of conducting a grounded theory analysis, one moves from lower-level concepts to higher-level theorizing:

Data go to concepts, and concepts get transcended to a core variable, which is the underlying pattern. Formal theory is on the fourth level, but the theory can be boundless as the research keeps comparing and trying to figure out what is going on and what the latent patterns are. (Glaser, 2000, p. 4)

Glaser (2000) worries that the popularity of grounded theory has led to a preponderance of lower-level theorizing without completing the job. Too many qualitative analysts, he warns, are satisfied to stop when they've merely generated "theory bits."

Theory bits are a bit of theory from a substantive theory that a person will use briefly in a sentence or so. . . .

Theory bits come from two sources. First, they come from generating one concept in a study and conjecturing without generating the rest of the theory. With the juicy concept, the conjecture sounds grounded, but it is not; it is only experiential. Second, theory bits come from a generated substantive theory. A theory bit emerges in normal talk when it is impossible to relate the whole theory. So, a bit with grab is related to the listener. The listener can then be referred to an article or a report that describes the whole theory. . . .

Grounded theory is rich in imageric concepts that are easy to apply "on the fly." They are applied intuitively, with no data, with a feeling of "knowing" as a quick analysis of a substantive incident or area. They ring true with great credibility. They empower conceptually and perceptually. They feel theoretically complete ("Yes, that accounts for it."). They are exciting handles of explanation. They can run way ahead of the structural constraints of research. They are simple one or two variable applications, as opposed to being multivariate and complex. . . . They are quick and easy. They invade social and professional conversations as colleagues use them to sound knowledgeable. . . . The danger, of course, is that they might be just plain wrong or irrelevant unless based in a grounded theory. Hopefully, they get corrected as more data come out. The grounded theorist should try to fit, correct, and modify them even as they pass his or her lips.

Unfortunately, theory bits have the ability to stunt further analysis because they can sound so correct. . . . Multivariate thinking stops in favor of a juicy single variable, a quick and sensible explanation. . . . Multivariate thinking can continue these bits to fuller explanations. This is the great benefit of trusting a theory that fits, works, and is relevant as it is continually modified. . . . But a responsible grounded theorist always should finish his or her bit with a statement to the effect that "Of course, these situations are very complex or multivariate, and without more data, I cannot tell what is really going on." (pp. 7–8)

As noted throughout this chapter in commenting on how to learn qualitative analysis, it is crucial to study examples. Bunch (2001) has published a grounded theory study about people living with HIV/AIDS. Glaser (1993) and Strauss and Corbin (1997) have collected together in edited volumes a range of grounded theory exemplars that include several studies of health (life after heart attacks, emphysema, chronic renal failure, chronically ill men, tuberculosis, and Alzheimer's disease), organizational headhunting, abusive relationships, women alone in public places, selfhood in women, prison time, and the characteristics of contemporary Japanese society. The journal *Grounded Theory Review* began publication in 2000.

The theory generated can take the form of a model and be told as a story based on the coding categories used to organize the data. So, for example, causal conditions will have been identified that produced the phenomenon that is being studied. Strategies show how these causal conditions operated in particular contexts. These strategies are mediated by intervening conditions and produce action and interactions that result in consequences. The model, which articulates a theory, also tells a causal story.

Narrative Analysis and Case Studies: Telling a Coherent Causal Story

Narrative analysis (see Chapter 3, pp. 128–131) interprets stories, life history narratives, historical memoirs, and creative nonfiction to reveal cultural and social patterns through the lens of individual experiences. Chapter 2 featured an interpretation of the significance of the story of Henrietta Lacks, whose cells were taken without her knowledge and used for medical research (pp. 78–80). Her story takes much of its meaning from what it reveals about the African American community at that time, the nature and norms of medical research, the policies of a major university, and the larger political, social, and economic context within which her story unfolded. The story explains why her cells were taken for medical research without her consent. The case study traces the consequences that resulted— consequences for the researchers, her family, the university, and people suffering from a great variety of diseases whose treatment came from what was learned from her cells. Case studies can be written in a variety of ways, one of which is "causal narratives" specifically constructed to "elucidate the processes at work in one case, or a small number of cases, using in-depth intensive analysis and a narrative presentation of the argument" (Maxwell & Mittapalli, 2008, p. 324).

Yin (2012) distinguishes descriptive case studies from explanatory case studies. Descriptive case studies generate "rich and revealing insights into the social world of a particular case" (p. 49). An explanatory case study, in contrast, "seeks to explain *how* and *why* a series of *events* occurred. In real-world settings, the explanations can be quite complex and can cover an extended period of time. Such conditions create the need for using the case study method rather than conducting, say, an experiment or a survey" (p. 89). Explanatory case studies use causal reasoning to create a coherent, data-based explanation of how one thing led to another.

The internal validity of an explanatory case study depends on the richness of the explanation, the detailed depiction of processes and actions that lead to observed outcomes and consequences, triangulation of data sources, and exploration of rival explanations to arrive at the one that best fits the data. Causal reasoning is the basis for organizing and making sense of a case study or narrative through systematic process tracing (see Exhibit 8.18).

Qualitative Comparative Analysis (QCA)

Qualitative comparative analysis, developed and championed by political sociologist Charles Ragin (1987, 2000), focuses on systematically making comparisons to generate explanations. He has developed a systematic approach for making comparisons of large case units, like nation-states and historical periods, or macrosocial phenomena, like social movements. He constructs and codes each case as a combination of causal and outcome conditions. These combinations can be compared with each other and then logically simplified through a bottom-up process of paired comparison. He aims to draw on the strength of holistic analysis manifest in context-rich individual cases while making possible systematic cross-case comparisons of relatively large numbers of cases, for example, 15 to 25 or more. Ragin (2000, 2008) draws on fuzzy set theory and calls the result "diversity-oriented research" because he systematically codes and takes into account case variations and uniquenesses as well as commonalities, thereby elucidating both similarities and differences. The comparative analysis involves constructing a "truth table" in which the analyst codes each case for the presence or absence of each attribute of interest (Fielding & Lee, 1998, pp. 158–159). The information in the truth table displays the different combinations of conditions that produce a specific outcome. To deal with the large number of comparisons needed, QCA is done using a customized software program.

Analysts conducting diversity-oriented research are admonished to assume maximum causal complexity by considering the possibility that no single causal condition may be either necessary or sufficient to explain the outcome of interest. Different combinations of causal conditions might produce the observed result, though singular causes can also be considered, examined, and tested. Despite reducing large amounts of data to broad patterns represented in matrices or some other form of shorthand, Ragin (1987) stresses repeatedly that these representations must ultimately

EXHIBIT 8.18	**Process Tracing for Causal Analysis**

Process tracing identifies, tests, elucidates, and validates causal mechanisms by describing the "causal chain between an independent variable (or variables) and the outcome of the dependent variable" (George & Bennett, 2005, p. 206–207).

Beach and Pedersen (2013) assert that process-tracing methods are arguably the only methods that allow us to study causal mechanisms:

Studying causal mechanisms with process-tracing methods enables the researcher to make strong within-case inferences about the causal process whereby outcomes are produced, enabling us to update the degree of confidence we hold in the validity of a theorized causal mechanism . . . [and] enabling us to open up the black box of causality using in-depth case study methods to make strong within-case inferences about causal mechanisms based on in-depth single-case studies that are arguably not possible with other social science methods. (pp. 1–2)

Beach and Pedersen (2013) differentiate three approaches to process tracing: (1) theory testing, (2) theory building, and (3) explaining outcomes:

Theory-testing process tracing deduces a theory from the existing literature and then tests whether evidence shows that each part of a hypothesized causal mechanism is present in a given case, enabling within-case inferences about whether the mechanism functioned as expected in the case and whether the mechanism as a whole was present. . . .

Theory-building process-tracing seeks to build a generalizable theoretical explanation from empirical evidence, inferring that a more general causal mechanism exists from the facts of a particular case. . . .

Finally, *explaining-outcome process-tracing* attempts to craft a minimally sufficient explanation of a puzzling outcome in a specific historical case. Here the aim is not to build or test more general theories but to craft a (minimally) sufficient explanation of the outcome of the case where the ambitions are more case-centric than theory-oriented. This distinction reflects the case-centric ambitions of many qualitative scholars. (p. 3)

be evaluated by the extent to which they enhance understanding of specific cases. A cause–consequence comparative matrix, then, can be thought of as a map providing guidance through the terrain of multiple cases and causes.

QCA [qualitative comparative analysis] seeks to recover the complexity of particular situations by recognizing the conjunctural and context-specific character of causation. Unlike much qualitative analysis, the method forces researchers to select cases and variables in a systematic manner. This reduces the likelihood that "inconvenient" cases will be dropped from the analysis or data forced into inappropriate theoretical moulds. . . .

QCA clearly has the potential to be used beyond the historical and cross-national contexts originally envisioned by Ragin. (Fielding & Lee, 1998, pp. 160, 161–162)

CROSS-CULTURAL CASE ANALYSIS COMPARABILITY

In cross-cultural research, the challenge of determining *comparable units of analysis* has created controversy. For example, when definitions of "family" vary dramatically, can one really do systematic comparisons? Are extended families in nonliterate societies and nuclear families in modern societies such different entities that, beyond the obvious surface differences, they cease to be comparable units for generating theory? "The main problem for ethnologists has been to define and develop adequate and equivalent cultural units for cross-cultural comparison" (De Munck, 2000, p. 279).

Comparative Pattern Analysis

"Your bright red skin is a delight to see. But the seeds inside are what make you like me."

"As a child I was told never to dare, an apple and orange— one could not compare.

To find we're alike I scarcely can bear. Don't tell me we also are like a green pear."

COMPARING APPLES AND ORANGES

Analytic Induction

The analysis of set relations is critically important to social research. . . . Qualitative analysis is fundamentally about set relations. Consider this simple example: if all (or almost all) of the anorectic teenage girls I interview have highly critical mothers (that is, the anorectic girls constitute a consistent subset of the girls with highly critical mothers), then I will no doubt consider this connection when it comes to explaining the causes and contexts of anorexia. This attention to consistent connections (e.g., causally relevant commonalities that are more or less uniformly present in a given set of cases) is characteristic of qualitative inquiry. It is the cornerstone of the technique commonly known as analytic induction. (Ragin, 2008, p. 2)

Analytic induction also involves cross-case analysis in an effort to seek explanations. Ragin's qualitative comparative analysis formalized and moderated the logic of analytic induction (Ryan & Bernard, 2000, p. 787), but it was first articulated as a method of "exhaustive examination of cases in order to prove universal, causal generalizations" (Peter Manning, quoted by Vidich & Lyman, 2000, p. 57). In Norman Denzin's sociological methods classic *The Research Act* (1978b), he identified analytic induction based on comparisons of carefully done case studies as one of the three primary strategies available for dealing with and sorting out rival explanations in generating theory; the other two are experiment-based inferences and multivariate analysis. Analytic induction as a comparative case method

was to be the critical foundation of a revitalized qualitative sociology. The claim to universality of the causal generalizations is . . . derived from the examination of a single case studied in light of a preformulated hypothesis that might be reformulated if the hypothesis does not fit the facts. . . . Discovery of a single negative case is held to disprove the hypothesis and to require its reformulation. (Vidich & Lyman, 2000, p. 57)

Over time, those using analytic induction have eliminated the emphasis on discovering universal causal generalizations and have, instead, emphasized it as a strategy for engaging in qualitative inquiry and comparative case analysis that includes examining preconceived hypotheses—that is, without the pretense of the mental blank slate advocated in purer forms of phenomenological inquiry and grounded theory.

In analytic induction, researchers develop hypotheses, sometimes rough and general approximations, prior

to entry into the field or, in cases where data already are collected, prior to data analysis. These hypotheses can be based on hunches, assumptions, careful examination of research and theory, or combinations. Hypotheses are revised to fit emerging interpretations of the data over the course of data collection and analysis. Researchers actively seek to disconfirm emerging hypotheses through negative case analysis, that is, analysis of cases that hold promise for disconfirming emerging hypotheses and that add variability to the sample. In this way, the originators of the method sought to examine enough cases to assure the development of universal hypotheses.

Originally developed to produce universal and causal hypotheses, contemporary researchers have de-emphasized universality and causality and have emphasized instead the development of descriptive hypotheses that identify patterns of behaviors, interactions and perceptions. . . . Bogdan & Biklen (1992) have called this approach modified analytic induction. (Gilgun, 1995, pp. 268–269)

Jane Gilgun (1995) used modified analytic induction in a study of incest perpetrators to test hypotheses derived from the literature on care and justice and to modify them to fit an in-depth subjective account of incest perpetrators. She used the literature-derived concepts to sensitize herself throughout the research while remaining open to discovering concepts and hypotheses not accounted for in the original formulations. And she did gain new insights:

Most striking about the perpetrators' accounts was that almost all of them defined incest as love and care. The types of love they expressed ranged from sexual and romantic to care and concern for the welfare of the children. These were unanticipated findings. I did not hypothesize that perpetrators would view incest as caring and as romantic love. Rather, I had assumed that incest represented lack of care and, implicitly, an inability to love [literature-derived hypotheses]. It did not occur to me that perpetrators would equate incest and romance, or even incest and feelings of sexualized caring. From previous research, I did assume that incest perpetrators would experience profound sexual gratification through incest. Ironically, their professed love of whatever type was contradicted by many other aspects of their accounts, such as continuing the incest when children wanted to stop, withholding permission to do ordinary things until the children submitted sexually, and letting others think the children were lying when the incest was disclosed. These perpetrators, therefore, did not view incest as harmful to victims, did not reflect on how they used their power and authority to coerce children to cooperate, and even interpreted their behavior in many cases as forms of care and romantic love. (p. 270)

Analytic induction reminds us that qualitative inquiry can do more than discover emergent concepts and generate new theory. A mainstay of science has always been examining and reexamining and reexamining yet again those propositions that have become the dominant belief or explanatory paradigm within a discipline or group of practitioners. Modified analytic induction provides a name and guidance for undertaking such qualitative inquiry and analysis.

Forensic Causal Analysis

On August 1, 2007, the major six-lane I-35 interstate highway bridge connecting Minneapolis and Saint Paul collapsed during rush hour. Thirteen people died, and another 145 were injured. The National Transportation Safety Board investigators studied evidence from the accident, including a security video camera that recorded the collapse. The investigators determined that the bridge's steel gusset plates were undersized and inadequate to support the load of the bridge, a load that had had increased substantially over time since the bridge's construction. During the recovery of the wreckage, investigators discovered gusset plates at eight different joint locations that were fractured. The investigation turned up photos from a June 2003 inspection of the bridge that showed gusset plate bowing. The investigation concluded that the primary cause of the collapse was the undersized gusset plates. Contributing to the collapse was the fact that 2 inches of concrete was added to the road surface over the years, increasing the dead load by 20%. Also contributing was the weight of construction equipment and material resting on the bridge just above its weakest point at the time of the collapse. That load was estimated at 578,000 pounds (262,000 kilograms), consisting of sand, water, and vehicles.

This kind of in-depth case analysis is mandatory for accidents of all kinds: vehicular accidents, fires, building collapses, and so on. Crimes are investigated. Deaths in hospitals are investigated. What all such investigations share are retrospective case study methods that consider possible causes and, based on the preponderance of evidence, the most probable cause. Exhibit 5.12 (p. 296) describes the rigorous forensic case analyses of railroad switching operations fatalities. As noted in the exhibit, 55 railroad employees died in switching yard accidents from 2005 to 2010. To analyze the causes of these deaths, investigators from the railroad industry, labor unions, locomotive engineers, and federal regulators worked together to look for patterns of causes of the accidents. They carefully reviewed every case, coding a variety of variables related to conditions, contributing factors, and kinds of operations involved. They then looked for patterns in the qualitative case data and correlations in the quantitative cross-case data. They spent three to four hours coding each case and hours analyzing patterns across cases.

They found that fatalities happen for a reason. Accidents are not random occurrences, unfortunate events, or just plain bad luck. The risks to employees engaged in switching operations are real, ever present, and preventable. The data showed patterns in why switching fatalities occur. Findings about the causes of railroad accidents accumulated through this rigorous analysis over time and across cases identified five common behaviors and 10 common hazards that contribute to fatal accidents.

General Elimination Method and Modus Operandi Analysis

The General Elimination Methodology (GEM) approach for case analysis involves identifying rival alternative causal explanations, comparing each with the evidence, and eliminating those that don't conform to the evidence, until that causal explanation remains that best fits the preponderance of the evidence (Scriven, 2008).

> To take an example from work in which I have been involved, when looking at the effect of aid given by Heifer or Gates to extremely poor farmers in East Africa, after determining that a substantial improvement in welfare has followed the arrival of aid, and has been sustained for a few years, we check for the presence of more than a dozen other possible causes of this observed subsequent increase in welfare, including: efforts by the country's government that have actually trickled down to the village level, analogous efforts by other philanthropies, self-help gains resulting from inspired leadership in the local communities, increased income from family members traveling to well-paid job openings elsewhere and remitting money back home, increased prices for milk or calves in the local markets, the beneficial results of a few years of good weather or of improved water supply, or of technology-driven improvements in the quality of available commercial feed, veterinary meds or services, or grass seed for improving pastures. This requires

considerable systematic effort, but no sophisticated experimental design, no sophisticated statistics or risk analysis. (Scriven, 2008, p. 22)

Modus operandi (or method of operating, MO) analysis was also conceptualized by evaluation theorist Michael Scriven (1976) as a way of inferring causality when experimental designs are impractical or inappropriate. The MO approach, drawing from forensic science, makes the inquirer a detective. Detectives match clues discovered at a crime scene with known patterns of possible suspects. Those suspects whose MO does not fit the crime scene pattern are eliminated from further investigation. Translated to research and evaluation, the inquirer/detective observes some pattern and makes a list of possible causes. Evidence from the inquiry is matched with the list of suspects (possible causes). Those possible causes that do not fit the pattern of evidence can be eliminated from further consideration. Following the autopsy-like logic of Occam's razor, as each possible cause is matched with

the evidence, that cause supported by the *preponderance of evidence* and offering the simplest causal inference among competing possibilities is chosen as most likely. This is also known as *inference to the best explanation*: If, based on the facts of the case and cumulative evidence, including interpretation, knowledge, and explanatory theory brought to the analysis, it is possible to identify one explanation as better than the others, then inferences based on that explanation will be warranted as the best explanation.

Forensic case analysis findings are accepted as evidence in courts of law for both criminal and civil proceedings. Clearly, such analyses and their use throughout the world demonstrate that thorough, systematic, and independent case analysis can identify causes at a reasonable level of confidence (preponderance of evidence) for individual cases. Cross-case analyses are used to generate regulations, policies, and operational procedures in fields as varied as transportation, hospitals, utilities, and construction, to name but a few examples.

SOURCE: Brazilian cartoonist Claudius Ceccon. Used with permission.

GENERAL ELIMINATION CASE STUDY METHOD EXAMPLE: EVALUATING AN ADVOCACY CAMPAIGN AIMED AT INFLUENCING A SUPREME COURT DECISION

Several foundations funded a campaign aimed at influencing a Supreme Court decision. The collaboration of foundations committed more than $2 million to a focused advocacy effort within a window of nine months to potentially influence the Court. The evaluation case study examined the following question: To what extent, if at all, did the final-push campaign influence the Supreme Court's decision?

The method we used in evaluating the Supreme Court advocacy campaign is what Scriven (2008) has called General Elimination Methodology, or GEM. It is a kind of "inverse epidemiological method." Epidemiology begins with an effect and searches for its cause. In this application of GEM, we had both an effect (the Supreme Court decision in favor of the campaign position) and an intervention (the advocacy campaign), and we were searching for connections between the two. In doing so, we conducted a retrospective case study. Using evidence gathered through fieldwork—interviews, document analysis, detailed review of the Court arguments and decision, news analysis, and the documentation of the campaign itself—we aimed to eliminate alternative or rival explanations until the most compelling explanation, supported by the evidence, remained. This is also called the forensic method or MO (modus operandi, or method of operating) approach. Scriven brought the concept into evaluation from detective work, in which a criminal's MO is established as a "signature trace" that connects the same criminal to different crimes (Davidson, 2005, p. 75). The modus operandi method works well in tracing the effects of interventions that have highly distinctive patterns of effects.

The evidence brought to bear in the evaluation of the judicial advocacy campaign was organized and presented as an in-depth case study of the campaign in four sections: (1) the litigation work; (2) the coordinated, targeted state organizing campaigns; (3) the communications and public education strategies; and (4) the overall coalition coordination. The case study involved detailed examination of campaign documents and interviews with 45 people directly involved in and knowledgeable about the campaign and/or the case, including the attorneys who argued both sides of the case before the Supreme Court. Several key people were interviewed more than once. The case also involved examining and analyzing hundreds of documents, including legal briefs, the Court's opinions, more than 30 other court documents, more than 20 scholarly publications and books about the Supreme Court, media reports on the case, and confidential campaign files and documents, including three binders of media clips from campaign files. The case also drew on reports and documents describing related cases, legislative activity, and policy issues. Group discussions with key campaign strategists and advocates were especially helpful in clarifying important issues in the case.

Given the multifaceted and omnibus nature of the total campaign, a particular value of constructing this kind of in-depth case study is that none of the informants completely knew the full story. And, of course, different informants about the same events and processes had varying perspectives about what occurred and what it meant. A case study, then, involves ongoing comparative analysis—the sorting out, comparing, and reporting of different perspectives.

The full case did not emerge all at once. Indeed, it took time, including follow-up interviews, rereading documents, and continuous fact checking, for the full story to emerge. In a retrospective case study of this kind, we are often talking to people about events that they have "moved beyond" in their busy lives. Documentation is useful in returning to the past, but the critical judgments and perceptions stored in the memories of key players often take time and care to reignite. Developing relationships with key players was critical to this process.

Evaluation Conclusion

Based on a thorough review of the campaign's activities, interviews with key informants and key knowledgeables, and careful analysis of the Supreme Court decision, we conclude that *the coordinated final-push campaign contributed significantly to the Court's decision.*

See Exhibit 8.24 (p. 614) in Module 73 for a graphic depiction of the finding about the factors that contributed to the campaign's success.

SOURCES: Patton (2008) and Patton and Sherwood (2007).

New Analysis Directions: Contribution Analysis, Participatory Analysis, and Qualitative Counterfactuals

> Look carefully for words and phrases that indicate attributes and various kinds of causal or conditional relations . . . .
>
> *Causal relations:* "because" and its variants, 'cause, 'cuz, as a result, since, and the like. For example, "Y'know, we always take [highway] 197 there 'cuz it avoids all that traffic at the mall." But notice the use of the word "since" in the following: "Since he got married, it's like he forgot his friends." Text analysis that involves the search for linguistic connectors like these requires very strong skills in the language of the text because you have to be able to pick out very subtle differences in usage.
>
> —Bernard and Ryan (2010, p. 60)
> *Analyzing Qualitative Data*

Contribution Analysis

One qualitative critique of traditional causal attribution approaches is that the language and concepts are overly deterministic. The word *cause* connotes singular, direct, and linear actions leading to clear, precise, and verifiable results: *X* caused *Y*. But such a direct, singular, linear causation is rare in the complex and dynamic interactions of human beings. More often, there are multiple causal influences and multiple outcomes.

Reframing Explanation From Cause and Effect to Influences and Interactions

Contribution analysis (Mayne, 2007, 2011, 2012; Patton, 2012b) was developed as an approach in program evaluation to examine a causal hypothesis (theory of change) against logic and evidence

to examine what factors could explain the findings. Attribution questions are different from contribution questions, as follows.

Traditional evaluation causality questions (attribution)

- Has the program caused the outcome?
- To what extent has the program caused the outcome?
- How much of the outcome is caused by the program?

Contribution questions

- Has the program made a difference? That is, has the program made an important contribution to the observed result? Has the program influenced the observed result?
- How much of a difference has the program made? How much of a contribution?

The result of a contribution analysis is not definitive proof that the intervention or program has made an important contribution but rather evidence and argumentation from which it is reasonable to draw conclusions about the degree and importance of the contribution, within some level of confidence. The aim is to get *plausible association* based on a preponderance of evidence, as in the judicial and forensic traditions. The question is whether a reasonable person would agree from the evidence and argument that the program has made an important contribution to the observed result. Contribution analysis can be used in impact evaluations for interventions in complex development situations with multiple actors (Stern et al., 2012).

A contribution analysis produces a *contribution story* that presents the evidence and other influences on program outcomes. A major part of that story may tell about behavioral changes that intended beneficiaries have made as a result of the intervention.

Attributes of a Credible Contribution Story

A credible statement of contribution would entail the following:

- A well-articulated context of the program, discussing other influencing factors

- A plausible theory of change (no obvious flaws) that is not disproven
- A description of implemented activities and resulting outputs of the program
- A description of the observed results
- The results of contribution analysis

  ○ The evidence in support of the assumptions behind the key links in the theory of change
  ○ Discussion of the roles of the other influencing factors

- A discussion of the quality of the evidence provided, noting weaknesses

Contribution analysis focuses on identifying *likely influences.* Such causes, which on their own are neither necessary nor sufficient, represent the kind of contribution role that many interventions play. Contribution analysis, like detective work, requires connecting the dots between what was done and what resulted, examining a multitude of interacting variables and factors, and considering alternative explanations and hypotheses, so that in the end, we can reach an independent, reasonable, and evidence-based judgment based on *the cumulative evidence.* That is what we did in evaluating the judicial advocacy campaign featured earlier in a sidebar (see p. 595). From a contribution perspective, the question became how much influence the campaign appeared to have had rather than whether the campaign directly and singularly produced the observed results.

Outcome mapping (IDRC, 2010) and outcome harvesting (Wilson-Grau & Britt, 2012) are well-developed frameworks that use contribution analysis for evaluating outcomes in complex dynamic systems characterized by multiple influences, multiple outcomes, and multiple interrelationships.

Collaborative and Participatory Causal Analyses

Collaborative and participatory approaches to qualitative inquiry include working with nonresearchers and nonevaluators not only in collecting data but also in analyzing data. When dealing with making judgments about the extent to which the preponderance of evidence supports certain causal conclusions, or what contributory factors explain the results, the people in the setting studied can serve as the equivalent of an inquiry jury, rendering their own interpretations and judgments about causality. In a major study of "the difficult methodological and theoretical challenges faced by those who wish to evaluate the impacts of

international development policies," aimed at "broadening the range of designs and methods for impact evaluations," participatory approaches were highlighted as an important design and analysis option.

> Impact Evaluation (IE) aims to demonstrate that development programmes lead to development results, that the intervention as *cause* has an *effect.* . . . On the basis of literature and practice, a basic classification of potential designs is outlined [with] . . . five design approaches identified—Experimental, Statistical, Theory-based, Case-based and Participatory. (Stern et al., 2012, pp. i–ii)

> Participatory approaches to causal inference do not see recipients of aid as passive recipients but rather as active "agents." Within this understanding, beneficiaries have "agency" and can help "cause" successful outcomes by their own actions and decisions. (Stern et al., 2012, p. 29)

Chapter 4 discussed collaborative and participatory approaches at some length, including Exhibit 4.13: Principles of Fully Participatory and Genuinely Collaborative Inquiry (page 222). Participatory approaches require special facilitation skills to help those involved adopt analytical thinking. Some of the challenges include the following:

- Deciding how much involvement nonresearchers will have, for example, whether they will simply react and respond to the researcher's analysis or whether they will be involved in the generative phase of analysis (Determining this can be a shared decision. "In participatory research, participants make decisions rather than function as passive subjects" [Reinharz, 1992, p. 185].)
- Creating an environment in which those collaborating feel that their perspective is genuinely valued and respected
- Demystifying research
- Combining training in how to do analysis with the actual work of analysis
- Managing the difficult mechanics of the process, especially where several people are involved
- Developing processes for dealing with conflicts in interpretations (e.g., agreeing to report multiple interpretations)
- Determining how to maintain confidentiality with multiple analysts

A good example of these challenges concerns how to help lay analysts deal with counterintuitive findings and counterfactuals—that is, data that don't fit primary patterns, negative cases, and data that oppose primary preconceptions or predilections. Morris (2000) found that shared learning, especially the capacity to deal with counterfactuals, was reduced when participants feared judgment by others, especially those in positions of authority.

In analyzing hundreds of open-ended interviews with parents who had participated in early-childhood parent education programs throughout the state of Minnesota, I facilitated a process of analysis that involved some 40 program staff. The staff worked in groups of two and three, each analyzing 10 pre and post paired interviews at a time. No staff analyzed interviews with parents from their own programs. The analysis included coding interviews with a framework developed at the beginning of the study as well as inductive, generative coding in which the staff could create their own categories. Following the coding, new and larger groups engaged in interpreting the results and extracting central conclusions. Everyone worked together in a large center for three days. I moved among the groups, helping resolve problems. Not only did we get the data coded, but also the process, as is intended in collaborative and participatory research processes, proved to be an enormously stimulating and provocative learning experience for the staff participants. The process forced them to engage deeply with parents' perceptions and feedback, as well as to engage with each other's reactions, biases, and interpretations. In that regard, the process also facilitated communication among diverse staff members from across the state, another intended outcome of the collaborative analysis process. Finally, the process saved thousands of dollars in research and evaluation costs, while making a staff and program development contribution. The results were intended primarily for internal program improvement use. As would be expected in such a nonresearcher analysis process, external stakeholders placed less value on the results than did those who participated in the process (Mueller, 1996; Mueller & Fitzpatrick, 1998; Program Evaluation Division, 2001). However, as participatory processes have become better facilitated and understood, external stakeholders are coming to appreciate and value them more.

Qualitative Counterfactuals

A central issue in establishing causality is addressing the counterfactual: What would have happened without the cause or intervention? A counterfactual analysis looks at the case that is counter to the facts actually observed. In experimental designs, the control group constitutes the counterfactual, providing evidence of what would happen without the treatment received

FRAMEWORK FOR CAUSAL ANALYSIS IN EVALUATION USING PROGRAM THEORY

SIDEBAR

A systematic approach to causal analysis for program theory evaluation consists of three components: (1) congruence, (2) comparisons, and (3) critical review.

1. *Congruence with program theory.* Do the observed and documented results match the program theory?

2. *Counterfactual comparisons.* What would have happened without the intervention? (See discussion of qualitative and scenario-based counterfactuals, p. 599.)

3. *Critical review.* Are there plausible explanations for the results?

SOURCE: Funnell and Rogers (2011, pp. 473–499).

by the treatment group. Because qualitative designs do not assign people to treatment and control groups, critics of qualitative inquiry assert that causal claims cannot be made. Yet in many cases, it is neither feasible nor ethical to conduct experiments. One qualitative solution is to construct a hypothetical counterfactual case.

Assessing the plausible outcome of a combination of conditions that does not exist and instead must be imagined may seem esoteric. However, this analytic strategy has a long and distinguished tradition in the history of social science. A causal combination that lacks empirical instances and therefore must be imagined is a *counterfactual case*; evaluating its plausible outcome is *counterfactual analysis*.

To some, counterfactual analysis is central to case-oriented inquiry because such research typically embraces only a handful of empirical cases. If only a few instances exist (e.g., of social revolution), then researchers must compare empirical cases to hypothetical cases. The affinity between counterfactual analysis and case-oriented research, however, derives not simply from its focus on small *N*s, but from its configurational nature. Case-oriented explanations of outcomes are often combinatorial in nature, stressing specific configurations of causal conditions. Counterfactual cases thus often differ from empirical cases by a single causal condition, thus creating a decisive, though partially imaginary, comparison. (Ragin, 2008, p. 150)

An example of a historical counterfactual case is Moore's (1966) creation of an alternative history of the United States in which the South, rather than the North, won the U.S. Civil War. He used this counterfactual creation to support his theory that a "revolutionary break with the past" is an essential part of becoming a modern democracy.

Such counterfactuals constitute *thought experiments.* German social science pioneer Max Weber is commonly credited with being the first to advocate the use of thought experiments in social research to gain insight into causal relationships. Ragin (2008) offers an extensive discussion of counterfactuals in his configurational framework of qualitative comparative analysis, discussed earlier. In qualitative comparative analysis, counterfactual cases are created and used as substitutes for matched empirical cases when the real world offers no matching case. The hypothetical matched cases are identified by their configurations of causal conditions to illuminate the comparative analysis and causal patterns. This moves the qualitative analyst from writing fieldwork-based empirical case studies to creating relevant comparison cases through thought experiments and alternative case study scenarios.

Scenario-Based Counterfactuals

One innovative and intriguing approach to constructing counterfactual comparisons is a collaborative and participatory process being applied in program evaluation. *Scenario-based counterfactuals* are negotiated alternative scenarios developed jointly with decision makers and stakeholders as part of a collaborative analysis process. The negotiated and constructed *scenario-based counterfactual* specifies reasonable projected alternative processes and any differences that would have occurred under the negotiated alternative scenario, for example, in timing or scale. This amounts to creating a *plausible alternative hypothesis* and examining the degree of plausibility.

- A community adopts an energy conservation plan built around a collaboration of churches, businesses, schools, nonprofits, and government units. How important was the collaboration to the adoption and implementation of the plan? Knowledgeable key informants construct the alternative scenario as a fabricated case study—no collaboration, business as usual—as a counterfactual.
- Exhibit 7.20 (pp. 511–516) is a case study of Thmaris, a homeless youth who received services and describes

the difference it had made to him. In the case, he asserts that he would have landed in jail given the path he was on before he received services. A full case study alternative scenario could be constructed to fill in the details of a possible counterfactual. The credibility and validity of the counterfactual scenario are negotiated in a participatory process with key stakeholders.

The negotiated scenario-based counterfactual is based on a constructed alternative to what was implemented. The negotiated alternative scenario is a plausible (to decision makers and stakeholders) alternative; it is feasible in the sense that there are no budgetary, timing, or technical reasons why it could not have occurred, and it is legal in the sense that the alternative represents one of the options available within current law or plausible changes to relevant law (Rowe, Colby, Hall, & Niemeyer, in press).

Overview and Summary: Causal Explanation Thorough Qualitative Analysis

This module has examined a variety of ways of approaching casual explanation in qualitative analysis. Exhibit 8.19 summarizes these different approaches.

I opened this module on causal analysis with observations from two eminent philosophers of science. First, Tom Schwandt (2001), University of Illinois, warned that "exactly how to define a causal relationship is one of the most difficult topics in epistemology and the philosophy of science. There is little agreement on how to establish causation" (p. 23). Then, Michael Scriven (2008), Claremont Graduate School, observed that "the causal wars are still raging, and the amount of collateral damage is increasing" (p. 11); he was referring to the gold standard debate that pits RCTs against all the alternatives. He concluded, as I do, that "there is absolutely nothing imperative, and nothing in general superior, about . . . RCT designs" (p. 23).

The overall conclusion I reach is that developments in qualitative methods and analysis over the past two decades have demonstrated that *qualitative analysis can yield causal explanations rigorously and credibly.*

That said, I close this module with a cautionary tale and a reminder that some question the whole notion of simple linear causality, doubting both its accuracy and its utility.

EXHIBIT 8.19 Twelve Approaches to Qualitative Causal Analysis

STANCE TOWARD CAUSALITY	INQUIRY ORIENTATION	FUNDAMENTAL PREMISES
1. *Indigenous focus.* Describing how people in a setting explain things: What are the indigenous causal explanations?	Studying and reporting causal explanations among those studied; reporting indigenous explanations	People and groups construct causal explanations, for example, "The gods are angry when it thunders." Qualitative inquiry can legitimately describe patterns of how those studied explain things, but it is beyond the scope of qualitative inquiry to independently generate causal explanations.
2. *Skeptics and doubters.* Qualitative data can describe but not explain	Upholding the belief that only experimental and quasi-experimental designs have high internal validity and can test causal hypotheses credibly	The "gold standard" for causal inference and attribution is randomized controlled trials. Manipulating independent and dependent variables, randomization of treatment and control groups, and counterfactuals are necessary to establish causality. Qualitative causal analysis is inferior and lacks both validity and credibility.
3. *Exploratory but not confirmatory.* Qualitative causal analysis can be useful but should be treated as speculative and exploratory—a source of causal hypotheses, not confirmed explanations	Accepting the argument of the skeptics and doubters (2) but looking to qualitative inquiry as an important exploratory method to generate propositions for experimental design testing	Qualitative inquiry is exploratory, has low internal validity, and is best suited for generating hypotheses to be tested using experimental designs. Experimental designs are confirmatory, with high internal validity.
4. *Direct observation of causes and consequences.* High-quality fieldwork, long-term participant observation, and in-depth interviews can provide detailed, contextually sensitive information to make valid causal inferences	Relatively direct, observable, and immediate causal chains, such as a fist to the jaw or knocking out a boxing opponent, constitute a simple causal analysis; long-term, in-depth case studies provide the details necessary for valid causal inferences	"The real 'gold standard' for causal claims is the same ultimate standard as for all scientific claims; it is critical observation. Causation can be directly observed, in lab or home or field, usually as one of many contextually embedded observations. And causation can also be inferred from non-causal direct observations with no experimentation" (Scriven, 2008, p. 18).
5. *Theory-based qualitative inquiry.* Integrating qualitative analysis with theory makes the resulting causal explanation valid and credible	Theory-based approaches include realist inquiry, grounded theory, analytic induction, qualitative comparative analysis, and theory-driven evaluation	Theory is necessary to appropriately interpret empirical findings and validate causal analysis. Analysis needs theory for explanation. Different theoretical orientations provide specific frameworks and analytic processes for arriving at valid explanations.
6. *Causal purposeful sampling.* Selecting cases for study that will specifically illuminate causal relationships	Cases are purposefully selected for in-depth study to describe, examine, and establish causal pathways	Causal processes can be traced and validated when cases are selected carefully because they are rich with information about casual patterns. There are many kinds of purposeful sampling. To do the best job of qualitative causal analysis, cases should be selected for study with causality as a criterion. A *pathway case* provides "uniquely

STANCE TOWARD CAUSALITY	INQUIRY ORIENTATION	FUNDAMENTAL PREMISES
		penetrating insight into causal mechanisms" (Gerring, 2007, p. 122). "Maximize leverage over the causal hypothesis" (George & Bennett, 2005, p. 172).
7. *Forensic-type analysis.* The General Elimination Methodology of fitting evidence to explanation and eliminating those explanations that do not fit the evidence	The conclusion is based on the preponderance of evidence; rigorously examining competing explanations against the evidence is key	Accident investigation, epidemiological studies, criminal investigations, inquiries into hospital deaths, and other fields routinely address the question of what caused an occurrence retrospectively through carefully assembled evidence.
8. *Existence proof.* Documentation that something is possible	Something unusual is observed, and so its existence is proof of the phenomenon even when it is not clear how it has occurred	Two Berlin patients living with HIV appear to have been cured. The case data provide solid documentation, though the precise causal mechanism remains under study. Proof of existence can point to breakthroughs in knowledge that justify ongoing causal research, for example, *How the Berlin Patients Defeated HIV and Forever Changed Medical Science* (Holt, 2014; see also Johnson, 2014a, 2014b).
9. *Proof of concept.* An intervention is tested on a small sample as a pilot test of whether the idea works	A well-documented pilot test of an intervention that is successful as hypothesized demonstrates "proof of concept" and justifies further research and replication studies	Mayo Clinic achieved the first complete remission of blood cancer on a single case through a massive blast of measles vaccine, providing "'proof of concept' that a single massive dose of intravenous viral therapy can kill cancer by overwhelming its natural defenses" (Dr Stephen Russell, Mayo Clinic, quoted by Browning, 2014, p. 1).
10. *Contribution analysis.* Replaces attribution with attention to multiple interacting influences and multiple effects	Qualitative inquiry is used to describe and map interrelationships and assess the degree to which various identified factors contributed to the observed results	"Most development interventions are 'contributory causes.' They 'work' as part of a causal package in combination with other 'helping factors' such as stakeholder behaviour, related programmes and policies, institutional capacities, cultural factors or socio-economic trends" (Stern et al., 2012).
11. *Participatory and collaborative analysis.* Involves people in the setting studied in analyzing causal explanations	Participatory analysis can be a natural concluding step in participatory designs	Involving the people studied in the analysis process, with training, coaching, and capacity building, can lead to deeper insights and make the findings more meaningful and useful to those involved. Outcome mapping and outcome harvesting combine contribution analysis (10) and participatory analysis.
12. *Constructed counterfactual scenario.* Using qualitative data and expert knowledge to create valid and credible counterfactuals to enhance comparative analysis	It is important to address the counterfactual condition: What would have happened in the absence of the cause?	Fabricated counterfactuals can be created with sufficient detail, depth, and validity to serve as a credible basis for comparative analysis in examining and elucidating causal explanations. Counterfactual thinking is quite common and well understood in everyday life when we ask the "what if" question.

From Linear Causality to Complex Interrelationships

> The law of causality, I believe, like much that passes muster among philosophers, is a relic of a bygone age, surviving, like the monarchy, only because it is erroneously supposed to do no harm.
>
> —Philosopher Bertrand Russell (1872–1970)
> *Selected Papers*

Simple causal explanations are alluring, even seductive. We seem unable to escape simple linear modeling. We fall back on the linear assumptions of much of quantitative analysis and specify isolated independent and dependent variables that are mechanically linked together out of context. In contrast, the challenge of qualitative inquiry involves portraying a *holistic picture* of what the phenomenon, setting, or program is like and struggling to understand the fundamental nature of a particular set of activities and people in relationship in a specific context. "Particularization is an important aim, coming to know the particularity of the case," as qualitative case study expert Bob Stake (1995) admonishes us to remember (p. 39). Simple statements of linear relationships may be more distorting than illuminating. The ongoing challenge of qualitative analysis is moving between the phenomenon of interest constantly moving between the phenomenon of interest and our abstractions of that phenomenon, between the descriptions of what has occurred and our interpretations of those descriptions, between the complexity of reality and our simplifications of those complexities, between the circularities and interdependencies of human activity and our need for linear, ordered statements of cause and effect.

Distinguished social scientist Gregory Bateson traced at least part of the source of our struggle to the ways in which we have been taught to think about things. We are told that a "noun" is the "name of a person, place, or thing." We are told that a "verb" is an "action word." These kinds of definitions, Bateson (1977) argues, were the beginning of teaching us that "the way to define something is by what it supposedly is in itself—not by its relations to other things."

> Today all that should be changed. Children could be told a noun is a word having a certain relationship to a predicate. A verb has a certain relationship to a noun, its subject, and so on. Relationship could now be used as a basis for definition, and any child could then see that there is something wrong with the sentence "'Go' is a verb." . . . We could have been told something about the pattern which connects: that all communication necessitates context, and that without context there is no meaning. (p. 13)

Without belaboring this point about the difference between linear causal analysis (*x* causes *y*) and a holistic perspective that describes the interdependence and interrelatedness of complex phenomena, I would simply offer the reader a Sufi story. I suggest trying to analyze the data represented by the story in two ways. First, try to isolate specific variables that are important in the story, deciding which are the independent variables and which the dependent variable, and then write a statement of the form "These things caused this thing." Then, read the story again. For the second analysis, try to distinguish among and label the different meanings of the situation expressed by the characters. Then, write a statement of the form "These things and these things came together to create _____." Don't try to decide that one approach is right and the other is wrong; simply try to experience and understand the two approaches.

> Walking one evening along a deserted road, Mulla Nasrudin saw a troop of horsemen coming towards him. His imagination started to work; he imagined himself captured and sold as a slave, robbed by the oncoming horsemen, or conscripted into the army. Fearing for his safety, Nasrudin bolted, climbed a wall into a graveyard, and lay down in an open tomb.
>
> Puzzled at this strange behavior, the men—honest travelers—pursued Nasrudin to see if they could help him. They found him stretched out in the grave, tense and quivering.
>
> "What are you doing in that grave? We saw you run away and see that you are in a state of great anxiety and fear. Can we help you?"
>
> Seeing the men up close, Nasrudin realized that they were honest travelers who were genuinely interested in his welfare. He didn't want to offend them or embarrass himself by telling them how he had misperceived them, so Nasrudin simply sat up in the grave and said, "You ask what I'm doing in this grave. If you must know, I can tell you only this: *I* am here because of you, and you are here because of me." (Adapted from Shah, 1972, p. 16)

> At one time, one blade of grass is as effective as a sixteen-foot golden statue of Buddha. At another time, a sixteen-foot golden statute of Buddha is as effective as a blade of grass.
>
> —Zen master Wumen Huikai (1183–1260)

Some reports are thin as a blade of grass; others feel 16 feet thick. Size, of course, is not the issue. Quality is. But given the volume of data involved in qualitative inquiry and the challenges of data reduction already discussed, reporting qualitative findings is the final step in data reduction, and size is a real constraint, especially when writing in forms other than research monographs and book-length studies, like journal articles and newsletter summaries. Each step in completing a qualitative project presents *quality* challenges, but the final step is completing a report so that others can know what you've learned and how you learned it. This means finding and writing the story that has emerged from your analysis. It also means dealing with what Lofland (1971) called the "the agony of omitting"—deciding what material to leave out of the story.

> It can happen that an overall structure that organizes a great deal of material happens also to leave out some of one's most favorite material and small pieces of analysis. . . . Unless one decides to write a relatively disconnected report, he must face the hard truth that no overall analytic structure is likely to encompass every small piece of analysis and all the empirical material that one has on hand. . . .
>
> The underlying philosophical point, perhaps, is that everything is related to everything else in a flowing, even organic fashion, making coherence and organization a difficult and problematic human task. But in order to have any kind of understanding, we humans require that some sort of order be imposed upon that flux. No order fits perfectly. All order is provisional and partial. Nonetheless, understanding requires order, provisional and partial as it may be. It is with that philosophical view that one can hopefully bring himself to accept the fact that he cannot write about everything that he has seen (or analyzed) and still write something with overall coherence or overall structure. (p. 123)

Purpose Guides Writing and Reporting

This chapter opened with the reminder that *purpose guides analysis.* Purpose also guides report writing and dissemination of findings. The key to all writing starts with (a) knowing your audience and (b) knowing what you want to say to them—a form of strategic communication (Weiss, 2001).

Dissertations have their own formats and requirements (Biklen & Casella, 2007; Bloomberg & Volpe, 2012; Heer & Anderson, 2005; Piantanida & Garman, 2009). Scholarly journals in various disciplines and applied research fields have their own standards and norms for what they publish. The best way to learn them is to read and study them, and study specialized qualitative methods journals like *Qualitative Inquiry, Qualitative Research, Field Methods, Symbolic Interaction, Journal of Contemporary Ethnography, Grounded Theory Review, Qualitative Health Research,* and *American Journal of Evaluation.* The format for evaluation reports is a matter of negotiation and is usually specified in the contract that commissions the evaluation. In all of these cases, the guiding principles remain (a) know your audience and (b) know what you want to say to them. (For in-depth guidance on writing the results of qualitative analysis, see Goodall, 2008, *Writing Qualitative Inquiry;* Holliday, 2007, *Doing and Writing Qualitative Research;* Wolcott, 2009, *Writing Up Qualitative Research.*)

Reflexivity and Voice

> I do not put that note of spontaneity that my critics like into anything but the fifth draft.
>
> —Economist John Kenneth Galbraith (1986)
> (In the interview "The Art of Good Writing")

In Chapter 2, when presenting the major strategic themes of qualitative inquiry, I included as one of the 12 primary themes that of "Voice, Perspective, and Reflexivity."

The qualitative analyst owns and is reflective about her or his own voice and perspective; a credible voice conveys authenticity and trustworthiness; the inquirer's focus becomes balance—understanding and depicting the world *authentically* in all its

PRESENTATIONS AND FEEDBACK BEFORE A FINAL REPORT OR PUBLICATION

You may be called on to present your findings to colleagues or at seminars or conferences before you've completed a final report or formally written up your results for publication. Such occasions can be helpful in testing out your conclusions and getting feedback about how they are received. Indeed, it is wise to seek such opportunities to help you get outside your own perspective and find out what questions your analysis raises in the minds of others. You can use what you learn to focus and fine-tune your report.

Program Evaluation Feedback

A different but related challenge arises in evaluation when, as is typical, intended users (especially program staff and administrators) want preliminary feedback while fieldwork is still under way or as soon as data collection is over. Providing preliminary feedback provides an opportunity to reaffirm with intended users the final focus of the analysis and nurture their interest in findings. Academic social scientists have a tendency to want to withhold their findings until they have polished their presentation. Use of evaluation findings, however, does not necessarily center on the final report, which should be viewed as one element in a total utilization process, sometimes a minor element, especially in formative and developmental evaluation (Patton, 2012a; Torres et al., 1996).

Evaluators who prefer to work diligently in the solitude of their offices until they can spring a final report on a waiting world may find that the world has passed them by. Feedback can inform ongoing thinking about a program instead of serving only as a one-shot information input for a single decision point. However, sessions devoted to reestablishing the focus of the evaluation analysis and providing initial feedback need to be handled with care. The evaluator will need to explain that analysis of qualitative data involves a painstaking process requiring long hours of careful work, going over notes, organizing data, looking for patterns, checking emergent patterns against the data, cross-validating data sources and findings, and making linkages among the various parts of the data and the emergent dimensions of the analysis. Thus, any early discussion of findings can only be preliminary, directed at the most general issues and the most striking, obvious results. If, in the course of conducting the more detailed and complete analysis of the data, the evaluator finds that statements made or feedback given during a preliminary session were inaccurate, evaluation users should be informed about the discrepancy at once.

complexity while being self-analytical, politically aware, and reflexive in consciousness. This reiterates that the qualitative inquirer is the instrument of inquiry. (See Exhibit 2.1, pp. 46–47.)

Analysis and reporting are where reflexivity comes to the fore. As discussed in Chapter 2, the term *reflexivity* has entered the qualitative lexicon as a way of emphasizing the importance of deep introspection, political consciousness, cultural awareness, and ownership of one's perspective. Reflexivity calls on us to think about how we think and inquire into our thinking patterns even as we apply thinking to making sense of the patterns we observe around us. Being reflexive involves self-questioning, self-understanding, and

> *interpretation of interpretation* and the launching of a critical self-exploration of one's own interpretations . . . , a consideration of the perceptual, cognitive, theoretical, linguistic, (inter)textual, political and cultural circumstances that form the backdrop to—as well as impregnate–the interpretations. (Alvesson & Sköldberg, 2009, p. 9)

Reflexivity reminds the qualitative inquirer to be attentive to and conscious of the cultural, political, social, linguistic, and economic origins of one's own perspective and voice as well as the perspective and voices of those one interviews and those to whom one reports. To be reflexive, then, is to undertake an ongoing examination of what I know and how I know it.

So I repeat—*analysis and reporting are where reflexivity comes to the fore.* Throughout analysis and reporting, as indeed throughout all of qualitative inquiry, questions of reflexivity and voice must be asked as part of the process of engaging the data and extracting findings. Triangulated reflexive inquiry involves three sets of questions (see Exhibit 2.1 in Chapter 2, pp. 46–47.):

1. **The Self-Reflexivity question** What do I know? How do I know what I know? What shapes and has shaped my perspective? How have my perceptions and my background affected the data I have collected and my analysis of those data? How do I perceive those I have studied? With what voice do I share my perspective? What do I do with what I found? These questions challenge the researcher to also be a learner, to reflect on our "personal epistemologies"—the ways we understand knowledge and the construction of knowledge (Rossman & Rallis, 1998, p. 25).

2. **Reflexive questions about those studied** How do those studied know what they know? What shapes

and has shaped their worldview? How do they perceive me, the inquirer? Why? How do I know?

3. ***Reflexivity about the audience*** How do those who receive my findings make sense of what I give them? What perspectives do they bring to the findings I offer? How to they perceive me? How do I perceive them? How do these perceptions affect what I report and how I report it?

REFLEXIVITY WHEN RESEARCHING TRAUMA

Connolly and Reilly (2007) examined the impact of conducting narrative research focused on trauma and healing. They recounted what they found through three voices: (1) the study participants, who experienced the trauma; (2) the researchers, who shared their personal experiences of conducting the inquiry; (3) and "an academic colleague who acted as a reflective echo making sense of and normalizing the researcher's experience" (p. 522). Issues that emerged included

harmonic resonance between the story of the participant and the life experiences of the researcher; emotional reflexivity; complex researcher roles and identities; acts of reciprocity that redress the balance of power in the research relationship; the need for compassion for the participants; and self-care for the researcher when researching trauma. (p. 522)

Based on their reflexive process, the researchers concluded that when researching trauma, the researcher is a member of both a scholarly community and a human community and that maintaining the stance of a member of the human community is an essential element of conducting trauma research.

Reflexivity in the Methods Section of a Report

Because the qualitative inquirer is the instrument of qualitative inquiry and analysis, especially analysis, the methods section of a qualitative report should include some degree of reflexive discussion to acknowledge the perspective, skills, and experiences the inquirer has brought to the work. Such reflexivity is rare to nonexistent in quantitative reports, but qualitative reporting brings a different voice and perspective to the work. The methods section should reflect that difference. That's also why qualitative reports more often use the first-person voice. (For a discussion on using the first-person, active voice in reporting versus the third-person, passive voice, see the section on reflexivity meets voice [pp. 70–74 in Chapter 2].)

Self-awareness, even a certain degree of self-analysis, has become a requirement of qualitative inquiry. As the reflexive questions above suggest, attention to voice applies not only to intentionality about the voice of the analyst but also to intentionality and consciousness about whose voices and what messages are represented in the stories and interviews we report. Qualitative data "can be used to relay dominant voices or can be appropriated to 'give voice' to otherwise silenced groups and individuals" (Coffey & Atkinson, 1996, p. 78). Eminent qualitative sociologist Howard Becker (1967) posed this classically as the question "Whose side are we on?" Societies, cultures, organizations, programs, and families are stratified. Power, resources, and status are distributed differentially. How we sample in the field, and then sample again during analysis in deciding who and what to quote, involves decisions about whose voices will be heard.

Finally, as we report findings, we need to anticipate how what we report will be heard and understood. We need strategies for thinking about the nature of the reporter–audience interaction, for example, understanding how "six basic tendencies of human behavior come into play in generating a positive response: reciprocation, consistency, social validation, liking, authority, and scarcity" (Cialdini, 2001, p. 76). Some writers eschew this responsibility, claiming that they write only for themselves. But researchers and evaluators have larger social responsibilities to present their findings for peer review and, in the cases of applied research, evaluation, and action research, to present their findings in ways that are understandable and useful.

Triangulated reflexive inquiry provides a framework for sorting through these issues during analysis and report writing—and then including in the methods section of your report how these reflections informed your findings. (For examples of qualitative writings centered on illuminating issues of reflexivity and voice, see Hertz, 1997.) We turn now to the content of writing and reporting.

Balancing Description and Interpretation

One of the major decisions that has to be made in reporting is how much description to include. Rich, detailed description and direct quotations constitute the foundation of qualitative inquiry. Sufficient description and direct quotations should be included to allow the reader to enter into the situation observed and the thoughts of the people represented in the report. Description should stop short, however, of becoming trivial and mundane. The reader does not have to know everything that was done or said.

Focus comes from having determined what's substantively significant (see p. 572) and providing enough detail and evidence to illuminate and make that substantive case.

Yet the description must not be so "thin" as to remove context or meaning. Qualitative analysis, remember, is grounded in "thick description":

> A thick description does more than record what a person is doing. It goes beyond mere fact and surface appearances. It presents detail, context, emotion, and the webs of social relationships that join persons to one another. Thick description evokes emotionality and self-feelings. It inserts history into experience. It establishes the significance of an experience, or the sequence of events, for the person or persons in question. In thick description, the voices, feelings, actions, and meanings of interacting individuals are heard. (Denzin, 1989c, p. 83)

From Thick Description to Thick Interpretation

Thick description sets up and makes possible thick interpretation. By "thick interpretation," Denzin (1989c) means, in part, connecting individual cases to larger public issues and to the programs that serve as the linkage between individual troubles and public concerns: "The perspectives and experiences of those persons who are served by applied programs must be grasped, interpreted, and understood if solid, effective, applied programs are to be put into place" (p. 105).

Description precedes and is then balanced by analysis and interpretation. Endless description becomes its own muddle. The purpose of analysis is to organize the description so that it is manageable. Description provides the skeletal frame for analysis that leads to interpretation. An interesting and readable report provides sufficient description to allow the reader to understand the basis for an interpretation and sufficient interpretation to allow the reader to appreciate the description.

Details of verification and validation processes (topics of the next chapter) are typically placed in a separate methods section of a report, but parenthetical remarks throughout the text about findings that have been validated can help readers value what they are reading. For example, if I describe some program process and then speculate on the relationship between that process and client outcomes, I may mention that (a) staff and clients agreed with this analysis when they read it, (b) I experienced this linkage personally as a participant-observer in the program, or (c) this connection was independently arrived at by two analysts looking at the data separately.

The report should help readers understand the different degrees of significance of various findings, if these exist. Since qualitative analysis lacks the parsimonious statistical significance tests of statistics, the qualitative analyst must make judgments that provide clues for the reader as to the writer's belief about variations in the credibility and importance of different findings: When are patterns "clear"? When are they "strongly supported by the data"? When are the patterns "merely suggestive"? Readers will ultimately make their own decisions and judgments about these matters based on the evidence you've provided, but your analysis-based opinions and speculations deserve to be reported and are usually of interest to readers given that you've struggled with the data and know them better than anyone else.

Exhibit 8.35, at the end of this chapter (pp. 643–649), presents portions of a report describing the effects on participants of their experiences in the wilderness education program. The data come from in-depth, open-ended interviews. This excerpt illustrates the centrality of quotations in supporting and explaining thematic findings.

Communicating With Metaphors and Analogies

> All perception of truth is the detection of an analogy.
>
> —Henry David Thoreau (1817–1862)

The museum study reported earlier in the discussion of analyst-generated typologies differentiated different kinds of visitors by using metaphors: the "commuter," the "nomad," the "cafeteria type," and the "VIP." In the dropout study, we relied on metaphors to depict the different roles we observed teachers playing in interacting with truants: the "cop," the "old-fashioned schoolmaster," and "the ostrich." Language not only supports communication but also serves as a form of representation, shaping how we perceive the world (Chatterjee, 2001; Patton, 2000).

Metaphors and analogies can be powerful ways of connecting with readers of qualitative studies. But some analogies offend certain audiences, so they must be selected with some sensitivity to how those being described would feel and how intended audiences will respond. At a meeting of the Midwest Sociological Society, distinguished sociologist Morris Janowitz was asked to participate in a panel on the question "What is the cutting edge of sociology?" Janowitz (1979), having written extensively on the sociology of the military, took offense at the "cutting edge" metaphor. He explained,

> Paul Fussell, the humanist, has prepared a powerful and brilliant sociological study of the literary works of the great wars of the 20th century which he entitled *The Great War and Modern Memory.* It is a work which all sociologists should read. His conclusion is that World War I and World War II, Korea and Vietnam have militarized our language. I agree and therefore do not like the question "Where is the cutting edge of sociology?" "Cutting Edge" is a military term. I am put off by the very term cutting edge. Cutting edge, like the parallel term breakthrough, are slogans which intellectuals have inherited from the managers of violence. Even if they apply to the physical sciences, I do not believe that they apply to the social sciences, especially sociology, which grows by gradual accretion. (p. 591)

Of particular importance, in this regard, is avoiding metaphors with possible racist and sexist connotations, for instance, "It's black and white." One external reviewer of this book felt that this point deserves special emphasis, so I yield the floor, so to speak:

> A lack of respect for people is conveyed in insensitive metaphors. So let's avoid metaphors that are insensitive to regional differences, health and mental health differences, sexual orientation, and so on and so forth. No need to list them all (who could?), but at least make the comment about being respectful of people rather than seeming to be specific to just two types of insensitivity [racist and sexist insensitivities]. It is just too easy to inadvertently fall into habits of speech that can be hurtful.

At an Educational Evaluation and Public Policy Conference sponsored by the Far West Laboratory for Educational Research and Development, the women's caucus expressed concern about the analogies used in evaluation and went on to suggest some alternatives:

> To deal with diversity is to look for new metaphors. We need no new weapons of assessment—the violence has already been done! How about brooms to sweep away the attic-y cobwebs of our male/female stereotypes? The tests and assessment techniques we frequently use are full of them. How about knives, forks, and spoons to sample the feast of human diversity in all its richness and color. Where are the techniques that assess the delicious-ness of response variety, independence of thought, originality, uniqueness?

(And lest you think those are female metaphors, let me do away with that myth—at our house everybody sweeps and everybody eats!) Our workgroup talked about another metaphor—the cafeteria line versus the smorgasbord banquet of styles of teaching/learning/assessing. Many new metaphors are needed as we seek clarity in our search for better ways of evaluating. To deal with diversity is to look for new metaphors. (Hurty, 1976, p. 1)

When employing a metaphor, it is important to make sure that it serves the data and not vice versa. Don't manipulate the data to fit the metaphor. Moreover, because metaphors carry implicit connotations, it is important to make sure that the data fit the most prominent of those connotations so that what is communicated is what the analyst wants to communicate. Finally, one must avoid reifying metaphors and acting as if the world were really the way the metaphor suggests it is.

> The metaphor is chiefly a tool for revealing special properties of an object or event. Frequently, theorists forget this and make their metaphors a real entity in the empirical world. It is legitimate, for example, to say that a social system is like an organism, but this does not mean that a social system is an organism. When metaphors, or concepts, are reified, they lose their explanatory value and become tautologies. A careful line must be followed in the use of metaphors, so that they remain a powerful means of illumination. (Denzin, 1978b, p. 46)

How "real" metaphors are may turn out to be a specific manifestation of Thomas's theorem: What is perceived as real is real in its consequences. Brain scans are revealing that when we read a detailed description, an evocative metaphor, or an emotional story, the brain is stimulated. A team of brain researchers from Emory University found that

> when subjects in their laboratory read a metaphor involving texture, the sensory cortex, responsible for perceiving texture through touch, became active. Metaphors like "The singer had a velvet voice" and "He had leathery hands" roused the sensory cortex, while phrases matched for meaning, like "The singer had a pleasing voice" and "He had strong hands," did not. (Paul, 2012, p. SR6)

This invites research into the larger question of what happens to the brain when one is analyzing qualitative data or reading a full qualitative case study. Stay tuned.

Creating and Incorporating Visuals

The raw data of qualitative inquiry take the form of words, narratives, recorded observations, documents, and stories. This chapter has discussed how these data are analyzed and interpreted through content, pattern, and theme analysis; constructing cases studies and cross-case analyses; creating typologies; and depicting causal connections and interrelationships. At the center of all these analytical approaches have been words. Now, we turn to those things that have legendarily and metaphorically been worth a thousand words: pictures, visuals, and graphics.

No trend is more pronounced in the past decade than the visualization of data and findings. The capability to create meaningful and powerful visuals is a skill that is likely to become increasingly important in our short-attention-span world. Visualization rules. But before we exalt its place in contemporary analysis and reporting, let's pause and give a nod to the visualization pioneers who couldn't just go on the Internet and download some impressive photos and graphics. The legendary nurse Florence Nightingale (1820–1910) gathered data over a period of a year showing deaths in the Crimean War (1855–1856) due to battle wounds compared with deaths due to infections and disease (a much larger number). She converted the data into a color-coded visual display that proved hugely influential in changing sanitation practices and paved the way for attention to preventing infections, which saved hundreds of thousands of lives (Magnello, 2012). In 1894, George Waring Jr. was appointed Street Commissioner of the City of New York and began a systematic sanitation program that cleared the streets of New York of shin-deep garbage and animal and human waste. He took and had published in the newspapers before and after photographs showing the visual difference his reforms brought (Waring, 1897; Wells, 2012). In 1896, New York City honored him with a parade of appreciation for his contributions to the city's quality of life, which included draining the wetlands of Manhattan Island to create Central Park. That is the vision of visualization we build on.

The best way to present qualitative data visualization is not to talk about such visuals but to provide actual illustrations. Exhibit 3.13 (pp. 142–143) in Chapter 3 presented a theory-of-change baseline systems graphic depicting the situation at the beginning of a light rail construction project; the key system actors, structures, and processes; and the lack of integration among these subsystem elements. The purpose of that visual graphic was to provide an example of how qualitative inquiry can be used to depict a system and support the conceptualization and evaluation of system change. Take another look at that graphic from the perspective of visually depicting findings, in that case the results of focus groups with key knowledgeables.

In this section, I'll present examples of visual displays of qualitative findings with a minimum of accompanying verbiage, inviting you instead to engage with each type of illustration as a way of stimulating your thinking about making visualizations a part of your qualitative reporting (see Exhibits 8.20–8.27).

SIDEBAR

PICTURING DISABILITIES

Qualitative sociologist Robert Bogdan (2012) has assembled, analyzed, and published more than 200 historical photographs of people with disabilities. Beginning in the 1860s, when photography was emerging as a commercial enterprise, up to the 1970s, when the disability rights movement forced change, he shows how people with disabilities were portrayed. In one photo, a young woman with no arms wears a sequined tutu and smiles for the camera as she holds a teacup with her toes. In another, a man holds up two prosthetic legs while his own legs are bared to the knees to show his missing feet. Such photos were used as promotional material for circus sideshows and charity drives and hung in art galleries. They were found on "begging cards" and in family albums.

Bogdan's (2012) analysis includes an inquiry into the perspective, role, and values of the photographers who took these photos and the contexts within which such photographs were created and people with disabilities were exploited. He examines a wide range of purposes and uses of disability photographs, from sideshow souvenirs to clinical photographs. The photographs are both data and visualization of findings.

EXHIBIT 8.20   Photos Before and After to Illustrate Change

Vietnam Helmet Law

- Road traffic injuries have long been a leading cause of death and disability in Vietnam.
- 60% of fatalities occur in motorcycle riders and passengers.

- Vietnam has had a partial motorcycle helmet legislation since 1995, However implementation and enforcement had been limited.

- On 15 December 2007, Vietnam first comprehensive mandatory helmet law came into effect, covering all riders and passengers on all roads nationwide. Penalties increased ten-fold and cohorts of police were mobilized for enforcement.

Photos on street corners the day before the law took effect

Photos on the same street corners the day after the law took effect

Results

The Asia Injury Prevention Foundation reported: "Nearly 100% of Vietnam's motorbike users left home wearing a helmet. It was an unbelievable sight with a near instantaneous effect. Major hospitals report the number of patients admitted for traumatic brain injuries in the two days after the law's enactment was much lower than on previous weekends. In Ho Chi Minh City alone, serious traffic accident injuries fell by almost 50 percent compared with pre-helmet weekends."

SOURCE: McDonnell, Tran, and McCoy (2010).

EXHIBIT 8.21   Immigration Roadmap Into the United States

Based on interviews, document analysis, case studies, and key informants' expertise, this diagram depicts the barriers to foreigners getting a green card in the United States as of 2009. Describing (in words) what is on this one-page

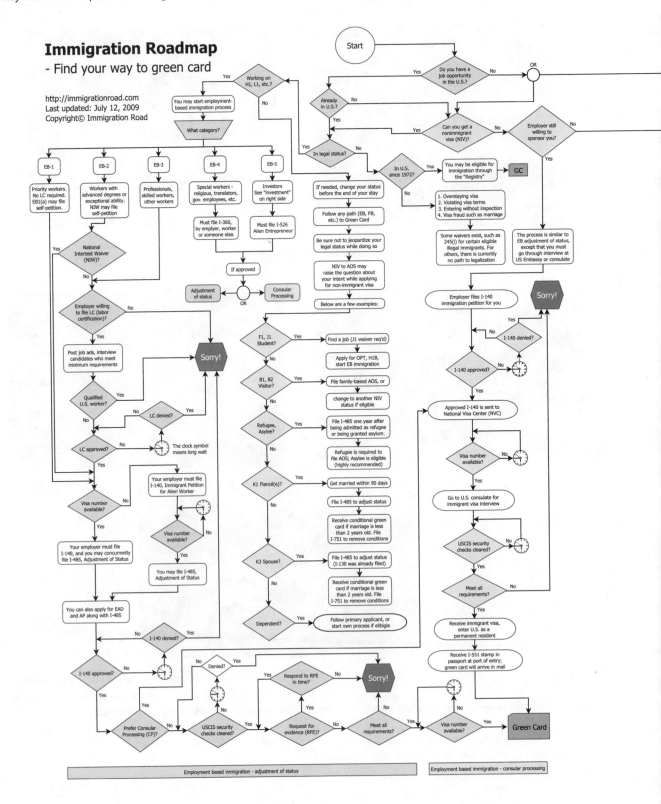

diagram would take 30 to 50 pages. The daunting nature of the visual and the seven "Sorry" hexagons that repre-

sent failure give a visceral feel to what it would be like to be caught somewhere in this maze.

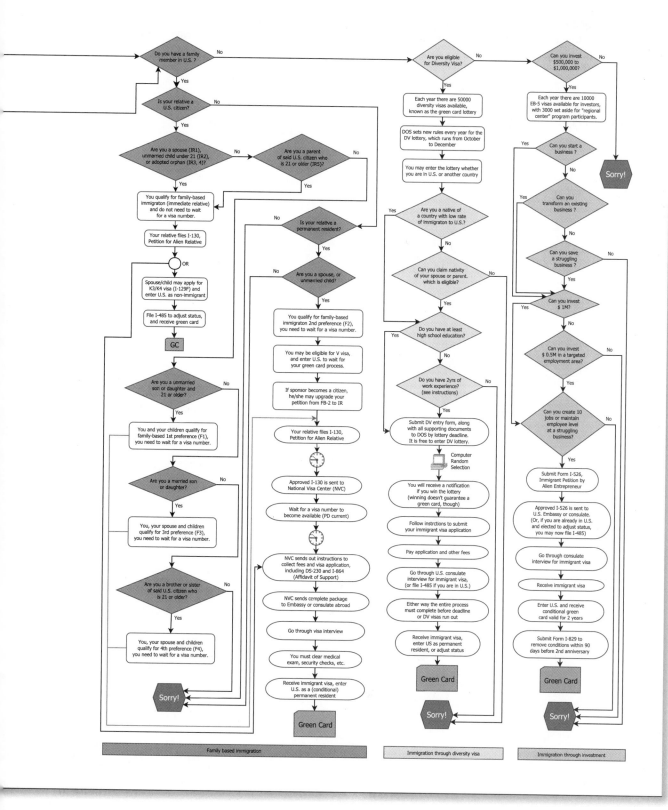

EXHIBIT 8.22   Interpersonal Systems

In a program for low-income, drug-addicted, pregnant teenagers, each young woman was asked to draw a map of her closest relationships as they were at the start of the program (baseline)—people who could support her during her pregnancy and after her baby was born. The program aimed to strengthen that network of support in keeping with the mantra that *it takes a village*. The interpersonal maps show the diversity of relationships these young women had, including one who was completely isolated and alone (Map 3). The program used the maps to help the program participants map the strength of their interpersonal support systems. The maps were used in the evaluation to graphically depict change, accompanied by a case study of each young woman.

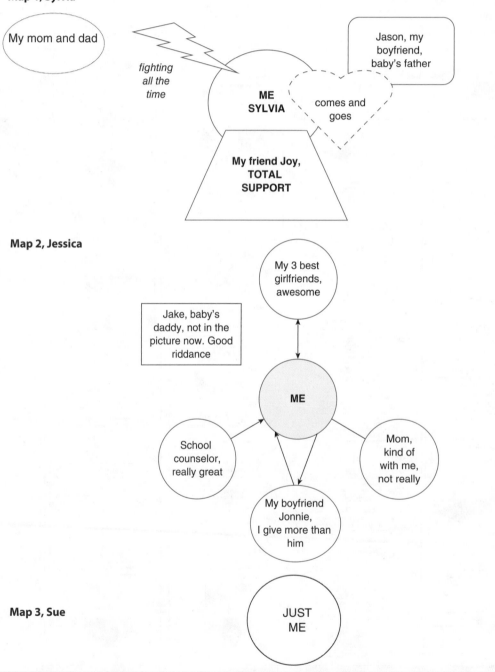

Map 1, Sylvia

My mom and dad

fighting all the time

ME SYLVIA

comes and goes

Jason, my boyfriend, baby's father

My friend Joy, TOTAL SUPPORT

Map 2, Jessica

My 3 best girlfriends, awesome

Jake, baby's daddy, not in the picture now. Good riddance

ME

School counselor, really great

My boyfriend Jonnie, I give more than him

Mom, kind of with me, not really

Map 3, Sue

JUST ME

SOURCE: McDonnell, Tran, and McCoy (2010).

EXHIBIT 8.23 Mountain of Accountability

I have worked with the The Blandin Foundation in Grand Rapids, Minnesota, on and off, over more than 30 years. In 2013, the senior staff undertook a strategic reflective practice exercise that included examining and making sense of the many different kinds of evaluation going on at the foundation. A lot was happening, but the pieces didn't seem to fit together. As we examined the various elements and what each did, we looked for a metaphor or graphic depiction that would bring order and cohesion to the messiness and multifaceted evaluation functions. What emerged was a Mountain of Accountability, based on Maslow's hierarchy framework, where basic necessities are at the bottom and organizational excellence is at the top. Information and patterns from lower levels inform higher levels, and understandings, lessons, and insights from higher levels flow back down to support lower levels and facilitate adaptation and ongoing learning.

(For the full explanation of the Mountain of Accountability, see the website http://blandinfoundation.org/who-we-are/accountability.php.)

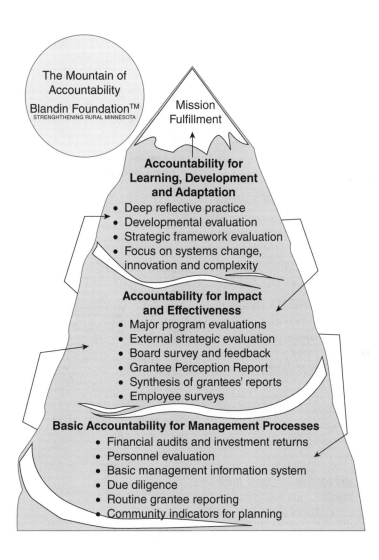

The Mountain of Accountability

Blandin Foundation™
STRENGHTENING RURAL MINNESOTA

Mission Fulfillment

Accountability for Learning, Development and Adaptation
- Deep reflective practice
- Developmental evaluation
- Strategic framework evaluation
- Focus on systems change, innovation and complexity

Accountability for Impact and Effectiveness
- Major program evaluations
- External strategic evaluation
- Board survey and feedback
- Grantee Perception Report
- Synthesis of grantees' reports
- Employee surveys

Basic Accountability for Management Processes
- Financial audits and investment returns
- Personnel evaluation
- Basic management information system
- Due diligence
- Routine grantee reporting
- Community indicators for planning

EXHIBIT 8.24   Depicting Interconnected Factors

In the discussion on causal attribution, I featured the example of how the GEM was used to determine how an advocacy campaign might have contributed to and influenced a U.S. Supreme Court decision. The preponderance of evidence pointed to some influence (see p. 595). Based on our synthesis of the findings, we generated a systems model depicting the interdependent elements of an integrated approach to policy reform that makes coalition building a centerpiece strategy. The model consists of six factors that, together, contribute to strengthening policy reform. There are six elements in the generic model:

1. **Strong, high-capacity coalitions:** Working through coalitions is a common centerpiece of advocacy strategy.

2. **Strong national–state grassroots coordination:** Effective policy change coalitions in the United States have to be able to work bottom up and top down, with national campaigns supporting and coordinating state and grassroots efforts while state efforts infuse national campaigns with local knowledge and grassroots energy. Strengthening strong national–state coordination is part of coalition development and field building.

3. **Disciplined and focused messages with effective communication:** Effective communication must occur within movements (message discipline) and to target audiences (focused messaging). Strengthening communication has been a key component of advocacy coalition building.

4. **Solid research and knowledge base:** The content of effective messages must be based on solid research and timely knowledge. In the knowledge age, policy coalitions must be able to marry their values with relevant research and real-time data about the dynamic policy environment.

5. **Timely, opportunistic lobbying and judicial engagement:** The evaluation findings emphasize that effective lobbying requires connections, skill, flexibility, coordination, and strategy.

6. **Collaborating funders engaged in strategic funding:** Effective funding involves not only financial support but also infusion of expertise and strategy as part of field building.

Overall Lesson for Effective Advocacy

In essence, *strong national–state–grassroots coordination* depends on having a *high-capacity coalition.* A *solid knowledge and research base* contributes to a *focused message and effective communication. Message discipline* depends on a strong coalition and national–state coordination, as *does timely and opportunistic lobbying and judicial engagement.* To build and sustain a high-capacity coalition, *funders must use their resources* and *knowledge to collaborate around shared strategies.* These factors *in combination* and as *mutual reinforcement* strengthen advocacy efforts. In classic systems framing, the whole is greater than the sum of the parts, and the optimal functioning of each part depends on the optimal integration and integrated functioning of the whole.

We needed a graphic depicting the dynamic nature of the relationship among these factors, basically depicting the interconnections among the factors as looking like a fluid spider web.

The interdependent system of factors that contribute to effective advocacy and change

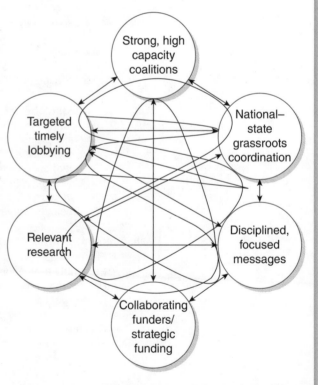

EXHIBIT 8.25 Depicting Organizational Tensions

An organizational development study revealed a number of fundamental tensions. The study was based on interviews, focus groups, and extensive review of documents, and observations of meetings and operational implementation and program delivery sessions. To report the findings, we created a set of graphic representations highlighting the tensions. Two ovals represent the two arenas of action that are in tension (e.g., strategy vs. execution).

Within each oval are listed key elements that define each strategic dimension (i.e., key elements of strategy in one oval and key elements of execution in the other oval). Connecting the two ovals is a large arrow that shows the two dimensions as interrelated. Highlighted between the two ovals are some of the principal tensions we identified between them.

STRATEGY versus EXECUTION

STRATEGY

- Articulate corporate values
- Focus strategic priorities
- Align perspective and position
- Strategy is concrete, explicit, and understood enough to guide action on the ground

- Strategy formulation competence versus execution competence
- Distinguishing strategy failure from execution failure
- Managing the discrepancy between implementation messiness versus strategy clarity
- Making partner agendas and concerns the focus versus our own strategic priorities and practices on the ground

EXECUTION

- Implement values in action
- Follow strategic priorities
- Deal with inevitable and inherent tensions in the field
- Competence to think and act under conditions of complexity and uncertainty where strategy execution has to be adapted to real, on-the-ground condition

(Continued)

(Continued)

MACRO versus MICRO

TENSIONS

MACRO

- Global program areas and problem definitions
- Relevance to the international community
- Participation by policy elites in countries and regions
- Long-term trends and impacts

TENSIONS

- Global problem definition versus local solutions
- Generalizable theory versus context sensitivity
- Top-down versus bottom-up strategy development and implementation
- Policy elites vs pluralism
- Big picture (from 10,000 meters) vs. on-the-ground view

MICRO

- Addressing local problems, politics, and context
- Participation by local partners
- Importance of specific, individual relationships at the local level
- Short-term deliverables and outcomes

EXHIBIT 8.26   Depicting Evaluation Tensions

No tension more deeply permeates evaluation than the admonition that evaluators should work closely with stakeholders to enhance mutual understanding, increase relevance, and facilitate use while maintaining independence to ensure credibility. Managing this tension was at the center of the way the evaluation of the Paris Declaration on Development Aid was structured, administered, and governed.

- At the country level, the national reference group in each country selected the evaluation team, based on published terms of reference and competitive processes; country evaluation teams were to operate independently in carrying out the evaluation.
- At the international level, the secretariat administered the evaluation, with an evaluation management group that selected the core evaluation team and provided oversight and guidance, while the

core evaluation team conducted the evaluation independently.

- The international reference group of country representatives, donor members, and international organization participants provided a stakeholder forum for engaging with the core evaluation team and the evaluation's draft findings.

Thus, at every level, the evaluation was structured to separate evaluation functions from political and management functions, while also providing forums and processes for meaningful interaction. Such structures and processes are fraught with tensions and risks. Yet, in the end, the credibility and integrity of the evaluation depended on doing both well: engaging key stakeholders to ensure relevance and buy-in while maintaining independence to ensure the credibility of findings. In our report, we created

TOP-DOWN VS. BOTTOM-UP EVALUATION PROCESSES

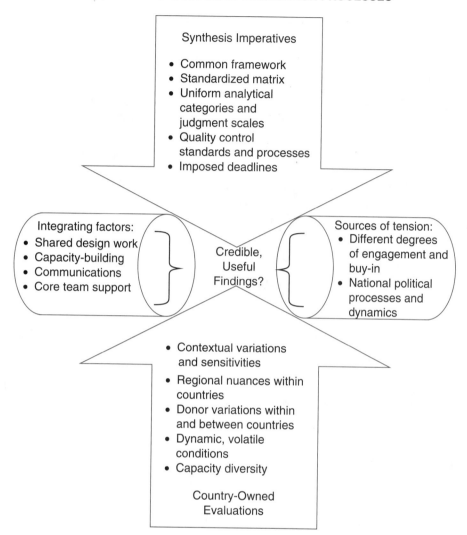

a graphic to depict this tension (Patton & Gornick, 2011a, p. 37).

The Paris Declaration evaluation team committed to a bottom-up design with country evaluations and donor studies feeding into a final synthesis of findings. But to make the synthesis manageable and facilitate aggregation of findings, the evaluation approach included building overarching, common elements:

- A common evaluation framework
- Common core questions
- A standardized data collection matrix
- Standardized scales for rendering judgments

- Standardized analytical categories for describing and interpreting progress: direction of travel, distance traveled, and pace
- A common glossary of terms

So the evaluation had a top-down, standardized framework to facilitate synthesis. The joint, collaborative, and participatory nature of the evaluation meant that these bottom-up and top-down processes had to be carefully managed. We found some confusion and tension about the relationship between country evaluations and the overall global synthesis, so we created a graphic to summarize, highlight, and depict the tensions that emerged in the data (Patton & Gornick, 2011a, p. 51).

(Continued)

(Continued)

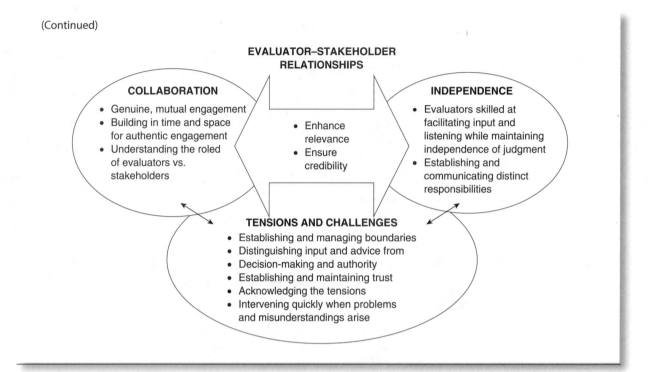

EVALUATOR–STAKEHOLDER RELATIONSHIPS

COLLABORATION
- Genuine, mutual engagement
- Building in time and space for authentic engagement
- Understanding the roled of evaluators vs. stakeholders

- Enhance relevance
- Ensure credibility

INDEPENDENCE
- Evaluators skilled at facilitating input and listening while maintaining independence of judgment
- Establishing and communicating distinct responsibilities

TENSIONS AND CHALLENGES
- Establishing and managing boundaries
- Distinguishing input and advice from
- Decision-making and authority
- Establishing and maintaining trust
- Acknowledging the tensions
- Intervening quickly when problems and misunderstandings arise

Strengths and Weaknesses of Visualization

The visualization examples I've shared are not meant to stand alone. They supplement and focus qualitative reporting but are not a substitute for narrative. Exhibit 8.27 concludes this section with a graphic overview of strengths and weaknesses in visualizing qualitative data and findings.

EXHIBIT 8.27   Visualization of Qualitative Data and Findings: Strengths and Weaknesses

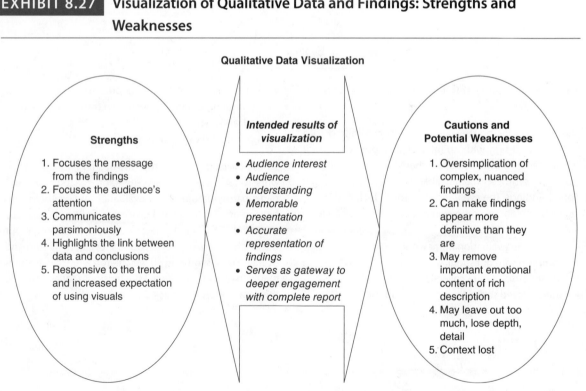

Qualitative Data Visualization

Strengths
1. Focuses the message from the findings
2. Focuses the audience's attention
3. Communicates parsimoniously
4. Highlights the link between data and conclusions
5. Responsive to the trend and increased expectation of using visuals

Intended results of visualization
- Audience interest
- Audience understanding
- Memorable presentation
- Accurate representation of findings
- Serves as gateway to deeper engagement with complete report

Cautions and Potential Weaknesses
1. Oversimplication of complex, nuanced findings
2. Can make findings appear more definitive than they are
3. May remove important emotional content of rich description
4. May leave out too much, lose depth, detail
5. Context lost

Resources for Qualitative Data Visualization

- Azzam and Evergreen (2013a, 2013b). *Data Visualization*
- Ball and Smith (1992). *Analyzing Visual Data*
- Banks (2007). *Using Visual Data in Qualitative Research*
- Bryson, Ackermann, and Eden (2014). *Visual Strategy: A Workbook for Strategy Mapping for Public and Nonprofit Organizations*
- Bryson, Ackerman, Eden, and Finn (2004). *Visible Thinking*
- Davis (2014). *Conversations About Qualitative Communication Research*
- Evergreen (2012). "Potent Presentations"
- Heath et al. (2010). *Video in Qualitative Research: Analyzing Social Interaction in Everyday Life*
- Henderson and Segal (2013). *Visualizing Qualitative Data in Evaluation*
- Kistler (2014). "Innovative Reporting"
- Margolis and Pauwels (2011). *The SAGE Handbook of Visual Research Methods*
- Mathison (2009). "Seeing Is Believing: The Credibility of Image-Based Research and Evaluation"
- Miles et al. (2014). *Qualitative Data Analysis: A Methods Sourcebook*

SOURCE: Brazilian cartoonist Claudius Ceccon. Used with permission.

MODULE

74

Special Analysis and Reporting Issues: Mixed Methods, Focused
Communication, Principles-Focused Report Exemplar, and Creativity

{

> For some critics, qualitative research has been getting better and more inventive. Even such a demanding veteran as Van Maanen suggests that ethnographers, especially those in the vanguard of the new ethnographic genres, are "learning to write better, less soothing, more faithful and ultimately more truthful accounts of their fellow human beings than ever before." And Denzin sees methodological blossoming in the cross-fertilization of new "epistemologies . . . and new genres."
>
> —Huberman and Miles (2002, pp. ix–x)
> *The Qualitative Researcher's Companion*

Integrating Qualitative and Quantitative Data

> Mixed methods researchers are extending our understandings of how to understand complex social phenomena, as well as how to use research to develop effective interventions to address complex social problems.
>
> —Donna M. Mertens (2013)
> Editor, *Journal of Mixed Methods Research*
> "Emerging Advances in Mixed Methods: Addressing Social Justice"

Exhibit 5.8, on purposeful sampling (pp. 266–272), includes several mixed-methods strategies for picking cases. Here, again, design frames analysis. Thus, a mixed-methods design anticipates a mixed-methods analysis. For example, a random, representative sample of recent immigrants in Minnesota was surveyed to determine their priority needs. A small number of respondents in different categories of need were then interviewed to understand in depth their situations and priorities and to make the statistical findings more personal through stories. The qualitative sample was also used to validate the accuracy and meaningfulness of the survey responses.

Mixed methods have increased in both importance and credibility as the strengths and weaknesses of both quantitative and qualitative approaches have become more evident. Quantitative data are strong in measuring how widespread a phenomenon is. Qualitative methods are strong in explaining what the phenomenon means. Combining methods would seem straightforward. It is not. Often, mixed-methods reports are written in two sections, a qualitative section and a quantitative section, and the two self-contained studies never engage each other. The two reports, often written by two different people, are like toddlers engaging in parallel play in a sandbox. They're old enough to recognize each other but not developed enough to interact. Ideally, however, studies are designed to be genuinely mixed (valuing both kinds of data) and, because the design makes it possible, the qualitative and quantitative data are integrated in analysis and reporting (Bergman, 2008; Greene, 2007; Mertens, 2013; Morgan, 2014). Exhibit 8.28 on pages 622–623 reviews challenges and solutions in mixed-methods analysis and reporting. But, first, to get into an appreciative mind-set for integrating qualitative and quantitative data, consider this intriguing example from the reflections of Abd Al-Rahman III, an emir and caliph of Córdoba in 10th-century Spain. He was an absolute ruler who lived in complete luxury. Here's how he assessed his life:

> I have now reigned above 50 years in victory or peace; beloved by my subjects, dreaded by my enemies, and respected by my allies. Riches and honors, power and pleasure, have waited on my call, nor does any earthly blessing appear to have been wanting to my felicity. I have diligently numbered the days of pure and genuine happiness which have fallen to my lot: They amount to 14. (Quoted by Brooks, 2014, p. SR1)

INSIGHTS FROM AN EXPERIENCED MIXED-METHODS EVALUATOR

Quantitative evidence is the bones; qualitative evidence is the flesh; and evaluative reasoning is the vital organs. If you are missing any of these you don't have the full evaluative picture.

I frame all my evaluation work around asking and answering important evaluative questions. These are the "what are we trying to find out from this whole evaluation?" questions (not the interview questions, survey questions, etc, which are much lower level).

I've yet to conduct an evaluation that didn't require some fairly substantial qualitative evidence to convincingly support the conclusions (answers to these high-level questions). In most cases I use mixed methods, but I have done a few evaluations that are 99–100% qualitative.

But more important than that, I use evaluation-specific methods. These are the ones that answer not just what's happening and what's changed, but the really tricky questions like whether the change is big enough, fast enough, worth the time, money and effort invested.

I also use qualitative methods a lot for causal inference. It's a rare case where quantitative data alone makes a convincing case for a causal contribution, and qualitative evidence can be used to substantially strengthen the case—and in some cases is enough on its own.

—Jane Davidson
*Evaluation Methodology Basics:
The Nuts and Bolts of Sound Evaluation* (2005)
*Actionable Evaluation Basics: Getting Succinct
Answers to the Most Important Questions* (2012)
Cofounder of Genuine Evaluation
blog (http://genuineevaluation.com/)

Focused Communications: Audience-Sensitive Presentations and Reports

I emphasized earlier that reports should be written to communicate what you want to say to specific audiences. The particular challenge of qualitative reporting is reducing the sheer volume of data into digestible morsels. Book-length monographs may be the only way to do justice to the rich descriptions and detail that undergird long-term fieldwork and the in-depth case studies that in-depth interviews yield. But even if you

get the opportunity to do a book, write a dissertation, or publish a full monograph, opportunities will arise for shorter and more focused articles and presentations. In such cases, focus is critical.

Even a comprehensive report will have to omit a great deal of information collected in a qualitative inquiry. If you try to include everything, you risk losing your readers or audience members in the sheer volume of the presentation. To enhance a report's coherence or a presentation's impact, follow the adage that *less is more*. This translates into covering a few key findings or conclusions well rather than lots of them poorly.

An evaluation report, for example, should address clearly each major evaluation question. The extent to which minor questions are addressed is a matter of negotiation with primary intended users. Sometimes, answers to minor questions are provided orally or in an appendix to keep the focus on major questions. For each one, present the descriptive findings, analysis, and interpretation succinctly. An evaluation report should be readable, understandable, and relatively free of academic jargon. The data should impress the reader beyond the academic credentials of the evaluator.

The advice I find myself repeating most often to students when they are writing articles or preparing

Oh good Bill! You're reading my report!

So, what do you think? Is the thick description thick enough?

Glad you like it. Since you don't have any recommended changes...

...I'll send it to the print shop for national distribution.

EXHIBIT 8.28   Mixed-Methods Challenges and Solutions

The column on the left presents common challenges in mixing methods (qualitative and quantitative) during analysis and reporting. As summary items in tabular format, these are admittedly overstated and oversimplified contrasts. Those who are sophisticated about mixing methods will certainly find the contrasts exaggerated. That said, the literature on mixed methods (e.g., the Journal of Mixed Methods *Research) and my own experience facilitating integration of different types of data lead me to emphasize the contrasts to highlight the potential and actual challenges of integration. Integrating different kinds of data is neither easy nor straightforward. That's the core message of this exhibit. The column on the right, therefore, offers integrative strategies and solutions.*

CHALLENGES IN MIXING METHODS (QUALITATIVE AND QUANTITATIVE) IN ANALYSIS AND REPORTING	INTEGRATIVE STRATEGIES AND SOLUTIONS
1. *Different design logics answering different questions.* Quantitative inquiry designs (measuring the amount of things and their relationships) versus qualitative inquiry designs (capturing perspectives and the meanings of things).	Understand what each design does. At the design stage, consider how the different questions derived from different methods will yield different perspectives on the same phenomenon. *Plan for integration at the design stage.*
2. *Different kinds of results.* Quantitative and qualitative designs do different things and produce different kinds of results. *Note:* Not just different results but *different kinds of results—numbers and statistical relationships versus words, narratives, and perceptions.*	Ask integrating questions of the data. How do the numbers inform the narrative, providing insight into the scope, scale, and extent of findings? How does the narrative give meaning to the numbers, explaining their substantive significance?
3. *Different quality standards.* Quantitative analysis emphasizes validity, reliability, and generalizability. Constructivist qualitative analysis emphasizes trustworthiness, authenticity, and transferability. Exhibit 9.7 (Chapter 9, pp. 680–681) discusses these varying, and sometimes conflicting, quality criteria in detail.	Design the inquiry, and collect data to meet quality standards appropriate to each inquiry tradition. A strong quantitative study integrates poorly with a weak qualitative study, and vice versa. Play to the strengths of each. *Worst-case scenario:* A weak qualitative study and a weak quantitative study create a really weak integrated design.
4. *Different knowledge and skill requirements.* Few researchers or evaluators are strong in both qualitative and quantitative methods.	Ensure that you have the expertise needed by filling whatever gaps there are in your own skill set, create an inquiry team of people with the needed range of skills, or bring in special expertise to supplement whatever expertise is missing.
5. *Different timelines and time commitments.* Quantitative measurement requires a lot of upfront work to ensure valid instruments and representative sampling, but the analysis is relatively standardized and quick. The openness of qualitative inquiry can make it possible to get into the field quickly, but data collection takes longer and analysis a lot longer.	Integrating qualitative and quantitative methods requires careful and thoughtful project planning and management. Schedule design, data collection, and analysis to support integration. Time phases of the work to support interaction and integration all the way along throughout the mixed-methods inquiry.
6. *Different data preferences.* A single inquirer is unlikely to value equally both qualitative and quantitative methods. Preferences derive from knowledge, expertise, experience, and training. In teams, different participants are likely to have different preferences. Preferring one kind of data over the other undermines integration.	Individuals using both methods will need to undertake serious reflexive and reflective analysis of how data preferences may be affecting mixed-methods integration. Teams may need special integrative coaching or a person with specialized "boundary-spanning expertise" to support and facilitate integration (Morgan, 2014, pp. 214–224).

CHALLENGES IN MIXING METHODS (QUALITATIVE AND QUANTITATIVE) IN ANALYSIS AND REPORTING	INTEGRATIVE STRATEGIES AND SOLUTIONS
7. *Conflicting or divergent results.* Qualitative and quantitative methods not only yield different kinds of results, the findings often also conflict or diverge. For example, quantitative data may show no statistically significant outcomes in an evaluation, while qualitative case studies may show substantial substantive significance, which may be the result of an evaluation of home visit programs (Sherwood, 2005).	Integration means interpreting different results and understanding how and why different methods yielded different results. Integrative analysis involves precisely this kind of interpretation, making sense of what each method produces and what that means for understanding the phenomenon of interest. Divergent results and different interpretations of significance are not necessarily conflicting results. Know the difference.
8. *Varying degrees of clarity from different methods.* Quantitative measures, by their nature, have an air of precision. Qualitative findings require more interpretation. But either method may produce ambiguous findings (and often does). If one method produces clear findings and the other is ambiguous, integration may fall a victim to highlighting the clear findings.	As with conflicting results, stay committed to understanding how and why different methods yielded different results with varying degrees of clarity and certainty. *Clear* does not mean more valid, more important, more useful, more credible, or more worthy of attention. It only means "clear." Weak and null results can be especially clear—and wrong. Likewise with strong results. Stay focused on what you can learn from each, understanding how and why different methods yielded different results.
9. *Writing involves different formats and styles.* Statistics and narratives have different norms for presenting data. Even voice can vary. Traditional quantitative research tends to employ the third-person, passive voice. Much of qualitative inquiry prefers the first-person, active voice for writing. (*Note:* These are writing tendencies, not rules.)	This is a particularly difficult challenge for integration. It's one of the reasons why quantitative and qualitative findings are often reported separately. Be creative about formatting and integration. Use visuals and graphics to integrate. And on voice, decide which voice will most resonate with your primary intended audience (10 on this list).
10. *Audiences bring personal expectations, methodological biases, and data preferences to their reading of reports.* Integrated mixed-methods reports are still so rare that some audiences haven't developed a taste or appreciation for them. And much of what they've seen may have been badly done. Moreover, people receiving reports are prone to their own biases about what kind of methods and data they prefer, so they may not be prepared to find value in balance and integration.	Explain carefully and fully in the report why you are taking an integrated mixed-methods approach. Acknowledge that it is not yet customary but it is the trend, especially in evaluation and policy research. Help readers understand what they will be looking at and provide guidance in how to make sense of mixed-methods findings. Repeat this explanation periodically as you present integrated, mixed-methods findings. In short, do some audience hand-holding throughout the report.
Resources for mixed-methods integration	Bergman (2008), Creswell and Clark (2011), Greene (2007), Mertens and Hesse-Biber (2013), Tashakkori and Teddlie (2003), and Teddlie and Tashakkori (2011)

presentations is FOCUS! FOCUS! FOCUS! The agony of omitting on the part of the qualitative researcher or evaluator is matched only by the readers' or listeners' agony in having to read or hear those things that were not omitted but should have been. Exhibit 8.29 presents an image of audience- and utilization-focused reporting.

Connecting With the Audience

An Evaluative Example Using Indigenous Typologies

Earlier in this chapter (pp. 546–548), I discussed analyzing findings through the framework of an

COLLABORATIVE OUTCOMES REPORTING

Collaborative Outcomes Reporting is a participatory approach to impact evaluation. Multiple kinds of evidence of outcomes are pulled together and presented as a "performance story." To ensure the validity and credibility of the performance story, it is independently reviewed by both technical experts and knowledgeable stakeholders, which may include both program participants and community members. A collaborative approach to developing the performance story and review by independent experts are the distinctive elements of this approach (Dart & Andrews, 2014).

Collaborative Outcomes Report Structure

The report aims to explore and report the extent to which a program has *contributed* to outcomes.

In Collaborative Outcomes Reporting, reports are short and generally structured in terms of the following sections:

1. A narrative section explaining the program context and rationale

2. A "results chart" summarizing the achievements of the program against a program logic model

3. A narrative section describing the implications of the results, for example, the achievements (expected and unexpected), the issues, and the recommendations

4. A section that provides a number of "vignettes" providing instances of significant change, usually first-person narratives

5. An index providing more detail on the sources of evidence

COR [Collaborative Outcomes Reporting] is based on the premise that the values of stakeholders, program staff, and key stakeholders are of highest importance in an evaluation. The evaluators attempt to "bracket off" their opinions and instead present a series of data summaries to panel and summit participants for them to analyze and interpret. Values are surfaced and debated throughout the process. Participants debate the value and significance of data sources and come to agreement on the key findings of the evaluation. In addition, qualitative inquiry is used to capture unexpected outcomes, and deliberative processes are used to make sense of the findings. (Dart & Johnson, 2014)

EXHIBIT 8.29   Audience- and Utilization-Focused Reporting

UNFOCUSED REPORTING	AUDIENCE- AND USE-FOCUSED REPORTING
Lots of side tracks; too much data	Presentation coheres in communicating primary findings and addressing priority concerns of those who can use the findings

indigenous typology—that is, using the language, concepts, and categories of the people studied to organize the data and make sense of the findings. Such an indigenous typology can also serve as an organizing framework for reports and presentations. Here's an example.

Our evaluation of a rural leadership program included participant observation. The program involved a six-day residential retreat experience. After six days and evenings of intense (and sometimes tense) participation, observation, interviewing participants, and taking extensive field notes, our team of three evaluators needed a framework for providing feedback to program staff about what we were finding *in a way that could be heard*. We knew that the staff were heavily ego involved in the program and would be very sensitive to an approach that might appear to substitute *our* concept of the program for theirs. Yet a major purpose of the evaluation was to help them identify and make explicit their operating assumptions as evidenced in what actually happened during the six-day retreat.

As our team of three accumulated more and more data, debriefing each night what we were finding, we became increasingly worried about how to focus feedback. The problem was solved the fifth night when we realized that we could use *their* frameworks for describing to them what we were finding. For example, a major component of the program was having participants work with the Myers-Briggs Type Indicator, an instrument that measures individual personality type based on the work of Carl Jung (see Berens & Nardi, 1999; Hirsh & Kummerow, 1987; Kroeger & Thuesen, 1988; Myers & Meyers, 1995). The Myers-Briggs Type Indicator gives individuals scores on four bipolar scales:

Extraversion (E)	Introversion (I)
Sensing (S)	Intuition (N)
Thinking (T)	Feeling (F)
Judging (J)	Perceiving (P)

In the feedback session, we began by asking the six staff members to characterize the overall retreat culture using the Myers-Briggs framework. The staff shared their separate ratings, on which there was no consensus, and then we shared our perceptions. We spent the whole morning discussing the database for and implications of each scale as a manifestation of the program's culture. We ended the session by discussing where the staff wanted the program to be on each dimension. Staff were able to hear what we said

without becoming defensive because we used *their* framework, a framework *they* defined, when giving individual feedback to participants, as nonjudgmental, facilitative, and developmental. Thus, they could experience our feedback as, likewise, nonjudgmental, facilitative, and developmental.

We formatted our presentation to staff using a distinction between "observations" and "perceived impacts" that program participants were taught as part of the leadership training. Observation: "You interrupted me in mid-sentence." Perceived impact: "I felt cut-off and didn't contribute after that." This simple distinction, aimed at enhancing interpersonal communication, served as a comfortable, familiar format for program staff to receive formative evaluation feedback. Our reporting, then, followed this format. Three of the 20 observations from the report are reproduced in Exhibit 8.30.

EXHIBIT 8.30   **Example of Reporting Feedback to Program Staff: Distinguishing Observations From Perceived Impacts Based on Their Indigenous Framework for Working With Participants in the Leadership Program**

OBSERVATIONS	PERCEIVED IMPACTS
1. The retreat setting, being away from the world, is culturally introverted.	1. There is deep bonding among group members, a sense of group as separate from the "real" world, though participants are expected to engage the "real" world after the retreat. This sets up a paradoxical tension.
2. The retreat is more conceptual and abstract in content than fact and skills oriented. It is primarily culturally intuitive (as opposed to step-by-step and practical).	2. Participants are conceptually stimulated and exposed to a variety of ideas. Some express uncertainty about what to do with the ideas (lack of practical applications).
3. The retreat culture is heavily affective—feelings oriented, not thinking oriented.	3. Highly emotional connections among participants. a. Participants are sensitized to how they feel about what they are experiencing and are explicitly encouraged to share feelings. b. Participants are affirmed as important; they feel special, cared about, and valued; it is a safe environment for learning. c. Participants are not stretched intellectually, logical distinctions are not made, and key concepts remain ambiguous. d. Affirming participants is clearly more important than challenging them; harmony is valued over clarity.

The critical point here is that we presented the findings using *their* categories and *their* frameworks. This greatly facilitated the feedback and enhanced the subsequent formative, developmental discussions. Capturing and using indigenous typologies can be a powerful analytical approach for making sense of and reporting qualitative data in evaluations.

The next example provides a quite different reporting format and illustrates a collaborative analysis and reporting effort.

Collaborative Reporting Example

A Principles-Focused Evaluation Report: An Example of the Flow of an Inquiry From Design to Data Collection and Analysis to Reporting

In 2012, six agencies serving homeless youth in Minneapolis and Saint Paul, Minnesota, began collaborating to evaluate the implementation and effectiveness of their shared principles. They first identified shared values and common principles and found that their work was undergirded and informed by eight principles. At the outset, these principles were just sensitizing concepts: labels and phrases without definition. Or where there was a descriptive definition, it varied among the agencies.

1. Trauma-informed care

2. Nonjudgmental engagement

3. Harm reduction

4. Trusting youth–adult relationships

5. Strengths-based approach

6. Positive youth development

7. Holistic approach

8. Collaboration

Working together, we designed a qualitative study to examine whether the principles actually guided the agencies' work with youth and whether the outcomes the youth achieved could be linked to the principles. Nora Murphy, a graduate student at the University of Minnesota, was commissioned to conduct the study, which also served as her doctoral dissertation (Murphy, 2014). Selecting a maximum variation sample of homeless youth who had been involved with several of the agencies, Murphy generated 14 case studies of homeless youth based on interviews with the youth and program staff and review of case records. The results of the case studies were then synthesized. The collaborative participated in reviewing every case study to determine which principles were at work and their effectiveness in helping the youth meet their needs and achieve their goals. The story of Thmaris (not his real name) is one of the cases (see Exhibit 7.20, pp. 511–516).

Coding the Case Studies and Interviews to Illuminate the Principles

The six agency directors participated in reviewing and validating the coding, as did the staff of the philanthropic foundation that funded the study. In pairs, they read through the case studies and identified passages, quotations, and events and experiences reported by the youth that seemed to illustrate a principle. Independently, as principal investigator, Murphy (2014) coded all the case studies against the sensitizing concept framework of the principles. Below, as an example, are three quotations from three different case studies (different youth) that were coded as illustrating the principle of *trusting youth–adult relationships*. (Pseudonym were selected by the youth.)

- *Pearl:* And you be like, "Okay, I have all this on my plate. I have to dig in and look into [the choices I'm making] to make my life more complete." And I felt that on my own, I really couldn't. Not even the strongest person on God's green Earth can do it. I couldn't do it. So I ended up reaching out to [the youth shelter], and they opened their arms. They were like just, "Come. Just get here," and they got me back on track.

- *Maria*: If I was to sit in a room and think about, like, everything that happened to me or I've been through, I'll get to cryin' and feelin' like I don't wanna be on Earth anymore—like I wanted to die. When I talk to somebody [at the youth program] about it, it makes me feel better. The people I talk to about it give me good advice. They tell me how much they like me and how [good] I'm doin'. They just put good stuff in my head, and then I think about it and realize I am a good person and everything's gonna work out better.

- *Thmaris*: [My case worker] is not going to send me to the next man, put me onto the next person's caseload. He just always took care of me. . . . I honestly feel like if I didn't have [him] in my corner, I would have been doing a whole bunch of dumb shit. I would have been right back at square one. I probably would have spent more time in jail than I did. I just felt like if it wasn't for him, I probably wouldn't be here right now, talking to you.

From Analysis to Reporting

Once all the case studies were coded against the principles, Murphy (2014) examined the data for additional principles. She found four possibilities and, in the end, confirmed one: being *journey oriented* in working with youth. In addition to the case studies, Murphy reviewed other available research (literature review for the dissertation) for definitions of and evidence of the effectiveness of each principle. Murphy used the cross-case analysis of the 14 interviews with the youth and the research literature to draft a definition for each principle. She and I then facilitated a process with the six agency directors to revise and finalize the definitions. In December 2013, the six agencies, acting together as a collaboration, formally adopted the nine principles (the original eight plus the one that emerged during the inquiry). Exhibit 8.31 presents the nine principles with the definitions developed and adopted. Next, the agency directors and Murphy (2014) worked together to summarize the evidence, explaining each principle in a two-page statement. The outline for the report the group produced is presented in Exhibit 8.31. One of the most frequent questions I'm asked is how to consolidate, synthesize, and summarize all the rich data that come from a qualitative inquiry into a useful report that can be communicated publicly and used to inform practice and policy. There is no recipe for doing so, but Exhibit 8.31 provides an example to show that it can be done. Even though heavily edited, this is a lengthy exhibit. I have included it because of the importance of having some sense of how one moves from analysis to reporting when the purpose of the inquiry includes communicating results to a larger audience, as in evaluation and policy studies.

EXHIBIT 8.31 **Principles-Focused Qualitative Evaluation Report Example**

Nine Evidence-Based, Guiding Principles to Help Youth Overcome Homelessness

The Homeless Youth Collaborative on Developmental Evaluation

Part 1. The report begins with one page on each of the six agencies in the collaboration organized by type of service provided to homeless youth:

Outreach: StreetWorks

Drop-in Centers: Face to Face: SafeZone and YouthLink

Shelters: Avenues for Homeless Youth; Catholic Charities: Hope Street; and The Salvation Army: Booth Brown House

Other contributors to the process and the report

Funder: The Otto Bremer Foundation, Saint Paul, Minnesota

Technical assistance: Nora Muphy (2013) and Michael Quinn Patton

Part 2. The nine evidence-based, guiding principles to help youth overcome homelessness are presented.

*The principles begin with the perspective that youth are on a journey; all of our interactions with youth are filtered through that **journey** perspective. This means we must be **trauma-informed, non-judgmental** and work to **reduce harm**. By holding these principles, we can build a **trusting relationship** that allows us to focus on **youths' strengths** and opportunities **for positive development**. Through all of this, we approach youth as **whole beings** through a youth-focused **collaborative** system of support.*

Journey Oriented: Interact with youth to help them understand the interconnectedness of past, present and future as they decide where they want to go and how to get there.

Trauma-Informed Care: Recognize that all homeless youth have experienced trauma; build relationships, responses and services on that knowledge.

Non-Judgmental Engagement: Interact with youth without labeling or judging them on the basis of background, experiences, choices or behaviors.

Harm Reduction: Contain the effects of risky behavior in the short-term and seek to reduce its effects in the long-term.

Trusting Youth-Adult Relationships: Build relationships by interacting with youth in an honest, dependable, authentic, caring and supportive way.

Strengths-based Approach: Start with and build upon the skills, strengths and positive characteristics of each youth.

Positive Youth Development: Provide opportunities for youth to build a sense of competency, usefulness, belonging and power.

(Continued)

(Continued)

Holistic: Engage youth in a manner that recognizes that mental, physical, spiritual and social health are interconnected and interrelated.

Collaboration: Establish a principles-based, youth-focused system of support that integrates practices, procedures and services within and across agencies, systems and policies.

Part 3. *Introduction to and explanation of principles-focused evaluation* (excerpts)

All homeless young people have experienced serious adversity and trauma. The experience of homelessness is traumatic enough, but most also have faced poverty, abuse, neglect or rejection. They have been forced to grow up way too early. Most have serious physical or mental health issues. Some are barely teenagers; others may be in their late teens or early twenties.

Some homeless youth have family connections, some do not; all crave connection and value family. They come from the big city, small towns and rural areas. Most are youth of color and have been failed by systems with institutionalized racism and programs that best serve the white majority. Homeless youth are straight, gay, lesbian, bisexual, transgender or questioning. Some use alcohol or drugs heavily. Some have been in and out of homelessness. Others are new to the streets.

The main point here is that, while all homeless youth have faced trauma, they are all unique. Each homeless youth has unique needs, experiences, abilities and aspirations. Each is on a personal journey through homelessness and, hopefully, to a bright future.

Because of their uniqueness, how we approach and support each homeless young person also must be unique. No recipe exists for how to engage with and support homeless youth. As homeless youth workers and advocates, we cannot apply rigid rules or standard procedures. To do so would result in failure, at best, and reinforce trauma in the young person, at worst. Rules don't work. We can't dictate to any young person what is best. The young people know what is best for their future and need the opportunity to engage in self-determination.

This is where Principles come in. Organizations and individuals that successfully support homeless youth take a principles-based approach to their work, rather than a rules-based approach. Principles provide guidance and direction to those working with homeless youth. They provide a framework for how we approach and view the

youth, engage and interact with them, build relationship with them and support them. The challenge for youth workers is to meet and connect with each young person where they are and build a supportive relationship from there. Principles provide the anchor for this relationship-building process.

Evaluation Design

Nora Murphy conducted 14 in-depth case studies of homeless youth who had interacted with the agencies, including interviews with the youth and program staff, and review of case records. The results of the case studies were then synthesized. The collaborative participated in reviewing every case study to determine which principles were at work and their effectiveness in helping the youth meet their needs and achieve their goals.

The results showed that *all the participating organizations were adhering to the principles and that the principles were effective (even essential) in helping them make progress out of homelessness.* While staff could not necessarily give a label to every principle at the time of the evaluation interviews, they clearly talked about and followed them in practice. In the interviews with the youth, their stories showed how the staffs' principles-based approaches made a critical difference in their journey through homelessness. (Reference for full dissertation: Murphy, 2013)

Part 4. *Statements of 2–3 pages in which each principle is defined and discussed.* For each principle, the summary statement explains why it matters, what key research shows, what the practice implications are, and concludes with illustrative quotes from the evaluation case studies.

Part 5. Following the principles statements are 2-page excerpts from each of the 14 case studies, excerpts that provide an overview of the outcomes the youth experienced through engagement with the youth-serving agencies.

Part 6. *One full 15-page case study*

Report Conclusion: Principles-Based Practice

Taken together, this set of principles provides a cohesive framework that guides practice in working with homeless youth.

For the full report, see Homeless Youth Collaborative on Developmental Evaluation (2014)

Nine Evidence-Based, Guiding Principles to Help Youth Overcome Homelessness (public report length, 71 pp.)

Use of the Report Findings and Case Studies

The six participating agencies are using the principles and case studies for program and staff development, recruitment and selection of new staff, policy advocacy on behalf of homeless youth, fund-raising, and expanded collaboration with other youth-serving organizations locally, statewide, and nationally. For further discussion of how and why qualitative inquiry is especially appropriate for evaluating principles-based programs and collaborations, see Chapters 4 (pp. 189–194) and 5 (p. 292)

Meeting the Challenge of Saying a Lot in a Small Space: The Executive Summary and Research Abstract

> The executive summary is a fiction.
>
> —Robert Stake (1998, p. 370)

The fact that qualitative reports tend to be relatively lengthy can be a major problem when busy decision makers do not have the time (or, more likely, will not take the time) to read a lengthy report. Stake's preference for insisting on telling the whole story notwithstanding is a preference I share, but my pragmatic, living-in-the-real-world side leads me to conclude that qualitative researchers and evaluators must develop the ability to produce an executive summary of one or two pages that presents the essential findings, conclusions, and reasons for confidence in the summary. The executive summary or research abstract is a dissemination document, a political instrument, and cannot be—nor is it meant to be—a full and fair representation of the study. An executive summary or abstract should be written in plain language, be highly focused, and state the core findings and conclusions. When writing the executive summary or research abstract, keep in mind that more people are likely to read that summary than any other document you produce.

So here are a couple of tips:

1. *Focus on findings not methods:* I see lots of summaries and abstracts where it's clear that the person reporting is more in love with his or her methods than his or her findings. But the reader is more likely to want to know about your findings than your methods. Report just enough about the methods to explain how the findings were generated, but highlight the most important findings.

2. *Pilot test your drafts:* Get some people you know to read your executive summary or abstract and tell you what it says to them.

3. *Take time to write a well-crafted executive summary or abstract:* French philosopher Blaise Pascal wrote a witty and oft-repeated apology in 1657: "I would have written a shorter letter if I had had more time."

Carpe Diem Briefings

As the hymn book is to the sound of music, the executive summary is to the oral briefing. Legendary are the stories of having spent a year of one's life gathering data, poring over it, and writing a rigorous and conscientious report and then encountering some "decision maker" (I use the term lightly here) who says, "Well, now, I know that you put a lot of work into this. I'm anxious to hear all about what you've learned. I've got about 10 minutes before my next appointment."

Should you turn heel and show him your backside? Not if you want your findings to make a difference. Use that 10 minutes well! Be prepared to make it count. *Carpe Diem.*

Chapter Summary and Conclusion, Plus Case Study Exhibits

The role of the qualitative researcher in research projects is often determined by the researcher's stance and intent. . . . All senses are used to understand the context of the phenomenon under study, the people who are participants in the study and their beliefs and behaviors, and of course the researcher's own orientation and purposes. . . . It is complicated, filled with surprises, and open to serendipity, and it often leads to something unanticipated in the original design of the research project. At the same time, the researcher works within the frame of a disciplined plan of inquiry, adheres to the high standards of qualitative inquiry, and looks for ways to complement and extend the description and explanation of the project through multiple methods of research, providing that this is done for a specific reason and makes sense. Qualitative researchers do not accept the misconception that more methods mean a better or richer analysis. Rather, the rationale for using selected methods is what counts. The qualitative researcher wants to tell a story in the best possible configuration.

—Valerie Janesick (2011, p. 176)
"Stretching" Exercises for Qualitative Researchers

Chapter Overview and Summary

This chapter opened by cautioning that no formula exists for transforming qualitative data into findings.

Purpose drives analysis. Design frames analysis. But even with a clear purpose and a strong, appropriate design, the challenge of qualitative analysis remains: making sense of massive amounts of data. This involves reducing the volume of raw information, sifting trivia from significant data, identifying significant patterns, distinguishing signal from noise, and constructing a framework for communicating the essence of what the data reveal.

Exhibit 8.1 (pp. 522–523) offered 12 tips for laying a strong foundation for qualitative analysis. Exhibit 8.2 (p. 528) provided guidance in connecting design and analysis by linking purposeful sampling and purpose-driven analysis. Part of that connection is that in qualitative inquiry, analysis begins during fieldwork. Quantitative research texts, focused on surveys, standardized tests, and experimental designs, typically make a hard-and-fast distinction between data collection and analysis. But the fluid and emergent nature of naturalistic inquiry makes the distinction between data gathering and analysis far less absolute. In the course of fieldwork, ideas about directions for analysis will occur. Patterns take shape. Signals start to emerge from the noise. Possible themes spring to mind. Thinking about implications and explanations deepens the final stage of fieldwork. While earlier stages of fieldwork tend to be generative and emergent, following wherever the data lead, later stages bring closure by moving toward confirmatory data collection—deepening insights into and confirming (or disconfirming) patterns that seem to have appeared. Indeed, sampling confirming and disconfirming cases requires a sense of what there is to be confirmed or disconfirmed. Thus, ideas for making sense of the data that emerge while still in the field constitute the beginning of analysis.

Once formal analysis begins, the first task is organizing the raw data for analysis; then coding and content analysis begin systematically and rigorously. Qualitative software can help manage the data, but sense making remains a quintessential human responsibility. Analysis includes identifying and reflecting on patterns and themes that emerge during analysis as well as being reflective and reflexive about the analysis process. How do you know what you know? How do you make sense of the perspectives of the people you've observed and interviewed? What do you know about the primary audience to whom your findings are addressed? These reflective questions are the doorway into reflexivity.

Constructing, analyzing, comparing, and interpreting case studies constitute core sense-making processes in many qualitative inquiries. Exhibit 8.33 (pp. 638–642), at the end of this chapter, presents an extensive example of an individual-level case study. Analytical approaches other than case studies are also part of qualitative analysis. Exhibit 8.10 (pp. 551–552) presented 10 types of qualitative analysis. Module 70 discussed interpreting findings, determining substantive significance, and elucidating phenomenological essence. Throughout the chapter, we've examined and discussed how and why analysis and reporting involve making distinctions between organizing data, presenting descriptive patterns and findings, interpreting themes, determining substantive significance, and, a particularly challenging challenge in qualitative inquiry, inferring causal relationships. Exhibit 8.19 (pp. 600–601) presents 12 approaches to qualitative causal analysis. The Halcolm graphic comic at the end of this chapter (pp. 635–637) offers an African fable to stimulate reflection on the nature of causal explanation.

Module 73 presented guidance on writing up and reporting findings, including incorporating creative visualizations that can powerfully and succinctly communicate core findings. Exhibits 8.20 to 8.27 presented an array of examples of visualization of findings. Exhibit 8.26 (pp. 616–618) summarized the strengths and weaknesses of visualization of qualitative data and findings. Exhibit 8.27 highlighted mixed-methods challenges and solutions, especially integrating qualitative and quantitative data and results. Exhibit 8.31 (pp. 627–628) presented an example of a qualitative report focused on principles for serving homeless youth. Exhibit 8.32 on the next page provides a checklist for data analysis, reporting, and interpretation that highlights some of the main points presented in this chapter. Exhibit 8.35 (pp. 643–649) presents excerpts from a qualitative report based on interviews.

Conclusion: The Creativity of Qualitative Inquiry

> Creativity will dominate our time after the concepts of work and fun have been blurred by technology.
>
> —Award-winning science fiction writer Isaac Asimov (1983, p. 42)

I have commented throughout this book that the human element in qualitative inquiry is both its strength and its weakness—its strength in allowing human insight and experience to blossom into new understandings and ways of seeing the world, its potential weakness in being so heavily dependent on the inquirer's skills, training, intellect, discipline, and creativity. Because the researcher is the instrument of qualitative inquiry, the quality of the result depends heavily on the qualities of that human being. Nowhere does this ring more true than in analysis. Being an empathetic interviewer or astute observer does not necessarily make one an insightful analyst—or a creative one. Creativity seems to be one of those special human qualities that play an especially important part in qualitative analysis, interpretation, and reporting. Therefore, I close this chapter with some observations on creativity in qualitative inquiry.

I opened this chapter by commenting on qualitative inquiry as both science and art, especially qualitative analysis. The scientific part demands systematic and disciplined intellectual work, rigorous attention to details within a holistic context, and a critical perspective in questioning emergent patterns even while bringing evidence to bear in support of them. The artistic part invites exploration, metaphorical flourishes, risk taking, insightful sense making, and imaginative connection making. While both science and art involve critical analysis and creative expression, science emphasizes critical faculties more, especially in analysis, while art encourages creativity. The critical thinker assumes a stance of doubt and skepticism: Things have to be proven. Faulty logic, slippery linkages, tautological theories, and unsupported deductions are targets of the critical mind. The critical thinker studies details and looks beyond appearances to find out what is really happening. Evaluators are trained to be rigorous and unyielding in critically thinking about and analyzing programs. Indeed, evaluation is built on the foundation of critical analysis.

Critical thinkers, however, tend *not* to be very creative. The creative mind generates new possibilities; the critical mind analyzes those possibilities, looking for inadequacies and imperfections. In summarizing research on critical and creative thinking, Barry F. Anderson (1980) has warned that the centrality of doubt in critical thinking can lead to a narrow, skeptical focus that hampers the creative ability to come up with innovative linkages or new insights.

EXHIBIT 8.32   Checklist for Qualitative Data Analysis, Interpreting Findings, and Reporting Results

1. *Keep analysis grounded in the data.* Return over and over to your field notes, interview transcripts, and the documents or artifacts you've collected that inform your analysis.

2. *Present actual data for readers to experience firsthand.* Use direct quotations, detailed observations, and document excerpts as the core of your reporting and as empirical support for your analysis and interpretations. "In this case, more is more. Convince your reader of your argument with evidence from transcripts, observations, reflective journals, and any other documentary evidence" (Janesick, 2011, p. 177).

3. *Distinguish description from interpretation and explanation.* Interpretation and explanation go beyond the data. Interpreting and explaining are part of the analyst's responsibility, but the first and foremost responsibility is organizing and presenting accurate and meaningful descriptive findings.

4. *Do justice to each case before doing cross-case analysis.* This admonition follows from the previous one. Well-constructed, well-written, thorough, and rich case studies are worthy in and of themselves. Don't let the allure of finding and presenting cross-case patterns and themes undercut full attention to the integrity and particularity of each case. And when distinguishing signal from noise, describe and interpret *both* the signal and the noise.

5. *Purpose drives analysis and interpretation.* If your purpose is to generate or test theory, then a significant part of your discussion will involve connecting your findings and analysis to inform and support theory-focused propositions. If your purpose is evaluative, to support program improvements and inform decision making about programs and policies, organize your findings to address core evaluation issues and provide actionable answers to evaluation questions.

6. *Throughout the study track, document, and report your inquiry methods and analytical procedures.* Findings flow from methods. Methods flow from inquiry questions. Inquiry questions flow from inquiry purpose. Connect these dots so that the linkages are clear. Report your methods and analytical process in sufficient detail that readers know from whence come your findings and interpretations and why you did what you did.

7. *Be reflective and reflexive.* As a qualitative inquirer, you are the instrument of the inquiry. Who you are, what brings you to this inquiry, your background, experience, knowledge, and training—*all these matter.* Report your reflexive inquiry in the methods section of your report (or elsewhere as appropriate). Acknowledge your prior beliefs.

You should work to reduce your biases, but to say you have none is a sign that you have many. To state your beliefs up front—to say "Here's where I'm coming from"—is a way to operate in good faith and to recognize that you perceive reality through a subjective filter. . . . Distinguishing the signal from noise requires both scientific knowledge and self-knowledge. (Silver, 2012, pp. 451, 453)

8. *Work at the writing.* How you present your findings will matter. Get editorial assistance as needed. Solicit feedback. Take the writing seriously. High school English teachers are right when they say that excellence in writing reflects excellence in thinking. All of your rigorous fieldwork and diligent analysis can come to naught if you present your findings poorly, incoherently, or in boring or error-plagued prose.

9. *Make the case for substantive significance.* Substantive significance is a matter of judgment. It's where you synthesize the evidence, consider its import, make transparent weaknesses and limitations in the data, balance strengths against weaknesses, weigh alternative possibilities, and *state why the findings are important and merit attention.* Don't overstate the case. Don't overgeneralize. But don't become timid and afraid of naysayers. You know the data better than anyone. You've immersed yourself in them, worked with them, and lived with them. Say what you've come to know and what you don't know, and provide the evidence for both judgments. Then, make the case for *substantive significance.* And, by all means, *stay qualitative.* Don't determine significance by the numbers of people who said something. It's not how many said something that matters. It's the import, wisdom, relevance, insightfulness, and applicability of what was said, by many or by a few. (See MQP Rumination #8, pp. 557–560.)

10. *Use both your critical and your creative faculties.* Qualitative analysis offers myriad opportunities for both critical and creative thinking. These are not antagonists. Both capabilities dwell within you. Draw on both. Critical thinking is manifest in rigorous attention to alternative interpretations and explanations and in settling on those interpretations and explanations that best fit the preponderance of evidence. Creative thinking is manifest in well-told stories, evocative case studies, astute selection and juxtaposition of quotes and observations, and powerful, captivating, and accurate visual representations of the findings.

The critical attitude and the creative attitude seem to be poles apart. . . . On the one hand, there are those who are always telling you why ideas won't work but who never seem able to come up with alternatives of their own; and, on the other hand, there are those who are constantly coming up with ideas but seem unable to tell good from the bad.

There are people in whom both attitudes are developed to a high degree . . . , but even these people say they assume only one of these attitudes at a time. When new ideas are needed, they put on their creative caps, and when ideas need to be evaluated, they put on their critical caps. (p. 66)

Qualitative inquiry draws on both critical and creative thinking—both the science and the art of analysis. But the technical, procedural, and scientific side of analysis is easier to present and teach. Creativity, while easy to prescribe, is harder to teach and perhaps harder to learn, but here's some guidance derived from research and training on creative thinking (De Bono, 1999; Kelley & Littman, 2001; Patton, 1987, pp. 247–248; Von Oech, 1998).

1. **Be open:** Creativity begins with openness to multiple possibilities.

2. **Generate options:** There's always more than one way to think about or do something.

3. **Diverge–converge–integrate:** Begin by exploring a variety of directions and possibilities before focusing on the details. Branch out, go on mental excursions, and brainstorm multiple perspectives before converging on the most promising.

4. **Use multiple stimuli:** Creativity training often includes exposure to many different avenues of expression: drawing, music, role-playing, story boarding, metaphors, improvisation, playing with toys, and constructing futuristic scenarios. Synthesizing through triangulation (see Chapter 9) promotes creative integration of multiple stimuli.

5. **Side track, zigzag, and circumnavigate:** Creativity is seldom a result of purely linear and logical induction or deduction. The creative person explores back and forth, round and about, in and out, over and under, and so on.

6. **Change patterns:** Habits, standard operating procedures, and patterned thinking pose barriers to creativity. Become aware of and change your patterned ways of thinking and behaving.

7. **Make linkages:** Many creative exercises include practice in learning how to connect the seemingly unconnected. The matrix approaches presented in this chapter push linkages. Explore linking qualitative and quantitative data. Work at creative syntheses of case studies and findings.

8. **Trust yourself:** Self-doubt short-circuits creative impulses. If you say to yourself, "I'm not creative," you won't be. Trust the process.

9. **Work at it:** Creativity is not all fun. It takes hard work, background research, and mental preparation.

10. **Play at it:** Creativity is not all work. It can and should be play and fun.

Drawing Conclusions and Closure

Certainty is false closure.

Ambiguity is honest closure.

Staying open eschews closure.

Enough is enough yields pragmatic closure.

Closure is hard. Not closing is harder.

Close . . . for now.

—From Halcolm's *Benedictions*

In his practical monograph *Writing Up Qualitative Research*, Wolcott (1990) considers the challenge of how to conclude a qualitative study. Purpose again rules in answering this question. Scholarly articles, dissertations, and evaluation reports have different norms for drawing conclusions. But Wolcott goes further by questioning the very idea of conclusions.

Give serious thought to dropping the idea that your final chapter must lead to a conclusion or that the account must build toward a dramatic climax. . . . In reporting qualitative work, I avoid the term *conclusion*. I do not want to work toward a grand flourish that might tempt me beyond the boundaries of the material I have been presenting or detract from the power (and exceed the limitations) of an individual case. (p. 55)

This admonition reminds us not to take anything for granted or fall into following some recipe for writing. Asking yourself, "When all is said and done, what conclusions do I draw from all this work?" can be a focusing question that forces you to get at the essence. Indeed, in most reports, you will be expected to include a final section on conclusions. But, as Wolcott (1990) suggests, it can be an unnecessary and inappropriate burden.

Or it can be a chance to look to the future. Spanish-born philosopher and poet George Santayana concluded thus when he retired from Harvard. Students and colleagues packed his classroom for his final appearance. He gave an inspiring lecture and was about to conclude when, in midsentence, he caught sight of a forsythia flower beginning to blossom in the melting snow outside the window. He stopped abruptly, picked up his coat, hat, and gloves, and headed for the door. He turned at the door and said emphatically, "Gentlemen, I should not be able to finish that sentence. I have just discovered that I have an appointment with April."

Or as Halcolm would say, *not concluding is its own conclusion.*

I close this chapter with a practical reminder that both the science and the art of qualitative analysis are constrained by limited time. Some people thrive under intense time pressure, and their creativity blossoms. Others don't. The way in which any particular analyst combines critical and creative thinking becomes partly a matter of style and partly a function of the situation, and it often depends on how much time can be found to play with creative possibilities. But exploring possibilities can also become an excuse for not finishing. There comes a time for bringing closure to analysis (or a book chapter) and getting on with other things. Taking too much time to contemplate creative possibilities may involve certain risks, a point made by the following story (to which you can apply both your critical and your creative faculties).

The Past and the Future: Deciding in Which Direction to Look

A spirit appeared to a man walking along a narrow road. "You may know with certainty what has happened in the past, or you may know with certainty what will happen in the future, but you cannot know both. Which do you choose?"

The startled man sat down in the middle of the road to contemplate his choices. "If I know with certainty what will happen in the future," he reasoned to himself, "then the future will soon enough become the past and I will also know with certainty what has happened in the past. On the other hand, it is said that the past is a prologue to the future, so if I know with certainty what has happened in the past, I will know much about what will happen in the future without losing the elements of surprise and spontaneity." Deeply lost to the present in the reverie of his calculations about the past and future, he was unaware of the sound of a truck approaching at great speed. Just as he came out of his trance to tell the spirit that he had chosen to know with certainty the future, he looked up and saw the truck bearing down on him, unable to stop its present momentum.

—From Halcolm's *Evaluation Parables*

EXHIBIT 8.33   Mike's Career Education Experience: An Illustrative Case Study

Background

Sitting in a classroom at Metro City High School was difficult for Mike. In some classes, he was way behind. In math, he was always the first to finish a test. "I loved math and could always finish a test in about 10 minutes, but I wasn't doing well in my other classes," Mike explained.

He first heard about Experience-Based Career Education (EBCE) when he was a sophomore. "I really only went to the assembly to get out of one of the classes I didn't like," Mike confessed.

But after listening to the EBCE explanation, Mike was quickly sold on the idea. He not only liked the notion of learning on the job but also thought the program might allow him to work at his own speed. The notion of no grades and no teachers also appealed to him.

Mike took some descriptive materials home to his parents, and they joined him for an evening session at the EBCE learning center to find out more about the program. Now, after two years in the program, Mike is a senior, and his parents want his younger brother to get into the program.

Early EBCE testing sessions last year verified the inconsistency of Mike's experiences in school. While his reading and language scores were well below the average scored by a randomly selected group of juniors at his school, he showed above-average abilities in study skills and demonstrated superior ability in math.

On a less tangible level, EBCE staff members early last school year described Mike as being hyperactive, submissive, lacking in self-confidence, and unconcerned about his health and physical appearance when he started the EBCE program. He was also judged to have severe writing deficiencies. Consequently, Mike's EBCE learning manager devised a learning plan that would build his communication skills (in both writing and interpersonal relations) while encouraging him to explore several career possibilities. Mike's job experiences and projects were designed to capitalize on his existing interests and broaden them.

First-Year EBCE Experiences. A typical day for Mike started at 8:00 a.m., just as in any other high school, but the hours in between varied considerably. On arriving at the EBCE learning center, Mike said he usually spent some time "fooling around" with the computer before he worked on projects under way at the center.

On his original application, Mike had indicated that his career preference would be computer operator. This led to an opportunity in the EBCE program to further explore that area and to learn more about the job. During April and May, Mike's second learning-level experience took place in the computer department of City Bank Services. He broke up his time there each day into morning and afternoon blocks, often arriving before his employer-instructor did for the morning period. Mike usually spent that time going through computer workbooks. When his employer-instructor arrived, they went over flow charts together and worked on computer language.

Mike returned to the high school for lunch and a German class he selected as a project. EBCE students seldom take classes at the high school, but Mike had a special interest in German since his grandparents speak the language.

Following German class, Mike returned to the learning center for an hour of work on other learning activities and then went back to City Bank. "I often stayed there until 5:00 p.m.," Mike said, even though high school hours ended at three.

Mike's activities and interests widened after that first year in the EBCE program, but his goal of becoming a computer programmer was reinforced by the learning experience at City Bank. The start of a new hobby—collection of computer materials—also occurred during the time he spent at City Bank. "My employer-instructor gave me some books to read that actually started the collection," Mike said.

Mike's interest in animals also was enhanced by his EBCE experience. Mike has always liked animals, and his family has owned a horse since he was 12 years old. By picking blueberries, Mike was able to save enough to buy his own colt two years ago. One of Mike's favorite projects during the year related to his horse. The project was designed to help Mike with basic skills and to improve his critical thinking skills. Mike read about breeds of horses and how to train them. He then joined a 4-H group with hopes of training his horse for a show.

Several months later, Mike again focused on animals for another EBCE project. This time he used the local zoo as a resource, interviewing the zoo manager and doing a thorough study of the Alaskan brown bear. Mike also joined an Explorer Scouting Club of volunteers to help at the zoo on a regular basis. "I really like working with the bears," Mike reflected. "They were really playful. Did you know when they rub their hair against the bars it sounds like a violin?" Evaluation of the zoo project, one of the last Mike completed during the year, showed much improvement. The learning manager commented to Mike, "You are getting your projects done faster, and I think you are taking more time than you did at first to do a better job."

Mike got off to a slow start in the area of life skills development. Like some of his peers, he went through a period described by one of the learning managers as "freedom shock," when removed from the more rigid structure normally experienced in a typical school setting. Mike tended to avoid his responsibility to the more "academic" side of his learning program. At first, Mike seldom followed up on commitments and often did not let the staff know what he was doing. By the end of the year, he had improved remarkably in both of these behavior areas.

Through the weekly writing required in maintaining his journal, Mike demonstrated a significant improvement in written communications, both in terms of presenting ideas and feelings and in the mechanics of writing. Mike also noted an interesting change in his behavior. "I used to watch a lot of TV and never did any reading." At the beginning of the following year, Mike said, "I read two books last year and have completed eight more this summer. Now I go to the book instead of the television." Mike's favorite reading material is science fiction.

Mike also observed a difference in his attitude to homework. "After going to school for six hours, I wouldn't sit down and do homework. But in the EBCE program, I wasn't sitting in a classroom, so I didn't mind going home with some more work on my journal or projects."

Mike's personal development was also undergoing change. Much of this change was attributed to one of his employer-instructors, an elementary school teacher, who told him how important it is in the work world to wear clean clothes. Both she and the project staff gave Mike much positive reinforcement when his dressing improved. That same employer also told Mike that she was really interested in what he had to say and therefore wanted him to speak slower so he could be understood.

Mike's school attendance improved while in the EBCE program. During the year, Mike missed only six days. This was better than the average absence for others in the program, which was found to be 12.3 days missed during the year, and much improved over his high school attendance.

Like a number of other EBCE students in his class, Mike went out on exploration-level experiences but completed relatively few other program requirements during the first three months of the school year. By April, however, he was simultaneously working on eight different projects and pursuing a learning experience at City Bank. By the time

Mike completed his junior year, he had finished 9 of the required 13 competencies, explored nine business sites, completed two learning levels, and carried through on 11 projects. Two other projects were dropped during the year, and 1 is uncompleted but could be finished in the coming year.

On a more specific level, Mike's competencies included transacting business on a credit basis, maintaining a checking account, designing a comprehensive insurance program, filing taxes, budgeting, developing physical fitness, learning to cope with emergency situations, studying public agencies, and operating an automobile.

Mike did not achieve the same level of success on all of his job sites. However, his performance consistently improved throughout the year. Mike criticized the exploration packages when he started them in the first months of the program and, although he couldn't pinpoint how, said they could be better. His own reliance on the questions provided in the package was noted by the EBCE staff, with a comment that he rarely followed up on any cues provided by the person he interviewed. The packets reflected Mike's disinterest in the exploration portion of EBCE work. They showed little effort and a certain sameness in the remarks about his impressions at the various sites.

Mike explored career possibilities at an automobile dealer, an audiovisual repair shop, a supermarket, an air control manufacturer, an elementary school, a housing development corporation, a city public works, a junior high school, and a bank services company.

Mike's first learning-level experience was at the elementary school. At the end of three and one-half months, the two teachers serving as his employer-instructors indicated concern about attendance, punctuality, initiative in learning, and the amount of supervision needed to see that Mike's time was used constructively. Mike did show significant improvement in appropriate dress, personal grooming, and quality of work on assignments.

Reports from the second learning-level experience—at the computer department of the bank services company—showed a marked improvement. The employer-instructor there rated Mike as satisfactory in all aspects and, by the time of the final evaluation, gave excellent ratings in 10 categories—(1) attendance/punctuality, (2) adhering to time schedules, (3) understanding and accepting responsibility, (4) observing employer rules, (5) showing interest and enthusiasm, (6) showing poise and self-confidence, (7) using initiative in seeking opportunities to learn, (8) using employer site learning resources,

(Continued)

(Continued)

(9) beginning assigned tasks promptly, and (10) completing the tasks assigned.

During the latter part of the school year, Mike worked on several projects at once. He worked on a project on basic electricity and took a course on "Beginning Guitar" for project credit.

To improve his communication skills, Mike also worked on an intergroup relations project. This project grew out of an awareness by the staff that Mike liked other students but seemed to lack social interaction with his peers and the staff. Reports at the beginning of the year indicated that he appeared dependent and submissive and was an immature conversationalist. In response to these observations, Mike's learning manager negotiated project objectives and activities with him that would help improve his communication skills and help him solve some of his interpersonal problems. At the end of the year, Mike noted a positive change related to his communication skills. "I can now speak up in groups," he said.

Mike's unfinished project related to his own experience and interests. He had moved to the Portland area from Canada 10 years previously and frequently returns to see relatives. The project was on immigration laws and regulations in the functional citizenship area. At the same time, it would help Mike improve his grammar and spelling. Since students have the option of completing a project started during their junior year when they are seniors, Mike had a chance to finish the project this year. Of the year, Mike said, "It turned out even better than I thought." Things he liked best about the new experience in EBCE were working at his own speed, going to a job, and having more freedom.

At the end of the year, Mike's tests showed significant increases in both reading and language skills. In the math and study skill areas, where he was already above average, only slight increases were indicated.

Tests on attitudes, given both at the beginning and at the end of the year, indicated positive gains in self-reliance, understanding of roles in society, tolerance for people with differences in background and ideas from his, and openness to change.

Aspirations did not change for Mike. He still wants to go into computer programming after finishing college. "When I started the year, I really didn't know too much about computers. I feel now that I know a lot and want even more to make it my career."

The description of Mike's second year in EBCE is omitted. We pick up the case study after his second-year experience.

Mike's Views of EBCE. Mike reported that his EBCE experiences, especially the learning levels, had improved all of his basic skills. He felt he had the freedom to do the kinds of things he wanted to do while at employer sites. These experiences, according to Mike, have strengthened his vocational choice in the field he wanted to enter and have caused him to look at educational and training requirements plus some other alternatives. For instance, Mike tried to enter the military, figuring it would be a good source of training in the field of computers, but was unable to because of a medical problem.

By going directly to job sites, Mike has gotten a feel for the "real world" of work. He said his work at computer repair–oriented sites furthered his conception of the patience necessary when dealing with customers and the fine degree of precision needed in the repair of equipment. He also discovered how a customer engineer takes a problem, evaluates it, and solves it.

When asked about his work values, Mike replied, "I figure if I get the right job, I'd work at it and try to do my best. . . . In fact, I'm sure that even though I didn't like the job I'd still do more than I was asked to. . . . I'd work as hard as I could." Although he has always been a responsible person, he feels that his experiences in EBCE have made him more trustworthy. Mike also feels that he is now treated more like an adult because of his own attitudes. In fact, he feels he understands himself a lot more now.

Mike's future plans concern trying to get a job in computer programming at an automobile dealership or computer services company. He had previously done some computer work at the automobile dealership in relationship to a project in Explorer Scouts. He also wants more training in computer programming and has discussed these plans with the student coordinator and an EBCE secretary. His attitude toward learning is that it may not be fun but it is important.

When asked in which areas he made less growth than he had hoped to, Mike responded, "I really made a lot of growth in all areas." He credits the EBCE program for this, finding it more helpful than high school. It gives you the opportunity to "get out and meet more people and get to be able to communicate better with people out in the community."

Most of Mike's experiences at the high school were not too personally rewarding. He did start a geometry class there this year but had to drop it as he had started late and could not catch up. Although he got along all right with the staff at the high school, in the past he felt the teachers there had a "barrier between them and the students." The EBCE staff "treat you on a more individual type circumstance.... [They] have the time to talk to you." In EBCE, you can "work at your own speed.... [You] don't have to be in the classroom."

Mike recommends the program to most of his friends, although some of his friends had already dropped out of school. He stated, "I would have paid to come into EBCE, I think it's really that good a program.... In fact, I've learned more in these two years in EBCE than I have in the last four years at the high school." He did not even ask for reimbursement for travel expenses because he said he liked the program so much.

Other Perspectives and Data

The Views of His Parents. When Mike first told his parents about the program, they were concerned about what was going to be involved and whether it was a good program and educational. When interviewed in March, they said that EBCE had helped Mike to become more mature and know where he is going.

Mike's parents said that they were well informed by the EBCE staff in all areas. Mike tended to talk to them about his activities in EBCE, while the only thing he ever talked about at the high school was photography. Mike's career plans have not really changed since he entered EBCE, and his parents have not tried to influence him, but EBCE has helped him to rule out mechanic and truck driving as possible careers.

Since beginning the EBCE program, his parents have found Mike to be more mature, dependable, and enthusiastic. He also became more reflective and concerned about the future. His writing improved, and he read more.

There are no areas where his parents felt that EBCE did not help him, and they rated the EBCE program highly in all areas.

Test Progress Measures on Mike. Although Mike showed great improvement in almost all areas of the Comprehensive Test of Basic Skills during the first year of participation, his scores declined considerably during the second year. Especially significant were the declines in Mike's arithmetic applications and study skills scores.

Mike's attitudinal scores all showed a positive gain over the total two-year period, but they also tended to decline during the second year of participation. On the semantic differential, Mike scored significantly below the EBCE mean at FY 75 posttest on the community resources, adults, learning, and work scales.

Mike showed continued growth over the two-year period on the work, self-reliance, communication, role, and trust scales of the Psychosocial Maturity Scale. He was significantly above the EBCE posttest means on the work, role, and social commitment scales and below average on only the openness to change scale. The openness to change score also showed a significant decline over the year.

The staff rated Mike on seven student behaviors. At the beginning of the year, he was significantly above the EBCE mean on "applies knowledge of his/her own aptitudes, interests, and abilities to potential career interests" and below the mean on "understands another person's message and feelings." At posttest time, he was still below the EBCE mean on the latter behavior, as well as on "demonstrates willingness to apply basic skills to work tasks and to vocational interests."

Over the course of the two years in the EBCE program, Mike's scores on the Self-Directed Search showed little change in pattern, although the number of interests and competencies did expand. Overall, realistic (R) occupations decreased and enterprising (E) occupations increased as his code changed from RCI (where C is conventional and I is investigative occupations) at pretest FY 74 to ICR at pretest FY 75 (a classification that includes computer operators and equipment repairers) to CEI at posttest FY 75. However, the I was only one point stronger than the R, and the CER classification includes data processing workers. Thus, Mike's Self-Directed Search codes appeared very representative of his desired occupational future.

Evaluators' Reflections. Mike's dramatic declines in attitudes and basic skills scores reflect behavior changes that occurred during the second half of his second year on the program and were detected by a number of people. In February, at a student staffing meeting, his learning manager reported of Mike that "no progress is seen in this zone with projects ... still elusive ... coasting right now ... may end up in trouble." The prescription was to "watch him—make him produce ... find out where he is." However, at the end of the next to last zone in mid-May, the report was still "the elusive butterfly! (Mike) needs to get himself in high gear to get everything completed on time!!!" Since the posttesting was completed before this time, Mike probably coasted through the posttesting as well.

(Continued)

(Continued)

Other data suggesting his lack of concern and involvement during the second half of his senior year was attendance. Although he missed only two days the first half of the year, he missed 13 days during the second half.

Mike showed a definite change in some of his personality characteristics over the two years he spent in the EBCE program. In the beginning of the program, he was totally lacking in social skills and self-confidence. By the time he graduated, he had made great strides in his social skills (although there was still much room for improvement). However, his self-confidence had grown to the point of overconfidence. Indeed, the employer-instructor on his last learning level spent a good deal of time trying to get Mike to make a realistic appraisal of his own capabilities.

When interviewed after graduation, Mike was working six evenings a week at a restaurant where he worked part-time for the last year. He hopes to work there for about a year, working his way up to cook, and then go to a business college for a year to study computers.

SOURCE: Fehrenbacher, Owens, and Haenn (1976, pp. 1, 17–21). Used by permission of Education Northwest.

EXHIBIT 8.34	Excerpts From Codebook for Use by Multiple Coders of Interviews With Decision Makers and Evaluators About Their Utilization of Evaluation Research

This codebook was developed from four sources: (1) the standardized open-ended questions used in interviewing; (2) review of the utilization literature for ideas to be examined and hypotheses to be reviewed; (3) our initial inventory review of the interviews, in which two of us read all the data and added categories for coding; and (4) a few additional categories added during coding when passages didn't fit well into the available categories.

Every interview was coded twice by two independent coders. Each individual code, including redundancies, was entered into our qualitative analysis database, so that we could retrieve all passages (data) on any subject included in the classification scheme, with brief descriptions of the content of those passages. The analyst could then go directly to the full passages and complete interviews from which these passages were extracted to keep quotations in context. In addition, the computer analysis permitted easy cross-classification and cross-comparison of passages for more complex analyses across interviews.

Characteristics of Program Evaluated

0101 Nature or kind of program

0102 Program relationship to government hierarchy

0103 Funding (source, amount, determination of, etc.)

0104 Purpose of program

0105 History of program (duration, changes, termination, etc.)

0106 Program effectiveness

Evaluator Role in Specific Study

0201 Evaluator's role in initiation and planning stage

0203 Evaluator's role in data collection stage

0204 Evaluator's role in final report and dissemination

0205 Relationship of evaluator to program (internal/external)

0206 Evaluator's organization (type, size, staff, etc.)

0207 Opinions/feelings about role in specific study

0208 Evaluator's background

0209 Comments on evaluator and evaluator process

Decision Maker's Role in Specific Study

0301 Decision maker's role in initiation and planning stage

0302 Decision maker's role in data collection stage

0303 Decision maker's role in final report and dissemination

0304 Relationship of decision maker to program

0305 Relationship of decision maker to other people or units in government

0306 Comments on decision maker and decision-making process (opinions, feelings, facts, knowledge, etc.)

Stakeholder Interactions

0501 Stakeholder characteristics

0502 Interactions during or about initiation of study

0503 Interactions during or about design of study

0504 Interactions during or about data collection

0505 Interactions during or about final report/findings

0506 Interactions during or about dissemination

Planning and Initiation Process of This Study (How and Who started)

0601 Initiator

0602 Interested groups or individuals

0603 Circumstances surrounding initiation

Purpose of Study (Why)

0701 Description of purpose

0702 Changes in purpose

Political Context

0801 Description of political context

0802 Effects on study

Expectations for Utilization

0901 Description of expectations

0902 Holders of expectations

0903 Effect of expectations on study

0904 Relationship of expectations to specific decisions

0905 Reasons for lack of expectations

0906 People mentioned as not having expectations

0907 Effect of lack of expectations on study

Data Collection, Analysis, Methodology

1001 Methodological quality

1002 Methodological appropriateness

1003 Factors affecting data collection and methodology

Findings, Final Report

1101 Description of findings/recommendations

1102 Reception of findings/recommendations

1103 Comments on final report (forms, problems, quality, etc.)

1104 Comments and description of dissemination

Impact of Specific Study

1201 Description of impacts on program

1202 Description of nonprogram impacts

1203 Impact of specific recommendations

Factors and Effects on Utilization

1301 Lateness

1302 Methodological quality

1303 Methodological appropriateness

1304 Positive/negative findings

1305 Surprise findings

1306 Central/peripheral objectives

1307 Point in life of program

1308 Presence/absence of other studies

1309 Political factors

1310 Interaction with evaluators

1311 Resources

1312 Most important factor

EXHIBIT 8.35 Excerpts From an Illustrative Interview Analysis: Reflections on Outcomes From Participants in a Wilderness Education Program

—Jeanne Campbell and Michael Patton

Experiences affect people in different ways. This experiential education truism means that the individual outcomes, impacts, and changes that result from participation in some set of activities are seldom predictable with any certainty. Moreover, the meaning and meaningfulness of such changes as do occur are likely to be highly specific to particular people in particular circumstances. While the individualized nature of learning is a fundamental tenet of experiential education, it is still important to stand back from those individual experiences in order to look at the patterns of change that cut across the specifics of person and circumstances. One of the purposes of the evaluation of the Learninghouse Southwest Field Training Project was to do just that—to document the experiences of individuals and then to look for the patterns that help provide an overview of the project and its impacts.

A major method for accomplishing this kind of reflective evaluation was the conduct of follow-up interviews with the 11 project participants. The first interviews were conducted at the end of October 1977, three weeks following the first field conference in the Gila Wilderness of New Mexico. The second interviews were conducted during the third week of February, three weeks after the wilderness experience in the Kofa Mountains of Arizona. The third and final interviews were conducted in early May, following the San Juan River conference in southern Utah. All interviews were conducted by telephone. An average interview took 20 minutes, with a range from 15 to 35 minutes. Interviews were tape-recorded and transcribed for analysis.

The interviews focused on three central issues: (1) How has your participation in the Learninghouse Project affected you personally? (2) How has your participation in the project affected you professionally? (3) How has your participation in the project affected your institution?

In the pages that follow, participant responses to these questions are presented and analyzed. The major purpose of the analysis was to organize participant responses in such a way that overall patterns would become clear. The emphasis throughout was on letting the participants speak for themselves. The challenge for the evaluators was to present participant responses in a cogent fashion that integrates the great variety of experiences and impacts recorded during the interviews.

Personal Change

"How has your participation in the Learninghouse Project affected you personally? What has been the impact of the project on you as a person?"

Questions about personal change generated more reactions from participants than subsequent questions about professional and institutional change. There is an intensity to these responses about individual change that makes it clear just how significant these experiences were in stimulating personal growth and development. Participants attempted throughout the interviews to indicate that they felt differently about themselves as persons because of their Learninghouse experiences. While such personal changes are often difficult to articulate, the interviews reflect a variety of personal impacts.

Confidence: A Sense of Self

During the three weeks in the wilderness, participants encountered a number of opportunities to test themselves. Can I carry a full pack day after day, uphill and downhill? Can I make it up that mountain? Do I have anything to contribute to the group? As participants encountered and managed stress, they learned things about themselves. The result was often an increase in personal confidence and a greater sense of self.

> It's really hard to say that LH did one thing or another. I think increased self-confidence has helped me do some things that I was thinking about doing. And I think that came, self-confidence came about largely because of the field experiences. I, right after we got back, I had my annual merit evaluation meeting with my boss, and at that I requested that I get a, have a change in title or a different title, and another title really is what it amounts to, and that I be given the chance for some other responsibilities that are outside the area that I work in. I want to get some individual counseling experience, and up to this point I have been kind of hesitant to ask for that, but I feel like I have a better sense of what I need to do for myself and that I have a right to ask for it least (Cliff, post-Kofas).

> I guess something that has been important to me in the last couple of trips and will be important in the

next one is just the outdoor piece of it. Doing things that perhaps I'd not been willing to attempt before whatever reason. And finding I'm better at it than expected. Before I was afraid (Charlene, post-Kofas).

The interviews indicate that increased confidence came not only from physical accomplishments but also—and especially—from interpersonal accomplishments.

After the Kofas I achieved several things that I've been working on for two years. Basically, the central struggle of the last two years of my life has been to no longer try to please people. No matter what my own feelings and needs are I try to please you. And in the past I had done whatever another person wanted me to do in spite of my own feelings and needs. And to have arrived at a point where I could tend to my own feelings and take care of what I needed to do for me is by far the most important victory I've won . . . a major one.

In the Kofas, I amazed myself that I didn't more than temporarily buy into how . . . I was being described . . . when I didn't recognize myself yet. And that's new for me. In the past I'd accept others' criticisms of me as if they were indeed describing me . . . and get sucked into that. And I felt that was an achievement for me to hold onto my sense of myself in the face of criticisms has long been one of my monsters I've been struggling with, so to hold onto me is, especially as I did, was definitely an achievement (Billie, post-Kofas).

I've been paying a lot of attention to not looking for validation from other people. Just sticking with whatever kinds of feelings I have and not trying to go outside of myself . . . and lay myself on a platter for approval. I think the project did have a lot to do with that, especially this second trip in the Kofas (Greg, post-Kofas).

I would say the most important thing that happened to me was being able to talk to other people quite honestly about, I think really about their problems more than mine. That's very interesting in that I think that I had, I think I had an effect upon Billie and Charlene both. As a result of that it gave me a lot more confidence and positive feelings. Do you follow that? Where rather than saying I had this problem and I talked to somebody and they solved it for me, it was more my helping other people to feel good about themselves that made me feel more adequate and better than myself (Rod, post-Gila).

Another element of confidence concerns the extent to which one believes in one's own ideas—a kind of intellectual confidence.

I think if I take the whole project into consideration. I think that I've gained a lot of confidence myself in some of the ideas that I have tried to use, both personally and let's say professionally. Especially in my teaching aspects, especially teaching at a woman's college where I think one of our roles is not only to teach women subject matter, but also to teach them to be more assertive. I think that's a greater component of our mission than normally would have it at most colleges. I think that a lot of the ideas that I had about personal growth and about my own interactions with people were maybe reinforced by the LH experience, so that I felt more confident about them, and as a result they have come out more in my dealings with people. I would say specifically in respect to a sort of a more humanistic approach to things (Rod, post-Kofas).

Increased confidence for participants was often an outcome of learning that they could do something new and difficult. At other times, however, increased confidence emerged as a result of finding new ways to handle old and difficult situations, for example, learning how to recognize and manage stress.

A change I've noticed most recently and most strongly is the ability to recognize stress. And also the ability to recognize that I can do a task without needing to make it stressful which is something I didn't know I did. So what I find I wind up doing, for example, is when I've had a number of things happen during the day and I begin to feel myself keying up I find myself very willing to say both to close friends and to people I don't know very well, I can't deal with this that you're bringing me. Can we talk about it tomorrow? This is an issue that really needs a lot of time and a lot of attention. I don't want to deal with it today, can we talk later . . . etc. So I'm finding myself really able to do that. And I'm absolutely delighted about it.

(Whereas before you just piled it on?)

Exactly. I'd pile it and pile it until I wouldn't understand why I was going in circles (Charlene, post-Kofas).

Personal Change—Overview

The personal outcomes cited by Learninghouse participants are all difficult to measure. What we have in the interviews are personal perceptions about personal

(Continued)

(Continued)

change. The evidence, in total, indicates that participants felt differently and, in many cases, behaved differently as a result of their project participation. Different participants were affected in different ways and to varying extent. One participant reported virtually no personal effects from the experiences.

> And as far as the effect it had on me personally, which was the original question, okay, to be honest with you, to a large degree it had very little effect, and that's not a dig on the program, because at some point in people's lives I think things start to have smaller effect, but they still have effect. So I think that for me, what it did have an effect on was tolerance. Because there were as lot of things that occurred on the trip that I didn't agree with. And still don't agree, but I don't find myself to be viciously in disagreement any longer, just plainly in disagreement. So it was kind of like before, I didn't want to listen to the disagreement, or I wanted to listen to it but resolve it. Now, you know, there's a third option, that I can listen to it, continue to disagree with it and not mind continuing to listen to it (Cory, post–San Juan).

The more common reaction, however, was surprise at just how much personal change occurred.

> My expected outcome was increase the number of contacts in the Southwest, and everyone of my expected outcomes were professional. That, you know, much more talk about potential innovations in education and directions to go, and you know, field-based education, what that's about, and I didn't expect at all, which may not be realistic on my part, but at least I didn't expect at all—the personal impact (Charlene, post-Gila).

For others, the year's participation in Learninghouse was among the most important learning experiences of a lifetime, precisely because the project embraced personal as well as professional growth.

> I've been involved in institutions and in projects as an educator, let's say, for 20 years I started out teaching in high school, going to the NSF institutions during the summertime and I've gone to a lot of Chautauqua things and a lot of conferences, you know, of various natures. And I really think that this project has by far the greatest. . . . has had by far the greatest impact on me. And I think that the reason is that in all the projects that I've had in the past . . . they've been all very specifically oriented towards one subject or toward

one . . . more of a, I guess, more of a science, more of a subject matter orientation to them. Whereas this having a process orientation has a longer effect. I mean a lot of the things I learn in these instances is out of date by now and you keep up with the literature, for example, and all that and maybe that stimulates you to keep up . . . but in reality as far as a growth thing on my part, I think on the part of other participants, I think that this has been phenomenal. And I just think that this is the kind of thing that we should be looking towards funding on any level, federal, or any level (Rod, post–San Juan).

We come now to a transition point in this report. Having reported participants' perceptions about personal change, we want to report the professional outcomes of the Learninghouse project. The problem is that in the context of a holistic experience like the Southwest Field Training Project, the personal–professional distinction becomes arbitrary. A major theme running throughout discussions during the conferences was the importance of reducing the personal–professional schism—the desirability of living an integrated life and being an integrated self. This theme is reflected in the interviews, as many participants had difficulty responding separately to questions about personal versus professional change.

Personal–Professional Change

Analytically, there is at least a connotative difference between personal and professional change. For evaluation purposes, we tried to distinguish one from the other as follows: personal changes concern the thoughts, feelings, behaviors, intentions, and knowledge people have about themselves; professional changes concern the skills, competences, ideas, techniques and processes people use in their work. There is, however, a middle ground. How does one categorize changes in thoughts, feelings, and intentions about competences, skills, and processes? There are changes in the person that affect that person's work. This section is a tribute to the complexity of human beings in defying the neat categories of social scientists and evaluators. This section reports changes that, for lack of a better nomenclature, we have called simply personal/professional impacts.

The most central and most common impact in this regard concerned changes in personal perspective that affected fundamental notions about and approaches to the world of work. The wilderness experiences and accompanying group processes permitted and/or forced many participants to stand back and take a look at themselves in relation to their work. The result was a changed perspective. The following four quotations are from interviews conducted after the first field conference in the Gila, a time when the contrasts provided by the first wilderness experience seemed to be felt most intensely.

The trip came at a real opportune time. I've been on this new job about 4–5 weeks and was really getting pretty thoroughly mired in it, kind of overwhelmed by it, and so it came after a particularly hellish week, so in that sense it was just a critical, really helpful time to get away. To feel that I had, to remember that I had some choices, both in terms of whether I stayed here or went elsewhere, get some perspective of what it was I actually wanted to accomplish in higher education rather than just surviving to keep my sanity. And it gave me some, it renewed some of my ability to think of doing what I wanted to do here at the University, or trying to, that there were things that were important for me to do rather than just handling the stuff that poured across my desk (Henry, post-Gila).

I think it's helped make me become more creative, and just, and that's kind of tied in with the whole idea of the theory of experiential education. And the way we approached it on these trips. And so for instance I'm talking with my wife the other night, after I got Laura's paper that she'd given in Colorado, and I said you oughta read this because you can go out and teach history and you know, experientially. Then I gave her an idea of how I would teach frontier history for instance, and I don't know beans about frontier history. But it was an idea which, then she told another friend about it, and this friend says oh, you can get a grant for that. You know. So that was just a real vivid example, and I feel like, it's, I've been able to apply, or be creative in a number of different situations, I think just because I give myself a certain freedom, I don't know, I can't quite pinpoint what brought it about, but I just feel more creative in my work (Cliff, post–San Juan).

You know my biggest problem is I've been trying to save the world, and what I'm doing is pulling back. Because, perhaps the way I've been going about it has been wrong or whatever, but at least my motives are clearer and I know much more directly what I need and what I don't need and so I'm more open but less, yeah, as I said, I've been in a let's save the world kind of thing, now I feel more realistic and honest (Charlene, post-Gila).

I've been thinking about myself and my relationship to men and my boss, and especially to ideas about fear and risk. . . . I decided that I needed to become a little more visible at the department. After the October experience, I just said I was a bit more ready to become visible at the department level. And I volunteered then to work on developing a department training policy and develop the plan and went down to the department and talked to the assistant about it and put myself in a consulting role while another person was assigned the actual job of doing it. And I think that I was ready to make that decision and act on it after I first of all got clear that I was working on male–female relationships. My department has a man, again, not a terribly easy one to know, so it's a risk for me to go talk with him and yet I did it. I was relatively comfortable and felt very good and very pleased with myself that I had done that and I think that's also connected (Billie, post-Kofas).

The connection between personal changes and professional activities was an important theme throughout the Learninghouse Project. The passages reported in this section illustrate how that connection took hold in the minds and lives of project participants. As we turn now to more explicit professional impacts, it is helpful to keep in mind the somewhat artificial and arbitrary nature of the personal–professional distinction.

(Omitted are sections on changed professional knowledge about experiential education, use of journals, group facilitation skills, individual professional skills, personal insights regarding work and professional life, and the specific projects the participants undertook professionally. Also omitted are sections on institutional impacts. We pick up the report in the concluding section.)

Final Reflections

Personal change . . . professional change . . . institutional change. . . . Evaluation categories aim at making sense out of an enormously complex reality. The reflections by participants throughout the interviews make it clear that most of them came away from the Learninghouse program feeling changes in themselves. Something had touched them. Sometimes it meant a change in perspective that would show up in completely unexpected ways.

For one thing, I just finished the purchase of my house. First of all, that's a new experience for me. I've never done it before. I've never owned a home and never

(Continued)

(Continued)

even wanted to. It seemed odd to me that my desire to "settle down" or make this type of commitment to a place occurred just right after the Gila trip. Just sort of one of those things that I woke up and went, "Wow, I want to stay here. I like this place. I want to buy it." And I had never in my life lived in a house or a place that I felt that way about. I thought that was kind of strange. And I do see that as a function of personal growth and stability. At least some kind of stability.

Other areas of personal growth: one has been, and this kind of crosses over I think into the professional areas, and that would be an ability to gain perspective. Certainly the trips I think . . . incredibly valuable for gaining perspective on what's happening in my home situation, my personal life, my professional life . . . the whole thing. And it has allowed me to focus on some priority types of things for me. And deal with some issues that I've been kind of dragging on for years and years and not really wanting to face up with them or deal with them. And I have been able to move on and move through those kinds of things in the last 6 or 9 months or so to a much greater extent than ever before (Tom, post–San Juan).

Other participants came away from the wilderness experiences with a more concrete orientation that they could apply to work, play, and life.

The thing that I realized as I was trying to make some connections between the river and raft trip, was that in some ways I can see the parallels of my life being kind of like our raft trip was, and the rapids, or the thrill ride, and they're a lot of fun, but it's nice to get out of them for a while and dry off. It's nice sometimes to be able to just drift along and not worry about things. But a lot of it also is just hard work. A lot of times I wish I could get out of it and go a different way, and that's been kind of a nice thing for me to think about and kind of a viewpoint to have whenever I see things in a lull or in a real high speed pace, that I can say, "Okay, I'm going to be in this for a while, but I'm going to come out of it and go into something else." And so that's kind of a metaphor that I use as somewhat of a philosophy or point of view that's helpful as I go from day to day (Cliff, post–San Juan).

A common theme that emerged as participants reflected on their year's involvement with Learninghouse was a new awareness of options, alternatives, and possibilities.

I would say that if I have one overall comment, the effect of the first week overall, is to renew my sense of the broader possibilities in my job and in my life. Opens things to me. I realize that I have a choice to be here and be myself. And since I have a choice, there are responsibilities. Which is a good feeling (Henry, post-Gila).

I guess to me what sticks out overall is that the experience was an opportunity for me to step out of the rest of my life and focus on it and evaluate it, both my personal life and my work, professional life aspect (Michael, post–San Juan).

As participants stood back and examined themselves and their work, they seemed to discover a clarity that had previously been missing—perspective, awareness, clarity—stuff of which personal/professional/institutional change is made.

I think I had a real opportunity to explore some issues of my own worth with a group of people who were willing to allow me to explore those. And it may have come later, but it happened then. On the Learninghouse, through the Learninghouse . . . and I think it speeded up the process of growing for me in that way, accepting my own worth, my own ideas about education, about what I was doing, and in terms of being a teacher it really aided my discussions of people and my interactions. It really gave me a lot of focus on what I was doing. I think I would've muddled around a long time with some issues that I was able to, I think, gain some clarity on pretty quickly by talking to people who were sharing their experience and were working towards the same goals, self-directed learning, and experiential education (Greg, post–San Juan).

I think what happened is that for me it served as a catalyst for some personal changes, you know, the personal, institutional, they're all wound up, bound up together. I think I was really wrestling with jobs and career and so on. For me the whole project was a catalyst, a kind of permission to look at things that I hadn't looked at before. One of the realizations, one of the insights that I had in the process was, kind of neat on my part, to become concrete, specific in my actions in my life, no matter whether that was writing that I was doing, or if it was in my job, or whatever it was. But to really pay attention to that. I think that's one of the things that happened to me (Peter, post–San Juan).

These statements from interviews do not represent a final assessment of the impacts of the Learninghouse Southwest Field Training Project. Several participants resisted the request to make summary statements about the effects and outcomes of their participation in the program because they didn't want to force premature closure.

(Can you summarize the overall significance of participation in the project?)

> I do want to make a summary, and I don't again. . . . It feels like the words aren't easy and for me being very much a words person, that's unusual. It's not necessarily that the impact hasn't been in the cognitive areas. There have been some. But what they've been, where the impact has been absolutely overwhelming is in the affective areas. Appreciation of other people, appreciation of this kind of education. Though I work in it, *I haven't done it before*! A real valuing of people, the profession, of my colleagues in a sense that I never had before. . . .
>
> The impact feels like it's been dramatic, and I'm not sure that I can say exactly how. I'm my whole . . . it all can be summarized perhaps by saying I'm much more in control. In a good kind of sense. In accepting risk and being willing to take it; accepting challenge and being willing to push myself on that; accepting and understanding more about working at the edge of my capabilities . . . what that means to me. Recognizing very comfortably what I can do and feeling good about that confidence, and recognizing that what I haven't yet done, and feeling okay about

trying it. The whole perception of confidence has changed (Charlene, post–San Juan).

The Learninghouse program was many things—the wilderness, a model of experiential education, stress, professional development—but most of all, the project was the people who participated. In response after response participants talked about the importance of the people to everything that happened. Because of the dominance of that motif throughout the interviews, we want to end this report with that highly personal emphasis.

> I said before I think that to know some people, that meant a lot to me, people who were also caring. And people who were also involved, very involved in some issues, philosophical and educational, that were pretty basic not only to education, but to living. Knowing these people has been really important to me. It's given me a kind of continuity and something to hold onto in the midst of a really frustrating, really difficult situation where I didn't have people where I could get much feedback from, or that I could share much thinking about, talking about, and working with. It's just kind of basic issues. That kind of continuity is real important to just my feelings, important to myself. Feeling like I have someplace to go. . . . Sometimes I feel funny about placing so much emphasis on the people. . . . But the people have really meant a lot to me as far as putting things together for myself. Being able to have my hands in something that might that really offers me a way to go (Greg, post–San Juan).

SOURCE: Patton (1978a, pp. 7–9).

APPLICATION EXERCISES

1. *Practicing cross-case analysis:* Three case studies are presented in different chapters of this book. The *story of Li* presents a Vietnamese woman's experience in an employment training program (Exhibit 4.3, pp. 182–183). *Thmaris* is a case study of a homeless youth and his journey to a more stable life (Exhibit 7.20, pp. 511–516). *Mike's story*, in this chapter, describes his experiences in a career education program (Exhibit 8.33, pp. 638–542). Conduct a cross-case analysis of these three cases. Identify patterns (descriptive similarities across cases) and themes (the labels you give to what the patterns mean). See page 541 for differentiation of patterns and themes.

2. *Conducting a causal analysis:* The module on causal analysis ends with a suggested assignment. Here, it is included again as a formal application exercise. In the Sufi story that follows, analyze the data represented by the story in two ways. (1) First, try to isolate specific variables that are important in the story, deciding which are the independent variables and which is the dependent variable, and then write a statement of the form "These things caused this thing." (2) Read the story again. For the second analysis, try to distinguish among and label the different meanings of the situation expressed by the characters observed in the story; then write a statement of the form "These things and these things came together to create _____." (3) Discuss the implications of these different interpretative and explanatory approaches. You aren't being asked to decide that one approach is right and the other is wrong. You are being asked to analyze the implications of each approach for interpreting qualitative case data. Here's the case data, otherwise known as a story.

    Walking one evening along a deserted road, Mulla Nasrudin saw a troop of horsemen coming toward him. His imagination started to work; he imagined himself captured and sold as a slave, robbed by the oncoming horsemen, or conscripted into the army. Fearing for his safety, Nasrudin bolted, climbed a wall into a graveyard, and lay down in an open tomb.

    Puzzled at this strange behavior, the men—honest travelers—pursued Nasrudin to see if they could help him. They found him stretched out in the grave, tense and quivering.

    "What are you doing in that grave? We saw you run away and see that you are in a state of great anxiety and fear. Can we help you?"

    Seeing the men up close, Nasrudin realized that they were honest travelers who were genuinely interested in his welfare. He didn't want to offend them or embarrass himself by telling them how he had misperceived them, so Nasrudin simply sat up in the grave and said, "You ask what I'm doing in this grave. If you must know, I can tell you only this: I am here because of you, and you are here because of me." (Adapted from Shah, 1972, p. 16)

3. *Exercise on diverse approaches to causal explanation:* The Halcolm graphic comic on causality (pp. 635–637) retells an African fable about causality. Exhibit 8.19 (pp. 600–601) presents 12 approaches to qualitative causal analysis. (a) Select three different approaches in Exhibit 8.19, and discuss those approaches

as applied to the Halcolm story about why dogs bark. (b) Identify which approaches in Exhibit 8.19 cannot be used with a single case story and why. What is needed to engage in certain kinds of causal explanatory analysis?

4. *Exercise on substantive significance:* Exhibit 8.35 (pp. 643–649) presents excerpts from an evaluation report. Add your own new section to the end of the report in which you discuss what you believe is *substantively significant* in the findings. Qualitative inquiry does not have statistical significance tests, so qualitative judgments must be rendered about what is substantively significant. Demonstrate that you can do this and understand what it means by identifying what is substantively significant in the report presented in Exhibit 8.35. Justify your judgments. (See pp. 643–649 for a discussion of substantive significance.)

5. *Exercise on reflexivity:* Select a topic, issue, or question on which you either have done qualitative inquiry or would like to do a qualitative inquiry. Present a brief outline of your methods, either actual or proposed. Now, write a *reflexive* analysis section. How does *who you are* come into play in this inquiry? Discuss yourself as the instrument of qualitative inquiry and analysis. What are your strengths and weaknesses that come into play for this particular actual or proposed inquiry? (See pp. 70–74 for a discussion of reflexivity.)

6. *The role of numbers in analysis:* MQP Rumination #8 argues that numbers (how many interviewees said what) should not be reported in a qualitative analysis; essentially, qualitative analysis should *stay qualitative* (see pp. 557–560). Make the opposite argument. Write your own rumination making the case that reporting how many people said what is altogether appropriate and useful.

7. *Evaluating qualitative analysis:* Locate a qualitative study on a topic of interest to you. How are the analysis process and approach described in the report? What questions do you have about the analysis that are left unanswered, if any? Is reflexivity addressed? Are descriptions clearly separated from interpretations? To what extent are causal explanations offered in the analysis? Are strengths and weaknesses in the data and the analytical process discussed? How is qualitative analysis software discussed, if at all? Use Exhibit 8.32 (pp. 631–632), the qualitative analysis checklist, to review and evaluate a qualitative study of your own choosing.

9 Enhancing the Quality and Credibility of Qualitative Studies

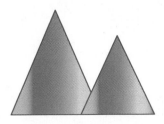

The Medieval alchemical symbol for fire was a single triangle, also the modern symbol for triangulation in geometry, trigonometry, and surveying: the process of locating an unknown point by measuring angles to it from known points. Triangulation in qualitative inquiry involves gathering and analyzing multiple perspectives, using diverse sources of data, and during analysis, using alternative frameworks.

The double-triangle symbol, shown here, represented *strong fire* in alchemy. Strong fire was needed to ensure that the transformative process would work. Building and sustaining a *strong fire* required quality materials, good ventilation, and ongoing monitoring. Using a strong fire required skill, experience, and rigorous implementation of the transformative process to achieve the desired effects. Strong fire produces both intense heat and bright illumination. Alchemists who could properly build, sustain, and appropriately use strong fire were held in high esteem, had great credibility, and produced much-valued products.

In another village you tell the hungry to give up their preoccupation with food. In yet another village you tell the people to pray for a richer harvest. In each village the problem is the same, but always your message is different. I can find no pattern of Truth in your teachings."

The Mulla looked piercingly at the young man.

"Truth? When you came here you did not tell me you wanted to learn Truth. Truth is like the Buddha. When met on the road it should be killed. If there were only one Truth to be applied to all villages, there would be no need of Mullahs to travel from village to village."

"When you first came to me you said you wanted to 'learn how to interpret' what you see as you travel through the world. Your confusion is simple. To interpret and to state Truths are two quite different things."

Having finished his story Halcolm smiled at the attentive youths. "Go, my children. Seek what you will, do what you must."

—From Halcolm's *Evaluation Parables*

Interpreting Truth

A young man traveling through a new country heard that a great Mulla, a Sufi guru with unequaled insight into the mysteries of the world, was also traveling in that region. The young man was determined to become his disciple. He found his way to the wise man and said, "I wish to place my education in your hands that I might learn to interpret what I see as I travel through the world."

After six months of traveling from village to village with the great teacher, the young man was confused and disheartened. He decided to reveal his frustration to the Mulla.

"For six months I have observed the services you provide to the people along our route. In one village you tell the hungry that they must work harder in their fields.

Chapter Preview

This chapter concludes the book by addressing ways to enhance the quality and credibility of qualitative analysis. Module 76 discusses and demonstrates analytical processes for enhancing credibility by systematically engaging and questioning the data. Module 77 presents four triangulation processes for enhancing credibility. Modules 78 and 79 present alternative and competing criteria for judging the quality of qualitative studies. Module 80 discusses how and why the credibility of the inquirer is critical to the overall credibility of qualitative findings. Module 81 examines core issues of generalizability, extrapolations, transferability, generating principles, and harvesting lessons. Module 82 concludes the chapter and the book by addressing philosophy of science issues related to the credibility and utility of qualitative inquiry.

Analytical Processes for Enhancing Credibility: Systematically Engaging and Questioning the Data

The credibility of qualitative inquiry depends on four distinct but related inquiry elements:

1. *Systematic, in-depth fieldwork* that yields high-quality data

2. *Systematic and conscientious analysis of data* with attention to issues of credibility

3. *Credibility of the inquirer*, which depends on training, experience, track record, status, and presentation of self

4. *Readers' and users' philosophical belief in the value of qualitative inquiry*—that is, a fundamental appreciation of naturalistic inquiry, qualitative methods, inductive analysis, purposeful sampling, and holistic thinking (indeed, all 12 core qualitative strategies presented in Exhibit 2.1, pp. 46–47)

The first of the elements that determine credibility, *systematic, in-depth fieldwork that yields high-quality data*, was covered in Chapter 5 (purposeful qualitative designs), Chapter 6 (in-depth fieldwork and rich observational data), and Chapter 7 (high-quality, skillful interviewing).

This module and the next focus on the remaining three elements of quality: systematic and conscientious analysis of data. Module 80 discusses credibility of the inquirer, and Module 82 examines readers' and users' philosophical belief in the value of qualitative inquiry.

Strategies for Enhancing the Credibility of Analysis

> ### Chance favors the prepared mind.
>
> —Louis Pasteur (1822–1895)
> French microbiologist (known as the "father of microbiology") who discovered the process for pasteurizing milk, named after him

Chapter 8 presented analytical strategies for coding qualitative data, identifying patterns and themes, creating typologies, determining substantive significance, and reporting findings. However, at the heart of much controversy about qualitative findings are doubts about the nature of qualitative analysis because it is so judgment dependent. Statistical analysis follows formulas and rules, while, at the core, qualitative analysis depends on the insights, conceptual capabilities, and integrity of the analyst. Qualitative analysis is driven by the capacity for astute pattern recognition from beginning to end. Staying open to the data, for example, involves aggregating and integrating the data around a particular expected pattern while also watching for unexpected patterns. This process is epitomized in health research by the scientist working on one problem who suddenly notices a pattern related to a quite different problem—and thus discovers *Viagra*; as Pasteur explained when he was asked how he happened to discover how to stop bacterial contamination of milk, "Chance favors the prepared mind." Here, then, are some techniques that prepare the mind for insight while also enhancing the credibility of the resulting analysis.

Integrity in Analysis: Generating and Assessing Alternative Conclusions and Rival Explanations

One barrier to credible qualitative findings stems from the suspicion that the analyst has shaped findings according to his or her predispositions and biases. Being able to report that you engaged in a systematic and conscientious search for alternative themes, divergent patterns, and rival explanations enhances credibility, not to mention that it is simply good analytical practice and the very essence of being rigorous in analysis. This can be done both inductively and logically. Inductively, it involves looking for other ways of organizing the data that might lead to different findings. Logically, it means thinking about other logical possibilities and then seeing if those possibilities can be supported by the data. When considering rival organizing schemes and competing explanations, your mind-set should not be one of attempting to disprove the alternatives; rather, *you look for data that support alternative explanations*.

In evaluation of a training program for chronically unemployed men of color, we conducted case studies of a group of successes. The program model was based on training in both hard skills (e.g., machine tooling, keyboarding, welding, and accounting) and soft skills (showing up to work on time, dressing appropriately, and respecting supervisors and coworkers). The cases studied validated the importance of both kinds of skills, but an additional explanation emerged in later cases, namely, that the program experience and peer support led to an identity shift: Successful trainees began to think of themselves as capable of holding a job. They were used to being labeled as "losers." The opportunity to think of themselves as "winners" involved more than acquiring "soft skills." It involved a shift in identity. We went back to earlier cases to find out if that phenomenon was evident there as well. It was, as was evidence for how that shift in identity occurred. Might this change be simply a function of participants being older by the time they entered this particular program (a maturation effect)? No, the change was evident in younger participants as well as older ones. We continued in this fashion, looking for alternative explanations and checking them out against the case data.

Failure to find strong supporting evidence for alternative ways of presenting data or contrary explanations helps increase confidence in the initial, principal explanation you generated. Comparing alternative patterns will not typically lead to clear-cut "yes there is support" versus "no there is no support" kinds of conclusions. You're searching for *the best fit*, the preponderance of evidence. This requires assessing the weight of evidence and looking for those patterns and conclusions that fit the preponderance of data. Keep track of and report alternative classification systems, themes, and explanations that you considered and "tested" during data analysis. This demonstrates intellectual integrity and lends considerable credibility to the final set of findings and explanations offered. Analysis of rival explanations in case studies is analogous to counterfactual analysis in experimental designs.

Searching for and Analyzing Negative or Disconfirming Evidence and *Cases*

Closely related to testing alternative constructs is the search for and analysis of *negative cases*. Where patterns and trends have been identified, our understanding of those patterns and trends is increased by considering the instances and cases that do not fit within the pattern. These may be exceptions that illuminate the boundaries of the pattern. They may also broaden understanding of the pattern, change the conceptualization of the pattern, or cast doubt on the pattern altogether.

> In qualitative analysis you need to keep analyzing the data to check any explanations and generalizations that

you wish to make, to ensure that you have not missed anything that might lead you to question their applicability. Essentially this means looking for negative or deviant cases—situations and examples that just do not fit the general points you are trying to make. However, the discovery of negative cases or counter-evidence to a hunch in qualitative analysis does not mean its immediate rejection. You should investigate the negative cases and try to understand why they occurred and what circumstances produced them. As a result, you might extend the idea behind the code to include the circumstances of the negative case and thus extend the richness of your coding. (Gibbs, 2007, p. 96)

In the Southwest Field Training Project involving wilderness education, virtually all participants reported significant "personal growth" as a result of their participation in the wilderness experiences; however, the two people who reported "no change" provided particularly useful insights into how the program operated and affected participants. These two had crises going on back home that limited their capacity to "get into" the wilderness experiences. The project staff treated the wilderness experiences as fairly self-contained, closed-system experiences. The two negative cases opened up thinking about "baggage carried in from the outside world," "learning-oriented mind-sets," and a "readiness" factor that subsequently affected participant selection and preparation.

Negative cases also provide instructive opportunities for new learning in formative evaluations. For example, in a health education program for teenage mothers where the large majority of participants complete the program and show knowledge gains, an important component of the analysis should include examination of reactions from dropouts, even if the sample is small for the dropout group. While the small proportion of dropouts may not be large enough to make a difference in a statistical analysis, qualitatively the dropout feedback may provide critical information about a niche group or a specific subculture, and/or clues to program improvement.

No specific guidelines can tell you how and how long to search for negative cases or how to find alternative constructs and hypotheses in qualitative data. Your obligation is to make an "assiduous search . . . until no further negative cases are found" (Lincoln & Guba, 1986, p. 77). You then report the basis for the conclusions you reach about the significance of the negative or deviant cases.

Readers of a qualitative study will make their own decisions about the plausibility of alternate explanations and the reasons why deviant cases do not fit within dominant patterns. But I would note that the

ADVOCACY–ADVERSARY ANALYSIS

In 1587, the Roman Catholic Church created advocacy–adversary roles to test the validity of evidence in support of the canonization process for elevating someone to sainthood. The Devil's Advocate *(Latin:* advocatus diaboli*) in this process (officially designated the Promoter of the Faith) was a canon lawyer whose job was to argue against the canonization by presenting doubts about or holes in the evidence, for example, to argue that any miracles attributed to the candidate were unsubstantiated or even fraudulent. The Devil's Advocate opposed God's Advocate, whose job was to present evidence supporting and make the argument in favor of canonization. This advocacy–adversary process endured until 1983, when it was abolished by Pope John Paul II as overly adversarial and contentious.*

Advocacy–Adversary Analysis in Evaluation

A formal and forced approach to engaging rival conclusions draws on the legal system's reliance on opposing perspectives battling it out in the courtroom. The advocacy-adversary model suggested by Wolf (1975) developed in response to concerns that evaluators could be biased in their conclusions. Also called the *Judicial Model of Evaluation* (Datta, 2005), to balance possible evaluator biases, two teams engage in debate. The *advocacy team* gathers and presents information that supports the proposition that the program is effective; the *adversary team* gathers information that supports the conclusion that the program ought to be changed or terminated.

Some years ago, I served as the judge for what would constitute admissible evidence in an advocacy–adversary evaluation of an innovative education program in Hawaii. The task of the advocacy team was to gather and present data supporting the proposition that the program was effective and ought to be continued. The adversaries were charged with marshalling all possible evidence demonstrating that the program ought to be terminated. When I arrived on the scene, I immediately felt the exhilaration of the competition. I wrote in my journal,

No longer staid academic scholars, these are athletes in a contest that will reveal who is best; these are lawyers prepared to use whatever means necessary to win their case. The teams have become openly secretive about

their respective strategies. These are experienced evaluators engaged in a battle not only of data but also of wits.

As the two teams prepared their final reports, a concern emerged among some about the narrow focus of the evaluation. The summative question concerned whether the program should be continued or terminated. Education officials were asking how to improve the program without terminating it. Was it possible that a great amount of time, effort, and money was directed at answering the wrong question? Was it appropriate to force the data into a simple save-it-or-scrap-it choice? In fact, middle-ground positions were more sensible. But the advocacy–adversary analytical process design obliged opposing teams to do battle on the unembellished question of whether to maintain or terminate a program. A systematic assessment of strengths and weaknesses, with ideas for improvement, gave way to an all-good, all-bad framing, and that's how the results were presented (Patton, 2008, pp. 142–143).

The weakness of the advocacy–adversary approach is that it emphasizes contrasts and opposite conclusions, to the detriment of appreciating and communicating nuances in the data and accepting and acknowledging genuine and meaningful ambiguities. *Advocacy–adversary analysis* forces data sets into combat with each other. Such oversimplification of complex and multifaceted findings is a primary reason why advocacy–adversary evaluation is rarely used (in addition to being expensive and time-consuming). Still, it highlights the importance of engaging in some systematic analysis of alternative and rival conclusions, and as one approach (but not the only one) to testing conclusions, it can be useful and revealing.

Practical Analytical Variations on a Theme

1. A variation of the overall advocacy–adversary approach would be to arbitrarily create advocacy and adversary teams *only* during the analysis stage so that both teams work with the same set of data but each team organizes and interprets those data to support different and opposite conclusions, including identifying ambiguous findings.

2. Another variation would be for a lone analyst to organize data systematically into *pro* and *con* sets of evidence to see what each yielded.

section of the report that involves exploration of alternative explanations and consideration of why certain cases do not fall into the main pattern can be among the most interesting sections of a report to read. When

well written, this section of a report reads something like a detective study in which the analyst (detective) looks for clues that lead in different directions and tries to sort out which direction makes the most sense given

ANALYTIC INDUCTION: HYPOTHESIS TESTING WITH NEGATIVE CASES

Analytic induction emphasizes giving special attention to negative or deviant cases for testing propositions that should, based on the theory being examined, apply to all cases that have been sampled in the design to manifest the phenomenon of interest. Analytic induction works through one case at a time. If the case data fit the hypothesis, the inductive analyst takes up the next case. If a case isn't consistent with the hypothesis—that is, it is a negative or deviant case—then the hypothesis is revised or the case is rejected as not actually relevant to the phenomenon being studied. The analytical focus is examining the extent to which *every case* confirms the hypothesis and to either refine the hypothesis or the statement of the problem to account for all cases. No cases can be ignored. All must be accounted for and used in the analysis.

Here's an example of testing a hypothesis about the effect of mother–daughter relationships on anorexia. The proposition being tested was "*If* mother was critical of daughter's body image *and* mother–daughter relationship was strained *and* daughter experiences weight loss, *then* count that as an example of mother's negative influence on daughter's self-image." Once particular interviews were identified as containing the codes identified in the hypothesis, the qualitative data from interviews and cases could be examined to determine whether support for this causal interpretation could be justified for each case (Hesse-Biber & Dupuis, cited in Silverman & Marvasti, 2008, p. 252). The rigor of this approach is that finding even a single disconfirming case disconfirms the hypothesis requiring either refinement or reformulation, for the goal is to identify and confirm a generalizable, universal, causal explanation for the phenomenon of interest (Flick, 2007a, p. 30; Schwandt, 2007, p. 6).

the clues (data) that are available. Such writing adds credibility by showing the analyst's authentic search for what makes most sense rather than marshalling all the data toward a single conclusion. Indeed, the whole tone of a report feels different when the qualitative analyst is willing to openly consider other possibilities than those finally settled on as most reasonable in accordance with the preponderance of evidence. Compare the approach of weighing alternatives with the report where all the data lead in a single-minded fashion, in a rising crescendo, toward an overwhelming presentation of a single point of view. Perfect patterns and omniscient explanations are likely to be greeted skeptically—and for good reason: The human world is not perfectly ordered, and human researchers are not omniscient. Humility can do more than certainty to enhance credibility. Dealing openly with the complexities and

dilemmas posed by negative cases is both intellectually honest and politically strategic.

Avoid the Numbers Game

Philosopher of science Thomas H. Kuhn (1970), having studied extensively the value systems of scientists, observed that "the most deeply held values concern predictions" and "quantitative predictions are preferable to qualitative ones" (pp. 184–185). The methodological status hierarchy in science ranks "hard data" above "soft data," where "hardness" refers to the precision of statistics. Qualitative data can carry the stigma of "being soft." This carries over into the public arena, especially in the media and among policymakers, creating what has been called the tyranny of numbers (Eberstadt, 1995).

How can one deal with a lingering bias against qualitative methods? A starting point is helping people understand that qualitative methods are not weaker or softer than quantitative approaches. Qualitative methods are *different*. Making the case for the value of qualitative inquiries involves being able to communicate the particular strengths of qualitative methods (Chapters 1 and 2) and the kinds of evaluation and other applications for which qualitative data are especially appropriate (Chapter 4). But those understandings can only open the door to dialogue. The fact is that numbers have a special allure in modern society. Statistics are seductive—so precise, so clear. Numbers convey that sense of precision and accuracy, even if the measurements that yielded the numbers are relatively unreliable, invalid, and meaningless (e.g., see Hausman, 2000; Silver, 2012).

Quantitizing

Quantitizing, commonly understood to refer to the numerical translation, transformation, or conversion of qualitative data, has become a staple of mixed-methods research (Sandelowski, Voils, &

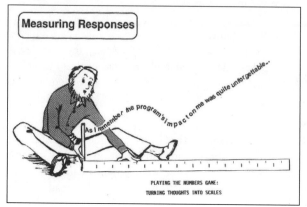

PLAYING THE NUMBERS GAME:
TURNING THOUGHTS INTO SCALES

Knafl, 2009, p. 208). Quantitized qualitative data are analyzed statistically, including using statistical significance tests (Collingridge, 2013).

There are different techniques by which quantitization may be achieved. Two common strategies are (1) dichotomizing and (2) counting. Dichotomizing refers to assigning a binary value (e.g., 0 and 1) to variables with two mutually exclusive and exhaustive categories, such as assigning "0" to participants who did not express a particular theme and "1" to participants who did express the theme. In contrast, counting involves calculating the number of themes expressed by each participant, as in the case of determining that a participant expressed two out of four themes in a study. Counting also includes calculating the number of qualitative codes assigned to specific themes, as in the case of determining that a participant expressed 10 qualitative codes associated with a theme (Collingridge, 2013, p. 82).

In Chapter 8, I devoted my MQP Rumination to why I consider this kind of quantizing to be generally a bad idea and advocated keeping qualitative analysis qualitative (see pp. 557–559). I won't repeat that argument here. Still, it strikes me as a worrisome trend. What's driving it? Partly, it's simply the cultural and political allure of numbers. But there's more.

> Pragmatic and ecumenical impulses, and the advent of computerized software programs to manage both qualitative and quantitative data, have served to promote a largely technical view of quantitizing. Moreover, the rhetorical appeal of numbers—their cultural association with scientific precision and rigor—has served to reinforce the necessity of converting qualitative into quantitative data.

A systematic literature review of quantitizing studies—that is, studies featuring quantitative analysis of qualitative interviews—shows the widespread nature of the phenomenon and some of the problems that arise, especially applying statistics to small sample sizes. Quantitative analyses of qualitative data are done to disaggregate results by background characteristics of participants (cross-tabs and correlations), to statistically test hypotheses, and to determine the prevalence of themes. But the overall problem is precisely what one would expect: "The conversion of the qualitative information to frequency counts has reduced the rich interpretation of people's experience that was expressed through their interviews" (Fakis, Hilliam, Stoneley, & Townend, 2014, p. 156). That is the crux of the issue, as is replacing a determination of substantive significance with the safe fallback position of replying on statistical significance.

Moreover, those engaged in quantiziting seem oblivious to the issues involved.

> Typically glossed, however, are the foundational assumptions, judgments, and compromises involved in converting qualitative into quantitative data and whether such conversions advance inquiry.... Such conversions "are by no means transparent, uncontentious, or apolitical" (Love, Pritchard, Maguire, McCarthy, & Paddock, 2005, p. 287; Sandelowski, Voils, & Knafl, 2009, p. 28).

Substantive Significance Trumps Statistical Significance

The point, however, is not to be anti-numbers. The point is to be *pro-meaningfulness.*

I'm not numbers phobic. I have used numbers regularly in titling exhibits throughout this book:

Exhibit 8.1 Twelve Tips for Ensuring a Strong Foundation for Qualitative Analysis (pp. 522–523).

Exhibit 8.10 Ten Types of Qualitative Analysis (see pp. 551–552).

Exhibit 9.1 Ten Systematic Analysis Strategies to Enhance Credibility and Utility (pp. 659–660).

Module 77 presents four triangulation processes for enhancing credibility.

When there is something meaningful to be counted, then count. As sample sizes increase, especially in mixed-methods studies, quantizing is likely to become even more pervasive. One study in the systematic review of quantizing articles had a sample size of 400 (Fakis et al., 2014, p. 146). Such studies will quantitize and do so appropriately. Weaver-Hightower (2014) studied political influence by reviewing public policy documents; from 1,459 transcript pages, he coded 2,294 unique arguments and relied heavily on quantitative analysis. That's understandable and appropriate, though reporting the results to two decimal places, "the average agreement score was 5.22%" (p. 125), illustrates the allure of pretentious precision. Or maybe just habit.

So while I advocate keeping qualitative analysis qualitative and focusing on substantive significance when interpreting findings, this is no hard-and-fast rule (my Chapter 8 MQP Rumination notwithstanding). Do what is appropriate. It doesn't make sense to report percentages in a sample of 10 interviewees; it does make sense with a sample of 400. By knowing the strengths and weaknesses of both quantitative and qualitative data, you can help those with whom

CONSTANT COMPARISON

A lot of qualitative analysis involves comparisons: comparing cases, comparing quotations, comparing observations, and comparing findings in others studies with your own findings.

The point about these comparisons is that they are constant; they continue throughout the period of analysis and are used not just to develop theory and explanations but also to increase the richness of description in your analysis and thus ensure that it closely captures what people have told you and what happened.

There are two aspects to this constant process:

1. Use the comparisons to check the *consistency and accuracy* of application of your codes, especially as you first develop them. Try to ensure that the passages coded the same way are actually similar. But at the same time, keep your eyes open for ways in which they are different. Filling out the detail of what is coded in this way may lead you to further codes and to ideas about what is associated with any variation. This can be seen as a circular or iterative process. Thus, develop your code, check for other occurrences in your data, compare these with the original, and then revise your coding (and associated memos) if necessary.

2. Look explicitly for *differences and variations* in the activities, experiences, actions and so on that have been coded.

In particular, look for variation across cases, settings and events (Gibbs, 2007, p. 96).

Constant comparison is an ongoing analysis of similarities and differences: What things go together in the data? What things are different? What explains these similarities and differences? What are the implications for your overall inquiry purpose and conclusions?

Design Checks: Keeping Methods and Data in Context

One issue that can arise during analysis is concerns about how design decisions affect results. For example, purposeful sampling strategies provide a limited number of cases for examination. When interpreting findings, it becomes important to reconsider how design constraints may have affected the data available for analysis. This means considering the rival methodological hypothesis that the findings are due to methodological idiosyncrasies.

By their nature, qualitative findings are highly context and case dependent. Three kinds of sampling limitations typically arise in qualitative research designs:

1. There are limitations in the situations (critical events or cases) that are sampled for observation (because it is rarely possible to observe all situations even within a single setting).

you dialogue focus on really important questions rather than, as sometimes happens, focusing primarily on how to generate numbers. The really important questions are about what the findings mean. A single illuminative case or interview may be more substantively meaningful and insightful than 20 routine cases. That 5% level of insight is not a reason to pay more attention to the 95% degree of mediocrity just because there's more of it. Information-rich cases stand out not because there are lots of them but precisely because they are so rare—and rich with revelation (the very definition of being information-rich). Rare, precious gems are valued over widely available (and less expensive), semiprecious stones for the same reason. Qualitative analysis must include the analytical insight to distinguish signal from noise and valuable insights from commonplace ones.

Summary of Strategies for Systematically Analyzing Qualitative Data to Enhance Credibility

Qualitative analysis aims to make sense of qualitative data: detecting patterns, identifying themes, answering the primary questions framing the study, and presenting substantively significant findings. In this chapter, we've been looking at ways of enhancing the credibility of findings by deepening the analysis, reexamining initial findings, and continuously working back and forth between the findings and the data to validate findings against data. Exhibit 9.1 summarizes the analytical techniques we've just covered and looks ahead to the four kinds of triangulation I'll present and discuss in the next module (Items 7–10 in Exhibit 9.1).

2. There are limitations from the time periods during which observations took place—that is, constraints of temporal sampling.

3. The findings will be limited based on selectivity in the people who were sampled for observations or interviews, or selectivity in document sampling.

In reporting how purposeful sampling decisions affect findings, the analyst returns to the reasons for having made the initial design decisions. Purposeful sampling involves studying information-rich cases in depth and detail to understand and illuminate important cases rather than generalizing from a sample to a population (see Chapter 5). For instance, sampling and studying highly successful and unsuccessful cases in an intervention yields quite different results from studying a "typical" case or a mix of cases. People unfamiliar with purposeful samples may think of small, purposeful samples as "biased," a perception that undermines credibility in their minds. In communicating findings, then, it becomes important to emphasize that the issue is not one of dealing with a distorted or biased sample but rather one of clearly delineating the purpose, strengths, and limitations of the sample studied—and therefore being careful about not inappropriately extrapolating the findings to other situations, other time periods, and other people—a caution we'll return to later in this chapter. Reporting both methods and results in their proper contexts will avoid many controversies that result from yielding to the temptation to overgeneralize from purposeful samples. *Keeping findings in context is a cardinal principle of qualitative analysis. Design decisions are context for analysis.*

The wise fool in Sufi tales, Mulla Nasrudin, was once called on to make this point to his monarch. Although he was supposed to be a wise man, Nasrudin was accused of being illiterate. Nagged to action by skeptics, the monarch decided to test him.

"Write something for me, Nasrudin," said the king.

"I would willingly do so, but I have taken an oath never to write so much as a single letter again," replied Nasrudin.

"Well, write something in the way in which you used to write before you decided not to write, so that I can see what it was like."

"I cannot do that, because every time you write something, your writing changes slightly through practice. If I wrote now, it would be something written for now."

"Then," addressing the crowd, the king commanded: "Bring me an example of Nasrudin's writing, anyone who has something he's written."

Someone brought a terrible scrawl that Nasrudin had once written to him.

"Is this your writing?" asked the monarch.

"No," said Nasrudin. "Not only does writing change with time, but reasons for writing change. You are now showing a piece of writing done by me to demonstrate to someone how he should *not* write." (Shah, 1973, p. 92)

EXHIBIT 9.1   Ten Systematic Analysis Strategies to Enhance Credibility and Utility

1. *Generate and assess alternative conclusions and rival explanations.* Don't settle quickly on initial conclusions. Go back to the data. What are other ways of explaining what you've found? Look for the explanation that best fits the preponderance of evidence.

2. *Advocacy–adversary analysis uses a debate format for testing the viability of conclusions.* What are the evidence and arguments that support your conclusions? What are the contrary evidence and counter-arguments? Get another analyst to play the "Devil's Advocate" role, or switch back and forth in advocacy and adversary roles yourself. The aim is to surface doubts and weaknesses as well as build on strengths and confirm solid conclusions.

3. *Search for and analyze negative or disconfirming evidence and cases.* There are "exceptions that prove the rule" and exceptions that question the rule. In either case, look for and learn from exceptions to the patterns you've identified.

4. *Make constant comparison your constant companion.* All analysis is ultimately comparative. You compare the data that fit into a category, pattern, or theme with the data that don't fit. You compare alternative explanations, conclusions, and chains of evidence. Compare and contrast. Then compare and contrast some more.

5. *Keep analysis connected to purpose and design.* When deeply enmeshed in cataloguing, classifying, and comparing the trees in your qualitative data—that is, the depth and details of rich, thick qualitative data—change perspectives now and again to see the forest—that is, reconnect with the big picture. Purpose drives design. Purpose and design drive data collection. Purpose, design, and the data collected, in combination, drive analysis. Make sure that your analysis is serving the purpose of the inquiry. A well-chosen, thoughtful design will have anticipated how analysis would unfold. Keep those linkages in mind so that analysis doesn't become isolated from the inquiry's overall purpose and context.

    Keeping findings in context is a cardinal principle of qualitative analysis.

6. *Keep qualitative analysis qualitative.* Paraphrasing poet Dylan Thomas, do not go gently into that numerical night. Quantitize thoughtfully, carefully, and even reluctantly. Do so when it's appropriate and enhances understanding, all the while aware of the allure of numbers and the danger of losing the richness of qualitative data in the parsimony of numerical reduction.

7. *Integrate and triangulate diverse sources of qualitative data: interviews, observations, document analysis.* Any single source of data has strengths and weaknesses. Consistency of findings across types of data increases confidence in the confirmed patterns and themes. Inconsistency across types of data invites questions and reflection about why certain methods produced certain findings.

8. *Integrate and triangulate quantitative and qualitative data in mixed-methods studies.* The logic of triangulation (see Item 7) applies in mixed-methods designs when the strengths and weaknesses of qualitative and quantitative data are used together to illuminate the inquiry.

9. *Triangulate analysts.* Having more than one pair of eyes look at and think about the data, identify patterns and themes, and test conclusions and explanations reduces concerns about the potential biases and selective perception of a single analyst.

10. *Undertake theory triangulation.* Look at the findings and conclusions through the lens of alternative theoretical frameworks. How would a symbolic interactionist interpret the data compared with a phenomenologist or realist? How would a behavioral psychologist interpret the findings compared with a humanistic psychologist? What does a mechanistic display reveal compared with a systems graphic? The point is not to conduct an endless set of such theoretical comparisons but to select only those theoretical frameworks most germane to your inquiry to see what the alternative perspectives yield by way of insight and explanation.

> By combining multiple observers, theories, methods and data sources, [researchers] can hope to overcome the intrinsic bias that comes from single-methods, single-observer, and single-theory studies.
>
> —Norman K. Denzin (1989c, p. 307)

Chapter 5 on design discussed the benefits of using multiple data-collection techniques, a form of triangulation, to study the same setting, issue, or program. You may recall from that discussion that the term *triangulation* is taken from land surveying. Knowing a single landmark only locates you somewhere along a line in a direction from the landmark, whereas with two landmarks you can take bearings in two directions and locate yourself at their intersection. The notion of triangulating also works metaphorically to call to mind the world's strongest geometric shape—the triangle, which in its double alchemical form serves as the symbol for this chapter. The logic of triangulation is based on the premise that no single method ever adequately solves the problem of rival explanations. Because each method reveals different aspects of empirical reality and social perception, multiple methods of data collection and analysis provide more grist for the analytical mill. Combinations of interviewing, observation, and document analysis are expected in most fieldwork. Mixed qualitative–quantitative studies are increasingly valued as more credible than single-method studies. Studies that use only one method are more vulnerable to errors linked to that particular method (e.g., loaded interview questions, biased or untrue responses) than studies that use multiple methods, in which different types of data provide cross-data consistency checks.

Four Kinds of Analytical Triangulation

It is in data analysis that the strategy of triangulation really pays off, not only in providing diverse ways of looking at the same phenomenon, but in adding to credibility by strengthening confidence in whatever conclusions are drawn. Four kinds of triangulation can contribute to the verification and validation of qualitative analysis:

1. *Triangulation of qualitative sources:* Checking out the consistency of different data sources within the same method (consistency across interviewees)

2. *Mixed qualitative–quantitative methods triangulation:* Checking out the consistency of findings generated by different data collection methods

3. *Analyst triangulation*: Using multiple analysts to review findings

4. *Theory/perspective triangulation*: Using multiple perspectives or theories to interpret data

By triangulating with multiple data sources, methods analysts, and/or theories, qualitative analysts can make substantial strides in overcoming the skepticism that greets singular methods, lone analysts, and single-perspective interpretations.

Interpreting Triangulation Results: Making Sense of Conflicting and Inconsistent Patterns

A common misconception about triangulation involves thinking that the purpose is to demonstrate that different data sources or inquiry approaches yield essentially the same result. The point is to *test for* such consistency. Different kinds of data may yield somewhat different results because different types of inquiry are sensitive to different real-world nuances. Thus, *understanding inconsistencies in findings across different kinds of data can be illuminative and important.* Finding such inconsistencies ought not to be viewed as weakening the credibility of results but, rather, as offering opportunities for deeper insight into the relationship between inquiry approach and the phenomenon under study. I'll comment briefly on each of the four types of triangulation.

1. Triangulation of Qualitative Data Sources

> Four kinds of persons: zeal without knowledge; knowledge without zeal; neither knowledge nor zeal; both zeal and knowledge.
>
> —Pascal, *Pensées*

> Four kinds of qualitative triangulation: interviews with observations; interviews with documents; observations with documents; and interviews from multiple sources with observations of diverse events and documents of many kinds.
>
> —Halcolm, *Qualitative Pensées*

Triangulation of data sources within and across different qualitative methods means comparing and cross-checking the consistency of information derived at different times and by different means from interviews, observations, and documents. It can include

- comparing observations with interviews;
- comparing what people say in public with what they say in private;
- checking for the consistency of what people say about the same thing over time;
- comparing the perspectives of people from different points of view—for example, in an evaluation, triangulating staff views, participants' views, funder views, and views expressed by people outside the program; and
- checking interviews against program documents and other written evidence that can corroborate what interview respondents report.

Quite different kinds of data can be brought together in a case study to illuminate various aspects of a phenomenon. In a classic evaluation of an innovative educational project, historical program documents, in-depth interviews, and ethnographic participant observations were triangulated to illuminate the roles of powerful actors in supporting adoption of the innovation (Smith & Kleine, 1986). The evaluation of the Paris Declaration on development aid triangulated interviews with a variety of key informants, government reports, donor agency reports, and observations of donor–recipient decision-making meetings (Wood et al., 2011).

Maxwell (2012) is especially insightful about the interrelationship of interview and observation data in qualitative inquiry and analysis.

> One belief that inhibits triangulation is the widespread (though often implicit) assumption that observation is mainly useful for describing behavior and events, while interviewing is mainly useful for obtaining the perspectives of actors. It is true that the immediate result of observation is description, but this is equally true of interviewing: The latter gives you a description of what the informant said, not a direct understanding of their perspective. Generating an interpretation of someone's perspective is inherently a matter of inference from descriptions of their behavior (including verbal behavior), whether the data are derived from observations, interviews, or some other source such as written documents.
>
> While interviewing is often an efficient and valid way of understanding someone's perspective, observation can enable you to draw inferences about this perspective that you couldn't obtain by relying exclusively on interview data. . . . For example, watching how a teacher responds to boys' and girls' questions in a science class may provide a much better understanding of the teacher's actual views about gender and science than what the teacher says in an interview.
>
> Conversely, although observation often provides a direct and powerful way of learning about people's behavior and the context in which this occurs, interviewing can also be a valuable way of gaining a description of actions and events—often the only way, for events that took place in the past or to which you can't gain observational access. Interviews can provide additional information that was missed in observation, and can be used to check the accuracy of the observations. However, in order for interviews to be useful for this purpose, you need to ask about *specific* events and actions rather than posing questions that elicit only generalizations or abstract opinions. . . . In both of these situations, triangulation of observations and interviews can provide a more complete and accurate account than either could alone. (pp. 106–107)

Triangulation of data sources within qualitative methods may not lead to a single, totally consistent picture. The point is to study and understand when and why differences appear. The fact that observational data produce different results from interview data does not mean that either or both kinds of data are "invalid," although that may be the case. More likely, it means that different kinds of data have captured different things and so the analyst attempts to understand the reasons for the differences. Either consistency in overall patterns of data from different sources or reasonable explanations for differences in data from divergent sources can contribute significantly to the overall credibility of findings.

ETHNOGRAPHIC TRIANGULATION

In ethnographic research practice, triangulation of data sorts and methods and of theoretical perspectives leads to extended knowledge potentials, which are fed by the convergences, and even more by the divergences, they produce.

As in other areas of qualitative research, triangulation in ethnography is a way of promoting quality of research.... Good ethnographies are characterized by flexible and hybrid use of different ways of collecting data and by a prolonged engagement in the field. As in other areas of qualitative research, triangulation can help reveal different perspectives on one issue in research such as knowledge about and practices with a specific issue. Thus, triangulation is again a way to promote quality of qualitative research in ethnography also and more generally a productive approach to managing quality in qualitative research. (Flick, 2007b, p. 89)

2. Mixed-Methods Triangulation: Integrating Qualitative and Quantitative Data

Tis not the many oaths that makes the truth,

But the plain single vow that is vow'd true.

—William Shakespeare (written 1604–1605)
Diana in *All's Wells That Ends Well*

Mixed-methods triangulation often involves comparing and integrating data collected through some kind of qualitative methods with data collected through some kind of quantitative method. Such efforts flow from a pragmatic approach to mixed-methods analysis that assumes potential compatibility and seeks to discover the degree and nature of such compatibility (Tashakkori & Teddlie, 1998; Teddlie & Tashakkori, 2003, 2011). This is seldom straightforward because certain kinds of questions lend themselves to qualitative methods (e.g., developing hypotheses or theory in the early stages of an inquiry, understanding particular cases in depth and detail, getting at meanings in context, and capturing changes in a dynamic environment), while other kinds of analyses lend themselves to quantitative approaches (e.g., generalizing from a sample to a population, testing hypotheses for statistical significance, and making systematic comparisons on standardized criteria). Thus, it is common that quantitative methods and qualitative methods are used in a complementary fashion to answer different questions that do not easily come together to provide a single, well-integrated picture of the situation.

Given the varying strengths and weaknesses of qualitative versus quantitative approaches, the researcher using different methods to investigate the same phenomenon should not expect that the findings generated by those different methods will automatically come together to produce some nicely integrated whole. Indeed, the evidence is that one ought to expect initial conflicts in findings from qualitative and quantitative data and expect those findings to be received with varying degrees of credibility. It is important, then, to consider carefully what each kind of analysis yields and thereby giving different interpretations the chance to arise, with each considered on its merits, before favoring one result over the other based on methodological biases.

Critical Multiplism as an Analytical Strategy

Critical multiplism is a research strategy that advocates designing packages of imperfect methods and theories in a manner that minimizes the respective and inevitable biases of each. Multiplism, applied to analysis, acknowledges that any analysis can usually be conducted in any one of several ways, but in many cases, no single way is known to be uniformly the best. Under such circumstances, a multiplist advocates making heterogeneous those aspects of analysis about which uncertainty exists, so that the task is conducted in several different ways, each of which is subject to different biases.

Critical refers to rational, empirical, and social efforts to identify the assumptions and biases present in the options chosen. Putting the two concepts together, we can say that the central tenet of critical multiplism is this: When it is not clear which of several defensible options for a scientific task is least biased, we should select more than one, so that our options reflect different biases, avoid constant biases, and leave no plausible bias overlooked. (Shadish, 1993, p. 18)

When multiple analytical approaches yield similar results across different analytical biases, confidence in the resulting findings is increased. If different results occur when the analysis is done in different ways, then we have to try to explain the differences.

Different Findings From Different Methods

In a classic article, Shapiro (1973) described in detail her struggle to resolve basic differences between qualitative data and quantitative data in her study of Follow Through Classrooms; she eventually concluded that some of the conflicts between the two kinds of data were the result of measuring different things, although the ways in which different things were measured were not immediately apparent until she worked to sort out the conflicting findings. She began with greater trust in the data derived from quantitative methods and ended by believing that the most useful information came from the qualitative data.

Another pioneering article, by M. G. Trend (1978) of ABT Associates, has become required reading for anyone becoming involved in a team project that will involve collecting and analyzing both qualitative and quantitative data, where different members of the team have responsibilities for different kinds of data. The Trend study involved an analysis of three social experiments designed to test the concept of using direct-cash housing allowance payments to help low-income families obtain decent housing on the open market. The analysis of qualitative data from a participant observation study produced results that were at variance with those generated by analysis of quantitative data. The credibility of the qualitative data became a central issue in the analysis.

> The difficulty lay in conflicting explanations or accounts, each based largely upon a different kind of data. The problems we faced involved not only the nature of observational versus statistical inferences, but two sets of preferences and biases within the entire research team. . . .
>
> Though qualitative/quantitative tension is not the only problem which may arise in research, I suggest that it is a likely one. Few researchers are equally comfortable with both types of data, and the procedures for using the two together are not well developed. The tendency is to relegate one type of analysis or the other to a secondary role, according to the nature of the research and the predilections of the investigators. . . . Commonly, however, observational data are used for "generating hypotheses," or "describing process." Quantitative data are used to "analyze outcomes," or "verify hypotheses." I feel that this division of labor is rigid and limiting. (Trend, 1978, p. 352)

Early Efforts at Quantitative–Qualitative Triangulation

Anthropologists participating in teams in which both quantitative and qualitative data were being

STRATEGY FOR ACHIEVING QUALITY IN MIXED-METHODS STUDIES

The quantitative researchers work side by side every step of the way as full members of the case study team, bringing the analytic rigor of their quantitative frameworks to bear on case study and observation design, data collection, analysis, integration with other methods, and reporting. The qualitative researchers, in turn, are full members of the quantitative team (analysis of administrative data, survey research, and time series assessments), bringing their own rigor to survey designs, data reduction decisions, and interpretations. As a result, assumptions are more rigorously examined, methodological lacunae more clearly (and early) identified, and the team leaders become sufficiently methodologically multilingual so that they can discuss both qualitatively and quantitatively based findings with equal confidence (Datta, 2006, p. 427).

collected applied their inquiry skills to examine the nature of the experience in the 1970s. The problems they have shared were stark evidence that qualitative methods at that time were typically perceived as exploratory and secondary when used in conjunction with quantitative/experimental approaches. When qualitative data supported quantitative findings, that was the icing on the cake. When qualitative data conflicted with quantitative data, the qualitative data have often been dismissed or ignored (Society of Applied Anthropology, 1980).

A strategy of methods triangulation, then, doesn't magically put everyone on the same page. While valuing and endorsing triangulation, Trend (1978) suggested that we give different viewpoints the chance to arise, and postpone the immediate rejection of information or hypotheses that seem out of joint with the majority viewpoint. Observationally derived explanations are particularly vulnerable to dismissal without a fair trial. (pp. 352–353)

From Separation to Integration

Qualitative and quantitative data can be fruitfully combined to elucidate complementary aspects of the same phenomenon. For example, a community health indicator (e.g., teenage pregnancy rate) can provide a general and generalizable picture of an issue, while case studies of a few pregnant teenagers can put faces on the numbers and illuminate the stories behind the quantitative data; this becomes even more powerful when the indicator is broken into categories (e.g., those under the age of 15, those 16 and above), with case studies illustrating the implications of and rationale for such categorization.

A STORY OF MIXED-METHODS TRIANGULATION: TESTING CONCLUSIONS WITH MORE FIELDWORK

Economists Lawrence Katz and Jeffrey Liebman of Harvard, and Jeffrey R. Kling of Princeton, were trying to interpret data from a federal housing experiment that involved randomly assigning people to a program that would help them get out of the slums. The evaluation focused on the usual outcomes of improved school and job performance. However, to get beyond the purely statistical data, they decided to conduct interviews with residents in an inner-city poverty community.

Professor Lieberman commented to a *New York Times* reporter,

I thought they were going to say they wanted access to better jobs and schools, and what we came to understand was their consuming fear of random crime; the need the mothers felt to spend every minute of their day making sure their children were safe. (Uchitelle, 2001, p. 4)

By adding qualitative, field-based interview data to their study, Kling, Liebman, and Katz (2001) came to a new and different understanding of the program's impacts and participants' motivations based on interviewing the people directly affected, listening to their perspectives, and including those perspectives in their analysis.

In essence, triangulation of qualitative and quantitative data constitutes a form of comparative analysis. The question is "What does each analysis contribute to our understanding?" Areas of convergence increase confidence in findings. Areas of divergence open windows to better understanding of the multifaceted, complex nature of a phenomenon. Deciding whether results have converged remains a delicate exercise subject to both disciplined and creative interpretation. Focusing on *the degree of convergence* rather than forcing a dichotomous choice—that the different kinds of data do or do not converge—yields a more balanced overall result.

Mixed-Methods Analysis and Triangulation in the Twenty-First Century

While difficulties still arise in triangulating and integrating qualitative and quantitative data, advances in mixed methods have propelled integrated analyses into the spotlight, especially in applied and interdisciplinary areas like policy analysis, program evaluation, environmental studies, international development, and global health. Where disciplinary barriers have yielded to genuine interdisciplinary engagement, traditional methodological divisions have yielded to collaboration and integration. Exhibit 8.27 (pp. 618–619) presented mixed-methods challenges and solutions. Exhibit 9.2 presents 10 developments that are making mixed-methods triangulation both valued and, increasingly, expected in applied social science.

3. Triangulation With Multiple Analysts

A third kind of triangulation is investigator or analyst triangulation—that is, using multiple as opposed to singular observers or analysts. This is the core of qualitative team research (Guest & MacQueen, 2008). Triangulating observers or using several interviewers helps reduce the potential bias that comes from a single person doing all the data collection and provides means of more directly assessing the consistency of the data obtained. Triangulating observers provides a check on potential bias in data collection.

A related strategy is *triangulating analysts*—that is, having two or more persons independently analyze the same qualitative data and compare their findings. In the traditional social science approach to qualitative inquiry, engaging multiple analysts and computing the interrater reliability among these different analysts is valued, even expected, as a means of establishing credibility of findings (Silverman & Marvasti, 2008, pp. 238–239).

EXHIBIT 9.2   Ten Developments Enhancing Mixed-Methods Triangulation

1. *Designs that are truly mixed-methods inquiries are demonstrating the value of systematic, planned triangulation.* Increased understanding of the strengths and weaknesses of qualitative and quantitative data has led to both the commitment and capacity to build on the strengths of each *at the design stage.*

2. *Asking integrating questions of the data supports triangulation.* Triangulation is most powerful when mixed-methods studies are designed for integration, which begins by asking the same questions of both methods and gathering both qualitative and quantitative data on those questions. That is happening at a level unprecedented in applied social science research and evaluation.

3. *Mixed-methods sampling strategies anticipate and facilitate triangulation.* Sampling with triangulation in mind is a collaborative strategy that anticipates and lays the foundation for mixed-methods analysis.

4. *Specific methods are incorporating mixed data intentionally to support triangulation.* Surveys ask both closed- and open-ended questions. Case studies collect both quantitative and qualitative data. Strong experimental designs gather both standardized intervention and quantitative effects data plus qualitative process data.

5. *Mixed methods are proving especially appropriate for studying complex issues.* Mixed-methods researchers are extending our understandings of how to understand complex social phenomena as well as how to use research to develop effective interventions to address complex social problems (Mertens, 2013; Patton, 2011).

6. *Team approaches are being created and implemented with mixed-methods skills and capabilities in mind.* High-quality mixed-methods designs often require teams because individuals lack the full skill set needed. Knowing how to form and manage such teams has advanced significantly as experience has accumulated about what to do—and what not to do (Guest & MacQueen, 2008; Morgan, 2014).

7. *Software supports mixed-methods data analysis and triangulation.* As data analysis software has become more sophisticated, flexible, and responsive to analysts' needs, techniques and processes for triangulation are becoming more common and easier to use.

8. *Resources available for mixed-methods designs and analysis have burgeoned.* The *Journal of Mixed Methods* began publishing in 2007, with an opening editorial by Abbas Taskhakkori and John Creswell proclaiming, "The New Era of Mixed Methods." This means that there are more outlets for publishing mixed-methods studies. *The Handbook of Mixed Methods* was published in 2003 (Tashakkori & Teddlie). Excellent mixed-methods texts provide guidance on the full process from designing mixed-methods studies to analyzing and triangulating mixed data (Bamberger, 2013; Bergman, 2008; Greene, 2007; Mertens, 1998; Mertens & Hesse-Biber, 2013; Morgan, 2014).

9. *Researchers are developing mixed skills, capabilities, and capacities—and being recognized and valued for their mixed-methods expertise.* In 2014, the International Association of Mixed Methods Research was launched and hailed as "a momentous development in mixed-methods research" (Mertens, 2014).

10. *Mixed-methods exemplars show what is possible.* Early experiences with qualitative–quantitative triangulation were mixed at best—and many were quite negative, as indicated in the cautionary tales reported preceding the exhibit. When I was doing earlier editions of this book, there were more bad examples and negative experiences than good and positive exemplars. That balance has shifted for all the reasons listed here. The momentum is building as funders of research and evaluation are coming to demand mixed-methods studies.

Here, however, is a perfect example of how different criteria for judging quality lead to different practices. In a lead editorial for the journal *Qualitative Health Research*, Janet Morse (1997) took on "the myth of inter-rater reliability" from a social constructionist perspective. She begins by distinguishing standardized interview formats from more flexible and open interview guide approaches.

She acknowledges that interrater reliability may be acceptable when everyone is asked the same question in the same way (the preferred interviewing approach to meet traditional social science concerns about validity and reliability), but in the more adaptive, personalized, individualized, and flexible approach of interview guides and conversational interviewing, what constitutes coherent passages for coding is more problematic and depends on the analyst's interpretive framework. Multiple analysts might still discuss what they see in the data, share insights, and consider what emerges from their different perspectives, but that's quite different from computing a statistical interrater reliability coefficient. (See the sidebar on "the myth of interrater reliability" for her full argument.)

PERFECTLY HEALTHY BUT DEAD: THE MYTH OF INTERRATER RELIABILITY

—Janet M. Morse (1997)

Qualitative researchers seem to have inherited a host of habits from quantitative researchers and have adopted them into the qualitative paradigm without considering the appropriateness of their purpose, rationale, or underlying assumptions. On the surface, these practices seem right, so they are unquestioningly maintained. One of these adopted habits is the practice of obtaining interrater reliability of coding decisions used in qualitative research when coding unstructured, interactive interviews.

The argument goes something like this: To be reliable, coding should be replicable. Replication is checked by duplication; if coding decisions are explicit and communicated to another researcher, that researcher should be able to make the same coding decisions as the first researcher. The result is reliable research. Right?

Wrong. Interrater reliability is appropriate with semistructured interviews, wherein all participants are asked the same questions, in the same order, and data are coded all at once at the end of the data collection period. But this does not hold for unstructured interactive interviews. Recall that unstructured, interactive interviews are used in research because the researcher does not know enough about the topic or its parameters to construct interview questions. With unstructured, interactive interviews, the researcher first assumes a listening stance and learns about the topic as she or he goes along. Thus, once the researcher has learned something about the phenomenon from the first few participants, the substance of the interview then changes and becomes targeted on another aspect of the phenomenon. Importantly, unlike semistructured interviews, all participants are not asked the same questions. Participants are used to verify the information learned in the first interviews and are encouraged both to speak from their own experience and to speak for others. Each interview may overlap with the others but may also have a slightly different focus and different content.

This notion, learning from participants as the study progresses, is crucial to the understanding of the fluid nature of coding unstructured interviews. Initially, coding decisions may be quite superficial—by topic, for instance—but later coding decisions are made with the knowledge of, and in consideration of, information gained from all the previously analyzed interviews. Such coding schemes are not superficial, and

in light of all the knowledge gained, small pieces of data may have monumental significance. The process is not necessarily superficially objective: It is conducted in light of comprehensive understanding of the significance of each piece of text. The coding process is highly interpretative.

This comprehensive understanding of data bits cannot be acquired in a few objective definitions of each category. Moreover, it cannot be conveyed quickly and in a few definitions to a new member of the research team who has been elected for the purpose of determining a percentage agreement score. This new coder does not have the same knowledge base as the researcher, has not read all the interviews, and therefore does not have the same potential for insight or depth of knowledge required to code meaningfully. Maintaining a simplified coding schedule for the purposes of defining categories for an interrater reliability check will maintain the coding scheme at a superficial level. It will simplify the research to such an extent that all of the richness attained from insight will be lost. Ironically, it forcibly removes each piece of data from the context in which each coding decision should be made. The study will become respectably reliable with an interrater reliability score, but this will be achieved at the cost of losing all the richness and creativity inherent in analysis, ultimately producing a superficial product.

The cost of such an endeavor is equivalent to Mrs. Frisby, who, when the farmer commented that the poisoned rat looked perfectly healthy, said sadly, "Perfectly healthy, but dead!" Your research will be perfectly reliable, but trivial.

There is often a shocked silence when I discuss this with students. But then I ask two questions: "How many of you have written a literature review lately?" Almost every hand is raised. I then ask, "How many of you took a second person to the library with you to make sure you interpreted each article in a manner that was replicable?" Not a single hand remains raised. "Aren't you concerned?" I ask, "How do you know that your analysis, your interpretation of those articles, was reliable?"

The analysis of unstructured, interactive interviews is exactly the same case. Researchers must learn to trust themselves and their judgments and be prepared to defend their interpretations and analyses. But it is death to one's study to simplify one's insights, coding, and analyses so that another person may place the same piece of datum in the same category.

Triangulation Through Distinct Evaluation Teams: The Goal-Free Approach

In program evaluation, an interesting form of team triangulation has been used. Michael Scriven (1972b) has advocated and used two separate teams, one that conducts a traditional goals-based evaluation (assessing the stated outcomes of the program) and a second that undertakes a "goal-free evaluation" in which the evaluators assess clients' needs and program outcomes without focusing on stated goals (see Chapter 4, p. 206). Comparing the results of the goals-based team with those of the goal-free team provides a form of analytical triangulation for determining program effectiveness (Youker & Ingraham, 2014).

Review by Inquiry Participants

Having those who were studied review the findings offers another approach to analytical triangulation. Researchers and evaluators can learn a great deal about the accuracy, completeness, fairness, and perceived validity of their data analysis by having the people described in that analysis react to what is described and concluded. To the extent that participants in the study are unable to relate to and confirm the description and analysis in a qualitative report, questions are raised about the credibility of the findings. In what became a classic study of how evaluations were used, key informants in each case study were asked for both verbal and written reactions to the accuracy and comprehensiveness of the cases. The evaluation report then included those written reactions (Alkin et al., 1979). In her study of homeless youth, Murphy (2014) met with each of the 14 youth to go over the details of the case study she created from their transcribed interviews to affirm accuracy, add additional details and reflections if they so desired, and choose a pseudonym that they wanted to be called in the study, if they had not already done so. (See Thmaris's case study example, pp. 511–516.)

> Obtaining the reactions of respondents to your working drafts is time-consuming, but respondents may (1) verify that you have reflected their perspectives; (2) inform you of sections that, if published, could be problematic for either personal or political reasons; and (3) help you to develop new ideas and interpretations. (Glesne, 1999, p. 152)

Different Purposes Drive Different Review Procedures

Different kinds of studies have different participant review processes, some none at all. Collaborative and participatory inquiry builds in participants' review of

cases, quotations, and findings as a matter of course; that's part of what collaboration and participation mean. However, investigative inquiries (Douglas, 1976) aimed at exposing what goes on beyond the public eye are often antagonistic to those in power, so their responses would not typically be used to revise conclusions but might be used to at least offer them an opportunity to provide context and an alternative interpretation. Some traditional social science researchers and evaluators worry that sharing findings with participants for their reactions will undermine the independence of their analysis. Others view it as an important form of triangulation. In an Internet listserv discussion of this issue, one researcher reported this experience:

> I gave both transcripts and a late draft of findings to participants in my study. I wondered what they would object to. I had not promised to alter my conclusions based on their feedback, but I had assured them that my aim was to be sure not to do them harm. My findings included some significant criticisms of their efforts that I feared/expected they might object to. Instead, their review brought forth some new information about initiatives that had not previously been mentioned. And their primary objection was to my not giving the credit for their successes to a wider group in the community. What I learned was not to make assumptions about participants' thinking.

Exhibit 9.3 summarizes three contrasting views of involving those studied in reviewing findings and conclusions.

Critical Friend Review

> A critical friend can be defined as a trusted person who asks provocative questions, provides data to be examined through another lens, and offers critiques of a person's work as a friend. A critical friend takes the time to fully understand the context of the work presented and the outcomes that the person or group is working toward. The friend is an advocate for the success of that work. (Costa & Kallick, 1993, p. 49)

Tessie Tzavaras Catsambasis is president of EnCompass LLC, an international evaluation research company. She is active in evaluation capacity building around the world, including leadership service with the *International Organization for Cooperation in Evaluation*. She also plays the role of critical friend with colleagues' projects and within her own organization. Here's an example she shared with me (and kindly gave permission to include here) that

EXHIBIT 9.3   **Different Perspectives on Triangulation by Those Who Were Studied**

PERSPECTIVES ON CHECKING IN WITH PARTICIPANTS TO REVIEW THEIR CASES AND OVERALL FINDINGS	RATIONALE
1. Against participants' reviews	a. May raise questions about the independence of findings: risks too much influence by participants on interpretation by the researcher
	b. Not sure what to do if participants and researchers disagree; whose opinion prevails?
2. Weigh pros and cons situationally: It depends	a. Takes time and resources
	b. Must be carefully planned and done well or could create problems in meeting deadlines; may not be worth the hassle
	c. Could be hard to get back to everyone, so some unfairness arises in who gets to review what
	d. Depends on how important and credible it is to the audiences who will receive findings
3. In favor of participant reviews	a. It's the ethical thing to do
	b. It's a chance to correct errors and inaccuracies so you end up with better data
	c. It's a chance to update the data

nicely illustrates the critical friend role as a form of analyst triangulation.

My team and I conducted an evaluation of a UN organization's Internet-based system that countries could download to track their own HIV/AIDS activities in any sector and area, nationally down to district level. A previous organizational review of the UN organization recommended discontinuing this program based on resource constraints and rumors about problems, but without looking at it closely. The department supporting this program decided to evaluate it first, because they had invested in it significantly and wanted to make a final decision based on evidence. The evaluation we conducted (country visits, focus groups, interviews, survey, benchmarking) revealed many, many problems. But interestingly, some 20 countries were using it (the tracking system). My colleagues who did the data collection were ready to push the button to kill it, citing all the problems we had found. I got involved at the last stage of the data analysis process.

I grilled my colleagues, asking them to justify every conclusion. Their perspective was clear: "This program has so many operational obstacles in the field, no Internet, low

capacity, we should recommend discontinuing it." Then, I asked what turned out to be the "turning point" questions: "If this program has so many problems, why are 20 countries choosing to use it?" and "How are those countries addressing the problems you have documented?"

This kind of question (at its best, how is it working?) is an appreciative analytical question asked as a critical friend from a systems dynamics perspective (change the shoes you are wearing, and from the perspective of a country, what do you see?). In response, my colleagues listed many innovations that countries were undertaking to make this tracking program work, and then they concluded, "It is the only option out there that they can control fully, and it is cheap." So "country controlled" was also important, and so was "low cost." Then, I asked them, "Imagine you hold the button to kill the program, do you push it?" They each said, "No, but we would . . ." and proceeded to give me three fabulous recommendations. Their responses enabled us to present to the client the findings, and engage the client in grappling with a tough decision.

Essentially, we said, "This program is filling a demand, and 20 counties are using it in spite of significant operational problems. We know you have resource constraints, and this system requires more technical assistance, but if you decide to stop supporting it, consider transferring the system's administration to another funding agency, and also consider certifying independent consultants as technical assistance providers, so countries can contract with them directly for help on the system. And, if you cannot even do that, think about how you will transition in a way that will not hurt countries."

From an evaluation point of view, two things are important: (1) if it were not for these two questions in the analysis, the team would have concluded something very different from the same data, and (2) asking these questions enabled us to facilitate the client to face these challenging findings, have an internal debate about what to do, and own the final decision.

Audience Review as Credibility Triangulation

Reflexive triangulation (Exhibit 2.5, p. 72) includes the audience's reactions to the triangulation mix: (1) the inquirer's reflexive perspective, (2) the perspectives of those studied, and (3) the perspectives of those who received the findings. The opening module of this chapter emphasized that different readers of qualitative reports will apply different criteria to judge quality and credibility. Audience reactions constitute additional data. Whenever possible, I prefer to present draft findings to multiple audiences to learn how they react, what they focus on, what is clear and unclear, and what questions are inadequately answered. In a sense, this is equivalent to theater or movie previews when producers and directors get to gauge audience reaction to a performance or film before it is released. Time and procedures for audience previews and reactions have to be planned in advance, but whenever I've done them, I've been glad I did.

In a study of a community development effort in an inner-city, low-income neighborhood, focus groups were done with diverse groups: African Americans, Native Americans, Hispanics, Hmong residents, and low-income whites. Age-based focus groups were also done: youth under the ages of 25, 25- to 55-year-olds, and those over 55 years. A community advisory group reviewed the study design and voiced no objections to focus groups done homogeneously by either ethnicity or age. In fact, they thought such focus groups were a good idea. But when the draft results were reported in a public meeting that included community people and public officials, the focus group results made it appear that there were great divisions and differences of perspectives among neighborhood ethnic and

age-groups. Audience members outside the community were especially focused in on conflicts and differences reported in the findings. Similarities and important areas of agreement got lost amid the reports' overemphasis on differences. Moreover, perspectives within ethnic group and age-group appeared much more monolithic and homogeneous than, in fact, they were. As a result of this feedback, we went back to the field and added heterogeneous focus groups to the data and then drafted a more balanced report. This was in no way undermining inquirer independence. It was making sure we got it right.

Evaluation Audiences and Intended Users

Program evaluation constitutes a particular challenge in establishing credibility because the ultimate test of the credibility of an evaluation report is the response of primary intended users and readers of that report. Their reactions often revolve around *face validity*. On the face of it, is the report believable? Are the data reasonable? Do the results connect to how people understand the world? In seriously soliciting intended users' reactions, the evaluator's perspective is joined to the perspective of the people who must use the findings. Evaluation theorist Ernie House (1977) has suggested that the more "naturalistic" (qualitative) the evaluation, the more it relies on its audiences to reach their own conclusions, draw their own generalizations, and make their own interpretations:

> Unless an evaluation provides an explanation for a particular audience, and enhances the understanding of that audience by the content and form of the argument it presents, it is not an adequate evaluation for that audience, even though the facts on which it is based are verifiable by other procedures. One indicator of the explanatory power of evaluation data is the degree to which the audience is persuaded. Hence, an evaluation may be "true" in the conventional sense but not persuasive to a particular audience for whom it does not serve as an explanation. *In the fullest sense, then, an evaluation is dependent both on the person who makes the evaluative statement and on the person who receives it* [italics added]. (p. 42)

Understanding the interaction and mutuality between the evaluator and the people who use the evaluation, as well as relationships with participants in the program, is critical to understanding the human side of evaluation. This is part of what gives evaluation—and the evaluator—situational and interpersonal "authenticity" (Lincoln & Guba, 1986). Exhibit 9.16, at the end of this chapter (pp. 736–741), provides an experiential account from an evaluator dealing with issues of credibility while building relationships with program participants and evaluation users; her reflections provide a personal, in-depth description of what *authenticity* is like from the perspective of one participant-observer.

Expert Audit Review

A final review alternative involves using experts to assess the quality of analysis or, where the stakes for external credibility are especially high, performing a meta-evaluation or process audit. An external audit by a disinterested expert can render judgment about the quality of data collection and analysis. "That part of the audit that examines the process results in a *dependability judgment* [italics added], while that part concerned with the product (data and reconstructions) results in a *confirmability judgment* [italics added]" (Lincoln & Guba, 1986, p. 77). Such an audit would need to be conducted according to appropriate criteria. For example, it would not be fair to audit an aesthetic and evocative qualitative presentation by traditional social science standards or vice versa. But within a particular framework, expert reviews can increase credibility for those who are unsure how to distinguish high-quality work. That, of course, is the role of the doctoral committee for graduate students and peer reviewers for scholarly journals. Problems arise when peer reviewers apply traditional scientific criteria to constructivist studies, and vice versa. In such cases, the review or audit itself lacks credibility. Exhibit 9.4 on the next page presents an example of an expert meta-evaluation (evaluation of the evaluation) to independently judge the quality and establish credibility for a high-stakes international mixed-methods evaluation.

The challenge of getting the right expert, one who can apply an appropriately critical eye, is wittily illustrated by a story about the great French artist Pablo Picasso. Marketing of fakes of his paintings plagued Picasso. His friends became involved in helping check out the authenticity of supposed genuine originals. One friend in particular became obsessed with tracking down frauds and brought several paintings to Picasso, all of which the master identified as fake. A poor artist who had hoped to profit from having obtained a Picasso before the great artist's works had become so valuable sent his painting for inspection via the friend. Again Picasso pronounced it a forgery.

"But I saw you paint this one with my very own eyes," protested the friend.

"I can paint false Picassos as well as anyone," retorted Picasso.

©2002 Michael Quinn Patton and Michael Cochran

4. Theory Triangulation

Greek legend tells of the fearsome hotelier Procrustes who would adjust his guests to match the length of his bed, stretching the short and trimming off the legs of the tall. Guides to program theory that are too prescriptive risk creating such a Procrustean bed. When the same approach to program theory is used for all types of interventions and all types of purposes, the risk is that the interventions will be distorted to fit into a preconceived format. Important aspects may be chopped off and ignored, and other aspects may be stretched to fit into preconceived boxes of a factory model, with inputs, processes, outcomes, and impacts.

Purposeful program theory requires thoughtful assessment of circumstances, asking in particular, "Who is going to use the program theory and for what purposes?" and "What is the nature of the intervention and the situation in which it is implemented?" It requires a wide repertoire, not a one-size-fits-all approach to program theory.

Purposeful program theory also requires attention to the limitations of any one program theory, which must necessarily be a simplification of reality and a willingness to revise it as needed to address emerging issues.

—Funnell and Rogers (2011, p. xxi)
Purposeful Program Theory

Having discussed triangulation of qualitative data sources, mixed-methods triangulation, and multiple analyst triangulation, we turn now to the fourth and final kind of triangulation: using different theoretical perspectives to look at the same data.

EXHIBIT 9.4   Metaevaluation: Evaluating the Evaluation of the Paris Declaration on Development Aid

It has become a standard in major high-stakes evaluations to commission an independent review to determine whether the evaluation meets generally accepted standards of quality and, in so doing, to identify strengths, weaknesses, and lessons (Stufflebeam & Shrinkfield, 2007, p. 649). The major addition to the Joint Committee Standards for Evaluation, when revised in 2010, was that of "Evaluation Accountability Standards" focused on meta-evaluation.

Evaluation Accountability Standards

E1 Evaluation documentation: Evaluations should fully document their negotiated purposes and implemented designs, procedures, data, and outcomes.

E2 Internal meta-evaluation: Evaluators should use these and other applicable standards to examine the accountability of the evaluation design, procedures employed, information collected, and outcomes.

E3 External meta-evaluation: Program evaluation sponsors, clients, evaluators, and other stakeholders should encourage the conduct of external meta-evaluations using these and other applicable standards (Joint Committee on Standards, 2010; Yarbrough, Shulha, Hopson, & Caruthers, 2010).

Evaluating the Evaluation of the Paris Declaration

Given the historic importance of the Evaluation of the Paris Declaration on Development Aid (Dabelstein & Patton, 2013b), the Management Group overseeing the evaluation commissioned an independent assessment of the evaluation. Prior to undertaking this review, we had no prior relationship with any members of the Management Group or the Core Evaluation Team. We had complete and unfettered access to any and all evaluation documents and data, and to all members of the International Reference Group, the Management group, the Secretariat, and the Core Evaluation Team. Our evaluation of the evaluation included reviewing data collection instruments, templates, and processes; reviewing the partner country and donor evaluation reports on which the synthesis of findings was based; directly observing two meetings of the International Reference Group where the evidence was examined and the conclusions refined

and sharpened accordingly; engaging International Reference Group participants in a reflective practice, lessons-learned session; surveying participants about the evaluation process and partner country evaluations; and interviewing key people involved in and knowledgeable about how the evaluation was conducted. The evaluation of the evaluation included assessing both the evaluation report's findings and the technical appendix that details how the findings were generated. The Development Assistance Committee (DAC) of the Organization for Economic Co-operation and Development (OECD) established international standards for evaluation in 2010, and those were the standards used for the meta-evaluation (OECD-DAC, 2010). A meta-evaluation audit statement confirming the quality, credibility, and usability of the evaluation was included as a preface to the full evaluation reports. The meta-evaluation report (Patton & Gornick, 2011a) was published and made available online two weeks after the Final Evaluation report was published. This timing was possible because the meta-evaluation began halfway through the Paris Declaration Evaluation and the meta-evaluation team had access to draft versions of the final report at each stage of the report's development. The process for conducting the meta-evaluation and its uses are discussed in detail in Patton (2013).

The Paris Declaration Evaluation received the 2012 American Evaluation Association (AEA) Outstanding Evaluation Award. At the award ceremony, the chair of the AEA Awards Committee, Frances Lawrenz (2013), summarized the merits of the evaluation that led to the award selection and recognition:

> The success of the Paris Declaration Phase 2 Evaluation required an unusually skilled, knowledgeable and committed evaluation team; a visionary, well-organized, and well-connected Secretariat to manage the logistics, international stakeholder meetings, and financial accounts; and a highly competent and respected Management Group to provide oversight and ensure the Evaluation's independence and integrity. This was an extraordinary partnership where all involved understood their roles, carried out their responsibilities fully and effectively, and respected the contributions of other members of the collaboration.

Chapter 3 presented a number of general theoretical frameworks derived from diverse intellectual and disciplinary traditions. More concretely, multiple theoretical perspectives can be brought to bear on specialized substantive issues. For example, one might examine interviews with therapy clients from different psychological perspectives: psychotherapy, Gestalt, Adlerian, and behavioral psychology. Observations of a group, community, or organization can be examined from a Marxian or Weberian perspective, a conflict or functionalist point of view. The point of theory triangulation is to understand how differing assumptions and premises affect findings and interpretations.

Examples of Theory Triangulation

Let's suppose we are studying famine in a drought-afflicted region of an African country. We have quantitative data on food production (sorghum and millet), nutrition data from household surveys, health data from clinics, rainfall data over many years, interviews with villagers (males and females), key informant interviews (e.g., government officials, agricultural experts, aid agency staff members, and village leaders), and case studies of purposefully sampled villages telling the story of their agricultural and nutritional situations and experiences before and during the famine. Put all of these data together and we have an in-depth description of the extent and nature of the famine, its effects on subsistence agriculture families, food and agricultural assistance provided, and the interventions of government and international agencies. We have (a) mixed-methods triangulation and (b) multiple sources of qualitative data (interviews, observations, case studies, documents), and (c) our team members have analyzed the patterns independently to confirm the findings as well as had the findings externally reviewed by experts. Thus, we can make a credible case for the nature, extent, and impacts of the famine. What does theory triangulation add?

When we move from description to interpretation, we need a framework to make sense of and explain the patterns in the data. Why is the region experiencing famine? Why aren't interventions more effective? Different theoretical frameworks emphasize different explanatory variables.

- Climate change theory would emphasize long-term weather and climate trends.
- Malthusian theory would emphasize overpopulation.
- Marxian theory would emphasize power dynamics

"I envy your confidence. Even after decades of evaluations, these metaevaluations still make me feel naked."

(Who controls the means of production? How do the powerful benefit from famine?).

- Weberian theory would emphasize organizational competence and incompetence (How does the functioning and activities of government and international agencies exacerbate or alleviate famine?).
- Ecological systems theory would call for examining the interactions between the ecosystem, farming practices, soil and water conditions, and markets.
- Cultural systems theory would emphasize the way in which cultural beliefs and norms affect the experience of and responses to famine by the people affected.
- Feminist theory would point to the role of women in the system as a factor in how the famine affects families and their responses to the crisis (Podems, 2014b).
- Cognitive theory would focus on how people make decisions in the face of changing conditions.

When designing the famine study, these various theoretical perspectives would inform the kinds of questions to be asked and data to be collected. When analyzing the findings and explaining results, these diverse theoretical perspectives provide competing interpretations for explaining the patterns and observed impacts. *Theory triangulation* involves examining the data through different theoretical lenses to see what theoretical framework (or combination) aligns most convincingly with the data (best fit).

Theory triangulation for evaluation can involve examining the data from the perspectives of various stakeholder positions. It is common for diverse stakeholders to disagree about program purposes, goals, and means of attaining goals. These differences represent different "theories of action" (Patton, 2012a) that can cast the same findings in different perspective-based lights. When we were seeking explanations to explain dropout rates for adult literacy programs in Minnesota, the predominant staff theory was that low-income people led chaotic lives and couldn't manage regular attendance and follow-through in a program. Political explanations included laziness, effects of multigenerational poverty, lack of good jobs to motivate participants to complete programs, and cultural deprivation theories. But the explanation that best fit the data (interviews with dropouts) was that the adult literacy programs were lousy learning experiences: large class sizes; disinterested and disrespectful teachers, poorly paid and exhausted from having already taught all day in their regular jobs; uninteresting and outdated curriculum materials; and an all-around depressing environment. Most

traditional explanations blamed the participants or the larger societal problems that affected the participants, but the actual data pointed to ineffective programs, something that was actionable. Changes were made, and dropout rates went down significantly.

Thoughtful, Systematic Triangulation

All four of these different types of triangulation—(1) mixed-methods triangulation, (2) triangulation of qualitative data sources, (3) analyst triangulation, and (4) theory or perspective triangulation—offer strategies for reducing systematic bias and distortion during data analysis, and thereby increasing credibility. In each case, the strategy involves checking findings against other sources and perspectives. Triangulation, in whatever form, increases credibility and quality by countering the concern (or accusation) that a study's findings are simply an artifact of a single method, a single source, or a single investigator's blinders. Exhibit 9.1 (p. 660) reviews and summarizes the four types of triangulation (items 7–10 in Exhibit 9.1).

Exhibit 9.5 presents a model for rigorous analysis that broadens and deepens triangulation processes in high-stakes, high-visibility situations. Eight attributes of a rigorous analysis process were identified by studying experienced intelligence analysts from multiple U.S. federal investigative agencies. The researchers used a cognitive systems approach in which professional intelligence analysts were engaged in going beyond assessment of the quality of an analysis based on product quality to examine the analytic processes necessary to generate a high-quality, credible, and useful product. The understanding of rigor that emerged was that it is not about following a standardized, highly prescribed analytical process (a formula or recipe) but, rather, "assessing the contextual sufficiency of many different aspects of the analytic process" (Zelik et al., 2007). The researchers posited that these dimensions could be relevant to any process where analysts must make sense of complex data, and the rigor and resulting credibility of their analytical process will affect the utility of the findings for decision making. Examine Exhibit 9.5 carefully and thoughtfully. There's a lot there pulled together in a comprehensive, coherent, and integrated triangulation model: The Rigor Attribute Model. What comes across most powerfully from the work that generated the model is that a product (report, findings, or presentation of results) cannot be assessed for quality and credibility without knowing the nature and rigor of the analytical process that generated the findings. That insight is consistent with the focus of my MQP Rumination, avoiding research rigor mortis, in this chapter (see pp. 701–703).

SIDEBAR

THEORY INTEGRATION MEETS THEORY TRIANGULATION

Different criteria for evaluating the quality of qualitative remain fluid as qualitative inquirers move back and forth among genres, ignoring the boundaries, much as birds ignore human fences—except to use them occasionally as convenient places to rest. Consider the reflections on working across and integrating multiple genres and theoretical orientations of self-described "critical educators" Patricia Burdell and Beth Blue Swadener (1999). They combine autobiographical narratives with a variety of theoretical perspectives, including critical, dialogic, phenomenological, feminist, and semiotic perspectives. They speculate that "it is perhaps both the intent and effect of many of these texts to broaden the 'acceptable' or give voice to the intellectual contradictions and tensions in everyday lives of scholar-teachers and researchers" (p. 23).

> Our research has used narrative inquiry, collaborative ethnography, and applied semiotics. Between us, we share an identity and scholarship in critical and feminist curriculum theory. We are frequent border-crossers. We seek texts that allow us to enter the world of others in ways that have us more present in their experience, while better understanding our own. (p. 23).

They call this border-crossing genre "critical personal narrative and autoethnography." The real world in which inquiry occurs is not a very neat and orderly place. Nor is it likely to become so. Theoretical and methodological border crossers are natural and determined *triangulators*.

EXHIBIT 9.5   Dimensions of Rigorous Analysis and Critical Thinking

The Rigor Attribute Model

Eight attributes of a rigorous analysis process were identified by studying experienced intelligence analysts from multiple U.S. federal investigative agencies. The researchers used a cognitive systems approach in which professional intelligence analysts were engaged in going beyond assessment of the quality of an analysis based on product quality to examine the analytic process that generated the product. The understanding of rigor that emerged was that it is not about following a standardized process but, rather, "assessing the contextual sufficiency of many different aspects of the analytic process" (Zelik et al., 2007). The researchers posited that these dimensions could be relevant to any process where analysts must make sense of complex data, and the rigor and resulting credibility of their analytical process will affect the utility of the findings for decision making.

Overview of the Eight Dimensions of Rigorous Analysis

RIGOR ATTRIBUTE	HIGH-RIGOR PROCESS	LOW-RIGOR PROCESS
1. *Hypothesis exploration.* Extent to which multiple hypotheses were seriously examined against the data	Test multiple hypotheses to identify the best, most probable explanations	Minimal weighing of alternatives
2. *Information search.* Depth and breadth of the search process used in collecting data	Comprehensively explore as much data as relevant to the inquiry (diligent, purposeful sampling)	Data collection limited to routine and readily available data sources (convenience sampling)
3. *Information validation.* Information sources are corroborated and cross-validated (triangulation)	Systematic verification and triangulation of information and sampling information-rich, trustworthy, and knowledgeable sources	Little effort made to triangulate (use converging evidence to verify source accuracy)
4. *Stance analysis.* Evaluation of data to identify and contextualize the perspective of the source	Investigate key informants' backgrounds to assess how their perspective might bias information they provide	Nothing is done when clear bias in a source is detected
5. *Sensitivity analysis.* The extent to which analysts consider, understand, and make explicit the assumptions, strengths, weaknesses, limitations, and gaps of their analysis	Systematic and strategic assessment of implications for interpretations, conclusions, and explanations if elements of the supporting sources and evidence prove invalid, inadequate, or otherwise problematic	Explanations accepted and reported if they seem appropriate and valid on a surface level. Emphasis on face validity
6. *Specialist collaboration.* The degree to which an analyst incorporates the perspectives of experts and key knowledgeables into their assessments	The analyst has talked to, or may be, a leading expert in the key content areas of the analysis. Seeks out independent, external expert peer review for high-stakes analysis	Little or no effort is made to seek out and incorporate independent, external expertise; no peer review before reporting
7. *Information synthesis.* Refers to how far beyond simply collecting, listing, and analyzing distinct data elements, sources, and cases an analyst went in the interpretive process	Extracted and integrated information with a thorough consideration of diverse interpretations of relevant data, noting both areas of consistency in findings and areas where different methods and data yield conflicting findings	Analyst simply complies and reports the relevant information in a sequential and compartmentalized form; little or no integration and synthesis

(Continued)

(Continued)

RIGOR ATTRIBUTE	HIGH-RIGOR PROCESS	LOW-RIGOR PROCESS
8. *Explanation critique.* A form of collaboration that engages different perspectives in examining the preponderance of evidence supporting primary conclusions	Peers and experts are involved in independently examining the interpretive chain of reasoning and inferences made, explicitly distinguishing which are stronger and which weaker	Little or no use of other analysts to give input on explanation quality

SOURCE: Adapted and revised from Zelik, Patterson, and Woods (2007).

SIDEBAR

INTERPRETING TRIANGULATION RESULTS: MAKING SENSE OF CONFLICTING AND INCONSISTENT CONCLUSIONS

A common misconception about triangulation involves think-ing that the purpose is to demonstrate that different data sources or inquiry approaches yield essentially the same result. The point is to test for such consistency. Different kinds of data may yield somewhat different results because different types of inquiry are sensitive to different real-world nuances. Different theoretical frameworks will likely foster different interpreta-tions of the same findings. Different analysts may well inter-pret the same patterns in different ways. Thus, *understanding inconsistencies in findings across different kinds of triangulation can be illuminative and important.* Finding such inconsistencies ought not to be viewed as weakening the credibility of results but, rather, as offering opportunities for deeper insight into the relationship between inquiry approach and the phenomenon under study.

Alternative and Competing Criteria for Judging the Quality of Qualitative Inquiries, Part 1

Universal Criteria, and Traditional Scientific Research Versus Constructivist Criteria

> Every way of seeing is also a way of not seeing.
>
> —David Silverman (2000, p. 825)

Judging Quality: The Necessity of Determining Criteria

It all depends on criteria. Judging quality requires criteria. Credibility flows from those judgments. Quality and credibility are connected in that judgments of quality constitute the foundation for perceptions of credibility.

Diverse approaches to qualitative inquiry—phenomenology, ethnomethodology, ethnography, hermeneutics, symbolic interaction, heuristics, critical theory, realism, grounded theory, and feminist inquiry, to name but a few—remind us that issues of quality and credibility intersect with audience and intended inquiry purposes. Research directed to an audience of independent feminist scholars, for example, may be judged by somewhat different criteria from research addressed to an audience of government economic policymakers. Formative research or action inquiry for program improvement involves different purposes and therefore different criteria of quality compared with summative evaluation aimed at making fundamental continuation decisions about a program or policy. Thus, it is important to acknowledge at the outset that particular philosophical underpinnings or theoretical orientations and special purposes for qualitative inquiry will generate different criteria for judging quality and credibility.

Despite this, efforts to generate universal criteria and checklists for quality abound. The results are as follows: multiple possibilities, no consensus, and ongoing debate.

A Review of Quality Assurance Recommendations for Qualitative Research

An interdisciplinary team of health researchers engaged in worldwide malaria prevention and treatment set out to identify quality criteria for qualitative research and evaluation (Reynolds et al., 2011). They found 93 papers published between 1994 and 2010 that offered and discussed quality criteria, 37 of which were sufficiently detailed to merit further analysis. (The 56 papers that were rejected focused only on review criteria for publication or guidance on a specific qualitative method or single stage of the research process, such as data analysis.) They found no consensus about how to ensure the quality of qualitative research. However, they were able to categorize approaches into two "narratives" about quality: (1) an output-oriented approach versus (2) a process-oriented approach:

1. The most dominant narrative detected was that of an *output-oriented approach*. Within this narrative, quality is conceptualized in relation to theoretical constructs such as validity or rigor, derived from the positivist paradigm, and is demonstrated by the inclusion of certain recommended methodological techniques: the use of triangulation, member (or participant) validation of findings, peer review of findings, deviant or negative case analysis, and multiple coders of data.

Strengths of the output-oriented approach for assuring quality of qualitative studies include the acceptability and credibility of this approach within the dominant positivist environment where decision making is based on "objective" criteria of quality. Checklists equip those unfamiliar with qualitative research with the means to assess its quality.

The weakness of this approach is that "following of check-lists does not equate with understanding of and commitment to the theoretical underpinnings of qualitative paradigms or what constitutes quality within the approach. The privileging of guidelines as a mechanism to demonstrate quality can mislead inexperienced qualitative researchers as to what constitutes good qualitative research. This runs the risk of reducing qualitative research to a limited set of methods, requiring little theoretical expertise and diverting attention away from the analytic content of research unique to the qualitative approach. Ultimately, one can argue that a solely output-oriented approach risks the values of qualitative research becoming skewed towards the demands of the positivist paradigm without retaining quality in the substance of the research process."

2. By contrast, the second, *process-oriented* narrative, presented conceptualizations of quality that were linked to principles or values considered inherent to the qualitative approach, to be understood and enacted throughout the research process. Six common principles were identified across the narrative: (1) reflexivity of the researcher's position, assumptions, and practice; (2) transparency of decisions made and assumptions held; (3) comprehensiveness of approach to the research question; (4) responsibility toward decision making acknowledged by the researcher; (5) upholding good ethical practice throughout the research; and (6) a systematic approach to designing, conducting, and analyzing a study.

Strengths of the process-oriented approach include the ability of the researcher to address the quality of their research in relation to the core principles or values of qualitative research. The core principles identified in this narrative also represent continuous, researcher-led activities rather than externally determined indicators such as validity, or endpoints. Reflexivity, for example, is an active, iterative process—*an attitude of attending systematically to the context of knowledge construction . . . at every step of the research process.* As such, this approach emphasises the need to consider quality throughout the whole course of research, and locates the responsibility for enacting good qualitative research practice firmly in the lap of the researcher(s).

Need for a Flexible Quality Framework

The review team (Reynolds et al., 2011) found that "there is an increasing demand for the qualitative research field to move forward in developing and establishing coherent mechanisms for quality assurance of qualitative research." They concluded with a recommendation for "the development of a flexible framework to help qualitative researchers to define, apply and demonstrate principles of quality in their research." They further recommended that "the strengths of both the output-oriented and process-oriented narratives be brought together to create guidance that reflects core principles of qualitative research but also responds to expectations of the global health field for explicitly assured quality in research."

We recommend the development of a framework that helps researchers identify their core principles, appropriate for their epistemological and methodological approach, and ways to demonstrate that these have been upheld throughout the research process. . . . We propose that this framework be flexible enough to accommodate different qualitative methodologies without dictating essential activities for promoting quality. (Reynolds et al., 2011)

This chapter addresses this recommendation, offering both a generic quality framework as well as specialized quality criteria for specific types of qualitative inquiry.

THE PURPOSE OF AND DEBATE ABOUT CRITERIA

Criteria are standards, benchmarks, norms, and, in some cases, regulative ideals that guide judgments about the goodness, quality, validity, truthfulness, and so forth of competing claims (or methodologies, theories, interpretations, etc.). . . .

Criteria that have been proposed for judging the processes and products of social inquiry include truth, relevance, validity, credibility, plausibility, generalizability, social action, and social transformation, among others. Some of these criteria are epistemic (i.e., concerned with justifying knowledge claims as true, accurate, correct), others are political (i.e., concerned with warranting the power, use, and effects of knowledge claims or the inquiry process more generally); still others are moral or ethical standards (i.e., concerned with the right conduct of the inquirer and the inquiry process in general). . . .

Poststructuralist and postmodernist approaches to qualitative inquiry are also shaping the way we conceive of criteria. Given the growing influence of narrative approaches and experimental texts in qualitative inquiry, it is becoming more common to find discussions of rhetorical and aesthetic criteria replacing discussions of epistemic criteria. Other scholars argue that epistemological criteria cannot be neatly decoupled from political and critical agendas and ethical concerns. Some scholars in qualitative inquiry have little patience for discussing criteria within different epistemological frameworks and theoretical perspectives and prefer to focus on the craft of using various methodological procedures for producing "quality" work.

—Schwandt (2007, pp. 49–50)
The Sage Dictionary of Qualitative Inquiry

Judging the Quality of Alternative Approaches to Qualitative Inquiry

There can be no universal, generic, standardized, and all-encompassing criteria for judging the quality of qualitative studies because qualitative inquiry is not monolithic, uniform, or standardized. It's as if someone set out to create a universal checklist for beauty that ignored culture, human variability, variety, and differences in taste, socialization, and values (oh yes, the "Miss Universe" and "Miss World" contests notwithstanding). The common core elements across all kinds of qualitative inquiry are attention to language, words, narrative, description, stories, cases, worldviews, and how people make sense of their worlds. Tracy (2010), for example, identified "eight 'big-tent' criteria for excellent qualitative research": (1) worthy topic, (2) rich rigor, (3) sincerity, (4) credibility, (5) resonance, (6) significant contribution, (7) ethics, and (8) meaningful coherence. But approaches to inquiring into and judging attainment of these general criteria are diverse and multifaceted, serve competing purposes, and are, ultimately, a matter of debate (Gordon & Patterson, 2013).

It is possible to specify quality criteria for research generally. These are not unique to qualitative inquiry but apply to scientific inquiries of all kinds. Exhibit 9.6 presents these general, science-based quality criteria.

EXHIBIT 9.6 General Scientific Research Quality Criteria

QUALITY CRITERIA	ELABORATION/EXAMPLES
1. Clarity of purpose	Basic research, applied research, and evaluation research, for example, serve different purposes and are judged by different standards. (See Exhibit 5.1, p. 250.)
2. Epistemological clarity	Inquiry traditions like positivism, naturalism, social construction, realism, pheneomenology, and pragmatism are based on different criteria about what constitutes knowledge, how it is acquired, and how it should be judged. (See Chapter, especially Exhibit 3.3, pp. 97–99.)
3. Questions and hypotheses flow from and are consistent with purpose and epistemology	Different purposes and iquiry traditions emphasize different priority questions. (See Exhibit 3.3, pp. 97–99.)
4. Methods, design, and data collection procedures are aprorpriate for the nature of the inquiry	Purpose, epistemology, and research questions, in combination, drive methods, design, and data collection decisions. Matching methods to questions and hypotheses, given constraints of time, resources, and access, is basic.
5. Data collection procedures are systematic and carefully documented	The foundation of all science is careful methodological documentation so that those reviewing findings can determine how they were produced.
6. Data analysis is appropriate for the kind of data collected	Matching analytical procedures to the nature and type of data collected is a basic standard. There may be disagreemnts about what is appropriate, but the researcher has the obligation to make the case for appropriateness and justify methodological and analytical decsions made.
7. Strengths and weaknesses are acknowledged and discussed	No studies are perfect. All have limitations. These should be acknowledged and their implications for interpreting findings discussed.
8. Findings should flow from the data and analysis	The connection between data collected, analysis undertaken, and findings (conclusions, explanations) should be clear and explained.
9. Research should be presented for review	A fundamental principle of science is openness to review by those in a position to judge quality.
10. Ethical reflection and disclosure	All scientific traditions and disciplines have ethical standards, like avoiding (or at least disclosing) conflicts of interest and treating human subjects with respect. Compliance with ethical standards should be discussed.

EXHIBIT 9.7   Alternative Sets of Criteria for Judging the Quality and Credibility of Qualitative Inquiry

1. **Traditional Scientific Research Criteria**

    1. Objectivity of the inquirer (minimize bias)
    2. Hypothesis generation and testing
    3. Validity of the data
    4. Interrater reliability of codings and pattern analyses
    5. Conclusions about the correspondence of findings to reality
    6. Generalizability (external validity)
    7. Strength of causal explanations (attribution analysis)
    8. Contributions to theory
    9. Independence of conclusions and judgments
    10. Credibility to knowledgeable disciplinary researchers (peer review)

(These criteria are explained and discussed on pp. 683–684.)

Fight
TRUTH
Decay

2. **Social Construction and Constructivist Criteria**

    1. Subjectivity acknowledged (discuss and take into account inquirer perspective)
    2. Trustworthiness and authenticity
    3. Interdependence: relationship based (intersubjectivity)
    4. Triangulation (capturing and respecting multiple perspectives)
    5. Reflexivity
    6. Particularity (doing justice to the integrity of unique cases)
    7. Enhanced and deepened understanding (*verstehen*)
    8. Contributions to dialogue
    9. Extrapolation and transferability
    10. Credible to and deemed accurate by those who have shared their stories and perspectives

(These criteria are explained and discussed on pp. 684–686.)

Deconstruct
TRUTHS

3. **Artistic and Evocative Criteria**

    1. Emotionally evocative: connects with and moves the audience
    2. Integrates science and art to open the world to us
    3. Creativity
    4. Aesthetic quality, artistic representation
    5. Interpretive vitality, sensuous
    6. Embedded in lived experience
    7. Stimulating and provocative
    8. Voice distinct, expressive
    9. *Feels* "true" or "authentic" or "real"
    10. Crystallization

(These criteria are explained and discussed on pp. 687–690.)

Create
TRUTHS

4. **Participatory and Collaborative Criteria**

    1. Genuine and significant participation from inquiry focus, through design, data collection, analysis, and reporting; participation is real
    2. Researchers and participants are co-inquirers, sharing power and decision making
    3. Interactive validity and interpersonal competence

4. Builds capacity through learning by doing
5. Mutual respect
6. Group reflexivity
7. Interdependence
8. Sense of group ownership ("We did this.")
9. Group accountability: negotiated trade-offs explicit and transparent
10. Credibility within the group the basis for external credibility

(These criteria are explained and discussed on pp. 690–691.)

Group-Sourcing TRUTH

5. **Critical Change Criteria**

1. Critical perspective: increases consciousness about injustices
2. Identifies nature and sources of inequalities and injustices
3. Represents the perspective of the less powerful
4. Makes visible the ways in which those with more power exercise and benefit from power
5. Engages those with less power respectfully and collaboratively
6. Builds the capacity of those involved to take action
7. Identifies potential change-making strategies
8. Praxis
9. Clear historical and values context
10. Consequential validity

(These criteria are explained and discussed on pp. 691–693.)

Speak TRUTH to Power

6. **Systems Thinking and Complexity Criteria**

1. Analyze and map systems of interests
2. Attend to interrelationships
3. Capture perspectives
4. Sensitive to and explicit about boundary implications
5. Capture emergence
6. Expect and document nonlinearities
7. Adapt inquiry in the face of uncertainties
8. Describe systems changes and their implications
9. Contribution analysis
10. Credible to systems thinkers

(These criteria are explained and discussed on pp. 693–695.)

Truth is COMPLEX

7. **Pragmatic, Utilization-Focused Criteria**

1. Focus inquiry on informing action and decisions
2. Identify intended uses and users
3. Interactive engagement with intended users to enhance relevance and use
4. Practical orientation throughout
5. Relevance to real-world issues and concerns
6. Time findings and feedback to support use
7. Understandable methods and findings
8. Actionable findings
9. Credible to primary intended users
10. What is useful is true
11. Extract lessons

What Is Useful Is TRUE

(These criteria are explained and discussed on pp. 695–697.)

From the General to the Particular: Seven Sets of Criteria for Judging the Quality of Different Approaches to Qualitative Inquiry

Once we move beyond general criteria for scientific inquiry (Exhibit 9.6) to address specific quality criteria for qualitative inquiry, we must move from the general to the particular and contextual. Exhibit 9.7 lists criteria that are embedded in and flow from distinct qualitative inquiry frameworks. The *traditional scientific research criteria* are embedded in and derived from what I discussed in Chapter 3 as *reality-testing inquiry* frameworks that include positivist, postpositivist, empiricist, and foundationalist epistemologies pp. 105–108. The *social construction criteria* are derived from the discussion of "constructivism" in Chapter 3 pp. 121–126. The *artistic and evocative criteria* are derived from the discussion of autoethnography and evocative forms of inquiry in Chapter 3, especially the criteria suggested by Richardson (2000b) for "creative analytic practice of ethnography." The fourth set of criteria, participatory and collaborative approaches, are based on traditions and approaches reviewed in Chapter 4 pp. 213–222. The fifth set of criteria, *critical change criteria*, flow from critical theory, feminist inquiry, activist research, and participatory research processes aimed at empowerment. The sixth set of criteria, *systems and complexity criteria*, are derived from the discussion in Chapter 3 pp. 139–151.The seventh and final set of criteria, *pragmatic and utilization-focused criteria*, are based on discussions in Chapters 3 and 4 pp. 152–157 as well as program evaluation standards and principles (Joint Committee and Standards, 2010) and "Guiding Principles for Evaluators" (AEA Task Force on Guiding Principles for Evaluators, 1995).

To some extent, all of the theoretical, philosophical, and applied orientations reviewed in Chapters 3 and 4 provide somewhat distinct criteria, or at least priorities and emphases, for what constitutes a quality contribution within those particular perspectives and concerns. I've chosen these seven broader sets of criteria to capture the primary debates that differentiate qualitative approaches and, more specifically, to highlight what seem to me to differentiate *reactions* to qualitative inquiry. In this chapter, we are primarily concerned with how others respond to our work. With what perspectives and by what criteria will our work be judged by those who encounter and engage it?

Some of the confusion that people have in assessing qualitative research stems from thinking it represents a uniform perspective, especially in contrast to quantitative research. This makes it hard for them to make sense of the competing approaches within qualitative inquiry. By understanding the criteria that others bring

DIFFERENT AUDIENCES INTERESTED IN AND INVOLVED IN ASSESSING THE QUALITY OF QUALITATIVE RESEARCH AND EVALUATION

Criteria of quality can and often do vary by audience (Flick, 2007b, pp. 3–8). Here are some questions to consider in thinking about the intersection of quality criteria and audience.

1. *Your criteria.* You, the inquirer, presumably have an interest in doing quality work. How do you decide what standards and criteria of quality you will adhere to?

2. *Primary users of your findings.* Others will read and potentially use your findings. Who are the intended users of what you generate, and what criteria will they apply in judging the quality and credibility of your work?

3. *Funders of your inquiry.* If your inquiry has been funded by a grant, an agency, an evaluation contract, or some other funding mechanism, funders will be judging whether what you produced was worth what it cost. How will they make that judgment?

4. *Publication reviewers.* You may want to publish your findings. How will journals, book editors, and peer reviewers judge your work?

You need not be passive about others' criteria and judgments. Indeed, you ought not to be passive. You should make explicit the quality criteria you have applied in designing and implementing your inquiry and invite readers, funders, and peer reviewers to join you in using your criteria. You may also add the caveat that if they apply different criteria, their judgments of quality may well differ from yours and from those who follow the criteria you're operating under. In all of this keep in mind that

> the question of how to ascertain the quality of qualitative research has been asked since the beginning of qualitative research and attracts continuous and repeated attention. However, answers to this question have not been found—at least not in a way that is generally agreed upon. (Flick, 2007b, p. 11)

to bear on our work, we can anticipate their reactions and help them position our intentions and criteria in relation to their own expectations and criteria. In terms of the Reflexive Triangulated Inquiry model presented in Chapter 2 as Exhibit 2.5 (see p. 72), we're dealing here with the intersection between the inquirer's perspective and the perspective of those receiving the study (the audiences).

Criteria Determine What We See: The Umpires' Perspectives

Different perspectives about things such as truth and the nature of reality constitute paradigms or worldviews based on alternative epistemologies and ontologies. People viewing qualitative findings through different paradigmatic lenses will react differently just as we, as researchers and evaluators, vary in how we think about what we do when we study the world. These differences are nicely illustrated by the classic story of three baseball umpires who, having retired after a game to a local establishment for the dispensing of reality-distorting but truth-enhancing libations, are discussing how they call balls and strikes.

"I call them as I see them," says the first.

"I call them as they are," says the second.

"They ain't nothing until I call them," says the third.

That's the classic version of the story. Now, thanks to high-speed camera technology, we can update the story.

As chance would have it, two management researchers, Brayden King of Northwestern University and Jerry Kim of Columbia Business School, happened to be in the same bar going over their research on the accuracy of umpires' calls. Overhearing the three umpires, they went up to them and said, "Fourteen percent of the time you call them wrong." Before the umpires could argue, they explained,

We analyzed more than 700,000 pitches thrown during the 2008 and 2009 seasons. In addition to an average error rate of 14%, we found that umpires tended to favor the home team and that umpires were more likely to make mistakes when the game was on the line. (Based on King & Kim, 2014, p. SR12)

The two researchers went on like this for some time, breaking down the error rates by innings, situation, pitcher and batter ethnicity and race, pitcher reputation, and so forth and so on, until finally the umpires together put up their hands and told them to stop.

The first umpire said, "Your criteria are based on a high-speed camera. We get that. You love your numbers. We get that. And your analysis is interesting, even fascinating. We get that. But during a game we don't use a camera. So I still call them as I see them," he reiterated.

"And I call them as they are," repeated the second.

"And they ain't nothing until I call them," concluded the third.

We turn now to discussion and elaboration of the seven alternative sets of criteria for judging the quality of qualitative work summarized in Exhibit 9.7.

1. Traditional Scientific Research Criteria

> The saddest aspect of life right now is that science gathers knowledge faster than society gathers wisdom.
>
> —Isaac Asimov (1920–1992)
> Science author and science fiction writer

One way to increase the credibility and legitimacy of qualitative inquiry among those who place priority on traditional scientific research criteria is to emphasize those criteria that have priority within that tradition. Science has traditionally emphasized objectivity, so qualitative inquiry within this tradition emphasizes procedures for minimizing investigator bias. Those working within this tradition will emphasize rigorous and systematic data collection procedures, for example, cross-checking and cross-validating sources during fieldwork. In analysis it means, whenever possible, using multiple coders and calculating intercoder consistency to establish the validity and reliability of pattern and theme analysis. Qualitative researchers working in this tradition are comfortable using the language of "variables" and "hypothesis testing" and striving for causal explanations and generalizability, especially in combination with quantitative data (e.g., Hammersley, 2008b). Qualitative approaches that manifest some or all of these characteristics include grounded theory (Glaser, 2000), qualitative comparative analysis (Ragin, 1987, 2000), and realism (Miles et al., 2014). Their common aim is to use qualitative methods to describe and explain phenomena as accurately and completely as possible so that their descriptions and explanations correspond as closely as possible to the way the world is and actually operates (Reynolds et al., 2011). Government agencies supporting qualitative research (e.g., the U.S. Government Accounting Office, the National Science Foundation, or the National Institutes of Health) usually operate within this traditional scientific framework.

THE ROOTS OF TRADITIONAL SOCIAL SCIENCE CRITERIA APPLIED TO QUALITATIVE INQUIRY

An emphasis on valid and reliable knowledge, as generated by neutral researchers utilizing the scientific method to discover universal Truth, reflects an epistemology commonly referred to as positivism. Historically, social scientists understood positivism as reflected in a "realist ontology, objective epistemology, and value-free axiology." Few, if any, qualitative researchers currently subscribe to an absolute faith in positivism, however. Many postpositivists, or researchers who believe that achievement of objectivity and value-free inquiry are not possible, nonetheless embrace the goal of production of generalizable knowledge through realist methods and minimization of researcher bias, with objectivity as a "regulatory ideal" rather than an attainable *goal*. In short, postpositivism does not embrace naive *belief* in pure scientific truth; rather, qualitative research conducted in a strict postpositivist tradition utilizes precise, prescribed processes and produces *social* scientific reports that enable researchers to make generalizable claims about the social phenomenon within particular populations under examination.

Postpositivists commonly utilize qualitative methods that bridge quantitative methods, in which researchers conduct an inductive analysis of textual data, form a typology grounded in the data (as contrasted with a preexisting, validated typology applied to new data), use the derived typology to sort data into categories, and then count the frequencies of each theme or category across data. Such research typically emphasizes validity of the coding schema, inter-coder reliability, and careful delineation of procedures, including random or otherwise *systematic sampling* of texts. Content analyses of media typify this approach. (Ellingson, 2011, pp. 596, 598; within-quote references omitted)

2. Social Construction and Constructivist Criteria

> What is perceived as real is real in its consequences.
>
> —The Thomas theorem

Social construction, constructivist, and "interpretivist" perspectives have generated new language and concepts to distinguish quality in qualitative research (e.g., Glesne, 1999, pp. 5–6). Lincoln and Guba (1986) proposed that constructivist inquiry demanded different criteria from those inherited from traditional social science. They suggested "credibility as an analog to internal validity, transferability as an analog to external validity, dependability as an analog to reliability, and confirmability as an analog to objectivity." In combination, they viewed these criteria as addressing "trustworthiness (itself a parallel to the term *rigor*)" (pp. 76–77). They went on to emphasize that naturalistic inquiry should be judged by dependability (a systematic process systematically followed) and authenticity (reflexive consciousness about one's own perspective, appreciation for the perspectives of others, and fairness in depicting constructions in the values that undergird them). They viewed the social world (as opposed to the physical world) as socially, politically, and psychologically constructed, as are human understandings and explanations of the physical world. They advocated triangulation to capture and report multiple perspectives rather than seek a singular truth. The team of researchers who reviewed approaches to assessing quality in qualitative research found that "the post-positivist criteria developed by Lincoln and Guba, based around the construct of 'trustworthiness,' were referenced frequently and appeared to be the basis upon which a number of authors made their recommendations for improving quality of qualitative research" (Reynolds et al., 2011).

Constructivists embrace subjectivity as a pathway deeper into understanding the human dimensions of the world in general as well as whatever specific phenomena they are examining (Peshkin, 1985, 1988, 2000a,b). They're more interested in deeply understanding specific cases within a particular context than in hypothesizing about generalizations and causes across time and space. Indeed, they are suspicious of causal explanations and empirical generalizations applied to complex human interactions and cultural systems. They offer perspective and encourage dialogue among perspectives rather than aiming at singular truths and linear predictions. Social constructivists' case studies, findings, and reports are explicitly informed by attention to praxis and reflexivity—that is, understanding how one's own experiences and background affect what one understands and how one acts in the world, including acts of inquiry. For an in-depth discussion of this perspective and its implications, see the *Handbook of Constructionist Research* (Holstein & Gubrium, 2008). Also see Chapter 3 pp. 121–126 for a much lengthier discussion of constructionism and constructivism.

Here are three examples of social construction as a framework for program evaluation.

1. The evaluation of a community development project in an ethnically and racially diverse neighborhood collected and reported stories from residents purposefully sampled to present a range of experiences and perspectives. The evaluation did not render judgments but was called a "multivocal evaluation" in which the diverse stories were used for dialogue and to enhance mutual understanding.

2. The evaluation of the international Paris Declaration on Development Aid included case studies that revealed the different perspectives and contexts within which aid is given and received. Donors (wealthier countries) and beneficiaries (poorer countries) experience different "realities."

One purpose of the evaluation was to capture those different realities, including diverse experiences with the Paris Declaration principles, to facilitate dialogue on future international development policies and practices.

3. Social constructivism was the foundation of Nora Murphy's (2014) study of homeless youth. The 14 case studies showed diverse experiences and perspectives on homelessness. The study concluded that the unique situation of each homeless youth meant that program responses needed to be socially constructed together with the youth to be meaningful to them and to build trusting adult–youth relationships. (For more on this evaluation, see pp. 194, 626–628.)

CONSTRUCTIVIST TRUSTWORTHINESS

The credibility of your findings and interpretations depends on your careful attention to establishing trustworthiness. Lincoln and Guba (1985) describe prolonged engagement (spending sufficient time at your research site) and persistent observation (focusing in detail on those elements that are most relevant to your study) as critical in attending to credibility. "If prolonged engagement provides scope, persistent observation provides depth" (p. 304). With each, time is a major factor in the acquisition of trustworthy data. Time at your research site, time spent interviewing, and time building sound relationships with respondents all contribute to trustworthy data. When a large amount of time is spent with your research participants, they less readily feign behavior or feel the need to do so; moreover, they are more likely to be frank and comprehensive about what they tell you. Lincoln and Guba posited four constructivist criteria as parallel to but distinct from traditional research criteria:

First, *credibility* (parallel to internal validity) addressed the issue of the inquirer providing assurances of the fit between respondents' views of their life ways and the inquirer's reconstruction and representation of same. Second, *transferability* (parallel to external validity) dealt with the issue of generalization in terms of case-to-case transfer. It concerned the inquirer's responsibility for providing readers with sufficient information on the case studied such that readers could establish the degree of similarity between the case studied and the case to which findings might be transferred. Third, dependability (parallel to reliability) focused on the process of the inquiry and the inquirer's responsibility for ensuring that the process was logical, traceable, and documented. Fourth, *confirmability* (parallel to objectivity) was concerned with establishing the fact that the data and interpretations of an inquiry were not merely figments of the inquirer's imagination. It called for linking assertions, findings, interpretations, and

so on to the data themselves in readily discernible ways. For each of these criteria, Lincoln and Guba also specified a set of procedures that could be used to meet the criteria. For example, auditing was highlighted as a procedure useful for establishing both dependability and confirmability, and member check and peer debriefing, among other procedures, were defined as most appropriate for credibility.

In *Fourth Generation Evaluation* (1989), Guba and Lincoln reevaluated this initial set of criteria. They explained that trustworthiness criteria were parallel, quasi-foundational, and clearly intended to be analogs to conventional criteria. Furthermore, they held that trustworthiness criteria were principally methodological criteria and thereby largely ignored aspects of the inquiry concerned with the quality of outcome, product, and negotiation. Hence, they advanced a second set of criteria called *authenticity criteria*, arguing that this second set was better aligned with the constructivist epistemology that informed their definition of qualitative inquiry. (Schwandt, 2007, pp. 299–300)

Continual alertness to your own biases and subjectivity (reflexivity) also assists in producing more trustworthy interpretations. Consider your subjectivity within the context of the trustworthiness of your findings. Ask yourself a series of questions: Whom do I not see? Whom have I seen less often? Where do I not go? Where have I gone less often? With whom do I have special relationships, and in what light would they interpret phenomena? What data-collecting means have I not used that could provide additional insight? Triangulated findings contribute to credibility. Triangulation may involve the use of multiple data collection methods, sources, investigators, or theoretical perspectives. To improve trustworthiness, you can also consciously search for negative cases.

Alternative Criteria Review

Exhibit 9.7 (pp. 680–681) presented seven different sets of criteria for judging the quality of qualitative studies. This module reviewed the first two sets of criteria: (1) traditional scientific research criteria versus (2) constructivist criteria. Constructivist criteria emerged from the critique that traditional scientific research criteria were based on quantitative and experimental design thinking that, by the very nature of using those criteria for defining quality, led to qualitative studies being judged inferior. The next module makes the issue of judging quality even more complicated by adding five more sets of alternative and competing criteria.

A Realist Views a Constructivist Proposal

Alternative and Competing Criteria, Part 2

Artistic, Participatory, Critical Change, Systems, Pragmatic, and Mixed Criteria

> The moral and social yearnings of fully realized human beings are not reducible to universal laws and cannot be studied like physics.
>
> —Brooks (2010, p. A27)

This module continues the presentation and discussion of seven alternative sets of criteria for judging the quality of qualitative studies. The previous module covered (1) traditional social science research criteria and (2) social construction and constructivist criteria. This module covers (3) artistic and evocative criteria, (4) participatory and collaborative criteria, (5) critical change criteria, (6) systems and complexity criteria, and (7) pragmatic and utilization-focused criteria. We'll then examine mixing criteria.

3. Artistic and Evocative Criteria

> TRUTH is visceral, palpable, sensuous, wrenching, hormonal, cognitive, cathartic, lyrical, contextual, awakening, fleeting, universal, and debatable. In other words, truth is art.
>
> —From Halcolm's *Ruminations*

Researchers and audiences operating from the perspective of traditional scientific research criteria emphasize the scientific nature of qualitative inquiry. Researchers and audiences who view the world through the lens of social construction emphasize qualitative inquiry as a particularly human form of understanding centered on the capacity of people and groups to construct meaning. That brings us to this third alternative, which emphasizes that human beings both think and feel. Traditional social science and constructivist inquiries focus on cognitive, logical, sense-making analyses. The artistic and evocative

approaches to qualitative inquiry want to bring forth our emotional selves and do so by integrating art and science. Science makes us think. Great art makes us feel. From the perspective of artistic and evocation qualitative inquirers, great qualitative studies should evoke both understandings (cognition) and feelings (emotions).

> Persons are moved by emotion. . . . People are their emotions. To understand who a person is, it is necessary to understand emotion. . . . Emotions cut to the core of people. Within and through emotion people come to define the surface and essential, or core, meanings of who they are. Emotions and moods are ways of disclosing the world for the person. (Denzin, 2009, pp. 1–2)

Artistic and evocative criteria focus on aesthetics, creativity, interpretive vitality, and expressive voice. Case studies become literary works. Poetry or performance art may be used to enhance the audience's direct experience of the essence that emerges from analysis. Artistically oriented qualitative analysts seek to engage those receiving the work, to connect with them, move them, provoke them, and stimulate them. Creative nonfiction and fictional forms of representation blur the boundaries between what is "real" and what has been created to represent the essence of a reality. A literal presentation of reality, real (scientific) or perceived (constructivism), yields to artistically created reality. The results may be called creative syntheses, ideal-typical case constructions, scientific poetics, or any number of phrases that suggest the artistic emphasis. Artistic expressions of qualitative analysis strive to provide an experience with the findings where "truth" or "reality" is understood to have *a feeling dimension* that is every bit as important as the cognitive dimension. Such qualitative inquiry is explicitly *sensuous* (Stoller, 2004) and emotional (Denzin, 2009).

The performance art of *The Vagina Monologues* (Ensler, 2001), based on interviews with women about their experiences of coming of age sexually, but presented as theater, offers a prominent example. The audience feels as much as knows the truth of the presentation because of the essence it reveals. In the artistic tradition, the analyst's interpretive and expressive voice, experience, and perspective may become as central to the work as depictions of others or the phenomenon of interest. Here are some examples of artistic and evocative approaches used in program evaluation.

- A program for low-income, pregnant, drug-addicted teenagers asked the young women to draw pictures of their hearts and tell what the pictures meant. The initial hearts were portrayed as wounded, knifed, torn, mangled, and tortured. Over a period of four months (four drawings and accompanying stories), the pictures showed some sunshine, flowers, rainbows, and, most striking, connections to other hearts. No perfect valentines. Not even close. But to look at those raw drawings was to see hearts healing.

- Theater for development is being used in Nigeria to engage community members in discussing preliminary results from an evaluation, with actors replaying scenarios relating to a program uncovered through fieldwork. By involving community members in role-play, the early findings can be verified or corrected based on the participants' experiences of the program, and new perspectives can be recorded, which might otherwise have remained dormant (Folorunsho, 2014).

- A team of educational evaluators concerted interviews with students and teachers about an international cross-cultural summer experience into a play dramatizing critical events and key learnings. The play was performed for the school board as the project's evaluation report.

- Photographs taken before and after Vietnam implemented a motorbike helmet law showed dramatic changes in compliance. (See Exhibit 8.20, pp. 609; for other examples of using visuals created to present evaluation findings, see Exhibits 8.21 through 8.27, pp. 610–619.)

Exhibit 9.8 shows how an interview transcript was converted into a poem when presenting the findings, all the better to give the reader a feel for what was said and the affect it carried.

EXHIBIT 9.8 **From Interview Transcript to Poem: An Artistic and Evocative Presentation**

In May 1994, Corrine Glesne (1997) interviewed Dona Juana, an 86-year-old professor in the College of Education at the University of Puerto Rico.

> That she chose a bird to represent her was no surprise. Standing 5 feet tall, very thin ("a problem all my life"), and with bright dark eyes, she was birdlike in appearance. Her office was a nest of books, papers, and folders in organized piles on her large desk, on the beige metal filing cabinets next to the door opposite her desk, in the wooden cabinet along the wall to the right of her desk, on the shelves below the window to her left, and on the two chairs before her desk. There was no sense of disorder, but rather an impression of an archive that would illuminate Dona Juana's 50 years in research and higher education. (p. 203)

Below is the poem Glesne (1997) created from the interview transcript, followed by a table showing the conversion from transcript to poem.

The Poem

That Rare Feeling

I am a flying bird

moving fast

seeing quickly

looking with the eyes of God,

from the tops of trees.

How hard for country people

picking green worms

from fields of tobacco,

sending their children to school,

not wanting them to suffer

as they suffer.

In the urban zone,

students worked at night

and so they slept in school.

Teaching was the real university.

So I came to study

to find out how I could help.

I am busy here at the university,

there is so much to do.

But the University

is not the Island.

I am a flying bird

moving fast, seeing quickly

so I can give strength,

so I can have that rare feeling

of being useful. (pp. 202–203)

Composing the Poem From the Interview Transcript

TRANSCRIPT	POETIC NARRATIVE CREATED
C (Corrine): If I asked you to use a metaphor to describe yourself as a professor, what would you say you were like? Someone I asked said that she was a bridge and then she told me why. What metaphor comes to mind for you?	*Version 1: Chronologically and linguistically faithful to the transcript* I would be a flying bird. I want to move so fast
J (Juana): I would be a flying bird.	so I can see quickly, everything. I wish I could look at the world
C: A flying bird. Tell me about it. How are you a flying bird?	with the eyes of God, to give strength to those that need . . .
J: Because I want to move so fast.	*Version 2: Draws from other sections* *of the interviews, takes more license with words.*
C: Mrn-hmmm. Cover a lot of territory.	
J: Yes. Yes.	I am a flying bird
C: Are you any kind of bird or just any bird?	moving fast, seeing quickly, looking with the eyes of God
J: Well, any bird because I don't want to mention some birds, some birds here are destructive.	from the tops of trees: How hard for country people
C: Are what?	picking green worms from fields of tobacco,
J: Are destructive. They destroy and I don't want to . . .	sending their children to school, not wanting them to suffer as they suffer
C: No, you don't want to be one of them. No. You're just a bird that moves fast.	In the urban zone, students worked at night
J: That moves fast and sees from the tops of trees. So I can see quickly.	and so they slept in school. Teaching was the real university.
C: See quickly, see everything.	So I came to study
J: Everything.	to find out how I could help. I am busy here at the university,
J: So you can see me?	there is so much to do. But the university is not the Island.
C: I can. I can see you, a flying bird.	
J: I wish I could look at the world with the eyes of God.	I am a flying bird moving fast, seeing quickly so I can give strength,
C: With the eyes of what?	so I could have that rare feeling of being useful. (Glesne, 1997, p. 207)
J: Of God, of that spiritual power that can give strength.	
C: That can give strength? Strength?	
J: Yes, to those that need.	

Crystallization

Sociologist Laurel Richardson (2000b) introduced *crystallization* as a criterion of quality in artistic and evocative qualitative inquiry, a replacement for triangulation as a criterion.

The scholar draws freely on his or her productions from literary, artistic, and scientific genres, often breaking the boundaries of each of those as well. In these productions, the scholar might have different "takes" on the same topic, what I think of as a postmodernist deconstruction of triangulation. . . . In postmodernist mixed-genre texts, we do not triangulate, we *crystallize*. . . . I propose that the central image for "validity" for postmodern texts is not the triangle—a rigid, fixed, two-dimensional object. Rather, the central imaginary is the crystal, which combines symmetry and substance with an infinite variety of shapes, substances, transmutations, multidimensionalities, and angles of approach. . . . Crystallization provides us with a deepened, complex, thoroughly partial, understanding of the topic. Paradoxically, we know more and doubt what we know. Ingeniously, we know there is always more to know. (p. 934)

Crystallization's roots can be traced to

the creative and courageous work of feminist methodologists who blasphemed the boundaries of art and science. . . .

Art and science do not oppose one another; they anchor ends of a continuum of methodology, and most of us situate ourselves somewhere in the vast middle ground. When scholars argue that we cannot include narratives alongside analysis or poems within grounded theory, they operate under the assumption that art and science negate one another and hence are incompatible, rather than merely differ in some dimensions. . . . My explanation of crystallization assumes a basic understanding of the complexities involved in combining methods and genres from across regions of the continuum. (Ellingson, 2009, pp. 3, 5)

4. Participatory and Collaborative Criteria

> To be human is to engage in interpersonal dynamics. *Inter:* between. *Personal:* people. *Dynamics:* forces that produce activity and change. Combining these definitions, interpersonal dynamics are the forces between people that lead to activity and change. Whenever and wherever people interact, these dynamics are at work.
>
> —King and Stevahn (2013, p. 2)
> *Interactive Evaluation Practice*

Participatory and collaborative qualitative inquiries have four purposes and justifications:

1. *Values premise:* The right way to inquire into a phenomenon of interest is to do it with the people involved and affected. This means doing research and evaluation *with* as opposed *to* people. It means engaging them as fellow inquirers and coresearchers rather than as research subjects.

RIGOR IN ARTISTIC AND EVOCATIVE CRYSTALLIZATION

One of the most helpful (albeit not foolproof) ways to enhance your account and ward off editorial defensiveness toward creative analytic work, in general, and crystallization, in particular, is to be absolutely clear about what you did (and did not do) in producing your manuscript. This includes data collection, analysis, and especially choices made about representation. . . . By explaining my process, I help alleviate suspicions that I took an "anything goes" sloppy attitude toward constructing my representation.

While some colleagues may not like or approve of what you did no matter how you explain it, concise, explicit details of your process make it more difficult for them to dismiss it as careless or random. Accounting for your process (even in an appendix or endnote) constitutes an important nod toward methodological rigor. As many have posited, engaging in creative analytic work should be no less rigorous, exacting, and subject to strict standards of peer evaluation. . . .   Moreover, such a roadmap assists others who may seek to follow your lead. . . .

Some suggestions on issues to own:

- Explain choices you made in composing narratives, poems, or other artistic work; in other words, how did you get from data to text?
- Describe your standpoint vis-à-vis your topic, not just what it is, but (at least some of) how it shapes your interactions with your data (e.g., I am a cancer survivor studying clinics so I tend to be more empathetic with patients than health care providers; I am a feminist so I pay a lot of attention to power dynamics).
- Indicate your awareness of and response to ethical considerations about voice, privacy, and responsibility to others. What steps did you take to ensure participant confidentiality? To privilege participants' voices? Consider how your work might be read in ways that do not reflect your intentions—for example, what quotes from participants could be taken out of context and used as justification for blaming the victim?—and surround vulnerable voices with preemptory statements that make it more difficult for oppositional forces to excerpt and reinterpret their meaning in regressive ways.
- Detail your analytic procedures. . . . Even if you construct a unique, outside-the-box artistic creation, you should explain your methodology and cite some sources to contextualize your work. Again, this need not interfere with your aesthetic goals; details should be concise and can be placed in an appendix, footnote, or even a separate piece altogether. The goal is to reveal crystallized projects as embodied, imperfect, insightful constructions rather than as immaculate end products. (Ellingson, 2009, pp. 199–120)

2. *Quality premise:* Data will be better when people who are the focus of the inquiry willingly participate, understand the nature of the inquiry, and agree with the importance of the study. Interviews will be richer and more detailed. Observations will be open and unguarded. Documents will be readily available. Data are better.

3. *Reciprocity premise:* Researchers get data, publications, knowledge, and career advancement from research and evaluation studies. Those who are the focus of inquiry should benefit as well. As coresearchers, through participation in the inquiry, they learn research skills, learn to think more systematically, and gain knowledge that they can use for their own purposes.

4. *Utility premise:* In program evaluation and action research inquiries, the findings are more likely to be useful—and actually used—when those who must act on the findings collaborate in generating and interpreting them.

From the classic articulation and justification of *Participatory Action Research* by William Foote Whyte (1989, 1991) to methods and facilitation guides on how to actually do it (Caister et al., 2011; Hacker, 2013; King & Stevahn, 2013; Pyrch, 2012; Taylor, Suarez-Balcazar, Forsyth, & Kielhofner, 2006), participatory and collaborative engagement has been a major approach to qualitative inquiry. When conducting research in a collaborative mode, professionals and nonprofessionals become coresearchers. Participatory action research encourages collaboration within a mutually acceptable inquiry framework to understand and/or solve organizational or community problems. Chapter 4 includes an in-depth discussion of participatory and collaborative approaches (pp. 213–222), including Exhibit 4.13, Principles of Fully Participatory and Genuinely Collaborative Inquiry (p. 222).

Here's an example of a participatory and collaborative qualitative inquiry. Robin Boylorn (2008) studied the experiences of black Southern women. She recruited a group of participants from the community in which she had grown up and invited them to share stories about their experiences and lives growing up and raising families in the rural South. She facilitated their interactions together as co-investigators so that they felt "equally invested and equally involved in the process of collecting, writing, interpreting, and editing the stories they wrote" (p. 600). She shared her experiences with the participants, and together they compared and contrasted their ideas and experiences.

Their involvement began during the early stages of recommending other participants and retelling stories in

INTERPERSONAL VALIDITY

Educational evaluator Karen Kirkhart (1995) coined the term *interpersonal validity*: the extent to which an evaluator is able to relate meaningfully and effectively to individuals in the evaluation setting. The interpersonal factor that undergirds interpersonal validity highlights the competence of a participatory evaluator or researcher do two things: (1) interact with people constructively throughout the framing and implementation of an inquiry and (2) create activities and conditions conducive to positive interactions among participants. The *interpersonal factor* is concerned with creating, managing, and ultimately mastering the interpersonal dynamics that make a collaborative inquiry possible and inform its findings. One is concerned with eventual use, the other with establishing buy-in among participants and a valid inquiry process. (King & Stevahn, 2013, p. 6)

individual and group settings to ensure adequate information was available. As co-investigators their stories were instrumental in establishing and representing a corporate set of themes and experiences. Though the co-researchers in this project were not involved in the writing stages, they did have the opportunity to respond to the stories the author wrote, offering their unique perspectives and feedback as participants in the research and characters in the stories. The resulting research project is a collaboration between the researcher and the researched, including participants as co-researchers. (p. 600)

5. Critical Change Criteria

We are distressed by underprivilege. We see gaps among privileged patrons and managers and staff and underprivileged participants and communities. . . . We are advocates of a democratic society.

—Robert Stake (2004, pp. 103–107)
Qualitative evaluation pioneer

How Far Dare an Evaluator Go Toward Saving the World?

Those engaged in qualitative inquiry as a form of critical analysis aimed at social and political change eschew any pretense of open-mindedness or objectivity; they take an activist stance. Critical change inquiry aims to critique existing conditions and through that critique bring about change. Critical change criterion is derived from *critical theory*, which frames and engages in qualitative inquiry with an explicit agenda of elucidating power, economic, and social inequalities. The "critical" nature of critical theory flows from a commitment to go beyond just studying society for the sake of increased understanding. Critical change researchers set out to use inquiry to critique society, raise consciousness, and change the balance of power in favor of those less powerful. Influenced by Marxism, informed by the presumption of the centrality of class conflict in understanding community and societal structures, and updated in the radical struggles of the 1960s, *critical theory* provides both philosophy and methods for approaching research and evaluation as fundamental and explicit manifestations of political praxis (connecting theory and action), and as change-oriented forms of engagement.

> Critical social science and critical social theory attempt to understand, analyze, criticize, and alter social, economic, cultural, technological, and psychological structures and phenomena that have features of oppression, domination, exploitation, injustice, and misery. They do so with a view to changing or eliminating these structures and phenomena and expanding the scope of freedom, justice, and happiness. The assumption is that this knowledge will be used in processes of social change by people to whom understanding their situation is crucial in changing it. (Bentz & Shapiro, 1998, p. 146; Kincheloe & McLaren, 2000)

Critical change has three interconnected elements: (1) inquiry into situations of social injustice, (2) interpretation of the findings as a critique of the existing situation, and (3) using the findings and critique to mobilize and inform change.

> Critical theory looks at, exposes, and questions hegemony—traditional power assumptions held about relationships, groups, communities, societies, and organizations—to promote social change. Combined with action research, critical theory questions the assumed power that researchers typically hold over the people they typically research. Thus, critical action research is based on the assumption that society is essentially discriminatory but is capable of becoming less so through purposeful human action.

> Critical action research also assumes that the dominant forms of professional research are discriminatory and must be challenged. Critical action research takes the

concept of knowledge as power and equalizes the generation of, access to, and use of that knowledge. Critical action research is an ethical choice that gives voice to, and shares power with, previously marginalized and muted people. (Davis, 2008, p. 140)

Critical change criteria apply to a number of specialized areas of qualitative inquiry (Given, 2008, pp. 139–179; Schwandt, 2007, pp. 50–55):

Critical ethnography	Critical discourse analysis	Critical realism
Critical education studies	Critical hermeneutics	Critical research
Critical arts-based inquiry	Critical humanism	Critical theory
Critical race theory	Critical pragmatism	Critical action research
Critical pedagogy	Critical social science	Critical systems analysis

In addition, *feminist inquiry* often includes an explicit agenda of bringing about social change (e.g., Benmayor, 1991; Brisolara, Seigart, & SenGupta, 2104; Hesse-Biber, 2013; Podems, 2014b). *Liberation research* and *empowerment evaluation* derive, in part, from Paulo Freire's philosophy of praxis and liberation education, articulated in his classics *Pedagogy of the Oppressed* (1970) and *Education for Critical Consciousness* (1973), still sources of influence and debate (e.g., Glass, 2001). Barone (2000) aspires to "emancipatory educational storysharing" (p. 247). Qualitative studies informed by critical change criteria range from largely intellectual and research-oriented approaches that aim to expose injustices to more activist forms of inquiry that actually engage in bringing about social change. Stephen Brookfield (2004) uses critical theory to illuminate adult education issues, trends, and inequities. Plummer (2011) integrates critical their and queer theory. Caruthers and Friend (2014) bring critical inquiry to online learning and engagement. Crave, Zaleski, and Trent (2014) emphasize the role of critical change in building a more equitable future through participatory program evaluation.

Here are two examples of critical change studies that would expect to be evaluated for quality by critical change criteria (Davis, 2008, p. 141):

1. Martin Diskin worked with policymakers and development agencies in Latin American studies to conduct what they called "power structure research," in which they exposed injustice as a strategy for building coalitions and motivating movements.

2. Christine Davis's ethnography of a children's mental health treatment team was an interdisciplinary research project involving the fields of communication studies, social work, and mental health. Conducted in partnership with community agencies, this research examined issues of power, marginalization, and control within these teams. It suggested a stance toward children and families that rejects the traditional hierarchical medical model of care and instead treats them as unique valuable humans and as equal partners in treatment.

Consequential validity as a critical change criterion for judging a research design or instrument makes the social consequences of its use a value basis for assessing its credibility and utility. Thus, standardized achievement tests are criticized because of the discriminatory consequences for minority groups of educational decisions made with "culturally biased" tests. Consequential validity asks for assessments of who benefits and who is harmed by an inquiry, measurement, or method (Brandon, Lindberg, & Wang, 1993; Messick, 1989; Shepard, 1993).

A QUALITATIVE MANIFESTO: A CALL TO ARMS

—Norman K. Denzin (2010)

The social sciences . . . should be used to improve quality of life. . . . For the oppressed, marginalized, stigmatized and ignored . . . and to bring about healing, reconciliation and restoration between the researcher and the researched.

—Stanfield (2006, p. 725)

Mills wanted his sociology to make a difference in the lives that people lead. He challenged persons to take history into their own hands. He wanted to bend the structures of capitalism to the ideologies of radical democracy. . . .

I want a critical methodology that enacts its own version of the sociological imagination. Like Mills, my version of the imagination is moral and methodological. And like Mills, I want a discourse that troubles the world, understanding that all inquiry is moral and political.

This book is an invitation and a call to arms. It is directed to all scholars who believe in the connection between critical inquiry and social justice (Denzin, 2010, p. 10).

Qualitative inquiry can contribute to social justice in the following ways:

1. It can help identify different definitions of a problem and/or a situation that is being evaluated with some agreement that change is required. It can show, for

example, how battered wives interpret the shelters, hotlines, and public services that are made available to them by social welfare agencies. Through the use of personal experience narratives, the perspectives of women and workers can be compared and contrasted.

2. The assumptions, often belied by the facts of experience, that are held by various interested parties—policy makers, clients, welfare workers, online professionals—can be located and shown to be correct, or incorrect (Becker, 1967, p. 239).

3. Strategic points of intervention into social situations can be identified. Thus, the services of an agency and a program can be improved and evaluated.

4. It is possible to suggest "alternative moral points of view from which the problem, the policy and the program can be interpreted and assessed" (see Becker, 1967, pp. 239–240). Because of its emphasis on experience and its meanings, the interpretive method suggests that programs must always be judged by and from the point of view of the persons most directly affected.

5. The limits of statistics and statistical evaluations can be exposed with the more qualitative, interpretive materials furnished by this approach. Its emphasis on the uniqueness of each life holds up the individual case as the measure of the effectiveness of all applied programs. (Denzin, 2010, pp. 24–25)

6. Systems Thinking and Complexity Criteria

In a finite game, it is easy to make sense. Everyone agrees on the goal; the rules are known; and the field of play has clear boundaries. Baseball, football, and bridge are examples of finite games. At one time in

the not-so-distant past we expected careers, marriages, parenthood, education, and citizenship to be finite games. When everyone agrees on the rules, and the consequences of our actions are undeniable, responsible people plan for what they want, take steps to achieve it, and enjoy the fruits of their labor. We know what it takes to make sense in a finite game.

Most of us realize we are playing in a very different game. We are playing in an infinite game. In which the boundaries are unclear or nonexistent, the scorecard is hidden, and the goal is not to win but to keep the game in play. There are still rules, but the rules can change without notice. There are still plans and playbooks, but many games are going on at the same time, and the winning plans can seem contradictory. There are still partners and opponents, but it is hard to know who is who, and besides that, the "who is who" changes unexpectedly.

—Glenda Eoyang and
Royce Holladay (2013, p. 4)

Adaptive Action: Leveraging Uncertainty in Your Organization

Studying "infinite games" in highly dynamic situations characterized by uncertainty and rapid change creates special challenges for qualitative inquiry. Systems thinking and complexity concepts offer a framework for studying such situations, tools for both inquiry and "coping with chaos" (Eoyang, 1997), and criteria for deciding whether such studies are of high quality. To be credible to systems thinkers and complexity scientists, the qualitative inquiry must capture, describe, map, and analyze, and map systems of interests; must attend to interrelationships, capture diverse perspectives, attend to emergence, and be sensitive to and explicit about boundary implications; and must document nonlinearities, adapt the inquiry in the face of uncertainties, and describe systems changes and their implications. In so doing, the explanatory approach moves from attribution to contribution analysis (see pp. 596–597 in Chapter 8).

Chapter 3 discussed systems theory and complexity theory as distinct, though intersecting, theoretical frameworks (see pp. 139–151). Exhibit 3.14 presents complexity theory concepts and qualitative inquiry implications (pp. 147–148). Exhibit 3.16 presents the relationship of systems theory to complexity theory (p. 150).

- **Systems theory inquiry questions:** How and why does this system function as it does? What are the system's boundaries and interrelationships, and how do these affect perspectives about how and why the system functions as it does?
- **Complexity theory inquiry question:** How can the emergent and nonlinear dynamics of complex adaptive systems be captured, illuminated, and understood?

For my purpose here, namely, differentiating distinct sets of criteria by which to judge the quality and credibility of various approaches to qualitative inquiry, the core systems and complexity dimensions can be integrated, as they are in Exhibit 9.7. That said, the systems field exemplifies the challenge of settling on some definitive set of quality criteria for judging qualitative inquiry, especially using a systems and complexity framing, because there are multiple approaches within the systems field (e.g., Hieronymi, 2013), each of which would assert and favor particular criteria unique to that perspective.

> Systemic inquiry covers a wide range methodologies, methods, and techniques with a strong focus on the behaviors of complex situations and the meanings we draw from those situations. It spans both the qualitative and quantitative research method domains but also includes approaches that fit neither category nor both categories....
>
> Any attempt to summarize a trans discipline like systemic inquiry is fraught with difficulties. Despite relatively simple origins, the field has sprawled many directions so that no single, universally accepted theory has emerged, and neither are there universally agreed definitions of basic concepts such as what is and what is not a system. Although we will find many definitions in the systems literature, many authors argue that single fixed definitions promote the kind of reductionist thinking that runs counter to systemic principles. Instead, they argue, the field should promote debates around methodological principles to create learning rather than fixed definitions—what Kurt Richardson calls "critical pluralism." (Williams, 2008, p. 858)

Thinking Systemically

So as not to get lost in or overwhelmed by different approaches to systems, let me close this section with an example grounded in the basics of attending to interrelationships, boundaries, perspectives, and emergence. An exemplar of applying systems thinking to understand an issue is the analysis done by Christopher Wells (2012) of the role and impact of the automobile in the United States. His analysis begins *before* there were automobiles (what complexity theorists call *initial conditions*). He examines emergent land use patterns in the nineteenth century, sanitation problems in cities, the development of agricultural markets, the role of horses in transportation, the influence of train routes, the challenges of riding bicycles on rutted and muddy roads, the function of farmers in maintaining roads along their farms, population growth, and many other factors that established the initial conditions that automobiles emerged into. To understand the automobile in American society, culture, politics, and economics, you must look at the systems before the automobile existed (transportation, commerce, public health, political jurisdictions, land use, and community values as starting places) and continue to examine those systems and their interactions through to the present day. The irony is that engaging and thinking through those complex interactions yields extraordinary clarity.

7. Pragmatic, Utilization-Focused Criteria

> *Usefulness!* It is not a fascinating word, and the quality is not one of which the aspiring spirit can dream o' nights, yet on the stage it is the first thing to aim at.
>
> —Dame Ellen Terry (1847–1928)
> Leading Shakespearean actress in Britain
>
> How use doth breed a habit in a man.
>
> —William Shakespeare

It is intriguing to find a great Shakespearean actress lauding *usefulness* as a matter of prime concern in her performances. Based on her musings about what she aspired to, usefulness concerned using anything and everything at her disposal to bring the play to life and connect with the audience. This is, perhaps, an artistic and evocative view of usefulness, but it also connotes a practical twist that makes for a provocative introduction to our final set of quality criteria: *pragmatic, utilization-focused criteria.*

- Observations of a high school cafeteria revealed substantial food waste. Interviews showed why the students were so dissatisfied with the food offered. The school had recently experienced an influx of immigrants from Asian countries, where people preferred rice rather than potatoes and bread. The results were used by school officials and the student council to advocate for more culturally appropriate food. Their efforts were successful.
- An early-childhood parent education program was experiencing a high dropout rate. Fewer than half the parents who started the program completed it. Interviews with the dropouts revealed that the program materials being used were academic and difficult for poorly educated parents to understand. Materials were revised to be more accessible and appropriate for parents with lower reading skills.
- The agricultural extension service serving a remote rural area in West Africa had very poor attendance at field trips aimed at helping farmers improve their basic growing practices for the subsistence crops sorghum and millet. Interviews with farmers revealed that they received no advance notice

DIVERSE METHODS BASED ON SYSTEMS AND QUALITY CONCEPTS

Systems and complexity concepts manifest nuances of difference under varying application frameworks:

- *System dynamics:* Focuses on the interrelationships between components of a situation, especially the consequences of feedback and delay
- *Viable systems:* Explores relationships that support an organization's viability within its environment
- *Soft systems methodology:* Looks at a situation from multiple viewpoints to understand and anticipate both interactions and unanticipated consequences
- *Critical systems heuristics:* Focuses on ethical issues, marginalization of people, and ideas of power and coercion
- *Activity systems:* Draws on cultural-historical activity theory to identify and track roles, tools, past features and dynamics, contradictions, tensions, conflicts, disturbances, innovations, processes, and learning opportunities
- *Complex adaptive systems:* Independent and interdependent elements or agents adapting to each other, self-organizing and emergent patterns, and nonlinear dynamics
- *Network analysis:* Examines dynamic interactions, connectivity, processes, and outcomes among a group or system of interconnected people or things. (Network Impact and Center for Evaluation Innovation, 2014)

SOURCES: Williams (2005) and Williams and Hummelbrunner (2011).

when the field training would occur. A radio news program agreed to announce extension field visits. Attendance increased significantly.

These are examples of simple inquiries aimed at providing practical and useful information to solve immediate problems. The pragmatic, utilization-focused criteria emphasize qualitative data generated to solve problems and inform decisions. This means focusing the inquiry on informing action and decisions. To be useful, specific intended users must be identified and their information needs met. Interactive engagement with intended users enhances relevance and use. Findings and feedback are timed to support use. Findings must be actionable and results understandable. The methods used need to be credible to those who will use the findings. Epistemologically, the orientation of pragmatic qualitative inquiry is that what is useful is true.

Pragmatic, utilization-focused inquiry begins with the premise that studies should be judged by their utility and actual use; therefore, evaluators and researchers should facilitate the inquiry process and design any study with careful consideration of how everything that is done, *from beginning to end*, will affect use. Use concerns how real people in the real world apply findings and experience the inquiry process. Therefore, the *focus* is on *intended use by intended users*. Since no study can be value-free, utilization-focused inquiry answers the question of whose values will frame the study by working with clearly identified, primary intended users who have the responsibility to apply findings and take action (Patton, 2008, 2012a).

PRAGMATIC EVALUATION STANDARDS

The evaluation profession has adopted standards that call for evaluations to be useful, practical, ethical, accurate, and accountable (Joint Committee on Standards, 2010). In the 1970s, as evaluation was just emerging as a field of professional practice, many evaluators took the position of traditional researchers that their responsibility was merely to design studies, collect data, and publish findings; what decision makers did with those findings was not their problem. This stance removed from the evaluator any responsibility for fostering use and placed all the "blame" for nonuse or underutilization on decision makers. Moreover, before the field of evaluation identified and adopted its own standards, criteria for judging evaluations could scarcely be differentiated from criteria for judging research in the traditional social and behavioral sciences, namely, technical quality and methodological rigor. Utility was largely ignored. Methods decisions dominated the evaluation design process. Validity, reliability, measurability, and generalizability were the dimensions that received the greatest attention in judging evaluation research proposals and reports. Indeed, evaluators concerned about increasing a study's usefulness often called for ever more methodologically rigorous evaluations to increase the validity of findings, thereby supposedly compelling decision makers to take findings seriously.

By the late 1970s, however, program staff and funders were becoming openly skeptical about spending scarce funds on evaluations that they couldn't understand and/or found irrelevant.

Evaluators were being asked to be "accountable," just as program staff were supposed to be accountable. The questions emerged with uncomfortable directness: Who will evaluate the evaluators? How will evaluation be evaluated? It was in this context that professional evaluators began discussing standards.

The most comprehensive effort at developing standards was hammered out over five years by a 17-member committee appointed by 12 professional organizations with input from hundreds of practicing evaluation professionals. Just prior to publication, Dan Stufflebeam, chair of the committee, summarized the results as follows:

> The standards that will be published essentially call for evaluations that have four features. These are *utility, feasibility, propriety* and *accuracy*. And I think it is interesting that the Joint Committee decided on that particular order. Their rationale is that an evaluation should not be done at all if there is no prospect for its being useful to some audience. Second, it should not be done if it is not feasible to conduct it in political terms, or practicality terms, or cost effectiveness terms. Third, they do not think it should be done if we cannot demonstrate that it will be conducted fairly and ethically. Finally, if we can demonstrate that an evaluation will have utility, will be feasible and will be proper in its conduct, then they said we could turn to the difficult matters of the technical adequacy of the evaluation. (Stufflebeam, 1980, p. 90)

High-Stakes Debate: What Counts as Credible Evidence, and by What Criteria Shall Credibility Be Judged?

The seven frameworks just reviewed show the range of criteria that can be brought to bear in judging a qualitative study. They can also be viewed as "angles of vision" or "alternative lenses" for expanding the possibilities available, not only for critiquing inquiry but also for undertaking it. What is most important to understand is that researchers and evaluators attending to and operating with any one of the seven different sets of quality criteria will (a) ask different questions, (b) use different methods, (c) follow different analytical processes, (d) report their findings in different ways, and (e) aim their claims of credibility to different audiences. These are not just academic distinctions. The differences are far from trivial. Quite the contrary, the different orientations have far-reaching implications for every aspect of inquiry. These different quality criteria constitute the underpinnings of significantly different ways of engaging in qualitative inquiry. At the heart of all scientific debate throughout history has been this burning question: *What counts as credible evidence and by what criteria shall credibility be judged?*

Nor is the debate about what counts as credible evidence just a matter of contention among scientists. Policymakers and politicians have gotten involved. It is down-and-dirty politics with millions of dollars in government-funded and philanthropic-sponsored research at stake (Denzin & Giardina, 2006, 2008; Donaldson, Christie, & Mark, 2008; Scriven, 2008). This means that advocates of qualitative inquiry must understand and be prepared to enter the debate about the politics of evidence (e.g., Eyben, 2013; Nutley et al., 2013; Schorr, 2012). In so doing, understanding the variety of approaches to qualitative inquiry, and which approaches are legitimate by what criteria, will become part of the debate.

So make no mistake about it, advocates of one particular set of criteria are likely to be vociferous critics of alternative criteria. Using the research process as an intervention to correct injustices and foment change is anathema to those who advocate traditional scientific research criteria as the only acceptable standards for judging quality. Those traditional criteria insist on a clear line of demarcation between studying a phenomenon (basic research and independent, external evaluation) versus engaging in change through the research process (advocacy). On the other hand, attempts to make traditional scientific research criteria the only legitimate approach to government-funded research are criticized as narrow-minded, self-serving political advocacy that constitutes "a conservative challenge to qualitative inquiry" (Denzin & Giardina, 2006, p. x). Constructivists generated their criteria of quality as a direct reaction to what they considered the gross inadequacies and methodological distortions of traditional scientific research criteria, which are essentially derived from the experimental/quantitative paradigm (see pp. 87–95). Thus, they systematically set out to replace traditional research criteria like validity and reliability with trustworthiness and authenticity (Lincoln & Guba, 1985, 1986). Advocates of artistic and evocative approaches attack both traditional research and constructivism as emotionally void. Traditional researchers have been disinclined to use participatory and collaborative approaches, sometimes believing that involving nonresearchers in research inevitably leads to poorer quality; in other cases, it's a matter of lacking incentives, capacity, or interest. Pragmatic, utilization-focused inquiries are attacked for being theoretically useless, unscholarly, and so practical as to be worthless for generating explanations or generalizations. Many traditional researchers don't even consider action research worthy of the name "research."

Choosing a Framework Within Which to Work

Which criteria you choose to emphasize in your work will depend on the purpose of your inquiry, the values and perspectives of the audiences for your work, and your own philosophical and methodological orientation. Operating within any particular framework and using any specific set of criteria will invite criticism from those who judge your work from a different framework and with different criteria. (For examples of the vehemence of such criticisms between those using traditional social science criteria and those using artistic narrative criteria, see Bochner, 2001; English, 2000.) Understanding that criticisms (or praise) flow from criteria can help you anticipate how to position your inquiry and make explicit what criteria to apply to your own work as well as what criteria to offer others given the purpose and orientation of your work.

The profession of program evaluation is a microcosm of these larger divisions. Program evaluation is a diverse, multifaceted profession manifesting many different models and approaches (Christie & Alkin, 2013; Fitzpatrick, Sanders, & Worthen, 2010; Funnell & Rogers, 2011; Patton, 2008; Stufflebeam, Madeus, & Kellaghan, 2000). All seven alternative quality criteria are advocated by various evaluation theorists, methodologists, and practitioners.

Any particular evaluation study has tended to be dominated by one set of criteria, with a second set as possibly secondary. For example, a primarily constructivist approach might add some artistic techniques as supporting methods. An evaluation dominated by the traditional scientific research approach might have a section dedicated to dealing with pragmatic issues. Exhibit 9.9 shows how the seven frameworks can be found in the approaches of various evaluation theorists, methodologists, and practitioners.

Clouds and Cotton: Mixing and Changing Perspectives

While each set of criteria manifest a certain coherence, many researchers mix and match approaches. The work of Tom Barone (2000), for example, combines aesthetic, political (critical change), and constructivist elements. Denzin's *Performance Ethnography* (2003) uses artistic and evocative approaches to foment and contribute to "radical social change, to economic justice, to a culture of politics that extends critical race theory and the principles of a radical democracy to all aspects of society" (p. 3). A team of evaluators collaborated to integrate constructivism, participatory evaluation, critical change, and a utilization focus (evaluations for improvement):

EXHIBIT 9.9 Alternative Quality Criteria Applied to Program Evaluation

Program evaluation is a diverse, multifaceted profession manifesting many different models and approaches (Alkin & Christie, 2013; Fitzpatrick, Sanders, & Worthen, 2010; Funnell & Rogers, 2011; Patton, 2008; Stufflebeam, Madeus, & Kellaghan, 2000). All seven alternative quality criteria are advocated by various evaluation theorists, methodologists, and practitioners.

QUALITY CRITERIA	PROGRAM EVALUATION FOCUS	LEADING CLASSIC TEXTS AND RESOURCES
1. Traditional scientific research criteria	Apply research methods to attribute documented outcomes to the intervention and generalize the findings.	Chen and Rossi (1987), Rossi, Lipsey, and Freeman (2004), and Silverman, Ricci, and Gunter (1990)
2. Social construction and constructivist criteria	Capture and report multiple perspectives on participants' experiences and diverse program outcomes.	Greene (1998, 2000), Guba and Lincoln (1981, 1989), Lincoln and Guba (1985), and Schwandt and Burgon (2006)
3. Artistic and evocative criteria	Connoisseurship evaluation: Use artistic representations to evoke participants' program experiences and judge a program's merit and worth.	Barone (2001, 2008), Eisner (1985, 1991), Knowles and Cole (2008), and Mathison (2009)
4. Participatory and collaborative criteria	Involve program staff and participants in evaluation to enhance use and build capacity for future evaluations.	Cousins and Chouinard (2012), Cousins and Earl (1992, 1995), King, Cousins, and Whitmore (2007), Greene (2006), and King and Stevahn (2013)
5. Critical change criteria	Use evaluation to address social justice, empower participants, and bring about change; support genuine democracy; and reduce power imbalances.	Fetterman (2000), Fetterman and Wandersman (2005), Fetterman, Kaftarian, and Wandersman (1996), Greene (2006), House and Howe (2000), Kirkhart (1995), and Mertens (1998, 1999, 2013)
6. Systems and complexity criteria	Understand programs through systems analysis and complexity concepts; support program innovation and adaptation; and evaluate systems change.	Eoyang and Holladay (2013), Jolley (2014), Mowles (2014), Patton (2011), Sterman (2006), Walton (2014), Williams and Hummelbrunner (2011), and Williams and Iman (2006)
7. Pragmatic, utilization-focused criteria	Get actionable answers to practical questions to support program improvement, guide problem solving, and enhance decision making, and ensure the utility and actual use of findings.	Alkin, Daillak, and White (1979), Davidson (2012), Patton (2008, 2012a), Rogers and Williams (2006), and Weiss (1977)

Evaluations for improvement, understanding lived experience, or advancing social justice are fundamentally participatory, involving key stakeholders in critical decisions about the evaluation's agenda, direction, and use. Such a principle is rooted epistemologically in the importance of understanding multiple perspectives and experiences in evaluation, and also politically in the importance of democratic inclusion.

—Whitmore et al. (2006, p. 341)

As an evaluator, I have worked with mixed criteria from all seven frameworks to match particular designs to the needs and interests of specific stakeholders and clients (Patton, 2008). But mixing and combining criteria means dealing with the tensions between them. After reviewing the tensions between traditional social science criteria and postmodern constructivist criteria, narrative researchers Lieblich, Tuval-Mashiach, and Zilber (1998) attempted "a middle course," but that middle course reveals the very tensions they were trying to supersede as they worked with one leg in each camp.

We do not advocate total relativism that treats all narratives as texts of fiction. On the other hand, we do not take narratives at face value, as complete and accurate representations of reality. We believe that stories are usually constructed around a core of facts or life events, yet allow a wide periphery for freedom of individuality and creativity in selection, addition to, emphasis on, and interpretation of these "remembered facts."...

Life stories are subjective, as is one's self or identity. They contain "narrative truth" which may be closely linked, loosely similar, or far removed from "historical truth."

Consequently, our stand is that life stories, when properly used, may provide researchers with a key to discovering identity and understanding it—both in its "real" or "historical" core, and as narrative construction. (p. 8)

Traditional scientific research criteria and critical change criteria are polar opposites. The same study cannot aspire to independence, objectivity, and a primary focus on contributing to theory while also being deeply engaged in using the inquiry process to foment change and ameliorate oppression. Mixing methods (qualitative and quantitative) is one thing. Mixing criteria of quality is a bit more challenging, one might even say daunting. Certainly, constructivist, artistic, and participatory criteria can be intermingled. But traditional scientific research criteria are less amenable to comingling.

The remainder of this chapter will elaborate some of the most prominent of these competing criteria that affect judgments about the quality and credibility of qualitative inquiry and analysis. But it's not always easy to tell whether someone is operating from a realist, constructionist, artistic, activist, or evaluative framework. Indeed, the criteria can shift quickly. Consider this example. My six-year-old son, Brandon, was explaining a geography science project he had done for school. He had created an ecological display out of egg cartons, ribbons, cotton, bottle caps, and styrofoam beads. "These are three mountains and these are four valleys," he said, pointing to the egg cup arrangement. "And is that a cloud?" I asked, pointing to the big hunk of cotton. He looked at me, disgusted, as though I've just said about the dumbest thing he's ever heard. "That's a piece of cotton, Dad."

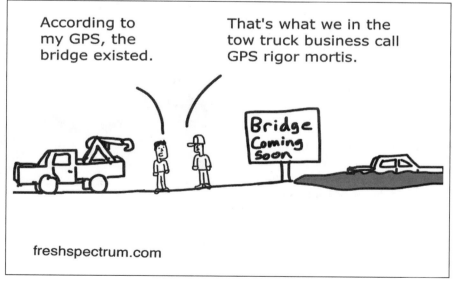

Foreshadowing research rigor mortis, MQP Rumination # 9 in the next module.

The previous modules in this chapter have reviewed strategies for enhancing the quality and credibility of qualitative analysis: selecting appropriate criteria for judging quality, searching for rival explanations, explaining negative cases, triangulation, and keeping data in context. Technical rigor in analysis is a major factor in the credibility of qualitative findings. This section now takes up the issue of how the credibility of the inquirer affects the way findings are received.

One barrier to credible qualitative findings stems from the suspicion that the analyst has shaped findings according to her or his predispositions and biases. Whether this may have happened unconsciously, inadvertently, or intentionally (with malice and forethought) is not the issue. The issue is how to counter such a suspicion before it takes root. One strategy involves discussing your predispositions and making biases explicit, to the extent possible. This involves systematic and studious reflexivity (see pp. 70–74). Another approach is engaging in mental cleansing processes (e.g., *epoche* in phenomenological analysis, p. 575). Or one may simply acknowledge one's orientation as a feminist researcher (Podems, 2014b) or critical theorist and move on from there. The point is that you have to address the issue of your credibility.

The Researcher as the Instrument in Qualitative Inquiry

Because the researcher is the instrument in qualitative inquiry, a qualitative report should include some information about you, the researcher. What experience, training, and perspective do you bring to the study? Who funded the study and under what arrangements with you? How did you gain access to the study site and the people observed and interviewed? What prior knowledge did you bring to the research topic and study site? What personal connections do you have to the people, program, or topic studied? For example, suppose the observer of an Alcoholics Anonymous program is a recovering alcoholic. This can either enhance or reduce credibility depending on how it has enhanced or detracted from data gathering and analysis. Either way, the analyst needs to deal with it in reporting findings. In a similar vein, it is only honest to report that the evaluator of a family counseling program was going through a difficult divorce at the time of fieldwork.

No definitive list of questions exists that must be addressed to establish investigator credibility. *The principle is to report any personal and professional information that may have affected data collection, analysis, and interpretation*—either negatively or positively—in the minds of users of the findings. For example, health status should be reported if it affected one's stamina in the field. Were you sick part of the time? Let's say that the fieldwork for evaluation of an African health project was conducted over three weeks, during which time the evaluator had severe diarrhea. Did that affect the highly negative tone of the report? The evaluator said it didn't, but I'd want to have the issue out in the open to make my own judgment. Background characteristics of the researcher (e.g., gender, age, race, and/or ethnicity) may be relevant to report in that such characteristics can affect how the researcher was received in the setting under study and what sensitivities the inquirer brings to the issues under study.

In preparing to interview farm families in Minnesota, I began building up my tolerance for strong coffee a month before the fieldwork. Being ordinarily not a coffee drinker, I knew my body would be jolted by 10 to 12 cups of coffee a day doing interviews in farm kitchens. In the Caribbean, I had to increase my tolerance for rum because some farmer interviews took place in rum shops. These are matters of personal preparation—both mental and physical—that affect perceptions about the quality of the study. Preparation and training for fieldwork, discussed at the beginning of Chapter 6, should be reported as part of the study's methodology.

Reflexivity and Intellectual Rigor

(Othello to Iago, interpreting what it means for someone to mutter something while sleeping)

> But this denoted a foregone conclusion.
>
> —William Shakespeare
> (Othello to Iago, interpreting what it means for someone to mutter something while sleeping)

The credibility of qualitative inquiry is so closely connected to the credibility of the person or team conducting the inquiry that the quality of reflexivity and reflectivity offered in a report is a window into the thinking processes that are the bedrock of qualitative analysis. Essentially, reflexivity involves turning qualitative analysis on yourself. Who are you, and how has how has who you are affected what you've found and reported in the study? This puts your intellectual rigor on display. The very notion of intellectual rigor connotes that as important as it is to employ systematic analytical strategies and techniques, the effectiveness and quality of those strategies and techniques depend on the quality of thinking that directs them. Which brings me to this chapter's rumination: Avoiding Research Rigor Mortis.

MQP Rumination # 9

Avoiding Research Rigor Mortis

I am offering one personal rumination per chapter. These are issues that have persistently engaged, sometimes annoyed, occasionally haunted, and often amused me over more than 40 years of research and evaluation practice. Here's where I state my case on the issue and make my peace.

Look for a pattern in what follows. See if you detect a theme.

Rigor (definition). Unyielding or inflexible; the quality of being extremely thorough, exhaustive, or accurate; being strict in conduct, judgment, and decision (*Oxford Dictionary*); scrupulous or inflexible accuracy or adherence (*Random House Dictionary*)

Measurement rigor. The underlying psychometric properties of a measure and its ability to fully and meaningfully capture the relevant construct; the fact that data have been collected in essentially the same manner, across time, the program, and jurisdictions, adds methodological rigor; the reliability and validity of instruments (Weitzman & Silver, 2012)

Research design rigor. The true experiment (randomized controlled trials) as the optimal (gold standard) design for developing evidence-based practice (Ross, Barkaoui, & Scott, 2007)

Methodological rigor. Design elements that support strong causal attributions and analytical generalization (Chatterji, 2007; Coryn, Schröter, & Hanssen, 2009)

Evaluation research rigor. Evidence testing the extent to which valid and reliable measures of program outcomes can be directly and confidently attributed to a standardized, high-fidelity, consistently implemented program intervention; the most rigorous evaluation is the randomized controlled trial (Chatterji, 2007; Henry, 2009; Ross et al., 2007; Rossi et al., 2004); "methodological rigor can be assessed from the evaluation plan and the quality of the evaluation's implementation" (Braverman, 2013, p. 101)

Analytical rigor. "Meticulous adherence to standard process . . .; scrupulous adherence to established standards for the conduct of work" (Zelik, Patterson, & Woods, 2007, p. 1)

Rigor mortis. Latin: *rigor* "stiffness," *mortis* "of death"—one of the recognizable signs of death, caused by chemical changes in the muscles after death, causing the limbs of the corpse to become stiff and difficult to move or manipulate

Research rigor mortis. Rigid designs, rigidly implemented, then rigidly analyzed through standardized, rigidly prescribed operating procedures and judged hierarchically by standardized, rigid criteria, thereby manifesting *rigorism* at every stage

Rigorism. Extreme strictness; no course may be followed that is contrary to doctrine (*Random House Dictionary*)

Research rigorism. Technicism—reducing research to "the application of techniques or the following of rules" (Hammersley, 2008b, p. 31)

Did you find the pattern? Did you detect a theme? Read on for the *countertheme.* (A countertheme is like a counterfactual: a theme that might be dominant, even should be dominant, in an alternate universe where the dominant theme is not so *dominant.*)

The Problem

"The Problem of Rigor in Qualitative Research"—that's the title of a classic article (Sandelowski, 1986) and a common refrain in textbooks about research methods. The "problem," it turns out, is that by traditional and dominant definitions of rigor, qualitative methods are inferior. But different criteria for what constitutes methodological quality lead to different judgments about rigor, the central point of this chapter. "The 'problem of multiple standards' describes the inherent difficulties in selecting which, among many viable candidates, is *the* standard process to which performance should be compared" (Zelik, Patterson, & Woods, 2007, p. 2). Rigor begets credibility. Different criteria for what constitutes methodological quality and rigor will yield different judgments about credibility. That much is straightforward.

The larger problem, it seems to me, is the focus on methods and procedures as the basis for determining quality and rigor. The notion that methods are more or less rigorous decouples methods from context and the thinking process that determined what questions to ask, what methods to use,

(Continued)

(Continued)

what analytical procedures to follow, and what inferences to draw from the findings. Avoiding *research rigor mortis* requires rigorous thinking.

Rigorous Thinking

> No problem can withstand the assault of sustained thinking.

> —Voltaire (1694–1778)
> French philosopher

Rigorous thinking combines (a) critical thinking, (b) creative thinking, (c) evaluative thinking, (d) inferential thinking, and (e) practical thinking. *Critical thinking* demands questioning assumptions; acknowledging and dealing with preconceptions, predilections, and biases; diligently looking for negative and disconfirming cases that don't fit the dominant pattern; conscientiously examining rival explanations; relentlessly seeking diverse perspectives; and analyzing what and how you think, why you think that way, and the implications for your inquiry (Kahneman, 2011; Klein, 2011; Loseke, 2013).

Creative thinking invites putting the data together in new ways to see the interactions among separate findings more holistically; synthesizing diverse themes in a search for coherence and essence while simultaneously developing comfort with ambiguity and uncertainty in the messy, complex, and dynamic real work; distinguishing signal from noise while also learning from the noise; asking wicked questions that enter into the intersections and tensions between the search for coherent meaning and persistent uncertainties and ambiguities; bringing artistic, evocative, and visualization techniques to data analysis and presentations; and inviting outside-the-box, off-the-wall, and beyond-the-ken perspectives and interpretations.

Evaluative thinking forces clarity about the inquiry purpose, who it is for, with what intended uses, to be judged by what quality criteria; it involves being explicit about what criteria are being applied in framing inquiry questions, making design decisions, determining what constitutes *appropriate* methods, and selecting and following analytical processes and being aware of and articulating values, ethical considerations, contextual implications, strengths and weaknesses of the inquiry, and potential (or actual) misinterpretations, misuses, and misapplications. In contrast with the perspective of rigor as strict adherence to a standardized process, evaluative thinking emphasizes the importance of understanding the sufficiency of rigor relative to context and situational factors (Clarke, 2005; Patton, 2012a).

Inferential thinking involves examining the extent to which the evidence supports the conclusions reached. Inferential

thinking can be deductive, inductive, or abductive—and often draws on and creatively integrates all three analytical processes—but at the core, it is a fierce examination of and allegiance to where the evidence leads.

> A rigorously conducted evaluation will be convincing as a presentation of evidence in support of an evaluation's conclusions and will presumably be more successful in withstanding scrutiny from critics. Rigor is multifaceted and relates to multiple dimensions of the evaluation.... The concept of rigor is understood and interpreted within the larger context of validity, which concerns the "soundness or trustworthiness of the inferences that are made from the results of the information gathering process" (Joint Committee on Standards for Educational Evaluation, 1994, p. 145).... There is relatively broad consensus that validity is a property of an inference, knowledge claim, or intended use, rather than a property either of a research or evaluation study from the study's findings. (Braverman, 2013, p. 101)

In reflecting on and writing about "what counts as credible evidence in applied research and evaluation practice," Sharon Rallis (2009), former president of the AEA and experienced qualitative researcher, emphasized rigorous reasoning: "I have come to see *a true scientist* [italics added], then, as one who puts forward her findings and the reasoning that led her to those findings for others to contest, modify, accept, or reject" (p. 171).

Practical thinking calls for assiduously integrating theory and practice, examining real-world implications of findings, inviting interpretations and applications from nonresearchers (e.g., community members, program staff, and participants) who can and will apply to the data what ordinary people refer to as "common sense"; and applying real-world criteria to interpreting the findings, criteria like understandability, meaningfulness, cost implications, and implications in addressing societal issues and problems.

What's at Stake?

> My words fly up, my thoughts remain below:

> Words without thoughts, never to heaven go.

> —William Shakespeare (1564–1616)
> The king in *Hamlet*

As I noted in Chapter 4, and is worth repeating here, philosopher Hannah Arendt (1968) concluded that to resist efforts by the powerful to deceive and control thinking, people need to practice thinking: "Experience in thinking . . . can be won, like all experience in doing something, only through practice, through exercises" (p. 4).

Regardless of what one thinks of the U.S. invasion of Iraq to depose Saddam Hussein in 2003, both those who supported the war and those who opposed it ultimately agreed that the intelligence used to justify the invasion was deeply flawed and systematically distorted (U.S. Senate Select Committee on Intelligence, 2004). Under intense political pressure to show sufficient grounds for military action, those charged with analyzing and evaluating intelligence data began doing what is sometimes called cherry-picking or stove-piping—selecting and passing on only those data that support preconceived positions and ignoring or repressing all contrary evidence (Hersh, 2003; Tan, 2014; Zelik et al., 2007). The failure of the intelligence community to appropriately and accurately assess whether Iraq had weapons of mass destruction was not a function of poor data but of weak analysis, political manipulation of the analysis process, and a fundamental failure to think critically, creatively, evaluatively, and practically. The generation of the Rigor Attribute Model to support more rigorous intelligence analysis and restore credibility to the intelligence community focuses on *rigorous thinking* (Zelik et al., 2007; see Exhibit 9.5, pp. 675–677).

> Despite the etymological implication that to be rigorous is to "be stiff," expert information analysis processes often are not rigid in their application of a standard process, but rather, flexible and adaptive to highly dynamic environments. In information analysis, judgment of rigor reflects a relationship in the appropriateness of fit between analytic processes and contextual requirements. Thus, as supported by this and other research, rigor is more meaningfully viewed as an assessment of degree of sufficiency, rather than degree of adherence to an established analytic procedure. (Zelik, Patterson, & Woods, 2007, p. 1)

The phrase "degree of sufficiency" as a criterion for assessing rigor refers to an evaluation of the extent to which a multidimensional, multiperspectival, and critical thinking process was followed determinedly to yield conclusions that best fit the data, and therefore findings that are credible to and inspire confidence among those who must use the findings.

Bottom-Line Conclusion

Methods do not ensure rigor. A research design does not ensure rigor. Analytical techniques and procedures do not ensure rigor. Rigor resides in, depends on, and is manifest in *rigorous thinking*—about everything, including methods and analysis.

The thread that runs through this rumination is the importance of intellectual rigor. There are no simple formulas or clear-cut rules about how to do a credible, high-quality analysis. The task is to do one's best to make sense of things. A qualitative analyst returns to the data over and over again to see if the constructs, categories, interpretations, and explanations make sense—if they sufficiently reflect the nature of the phenomena studied. Creativity, intellectual rigor, perseverance, insight—these are the intangibles that go beyond the routine application of scientific procedures. It is worth quoting again Nobel prize–winning physicist Percy Bridgman: "There is no scientific method as such, but the vital feature of a scientist's procedure has been merely to do his utmost with his mind, *no holds barred*" (quoted in Waller, 2004, p. 106).

Varieties of and Concerns About Reactivity: How What We See and Do Affects What Is Seen and Done

Nasrudin denied that he was a fisherman. From a passing tourist he had heard of something called philanthropy and, feeling transformed by what he had learned, he instantly adopted the moniker for himself. He explained to his fellow villagers: "When we see a problem that needs solving, it is wrong to just stand by and observe as scholars are wont to do. We must react. It is wrong to remain passive and detached in the face of need and noble to render help."

"I am a philanthropist. Each day I strive to help fish that are drowning in the lake. I save them. I throw out my net and the fish rush in. I quickly put the many fish I've rescued on the dry ground, where they dance about in joy. But the dancing soon exhausts them and before long they cease to move. Alas, they dance themselves to death."

"It is sad, but it is also wrong not to honor their struggle. So I take the dead fish to market where people contribute money to my effort to save more fish in exchange for my gifts to them of those fish who have lost the struggle. With the financial tokens of appreciation I receive for my charitable work, I purchase more nets so I can rescue more fish."

—From Halcolm's *Chronicles of Lessons Learned: Teach a Man to Fish*

IN-DEPTH REFLEXIVITY: GUIDELINES FOR QUALITY IN AUTOBIOGRAPHICAL FORMS OF SELF-STUDY RESEARCH

- Autobiographical self-studies should ring true and enable connection.
- Self-studies should promote insight and interpretation.
- Autobiographical self-study research must engage history forthrightly, and the author must take an honest stand.
- Authentic voice is a necessary but not sufficient condition for the scholarly standing of a biographical self-study.
- The autobiographical self-study researcher has an ineluctable obligation to seek to improve the learning situation not only for the self but also for the other.
- Powerful autobiographical self-studies portray character development and include dramatic action: Something genuine is at stake in the story.
- Quality autobiographical self-studies attend carefully to persons in context or setting.
- Quality autobiographical self-studies offer fresh perspectives on established truths.
- To be scholarship, edited conversation or correspondence must not only have coherence and structure but that coherence and structure should also provide argumentation and convincing evidence.
- Interpretations made of self-study data should not only reveal but also interrogate the relationships, contradictions, and limits of the views presented (adapted from Bullough & Pinnegar, 2001, pp. 13–21).

Considering and Reporting Investigator Effects: Varieties of Reactivity

Reflectivity includes considering and reporting how your presence as an observer or evaluator may have affected what you observed. There are four primary ways in which the presence of an outside observer, or the fact that an evaluation is taking place, can affect, and possibly distort, the findings of a study, namely,

1. reactions of those in the setting (e.g., program participants and staff) to the presence of the qualitative fieldworker;

2. changes in you, the fieldworker (the measuring instrument), during the course of the data collection or analysis—that is, what has traditionally been called instrumentation effects in quantitative measurement;

3. the predispositions, selective perceptions, and/or biases you might bring to the inquiry that become evident to others during data collection; and

4. researcher incompetence (including lack of sufficient training or preparation).

Reactivity

> All accounts produced by researchers must be interpreted within the context in which they were generated. Interpretations must examine, as carefully as possible, how the presence of the researcher, the context in which data were obtained, and so on shaped the data.
>
> —Schwandt (2007, p. 256)

Problems of reactivity are well documented in the anthropological literature, which is one of the prime reasons why qualitative methodologists advocate long-term observations that permit an initial period during which observers and the people in the setting being observed get a chance to get used to each other. This increases trustworthiness, which supports credibility both within and outside the study setting.

> The credibility of your findings and interpretations depend upon your careful attention to establishing trustworthiness. . . . Time is a major factor in the acquisition of trustworthy data. Time at your research site, time spent interviewing, and time building sound relationships with respondents all contribute to trustworthy data. When a large amount of time is spent with your research participants, they less readily feign behavior or feel the need to do so; moreover, they are more likely to be frank and comprehensive about what they tell you. (Glesne, 1999, p. 151)

On the other hand, prolonged engagement may actually increase reactivity as the researcher becomes more a part of the setting and begins to affect what goes on through prolonged engagement. Thus, whatever the length of inquiry or method of data collection, researchers have an obligation to examine how their presence affects what goes on and what is observed.

> It is axiomatic that observers must record what they perceive to be their own reactive effects. They may treat this reactivity as bad and attempt to avoid it (which is impossible), or they may accept the fact that they will have a reactive effect and attempt to use it to advantage. . . . The reactive effect will be measured by daily

field notes, perhaps by interviews in which the problem is pointedly inquired about, and also in daily observations. (Denzin, 1978b, p. 200)

Anxieties that surround an evaluation can exacerbate reactivity. The presence of an evaluator can affect how a program operates as well as its outcomes. The evaluator's presence may, for example, create a halo effect so that staff perform in an exemplary fashion and participants are motivated to "show off." On the other hand, the presence of the evaluator may create so much tension and anxiety that performances are below par. Some forms of program evaluation, especially "empowerment evaluation" and "intervention-oriented evaluation," (Patton, 2008, chap. 5) turn this traditional threat to validity into an asset by designing data collection to enhance achievement of the desired program outcomes. For example, at the simplest level, the observation that "what gets measured gets done" suggests the power of data collection to affect outcomes attainment. A leadership program, for example, that includes in-depth interviewing and participant journal writing as ongoing forms of evaluation data collection may find that participating in the interviewing and writing reflectively have effects on participants' learning and program outcomes. Likewise, a community-based AIDS awareness intervention can be enhanced by having community participants actively engaged in identifying and doing case studies of critical community incidents. In short, a variety of reactive responses are possible, some that support program processes, some that interfere, and many that have implications for interpreting findings. Thus, the evaluator has a responsibility to think about the problem, make a decision about how to handle it in the field, attempt to monitor evaluator/observer effects, and reflect on how reactivities may have affected the findings.

Evaluator effects can be overrated, particularly by evaluators. There is more than a slight touch of self-importance in some concerns about reactivity. Lillian Weber, director of the Workshop Center for Open Education, City College School of Education, New York, once set me straight on this issue, and I pass her wisdom on to my colleagues. In doing observations of open classrooms, I was concerned that my presence, particularly the way kids flocked around me as soon as I entered the classroom, was distorting the evaluation to the point where it was impossible to do good observations. Lillian laughed and suggested to me that what I was experiencing was the way those classrooms actually were. She went on to note that this was common among visitors to schools; they were always concerned that the teacher, knowing visitors were coming, whipped the kids into shape for those visitors. She suggested that under the best of

circumstances a teacher might get kids to move out of habitual patterns into some model mode of behavior for as much as 10 or 15 minutes but that, habitual patterns being what they are, the kids would rapidly revert to normal behaviors and whatever artificiality might have been introduced by the presence of the visitor would likely become apparent.

Evaluators and researchers should strive to neither overestimate nor underestimate their effects but to take seriously their responsibility to describe and study what those effects are.

Effects on the Inquirer of Being Engaged in the Inquiry

A second form of reactivity arises from the possibility that the researcher or evaluator changes during the course of the inquiry. In Chapter 7, on interviewing, I offered several examples of this, including how in a study of child sexual abuse, those involved were deeply affected by what they heard. One of the ways this sometimes happens in anthropological research is when participant observers "go native" and become absorbed into the local culture. The epitome of this in a short-term observation is the legendary story of the student observers who became converted to Christianity while observing a Billy Graham evangelical crusade (Lang & Lang, 1960). Evaluators sometimes become personally involved with program participants or staff and therefore lose their sensitivity to the full range of events occurring in the setting.

Johnson (1975) and Glazer (1972) have reflected on how they and others have been changed by doing field research. The consensus of advice on how to deal with the problem of changes in observers as a result of involvement in research is similar to advice about how to deal with the reactive effects created by the presence of observers.

> It is central to the method of participant observation that changes will occur in the observer; the important point, of course, is to record these changes. Field notes, introspection, and conversations with informants and colleagues provide the major means of measuring this dimension, . . . for to be insensitive to shifts in one's own attitudes opens the way for placing naive interpretations on the complex set of events under analysis. (Denzin, 1978b, p. 200)

Inquirer-Selective Perception and Predispositions

The third concern about inquirer effects related to credibility has to do with the extent to which the predispositions or biases of the inquirer may affect data analysis and interpretations. This issue carries mixed messages because, on the one hand, rigorous data collection and analytical procedures, like triangulation, are aimed at substantiating the credibility of the findings and minimizing inquirer biases and, on the other, the interpretative and constructivist perspectives remind us that data from and about humans inevitably represent some degree of perspective rather than absolute truth. Getting close enough to the situation observed to experience it firsthand means that researchers can learn from their experiences, thereby generating personal insights; but that closeness makes their objectivity suspect. "For social scientists to refuse to treat their own behavior as data from which one can learn is really tragic" (Scriven, 1972a, p. 99). In effect, all of the procedures for validating and verifying analysis that have been presented in this chapter are aimed at reducing distortions introduced by inquirer predisposition. Still, people who use different criteria in determining evidential credibility will come at this issue from different stances and end up with different conclusions.

Consider the interviewing stance of *emphatic neutrality* introduced in Chapter 2 and elaborated in Chapter 7. An emphatically neutral inquirer will be perceived as caring about and interested in the people being studied but neutral about the content of what they reveal. House (1977) balances the caring, interested stance against independence and impartiality for evaluators, a stance that also applies to those working according to the standards of traditional science.

> The evaluator must be seen as caring, as interested, as responsive to the relevant arguments. He must be impartial rather than simply objective. The impartiality of the evaluator must be seen as that of an actor in events, one who is responsive to the appropriate arguments but in whom the contending forces are balanced rather than non-existent. The evaluator must be seen as not having previously decided in favor of one position or the other. (pp. 45–46)

But neutrality and impartiality are not easy stances to achieve. Denzin (1989b) cites a number of scholars who have concluded, as he does, that every researcher brings preconceptions and interpretations to the problem being studied, regardless of the methods used.

> All researchers take sides, or are partisans for one point of view or another. Value-free interpretive research is impossible. This is the case because every researcher brings preconceptions and interpretations to the problem being studied. The term *hermeneutical circle or situation* refers to this basic fact of research. All scholars are caught in the circle of interpretation. They can never

be free of the hermeneutical situation. This means that scholars must state beforehand their prior interpretations of the phenomenon being investigated. Unless these meanings and values are clarified, their effects on subsequent interpretations remain clouded and often misunderstood. (p. 23)

Earlier I presented seven sets of criteria for judging the quality of qualitative inquiry (Exhibit 9.7, pp. 680–681). Those varying and competing frameworks offer different perspectives on how inquirers should deal with concerns about bias. Neutrality and impartiality are expected when qualitative work is being judged by traditional scientific criteria or by evaluation standards, thus the source of House's (1977) admonition quoted above. In contrast, constructivist analysts are expected to deal with these issues through conscious and committed reflexivity—entering the *hermeneutical circle of interpretation* and therein reflecting on and analyzing how their perspective interacts with the perspectives they encounter. Artistic inquirers often deal with issues of how they personally relate to their work by invoking aesthetic criteria: Judge the work on its *artistic merits*. Participatory and collaborative inquiries encourage the formation of meaningful and trusting relationships between researchers and those participating in the inquiry. When critical change criteria are applied in judging reactivity, the issue becomes whether, how, and to what extent the inquiry furthered the cause or enhanced the well-being of those involved and studied; neutrality is eschewed in favor of explicitly using the inquiry process to facilitate change, or at least illuminate the conditions needed for change.

Inquirer Competence

Concerns about the extent to which the inquirer's findings can be trusted—that is, trustworthiness—can be understood as one dimension of perceived methodological rigor. But ultimately, for better or worse, the trustworthiness of the data is tied directly to the trustworthiness of those who collect and analyze the data—and their demonstrated competence. Competence is demonstrated by using the verification and validation procedures necessary to establish the quality of analysis and thereby building a "track record" of quality work. As Exhibit 9.10 shows, inquirer competence includes not just systematic inquiry knowledge and skill but also interpersonal competence, reflective practice skills, situational analysis, professional practice competence, and project management. This array of competencies is being acknowledged and certified by professional evaluation associations around the world (King & Podems, 2014; Podems, 2014a). Consistent with the overall message of this chapter, especially my MQP Rumination on avoiding research rigor mortis, thinking skills also need ongoing development. An excellent resource in that regard is the *Critical Evaluation Skills Toolkit* (Crebert, Patrick, Cragnolini, Smith, Worsfold, & Webb, 2011).

EXHIBIT 9.10 **The Multiple Dimensions of Program Evaluator Competence**

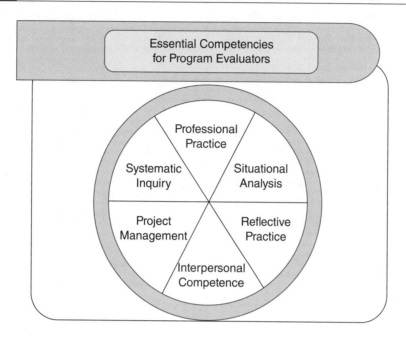

SOURCES: Ghere, King, Stevahn, and Minnema (2006) and King, Stevahn, Ghere, and Minnema (2001).

The principle for dealing with inquirer competence is this: Don't wait to be asked. Anticipate competence as an issue. Address the issue of competence proactively, explicitly, and multidimensionally. With quantitative methods, validity and reliability reside in tools, instruments, design parameters, and procedures. In qualitative inquiry, the competency stakes are greater because the inquirer is the instrument. Trustworthiness and authenticity are functions of systematic inquiry procedures, interpersonal (relational) dynamics in the field, and competency to engage in the challenges and deal with the ambiguities of qualitative inquiry.

Review: The Credibility of the Inquirer

Because the researcher is the instrument in qualitative inquiry, the credibility of the inquirer is central to the credibility of the study. Exhibit 9.11 on the next page summarizes the issues that arise in establishing and judging the credibility of the inquirer.

EXHIBIT 9.11   The Credibility of the Inquirer: Issues and Solutions

CREDIBILITY CONCERN	WHAT'S THE ISSUE?	WAYS TO ADDRESS THE ISSUES AND ENHANCE CREDIBILITY	EXAMPLES FROM NORA MURPHY'S (2014) STUDY OF HOMELESS YOUTH
1. Who did the study? Who is the inquirer?	Because the researcher is the instrument in qualitative inquiry, who did the work and carried out the analysis matters.	The methodology section of the report should present not only the usual design and data collection details and rationale but also description of the inquirer's relevant experiences, training, perspective, competence, and purpose.	A University of Minnesota doctoral student on the staff of the Minnesota Evaluation Studies Institute, who has completed doctoral studies in evaluation theory, methods, and practice and is being supervised by experienced and knowledgeable researchers and evaluators, did the study.
2. Reflexivity	How has the inquirer's background and perspective affected the findings?	Reflexivity goes beyond reporting background, experience, and training; it involves reflecting on and reporting your reflexive process and the answers to reflexive questions: How do you know what you know? What shapes and has shaped your perspective? (See Exhibit 2.5, p. 72.)	"I am a constructivist working from a systems perspective. My evaluation approach is *utilization focused*, and I'm using *developmental evaluation* because it fits the dynamics, complexities, and developmental nature of the initiative being evaluated. I have worked as a teacher and program staff member with disadvantaged youth."
3. Potential inquirer bias	How might the findings be a function of the inquirer's selective perception, predispositions, and bias? What steps have been taken to deal with potential bias?	*Options (not mutually exclusive)* a. Acknowledge potential sources of bias: What brought you to this inquiry? Why do you care about what you're studying? What are the implications of caring? How do you deal with concerns about bias (which you acknowledge as legitimate)? b. Acknowledge your perspective and present it as a strength: "I am a constructivist and view getting close enough to people to experience empathy as a strength of in-depth fieldwork and interviewing." c. Describe your process for surfacing and setting aside any preconceptions (e.g., *epoche* in phenomenological analysis). d. Subject your analysis to independent review (analyst triangulation, peer review, or external audit).	"I care about homeless youth and believe they deserve an opportunity to move on in their life's journey past their period of homelessness. I believe in and subscribe to the values and principles expressed by the programs participating in the evaluation. I want to help them better elucidate and implement those principles. I also want evaluation to be a vehicle for giving voice to homeless young people, to honor their stories, and help them articulate what they've experienced." The study is being done collaboratively with six youth-serving agencies, which have monitored the appropriateness and integrity of the data collection and analysis. The study is supervised by experienced researchers and evaluators who monitor the integrity of the methods and analysis.

CREDIBILITY CONCERN	WHAT'S THE ISSUE?	WAYS TO ADDRESS THE ISSUES AND ENHANCE CREDIBILITY	EXAMPLES FROM NORA MURPHY'S (2014) STUDY OF HOMELESS YOUTH
4. Reactivity	How has the inquiry affected the people in the setting studied?	*Options (not mutually exclusive)* Demonstrate awareness of the issue and take it seriously: a. Keep field notes on your observations about how your presence may have, or actually did appear, to affect things. Describe effects and their implications for your findings. b. Gather data about reactions; ask key informants how your presence affected the setting observed and people interviewed.	a. Youth interviewed were compensated for their time. They reviewed and approved the case studies created from their interview transcripts. They expressed appreciation for the opportunity to tell their stories. b. The six participating agencies report having strengthened their collaboration with each other as a result of being part of this study (which was the intent). They reported learning from the experience, feeling that their work and approach was validated, and are using the findings for staff development in their agencies.
5. Effects on the inquirer of involvement in the inquiry	How were you affected or changed by engaging in this inquiry?	Reflect on and report what you've seen and how it has affected you. Acknowledge emotional responses and any actions taken. Acknowledge that as the research instrument, you are also a human being—and report honestly your human responses.	"This study took a personal toll. There were times when I went home and cried. I felt guilt that I could not help them more and fear that I was exploiting them. What helped me was that listening seemed to help them. I still carry some of the sadness that I experienced as I sat with the youth and their telling of their lives, but I also carry the hope that I felt when I experienced their optimism and their strength."
6. Competence	How can I, the reader and user of your findings, be assured of your competence to undertake this inquiry?	Acknowledge the importance of competence and its multiple dimensions (see Exhibit 9.10), and report on your competence in these areas.	The entire collaboration engaged in reflective practice together. Confidentiality, rapport, and trust were essential in interviewing the youth. Sensitivity to race, gender orientation, and the trauma experienced by homeless youth were monitored by the participating agencies. The methods section of the study addresses these and related issues in depth.

Generalizations, Extrapolations, Transferability,
Principles, and Lessons Learned

> The trouble with generalizations is that they don't apply to particulars.
>
> —Lincoln and Guba (1985, p. 110)

Credibility and utility are linked. What can one do with qualitative findings? The results illuminate a particular situation or small number of cases. But can qualitative findings be generalized? Here, again, different qualitative frameworks based on different criteria offer different answers. The traditional scientific research criteria include generalizability. Constructivist criteria, in contrast, emphasize particularity; constructivists generally eschew, and are skeptical about, generalizability. They offer *extrapolations* and *transferability* instead. So let's see if we can sort out these different perspectives and their implications.

Purposeful Sampling and Generalizability

Chapter 5 discussed the logic and value of purposeful sampling with small but carefully selected information-rich cases. Certain kinds of small samples and qualitative studies are designed for generalizability and broader relevance: a critical case, an index case, a causal pathway sample, a positive deviance case, and a qualitative synthesis review are examples (see Exhibit 5.8, pp. 266–272). Other sampling strategies, for example, outlier cases (exemplars of excellence or failure), a high-impact case, sensitizing concept exemplars, and principles-focused sampling, aim to yield insights about principles that might be adapted for application elsewhere. In short, the conditions for, possibility of, and relative importance attached to generalizability are determined at the design stage. *To review:* Purpose drives design. Design drives data collection. Data drive analysis. Purpose, design, data, and analysis, in combination, determine generalizability.

Principles of Generalizability

Shadish (1995a) has made the case that certain core principles of generalization apply to both experiments and ethnographies (or qualitative methods generally). Both experiments and case studies share the problem of being highly localized. Findings from a study,

experimental or naturalistic in design, can be generalized according to five principles:

1. *The principle of proximal similarity:* We generalize most confidently to applications where treatments, settings, populations, outcomes, and times are most similar to those in the original research. . . .

2. *The principle of heterogeneity of irrelevancies:* We generalize most confidently when a research finding continues to hold over variations in persons, settings, treatments, outcome measures, and times that are presumed to be conceptually irrelevant. The strategy here is identifying irrelevancies, and where possible including a diverse array of them in the research so as to demonstrate generalization over them. . . .

3. *The principle of discriminant validity:* We generalize most confidently when we can show that it is the target construct, and not something else, that is necessary to produce a research finding. . . .

4. *The principle of empirical interpolation and extrapolation:* We generalize most confidently when we can specify the range of persons, settings, treatments, outcomes, and times over which the finding holds more strongly, less strongly, or not all. The strategy here is empirical exploration of the existing range of instances to discover how that range might generate variability in the finding for instances not studied. . . .

5. *The principle of explanation:* We generalize most confidently when we can specify completely and exactly (a) which parts of one variable (b) are related to which parts of another variable (c) through which mediating processes (d) with which salient interactions, for then we can transfer only those essential components to the new application to which we wish to generalize. The strategy here is breaking down the finding into component parts and processes so as to identify the essential ones. (pp. 424–426)

Generalizability Versus Contextual Particularity

Deep philosophical and epistemological issues are embedded in concerns about generalizing. What's desirable or hoped for in science (generalizations across

CULTURAL LIMITS ON GENERALIZABILITY

Psychological experiments have been used to study how people react to things like negotiating rewards and perceptions of whether two lines are of equal length when phony participants in the experiment say that the shorter line is really longer. The results of most such laboratory research have been interpreted as showing "evolved psychological traits common to all humans" (Watters, 2013, p. 1). However, when such experiments are repeated in other cultures, the ways in which Americans respond can be quite different from how nonliterate peoples respond. What were once thought to be tests of basic perception (how the brain works) have turned out to be culturally determined. Social scientists had assumed that lab experiments studied

> the human mind stripped of culture, [that] the human brain is genetically comparable around the globe, it was agreed, so human hardwiring for much behavior, perception, and cognition should be similarly universal. No need, in that case, to look beyond the convenient population of undergraduates for test subjects. A 2008 survey of the top six psychology journals dramatically shows how common that assumption was: more than 96 percent of the subjects tested in psychological studies from 2003 to 2007 were Westerners—with nearly 70 percent from the United States alone. Put another way: 96 percent of human subjects in these studies came from countries that represent only 12 percent of the world's population. (Watters, 2013, p. 1)

Cross-cultural research is now revealing that

the mind's capacity to mold itself to cultural and environmental settings was far greater than had been assumed. The most interesting thing about cultures may not be in the observable things they do—the rituals, eating preferences, codes of behavior, and the like—but in the way they mold our most fundamental conscious and unconscious thinking and perception. (Watters, 2013, p. 1)

Moreover, the experiments done on American undergraduate students may be especially prone to inappropriate overgeneralizations.

> It is not just our Western habits and cultural preferences that are different from the rest of the world, it appears. The very way we think about ourselves and others—and even the way we perceive reality—makes us distinct from other humans on the planet, not to mention from the vast majority of our ancestors. Among Westerners, the data showed that Americans were often the most unusual, leading the researchers to conclude that "American participants are exceptional even within the unusual population of Westerners—outliers among outliers."

> Given the data, they concluded that social scientists could not possibly have picked a worse population [American undergraduate students)] from which to draw broad generalizations. Researchers had been doing the equivalent of studying penguins while believing that they were learning insights applicable to all birds. (Watters, 2013, p. 1)

time and space) runs into real-world considerations about what's possible. Lee J. Cronbach (1975), one of the major figures in psychometrics and research methodology in the twentieth century, devoted considerable attention to the issue of generalizations. He concluded that social phenomena are too variable and context bound to permit very significant empirical generalizations. He compared generalizations in natural sciences with what was likely to be possible in behavioral and social sciences. His conclusion was that "generalizations decay. At one time a conclusion describes the existing situation well, at a later time it accounts for rather little variance, and ultimately is valid only as history" (p. 122). Cronbach (1975) offers an alternative to generalizing that constitutes excellent advice for the qualitative analyst:

> Instead of making generalization the ruling consideration in our research, I suggest that we reverse our priorities. An observer collecting data in a particular situation

is in a position to appraise a practice or proposition in that setting, observing effects in context. In trying to describe and account for what happened, he will give attention to whatever variables were controlled, but he will give equally careful attention to uncontrolled conditions, to personal characteristics, and to events that occurred during treatment and measurement. As he goes from situation to situation, his first task is to describe and interpret the effect anew in each locale, perhaps taking into account factors unique to that locale or series of events.... When we give proper weight to local conditions, any generalization is a working hypothesis, not a conclusion. (pp. 124–125)

Robert Stake (1978, 1995, 2000, 2006, 2010), master of the case study, concurs with Cronbach that the first priority is to do justice to the specific case, to do a good job of "particularization" before looking for patterns across cases. He quotes William Blake on the subject:

To generalize is to be an idiot. To particularize is the lone distinction of merit. General knowledges are those that idiots possess.

Stake (1978) continues,

Generalization may not be all that despicable, but particularization does deserve praise. To know particulars fleetingly, of course, is to know next to nothing. What becomes useful understanding is a full and thorough knowledge of the particular, recognizing it also in new and foreign contexts. That knowledge is a form of generalization too, not scientific induction but naturalistic generalization, arrived at by recognizing the similarities of objects and issues in and out of context and by sensing the natural covariations of happenings. To generalize this way is to be both intuitive and empirical, and not idiotic. (p. 6)

Stake (2000) extends *naturalistic generalizations* to include the kind of learning that readers take from their encounters with specific case studies. The "vicarious experience" that comes from reading a rich case account can contribute to the social construction of knowledge, which, in a cumulative sense, builds general, if not necessarily generalizable, knowledge.

Readers assimilate certain descriptions and assertions into memory. When researcher's narrative provides opportunity for vicarious experience, readers extend their memories of happenings. Naturalistic, ethnographic case materials, to some extent, parallel actual experience, feeding into the most fundamental processes of awareness and understanding . . . [to permit] *naturalistic generalizations.* The reader comes to know some things told, as if he or she had experienced it. Enduring meanings come from encounter, and are modified and reinforced by repeated encounter.

In life itself, this occurs seldom to the individual alone but in the presence of others. In a social process, together they bend, spin, consolidate, and enrich their understandings. We come to know what has happened partly in terms of what others reveal as their experience. The case researcher emerges from one social experience, the observation, to choreograph another, the report. Knowledge is socially constructed, so we constructivists believe, and, in their experiential and contextual accounts, case study researchers assist readers in the construction of knowledge. (p. 442)

Guba (1978) considered three alternative positions that might be taken in regard to the generalizability of naturalistic inquiry findings:

1. Generalizability is a chimera; it is impossible to generalize in a scientific sense at all. . . .

2. Generalizability continues to be important, and efforts should be made to meet normal scientific criteria that pertain to it. . . .

3. Generalizability is a fragile concept whose meaning is ambiguous and whose power is variable. (pp. 68–70)

Having reviewed these three positions, Guba (1978) proposed a resolution that recognizes the diminished value and changed meaning of generalizations and echoes Cronbach's emphasis, cited above, on treating conclusions as hypotheses for future applicability and testing rather than as definitive.

The evaluator should do what he can to establish the generalizability of his findings. . . . Often naturalistic inquiry can establish at least the "limiting cases" relevant to a given situation. But in the spirit of naturalistic inquiry he should regard each possible generalization only as a working hypothesis, to be tested again in the next encounter and again in the encounter after that. For the naturalistic inquiry evaluator, premature closure is a cardinal sin, and tolerance of ambiguity a virtue. (p. 70)

Guba and Lincoln (1981) emphasized appreciation of and attention to context as a natural limit to naturalistic generalizations. They ask, "What can a generalization be except an assertion that is context free? [Yet] *it is virtually impossible to imagine any human behavior that is not heavily mediated by the context in which it occurs*" (p. 62). They proposed substituting the concepts "transferability" and "fittingness" for generalization when dealing with qualitative findings:

The degree of *transferability* is a direct function of the *similarity* between the two contexts, what we shall call "*fittingness*." Fittingness is defined as degree of congruence between sending and receiving contexts. If context A and context B are "sufficiently" congruent, then working hypotheses from the sending originating context may be applicable in the receiving context. (Lincoln & Guba, 1985, p. 124)

Cronbach (1980) offered a middle ground in the debate over generalizability. He found little value in experimental designs that are so focused on carefully controlling cause and effect (internal validity) that the findings are largely irrelevant beyond that highly controlled experimental situation (external validity). On the other hand, he was equally concerned about entirely idiosyncratic case studies that yield little of use beyond the case study setting. He was also skeptical that highly specific empirical findings would be meaningful under new conditions. He suggested instead that designs balance depth and breadth, realism and control so as to permit reasonable "extrapolation" (pp. 231–235).

TESTING THEORY FROM A PURPOSEFUL SAMPLE OF QUALITATIVE CASES TO GENERALIZE: A CLASSIC CASE EXAMPLE

Sociologist Alfred Lindesmith (1905–1991), Indiana University, wanted to test his theory about addiction to opiate drugs. The theory posited that people became addicted to opium, morphine, or heroin when they took the drug often enough and in sufficient quantity to develop physical withdrawal. But Lindesmith had observed that people become habituated to opiates in a hospital when medicated for pain and manifest junkie behavior of compulsively searching for drugs at almost any cost after hospitalization. He hypothesized that two other things had to happen: Having become habituated, the potential addict now had to (1) stop using drugs and experience the painful withdrawal symptoms that resulted and (2) consciously connect withdrawal distress with ceasing drug use, a connection not everyone made. Junkies, unlike former hospital patients, then had to act on that realization and take more drugs to relieve the symptoms. Those steps, taken together and taken repeatedly, create the compulsive activity that is addiction.

A well-known statistician criticized Lindesmith's sample because he had generalized to a large population (all the addicts in the United States or in the world) from a small, purposefully selected sample rather than studying a random sample. Lindesmith replied that the purpose of random sampling was to ensure that every case had a known probability of being drawn for a sample and that researchers randomize to permit generalizations about distributions of some phenomenon in a population and in subgroups in a population. But, he argued, random sampling was irrelevant to his research on addicts because he was interested not in distributions but in a universal process—how one became and remained an addict. He didn't want to know the probability that any particular case would be chosen for his sample. He wanted to maximize the probability of finding a negative case so as all the better to test the theory. Not finding disconfirming cases strengthened his confidence in generalizing his findings.

—Adapted from Becker (1998, pp. 86–87)

Extrapolation

Unlike the usual meaning of the term *generalization*, an *extrapolation* clearly connotes that one has gone beyond the narrow confines of the data to *think about other applications of the findings*. Extrapolations are modest speculations on the likely applicability of findings other situations under similar, but not identical, conditions. Extrapolations are logical, thoughtful, case derived and problem oriented rather than statistical and probabilistic.

Distinguished methodologist Thomas D. Cook (2014) has explained the nature and significance of extrapolation.

> Informing future policy decisions also requires justified procedures for extrapolating past findings to future periods when the populations of treatment providers and recipients might be different, when adaptations of a previously studied treatment might be required, when a novel outcome is targeted, when the application might be to situations different from earlier, and when other factors affecting the outcome are novel too. We call this the *extrapolation function* since inferences are required about populations and categories that are now in some ways different from the sampled study particulars. Sampling theory cannot even pretend to deal with the framing of causal generalization as extrapolation since the emphasis is on taking observed causal findings and projecting them beyond the observed sampling specifics.

> We argue here that both representation and extrapolation are part of a broad and useful understanding of external validity; that each has been quite neglected in the past relative to internal validity—namely, whether the link between manipulated treatments and observed effects is plausibly causal; that few practical methods exist for validly representing the populations and other constructs sampled in the existing literature; and that even fewer such methods exist for extrapolation. Yet, causal extrapolation is more important for the policy sciences, I argue, than is causal representation. (p. 527)

Extrapolations can be particularly useful when based on information-rich samples and designs—that is, studies that produce relevant information carefully targeted to specific concerns about both the present and the future. Users of evaluation, for example, will usually expect evaluators to thoughtfully extrapolate from their findings in the sense of pointing out *lessons learned* and potential applications to future efforts. Sampling strategies in qualitative evaluations can be planned with the stakeholders' desire for extrapolation in mind.

High-Quality Lessons Learned

The notion of identifying and articulating "lessons learned" has become popular as a way of extracting useful and actionable knowledge from cross-case analyses. Rather than being stated in the form of traditional scientific empirical generalizations, lessons learned take the form of *principles of practice* that must be adapted to particular settings in which the principle is to be applied. For example, a lesson learned from research on evaluation use is that evaluation use will likely be enhanced by designing an evaluation to answer the focused questions of specific primary intended users (Cousins & Bourgeois, 2014; Patton, 2008).

Ricardo Millett, former Director of Evaluation at the W. K. Kellogg Foundation, and I analyzed the lessons-learned sections of grantee evaluation reports. What we found was massive confusion and inconsistency. Listed under the heading "lessons" were findings, opinions, ideas, visions, and recommendations—but seldom lessons. Exhibit 9.12 provides examples of what we found.

EXHIBIT 9.12   Confusion About What Constitutes *a Lesson Learned*

A lesson, in the context of extracting useable knowledge from findings, takes the form of an *if . . . then* proposition that provides direction for future action in the real world.

Lesson about evaluation use. If you actively involve intended users in designing an evaluation to ensure its relevance, they are more likely to be interested in and actually use the findings.

This lesson meets two criteria: (1) it is based on evidence from studies of evaluation use (Cousins & Bourgeois, 2014; Patton, 2008) and (2) it provides guidance for future action (an extrapolation from past evidentiary patterns to future desired outcomes). A lesson provides guidance, but it is different from a law, a recipe, or a theoretical proposition.

A physical law. If you heat water to 100 degrees Celsius at sea level, it will boil.

A recipe. Place a cup of oats in two cups of water, add a pinch of salt, and boil for five minutes. Remove from heat, and leave covered for two minutes. It is then ready to serve.

A theoretical proposition. It describes how the world works, as with *natural selection*: If a mutation provides a reproductive advantage that is heritable, over many generations that trait will become dominant in the population.

Using the definition of lesson and these distinctions, here is a sample of statements from evaluation reports illustrating confusion about what constitutes a lesson—and a lesson learned.

STATEMENT REPORTED UNDER THE HEADING "LESSONS" IN EVALUATION REPORTS	WHAT THE STATEMENTS ACTUALLY ARE
1. "Students whose parents helped them with homework got higher grades than those who did not get such help at home."	This is *a finding*. The lesson remains implicit and unexpressed.
2. "One size doesn't fit all."	This is *a conclusion* (based on findings that different people in a program wanted and needed different things), but the conditions to which this conclusion applies (the "if" statement) and what will result (the "then" statement) are implicit.
3. "There are no workarounds powerful enough to compensate for a failing educational system."	This is *an opinion* based on negative findings from an evaluation about a single program. It is a gross overgeneralization born of frustration and skepticism, but it lacks both supporting evidence and guidance about what to do in any applicable and useful manner.
4. "Be sure to provide daycare when you hold community meetings."	This is *a recommendation*. It prescribes a quite specific action, but both the basis for the recommendation and the outcome that will follow its implementation are implicit.

STATEMENT REPORTED UNDER THE HEADING "LESSONS" IN EVALUATION REPORTS	WHAT THE STATEMENTS ACTUALLY ARE
5. "Be prepared: By failing to prepare, you are preparing to fail."	This is *an aphorism*, no doubt wise, and certainly oft cited since published by Ben Franklin and adopted by the Boy Scouts, but reporting it as a central lesson in an evaluation report might at least include some acknowledgment that the observation has a long and distinguished history. (The report in which this appeared had nine others of like sentiment and no lessons original to or grounded in the actual evaluation done.)
6. "Lesson learned: Take time to do reflective practice."	This, too, is *a recommendation*, but it invites introduction of a useful distinction between "lessons" and "lessons learned." A lesson is a cognitive insight or understanding that if you do a certain thing, a certain result is likely to follow. A lesson is not "learned" until it is put into practice (behavioral change).
7. "We will never stop working to make the world a better place."	This is a *visionary promise*, an organizational commitment, and an inspirational reassurance to actual or potential funders. It is not a lesson.
8. If you want to formulate a meaningful and useful lesson that provides guidance for future action, then learn what a lesson is (as distinct from a finding, conclusion, opinion, recommendation, aphorism, or vision).	*That's a lesson.* If you put that lesson into action, you will have *a lesson learned.*

High-Quality Lessons

As we looked at examples of "lessons" listed in a variety of evaluation reports, it became clear that the label was being applied to any kind of insight, evidentially based or not. We began thinking about what would constitute "high-quality lessons" and decided that one's confidence in the transferability or extrapolated relevance of a supposed lesson would increase to the extent towhich it was supported by multiple sources and types of learnings (triangulation). Exhibit 9.13 on the next page presents a list of kinds of evidence that could be accumulated to support a proposed lesson, making it more worthy of application and adaptation to new settings if it has triangulated support from a variety of perspectives and data sources. Questions for generating lessons learned are also listed. Thus, for example, the lesson that designing an evaluation to answer the focused questions of specific primary intended users enhances evaluation use is supported by research on use, theories about diffusion of innovation and change, practitioner wisdom, cross-case analyses of use, the profession's articulation of standards, and expert testimony. *High-quality lessons*, then, constitute guidance extrapolated from multiple sources and independently triangulated to increase transferability as cumulative knowledge and working hypotheses that can be adapted and applied to new situations. This is a form of pragmatic utilitarian generalizability, if you will. The pragmatic bias in this approach reflects the wisdom of Samuel Johnson: "As gold which he cannot spend will make no man rich, so knowledge which he cannot apply will make no man wise."

Principles

Principles are lessons expressed more generically, taken to a higher level of generalizability, and stated in a more direct and less contingent manner.

Lesson about evaluation use: If you actively involve intended users in designing an evaluation to ensure its relevance, they are more likely to be interested in and actually use the findings.

Principle to enhance evaluation use: Form and nurture a relationship with primary intended users built around their information needs and intended uses of the evaluation.

Principles are built from lessons that are based on evidence about how to accomplish some desired result. Qualitative inquiry is an especially productive way

EXHIBIT 9.13   High-Quality Lessons Learned

High-quality lessons learned. Knowledge that can be applied to future action and derived from multiple sources of evidence (triangulation)

1. Evaluation findings—patterns across programs

2. Basic and applied research

3. Practice wisdom and experience of practitioners

4. Experiences reported by program participants/clients/ intended beneficiaries

5. Expert opinion

6. Cross-disciplinary findings and patterns

The idea is that the greater the number and quality of supporting sources for a "lesson," the more rigorous the supporting evidence, and the greater the *triangulation of supporting sources*, the more confidence one has in the significance and meaningfulness of the lesson. Lessons promulgated with only one type of supporting evidence would be considered a "lessons" hypothesis. Nested within and cross-referenced to lessons should be the actual cases from which practice wisdom and evaluation findings have been drawn. A critical principle here is to maintain the contextual frame for lessons— that is, to keep lessons grounded in their context.

For ongoing learning, the trick is to follow future-supposed applications of lessons to test their wisdom and relevance over time in action in new settings. If implemented and validated, they become *high-quality lessons learned.*

Questions for Generating High-Quality Lessons Learned

1. What is meant by a "*lesson*"?

2. What is meant by "*learned*"?

3. By whom was the lesson learned?

4. What's the evidence supporting each lesson?

5. What's the evidence the lesson was learned?

6. What are the contextual boundaries around the lesson (i.e., under what conditions does it apply)?

7. Is the lesson specific, substantive, and meaningful enough to guide practice in some concrete way?

8. Who else is likely to care about this lesson?

9. What evidence will they want to see?

10. How does this lesson connect with other "lessons"?

---

to generate lessons and principles precisely because purposeful sampling of information-rich cases, systematically and diligently analyzed, yields rich, contextually sensitive findings. This combination of qualitative elements constitutes the intellectual farming system from which nutritious lessons and principles grow and thrive. I have discussed principles-focused qualitative inquiry throughout this book.

- *Chapter 1:* Examples of principles as both a focus of inquiry (Paris Declaration Principle for Development Aid, p. 10) and the result of comparative case study analysis (principles that distinguish great from good organizations, Collins, 2001a; adaptive from nonadaptive companies, Collins & Hansen, 2011)
- *Chapter 2:* Strategic principles for qualitative inquiry (Exhibit 2.1, pp. 46–47)

- *Chapter 3:* Principles that undergird and guide various theoretical perspectives: constructivism, hermeneutics, pragmatism
- *Chapter 4:* Practical qualitative inquiry principles to get actionable answers (Exhibit 4.1, pp. 172–173); principles of fully participatory and genuinely collaborative inquiry (p. 222); and principles-focused evaluation (p. 194)
- *Chapter 5:* Principles-focused purposeful sampling (p. 292)
- *Chapter 6:* Principles for engaging in qualitative fieldwork (pp. 415–416)
- *Chapter 7:* Ten interview principles and skills (Exhibit 7.2, p. 428)
- *Chapter 8:* A principles-focused evaluation report (pp. 627–528)
- *Chapter 9:* Rigor attribute analysis principles (pp. 675–676)

How to Extract Credible and Useful Principles: A Case Example

> *Scaling Up Excellence* tackles a challenge that confronts every leader and organization—spreading constructive beliefs and behavior from the few to the many. This book shows what it takes to build and uncover pockets of exemplary performance, spread those splendid deeds, and as an organization grows bigger and older—rather than slipping toward mediocrity or worse—recharge it with better ways of doing the work at hand.
>
> —Sutton and Rao (2014, p. 1)

This is how Robert Sutton and Huggy Rao (2014) open their influential book *Scaling Up Excellence*. Scaling is an applied version of the challenge of generalization. Scholars worry about generalizing findings. Philanthropic foundations, policymakers, and social innovators worry about spreading effective programs. Sutton and Rao identify five principles to guide scaling. How did they do it?

Sutton and Rao (2014) focused on two goals:

> Uncovering the most *rigorous* evidence and theory we could find and generating observations and advice that were *relevant* to people who were determined to scale up excellence.

This meant bouncing back and forth between

> the clean, careful, and orderly world of theory and research—that rigor we love so much as academics—and the messy problems, crazy constraints, and daily twists and turns that are relevant to real people as they strive and struggle to spread excellence to those who need it. (p. 298)

Seven Years of Inquiry

Sutton and Rao (2014) report that they began by gathering ideas and evidence, a process that took years.

> We did case studies, reviewed theory and research, and huddled to develop insights about scaling challenges

and how to overcome them. Little by little, this process changed from a private conversation between the two of us to ongoing conversations about scaling with an array of smart people. We were at the center of this process: making decisions about which leads, stories, and evidence to pursue; choosing which to keep, discard, or save for later; and weaving them together into (we hope) a coherent form. (p. 299)

Sutton and Rao (2014) then analyzed the evidence to reach preliminary conclusions. As conclusions emerged, they presented what they had found to people who had read their prior publications and/or attended their classes and speeches. They recruited knowledgeable and thoughtful people to review, question, and enhance their work.

> This book is best described as the product of years of give-and-take between us and many thoughtful people, not as an integrated perspective that we constructed in private and are now unveiling for the first time. Hundreds of people played direct roles in helping us, and thousands more played indirect roles—even if they didn't realize it. (p. 299)

To speak to issues of rigor and credibility, Sutton and Rao (2014) have distilled their inquiry process into seven core methods, each of which they elaborate in the methodological appendix of the book.

1. Combing through research from the behavioral sciences and beyond

2. Conducting and gathering detailed case studies

3. Brief examples from diverse media sources

4. Targeted interviews as unplanned conversations

5. Presenting emerging scaling ideas to diverse audiences

6. Teaching a "*Scaling Up Excellence*" class to Stanford graduate students

7. Participation in and observation of scaling at the *Stanford* school (an executive professional development program) (pp. 301–306)

What emerges from their description of their inquiry methods is a portrayal of an ongoing, generative, and iterative process of integrating theory, research, and practice around gathering and making sense of the evidence and, ultimately, distilling what they found into principles. The principles constitute a form of generalized guidance derived from and based on lessons. Remember, earlier I postulated that lessons lead to principles. The book opens with the four lessons they identified that became the basis for formulating their five scaling principles. Here's how Sutton and Rao (2014) describe that connection and the first lesson, which is the basis for treating the principles as generalizations.

> Our first big lesson is that, although the details and daily dramas vary wildly from place to place, the similarities among scaling challenges are more important than the differences. The key choices that leaders face and the principles that help organizations scale up without screwing up are strikingly consistent. (p. xi)

Why Principles?

The seven years of inquiry described by Sutton and Rao (2014) generated five principles. Why principles? Because people engaged in a scaling initiative cannot simply look up some right answers and apply them. There is no recipe.

> In the case of scaling, there are so many different aspects of the challenge, and the right answers vary so much across teams, organizations, and industries (and even across challenges faced by a single team or organization), that it is impossible to develop a useful "paint by numbers" approach. Regardless of how many cases, studies, and books (including this one) you read, success at scaling will always depend on making constantly shifting, complex, and not easily codified judgments. (p. 298)

Principles guide judgment. Context informs judgment. Qualitative inquiry generates principles, and then further qualitative inquiry, in a specific context, illuminates that context so that the principles can be interpreted and applied appropriately within that particular context. That process involves both extrapolation and assessing transferability, the qualitative approach to the challenge of generalizing.

Perspectives on Generalizability: A Review

Four core epistemological issues are at the center of debates about the credibility and utility of qualitative inquiry: (1) judging the quality of findings, (2) inferring causality (the challenge of attribution), (3) the validity of generalizations, and (4) determining what is true.

FROM LESSONS TO PRINCIPLES: SCALING THE TRANSFORMATIVE CHANGE INITIATIVE

Started in 2012, the Transformative Change Initiative (TCI) assists community colleges in scaling up innovation: "evidence-based strategies to improve student outcomes and program, organization, and system performance." The TCI evaluation team reviewed case studies of effective innovations, extracted themes and lessons from those separate evaluations of diverse programs, and generated seven principles to guide the next stage of innovation.

The TCI Framework presents the rationale and guiding principles for scaling innovation in the community college context. It is important to link scaling to guiding principles because principles provide direction rather than prescription. They represent the intentionality of the innovation in ways that often allow for multiple actions (practices) to take place. Principles provide "guidance for action in the face of complexity" so that adaptation can occur in ways that achieve the intended outcome.

The theory of change for TCI suggests scaling happens most successfully when practitioners apply guiding principles to their implementation and scaling efforts. In this view, scaling is not so much about replicating what others assert is good practice, which is a classic theory of scaling, but about practitioners and stakeholders becoming instrumental to the scaling process by igniting a chain of actions, reactions, and outcomes that reflect and ultimately reshape the context. To make this happen, practitioners need to

- be aware of the principles that guide the changes they are making to their practice,
- reflect those principles in implementation over time, and
- measure and assess whether the changes are producing the intended improved performance.

—Bragg et al. (2014, p. 6)
Transformative Change Initiative

The first part of this chapter dealt with the issue of quality by examining alternative criteria for judging quality (Modules 76 and 77). Chapter 8 included an extensive discussion of causal inference (pp. 582–595). This module has been examining perspectives on making generalizations. The next and final module will take up the issue of determining what is true. This

EXHIBIT 9.14   Twelve Perspectives on and Approaches to Generalization of Qualitative Findings

INQUIRY PERSPECTIVE	APPROACH TO GENERALIZATION	ELABORATION
1. Traditional scientific research approaches: • Grounded theory • Analytic induction • Qualitative comparative analysis	Generalizations must be theory based: rigorous and systematic comparisons of observed patterns with theoretical propositions.	Qualitative inquiry can contribute generalizable knowledge by generating, testing, and validating theory.
2. Realism	Generalizations depend on purposeful theoretical sampling.	"By linking decisions about whom or what to sample both empirical and theoretical considerations are combined and claims can be made about how the chosen sample relates to a wider universe or population" (Emmel, 2013, p. 60).
3. Constructivism	*Transferability* of findings from particular cases to others based on similarity of context and conditions—also called *inferential generalization* (Lewis & Ritchie, 2003)	Eschew "generalization" in favor of assessing transferability based on in-depth knowledge about the cases studied that provides a basis for assessing the relevance of findings to other similar cases (Lincoln & Guba, 1985).
4. In-depth case study particularity	Focus first on in-depth particularity. Do justice to the case. The issue here is *internal generalization*: "generalizing within the setting . . . studied to people, events, and settings that were not directly observed or interviewed . . . ; the extent to which the times and places observed may differ from those that were not observed, either because of sampling or because of the observation itself" (Maxwell, 2012, p. 142).	"Particularization does deserve praise. . . . What becomes useful understanding is a full and thorough knowledge of the particular, recognizing it also in new and foreign contexts. That knowledge is a form of generalization too . . . , arrived at by recognizing the similarities of objects and issues in and out of context and by sensing the natural covariations of happenings. To generalize this way is to be both intuitive and empirical" (Stake, 1978, p. 6).
5. Social construction	Reflective, experiential, and socially shared generalizations: People naturally make comparisons, which become a "natural" form of generalizing among a group of people. Qualitative cases can enhance those comparisons through the depth and detail that enhances understanding.	*Naturalistic generalizations.* The kind of learning that ordinary people take from their encounters with specific case studies: "The 'vicarious experience' that comes from reading a rich case account can contribute to the social construction of knowledge which, in a cumulative sense, builds general, if not necessarily generalizable knowledge" (Stake, 1995, p. 38).

(Continued)

(Continued)

INQUIRY PERSPECTIVE	APPROACH TO GENERALIZATION	ELABORATION
6. Phenomenology	*Essence.* "A unified statement of the essences of the experience of the phenomenon as a whole. . . . Essence . . . means that which is common or universal, the condition or quality without which a thing would not be what it is" (Moustakas, 1994, p. 100).	Essence emerges from a synthesis of meanings, a reduction of variation to what is essential. Essence integrates and supersedes individual experiences. "The essences of any experience are never totally exhausted. The fundamental textural-structural synthesis represents the essences at a particular time and place from the vantage point of an individual researcher following an exhaustive imaginative and reflective study of the phenomenon" (Moustakas, 1994, p. 100).
7. Ethnography	*Conceptual generalization*: describing both the common and variable meanings and manifestations of universal concepts across cultures, e.g., kinship, conflict, religion, coming of age, etc.	Connecting the microscopic or situation-specific findings of a particular ethnography to more general understandings of culture (Geertz, 1988, 2001) constitutes a form of ethnographic generalization. It involves seeking "the pattern that connects" (Bateson, 1977, 1988).
8. Pragmatism	*Extrapolations.* Modest, practical speculations on the likely applicability of findings to future times and other situations under similar, but not identical, conditions; allows for more interpretive flexibility than a direct transferability assessment; as interested in temporal application of findings (applying what was learned to the future) as in applications in other places.	Extrapolations are logical, thoughtful, case derived, problem oriented, and future oriented rather than statistical and probabilistic: what of practical value has been learned that can be extrapolated to guide future actions, whether in the same place or a different one. Extrapolations can be particularly useful when based on information-rich samples targeted to specific concerns (Cronbach, 1980).
9. Program evaluation and policy analysis	*Lessons.* Qualitative evaluation's contribution to general knowledge takes the form of lessons identified in one evaluation (or a cluster of evaluations) that are offered for application to other places and future programs.	Patterns of effectiveness identified from cross-case analysis of different programs are analyzed to extract common lessons. A classic example is Schorr's (1988) "Lessons of Successful Programs," serving high-risk children and families in poverty. Here is a policy example: *Learning From Iraq* (Special Inspector General for Iraq Reconstruction, 2013).
10. Systems and complexity	*Principles.* Dynamic and complex systems defy simple empirical generalizations. Instead, principles can be identified that inform future systems analyses and guide innovation in complex situations.	Case-based principles provide guidance for adaptive action in the face of complexity. Adaptive action through principles contrasts high-fidelity replication of standardized models. Principles emphasize contextual sensitivity and situational analysis (Eoyang & Holladay, 2013; Patton, 2011).

INQUIRY PERSPECTIVE	APPROACH TO GENERALIZATION	ELABORATION
11. Artistic, evocative representations	Emotional connections and empathy are a form of human generalizability based on shared feelings. *Interpretive interactionism* (Denzin, 1989b) involves intersubjective understandings and feelings.	Finding shared meaning in stories and artistic works (as representations of qualitative findings) moves people from their isolated, particular experience to a more general experience and understanding of the human condition. Emotional resonance among humans (Denzin, 2009) is a form of empathic generalization.
12. Postmodernism	All knowledge is local, specific, and immediate. Generalizations, either empirical or theoretical, are impossible and undesirable.	"Postmodernism is characterized by its distrust of and incredulity toward all 'totalizing' discourses or metanarratives" (Schwandt, 2007, p. 235). This discounts the validity of generalizations, theories, and predictions of any kind, "alerting us to postmodernism's nihilistic tendencies" (Gubrium & Holstein, 2003, p. 5).

"ALL GENERALIZATIONS ARE FALSE."

"All generalizations are false, including this one"—doesn't clarify much of anything.

"All generalizations are false, including this one" leads logically to "Some generalizations are true." If you wish to trace this error back, consider "All Cretans lie," uttered by a Cretan. It can't be true, but it can be false. What's interesting is that it not only leads to "Some Cretans tell the truth," but it also leads to the conclusion that the Cretan speaking is not one of them.

—Errol Morris (2014)
Documentary filmmaker and philosopher

Enhancing the Credibility and Utility of Qualitative
Inquiry by Addressing Philosophy of Science Issues

module concludes with a summary of perspectives on and approaches to generalization in Exhibit 9.14.

We come now to the fourth and final dimension of credibility. Let's review. The first dimension is *systematic, in-depth fieldwork that yields high-quality data*. The second dimension that informs judgments of credibility is *systematic and conscientious analysis*. The third concerns judgments about the *credibility of the researcher*, which depends on training, experience, track record, status, and presentation of self. Now, to

conclude, we take up the issue of *philosophical belief in the value of qualitative inquiry*, that is, a fundamental appreciation of naturalistic inquiry, qualitative methods, inductive analysis, purposeful sampling, and holistic thinking. Exhibit 9.15 graphically depicts these four dimensions of credibility. In the center of the graphic are the alternative criteria for judging quality that opened this chapter: traditional scientific research criteria, constructivist and social construction criteria, artistic and evocative criteria, participatory

EXHIBIT 9.15 Criteria for Judging Quality

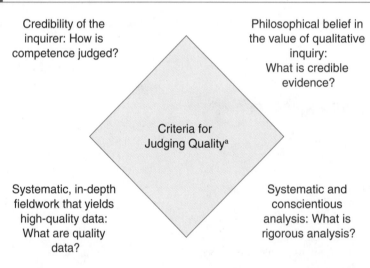

Credibility of the inquirer: How is competence judged?

Philosophical belief in the value of qualitative inquiry: What is credible evidence?

Criteria for Judging Quality[a]

Systematic, in-depth fieldwork that yields high-quality data: What are quality data?

Systematic and conscientious analysis: What is rigorous analysis?

a. Traditional research criteria, constructivist and social construction criteria, artistic and evocative criteria, participatory and collaborative criteria, critical change criteria, systems and complexity criteria, and pragmatic criteria.

and collaborative criteria, critical change criteria, systems and complexity criteria, and pragmatic criteria.

Philosophical belief in the value of qualitative inquiry is a prime determinant of credibility—and a matter of debate and controversy. Given the often-controversial nature of qualitative findings and the necessity, on occasion, to be able to explain and

even defend the value and appropriateness of qualitative inquiry, this module will briefly discuss some of the most contentious issues. The selection of which philosophy of science issues to address in this closing section of the book is based on the workshops I regularly teach on qualitative evaluation methods. In those two- and three-day courses, which typically include participants from around the world, I reserve the final afternoon for open-ended exchanges about whatever matters of interest and concern participants want to raise. By then, we have covered types and applications of qualitative inquiry, design options, purposeful sampling approaches, fieldwork techniques, observational methods, interviewing skills, how to do systematic and rigorous analysis, and ethical standards. Inevitably, questions come pouring forth about the paradigms debate, political considerations, and fundamental doubts participants encounter about the legitimacy of qualitative inquiry. I'll reproduce the questions that arise and offer my responses.

Paradigms question: Why are qualitative methods so controversial? I just want to interview people, see what they say, analyze the patterns, and report my findings? I don't want to debate paradigms. Do we really have to deal with paradigms stuff?

You have to deal with what constitutes credible evidence. What constitutes credible evidence is a matter of debate among both scientists and nonscientists. While not always framed as a paradigms debate, and there are disagreements about what a paradigm is and whether it's a useful concept, I think framing the controversy as a paradigms debate is both accurate and illuminating. In Chapter 3, I discussed the qualitative/quantitative paradigms debate at some length (see pp. 87–95), including an MQP Rumination against designating randomized controlled trials as the "gold standard." In this module, I'm going to focus specifically on how that debate affects credibility and utility.

Paradigms are a way of distinguishing different perspectives in science about how best to study and understand the world. The debate sometimes takes the form of natural science versus social science, qualitative versus quantitative methods, behavioral psychology versus phenomenology, positivism versus constructivism, or realism versus interpretivism. How the debate is framed depends on the perspectives that people bring to it and the language available to them to talk about it. Whatever the terminology and labels for contrasting points of view, the debate is rooted in philosophical differences about the nature of reality and epistemological differences in what constitutes knowledge and how it is created. The *paradigms debate,* whatever form it takes, affects credibility and utility when particular worldviews are pitted against one another at the intersection of philosophy and methods to determine what kinds of evidence are acceptable, believable, and useful.

You may be able to carry out a qualitative study without ever addressing the issue of paradigms. But you ought to know enough about the debate and its implications, it seems to me, to address the issue if it comes up. I would alert those new to the debate that

it has been and can be intense, divisive, emotional, and rancorous. And to those experienced in and tired of the debate, let me say that I've followed it, and been personally engaged in it, for more than 40 years. I've watched the debate ebb and flow, take on new forms, and attract new advocates and adversaries. But it doesn't go away. The paradigms debate is an epistemological phoenix that emerges anew when fires of dissent mellow into become dying embers only to flame again on new winds of contention. I doubt that you can use qualitative methods without encountering and needing to deal with some aspects of the debate. As I have illustrated throughout this chapter, both scientists and nonscientists hold strong opinions about what constitutes credible evidence. Those opinions are paradigm derived and paradigm dependent because a paradigm constitutes a worldview built on epistemological assumptions, preferred definitions of key concepts, comfortable habits, entrenched values defended as truths, and beliefs offered up as evidence. As such, paradigms are deeply embedded in the socialization of adherents and practitioners, telling them what is important, legitimate, and reasonable.

So be prepared to address controversies and competing perspectives about what constitutes credible evidence even if it doesn't come cloaked in the guise of a paradigms debate. Moreover, these are not simply matters of academic debate. They have entered the public policy arena as matters of political debate.

Politics of evidence question: What makes research methods a matter of concern for politics and politicians?

In the public policy arena, advocates of randomized control trials are organized and funded to lobby the U.S. Congress to put their paradigm preferences into legislation (Coalition for Evidence-Based Policy, 2014). They have communications experts who supply reporters with positive news accounts (e.g., Keating, 2014; Kolata, 2013, 2014). On the other side, there are strong political advocacy statements for alternative paradigms: *The Qualitative Manifesto* (Denzin, 2010), *Qualitative Inquiry and the Conservative Challenge* (Denzin & Giardina, 2006), and *Qualitative Inquiry and the Politics of Evidence* (Denzin & Giardina, 2008). Ray Pawson (2013) has produced *A Realist Manifesto.* But there is no organized and funded lobbying effort on behalf of qualitative, mixed-methods, and/or realist approaches. So guess which group is successful in getting its paradigm legitimated and funded in legislation? *Hint:* It's not the qualitative manifesto.

Objectivity question: Doesn't the paradigms debate come down to objectivity versus subjectivity?

French philosopher Jean-Paul Sartre once observed that "words are loaded pistols." The words "objectivity" and "subjectivity" are bullets people arguing fire

DIFFERENT MEANINGS AND USES OF OBJECTIVITY

1. *Objective person.* Unbiased, open-minded, and neutral

2. *Objective process.* Follow, document, and report procedures that do not predetermine results

3. *Objective statement.* Just the facts, unvarnished, put forward by an objective person following an objective process

4. *Objective reality.* Belief that there is knowable, absolute reality

5. *Objective scientific claim.* Findings subjected to scientific peer review by members of a discipline capable of judging the extent to which a claim has been produced by appropriate scientific methods and analysis

6. *Objective methods.* A design, data collection procedures, and analysis that follow accepted inquiry norms of a scientific discipline

7. *Objective measure.* The extent to which a given number can be interpreted as indicating the same amount of the thing measured, across persons or thing measured, using a validated and reliable instrument

8. *Objective decisions.* Fair and balanced judgment based on preponderance of evidence presented and explicit; transparent criteria for weighing the evidence

SOURCE: © Chris Lysy—freshspectrum.com

controlled experimental designs. Yet the ways in which measures are constructed in psychological tests, questionnaires, cost–benefit indicators, and routine management information systems are no less open to the intrusion of biases than making observations in the field or asking questions in interviews. Numbers do not protect against bias; they merely disguise it. All statistical data are based on *someone's* definition of what to measure and how to measure it. An "objective" statistic like the consumer price index is really made up of very subjective decisions about what consumer items to include in the index. Periodically, government economists change the basis and definition of such indices.

Philosopher of science Michael Scriven (1972a) has insisted that quantitative methods are no more synonymous with objectivity than qualitative methods are synonymous with subjectivity:

> Errors like this are too simple to be explicit. They are inferred confusions in the ideological foundations of research, its interpretations, its application. . . . It is increasingly clear that the influence of ideology on methodology and of the latter on the training and behavior of researchers and on the identification and disbursement of support is staggeringly powerful. Ideology is to research what Marx suggested the economic factor was to politics and what Freud took sex to be for psychology. (p. 94)

Scriven's (1972a) lengthy discussion of objectivity and subjectivity in educational research deserves careful reading by students and others concerned by

at each other. It's true that objectivity is held in high esteem. Science aspires to objectivity and a primary reason why decision makers commission an evaluation is to get objective data from an independent source external to the program being evaluated. The charge that qualitative methods are inevitably "subjective" casts an aspersion connoting the very antithesis of scientific inquiry. Objectivity is traditionally considered the sine qua non of the scientific method. To be subjective means to be biased, unreliable, and irrational. Subjective data imply opinion rather than fact, intuition rather than logic, impression rather than confirmation. Chapter 2 briefly discussed concerns about objectivity versus subjectivity, but I return to the issue here to address how these concerns affect the credibility and utility of qualitative analysis.

Let's take a closer look at the objective/subjective distinction. The conventional means for controlling subjectivity and maintaining objectivity are the methods of quantitative social science: distance from the setting and people being studied, standardized quantitative measures, formal operational procedures, manipulation of isolated variables, and randomized this distinction. He skillfully detaches the notions of objectivity and subjectivity from their traditionally narrow associations with quantitative and qualitative methodology, respectively. He presents a clear explanation of how objectivity has been confused with consensual validation of something by multiple observers. Yet a little research will yield many instances of "scientific blunders" (Dyson, 2014; Livio, 2013; Youngson, 1998) where the majority of scientists were factually wrong while *one* dissenting observer described things as they really were (Kuhn, 1970).

Qualitative rigor has to do with the quality of the observations made by an inquirer. Scriven (1972a) emphasizes the importance of being factual about observations rather than being distant from the phenomenon being studied. *Distance does not guarantee objectivity; it merely guarantees distance.* Nevertheless, in the end, Scriven (1998) still finds the ideal of objectivity worth striving for as a counter to bias, and he continues to find the language of objectivity serviceable.

In contrast, Lincoln and Guba (1986), as noted earlier, have suggested replacing the traditional mandate to be objective with an emphasis on *trustworthiness* and *authenticity* by being balanced, fair, and conscientious in taking account of multiple perspectives, multiple interests, multiple experiences, and diverse constructions of realities. Guba (1981) suggested that researchers and evaluators can learn something about these attributes from the stance of investigative journalists.

Journalism in general and investigative journalism in particular are moving away from the criterion of objectivity to an emergent criterion usually labeled "fairness" . . . Objectivity assumes a single reality to which the story or evaluation must be isomorphic; it is in this sense a one-perspective criterion. It assumes that an agent can deal with an objective (or another person) in a nonreactive and noninteractive way. It is an absolute criterion.

Journalists are coming to feel that objectivity in that sense is unattainable. . . .

Enter "fairness" as a substitute criterion. In contrast to objectivity, fairness has these features:

- It assumes multiple realities or truths—hence a test of fairness is whether or not "both" sides of the case are presented, and there may even be multiple sides.

- It is adversarial rather than one-perspective in nature. Rather than trying to hew the line with the truth, as the objective reporter does, the fair reporter seeks to present each side of the case in the manner of an advocate—as, for example, attorneys do in making a case in court. The presumption is that the public, like a jury, is more likely to reach an equitable decision after having heard each side presented with as much vigor and commitment as possible.

- It is assumed that the subject's reaction to the reporter and interactions between them heavily determines what the reporter perceives. Hence one test of fairness is the length to which the reporter will go to test his own biases and rule them out.

- It is a relative criterion that is measured by *balance* rather than by isomorphism to enduring truth. (pp. 76–77)

But times change, and Guba would be unlikely to use the language of "fairness and balance" now that the most politically conservative and deliberately biased American television channel has adopted that phrase as its brand. *Fairness and balance* has become a euphemism for *prejudiced and one-sided.* Objectivity has also taken on unfortunate political and cultural connotations in some quarters, meaning uncaring, unfeeling, disengaged, and aloof. What about subjectivity, the constructivist badge of honor?

Subjectivity Deconstructed

In public discourse, it is not particularly helpful to know that philosophers of science now typically doubt the possibility of anyone or any method being totally "objective." But subjectivity fares even worse. Even if acknowledged as inevitable (Peshkin, 1988), or valuable as a tool to understanding (Soldz & Andersen,

2012), subjectivity carries such negative connotations at such a deep level and for so many people that the very term can be an impediment to mutual understanding. For this and other reasons, as a way of elaborating with any insight the nature of the research process, the notion of subjectivity may have become as useless as the notion of objectivity.

> The death of the notion that objective truth is attainable in projects of social inquiry has been generally recognized and widely accepted by scholars who spend time thinking about such matters. . . . I will take this recognition as a starting point in calling attention to a second corpse in our midst, an entity to which many refer as if it were still alive. Instead of exploring the meaning of subjectivity in qualitative educational research, I want to advance the notion that following the failure of the objectivists to maintain the viability of their epistemology, the concept of subjectivity has been likewise drained of its usefulness and therefore no longer has any meaning. Subjectivity, I feel obliged to report, is also dead. (Barone, 2000, p. 161)

But for other qualitative researchers, subjectivity is not so much about philosophy of science as it is about using one's own experience to make sense of the world through reflexivity (Connolly & Reilly, 2007). That perspective, once entertained, can lead from a focus on the researcher's subjectivity as a window into sense making to *shared meaning making*: intersubjectivity.

Intersubjectivity

> "Subjective" versus "objective" no longer makes sense, since everyone involved is a subject. . . . Human Social Research is *intersubjective* . . . built from encounters among subjects, including researchers who, like it or not, are also subjects. (Agar, 2013, pp. 108–109)

Eschewing both objectivity and subjectivity, intersubjectivity focuses on knowledge as socially constructed in human interactions. Human science research, what anthropologist Michael Agar (2013) calls the *Lively Science*, requires "human social relationships in order to happen at all. They are *intersubjective* sciences. They require social relationships with those who support the science, those who do it, those who serve as subjects of it, and those who consume it" (p. 215).

> The difficult judgment call for the researcher is this: To some extent he or she *should* translate his or her own framework and jointly build a framework for communication with subjects of all those different types. . . . The bedrock of intersubjective research isn't to preach or to

lecture, but rather to learn and to communicate the results, though not at the price of abandoning the core principles of the science. The pressure always exists to achieve a balance, and a researcher always has to make the call of how much and in what way to handle it.

> *This fact has to be part of the* science, *not to mention a central part of training for human social researchers.* How to navigate this ambiguous territory with professional integrity and product quality is a neglected topic, a neglect understandable in light of academic traditions where one could assume that whatever the dissertation committee or disciplinary peers would like was the right thing to do. That isolation is no longer possible. In my view, taking human social research out into the world makes it more difficult, more interesting, more intellectually challenging, and of higher moral value than it has ever been. (pp. 215–216)

Empathic Neutrality

No consensus about substitute terminology has emerged. I prefer empathic neutrality, one of the 12 qualitative themes that I presented in Chapter 2.

> While empathy describes a stance toward the people we encounter in fieldwork, calling on us to communicate interest, caring, and understanding, neutrality suggests a stance toward their thoughts, emotions, and behaviors, a stance of being nonjudgmental. Neutrality can actually facilitate rapport and help build a relationship that supports empathy by disciplining the researcher to be open to the other person and nonjudgmental in that openness.

(See pp. 57–62 for the full discussion of empathic neutrality.)

Open-Mindedness and Impartiality

I have evaluation colleagues who simply describe themselves as open-minded, which seems to satisfy most lay people. The political nature of evaluation means that individual evaluators must make their own peace with how they are going to describe what they do. The meaning and connotations of words like *objectivity*, *subjectivity*, *neutrality*, and *impartiality* will have to be worked out with particular stakeholders in specific evaluation settings. In her leadership role in evaluation in the U.S. federal government, former AEA president Eleanor Chelimsky emphasized her unit's independence and impartiality. The perception of impartiality, she has explained, is at least as important as methodological rigor in highly political environments. Credibility, and therefore utility, are affected by "the steps we take to make and explain our evaluative decisions, [and] also intellectually, in the effort we put forth to look at all sides and all stakeholders of an evaluation" (Chelimsky, 1995, p. 219; see also Chelimsky, 2006).

I think it is worth noting that the official Program Evaluation Standards (Joint Committee on Standards, 2010) do not call for objectivity. The standards have been guiding evaluation practice for nearly four decades. They were originally formulated by social scientists and evaluators representing all the major disciplinary associations. They have twice gone through major review processes. The language used, therefore, has been thoroughly vetted. The standards call for evaluations to be credible, systematic, accurate, useful, accurate, and dependable, but not objective. The term *objectivity* has become a lightning rod attracting epistemological paradigms debate and therefore not useful as a standard for evaluation in the American context. In contrast, the international Quality Standards for Development Evaluation define evaluation as "objective assessment" (OECD-DAC, 2010, p. 5). Different context, different language.

Given the seven different sets of criteria for judging the quality of qualitative inquiry I identified at the beginning of this chapter, and the terms associated with each, it seems unlikely that a consensus about terminology is on the horizon. The methodological and scientific Tower of Babel stands tall and casts a long shadow. But the different perspectives on and uses of terms can be liberating because they opens up the possibility of getting beyond the meaningless abstractions and heavy-laden connotations of objectivity and subjectivity to move instead toward carefully selecting *descriptive* methodological language that best describes your own inquiry processes and procedures. That is, don't label those processes as "objective," "subjective," "intersubjective," "trustworthy," or "authentic." Instead, eschew overarching labels. Describe how you approach your inquiry, what you bring to your work, and how you've reflected on what you do, and then let the reader be persuaded, or not, by the intellectual and methodological rigor, meaningfulness, value, and utility of the result. In the meantime, be very careful how you use particular terms in specific contexts. Words are bullets. They are also landmines. I end this diatribe with a cautionary tale about being sensitive to the cultural context within which terms are used.

During a tour of America, former British prime minister Winston Churchill attended a buffet luncheon at which chicken was served. As he returned to the buffet for a second helping he asked, "May I have some more breast?"

His hostess, looking embarrassed, explained that "in this country we ask for white meat or dark meat."

Churchill, taking the white meat he was offered, apologized and returned to his table.

The next morning the hostess received a beautiful orchid from Churchill with the following card: "I would be most obliged if you would wear this on your white meat."

A REALIST PERSPECTIVE ON OBJECTIVITY

Evaluation cannot hope for perfect objectivity but neither does this mean it should slump into rampant subjectivity. We cannot hope for absolute cleanliness but this does not require us to enjoy a daily roll in the manure. The alternative to these two termini is for evaluation to embrace the goal of being "validity increasing" . . .

Skepticism . . . , in its English spelling, . . . constitutes the final desideratum of evaluation science.

Organised scepticism means that any scientific claim must be exposed to critical scrutiny before it becomes accepted. . . . What counts is the depth of critical scrutiny applied to the inferences drawn from any inquiry. And this level of attention depends, in turn, on the presence of a collegiate group of stakeholders and their willingness to put each other's work under the microscope.

—Ray Pawson (2013, p. 107)
*The Science of Evaluation:
A Realist Manifesto*

Truth and reality question: *"I don't understand this talk about multiple realities and different truths for different people. If research is anything, it ought to be about getting at true reality. I know you like quotes, so here's one of my favorite quotes for you, from George Orwell: 'In a time of universal deceit—telling the truth is a revolutionary act.' I think we ought to be research revolutionaries and speak the truth. In fact, the mantra of evaluation is: Speak truth to power. So, truth or not truth?"*

It's an important question. Certainly, there are a lot of quotes about truth. This is a thick book, and it could contain nothing but quotes about truth, which would serve to illustrate its evasiveness. Let me offer a quote from the great comedian Lily Tomlin, who, playing the character of a little girl accused by a scolding adult of making things up, responded thus:

Lady, I do not make up things. That is lies. Lies are not true. But the truth could be made up if you know how. And that's the truth.

Or consider this observation by Thomas Schwandt, a philosopher of science and professional evaluator, who has spent much of a distinguished career grappling with this very issue. His conclusion:

> **TRUTH** is one of the most difficult of all philosophical topics, and controversies surrounding the nature of truth lie at the heart of both apologies for and criticisms of varieties of qualitative work. Moreover, truth is intimately related to questions of *meaning*, and establishing the nature of that relationship is also complicated and contested.
>
> There is general agreement that *what* is true or what carries truth are statements, propositions, beliefs, and assertions, but *how* the truth of same is established is widely debated. (Schwandt, 2007, p. 300)

Schwandt presents 10 different philosophical orientations to and theories about truth: (1) correspondence, (2) consensus, (3) coherence, (4) contextualist, (5) pragmatic, (6) hermeneutic, (7) critical theory (Foucault), (8) realist, (9) constructivist, and (10) objectivist theory. Pick your poison—or truth. We won't resolve the debate here. Not even close. Nor will others for, to add yet another quote to the collection, here's cynic Ambrose Bierce's (1999) assessment:

> Discovery of truth is the sole purpose of philosophy, which is the most ancient occupation of the human mind and has a fair prospect of exiting with increasing activity to the end of time. (p. 201)

Since we can't resolve the nature of truth, indulge me in a story that illustrates why it may be important to have figured out where you, yourself, stand on matters of truth. Following a presentation of evaluation findings at a public school board meeting, I was asked by the school district's internal evaluator, "Do you, as a qualitative researcher, swear to tell the truth, the whole truth and nothing but the truth?" The question was meant to embarrass me. The researcher had an article I had written attacking overreliance on standardized tests for school evaluations and another advocating soliciting multiple perspectives from parents, teachers, students, and community members about their experiences with the school district to document diverse perspectives. In that article, and earlier editions of this book, I had expressed doubt about the utility of

truth as a criterion of quality and I suspected that he hoped to lure me into an academic-sounding, arrogant, and philosophical discourse on the question "What is truth?" in the expectation that the public officials present would be alienated and dismiss my presentation. So when he asked, "Do you, as a qualitative researcher, swear to tell the truth, the whole truth and nothing but the truth?" I did *not* reply, "That depends on what *truth* means." I said simply, "Certainly I promise to respond honestly." Notice the shift from truth to honesty.

The researcher applying traditional social science criteria might respond, "I can show you truth insofar as it is revealed by the data."

The constructivist might answer, "I can show you multiple truths."

The artistically inclined might suggest that "beauty is truth." And "fiction often reveals truth better than nonfiction."

The critical theorist could explain that "truth depends on one's consciousness."

The participatory qualitative inquirer would say, "We create truth together."

The critical change activist might say, "I offer you praxis. Here is where I take my stand. This is true for me."

The pragmatic evaluator might reply, "I can show you what is useful. What is useful is true."

Indeed, in this vein, Exhibit 9.7, in presenting the seven sets of criteria for judging quality, offers a political campaign button about TRUTH for each (pp. 680–681).

By the way, I noted earlier that the Program Evaluation Standards do not use the language of objectivity, but the "the Accuracy Standards are intended to increase the dependability and *truthfulness* [italics added] of evaluation representations" (Joint Committee on Standards, 2010). *Note: Truthfulness* is not TRUTH. You could do a little hermeneutic work on that distinction, should you be so inclined.

Ironically, it is sometimes easier to determine what is false than what is true. For insights into how the academic peer review process has been distorted and corrupted to generate invalid and untrustworthy results, see the widely cited and influential analysis by Professor of Health Research and Policy at Stanford School of Medicine, John P. A. Ioannidis (2005) "Why Most Published Research Findings Are False."

Truth Tests and Utility Tests

Previously I have cited the influential research by Weiss and Bucuvalis (1980) that decision makers apply both "truth" tests and "utility" tests to evaluation. "Truth," in this case, however, means reasonably accurate and credible data (the focus of the program evaluation standards) rather than data that are true in some absolute sense. Savvy policymakers know better than most the context and perspective-laden nature of competing truths. Qualitative inquiry can present accurate data on various perspectives, including the evaluator's perspective, without the burden of determining that only one perspective must be true.

Evaluation theorist and methodologist Nick Smith (1978), pondering these questions, has noted that to act in the world we often accept either approximations to truth or even untruths.

> For example, when one drives from city to city, one acts as if the earth is flat and does not try to calculate the earth's curvature in planning the trip, even though acting as if the earth is flat means acting on an untruth. Therefore, in our study of evaluation methodology, two criteria replace exact truth as paramount: practical utility and level of certainty. The level of certainty required to make an adequate judgment under the law differs depending on whether one is considering an administrative hearing, an inquest, or a criminal case. Although it seems obvious that much greater certainty about the nature of things is required when legislators set national and educational policy than when a district superintendent decides whether to continue a local program, the rhetoric in evaluation implies that the same high level of certainty is required of both cases. If we were to first determine the level of certainty desired in a specific case, we could then more easily choose appropriate methods. Naturalistic descriptions give us greater certainty in our understanding of the nature of an educational process than randomized, controlled experiments do, but less certainty in our knowledge of the strength of a particular effect.... Our first concern should be the practical utility of our knowledge, not its ultimate truthfulness. (p. 17)

In studying evaluation use (Patton, 2008), I found that decision makers did not expect evaluation reports to produce "TRUTH" in any fundamental sense. Rather, they viewed evaluation findings as additional information that they could and did combine with other information (political, experiential, other research, colleague opinions, etc.), all of which fed into a slow, evolutionary process of incremental decision making. Kvale (1987) echoed this interactive and contextual approach to truth in emphasizing the "pragmatic validation" of findings in which the results of qualitative analysis are judged by their relevance to and use by those to whom findings are presented.

This *criterion of utility* can be applied not only to evaluation but also to qualitative analyses of all kinds, including textual analysis. Barone (2000), having rejected objectivity and subjectivity as meaningless criteria in the postmodern age, makes the case for pragmatic utility:

> If all discourse is culturally contextual, how do we decide which deserves our attention and respect? The pragmatists offer the criterion of usefulness for this purpose.... An idea, like a tool, has no intrinsic value and is "true" only in its capacity to perform a desired service for its handler within a given situation. When the criterion of usefulness is applied to context-bound, historically situated transactions between itself and a text, it helps us to judge which textual experiences are to be valued.... The gates are opened for textual encounters, in any inquiry genre or tradition, that serve to fulfill an important human purpose. (pp. 169–170)

Focusing on the connection between truth tests and utility tests shifts attention back to credibility and quality, not as absolute generalizable judgments but as contextually dependent on the needs and interests of those receiving our analysis. This obliges researchers and evaluators to consider carefully how they present their work to others, with attention to the purpose to be fulfilled. That presentation should include reflections on how your perspective affected the questions you pursued in fieldwork, careful documentation of all procedures used so that others can review your methods for bias, and being open in describing the limitations of the perspective presented. Exhibit 9.16, at the end of this chapter (pp. 736–741), offers an in-depth description of how one qualitative inquirer dealt with these issues in a long-term participant–observer relationship. The exhibit, titled A Documenter's Perspective, is based on her research journal and field notes. It moves the discussion from abstract philosophizing to day-to-day, in-the-trenches fieldwork encounters aimed at sorting out what is true (small *t*) and useful.

Finding TRUTH can be a heavy burden. I once had a student who was virtually paralyzed in writing an evaluation report because he wasn't sure if the patterns he thought he had uncovered were really true. I suggested that he not try to convince himself or others that his findings were true in any absolute sense but, rather, that he had done the best job he could in describing the patterns that appeared to him to be present in the data and that he present those patterns as *his* perspective based on his analysis and interpretation of the data he had collected. Even if he believed that what he eventually produced was Truth, any sophisticated person reading the report would know that what he presented was no more than his perspective, and they would judge that perspective by their own commonsense understandings and use the

TRUTH VERSUS RELATIVISM

Postmodern work is often accused of being relativistic (and evil) since it does not advocate a universal, independent standard of truth. In fact, relativism is only an issue for those who believe there is a foundation, a structure against which other positions can be objectively judged. In effect, this position implies that there is no alternative between objectivism and relativism. Postmodernists dispute the assumptions that produce the objectivism/relativism binary since they think of truth as multiple, historical, contextual, contingent, political, and bound up in power relations. Refusing the binary does not lead to the abandonment of truth, however, as Foucault emphasizes when he says, "I believe too much in truth not to suppose that there are different truths and different ways of speaking the truth."

Furthermore, postmodernism does not imply that one does not discriminate among multiple truths, that "anything goes.". . . If there is no absolute truth to which every instance can be compared for its truth-value, if truth is instead multiple and contextual, then the call for ethical practice shifts from grand, sweeping statements about truth and justice to engagements with specific, complex problems that do not have generalizable solutions. This different state of affairs is not irresponsible, irrational, or nihilistic. . . . As with truth, postmodern critiques argue for multiple and historically specific forms of reason. (St. Pierre 2000, p. 25)

information according to how it contributed to their own needs.

As one additional source of reflection on these issues, perhaps the following Sufi story will provide some guidance about the difference between truth and perspective. Sagely, in this encounter, Nasrudin gathers data to support his proposition about the nature of truth. Here's the story.

Mulla Nasrudin was on trial for his life. He was accused of no less a crime than treason by the king's ministers, wise men charged with advising on matters of great import. Nasrudin was charged with going from village to village inciting the people by saying, "The king's wise men do not speak truth. They do not even know what truth is. They are confused." Nasrudin was brought before the king and the court. "How do you plead, guilty or not guilty?"

"I am both guilty and not guilty," replied Nasrudin.

"What, then, is your defense?"

Nasrudin turned and pointed to the nine wise men who were assembled in the court. "Have each sage write an answer to the following question: 'What is water?'"

The king commanded the sages to do as they were asked. The answers were handed to the king, who read to the court what each sage had written.

The first wrote, "Water is to remove thirst."

The second, "It is the essence of life."

The third, "Rain."

The fourth, "A clear, liquid substance."

The fifth, "A compound of hydrogen and oxygen."

The sixth, "Water was given to us by God to use in cleansing and purifying ourselves before prayer."

The seventh, "It is many different things—rivers, wells, ice, lakes, so it depends."

The eighth, "A marvelous mystery that defies definition."

The ninth, "The poor man's wine."

Nasrudin turned to the court and the king: "I am guilty of saying that the wise men are confused. I am not, however, guilty of treason because, as you see, the wise men are confused. How can they know if I have committed treason if they cannot even decide what water is? If the sages cannot agree on the truth about water, something which they consume every day, how can one expect that they can know the truth about other things?"

The king ordered that Nasrudin be set free.

TRUE FACTS VERSUS TRUE THEORIES

Facts and theories are born in different ways and are judged by different standards. Facts are supposed to be true or false. They are discovered by observers or experimenters. A scientist who claims to have discovered a fact that turns out to be wrong is judged harshly. One wrong fact is enough to ruin a career.

Theories have an entirely different status. They are free creations of the human mind, intended to describe our understanding of nature. Since our understanding is incomplete, theories are provisional. Theories are tools of understanding; and a tool does not need to be precisely true in order to be useful. Theories are supposed to be more-or-less true, with plenty of room for disagreement. A scientist who invents a theory that turns out to be wrong is judged leniently. Mistakes are tolerated, so long as the culprit is willing to correct them when nature proves them wrong.

—Physicist Freeman Dyson (2014, p. 4)
Institute for Advanced Studies, Princeton

Enhanced Credibility and Increased Legitimacy for Qualitative Methods: Looking Back and Looking Ahead

> The distinction between the past, present, and future is only a stubbornly persistent illusion.
>
> —Physicist Albert Einstein
> *Theory of Relativity*

Chapter Summary

This chapter has reviewed ways of enhancing the quality, credibility, and utility of qualitative analysis by dealing with four distinct but related inquiry concerns:

- Rigorous methods for doing fieldwork that yield high-quality data
- Systematic and conscientious analysis with attention to issues of credibility
- The credibility of the researcher, which depends on training, experience, track record, status, and presentation of self
- Philosophical belief in the value of qualitative inquiry—that is, a fundamental appreciation of naturalistic inquiry, qualitative methods, inductive analysis, purposeful sampling, and holistic thinking.

Exhibit 9.15 presented a graphic depicting these four dimensions, with criteria for judging quality in the center.

Conclusion: Beyond the Qualitative/Quantitative Debate

Question: What's the status of the qualitative/quantitative debate today? From your perspective, what does the future look like for qualitative inquiry?

The debate between qualitative and quantitative methodologists was often strident historically, but in recent years the debate has mellowed. A consensus has gradually emerged that the important challenge is to appropriately match methods to purposes and inquiry questions, not to universally and unconditionally advocate any single methodological approach for all inquiry situations. Indeed, eminent methodologist Thomas Cook, one of evaluation's luminaries, pronounced in his keynote address to the 1995 International Evaluation Conference in Vancouver that "qualitative researchers have won the qualitative/quantitative debate."

Won in what sense?

Won acceptance.

The validity of experimental methods and quantitative measurement, appropriately used, was never in doubt. Now, qualitative methods have ascended to a level of parallel respectability. I have found increased interest in and acceptance of qualitative methods in particular and multiple methods in general. Especially in evaluation, a consensus has emerged that researchers and evaluators need to know and use a variety of methods in order to be responsive to the nuances of particular empirical questions and the idiosyncrasies of specific stakeholder needs. The debate has shifted from quantitative versus qualitative to strong differences of opinion about how to establish causality (the attribution and so-called gold standard debate discussed in Chapters 3 and 8). While related, that's a narrower issue.

The credibility and respectability of qualitative methods varies across disciplines, university departments, professions, time periods, and countries. In the field I know best, program evaluation, the increased legitimacy of qualitative methods is a function of more examples of useful, high-quality evaluations employing qualitative methods and an increased commitment to providing useful and understandable information based on stakeholders' concerns. Other factors that contribute to increased credibility include more and higher-quality training in qualitative methods and the publication of a substantial qualitative literature.

The history of the paradigms debate parallels the history of evaluation. The earliest evaluations focused largely on quantitative measurement of clear, specific goals and objectives. With the widespread social and educational experimentation of the 1960s and early 1970s, evaluation designs were aimed at comparing the effectiveness of different programs and treatments through rigorous controls and experiments. This was the period when the quantitative/experimental paradigm dominated. By the middle 1970s, the paradigms debate had become a major focus of evaluation discussions and writings. By the late 1970s, the alternative qualitative/naturalistic paradigm had been fully articulated (Guba, 1978; Patton, 1978; Stake, 1975, 1978). During this period, concern about finding ways to increase use became predominant in evaluation, and evaluators began discussing standards. A period of pragmatism and dialogue followed, during which calls for and experiences with multiple methods and a synthesis of paradigms became more common. The advice of Cronbach (1980), in his important book on reform of program evaluation, was widely taken to heart: "The evaluator will be wise not to declare allegiance to either a quantitative–scientific–summative methodology or a qualitative–naturalistic–descriptive methodology" (p. 7).

Signs of detente and pragmatism now abound. Methodological tolerance, flexibility, eclecticism, and concern for appropriateness rather than orthodoxy now characterize the practice, literature, and discussions of evaluation. Several developments seem to me to explain the withering of the methodological paradigms debate.

1. The articulation of professional standards has emphasized methodological appropriateness rather than paradigm orthodoxy (Joint Committee, 2010; OECD-DAC, 2010). Within the standards as context, the focus on conducting evaluations that are useful, practical, ethical, accurate, and accountable have reduced paradigms polarization.

2. The strengths and weaknesses of both quantitative/experimental methods and qualitative/naturalistic methods are now better understood. In the original debate, quantitative methodologists tended to attack some of the worst examples of qualitative evaluations while the qualitative evaluators tended to hold up for critique the worst examples of quantitative/experimental approaches. With the accumulation of experience and confidence, exemplars of both qualitative and quantitative approaches have emerged with corresponding analyses of the strengths and weaknesses of each. This has permitted more balance and a better understanding of the situations for which various methods are most appropriate as well as grounded experience in how to combine methods.

3. A broader conceptualization of evaluation, and of evaluator training, has directed attention to the relation of methods to other aspects of evaluation, like use, and has therefore reduced the intensity of the methods debate as a topic unto itself.

4. Advances in methodological sophistication and diversity within both paradigms have strengthened diverse applications to evaluation problems. The proliferation of books and journals in evaluation, including but not limited to methods contributions, has converted the field into a rich mosaic that cannot be reduced to quantitative versus qualitative in primary orientation. Moreover, the upshot of all the developmental work in qualitative methods is that, as documented in Chapter 3, today there is as much variation among qualitative researchers as there is between qualitatively and quantitatively oriented scholars.

5. Support for methodological eclecticism from major figures and institutions in evaluation increased methodological tolerance. When eminent measurement and methods scholars like Donald Campbell and Lee J. Cronbach began publicly recognizing the contributions that qualitative methods could make, the acceptability of qualitative/naturalistic approaches was greatly enhanced. Another important endorsement of multiple methods came from the Program Evaluation and Methodology Division of the U.S. General Accounting Office (GAO), which arguably did the most important and influential evaluation work at the national level. Under the leadership of Assistant Comptroller General and former AEA president (1995) Eleanor Chelimsky, GAO published a series of methods manuals, including *Case Study Evaluations* (GAO, 1987), *Prospective Evaluation Methods* (GAO, 1989), and *The Evaluation Synthesis* (GAO, 1992). The GAO manual *Designing Evaluations* put the paradigms debate to rest as it described what constituted a "strong evaluation."

Strength is not judged by adherence to a particular paradigm. It is determined by use and technical adequacy, whatever the method, within the context of purpose, time, and resources.

Strong evaluations employ methods of analysis that are appropriate to the question; support the answer with evidence; document the assumptions, procedures, and modes of analysis; and rule out competing evidence. Strong studies pose questions clearly, address them appropriately, and draw inferences commensurate with the power of the design and the availability, validity, and reliability of the data. Strength should not be equated with complexity. Nor should strength be equated with the degree of statistical manipulation of data. Neither infatuation with complexity nor statistical incantation makes an evaluation stronger.

The strength of an evaluation is not defined by a particular method. Longitudinal, experimental, quasi-experimental, before-and-after, and case study evaluations can be either strong or weak. . . . That is, the strength of an evaluation has to be judged within the context of the question, the time and cost constraints, the design, the technical adequacy of the data collection and analysis, and the presentation of the findings. A strong study is technically adequate and useful—in short, it is high in quality. (GAO, 1991, pp. 15–16)

6. Evaluation professional societies have supported exchanges of views and high-quality professional practice in an environment of tolerance and eclecticism. The evaluation professional societies

and journals serve a variety of people from different disciplines who operate in different kinds of organizations at different levels, in and out of the public sector, and in and out of universities. This diversity, and opportunities to exchange views and perspectives, has contributed to the emergent pragmatism, eclecticism, and tolerance in the field. A good example was the appearance two decades ago of a volume of New Directions for Program Evaluation on *The Qualitative-Quantitative Debate: New Perspectives* (Reichardt & Rallis, 1994). The tone of the eight distinguished contributions in that volume is captured by phrases such as "peaceful coexistence," "each tradition can learn from the other," "compromise solution," "important shared characteristics," and "a call for a new partnership."

7. There is increased advocacy of and experience in combining qualitative and quantitative approaches. The Reichardt and Rallis (1994) volume just cited also included these themes: "blended approaches," "integrating the qualitative and quantitative," "possibilities for integration," "qualitative plus quantitative," and "working together." Exhibit 9.2 presented 10 developments enhancing mixed-methods triangulation (p. 666).

Matching Claims and Criteria

The withering of the methodological paradigms debate holds out the hope that studies of all kinds can be judged on their merits according to the claims they make and the evidence marshaled in support of those claims. The thing that distinguishes the seven sets of criteria for judging quality introduced in this chapter (Exhibit 9.7) is that they support different kinds of claims. Traditional scientific claims, constructivist claims, artistic claims, participatory inquiry claims, critical change claims, systems claims, and pragmatic claims will tend to emphasize different kinds of conclusions with varying implications. In judging claims and conclusions, the validity of the claims made is only partly related to the methods used in the process.

Validity is a property of knowledge, not methods. No matter whether knowledge comes from an ethnography or an experiment, we may still ask the same kind of questions about the ways in which that knowledge is valid. To use an overly simplistic example, if someone claims to have nailed together two boards, we do not ask if their hammer is valid, but rather whether the two boards are now nailed together, and whether the claimant was, in fact, responsible for that result. In fact, this particular claim may be valid whether the nail was set in place by a hammer, an airgun, or the butt of a screwdriver. A hammer does not guarantee successful nailing, successful nailing does not require a hammer, and the validity of the claim is in principle separate from which tool was used. The same is true of methods in the social behavioral sciences. (Shadish, 1995a, p. 421)

This brings us back to a pragmatic focus on the utility of findings as a point of entry for determining what's at stake in the claims made in a study and therefore what criteria to use in assessing those claims. As I noted in opening this chapter, judgments about credibility and quality depend on criteria. And though this chapter has been devoted to ways of enhancing quality and credibility, all such efforts ultimately depend on the willingness of the inquirer to weigh the evidence carefully and be open to the possibility that what has been learned most from a particular inquiry is how to do it better next time.

Canadian-born bacteriologist Oswald Avery, discoverer of DNA as the basic genetic material of the cell, worked for years in a small laboratory at the hospital of the Rockefeller Institute in New York City. Many of his initial hypotheses and research conclusions turned out, on further investigation, to be wrong. His colleagues marveled that he never turned argumentative when findings countered his predictions and never became discouraged. He was committed to learning and was often heard telling his students, "Whenever you fall, pick up something."

A final Halcolm story on the nature of journeys ends this chapter—and this book.

EXHIBIT 9.16   A Documenter's Perspective

by Beth Alberty

Introduction

This exhibit provides a reflective case study of the struggle experienced by one internal, formative program evaluator of an innovative school art program as she tried to figure out how to provide useful information to program staff from the voluminous qualitative data she collected. Beth begins by describing what she means by "documentation" and then shares her experiences as a novice in analyzing the data, a process of moving from a mass of documentary material to a unified, holistic document.

Documentation

Documentation, as the word is commonly used, may refer to "slice of life" recordings in various media or to the marshalling of evidence in support of a position or point of view. We are familiar with "documentary" films; we require lawyers or journalists to "document" their cases. Both meanings contribute to my view of what documentation is, but they are far from describing it fully. Documentation, to my mind, is the interpretive reconstitution of a focal event, setting, project, or other phenomenon, based on observation and on descriptive records set in the context of guiding purposes and commitments.

I have always been a staff member of the situations I have documented, rather than a consultant or an employee of an evaluation organization. At first this was by accident, but now it is by conviction: My experience urges that the most meaningful evaluation of a program's goals and commitments is one that is planned and carried out by the staff and that such an evaluation contributes to the program as well as to external needs for information. As a staff member, I participate in staff meetings and contribute to decisions. My relationships with other staff members are close and reciprocal. Sometimes I provide services or perform functions that directly fulfill the purposes of the program—for example, working with children or adults, answering visitor's questions, and writing proposals and reports. Most of my time, however, is spent planning, collecting, reporting, and analyzing documentation.

First Perceptions

With this context in mind, let me turn to the beginning plunge. Observing is the heart of documenting, and it was into observing that I plunged, coming up delighted at the apparent ease and swiftness with which I could fish insight and ideas from the ceaseless ocean of activity around me. Indeed, the fact that observing (and record keeping) does generate questions, insight, and matters for discussion is one of many reasons why records for any documentation should be gathered by those who actually work in the setting.

My observing took many forms, each offering a different way of releasing questions and ideas—interactive and noninteractive observations were transcribed or discussed with other staff members and thereby rethought; children's writing was typed out, the attention to every detail involving me in what the child was saying; notes of meetings and other events were rewritten for the record; and so on. Handling such detail with attention, I found, enabled me to see into the incident or piece of work in a way I hadn't on first look. Connections with other things I knew, with other observations I made, or questions I was puzzling over seemed to proliferate during these processes; new perceptions and new questions began to form.

I have heard others describe similarly their delighted discovery of the provocativeness of record-keeping processes. The teacher who begins to collect children's art, without perhaps even having a particular reason for the collecting, will, just by gathering the work together, begin to notice things about them that he or she had not seen before—how one child's work influences another's, how really different (or similar) are the trees they make, and so on. The in-school advisor or resource teacher who reviews all his or her contacts with teachers—as they are recorded or in a special meeting with his or her colleagues—may begin, for example, to see patterns of similar interest in the requests he or she is getting and thus become aware of new possibilities for relationships within the school.

My own delight in this apparently easy access to a first level of insight made me eager to collect more and more, and I also found the sheer bulk of what I could collect satisfying. As I collected more records, however, my enthusiasm gradually changed to alarm and frustration. There were so many things that could be observed and recorded, so many perspectives, such a complicated history! My feelings of wanting more changed to a feeling of needing to get everything. It wasn't enough for me to know how the program worked now—I felt I needed to know how it got started and how the present workings had evolved. It wasn't enough to know how the central part of the program worked—I felt I had to know about all its spinoff activities and from all points of view. I was quickly drawn into a fear of losing something significant, something I might need later on. Likewise, in my early observations of class sessions, I sought to write down everything I saw. I have had this experience of wanting to get everything in every setting in which I have documented, and I think it is not unique.

I was fortunate enough to be able to indulge these feelings and to learn from where they led me. It did become clear to me after a while that my early ambitions for documenting everything far exceeded my time and, indeed, the needs of the program. Nevertheless, there was a sense to them. Collecting so much was a way of getting to know a new setting, of orienting myself. And, not knowing the setting, I couldn't know what would turn out to be important in "reconstituting" it; also, the purpose of "reconstituting" it was sufficiently broad to include any number of possibilities from which I had not yet selected. In fact, I found that the first insights, the first connections that came from gathering the records were a significant part of the process of determining what would be important and what were the possibilities most suited to the purposes of the documentation. The process of gathering everything at first turned out to be important and, I think, needs to be allowed for at the beginning of any documenting effort. Even though much of the material so gathered may remain apparently unused, as it was in my documenting, in fact it has served its purpose just in being collected. A similar process may be required even when the documenter is already familiar with the setting, since the new role entails a new perspective.

The first connections, the first patterns emerging from the accumulating records were thus a valuable aspect of the documenting process. There came a moment, however, when the data I had collected seemed more massive than was justified by any thought I'd had as a result of the collecting. I was ill at ease because the first patterns were still fairly unformed and were not automatically turning into a documentation in the full sense I gave earlier, even though I recognized them as part of the documentary data. Particularly, they did not function as "evaluation." Some further development was needed, but what? "What do I do with them now?" is a cry I have heard regularly since then from teachers and others who have been collecting records for a while.

I began with the relatively simple procedure of rereading everything I had gathered. Then, I returned to rethink what my purposes were and sought out my original resources on documentation. Rereading qualitative references, talking with the staff of the school and with my staff colleagues, I began to imagine a shape I could give to my records that would make a coherent representation of the program to an outside audience.

At the same time, I began to rethink how I could make what I had collected more useful to the staff. Conceiving an audience was very important at this stage. I will be returning to this moment of transition from initial collecting to rethinking later, to analyze the entry into interpretation that it entails. Descriptively, however, what occurred was that I began to see my observations and records as a body with its own configurations, interrelationships, and possibilities, rather than simply as excerpts of the larger program that related only to the program. Obviously, the observations and records continued to have meaning through their primary relationship to the setting in which they were made; but they also began to have meaning through their secondary relationships to each other.

These secondary relationships also emerge from observation as a process of reflecting. Here, however, the focus of observation is the setting as it appears in and through the observations and records that have accumulated, with all their representation of multiple perspectives and longitudinal dimensions. These observations in and through records —"thickened observations"—are of course confirmed and added to by continuing direct observation of the setting.

Beginning to see the records as a body and the setting through thickened observation is a process of integrating data. The process occurs gradually and requires a broad base of observation about many aspects of the program over some period of time. It then requires concentrated and systematic efforts to find connections within the data and weave them into patterns, to notice changes in what is reported, and find the relationship of changes to what remains constant. This process is supported by juxtaposing the observations and records in various ways as well as by continual return to reobserve the original phenomenon. There is, in my opinion, no way to speed up the process of documenting. Reflectiveness takes time.

In retrospect, I can identify my own approach to an integration of the data as the time when I began to give my opinions on long-range decisions and interpretations of daily events with the ease of any other staff member. Up to the moment of transition, I shared specific observations from the records and talked them over as a way of gathering yet more perspectives on what was happening. I was aware, however, that my opinions or interpretations were still personal. They did not yet represent the material I was collecting.

(Continued)

(Continued)

Thus, it may be that integration of the documentary material becomes apparent when the documenter begins to evince a broad perspective about what is being documented, a perspective that makes what has been gathered available to others without precluding their own perceptions. This perspective is not a fixed-point view of a finished picture, both the view and the picture constructed somehow by the documenter in private and then unveiled with a flourish. It is also not a personal opinion; nor does it arise from placing a predetermined interpretive structure or standard on the observations. The perspective results from the documenter's own current best integration of the many aspects of the phenomenon, of the teachers' or staff's aims, ideas, and current struggles, and of their historical development as these have been conveyed in the actions that have been observed and the records that have been collected.

As documenter, my perspective of a program or a classroom is like my perspective of a landscape. The longer I am in it, the sharper defined become its features, its hills and valleys, forests and fields, and the folds of distance; the more colorful and yet deeply shaded and nuanced in tone it appears; the more my memory of how it looks in other weather, under other skies, and in other seasons, and my knowledge of its living parts, its minute detail, and its history deepen my viewing and valuing of it at any moment. This landscape has constancy in its basic configurations, but is also always changing as circumstances move it and as my perceptions gather. The perspective the documenter offers to others must evoke the constancy, coherence, and integrity of the landscape, and its possibilities for changing its appearance. Without such a perspective, an organization or integration that is both personal and informed by all that has been gathered by myself and by others in the setting—others could not share what I have seen—could not locate familiar landmarks and reflect on them as they exhibit new relationships to one another and to less familiar aspects. All that material, all those observations and records, would be a lifeless and undoubtedly dusty pile.

The process of forming a perspective in which the data gathered are integrated into an organic configuration is obviously a process of interpretation. I had begun documenting, however, without an articulated framework for interpretation or a format for representation of the body of records, like the theoretical framework researchers bring to their data. Of course, there was a framework. Conceptions of artistic process, of learning and development, were inherent in the program; but these were not explicit in its goals as a program to provide certain kinds

of service. The plan of the documentation had called for certain results, but there was no specified format for presentation of results. Therefore, my entry into interpretation became a struggle with myself over what I was supposed to be doing. It was a long internal debate about my responsibilities and commitments.

When I began documenting this particular school's art program, for example, I had priorities based on my experience and personal commitments. It seemed to me self-evidently important to provide art activities for children and to try and connect these to other areas of their learning. I knew that art was not something that could be "learned" or even experienced on a once-a-week basis, so I thought it was important to help teachers find various ways of integrating art and other activities into their classrooms. I had already made a personal estimate that what I was documenting was worthwhile and honest. I had found points of congruence between my priorities and the program. I could see how the various structures of the program specified ways of approaching the goals that seemed possible and that also enabled the elaboration of the goals.

This initial commitment was diffuse; I felt a kind of general enthusiasm and interest for the efforts I observed and a desire to explore and be helpful to the teachers. In retrospect, however, the commitment was sufficiently energizing to sustain me through the early phases of collecting observations and records, when I was not sure what these would lead to. Rather than restricting me, the commitment freed me to look openly at everything (as reflected in the early enthusiasm for collecting everything). Obviously, it is possible to begin documenting from many other positions of relative interest and investment, but I suspect that even if there is no particular involvement in program content on the part of the documenter, there must be at least some idea of being helpful to its staff. (Remember, this was a formative evaluation.) Otherwise, for example, the process of gathering data may be circumscribed.

At the point of beginning to "do something" with the observations and records, I was forced to specify the original commitment, to rethink my purposes and goals. Rereading the observations and records as a preliminary step in reworking to address different audiences, I found myself at first reading with an idea of "balancing" success and failure, an idea that constricted and trivialized the work I had observed and recorded. Thankfully, it was immediately evident from the data itself that such balance was not possible. If, during 10 days of observation, a child's experience was intense 1 day and characterized by rowdy socializing the other 9, a simple weigh-off would not establish the success or failure of the child's experience. The idea was ludicrous. Similarly, the staff might be thorough in its planning and follow-through on one day and disorganized on another day, but organization and planning were clearly not the totality of the experience for children.

Such trade-offs implied an external, stereotyped audience awaiting some kind of quantitative proof, which I was supposed to provide in a disinterested way, like an external, summative evaluator. The "balanced view" phase was also like my early record gathering of everything. What I was documenting was still in fragments for me, and my approach was to the particulars, to every detail.

A second approach to interpreting, also brief, took a slightly broader view of the data, a view that acknowledged my original estimate of program value and attempted to specify it. Perceiving through the data the landscape-like configurations of program strengths, I made assessments that included statements of past mistakes or inadequacies like minor "flaws" in the landscape (e.g., a few odd billboards and a garbage dump in one of Poussin's dreams of classical Italy) rather than debits on a balance sheet. Here again, the implication was of an external audience, expecting some absolute of accomplishment. The "flaws" could be "minor" only by reference to an implied major flaw—that of failing to carry out the program goals altogether.

The formulation of strength subsuming weakness could not withstand the vitality of the records I was reading. The reality the data portrayed became clearer as the inadequacy of my first formulations of how to interpret the documentary material was revealed. Similarly, the implications of external audience expectations were not justified by the actuality of my relationship to the program and staff. My stated goal as documenter had been originally to set up record-keeping procedures that would preserve and make available to staff and to other interested persons aspects of the beginnings and workings of the program, and to collect and analyze some of the material as an assessment of what further possibilities for development actually existed. My goals had not been to evaluate in the sense of an external judgment of success or failure.

Thinking over what other approaches to interpretation were possible, I recalled that I had gathered documentary materials quite straightforwardly as a participant, whose engagement was initially through recognition of shared convictions and points of congruence with the program. Perhaps, I decided, I could share my viewpoint of the observations just as straightforwardly, as a participant with

a particular point of view. In examining this possibility, I came to a view of interpreting observational data as a process of "rendering," much as a performer renders a piece of classical music. The interpretation follows a text closely—as a scientist might say, it sticks closely to the facts. But it also reflects the performer, specifically the performer's particular manner of engagement in the enterprise shared by text and performer, the enterprise of music. The same relationship could exist, it seemed to me, between a body of observations and records gathered participatively and as documenter. The relationship would allow my personal experience and viewpoint to enhance rather than distort the data. Indeed, I would become their voice.

Through this relationship I could make the observations available to staff and to other audiences in a way that was flexible and responsive to *their* needs, purposes, and standards. In so doing, of course, the framework of inherent conceptions underlying the work of the program would be incorporated. Thus, to interpret the observational data I had gathered, I had to reaffirm and clarify my relationship, my attachment to and participation in the program.

My initial engagement, with its strong coloring of prior interests and ideas, had never meant that I understood or was sympathetic with every goal or practice of every participant of the program all the time. In any joint enterprise, such as a school or program, there are diverse and multiple goals and practices. Part of the task of documenting is to describe and make these various understandings, points of view, and practices visible so that participants can reflectively consider them as the basis for planning. No participant agrees on all issues and points of practice. Part of being a participant is exploring differences and how these illuminate issues or contribute to practice. My participation allowed me to examine and extend the interests and ideas I came with as well as observing and recording those other people brought. In this process, my engagement was deepened, enabling me to make assessments closer to the data than my first readings brought. These assessments are evaluation in its original sense of "drawing-value-from," an interactive process of valuing, of giving weight and meaning.

In the context of renewed engagement and deepened participation, assessments of mistakes or inadequacies are construed as discrepancies between a particular practice and the intent behind it, between immediate and long-range purposes. The discrepancy is not a flaw in an otherwise perfect surface, but—like the discrepancy in a child's understanding that stimulates new learning—is the

(Continued)

(Continued)

occasion for growth. It is a sign of life and possibility. The burden of the discrepancy can lie either with the practice or with the intent, and that is the point for further examination. Assessment can also occur through the observation of and search for underlying themes of continuity between present and past intent and practice, and the point of change or transformation in continuity. Whereas discrepancy will usually be a more immediate trigger to evaluation, occasions for the consideration of continuity may tend to be longer-range planning for the coming year, contemplating changes in staff and function, or commemorating an anniversary.

I have located the documenter as participant, internal to the program or setting, gathering and shaping data in ways that make them available to participants and potentially to an external audience. Returning to the image of a landscape, let me comment on the different forms availability assumes for these different audiences.

Participant access to the landscape through the documenter's perspective cannot be achieved through ponderous written descriptions and reports on what has been observed but must be concentrated in interaction. Sometimes this may require the development of special or regular structures—a series of short-term meetings on a particular issue or problem; an occasional event that sums up and looks ahead; a regular meeting for another kind of planning. But many times the need is addressed in very slight forms, such as a comment in passing about something a child or adult user is doing, about the appearance of a display, or the recounting of another staff member's observation. I do not mean that injecting documentation into the self-assessment process is a juggling act or some feat of manipulation; merely that the documenter must be aware that his or her role is to keep things open and that, while the observations and records are a resource for doing this, a sense of the whole they create is also essential. The landscape is, of course, changed by the new observations offered by fellow viewers.

The external audience places different requirements on the documenter who seeks to represent to it the documentary perspective. By external audience I refer to funding agencies, supervisors, school boards, institutional hierarchies, and researchers. Proposals, accounts, and reports to these audiences are generally required. They can be burdensome because they may not be organically related to the process of internal self-reflection and because the external audience has its own standards, purposes, and questions; it is unfamiliar with the setting and with the documenter, and it needs the time offered by written accounts to return and review the material. The external

audience will need more history and formal description of the broad aspects than the internal audience, with commentary that indicates the significance of recent developments. This need can be met in the overall organization, arrangement, and introduction of documents, which also convey the detail and vividness of daily activity.

To limit the report to conventional format and expectations would probably misrepresent the quality of thought, of relating, of self-assessment that goes into developing the work. If there is intent to use the occasion of a report for reflection—for example, by including staff in the development of the report—the reporting process can become meaningful internally while fulfilling the legitimate external demands for accounting. Naturally, such a comment engages the external audience in its own evaluative reflections by evoking the phenomenon rather than reducing it.

In closing, I return to what I see as the necessary engaged participation of the documenter in the setting being documented, not only for data gathering but for interpretation. Whatever authenticity and power my perspective as documenter has had has come, I believe, from my commitment to the development of the setting I was documenting and from the opportunities in it for me to pursue my own understanding, to assess and reassess my role, and to come to terms with issues as they arose.

We come to new settings with prior knowledge, experience, and ways of understanding, and our new perceptions and understandings build on these. We do not simply look at things as if we had never seen anything like them before. When we look at a cluster of light and dark greens with interstices of blue and some of deeper browns and purples, what we identify is a tree against the sky. Similarly, in a classroom we do not think twice when we see, for example, a child scratching his head, yet the same phenomenon might be more strictly described as a particular combination of forms and movements. Our daily functioning depends on this kind of apparently obvious and mundane interpretation of the world. These interpretations are not simply personal opinions—though they certainly may be unique—nor are they made up. They are instead organizations of our perceptions as "tree" or "child scratching" and they correspond at many points with the phenomena so described.

It is these organizations of perception that convey to someone else what we have seen and that make objects available for discussion and reflection. Such organizations need not exclude our awareness that the tree is also a cluster of colors or that the child scratching his head is also a small human form raising its hand in a particular way. Indeed, we know that there could be many other ways to describe the same phenomena, including some that would be completely numerical—but not necessarily more accurate, more truthful, or more useful! After all, we organize our perceptions in the context of immediate purposes and relationships. The organizations must correspond to the context as well as to the phenomenon.

Facts do not organize themselves into concepts and theories just by being looked at; indeed, except within the framework of concepts and theories, there are no scientific facts but only chaos. There is an inescapable a priori element in all scientific work. Questions must be asked before answers can be given. The questions are all expressions of our interest in the world; they are at bottom valuations. Valuations are thus necesssarily involved already at the stage when we observe facts and carry on theoretical analysis and not only at the stage when we draw political inferences from facts and valuations (Myrdal, 1969, p. 9).

My experience suggests that the situation in documenting is essentially the same as what I have been describing with the tree and the child scratching and what Myrdal describes as the process of scientific research. Documentation is based on observation, which is always an individual response both to the phenomena observed and to the broad purposes of observation. In documentation, observation occurs both at the primary level of seeing and recording phenomena and at secondary levels of re-observing the phenomena through a volume of records and directly, at later moments. Since documentation has as its purpose to offer these observations for reflections and evaluation in such a way as to keep alive and open the potential of the setting, it is essential that observations at both primary and secondary levels be interpreted by those who have made them. The usefulness of the observations to others depends on the documenter's rendering them as finely as he or she is able, with as many points of correspondence to both the phenomena and the context of interpretation as possible. Such a rendering will be an interpretation that preserves the phenomena and so does not exclude but rather invites other perspective.

Of course, there is a role for the experienced observer from outside who can see phenomenon freshly; who can suggest ways of obtaining new kinds of information about it, or, perhaps more important, point to the significance of already existing procedures or data; who can advise on technical problems that have arisen within a documentation; and who can even guide efforts to interpret and integrate documentary information. I am stressing, however, that the outside observer in these instances provides support, not judgment or the criteria for judgment.

The documenter's obligation to interpret his or her observations and those reflected in the records being collected becomes increasingly urgent, and the interpretations become increasingly significant, as all the observers in the setting become more knowledgeable about it and thus more capable of bringing range and depth to the interpretation. Speaking of the weight of her observations of the Manus over a period of some 40 years to great change, Margaret Mead clarifies the responsibility of the participant–observer to contribute to both people studied and to a wider audience the rich individual interpretation of his or her own observations:

> Uniqueness, now, in a study like this (of people who have come under the continuing influence of contemporary world culture), lies in the relationships between the fieldworker and the material. I still have the responsibility and incentives that come from the fact that because of my long acquaintance with this village I can perceive and record aspects of this people's life that no one else can. But even so, this knowledge has a new edge. This material will be valuable only if I myself can organize it. In traditional fieldwork, another anthropologist familiar with the area can take over one's notes and make them meaningful. But here it is my individual consciousness that provides the ground on which the lives of these people are figures. (Mead, 1977, pp. 282–283)

In documenting, it seems to me the contribution is all the greater, and all the more demanded, because what is studied is one's own setting and commitment.

APPLICATION EXERCISES

1. Locate a published qualitative study on a subject of interest to you. How does the study address and establish the credibility of qualitative inquiry? Use Exhibits 9.4 and 9.15 to review and critique how credibility is addressed in the study you've chosen. What questions are left unanswered in the study you're reviewing that, from your perspective, if answered, would enhance credibility?

2. Locate a study that highlights use of mixed methods. What was the nature of the mix? What rationale was used for mixing methods? To what extent was *triangulation* an explicit justification for mixing methods? How integrated was the analysis of qualitative and quantitative data? Based on your review, what are the strengths and weaknesses of the mixed-methods design and analysis you reviewed.

3. Exhibit 9.11 (pp. 707–709) provides the framework for establishing the credibility of a qualitative inquirer. If you have conducted a qualitative study, or been part of one, complete that table using your own experience (fill in the column for yourself that reports Nora Murphy's experiences, perspectives, reactions, and competence in Exhibit 9.11). If you haven't done a qualitative study, imagine one and complete the table for a qualitative scenario that you construct. The purpose is to practice being reflexive and addressing inquirer credibility.

4. The discussion on objectivity considers a number of alternative ways of describing an inquirer's stance and philosophy (pp. 723–728). What is your preferred terminology? Write a statement describing your paradigm stance, a statement that you could give someone who was considering funding you to do a qualitative study. Describe a scenario or situation where you would need to explain your stance—and then do so. (You don't have to be limited to the language options discussed here.)

5. a. As an exercise in distinguishing quality criteria frameworks, try matching the three umpires' perspectives (p. 683) to the frameworks in Exhibit 9.7 (pp. 680–881). Explain your choices.

   b. What would a systems-oriented umpire say about umpiring? (Explain).

   c. What would an artistic-evocative-oriented umpire say about umpiring (Explain).

   d. What would a critical change umpire say?

6. (Advanced application) On the next page is a description of an edited volume of qualitative inquiries into the nature of family. Use the criteria for autoethnography in Chapter 3 (pp. 102–104) and the sets of criteria in Exhibit 9.7. Create your own set of 10 criteria for judging the methodological quality of this book by selecting criteria that seem especially relevant given the description of the book's approach. Use this example to discuss the nature of quality criteria in judging the quality of qualitative inquiries.

From *On (Writing) Families: Autoethnographies of Presence and Absence, Love and Loss* (Wyatt & Adams, 2014):

Who are we with—and without—families? How do we relate as children to our parents, as parents to our children? How are parent–child relationships—and familial relationships in general—made and (not) maintained?

Informed by narrative, performance studies, poststructuralism, critical theory, and queer theory, contributors to this collection use autoethnography—a method that uses the personal to examine the cultural—to interrogate these questions. The essays write about/around issues of interpersonal distance and closeness, gratitude and disdain, courage and fear, doubt and certainty, openness and secrecy, remembering and forgetting, accountability and forgiveness, life and death.

Throughout, family relationships are framed as relationships that inspire and inform, bind and scar—relationships replete with presence and absence, love and loss (p. 1).

7. (Advanced application) Martin Rees, astronomer, former Master of Trinity College, and ex-president of the Royal Society of Astronomy said, "Ultimately, I don't think there's anything special in the scientific method that goes beyond what a detective does" (quoted by Morris, 2014). Imagine that you are using this quotation to support the credibility of qualitative inquiry. Discuss how this quotation applies to each of the four dimensions of credibility discussed in this chapter (see Exhibit 9.15, p. 722)

References

Abbey, E. (1968). *Desert solitaire: A season in the wilderness*. New York, NY: McGraw-Hill.

Abdul-Quader, A. S., Heckathorn, D. D., Sabin, K., & Saidel, T. (2006). Implementation and analysis of respondent driven sampling: Lessons learned from the field. *Journal of Urban Health: Bulletin of the New York Academy of Medicine, 83*(7), i1–i5.

Abramovitch, H. (2014). *The anthropology of death*. Retrieved from http://henry-a.com/apage/115917.php

Abrams, L. (2010). *Oral history theory*. New York, NY: Routledge.

Academy of Achievement. (2005). *Barry Marshall biography*. Retrieved from http://www.achievement.org/autodoc/page/mar1bio-1

Achebe, C. (1994). *Things fall apart*. New York, NY: Anchor Books.

Ackerly, B., & True, J. (2010). *Doing feminist research in political and social science*. New York, NY: Palgrave Macmillan.

Ackoff, R. L. (1999a). *Ackoff's best: His classic writings on management*. New York, NY: Wiley.

Ackoff, R. L. (1999b). *Re-creating the corporation: A design of organizations for the 21st century*. New York, NY: Oxford University Press.

Ackoff, R. L., & Emery, F. E. (2005). *On purposeful systems: An interdisciplinary analysis of individual and social behavior as a system of purposeful events*. New York, NY: Transaction.

Active Learning Network for Accountability and Performance in Humanitarian Action. (2014, March). Engaging disaster-affected people in humanitarian action. Paper presented at the annual conference, Addis Ababa, Ethiopia. Retrieved from http://www.alnap.org/

Adams, C., & Van Manen, M. (2008). Phenomenology. In L. M. Given (Ed.), *The SAGE encyclopedia of qualitative research methods* (Vol. 2, pp. 614–619). Thousand Oaks, CA: Sage.

Adams, T. E. (2011). *Narrating the closet: An autoethnography of same-sex attraction*. Walnut Creek, CA: Left Coast Press.

Agafanoff, N. (2014). *Real ethnography*. Retrieved from http://www.realethnography.com.au/

Agar, M. (1999). Complexity theory: An exploration and overview based on John Holland's work. *Field Methods, 11*(2), 99–120.

Agar, M. (2000). Border lessons: Linguistic "rich points" and evaluative understanding. *New Directions for Evaluation, 86*, 93–109.

Agar, M. (2013). *The lively science: Remodeling human social research*. Minneapolis, MN: Mill City Press.

Albert, S. (2013). *When: The art of perfect timing*. San Francisco, CA: Jossey-Bass.

Alderson, P. (2001). Research by children. *International Journal of Social Research Methodology, 4*(2), 139–153.

Alkin, M. C. (1972). Wider context goals and goals-based evaluators. *Evaluation Comment: The Journal of Educational Evaluation, 3*(4), 10–11.

Alkin, M. C. (1975). Evaluation: Who needs it? Who cares? *Studies in Educational Evaluation, 1*, 201–212.

Alkin, M. C., Daillak, R., & White, P. (1979). *Using evaluations: Does evaluation make a difference?* Beverly Hills, CA: Sage.

Alkin, M. C., & Patton, M. Q. (1987). Working both sides of the street. *New Directions for Program Evaluation, 36*, 19–32.

Alkin, M. C., Vo, A. T., & Hanson, M. (Eds.). (2013). Using logic models to facilitate comparisons of evaluation theory. *Evaluation and Program Planning, 38*, 33.

Allen, C. (1997). Spies like us: When sociologists deceive their subjects. *Lingua Franca, 7*, 31–38.

Allen, D. (2013). "Just a typical teenager": The social ecology of "normal adolescence"—insights from diabetes care. *Symbolic Interaction, 36*(1), 40–59.

Allison, M. (2000). Enriching your practice with complex systems thinking. *OD Practitioner, 32*(3), 11–22.

Allman, W. F. (2012). The accidental history of the @ symbol. *Smithsonian Magazine, September*, p. 63.

Altheide, D. L., & Johnson, J. M. (2011). Reflections on interpretive adequacy in qualitative research. In N. K. Denzin & Y. S. Lincoln (Eds.), *The SAGE handbook of qualitative research* (4th ed., pp. 581–594). Thousand Oaks, CA: Sage.

Alvesson, M., & Sköldberg, K. (2009). *Reflexive methodology: New vistas for qualitative research* (2nd ed.). London, England: Sage.

American Association for Public Opinion Research. (2013). *Non-probability sampling* (Task Force report on Non-Probability Sampling). Retrieved from http://bit.ly/10bimiK

American Evaluation Association. (2003). *Scientifically based evaluation methods*. Retrieved from http://www.eval.org/p/cm/ld/fid=95

American Evaluation Association. (2013). *Statement on cultural competence in evaluation*. Retrieved from http://www.eval.org/p/cm/ld/fid=92

American Evaluation Association. (2014). *Statement on what is evaluation*. Retrieved from http://www.eval.org/p/bl/et/blogid=2&blogaid=4#a_comm_3

American Evaluation Association Task Force on Guiding Principles for Evaluators. (1995). Guiding principles for evaluators. *New Directions for Program Evaluation, 66*, 19–34.

American Psychiatric Association. (2013). *Autism spectrum disorder*. Retrieved from http://www.dsm5.org/Documents/Autism%20Spectrum%20Disorder%20Fact%20Sheet.pdf

Anderson, B. F. (1980). *The complete thinker: A handbook of techniques for creative and critical problem solving*. Englewood Cliffs, NJ: Prentice Hall.

Anderson, R. A., Crabtree, B. F., Steele, D. J., & McDaniel, R. R., Jr. (2005). Case study research: The view from complexity science. *Qualitative Health Research, 15*(5), 669–685.

Anderson, R. B. (1977, April). *The effectiveness of follow through: What have we learned?* Speech presented at the annual meeting of the American Educational Research Association, New York.

Anderson, V., & Johnson, L. (1997). *Systems thinking basics: From concepts to causal loops*. Cambridge, MA: Pegasus Communications.

Andrews, M., Squire, C., & Taboukou, M. (2008). *Doing narrative research*. London, England: Sage.

Angrosino, M., & Rosenberg, J. (2011). Observations on observations: Continuities and challenges. In N. Denzin & Y. Lincoln (Eds.), *The SAGE handbook of qualitative research* (4th ed., pp. 467–478). Thousand Oaks, CA: Sage.

Arcana, J. (1981). *Our mothers' daughters*. Berkeley, CA: Women's Press.

Arcana, J. (1983). *Every mother's son: The role of mothers in the making of men*. London, England: Women's Press.

Arditti, R. (1999). *Searching for life: The grandmothers of the Plaza de Mayo and the disappeared children of Argentina*. Berkeley: University of California Press.

Arendt, H. (1968). *Between past and future: Eight exercises in political thought*. New York, NY: Viking Press.

Argyris, C. (1982). *Reasoning, learning, and action: Individual and organizational*. San Francisco, CA: Jossey-Bass.

Argyris, C., Putnam, R., & Smith, D. M. (1985). *Action science*. San Francisco, CA: Jossey-Bass.

Argyris, C., & Schön, D. A. (1978). *Organizational learning: A theory of action perspective*. Reading, MA: Addison-Wesley.

Armstrong, D. M. (1992). *Managing by storying around*. New York, NY: Doubleday.

Armstrong, E. A., & Hamilton, L. T. (2013). *Paying for the party: How college maintains inequality*. Cambridge, MA: Harvard University Press.

Arthur, G. F. K. (2001). *Cloth as metaphor: Reading the Adinkra cloth symbols of the Akan of Ghana*. Accra, Ghana: Centre for Indigenous Knowledge Systems.

Asimov, I. (1983). Creativity will dominate our time after the concepts of work and fun have been blurred by technology. *Personal Administrator, 28*(2), 42–46.

Astbury, B. (2013). Some reflections on Pawson's science of evaluation: A realist manifesto. *Evaluation, 19*(4), 383–401.

Atkinson, R. (1998). *The life story interview.* Thousand Oaks, CA: Sage.

Atkinson, R. (2002). The life story interview. In J. F. Gubrium & J. A. Holstein (Eds.), *Handbook of interview research: Context and method* (pp. 121–140). Thousand Oaks, CA: Sage.

Atkinson, R. (2012). The life story interview as a mutually equitable relationship. In J. F. Gubrium, J. A. Holstein, A. B. Marvasti, & K. D. McKinney (Eds.), *The SAGE handbook of interview research: The complexity of the craft* (2nd ed., pp. 115–128). Thousand Oaks, CA: Sage.

Aubrey, R., & Cohen, P. M. (1995). *Working wisdom: Timeless skills and vanguard strategies for learning organizations.* San Francisco, CA: Jossey-Bass.

Australian Institute of Aboriginal and Torres Strait Islander Studies. (2011). *Guidelines for ethical research in Australian indigenous studies* (2nd ed.). Canberra, Australia: Author.

Avery, M. L. (2013, January 10). *IFTF remembers Robert B. Textor, anticipatory anthropologist* [Future Now blog]. Retrieved from http://www.iftf.org/future-now/article-detail/iftf-remembers-robert-b-textor-anticipatory-anthropologist/

Azumi, K., & Hage, J. (Eds.). (1972). Towards a synthesis: A system's perspective. In *Organizational systems: A text-reader in the sociology of organizations* (pp. 511–522). Lexington, MA: D. C. Heath.

Azzam, T., & Evergreen, S. (Eds.). (2013a). *Data visualization: Part 1* (*New Directions for Evaluation*, No. 139). Hoboken, NJ: Wiley Periodicals.

Azzam, T., & Evergreen, S. (Eds.). (2013b). *Data visualization: Part 2* (*New Directions for Evaluation*, No. 140). Hoboken, NJ: Wiley Periodicals.

Bach, H. (1998). *A visual narrative concerning curriculum, girls, photography etc.* Walnut Creek, CA: Left Coast Press.

Baert, P. (1998). *Social theory in the twentieth century.* New York: New York University Press.

Bailey, K. D. (1994). *Typologies and taxonomies: An introduction to classification techniques.* Thousand Oaks, CA: Sage.

Baker, A., & Sabo, K. (2004). *Participatory evaluation essentials: A guide for nonprofit organizations and their evaluation partners.* Cambridge, MA: Bruner Foundation.

Baldwin, J. (1990). *Notes of a native son.* Boston, MA: Beacon Press.

Ball, M. S., & Smith, G. W. H. (1992). *Analyzing visual data* (Qualitative Research Methods Series No. 24). London, England: Sage.

Ball, S. J. (2013). *Foucault, power and education.* New York, NY: Routledge.

Bamberger, M., & Podems, D. (2002). Feminist evaluation: Explorations and experiences. *New Directions for Evaluation, 96*, 83–96.

Bamberger, M. (2012, July). *Unanticipated consequences of development interventions: A blind spot for evaluation theory and practice.* Paper presented at the International Program for Development Evaluation Training, Carleton University, Ottawa, Ontario, Canada.

Bamberger, M. (2013). *Introduction to mixed-methods in impact evaluation.* Washington, DC: InterAction. Retrieved from http://www.interaction.org/document/guidance-note-3-introduction-mixed-methods-impact-evaluation

Bamberger, M. (2014). *Calendrier de la Femme Marocaine.* Retrieved from http://michaelbambergersculpture.com/wallcalendrier.htm

Bamberger, M., Rugh, J., & Mabry, L. (2011). *Real world evaluation: Working under budget, time, data, and political constraints* (2nd ed.). Thousand Oaks, CA: Sage.

Bandler, R., & Grinder, J. (1975). *The structure of magic* (Vols. 1 & 2). Palo Alto, CA: Science and Behavior Books.

Banks, M. (2007). *Using visual data in qualitative research.* Thousand Oaks, CA: Sage.

Barbour, R. (2007). *Doing focus groups.* London, England: Sage.

Barbour, R. S. (2001). Checklists for improving rigour in qualitative research: A case of the tail wagging the dog? *British Medical Journal, 322*(7294), 1115–1117.

Bare, J. (2013). Evaluation, accountability, and social change. *Foundation Review, 1*(4), 84–104.

Barnett, M. (2005). Chronic obstructive pulmonary disease: A phenomenological study of patients' experiences. *Journal of clinical nursing, 14*(7), 805–812.

Barnes, L. B., Christensen, C. R., & Hansen, A. J. (1994). *Teaching and the case method: Text, cases, and readings.* Cambridge, MA: Harvard University Press.

Barone, T. (2000). *Aesthetics, politics, and educational inquiry: Essays and examples.* New York, NY: Peter Lang.

Barone, T. (2001). *Touching eternity: The enduring outcomes of teaching.* New York, NY: Teachers College Press.

Barone, T. (2008). Arts-based research. In L. M. Given (Ed.), The *SAGE encyclopedia of qualitative research methods* (Vol. 1, pp. 29–36). Thousand Oaks, CA: Sage.

Barrett, K., & Greene, R. (2013, August 22). New Yorkers lose homes to the homeless. *B&G Report, Governing,* p. 1.

Barrington, G., Shimoni, R., & Legaspi, A. (2014). Case study site selection: Using an evidence-based approach in health care settings. *Sage Research Methods Cases.* Retrieved from http://srmo.sagepub.com/view/methods-case-studies-2013/n152.xml

Barton, D., Hamilton, M., & Ivanič, R. (2000). *Situated literacies: Reading and writing in context.* London, England: Routledge.

Bartunek, J., & Louis, M. R. (1996). *Insider/outsider team research: Vol. 40. Qualitative research methods.* Thousand Oaks, CA: Sage.

Bateson, G. (1977). The pattern which connects. *CoEvolution Quarterly, Summer,* 5–15.

Bateson, G. (1988). *Mind and nature.* New York, NY: Bantam Books.

Bateson, M. C. (2000). *Full circles, overlapping lives: Culture and generation in transition.* New York, NY: Random House.

BBC. (2004). *Lorenzo's oil: The full story.* Retrieved from http://news.bbc.co.uk/2/hi/health/3907559.stm

Beach, D., & Pedersen, R. B. (2013). *Process-tracing methods.* Ann Arbor: University of Michigan Press.

Beall, A. E. (2010). *Strategic market research: A guide to conducting research that drives businesses.* Bloomington, IN: iUniverse.

Beaudoin, C. E., & Thorson, E. (2004). Social capital in rural and urban communities: Testing differences in media effects and models. *Journalism & Mass Communication Quarterly, 81*, 378–399.

Beaver, D. (2012). *Public ethnography.* Retrieved from http://publicethnography.net/home

Becker, H. S. (1953). Becoming a Marihuana user. *American Journal of Sociology, 59*, 235–242.

Becker, H. S. (1967). Whose side are we on? *Social Problems, 14*(3), 239–247.

Becker, H. S. (1970). *Sociological work: Method and substance.* Chicago, IL: Aldine.

Becker, H. S. (1985). *Outsiders: Studies in the sociology of deviance.* New York, NY: Free Press.

Becker, H. S. (1998). *Tricks of the trade: How to think about your research while you're doing it.* Chicago, IL: University of Chicago Press.

Becker, H. S., & Geer, B. (1970). Participant observation and interviewing: A comparison. In W. J. Filstead (Ed.), *Qualitative methodology: Firsthand involvement with the social world* (pp. 133–159). Chicago, IL: Markham.

Becker, H. S., Geer, B., Hughes, E. C., & Strauss, A. (1961). *Boys in white: Student culture in medical school.* New Brunswick, NJ: Transaction.

Beebe, J. (2001). *Rapid assessment process: An introduction.* Walnut Creek, CA: AltaMira Press.

Beebe, J. (2008). Rapid assessment process. In L. M. Given (Ed.), *The SAGE encyclopedia of qualitative research methods* (Vol. 2, pp. 726–728). Thousand Oaks, CA: Sage.

Beh, H. G., & Diamond, M. (2006). *The failure of abstinence-only education: Minors have a right to honest talk about sex.* Manoa, Hawaii: University of Hawaii, Pacific Center for Sex and Society. Retrieved from http://www.hawaii.edu/PCSS/biblio/articles/2005to2009/2006-failure-of-abstinence-only-education.html

Belenky, M. F., Clinchy, B. M., Goldberger, N. R., & Tarule, J. M. (1986). *Women's way of knowing: The development of self, voice, and mind.* New York, NY: Basic Books.

Bell, J. S. (2002). Narrative inquiry: More than just telling stories. *TESOL Quarterly, 36*(2), 207–213.

Benedict, R. (1934). *Patterns of culture.* New York, NY: Houghton Mifflin.

Benko, S. S., & Sarvimaki, A. (2000). Evaluation of patient-focused health care from a systems perspective. *Systems Research and Behavioral Science, 17*(6), 513–525.

Benmayor, R. (1991). Testimony, action research, and empowerment: Puerto Rican women and popular education. In S. B. Gluck & D. Patai (Eds.), *Women's words: The feminist practice of oral history* (pp. 159–174). New York, NY: Routledge.

Benson, O. (2013). Whose pattern? *Free Inquiry, 33*(6), 11.

Bentz, V. M., & Shapiro, J. J. (1998). *Mindful inquiry in social research.* Thousand Oaks, CA: Sage.

Berens, L. V., & Nardi, D. (1999). *The 16 personality types: Description for self discovery.* New York, NY: Telos.

Berger, A. A. (2010). *The cultural theorist's book of quotations.* Walnut Creek, CA: Left Coast Press.

Berger, A. A. (2014). *What objects mean: An introduction to material culture* (2nd ed.). Walnut Creek, CA: Left Coast Press.

Berger, J. (1972). *Ways of seeing.* New York, NY: Penguin Books.

Berger, P. L., & Luckmann, T. (1967). *The social construction of reality: A treatise in the sociology of knowledge.* Garden City, NJ: Doubleday.

Bergman, M. M. (Ed.). (2008). *Advances in mixed methods research.* London, England: Sage.

Bering, J. (2012). *Why is the penis shaped like that? And other reflections on being human.* New York, NY: Farrar, Straus & Giroux.

Bering, J. (2013). *PERV: The sexual deviant in all of us.* New York, NY: Farrar, Straus & Giroux.

Berkes, F., & Folke, C. (Eds.). (2000). *Linking social and ecological systems: Management practices and social mechanisms for building resilience.* Cambridge, England: Cambridge University Press.

Berland, J. (1997). Nationalism and modernist legacy: Dialogues with Innis. *Culture and Policy, 8*(3), 9–39.

Bernard, H. R. (1994). *Research methods in anthropology: Qualitative and quantitative approaches.* Thousand Oaks, CA: Sage.

Bernard, H. R. (1998). *Handbook of methods in cultural anthropology.* Walnut Creek, CA: AltaMira Press.

Bernard, H. R. (2000). *Social research methods: Qualitative and quantitative approaches.* Thousand Oaks, CA: Sage.

Bernard, H. R. (2013). *Social research methods: Qualitative and quantitative approaches* (2nd ed.). Thousand Oaks, CA: Sage.

Bernard, H. R., & Ryan, G. W. (2010). *Analyzing qualitative data: Systematic approaches.* Thousand Oaks, CA: Sage.

Berry, D. M. (2004). Internet research: Privacy, ethics and alienation. *Internet Research, 14*(4), 323–332.

Best Practices, LLC. (2014). *Knowledge management of internal best practices.* Retrieved from http://www.best-in-class.com/

Better Evaluation. (2013). *Positive deviance approach.* Retrieved from http://betterevaluation.org/plan/approach/positive_deviance

Better Evaluation. (2014). *Collaborative outcomes reporting.* Retrieved from http://betterevaluation.org/plan/approach/cort

Beynon, P., & Fisher, C. (2013, February). *Learning about reflecting on knowledge: An approach for embedding reflective practice in an action research team* (IDS Practice Paper, Brief No. 9). Brighton, England: Institute of Development Studies. Retrieved from http://www.ids.ac.uk/publication/learning-about-reflecting-on-knowledge-an-approach-for-embedding-reflective-practice-in-an-action-research-team

Bhaskar, R. (1975). *A realist theory of science.* Leeds, England: Leeds Books.

Bierce, A. (1999). *The devil's dictionary.* New York, NY: Oxford University Press.

Biklen, S. K., & Casella, R. (2007). *A practical guide to the qualitative dissertation.* New York, NY: Teachers College Press.

Binnendijk, A. L. (1986). *AID's experience with contraceptive social marketing: A synthesis of project evaluation findings.* Washington, DC: U.S. Agency for International Development.

Blaikie, N. W. H. (1991). A critique of the use of triangulation in social research. *Quality & Quantity, 25*(2), 115–136.

Blater, J. K. (2008). Case study. In L. M. Given (Ed.), *The SAGE encyclopedia of qualitative research methods* (Vol. 1, pp. 68–71). Thousand Oaks, CA: Sage.

Blewett, M. H. (2009). *The Yankee Yorkshireman: Migration lived and imagined.* Urbana: University of Illinois Press.

Bloomberg, L. D., & Volpe, M. F. (2012). *Completing your qualitative dissertation: A road map from beginning to end.* Thousand Oaks, CA: Sage.

Blumenthal, D. S., DiClemente, R. J., Braithwaite, R., & Smith, S. (2013). *Community-based participatory health research: Issues, methods, and translation to practice* (2nd ed.). New York, NY: Springer.

Blumer, H. (1954). What is wrong with social theory? *American Sociological Review, 19,* 3–10.

Blumer, H. (1969). *Symbolic interactionism: Perspective and method.* Englewood Cliffs, NJ: Prentice Hall.

Blumer, H. (1978). Methodological principles of empirical science. In N. K. Denzin, (Ed.), *Sociological methods: A sourcebook* (pp. 20–41). New York, NY: McGraw-Hill.

Boas, F. (1943). Recent Anthropology. *Science, 98,* 311–337.

Bochner, A. P. (2001). Narrative's virtues. *Qualitative Inquiry, 7*(2), 131–157.

Bochner, A. P. (2014). *Coming to narrative: A personal history of paradigm change in the human sciences.* Walnut Creek, CA: Left Coast Press.

Bochner, A. P., Ellis, C., & Tillman-Healy, L. (1997). Relationships as stories. In S. Duck (Ed.), *Handbook of personal relationships: Theory, research and interventions* (2nd ed., pp. 307–324). New York, NY: Wiley.

Boeije, H. (2010). *Analysis in qualitative research.* London, England: Sage.

Boellstorff, T., Nardi, B., & Pearce, C. (2012). *Ethnography and virtual worlds: A handbook of method.* Princeton, NJ: Princeton University Press.

Boeri, M. (2002). Women after the utopia: The gendered lives of former cult members. *Journal of Contemporary Ethnography, 31*(3), 323–360.

Bogdan, R. (2012). *Picturing disability: Beggar, freak, citizen and other photographic rhetoric.* New York: Syracuse University Press.

Bogdan, R., & Biklen, S. K. (1992). *Qualitative research for education: An introduction to theory and methods.* Boston, MA: Allyn & Bacon.

Bold, C. (2012). *Using narrative in research.* Thousand Oaks, CA: Sage.

Bonnet, D., Ranney, J., Snow, J., Coplen, M., & Patton, M. Q. (2009). *Evaluation of the switching operations fatality analysis 2010 working groups processes.* Cambridge, MA: John A. Volpe National Transportation.

Borer, M. I., & Fontana, A. (2012). Postmodern trends: Expanding the horizons of interviewing practices and epistemologies. In J. F. Gubrium, J. A. Holstein, A. B. Marvasti, & K. D. McKinney (Eds.), *The SAGE handbook of interview research: The complexity of the craft* (2nd ed., pp. 45–60). Thousand Oaks, CA: Sage.

Bornat, J. (2004). The interview and its analysis: Oral history. In S. S. Lewis-Beck, A. Bryman, & T. F. Liao (Eds.), *The SAGE encyclopedia of social science methods* (Vol. 2, pp. 770–772). Thousand Oaks, CA: Sage.

Boston Women's Teachers' Group. (1986). *The effect of teaching on teachers.* Grand Forks: University of North Dakota, North Dakota Study Group on Evaluation.

Boulding, K. E. (1985). *Human betterment.* Beverly Hills, CA: Sage.

Box, G. E., & Draper, N. R. (1987). *Empirical model-building and response surfaces.* New York, NY: Wiley.

Boxill, N. A. (Ed.). (1990). *The watchers and waiters: America's homeless children.* Pittsburgh, PA: Haworth.

Boyatzis, R. E. (1998). *Transforming qualitative information: Thematic analysis and code development.* Thousand Oaks, CA: Sage.

Boyd, D. (2014). *It's complicated: The social lives of networked teens.* New Haven, CT: Yale University Press.

Boylorn, R. (2008). Participants as co-researchers. In L. M. Given (Ed.), *The SAGE encyclopedia of qualitative research methods* (Vol. 2, pp. 599–601). Thousand Oaks, CA: Sage.

Brady, I. (2000). Anthropological poetics. In N. K. Denzin & Y. S. Lincoln (Eds.), *Handbook of qualitative research* (pp. 949–979). Thousand Oaks, CA: Sage.

Bragg, D. D., Kirby, C., Witt, M. A., Richie, D., Mix, S., Feldbaum, M., & Mason, M. (2014). *Transformative change initiative.* Urbana-Champaign, IL: University of Illinois, Office of Community College Research and Leadership. Retrieved from http://occrl.illinois.edu/files/Projects/CCTCI/2014-tci-booklet.pdf

Brajuha, M., & Hallowell, L. (1986). Legal intrusion and the politics of fieldwork: The impact of the Brajuha case. *Urban Life, 1*(4), 454–478.

Brandon, P. R., & Ah Sam, A. L. (2014). Program evaluation. In P. Leavy (Ed.), *The Oxford handbook of qualitative research* (pp. 471–497). New York, NY: Oxford University Press.

Brandon, P. R., & Fukunaga, L. L. (2014). The state of empirical research literature on stakeholder involvement in program evaluation. *American Journal of Evaluation, 25*(1), 26–44.

Brandon, P. R., Lawton, B. E., & Harrison, G. M. (2014). Issues of rigor and feasibility when observing the quality of program implementation: A case study. *Evaluation and Program Planning, 44,* 75–80.

Brandon, P. R., Lindberg, M. A., & Wang, Z. (1993). Involving program beneficiaries in the early stages of evaluation: Issues of consequential validity and influence. *Educational Evaluation and Policy Analysis, 15*(4), 420–428.

Braverman, M. T. (2013). Negotiating measurement methodological and interpersonal considerations in the choice and interpretation of instruments. *American Journal of Evaluation, 34*(1), 99–114.

Breustedt, S., & Puckering, C. (2013). A qualitative evaluation of women's experiences of the Mellow Bumps antenatal intervention. *British Journal of Midwifery, 21*(3), 187–194.

Brewer, J., & Hunter, A. (1989). *Multimethod research: A synthesis of styles.* Newbury Park, CA: Sage.

Brinkerhoff, R. O. (2003). *The success case method: Find out quickly what's working and what's not.* San Francisco, CA: Berrett-Koehler.

Brinkley, D. (1968). *Public broadcasting laboratory.* [Interview on public broadcasting service]. *The New Yorker,* August 8, p. 41.

Brisolara, S., Seigart, D., & SenGupta, S. (Eds.). (2014). *Feminist evaluation and research: Theory and practice.* New York, NY: Guilford Press.

Brizuela, B. M., Stewart, J. P., Carrillo, R. G., & Berger, J. G. (2000). *Acts of inquiry in qualitative research* (Reprint Series No. 34). Cambridge, MA: Harvard Educational Review.

Brookfield, S. D. (2004). *The power of critical theory: Liberating adult learning and teaching.* San Francisco: Jossey-Bass.

Brooks, A. C. (2014, July 20). Love people, not pleasure. *New York Times Sunday Review,* p. 1.

Brooks, D. (2010, March 26). The return of history. *The New York Times,* p. A27.

Brooks, D. (2011). *The social animal: The hidden sources of love, character, and achievement.* New York, NY: Random House.

Brooks, D. (2014, May 6). The streamlined life. *The New York Times,* p. A25.

Brown, J. R. (1996). *The I in science: Training to utilize subjectivity in research.* Oslo, Norway: Scandinavian University Press.

Brown, L. M., Tappan, M. B., Gilligan, C., Miller, B. A., & Argyris, D. E. (1989). Reading for self and moral voice: A method for interpreting narratives of real-life moral conflict and choice. In M. J. Packer & R. B. Addison (Eds.), *Entering the circle: Hermeneutic investigation in psychology* (pp. 141–164). Albany: State University of New York Press.

Brown, T. (2014, April 13). Providing the balm of truth. *The New York Times,* p. SR8.

Browne, A. (1987). *When battered women kill.* New York, NY: Free Press.

Browne, K., & Nash, C. J. (Eds.). (2010). *Queer methods and methodologies.* Burlington, VT: Ashgate.

Browning, D. (2014, May 14). Massive blast of measles vaccine wipes out cancer. *Minneapolis Star Tribune,* p. 1

Brunton, G., Stansfield, C., & Thomas, J. (2012). Finding relevant studies. In D. Gough, S. Oliver, & J. Thomas (Eds.). (2012). *An introduction to systematic reviews* (pp. 107–134). Thousand Oaks, CA: Sage.

Bruyn, S. (1963). The methodology of participant observation. *Human Organization, 21,* 224–235.

Bruyn, S. T. (1966). *The human perspective in sociology: The methodology of participant observation.* Englewood Cliffs, NJ: Prentice Hall.

Bryant, A., & Charmaz, K. (Eds.). (2010). *The SAGE handbook of grounded theory.* London, England: Sage.

Brydon-Miller, M., Kral, M., Maguire, P., Noffke, S., & Sabhlok, A. (2011). Jazz and the banyan tree: Roots and riffs on participatory action research. In N. K. Denzin & Y. S. Lincoln (Eds.), *The SAGE handbook of qualitative research* (4th ed., pp. 387–400). Thousand Oaks, CA: Sage.

Bryson, J. M., Ackerman, F., Eden, C., & Finn, C. B. (2004). *Visible thinking.* San Francisco, CA: Wiley.

Bryson, J. M., & Patton, M. Q. (2010). Analyzing and engaging stakeholders. In J. S. Wholey, H. P. Hatry, & K. E. Newcomer (Eds.), *Handbook of practical program evaluation* (3rd ed., pp. 30-54). San Francisco: Jossey-Bass.

Bubaker, S., Balakrishnan, P., & Bernadine, C. B. (2013). *Qualitative case study research in Africa and Asia: Challenges and prospects.* Retrieved from http://ro.uow.edu.au/cgi/viewcontent.cgi?article=2619&context =commpapers&sei-redir=1&referer=http%3A%2F%2Fscholar.google .com%2Fscholar%3Fhl%3Den%26q%3DBubaker.%2BBala%2B% 2526%2BBernadine%2BQualitative%26btnG%3D%26as_ sdt%3D1%252C24%26as_sdtp%3D%23search=%22Bubaker.%20 Bala%20%26%20Bernadine%20Qualitative%22

Buckholt, M. (2001). *Women's voices of resilience: Female adult abuse survivors define the phenomenon* (Unpublished doctoral dissertation). Union Institute, Cincinnati, OH.

Bullough, R. V., & Pinnegar, S. (2001). Guidelines for quality in autobiographical forms of self-study research. *Educational Researcher, 30*(3), 13–21.

Bunch, E. H. (2001). Quality of life in people with advanced HIV/AIDS in Norway. *Grounded Theory Review, 2,* 30–42.

Bunge, M. (1959). *Causality: The place of the causal principle in modern science.* Cambridge, MA: Harvard University Press.

Burdell, P., & Swadener, B. B. (1999). Critical personal narrative and autoethnography in education: Reflections on a genre. *Educational Researcher, 28*(6), 21–26.

Burns, D. (2007). *Systemic action research: A strategy for whole systems change.* Bristol, England: Policy Press.

Burns, T., & Stalker, G. M. (1972). Models of mechanistic and organic structure. In K. Azumi & J. Hage (Eds.), *Organizational systems: A text-reader in the sociology of organizations* (pp. 240–255). Lexington, MA: D. C. Heath.

Bushkin, H. (2013). *Johnny Carson.* New York, NY: Houghton Mifflin Harcourt.

Butler, L. M. (1995). The Sondeo: A rapid reconnaissance approach for rapid assessment. *Western Region Extension Publication, WREP0127.* Retrieved from http://wrdc.usu.edu/files/publications/publication/pub__1758647.pdf

Buvala, K. S. (2013). *Home page.* Retrieved from http://storyteller.net/

Buxton, A. (1982). *Children's journals: Further dimensions of assessing language development* (North Dakota Study Group on Evaluation). Grand Forks: University of North Dakota.

Caelli, K., Ray, L., & Mill, J. (2003). Clear as mud: Towards a greater clarity in generic qualitative research. *International Journal of Qualitative Methods, 2*(2), 1–23.

Caister, K., Green, M., & Worth, S. (2011). Learning how to be participatory: An emergent research agenda. *Action Research, 10*(1), 22–39.

Cameron, W. B. (1963). *Informal sociology: A casual introduction to sociological thinking.* New York, NY: Random House.

Campbell, D. T. (1999a). Legacies of logical positivism and beyond. In D. T. Campbell & M. J. Russo (Eds.), *Social experimentation* (pp. 131–144). Thousand Oaks, CA: Sage.

Campbell, D. T. (1999b). On the rhetorical use of experiments. In D. T. Campbell & M. J. Russo (Eds.), *Social experimentation* (pp. 149–158). Thousand Oaks, CA: Sage.

Campbell, D. T., & Russo, M. J. (Eds.). (1999c). *Social experimentation.* Thousand Oaks, CA: Sage.

Campbell, D. T., & Stanley, J. C. (1963). *Experimental and quasi-experimental designs for research.* Chicago, IL: Rand McNally.

Campbell, J. L. (1983). *Factors and conditions influencing usefulness of planning, evaluation and reporting in schools* (Unpublished doctoral dissertation). University of Minnesota, Minneapolis.

Campbell, M., Patton, M. Q., & Patrizi, P. (2003). *Changing stakeholder needs and changing evaluator roles: The Central Valley Partnership of the James Irvine Foundation.* Evaluation Roundtable. Retrieved from http://www.evaluationroundtable.org/documents/cs-changing -stakeholder-needs.pdf

CAQDAS Networking Project. (2014). *Practical support, training and information in the use of a range of software programs designed to assist qualitative data analysis.* Retrieved from http://www.surrey .ac.uk/sociology/research/researchcentres/caqdas/

Carden, F. (2009). *Knowledge to policy: Making the most of development research.* Ottawa, Ontario, Canada: International Development Research Centre.

Carey, M. A., Asbury, J., & Tolich, M. (2012). *Focus group research.* Walnut Creek, CA: Left Coast Press.

Carini, P. F. (1975). *Observation and description: An alternative method for the investigation of human phenomena* (North Dakota Study Group on Evaluation Monograph Series). Grand Forks: University of North Dakota.

Carini, P. F. (1979). *The art of seeing and the visibility of the person.* Grand Forks: University of North Dakota.

Carlin, G. (1997). *Brain droppings.* New York, NY: Hyperion.

Carr, N. G. (2010). *The shallows: What the internet is doing to our brains.* New York, NY: W. W. Norton.

Carr, N. G. (2014). *Anti-anecdotalism.* Retrieved from http://www.edge .org/responses/what-scientific-idea-is-ready-for-retirement

Caruthers, L., & Friend, J. (2014). Critical pedagogy in online environments as thirdspace: A narrative analysis of voices of candidates in educational preparatory programs. *Educational Studies, 50*(1), 8–35.

Castaneda, C. (1972). *Journey to Ixtlan: The lessons of Don Juan.* New York, NY: Simon & Schuster.

CBS News. (2007, February 25). Brundibar: How the Nazis conned the world. *60 Minutes*. Retrieved from www.cbsnews.com/stories/2007/02/23/60minutes/main2508458_page2.shtml

Center for Earth and Environmental Science. (2014). *Water quality studies*. Indianapolis: Indiana and Purdue Universities. Retrieved from http://www.cees.iupui.edu/DSE/water-quality-studies

Centers for Disease Control and Prevention. (2008). *Introduction to process evaluation in tobacco use prevention and control*. Atlanta, GA: Author. Retrieved from http://www.cdc.gov/tobacco/tobacco_control_programs/surveillance_evaluation/process_evaluation/

Cernea, M. M., & Guggenheim, S. E. (1985). *Is anthropology superfluous in farming systems research?* Washington, DC: World Bank.

Cervantes, S. M. (1964). *Don Quixote*. New York, NY: Signet Classics.

Chagnon, N. A. (1992). *Yanomamö: The last days of Eden*. New York, NY: Harcourt Brace.

Chamberlayne, P., Bornat, J., & Wengraf, T. (2000). *The turn to biographical methods in social science: Comparative issues and examples*. London, England: Routledge.

Chambers, E. (2000). Applied ethnography. In N. K. Denzin & Y. S. Lincoln (Eds.), *The handbook of qualitative research* (2nd ed., pp. 851–869). Thousand Oaks, CA: Sage.

Chambers, R., & Carruthers, I. D. (1986). *Rapid appraisal to improve canal irrigation performance: Experience and options* (IMI Research Paper No. 3). Colombo, Sri Lanka: International Irrigation Management Institute.

Chan, A. W., Hrobjartsson, A., Haahr, M. T., Gotzsche, P. C., & Altman, D. G. (2004). Empirical evidence for selective reporting of outcomes in randomized trials: Comparison of protocols to published articles. *Journal of the American Medical Association, 291*, 2457–2465.

Chandler, D. (2007). *Semiotics: The basics* (2nd ed.). New York, NY: Routledge.

Chang, H. (2008). *Autoethnography as method*. Walnut Creek, CA: Left Coast Press.

Chang, H., Wambura, F., & Hernandez, K. C. (2013). *Collaborative autoethnography*. Walnut Creek, CA: Left Coast Press.

Changing minds. (2013). *Snowball sampling*. Retrieved from http://changingminds.org/explanations/research/sampling/snowball_sampling.htm

Chapman, J. (2004). *System failure: Why governments must learn to think differently*. London, England: Demos.

Charmaz, K. (2000). Grounded theory: Objectivist and constructivist methods. In N. K. Denzin & Y. S. Lincoln (Eds.), *Handbook of qualitative research* (2nd ed., pp. 509–535). Thousand Oaks, CA: Sage.

Charmaz, K. (2009). Constructivist grounded theory. In J. M. Morse, P. N. Stern, J. Corbin, B. Bowers, K. Charmaz, & A. E. Clarke (Eds.), *Developing grounded theory: The second generation* (pp. 127–193). Walnut Creek, CA: Left Coast Press.

Charmaz, K. (2011). Grounded theory methods in social justice research. In N. K. Denzin & Y. S. Lincoln (Eds.). *The SAGE handbook of qualitative research methods* (pp. 359–380). Thousand Oaks, CA: Sage.

Charmaz, K., & Bryant, A. (2008). Grounded theory. In L. M. Given (Ed.), *The SAGE encyclopedia of qualitative research* (Vol. 2, pp. 374–377). Thousand Oaks, CA: Sage.

Charon, R., Greene, M. G., & Adelman, R. D. (1998). Qualitative research in medicine and health care: Questions and controversy, a response. *Journal of General Internal Medicine, 13*(January), 67–68.

Chase, S. E. (2008). Narrative inquiry. In L. M. Given (Ed.), *The SAGE encyclopedia of qualitative research methods* (Vol. 2, pp. 421–434). Thousand Oaks, CA: Sage.

Chatterjee, A. (2001). Language and space: Some interactions. *Trends in Cognitive Science, 5*(2), 55–61.

Chatterji, M. (2007). Grades of evidence variability in quality of findings in effectiveness studies of complex field interventions. *American Journal of Evaluation, 28*(3), 239–255.

Chazdon, S., & Alviz, K. (2013, May). *Ripple effect mapping: A tool for evaluating the impacts of complex interventions*. Paper presented at the University of Minnesota Evaluation Studies Institute Conference, St. Paul, MN. Retrieved from http://evaluation.umn.edu/wp-content/uploads/MESI13.Ripple-Effects-handouts.pdf

Checkland, P. (1999). *Systems thinking, systems practice: A 30-year retrospective*. New York, NY: Wiley.

Checkland, P., & Poulter, J. (2006). *Learning for action: A short definitive account of soft systems methodology and its use for practitioner, teachers, and students*. Chichester, NY: Wiley.

Chelimsky, E. (1983). Improving the cost effectiveness of evaluation. In M. C. Alkin & L. C. Solomon (Eds.), *The costs of evaluation* (pp. 149–170). Beverly Hills, CA: Sage.

Chelimsky, E. (1995). The political environment of evaluation and what it means for the development of the field. *Evaluation Practice, 16*(3), 215–225.

Chelimsky, E. (2006). The purposes of evaluation in a democratic society. In I. F. Shaw, J. C. Greene, & M. M. Mark (Eds.), The *SAGE handbook of evaluation: Policies, programs and practices* (pp. 33–55). Thousand Oaks, CA: Sage.

Chelimsky, E. (2007). Factors influencing the choice of methods in federal evaluation practice. *New Directions for Evaluation, 113*, 13–33.

Chen, H. T. (2014). *Practical program evaluation: Theory-driven evaluation and the integrated evaluation perspective*. Thousand Oaks, CA: Sage.

Chen, H. T., Donaldson, S., & Mark, M. (Eds.). (2011). *Advancing validity in outcomes evaluation: Theory and practice* (*New Directions for Evaluation*, No. 130). San Francisco, CA: Jossey-Bass.

Chen, H., & Rossi, P. H. (1987). The theory-driven approach to validity. *Evaluation and Program Planning, 10*(1), 95–103.

Chevalier, J. M., & Buckles, D. J. (2013). *Participatory action research: Theory and methods for engaged inquiry*. New York, NY: Routledge.

Chew, S. T. (1989). *Agroforestry projects for small farmers: A project manager's reference*. Washington, DC: U.S. Agency for International Development.

Chilisa, B. (2012). *Indigenous research methodologies*. Thousand Oaks, CA: Sage.

Christians, C. G. (2000). Ethics and politics in qualitative research. In N. K. Denzin & Y. S. Lincoln (Eds.), *Handbook of qualitative research* (2nd ed., pp. 133–155). Thousand Oaks, CA: Sage.

Christiansen, O., Scott, H., & Sørensen, S. E. (2013). A partial application of classic grounded theory in a study of poverty in Greenland. *Grounded Theory Review, 12*(2), 1–6.

Christie, C. A., & Alkin, M. C. (2013). An evaluation theory tree. In M. C. Alkin (Ed.), *Evaluation roots: A wider perspective of theorists' views and influences* (2nd ed.). Thousand Oaks, CA: Sage.

Christoplos, I. (2006). *Links between relief, rehabilitation and development in the tsunami response*. London, England: Tsunami Evaluation Coalition.

Cialdini, R. B. (2001). The science of persuasion. *Scientific American, 284*(2), 76–81.

Clandinin, D. J. (2013). *Engaging in narrative inquiry*. Walnut Creek, CA: Left Coast Press.

Clarke, A. E. (2005). *Situational analysis: Ground theory after the postmodern turn*. Thousand Oaks, CA: Sage.

Clarke, S., & Long, M. (2013). *Using cognitive interviewing to identify and resolve data collection problems* (American Evaluation Association Annual Conference Session No. 67). Washington, DC: American Evaluation Association.

Cleveland, H. (1985). *The knowledge executive: Leadership in an information society*. New York, NY: Dutton.

Clifford, J. (1988). *The predicament of culture*. Cambridge, MA: Harvard University Press.

Coalition for Evidence-Based Policy. (2014). *Evidence-based reform*. Retrieved from http://coalition4evidence.org/mission-activities/

Coffey, A., & Atkinson, P. (1996). *Making sense of qualitative data: Complementary research strategies*. Thousand Oaks, CA: Sage.

Coles, R. (1989). *The call of stories: Teaching and the moral imagination*. Boston, MA: Houghton Mifflin.

Coles, R. (1990). *The spiritual life of children*. Boston, MA: Houghton Mifflin.

Colic-Peisker, V. (2008). *Class and transnational identities: Croatians in Australia and America*. Urbana: University of Illinois Press.

Collier, D., Seawright, J., & Munck, G. L. (2010). The quest for standards. In H. E. Brady & D. Collier (Eds.), *Rethinking social inquiry: Diverse tools, shared standards* (pp. 33–63). New York, NY: Rowman & Littlefield.

Collingridge, D. S. (2013). A primer on quantitized data analysis and permutation test. *Journal of Mixed Methods Research, 7*(1), 81–97.

Collins, J. (2000, August). *The timeless physics of great companies*. Retrieved from http://www.jimcollins.com/article_topics/articles/the-timeless-physics.html

Collins, J. (2001a). *Good to great: Why some companies make the leap . . . and others don't*. New York, NY: HarperBusiness.

Collins, J. (2001b). Level 5 Leadership: The triumph of humility and fierce resolve. *Harvard Business Review, 79*(1), 67–76, 175.

Collins, J. (2009). *How the mighty fall: And why some companies never give in*. New York, NY: CollinsBusiness Book.

Collins, J. (2012). *It's in the research*. Retrieved from http://www .jimcollins.com/books/research.html

Collins, J., & Hansen, M. T. (2011). *Great by choice: Uncertainty, chaos, and luck—Why some thrive despite them all*. New York, NY: HarperBusiness.

Collins, J., & Porras, J. I. (2004). *Built to last: Successful habits of visionary companies*. New York, NY: HarperBusiness.

Columbia University Medical School. (2013). *Program in narrative medicine*. Retrieved from http://www.narrativemedicine.org/

Conner, R. F., Fitzpatrick, J. L., & Rog, D. J. (2012). A first step forward: Context assessment. In D. J. Rog, J. L. Fitzpatrick, & R. F. Conner (Eds.), *Context: A framework for its influence on evaluation practice* (*New Directions for Evaluation*, No. 135, pp. 89–105). Hoboken, NJ: Wiley Periodicals.

Connolly, D. R. (2000). *Homeless mothers: Face to face with women and poverty*. Minneapolis: University of Minnesota Press.

Connolly, K., & Reilly, R. C. (2007). Emergent issues when researching trauma: A Confessional tale. *Qualitative Inquiry, 13*(4), 522–540.

Connor, R. (1985). International and domestic evaluation: Comparisons and insights. In M. Q. Patton (Ed.), *Culture and evaluation* (pp. 19–28). San Francisco, CA: Jossey-Bass.

Conrad, J. (1960). *Heart of darkness*. New York, NY: Knopf.

Conroy, D. L. (1987). *A phenomenological study of police officers as victims* (Unpublished doctoral dissertation). Union Institute, Cincinnati, OH.

Constas, M. A. (1998). Deciphering postmodern educational research. *Educational Researcher, 27*(9), 36.

Cook, T. D.(2014). Generalizing causal knowledge in the policy sciences: External validity as a task of both multi-attribute representation and multi-attribute extrapolation. *Journal of Policy Analysis and Management,33(2),* 527–536.

Cook, T. D., Leviton, C., & Shadish, W. R. (1985). Program evaluation. In G. Lindzey & E. Aronson (Eds.), *Handbook of social psychology, theory and method* (3rd ed., Vol. 1, pp. 699–777). New York, NY: Random House.

Cooper, K., & White, R. E. (2012). *Qualitative research in the post-modern era*. New York, NY: Springer.

Cooper, S., & Endacott, R. (2007). Generic qualitative research: A design for qualitative research in emergency. *Emergency Medical Journal, 24*(12), 816–819.

Coryn, C. L. S., Schröter, D. C., & Hanssen, C. E. (2009). Adding a time-series design element to the success case method to improve methodological rigor: An application for nonprofit program evaluation. *American Journal of Evaluation, 30*(1), 80–92.

Costa, A., & Kallick, B. (1993). Through the lens of a critical friend. *Educational Leadership, 51*(2), 49–51.

Coulon, A. (1995). *Ethnomethodology* (Qualitative Research Methods Series No. 36). Thousand Oaks, CA: Sage.

Coursen, D. (2014, April 23). *The varieties of holistic human service information* [Human Service Infomatics blog]. Retrieved from http:// humanserviceinformatics.wordpress.com/2014/04/23/the-three -patterns-of-holistic-human-service-information/

Cousins, J. B., & Bourgeois, I. (Eds.). (2014). *Organizational capacity to do and use evaluation* (New Directions for Evaluation, No. 141). Hoboken, NJ: Wiley Periodicals.

Cousins, J. B., & Chouinard, J. A. (2012). *Participatory evaluation up close: An integration of research-based knowledge*. Charlotte, NC: Information Age.

Cousins, J. B., & Earl, L. M. (1992). The case for participatory evaluation. *Educational Evaluation and Policy Analysis, 14*(4), 397–418.

Cousins, J. B., & Earl, L. M. (1995). *Participatory evaluation in education: Studies in evaluation use and organizational learning*. London, England: Falmer Press.

Cousins, J. B., Whitmore, E., & Shulha, L. (2014). Let there be light. *American Journal of Evaluation, 35*(1), 149–153.

Covey, S. R. (2013). *The 7 habits of highly effective people: Powerful lessons in personal change*. New York, NY: Simon & Schuster.

Cowden, K. J., McDonald, L., & Littlefield, R. (2012). Using community based participatory action research as service-learning for tribal

college students. *Journal of Indigenous Research, 2*(1). Retrieved from http://digitalcommons.usu.edu/kicjir/vol2/iss1/1

Cowell, A. (2011, June 11). Germany says bean sprouts are likely *E. coli* source. *The New York Times*, p. A5. Retrieved from http://www .nytimes.com/2011/06/11/world/europe/11ecoli.html?hp

Cowie, B., Hipkins, R., Boyd, S., Bull, A., Keown, P., McGee, C., . . . Yates, R. (2009). *Curriculum implementation exploratory studies: Final report*. Wellington, New Zealand: Learning Media. Retrieved from http:// www.educationcounts.govt.nz/publications/curriculum/57760/4

Cox, A. (2005). What are communities of practice? A comparative review of four seminal works. *Journal of Information Science, 31*(6), 527–540.

Craig, P. (1978). *The heart of a teacher: A heuristic study of the inner world of teaching*. Boston, MA: Boston University Press.

Crave, M., Zaleski, K., & Trent, T. (2014). *Do your participatory methods contribute to an equitable future?* Retrieved from http://aea365.org/ blog/pd-presenters-week-mary-crave-kerry-zaleski-and-tererai-trent -on-do-your-participatory-methods-contribute-to-an-equitable -future/?utm_source=feedburner&utm_medium=email&utm_ campaign=Feed%3A+aea365+%28AEA365%29

Crawford, K. (2013). The hidden biases in big data. *HBR Blog Network*, p. 1.

Crebert, G., Patrick, C.-J., Cragnolini, V., Smith, C., Worsfold, K., & Webb, F. (2011). *Critical evaluation skills toolkit*. Retrieved from http://www .griffith.edu.au/__data/assets/pdf_file/0004/290659/Critical- evaluation-skills.pdf

Cressey, D. R. (1976). Intensive interviews. In W. B. Sanders (Ed.), *The sociologist as detective* (2nd ed., pp. 208–221). New York, NY: Praeger.

Creswell, J. W. (1998). *Qualitative inquiry and research design: Choosing among five*. Thousand Oaks, CA: Sage.

Creswell, J. W. (2009). *Research design: Qualitative, quantitative, and mixed methods approaches* (3rd ed.). Thousand Oaks, CA: Sage.

Creswell, J. W. (2012). *Qualitative inquiry and research design: Choosing among five traditions* (3rd ed.). Thousand Oaks, CA: Sage.

Creswell, J. W., & Clark, V. L. P. (2011). *Designing and conducting mixed methods research*. Thousand Oaks, CA: Sage.

Cronbach, L. J. (1975). Beyond the two disciplines of scientific psychology. *American Psychologist, 30*(2), 116–127.

Cronbach, L. J. (1980). *Toward reform of program evaluation*. San Francisco, CA: Jossey-Bass.

Cronbach, L. J. (1982). *Designing evaluations of educational and social programs*. San Francisco, CA: Jossey-Bass.

Crosby, P. B. (1979). *Quality is free: The art of making quality certain*. New York, NY: McGraw-Hill.

Crotty, M. (1998). *The foundations of social research: Meaning and perspective in the research process*. London, England: Sage.

Crow, G. (2008). Reciprocity. In L. M. Given (Ed.), *The SAGE encyclopedia of qualitative research methods* (Vol. 2, pp. 739–740). Thousand Oaks, CA: Sage.

Cueva, M., Dignan, M., Lanier, A., & Kuhnley, R. (2013). Qualitative evaluation of a colorectal cancer education CD-ROM for community health aides/practitioners in Alaska. *Journal of Cancer Education, November*. Retrieved from http://link.springer.com/ article/10.1007/s13187-013-0590-x

Culver, C. L. (2013). *The cognitive interview: A witness interview technique for the field researcher*. Hiawassee, GA: Enigma Research group. Retrieved from http://www.enigmaresearchgroup.com/article27.htm

Cumpa, J., & Tegtmeier, E. (Eds.). (2009). *Phenomenological realism versus scientific realism*. Frankfurt, Germany: Ontos Verlag.

Curry, C. (1995). *Silver rights*. Chapel Hill, NC: Algonquin Books of Chapel Hill.

Czarniawska, B. (1998). *A narrative approach to organization studies* (Qualitative Research Methods Series No. 43). Thousand Oaks, CA: Sage.

Dabelstein, N., & Patton, M. Q. (2013a). Lessons learned and the contributions of the Paris Declaration Evaluation to evaluation theory and practice. *Canadian Journal of Program Evaluation, 27*(3), 173–200. Retrieved from http://evaluationcanada.ca/secure/27-3-019.pdf

Dabelstein, N., & Patton, M. Q. (2013b). The Paris Declaration on aid effectiveness: History and significance. *Canadian Journal of Program Evaluation, 27*(3), 19–36. Retrieved from http://evaluationcanada.ca/ site.cgi?s=4&ss=21&_lang=en&volume=2012//3

Dahler-Larsen, P., & Schwandt, T. A. (2012). Political culture as context for evaluation. In D. J. Rog, J. L. Fitzpatrick, & R. F. Conner (Eds.), *Context: A framework for its influence on evaluation practice* (New

Directions for Evaluation, No. 135, pp. 75–87). Hoboken, NJ: Wiley Periodicals.

Dallaire, R. A., & Beardsley, B. (2004). *Shake hands with the devil: The failure of humanity in Rwanda*. Toronto, Ontario, Canada: Random House.

Dart, J. J., Drysdale, G., Cole, D., and Saddington, M. (2000). The most significant change approach for monitoring an Australian extension project. *PLA Notes, 38,* 47–53, International Institute for Environment and Development, London.

Dart, J., & Roberts, M. (2014). Collaborative outcomes reporting. *Better Evaluation.* Retrieved from http://betterevaluation.org/plan/approach/cort

Dart, J., Drysdale, G., Cole, D., & Saddington, M. (2000). The Most Significant Changes approach for monitoring an Australian extension project. *PLA Notes, 38,* 47–53.

Datta, L. (1997). A pragmatic basis for mixed-method designs. *New Directions for Evaluation, (74),* 33–46.

Datta, L. (2005). Judicial model of evaluation. In S. Mathison (Ed.), *Encyclopedia of evaluation* (pp. 214–217). Thousand Oaks, CA: Sage.

Datta, L. (2006). The practice of evaluation: Challenges and new directions. In I. Shaw, J. Greene, & M. Mark (Eds.), *The SAGE handbook of evaluation* (pp. 419–438). Thousand Oaks, CA: Sage.

Davidson, E. J. (2005). *Evaluation methodology basics: The nuts and bolts of sound evaluation.* Thousand Oaks, CA: Sage.

Davidson, E. J. (2012). *Actionable evaluation basics: Getting succinct answers to the most important questions.* Auckland, New Zealand: Real Evaluation. Retrieved from http://genuineevaluation.com/the-worlds-first-evaluation-e-minibook-actionable-evaluation-basics/

Davidson, E. J. (2014, June 2). Anecdata. In *Genuine evaluation* [Web log post]. Retrieved from http://genuineevaluation.com/anecdata/?utm_source=feedburner&utm_medium=email&utm_campaign=Feed%3A+GenuineEvaluation+%28Genuine+Evaluation%29

Davidson, J., & di Gregorio, S. (2011). Qualitative research and technology: In the midst of a revolution. In N. K. Denzin & Y. S. Lincoln (Eds.), *The SAGE handbook of qualitative research* (4th ed., 627–643). Thousand Oaks, CA: Sage.

Davies, C. (1989). Goffman's concept of the total institution: Criticisms and revisions. *Human Studies, 12*(1), 77–95.

Davies, R. (1998a). An evolutionary approach to organisational learning: An experiment by an NGO in Bangladesh (aka Most Significant Changes approach to monitoring). *Impact Assessment and Project Appraisal, 16*(3), 243–250.

Davies, R. (1998b). *Order and diversity: Representing and assisting organisational learning in non-government aid organisations* (Doctoral thesis). Centre for Development Studies, University of Wales, Swansea, Wales.

Davies, R. J. (1996). *An evolutionary approach to facilitating organizational learning: An experiment by the Christian Commission for Development in Bangladesh.* Swansea, England: Center for Development Studies.

Davies, R. J. (2013, August). *Planning evaluability assessments: A synthesis of the literature with recommendations* (DFID Working Paper No. 40). London, England. Retrieved from http://r4d.dfid.gov.uk/Project/61141/%5C

Davies, R. J., & Dart, J. (2005). *The "Most Significant Change" (MSC) technique: A guide to its use.* Retrieved from http://www.mande.co.uk/docs/MSCGuide.pdf

Davis, C. S. (2008). Critical action research. In L. M. Given (Ed.), *The SAGE encyclopedia of qualitative research methods* (Vol. 1, pp. 139–142). Thousand Oaks, CA: Sage.

Davis, C. S. (2014). *Conversations about qualitative communication research.* Walnut Creek, CA: Left Coast Press.

Davis, K. (1940). Extreme social isolation of a child. *American Journal of Sociology, 45*(4), 554–565.

Davis, K. (1947). Final note on a case of extreme social isolation. *American Journal of Sociology, 52*(March), 432–437.

Danida (2005). *Lessons from Rwanda.* Copenhagen, Denmark: Danish Ministry of Foreign Affairs, Evaluation Department.

de Bono, E. (1985). *Six thinking hats.* New York, NY: Little, Brown.

de Bono, E. (1999). *Six thinking hats* (2nd ed.). New York, NY: Little, Brown.

de Bono, E. (2009). *Think! Before it's too late.* New York, NY: Ebury.

De Munck, V. (2000). Introduction: Units for describing and analyzing culture and society. *Ethnology, 39*(4), 279–292.

DeCramer, G. (1997a). *Minnesota's district/area transportation partnership process: Vol. 1. Cross-case analysis.* St. Paul, MN: Center for Transportation Studies.

DeCramer, G. (1997b). *Minnesota's district/area transportation partnership process: Vol. 2. Case studies and other perspectives.* St. Paul, MN: Center for Transportation Studies.

Deming, W. E. (2000). *The new economics for industry, government, education* (2nd ed.). Cambridge: MIT Center for Advanced Educational Services.

Denning, S. (2001). *The springboard: How storytelling ignites action in knowledge-era organizations.* Boston, MA: Butterworth-Heinemann.

Denyer, D., & Tranfield, D. (2006). Using qualitative research synthesis to build an actionable knowledge base. *Management Decision, 44*(2), 213–227.

Denzin, N. K. (1978a). The logic of naturalistic inquiry. In *Sociological methods: A sourcebook* (pp. 6–29). New York, NY: McGraw-Hill.

Denzin, N. K. (1978b). *The research act: A theoretical introduction to sociological methods* (2nd ed.). New York, NY: McGraw-Hill.

Denzin, N. K. (1983). Interpretive interactionism. In G. Morgan (Ed.), *Beyond method: Strategies for social research* (pp. 129–146). Beverly Hills, CA: Sage.

Denzin, N. K. (1989a). *Interpretive biography* (Qualitative Research Methods Series No. 17). Newbury Park, CA: Sage.

Denzin, N. K. (1989b). *Interpretive interactionism.* Newbury Park, CA: Sage.

Denzin, N. K. (1989c). *The research act: A theoretical introduction to sociological methods.* Englewood Cliffs, NJ: Prentice Hall.

Denzin, N. K. (1991). *Images of postmodern society: Social theory and contemporary cinema.* London, England: Sage.

Denzin, N. K. (1997a, November). Coffee with Anselm. *Qualitative Family Research, 11*(1–2), 11–18.

Denzin, N. K. (1997b). *Interpretive ethnography: Ethnographic practices for the 21st century.* Thousand Oaks, CA: Sage.

Denzin, N. K. (2001). *Interpretive interactionism* (2nd ed.). Thousand Oaks, CA: Sage.

Denzin, N. K. (2003). *Performance ethnography: Critical pedagogy and the politics of culture.* Thousand Oaks, CA: Sage.

Denzin, N. K. (2009). *On understanding emotion.* New Brunswick, NJ: Transaction.

Denzin, N. K. (2010). *The qualitative manifesto: A call to arms.* Walnut Creek, CA: Left Coast Press.

Denzin, N. K., & Lincoln, Y. S. (Eds.) (2000). *Handbook of qualitative research.* Thousand Oaks, CA: Sage.

Denzin, N. K., & Giardina, M. D. (Eds.). (2006). *Qualitative inquiry and the conservative challenge.* Walnut Creek, CA: Left Coast Press.

Denzin, N. K., & Giardina, M. D. (Eds.). (2008). *Qualitative inquiry and the politics of evidence.* Walnut Creek, CA: Left Coast Press.

Denzin, N. K., & Lincoln, Y. S. (Eds.). (2002). *The qualitative inquiry reader.* Thousand Oaks, CA: Sage.

Denzin, N. K., & Lincoln, Y. S. (Eds.). (2003). *Turning points in qualitative research.* Lanham, MD: AltaMira Press.

Denzin, N. K., & Lincoln, Y. S. (Eds.). (2011). *The SAGE handbook of qualitative research* (4th ed.). Thousand Oaks, CA: Sage.

Denzin, N. K., Lincoln, Y. S., & Smith. L. T. (Eds.). (2008). *Handbook of critical and indigenous methodologies.* Thousand Oaks, CA: Sage.

Department for International Development. (2012, April). *Broadening the range of designs and methods for impact evaluations* (Working Paper No. 38). London, England: Author.

Deutsch, C. H. (1998, November 15). The guru of doing it right still sees much work to do. *The New York Times,* p. BU-5.

Deutscher, I. (1970). Words and deeds: Social science and social policy. In W. Filstead (Ed.), *Qualitative methodology* (pp. 27–51). Chicago, IL: Markham.

Development Assistance Committee. (2010). *Quality standards for development evaluation* (DAC Guidelines and Reference Series). Paris, France: Organisation for Economic Co-operation and Development. Retrieved from http://www.oecd.org/dataoecd/55/0/44798177.pdf

Devereux, S., & Roelen, K. (2013). *Evaluating outside the box: Mixing methods in analysing social protection programmes.* London, England: Institute of Development Studies, Center for Social Protection. Retrieved from http://www.uea.ac.uk/documents/439774/3502783/Roelen+and+Devereux+-+Evaluating+outside+the+box.pdf/3e18fd9c-5fb1-4a6c-88e1-a6b4cc28e003

Dewey, J. (1956). *The child and the curriculum, and the school and society.* Chicago, IL: University of Chicago Press.

Dick, B. (2009a). *Action research resources.* Retrieved from http://www .scu.edu.au/schools/gcm/ar/arhome.html

Dick, B. (2009b). *What is action research?* Retrieved from http://www.scu .edu.au/schools/gcm/ar/whatisar.html

Djuraskovic, I., & Arthur, N. (2010). Heuristic inquiry: A personal journey of acculturation and identity reconstruction. *Qualitative Report, 15*(6), 1569–1593. Retrieved from http://www.nova.edu/ssss/QR/QR15-6/ djuraskovic.pdf

Domaingue, R. (1989). Community development through ethnographic futures research. *Journal of Extension, Summer,* 22–23.

Donaldson, S. I., Christie, C. A., & Mark, M. M. (Eds.). (2008). *What counts as credible evidence in applied research and evaluation practice?* Thousand Oaks, CA: Sage.

Douglas, J. D. (1976). *Investigative social research: Individual and team field research.* Beverly Hills, CA: Sage.

Douglas, J. D. (1985). *Creative interviewing.* Beverly Hills, CA: Sage.

Douglass, B. G., & Moustakas, C. (1985). Heuristic inquiry: The internal search to know. *Journal of Humanistic Psychology, 25*(3), 39–55.

Downing-Matibag, T. M., & Geisinger, B. (2009). Hooking up and sexual risk taking among college students: A health belief model perspective. *Qualitative Health Research,* 19(9), 1196–1209.

Duhigg, C. (2013). *The power of habit. Why we do what we do and how to change.* New York, NY: Random House.

Dunn, S. (2000, April 10). Empathy. *The New Yorker, 10,* p. 62.

Dunning, J., & Kruger, D. (1999). Unskilled and unaware of it: How difficulties in recognizing one's own incompetence lead to inflated self-assessments. *Journal of Personality and Social Psychology, 77*(6), 1121–1134.

Dupret, B. (2011). *Adjudication in action: An ethnomethodology of law, morality and justice.* Farnham, England: Ashgate.

Durkin, T. (1997). Using computers in strategic qualitative research. In G. Miller & R. Dingwall (Eds.), *Context and method in qualitative research* (pp. 92–105). Thousand Oaks, CA: Sage.

Dyson, F. (2014). The case for blunders. *New York Review of Books, 61*(4), 4–8.

Eberstadt, N. (1995). *The tyranny of numbers: Mismeasurement and misrule.* Washington, DC: AEI Press.

Ebert, R. (1996, March 21). On watching a movie multiple times. *Fresh Air, National Public Radio.* Retrieved from http://www.npr.org/templates/ story/story.php?storyId=129161766&sc=emaf

Eco, U. (1976). *A theory of semiotics.* Bloomington: Indiana University Press.

Eder, D., & Fingerson, L. (2002). Interviewing children and adolescents. In J. F. Gubrium & J. A. Holstein (Eds.), *Handbook of interview research: Context and method* (pp. 181–201). Thousand Oaks, CA: Sage.

Eder, D., & Fingerson, L. (2003). Interviewing children and adolescents. In J. A. Holstein & J. F. Gubrium (Eds.), *Inside interviewing: New lenses, new concerns* (pp. 33–54). Thousand Oaks, CA: Sage.

Edmonson, A. C. (2011, April). Strategies for learning from failure. *Harvard Business Review Magazine.* Retrieved from http://hbr .org/2011/04/strategies-for-learning-from-failure/ar/

Ehrlinger, J., Johnson, K., Banner, M., Dunning, D., & Kruger, J. (2008). Why the unskilled are unaware: Further explorations of (absent) self-insight among the incompetent. *Organizational Behavior and Human Decision Processes, 105*(1), 98–121.

Eichelberger, R. T. (1989). *Disciplined inquiry: Understanding and doing educational research.* New York, NY: Longman.

Eichengreen, B., & Temin, P. (1997). *The gold standard and the Great Depression* (NBER Working Paper No. 6060). Cambridge, MA: National Bureau Of Economic Research.

Eisner, E. W. (1985). *The art of educational evaluation: A personal view.* London, England: Falmer Press.

Eisner, E. W. (1991). *The enlightened eye: Qualitative inquiry and the enhancement of educational practice.* New York, NY: Macmillan.

Ellingson, L. L. (Ed.). (2009). *Engaging crystallization in qualitative research: An introduction.* London, England: Sage.

Ellingson, L. L. (2011). Analysis and representation across the continuum. In N. K. Denzin & Y. S. Lincoln (Eds.), *The SAGE handbook of qualitative research* (4th ed., pp. 595–610). Thousand Oaks, CA: Sage.

Elliott, J. (1976). *Developing hypotheses about classrooms from teachers' practical constructs: An account of the work of the Ford Teaching Project.* Grand Forks: University of North Dakota.

Elliott, J. (2005). Action research. In S. Mathison (Ed.), *Encyclopedia of evaluation* (pp. 8–10). Thousand Oaks, CA: Sage.

Ellis, C. (1986). *Fisher folk: Two communities on Chesapeake Bay.* Lexington: University Press of Kentucky.

Ellis, C. S., & Brochner, A. (2000). Autoethnography, personal narrative, reflexivity: Researcher as subject. In N. Denzin & Y. Lincoln (Eds.), *The handbook of qualitative research* (2nd ed., pp. 733–768). Thousand Oaks, CA: Sage.

Elmore, R. F. (1976). Follow through planned variation. In W. Williams & R. F. Elmore (Eds.), *Social program implementation* (pp. 101–123). New York, NY: Academic Press.

Ember, C. R., & Ember, M. (2009). *Cross-cultural research methods* (2nd ed.). Lanham, MD: AltaMira Press.

Emmel, N. (2013). *Sampling and choosing cases in qualitative research: A realist approach.* London, England: Sage.

Engberg, K. (2013). *The EU and military operations: A comparative analysis.* New York, NY: Routledge.

England, K. V. L. (1994). Getting personal: Reflexivity, positionality, and feminist research. *The Professional Geographer, 46*(1), 80–89.

English, F. W. (2000). A critical appraisal of Sara Lawrence-Lightfoot's "Portraiture" as a method of educational research. *Educational Researcher, 29*(7), 21–26.

Ensler, E. (2001). *The vagina monologues.* New York, NY: Villard.

Eoyang, G. H. (1997). *Coping with chaos: Seven simple tools.* Cheyenne, WY: Lagumo.

Eoyang, G. H., & Holladay, R. (2013). *Adaptive action: Leveraging uncertainty in your organization.* Stanford, CA: Stanford University Press.

Epstein, M. (2012). Introduction to the philosophy of science. In C. Seale (Ed.), *Research society and culture* (3rd ed., pp. 7–28). Thousand Oaks, CA: Sage.

Erber, J. T. (2010). *Aging and older adulthood* (2nd ed.). London, England: Wiley-Blackwell.

Erickson, F. (1973). What makes school ethnography "ethnographic"? *Anthropology and Education Quarterly, 4*(2), 10–19.

Ericsson, K. A., & Simon, H. A. (1993). *Protocol analysis: Verbal reports as data.* Cambridge: MIT Press.

Eriksson, P., & Kovalainen, A. (2008). *Qualitative methods in business research.* London, England: Sage.

Ess, C. (2014). *Digital media ethics.* Cambridge, England: Polity Press.

European Evaluation Society. (2007). *The importance of a methodologically diverse approach to impact evaluation.* Retrieved from http://europeanevaluation.org/sites/default/files/EES%20 Statement_0.pdf

EvalPartners. (2014). *2015 declared as the International Year of Evaluation.* Retrieved from http://mymande.org/evalyear/Declaring_2015_as_ the_International_Year_of_Evaluation

Evaluation Exchange. (2010). Resources on scaling impact. *Evaluation Exchange, 15*(1), 1. Retrieved from http://www.hfrp.org/evaluation/the -evaluation-exchange/current-issue-scaling-impact/new-noteworthy

Evaluation Roundtable. (2014). *Case studies.* Retrieved from http://www .evaluationroundtable.org/publications.html

Evashwick, C. (1989). Creating the continuum of care. *Health Matrix, 7*(1), 30–39.

Evergreen, S. (2012, June 16). Potent presentations [AEA365 blog]. Retrieved from http://aea365.org/blog/

Eyben, R. (2013). *Uncovering the politics of "evidence" and "results": A framing paper for development practitioners.* Retrieved from http:// bigpushforward.net/wp-content/uploads/2011/01/The-politics-of -evidence-11-April-20133.pdf

Eynon, R., Fry, J., & Schroeder, R. (2008). Ethics of internet research. In N. G. Fielding, R. M. Lee, & G. Blank (Eds.). (2008). *The SAGE handbook of online research methods* (pp. 23–41). Thousand Oaks, CA: Sage.

Eysenbach, G., & Till, J. E. (2001). Ethical issues in qualitative research on internet communities. *British Medical Journal.* Retrieved from http:// www.bmj.com/content/323/7321/1103

Fadiman, C. (2000). *Bartlett's book of anecdotes* (Rev. ed.). Boston, MA: Little, Brown.

Farhi, P. (2013, November 8). "60 Minutes" retracts, apologizes for Benghazi report: CBS says it was misled by a source. *The Washington Post.* Retrieved from http://www.washingtonpost .com/lifestyle/style/60-minutes-apologizes-for-benghazi-report/ 2013/11/08/6e7b6b9a-487e-11e3-a196-3544a03c2351_story.html

Farming Systems Support Project. (1986). *Diagnosis in farming systems research and extension* (Vol. 1). Gainesville: University of Florida Institute of Food and Agricultural Sciences.

Farming Systems Support Project. (1987). *Bibliography of readings in farming systems* (Vol. 4). Gainesville: University of Florida Institute of Food and Agricultural Sciences.

Fakis, A., Hilliam, R., Stoneley, H., & Townend, M. (2014). Quantitative Analysis of Qualitative Information From Interviews A Systematic Literature Review. *Journal of Mixed Methods Research, 8*(2), 139–161.

Fehrenbacher, H. L., Owens, T. R., & Haenn, J. F. (1976). *The use of student case study methodology in program evaluation* (Research Evaluation Development Paper No. 10). Portland, OR: Northwest Regional Educational Laboratory.

Feiler, B. (2013a). *The secrets of happy families: How to improve your morning, rethink family dinner, fight smart, go out and play, and much more.* New York, NY: William Morrow.

Feiler, B. (2013b). The stories that bind us. *The New York Times*. Retrieved from http://www.nytimes.com/2013/03/17/fashion/the-family -stories-that-bind-us-this-life.html?smid=pl-share&_r=0

Feiman, S. (1977). Evaluating teacher centers. *The School Review, 85*(3), 395–411.

Ferguson, C. (1989). *The use and impact of evaluation by decision makers: Four Australian case studies* (Unpublished doctoral dissertation). Macquarie University, Sydney, New South Wales, Australia.

Festinger, L., Riecken, H. W., & Schachter, S. (1956). *When prophecy fails: A social and psychological study of a modern group that predicted the destruction of the world.* New York, NY: Harper Torchbooks.

Fetterman, D. M. (1989). *Ethnography step by step.* Thousand Oaks, CA: Sage.

Fetterman, D. M. (2000). *Foundations of empowerment evaluation: Step by step.* Thousand Oaks, CA: Sage.

Fetterman, D. M. (2008a). Ethnography. In L. M. Given (Ed.), *The SAGE encyclopedia of qualitative research methods* (Vol. 1, pp. 288–292). Thousand Oaks, CA: Sage.

Fetterman, D. M. (2008b). Key informants. In L. M. Given (Ed.), *The SAGE encyclopedia of qualitative research methods* (Vol. 1, pp. 477–479). Thousand Oaks, CA: Sage.

Fetterman, D. M. (2009). *Ethnography: Step by step* (Rev. ed.). Thousand Oaks, CA: Sage.

Fetterman, D. M. (2013). *Empowerment evaluation in the digital villages: Hewlett-Packard's $15 million race toward social justice.* Palo Alto, CA: Stanford University Press.

Fetterman, D. M., Kaftarian, S. J., & Wandersman, A. (1996). *Empowerment evaluation: Knowledge and tools for self-assessment & accountability.* Thousand Oaks, CA: Sage.

Fetterman, D. M., Rodríguez-Campos, L., Wandersman, A., & O'Sullivan, R. G. (2014). Collaborative, participatory, and empowerment evaluation: Building a strong conceptual foundation for stakeholder involvement approaches to evaluation. *American Journal of Evaluation, 35*(1), 144–148.

Fetterman, D. M., & Wandersman, A. (2005). *Empowerment evaluation principles and practice.* New York, NY: Guilford Press.

Fielding, N. G., & Lee, R. M. (1998). *Computer analysis and qualitative research.* Thousand Oaks, CA: Sage.

Fielding, N. G., Lee, R. M., & Blank, G. (Eds.). (2008). *The SAGE handbook of online research methods.* Thousand Oaks, CA: Sage.

Fiester, L. (2010). *Measuring change while changing measures: Learning in, and from, the evaluation of making connections: Evaluation roundtable teaching forum.* Retrieved from http://www.evaluationroundtable .org/documents/cs-measuring-change.pdf

Filstead, W. J. (1970a). *Qualitative methodology.* Chicago, IL: Markham.

Finlayson, K. W., & Dixon, A. (2008). Qualitative meta-synthesis: A guide for the novice. *Nurse Researcher, 15*(2), 59.

Firestein, S. (2012). *Ignorance: How it drives science.* New York, NY: Oxford University Press.

Fitz-Gibbon, C. T., & Morris, L. L. (1987). *How to design a program evaluation.* Newbury Park, CA: Sage.

Fitzpatrick, J. L., Sanders, J. R., & Worthen, B. R. (2010). *Program evaluation: Alternative approaches and practical guidelines* (4th ed.). Boston, MA: Pearson.

Flick, U. (2000). Qualitative inquiries into social representations of health. *Journal of Health Psychology, 5*, 309–318.

Flick, U. (2007a). *Designing qualitative research.* London, England: Sage.

Flick, U. (2007b). *Managing quality in qualitative research.* London, England: Sage.

Flick, U. (2009). *An introduction to qualitative research* (4th ed.). Thousand Oaks, CA: Sage.

Flick, U., von Kardoff, E., & Steinke, I. (Eds.). (2004). *A companion to qualitative research.* Thousand Oaks, CA: Sage.

Flyvbjerg, B. (2011). Case study. In N. Denzin & Y. Lincoln (Eds.), *The SAGE handbook of qualitative research* (4th ed., pp. 301–316). Thousand Oaks, CA: Sage.

Folorunsho, M. (2014). *Theatre for development.* Yaoundé, Cameroons: African Evaluation Association.

Fontana, A., & Frey, J. H. (2000). The interview: From structured questions to negotiated text. In N. K. Denzin & Y. S. Lincoln (Eds.), *Handbook of qualitative research* (2nd ed., pp. 645–672). Thousand Oaks, CA: Sage.

Foucault, M. (2013). *Foucault.* Retrieved from http://en.wikipedia.org/ wiki/Michel_Foucault

Fox, N. J. (2008). Leaving the field. In L. M. Given (Ed.), *The SAGE encyclopedia of qualitative research methods* (Vol. 1, pp. 483–484). Thousand Oaks, CA: Sage.

Frake, C. (1962). The ethnographic study of the cognitive system. In T. Gladwin & W. Sturtlevant (Eds.), *Anthropology and human behavior* (pp. 28–41). Washington, DC: Anthropology Society of Washington.

Frank, A. W. (1995). *The wounded storyteller: Body, illness, and ethics.* Chicago, IL: University of Chicago Press.

Frank, A. W. (2000). Illness and autobiographical work. *Qualitative Sociology, 23*, 135–156.

Frankl, V. (2006). *Man's search for meaning: An introduction to logotherapy.* Boston, MA: Beacon Press.

Freire, P. (1970). *Pedagogy of the oppressed.* New York, NY: Seabury Press.

Freire, P. (1973). *Education for critical consciousness.* New York, NY: Seabury Press.

Frey, J. H. (2004). Open-ended questions. In S. S. Lewis-Beck, A. Bryman, & T. F. Liao (Eds.), *The SAGE encyclopedia of social science methods* (Vol. 2, p. 768). Thousand Oaks, CA: Sage.

Fujita, N. (2010). *Beyond log frame: Using systems concepts in evaluation.* Tokyo, Japan: Foundation for Advanced Studies on International Development.

Fuller, S. (2000). *Thomas Kuhn: A philosophical history for our times.* Chicago, IL: University of Chicago Press.

Funnell, S. C., & Rogers, P. J. (2011). *Purposeful program theory: Effective use of logic models and theories of change.* San Francisco, CA: Jossey-Bass.

Galbraith, J. K. (1986). *The art of good writing: Interview with Harry Kreisler.* Retrieved from http://globetrotter.berkeley.edu/conversations/ Galbraith/galbraith2.html

Gale, J. E., & Dolbin-MacNab, M. L. (2014). Qualitative research for family therapy. In R. B. Miller & L. N. Johnson (Eds.), *Advanced methods in family therapy research: A focus on validity and change* (pp. 247–265). New York, NY: Routledge.

Galea, S., Tracy, M., Hoggatt, K. J., DiMaggio, C., & Karpati, A. (2011). Estimated deaths attributable to social factors in the United States. *American Journal of Public Health, 101*(8), 1456–1465.

Galman, S. C. (2013). *The good, the bad, and the data: Shane the lone ethnographer's basic guide to qualitative data analysis.* Walnut Creek, CA: Left Coast Press.

Galt, D., & Mathema, B. (1987). Farmer participation in farming systems research. *Farming Systems Support Project Newsletter, 5*(1), 1–20. Retrieved from http://pdf.usaid.gov/pdf_docs/PNAAW987.pdf

Garcia, S. E. (1989). *Alexander the Great: A strategy review.* Report number 84-0960, Maxwell Air Force Base. Alabama: Air Command and Staff College.

Garber, A. R. (2001). Death by powerpoint. *Small Business Computing Staff Journal.* Retrieved from http://www.smallbusinesscomputing .com/biztools/article.php/684871/Death-By-Powerpoint.htm

Garfinkel, H. (1967). *Studies in ethnomethodology.* Englewood Cliffs, NJ: Prentice Hall.

Gauthier, B. (2013, March). Face of AEA: Meet Benoît Gauthier. *American Evaluation Association Newsletter, 13*(3).

Gawande, A. (2002). *Complications: A surgeon's notes on an imperfect science.* New York, NY: Metropolitan Books.

Gawande, A. (2004, January 12). Medical dispatch: The mop-up. *The New Yorker*, pp. 34–40.

Gawande, A. (2007, May 1). The power of negative thinking. *The New York Times*, p. A23.

Gawande, A. (2010). *The checklist manifesto: How to get things right*. New York, NY: Henry Holt.

Geertz, C. (1973). Deep play: Notes on the Balinese cockfight. In *The interpretation of cultures: Selected essays* (pp. 412–453). New York, NY: Basic Books.

Geertz, C. (1986). Making experiences, authoring selves. In V. W. Turner & E. Bruner (Eds.), *The anthropology of experience* (pp. 373–380). Urbana: University of Illinois Press.

Geertz, C. (1988). Thick description: Toward an interpretation of the theory of culture. In R. Emerson (Ed.), *Contemporary field research: A collection of readings* (pp. 37–59). Prospect Heights, IL: Waveland Press.

Geertz, C. (2001). Life among the anthros. *New York Review of Books, 48*(2), 18–22.

General Accounting Office. (1987). *Case study evaluations* (Transfer Paper No. 9). Washington, DC: Author.

General Accounting Office. (1989). *Prospective methods: The prospective evaluation synthesis*. Washington, DC: Author.

General Accounting Office. (1991). *Designing evaluations*. Washington, DC: Author.

General Accounting Office. (1992). *The evaluation synthesis*. Washington, DC: Author.

Genomics. (2010). *Genetic anthropology, ancestry, and ancient human migration* (Human Genome Program). Washington, DC: U.S. Department of Energy Office of Science, Office of Biological and Environmental Research. Retrieved from http://www.ornl.gov/sci/techresources/Human_Genome/elsi/humanmigration.shtml

George, A. L., & Bennett, A. (2005). *Case studies and theory development in the social sciences*. Cambridge, MA: Harvard University Press.

Gergen, K. J., & Gergen, M. M. (2008). Social construction and psychological inquiry. In J. A. Holstein & J. F. Gubrium (Eds.), *Handbook of constructionist research* (pp. 171–188). New York, NY: Guilford Press.

Gergen, M. M., & Gergen, K. J. (2000). Qualitative inquiry: Tensions and transformation. In N. K. Denzin & Y. S. Lincoln (Eds.), *Handbook of qualitative research* (2nd ed., pp. 1025–1046). Thousand Oaks, CA: Sage.

Gerring, J. (2007). *Case study research: Principles and practices*. Cambridge, England: Cambridge University Press.

Gharajedaghi, J. (1985). *Toward a systems theory of organization*. Seaside, CA: Intersystems.

Gharajedaghi, J., & Ackoff, R. L. (1985). Toward systemic education of systems scientists. *Systems Research, 2*(1), 21–27.

Ghere, G., King, J. A., Stevahn, L., & Minnema, J. (2006). A professional development unit for reflecting on program evaluator competencies. *American Journal of Evaluation, 27*(1), 108–123.

Gibbs, G. (2007). *Analyzing qualitative data*. London, England: Sage.

Gilgun, J. F. (1991). The resilience and intergenerational transmission of child sexual abuse. In M. Q. Patton (Ed.), *Family sexual abuse* (pp. 93–105). Thousand Oaks, CA: Sage.

Gilgun, J. F. (1994). Avengers, conquerors, playmates, and lovers: A continuum of roles played by perpetrators of child sexual abuse. *Families in Society, 75,* 467–480.

Gilgun, J. F. (1995). We shared something special: The moral discourse of incest perpetrators. *Journal of Marriage and the Family, 57*(2), 265–281.

Gilgun, J. F. (1996). Human development and adversity in ecological perspective, Part 2: Three patterns. *Families in Society, 77,* 459–526.

Gilgun, J. (1999). Fingernails painted red. A feminist semiotic analysis of "hot text." *Qualitative inquiry,* 5(2), 181–207.

Gilgun, J. F., & Connor, T. M. (1989). How perpetrators view child sexual abuse. *Social Work, 34*(3), 249–251.

Gilgun, J. F., & McLeod, L. (1999). Gendering violence. *Studies in Symbolic Interaction, 22,* 167–193.

Gilligan, C. (1982). *In a different voice: Psychological theory and women's development*. Cambridge, MA: Harvard University Press.

Giorgi, A. (1971). Phenomenology and experimental psychology. In A. Giorgi, W. F. Fischer, & E. R. Von (Eds.), *Duquesne studies in phenomenological psychology* (pp. 245–277). Pittsburgh, PA: Duquesne University Press.

Giorgi, A. (1985). *Phenomenology and psychological research*. Pittsburgh, PA: Duquesne University Press.

Giorgi, A. (2006). Difficulties encountered in the application of the phenomenological method in the social sciences. *Análise Psicológica, 24*(3), 353–361.

Given, L. M. (Ed.). (2008). *The SAGE encyclopedia of qualitative research methods*. Thousand Oaks, CA: Sage.

Gladwell, M. (2000, March 13). Annals of medicine. *The New York Times*, pp. 55–56.

Gladwell, M. (2002). *The tipping point: How little things can make a big difference*. Boston, MA: Little, Brown.

Gladwell, M. (2008). *Outliers: The story of success*. New York, NY: Little, Brown.

Gladwell, M. (2013). *David and Goliath: Underdogs, misfits, and the art of battling giants*. New York, NY: Little, Brown.

Glaser, B. G. (1978). *Theoretical sensitivity: Advances in the methodology of grounded theory*. Mill Valley, CA: Sociology Press.

Glaser, B. G. (Ed.). (1993). *Examples of grounded theory: A reader*. Mill Valley, CA: Sociology Press.

Glaser, B. G. (2000). The future of grounded theory. *Grounded Theory Review, 1,* 1–18.

Glaser, B. G. (2001). Doing grounded theory. *Grounded Theory Review, 1,* 1–8.

Glaser, B. G. (2012). *STOP, WRITE! Writing grounded theory*. Mill Valley, CA: Sociology Press.

Glaser, B. G. (2014a). *Memoing: A vital grounded theory procedure*. Mill Valley, CA: Sociology Press.

Glaser, B. G. (2014b). No preconceptions: The grounded theory dictum. Mill Valley, CA: Sociology Press.

Glaser, B. G., & Strauss, A. L. (1965). *Awareness of dying*. Chicago, IL: Aldine.

Glaser, B. G., & Strauss, A. L. (1967). *The discovery of grounded theory: Strategies for qualitative research*. Chicago, IL: Aldine.

Gläser, J., & Laudel, G. (2013). Life with and without coding: Two methods for early-stage data analysis in qualitative research aiming at causal explanations (FSG: Forum). *Qualitative Social Research, 14*(2). Retrieved from http://www.qualitative-research.net/index.php/fqs/article/view/1886/3528

Glass, R. (2001). On Paulo Freire's philosophy of praxis and the foundations of liberation education. *Educational Researcher, 30*(2), 15–25.

Glazer, M. (1972). *The research adventure, promise and problems of field work*. New York, NY: Random House.

Gleick, J. (1987). *Chaos: Making a new science*. New York, NY: Penguin Books.

Glesne, C. (1997). Rare feeling: Representing research through poetic transcription. *Qualitative Inquiry, 3*(June), 202–221.

Glesne, C. (1999). *Becoming qualitative researchers: An introduction* (2nd ed.). New York, NY: Longman.

Glesne, C. (2010). *Becoming qualitative researchers: An introduction* (4th ed.). New York, NY: Pearson.

Gobo, G. (2011). Ethnography. In D. Silverman (Ed.), *Qualitative research* (3rd ed., pp. 15–34). Thousand Oaks, CA: Sage.

Goffman, E. (1961). *Asylums: Essays on the social situation of mental patients and other inmates*. Garden City, NY: Anchor Books.

Goffman, E., & Helmreich, W. B. (1961). *Asylums: Essays on the social situation of mental patients and other inmates* (Vol. 277). New York, NY: Anchor Books.

Goleman, D. (2013, October 6). Rich people just care less. *The New York Times*, p. SR12. Retrieved from http://opinionator.blogs.nytimes.com/2013/10/05/rich-people-just-care-less/?hp&_r=0

Golembiewski, B. (2000). Three perspectives on appreciative inquiry. *OD Practitioner, 32*(1), 53–58.

Goodall, H. L., Jr. (2000). *Writing the new ethnography*. Walnut Creek, CA: AltaMira Press.

Goodall, H. L., Jr. (2008). *Writing qualitative inquiry*. Walnut Creek, CA: Left Coast Press.

Goodenough, W. H. (1971). *Culture, language, and society*. Reading, MA: Addison-Wesley.

Goodyear, L., Jewiss, J., Usinger, J., & Barela, E. (Eds.). (2014). *Qualitative inquiry in the practice of evaluation*. San Francisco, CA: Jossey-Bass.

Golden-Biddle, K., & Locke, K. (2007). *Composing qualitative research*. Sage.

Gopalakrishnan, S., Preskill, P., & Lu, S. J. (2013). *Next generation evaluation: Embracing complexity, connectivity, and change*. Boston, MA: Foundation Strategy Group.

QUALITATIVE RESEARCH & EVALUATION METHODS

Gordon, J., & Patterson, J. A. (2013). Response to Tracy's under the "Big Tent": Establishing Universal criteria for evaluating qualitative research. *Qualitative Inquiry, 19*(9), 689–695.

Gosling, S. D., & Johnson, J. A. (2013). *Advanced methods for conducting online behavioral research* (Electronic ed.). Washington, DC: American Psychological Association.

Gottlieb, M., & Ernst-Slavit, G. (Eds.). (2013). *Academic language in diverse classrooms: Promoting content and language learning.* Newbury Park, CA: Corwin Press.

Gottschall, J. (2012). *The storytelling animal: How stories make us human.* Boston, MA: Houghton Mifflin Harcourt.

Gough, D., Oliver, S., & Thomas, J. (Eds.). (2012). *An introduction to systematic reviews.* Thousand Oaks, CA: Sage.

Graham, R. J. (1993). Decoding teaching: The rhetoric and politics of narrative form. *Journal of Natural Inquiry, 8*(1), 30–37.

Graue, M. E., & Walsh, D. J. (1998). *Studying children in context: Theories, methods, and ethics.* Thousand Oaks, CA: Sage.

Green, J. A., Whitney, P. G., & Potegal, M. (2011). Screaming, yelling, whining, and crying: Categorical and intensity differences in vocal expressions of anger and sadness in children's tantrums. *Emotion, 11*(5), 1124–1133.

Greene, J. (2006). Evaluation, democracy, and social change. In I. Shaw, J. Greene, & M. Mark (Eds.), *The SAGE handbook of evaluation* (pp. 118–140). Thousand Oaks, CA: Sage.

Greene, J. C. (1998). Balancing philosophy and practicality in qualitative evaluation. In R. Davis (Ed.), *Proceedings of the Stake symposium on educational evaluation* (pp. 35–49). Urbana: University of Illinois Press.

Greene, J. C. (2000). Understanding social programs through evaluation. In Y. S. Lincoln & N. K. Denzin (Eds.), *Handbook of qualitative research* (pp. 35–49). Thousand Oaks, CA: Sage.

Greene, J. C. (2007). *Mixed methods in social inquiry.* San Francisco, CA: Jossey-Bass.

Greig, A., & Taylor, J. (1998). *Doing research with children.* Thousand Oaks, CA: Sage.

Griffiths, T. (2014). *What scientific idea is ready for retirement? Bias is always bad.* Retrieved from http://www.edge.org/responses/what-scientific-idea-is-ready-for-retirement

Grinder, J., DeLozier, J., & Bandler, R. (1977). *Patterns of the hypnotic techniques of Milton Erickson, M.D.* (Vol. 2). Cupertino, CA: Meta.

GRM International. (2013). *Research into the long term impact of development interventions in the Koshi Hills of Nepal: Reality check approach study report.* In association with the Effective Development Group and the Foundation for Development Management, Government of Nepal, and UK AID (DFID). Retrieved from http://www.edgroup.com.au/wp-content/uploads/2013/09/NPRKH_RCA-Report.pdf

Groupman, J. (2008). *How doctors think.* New York, NY: Houghton Mifflin.

Guba, E. G. (1978). *Toward a methodology of naturalistic inquiry in educational evaluation* (CSE Monograph Series in Evaluation No. 8). Los Angeles: University of California, Center for the Study of Evaluation, UCLA Graduate School of Education.

Guba, E. G. (1981). Investigative reporting. In N. L. Smith (Ed.), *Metaphors for evaluation* (pp. 67–86). Beverly Hills, CA: Sage.

Guba, E. G., & Lincoln, Y. S. (1981). *Effective evaluation: Improving the usefulness of evaluation results through responsive and naturalistic approaches.* San Francisco, CA: Jossey-Bass.

Guba, E. G., & Lincoln, Y. S. (1988). Do inquiry paradigms imply inquiry methodologies. In D. M. Fetterman (Ed.), *Qualitative approaches to evaluation in education: The silent scientific revolution* (pp. 89–115). New York, NY: Praeger.

Guba, E. G., & Lincoln, Y. S. (1989). *Fourth generation evaluation.* Newbury Park, CA: Sage.

Guba, E. G., & Lincoln, Y. S. (1990). Can there be a human science. *Person Centered Review, 5*(2), 130–154.

Gubrium, A., & Harper, K. (2013). *Participatory visual and digital methods.* Walnut Creek, CA: Left Coast Press.

Gubrium, J. F., & Holstein, J. (2000). Analyzing interpretive practice. In N. K. Denzin & Y. S. Lincoln (Eds.), *Handbook of qualitative research* (2nd ed., pp. 487–508). Thousand Oaks, CA: Sage.

Gubrium, J. F., & Holstein, J. A. (Eds.). (2003). *Postmodern interviewing.* Thousand Oaks, CA: Sage.

Gubrium, J. F., & Holstein, J. A. (2009). *Analyzing narrative reality.* Thousand Oaks, CA: Sage.

Guerrero, S. H. (Ed.). (1999a). *Gender-sensitive & feminist methodologies: A handbook for health and social researchers.* Quezon City, Philippines: University of the Philippines, Center for Women's Studies.

Guerrero, S. H. (Ed.). (1999b). *Selected readings on health & feminist research: A sourcebook.* Quezon City, Philippines: University of the Philippines, Center for Women's Studies.

Guerrero-Manalo, S. (1999). Child sensitive interviewing: Pointers in interviewing child victims of abuse. In S. H. Guerrero (Ed.), *Gender-sensitive and feminist methodologies: A handbook for health and social researchers* (pp. 195–203). Quezon City, Philippines: University of the Philippines, Center for Women's Studies.

Guest, G., & MacQueen, K. M. (2008). *Handbook for team-based qualitative research.* Lanham, MD: AltaMira Press.

Guest, G. S., MacQueen, K. M., & Namey, E. E. (2012). *Applied thematic analysis.* Thousand Oaks, CA: Sage.

Gunderson, L. H., & Holling, C. S. (Eds.). (2002). *Panarchy: Understanding transformations in human and natural systems.* Washington, DC: Island Press.

Gustason, C. (2012, March 12). Closure [Electronic mailing list message]. QUALRS-L@LISTSERV.

Gutin, S. A., Cummings, B., Jaiantilal, P., Johnson, K., Mbofana, F., & Rose, C. D. (2014). Qualitative evaluation of a positive prevention training for health care providers in Mozambique. *Evaluation and Program Planning, 43,* 38–47.

Hacker, K. A. (2013). *Community-based participatory research.* Thousand Oaks, CA: Sage.

Hacking, I. (2000). *The social construction of what?* Cambridge, MA: Harvard University Press.

Hall, N. (Ed.). (1993). *Exploring chaos: A guide to the new science of disorder.* New York, NY: W. W. Norton.

Hall, S. S. (2007, May 6). The older-and-wiser hypothesis. *Sunday New York Times Magazine.* Retrieved from http://www.nytimes.com/2007/05/06/magazine/06Wisdom-t.html?pagewanted=all&module=Search&mabReward=relbias%3Ar

Halliday, M. A. K. (1978). *Language as social semiotic: The social interpretation of language and meaning.* Baltimore, MD: University Park Press.

Hamera, J. (2011). Peformative enthnography. In N. Denzin & Y. Lincoln (Eds.), *The SAGE handbook of qualitative research* (4th ed., pp. 317–329). Thousand Oaks, CA: Sage.

Hammer, L. J. (2011a). *Me, my intimate partner, and HIV: Fijian self-assessments of transmission risks.* Suva, Fiji: United Nations Development Programme Pacific Centre.

Hammer, L. J. (2011b, December 13). Me, my intimate partner, and HIV [Electronic mailing list message]. EvalTalk Internet listserv of the American Evaluation Association.

Hammersley, M. (2008a). *Questioning qualitative inquiry: Critical essays.* Thousand Oaks, CA: Sage.

Hammersley, M. (2008b). Troubles with triangulation. In M. M. Bergman (Ed.), *Advances in mixed methods* (pp. 22–36). Thousand Oaks, CA: Sage.

Hamon, R. R. (1996). Bahamian life as depicted by wives' tales and other old sayings. In M. B. Sussman & J. F. Gilgun (Eds.), *The methods and methodologies of qualitative family research* (pp. 57–88). New York, NY: Haworth Press.

Hancock, J. (2003). *Scaling-up the impact of good practices in rural development: A working paper to support implementation of the world bank's rural development strategy.* Washington, DC: World Bank.

Handwerker, W. P. (2001). *Quick ethnography.* Walnut Creek, CA: AltaMira Press.

Handwerker, W. P. (2009). *The origin of cultures: How individual choices make cultures change.* Walnut Creek, CA: Left Coast Press.

Hannes, K., & Lockwood, C. (2012). *Synthesizing qualitative research: Choosing the right approach.* London, England: Sage.

Harden, A., & Gough, D. (2012). Quality and relevance appraisal. In D. Gough, S. Oliver, & J. Thomas (Eds.), *An introduction to systematic reviews* (pp. 153–178). Thousand Oaks, CA: Sage.

Hargreaves, M. (2013). *Rapid evaluation approaches for complex initiatives: White paper.* Cambridge, MA: Mathematica Policy Research.

Hargreaves, M., & Paulsell, D. (2009). *Evaluating systems change efforts to support evidence-based home visiting: Concepts and methods.* Cambridge, MA: Mathematica Policy Research.

Harper, D. (2000). Reimagining visual methods. In N. K. Denzin & Y. S. Lincoln (Eds.), *Handbook of qualitative research* (2nd ed., pp. 717–732). Thousand Oaks, CA: Sage.

Harper, D. (2008). What's new visually. In N. K. Denzin & Y. S. Lincoln (Eds.), *Collecting and interpreting qualitative materials* (pp. 185–204). Thousand Oaks, CA: Sage.

Harvey, C., & Denton, J. (1999). To come of age: The antecedents of organizational learning. *Journal of Management Studies, 36*(7), 897–918.

Harvey, J. H., Turnquist, D. C., & Agostinell, G. (1998). Identifying attributions in oral and written explanations. In C. Antaki (Ed.), *Analyzing everyday explanation* (pp. 32–42). London, England: Sage.

Harwood, R. R. (1979). *Small farm development: Understanding and improving farming systems in the humid tropics.* Boulder, CO: Westview Press.

Hastings, G., & Domegan, C. (2014). Research and the art of storytelling. *Social Marketing: From tunes to symphonies* (chap. 5). New York, NY: Routledge.

Hausman, C. (2000). *Lies we live by.* New York, NY: Routledge.

Hayano, D. M. (1979). Autoethnography: Paradigms, problems, and prospects. *Human Organization, 38,* 113–120.

Hayes, T. A. (2000). Stigmatizing indebtedness: Implications for labeling theory. *Symbolic Interaction, 23*(1), 29–46.

Hays, D. G., & Singh, A. A. (2012). *Qualitative inquiry in clinical and educational settings.* New York, NY: Guilford Press.

Headland, T. N., Pike, K. L., & Harris, M. (Eds.). (1990). *Emics and etics: The insider/outsider debate.* Newbury Park, CA: Sage.

Heath, C., Hindmarsh, J., & Luff, P. (2010). *Video in qualitative research: Analyzing social interaction in everyday life.* Thousand Oaks, CA: Sage.

Heckathorn, D. D. (1997). Respondent-driven sampling: A new approach to the study of hidden population. *Social Problems, 44*(2), 174–199.

Heer, K. G., & Anderson, G. L. (2005). *The action research dissertation: A guide for students and faculty.* Thousand Oaks, CA: Sage.

Heinlein, R. A. (1978). *The notebooks of Lazarus Long.* New York, NY: Putnam.

Helsing, D., Howell, A., Kegan, R., & Lahey, L. (2008). Putting the "development" in professional development: Understanding and overturning educational leaders' immunities to change. *Harvard Educational Review, 78*(3), 437–465.

Henderson, S., & Segal, E. H. (2013). Visualizing qualitative data in evaluation. *New Directions for Evaluation, 139,* 53–72.

Henige, D. (1988). *Oral historiography.* London, England: Longman.

Henry, G. (2009). When getting it right matters: The case for high-quality policy and program impact evaluations. In S. I. Donaldson, C. A. Christie, & M. Mark (Eds.), *What counts as credible evidence in applied research and evaluation practice* (pp. 32–50). Thousand Oaks, CA: Sage.

Henry, S. G., & Fetters, M. D. (2012). Video elicitation interviews: A qualitative research method for investigating physician-patient interactions. *Annals of Family Medicine, 10*(2), 118–125.

Hersh, S. M. (2003, October 27). The stovepipe: How conflicts between the Bush administration and the intelligence community marred the reporting on Iraq's weapons. *The New Yorker.* Retrieved from http://www.newyorker.com/archive/2003/10/27/031027fa_fact?currentPage=all

Hertz, R. (Ed.). (1997). *Reflexivity & voice.* Thousand Oaks, CA: Sage.

Hesse-Biber, S. J. N. (Ed.). (2011). *Handbook of feminist research: Theory and praxis.* Thousand Oaks, CA: Sage

Hesse-Biber, S. J. N. (Ed.). (2013). *Feminist research practice: A primer* (2nd ed.). Thousand Oaks, CA: Sage.

Hesse-Biber, S. J. N., & Leavy, P. L. (2006a). *Emergent methods in social research.* Thousand Oaks, CA: Sage.

Hesse-Biber, S. J. N., & Leavy, P. L. (Eds.). (2006b). *Feminist research practice: A primer.* Thousand Oaks, CA: Sage.

Hesse-Biber, S. J. N., & Leavy, P. L. (2008). *Handbook of emergent methods.* New York, NY: Guilford Press.

Hesse-Biber, S. J. N., & Leavy, P. L. (2011). *The practice of qualitative research* (2nd ed.). Thousand Oaks, CA: Sage

Hieronymi, A. (2013). Understanding systems science: A visual and integrative approach. *Systems Research and Behavioral Science, 30*(5), 580–595. Retrieved from http://onlinelibrary.wiley.com/doi/10.1002/sres.2215/pdf

Higginbotham, J. B., & Cox, K. K. (1979). *Focus group interviews: A reader.* Chicago, IL: American Marketing Association.

Hiles, D. (2001). *Heuristic inquiry and transpersonal research.* Leicester, England: De Montfort University, Psychology Department. Retrieved from http://psy.dmu.ac.uk/drhiles/HIpaper.htm

Hilliard, A. G. (1989). Kemetic (Egyptian) historical revision: Implications for cross-cultural evaluation and research in education. *American Journal of Evaluation, 10*(2), 7–23.

Hinton, B. (1988). Audit tales: Kansas intrigue. *Legislative Program Evaluation Society (LPES) Newsletter, Spring,* 3.

Hipkins, R., Cowie, B., Boyd, S., Keown, P., & McGee, C. (2011). *Curriculum implementation exploratory studies 2: Final report.* Wellington, New Zealand: Learning Media. Retrieved from http://www.educationcounts.govt.nz/publications/curriculum/curriculum-implementation-exploratory-studies-2

Hird, M. J., & Abshoff, K. (2000). Women without children: A contradiction in terms? *Journal of Comparative Family Studies, 31*(3), 347–366.

Hirsh, S. K., & Kummerow, J. M. (1987). *Introduction to type in organizational settings.* Palo Alto, CA: Consulting Psychologists Press.

Hodder, I. (2000). The interpretation of documents and material culture. In N. K. Denzin & Y. S. Lincoln (Eds.), *Handbook of qualitative research* (pp. 703–713). Thousand Oaks, CA: Sage.

Hoevemeyer, V. A. (2005). *High-impact interview questions.* New York, NY: American Management Association.

Hoffman, L. (1981). *Foundations of family therapy: A conceptual framework for systems change.* New York, NY: Basic Books.

Hogle, J. A., & Moberg, D. P. (2013). Success case studies contribute to evaluation of complex research infrastructure. *Evaluation & the Health Professions, 37*(1), 98–113. Retrieved from http://ehp.sagepub.com/content/early/2013/08/07/0163278713500140.abstract

Hohman, M. (2011). *Motivational interviewing in social work practice.* New York, NY: Guilford Press.

Holbrook, T. L. (1996). Document analysis: The contrast between official case records and the journal woman on welfare. In M. B. Sussman & J. F. Gilgun (Eds.), *The methods and methodologies of qualitative family research* (pp. 41–56). New York, NY: Haworth Press.

Holley, H., & Arboleda-Florez, J. (1988). Utilization isn't everything. *Canadian Journal of Program Evaluation, 3*(2), 93–102.

Holliday, A. (2007). *Doing and writing qualitative research* (2nd ed.). Thousand Oaks, CA: Sage.

Hollinger, D. (2000). Paradigms lost (J. A. Holstein & J. F. Gubrium, Eds.). *The New York Times Book Review,* p. 23.

Holmes, R. M. (1998). *Fieldwork with children.* Thousand Oaks, CA: Sage.

Holstein, J. A., & Gubrium, J. F. (1995). *The active interview: Vol. 37. Qualitative research methods.* Thousand Oaks, CA: Sage.

Holstein, J. A., & Gubrium, J. F. (2003). Active interviewing. In J. F. Gubrium & J. A. Holstein (Eds.), *Postmodern interviewing* (pp. 67–80). Thousand Oaks, CA: Sage.

Holstein, J. A., & Gubrium, J. F. (Eds.). (2008). *Handbook of constructionist research.* New York, NY: Guilford Press.

Holstein, J. A., & Gubrium, J. F. (2011). The constructive analytics of interpretive practice. In N. K. Denzin & Y. S. Lincoln (Eds.), *The SAGE handbook of qualitative research* (4th ed., pp. 341–358). Thousand Oaks, CA: Sage.

Holt, N. (2014). *Cured: How the Berlin patients defeated HIV and forever changed medical science.* New York, NY: Dutton.

Holte, J. (Ed.). (1993). *Chaos: The new science: Nobel Conference XXVI.* St. Peter, MN: Gustavus Adolphus College.

Holtzman, J. S. (1986). *Rapid reconnaissance guidelines for agricultural marketing and food systems research in developing countries* (Working Paper No. 30). Ann Arbor: Michigan State University, Department of Agricultural Economics.

Homeless Youth Collaborative on Developmental Evaluation. (2014). *Nine evidence-based, guiding principles to help youth overcome homelessness.* St. Paul, MN: Otto Bremer Foundation. Retrieved from http://www.terralunacollaborative.com/wp-content/uploads/2014/03/9-Evidence-Based-Principles-to-Help-Youth-Overcome-Homelessness-Webpublish.pdf

Hoorens, V. (1993). Self-enhancement and superiority biases in social comparison. *European Review of Social Psychology, 4*(1), 113–139.

Hopson, R. K. (Ed.). (2000). *How and why language matters in evaluation* (*New Directions for Evaluation*, No. 86). San Francisco, CA: Jossey-Bass.

House, E. R. (1977). *The logic of evaluative argument: Vol. 7. CSE monograph series in evaluation.* Los Angeles, CA: University of

REFERENCES

California, UCLA Graduate School of Education, Center for the Study of Evaluation.

House, E. R. (1991). Confessions of a responsive goal-free evaluation. *Evaluation Practice, 12*(1), 109-113.

House, E. R. (1994). Integrating the qualitative and quantitative. *New Directions for Program Evaluation, 6,* 13–22.

House, E. R., & Howe, K. R. (2000). Deliberative democratic evaluation. *New Directions for Evaluation, 85,* 3–12.

Huberman, A. M., & Miles, M. B. (2002). *The qualitative researcher's companion.* Thousand Oaks, CA: Sage.

Huffington Post. (2013, August 17). *"Typhoid Mary" mystery may have been solved at last, scientists say.* Retrieved from http://www.huffingtonpost.com/2013/08/17/typhoid-mary-mystery-solved_n_3762822.html

Hughes, C. (2002). *Key concepts in feminist theory and research.* London, England: Sage.

Hull, B. (1978). *Teachers' seminars on children's thinking* (North Dakota Study on Group Evaluation). Grand Forks: University of North Dakota.

Humphreys, L. (1970). *Tearoom trade: Impersonal sex in public places.* New York, NY: Aldine.

Hunt, S. A., & Benford, R. D. (1997). Dramaturgy and methodology. In G. Miller & R. Dingwall (Eds.), *Context and method in qualitative research* (pp. 106–118). Thousand Oaks, CA: Sage.

Hurty, K. (1976). Report by the women's caucus. In *Proceedings: Educational evaluation and public policy, a conference.* San Francisco, CA: Far West Laboratory for Educational Research and Development.

Husserl, E. (1954). *Ideas.* New York, NY: Collier. (Original work published 1913, London, England: Allen & Unwin)

Husserl, E. (1967). The thesis of the natural standpoint and its suspension. In J. Kockelmans (Ed.), *Phenomenology* (pp. 68–79). Garden City, NY: Doubleday.

Husserl, E., & Moran, D. (2012). *Ideas: General introduction to pure phenomenology.* New York, NY: Routledge.

Huxley, A. (1937). *Ends and means.* London, England: Chatto & Windus.

Hyman, H. H. (1954). *Interviewing in social research* (Original work published 1913). Chicago, IL: University of Chicago Press.

Ihde, D. (1977). *Experimental phenomenology.* New York: G. P. Putnams's Sons.

ImmigrationRoad.com. (2013). *Immigration roadmap.* San Diego, CA: Author.

International Development Research Centre. (2010). *Outcome mapping.* Ottawa, Ontario, Canada: Author. Retrieved from http://www.idrc.ca/EN/Resources/Publications/Pages/ArticleDetails.aspx?PublicationID=1004

Ioannidis, J. P. A. (2005). Why most published research findings are false. *PLoS Medicine, 2*(8). Retrieved from http://www.plosmedicine.org/article/info:doi/10.1371/journal.pmed.0020124

Iorio, S. H. (Ed.). (2009). *Qualitative research in journalism: Taking it to the streets.* New York, NY: Taylor & Francis.

Ironside, W. (1993). Cade, John Frederick Joseph (1912–1980). In *Australian dictionary of biography.* Retrieved from http://adb.anu.edu.au/biography/cade-john-frederick-joseph-9657

Israel, B. A., Eng, E., Schulz, A. J., & Parker, E. A. (2013). *Methods for community-based participatory research for health.* San Francisco, CA: Wiley.

Jacob, E. (1987). Qualitative research traditions: A review. *Review of Educational Research, 57*(1), 1–50.

Jacob, S., & Desautels, G. (2013). Evaluation of Aboriginal programs: What place is given to participation and cultural sensitivity? *International Indigenous Policy Journal, 4*(2), 1–29.

James, N., & Busher, H. (2012). Internet interviewing. In J. F. Gubrium, J. A. Holsten, A. B. Marvasti, & K. D. McKinney (Eds.), *The SAGE handbook of interview research: The complexity of the craft* (2nd ed., pp. 177–192). Thousand Oaks, CA: Sage.

Jamison, L. (2014). *The empathy exams.* Minneapolis, MN: Graywolf Press.

Janesick, V. J. (2000). The choreography of qualitative research design: Minuets, improvisations, and crystallization. In N. K. Denzin & Y. S. Lincoln (Eds.), *Handbook of qualitative research* (2nd ed., pp. 379–399). Thousand Oaks, CA: Sage.

Janesick, V. J. (2010). *Oral history for the qualitative researcher.* New York, NY: Guilford Press.

Janesick, V. J. (2011). *Stretching exercises for qualitative researchers* (3rd ed.). Thousand Oaks, CA: Sage.

Janowitz, M. (1979). Where is the cutting edge of sociology? *Sociological Quarterly, 20*(4), 591–593.

Jarvis, S. (2000). *Getting the log out of our own eyes: An exploration of individual and team learning in a public human services agency.* Unpublished Doctoral Dissertation, The Union Institute Graduate College, Cincinnati, Ohio.

Jenkings, K. N. (2013). Adjudicating accounts: A review of Dupret's ethnomethodological studies of Arab practices. *Symbolic Interaction, 36*(1), 99–103.

Johnson, A., & Sackett, R. (1998). Direct systematic observation of behavior. In H. R. Bernard (Ed.), *Handbook of methods in cultural anthropology* (pp. 301–331). Walnut Creek, CA: AltaMira Press.

Johnson, G. (2014a, January 20). New truths that only one can see. *The New York Times,* p. D1.

Johnson, G. (2014b, May 11). Patients and fortitude. *The New York Times Sunday Book Review,* p. 30.

Johnson, J. M. (1975). *Doing field research.* Beverly Hills, CA: Sage.

Johnson, S. (2001). *Emergence: The connected lives of ants, brains, cities, and software.* New York, NY: Scribner.

Johnston, L. G., & Sabin, K. (2010). Sampling hard-to-reach populations with respondent driven sampling. *Methodological Innovations Online, 5*(2), 38–48. Retrieved from http://www.pbs.plym.ac.uk/mi/pdf/05-08-10/5.%20Johnston%20and%20Sabin%20English%20(formatted).pdf

Joint Committee on Standards. (2010). *The program evaluation standards.* Thousand Oaks, CA: Sage. Retrieved from http://www.jcsee.org/

Joint Committee on Standards for Educational Evaluation. (1994). *The standards for program evaluation.* Thousand Oaks, CA: Sage.

Jolley, G. (2014). Evaluating complex community-based health promotion: Addressing the challenges. *Evaluation and Program Planning, 45*(1), 71–81.

Jones, J. H. (1993). *Bad blood: The Tuskegee syphilis experiment.* New York, NY: Free Press.

Jones, M. O. (1996). *Studying organizational symbolism: What, how, why?* (Qualitative Research Methods Series No. 39). Thousand Oaks, CA: Sage.

Jones, S. H., Adams, T. E., & Ellis, C. (Eds.). (2013). *Handbook of autoethnography.* Walnut Creek, CA: Left Coast Press.

Jordan, B. (Ed.). (2012). *Advancing ethnography in corporate environments.* Walnut Creek, CA: Left Coast Press.

Josselson, R. (2013). *Interviewing for qualitative inquiry: A relational approach.* New York, NY: Guilford Press.

Joyce, C. (2012, August 23). *How the Smokey Bear effect led to raging wildfires* (Morning edition radio). Washington, DC: NPR. Retrieved from http://www.npr.org/2012/08/23/159373691/how-the-smokey-bear-effect-led-to-raging-wildfires?sc=emaf

Joyce, P. G. (2011). *The Congressional Budget office: Honest numbers, power, and policymaking.* Washington, DC: Georgetown University Press.

Julnes, G. (Ed.). (2012). *Promoting valuation in the public interest: Informing policies for judging value in evaluation* (New Directions for Evaluation, No. 133). Hoboken, NJ: Wiley Periodicals.

Julnes, G., & Rog, D. (Eds.). (2007). *Informing federal policies on evaluation methodology: Building the evidence base for method choice in government sponsored evaluation* (New Directions for Evaluation, No. 113). San Francisco, CA: Jossey-Bass.

Jung, C. G. (1902). *Collected works of C. G. Jung* (2 vols.). Princeton, NJ: Princeton University Press.

Junker, B. H. (1960). *Field work: An introduction to the social sciences.* Chicago, IL: University of Chicago Press.

Juran, J. M. (1951). *Quality control handbook.* New York, NY: McGraw-Hill.

Juster, N. (1961). *The phantom tollbooth.* New York, NY: Alfred A. Knopf.

Kaczynski, D., Salmona, M., & Smith, T. (2014). Qualitative research in finance. *Australian Journal of Management, 39*(1), 127–135.

Kahane, A. (2012). *Transformative scenario planning: Working together to change the future.* San Francisco, CA: Barrett-Koehler.

Kahn, A. (1990). The hermeneutics of sexual order. *Santa Clara Law Review, 31*(1), 47–102.

Kahneman, D. (2011). *Thinking, fast and slow.* New York, NY: Farrar, Straus & Giroux.

Kahneman, D., & Krueger, A. B. (2006). Developments in the measurement of subjective well-being. *Journal of Economic Perspectives, 20*(1), 3–24.

Kalamazoo schools. (1974). *American School Board Journal, April,* 32–40.

Kamberelis, G., & Dimitriadis, G. (2011). Focus groups: Contingent articulations of pedagogy, politics, and inquiry. In N. K. Denzin & Y. S. Lincoln (Eds.), *The SAGE handbook of qualitative research* (4th ed., pp. 545–561). Thousand Oaks, CA: Sage.

Kanter, R. M. (1983). *The change masters: Innovations for productivity in the American corporation.* New York, NY: Simon & Schuster.

Kaplan, A. (1973). *The conduct of inquiry.* New York, NY: Transaction.

Kaplan, L. (2013, September 17). Storytelling is not a magic bullet. *Nonprofit Quarterly Newswire.* Retrieved from http://www.nonprofitquarterly.org/management/22914-storytelling-is-not-a-magic-bullet.html

Kaplowitz, M. D. (2000). Statistical analysis of sensitive topics in group and individual interviews. *Quality & Quantity, 34*(4), 419–431.

Karpiak, I. E. (2006). Chaos and complexity: A framework for understanding social workers in midlife. In V. A. Anfara Jr. & N. T. Mertz (Eds.), *Theoretical frameworks in qualitative research* (pp. 85–108). Thousand Oaks, CA: Sage.

Kartunnen, L. (1973, March). *Remarks on presuppositions.* Paper presented at the Texas Conference on Performances, Conversational Implicature and Presuppositions, University of Texas, Austin.

Katz, L. (1987). *The experience of personal change* (Doctoral dissertation). The Union for Experimenting Colleges and Universities, Cincinnati, OH.

Katzer, J., Cook, K. H., & Crouch, W. W. (1998). *Evaluating information: A guide for users of social science research.* Boston, MA: McGraw-Hill.

Kavanagh, D. (2013). Children: Their place in organization studies. *Organization Studies, 34*(1), 1487–1503.

Kean, S. (2014, May 3). Beyond the damaged brain. *The New York Times Sunday Review,* SR8.

Keating, J. (2014). Random acts: What happens when you approach global poverty as a science experiment? *Slate.* Retrieved from http://www.slate.com/articles/business/crosspollination/2014/03/randomized_controlled_trials_do_they_work_for_economic_development.1.html

Kegan, J. (1982). *The evolving self: Problem and process in human development.* Cambridge, MA: Harvard University Press.

Kellehear, A. (1993). *The unobtrusive researcher.* Sydney, New South Wales, Australia: Allen & Unwin.

Kelley, T., & Littman, J. (2001). *The art of innovation: Lessons in creativity from IDEO, America's leading design firm.* New York, NY: Currency/Doubleday.

Kelso, J. A. S., & Engstrom, D. A. (2006). *The complementary nature.* Cambridge: MIT Press.

Kemmis, S., & McTaggart, R. (2000). Participatory action research. In N. K. Denzin & Y. S. Lincoln (Eds.), *Handbook of qualitative research* (2nd ed., pp. 567–606). Thousand Oaks, CA: Sage.

Kemper, E. A., Stringfield, S., & Teddlie, C. (2003). Mixed methods sampling strategies in social science research. In A. Tashakkori & C. Teddlie (Eds.), *Handbook of mixed methods in social and behavioral research* (pp. 273–296). Thousand Oaks, CA: Sage.

Khagram, S., & Thomas, C. W. (2010). Toward a platinum standard for evidence-based assessment by 2020. *Public Administration Review, 70,* S100-S106.

Khan, I. (2013, June 16). Polio workers killed in Pakistan. *The New York Times.* Retrieved from http://www.nytimes.com/2013/06/17/world/middleeast/2-polio-workers-killed-in-pakistan.html?_r=0

Kibel, B. M. (1999). *Success stories as hard data: An introduction to results mapping.* New York, NY: Kluwer Academic/Plenum Press.

Kim, D. H. (1993). *Systems archetypes I.* Williston, VT: Pegasus Communications.

Kim, D. H. (1994). *Systems archetypes II: Using systems archetypes to take effective action.* Williston, VT: Pegasus Communications.

Kim, D. H. (1999). *An introduction to systems thinking.* Williston, VT: Pegasus Communications.

Kincheloe, J. L., & McLaren, P. (2000). Rethinking critical theory and qualitative research. In N. K. Denzin & Y. S. Lincoln (Eds.), *Handbook of qualitative research* (pp. 279–313). Thousand Oaks, CA: Sage.

King, B., & Kim, J. (2014, March 30). What umpires get wrong. *The New York Times Sunday Review,* p. SR12.

King, G., Keohane, R. O., & Verba, S. (1994). *Designing social inquiry: Scientific inference in qualitative research.* Princeton, NJ: Princeton University Press.

King, J. A. (1995). Involving practitioners in evaluation studies: How viable is collaborative evaluation in schools. In J. B. Cousins & L. M. Earl (Eds.), *Participatory evaluation in education: Studies in evaluation use and organizational learning* (pp. 86–102). London, England: Falmer Press.

King, J. A. (2005). Participatory evaluation. In S. Mathison (Ed.), *Encyclopedia of evaluation* (pp. 291–294). Thousand Oaks, CA: Sage.

King, J. A., Cousins, J. B., & Whitmore. E. (2007). Making sense of participatory evaluation: Framing participatory evaluation. In S. Mathison (Ed.), *Enduring issues in evaluation: The 20th anniversary of the collaboration between NDE and AEA* (*New Directions for Evaluation,* No. 114, pp. 83–105). Hoboken, NJ: Wiley Periodicals.

King, J. A., Morris, L. L., & Fitz-Gibbon, C. T. (1987). *How to assess program implementation.* Newbury Park, CA: Sage.

King, J. A., & Pechman, E. (1982). *Improving evaluation use in local schools.* Washington, DC: National Institute of Education.

King, J. A., & Podems, D. (Eds.). (2014). Professionalizing evaluation: A global perspective on evaluator competencies [Special issue]. *Canadian Journal of Program Evaluation, 28*(3).

King, J. A., & Stevahn, L. (2013). *Interactive evaluation practice: Mastering the interpersonal dynamics of program evaluation.* Thousand Oaks, CA: Sage.

King, J. A., Stevahn, L., Ghere, G., & Minnema, J. (2001). Toward a taxonomy of essential evaluator competencies. *American Journal of Evaluation, 22*(2), 229–247.

King, N. (2004). Using templates in the thematic analysis of texts. In C. Cassell & G. Symon (Eds.), *Essential guide to qualitative methods in organizational research* (pp. 256–270). London, England: Sage.

King, N. (2012). Doing template analysis. In C. Cassell & G. Symon (Eds.), *Qualitative organizational research: Core methods and current challenges* (pp. 426–450). London, England: Sage.

King, N., & Horrocks, C. (2012). *Interviews in qualitative research.* London, England: Sage.

King, S. A. (1996). Researching Internet communities: Proposed ethical guidelines for the reporting of results. *Information Society, 12*(2), 119–128.

Kinn, L. G., Holgersen, H., Ekeland, T., & Davidson, L. (2013). Metasynthesis and bricolage. *Qualitative Health Research, 23*(9), 1285–1292.

Kirk, J., & Miller, M. L. (1986). *Reliability and validity in qualitative research.* Beverly Hills, CA: Sage.

Kirkhart, K. E. (1995). Evaluation and social justice seeking multi-cultural validity. *American Journal of Evaluation, 16*(1), 1–12.

Kirkman, J. (2013). *I can barely take care of myself: Tales from a happy life without kids.* New York, NY: Simon & Schuster.

Kistler, S. (2014, May 13). Innovative reporting [AEA365 blog]. Retrieved from http://aea365.org/blog/

Klein, G. (2011). Critical thinking. *Theoretical Issues in Ergonomics Science, 12*(3), 210–224.

Klein, G. (2014). *Evidence-based medicine.* Retrieved from http://www.edge.org/responses/what-scientific-idea-is-ready-for-retirement

Kling, J. R., Liebman, J. B., & Katz, L. F. (2001, January). *Bullets don't got no name: Consequences of fear in the ghetto.* Paper presented at the Conference on Mixed Methods sponsored by the Macarthur Network on Successful Pathways Through Middle Childhood, Santa Monica, CA.

Kneller, G. F. (1984). *Movements of thought in modern education.* New York, NY: Wiley.

Knowles, M. (1989). *The making of an adult educator.* San Francisco, CA: Jossey-Bass.

Knowles, J. G., & Cole, A. L. (Eds.). (2008). *Handbook of the arts in qualitative research.* Thousand Oaks, CA: Sage.

Knox, R. (2013, March). Scientists report first cure of HIV in a child, say it's a game-changer. *National Public Radio.* Retrieved from http://www.npr.org/blogs/health/2013/03/04/173258954/scientists-report-first-cure-of-hiv-in-a-child-say-its-a-game-changer

Knutson, B. (2014). *Scientific idea ready for retirement: Emotion is peripheral.* Retrieved from http://www.edge.org/responses/what-scientific-idea-is-ready-for-retirement

Kolata, G. (2013, September 2). Guesses and hype give way to data in study of education. *The New York Times,* p. D1. Retrieved from http://

www.nytimes.com/2013/09/03/science/applying-new-rigor-in
-studying-education.html

Kolata, G. (2014, February 2). Method of study is criticized in group's health policy tests. *The New York Times*, p. A1. Retrieved from http://www.nytimes.com/2014/02/03/health/effort-to-test-health-policies-is-criticized-for-study-tactics.html?ref=ginakolata

Kolkey, C. (2011). *The immigrant advantage: What we can learn from newcomers to America about health, happiness and hope*. New York, NY: Free Press.

Koro-Ljungberg, M. (2008). A social constructionist framing of the research interview. In J. A. Holstein & J. F. Gubrium (Eds.), *Handbook of constructionist research* (pp. 429–444). New York, NY: Guilford Press.

Kovach, M. (2009). *Indigenous methodologies: Characteristics, conversations, and context*. Toronto, Ontario, Canada: University of Toronto Press.

Krahmer, E., & Ummelen, N. (2004). Thinking about thinking aloud: A comparison of two verbal protocols for usability testing. *IEEE Transactions on Professional Communication, 47*(2), 105–117.

Krishnamurti, J. (1964). *Think on these things*. New York, NY: Harper & Row.

Kroeger, O. & Thuesen, J. M. (1988). *Type talk*. New York: Delacorte Press.

Krueger, R. A. (1994). *Focus groups: A practical guide for applied research* (2nd ed.). Thousand Oaks, CA: Sage.

Krueger, R. A. (1997). *The focus group kit* (1–6 vols.). Thousand Oaks, CA: Sage.

Krueger, R. A. (2010). Using stories in evaluation. In J. Wholey, H. Hatry, & K. Newcomer (Eds.), *Handbook of practical program evaluation* (3rd ed., pp. 404–423). San Francisco, CA: Jossey-Bass.

Krueger, R. A., & Casey, M. A. (2000). *Focus groups: A practical guide for applied research* (3rd ed.). Thousand Oaks, CA: Sage.

Krueger, R. A., & Casey, M. A. (2008). *Focus groups: A practical guide for applied research* (4th ed.). Thousand Oaks, CA: Sage.

Krueger, R. A., & Casey, M. A. (2012, October 26). *Internet focus groups: Lessons learned and important considerations*. Paper presented at the annual meeting panel presentation of the American Evaluation Association, Minneapolis, MN.

Krueger, R. A., & King, J. A. (1997). *Involving community members in focus groups: Vol. 4. The focus group kit*. Thousand Oaks, CA: Sage.

Kruska, W., & Mosteller, F. (1980). Representative sampling, IV: The history of the concept is statistics, 1895–1939. *International Statistical Review, 48*, 169–195.

Kuhn, M. H. (1964). Major trends in symbolic interaction theory in the past twenty-five years. *Sociological Quarterly, 5*(1), 61–84.

Kuhn, T. S. (1970). *The structure of scientific revolutions*. Chicago, IL: University of Chicago Press.

Kulish, N. (2001, March 12). Ancient split of Assyrians and Chaldeans leads to modern-day battle over census. *The Wall Street Journal*, p. 1.

Kurth, E., Kennedy, H. P., Stutz, E. Z., Kesselring, A., Fornaro, I., & Spichiger, E. (2013). Responding to a crying infant—you do not learn it overnight: A phenomenological study. *Midwifery*. Retrieved from http://www.midwiferyjournal.com/article/S0266-6138(13)00199-X/abstract

Kurtzman, J., & Goldsmith, M. (2010). *Common purpose: How great leaders get organizations to achieve the extraordinary*. San Francisco, CA: Jossey-Bass.

Kurylo, E. (2013). *Family storytelling may be key to resilience in children*. Retrieved from http://www.marial.emory.edu/faculty/profiles/duke.html

Kushner, S. (2000). *Personalizing evaluation*. London, England: Sage.

Kusters, C. S. L., van Vugt, S. M., Wigboldus, S. A., Williams, B., & Woodhill, A. J. (2011). *Making evaluations matter: A practical guide for evaluators*. Wageningen, Netherlands: Wageningen University, Centre for Development Innovation.

Kuzmin, A. (2005). *Exploration of factors that affect the use of evaluation training in evaluation capacity development* (Doctoral dissertation). Union Institute and University, Cincinnati, OH.

Kvale, S. (1987). Validity in the qualitative research interview. *Methods: A Journal for Human Science, 1* (2, Winter), 37–72.

Kvale, S. (1996). *Interviews. An introduction to qualitative research interviewing*. Thousand Oaks, CA: Sage.

Kvale, S. (2007). *Doing interviews*. London, England: Sage.

Kvale, S., & Brinkmann, S. (2008). *Interviews: Learning the craft of qualitative research interviewing* (2nd ed.). Thousand Oaks, CA: Sage.

Ladner, S. (2014). *Practical ethnography: A guide to doing ethnography in the private sector*. Walnut Creek, CA: Left Coast Press.

Lalonde, B. I. D. (1982). Quality assurance. In M. J. Austin & W. E. Hershey (Eds.), *Handbook on mental health administration* (pp. 352–375). San Francisco, CA: Jossey-Bass.

Lambert, V., Glacken, M., & McCarron, M. (2013). Using a range of methods to access children's voices. *Journal of Research in Nursing, 18*(7), 601–616.

Lang, K., & Lang, G. E. (1960). Decisions for Christ: Billy Graham in New York City. In M. Stein, A. J. Vidich, & D. M. White (Eds.), *Identity and anxiety* (pp. 415–425). New York, NY: Free Press.

Lawrence-Lightfoot, S. (1997). Illumination: Framing the terrain. In S. Lawrence-Lightfoot & J. H. Davis (Eds.), *The art and science of portraiture* (pp. 41–59). San Francisco, CA: Jossey-Bass.

Lawrence-Lightfoot, S. (2000). *Respect: An exploration*. Cambridge, MA: Perseus Books.

Lawrence-Lightfoot, S. (2012). *Exit: The endings that set us free*. New York, NY: Farrar, Straus & Giroux.

Lawrence-Lightfoot, S., & Hoffman Davis, J. (1997). *The art and science of portraiture*. San Francisco, CA: Jossey-Bass.

Lawrenz, F. (2013). *Outstanding evaluation of 2012*. Minneapolis, MN: American Evaluation Association Awards Luncheon.

Leavy, P. (Ed.). (2014). *The Oxford handbook of qualitative research*. New York, NY: Oxford University Press.

Lee, M. G., & O'Brien, M. K. (2012, October 26). *Internet focus groups: Lessons learned and important considerations*. Paper presented at the annual meeting panel presentation of the American Evaluation Association, Minneapolis, MN.

Lee, P. (1996). *The Whorf theory complex: A critical reconstruction: Vol. 81. Studies in the history of the language sciences*. Amsterdam, Netherlands: John Benjamins.

Leeuw, F., Rist, R., & Sonnichsen, R. (Eds.). (1993). *Comparative perspectives on evaluation and organizational learning*. New Brunswick, NJ: Transaction.

Leibovich, M. (2013, December 23). Behind the cover story. *The New York Times*. Retrieved from http://6thfloor.blogs.nytimes.com/2013/12/23/behind-the-cover-story-mark-leibovich-on-checking-in-with-john-mccain/?hp&_r=0

Leinicke, L. M., Ostrosky, J. A., Rexroad, W. M., Baker, J. R., & Beckman, S. (2005). Interviewing as an auditing tool. *The CPA Online Journal*. Retrieved from http://www.nysscpa.org/cpajournal/2005/205/essentials/p34.htm

Lennie, J. (2011). The most significant change technique. Retrieved from http://betterevaluation.org/sites/default/files/EA_PM%26E_toolkit_MSC_manual_for_publication.pdf

Lennie, J., & Tacchi, J. (2013). *Evaluating communication for development: A framework for social change*. New York, NY: Routledge.

Leonard, E. (2001, March 11). Anecdotes. *Week in Review, The New York Times*, p. 7.

Lesser, E. L., & Stork, J. (2001). Communities of practice and organizational performance. *IBM Systems Journal, 40*(4), 831–841.

Levin, B. (1993). Collaborative research in and with organizations. *Qualitative Studies in Education 6*(4), 331–340.

Levin-Rozalis, M. (2000). Abduction: A logical criterion for programme and project evaluation. *Evaluation, 6*(4), 415–432.

Levi-Strauss, C. (1966). *The savage mind* (2nd ed.). Chicago, IL: University of Chicago Press.

Levitt, N. (1998). Why professors believe weird things. *SKEPTIC, 6*(3), 28–35.

Levy, P. F. (2001). The nut island effect: When good teams go wrong. *Harvard Business Review, 79*(3), 51–59, 163.

Lewis, C. (1929). *Mind and the world order*. New York, NY: Scribner.

Lewis, J., & Ritchie, J. (2003). Generalising from qualitative research. In J. Ritchie & J. Lewis (Eds.), *Qualitative research practice* (pp. 263–286). London, England: Sage.

Lewis, M., & Staehler, T. (2010). *Phenomenology: An introduction*. New York, NY: Continuum.

Lewis, P. J. (2001). The story of I and the death of a subject. *Qualitative Inquiry, 7*(1), 109–128.

Lie, J. (2013). Challenging anthropology: Anthropological reflections on the ethnographic turn in international relations. *Millennium: Journal of International Studies, 41*(2), 201–220.

Lieblich, A., Tuval-Mashiach, R., & Zilber, T. (1998). *Narrative research: Reading, analysis, and interpretation*. Thousand Oaks, CA: Sage.

Liebow, E. (1967). *Tally's corner*. Boston, MA: Little, Brown.

Lincoln, Y. S. (1985). *Organizational theory and inquiry: The paradigm revolution*. Beverly Hills, CA: Sage.

Lincoln, Y. S. (1990). Toward a categorical imperative for qualitative research. In E. Eisner & A. Peshkin (Eds.), *Qualitative inquiry in education: The continuing debate* (pp. 277–295). New York, NY: Teachers College Press.

Lincoln, Y. S., & Guba, E. G. (1985). *Naturalistic inquiry*. Beverly Hills, CA: Sage.

Lincoln, Y. S., & Guba, E. G. (1986). But is it rigorous? Trustworthiness and authenticity in naturalistic evaluation. In D. D. Williams (Ed.), *Naturalistic evaluation* (New Directions for Evaluation, No. 30, pp. 73–84). San Francisco, CA: Jossey-Bass.

Lincoln, Y. S., & Guba, E. G. (2013). *The constructivist credo*. Walnut Creek, CA: Left Coast Press.

Lincoln, Y. S., Lynham, S. A., & Guba, E. C. (2011). Paradigmatic controversies, contradictions, and emerging confluences revisited. In N. K. Denzin & Y. S. Lincoln (Eds.), *The SAGE handbook of qualitative research* (4th ed., pp. 97–128). Thousand Oaks, CA: Sage.

Lindgren, M., & Bandhold, H. (2009). *Scenario planning: The link between future and strategy* (Rev. ed.). New York, NY: Palgrave Macmillan.

Lipsey, M., & Wilson, D. (2001). *Practical meta-analysis*. Thousand Oaks, CA: Sage.

Livio, M. (2013). *Brilliant blunders: From Darwin to Einstein*. New York, NY: Simon & Schuster.

Loevinger, J. (1976). *Ego development*. San Francisco, CA: Jossey-Bass.

Lofland, J. (1971). *Analyzing social settings*. Belmont, CA: Wadsworth.

Lofland, J., & Lofland, L. H. (1984). *Analyzing social settings*. Belmont, CA: Wadsworth.

Long, L. N., & Christensen, W. F. (2013). When justices (subconsciously) attack: The theory of argumentative threat and the Supreme Court. *Oregon Law Review, 91*, 933–960.

Lonner, W. J., & Berry, J. W. (1986). *Field methods in cross-cultural research*. Beverly Hills, CA: Sage.

Lorenz, L. S. (2011). A way into empathy: A "case" of photo-elicitation research. *Health: An Interdisciplinary Journal for the Social Study of Health, Illness and Medicine, 15*(2), 259–276.

Loseke, D. (2013). *Methodological thinking: Basic principles of social research design*. Thousand Oaks, CA: Sage.

Love, K., Pritchard, C., Maguire, K., McCarthy, A., & Paddock, P. (2005). Qualitative and quantitative approaches to health impact assessment: An analysis of the political and philosophical milieu of the multimethod approach. *Critical Public Health, 15*, 275–289.

Lund, A., Heggen, K., & Strand, R. (2013). Knowledge and power: Exploring unproductive interplay between quantitative and qualitative researchers. *Journal of Mixed Methods Research, 7*(2), 197–210.

Mabry, L. (Ed.). (1997). *Evaluation and the postmodern dilemma: Vol. 3. Advances in program evaluation* (R. E. Stake, Series Ed.; L. Mabry, Vol. Ed.). Greenwich, CT: JAI Press.

MacCurdy, J. T. (1943). *The structure of morale*. Cambridge, England: Cambridge University Press.

Mack, J. E. (2007). *Abduction: Human encounters with aliens*. New York, NY: Scribner.

Mack, J. E. (2010). *Passport to the cosmos: Human transformation and alien encounters*. Guildford, England: White Crow Books.

Mackay, N., Quinlan, M. K., & Sommer, B. W. (2013). *Community oral history toolkit* (5 Vols.). Walnut Creek, CA: Left Coast Press.

Macmillan. (2014). Anecdata. In *BuzzWord* (Macmillan Dictionary). Retrieved from http://www.macmillandictionary.com/buzzword/entries/anecdata.html

MacQueen, K. M., & Milstein, B. (1999). A systems approach to qualitative data management and analysis. *Field Methods, 11*(1), 27–39.

Madill, A. (2008). Realism. In L. M. Given (Ed.), *The SAGE encyclopedia of qualitative research* (Vol. 2, pp. 731–735). Thousand Oaks, CA: Sage.

Madriz, E. (2000). Focus groups in feminist research. In N. K. Denzin & Y. S. Lincoln (Eds.), *Handbook of qualitative research* (2nd ed., pp. 835–850). Thousand Oaks, CA: Sage.

Magnello, M. E. (2012). Victorian statistical graphics and the iconography of Florence Nightingale's polar area graph. *BSHM Bulletin: Journal of the British Society for the History of Mathematics, 27*(1), 13–37.

Magolda, B., & King, P. M. (2012). *Assessing meaning making and self-authorship: Theory, research, and application* (ASHE Higher Education Rep., Vol. 38, No. 3). Hoboken, NJ: Wiley Periodicals.

Manderson, L. (2011). *Surface tensions: Surgery, bodily boundaries and the social self*. Walnut Creek, CA: Left Coast Press.

Manning, P. K. (1987). *Semiotics and fieldwork* (Qualitative Research Methods Series No. 7). Newbury Park, CA: Sage.

Mansuri, G., & Vijayendra, R. (2012). *Localizing development: Does participation work?* Washington, DC: World Bank.

Mantel, H. (Ed.). (2013). *No kidding: Women writers on bypassing parenthood*. Berkeley, CA: Perseus Books.

Manturuk, K. (2013, April 15). Cultural concerns with the household food insecurity access scale [Electronic mailing list message]. EvalTalk Internet listserv of the American Evaluation Association.

Marcotty, J. (2013, October 20). Saving the great north woods. *Minneapolis Star Tribune*, pp. 1, A10–A12. Retrieved from http://www.startribune.com/local/228250501.html

Marczak, M., & Sewell, M. (2013). Using focus groups for evaluation. *CYFERnet-Evaluation*. Retrieved from http://ag.arizona.edu/sfcs/cyfernet/cyfar/focus.htm

Margolis, E., & Pauwels, L. (Eds.). (2011). *The SAGE handbook of visual research methods*. Thousand Oaks, CA: Sage.

Marion, J. S., & Crowder, J. W. (2013). *Visual research: A concise introduction to thinking visually*. New York, NY: Bloomsbury Academic.

Mark, M. M., Henry, G. T., & Julnes, G. (2000). *Evaluation: An integrated framework for understanding, guiding, and improving public and nonprofit policies and programs*. San Francisco, CA: Jossey-Bass.

Marshall, C., & Rossman, G. B. (2011). *Designing qualitative research* (5th ed.). Thousand Oaks, CA: Sage.

Martin, K. (2007). *When a baby dies of SIDS: The parents' grief and search for reason*. Edmonton, Alberta, Canada: Qual Institute Press.

Maslow, A. H. (1954). *Motivation and personality*. New York, NY: Harper.

Maslow, A. H. (1956). Toward a humanistic psychology. *ETC: A Review of General Semantics, 14*, 10–22.

Maslow, A. H. (1966). *The psychology of science*. New York, NY: Harper & Row.

Mason, M. (2010). Sample size and saturation in PhD studies using qualitative interviews. Forum: Qualitative Social Research, 11(3). Retrieved from http://www.qualitative-research.net/index.php/fqs/article/view/1428/3027

Mathews, J. K., Raymaker, J., & Speltz, K. (1991). Effects of reunification on sexually abusive families. In M. Q. Patton (Ed.), *Family sexual abuse: Frontline research and evaluation* (pp. 147–161). Newbury Park, CA: Sage.

Mathews, R., Matthews J. K., & Speltz, K. (1989). *Female sexual offenders*. Orwell, VT: Safer Society Press.

Mathison, S. (Ed.). (2005). *Encyclopedia of evaluation*. Thousand Oaks, CA: Sage.

Mathison, S. (2009). Seeing is believing: The credibility of image-based research and evaluation. In S. I. Donaldson, C. A. Christie, & M. Mark (Eds.), *What counts as credible evidence in applied research and evaluation practice* (pp. 181–196). Thousand Oaks, CA: Sage.

Maxcy, S. J. (2003). Pragmatic threads in mixed method research in the social sciences: The search for multiple modes of inquiry and the end of the philosophy of formalism. In A. Tashakkori & C. Teddlie (Eds.), *Handbook of mixed methods in the social and behavioral research* (pp. 51–90). Thousand Oaks, CA: Sage.

Maxson, M. (2014). *Lessons from four years of story-based monitoring*. Retrieved from http://dmeforpeace.org/discuss/lessons-four-years-story-based-monitoring

Maxwell, J. A. (2004). Causal explanation, qualitative research, and scientific inquiry in education. *Educational Researcher, 33*(2), 3–11.

Maxwell, J. A. (2012). *A realist approach for qualitative research*. Thousand Oaks, CA: Sage.

Maxwell, J. A., & Mittapalli, K. (2008). Explanation, and explanatory research. In L. M. Given (Ed.), *The SAGE encyclopedia of qualitative research methods* (pp. 322–325). Thousand Oaks, CA: Sage.

Mayne, J. (2007). *Contribution analysis: An approach to exploring cause-effect*. (ILAC [Institutional Learning and Change] Brief 16). Retrieved from http://www.outcomemapping.ca/download.php?file=/resource/files/csette_en_ILAC_Brief16_Contribution_Analysis.pdf

Mayne, J. (2011). Exploring cause-effect questions using contribution analysis. In K. Forss, M. Mara, & R. Schwartz (Eds.), *Evaluating the complex: Attribution, contribution and beyond* (chap. 3). New Brunswick, NJ: Transaction Books.

Mayne, J. (2012). Contribution analysis: Coming of age? *Evaluation: The International Journal, 18*(3), 270–280.

McClure, G. (1989). *Organizational culture as manifest in critical incidents: A case study of the Faculty of Agriculture, University of the West Indies* (Unpublished doctoral dissertation). University of Minnesota, Minneapolis.

McConaughy, S. H. (2013). *Clinical interviews for children and adolescents: Assessment to intervention* (2nd ed.). New York, NY: Guilford Press.

McCracken, G. (1988). *The long interview* (Qualitative Research Methods Series No. 13). Newbury Park, CA: Sage.

MacDonald, B. (1987). *Evaluation and control of education*. In R. Murphy & H. Torrance (Eds.), *Issues and methods in evaluation* (pp. 36–48). London, England: Paul Chapman.

McDonnell, M. B., Tran, V. B. T., & McCoy, N. R. (2010). *Helmet day: Lessons learned on Vietnam's road to healthy behavior*. Brooklyn, NY: Social Science Research Council.

McGarvey, C. (2007). *Participatory action research*. New York, NY: GrantCraft, Ford Foundation.

McGrory, K., & Solochek, J. S. (2013, August 2). Amid school grading controversy, Florida education chief Tony Bennett Resigns. *Miami Herald*. Retrieved from http://www.miamiherald .com/2013/08/01/3535902/amid-grading-controversy-florida.html

McInnes, R. J., & Chambers, J. A. (2008). Supporting breastfeeding mothers: Qualitative synthesis. *Journal of Advanced Nursing, 62*(4), 407–427.

McIntyre, A. (2007). *Participatory action research*. Thousand Oaks, CA: Sage.

McKechnie, L. E. F. (2008). Reactivity. In L. M. Given (Ed.), *The SAGE encyclopedia of qualitative research methods* (Vol. 2, pp. 729–730). Thousand Oaks, CA: Sage.

McKegg, K., Wehipeihana, N., Pipi, K., & Thompson, V. (2013). *He Oranga Poutama: What have we learned?* A report on the developmental evaluation of *He Oranga Poutama*. Wellington, New Zealand: Sport New Zealand. Retrieved from http://www.outdoorsnz.org.nz/news/ study-he-oranga-poutama-what-have-we-learned-report -developmental-evaluation-he-oranga-poutama

McLaughlin, M. (1976). Implementation as mutual adaptation. In W. Williams & R. F. Elmore (Eds.), *Social program implementation* (pp. 167–180). New York, NY: Academic Press.

McLeod, J. (2001). *Qualitative research in counseling and psychotherapy*. Thousand Oaks, CA: Sage.

McNabb, M. (2013, May). On-line "snowball" sample surveys [Electronic mailing list message]. EvalTalk Internet listserv of the American Evaluation Association.

McNamara, C. (1996). *Evaluation of a group-managed, multi-technique management development program that includes action learning* (Unpublished doctoral dissertation). Union Institute, Cincinnati, OH.

McPhee, J. (2014, April 7). Elicitation. *The New Yorker*, pp. 50–57.

Mead, G. H. (1934). *Mind, self and society*. Chicago, IL: University of Chicago Press.

Mead, G. H. (1936). *Movements of thought in the nineteenth century*. Chicago, IL: University of Chicago Press.

Mead, M. (1977). *Letters from the field, 1925–1975*. New York, NY: Harper & Row.

Memon, A., Meissner, C. A., & Fraser, J. (2010). The cognitive interview: A meta-analytic review and study space analysis of the past 25 years. *Psychology, Public Policy, and Law, 16*(4), 340–372.

Merleau-Ponty, M. (1962). *The phenomenology of perception*. London, England: Routledge.

Merriam, S. (1997). *Qualitative research and case study applications in education*. San Francisco, CA: Jossey-Bass.

Mertens, D. M. (1998). *Research methods in education and psychology: Integrating diversity with quantitative and qualitative approaches*. Thousand Oaks, CA: Sage.

Mertens, D. M. (1999). Inclusive evaluation: Implications of transformative theory for evaluation. *American Journal of Evaluation, 20*(1), 1–14.

Mertens, D. M. (2005). Feminism. In S. Mathison (Ed.), *Encyclopedia of evaluation* (p. 154). Thousand Oaks, CA: Sage.

Mertens, D. M. (2013). Emerging advances in mixed methods. *Journal of Mixed Methods Research, 7*(3), 215–218.

Mertens, D. M. (2014). A momentous development in mixed methods research. *Journal of Mixed Methods Research, 8*(1), 3–5.

Mertens, D. M., Cram, F., & Chilisa, B. (2013). *Indigenous pathways into social research*. Walnut Creek, CA: Left Coast Press.

Mertens, D. M., & Hesse-Biber, S. (Eds.). (2013). *Mixed methods and credibility of evidence in evaluation* (*New Directions for Evaluation*, No. 138). Hoboken, NJ: Wiley Periodicals.

Mertens, D. M., & Wilson, A. T. (2012). *Program evaluation theory and practice: A comprehensive guide*. New York, NY: Guilford Press.

Merton, R. K. (1936). The unintended consequences of purposive social action. *American Sociological Review, 1*(6), 894–904.

Merton, R. K. (1968a). On sociological theories of the middle range. In *Social theory and social structure* (pp. 39–72). London, England: Free Press.

Merton, R. K. (1968b). The self-fulfilling prophecy. In *Social theory and social structure* (pp. 475–490). London, England: Free Press.

Merton, R. K. (1973). The normative structure of science. In *The sociology of science* (pp. 267–278). Chicago, IL: University of Chicago.

Merton, R. K. (1976). Social knowledge and public policy. In *Sociological ambivalence* (pp. 156–179). New York, NY: Free Press.

Merton, R. K. (1995). The Thomas theorem and the Matthew effect. *Social Forces, 74*(2), 379–424.

Merton, R. K., Riske, M., & Kendall, P. L. (1956). *The focused interview*. New York, NY: Free Press.

Messick, S. (1989). Validity. In R. L. Linn (Ed.), *Educational measurement* (3rd ed., pp. 13–103). New York, NY: American Council on Education.

Method of study is criticized in group's health policy tests. (2014, February 3). *The New York Times*, p. A1. Retrieved from http://www .nytimes.com/2014/02/03/health/effort-to-test-health-policies-is -criticized-for-study-tactics.html?hp&_r=0

Michon, G. P. (2013). *Escutcheons of science: Armorial of Scientists: Numericana*. Retrieved from http://www.numericana.com/arms/ index.htm#newton

Micklewright, M. (2013). So God made a quality manager. *Quality Digest*. Retrieved from http://www.youtube.com/ watch?v=Y95o94aVIBs&feature=youtu.be

Miles, M. B., & Huberman, A. M. (1984). *Qualitative data analysis: A sourcebook of new methods*. Beverly Hills, CA: Sage.

Miles, M. B., & Huberman, A. M. (1994). *Qualitative data analysis: An expanded sourcebook* (2nd ed.). Thousand Oaks, CA: Sage.

Miles, M. B., Huberman, A. M., & Saldaña, (2014). *Qualitative data analysis: A methods sourcebook* (3rd ed.). Thousand Oaks, CA: Sage.

Milgram, S. (1974). *Obedience to authority*. New York, NY: Harper & Row.

Milius, S. (1998). When worlds collide. *Science News, 154*(6), 92–93.

Miller, D. (1981). *The book of jargon*. New York, NY: Macmillan.

Miller, R. L., & Campbell, R. (2006). Taking stock of empowerment evaluation: An empirical review. *American Journal of Evaluation, 27*(3), 296–319.

Miller, W. R., & Crabtree, B. (2000). Clinical research. In N. K. Denzin & Y. S. Lincoln (Eds.), *Handbook of qualitative research* (2nd ed., pp. 607–632). Thousand Oaks, CA: Sage.

Miller, W. R., & Rollnick, S. (2012). *Motivational interviewing: Helping people change* (3rd ed.). New York, NY: Guilford Press.

Miller, R. B. & Johnson, L. N. (Eds.) (2014). Advanced methods in family therapy research. New York: Routledge.

Mills, A. J., Durepos, G., & Wiebe, E. (Eds.). (2010). *Encyclopedia of case study research*. Thousand Oaks, CA: Sage.

Mills, C. W. (2000). *The sociological imagination*. Oxford, England: Oxford University Press. (Original work published 1959)

Minkler, M., & Wallerstein, N. (2008). *Community-based participatory research for health: From process to outcomes*. San Francisco, CA: Wiley.

Minnesota Biological Survey. (2013). *Areas of biodiversity significance in Minnesota*. St. Paul: Minnesota Department of Natural Resources. Retrieved from http://files.dnr.state.mn.us/eco/mcbs/maps/areas_ of_biodiv_sig_statewide_2012_06_12_8x11.pdf

Minnesota Department of Health. (2013). *Youth photovoice project*. Retrieved from http://www.health.state.mn.us/healthreform/ship/ events/policyconference/photovoice.pdf

Minnich, E. (2004). *Transforming knowledge* (2nd ed.). Philadelphia, PA: Temple University Press.

Mintzberg, H. (2007). *Tracking strategies*. New York, NY: Oxford University Press.

Mintzberg, H., Lampel, J., & Ahlstrand, B. (2008). *Strategy safari: The complete guide through the wilds of strategic management* (2nd ed.). New York, NY: Free Press.

MIT Media Lab. (2014). *The human speechome project*. Cambridge: MIT Press. Retrieved from http://www.media.mit.edu/cogmac/projects/hsp.html

Mitchell, M. (2002). Exploring the future of the digital divide through ethnographic futures research. *First Monday Internet Journal, 7*(11). Retrieved from http://firstmonday.org/ojs/index.php/fm/article/view/1004

Mitchell, M. (2009). *Complexity: A guided tour*. New York, NY: Oxford University Press.

Mitchell, R. G., Jr. (1993). *Secrecy and fieldwork* (Qualitative Research Methods Series No. 29). Newbury Park, CA: Sage.

Molenaar, P. C. M., Lerner, R. M., & Newell, K. M. (Eds.). (2014). *Handbook of developmental systems theory and methodology*. New York, NY: Guilford Press.

Montgomery, J., & Fewer, W. (1988). *Family systems and beyond*. New York, NY: Human Science Press.

Montori, V. (2014, February 10). Minimally disruptive medicine [Healthcare PlexusCall]. *Plexus Institute News*. Retrieved from http://www.plexusinstitute.org/event/id/397214/Healthcare-PlexusCall-Minimally-Disruptive-Medicine.htm

Moore, B., Jr. (1966). *The social origins of dictatorship and democracy*. Boston, MA: Beacon Press.

Moore, R. (2011). *Semiotics: The key* (Kindle ed.). Amazon Digital Services.

Moos, R. (1975). *Evaluating correctional and community settings*. New York, NY: Wiley Interscience.

Morell, J. A. (2010). *Evaluation in the face of uncertainty: Anticipating surprise and responding to the inevitable*. New York, NY: Guilford Press.

Morgan, D. L. (1988). *Focus groups as qualitative research* (Qualitative Research Methods Series No. 16). Beverly Hills, CA: Sage.

Morgan, D. L. (1997a). *The focus group guidebook: Vol. 1. The focus group kit*. Thousand Oaks, CA: Sage.

Morgan, D. L. (1997b). *Planning focus groups: Vol. 2. The focus group kit*. Thousand Oaks, CA: Sage.

Morgan, D. L. (2007). Paradigms lost and pragmatism regained: Methodological implications of combining qualitative and quantitative methods. *Journal of Mixed Methods Research, 1*(1), 48–76.

Morgan, D. L. (2008a). Focus groups. In L. M. Given (Ed.), *The SAGE encyclopedia of qualitative research methods* (Vol. 1, pp. 352–354). Thousand Oaks, CA: Sage.

Morgan, D. L. (2008b). Quota sampling. In L. M. Given (Ed.), *The SAGE encyclopedia of qualitative research methods* (Vol. 2, pp. 722–723). Thousand Oaks, CA: Sage.

Morgan, D. L. (2012, March 14). *Closure* [Electronic mailing list message]. QUALRS-L@LISTSERV.

Morgan, D. L. (2014). *Integrating qualitative and quantitative methods: A pragmatic approach*. Thousand Oaks, CA: Sage.

Morgan, D. L., Ataie, J., Carder, P., & Hoffman, K. (2013). Introducing dyadic interviews as a method for collecting qualitative data. *Qualitative Health Research, 23*(9), 1276–1284.

Morgan, G. (Ed.). (1983). *Beyond methods: Strategies for social research*. Beverly Hills, CA: Sage.

Morgan, G. (1986). *Images of organizations*. Beverly Hills, CA: Sage.

Morgan, G. (1989). *Creative organizational theory: A resource book*. Newbury Park, CA: Sage.

Morris, E. (Director). (2000). *Mr. Death: The rise & fall of Fred A. Leuchter Jr.* [Documentary]. Hollywood, CA: Universal Studios.

Morris, E. (Director). (2003). *The fog of war: Eleven lessons from the life of Robert S. McNamara* [Documentary]. New York, NY: Sony Pictures Classic.

Morris, E. (2011). *Believing is seeing: Observations on the mysteries of photography*. New York, NY: Penguin Books.

Morris, E. (Director). (2012). *The unknown known: The life and times of Donald Rumsfeld* [Documentary]. Hollywood, CA: Warner.

Morris, E. (2013, December 3). The interminable, everlasting Lincolns. *The New York Times*. Retrieved from http://opinionator.blogs.nytimes.com/2013/12/02/the-interminable-everlasting-lincolns-part-1/

Morris, E. (2014, March 28). The certainty of Donald Rumsfeld (Part 4). *The New York Times*. Retrieved from http://opinionator.blogs.nytimes.com/2014/03/28/the-certainty-of-donald-rumsfeld-part-4/?_php=true&_type=blogs&_php=true&_type=blogs&hp&rref=opinion&_r=1

Morris, Z. S. (2009). The truth about interviewing elites. *Politics, 29*(3), 209–217.

Morrison, D. (1999). The role of observation. *Skeptical Briefs, 9*(1), 8.

Morrison, J. (2014). *Diagnosis made easier: Principles and techniques for mental health clinicians* (2nd ed.). New York, NY: Guilford Press.

Morse, J. M. (1997). *Completing a qualitative project*. Thousand Oaks, CA: Sage.

Morse, J. M. (2000). Determining sample size. *Qualitative Health Research, 10*(1), 3–5.

Morse, J. M. (2009). Tussles, tensions, and resolutions. In J. M. Morse, P. N. Stern, J. Corbin, B. Bowers, K. Charmaz, & A. E. Clarke (Eds.). *Developing grounded theory: The second generation* (pp. 13–19). Walnut Creek, CA: Left Coast Press.

Morse, J. M. (2010). Sampling in grounded theory. In A. Bryant & K. Charmaz (Eds.), *The SAGE handbook of grounded theory* (pp. 229–244). London, England: Sage.

Morse, J. M., Stern, P. N., Corbin, J., Bowers, B., Charmaz, K., & Clarke, A. E. (Eds.). (2009). *Developing grounded theory: The second generation*. Walnut Creek, CA: Left Coast Press.

Mossman, E., & Mayhew, P. (2007). *Key informant interviews: Review of the Prostitution Reform Act 2003*. Wellington, New Zealand: Victoria University, Crime and Justice Research Centre. Retrieved from http://www.justice.govt.nz/policy/commercial-property-and-regulatory/prostitution/prostitution-law-review-committee/publications/key-informant-interviews/documents/report.pdf

Moustakas, C. (1961). *Loneliness*. Englewood Cliffs, NJ: Prentice Hall.

Moustakas, C. (1972). *Loneliness and love*. Englewood Cliffs, NJ: Prentice Hall.

Moustakas, C. (1975). *The touch of loneliness*. Englewood Cliffs, NJ: Prentice Hall.

Moustakas, C. (1988). *Phenomenology, science and psychotherapy*. Sydney, Nova Scotia, Canada: University College of Cape Breton, Family Life Institute.

Moustakas, C. (1990). *Heuristic research: Design, methodology, and applications*. Newbury Park, CA: Sage.

Moustakas, C. (1994). *Phenomenological research methods*. Thousand Oaks, CA: Sage.

Moustakas, C. (1995). *Being-in, being-for, being-with*. Northvale, NJ: Jason Aronson.

Mowles, C. (2014). Complex, but not quite complex enough: The turn to the complexity sciences in evaluation scholarship. *Evaluation, 20*(2), 160–175.

Mueller, C. W. (2004). Conceptualization, operationalization, and measurement. In M. S. Lewis-Beck, A. Bryman, & T. Futing Liao (Eds.), *The SAGE encyclopedia of social science research methods* (pp. 161–165). Thousand Oaks, CA: Sage.

Mueller, M. R. (1996). *Immediate outcomes of lower-income participants in Minnesota's universal access early childhood fairly education*. St. Paul: Minnesota Department of Children, Families, and Learning.

Mueller, M. R., & Fitzpatrick, J. (1998). Dialogue with Marsha Mueller. *American Journal of Evaluation, 19*(1), 97–98.

Muise, A., Impett, E. A., & Desmarais, S. (2013). Getting it on versus getting it over with: Sexual motivation, desire, and satisfaction in intimate bonds. *Personality and Social Psychology Bulletin, 20*(10), 1–13.

Mukherjee, A. (2010). *The emperor of all maladies: A biography of cancer*. New York, NY: Scribner.

Muncey, T. (2010). *Creating autoethnographies*. Thousand Oaks, CA: Sage.

Munk, N. (2013). *The idealist: Jeffrey Sachs and the quest to end poverty*. New York, NY: Random House.

Murali, M. L. (Ed.). (1995). *Chaos in nonlinear oscillators: Controlling and synchronization* (World Scientific Series on Nonlinear Science, Series A: Monographs and Treatises). New York, NY: World Scientific.

Murchison, J. (2010). *Ethnography essentials: Designing, conducting, and presenting your research*. San Francisco, CA: Jossey-Bass.

Murphy, N. F. (2014). *Developing evidence-based effective principles for working with homeless youth: A developmental evaluation of the Otto Bremer Foundation's support for Collaboration Among Agencies Serving Homeless Youth* (Unpublished doctoral dissertation). University of Minnesota, Minneapolis.

REFERENCES

Mutua, K., & Swadener, B. B. (Eds.). (2011). *Decolonizing research in cross-cultural contexts: Critical personal narratives.* Albany: State University of New York Press.

Mwaluko, G. S., & Ryan, T. B. (2000). The systemic nature of action learning programmes. *Systems Research and Behavioral Science, 17*(4), 393–401.

Myers, I. B. (with Meyers, P.). (1995). *Gifts differing.* Palo Alto, CA: Consulting Psychologists Press.

Myrdal, G. (1969). *Objectivity in social research.* New York, NY: Random House/Pantheon.

Nadel, L., & Stein, D. (Eds.). (1995). *The 1993 lectures in complex systems* (Santa Fe Institute Studies in the Sciences of Complexity, Lectures, Vol. 6). Boulder, CO: Perseus Press.

Nagel, E. (1961). *The structure of science.* New York, NY: Harcourt, Brace & World.

Nagel, T. (2013). Ronald Dworkin: The moral quest. *New York Review of Books, 60*(18), 56–58.

Narayan, D., Chambers, R., Shah, M. K., & Petesch, P. (2000). *Crying out for change: Voices of the Poor.* Washington, DC: World Bank.

Narayan, D., Patel, R., Schafft, K., Rademacher, A., & Koch-Schulte, S. (2000). *Can anyone hear us? Voices of the poor.* Washington, DC: World Bank.

Narayan, D., & Petesch, P. (Eds.). (2002). *Voices of the poor: From many lands.* Washington, DC: World Bank.

Narayan, K., & George, K. M. (2003). Personal and folk narrative as cultural representation: Stories about getting stories. In J. A. Holstein & J. F. Gubrium (Eds.), *Inside interviewing: New lenses, new concerns* (pp. 449–465). Thousand Oaks, CA: Sage.

Nash, R. (1986). *Wilderness and the American mind.* New Haven, CT: Yale University Press.

National Honey Bee Health Stakeholder Conference Steering Committee. (2012). *Report on the national stakeholders conference on honey bee health.* Washington, DC: U.S. Department of Agriculture. Retrieved from http://www.usda.gov/documents/ReportHoneyBeeHealth.pdf

National Resource Center for Community-Based Child Abuse Prevention. (2009). *Using qualitative data in program evaluation: Telling the story of a prevention program.* Washington, DC: FRIENDS National Resource Center for Community-Based Child Abuse Prevention. Retrieved from http://friendsnrc.org/using-qualitative-in-program-evaluation

Network Impact and Center for Evaluation Innovation. (2014). *Part 1, Guide to Network Evaluation. Framing Paper: The State of Network Evaluation.* Retrieved from http://www.evaluationinnovation.org/sites/default/files/NetEval1_Framing.pdf
Part 2, Guide to Network Evaluation. Evaluating Networks for Social Change: A casebook. Retrieved from http://www.evaluationinnovation.org/sites/default/files/NetEval2_Casebook.pdf

Neuendorf, K. A. (2002). *The content analysis guidebook.* Thousand Oaks, CA: Sage.

Nhat Hahn, T. (1999). *Call me by my true names: The collected poems of Thich Nhat Hanh.* Berkeley, CA: Parallax Press.

Noblit, G. W., & Hare, R. W. (1988). *Meta-ethnography: Synthesizing qualitative studies.* Newbury Park, CA: Sage.

Norris, J., Sawyer, R. D., & Lund, D. E. (Eds.). (2012). *Duoethnography: Dialogic methods for social, health, and educational research.* Walnut Creek, CA: Left Coast Press.

Northcutt, N., & McCoy, D. (2004). *Interactive qualitative analysis: A systems method for qualitative research.* Thousand Oaks, CA: Sage.

Norum, K. E. (2008). Reality and multiple realities. In L. M. Given (Ed.), *The SAGE encyclopedia of qualitative research* (Vol. 2, pp. 737–739). Thousand Oaks, CA: Sage.

Nutley, S., Powell, A., & Davies, H. (2013). *What counts as good evidence.* Scotland, England: University of St Andrews, Research Unit for Research Utilisation, School of Management.

Oakley, A. (1981). Interviewing women: A contradiction in terms. In H. Roberts (Ed.), *Doing feminist research* (pp. 30–61). London, England: Routledge.

Organisation for Economic Co-operation and Development. (2005). *Paris Declaration principles.* Retrieved from http://www.oecd.org/dac/effectiveness/parisdeclarationandaccraagendaforaction.htm

Organisation for Economic Co-operation and Development. (2005). *Paris Declaration and Accra agenda for action.*

Retrieved from http://www.oecd.org/dac/effectiveness/parisdeclarationandaccraagendaforaction.htm

Organisation for Economic Co-operation and Development, and Development Assistance Committee. (2010). *Quality standards for development evaluation.* Retrieved from http://www.oecd.org/document/30/0,3746,en_21571361_34047972_38903582_1_1_1_1,00.html

Ogbor, J. O. (2000). Mythicizing and reification in entrepreneurial discourse: Ideology-critique of entrepreneurial studies. *Journal of Management Studies, 37*(5), 605–635.

Olesen, V. L. (2000). Feminisms and qualitative research at and into the millennium. In N. K. Denzin & Y. S. Lincoln (Eds.), *Handbook of qualitative research* (2nd ed., pp. 215–256). Thousand Oaks, CA: Sage.

Olesen, V. L. (2011). Feminist qualitative research in the millennium's first decade. In N. K. Denzin & Y. S. Lincoln (Eds.), *The SAGE handbook of qualitative research* (4th ed., pp. 129–148). Thousand Oaks, CA: Sage.

Onwuegbuzie, A. J. (2012). Putting the MIXED back into quantitative and qualitative research in educational research and beyond: Moving toward the radical middle. *International Journal of Multiple Research Approaches, 6*(3), 192–219.

Oppenheimer, S. (2003). *The real eve: Modern man's journey out of Africa.* New York, NY: Basic Books.

O'Reilly, K. (2012a). *Ethnographic methods* (2nd ed.). New York, NY: Routledge.

O'Reilly, K. (2012b). *International migration and social theory.* London, England: Palgrave Macmillan.

O'Reilly, K. (2013). *Practice stories.* Retrieved from http://karenoreilly.wordpress.com/international-migration-and-social-theory/practice-stories/

Ormerod, P. (2001). *Butterfly economics: A new general theory of social and economic behavior.* New York, NY: Basic Books.

Ornish, D. (2014). *Scientific idea ready for retirement: Large randomized controlled trials.* Retrieved from http://www.edge.org/responses/what-scientific-idea-is-ready-for-retirement

O'Toole, G. (2010). *Not everything that counts can be counted, and not everything that can be counted counts.* Retrieved from http://quoteinvestigator.com/2010/05/26/everything-counts-einstein/

Owen, J. A. T. (2008). Naturalistic inquiry. In L. M. Given (Ed.), *The SAGE encyclopedia of qualitative research methods* (Vol. 2, pp. 547–550). Thousand Oaks, CA: Sage.

Owens, T., Haenn, J. F., & Fehrenbacher, H. L. (1976). *The use of multiple strategies in the evaluation of an experience-based career education program* (Research Evaluation Development Paper Series No. 9). Portland, OR: Northwest Regional Educational Laboratory.

Packer, M., & Addison, R. (1989). *Entering the circle: Hermeneutic investigation in psychology.* Albany: State University of New York Press.

Padgett, D. (2004). *The qualitative research experience.* Stamford, CT: Wadsworth/Thomson Learning.

Pagels, H. R. (1988). *The dreams of reason: The rise of the sciences of complexity.* New York, NY: Simon & Schuster.

Palmer, L. (1988). *Shrapnel in the heart.* New York, NY: Vintage Books.

Palmer, R. E. (1969). *Hermeneutics.* Evanston, IL: Northwestern University Press.

Panati, C. (1987). *Extraordinary origins of everyday things.* New York, NY: Harper & Row.

Pant, L. P., & Odame, H. H. (2009). The promise of positive deviants: Bridging divides between scientific research and local practices in smallholder agriculture. *Knowledge Management for Development Journal, 5*(2), 138–150.

Parameswaran, R. (2001). Feminist media ethnography in India: Exploring power, gender, and culture in the field. *Qualitative Inquiry, 7*(1), 69–103.

Park, C. C. (with Sacks, O.). (2001). *Exiting nirvana: A daughter's life with autism.* Boston, MA: Little, Brown.

Parlett, M., & Hamilton, D. (1976). Evaluation as illumination: A new approach to the study of innovatory programs. In G. V. Glass (Ed.), *Evaluation studies review annual* (Vol. 1, pp. 140–157). Beverly Hills, CA: Sage.

Partnow, E. (1978). *The quotable woman, 1800-on.* Garden City, NY: Anchor Books.

REFERENCES

Pascale, R., Sternin, J., & Sternin, M. (2010). *The power of positive deviance: How unlikely innovators solve the world's toughest problems.* Cambridge, MA: Harvard Business Press.

Patrizi, P., & Patton, M. Q. (Eds.). (2010). *Evaluating strategy* (New Directions for Evaluation, No. 128). Hoboken, NJ: Wiley Periodicals.

Patrizi, P., Thompson, E. H., Coffman, J., & Beer, T. (2013). Eyes wide open: Learning as strategy under conditions of complexity and uncertainty. *Foundation Review, 5*(3), 50–65. Retrieved from http://dx.doi.org/10.9707/1944–5660.1170

Patrizi, P., Thompson, E. H., & Spector, A. (2008). *Death is certain: Strategy isn't. Assessing RWJF's end-of-life grantmaking: Field building in end of life.* Princeton, NJ: Robert Wood Johnson Foundation. Retrieved from http://www.evaluationroundtable.org/documents/cs-death-is-certain.pdf

Patton, M. Q. (1970, August). *British colonial settlement schemes aimed at the Wagogo people of the Central Dodoma Region, Tanzania: An inventory and assessment.* Report submitted to the Tanzania Ministry of Agriculture, Dar es Salaam, Tanzania.

Patton, M. Q. (1978a). *Evaluation of Southwest Field Training Project.* Unpublished evaluation report, Saint Paul, MN.

Patton, M. Q. (1978b). *Utilization-focused evaluation* (1st ed.). Newbury Park, CA: Sage.

Patton, M. Q. (1980). *Qualitative evaluation methods.* Beverly Hills, CA: Sage.

Patton, M. Q. (1981). *Practical evaluation.* Beverly Hills, CA: Sage.

Patton, M. Q. (1985). Logical incompatibilities and pragmatism. *Evaluation and Program Planning, 8,* 307–308.

Patton, M. Q. (1987). *Creative evaluation* (2nd ed.). Newbury Park, CA: Sage.

Patton, M. Q. (1988a). Integrating evaluations into a program for increased utility and cost-effectiveness. In J. A. McLaughlin, L. J. Weber, R. W. Covert, & R. B. Ingle (Eds.), *Evaluation utilization* (New Directions for Evaluation, No. 39, pp. 85–94). San Francisco, CA: Jossey-Bass.

Patton, M. Q. (1988b). Paradigms and pragmatism. In D. M. Fetterman (Ed.), *Qualitative approaches to evaluation in education: The silent scientific revolution* (pp. 116–137). New York, NY: Praeger.

Patton, M. Q. (1990). Humanistic psychology and qualitative research: Shared principles and processes. *Person Centered Review, 5*(2), 191–202.

Patton, M. Q. (Ed.). (1991). *Family sexual abuse: Frontline research and evaluation.* Newbury Park, CA: Sage.

Patton, M. Q. (1993). *The aid to families in poverty program: A synthesis of themes, patterns and lessons learned.* Minneapolis, MN: McKnight Foundation.

Patton, M. Q. (1997). *Utilization-focused evaluation: The new century text* (3rd ed.). Thousand Oaks, CA: Sage.

Patton, M. Q. (1999). *Grand Canyon celebration: A father-son journey of discovery.* Amherst, NY: Prometheus Books.

Patton, M. Q. (2000). Language matters. In H. Rodney (Ed.), *How and why language matters in evaluation* (New Directions for Evaluation, No. 86, pp. 5–16). San Francisco, CA: Jossey-Bass.

Patton, M. Q. (with Patton, B. Q. T.). (2001). What's in a name? Heroic nomenclature in the Grand Canyon. *Plateau Journal, 4*(2), 16–29.

Patton, M. Q. (2002). *Qualitative research and evaluation methods* (3rd. ed.). Thousand Oaks, CA: Sage.

Patton, M. Q. (2008). Advocacy impact evaluation. *Journal of MultiDisciplinary Evaluation, 5*(9), 1–10. Retrieved from http://survey.ate.wmich.edu/jmde/index.php/jmde_1/issue/view/25

Patton, M. Q. (2011). *Developmental evaluation: Applying complexity concepts to enhance innovation and use.* New York, NY: Guilford Press.

Patton, M. Q. (2012a). *Essentials of utilization-focused evaluation* (4th ed.). Thousand Oaks, CA: Sage.

Patton, M. Q. (2012b). A utilization-focused approach to contribution analysis. *Evaluation: The International Journal, 18*(3), 364–377.

Patton, M. Q. (2013). Meta-evaluation: Evaluating the evaluation of the Paris Declaration. *Canadian Journal of Program Evaluation, 27*(3), 147–171. Retrieved from http://evaluationcanada.ca/site.cgi?s=4&ss=21&_lang=en&volume=2012//3

Patton, M. Q., & Gornick, J. K. (2011a). *Evaluation of the Phase 2 evaluation of the Paris Declaration: An independent review of strengths, weaknesses, and lessons.* Retrieved from http://www.oecd.org/dataoecd/37/3/48620425.pdf

Patton, M. Q., & Gornick, J. K. (2011b). Qualitative approaches to outcomes evaluation. In J. L. Magnabosco & R. W. Mandersched (Eds.), *Outcomes measurement in the human services* (2nd ed., pp. 31–45). Washington, DC: National Association of Social Workers.

Patton, M. Q., & Patrizi, P. (Eds.). (2005). *Teaching evaluation using the case method* (New Directions for Evaluation, No. 105). Hoboken, NJ: Wiley Periodicals.

Patton, M. Q., & Sherwood, K. (2007). *Kids are different: The successful campaign to overturn the juvenile death penalty in America. Case study and retrospective evaluation of the Roper v. Simmons Supreme Court case that ended the juvenile death penalty in the United States.* New York, NY: Atlantic Philanthropies.

Patton, M. Q., & Stockdill, S. (1987). *Summative evaluation of the technology for literacy center.* St. Paul, MN: Saint Paul Foundation.

Paul, A. M. (2012, March 17). Your brain on fiction. *New York Times,* p. SR6.

Paulson, R. G. (1990). From paradigm wars to disputatious community. *Comparative Education Review, 34*(3), 395–400.

Pawson, R. (1989). *Measure for measure: A manifesto for empirical sociology.* London, England: Routledge.

Pawson, R. (2013). *The science of evaluation: A realist manifesto.* London, England: Sage.

Pawson, R., & Tilley, N. (1997). *Realistic evaluation.* London, England: Sage.

Payne, S. L. (1951). *The art of asking questions.* Princeton, NJ: Princeton University Press.

Pedler, M. (Ed.). (1991). *Action learning in practice.* Aldershot Hauts, England: Gower.

Pedler, M. (2008). *Action learning for managers.* Burlington, VT: Glower.

Peirce, C. S. (1878). How to make our ideas clear. *Popular Science Monthly, 12*(1), 286–302.

Pelto, P. J. (2013). *Applied ethnography: Guidelines for field research.* Walnut Creek, CA: Left Coast Press.

Pelto, P. J., & Pelto, G. H. (1978). *Anthropological research: The structure of inquiry.* Cambridge, England: Cambridge University Press.

Perls, F. (1973). *The Gestalt approach and eye witness to therapy.* Palo Alto, CA: Science and Behavior Books.

Perrone, V. (Ed.). (1985). *Portraits of high schools* (The Carnegie Foundation for the Advancement of Teaching). Lawrenceville, NJ: Princeton University Press.

Perrone, V., & Patton, M. Q. (with French, B.). (1976). *Does accountability count without teacher support?* Minneapolis: University of Minnesota, Minnesota Center for Social Research.

Perry, B., & Szalavitz, M. (2006). *The boy who was raised as a dog and other stories from a child psychiatrist's notebook.* New York, NY: Basic Books.

Peshkin, A. (1985). Virtuous subjectivity: In the participant-observer's I's. In D. Berg & K. Smith (Eds.), *Exploring clinical methods for social research* (pp. 267–268). Beverly Hills, CA: Sage.

Peshkin, A. (1986). *God's choice: The total world of a fundamentalist Christian school.* Chicago, IL: University of Chicago Press.

Peshkin, A. (1988). In search of subjectivity: One's own. *Educational Researcher, 29*(9), 5–9.

Peshkin, A. (1997). *Places of memory: Whiteman's schools and Native American communities* (Sociocultural, political, and historical studies in education). Mahwah, NJ: Lawrence Erlbaum.

Peshkin, A. (2000a). The nature of interpretation in qualitative research. *Educational Researcher, 17*(7), 17–22.

Peshkin, A. (2000b). *Permissible advantage? The moral consequences of elite schooling.* Mahwah, NJ: Lawrence Erlbaum.

Peters, T. J., & Waterman, R. H., Jr. (1982). *In search of excellence: Lessons from America's best-run companies.* New York, NY: Harper & Row.

Pettigrew, A. M. (1983). On studying organizational cultures. In J. Van Maanen (Ed.), *Qualitative methodology* (pp. 87–104). Beverly Hills, CA: Sage.

Pew. (2013). *Focus groups.* Washington, DC: Pew Internet and American Life Project. Retrieved from http://www.pewinternet.org/Reports/2013/Teens-Social-Media-And-Privacy/Methods/Focus-Groups.aspx

Pfahl, N. L. (2011). Using narrative inquiry and analysis of life stories to advance elder learning. In G. Boulton-Lewis & M. Tam (Eds.), *Active ageing, active learning: Issues and challenges* (chap. 5; pp. 67–87). New York, NY: Springer.

Philliber, S. (1989, March). *Workshop on evaluating adolescent pregnancy prevention programs.* Washington, DC: Children's Defense Fund Conference.

Phillips, D. C. (2000). *The expanded social scientist's bestiary: A guide to fabled threats to, and defences of, naturalistic social science.* Lanham, MD: Rowman & Littlefield.

Piantanida, M., & Garman, N. B. (2009). *The qualitative dissertation: A guide for students and faculty* (2nd ed.). Thousand Oaks, CA: Corwin Press.

Pietro, D. S. (1983). *Evaluation sourcebook for private and voluntary organizations.* New York, NY: American Council of Voluntary Agencies for Foreign Service.

Pike, K. (1954). *Language in relation to a unified theory of the structure of human behavior* (Vol. 1). Glendale, CA: Summer Institute of Linguistics. (Republished in 1967, The Hague, Netherlands: Mouton)

Pillemer, K. (2011). *30 lessons for living: Tried and true advice from the wisest Americans.* New York, NY: Penguin Books.

Pillow, W. S. (2000, June–July). Deciphering attempts to decipher postmodern educational research. *Educational Researcher, 29*(5), 21–24.

Pincus, H. A., Abedin, Z., Blank, A. E., & Mazmanian, P. E. (2013). Evaluation and the NIH clinical and translational science awards: A "top ten" list. *Evaluation of Health Professions, 36*(4), 411–431.

Plummer, K. (2011). Critical humanism and queer theory. In N. K. Denzin & Y. S. Lincoln (Eds.), *The SAGE handbook of qualitative research* (4th ed., pp. 195–207). Thousand Oaks, CA: Sage.

Podems, D. (2010). Feminist evaluation and gender approaches: There's a difference. *Journal of MultiDisciplinary Evaluation, 6*(14). Retrieved from file:///C:/Documents%20and%20Settings/quads/My%20 Documents/199–943–1-PB.pdf

Podems, D. (2013). *Feminist evaluation for non-feminist evaluators* (AEA365 blog post), September 17. Retrieved from http://aea365. org/blog/fiemme-week-donna-podems-on-applying-feminist -evaluation-for-non-feminist-evaluators/

Podems, D. (2014a). Evaluator competencies and professionalizing the field: Where are we now? *Canadian Journal of Program Evaluation, 28*(3), 127–136.

Podems, D. (2014b). Feminist evaluation for nonfeminists. In S. Brisolara, D. Seigart, & S. SenGupta (Eds.), *Feminist evaluation and research: Theory and practice* (chap. 5, pp. 113–142). New York, NY: Guilford Press.

Polanyi, M. (1962). *Personal knowledge.* Chicago, IL: University of Chicago Press.

Polanyi, M. (1967). *The tacit dimension.* Magnolia, MA: Peter Smith. (Reprinted 1983)

Pollack, A., & McNeil, D. G., Jr. (2013, March 3). A medical first, a baby with H.I.V. is deemed cured. *The New York Times,* p. A1. Retrieved from http://www.nytimes.com/2013/03/04/health/for-first-time-baby -cured-of-hiv-doctors-say.html?ref=health?src=dayp&_r=1&

Pomeroy, R. (2013, April 19). Creatures of coherence: Why we're so obsessed with causality. *Pacific Standard.* Retrieved from http://www .psmag.com/science/creatures-of-coherence-why-were-so-obsessed -with-causation-55801/

Porter, S. E., & Robinson, J. C. (2011). *Hermeneutics: An introduction to interpretive theory.* Grand Rapids, MI: Wm. B. Eerdmans.

Powdermaker, H. (1966). *Stranger and friend.* New York, NY: W. W. Norton.

Power, K. A. (2012). *Hooking up: An ecological psychology perspective on the sexual behaviors of women and potential effects on sexually transmitted infections* (Unpublished MA thesis). Boston University School of Public Health, MA.

Poynter, R. (2010). *The handbook of online and social media research.* Chichester, England: Wiley.

Preskill, H. (2005). Appreciative inquiry. In S. Mathison (Ed.), *Encyclopedia of evaluation* (pp. 18–19). Thousand Oaks, CA: Sage.

Preskill, H., & Beer, T. (2013). *Evaluating social innovation.* Washington, DC: Center for Evaluation Innovation. Retrieved from http://www .fsg.org/Portals/0/Uploads/Documents/PDF/Evaluating_Social_ Innovation.pdf

Preskill, H., & Catsambas, T. T. (2006). *Reframing evaluation through appreciative inquiry.* Thousand Oaks, CA: Sage.

Preskill, H., & Coghlan, A. T. (Eds.). (2003). *Using appreciative inquiry in evaluation* (New Directions for Evaluation, No. 100). San Francisco, CA: Jossey-Bass.

Preskill, H., & Mack, K. (2013). *Building a strategic learning and evaluation system for your organization.* Boston, MA: Foundation Strategy Group.

Preskill, S., & Jacobvitz, R. S. (2000). *Stories of teaching: A foundation for educational renewal.* Englewood Cliffs, NJ: Prentice Hall.

Pressley, M., & Afflerbach, P. (1995). *Verbal protocols of reading: The nature of constructively responsive reading.* Mahwah, NJ: Lawrence Erlbaum.

Pressman, J. L., & Wildavsky, A. (1984). *Implementation: How great expectations in Washington are dashed in Oakland: Or, why it's amazing that federal programs work at all, this being a saga of the Economic Development Administration as told by two sympathetic observers who seek to build morals on a foundation of ruined hopes.* Berkeley: University of California Press.

Prigogine, I. (2009). Complexity theory. In M. Ramage & K. Shipp (Eds.), *Systems thinkers* (pp. 224–239). New York, NY: Springer.

Private Agencies Collaborating Together. (1986). *Participatory evaluation.* New York, NY: Author.

Program Evaluation Division. (2001). *Early childhood education programs: Program evaluation report* (Report No. 01–01). St. Paul, MN: Office of the Legislative Auditor.

Prosser, J. (2011). Visual methodology: Toward a more seeing research. In N. K. Denzin & Y. S. Lincoln (Eds.), *The SAGE handbook of qualitative research* (4th ed., pp. 479–495). Thousand Oaks, CA: Sage.

Public Radio Program Directors Association. (2008). *Public radio talk show handbook.* Retrieved from http://www.prpd.org/ knowledgebase/talkshow_handbook_intro/Talkshow_handbook_ onair/Talkshow_handbook_interviewing.aspx

Punch, M. (1985). *Conduct unbecoming: Police deviance and control.* London, England: Tavistock.

Punch, M. (1986). *The politics and ethics of fieldwork* (Qualitative Research Methods Series No. 3). London, England: Sage.

Punch, M. (1989). Researching police deviance: A personal encounter with the limitations and liabilities of fieldwork. *British Journal of Sociology, 40*(2), 177–204.

Punch, M. (1997). *Dirty business: Exploring corporate misconduct.* London, England: Sage.

Putnam, H. (1987). *The many faces of realism.* LaSalle, IL: Open Court.

Putnam, H. (1990). *Realism with a human face.* Cambridge, MA: Harvard University Press.

Pyrch, T. (2012). *Breaking free: A facilitator's guide to participatory action research practice.* Retrieved from http://www.lulu.com/shop/timothy -pyrch/breaking-free-a-facilitators-guide-to-participatory-action -research-practice/paperback/product-20113382.html

Qualia. (2013). *Wikipedia.* Retrieved from http://en.wikipedia.org/wiki/ Qualia

Quinion, M. (1999). More than one way to skin a cat. *World Wide Words.* Retrieved from http://www.worldwidewords.org/qa/qa-mor1.htm

Radavich, D. (2001, January). On poetry and pain. *A View From the Loft, 24*(6), 3–6, 17.

Ragin, C. C. (1987). *The comparative method: Moving beyond qualitative and quantitative strategies.* Berkeley: University of California Press.

Ragin, C. C. (1992). Introduction: Cases of "What is a case?" In C. C. Ragin & H. S. Becker (Eds.), *What is a case?* (pp. 1–18). Cambridge, England: Cambridge University Press.

Ragin, C. C. (2000). *Fuzzy-set social science.* Chicago, IL: University of Chicago Press.

Ragin, C. C. (2008). *Designing social inquiry.* Chicago, IL: University of Chicago Press.

Ragin, C. C., & Becker, H. S. (Eds.). (1992). *What is a case? Exploring the foundations of social inquiry.* Cambridge, England: Cambridge University Press.

Rain, C. B. (2012). *Grasshopper moments: The Kung-Fu Masters of process evaluation* [AEA365 blog]. Retrieved from http://aea365.org/blog/?s= Grasshopper+moments%3A+The+Kung-Fu+Masters+of+process+e valuation&submit=Go

Rallis, S. F. (2009). Reasoning with rigor and probity: Ethical premises for credible evidence. In S. I. Donaldson, C. A. Christie, & M. Mark (Eds.), *What counts as credible evidence in applied research and evaluation practice* (pp. 168–180). Thousand Oaks, CA: Sage.

Rallis, S., & Rossman, G. (2003). Mixed methods in evaluation contexts: A pragmatic framework. In A. Tashakkori & C. Teddlie (Eds.), *Handbook of mixed methods in the social and behavioral sciences* (pp. 491–512). Thousand Oaks, CA: Sage.

Ramachandran, V. S., & Blakeslee, S. (1998). *Phantoms in the brain: Probing the mysteries of the human mind*. London, England: Fourth Estate.

Ramalingam, B., & Jones, H., with Reba, T. and Young, J. (2008) *Exploring the science of complexity: Ideas and implications for development and humanitarian efforts*. ODI Working Paper 285. London: ODI.

Ramalingam, B. (2013). *Aid on the edge of chaos: Rethinking international cooperation in a complex world*. Oxford, England: Oxford University Press.

Ramsey, K., & Peale, C. (2010, March 29). *First-generation college students stay the course*. USA Today. Retrieved from http://www.usatoday.com/news/education/2010-03-30-FirstGenDorm30_ST_N.htm

Rawls, A. W. (Ed.). (2002). *Ethnomethodology's program: Working out Durkeim's aphorism*. Oxford, England: Rowman & Littlefield.

Rawls, A. W. (2008). Harold Garfinkel, ethnomethodology and workplace studies. *Organization Studies, 29*(5), 701–732.

Redfield, R. (1953). *The primitive world and its transformations*. Ithaca, NY: Cornell University Press.

Reichardt, C. S., & Rallis, S. F. (Eds.) (1994). *The qualitative-quantitative debate: New perspectives* (*New Directions for Program Evaluation* No. 61). San Francisco: Jossey-Bass.

Reichertz, J. (2004). Abduction, deduction and induction in qualitative research. In U. Flick, E. von Kardoff, & I. Steinke (Eds.), *A companion to qualitative research* (pp. 159-164). Thousand Oaks, CA: Sage.

Reinharz, S. (1992). *Feminist methods in social research*. New York, NY: Oxford University Press.

Restivo, S., & Croissant, J. (2008). Social constructionism in science and technology studies. In J. A. Holstein & J. F. Gubrium (Eds.), *Handbook of constructionist research* (pp. 213–229). New York, NY: Guilford Press.

Rettig, K., Chiu-Wan Tam, V., & Maddock, M. B. (1996). Using pattern matching and modified analytic induction in examining justice principles in child support guidelines. In M. B. Sussman & J. F. Gilgun (Eds.), *The methods and methodologies of qualitative family research* (pp. 193–222). New York, NY: Haworth Press.

Reynolds, J., Kizito, J., Ezumah, N., Mangesho, P., Allen, E., & Chandler, C. (2011). Quality assurance of qualitative research: A review of the discourse. *Health Research Policy and Systems, 9*, 43. Retrieved from http://www.health-policy-systems.com/content/9/1/43

Rheingold, H. (1988). *They have a word for it: A lighthearted lexicon of untranslatable words and phrases* (1st ed.). Los Angeles, CA: Tarcher.

Rheingold, H. (2000). *They have a word for it: A lighthearted lexicon of untranslatable words and phrases* (2nd ed.). Louisville, KY: Sarabande Books.

Rhodes, T., & Treloar, C. (2008). The social production of hepatitis C risk among injecting drug users: A qualitative synthesis. *Addiction, 103*(10), 1593–1603.

Rice, G. (2009). Reflections on interviewing elites. *AREA: Royal Geographical Society, 42*(1), 70–75.

Richardson, J. T. (1998). Religious conversion. In W. H. Swatos Jr. (Ed.), *Encyclopedia of religion and society* (pp. 119–120). Walnut Creek, CA: AltaMira Press. Retrieved from http://hirr.hartsem.edu/ency/conversion.htm

Richardson, L. (2000a, June). Evaluating ethnography. *Qualitative Inquiry, 6*(2), 253–255.

Richardson, L. (2000b). Writing: A method of inquiry. In N. K. Denzin & Y. S. Lincoln (Eds.), *Handbook of qualitative research* (2nd ed., pp. 923–948). Thousand Oaks, CA: Sage.

Riessman, C. K. (1993). *Narrative analysis*. Newbury Park, CA: Sage.

Riessman, C. K. (2003). Analysis of personal narratives. In J. A. Holstein & J. F. Gubrium (Eds.), *Inside interviewing: New lenses, new concerns* (pp. 331–346). Thousand Oaks, CA: Sage.

Riessman, C. K. (2008). *Narrative methods for the human sciences*. Thousand Oaks, CA: Sage.

Rihoux, B., & Ragin, C. C. (Eds.). (2009). *Configurational comparative methods: Qualitative comparative analysis (QCA) and related techniques*. Thousand Oaks, CA: Sage.

Robert Wood Johnson Foundation. (2008). *Sampling* (Qualitative Research Guidelines Project). Retrieved from http://www.qualres.org/HomeSamp-3702.html

Roberts, L. C. (1997). *From knowledge to narrative: Educators and the changing museum*. Washington, DC: Smithsonian Institution Press.

Robertson, L., & Hale, B. (2011). Interviewing older people: Relationships in qualitative research. *Internet Journal of Applied Health Sciences and Practice, 9*(3), 1540–1580. Retrieved from http://ijahsp.nova.edu/articles/Vol9Num3/pdf/Robertson.pdf

Robinson, C. A., Jr. (1949). *Alexander the great, the meeting of east and west in world government and brotherhood*. New York, NY: Dutton.

Robinson, S. B. (2013). Using a grounded theory approach to evaluate professional development. *Sage Research Methods Cases*. Retrieved from http://srmo.sagepub.com/view/methods-case-studies-2013/n102.xml?rskey=bUQCAe&row=3

Rog, D. (2009, November 9–14). *Context and evaluation* (Presidential address). Evaluation 2009: 23rd annual conference of the American Evaluation Association, Orlando, FL. Retrieved from http://archive.eval.org/search09/session.asp?sessionid=8427&presenterid=2017

Rog, D. J. (2012). When background becomes foreground: Toward context-sensitive evaluation practice. In D. J. Rog, J. L. Fitzpatrick, & R. F. Conner (Eds.), *Context: A framework for its influence on evaluation practice* (*New Directions for Evaluation*, No. 135, pp. 25–40). Hoboken, NJ: Wiley Periodicals.

Rogers, B. L., & Wallerstein, M. B. (1985). *PL 480 Title I: Impact evaluation results and recommendations* (A.I.D. Program Evaluation Report No. 13). Washington, DC: U.S. Agency for International Development.

Rogers, C. (1961). *On becoming a person*. Boston, MA: Houghton Mifflin.

Rogers, C. (1969). Toward a science of the person. In A. Sutich & M. Vich (Eds.), *Reading in humanistic psychology* (pp. 21–50). New York, NY: Free Press.

Rogers, C. (1977). *On personal power*. New York, NY: Delacorte Press.

Rogers, E. (1962). *Diffusion of innovations*. New York, NY: Free Press.

Rogers, P. J. (2008). Using programme theory to evaluate complicated and complex aspects of interventions. *Evaluation, 14*(1), 29–48.

Rogers, P. J. (2012). *Introduction to impact evaluation*. Washington, DC: InterAction. Retrieved from http://www.interaction.org/document/introduction-impact-evaluation

Rogers, P. J. (2014). Realist synthesis. *Better Evaluation*. Retrieved from http://betterevaluation.org/evaluation-options/realistsynthesis

Rogers, P. J., & Fraser, D. I. (2014). Development evaluation. In B. Currie-Alder, R. Kanbur, D. M. Malone, & R. Medhora (Eds.), *International development: Ideas, experience, and prospects* (pp. 151–167). Oxford, England: Oxford University Press.

Rogers, P. J., & Williams, B. (2006). Evaluation for practice improvement and organizational learning. In I. Shaw, J. Greene, & M. Mark (Eds.), *The SAGE handbook of evaluation* (pp. 76–97). Thousand Oaks, CA: Sage.

Roholt, R. V., Hildreth, R. W., & Baizerman, M. (2009). *Becoming citizens: Deepening the craft of youth civic engagement*. New York, NY: Routledge.

Rorty, R. (1994). Method, social science, and social hope. In S. Seidman (Ed.), *The postmodern turn: New perspectives on social theory* (pp. 46–64). Cambridge, England: Cambridge University Press.

Rosenthal, R. (1994). *Homeless in paradise: A map of the terrain*. Philadelphia, PA: Temple University Press.

Ross, J. A., Barkaoui, K., & Scott, G. (2007). Evaluations that consider the cost of educational programs the contribution of high-quality studies. *American Journal of Evaluation, 28*(4), 477–492.

Rossi, P. H., Lipsey, M. W., & Freeman, H. E. (2004). *Evaluation: A systematic approach* (7th ed.). Thousand Oaks, CA: Sage.

Rossi, P. H., & Williams, W. (Eds.). (1972). *Evaluating social programs: Theory, practice, and politics*. New York, NY: Seminar Press.

Rossman, G. B., & Rallis, S. F. (1998). *Learning in the field: An introduction to qualitative research*. Thousand Oaks, CA: Sage.

Rossman, G. B., & Rallis, S. F. (2011). *Learning in the field: An introduction to qualitative research*. Thousand Oaks, CA: Sage.

Roulston, K. J. (2010). *Reflective interviewing*. Thousand Oaks, CA: Sage.

Rowe, A., Colby, B., Hall, W., & Niemeyer, M. (in press)**.** *Rapid impact evaluation*.

Roy, D. (2009). New horizons in the study of child language acquisition. In Proceedings of Interspeech. Retrieved from http://www.media.mit.edu/cogmac/publications/Roy_interspeech_keynote.pdf

Rubin, H. J., & Rubin, I. S. (1995). *Qualitative interviewing: The art of hearing data*. Thousand Oaks, CA: Sage.

Rubin, H. J., & Rubin, I. S. (2012). *Qualitative interviewing: The art of hearing data* (3rd ed.). Thousand Oaks, CA: Sage.

Russell, B. (1959). *Wisdom of the west*. London, England: Rathbone.

Ryan, G. W., & Bernard, H. R. (2000). Data management and analysis methods. In N. K. Denzin & Y. S. Lincoln (Eds.), *Handbook of qualitative research* (2nd ed., pp. 769–802). Thousand Oaks, CA: Sage.

Ryan, K. E., & Cousins, J. B. (Eds.). (2009). *The SAGE international handbook of educational evaluation*. Thousand Oaks, CA: Sage.

Ryen, A. (2011). Ethics and qualitative research. In D. Silverman (Ed.), *Qualitative research* (3rd ed., pp. 416–438). Thousand Oaks, CA: Sage.

Sachs, J. (2005). The end of poverty: Economic possibilities for our time. New York, NY: Penguin Books.

Sacks, O. (1985). *The man who mistook his wife for a hat and other clinical tales.* New York, NY: Summit Books. Retrieved from http://www .oliversacks.com/books/man-who-mistook-his-wife/

Sacks, O. (2012). *Hallucinations.* New York, NY: Alfred A. Knopf.

Safire, W. (2007, May). Halfway humanity: On language. *Sunday New York Times Magazine.* Retrieved from http://www.nytimes .com/2007/05/06/magazine/06wwln-safire-t.html

Safire, W., & Safire, L. (1991). *Leadership.* New York, NY: Fireside.

Sagan, C. (1987). The burden of skepticism. *Skeptical Inquirer, 12*(1), 38–46. Retrieved from http://www.csicop.org/si/show/burden_of_skepticism

Saini, M., & Shlonsky, A. (2012). *Systematic synthesis of qualitative research.* New York, NY: Oxford University Press.

Saldaña, J. (2009). *The coding manual for qualitative researchers.* Thousand Oaks, CA: Sage.

Saldaña, J. (2011). *Fundamentals of qualitative research.* New York, NY: Oxford University Press.

Salmen, L. F. (1987). *Listen to the people: Participant-observer evaluation of development projects.* New York, NY: Oxford University Press for the World Bank.

Salmen, L. F., & Kane, E. (2006). *Bridging diversity: Participatory learning for responsive development.* Washington, DC: World Bank.

Salmons, J. (2010). *Online interviews in real time.* Thousand Oaks, CA: Sage.

Salmons, J. (Ed.). (2012). *Cases in online interview research.* Thousand Oaks, CA: Sage.

Sandage, S. (2014, April 6). Epic fail. *The New York Times Sunday Book Review,* p. 18.

Sandelowski, M. (1986). The problem of rigor in qualitative research. *Advances in Nursing Science, 8*(3), 27–37.

Sandelowski, M., Voils, C. I., & Knafl, G. (2009). On quantitizing. *Journal of Mixed Methods Research, 3*(3), 12–28.

Sanders, W. (1976). *The sociologist as detective* (2nd ed.). New York, NY: Praeger.

Sands, D. M. (1986). Farming systems research: Clarification of terms and concepts. *Experimental Agriculture, 22,* 87–104.

Sands, G. (2000). *A principal at work: A story of leadership for building sustainable capacity of a school* (Doctoral thesis). Queensland University of Technology, Brisbane, Queensland, Australia.

Sangi, S. (2009). *Documentation.* Retrieved from http://www.slideshare .net/sangi/what-is-documentation-and-its-techniques

Saul, J. E., Willis, C. D., Bitz, J., & Best, A. (2013). A time-responsive tool for informing policymaking: Rapid realist review. *Implementation Science, 8,* 103. Retrieved from http://www.implementationscience .com/content/8/1/103

Saumure, K., & Given, L. M. (2008). Convenience sample. In L. M. Given (Ed.), *The SAGE encyclopedia of qualitative methods* (Vol. 1, pp. 124– 125). Thousand Oaks, CA: Sage.

Schell, J. (2008). *The art of game design: A book of lenses.* Burlington, MA: Morgan Kaufman.

Schensul, J., & LeCompte, M. D. (Eds.). (2010). *Designing and conducting ethnographic research: Ethnographer's toolkit* (2nd ed.). Walnut Creek, CA: AltaMira Press.

Schlangen, R. (2014). *Monitoring and evaluation for human rights organizations: Three case studies.* Washington, DC: Center for Evaluation Innovation. Retrieved from http://www.evaluationinnovation .org/sites/default/files/CEI%20HR%20Case%20Studies.pdf

Schlechty, P., & Noblit, G. (1982). Some uses of sociological theory in educational evaluation. In R. Corwin (Ed.), *Policy research* (pp. 283–306). Greenwich, CT: JAI Press.

Schon, D. A. (1983a). Organizational learning. In G. Morgan (Ed.), *Beyond method: Strategies for social research* (pp. 114–128). Newbury Park, CA: Sage.

Schon, D. A. (1983b). *The reflective practitioner: How professionals think in action.* New York, NY: Basic Books.

Schon, D. A. (1987). *Educating the reflective practitioner: Toward a new design for teaching and learning in the professions.* San Francisco, CA: Jossey-Bass.

Schorr, L. B. (1988). *Within our reach: Breaking the cycle of disadvantage.* New York, NY: Doubleday.

Schorr, L. B. (2012). Broader evidence for bigger impact. *Stanford Social Innovation Review.* Retrieved from http://www.ssireview.org/articles/ entry/broader_evidence_for_bigger_impact

Schultz, E. (Ed.). (1991). *Dialogue at the margins: Whorf, bakhtin, and linguistic relativity.* Madison: University of Wisconsin Press.

Schultz, S. J. (1984). *Family systems therapy: An integration.* Northvale, NJ: Aronson.

Schulz, K. (2014, April 7). Final forms: What death certificates can tell us, and what they can't. *The New Yorker,* pp. 32–37.

Schutz, A. (1967). *The phenomenology of the social world.* Evanston, IL: Northwestern University Press.

Schutz, A. (1970). *On phenomenology and social relations.* Chicago, IL: University of Chicago Press.

Schutz, A. (1977). Concepts and theory formation in the social sciences. In F. R. Pallmayr & T. A. McCarthy (Eds.), *Understanding and social inquiry.* Notre Dame, IN: University of Notre Dame Press.

Schwandt, T. A. (1989). Recapturing moral discourse in evaluation. *Educational Researcher, 18*(8), 11–16, 34.

Schwandt, T. A. (1997). *Qualitative inquiry: A dictionary of terms.* Thousand Oaks, CA: Sage.

Schwandt, T. A. (2000). Three epistemological stances for qualitative inquiry: Interpretivism, hermeneutics, and social constructivism. In N. K. Denzin & Y. S. Lincoln (Eds.), *Handbook of qualitative research* (2nd ed., pp. 189–214). Thousand Oaks, CA: Sage.

Schwandt, T. A. (2001). *Dictionary of qualitative inquiry* (2nd Rev. ed.). Thousand Oaks, CA: Sage.

Schwandt, T. A. (2002). *Evaluation practice reconsidered.* New York, NY: Peter Lang.

Schwandt, T. A. (2007). *The SAGE dictionary of qualitative inquiry* (3rd ed.). Thousand Oaks, CA: Sage.

Schwandt, T. A., & Burgon, H. (2006). Evaluation and the study of lived experience. In I. Shaw, J. Greene, & M. Mark (Eds.), *The SAGE handbook of evaluation* (pp. 98–117). Thousand Oaks, CA: Sage.

Schwartz, J. (2010, September 14). Confessing to crime, but innocent. *The New York Times,* p. 14.

Schwartzman, H. B. (1993). *Ethnography in organizations* (Qualitative Research Methods Series No. 27). Newbury Park, CA: Sage.

Schwed, F. (2014, January 1). The best financial advice I ever got (or gave). *Yahoo News.* Retrieved from http://finance.yahoo.com/news/ best-financial-advice-ever-got-231000077.html

Scott, B. (2011, August). *Stay interviews.* Innovation Partners International newsletter. Retrieved from http://www .innovationpartners.com/

Scriven, M. (1970). Some comparisons: Methods of reasoning and justification in social science and law. *Journal of Legal Education, 23,* 189.

Scriven, M. (1972a). Objectivity and subjectivity in educational research. In L. G. Thomas (Ed.), *Philosophical redirection of educational research: The seventy-first yearbook of the national society for the study of education* (pp. 94–142). Chicago, IL: University of Chicago Press.

Scriven, M. (1972b). Pros and cons about goal-free evaluation. *Evaluation Comment, 3,* 1–7.

Scriven, M. (1976). Maximizing the power of causal investigations: The modus operandi method. *Evaluation Studies Review Annual, 1,* 101–118.

Scriven, M. (1993). *Hard-won lessons in program evaluation* (*New Directions for Evaluation,* No. 58). San Francisco, CA: Jossey-Bass.

Scriven, M. (1998). The meaning of bias. In R. Davis (Ed.), *Proceedings of the Stake symposium on educational evaluation* (pp. 13–24). Urbana: Urbana University Press.

Scriven, M. (2008). A summative evaluation of RCT methodology: An alternative approach to causal research. *Journal of MultiDisciplinary Evaluation, 5*(9), 11–24.

Scruton, R. (2009). *I drink therefore I am: A philosopher's guide to wine.* London, England: Continuum International.

Seale, C. (2012). *Researching society and culture* (3rd ed.). Thousand Oaks, CA: Sage.

Searle, B. (Ed.). (1985). *Evaluation in World Bank education projects: Lessons from three case studies* (Report No. EDT 5). Washington, DC: World Bank.

Senge, P. M. (1990). *The fifth disciple: The art and practice of the learning organization.* New York, NY: Doubleday.

A sentimental education. (2014, February 9). *New York Times Book Review*, 16–19. Retrieved from http://www.nytimes.com/2014/02/09/books/review/a-sentimental-education.html?ref=books&_r=1

Shadish, W. R. (1993). Critical multiplism: A research strategy and its attendant tactics. In L. Sechrest (Ed.), *Program evaluation: A pluralistic enterprise* (*New Directions for Evaluation*, No. 60, pp. 13–57). San Francisco, CA: Jossey-Bass.

Shadish, W. R. (1995a). The logic of generalization: Five principles common to experiments and ethnographies. *American Journal of Community Psychology, 23*(3), 419–428.

Shadish, W. R. (1995b). Philosophy of science and the quantitative-qualitative debates: Thirteen common errors. *Evaluation and Program Planning, 18*(1), 63–75.

Shadish, W. R., Cook, T., & Campbell, D. (2001). *Experimental and quasi-experimental designs for generalized causal inference*. Independence, KY: Cengage Learning.

Shah, I. (1972). *The exploits of the incomparable Mulla Nasrudin*. New York, NY: Dutton.

Shah, I. (1973). *The subtleties of the inimitable Mulla Nasrudin*. New York, NY: Dutton.

Shaner, W. W., Philipp, P. F., & Schmehl, W. R. (1982a). *Farming systems research and development: Guidelines for developing countries*. Boulder, CO: Westview Press.

Shaner, W. W., Philipp, P. F., & Schmehl, W. R. (1982b). *Readings in farming systems research and development*. Boulder, CO: Westview Press.

Shank, G. (2008). Semiotics. In L. M. Given (Ed.), *The SAGE encyclopedia of qualitative research* (Vol. 2, pp. 806–810). Thousand Oaks, CA: Sage.

Shapiro, E. (1973). Educational evaluation: Rethinking the criteria of competence. *School Review, 81*, 523–549.

Shaw, G., Brown, R., & Bromiley, P. (1998, May–June). Strategic stories: How 3M is rewriting business planning. *Harvard Business Review, 76*(3), 41–50.

Sheehan, C., De Cieri, H., Cooper, B., & Brooks, R. (2014). Exploring the power dimensions of the human resource function. *Human Resource Management Journal*. doi:10.1111/1748-8583.12027

Sheehy, G. (2006). *Passages: Predictable crises of adult life*. New York, NY: Ballantine.

Shepard, L. (1993). Evaluating test validity. *Review of Research in Education, 19*, 405–450.

Sherwood, K. (2005). Evaluating home visitation: A case study of evaluation at the David and Lucille Packard Foundation. *New Directions for Evaluation, 105*, 59–82.

Shils, E. A. (1959). Social inquiry and the autonomy of the individual. In D. Lerner (Ed.), *The human meaning of the social sciences* (pp. 114–158). Cleveland, OH: Meridian.

Shopes, L. (2011). Oral history. In N. Denzin & Y. Lincoln (Eds.), *The SAGE handbook of qualitative research* (4th ed., pp. 451–465). Thousand Oaks, CA: Sage.

Shulman, L. (1986). Those who understand: Knowledge growth in teaching. *Educational Researcher, 15*(2), 4–14.

Silver, N. (2012). *The signal and the noise: Why so many predictions fail—but some don't*. New York, NY: Penguin Books. Retrieved from http://www.fastcompany.com/3009258/most-creative-people-2013/1-nate-silver

Silverman, D. (2000). Analyzing talk and text. In N. K. Denzin & Y. S. Lincoln (Eds.), *Handbook of qualitative research* (2nd ed., pp. 821–834). Thousand Oaks, CA: Sage.

Silverman, D., & Marvasti, A. (2008). *Doing qualitative research: A comprehensive guide*. Thousand Oaks, CA: Sage.

Silverman, M., Ricci, E. M., & Gunter, M. J. (1990). Strategies for increasing the rigor of qualitative methods in evaluation of health care programs. *Evaluation Review, 14*(1), 57–74.

Simic, C. (2000). Tragicomic soup. *New York Review of Books, 47*(9), 8–11.

Simmons, R. (1985). *Farming systems research: A review* (World Bank Technical Paper No. 43). Washington, DC: World Bank.

Simon, H. A. (1971). Designing organizations for an information-rich world. In M. Greenberger (Ed.), *Computers, communication, and the public interest* (pp. 37–72). Baltimore, MD: Johns Hopkins Press.

Sims, C. (2001, March 11). Stone age ways surviving, barely: Indonesian village is caught between worlds very far apart. *The New York Times*, p. 6.

Singhal, A. (2013). *A new normal for social change interventions: Focus on the positive deviants*. Retrieved from http://www.t2w2.org/which-way-is-north-2/

Singleton, R. A., Jr., & Straits, B. C. (2012). Survey interviewing. In J. F. Gubrium, J. A. Holstein, A. B. Marvasti, & K. D. McKinney (Eds.), *The SAGE handbook of interview research: The complexity of the craft* (2nd ed., pp. 77–98). Thousand Oaks, CA: Sage.

Skinner, J. (Ed.). (2013). *The interview: An ethnographic approach*. New York, NY: Bloomsbury.

Skloot, R. (2010). *The immortal life of Henrietta Lacks*. New York, NY: Broadway Books.

Slaughter, A. (2011). Preface. In Mr. Y. (Ed.), *A national strategic narrative* (pp. 2–4). Princeton, NJ: Woodrow Wilson Center. Retrieved from http://theglobalrealm.com/2011/04/27/a-national-strategic-narrative/

Smith, E. E. (2012, November 5). A plan to reboot dating. *The Atlantic*. Retrieved from http://www.theatlantic.com/sexes/archive/2012/11/a-plan-to-reboot-dating/264184/

Smith, G. C. S., & Pell, J. P. (2003). Parachute use to prevent death and major trauma related to gravitational challenge: Systematic review of randomised controlled trials. *British Medical Journal, 327*, 1459–1461.

Smith, J. M. (2000). The Cheshire cat's DNA: Review of the century of the gene. *New York Review of Books, 47*(20), 43–46.

Smith, L. M., & Kleine, P. F. (1986). Qualitative research and evaluation: Triangulation and multimethods reconsidered. In D. D. Williams (Ed.), *Naturalistic evaluation* (*New Directions for Evaluation*, No. 30, pp. 55–72). San Francisco, CA: Jossey-Bass.

Smith, L. T. (2012). *Decolonizing methodologies: Research and indigenous peoples* (2nd ed.). New York, NY: Zed Books.

Smith, M. K. (2005). Elliot W. Eisner, connoisseurship, criticism and the art of education. The encyclopedia of informal education. Retrieved from http://infed.org/mobi/elliot-w-eisner-connoisseurship-criticism-and-the-art-of-education/

Smith, N. (1978). Truth, complementarity, utility, and certainty. *CEDR Quarterly, 11*, 16–17.

Smith, N. L. (2013). How micro-cultures of evaluation practice shape evaluation design. *Journal of MultiDisciplinary Evaluation, 9*(21), 21–26. Retrieved from http://journals.sfu.ca/jmde/index.php/jmde_1/article/view/385/374

Smith-Acuña, S. (2011). *Systems theory in action: Applications to individual, couple, and family therapy*. Hoboken, NJ: Wiley.

Snow, D. A. (1980). The disengagement process: A neglected problem in participant observation research. *Qualitative Sociology, 3*(2), 100–122.

Snow, D. A. (2001). Extending and broadening Blumer's conceptualization of symbolic interactionism. *Symbolic Interaction, 24*(3), 367–377.

Snowden, D. J., & Boone, M. E. (2007). A leader's framework for decision making. *Harvard Business Review, 85*(11), 68–77.

Social Impact. (2013). Rapid appraisal methods. *Evaluation: Some tools, methods and approaches* (pp. 31–33). Washington, DC: Author.

Society of Applied Anthropology. (1980). *Symposium on integrating qualitative and quantitative data*, Denver, CO. Retrieved from http://sfaa.metapress.com/home/main.mpx

Soeffner, H. (2010). Social science hermeneutics. In U. Flick, E. von Kardoffr, & I. Steinke (Eds.), *A companion to qualitative research* (3rd ed., pp. 95–100). Thousand Oaks, CA: Sage.

Soldz, S., & Andersen, L. L. (2012). Expanding subjectivities: Introduction to the special issue on new directions in psychodynamic research. *Journal of Research Practice, 8*(2), 1–8.

Solomon, A. (2012). *Far from the tree: Parents, children, and the search for identity*. New York, NY: Scribner.

Sonnemann, U. (1954). *Existence and therapy: An introduction to phenomenological psychology and existential analysis*. New York, NY: Grune & Stratton.

Sources of Insight. (2009). *Lessons learned from Peter Drucker*. Retrieved from http://sourcesofinsight.com/lessons-learned-from-peter-drucker/

Special Inspector General for Iraq Reconstruction. (2013). *Learning from Iraq: Final report of the Special Inspector General for Iraq Reconstruction*. Washington, DC: U.S. Congress.

Spry, T. (2011). *Body, paper, stage: Writing and performing autoethnography*. Walnut Creek, CA: Left Coast Press.

Stacey, R. D. (2001). *Complex responsive processes in organizations: Learning and knowledge creation*. New York, NY: Routledge.

Stacey, R. D. (2007). *Strategic management and organisational dynamics: The challenge of complexity to ways of thinking about organizations*. New York, NY: Prentice Hall Financial Times.

REFERENCES

Stake, R. E. (1975). *Evaluating the arts in education: A responsive approach.* Columbus, OH: Charles E. Merrill.

Stake, R. E. (1978). The case study method in a social inquiry. *Educational Researcher, 7,* 5–8.

Stake, R. E. (1995). *The art of case study research.* Thousand Oaks, CA: Sage.

Stake, R. E. (1998). Hoax? In R. Davis (Ed.), *Proceedings of the Stake symposium on educational evaluation* (p. 363). Urbana: University of Illinois Press.

Stake, R. E. (2000). Case studies. In N. K. Denzin & Y. S. Lincoln (Eds.), *Handbook of qualitative research* (2nd ed., pp. 435–454). Thousand Oaks, CA: Sage.

Stake, R. E. (2004). How far dare an evaluator go toward saving the world? Beyond neutrality: What evaluators care about. *American Journal of Evaluation, 25*(1), 103–107.

Stake, R. K. (2006). *Multiple case study analysis.* New York, NY: Guilford Press.

Stake, R. K. (2010). *Qualitative research: How things work.* New York, NY: Guilford Press.

Stake, R. K., Bresler, L., & Mabry, L. (1991). *Custom and cherishing: The arts in elementary schools.* Urbana: University of Illinois, Council for Research in Music Education.

Standing, G. (2011). *The precariat: The new dangerous class.* New York, NY: Bloomsbury.

Stanfield, J. H., II. (2006). The possible restorative justice functions of qualitative research. *International Journal of Qualitative Studies in Education, 19*(6), 723–727.

Starr, D. (2013, December 9). The interview: Do police interrogation techniques produce false confessions? *The New Yorker.* Retrieved from http://www.newyorker.com/reporting/2013/12/09/131209fa_fact_starr

Steckler, A., & Linnan, L. (Eds.). (2002). *Process evaluation for public health interventions and research.* San Francisco, CA: Wiley.

Steinberg, D. I. (1983). *Irrigation and AID's experience: A consideration based on evaluation* (A.I.D. Program Evaluation Report No. 8). Washington, DC: U.S. Agency for International Development.

Stenhouse, L. (1977). *Case study as a basis for research in a theoretical contemporary history of education.* East Anglia, England: University of East Anglia, Centre for Applied Research in Education.

Sterman, J. D. (2006). Learning from evidence in a complex world. *American Journal of Public Health, 96,* 505–514.

Stern, E., Stame, N., Mayne, J., Forss, K., Davies, R., & Befani, B. (2012). *Broadening the range of designs and methods for impact evaluations.* A report of a study commissioned by the Department for International Development (Working Paper No. 38). London, England: Department for International Development.

Stern, M. J., Powell, R. B., & Hill, D. (2013). Environmental education program evaluation in the new millennium: What do we measure and what have we learned? *Environmental Education Research.* Retrieved from http://www.tandfonline.com/doi/full/10.1080/13504622.2013.838749

Stewart, J. A. (2009). *Indigenous narratives of success.* Flaxton, Queensland, Australia: Post Pressed.

Stewart, J. B. (2009, September 21). Eight days. *The New Yorker,* p. 59. Retrieved from http://www.newyorker.com/reporting/2009/09/21/090921fa_fact_stewart#ixzz0uT8hal2Q

Stewart, R., & Oliver, S. (2012). Making a difference with systematic reviews. In D. Gough, S. Oliver, & J. Thomas (Eds.), *An introduction to systematic reviews* (pp. 227–244). Thousand Oaks, CA: Sage.

Stockdill, S. H., Duhon-Sells, R. M., Olson, R. A, & Patton, M. Q. (1992). Voices in the design and evaluation of a multicultural education program: A developmental approach. *New Directions for Program Evaluation, 53,* 17–34.

Stoller, P. (2004). Sensuous ethnography, African persuasions, and social knowledge. *Qualitative Inquiry, 10*(6), 817–835.

Storey, D. (2006, May). *Evaluating business entrepreneurship programs.* Paper presented at the International Finance Corporation Technical Assistance Programs Monitoring and Evaluation Meeting, "Results Measurement For Technical Assistance," Washington, DC.

Storm, J., & Vitt, M. (2000). *Master of creative philanthropy: The story of Russ Ewald.* Minneapolis, MN: Philanthropoid Press.

Stoto, M., Nelson, C., & Klaiman, T. (2013). Getting from what to why: Using qualitative methods in public health systems research.

Academy Health Brief. Retrieved from http://www.academyhealth.org/files/publications/QMforPH.pdf

St. Pierre, E. A. (2000). The call for intelligibility in postmodern education. *Educational Researcher, 29*(5), 25–28.

Strauss, A., & Corbin, J. (1990). *Basics of qualitative research: Grounded theory procedures and techniques.* Newbury Park, CA: Sage.

Strauss, A., & Corbin, J. (Eds.). (1997). *Grounded theory in practice.* Thousand Oaks, CA: Sage.

Strauss, A., & Corbin, J. (1998). *Basics of qualitative research: Techniques and procedures for developing grounded theory* (2nd ed.). Thousand Oaks, CA: Sage.

Strickhouser, S. M., & Wright, J. D. (2014). Agency resistance to outcome measurement. *Journal of Applied Social Science.* Retrieved from http://jax.sagepub.com/content/early/2014/02/18/1936724414523966.abstract

Stringer, E. T. (2007). *Action research* (3rd. ed.). Thousand Oaks, CA: Sage.

Strong, P. (1979). *The ceremonial order of the clinic: Parents, doctors and medical bureaucracies.* London, England: Routledge & Kegan Paul.

Stufflebeam, D. L. (1980). An interview with Daniel L. Stufflebeam. *Educational Evaluation and Policy Analysis, 2*(4), 90–92.

Stufflebeam, D. L., Madeus, G. F., & Kellaghan, T. (Eds.). (2000). *Evaluation models: Viewpoints on educational and human services evaluation* (2nd ed.). Boston, MA: Kluwer.

Stufflebeam, D., & Shinkfield, A. (2007). *Metaevaluation: Evaluating evaluations. Evaluation theory, models, and applications* (chap. 27). San Francisco, CA: Jossey-Bass.

Sullivan, J. (2013). *Stay interviews: An essential tool for winning "the war to keep your employees."* Retrieved from http://www.ere.net/2013/12/02/stay-interviews-an-essential-tool-for-winning-the-war-to-keep-your-employees/

Sussman, M. B., & Gilgun, J. F. (Eds.). (1996). *The methods and methodologies of qualitative family research.* New York, NY: Haworth Press.

Sutherland, P. L., & Denny, R. M. (2007). *Doing anthropology in consumer research.* Walnut Creek, CA: Left Coast Press.

Sutton, R. I., & Rao, H. (2014). *Scaling up excellence: Getting to more without settling for less.* New York, NY: Crown Business.

Sweeney, C., & Gosfield, J. (2013). *The art of doing: How super achievers do what they do and how they do it so well.* New York, NY: Penguin Books.

Switching Operations Fatality Analysis Working Group. (2010). *Railroad industry collaboration enters new phase in working to prevent employee switching fatalities.* Washington, DC: Author. Retrieved from https://www.fra.dot.gov/eLib/details/L01005

Symbols. (2006). Cross with arms of equal length. *Online encyclopedia of symbols and ideograms.* Stockholm, Sweden: HME. Retrieved from http://www.symbols.com/encyclopedia/09/091.html

Taleb, N. N. (2005). *Fooled by randomness: The hidden role of chance in life and the markets.* New York, NY: Random House.

Taleb, N. N. (2007). *The black swan: The impact of the highly improbable.* New York, NY: Random House.

Taleb, N. N. (2012). *Antifragile: Things that gain from disorder.* New York, NY: Random House.

Tallmadge, J. (1997). *Meeting the tree of life: A teachers' path.* Salt Lake City: University of Utah Press.

Tan, C. (2014). Reflective thinking for intelligence analysis using a case study. *Reflective Practice, 15*(2), 218–231.

Tashakkori, A., & Teddlie, C. (1998). *Mixed methodology: Combining qualitative and quantitative approaches.* Thousand Oaks, CA: Sage.

Tashakkori, A., & Teddlie, C. (Eds.). (2003). *Handbook of mixed methods in the social and behavioral research.* Thousand Oaks, CA: Sage.

Taylor, R. R., Suarez-Balcazar, Y., Forsyth, K., & Kielhofner, G. (2006). Participatory research in occupational therapy. In G. Kielhofner (Ed.), *Research in occupational therapy: Methods of inquiry for enhancing practice* (pp. 548–564). Philadelphia, PA: F. A. Davis.

Taylor, S. J., & Bogdan, R. (1984). *Introduction to qualitative research methods. The search for meaning* (2nd ed.). New York, NY: Wiley.

Teddlie, C., & Tashakkori, A. (2003). Major issues and controversies in the use of mixed methods in the social and behavioral sciences. In A. Tashakkori & C. Teddlie (Eds.), *Handbook of mixed methods in the social and behavioral sciences* (pp. 3–50). Thousand Oaks, CA: Sage.

Teddlie, C., & Tashakkori, A. (2011). Mixed methods research. In N. Denzin & Y. Lincoln (Eds.), *The SAGE handbook of qualitative research* (4th ed., pp. 285–299). Thousand Oaks, CA: Sage.

Teddlie, C., & Yu, F. (2007). Mixed methods sampling: A typology with examples. *Journal of Mixed Methods Research, 1*(1), 77–100.

Tedlock, B. (2000). Ethnography and ethnographic representation. In N. K. Denzin & Y. S. Lincoln (Eds.), *The handbook of qualitative research* (2nd ed., pp. 455–486). Thousand Oaks, CA: Sage.

Tesch, R. (1990). *Qualitative research: Analysis types and software tools.* New York, NY: Falmer.

Textor, R. (1980). *A handbook on ethnographic futures research.* Stanford, CA: Stanford University Cultural and Educational Future Research Project.

Thiselton, A. C. (2009). *Hermeneutics: An introduction.* Grand Rapids, MI: Wm. B. Eerdmans.

Thomas, J. (1993). *Doing critical ethnography* (Qualitative Research Methods Series No. 26). Newbury Park, CA: Sage.

Thomas, W. I., & Thomas, D. S. (1928). *The child in America: Behavior problems and programs.* New York, NY: Alfred A. Knopf.

Thompson, A. I. (2013). "Sometimes, I think I might say too much": Dark secrets and the performance of inflammatory bowel disease. *Symbolic Interaction, 36*(1), 21–39.

Tierney, P. (2000a). *Darkness in El Dorado: How scientists and journalists devastated the Amazon.* New York, NY: W. W. Norton.

Tierney, P. (2000b, October 9). The fierce anthropologist. *The New Yorker,* pp. 50–61.

Tierney, W. (2000). Undaunted courage: Life history and the postmodern challenge. In N. K. Denzin & Y. S. Lincoln (Eds.), *Handbook of qualitative research* (2nd ed., pp. 537–565). Thousand Oaks, CA: Sage.

Tikunoff, W. (with Ward, B.). (1980). *Interactive research and development on teaching.* San Francisco, CA: Far West Laboratory for Educational Research and Development.

Tilney, J. S., Jr., & Riordan, J. (1988). *Agricultural policy analysis and planning: A summary of two recent analysis of A.I.D.—supported projects worldwide* (A.I.D. Evaluation Special Study No. 55). Washington, DC: U.S. Agency for International Development.

Timmermans, S., & Berg, M. (2003). *"The gold standard" The challenge of evidence-based medicine and standardization in health care.* Philadelphia, PA: Temple University Press.

To, S. (2013). Understanding *Sheng Nu* ("leftover women"): The phenomenon of late marriage among Chinese professional women. *Symbolic Interaction, 36*(1), 1–20.

Tomich, T. P., Brodt, S., Ferris, H., Galt, R., Horwath, W. R., Kebreab, E., . . . Yang, L. (2011). Agroecology: A review from a global-change perspective. *Annual Review of Environment and Resources, 36,* 193–222.

Torres, R., Preskill, H., & Piontek, M. (1996). *Evaluation strategies for communicating and reporting: Enhancing learning in organizations.* Thousand Oaks, CA: Sage.

Tötösy de Zepetnek, S. (1993). *The social dimensions of fiction: On the rhetoric and function of prefacing novels in nineteenth-century Canada.* Braunschweig, Germany: Vieweg & Sohn. Retrieved from http://docs .lib.purdue.edu/clcweblibrary/totosystudyofpreface/

Tourigny, S. C. (1998). Some new dying trick: African American youths choosing HIV/AIDS. *Qualitative Health Research, 8*(2), 149–167.

Tracy, S. J. (2010). Qualitative quality: Eight "big-tent" criteria for excellent qualitative research. *Qualitative inquiry, 16*(10), 837–851.

Tremmel, R. (1993). Zen and the art of reflective practice in teacher education. *Harvard Educational Review, 63*(4), 434–458.

Trend, M. G. (1978). On the reconciliation of qualitative and quantitative analyses: A case study. *Human Organization, 37,* 345–354.

Trow, M. (1970). Comment on participant observation and interviewing: A comparison. In W. J. Filstead (Ed.), *Qualitative methodology* (pp. 79–83). Chicago, IL: Markham.

Tucker, E. (1977, April). *The follow through planned variation experiment: What is the pay-off?* Paper presented at the annual meeting of the American Educational Research Association, New York City.

Tufte, E. (2003). *PowerPoint is evil: Power corrupts: PowerPoint corrupts absolutely. WIRED.* Retrieved from http://www.wired.com/wired/ archive/11.09/ppt2.html

Tufte, E. (2006). *The cognitive style of PowerPoint: Pitching out corrupts within* (2nd ed.). Cheshire, CT: Graphics Press.

Turner, A. (2000). Embodied ethnography. Doing culture. *Social Anthropology, 8*(1), 51–60.

Turner, J. H. (1998). *The structure of sociological theory.* Belmont, CA: Wadsworth.

Turner, R. (Ed.). (1974). *Ethnomethodology: Selected readings.* Baltimore, MD: Penguin Books.

Turner, T. S. (2004). *Behavioral interviewing guide: A practical, structured approach for conducting effective selection interviews.* Victoria, British Columbia, Canada: Trafford.

Tye, M. (2013). *Qualia* (Stanford Encyclopedia of Philosophy). Retrieved from http://plato.stanford.edu/entries/qualia/

Uchitelle, L. (2001, February 18). By listening, three economists show slums hurt the poor. *New York Times Business,* p. 4.

United Nations. (1996). *The United Nations and Rwanda 1993–1996.* New York, NY: Author.

The United Nations Children's Fund. (2008). *2004 Indian Ocean earthquake and tsunami: Lessons learned.* New York, NY: United Nations. Retrieved from http://www.unicef.org/har08/index_ tsunami.html

Uphoff, N. (1991). A field methodology for participatory self-evaluation. In *Evaluation of Social Development Projects* [Special issue]. *Community Development Journal, 26*(4), 271–285.

U.S. Agency for International Development. (2010). *Using rapid appraisal methods: Performance monitoring and evaluation tips, No. 5.* Retrieved from http://www.seachangecop.org/sites/default/files/ documents/2010%2003%20USAID%20-%20TIPS%2005_Using%20 Rapid%20Appraisal%20Methods.pdf

U.S. Agency for International Development. (2011). *Rapid data collection methods* (Online course). Retrieved from http://manage.c2ti.com/ review/USAID/content/rapiddatacollectionmethods/menu_a. html?id=6397076589055359

U.S. Conference of Mayors. (2014). *Best practices.* Retrieved from http:// www.usmayors.org/bestpractices/

U.S. Senate Select Committee on Intelligence. (2004*). Report of the Select Committee on Intelligence on the U.S. intelligence community's prewar intelligence assessments on Iraq.* Washington, DC: Author.

Vaillant, G. E. (2013). *Triumphs of experience: The men of the Harvard Grant Study.* Cambridge, MA: Belknap Press of Harvard University Press.

Van den Hoonaard, W. C. (1997). *Working with synthesizing concepts: Analytical field research* (Qualitative Research Methods Series No. 41). Thousand Oaks, CA: Sage.

Van Maanen, J. (1988). *Tales of the field: On writing ethnography.* Chicago, IL: University of Chicago Press.

Van Maanen, J. (2011). *Tales of the field: On writing ethnography* (2nd ed.). Chicago, IL: University of Chicago Press.

Van Manen, M. (1990). *Researching lived experience: Human science for an action sensitive pedagogy.* New York: State University of New York Press.

Van Manen, M. (2014). *Phenomenology of practice.* Walnut Creek, CA: Left Coast Press.

Vidich, A. J., & Lyman, S. M. (2000). Qualitative methods: Their history in sociology and anthropology. In N. K. Denzin & Y. S. Lincoln (Eds.), *Handbook of qualitative research* (2nd ed., pp. 37-84). Thousand Oaks, CA: Sage.

Von Bertalanffy, L. (1976). *General system theory: Foundations, development, applications.* New York, NY: George Braziller.

Von Oech, R. (1998). *A whack on the side of the head: How you can be more creative.* New York, NY: Warner Books.

Wadsworth, Y. (1984). *Do it yourself social research.* Melbourne, Victoria, Australia: Victorian Council of Social Service and Melbourne Family Care Organization in association with Allen & Unwin.

Wadsworth, Y. (1993). *What is participatory action research?* Melbourne, Victoria, Australia: Action Research Issues Association.

Wadsworth, Y. (2008a). Is it safe to talk about systems again yet? Self-organising processes for complex living systems and the dynamics of human inquiry. *Systematic Practice and Action Research, 21*(2), 153–170.

Wadsworth, Y. (2008b). Systemic human relations in dynamic equilibrium. *Systematic Practice and Action Research, 21*(1), 15-34.

Wadsworth, Y. (2010). *Building in research and evaluation: Human inquiry for living systems.* Melbourne, Victoria, Australia: Allen & Unwin.

Wadsworth, Y. (2011a). *Building in research and evaluation: Human inquiry for living systems.* Walnut Creek, CA: Left Coast Press.

Wadsworth, Y. (2011b). *Do it yourself social research* (3rd ed.). Walnut Creek, CA: Left Coast Press.

Wagner, E. (2011). Viewing illegal immigration through desert debris. *Pacific Standard.* Retrieved from http://www.psmag .com/?s=Viewing+illegal+immigration+through+desert+debris

REFERENCES

Walcott, H. (2010). *Ethnography lessons: A primer.* Walnut Creek, CA: Left Coast Press.

Waldrop, M. M. (1992). *Complexity: The emerging science at the edge of order and chaos.* New York, NY: Simon & Schuster.

Walker, J. (1996). Letters in the attic: Private reflections of women, wives and mothers. In M. B. Sussman & J. F. Gilgun (Eds.), *The methods and methodologies of qualitative family research* (pp. 9–40). New York, NY: Haworth Press.

Wallace, R. A., & Wolf, A. (1980). *Contemporary sociological theory.* Englewood Cliffs, NJ: Prentice Hall.

Waller, J. (2004). *Fabulous science: Fact and fiction in the history of scientific discovery.* New York, NY: Oxford University Press.

Wallerstein, I. (2004). *World-systems analysis: An introduction.* Durham, NC: Duke University Press.

Wallerstein, I. (2011). *The modern world-system I: Capitalist agriculture and the origins of the European world-economy in the sixteenth century* (with a new prologue; Vol. 1). Berkeley: University of California Press.

Walsh, S. M. (2009). *Linking coral reef health and human welfare* (Doctoral dissertation in marine biology). University of California, San Diego.

Walston, J. T., & Lissitz, R. W. (2000). Computer-mediated focus groups. *Evaluation Review, 5*(October), 457–483.

Walton, M. (2014). Applying complexity theory: A review to inform evaluation design. *Evaluation and Program Planning, 45*(1), 119–126.

Wang, C., & Burris, M. A. (1994). Empowerment through photo novella: Portraits of participation. *Health Education & Behavior, 21*(2), 171–186.

Waring, G. E., Jr. (1897). *Street-cleaning and the disposal of a city's wastes.* New York, NY: Doubleday & McClure.

Waring, T., & Wainwright, D. (2008). Issues and challenges in the use of template analysis. *Journal of Business Research Methods, 6*(1), 85–93.

Warren, M. K. (1984). *AID and education: A sector report on lessons learned* (A.I.D. Program Evaluation Report No. 12). Washington, DC: U.S. Agency for International Development.

Waskul, D., Douglass, M., & Edgley, C. (2000). Cybersex: Outercourse and the enselfment of the body. *Symbolic Interaction, 23*(4), 375–397.

Wasserman, G., & Davenport, A. (1983). *Power to the people: Rural electrification sector summary report* (A.I.D. Program Evaluation Report No. 11). Washington, DC: U.S. Agency for International Development.

Watkins, J. M., & Cooperrider, D. (2000). Appreciative inquiry: A transformative paradigm. *OD Practitioner, 32*(1), 6–12.

Watkins, K. E., & Marsick, V. J. (1993). *Sculpting the learning organization.* San Francisco, CA: Jossey-Bass.

Watson, G., & Goulet, J. G. (1998, March). What can ethnomethodology say about power? *Qualitative Inquiry, 4*(1), 96–113.

Watters, E. (2013). We aren't the world. *Pacific Standard: The Science of Society.* Retrieved from http://www.psmag.com/magazines/magazine-feature-story-magazines/joe-henrich-weird-ultimatum-game-shaking-up-psychology-economics-53135/

Wax, R. H. (1971). *Doing fieldwork: Warnings and advice.* Chicago, IL: University of Chicago Press.

Weaver-Hightower, M. B. (2014). A mixed methods approach for identifying influence on public policy. *Journal of Mixed Methods Research, 8*(2), 115–138.

Webb, E. J., Campbell, D. T., Schwartz, R., & Sechrest, L. (1966). *Unobtrusive measures: Nonreactive research in the social sciences.* Chicago, IL: Rand McNally.

Webb, E. J., Donald, T., Campbell, D. T., Schwartz, R., & Sechrest, L. (1999). *Unobtrusive measures: Nonreactive research in the social sciences* (Rev. ed.). Thousand Oaks, CA: Sage.

Webb, E. J., & Weick, K. E. (1983). Unobtrusive measures in organizational theory: A reminder. In J. Van Maanen (Ed.), *Qualitative methodology* (pp. 209–224). Beverly Hills, CA: Sage.

Weick, K., & Sutcliffe, K. (2001). *Managing the unexpected: Assuming high performance in an age of complexity.* San Francisco, CA: Jossey-Bass.

Weiss, C. H. (Ed.). (1977). *Using social research in public policy making.* Lexington, MA: D. C. Heath.

Weiss, C. H. (2002). "Lessons learned:" A comment. *American Journal of Evaluation, 23*(2), 229–230.

Weiss, C. H. (with the Oral History Project Team). (2006). The Oral History of Evaluation, Part 4: The professional evolution of Carol H. Weiss. *American Journal of Evaluation, 27*(4), 475–484.

Weiss, C. H., & Bucuvalas, M. (1980). Truth test and utility test: Decision makers' frame of reference for social science research. *American Sociological Review, April,* 302–313.

Weiss, H. B. (2001). Strategic communications: From the director's desk. *Evaluation Exchange, 7*(1), 1.

Weiss, H. B., & Greene, J. C. (1992). An empowerment partnership for family support and education programs and evaluations. *Family Science Review, 5*(1–2), 145–163.

Weiss, R. S. (1994). *Learning from strangers: The art and method of qualitative interview studies.* New York, NY: Free Press.

Weiss, W. M., Bolton, P., & Shankar, A. V. (2000). *Rapid assessment procedures: Addressing the perceived needs of refugees and internally displaced persons through participatory learning and action.* Baltimore, MD: Johns Hopkins University, Center for Refugee and Disaster Studies. Retrieved from http://www.jhsph.edu/research/centers-and-institutes/center-for-refugee-and-disaster-response/publications_tools/publications/rap.html

Weitzman, B. C., & Silver, D. (2012). Good evaluation measures: More than their psychometric properties. *American Journal of Evaluation, 34*(1), 115–119.

Weitzer, R. (2007a). Prostitution as a form of work. *Sociology Compass, 1*(1), 143–155.

Weitzer, R. (2007b). The social construction of sex trafficking: Ideology and institutionalization of a moral crusade. *Politics & Society, 35*(3), 447–475.

Wells, C. W. (2012). *Car country: An environmental history.* Seattle: University of Washington Press.

Wenger, E. (1998). *Communities of practice: Learning, meaning, and identity.* Cambridge, England: Cambridge University Press.

Wenger, E., McDermott, R. A., & Snyder, W. M. (2002). *Cultivating communities of practice: A guide to managing knowledge.* Boston, MA: Harvard Business School.

Wenger, E., Trayner, B., & de Laat, M. (2011). *Promoting and assessing value creation in communities and networks: A conceptual framework* (Report 18). Heerlen, Netherlands: Ruud de Moor Centrum, Open Universiteit.

Wenger, G. C. (2003). Interviewing older people. In J. A. Holstein & J. F. Gubrium (Eds.), *Inside interviewing: New lenses, new concerns* (pp. 111–130). Thousand Oaks, CA: Sage.

Wengraf, T. (2001). *Qualitative research interviewing: Biographic narrative and semi-structured methods.* Thousand Oaks, CA: Sage.

Wengraf, T. (2013). *Interviewing for life-histories, lived periods and situations, and ongoing personal experiencing using the Biographic-Narrative Interpretive Method (BNIM) BNIM Short Guide bound with the BNIM Detailed Manual* (Unpublished manuscript).

Wertsch, J. V. (1991). *Voices of the mind: A sociocultural approach to mediated action.* Cambridge, MA: Harvard University Press.

Westhorp, G. (2013). Developing complexity-consistent theory in a realist investigation. *Evaluation, 19*(4), 364–382.

Westley, F., Zimmerman, B., & Patton, M. Q. (2006). *Getting to maybe: How the world is changed.* Toronto, Ontario, Canada: Random House.

Westra, H. A. (2012). *Motivational interviewing in the treatment of anxiety.* New York, NY: Guilford Press.

Wheatley, M. (1992). *Leadership in the new science.* San Francisco, CA: Berrett-Koehler.

White, G. G., Lansky, A., Goel, S., Wilson, D., Hladik, W., Hakim, A., & Fraost, S. D. W. (2012). Respondent driven sampling: Where we are and where should we be going? *Sexually Transmitted Infection, 88*(6), 397–399.

White, M., & Epston, D. (1990). *Narrative means to therapeutic ends.* New York, NY: W. W. Norton.

White, R. C. (2014). *Global case studies in maternal and child health.* Burlington, MA: Jones & Bartlett.

Whitehead, A. N. (1958). *Modes of thought.* New York, NY: Capricorn.

Whiting, R. (1990). *You gotta have wa.* New York, NY: Vintage Books.

Whitmore, E., Guijt, I., Mertens, D. M., Imm, P. S., Chinman, M., & Wandersman, A. (2006). Embedding improvements, lived experience, and social justice in evaluation practice. In Shaw, I., Greene, J.C., & Mark, M.(Eds.), *The SAGE Handbook of Evaluation,* pp. 340–359. Thousand Oaks, CA: Sage.

Whyte, W. F. (1943). *Street corner society.* Chicago, IL: University of Chicago Press.

Whyte, W. F. (1984). *Learning from the field: A guide from experience.* Beverly Hills, CA: Sage.

Whyte, W. F. (Ed.). (1989, May–June). Action research for the twenty-first century: Participation, reflection, and practice [Special issue]. *American Behavioral Scientist, 32*(5).

Whyte, W. F. (Ed.). (1991). *Participatory action research.* Newbury Park, CA: Sage.

Wiebeck, V., & Dahlgren, M. (2007). Learning in focus groups: An analytical dimension for enhancing focus group research. *Qualitative Research, 7*(2), 249–267.

Wikimedia Commons. (2012). *Universal symbol of danger.* Retrieved from http://commons.wikimedia.org/wiki/File:Universal_Symbol_of_Danger.jpg

Wildavsky, A. (1985). The self-evaluating organization. In E. Chelimsky (Ed.), *Program evaluation: Patterns and directions* (pp. 246–265). Washington, DC: American Society for Public Administration.

Wilkinson, A. (1999, February 15). Notes left behind. *The New Yorker,* pp. 44–49.

Williams, B. (2005). Systems and systems thinking. In S. Mathison (Ed.), *Encyclopedia of evaluation* (pp. 405–412). Thousand Oaks, CA: Sage.

Williams, B. (2008). Systemic inquiry. In L. M. Given (Ed.), *The SAGE encyclopedia of qualitative research methods* (Vol. 2, pp. 854–859). Thousand Oaks, CA: Sage.

Williams, B., & Hummelbrunner, R. (2011). *Systems concepts in action: A practitioner's toolkit.* Stanford, CA: Stanford University Press.

Williams, B., & Iman, I. (2006). *Systems concepts in evaluation: An expert anthology* (American Evaluation Association Monograph No. 6). Point Reynes, CA: EdgePress.

Williams, B. F. (1991). *Stains on my name, war in my veins: Guyana and the politics of cultural struggle.* Durham, NC: Duke University Press.

Williams, M. (2004). Operationism/operationalism. In M. S. Lewis-Beck, A. Bryman, & T. Futing Liao (Eds.), *The SAGE encyclopedia of social science research methods* (pp. 768–769). Thousand Oaks, CA: Sage.

Willis, B. (1998). *The Adinkra dictionary : A visual primer on the language of Adinkra.* Washington, DC: Pyramid Complex.

Willis, G. B. (1999). *Cognitive interviewing: A how-to guide.* Research Triangle Park, NC: Research Triangle Institute. Retrieved from http://www.hkr.se/pagefiles/35002/gordonwillis.pdf

Wilson, S. (2000). *Construct validity and reliability of a performance assessment rubric to measure student understanding and problem solving in college physics: Implications for public accountability in higher education* (Doctoral dissertation, Dissertation Abstracts International, AAT 9970526). San Francisco, CA. University of San Francisco.

Wilson, S. (2009). *Research is ceremony: Indigenous research methods.* Winnipeg, Manitoba, Canada: Fernwood.

Wilson-Grau, R., & Britt, H. (2012). *Outcome harvesting.* Cairo, Egypt: Ford Foundation Middle East and North Africa Office. Retrieved from http://www.outcomemapping.ca/resource/resource.php?id=374

Wolcott, H. F. (1990). *Writing up qualitative research* (Qualitative Research Methods Series No. 20). Newbury Park, CA: Sage.

Wolcott, H. F. (1992). Posturing in qualitative inquiry. In M. D. LeCompte, W. L. Milroy, & J. Preissle (Eds.), *The handbook of qualitative research in education* (pp. 3–52). New York, NY: Academic Press.

Wolcott, H. F. (2008). *Ethnography: A way of seeing.* Walnut Creek, CA: AltaMira Press.

Wolcott, H. F. (2009). *Writing up qualitative research* (3rd ed.). Thousand Oaks, CA: Sage.

Wolf, R. L. (1975). Trial by jury: A new evaluation method. *Phi Delta Kappan, 3*(57), 185–187.

Wolf, R. L., & Tymitz, B. (1978). *Whatever happened to the giant wombat: An investigation of the impact of the ice age mammals and emergence of man exhibit.* Washington, DC: National Museum of Natural History, Smithsonian Institutes.

Wood, B., Betts, J., Etta, F., Gayfer, J., Kabell, F. D., Ngwira, N., . . . Samaranayake, M. (2011). *The evaluation of the Paris Declaration, Phase 2 final report.* Copenhagen, Denmark: Danish Institute for International Studies. Retrieved from http://pd-website.inforce.dk/content/pdf/PD-EN-web.pdf

Work Institute. (2013). *Essential guide to exit interviews.* Retrieved from http://workinstitute.com/Essential-Guide-to-Exit-Interviews#.UrcohvRDs08

Wright, B. (2013). *Reuse, recycle: Rethink research. Questions to ask in reviewing research evidence.* Retrieved from http://betterevaluation.org/resources/guide/questions_to_ask_in_reviewing_research_evidence

Writers beta. (2012). *Is being paid by the word/page making American books longer?* Retrieved from http://writers.stackexchange.com/questions/5503/is-being-paid-by-the-word-page-making-american-books-longer

Wyatt, J., & Adams, T. (Eds.). (2014). *On (writing) families: Autoethnographies of presence and absence, love and loss.* Rotterdam, Netherlands: Sense.

Yanarella, E. J. (1975). "Reconstructed logic" & "logic-in-use" in decision-making analysis. *Polity, 8*(1), 156–172.

Yarbrough, D. B., Shulha, L. M., Hopson, R. K., & Caruthers, F. A. (2010). *The program evaluation standards: A guide for evaluators and evaluation users* (3rd ed.). Thousand Oaks, CA: Sage.

Yin, R. K. (Ed.). (2004). *The case study anthology.* Thousand Oaks, CA: Sage.

Yin, R. K. (2009). *Case study research: Design and methods* (4th ed.). Los Angeles, CA: Sage.

Yin, R. K. (2011). *Qualitative research from start to finish.* New York, NY: Guilford Press.

Yin, R. K. (2012). *Applications of case study research.* Thousand Oaks, CA: Sage.

Youker, B. W., & Ingraham, A. (2014). Goal-free evaluation: An orientation for foundations' evaluations. *Foundation Review, 5*(4), 51–61.

Youngson, R. (1998). *Scientific blunders: A brief history of how wrong scientists can sometimes be.* New York, NY: Caroll & Graf.

Youth Homelessness Initiative Collaboration. (2014). *Nine guiding principles to help youth overcome homelessness: A developmental evaluation.* Minneapolis, MN: Author.

Zaner, R. M. (1970). *The way of phenomenology: Criticism as a philosophical discipline.* New York, NY: Pegasus Communications.

Zelik, D. J., Patterson, E. S., & Woods, D. D. (2007, October). Judging sufficiency: How professional intelligence analysts assess analytical rigor. In *Proceedings of the Human Factors and Ergonomics Society annual meeting* (Vol. 51, No. 4, pp. 318–322). Thousand Oaks, CA: Sage.

Zhang, X. (2009). *Earnings of women with and without children.* Statistics Canada. Retrieved from http://www.statcan.gc.ca/pub/75–001-x/2009103/pdf/10823-eng.pdf

Zimmerman, R. (2014). *The seven deadly sayings of nonprofit leaders . . . and how to avoid them.* Rancho Santa Margarita, CA: Charity Channel Press.

Zuber-Skerritt, O. (Ed.). (2009). *Action learning and action research.* Rotterdam, Netherlands: Sense.

Zulawski, D. E., & Wicklander, D. E. (1998). *Practical aspects of interview and interrogation.* Ann Arbor, MI: CRC Press.

Zuschlag, M., Ranney, J., Coplen, M., & Harnar, M. (2012). *Transformation of safety culture on the San Antonio service unit of Union Pacific Railroad.* Washington, DC: U.S. DOT Federal Railroad Administration. DOT-FRA-ORD-12–16. Retrieved from http://www.fra.dot.gov/eLib/details/L04121

Author Index

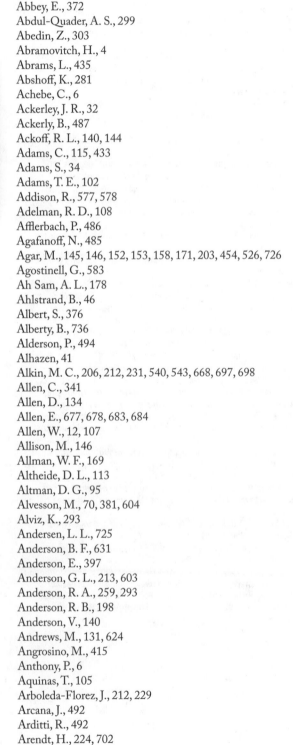

AUTHOR INDEX

Subject Index

SUBJECT INDEX

$SAGE researchmethods

The essential online tool for researchers from the world's leading methods publisher

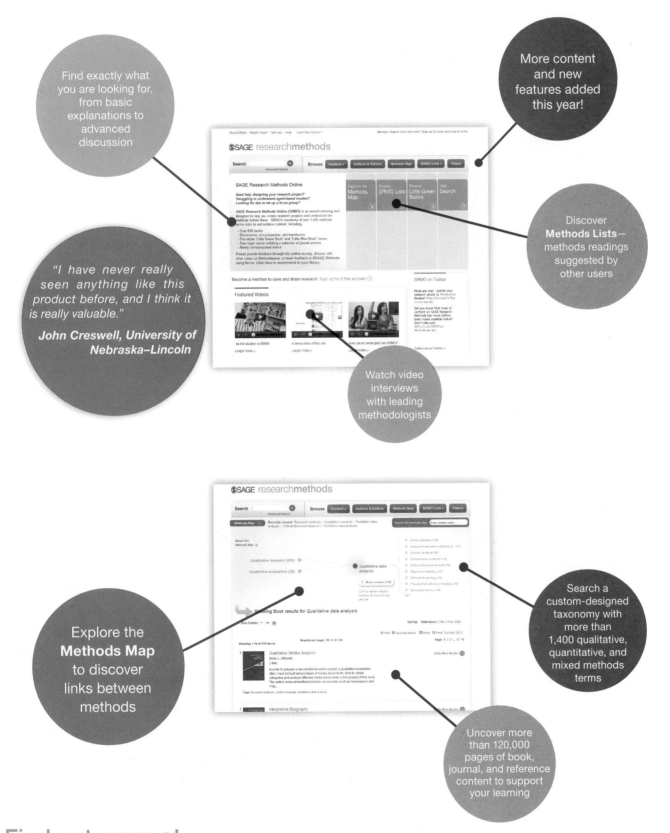

Find exactly what you are looking for, from basic explanations to advanced discussion

More content and new features added this year!

"I have never really seen anything like this product before, and I think it is really valuable."

John Creswell, University of Nebraska–Lincoln

Discover **Methods Lists**— methods readings suggested by other users

Watch video interviews with leading methodologists

Explore the **Methods Map** to discover links between methods

Search a custom-designed taxonomy with more than 1,400 qualitative, quantitative, and mixed methods terms

Uncover more than 120,000 pages of book, journal, and reference content to support your learning

Find out more at
www.sageresearchmethods.com